Foreword

The *Journal of Physical and Chemical Reference Data* is published jointly by the American Institute of Physics and the American Chemical Society for the National Institute of Standards and Technology (NIST). Its objective is to provide critically evaluated physical and chemical property data, fully documented as to the original sources and the criteria used for evaluation. One of the principal sources of material for the journal is the NIST Standard Reference Data Program, a program promoting the compilation and critical evaluation of property data.

The regular issues of the *Journal of Physical and Chemical Reference Data* are published bimonthly and contain compilations and critical data reviews of moderate length. Longer works, volumes of collected tables, and other material unsuited to a periodical format have previously been published as *Supplements to the Journal*. Beginning in 1989 the generic title of these works was changed to *Monograph*, which reflects their character as independent publications. This volume, ''Heat Capacity of Liquids: Volumes I and II. Critical Review and Recommended Values'' by Milan Zábranský, Vlastimil Růžička Jr., Vladimír Majer, and Eugene S. Domalski is the first of two volumes of *Monograph No. 6* of the *Journal of Physical and Chemical Reference Data.*

Jean W. Gallagher, Editor
Journal of Physical and Chemical Reference Data

Journal of
**Physical and
Chemical
Reference Data**

Monograph No. 6

Heat Capacity of Liquids: Volume I
Critical Review and Recommended Values

Milan Zábranský and Vlastimil Růžička Jr.
Department of Physical Chemistry
Institute of Chemical Technology
166 28 Prague 6, Czech Republic

Vladimír Majer
Laboratory of Thermodynamics and Chemical Engineering
University Blaise Pascal & CNRS
63177 Aubière, France

Eugene S. Domalski
Chemical Kinetics and Thermodynamics Division
National Institute of Standards and Technology
Gaithersburg, Maryland 20899-0001, USA

Published by the **American Chemical Society**
and the **American Institute of Physics** for
the **National Institute of Standards and Technology**

In association with the
International Union of Pure and Applied Chemistry
Commission on Thermodynamics

International Standard Book Number
1-56396-600-X

American Institute of Physics
500 Sunnyside Boulevard
Woodbury, New York 11797-2999

Printed in the United States of America

Contents

Volume I

General

Indices

Data

Volume II

Data

Preface 1

The Commission I.2 on Thermodynamics of the International Union of Pure and Applied Chemistry is responsible, through its Subcommittee on Thermodynamic Data, for the dissemination of internationally-approved thermodynamic tables. The projects are conveniently discussed under two headings. Firstly, there is the compilation of internationally-agreed values for the equilibrium thermodynamic properties of individual substances of special interest and importance to scientists and technologists. This work is carried out at the IUPAC Thermodynamic Tables Project Centre at Imperial College of Science, Technology and Medicine, London with publication in the series International Thermodynamic Tables of the Fluid State, now published by Blackwell Scientific Publications, Oxford. The second type of project is the critical compilation of values for individual equilibrium properties for a large group of compounds, which is carried out through international collaboration. The present publication on Heat Capacities of Liquids is the result of the first of these projects to be approved, because of the importance of this property in thermochemistry, in thermodynamics and in chemical engineering generally. The authors are to be congratulated on their careful assessment of all the experimental data and the clear presentation of their results: the summary of the experimental data, the selected and rejected sets, the coefficients in the correlating equations, the recommended data and the deviation plots. The tables given here should be of considerable value to all thermodynamicists. The emphasis on the accuracy of the values will be of particular importance to the user.

J. H. Dymond
Chairman
Subcommittee on Thermodynamic Data
Commission I.2 on Thermodynamics
International Union of Pure and Applied Chemistry

Preface 2

This monograph is a result of an IUPAC data project carried out in cooperation with the Department of Physical Chemistry at the Institute of Chemical Technology, Prague (Czech Republic), the Laboratory of Thermodynamics and Chemical Engineering at the University Blaise Pascal in Clermont-Ferrand (France) and the Chemical Thermodynamics and Kinetics Division at the National Institute of Standards and Technology (NIST), Gaithersburg, MD (U.S.A). The authors started cooperation on data collection and evaluation of liquid heat capacities in the mid-1980s, and this endeavor was accepted as an official project of the IUPAC Subcommittee on Thermodynamic Data (STD) of the Commission on the Chemical Thermodynamics in 1987 at the 34th IUPAC General Assembly in Boston. A specimen copy including all introductory material and representative examples of tables was prepared in the spring of 1991 and sent for review to experts from all over the world selected by STD. Altogether eleven reviews were received, providing constructive criticisms and suggestions. As a collective group, they accepted the project output in a positive manner without any serious objections. Most reports from reviewers were available before the 36th IUPAC General Assembly held in Hamburg in 1991 where all the input from the review process was discussed and minor modifications to the text and data presentation were decided upon. The manuscript was corrected accordingly and final approval for publication was given by the IUPAC Commission on Chemical Thermodynamics in the summer of 1992 during its meeting at Snowbird, Utah. Although STD and the authors were approached by several scientific book publishers, the *Journal of Physical and Chemical Reference Data* Monograph Series was selected as the best possible choice to publish the results of the IUPAC data project. This Monograph Series was selected because the NIST Standard Reference Data Program (SRDP) from which it originates has considerable experience in the dissemination of evaluated data and also because it is in charge of distributing a computerized database on liquid heat capacities which is closely connected with this data project. All the materials for publishing the monograph were submitted electronically to the SRDP staff during 1994.

The objective of the data project was to collect and critically assess all calorimetric, isobaric, and saturation heat capacities for well-defined organic and inorganic compounds in the liquid state which were published in the primary literature between 1862 and 1994. Older data were considered in exceptional cases where no other literature sources were found. Only pure chemicals with melting temperature below 573 K were considered. High melting chemicals such as most metals, inorganic compounds, and some organic compounds, were not included. The collected heat capacity data were related mainly to atmospheric pressure or saturated vapor pressure while investigations which focused on heat capacities as a function of pressure were considered beyond the scope of this project. The main outputs of the project are the recommended heat capacities provided as a function of temperature along the saturation line in the form of correlation equation(s) parameters and tabulated values with the expected uncertainties. Detailed characterization of all literature sources of experimental data (both included and rejected in the final selection) is also given.

The authors have followed, in general, the terminology and symbols recommended by IUPAC as well as the values for fundamental constants and molar masses. With respect to nomenclature, the authors followed the system used by Chemical Abstracts Service because it was generally an unambiguous and computer readable scheme.

At this point, we wish to thank all those who contributed to the success of the project in its different stages.

First of all we are grateful to the successive Chairmen at the Department of Physical Chemistry, Institute of Chemical Technology, Prague — J. Pick, R. Holub and J. Matouš, for allowing us to use Department facilities for most of the practical work of the project. We thank the Chief of the Standard Reference Data Program, M. W. Chase, Jr., and the Editor of the Journal of Physical and Chemical Reference Data, J. W. Gallagher, for their assistance with arrangements in the final publication phase of the project. The Chairman of the Subcommittee on Thermodynamic Data, J. H. Dymond, was always instrumental and ready to help when problems had to be overcome.

Our sincere thanks go to all the reviewers who devoted their valuable time to reading the specimen copy. Their ideas and remarks certainly improved the manuscript and affected the final version of the book in a positive manner. The individual reviewers were: R. C. Chirico from NIPER, Bartlesville, OK (U.S.A.), G. T. Furukawa from NIST, Gaithersburg, MD (U.S.A.), A. I. Johns from the National Engineering Laboratory, Glasgow (United Kingdom), B. V. Lebedev from the Lobachevsky State University, Nizhnii Novgorod (Russia), P. Liley from Purdue University, IN (U.S.A.), K. N. Marsh from The Texas A&M University System, TX (U.S.A.), P. A. G. O'Hare from NIST, Gaithersburg, MD (U.S.A.), H. Suga from the University of Osaka (Japan), I. Wadsö from the University of Lund (Sweden), E. F. Westrum, Jr. from the University of Michigan, Ann Arbor, MI (U.S.A.) and E. Wilhelm from the Vienna University (Austria). Reviews by P. A. G. O'Hare and E. F. Westrum, Jr., were especially detailed and helpful, once again many thanks.

Regarding problems with nomenclature, our valuable advisors were F. Liška and M. Ferles, professors of organic chemistry at the Institute of Chemical Technology, Prague. The program for cubic spline fitting was made available to us through the courtesy of K. Kollár-Hunek from the Budapest Technical University. Deviation plots were generated by using a library of Fortran subroutines known as VolksGrapher which were kindly provided to us by W. E. Anderson from NIST, Gaithersburg, MD. All figures for the monograph were carefully prepared by J. Margoliusová from the Institute of Chemical Technology in Prague. We also want to acknowledge help in the literature search given to us by all librarians in the Central Library (directed at the time by K. Sgallová) at the Institute of Chemical Technology, Prague as well as help from many unnamed staff persons in more than ten scientific libraries in the Czech Republic and abroad. Last but not least, we are grateful to Ilse Putman, Constance Seymour, and Judy Calabrese, staff persons in the NIST Standard Reference Data Program, for their efforts in the computer transfer and format manipulation of this monograph from the submitted computer files.

Milan Zábranský
Vlastimil Růžička Jr.
Vladimír Majer
Eugene S. Domalski

Heat Capacity of Liquids: Volume 1
Critical Review and Recommended Values

Heat Capacity of Liquids: Volume I
Critical Review and Recommended Values

Milan Zábranský, and Vlastimil Růžička Jr.,
Institute of Chemical Technology, Prague, Czech Republic

Vladimír Majer,
Université Blaise Pascal & CNRS, Aubière, France

and

Eugene S. Domalski
National Institute of Standards and Technology, Gaithersburg, Maryland 20899-0001, USA

This monograph contains recommended data on liquid heat capacities for over 1600 pure, mostly organic compounds that have melting temperatures below 573 K. The data were obtained by critical evaluation of calorimetrically measured heat capacities published in the literature. The introductory part of the monograph includes a discussion of the liquid heat capacity as a thermodynamic quantity, a brief survey of relevant calorimetric techniques, and a description of a procedure adopted for data treatment. The bulk of the monograph consists of tables for individual compounds which include information on experimental sources, the selection process, quality of correlation, and recommended data. Recommended data are presented as parameters of a correlating equation that expresses the dependence of the heat capacity on temperature, and in tables at several equally spaced temperatures. An estimate of the overall uncertainty of the recommended data is also given. At the end of the monograph, a formula and name index is provided as well as a bibliography.

Key words: bibliography; correlating equations; critically evaluated data; heat capacity; liquids; recommended values.

1. Introduction

Heat capacities belong among the basic thermophysical and thermodynamic properties which characterize a liquid. They are directly linked with temperature derivatives of basic thermodynamic functions and are therefore indispensable for the calculation of differences in these functions between different temperatures. This information is widely used in chemical engineering for establishing energy balances, in thermodynamics for obtaining entropy and enthalpy values, and in thermochemistry for calculating changes in reaction enthalpies with temperature. A knowledge of heat capacities is also required for evaluating the effect of temperature on phase and reaction equilibria. Variations in heat capacities serve as a sensitive indicator of phase transitions and are an important tool for understanding changes in the structure of liquid solutions.

1.1. Heat Capacity Calorimetry for Liquids; from the Early Measurements to the Present Status

Heat capacities were among the first physico-chemical quantities obtained by calorimetric measurements. For a number of chemicals, the first determinations were made in the 19th century using a Dewar flask as a simple drop calorimeter. As the temperatures of an ampoule with a sample and the water inside the flask differed considerably, this technique yielded enthalpy differences rather than values corresponding to true heat capacities. Other approaches were also considered, but these suggestions were more of a theoretical than practical character; the technical means available in the 19th century did not allow for the development of more sophisticated methods. The principle of relative calorimetric measurements was first mentioned as early as 1845 when Joule proposed the comparison of a sample under investigation with a well known reference substance (*45JOU) (See Sec. 6.3.1. for definition of reference format.). The first isoperibol calorimeters appeared after 1900 (02GAE); major contributions were due to Eucken (09EUC) and Nernst (10NER, 11NER), Nernst was the first to use the isothermal shield (14NER/SCH). This type of instrument (''Nernst calorimeter'') made it possible to perform measurements with temperature increments of several kelvin. These were the first results which could be considered as true heat capacity values.

The most important step in the development of instruments for heat capacity measurement was the introduction of adiabatic calorimetry. This technique became widely utilized shortly after the description of the first instrument using an adiabatic jacket (20COH/MOE). Since then, a great number of such instruments have been constructed. This method achieved a high degree of sophistication and became the most important and accurate tool for heat capacity determination over a wide range of temperatures. The years between 1935 and 1970 were the most fruitful concerning both the quantity and quality of the published data; the error of high-precision measurements decreased to 0.05 percent. Most determinations were performed for condensed phases over the temperature range from several kelvin to room temperature and permitted the calculation of the absolute entropy, enthalpy, and Gibbs function at 298.15 K.

Although the accuracy of adiabatic calorimetry has not been surpassed so far, several other approaches to heat capacity measurement have been developed which are less costly, faster, and easier to use. Calculations of heat capacity from heating or cooling curves became wide-spread especially in the high-temperature region; an alternative approach was the introduction of the differential thermoconducting calorimeters. The latter concept was applied successfully in the development of commercial instruments for differential thermal analysis and differential scanning calorimetry, and has been routinely used for determining heat capacities. Recently, a significant amount of data have been generated from differential flow calorimeters which were used extensively for measurements near room temperature. Adiabatic calorimetry has not undergone any major changes or experienced new developments. The main area of change or progress has been in the high degree of automation of process control and in the introduction of microcomputers for logging, retrieval, and processing of data.

Over the past twenty years, heat capacities of liquid organic mixtures and of electrolyte and nonelectrolyte aqueous solutions have been intensely investigated near ambient temperatures; the number of systems studied largely outnumbered measurements with pure substances. This occurred due to the arrival of the Picker differential flow calorimeter (71PIC/LED) which permits very fast data acquisition over the whole concentration range; this calorimeter is available commercially. After 1980, this principle was also successfully used for investigating liquid mixtures at high temperatures and pressures.

In recent years, little experimental activity was oriented specifically towards heat capacities of pure liquids. Low-temperature measurements have been performed primarily for solids, and often in connection with superconductivity research. Systematic studies for liquid compounds contained in coal and petroleum have been performed at the National Institute for Petroleum and Energy Research (NIPER), Oklahoma (USA) and for liquid macromolecular and organometallic compounds at the Gorkii University (USSR, now Russia). Occasionally, data for various liquids have been measured in a limited temperature interval using commercial differential scanning calorimeters. Frequent measurements on mixtures near ambient conditions yield heat capacity data for pure substances as values at the limit of a concentration range.

Several reasons led to a considerable decline in measurements on pure liquids. First, low-temperature heat capacities and the derived quantities resulting from the third law thermodynamics (S, H, and G functions) are now available for many pure liquids of theoretical and/or practical interest. Funding for low-temperature measurements which are time-consuming, laborious, and require expensive and complex instrumentation are generally scarce. Second, the available data base of experimental heat capacity values is often used for the development of group contribution methods and other estimation techniques which allow prediction of thermophysical properties for a wide variety of organic structures with a typical accuracy of 5 to 10 percent. This is satisfactory for many industrial and technological applications. Because group contribution predictions are less successful for solids, thermal properties are more often measured on them. Third, attention is now primarily focused on determining heat capacities of mixtures which offer many more possibilities for theoretical interpretation using different statistical thermodynamic models.

1.2. Scope of the Project

The project was started under auspices of the IUPAC Subcommittee on Thermodynamic Data. Its purpose was to review the state of available knowledge on heat capacities of pure substances in the liquid state and to produce the recommended values based on the evaluation of all available experimental data. The project covered organic and inorganic compounds. The criterion for inclusion of a substance in the compilation required a substance to have a melting point below 573 K. Only well-defined compounds in the isotropic liquid state were considered. Data on compounds which existed as undercooled liquids were excluded.

In order to produce thermodynamically well-defined data, we gave considerable attention to the origin of experimental values and to the conditions of their measurements. Only calorimetric heat capacities were compiled; primarily the values measured at (or close to) the saturation line were considered. Neither enthalpic measurements on reference-fluid boil-off calorimeters nor heat capacities determined by indirect techniques (see below) were considered. In general, the raw data base comes directly from experimental measurements which have been reported in the literature. Except for a few compounds for which no other data are available, comprehensive tables, calculated values, and other secondary data sources were not included. The bulk of the literature used has been published after 1920, which marks the beginning of the development of modern low-temperature calorimetry. Measurements published previous to this time were included only if no other data sources were available in the corresponding temperature interval or if the later data were obviously doubtful. All the sources of heat capacities referenced in Chemical Abstracts up to the end of 1992 were considered. Moreover the most important journals were checked for the data up to September, 1993.

Individual steps in the project were as follows:

1. Compilation of all available heat capacity data following the criteria described above.
2. Critical evaluation of the data.
3. Correlation of the selected data as a function of temperature.
4. Generation of sets of recommended values and estimation of their accuracy.

The main output of the project is the recommended isobaric and saturation heat capacities for 1624 substances. They are presented as parameters of two different correlating equations expressing heat capacity as a function of temperature for all substances (about 80 percent) for which the values are available at several temperatures. One equation serves for accurate representation of heat capacity within the temperature limits of experimental values. Discrete heat capacity values calculated from this equation are also tabulated. Parameters of the second correlating equation describe heat capacity over a larger temperature range allowing meaningful extrapolation in the direction of the critical point. Estimates of uncertainty are given for all recommended data.

1.3. Organization of the Monograph

The monograph consists of three parts: General, Indices, and Data.

The *General* part consists of several chapters. The heat capacity of liquids is discussed as a thermodynamic quantity, the result of calorimetric measurements and object of compilation efforts; the procedure adopted in this project is described and the presentation of data in this monograph is explained.

Chapter 2 defines different types of heat capacities, demonstrates their interconnection, and gives the basic thermodynamic background. Different calorimetric approaches to the heat capacity determination are analyzed in light of the thermodynamic interpretation of results. The variation of heat capacity with temperature and possible approaches to its correlation are examined.

A brief analysis of calorimetric techniques for determining the heat capacities of liquids is given in *Chapter 3*. Terminology and criteria for classifying experimental methods adopted in this project are introduced. Thereafter, a critical outline of the main experimental approaches with examples of representative calorimeters is provided. At the end of the chapter, a table summarizes all of the calorimeters referenced in the data sources which were utilized in this compilation.

Chapter 4 is an annotated bibliography of the previous compilations and evaluations regarding heat capacities of liquids and gives some description of all important secondary sources.

All the facts on the methodology of the project and handling of data are in *Chapter 5*. Rules for compiling and evaluating information from the original sources are described together with the techniques used for selection and correlation of data. Specifications are given concerning the types of recommended data and the methods used for their determination.

Chapter 6 defines the division of substances into groups for presentation purposes; the order in which the substances are listed is specified. A guide to tables gives detailed information as to how the data and accompanying information are presented and how the tables should be read.

The part *Indices* includes one literature index denoted *References* presenting all references to literature sources. For locating a substance according to its code number, the *Formula index* and *Name index* at the end of the monograph can be used.

The bulk of the monograph is the *Data* part consisting of tables for individual substances which include information on experimental sources, selection process, quality of correlation and recommended data. All of the data for one substance are presented together in several tables along with a deviation plot. The substances are ordered into groups according to chemical structure; short introductory discussions precede each group. Each substance presented is assigned a characteristic code which determines its position inside the tables.

1.4. Summary of Notation

Note: Symbols which occur only once in the text are not listed

Symbols:

Roman letters:

A	Helmholtz energy
A_i	correlation parameters
A'	time integral of the thermopile voltage
C	heat capacity
c	specific heat capacity
d	deviation
F	volumetric flow rate
f	correction factor
G	Gibbs energy
H	enthalpy
k_e	constant of proportionality
k_{ht}	heat transfer coefficient
L	heat loss
m	number of adjustable parameters
N	amount of substance (moles)
n	number of experimental points
P	power input
p	pressure
Q	heat
R	universal gas constant
S	minimized objective function
S	entropy
s	deviation
T	temperature (kelvin)
t	temperature (°C)
U	internal energy
U_e	voltage
V	volume

Greek letters:

α	thermal expansivity
β	isothermal compressibility
γ	thermal pressure
Δ	difference function
δ	deviation
ξ	function $(1 - T/T_c)$
ρ	specific density
Σ	sum
σ	error
σ^2	variance
τ	time

Superscripts:

g	gaseous
l	liquid

Subscripts:

avg	average
amp	ampoule
b	bias
c	critical point
cv	calorimeter vessel
exp	experimental (raw) value
fi	final
i	general index
in	initial
j	general index
m	molar
nb	normal boiling temperature (p_{sat} = 101325 Pa)
p	constant pressure
r	relative; reduced
re	reference
rec	receiver
s	sample
sat	saturation curve
sm	fitted (smoothed) value
su	surroundings
T	constant temperature
t	two-phase equilibrium system
V	constant volume
v	vaporization
w	weighted

2. Heat Capacities of Liquids—Thermodynamic Background

2.1. Definitions, Basic Relationships

Heat capacity C is defined by the differential equation

$$C = (\partial Q/\partial T)_x \qquad \text{(II-1)}$$

where Q indicates the amount of heat exchanged between the system studied and the surroundings when the temperature T changes under conditions specified by x. These conditions have to be specified as the heat Q is not a state function or quantity, and C varies according to different paths along which Q is exchanged. The corresponding intensive quantities are related to the unit amount of mass — *specific* heat capacity c, or to one mole — *molar* heat capacity C_m. Sometimes, heat capacities are also denoted as *thermal coefficients*.

Several different temperature gradients of thermodynamic functions are encountered in the literature: the *isochoric heat capacity* or *heat capacity at constant volume*

$$C_V = T(\partial S/\partial T)_V = (\partial U/\partial T)_V = -T(\partial^2 A/\partial T^2)_V, \qquad \text{(II-2)}$$

the *isobaric heat capacity* or *heat capacity at constant pressure*

$$C_p = T(\partial S/\partial T)_p = (\partial H/\partial T)_p = -T(\partial^2 G/\partial T)_p\,, \qquad \text{(II-3)}$$

the *saturation heat capacity*

$$C_{sat} = T(\partial S/\partial T)_{sat}\,, \qquad \text{(II-4)}$$

and $(\partial H/\partial T)_{sat}$ and $(\partial U/\partial T)_{sat}$ which do not have any specific designation and are less common. The thermodynamic functions U, H, S, A, and G are internal energy, enthalpy, entropy, Helmholtz function, and Gibbs function, respectively. The subscript "sat" denotes derivatives along the saturation (*orthobaric*) curve. For simplification of the notation, we use throughout this monograph the symbols C_p, C_V, and C_{sat}; they are also used for denoting molar heat capacities (omitting subscript m). The connection of individual coefficients with entropy and enthalpy is illustrated schematically in Fig. 1. For plots of S versus the logarithm of T and of H versus T the slopes of isochores, isobars, and saturation line correspond directly to the quantities introduced above. While the heat capacities defined by equations (II-2) and (II-3) can relate to any combination of temperature and pressure, the other three temperature gradients are by definition confined to the saturation curve (T, p_{sat}). It should be stressed that C_{sat} is *not* identical to $(\partial H/\partial T)_{sat}$ as is sometimes expected on the basis of the mistaken analogy with equation (II-3).

Using well known thermodynamic relationships, one can derive conversion equations between the individual thermal coefficients (56HAA, 82ROW/SWI, 89MAJ/SVO):

$$\begin{aligned} C_p - C_V &= T(\partial p/\partial T)_V(\partial V/\partial T)_p \\ &= -T(\partial V/\partial T)_p^2/(\partial V/\partial p)_T \\ &= -T(\partial p/\partial T)_V^2/(\partial p/\partial V)_T \end{aligned} \qquad \text{(II-5)}$$

$$C_p - C_{sat} = T(\partial V/\partial T)_p/(\partial p/\partial T)_{sat} \qquad \text{(II-6)}$$

$$C_V - C_{sat} = -T(\partial p/\partial T)_V/(\partial V/\partial T)_{sat} \qquad \text{(II-7)}$$

$$C_p - (\partial H/\partial T)_{sat} = -[V - T(\partial V/\partial T)_p](\partial p/\partial T)_{sat} \qquad \text{(II-8)}$$

$$C_V - (\partial U/\partial T)_{sat} = [p - T(\partial p/\partial T)_V](\partial V/\partial T)_{sat} \qquad \text{(II-9)}$$

$$C_{sat} - (\partial H/\partial T)_{sat} = -V(\partial p/\partial T)_{sat} \qquad \text{(II-10)}$$

$$C_{sat} - (\partial U/\partial T)_{sat} = p(\partial V/\partial T)_{sat} \qquad \text{(II-11)}$$

where

$$(\partial V/\partial T)_{sat} = (\partial V/\partial T)_p + (\partial V/\partial p)_T(\partial p/\partial T)_{sat}\,. \qquad \text{(II-12)}$$

The partial derivatives on the right-hand side of equations (II-5) to (II-12) are often expressed in terms of mechanical coefficients $\alpha_p = (\partial V/\partial T)_p/V$ (*thermal expansivity*), $\beta_T = -(\partial V/\partial p)_T/V$ (*isothermal compressibility*) and $\gamma_V = (\partial p/\partial T)_V$ (*thermal pressure*) which are available for many pure liquids. The coefficient $(\partial p/\partial T)_{sat}$ is easily accessible from the temperature dependence of the vapor pressure. Below the normal boiling temperature, the term $(\partial V/\partial p)_T(\partial p/\partial T)_{sat}$ in equation (II-12) is negligible compared with $(\partial V/\partial T)_p$. Except for γ_V, all coefficients increase with

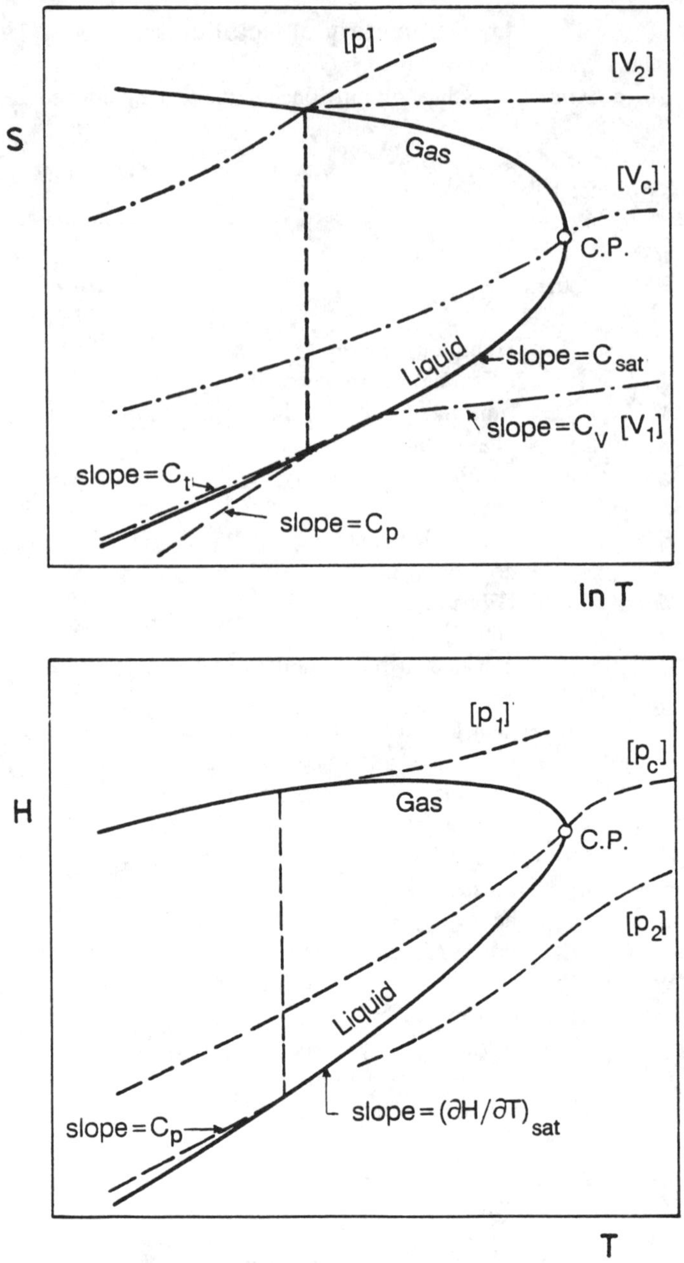

FIG. 1. Schematic plot for entropy and enthalpy of a fluid as a function of the logarithm of temperature and temperature, respectively; C_p—isobaric heat capacity, C_V—isochoric heat capacity, C_{sat}—saturation heat capacity, C_t—isochoric heat capacity of a two phase system; full lines—saturation curve, dashed lines—isobars $p_1<p_c<p_2$, dashed-dotted lines—isochores $V_1<V_c<V_2$, C.P.—critical point.

increasing temperature; α_p and β_T are equal to plus infinity at the critical point. For example, for liquid benzene at the normal boiling temperature ($T_{nb} = 353.3$ K), the values of the coefficients are as follows: $V_m = 9.59 \cdot 10^{-5}\ m^3 \cdot mol^{-1}$, $\alpha_p = 1.35 \cdot 10^{-3} K^{-1}$, $\beta_T = 1.52 \cdot 10^{-4} kPa^{-1}$, $\gamma_V = 8.9\ kPa \cdot K^{-1}$, and $(\partial p/\partial T)_{sat} = 3.13\ kPa \cdot K^{-1}$.

It holds for the saturated liquid that (82ROW/SWI):

$$(\partial H/\partial T)_{sat} > C_p > C_{sat} > (\partial U/\partial T)_{sat} > C_V. \qquad \text{(II-13)}$$

The first nonequality applies only when $T\alpha_p < 1$, which is the case for most liquids. The differences between the individual coefficients increase with increasing temperature and all five are equal to infinity at the critical point. The first four quantities usually agree below the normal boiling temperature T_{nb} to within several tenths of one percent and divergence becomes more important only above that temperature. For liquids, therefore, it is not necessary to make any clear distinction below $\sim 0.9\ T_{nb}$ between C_p at the saturation curve and C_{sat} as the differences are less than the uncertainty obtainable in high precision measurements. For instance, a difference of 0.1 percent between C_{sat} and C_p is observed for benzene at 333 K, and the same difference for n-heptane and water, is observed at 358 and 403 K ($T_{nb} = 371.6$ and 373.2 K), respectively.

The differences between C_V and the other coefficients are more important at all temperatures. For example, for liquid benzene and n-heptane, C_p and C_V differ by 40 and 50 $J \cdot K^{-1} \cdot mol^{-1}$ near their respective T_{nb}. The typical variation of C_p, C_V, and C_{sat} with temperature for n-heptane is depicted in Fig. 2. It should be stressed that the above inequality does not generally apply to the saturated gas where C_{sat} can assume negative values for small molecules. Corresponding heat capacities for gases are always lower when compared with those for liquids.

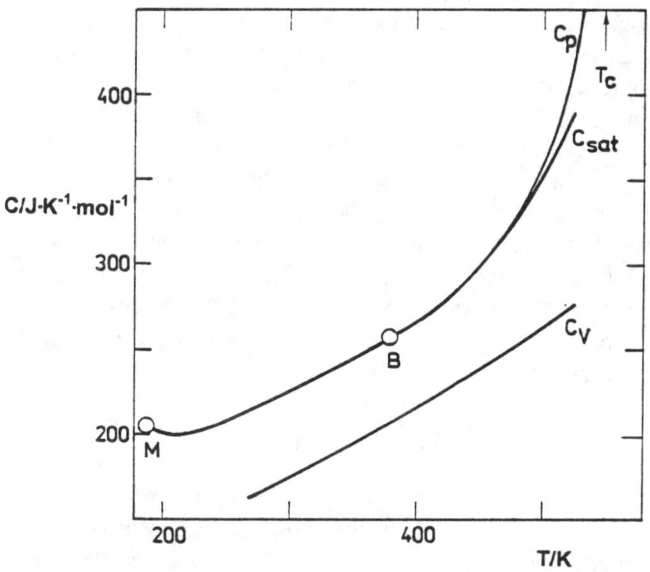

FIG. 2. Temperature variation of C_V, C_p, C_{sat} for n-heptane (according to 82ROW/SWI) M—melting temperature, B—normal boiling temperature, T_c—critical temperature, C_p—isobaric heat capacity, C_V—isochoric heat capacity, C_{sat}—saturation heat capacity.

Data on C_p and C_V are often available at conditions removed from the saturation curve. The variation of these coefficients with pressure and volume, respectively, is expressed by the differential equations:

$$\left(\frac{\partial C_p}{\partial p}\right)_T = -T\left(\frac{\partial^2 V}{\partial T^2}\right)_p \tag{II-14}$$

$$\left(\frac{\partial C_V}{\partial V}\right)_T = T\left(\frac{\partial^2 p}{\partial T^2}\right)_V \tag{II-15}$$

which can be used for the conversion of the data to saturation conditions. By integration, one obtains the relationships:

$$C_p(T, p_{sat}) - C_p(T, p) = -T\int_p^{p_{sat}} (\partial^2 V/\partial T^2)_p\, dp \tag{II-16}$$

$$C_V(T, V_{sat}) - C_V(T, V) = T\int_V^{V_{sat}} (\partial^2 p/\partial T^2)_V\, dV \tag{II-17}$$

where V_{sat} is the molar volume at T and p_{sat}. The conversion between the two states can be performed provided the pVT data for the liquid are available.

There are two principal methods for the determination of the heat capacities of liquids. The first method is an *indirect* approach and consists in the calculation from the other properties which are available experimentally. This approach utilizes the thermodynamic relationship of heat capacities to the combinations of different mechanical and adiabatic coefficients which can be obtained from pVT and speed of sound measurements. The calculations are easy, but not very accurate, and results with an error below one percent can rarely be obtained. Indirect determinations have not yielded a significant number of results because the input data are available only for a limited number of substances. This method is described by Cruickshank and coauthors in 68MCC/SCO. In compiling data for this monograph, we did not consider any heat capacities obtained by indirect methods.

The second method consists of *direct* determination by a calorimetric measurement. The thermodynamic and methodological aspects of this approach are discussed briefly below.

2.2. True and Average Heat Capacity

In most calorimetric measurements, a sample under investigation is heated or cooled and the quantity measured is the amount of exchanged heat corresponding to a finite temperature change. The *true* heat capacity is, however, a differential quantity and the experimental value obtained as the ratio of heat and temperature increment cannot be automatically considered as C in equation (II-1). It is rather the *average* heat capacity, C_{avg}, defined as the amount of heat necessary to change the temperature of a system by one kelvin. The average heat capacity is expressed by:

$$C_{avg} = \frac{Q}{(T_2 - T_1)} \tag{II-18}$$

in an interval defined by the temperatures T_1 and T_2. This value is related to the average temperature, $T_{avg} = (T_1 + T_2)/2$.

The relationship between C and C_{avg} is obtained by introduction of the integrated equation (II-1) into equation (II-18):

$$C_{avg} = \frac{1}{(T_2 - T_1)}\int_{T_1}^{T_2} C\,dT. \tag{II-19}$$

The heat capacity of a liquid always changes with temperature and C and C_{avg} are identical at T_{avg} only when heat capacity is a linear function of temperature. The two quantities approach each other when the temperature interval over which the average heat capacity is determined converges and they are equal to each other in the limit:

$$C = \lim_{T_1 \to T_2} C_{avg}. \qquad \text{(II-20)}$$

We find that the heat capacity of most liquids changes moderately with temperature without a strong curvature (except in the region of high vapor pressures) and the true heat capacity agrees in most cases with the experimental value within the error of measurement provided the temperature interval is not too large (intervals up to 10 K are usually acceptable). Sometimes it is preferable to proceed with a wider temperature increment in order to reduce the time of experiments.

When the $C = C(T)$ function is nonlinear and the temperature steps are too large, it is possible to apply a *curvature correction*. Three methods have been described in the literature (68MCC/SCO):

a. The heat capacity can be expanded around T_{avg} by means of a Taylor series using the terms up to T^3.

$$C = C(T_{avg}) + \left(\frac{\partial C}{\partial T}\right)\delta T + \left(\frac{\partial^2 C}{\partial T^2}\right)\frac{\delta T^2}{2} + \left(\frac{\partial^3 C}{\partial T^3}\right)\frac{\delta T^3}{6}, \qquad \text{(II-21)}$$

where all derivatives are taken at T_{avg} and $\delta T = T - T_{avg}$. After introduction of (II-21) into (II-19) and transformation of the integration variable we obtain:

$$C_{avg} = \frac{1}{(T_2 - T_1)} \int_{\frac{1}{2}(T_1 - T_2)}^{\frac{1}{2}(T_2 - T_1)} C \mathrm{d}(\delta T) = C(T_{avg}) + \left(\frac{\partial^2 C}{\partial T^2}\right)\frac{(T_2 - T_1)^2}{24}. \qquad \text{(II-22)}$$

The second temperature derivative of C can be obtained in an iterative way using the $C_{avg} = C_{avg}(T)$ function as the first approximation and repeating the calculation until the $C = C(T)$ values do not change.

b. The difference between C and C_{avg} is usually small and with minimal temperature dependence over the interval between T_1 and T_2. Hence:

$$(C - C_{avg})(T_2 - T_1) = \int_{T_1}^{T_2} (C - C_{avg})\mathrm{d}T =$$

$$C_{avg}(T_2 - T_1) - \int_{T_1}^{T_2} C_{avg}\, \mathrm{d}T \qquad \text{(II-23)}$$

where values of $C_{avg}(T_1)$ and $C_{avg}(T_2)$ can be obtained numerically or graphically from the $C_{avg} = C_{avg}(T)$ curve at the temperatures T_1 and T_2. This procedure can serve the experimentalist as a quick test to find out if the correction for curvature is needed.

c. If the average heat capacities are available at several temperatures and the true heat capacity as a function of temperature is described by a polynomial then by combining Eqns. (II-19) and (II-28) we obtain:

$$C_{avg} = \frac{1}{(T_2 - T_1)} \sum_{j=0}^{m} \frac{A_j}{j+1}\left(T_2^{j+1} - T_1^{j+1}\right), \qquad \text{(II-24)}$$

where parameters A_j are adjusted by least squares fitting. This method was used for correcting average heat capacity data by Wilhoit and coworkers (85WIL/CHA).

2.3. Calorimetric Determination of Heat Capacities

From the experimental point of view the individual heat capacities defined by equations (II-2) to (II-4) differ by the paths along which the heat is exchanged between the sample and its surroundings. C_V and C_p can be obtained directly by measuring with a liquid sample at constant volume and pressure, respectively. C_{sat} can be derived from measurements at constant volume using a sample in a vessel containing small vapor space. These experiments are briefly characterized below.

2.3.1. Isochoric Measurements Without Vapor Space

Isochoric measurements with liquids are difficult to carry out at temperatures far below the critical temperature where the thermal pressure coefficient, γ_V, of the liquid is large and when it is difficult to prevent any changes in the volume of the vessel. The vessel has to be constructed from a highly rigid material with a considerable wall thickness. This latter requirement has as a consequence an unfavorable ratio in the overall heat capacities of the sample and the vessel which leads to a decrease in sensitivity. The measurements at low reduced temperatures are, therefore, rare and C_V is usually calculated from the other types of heat capacity measurements via equations (II-5) and (II-7). Direct determination of isochoric heat capacities is feasible close to the critical region where γ_V is substantially smaller than that in the region remote from the critical point (61GOO).

2.3.2. Measurements at Constant Pressure

The isobaric heat capacity of a liquid can be measured directly at pressures above the saturation vapor pressure. Although the direct determination of C_p at the saturation line is not feasible, it is possible to obtain experimental values close to the conditions of vapor-liquid equilibria and to correct them by equation (II-16). Direct measurements are easy to carry out in a flow calorimeter where the liquid is pumped through the system under a pressure maintained constant by means of a controller connected in series after the instrument. This approach was recently used extensively for studying isobaric heat capacities of mixtures in a wide range of temperatures and pressures.

Determination of C_p in a batch calorimeter is much less convenient; it is necessary to conceive of a calorimetric vessel which would allow its volume to change continuously during heating in order to allow for expansion of the liquid sample under isobaric conditions. Several variations of batch calorimeters were reported in the literature; due to numerous technical difficulties, only a limited amount of data was obtained by this technique. Aqueous solutions at elevated pressures were investigated by using a cylindrical vessel closed with a sliding piston (55GUC/CHR). Another experimental approach was the placement of a flexible metallic diaphragm at the top and/or bottom of the vessel (61BRO/FOC, 55TOD); a limited change of the internal volume is permitted in this way. When expansion occurs the pressure is not quite constant and correction *via* equation (II-14) is necessary if the temperature range of measurement with one filling is important. A more dependable alternative is to accommodate the interior of the calorimeter vessel with a highly compressible capsule as a reentrant well (64TIM, 68REC1); for instance, a flexible metal bellows made of brass or phosphoric bronze can be used. In this way, the internal volume of the vessel is substantially reduced and measurements over as much as 100 K can be performed with one filling without any serious pressure increase. Temperature inhomogeneities due to insufficient mixing of the liquid can arise. Rastorguev et al. (75RAS/GRI) used a steel calorimetric vessel connected through a metallic capillary to a large volume tank filled with argon for maintaining isobaric conditions. The effective quantity of sample in an experiment was calculated by considering the amount of liquid expelled from the calorimeter and the thermal expansion of the calorimeter vessel. A similar approach was used by Coxam and coworkers (91COX/QUI).

2.3.3. Isochoric Measurements in a Two-Phase System

Examination of the plot of the entropy versus the logarithm of temperature (Fig. 1) shows that the lowest dashed-dotted line represents an entropic isochore corresponding to a subcritical molar volume. The slope of this line on the right of the intersection with the liquid saturation curve represents C_V of a liquid and on the left of the intersection represents the isochoric heat capacity of a two-phase equilibrium system C_t for which the volume of the vapor space becomes zero when the isochore crosses the saturation curve. It is clear that except for the vicinity of the critical point, C_{sat} for the liquid must be closely related to the isochoric heat capacity of this two-phase system, with the relative mass of vapor being much smaller compared with that of the liquid. This is schematically illustrated in the entropy diagram by the proximity of the two-phase isochore and the saturation curve. From the experimental point of view, it means that the saturation heat capacity is accessible through a calorimetric measurement with a vessel filled under vacuum with a liquid which fills most of the inner space and is in equilibrium with its vapor. Then, only a small correction has to be applied in order to obtain the true value of C_{sat}. This experiment is also the most effective way to determine C_p at the saturation line (correction using equation (II-6) has to be applied when necessary). Isochoric measurement with a two-phase system is much easier to perform compared with isochoric or isobaric measurements with the liquid phase only. Most heat capacities for pure liquid substances reported in the literature were determined by this procedure. The overall heat capacity of such a two-phase system C_t consists of the contribution from the saturation heat capacities of the liquid and vapor and further reflects the heat absorbed or evolved during vaporization or condensation inside the vessel. When the temperature increases, the liquid expands and the mass transfer between the coexisting phases is necessary for maintaining equilibrium conditions. Typically, the amount of substance in the vapor space increases first, attains a maximum, and then decreases finally towards zero (under the condition that the vessel was filled in such a way that there is no vapor space at the maximum temperature).

The relationship between C_{sat} and C_t can be derived as follows (46HOG, 82ROW/SWI). The differential of the entropy of a two-phase equilibrium system S_t containing N moles of a sample can be written as:

$$\mathrm{d}S_t = N\mathrm{d}S_m^l + \mathrm{d}\left[N^g(S_m^g - S_m^l)\right]. \qquad \text{(II-25)}$$

S_m^g and S_m^l are the molar entropies of the coexisting phases and N^g is the amount of substance in the vapor space which can be expressed from the total vessel volume V and molar volumes of the coexisting phases as $N^g = (V - NV_m^l)/(V_m^g - V_m^l)$. By taking the temperature derivative of equation (II-25), the saturation heat capacity of the liquid can be expressed as:

$$\frac{C_t}{N} = C_{sat} + \frac{T}{N}\left[\frac{\partial\left(\frac{\Delta S_v}{\Delta V_v}[V - NV_m^l]\right)}{\partial T}\right]_{sat} \qquad \text{(II-26)}$$

where $\Delta S_v = S_m^g - S_m^l$ and $\Delta V_v = V_m^g - V_m^l$ are the molar entropy and molar volume of vaporization, respectively. Equation (II-26) can be rearranged with use of the Clapeyron equation $(\mathrm{d}p/\mathrm{d}T)_{sat} = \Delta S_v/\Delta V_v$ to a more convenient form:

$$C_{sat} = \frac{C_t}{N} + T\left(\frac{\partial p}{\partial T}\right)_{sat}\left(\frac{\partial V_m^l}{\partial T}\right)_{sat} - T\left(\frac{V}{N} - V_m^l\right)\left(\frac{\partial^2 p}{\partial T^2}\right)_{sat}. \qquad \text{(II-27)}$$

The correction on the right hand side of this relationship can be easily calculated from the vapor pressure equation and pVT properties of the liquid (except when in or near the critical region, $(\partial V_m^l/\partial T)_{sat}$ is close to the isobaric thermal expansivity α_p – see equation (II-12)). The temperature variation of the two terms of equation (II-27) is illustrated for benzene in Table 1. The first term must always be positive and increase with temperature; the second term has the opposite sign as the vapor pressure is a convex function of temperature and $V/N \geqslant V_m^l$. The variation of the latter term with temperature is more complex in that near and below the normal boiling temperature its absolute value increases with temperature when the vapor space volume is appreciable. Thereafter, it passes through a maximum, and finally drops rapidly towards zero when the vapor space vanishes. In this limit, equation (II-27) becomes identical with equation (II-7). If the vessel is completely filled with a liquid near the upper temperature limit there is typically a temperature at which the correction terms compensate and $C_t = C_{sat}$. Table 1 shows that at temperatures below T_{nb} the overall correction never exceeds 0.1 percent and it should be considered only for high precision measurements. It can play, however, quite an important role during high pressure measurements at saturation conditions.

TABLE 1. Correction terms in equation (II-27) for benzene; $V_{vap} = 100(V - NV_m^1)/V$ denotes the percentage of the vessel volume occupied by the vapor, $I = T(\partial p/\partial T)_{sat}(\partial V_m^1/\partial T)_{sat}$, $II = -T(V/N - V_m^1)(\partial^2 p/\partial T^2)_{sat}$, $Corr = 100(C_{sat} - C_t/N)/C_{sat}$ denotes the overall percentage difference between C_{sat} and C_t/N

T / K	V_{vap} / %	C_{sat}[a]	I	II	Corr / %
		J · K^{-1} · mol^{-1}			
278. 15[b]	11.4	131.88	0.007	–0.035	–0.02
293. 15	9.8	134.87	0.015	–0.054	–0.03
313. 15	7.6	139.27	0.036	–0.079	–0.03
333. 15	5.1	144.22	0.075	–0.088	–0.01
353. 15	2.7	149.31	0.139	–0.069	0.05

[a] recommended values of C_{sat} from this compilation
[b] extrapolation outside the temperature interval of experimental data using the quasipolynomial equation (V-12).

2.4. Variation of Heat Capacity With Temperature, Correlation Equations

Thermodynamics is limited in the exact information it provides to the temperature dependence of the heat capacity for liquids; the examples of $C=C(T)$ curves are shown in Fig. 3. Typically, the heat capacity is considered to be an increasing function of temperature though there are some substances for which the heat capacity can decrease over a certain temperature range. Heat capacities, however, always increase progressively when approaching the critical region where they are unbounded. Isobaric measurements performed at supercritical pressures by flow calorimetry indicate sharp maxima on the heat capacity curve above the critical temperature. With increasing pressure, these peaks for the supercritical fluid become less pronounced and shift towards higher temperatures.

For many substances, the heat capacity curves are flat above the solid-liquid phase transition and some of them exhibit a shallow minimum in the region above their melting temperature. This is the case for many simple inorganic compounds (e.g. AsH_3, $COCl_2$, SOF_2, PH_3, SO_2, H_2O, halogens) and some organic compounds having up to five carbon atoms in the molecule (e.g., alkenes, halogenated alkanes and alkenes, oxygenated heterocyclic compounds, sulfides, thiols). This minimum is attributed to changes in the structure of the liquid near melting conditions and is not usually pronounced. At higher temperatures, a typical monotonic increase of heat capacity is observed for most of these substance.

The temperature variation of the heat capacity can become rather complex for substances exhibiting some ordering in the liquid phase (hydrogen bonded compounds). The degree of association usually decreases with increasing temperature and this association is reflected in the shape of the heat capacity curve which can have an S form with an inflection point (90ZAB/RUZ). For 1-alkanols with more than two carbon atoms and some of their branched isomers, the second temperature derivative of the heat capacity is positive above the triple point (C curve is convex), it becomes negative in the middle part of the temperature interval (curve is concave) and finally must again become positive when the critical point is approached.

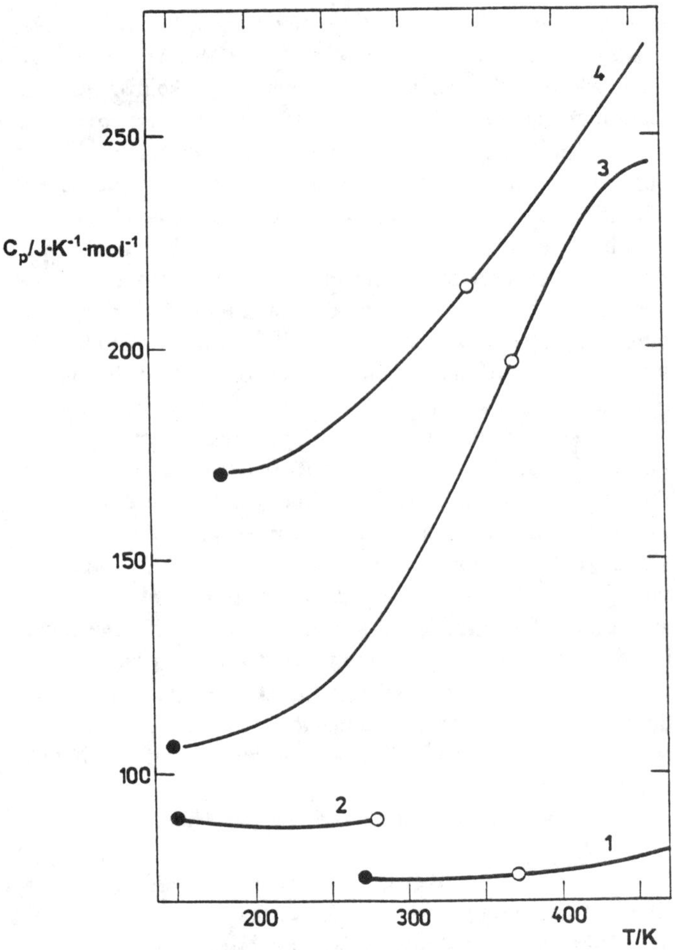

FIG. 3. Typical heat capacity curves for liquids: 1—water, temperature dependence typical for metals and simple inorganic substances, 2—methanethiol, 3—1-propanol, variation typical for 1-alkanols C_3 and higher, 4—n-hexane ● triple point temperature, ○ normal boiling temperature.

For several compounds, discontinuities (called λ-peak because of their shape) have been observed on the heat capacity curve. Such behavior was reported for instance for sulfur at 435.2 K (54BRA/MOL, 59WES1, 73KOM/MIL) and for several aldehydes (acetaldehyde at 241.9 K, 88LEB/VAS; propanal at 243 and 286 K, 77KOR/VAS; butanal at 198.2 and 284.8 K, 89VAS/LEB; hexanal at 243 K and 298 K, 91VAS/BYK). These transitions, however, did not seem to be the typical second-order phase transitions as heat exchange was observed. Little attention was paid to the explanation of these anomalies in the literature; the above cases of atypical heat capacity curves were reviewed and discussed by Zábranský at al. (93ZAB/BUR).

Empirical equations are used for correlating experimental heat capacity data. Two main criteria should be considered in selecting a relationship for fitting such experimental data:

1. The ability to describe reasonably the experimental data over the whole temperature range with a relatively small number of parameters is required.

2. The possibility of a meaningful extrapolation outside the temperature limits of experimental data is desired.

A *polynomial* seems to be the most dependable expression to describe heat capacity as a function of temperature and has been used most often. The general form of a polynomial expression is:

$$C = \sum_{j=0}^{m} A_{j+1} T^j, \qquad \text{(II-28)}$$

where A_j are adjustable parameters and, usually, $3 \geqslant m \geqslant 1$. However, for some compounds up to eight parameters would have to be used for a satisfactory description of C over the whole temperature range where experimental data are available. A substantial drawback of a high-degree polynomial is the possibility of oscillations.

A reasonable alternative is the use of cubic splines which prove to be valuable in describing complex functions in a number of applications. Cubic splines are used in this monograph for fitting the data for some compounds. They are as flexible as high-degree polynomials without having the risk of numerical instabilities.

The principle of interpolation by cubic splines is as follows. The overall interval of the independent variable (temperature) $<T_{\min}, T_{\max}>$ is divided into k-1 subintervals which are delimited by temperatures $T_{\min} = T_1, \ldots, T_j, \ldots, T_k = T_{\max}$ called *knots*. The heat capacity as a function of temperature is approximated between the neighboring knots by cubic polynomials in temperature. The basic property of the splines is that the j-1th and the jth polynomials must have at the temperature T_j not only the same values with respect to heat capacity but also must have identical values for first and second temperature derivatives. As the overall temperature range is described by k-1 cubic polynomials (each having four parameters) and the above three constraints apply to k-2 knot temperatures T_j, there are $4(k-1) - 3(k-2) = k + 2$ independent adjustable parameters. They are expressed as heat capacity values at k knot temperatures and two further constraints, typically taken as boundary conditions at T_1 and T_k. The end conditions that require the second derivative to be zero are the most common (*natural* spline).

When using cubic splines for smoothing experimental data, the knot temperatures have to be chosen first and then the heat capacity values at the respective knots are adjusted by a spline correlation routine in order to minimize the sum for squares of deviations between experimental and calculated values. The flexibility can be increased by considering the boundary conditions as two additional adjustable parameters instead of using the natural spline. It was shown (88MAJ/GAT) that spline fitting with an adjustable second-derivative end condition is equivalent to adding an arbitrary knot beyond both end knots and fitting the data with a natural boundary condition. This simplifies substantially the minimization algorithm. More specific information regarding use of splines in this monograph can be found in Section 5.3.1.

The cubic splines are very flexible when describing the temperature dependence of the heat capacity inside the limits of the temperature range of experimental data, i.e. they are excellent for interpolation. However, as with conventional polynomials, the cubic splines cannot be used for meaningful extrapolation. Thus, other formulations were used to provide reasonable extrapolations, especially towards the critical point. Engineering Science Data Unit, ESDU, Great Britain represents in its publications the heat capacities along the saturation line by expansions of the type (79ENG, 86ENG2):

$$C = \sum_{j=-1}^{2} A_j \xi^j, \qquad \text{(II-29)}$$

$$C = \sum_{j=-2}^{1} A_j \xi^{j/3}, \qquad \text{(II-30)}$$

where $\xi = (1 - T/T_c)$. Although the C values obtained from these relationships are always infinite at the critical point, there is no assurance against unrealistic behavior between the upper temperature limit of experimental data and T_c.

In this project we have used a modified version of a new type of relationship proposed by Zábranský et al. (90ZAB/RUZ) which is called the *quasi-polynomial equation*. The modified quasi-polynomial equation, even though it contains less adjustable parameters, is more flexible than the original relationship. Contrary to the original quasi-polynomial equation, the modified equation gives a positive second derivative of heat capacity with respect to temperature for the entire range from the lower temperature limit to the critical temperature. A flat area on the heat capacity curve $C = C(T)$ located above the upper limit of experimental data, that was observed for some compounds when using the original equation, has not been encountered for the modified equation.

The derivation of a quasi-polynomial equation was based on the assumption that the heat capacity is always an increasing function of temperature. As attention was focused on the extrapolation capability towards high temperatures rather than on a perfect fit of experimental data, the possible presence of a minimum on the heat capacity curve was ignored. Let us suppose that the first derivative of the heat capacity with respect to the variable $T_r = T/T_c$ has the form:

$$\frac{dC}{dT_r} = R\left(\frac{a_2}{1-T_r} + \sum_{i=0}^{m_1} a_{i+3} T_r^i\right)^2. \qquad \text{(II-31)}$$

The expression on the right-hand side of this equation is always positive (C must increase with increasing T_r) and can be integrated to obtain:

$$\frac{C}{R} = A_1 \ln(1-T_r) + \frac{A_2}{1-T_r} + \sum_{j=0}^{m} A_{j+3} T_r^j, \qquad \text{(II-32)}$$

A_j and a_i are adjustable parameters; A_j's are automatically constrained by equation (II-31) in such a way that the temperature derivative of the heat capacity must always be positive and equal to plus infinity at the critical temperature. When m_1 is equal to 0, then $m = 1$, and three parameters are independent; when m_1 is equal to 1, then $m = 3$, and four parameters are independent. More specific information on the use of this correlation equation can be found in Section 5.3.2.

3. Calorimetric Techniques for Determining Heat Capacities of Liquids

Calorimetry is the most frequently used method for determining heat capacities of substances in the condensed state. Several publications describe heat capacity calorimeters and attempt to introduce some system of classification. A general overview of calorimetric techniques for investigation of non-reacting systems is given in an IUPAC monograph edited by McCullough and Scott (68MCC/SCO) where the emphasis is on heat capacity measurements. The review articles by Wadsö (70WAD, 85WAD) present microcalorimetric techniques and those by Lakshmikumar and Gopal (81LAK/GOP, 82LAK/GOP) describe new modern trends in determination of heat capacities; low-temperature calorimetry was reviewed by Gmelin (79GME, 87GME). Heat capacities can be determined in a variety of calorimeters which are not specifically designed for this purpose. The majority of heat capacity calorimeters for measurements with liquids can be used also for determination with solids. On the other hand, only a small number of instruments (flow calorimeters) permit measurements with both liquids and gases (or vapors).

3.1. Terminology and Criteria for Classification of Calorimeters

There is no officially established nomenclature for describing different types of calorimeters; an attempt was made by Rouquerol and Zielenkiewicz (86ROU/ZIE) to introduce a classification of calorimeters based on heat exchange considerations, paying special attention to the dynamic behavior of calorimeters. A somewhat different terminology used in this monograph is described in this section.

That part of the calorimeter which accommodates the sample will be called the *calorimeter vessel* or *cell*. In most cases, it also contains a temperature sensing element and a resistance heater which serves either in the measurements directly or during calibration. Adjacent parts of the calorimeter which surround the vessel (jackets, shields, calorimeter block etc.) will be called the *surroundings*. The heat transfer between these two parts is, in most cases, proportional to the difference in average temperatures of the heat exchanging surfaces; they are denoted below as T_{cv} (corresponding to the calorimeter vessel) and T_{su} (corresponding to the surroundings). Heat capacity determination is usually based on recording a temperature change connected with either heating or cooling of the sample. This change can be either *intermittent* (determination of temperature increment ΔT_{cv} corresponding to the exchange of a certain finite amount of heat Q) or *continuous* (measuring the heating rate $dT_{cv}/d\tau$ corresponding to the power input $P = dQ/d\tau$). Highly automated instruments which are capable of performing measurements over a wide temperature range are called *scanning* calorimeters. For intermittent heating, the heat capacity of the calorimetric vessel with its contents C_{cv} is obtained from the relationship:

$$Q = C_{cv}\,\Delta T_{cv} + \int L\,d\tau \qquad \text{(III-1)}$$

where τ denotes the heating period, and for continuous heating, from the equation:

$$P = C_{cv}\,(dT_{cv}/d\tau) + L. \qquad \text{(III-2)}$$

The first term on the right hand side of the equations expresses the accumulation of heat in the vessel and the second term corresponds to the unaccounted heat transfer between the vessel and its surroundings. The *heat loss (leak)* L stands for the amount of energy lost per unit time. In its simplest form, the equation can be expressed as:

$$L = k_{ht}(T_{cv} - T_{su}), \qquad \text{(III-3)}$$

where the *heat transfer coefficient* k_{ht} quantifies the heat exchange across the space separating the vessel from its surroundings. In some cases, this expression can become complex because terms other than $T_{cv} - T_{su}$ can also govern the heat losses (see Section 3.2.3).

Measurements are considered as *absolute* when the amount of energy exchanged with the sample is quantified, or *comparative* (*relative*) when *calibration* is necessary for determining the value of heat capacity. Calibration can be either *chemical* using a *reference substance* of well-known heat capacity or can be *electrical* using the Joule effect to determine the response of the instruments to the introduction of an exactly known amount of heat. Corrections for heat losses are established by specific procedures that depend on the design of the instrument and the technique adopted for measuring heat capacity.

Heat capacity calorimeters can be classified according to four criteria:

1. *motion of the sample inside the instrument* (batch, flow, and drop calorimeters)
2. *mode of measurement* (stationary or dynamic conditions)
3. *temperature relationship between the calorimeter vessel and its surroundings – mode of heat transfer* (adiabatic, isoperibol, isothermal, and conduction calorimeters)
4. *number of calorimeter vessels* (single vessel and twin vessel calorimeters)

TABLE 2. Criteria codes used for classification of calorimeters

Motion of sample 1		Mode of measurement 2		Temperature relation of vessel to surroundings 3		Number of vessels 4	
batch	B	stationary	S	adiabatic	A	one vessel	O
flow	F	dynamic	D	isoperibol	I	two vessels	T
drop	D			isothermal	T		
				conduction	H		
				thermopile conduction	C		

Table 2 lists individual criteria with the codes used for characterizing calorimeters in Tables 3 and 4 and in tables with experimental values (see Section 6.3.1.). In the subsequent text, the above characteristics are briefly described. It should be kept in mind that the codes serve for a concise categorization of instruments and the adopted system of classification does not pretend to be exhaustive.

TABLE 3. Types of calorimeters used for determining heat capacity of pure liquids

Type of calorimeter	Classification 1	2	3	4	Section
Adiabatic batch calorimeters with intermittent heating	B	S	A	O	3.2.1.
Adiabatic batch calorimeters with continuous heating	B	D	A	O	3.2.1.
Isoperibol batch calorimeters	B	S	I	O/T	3.2.2.
Isoperibol flow calorimeters	F	S	I	O/T	3.2.3.
Isoperibol drop calorimeters	D	S	I	O	3.2.4.
Adiabatic drop calorimeters	D	S	A	O	3.2.4.
Isothermal drop (phase change) calorimeters	D	S	T	O	3.2.4.
Thermopile conduction drop calorimeters	D	D	C	T	3.2.4.
Measurement of heating and/or cooling curves	B	D	H	O	3.2.5.
Differential heat conduction calorimeters (TA, DTA, DSC)	B	D	H	T	3.2.6.
Differential thermopile-conduction calorimeters	B	D	C	T	3.2.6.
Special dynamic methods using pulse heating	B	D	H	O	3.2.7.

TABLE 4. Survey of calorimeters for determining heat capacity of liquids

Reference	Temp. range	Measur. accuracy	Note
Adiabatic batch calorimeters with intermittent heating			
20COH/MOE	room	medium	
24WIL/DAN	300-360 K	medium	
30SOU/AND	low	0.50 %	
33SOU/BRI	low	medium	
35AOY/KAN	low	medium	
35SAG/LAC	294-570 K	low	
37OSB/STI	373-647 K	high	also for heat of vaporization
38EUC/SCH	low	medium	
39AST/EID1	low	high	
39OSB/STI	273-373 K	high	constructed for measurement of water only
39SAG/EVA	300-370 K	2.00 %	high pressure, also for heat of vaporization
40AWB/GRI	low	medium	
41YOS/GAR	low	0.20 %	
43RUE/HUF	low	0.10 %	
44BAI/TOD	80-340 K	1.00 %	
45GUT/PIT	low	0.20 %	
45SCO/MEY	low	0.10 %	
47AST/SZA	low	high	measurement error 0.1 and 0.2 % below and above 200 K, respectively
47HUF	low	0.10 %	detailed description of improvements in calorimeter, 43RUE/HUF
47OSB/GIN	medium	0.10 %	also for heat of vaporization
47SKU	room	0.30 %	

TABLE 4. Survey of calorimeters for determining heat capacity of liquids — (Continued)

Reference	Temp. range	Measur. accuracy	Note
49HOR/STO	300-500 K	0.50 %	
49STA/GUP	low	medium	
50SAG/HOU	medium and high	1.00 %	high pressure (up to 6.9 MPa)
51AST/FIN	15-300 K	0.20%	
51DOL/HET	253-573 K	low	
51EUC/EIG	medium	medium	
52ADA/JOH	low and medium	high	
53WES/HAT	250-550 K	0.15 %	
54BRA/MOE	medium and high	0.50 %	
54STR/ICK	low	0.30 %	
55DAU/MAR	20- 330 K	0.20 %	
55MAR/SMI	253-553 K	0.25 %	measurement error of 0.5 % near upper temperature limit
55PAC/PIE	low	0.10 %	also for heat of adsorption
55STA/TUP	283-373 K	1.00 %	
56COO/BAL	room	high	also for heat of solution
56POP/KOL	60-300 K	0.50 %	
57PIL	low and medium	medium	
58HIL/KRA	low	0.30 %	
58WES/GIN	303-773 K	0.10 %	
59COX/SMI	273-303 K	low	
59ONK	293-373 K	high	
60BRA/MYE	low	medium	
61FLU/LEA	2-300 K	0.50 %	also for heat of vaporization
61GOO	low	high	high pressure
61ROU	283-323 K	medium	
62KOL/SER	12-340 K	0.50 %	
63AND/COU1	low	high	
63BEN	273-343 K	high	
63FEH/SEY	medium	2.00 %	
63VAR/KOP	medium and high	1.50 %	
63ZIE/MUL	20-300 K	0.50 %	
64ARN1	293-453 K	0.30 %	
65STE/BLA	10-360 K	0.10 %	
65SUG/SEK	15-310 K	0.30 %	
66DWO/GUI	low	2.00 %	
66FIE	low	medium	
66KLE	293-343 K	high	
66NIK/LEB	60-300 K	0.30 %	
66SHI/ATA	low	high	
67AND/COU	300-450 K	medium	
67GRO	293-1048 K	0.30 %	
67RAS/GAN	298-473 K	0.50 %	
68BAG/KUC	2-300 K	1.00 %	measurement error: 4 % at 2.6 K, 2 % at 4 K, 1 % at higher temperatures
68CLE/MEL	80-320 K	high	
68LEA	298-773 K	high	measurement error 0.1 and 0.2 % below and above 673 K, respectively
68REC1	room	medium	
68WES/FUR	low	high	
68WES/WES	300-800 K	high	
69PAU/LAV2	low	high	
71GOP/GAM	303-343 K	1.00 %	
71MAY/WAL	77-400 K	medium	
72VAN	123-373 K	high	
73SHI/ENO	low	high	

TABLE 4. Survey of calorimeters for determining heat capacity of liquids — (Continued)

Reference	Temp. range	Measur. accuracy	Note
74ATA/CHI	2-300 K	high	
74DIA/REN	medium	0.30 %	
74MOS/MOU	273-346 K	low	
75MON	12-433 K	0.20 %	
76DWO/FIG	low	medium	
76LEB/LIT	5-340 K	0.30 %	measurement error: 3-5 % below 20 K, 0.5-1 % between 20-60 K, 0.15-0.3 % above 60 K
77KU /COM	low	1.00 %	
77NAZ/MUS	293-620 K	1.50 %	high pressure (up to 50 MPa)
77VOR/PRI	room	high	
78SOL/SUK	123-283 K	medium	also for heat of vaporization
79AMI/LEB	5-300 K	high	
79SCH/OFF	90-350 K	high	
79VES/ZAB	293-318 K	0.50 %	
79ZHO/KOS	9-300 K	0.40 %	
80GUR/GAV	5-300 K	0.20 %	measurement error: 1.5 % below 12 K, 0.5 % between 12-30 K and 0.2 % above 30 K
80KAL/JED	90-300 K	0.10 %	
80SHA/LYU	15-330 K	0.50 %	
80VAS/TRE	313-623 K	low	
82KAR/IGA	medium	0.50 %	
82OGU/WAT	100-370 K	0.50 %	high pressure (up to 110 MPa)
83ANI/VOR	298-393 K	medium	
83KUK/KOR	263-353 K	0.20 %	
83TAN/ZHO	medium	high	
83YOS/SOR1	15-390 K	0.20 %	
84POD/RAC	medium	medium	
85BOU/DEL	medium and high	medium	
85RAB/SHE	low	0.30 %	measurement error: 1.8 % below 10 K, 0.8 % between 10-25 K, 0.3 % above 30 K
86DEV/GUS	low	0.20 %	measurement error: 2 % between 5-25 K, 0.5 % between 25-50 K, 0.2 % above 50 K
87KHO/BUG	low	1.00 %	soviet commercial calorimeter produced at VNIIFTRI
87VAN/VAN	10-350 K	0.20 %	
87ZAB/HYN	290-373 K	0.50 %	
88ZHA/ZOU	medium	medium	
89KHO/PUL	low	medium	
89KON/STR	low	medium	
91SVO/ZAB1	293-353 K	0.30 %	also for heat of vaporization
Adiabatic batch calorimeters with continuous heating			
31DEE	low	medium	
51POP/GAL	300-1000 K	medium	64VAS2203-423 K0.50 %
65FIN/GRU	room	0.40 %	
71MUS	293-773 K	2.50 %	high pressure (up to 50 MPa)
75RAS/GRI	medium and high	1.00 %	high pressure
87OKH/RAZ	medium	0.30 %	
88LUS/RUB	14-300 K	0.05 %	
Isoperibol batch calorimeters			
07BAT	low	low	
09SCH	medium	low	
12SCH1	medium	low	
14GRI	medium	low	
16EUC	low	low	
19DEJ	medium	low	

TABLE 4. Survey of calorimeters for determining heat capacity of liquids — (Continued)

Reference	Temp. range	Measur. accuracy	Note
20GIB/LAT	low and medium	low	
21JAE/VON	room	medium	
23SIM/LAN	low	medium	
24EUC/KAR	low	0.60 %	
24KEY/BEA	low and medium	1.00 %	also for heat of vaporization
25DRU/WEI	room	low	
25PAR	low	0.50 %	
26DAN/JEN	low	2.00 %	
27FER	medium	low	
28EUC/HAU	low	low	
28GIA/WIE1	low	0.20 %	
28LAT/GRE	low	medium	
29CLU1	low	medium	calorimeter vessels, one glass and one metal
29KEL1	low	medium	
29MIT/HAR1	low	medium	
30CAR/STO	303-773 K	medium	
30WIE/HUB	low	medium	
31BLA/LEI	medium and high	3.00 %	
31FIO/GIN	medium and high	high	also for heat of vaporization
31FOR/BRU	medium and high	low	
32NEU	medium	medium	
33LEB/MOE	room	medium	
33POH/MEH	medium	low	
34KOL/UDO2	room	low	differential instrument
35JAC	medium	medium	
35KEE/KEE	low	medium	
36AST/MES	low	medium	
36CLU/GOL	low	medium	
36GIA/STO	low	0.20 %	
36PEA/BAK	low	medium	
37ELL	medium	medium	
37GIA/EGA	low	high	measurement error: 3 % at 15 K, 1 % at 20 K, 0.2 % above 35 K
37KAN	low	medium	
37PER1	low	0.40 %	
37VOL	room	low	
38HIC	low	0.30 %	
38KAI	low	0.40 %	
39BYK	298-305 K	low	differential instrument
39MAZ3	low and medium	low	
39RIE3	low	1.00 %	
40BEN/MCH	medium	low	
40CLU/POP	low	medium	
41GIA/MEA	low	high	
41NEL/NEW	273-333 K	0.50 %	
43BAC/PER	medium	medium	
48TSC1	room	low	also for heat of mixing
49TSC/RIC1	room	medium	also for heat of mixing
49WEI	room	low	
50JOH/CLA	low	high	
50URA/SID	medium	1.00 %	
51HUL/WIL	low	0.50 %	
51PAR/SIM	low	medium	adiabatic conditions below 190 K
52STA/AMI	medium	low	also for heat of vaporization
53LLE	medium	1.00 %	
53RIF/KER	low	medium	high pressure (up to 3.5 MPa)

TABLE 4. Survey of calorimeters for determining heat capacity of liquids — (Continued)

Reference	Temp. range	Measur. accuracy	Note
54OSB/ABR	low	5.00 %	
54ROB/SYD	low	high	
55HUT/MAN	297-299 K	medium	
55RUI1	room	medium	also for heat of mixing
55TAY/JOH	low	0.30 %	
56MUR/VAN	room	2.35 %	
57CRU/JOS	medium	2.00 %	
57HAR/MOE	medium	0.50 %	
57HIL/LOU	low	1.00 %	
57KEN	medium	1.00 %	also for heat of mixing
58CHE	low and medium	low	
59BAK	300-353 K	high	
60BAR/BOL	293-533 K	5.00 %	
61EGA/LUF	room	medium	
62KAT	medium	medium	
64KRO/VEN	medium	medium	
64MOE/THO	293-313 K	low	
64RAS/BAS	medium	1.50 %	
66DRA/LAN	room	medium	
66SAV	293-333 K	1.20 %	
67GRA	303-338 K	1.00 %	
69STO/MAR	room	low	for heat of mixing
69TOM/LIN	room	low	
70LKB/COM	278-333 K	medium	
70REC	room	medium	
75PED/KAY	medium	1.00 %	
76BON/CER	room	medium	
76GIE/WOL	low	3.00 %	
78RYB/EME	medium	low	
79CZA	room	low	high pressure (up to 1000 MPa)
80FUC	293-300 K	0.50 %	
86KAU/HEI	room	low	
88ROD/MAR	room	medium	commercial instrument Tronac model 458
89VAS/NOV	medium	low	
Isoperibol flow calorimeters			
23JEN/SHO	medium	medium	
28LAN	medium	medium	
31FOR/BRU	high	2.00 %	
59RIB/EGO	293-573 K	medium	high pressure (up to 25 MPa)
65KAU/BIT	293-350 K	1.00 %	
68AHL/YOU	medium	2.00 %	heat exchanger (cooling water)
71PIC/LED	room	medium	diff. instrument, comm. version SETARAM, SODEV
75SAF/GER	medium and high	0.60 %	high pressure (up to 50 MPa)
75SAN	293-573 K	0.60 %	high pressure (up to 5 MPa)
79VAN/ZEL1	100-300 K	1.00 %	high pressure (up to 7 MPa)
83GOR/SIM	medium	medium	
85OGA	room	0.10 %	
87LAN/CRI	high	medium	
Isoperibol drop calorimeters			
*79BER	medium	low	
*81VON	medium	medium	
*85STO/WIL	medium	low	
*86LUD	medium	low	
*86SCH	medium	medium	
*92STO	medium	low	
*98LOU	medium	medium	
01KAH	medium	low	
07GUI1	high	low	
07GUI2	high	medium	
07WAL	room	low	pure solid metal dropped into liquid sample
08BOG/WIN	medium	low	
09GOO/KAL	high	medium	
11LEW/RAN	293-663 K	low	
11POM	medium	low	
12LUS	medium	low	
13REN	288-373 K	low	
18NAR	medium	low	
20MAA/HAT	90-298 K	2.00 %	
22HER/SCH	room	low	
24GAR/RAN	medium	low	
26AND/LYN	low and medium	low	
26AUE	medium	low	
26AWB/GRI	high	low	
26MON	290-553 K	low	
26UMI	373-1523 K	low	
30WAS1	medium	2.00 %	
33ROT/MEY2	273-628 K	medium	
36KHO/KAL	298-423 K	low	
36KUR/VOS	room	low	
36NEG	medium	medium	
38FRE/HIL1	medium	1.00 %	
40KUB/SCH	high	low	
42YOU/HIL	medium	medium	
45DAV/WIE	medium	low	
47KUR	medium	low	estimated measurement error about 5 %
47PUS/FED	room	low	
50EGA/WAK	high	medium	
56SCH/HIL	high	low	
56WAL/GRA	high	medium	
58SWI/ZIE1	medium	low	
62STR/BAR	298-323 K	low	
63GAT/KRE	low and medium	low	
64CAM/NAG	medium and high	medium	
64EDI/CUB	high	medium	
69SAD/STE	medium and high	low	
69SMI/MAT	298-373 K	medium	
71BEL/HUL	high	medium	pure solid metal dropped into the same liquid metal
71MAR/CIO	300-1300 K	low	
76LEB/RYA	298-473 K	1.00 %	
Adiabatic drop calorimeters			
30BAR/MAA	medium	1.00 %	
62LEV	1273-2773 K	0.70 %	
Isoperibol drop (phase change) calorimeters			
05DIE	273-573 K	medium	Bunsen ice calorimeter
06BER	273-430 K	low	Bunsen ice calorimeter
16BRA	medium	0.40 %	Bunsen ice calorimeter
28FIS/BIL	medium and high	low	Bunsen ice calorimeter
35BAR/CLU	medium	low	Bunsen ice calorimeter
50GIN/DOU	273-1173 K	0.35 %	Bunsen ice calorimeter
51FOL/GIG	room	0.25 %	Bunsen ice calorimeter
52RED/LON	high	0.50 %	Bunsen ice calorimeter
55GIG/MOR1	medium	medium	diphenyl ether calorimeter

TABLE 4. Survey of calorimeters for determining heat capacity of liquids — (Continued)

Reference	Temp. range	Measur. accuracy	Note
Thermopile conduction drop calorimeters			
71KON/SUU	273-343 K	0.10 %	
74SUU/WAD	273-343 K	high	
85WAD	medium	high	
Measurement of heating and/or cooling curve			
*67HIR	medium and high	low	
*90PIC	medium	low	
24SHO	233-313 K	low	
26AND	low and medium	2.00 %	
31THO/PAR	293-773 K	low	
33FER/MIL	293-323 K	1.00 %	
33STR/MAL	medium	low	
37STU	90-320 K	1.00 %	
40COC/FER	room	low	
40TUR/BAR	298-1273 K	low	
41ZIE/MES	medium	2.00 %	
49WUY/JUN	medium	medium	
50KUS/CRO	medium	1.50 %	dielectric constant measured simultaneously
50YAG/UNT	high	10.0 %	
52HOF	273-523 K	1.00 %	
55SEK/MOM	medium	medium	
58LUT/PAN	323-380 K	0.70 %	
59ABA/MUS	room	low	
64VUK/RAS	medium	3.00 %	
73KOM/MIL	373-903 K	high	
79TAK/YOK	80-1100 K	1.00 %	
81ATA/ELS	room	low	
84FIL/LAU	medium	2.00 %	
86NAZ/BAS1	303-523 K	2.00 %	high pressure (up to 50 MPa)
Differential heat conduction calorimeters (TA, DTA, DSC)			
31SWI/RYB2	room	0.50 %	
38KRE	293-453 K	low	
58BRA/WAT	low	low	
59BEN/THO	room	low	
63GUD/CAM	medium	low	
65GOD/BAR	80-623 K	2.50 %	
66PER/COM	340-510 K	1.00 %	
68OST/DOB	290-473 K	medium	
69PER/COM	medium and high	low	measurement error 1 % above 200 K
71DU /COM	room	low	
73PER/COM	medium and high	low	
74DSM/COM	173-773 K	1.00 %	
75CUC1	medium and high	low	commercial instrument Du Pont 990 Thermal Analyzer with DSC module
76MET/COM	medium and high	5.00 %	
77LAG/PIE	2-300 K	1.00 %	
79DU /COM	medium and high	low	
81ARU	medium and high	1.50 %	high pressure (up to 60 MPa)
84GUS/MIR	303-523 K	2.00 %	high pressure
86CDA/COM	medium	2.00 %	
87PER/COM	medium and high	1.50 %	
89KNI/ARC	high	1.00 %	commercial instrument Perkin-Elmer DSC-2
89PRA/RAJ	318-333 K	3.00 %	
90JIN/WUN	180-750 K	1.00 %	commercial instrument TA 2100 system; error between 180-370 K is 3 %
92BAO/CAC	medium and high	low	commercial instrument TA 4000 Mettler
92KAB/KOZ	340-520 K	1.50 %	triple-heat bridge method
Differential thermopile conduction calorimeters			
70PAZ/PAZ	medium	medium	modified commercial SETARAM microcalorimeter
74PET/TER	298-475 K	1.00 %	
76CON/GIA	medium	medium	
78BYV/JAS	medium	2.00 %	
83ROU/ROU	278-368 K	high	modified commercial SETARAM microcalorimeter DSC
84HAN/HAW	85-270 K	1.00 %	modified commercial SETARAM calorimeter (model BT)
86MER/BEN	150-1100 K	medium	commercial instrument SETARAM DSC 111
89BRE/LIC	170-370 K	2.00 %	commercial SETARAM micro-calorimeter DSC 111G
91BAN/GAR	298-573 K	medium	commercial instrument SETARAM model C-80

3.1.1. Motion of the Sample in the Calorimeter

In a *batch calorimeter* the liquid is contained in a vessel anchored in the core of the calorimeter. For loading the sample, either the vessel has to be removed or a special filling device must be used; the mass of liquid is in most cases determined by differential weighing. The vessel has to provide for the expansion of the sample or a small vapor space must be preserved. In order to increase the sensitivity of the measurements, the ratio of sample and vessel heat capacities should be as large as possible. To prevent temperature gradients construction must be such that fast and uniform distribution of heat inside the vessel is achieved. The batch arrangement is the one most frequently used with instruments operating at low and ambient temperatures.

In a *flow calorimeter*, a fluid under investigation flows at a constant rate through a heated tube; heating wire is attached inside the tube along its axis or, more often, the wire is wound around the outer surface of the tube. The increase in temperature, ΔT_{cv}, is determined by comparing the readings of the thermometers before and after the heater. The *specific density* of the sample, ρ, has to be known as only the volumetric heat capacity is otherwise obtained. From measurements of the *volumetric flow rate*, $F = dV/d\tau$, and the *heating input*, $P = dQ/d\tau$, the specific isobaric heat capacity c_p can be calculated directly using the equation:

$$P = c_p \rho F \Delta T_{cv} + L, \tag{III-4}$$

where the product $c_p \rho F$ is called *heat capacity flux*. In a batch calorimeter, T_{cv} increases with time and becomes more or less uniform all over the vessel. In a flow calorimeter, the temperature in the heating zone increases in the direction of the fluid flow and the temperature profile is independent of time (provided the stationary state was achieved). Problems might occur due to temperature gradients in the liquid, fluctuations in the flow rate, and undetected heat leaks. In differential flow calorimeters, these problems are eliminated.

In a *drop calorimeter*, a sealed ampoule containing a sample is heated to a well defined temperature T_{in} outside the calorimeter and is dropped subsequently into a *receiver* (which replaces the calorimeter vessel) at the lower temperature T_{rec}. Depending on the design of the calorimeter, either the temperature of the calorimeter vessel T_{rec} remains constant (i.e. all the heat is transported to the surroundings), or it changes only moderately since the heat capacity of the receiver is substantially larger compared with the ampoule containing the sample. For the first case in which T_{rec} is constant:

$$C_{amp}(T_{in}-T_{rec}) = Q + \int L\,d\tau, \qquad \text{(III-5)}$$

and for the second case in which T_{rec} changes, it holds that:

$$C_{amp}(T_{in}-T_{fi}) = C_{rec}(T_{fi}-T_{rec}) + \int L\,d\tau, \qquad \text{(III-6)}$$

where C_{amp} and C_{rec} denote the average heat capacities of the ampoule containing a sample and of the receiver, respectively. The temperature T_{fi} is the final equilibrium temperature after the ampoule has been dropped. In many drop experiments, the difference between T_{in} and T_{rec} was considerable (several tens of degrees) and, therefore, only average heat capacities were usually determined; the method is not easily applicable to sluggish processes and in situations where a phase change can occur.

3.1.2. Mode of Measurement

During an experiment in the *stationary* mode, the heat capacity is determined from the electrical energy supplied and the corresponding temperature change, ΔT_{cv}, by comparison of the status of the system before and after heating. It is not necessary to monitor the sample temperature and/or the heat transfer to the vessel as a function of time. On the other hand, the measurements in the *dynamic* mode require careful monitoring of the temperature and of the heat exchanged over the whole time period of the experiment. It is apparent that all methods using continuous heating are dynamic, while methods using intermittent heating can have either stationary or dynamic character.

3.1.3. Temperature Relationship Between the Calorimeter Vessel and its Surroundings

This criterion is very important as it characterizes the mode of heat transfer between the calorimeter vessel and its surroundings. The schematic of the four basic arrangements is depicted in Fig. 4. In an ideal *adiabatic calorimeter* there is no heat exchange between the vessel and its surroundings. All the electrical energy introduced into the vessel from its resistance heater is fully contained, however, under real conditions, corrections for small deviations from the strictly adiabatic regime are usually needed. The measurements are absolute in character and calibration is not necessary; however, the heat capacity of the empty vessel has to be determined separately or two experiments (with low and high filling) have to be run. In order to maintain the adiabatic regime, the vessel is surrounded by an adiabatic jacket whose temperature follows exactly that of the vessel surface ($T_{cv} = T_{su}$): see Fig. 4a. The resistance heater of this jacket is usually controlled automatically using information from the sensor of the temperature difference $T_{cv} - T_{su}$. This temperature difference is monitored by a thermopile or two platinum resistance thermometers connected in two arms of a Wheatstone bridge.

It is difficult to keep strict adiabatic conditions over an extended period of time; during a heat capacity experiment, the temperature changes and the gradients across the heat exchanging surfaces cannot be avoided. Sometimes the temperature of the vessel is adjusted so that it is slightly higher than that of the surroundings in order to maintain a small controlled heat transfer between the vessel and its surroundings (quasi-adiabatic regime). This heat leak is determined before and after the actual temperature increase by measuring the amount of energy necessary to be supplied to the vessel to maintain a constant temperature (79VES/ZAB, 87ZAB/HYN).

In order to minimize the heat leaks, every effort is made to reduce the value of the heat transfer coefficient k_{ht} which includes contributions to heat conduction, convection, and radiation. The latter mode of heat transfer is predominant at high temperatures while the former two play important roles at low temperatures. Calorimeters are evacuated in order to suppress conduction and convection from gas molecules. The diameters of electric leads and supports are minimized to reduce the heat conduction through the solid connections and shields are used to reduce radiation. Calorimeters equipped with a system of heated and/or unheated radiation shields can be used up to a temperature of around 800 K (58WES/GIN). Discussion of the heat transfer in calorimeters can be found in the literature (68MCC/SCO). Adiabatic calorimeters are typically single-vessel instruments of the batch type and can work both in a stationary or dynamic mode.

In an *isoperibol calorimeter* (see Fig. 4b) the temperature of the surroundings is kept constant (T_{su} = const) or close to constant and the experiment is conducted in such a manner that the difference between T_{cv} and T_{su} is kept small at any time during an experiment (usually below 5 K). The vessel is insulated from its surroundings, thus enabling most of the heat introduced from the vessel heater to be contained. By measuring the drift in the temperature of the vessel before and after heating, the correction for the heat exchange with the surroundings can be calculated. Compared with the arrangement of adiabatic calorimeters, isoperibol calorimeters are easier to construct and to operate. Isoperibol instruments were used successfully, especially at low temperatures; when the inner space of the calorimeter is evacuated, the inequality of T_{cv} and T_{su} does not lead to serious heat leaks as radiation is very low in this region. A simple isoperibol arrangement is a Dewar flask with a Beckman thermometer and a heater, placed in a thermostatted room; such a rudimentary instrument can yield data with an accuracy of around 3 percent at room temperature. Isoperibol calorimeters are typically single vessel instruments of the batch type working in a stationary regime.

In an *isothermal calorimeter*, both the calorimeter vessel and its surroundings are always at the same constant temperature ($T_{cv} = T_{su}$). In these instruments, all the heat effect in the vessel is exactly compensated by introduction or removal of the same amount of energy. Supplying energy into the vessel by *Joule heating* is used for endothermic processes, *Peltier cooling* is used for exothermic processes. In older calorimeters, the heat was absorbed by melting or vaporizing a suitable substance; the actual amount of heat was determined from the amount of substance which underwent the phase transformation. The Bunsen ''ice'' calorimeter is a typical example of a *phase change calorimeter* where the difference in densities of ice and liquid water serves to quantify the heat effect (see Fig. 4c). The advantage of such a calorimeter is its simplicity of construction and no need for calibration; the substantial drawback is the limitation of the temperature of the phase change. The isothermal technique has limited use in heat capacity determinations as it can be used only for measuring average heat capacities in a drop calorimeter.

As *heat conduction calorimeters* are denoted different instruments in which the heat transfer between the calorimeter vessel and its surroundings plays a crucial role in the heat capacity determination. The experiment is based on recording the change of temperature with time after introduction of a certain amount of heat to the vessel or to the surroundings. Alternatively, the energy input needed to achieve a desired rate of temperature increase can be recorded. Heating can be continuous or intermittent, heat transfer between the sample and the surroundings can be facilitated by various types of environment. Measurements are performed in a dynamic mode and are always relative; they can be carried out in a twin calorimeter or by comparing two experiments performed subsequently under identical conditions in a single vessel calorimeter. All the instruments for heat capacity determination based on recording of *heating* and/or *cooling* curves can be considered as heat conduction calorimeters.

Thermopile conduction calorimeters are a special type of heat conduction calorimeter (see Fig. 4d) based on an idea of Tian (24TIA) and developed later by Calvet and other researchers. A thermopile consisting of a large number of thermocouples (up to several hundred) permits a *controlled* and rapid transfer of heat between the vessel and a massive *calorimeter block*. As its heat capacity is, in general, much larger compared with the thermal effect in the vessel the conditions are close to isothermal (T_{su} = const., $T_{cv} - T_{su} \to 0$). The transferred heat is proportional to the voltage U_e measured at the ends of the thermopile and its recording as a function of time τ permits one to determine the overall thermal effect from the equation:

$$Q = k_e \int U_e \mathrm{d}\tau = k_e A', \qquad \text{(III-7)}$$

where A' denotes the total area under the recorded curve representing U_e versus τ. A constant of proportionality k_e is determined through electric or chemical calibration. In the first case a heater inserted into an empty calorimeter vessel is used to determine the relationship between the recorded signal and the amount of the heat introduced. In the second case an experiment is performed with a substance (or substances) the thermal properties of which are well known. Most calorimeters use two cells housed in separate wells inside the calorimeter block. Differential connection of their thermopiles allows one to filter out most of the heat transfer which is not associated with the measured effect. This arrangement is well suited especially for studying slow or sluggish processes. Heat capacity measurements in a classical Calvet calorimeter are not very convenient as a combination of several experiments is necessary to obtain the final value of the heat capacity (56CAL/PRA). The principle of conduction through thermopiles is, however, utilized in several commercial instruments which can serve satisfactorily for heat capacity measurements (see Section 3.2.6).

3.1.4. Number of Calorimeter Vessels

Heat capacity determinations are performed either in *single vessel* or in *two vessel* (*twin*) instruments which are also denoted often as *differential calorimeters*. The single arrangement is simpler and less expensive as it is not necessary to manufacture two identical vessels. Beside the *measuring* cell where the actual thermal effect is measured, twin calorimeters have a *reference* cell which can be empty or filled with another substance; the difference in the thermal response of the two cells is directly recorded. The main advantage of twin calorimeters is the elimination of most errors due to heat losses and the minimization of perturbations which affect both cells in a similar way; this is important particularly for working with small samples and when measuring weak thermal effects (*microcalorimetry*) as well as for investigation of slow processes. An arrangement with two vessels is also more versatile as different kinds of experiments can be designed giving directly information on differences between the heat effects in the two vessels. Absolute measurements without calibration are performed more often in one-vessel instruments (e.g. adiabatic batch calorimeters) while the twin arrangement is typically used for comparative measurements (thermopile conduction calorimeters, Picker-type flow calorimeters).

However, it should be kept in mind that the reference vessel of a twin calorimeter should not be automatically associated with a reference substance. The use of the same word reference in the two contexts is only a terminological coincidence. Actually, in many differential experiments, the measuring cell is consecutively filled with a sample and with a reference substance while the contents of a reference cell does not play any important role. A reference fluid can be used also in experiments with single-vessel instruments for calibration purposes; on the other hand, an experiment can, in principal, be performed in a differential calorimeter without a reference fluid.

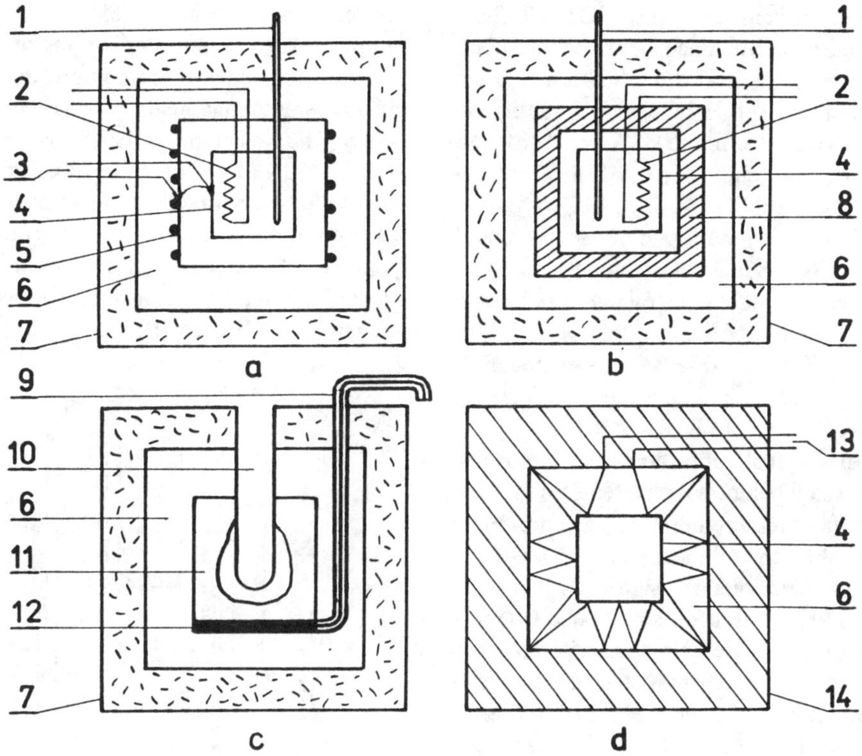

FIG. 4. Four main arrangements of calorimeters illustrating the heat transfer between the vessel and its surroundings: a. adiabatic, b. isoperibol, c. isothermal phase-change, d. thermopile conduction. 1—thermometer, 2—electric heater, 3—thermopile, 4—calorimeter vessel, 5—adiabatic jacket, 6—evacuated space, 7—outer insulation, 8—massive isothermal shield, 9—capillary filled with mercury, 10—space for dropping a sample (receiver), 11—solid-liquid equilibrium mixture of the substance used for quantifying the heat effect, 12—mercury in contact with the melting substance, 13—leads of the thermopile between the vessel and the thermostatted block, 14—thermostatted massive metallic block.

3.2. Main Experimental Approaches to Determination of Heat Capacity for Liquids

This section presents a concise overview of the most important calorimetric methods for determining the heat capacities of pure liquids with examples of several representative calorimetric designs. Most of the content of the well-known IUPAC monograph on the calorimetry of non-reacting systems (68MCC/SCO) is devoted to classical techniques of heat capacity measurements with pure condensed phases. Therefore, we shall pay more attention to the developments over the last few decades and mention some modern techniques which were not primarily designed for the investigation of pure liquids, but present an important aspect of progress in the evolution of heat capacity calorimetry.

3.2.1. Adiabatic Single Vessel Batch Calorimeters

Adiabatic instruments with intermittent heating represent the most accurate tool for determining the absolute heat capacity of substances in the condensed phases in the range from temperatures close to absolute zero up to ambient conditions. High-precision instruments can yield data with an error between 0.05 and 0.1 percent. These instruments are not, however, available commercially and the construction is both complicated and time-consuming.

With the first calorimeters of this type (20COH/MOE, 24WIL/DAN), adiabatic conditions were realized using a metallic vessel immersed in a bath whose temperature was controlled (wet shield). The advantage of such an arrangement was the possibility of active cooling. With such an arrangement the control system was, however, slow in response and was subsequently used only occasionally (35SAG/LAC, 50SAG/HOU). Dry shields made from highly conductive material (copper and brass are common) with a heating wire coiled on the outer surface of the shield were introduced by Andrews (29AND, 30SOU/AND); these have been used most often for maintaining adiabatic conditions.

Referenced below are some of the laboratories and adiabatic instruments which made a significant contribution to the heat capacity determination in the range between 5 and 320 K regarding both the quantity and quality of data (typical accuracy 0.05 to 0.3 percent above 20 K and 3.0 to 5.0 percent below that temperature): the former National Bureau of Standards (today's NIST), USA (37OSB/STI, 45SCO/MEY, 47OSB/GIN), the former laboratory of the Bureau of Mines at Bartlesville (today's NIPER), USA (43RUE/HUF, 47HUF), Pennsylvania State University, USA (39AST/EID1, 47AST/SZA, 51AST/FIN, 53AST/WOO), National Physical Laboratory in Teddington, Great Britain (63AND/COU1, 67AND/COU, 73MAR), University of Michigan, Ann Arbor, USA (68WES/FUR) and University at Gorkii, USSR, now Russia (66NIK/LEB, 76LEB/LIT). While few new adiabatic calorimeters have been constructed in the USA and Europe in recent years, there has been recently a good deal of experimental activity in Japan (66SHI/ATA, 74ATA/CHI; a calorimeter has been described for measurements on liquid crystals up to 390 K 83YOS/SOR1). Instruments allowing measurements at high pressures were constructed by Goodwin (61GOO); Rastorguev et al. (75RAS/GRI) and Oguni et al. (82OGU/WAT). Calorimeters with a small controlled heat leak (see Section 3.1.3) have an accuracy of heat capacity determination 0.3 to 0.5 percent (79VES/ZAB, 87ZAB/HYN).

Two examples of typical adiabatic heat capacity calorimeters (one for low-temperature measurements and the second for high temperature measurements) are given in Figs. 5 and 6. The instrument of Ruehrwein and Huffman (43RUE/HUF) is a low-temperature calorimeter designed for measurements between 10 and 320 K with accuracy around 0.1 percent. The calorimeter vessel is made of copper with an inner volume of 56 cm^3. The resistance thermometer and the resistance heater resistor are housed in the well. The vessel is suspended on thin wires of low conductivity inside an evacuated space. The temperature difference between the vessel and the brass adiabatic jacket is detected by three thermocouples. Liquid hydrogen and liquid air in the reservoirs serve for cooling down the interior of the instrument; all electrical leads are wound around the floating ring to minimize heat leaks by conduction. This calorimeter yielded an impressive number of accurate data over a period of 50 years, and a number of improvements regarding process control and retrieval of data were reported.

The instrument of West and Ginnings (58WES/GIN, Fig. 6) was designed for experiments between room temperature and 773 K. In order to prevent temperature gradients, a cylindrical aluminum block with a system of coaxial channels served as an unconventional calorimeter vessel. The central channel houses the platinum resistance standard, three other channels accommodate heating resistors and the rest contain the sample (the overall interior free volume is 70 cm^3, the diameter of the channels is 2 to 5 mm). The vessel is fixed in a heated silver ring and surrounded by a system of thin silver lids and shields which serve to minimize the radiation losses. Chromel-aluminum thermocouples are used as sensors of the temperature difference between the vessel and adiabatic surroundings which consist also of several heated silver shields and lids. Temperature is controlled automatically such that the differences between individual parts inside the calorimeter are below 0.001 K. A homogeneous atmosphere inside the calorimeter is maintained by a slow flow of carbon dioxide ($1\ cm^3 \cdot min^{-1}$).

Continuous heating allows for the measurements to be carried out rapidly compared with the experiments using a classical adiabatic calorimeter; the measurements are, however, usually less accurate. In order to avoid systematic errors due to temperature gradients, it is imperative to have a very fast temperature response of the entire calorimeter vessel to heating. Also, recording or logging of the primary experimental output becomes more complicated (see Eqn. III-2) since the temperature and heating rate have to be monitored continuously. The first calorimeter of this type was reported as early as the 1930's (31DEE), but significant progress was attained only with the arrival of automated techniques for temperature control and the processing of the primary output by computer methods (88LUS/RUB). A continuous heating method adapted for high pressures was reported by Guseinov and Mirzaliev (84GUS/MIR).

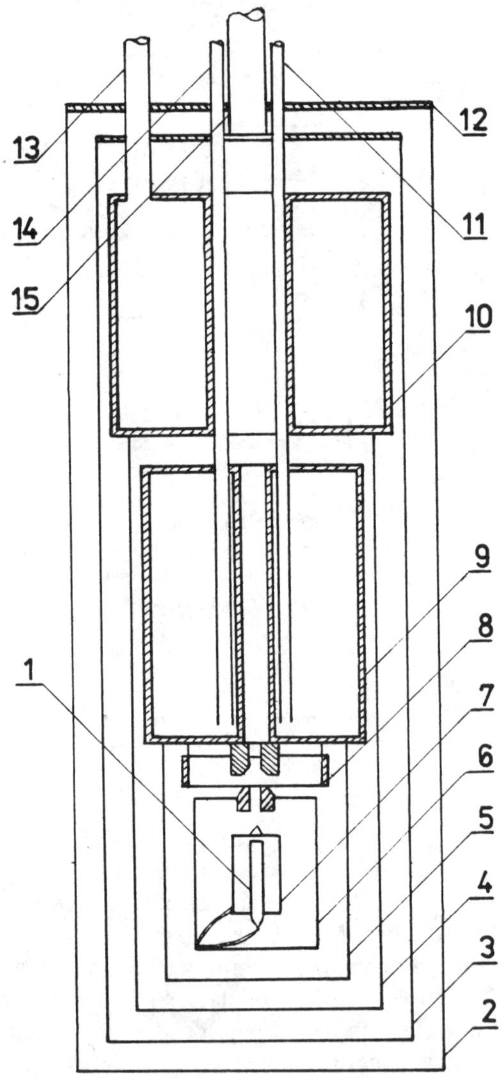

FIG. 5. Adiabatic calorimeter for low temperature measurements (43RUE/HUF)
1—well housing a heater and thermometer, 2—outer can, 3,4—copper radiation shields, 5—copper shield in contact with liquid hydrogen, 6—adiabatic jacket, 7—calorimeter vessel, 8—floating ring, 9—liquid hydrogen reservoir, 10—liquid air reservoir, 11—filling tube (hydrogen reservoir), 12—vacuum tight joint, 13—filling tube (liquid air reservoir), 14—emptying tube, 15—tube for electrical leads and connection to a vacuum line.

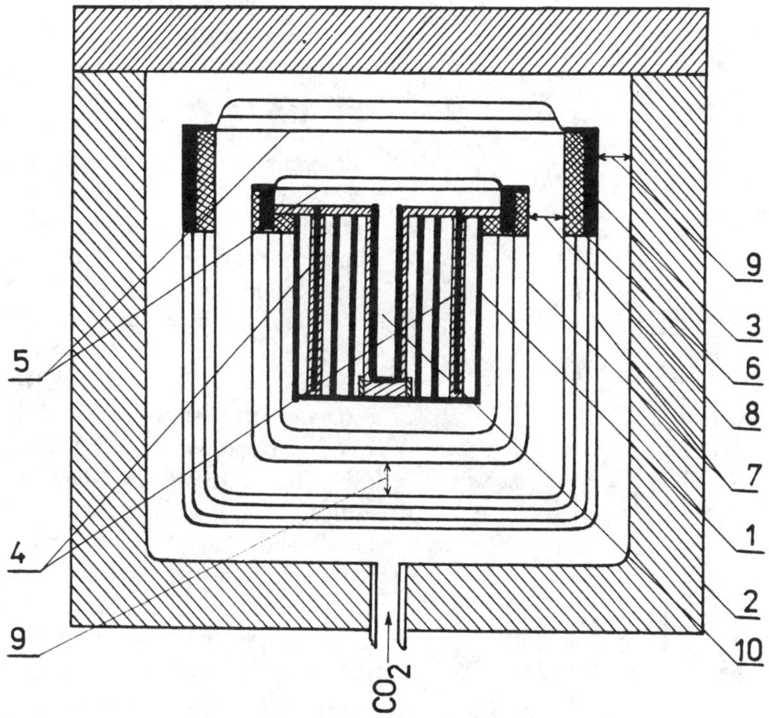

FIG. 6. Adiabatic calorimeter for high temperature measurements (58WES/GIN) 1—calorimeter vessel, 2—outer insulation, 3—heater of the adiabatic jacket, 4—vessel heaters, 5—radiation lids, 6—adiabatic ring supporting lids and shields, 7—radiation shields, 8, 9—thermocouples, 10—thermometer well.

3.2.2. Isoperibol Single Vessel Batch Calorimeters

Isoperibol calorimeters with a massive isothermal shield were frequently used for low-temperature heat capacity determination before 1940; the best measurements had an accuracy of 0.2 – 1 percent above 35 K and 5 percent below that temperature. Most of the measurements were, however, performed in simple isoperibol instruments without massive isothermal shields. Usually a Dewar vessel equipped with a thermometer and an electric heater was used (49WEI); in this case, the best accuracy which can be achieved is about 1 percent (80FUC).

Important laboratories which contributed to the development of this technique in a significant manner were University of California at Berkeley, USA (28GIA/WIE1, 37GIA/EGA) and Berlin University (29CLU1, 40CLU/POP). A typical example of this class of calorimeter is the instrument of Giauque and Egan (37GIA/EGA); its simplified scheme is depicted in Fig. 7. The calorimeter vessel is made of gold, its mass is about 500 g and the inner volume is near 170 cm^3. The platinum resistance thermometer which is calibrated using a standardized copper-constantan thermocouple is in the central well; the resistance heater is wound on the outer surface of the vessel. The measurement error is 0.2 percent except in the region of very low temperatures (see Table 4). This approach is, however, rarely used at present and is mainly of historical interest.

3.2.3. Isoperibol Flow Calorimeters

Most flow calorimeters have isoperibol character as the temperature of the surroundings is maintained constant close to the average temperature of the heated fluid. The ability to maintain a stable flow rate is the primary condition for the proper functioning of a flow calorimeter with a single cell. The first measurements on these calorimeters were, therefore, not of high accuracy (28LAN, 31FOR/BRU) and progress was achieved only with the advent of modern pumping devices. This type of instrument was successfully used for measurements at elevated pressures (75SAF/GER, 77HOF/SAN).

After the introduction of the differential arrangement was first proposed by Picker (71PIC/LED), flow calorimeters of this kind became widely utilized for measuring heat capacities of mixtures. The commercial versions of isoperibol flow calorimeters were distributed by SETARAM in Europe and by SODEV in America. Within the two last decades, a large number of measurements have been made using these calorimeters yielding much data near ambient conditions.

In a Picker calorimeter (see Fig. 8), a liquid passes through two identical cells connected in series inside a thermostatted evacuated chamber. The heating input is controlled in such a way that the temperature increase in the two cells is always identical. First the reference fluid is pumped into the system followed immediately by the sample; heating input to

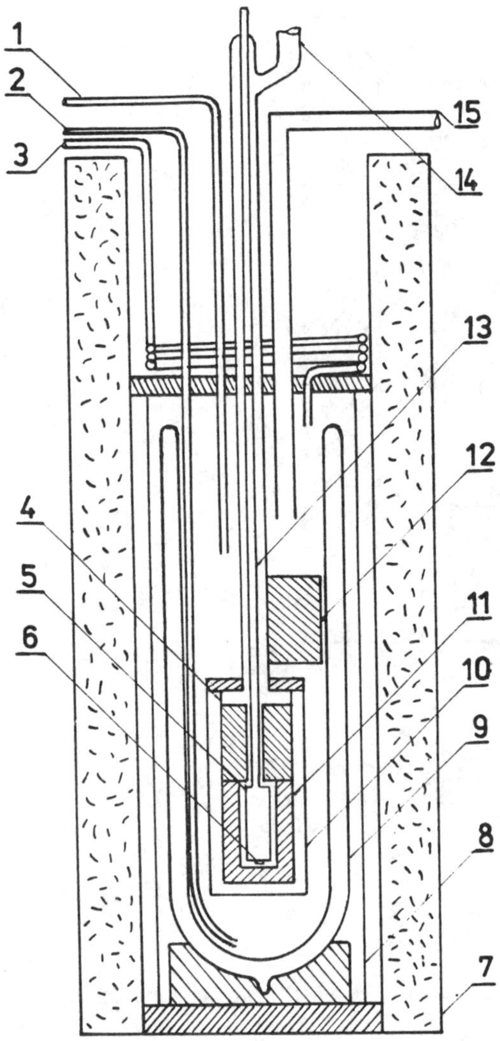

FIG. 7. Isoperibol calorimeter for low temperature measurements (37GIA/EGA)
1—transfer tube, 2—tube for removing liquid from Dewar vessel, 3—tube for introducing liquid hydrogen, 4—suspension cords for protective cylinder, 5—calorimeter vessel, 6—thermometer well, 7—outer insulation case, 8—inner case, 9—silvered Pyrex Dewar flask, 10—container of the vessel, 11—protective cylinder, 12—block for thermostatting of leads, 13—filling tube of the vessel, 14,15—vacuum lines.

the measuring cell is registered during the period when the reference fluid is present in both cells (P_{re}) and subsequently when the sample passes through the first cell and the reference fluid through the second cell (P_s). It is obvious that in such an arrangement the uncertainty due to instabilities in the flow rate are at a minimum. Experiments are rapid (about two data points in one hour) and fairly accurate (typical errors around 0.5 percent). The small diameter of the tube (inner diameter near 1 mm) reduces the risk of temperature gradients in the liquid and the amount of sample needed for an experiment (around 5 cm^3). The Picker calorimeter is not suitable for working with viscous liquids. In addition, the sample fluid must be miscible with the reference liquid of well-known heat capacity and density data for the ratio ρ_s/ρ_{re} must be available, otherwise, only heat capacities related to the unit volume can be obtained (see Eqn. III-8). Water is used as a reference for aqueous solutions and heptane is used in most cases for measurements with organic liquids.

Differential flow calorimeters are especially useful for measurements at high temperatures and pressures. It is not necessary to cool down the instrument and release the pressure for refilling. The instrument time constant is small since the volume of the cells is small, and thin-walled tubing can sustain considerable pressure. Isobaric heat capacities are directly obtained from the experiment. The absence of a vapor space or any other correction is a particular asset at superambient conditions where the correction terms are large and difficult to estimate. The first instrument of this type proposed by Smith-Magowan and Wood (81SMI/WOO) was further improved (88WHI/WOO, 91CAR/WOO) to allow measurements with corrosive fluids up to 723 K and 40 MPa (see Fig. 9). Reference liquid (water) flows first through the reference side of the calorimeter towards the six port valve. Depending on the valve's position, the reference liquid either by-passes the loop entering directly into the measuring side (base line determination) or penetrates into the loop forcing the sample into the instrument (plateau determination). Temperature equilibration to 0.01 K is secured by extensive heat exchange and a preheating system. The pressure is maintained constant to 0.02 MPa by means of two back pressure regulators connected in series. The pressure between the two regulators is utilized before the sample injection for prepressurizing the loop which is thermostatted at a well defined temperature (usually 298.15 K). This provision allows one to substitute at any temperature the ratio of sample and reference densities in Eqn. (III-8) for the density ratio at the thermostatting temperature; thus, in this arrangement, knowledge of density at only one temperature is satisfactory. During the calibration, an auxiliary pump allows the addition or withdrawal of some reference liquid before its passage through the measuring side (see below).

When Eqn. (III-4) is written for the measuring cell of a flow calorimeter for sample and reference conditions, it is easy to derive the relationship (76DES/DEV):

$$1+f\left(\frac{P_s-P_{re}}{P_{re}}\right)=\frac{(c_p\rho)_s}{(c_p\rho)_{re}}. \tag{III-8}$$

This equation allows one to calculate the heat capacity of the sample provided the *correction factor f*, which accounts for heat losses during experiment, is available. By comparing Eqns. (III-4) and (III-8), one finds that:

$$f=\frac{1-\dfrac{L_s-L_{re}}{P_s-P_{re}}}{1-\dfrac{L_{re}}{P_{re}}} \tag{III-9}$$

where L_s and L_{re} denote heat leaks from the measuring cell.

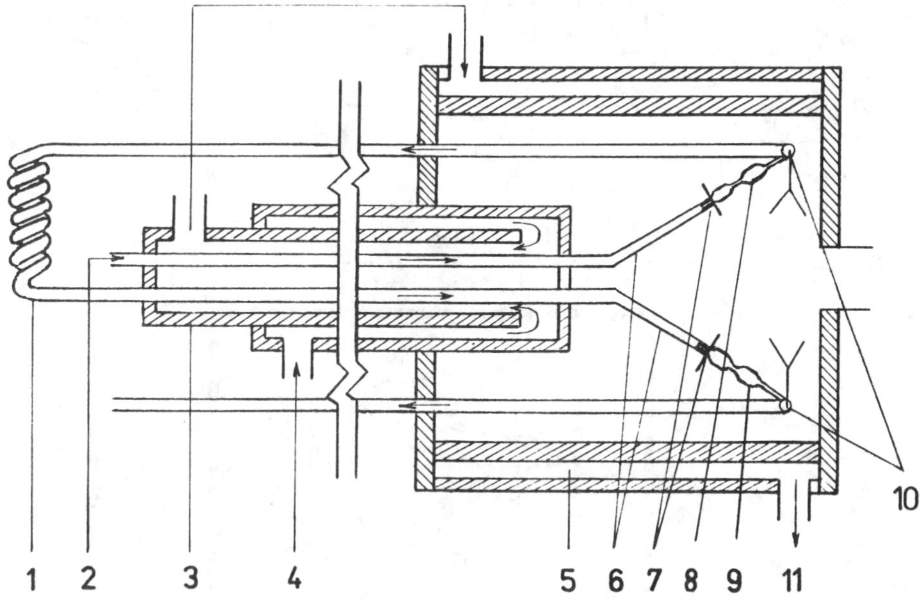

FIG. 8. Picker differential flow calorimeter (71PIC/LED)
1—external stainless steel coil, 2—injection system, 3—thermostat for injecting fluid, 4—water inlet to thermostat, 5—thermostatted water jacket, 6—stainless steel tube, 7—heaters, 8—reference cell, 9—measuring cell, 10—detector assemblies with thermistors, 11—water outlet from thermostat.

There are two methods for determining the correction factor f which in the ideal case is equal zero. The first is the measurement of ΔT_{cv}, F, and P for a fluid of well-known heat capacity and the calculation of L from Eqn. (III-4) (76FOR/BEN2, 81ROG/PIT). The second, more widespread method (71PIC/LED, 76FOR/BEN2, 81SMI/WOO, 91CAR/WOO), uses the change in heat capacity flux. Here, the change in C_p is mimicked by altering the flow rate through the measuring cell while keeping the flow rate of the same liquid through the reference cell constant. The factor f is calculated from:

$$f = \frac{P\,\Delta F}{F\,\Delta P} \qquad \text{(III-10)}$$

where ΔF and ΔP denote the change in flow rate and the corresponding adjustment of the electrical input, respectively.

In the case of a flow calorimeter, the heat losses depend on several factors; L can be expressed (82WHI/WOO, 91CAR/WOO) as:

$$L = k_{ht1}(T_{cv} - T_{su}) + k_{ht2}\frac{(T_{cv} - T_{su})}{c_p \rho F} + k_{ht3}P \qquad \text{(III-11)}$$

where the first term has the same meaning as in Eqn. (III-3). The second term expresses the heat loss due to the gradient along the tube; since temperature sensors are located at some distance from the heater, the measured ΔT_{cv} is different from the real temperature rise in the liquid due to heat exchange between the tube and its surroundings. This distortion decreases with increasing heat capacity flux. The third term represents heat losses due to the overheating of the heater on the tube surface. This term is obviously proportional to the energy input. An exhaustive analysis of heat losses and methods for determining the f factor in high-temperature flow calorimeters was published by Carter and Wood (91CAR/WOO).

3.2.4. Drop Calorimeters

In the simplest version of this technique, the interior of the Dewar flask filled with water serves as a receiver. An increase in the water temperature is measured after dropping a sealed ampoule with a sample, preheated in a furnace above the calorimeter. The actual thermal effect is calculated from Eqn. (III-6) using the calibration constant C_{rec} determined by the Joule effect or by measurement with a sample of the well known heat capacity. This rudimentary isoperibol calorimeter used to be the most widespread tool for heat capacity measurements in the 19th century; certain results were surprisingly accurate (around 1 percent, *81VON) and retain some value today. This type of experiments is denoted sometimes in the literature as the method of mixtures to express that the sample and water at two different temperatures are "mixed" to equilibrate at the third temperature. The heat capacity of the receiver is much larger compared with that of the ampoule and sample; the difference between temperatures of the furnace and the vessel must be considerable to achieve sufficient sensitivity. Therefore, "mixing" experiments give actually enthalpic differences rather than heat capacity values. Currently, this method is used only for measurements at high temperature where the heat capacities are large and do not change significantly with temperature; in this case, melted metal is used instead of water (71BEL/HUL).

Drop calorimeters were rarely run under adiabatic conditions (30BAR/MAA, 62LEV). The isothermal arrangement, in

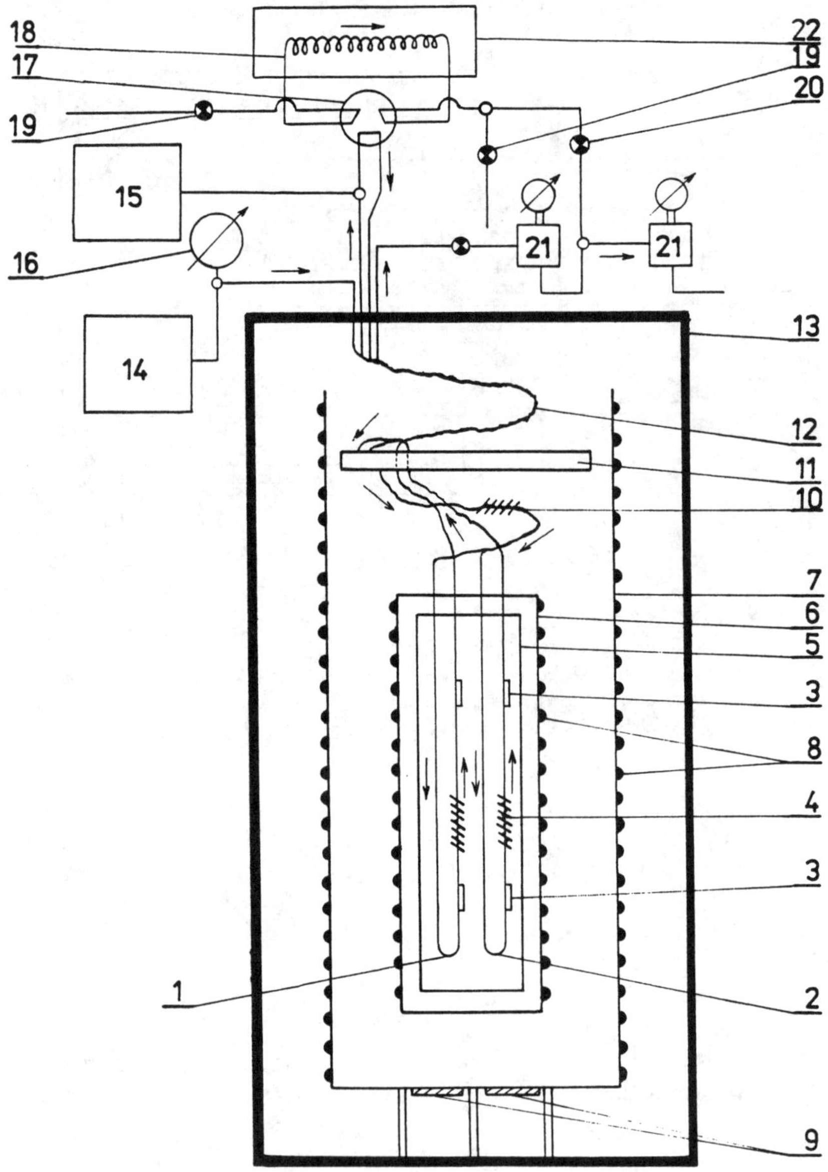

FIG. 9. Scheme of a differential flow calorimeter for measurements at high temperatures and pressures (91CAR/WOO), direction of flow is indicated by arrows
1—reference side, 2—measuring (sample) side, 3—platinum resistance thermometers, 4—heater, 5—calorimeter block, 6—inner isothermal jacket, 7—outer isothermal jacket, 8—coaxial heater cable, 9—strip heaters, 10—coaxial preheater, 11—preheater lid, 12—counter-current heat exchanger, 13—vacuum can, 14—HPLC pump, 15—auxiliary pump used for calibration, 16—pressure gauge, 17—HPLC six port valve, 18—thermostatted sample loop, 19—valves for filling the loop, 20—prepressurization valve, 21—back pressure regulators, 22—thermostat of the sample loop.

which all the heat transferred from the ampoule is "pumped out" from the receiver keeping T_{rec} constant (see Eqn. III-5), seems to be the most convenient for drop experiments. Phase-change calorimeters were often used as they are easy to construct and allow experiments to be performed under isothermal conditions without any electronic equipment. Measurements were carried out for example at the National Bureau of Standards, USA in a Bunsen ice calorimeter on a number of substances between 273 and 1173 K (47GIN/COR, 56FUR/DOU). Drawbacks of using the water-to-ice phase transition as a heat exchange indicator are the low sensitivity due to the high value of the specific enthalpy of melting for ice and the negative sign of volumetric change accompanying the melting ice. In this case, as shown in Fig. 4c, mercury is sucked inside the instrument. Diphenylether (melting temperature 300.05 K) is more suitable for use in phase change calorimeters (55GIG/MOR1), because its volumetric change at melting is positive and has a specific enthalpy of melting almost four times lower than water.

A thermopile conduction drop calorimeter for measurement with small samples below 1 g in the temperature range from 273 to 343 K was developed by Wadsö and coworkers (71KON/SUU, 74SUU/WAD, see Fig 10). Two identical ampoules can be preheated in a furnace and dropped simultaneously into the receivers of a calorimeter which is thermostatted at a working temperature about 10 K lower; the recorded

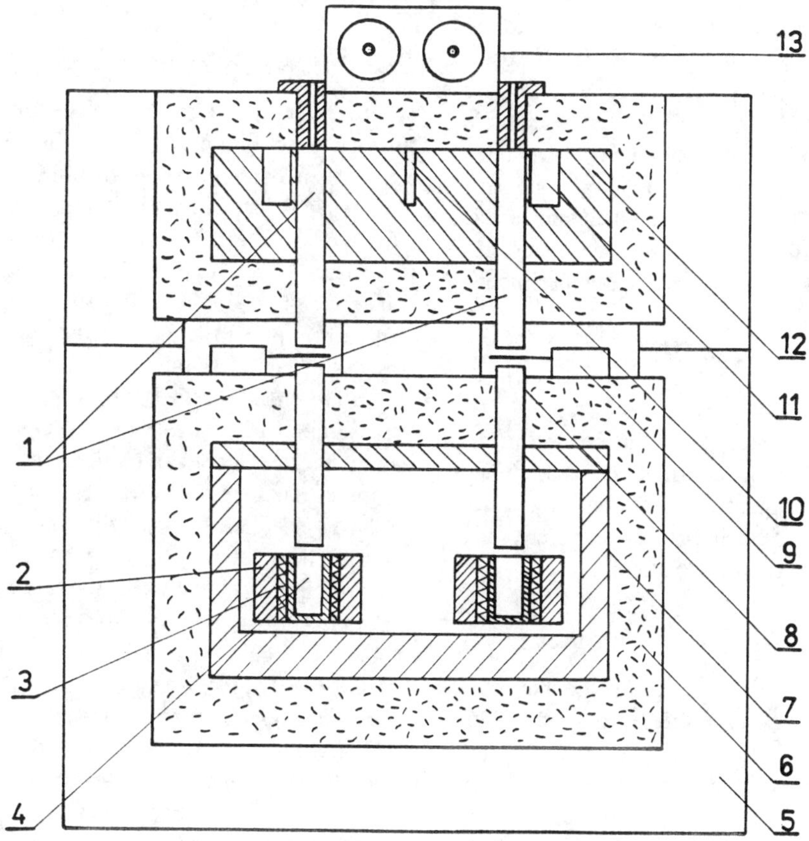

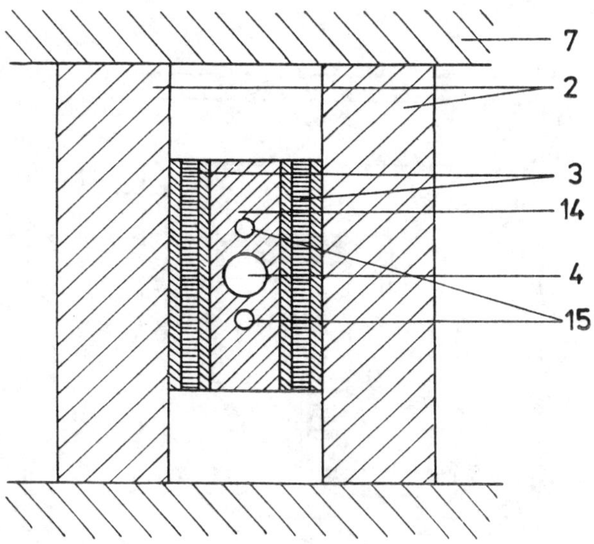

FIG. 10. Drop thermopile-conduction calorimeter for measurements near ambient temperatures (74SUU/WAD)
a. Schematic view of the calorimeter assembly
b. Calorimetric unit seen from above
1—holes for temperature equilibration of the ampoules containing a sample and a reference substance, 2—aluminum block, 3—thermopile plate, 4—receiver, 5—water thermostat, 6—thermal insulation, 7—main heat sink, 8—guiding tube, 9—electromechanical shutter, 10—cavity for temperature controlling element (thermistor), 11—cavity for temperature measuring element (quartz probe), 12—furnace, 13—mechanical lift, 14—aluminum plate, 15—cavities for calibration heaters.

differential signal from the thermopiles is proportional to the difference in the heat capacities of the ampoules. Two experiments are performed; the ampoule dropped on the measuring side of the calorimeter is either empty or filled with a sample and the contents of the ampoule on the reference side remains constant and does not play any role. Using the notation introduced in Eqns. (III-5) and (III-7), the sample heat capacity C is obtained by using the differential thermopile signals A'_{d1} and A'_{d2} corresponding to the first and second experiments, respectively:

$$C = k_e\left(\frac{A'_{d2}}{T_{in2}-T_{rec}} - \frac{A'_{d1}}{T_{in1}-T_{rec}}\right) \qquad \text{(III-12)}$$

where T_{in1} and T_{in2} are the temperatures of the furnace in the two experiments. The calibration constant, k_e, is obtained through electric calibration or measurement with a reference substance of well-known heat capacity. Experiments are fully automated with results accurate to 0.1 percent.

3.2.5. Analysis of Heating and/or Cooling Curves

A variety of techniques was reported in the literature which allow one to calculate heat capacities through the analysis of heating and/or cooling curves. These methods are always comparative and have usually been performed in single-vessel instruments. For instance, the ratio of the heat capacity of the sample C_s and that of a reference substance C_{re} can be obtained by measuring consecutively the rate of temperature change for the two substances:

$$\frac{C_s}{C_{re}} = \frac{(dT_{cv}/d\tau)_{re}}{(dT_{cv}/d\tau)_s}, \qquad \text{(III-13)}$$

where the temperature derivatives are obtained by analyzing heating or cooling curves. Alternatively, the heating times or amounts of heat necessary to produce identical temperature changes in the studied and reference substances are compared. This type of determination was very popular for its simplicity and many values reported between 1920 and 1950 were obtained by this technique. Because of its low accuracy (typically errors of 5 to 10 percent), it is now only used occasionally at high temperatures.

References (33FER/MIL, 41ZIE/MES, 50KUS/CRO) are examples of determinations where the heating times necessary for the identical increase in the temperature of the reference and test substances were measured. Hoffman (52HOF) used constant heating input and maintained a constant gradient between the vessel and its surroundings. A similar principle, with the use of both heating and cooling with a constant temperature gradient, was employed for measurements with liquid sulphur in the temperature range 393-600 K (73KOM/MIL); the instrument is depicted in Fig. 11. Both the quartz cell containing the substance and the reference shield are housed inside a furnace; the difference between these two parts is kept constant during all the experiments. The heat capacity of the substance is obtained by combining the heating times necessary to heat the vessel when it is empty or when filled with sample or when it contains reference substance (copper).

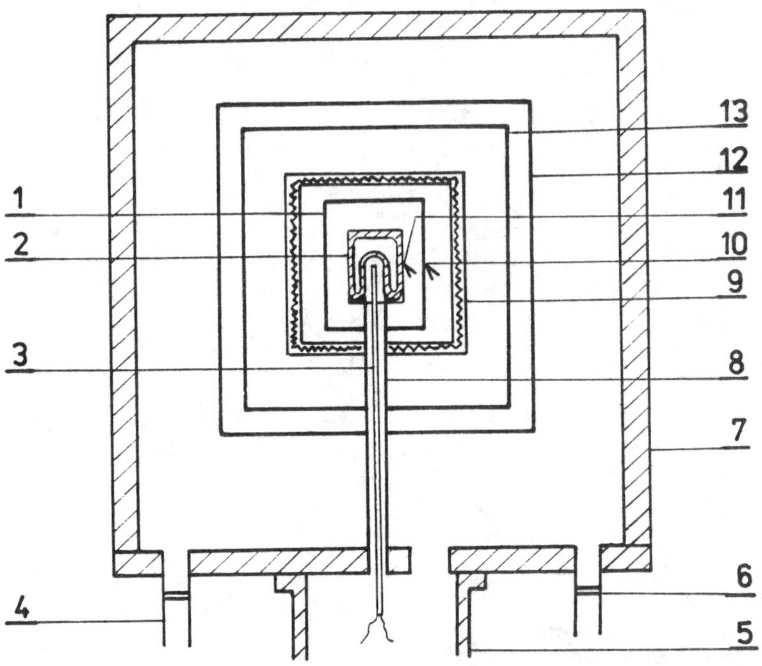

FIG. 11. Constant temperature gradient calorimeter for high temperature measurements (73KOM/MIL)
1—reference jacket, 2—calorimeter vessel (quartz cell inside a metallic shell), 3—thermocouple for measuring temperature, 4—vacuum tight connections for heater leads, 5—vacuum line, 6—vacuum tight connections for thermocouple leads, 7—brass can, 8—supporting tube of quartz, 9—furnace, 10,11—thermocouples for controlling the temperature difference between the reference jacket and calorimeter vessel, 12,13—radiation shield.

3.2.6. Differential Scanning Techniques

As differential scanning techniques, we denote the approaches using the twin vessel arrangement where the temperature of measuring and reference cells are increased simultaneously over a wide temperature range and the effect corresponding to the difference in the thermal response of the two cells is measured directly. These techniques can be performed in commercially available instruments which, in most cases, allow one to work with small samples over a wide temperature range; heating is often continuous but can be also intermittent.

Most instruments work as heat conduction calorimeters in the dynamic mode; the heat is not introduced in some cases (thermopile conduction instruments) into the cell directly but rather into its surroundings. Since the heat capacity of the cells is usually very small compared to that of the surroundings, the T_{cv}, and, T_{su}, are practically identical. Heat leaks from both cells can be considered practically identical and play only a minor role in the heat capacity determination. For a comparison of the thermal response in the two cells, two approaches were developed commercially:

1. *differential thermal analysis* (*DTA*); the temperature difference is measured between the two cells which are heated with identical electrical input energy.
2. *differential scanning calorimetry* (*DSC*); the heating inputs into the measuring and reference cells are adjusted such that the temperature increase in both cells are identical.

DTA and DSC are widely used for investigations with solids (84SES); they are sometimes employed for the measurement of the heat capacities of liquids with an average accuracy of 2 to 5 and 1 to 3 percent, respectively. Examples of the utilization of commercial instruments are the measurements performed on the Dupont 900DTA (67PAC), DTA Mettler TA200 (76ADE/SIM), Dupont 990 DSC (75CUC1), and Perkin-Elmer DSC-2C (87MIL/FEN2) calorimeters.

Mraw and Naas (79MRA/NAA) developed a method for measuring heat capacities of solids using the Perkin-Elmer DSC-2C calorimeter with an error around 1%. This approach was modified by Knipmeyer et al. (89KNI/ARC) in order to allow measurements with liquids up to the vicinity of the critical temperature.

Heat conduction through a thermopile is utilized in several commercial instruments produced by SETARAM. Fig. 12 shows the differential scanning calorimeter DSC111. A thermostatted metallic block can be operated in the temperature range between 150 and 1100 K. The measuring and reference cells are inserted separately into the central parts of two coaxial tubes made of ceramic material (only one is visible in the figure). Thermopiles are housed in the cavities of the metallic block which surround the central parts of the tubes and serve to conduct most of the heat exchanged between the vessel and the surroundings. When the metallic block is heated, the differential signal from the thermopiles reflects the difference in the amounts of heat transferred into the cells, which is proportional to the heat capacity difference between the measuring and reference cells.

A similar principle is used for the measurement of heat capacities of liquids up to 573 K in a modified commercial SETARAM C-80 calorimeter (91COX/QUI). The entire differential assembly is controlled by a computer while the experiments and processing of data are fully automated over the whole operating range of the calorimeter. Typically, the temperature is increased by 10 K steps symmetrical around the temperature plateaus at intervals about 25 K. The procedure consists of three experiments in which the reference vessel is always filled with water and the measuring cell is initially empty, then filled with water, and finally filled with the actual sample. The corresponding differential signals from the thermopiles are denoted as A'_{d1}, A'_{d2}, and A'_{d3}, respectively. As the temperature programming in these three experiments is kept identical, it holds for each temperature increment that:

$$\left[\int_{T_1}^{T_2} c_p\rho \mathrm{d}T\right]_{\mathrm{s}} = \frac{A'_{\mathrm{d3}} - A'_{\mathrm{d1}}}{A'_{\mathrm{d2}} - A'_{\mathrm{d1}}}\left[\int_{T_1}^{T_2} c_p\rho \mathrm{d}T\right]_{\mathrm{re}} \qquad \text{(III-14)}$$

where the terms in brackets on the left- and right-hand sides relate to the sample and reference substance, respectively, and T_1 and T_2 delimit the temperature increment. In this method, ρ and c_p for sample and reference fluids are expressed by polynomials in temperature whose parameters must be known over the whole temperature range of measurement except for the polynomial for the heat capacity of the sample. Its parameters can be obtained after integration by a least squares minimization, thus combining data for all temperature increments. This technique is rapid, yielding isobaric heat capacity directly; the pressure remains constant throughout the measurement as liquid can expand from the vessel. In order to avoid heat leaks in the vertical direction, the calorimeter is operated in an upside down position. The errors are generally below 1 percent. A disadvantage is, however, the necessity to know the density of both the sample and reference fluids over the whole temperature range.

3.2.7. Special Dynamic Methods Using Pulse Heating

Under this title, we group several techniques which use the periodic introduction of heating pulses into the sample and recording of the corresponding temperature oscillations as the principle of measurement. This type of determination is somewhat similar to an analysis of heating or cooling curves described in Section 3.2.5. In most cases, beside heat capacities, the measurements yield thermal conductivities. These techniques are very rapid but lack accuracy (error 2 to 5 percent). They are used most often for very small solid samples and occasionally for liquids usually at extreme conditions (low temperatures, high pressures, experiments in an electromagnetic field). Development of these techniques has been connected with research on superconductors.

Here, we mention briefly only three techniques referenced in Table 4 which summarize the calorimeters; a more detailed overview can be found in several review articles (81LAK/GOP, 82LAK/GOP, 79GME, 87GME). In a *laser-flash method*, a laser beam is used for pulse heating of a sample (79TAK/YOK). Electrical pulses are sent to a heating wire passing through a vessel completely filled with a sample (*hot-wire method*). The same wire serves simulta-

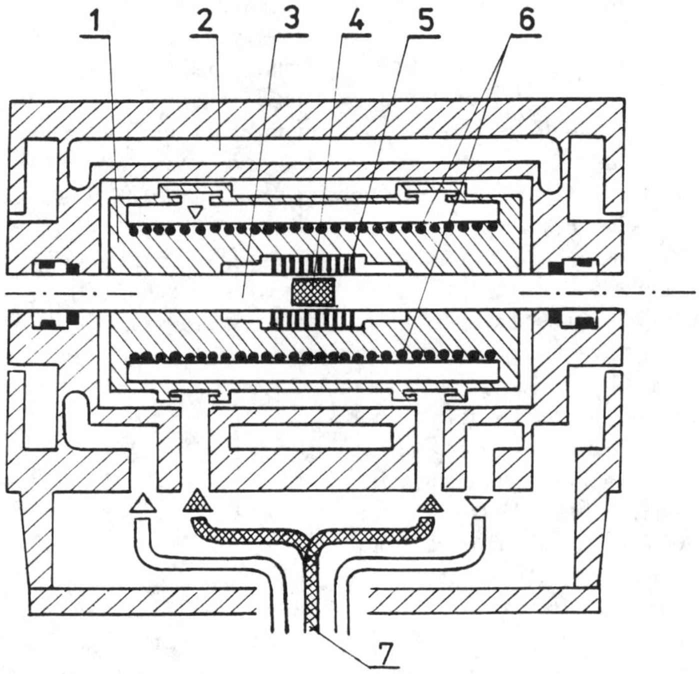

FIG. 12. Scheme of differential scanning calorimeter SETARAM DSC111 (86MER/BEN)
1—metallic bloc, 2—thermostatting unit, 3—one of the two central tubes in ceramics, 4—cell with the sample, 5—thermopile, 6—heating elements, 7—thermostatting system

neously as a temperature element for recording the induced temperature changes (81ATA/ELS). A similar scheme is used in *regular pulsing techniques* (86NAZ/BAS1). A narrow tube with a heating wire at its center passes through the axis of a cylindrical vessel filled with a sample. Periodically either electrical pulses are supplied to the heater or the gaseous cooling agent flows through the tube and temperature oscillations inside the vessel are observed.

3.3. Survey of Calorimeters Used for Determination of Heat Capacities of Pure Liquids

Table 3 divides techniques used for determining heat capacities of pure liquids into groups according to the classification introduced in the preceding sections. The codes defined in Table 2 are used to characterize individual methods and the section numbers refer to the description of the technique in the preceding text.

Table 4 surveys all the calorimeters cited in the original sources of experimental heat capacities which are referenced in this monograph and serve for establishing the data base of raw values. References to instruments with a brief characterization are listed in chronological order in groups defined in Table 3. This division of calorimeters according to the adopted categorization is necessarily a simplification of reality. With respect to the diversity of experimental designs, it is not always possible to associate an instrument unambiguously with a particular group or category.

The terms low, medium, and high are used for indicating the accuracy of measurements with a certain calorimeter in those cases when no quantitative indication was available in the original literature and estimates were made in most cases by the evaluator. For the column *Temperature Range* the meaning is as follows: "low" the lower temperature limit of measurements below room temperature, "medium" 270 - 370 K, "high" the upper temperature limit of measurements above 370 K. For the column *Measurement accuracy* the meaning is as follows: "low" the error is above one percent, "medium" the error is between 0.3 and 1 percent, "high" the error is below 0.3 percent.

4. Previous Compilations of Heat Capacity Data on Liquids

An annotated bibliography is provided which describes the publication of compilations containing reported values for the heat capacities of organic compounds (and some inorganic compounds) from the 19th century through the beginning of 1991. Some compilations provide a thorough coverage of the literature up to about the date of their publication for either organic compounds as an entire class of substances or the coverage may be limited to the gas phase or the condensed phase. These comprise the majority of the compilations listed in this chapter. Other compilations deal with selected classes of organic compounds, such as n-alkanes, 1-alkanols, or organic oxygen compounds, but provide a very detailed evaluation of the experimental data and select optimal heat capacity values for the corresponding class of compounds. Examples of the latter type are: 82WAG/EVA, 85WIL/CHA, 90CHA/GAD, 90ZAB/RUZ, 91RUZ/ZAB and any of the ESDU Series outputs. Some compilations provide a detailed description of the evaluation of the experimental data, such as 70TOU/MAK,

85WIL/CHA, 90ZAB/RUZ, or 91RUZ/ZAB, while others offer only a limited evaluation or quality rating, such as 76TOU/MAK, 84DOM/EVA, or 90DOM/HEA.

Examination of the experimental calorimetric methods from the 19th century to the present reveals improvements not only in experimental techniques resulting in enhanced precision of the data, but shows deliberate and significant attention being directed toward sample purity and sample characterization which assures that the data are also of enhanced accuracy.

Compilations are listed below in chronological order which report data on the heat capacities of liquids. The sustained interest and need for data on heat capacities is exemplified by the growth, continuation, and specificity of these compilations with time.

1897 Berthelot M.P.E., *Thermochimie*, Vol. I and II (*97BER).

Thermochemical (thermodynamic) properties of inorganic and organic compounds are tabulated and the corresponding references are provided whenever possible. The second part of volume II (pp. 393-693) is devoted to the thermochemistry of several hundred organic compounds. Reactions and processes, such as formation, combustion, isomerization, vaporization, fusion, and solution are examined in detail along with the enthalpies for the processes and heat capacities for the various classes of organic compounds (hydrocarbons, alcohols, acids, etc.) in the gas, liquid, solid, and dissolved phases. References are provided with the enthalpy and heat capacity data. This is an excellent summary of 19th century thermochemical research.

1912 *Landolt-Börnstein Physikalisch-Chemische Tabellen*, 4 Auflage, Börnstein R., Roth W.A. (Eds) (12LAN/BOR).

A section on the specific heats of organic compounds (pp. 766-772) lists the heat capacities for about 150 organic compounds in the liquid and solid phases. The data given are compound name, formula, state, temperature or temperature range, specific heat capacity in cal · °C^{-1} · g^{-1} and reference. The precision and accuracy of heat capacity measurements was limited during the 19th century as well as during beginning of the 20th century. Hence, these tables of heat capacity data are more representative of reliable collections of information from the literature but do not appear to be critically evaluated.

1923 *Landolt-Börnstein Physikalisch-Chemische Tabellen*, 5 Auflage, Vol 2., Börnstein R., Roth W.A. (Eds) (23LAN/BOR).

As an up-date from 12LAN/BOR, another section on the specific heat capacities of organic compounds (pp. 1265-1273) lists the values for about 200 organic compounds in the liquid and solid phases. The data provided is similar to that in 12LAN/BOR, also with references.

1927 *Landolt-Börnstein Physikalisch-Chemische Tabellen*, 5 Auflage, Suppl. 1, Börnstein R., Roth W.A. (Eds) (27LAN/BOR).

This collection of heat capacity data on organic substances in the solid and liquid states consists of tables containing about 80 compounds (pp. 692-696) which supplements previous data found in 23LAN/BOR. The supplemental data were extracted from papers published primarily in the 1920's.

1929 *International Critical Tables of Numerical Data, Physics, Chemistry, and Technology*, Vol. V, Washburn E.W. (Ed) (29WAS).

Critically evaluated data on the heat capacities of organic compounds are provided in volume V of the International Critical Tables. Two tables separate organic compounds in the solid state (pp. 101-105, about 125 compounds) from organic compounds in the liquid state (pp. 107-113, about 380 compounds). The data listed are formula, compound name, temperature or temperature range, specific heat capacity in J · °C^{-1} · g^{-1}, percent uncertainty, coefficients in an equation expressing specific heat capacity c_p as a function of temperature, temperature range for the equation, and references.

1931 *Landolt-Börnstein Physikalisch-Chemische Tabellen*, 5 Auflage, Suppl. 2, Vol. II, Roth W.A., Scheel K. (Eds) (31LAN/BOR).

Data on the specific heat capacity, c_p, and heat capacity, C_p, of organic solids and liquids are tabulated (pp. 1194-1217) in a different format with emphasis on providing recent and precise experimental data for specific temperatures and giving a plot of C_p versus temperature for good quality data on about 30 organic compounds. References are primarily to papers published in the 1920's and some during 1930.

1936 *Landolt-Börnstein Physikalisch-Chemische Tabellen*, 5 Auflage, Suppl. 3, Vol. III, Roth W.A., Scheel K. (Eds) (36LAN/BOR).

Data on the specific heat capacity, c_p, and heat capacity, C_p, of organic substances (pp. 2291-2315) are shown as a supplement to the data found in 31LAN/BOR. Experimental data having significant precision are tabulated at specific temperatures, plots of C_p versus temperature are shown for 4 organic compounds as well as a large number of equations representing C_p as a function of temperature.

1950 Timmermans J., *Physicochemical Constants of Pure Organic Compounds*, Vol. I (50TIM).

A thorough search of the chemical literature up to January 1, 1950 was conducted and compiled for physical property data on organic compounds. Particular emphasis was placed upon data for which the purity and characterization of the compounds received attention. Data on the heat capacities of organic compounds was limited to careful measurements for which such data were reported at specified temperatures. Average heat capacity data between two temperatures were rejected.

1952 Rossini F.D., Wagman D.D., Evans W.H., Levine S., Jaffe I., *Selected Values of Chemical Thermodynamic Properties*, National Bureau of Standards, Circular 500 (52ROS/WAG).

Tables of thermodynamic properties of inorganic and one and two carbon atom organic compounds were given after a thorough search, collection, and evaluation of the literature. Tables are divided into Series I and Series II. In Series I, properties listed are heat of formation, free energy of formation, logarithm of the equilibrium constant, entropy, and heat capacity at 298.16 K, and the heat of formation at 0 K. In Series II, phase transition processes, such as fusion, vaporization, sublimation, and solid-solid transitions are described, including the changes in heat, entropy, and heat capacity at the temperature and pressure corresponding to the transition.

1961 *Landolt-Börnstein Physikalisch-Chemische Tabellen*, 6 Auflage, Zahlenwert und Funktionen aus Natur-Wissenschaften und Technik, Vol. II, Part 4, Kalorische Zustandsgrossen (61LAN/BOR).

Separate tables provide data on the thermodynamic properties of organic compounds (pp. 262-372). In these tables, heat capacities are listed for about 550 compounds of which half are in the gas phase. The data provided are formula, compound name, physical state, heat capacity in $J \cdot K^{-1} \cdot mol^{-1}$ at 298 K, and the corresponding reference(s).

1965 Timmermans J., *Physicochemical Constants of Pure Organic Compounds*, Vol. II (65TIM).

Volume II of the Timmermans' compilation is an extension of Volume I (50TIM). The chemical literature is up-dated to 1964. The criteria for choosing physical property data for pure and well-characterized organic compounds is continued. Additional heat capacity data are reported.

1967 Tamplin W.S., Zuzic D.A., *Specific Heat of Organic Hydrocarbons* (67TAM/ZUZ).

After a critical evaluation of the heat capacity data for 83 hydrocarbons, constants are tabulated for an equation of the type: $C_p = A+Bt+Ct^2 \pm D$, where C_p is the heat capacity in $cal \cdot °C^{-1} \cdot mol^{-1}$, t is the temperature in Celsius, and D is the average difference between the calculated and selected values over the considered temperature range. A total of 84 references are listed.

1968 Wagman D.D., Evans W.E., Parker V.B., Halow I., Bailey S.M., Schumm R.H., *Selected Values of Chemical Thermodynamic Properties*, NBS Technical Note 270-3 (68WAG/EVA).

Selected values were made from the evaluation of available data for the enthalpy and Gibbs function of formation, entropy, and heat capacity at 298.15 K, the enthalpy difference between 298.15 K and 0 K, the enthalpy of formation at 0 K for the first 34 elements and their compounds from the standard order of arrangement. Data for organic compounds were limited to one and two carbon atom compounds. Physical states include crystalline solid, glassy or amorphous solid, liquid, gas, aqueous solution, and some data for non-aqueous solutions. In general, uncertainties are such that they lie between 2 and 20 units of the last figure. No bibliography or commentary are provided. NBS Technical Note 270-3 combines information provided in NBS Technical Notes 270-1 and 270-2, and adds data for two additional elements, zinc and cadmium, and their compounds.

1970 Touloukian Y.S., and Makita T., *Specific Heat of Nonmetallic Liquids and Gases*, Vol. 6 in Thermophysical Properties of Matter (70TOU/MAK).

A careful and critical evaluation is provided for 55 nonmetallic compounds in the gaseous and/or liquid states. References to the reported experimental heat capacity data are given along with information on sample purity and estimated uncertainty in the reported values. A list of recommended C_p values and a polynomial equation for C_p as a function of temperature are also given.

1975 Engineering Science Data Unit Series, *Heat Capacity and Enthalpy of Liquids. I: Aromatic Hydrocarbons*, Vol. 4 (Item No. 75015) (75ENG).

Heat capacity and enthalpy data for 34 aromatic hydrocarbons have been evaluated and are presented in tabular form as well as in equations as a function of temperature. The data on which the values are based are the most reliable available and for most compounds cover the range from the melting temperature to the normal boiling temperature or to the limit of the experimental data whichever is the higher. A total of 36 references is provided.

1975 Horvath A.L., *Physical Properties of Inorganic Compounds* (75HOR).

A critical review was conducted for the physical, thermodynamic, and transport properties of 31 inorganic compounds among which were CO, CO_2, and HCN. Heat capacities are plotted as a function of temperature for the gaseous and condensed phases. Properties at specific temperatures are listed, some of which are: heat of fusion at the melting temperature, heat of vaporization at the boiling temperature, and C_p/C_V for gases at 25 °C. References are provided for each compound listed.

1976 Engineering Science Data Unit Series, *Heat Capacity and Enthalpy of Liquids. II: Halogenated Methanes*, Vol. 4 (Item No. 76010) (76ENG).

Heat capacity and enthalpy data for 34 halogenated methanes have been evaluated and are presented in tabular form as well as in equations as a function of temperature. The data on which the values are based are the most reliable available and for most compounds cover the range from the melting temperature to the normal boiling temperature or to the limit of the experimental data whichever is the higher. A total of 24 references is provided.

1976 Miller J.W., Jr., Schorr G.R., Yaws C.L., *Heat Capacities of Liquids* (76MIL/SCH).

Heat capacity data have been correlated for 23 inorganic and 39 organic compounds using a series expansion in temperature of the type: $C_p = A+BT+CT^2+DT^3$, where C_p is the heat capacity of the saturated liquid in cal · K^{-1} · g^{-1} , A,B,C,D, are correlation constants for a specific compound, and T is the temperature in kelvin. In addition to the tabulated correlation constants, the C_p value at 25 °C, the temperature range, and references are given. The correlated data are not recommended for use in the immediate vicinity of the critical temperature, T_c.

1976 Raznjevic K., *Handbook of Thermodynamic Tables and Charts* (76RAZ).

A table is given which lists the specific heats at 20 °C for 22 common solid organic compounds. Another similar table provides specific heats for 82 liquid substances which include organic compounds, inorganic compounds, aqueous solutions, and mixtures. The specific heat data are given, in most cases, at several temperatures. The data do not appear to be evaluated. No references are listed for any property data in this handbook.

1976 Touloukian Y.S. and Makita T., *Specific Heat of Nonmetallic Liquids and Gases*, Suppl. to Vol. 6 in Thermophysical Properties of Matter (76TOU/MAK).

Heat capacities for 210 organic compounds in the gaseous and/or liquid states have been compiled as a by-product publishing 70TOU/MAK. This supplement does not appear to contain the careful and critical evaluation and commentary which was provided in 70TOU/MAK. Tables contain the following: compound name, formula, indication of sample purity, temperature in K, C_p in kJ · K^{-1} · kg^{-1}, pressure, method used, reported accuracy in %, and an internal TPRC accession number. References to original sources for gaseous and liquid heat capacities are frequently not correct. Only about 25% of the references to liquid heat capacity data appear to be satisfactory.

1977 Engineering Science Data Unit Series, *Heat Capacity and Enthalpy of Liquids. III: Halogenated Ethanes and Ethylenes*, Vol. 4 (Item No. 77007), (77ENG).

Heat capacity and enthalpy data have been evaluated and are presented in tabular form as well as in equations as a function of temperature. The data on which the values are based are the most reliable available and for most compounds cover the range from the melting temperature to the normal boiling temperature or to the limit of the experimental data whichever is the higher. A total of 42 references is provided.

1979 Engineering Science Data Unit Series, *Heat Capacity and Enthalpy of Liquids: Aliphatic Alcohols*, Vol. 4 (Item No. 79028) (79ENG).

Heat capacity and enthalpy data for 26 aliphatic alcohols have been evaluated and are presented in tabular form as well as in equations as a function of temperature. The data on which the values are based are the most reliable available and for most compounds cover the range from the melting temperature, temperature of glass transition, or pour point to a value of the reduced temperature that exceeds 0.5. A total of 42 references is provided.

1979 Herington E.F.G., *Physical Constants of Some Constituents of Coal Tar* (79HER).

Heat capacity data on 80 organic compounds are given at a specified temperature or for a temperature range in J · K^{-1} · g^{-1} . A bibliography and an index of precise C_p measurements which were carried out at NPL are available from another cited NPL Report.

1981 Engineering Science Data Unit Series, *Heat Capacity and Enthalpy of Liquids: Aliphatic Ketones*, Vol. 4 (Item No. 81030) (81ENG).

Heat capacity and enthalpy data for 20 aliphatic ketones have been evaluated and are presented in tabular form as well as in equations as a function of temperature. The data on which the values are based are the most reliable available and for most compounds cover the range from the melting temperature to a reduced temperature of about 0.8. A total of 16 references is provided.

1982 Wagman D.D., Evans W.H., Parker V.B., Schumm R.H., Halow I., Bailey S.M., Churney K.L., Nuttall R.L., *The NBS Tables of Chemical Thermodynamic Properties, Selected Values for Inorganic and* C_1 *and* C_2 *Organic Substances in SI Units* (82WAG/EVA).

Reliable values are provided as a result of a thorough evaluation of the available data for the chemical thermodynamic properties of inorganic compounds and organic compounds containing one or two carbon atoms. The properties tabulated are the enthalpy of formation, Gibbs function of formation, entropy, and heat capacity at 298.15 K, the enthalpy difference between 298.15 K and 0 K, and the enthalpy of formation at 0 K. Thermodynamic data for approximately 14 300 substances, in different phases and at different concentrations, are listed. This publication is the combination and culmination of NBS Technical Note 270-1 (October 1965) through NBS Technical Note 270-8 (May 1981). No bibliography or commentary are provided.

1984 Domalski E.S., Evans W.H., Hearing E.D., *Heat Capacities and Entropies of Organic Compounds in the Condensed Phase* (84DOM/EVA).

Heat capacity and entropy data have been compiled for approximately 1400 organic compounds in the liquid and solid phases. Values for the enthalpies and entropies of phase transitions, i.e., solid phase, fusion, vaporization, which were encountered as part of this evaluation, are included. The published literature from about 1925 to 1982 is covered as well as some earlier published papers. The bibliography contains over 850 references. Each referenced article is accompanied by a rating which indicates the quality of the reported data.

1984 Vasil'ev I.A., Petrov V.M., *Thermodynamic Properties of Oxygen-Containing Organic Compounds, Handbook* (84VAS/PET).

Selected thermodynamic properties for 271 organic oxygen compounds have been tabulated from the data available from the literature over the past 60 years. Heat capacity data are among the thermodynamic properties given which have been measured directly or correlated as a function of temperature. The compilation is valuable primarily as a source of data from the Russian literature which are not readily accessible elsewhere. A total of 217 references is listed.

1985 Engineering Science Data Unit Series, *Heat Capacity and Enthalpy of Liquids: Organic Sulphur Compounds (Thiols and Sulphides)*, Vol. 4 (Item No. 85024) (85ENG).

Heat capacity and enthalpy data for 34 organic sulfur compounds have been evaluated and are presented in tabular form as well as in equations as a function of temperature. The data on which the values are based are the most reliable available and for most compounds cover the range from the melting temperature to a reduced temperature of 0.86. A total of 24 references is provided.

1985 Wilhoit R.C., Chao J., Hall K.R., *Thermodynamic Properties of Key Organic Oxygen Compounds in the Carbon Range* C_1 *to* C_4. *I. Properties of Condensed Phase* (85WIL/CHA).

Data for heat capacities, enthalpies, and phase transitions which have been obtained from calorimetric measurements were examined and critically evaluated on the crystal, glass, and liquid phases for organic oxygen compounds containing one to four carbon atoms per molecule. Equations for the heat capacities expressed as polynomial functions of temperature were fitted to selected data by a least squares procedure. Tables of smoothed values of thermodynamic properties, derived from these functions, are given for 38 organic oxygen compounds. A total of 327 references comprise the bibliography.

1986 Engineering Science Data Unit Series, *Heat Capacity and Enthalpy of Liquids: Alkanes*, Vol. 4 (Item No. 86007) (86ENG1).

Heat capacity and enthalpy data for 70 alkanes have been evaluated and are presented in tabular form as well as in equations as a function of temperature. The data on which the values are based are the most reliable available and cover the range from the melting temperature to a reduced temperature of 0.96. A total of 80 references is provided.

1986 Engineering Science Data Unit Series, *Heat Capacity and Enthalpy of Liquids: Alkenes and Alkadienes*, Vol. 4 (Item No. 86023) (86ENG2).

Heat capacity and enthalpy data for 34 alkenes and 9 alkadienes have been evaluated and are presented in tabular form as well as in equations as a function of temperature. The data on which the values are based are the most reliable available and for most compounds cover the range from the melting temperature to a reduced temperature of 0.96. A total of 42 references is provided.

1988 Liley P.E., Makita T., Tanaka Y., *Properties of Inorganic and Organic Fluids*, CINDAS Vol. V-1 (88LIL/MAK).

Thermophysical properties, including heat capacities, are given for 10 inorganic and 18 organic fluids as a function of temperature and pressure. Chapters are devoted to individual fluids (i.e., liquid), some of which are: ammonia, chloroform, methane, benzene, sulfur dioxide, methanol, ethanol, water, ethylene glycol, glycerol, toluene, acetylene, and C_5 to C_{10} *n*-alkanes. Data are provided as tables and also in graphical form. An evaluation of the available data is given for each fluid along with the corresponding references. The selected data were used to develop a polynomial expression for the heat capacity of each of the fluids as a function of temperature.

1989 Daubert T.E., Danner R.P., *Physical and Thermodynamic Properties of Pure Chemicals: Data Compilation, DIPPR® Database*, 4 Volumes (89DAU/DAN).

A compilation containing critically evaluated data on the physical, thermodynamic, and transport properties of over 900 organic compounds has been prepared under the auspices of the Design Institute for Physical Property Research (DIPPR®), which is an AIChE activity. Heat capacity is one of 38 properties of which 13 are, like C_p, temperature dependent. Original or raw experimental data are not provided. Instead, coefficients are given for a specific equation which correlates selected data for a compound whose accuracy has been assessed and identified. An indication is offered as to which data (and references) were used and which were not. The units of C_p in the correlating equations are in $J \cdot K^{-1} \cdot kmol^{-1}$. Specific comments are provided for compounds when necessary. This data compilation is dynamic because it is loose-leaf so that new compounds may be added and data for others may be revised. A bibliography contains 4181 references which covers articles for all 38 properties.

1990 Chao J., Gadella N.A.M., Gammon B.E., Marsh K.N., Rodgers A.S., Somayajulu G.R., Wilhoit R.C., *Thermodynamic and Thermophysical Properties of Organic Nitrogen Compounds. Part I. Methanamine, Ethanamine, 1- and 2-Propanamine, Benzenamine, 2-, 3-, and 4-Methylbenzenamine* (90CHA/GAD).

The thermodynamic and thermophysical properties of eight primary amines have been examined and evaluated. Recommended values are given for the following properties: normal boiling temperature, freezing and triple point temperatures, critical constants, thermodynamic properties in the liquid and solid phases, vapor pressure, enthalpy of vaporization, density, second virial coefficients, and enthalpy of combustion. Ideal gas thermodynamic properties were calculated using statistical mechanical procedures. A bibliography contains about 480 references.

1990 Domalski E.S., Hearing E.D., *Heat Capacities and Entropies of Organic Compounds in the Condensed Phase*, Vol. II (90DOM/HEA).

This compilation of data on the heat capacities, entropies, and phase transitions of organic compounds in the condensed phase supplements the work published earlier on this subject (84DOM/EVA). Data on approximately 1300 organic compounds are provided. A total of 565 articles have been examined and evaluated. Each referenced article is provided with a rating which indicates the quality of the reported data. About half of the articles examined contain data published prior to 1982. Most of the 1990 literature is covered.

1990 Zábranský M., Růžička V., Jr., Majer V., *Heat Capacities of Organic Compounds in the Liquid State. I. C_1 to C_{18} 1-Alkanols* (90ZAB/RUZ).

Heat capacities of liquid C_1 to C_{18} 1-alkanols which were determined by calorimetric methods have been compiled and critically evaluated. The selected experimental data were fitted as a function of temperature with cubic splines using weighted least squares minimization. The parameters of the cubic spline polynomials and the recommended values for heat capacities are provided. An assessment of the quality of the recommended data is also given. A new quasi-polynomial equation was derived which permits extrapolation of heat capacities outside of the temperature range of experimental values. The parameters for this equation for C_1 to C_{10} 1-alkanols are presented. Over 140 references are given at the end of the paper. The results of the critical evaluation on C_1 to C_{18} 1-alkanols by 90ZAB/RUZ are part of this compilation.

1991 Marsh K. N. et al., *Selected Values of Thermodynamic Properties of Organic Compounds, TRC Data Project* (91MAR).

Tables of selected values of physical and thermodynamic properties of organic compounds are provided as a function of temperature, separately for hydrocarbons and non-hydrocarbons, in loose-leaf form. The tables are prepared only after an extensive evaluation and correlation of the available data from the literature. This work is a continuation and expansion of 47ROS/PIT and 53ROS/PIT. The expansion has led to an increase in the tabulation of data for the condensed phase over the years. A set of eleven (loose-leaf) volumes constitutes the physical and thermodynamic properties of hydrocarbons while similar information for non-hydrocarbon compounds is found in a set of eight (loose-leaf) volumes. Up-dates to both sets are available twice per year.

1991 Růžička V. Jr., Zábranský M., and Majer V., *Heat Capacities of Organic Compounds in the Liquid State. II. C_1 to C_{18} n-Alkanes* (91RUZ/ZAB).

Heat capacities of liquid C_1 to C_{18} *n*-alkanes which were determined by calorimetric methods have been compiled and critically evaluated. The selected experimental data were fitted as a function of temperature with cubic splines using weighted least squares minimization. The parameters of the cubic spline polynomials and the recommended values for heat capacities are given. Heat capacities were also fitted by a quasi-polynomial equation permitting extrapolation of heat capacities outside of the temperature range of the experimental values. About 150 references are provided at the end of the paper. The results of the critical evaluation on C_1 to C_{18} *n*-alkanes by 91RUZ/ZAB are part of this compilation.

5. Methodology of Data Treatment; Establishment of Recommended Data

5.1. Data Base of Raw Values

Section 1.2. specifies which liquids and what types of heat capacity data (regarding experimental techniques and literature sources) are covered in this compilation. The experimental values extracted from the literature together with auxiliary data were stored in the data base of raw experimental values. This data base covered 1624 substances and consisted of 3389 data sets (set of experimental values from one calorimeter reported by an author for one substance in one original source). Only the parameters for smoothing equations were available for 223 data sets (no discrete data points were reported in these sources). The data sets of discrete values contained 27430 raw data points. The auxiliary data comprised information on the quality of experimental values, sample purity and its determination, the nature of the experimental procedure (calorimetric technique and type of resulting heat capacity) and units used in the original source. This information was used later as a guide in the selection of data sets for the final correlation. Most auxiliary data are summarized for individual substances in tables of experimental heat capacities.

According to the information in the literature sources, we distinguished in the data base of raw values three types of heat capacities: *isobaric heat capacity* C_p, *saturation heat capacity* C_{sat}, and the *average heat capacity* C_{avg}. These three quantities are defined by equations (II-3), (II-4), and (II-18), respectively, and their thermodynamic background is described in Chapter 2.

Below the normal boiling temperature T_{nb} the measured isobaric heat capacity usually relates to atmospheric pressure though this is not frequently specified in the original source. C_p has limited dependence on pressure at low and moderate vapor pressures, and its effect can be neglected unless the pressure change is large. Below the normal boiling temperature no distinction was made between the literature C_p data relating to the saturation line, to atmospheric conditions, or to 101.325 kPa; these were considered as identical. Above the normal boiling temperature, the saturation heat capacities C_{sat} rather than C_p data are reported in the original literature.

Measurements at pressures remote from the vapor pressure are not numerous. They are of interest in studies dealing with mixtures and, hence, no particular attention has been paid to them in the context of this compilation. They were considered only in those cases when they were determined at pressures not far removed from the saturation line or when they extended above the temperature range of the experimental data. They were included only when it was possible to determine with fair accuracy the appropriate correction for their conversion to saturation conditions.

In most calorimetric determinations of heat capacities, the temperature change does not exceed several degrees (temperature steps between 3 and 8 degrees are the most frequent); then, in most cases, the true heat capacity equals the experimental value within the error of measurement. When the temperature change in the experiment is above 10 K, the value obtained is denoted in the following tables as C_{avg}; an exception was made when this type of heat capacity was used for establishing recommended values for several old sources when no other data were available in the corresponding temperature range. In this case Eqn. (II-24) was used for correcting the data to the true heat capacities. Several older data sources (*81VON, *86SCH, *87SCH, 18NAR) were treated in this way and the corresponding deviations in the tables of Correlated Heat Capacities equal zero.

5.2 Evaluation and Selection Process

All available sources were critically assessed. The important part of the selection process was the simultaneous correlation of all experimental data which served to test the consistency and helped to reveal systematic errors. The correlation was performed by the weighted least-squares method; the *minimized objective function* S had the form:

$$S = \sum_{i=1}^{n} \frac{(C_{sm} - C_{exp})^2}{\sigma^2 C_i} \qquad \text{(V-1)}$$

where the summation is over all the values included in the correlation and the subscript "exp" and "sm" relate to experimental and smoothed values, respectively. The reciprocal of the variance, $\sigma^2 C$, is the *weighting factor*. It was estimated for each value on the basis of the assumed experimental error of the data set used in the correlation. The input information was the percentage error of the experimental data, $\sigma_r C$, as given by the author for the whole data set or estimated by the evaluator in cases where no information was available in the literature. Thus, the variance of the i-th data point was expressed as:

$$\sigma^2 C_i = \left(\frac{C_i \sigma_r C}{100}\right)^2. \qquad \text{(V-2)}$$

The following criteria were observed in the selection process:

1. accuracy of the experimental technique claimed by the author,
2. performance history of the laboratory,
3. consistency of the data with the values from other sources (if available),
4. purity of the substance,
5. type of the calorimeter,
6. time of data origin (measurements published after the year 1940 were preferred),
7. scatter of the data.

It is not, however, possible to apply the above criteria for selecting compounds in a rigorous manner following strictly a certain policy. In many cases, the error of measurement claimed by the author is too optimistic and sometimes does not correspond to reality at all. The evaluating reputation of different laboratories is also a somewhat subjective enterprise. According to our experience the most reliable data were reported from the following laboratories which were preferred in the data selection: BMB, NIPER, NBSW, NBSB, NPLT, and UMAA (the list of acronyms is given in the Section 6.2.1., see below).

Selection and correlation of data for each compound was carried out simultaneously and involved the following steps:

(a) All available heat capacity data sets along with the information on their accuracies were read in and a preliminary joint correlation was performed. If necessary, experimental data were converted to the same type of heat capacity — C_p or C_{sat} (see Section 5.4. for details). Prior to the correlation, data sets considered a priori as unreliable were discarded if more accurate data were available for the same temperature range. However, deviations of the discarded data from the final values were always determined within the temperature range of the recommended values. For those sources where only parameters of a smoothing equation were available, pseudo-discrete data were generated in the temperature range of the parameters' validity. When the number of the experimental data points was not indicated in the original source, a temperature step of 5 to 10 K was used according to the length of the temperature interval and density of data points from other sources.

(b) In the next steps it was possible to make the following tentative changes: to discard individual values, to reject parts of or whole data sets that showed little consistency with the other data, to change weights of whole data sets by altering the expected percent error of measurement, to modify the number of generated pseudo-discrete data points, and to change the temperature limits within which the data sets are considered.

(c) The correlation was repeated several times until the final fit with the selected data was obtained where differences between experimental and smoothed values were roughly equal to or smaller than expected experimental errors. The final correlation of the selected experimental data was assigned a level of accuracy (I to VI, for the definition of levels of accuracy see Section 6.3.3) according to the quality of the correlated data.

The main criterion for judging the quality of the correlation was the *standard weighted deviation*, s_w:

$$s_w = \left(\frac{S_{min}}{n-m}\right)^{1/2}, \qquad \text{(V-3)}$$

where S_{min} is the value of the objective function at its minimum, n is the overall number of the fitted data points, and m is number of independent adjustable parameters in a correlation equation. When experimental data are consistent within the expected error limits, s_w should be close to unity. In addition, use of the following statistical criteria was made:

the *standard deviation*

$$s = \left(\frac{\sum_{i=1}^{n}(C_{sm} - C_{exp})_i^2}{n-m}\right)^{1/2} \qquad \text{(V-4)}$$

the *percentage standard deviation*

$$s_r = \left(\frac{\sum_{i=1}^{n}[(C_{sm} - C_{exp})/C_{exp}]_i^2}{n-m}\right)^{1/2} 100, \qquad \text{(V-5)}$$

the *bias*

$$s_b = \frac{\sum_{i=1}^{n}(C_{sm} - C_{exp})_i}{n}, \qquad \text{(V-6)}$$

and the difference between the number of experimental points with positive and negative deviation from the smoothed values (denoted in tables as +/–).

In order to get information on how data from individual sources compare with the final correlation, the deviations d_w, d, d_r, and d_b defined analogously to s_w, s, s_r and s_b were calculated for both included and rejected data sets. The definitions are as follows:

the *average weighted deviation*

$$d_w = \left(\frac{\sum_{i=1}^{n_1}[(C_{sm} - C_{exp})^2/\sigma^2 C]_i}{n_1}\right)^{1/2}, \quad (V\text{-}7)$$

the *average deviation*

$$d = \left(\frac{\sum_{i=1}^{n_1}(C_{sm} - C_{exp})_i^2}{n_1}\right)^{1/2} \quad (V\text{-}8)$$

the *average percentage deviation*

$$d_r = \left(\frac{\sum_{i=1}^{n_1}[(C_{sm} - C_{exp})/C_{exp}]_i^2}{n_1}\right)^{1/2} 100, \quad (V\text{-}9)$$

the *bias* of a data set

$$d_b = \frac{\sum_{i=1}^{n_1}(C_{sm} - C_{exp})_i}{n_1}, \quad (V\text{-}10)$$

where n_1 denotes the number of data points in one data set.

When heat capacity was measured at one temperature only and several literature sources were available, simple selection of the best value or averaging of several measurements was performed.

5.3. Temperature Correlation and Establishment of Recommended Values

Two different approaches were adopted for correlating heat capacities as a function of temperature. *Polynomials* or *cubic spline polynomials* were used for accurately fitting heat capacity inside the temperature limits of experimental values. A *quasi-polynomial equation* was also used when it was appropriate (see below); this equation enabled one to describe the temperature dependence by one set of parameters which, unlike polynomial equations, allows a meaningful extrapolation above the upper temperature limit of the experimental data. Both correlation approaches are described in Section 2.4; here we give only the specifics regarding their use in this project.

5.3.1. Representation of Data by Polynomial(s)

When data are available only in a limited temperature range, one polynomial of third or lower degree may be sufficient for fitting the data successfully. In all cases, when such a condition was probable, the correlation with one polynomial was tried first. Its degree was gradually increased to 3, and the statistical F test was used to determine when the addition of higher terms was not significant any longer.

Fitting with cubic splines was used only when the F test indicated need for a higher than third degree polynomial. We used the correlation algorithm developed at the Technical University of Budapest (79KOL) in which the two boundary conditions are considered during minimization as two additional adjustable parameters. This variant is more flexible than the natural spline. In the first step, three knot temperatures were used (two at the limits and one inside the temperature interval), and if necessary additional knots were added. Usually three and sometimes four knots (i.e. five and six independent parameters, respectively) were necessary for the satisfactory fit of the data within the expected experimental error limits. For exceptional cases, five knots (i.e. seven independent parameters) had to be used for the description of data covering a wide temperature range; in situations where the data sets reached the vicinity of the critical point, the heat capacity increased dramatically.

For many substances, the interior temperature knots were positioned such that the individual subintervals had about the same length. Adjustments were made when the density of experimental points changed significantly with temperature. This was often the case as high-temperature values were usually less numerous than low-temperature measurements. Situations in which one subinterval is almost "empty" are to be avoided. Shorter distances between knot temperatures were also needed when the shape of the heat capacity curve was changing considerably, especially close to the critical point.

The results of spline correlation are usually reported in terms of the knot values and two boundary conditions; the spline routine is then necessary for generating the recommended data. Since such a routine is not available to all users of the data, we preferred to tabulate directly the parameters of the cubic polynomials which are easier to use. This presentation is also consistent with the tabulation when only one polynomial is used. The tabulated adjustable parameters, A_j, relate to the equation expressing the dimensionless quantity C/R as a function of the scaled temperature $T/100$:

$$\frac{C}{R} = \sum_{j=0}^{m} A_{j+1}\left(\frac{T}{100}\right)^j, \quad (V\text{-}11)$$

where R is the gas constant; scaling was made to improve the numerical stability of the fitting procedure. The upper limit of the summation m is equal to 3 in all cases where the overall temperature range is subdivided (cubic splines fitting) or can be lower when only one polynomial is used to describe the data.

Tables of discrete values were generated using a polynomial representation of the recommended data.

5.3.2. Representation of Data by a Quasi-Polynomial Equation

The quasi-polynomial equation was used in the form:

$$\frac{C}{R} = A_1 \ln(1 - T_r) + \frac{A_2}{1 - T_r} + \sum_{j=0}^{m} A_{j+3} T_r^j, \qquad \text{(V-12)}$$

where $T_r = T/T_c$ and m equals 3 for about 50 % of substances. By combining Eqns. (II-31) and (II-32), it can be shown that the tabulated parameters A_j are constrained as follows: $A_1 = -2(a_2a_3+a_2a_4)$, $A_2 = a_2^{\,2}$, $A_3 = a_1+2a_2a_4$, $A_4 = a_3^{\,2} - 2a_2a_4$, $A_5 = a_3a_4$, $A_6 = a_4^{\,2}/3$. Parameters a_1, a_2, a_3 and a_4 relate to the Eqn. (II-31) with $m_1 = 1$. For the remaining 50 % of substances Eqn. (V-12) with 4 parameters A_j (m=1) was sufficient. In this case parameters A_5 and A_6 equal zero as a_4 equals zero and the remaining constraints for A_j are modified appropriately.

Under the above conditions, the temperature derivative of the heat capacity is always positive. For some substances, the heat capacity curve has a shallow minimum above the triple point temperature (see Fig.3); in those cases, the nearest experimental temperature above this minimum was selected as the lower limit of the correlation and the data below that temperature were not included. The quasi-polynomial equation was not used when fewer then seven experimental data points were available and the temperature range of experimental data was below 50 K, or when there was no possibility of obtaining the critical temperature.

The variation of heat capacity in the high-temperature range differs considerably from one compound to another. It should be kept in mind that the quasi-polynomial equation is mainly empirical and extrapolation above the upper temperature limit of experimental data is correct only in a qualitative manner. No guarantee can be given that the extrapolation will describe quantitatively the real heat capacity in this region except for the value at the critical point which is always correct and corresponds to plus infinity.

For organic substances, the critical temperatures from the data bank of the Prague Institute of Chemical Technology (91CDA) or from the TRC tables (91MAR) were used; for substances where T_c values were not found, estimates were performed according to Lydersen (55LYD). The compilation by Morachevskii and Sladkov (87MOR/SLA) was used as the main source of critical temperatures for inorganic substances.

The following expressions were derived for the calculation of the enthalpy and entropy differences when using the quasi-polynomial equation:

$$\frac{\Delta H}{RT_c} = A_1[(T_{r2}-1)\ln(1-T_{r2})-(T_{r1}-1)\ln(1-T_{r1})-(T_{r2}-T_{r1})]$$
$$-A_2[\ln(1-T_{r2})-\ln(1-T_{r1})] + A_3(T_{r2}-T_{r1}) \qquad \text{(V-13)}$$
$$+\frac{A_4}{2}(T_{r2}^2 - T_{r1}^2) + \frac{A_5}{3}(T_{r2}^3 - T_{r1}^3) + \frac{A_6}{4}(T_{r2}^4 - T_{r1}^4)$$

$$\frac{\Delta S}{R} = A_1[\mathrm{dilog}(1-T_{r1})-\mathrm{dilog}(1-T_{r2})]$$
$$+A_2[\ln(T_{r2})-\ln(1-T_{r2})-\ln(T_{r1})+\ln(1-T_{r1})] \qquad \text{(V-14)}$$
$$+A_3(\ln T_{r2}-\ln T_{Nr1}) + A_4(T_{r2}-T_{r1})$$
$$+\frac{A_5}{2}(T_{r2}^2 - T_{r1}^2) + \frac{A_6}{3}(T_{r2}^3 - T_{r1}^3)$$

where dilog (x) can be approximated by the following function

$$\mathrm{dilog}(x) = \frac{D_1+D_2x+D_3(x-0.1)^2+D_4(x-0.1)^3+D_5(x-0.1)^4+D_6(x-0.1)^5}{D_7+D_8x+D_9(x-0.1)^2+D_{10}(x-0.1)^3+D_{11}(x-0.1)^4+D_{12}(x-0.1)^5} \qquad \text{(V-15)}$$

where

D_1	=	−1.168876770,	D_2	=	24.68591493,
D_3	=	145.8845448,	D_4	=	242.8099791,
D_5	=	−178.5653452,	D_6	=	−338.8035369,
D_7	=	−1.096178670,	D_8	=	20.96178670,
D_9	=	150.3249552,	D_{10}	=	427.9410776,
D_{11}	=	414.7775590,	D_{12}	=	69.55852420.

5.4. Types of Heat Capacities Presented

When establishing the recommended values no distinction was made between the saturation heat capacity C_{sat} and the isobaric heat capacity C_p when their difference was smaller than the expected error of experimental data over the whole temperature range of their validity. This was the case for practically all substances where data did not reach above the normal boiling temperature. Then only one set of parameters was presented for each of the correlation equations and only one set of discrete values was generated.

For the compounds where experimental values of reasonable accuracy are available above the normal boiling temperature, distinct recommended data sets are presented for C_p and C_{sat}. In this case, all raw data were first converted to one type of heat capacity which was used in the evaluation and selection process. After the experimental values for the final correlation had been set, they were expressed as C_p and C_{sat} and both sets were correlated separately under identical conditions using the polynomial or cubic spline representation. Only isobaric heat capacity C_p was correlated by the quasi-polynomial equation.

Recommended C_{sat} data relate by definition to the saturation line. Recommended C_p data relate above the normal boiling temperature to the saturation line. Below T_{nb}, they represent both the heat capacity at standard pressure 101.325 kPa and the saturation pressure as their difference is smaller then the accuracy of the best literature data.

Thermodynamic relationships between isobaric and saturation heat capacities are discussed in detail in Section 2.1. When necessary, conversion between C_p and C_{sat} was performed using Eqn. (II-6). As there are not enough data on expansivities, the term $(\partial V/\partial T)_p$ was approximated by $(\partial V/\partial T)_{sat}$; the difference between them becomes important only far above the normal boiling temperature. The latter term was calculated from the temperature correlation of densities along the saturation line using a modified Rackett equation (85CAM/THO) or the Francis equation (86SMI/SRI). The derivative $(dp/dT)_{sat}$ was calculated from one of the following vapor pressure equations: Wagner, Frost-Kalkwarf, Cox or Antoine using parameters from the references 83MCG, 79DYK/REP, 90CHI/GAM, 92STE/CHI2 respectively. More specific information regarding individual substances is given when appropriate in the introductory discussions before the tables.

5.5. The Influence of Changes in Atomic Masses on the Heat Capacity Value

The values of atomic masses of elements are being continuously refined. When converting the primary measured data to the heat capacities the individual authors were using molar masses valid at the time of experiment. Thus the later modifications of molar masses necessarily also affect the heat capacity values; logically the most influenced are the old measurements. We have analyzed and evaluated the effect of the change in atomic masses on the heat capacity for three compounds, n-heptane, ethanol, and triethylamineborane, for which accurate raw data are available. The results of the evaluation are given in Table 5. The systematic error in the heat capacity due to the change in molar mass is equal to the percent difference in the original and the reference molar masses. It is apparent from the table that this error is at least one order of magnitude smaller when compared to the error in the data. The atomic masses recommended by the IUPAC Commission on Atomic Weights and Isotopic Abundances (91COM) were used as reference values throughout the publication.

TABLE 5. Examples of the systematic error Δ in the heat capacities measured over the last 60 years due to the change in atomic masses of elements.

Data source	Year	$\sigma_r{}^a/\%$	Molar mass	$\Delta^b/\%$
		n-heptane		
47OSB/GIN	1947	0.1	100.198	6.10^{-3}
54DOU/FUR	1954	0.1	100.205	-1.10^{-3}
61HUF/GRO	1961	0.2	100.20557	-2.10^{-3}
79SCH/OFF	1979	0.1	100.2034	-6.10^{-4}
		ethanol		
31FIO/GIN	1931	0.2	46.0468	5.10^{-2}
68PAZ/REC	1968	0.1	46.0694	-8.10^{-4}
76FOR/BEN1	1976	0.3	46.0688	5.10^{-4}
79BRO/ZIE	1979	0.2	46.0688	5.10^{-4}
		triethylamineborane		
67SMI/GOO1	1967	0.2	115.0267	-7.10^{-5}
70FIN/TOD	1970	0.2	115.0249	1.10^{-3}

[a] measurement error of the original data

[b] $\Delta = 100 \cdot (\text{molar mass}_{1991} - \text{molar mass}_{year}) / \text{molar mass}_{1991}$

5.6. Conversion of Temperatures and Original Heat Capacity Data to the International Temperature Scale of 1990

Most experimental data which served for producing the recommended data in this compilation were published over the last 60 years. During this period three different temperature scales (ITS-27, IPTS-48 and IPTS-68) were in use. Since the beginning of 1990, ITS-90 has been the standard. We have closely examined the effect which the introduction of the new temperature scale has to the published heat capacity values. Main points are as follows.

a. *What are the differences between the different temperature scales?*

Table 6 lists the differences between ITS-90 and the previous two temperature scales in the temperature range between −180 and 300 °C where most data compiled here were reported (data at higher temperatures are always subject to errors which are much larger than any changes due to a scale conversion). ITS-48 is identical with ITS-27 below 630 K. Temperature derivatives of the differences are also listed; they play a more important role in conversion than the shifts in temperature (see below). Note that starting with 473 K the differences between IPTS-68 and IPTS-48 are greater than between ITS-90 and the previous two scales (see Table 6).

b. *How to correct heat capacities for a change in the temperature scale*

When converting heat capacity values to ITS-90 two effects have to be considered.

Firstly, the temperatures, to which the experimental heat capacities are related, have to be converted. This procedure is trivial and has little effect upon the heat capacity except in the region close to the critical point where the variation with temperature is steep. In other cases, the correction in the heat capacity due to temperature shift rarely exceeds 0.02 percent and is typically one order of magnitude lower.

Secondly, the heat capacities are usually determined as a ratio of energy to the corresponding temperature increment ΔT. This increment can be affected by the scale change in a more substantial manner than T itself, especially at temperatures where $d((\delta T)/dT)$ has a large value (see Table 6); the changes in ΔT are typically several hundredths of a percent and attain in the most unfavorable cases as much as 0.09 and 0.03 percent for conversion from IPTS-48 and IPTS-68, respectively, for the compounds and temperature ranges considered herein. The percent change in the heat capacity is then identical.

It is clear that both effects are interrelated and can add to or compensate for each other depending on the shape of the δT versus T function. Specific examples in Tables 7 to 12 demonstrate the effects of conversion from IPTS-48 and IPTS-68 for the heat capacities of organic compounds for a wide temperature range.

c. *Is conversion of heat capacities C_{sat} and C_p to ITS-90 feasible?*

For converting the data to ITS-90 scale, it is necessary to have the following information:

— temperature scale used; this information is not often given explicitly in the original literature but can be guessed from the year of publication

— temperature increments for individual data points; this information is unfortunately missing in most data sources. It is systematically given only in the articles from the former National Bureau of Standards, Washington D.C. (now called NIST) for a very limited number of compounds and for a part of data reported by the former National Bureau of Mines, Bartlesville, Oklahoma (now called NIPER). For extending the temperature range of recommended values, the above sources are often combined with other data for which conversion is not possible. A partial conversion of data serving to produce the recommended values would mean the introduction of certain inconsistencies.

TABLE 6. Approximate differences $\delta = (t_{90} - t_{48})$/°C and $\delta = (t_{90} - t_{68})$/°C and derivatives of these differences (based on values reported by 92GOL/WEI)

	IPTS-48			IPTS-68	
t_{48}/°C	δ	$(d\delta/dT)\cdot 10^5$	t_{68}/°C	δ	$(d\delta/dT)\cdot 10^5$
−180	0.020	00	−180	0.008	20
−170	0.017	−46	−170	0.010	20
−160	0.007	−82	−160	0.012	13
−150	0.000	00	−150	0.013	10
−140	0.001	18	−140	0.014	00
−130	0.008	79	−130	0.014	00
−120	0.017	90	−120	0.014	00
−110	0.026	90	−110	0.013	00
−100	0.035	72	−100	0.013	00
−90	0.041	48	−90	0.012	00
−80	0.045	00	−80	0.012	00
−70	0.045	00	−70	0.011	−10
−60	0.042	−34	−60	0.010	−10
−50	0.038	−48	−50	0.009	−10
−40	0.032	−69	−40	0.008	−13
−30	0.024	−80	−30	0.006	−20
−20	0.016	−80	−20	0.004	−20
−10	0.008	−80	−10	0.002	−20
0	0.0	−69	0	0.0	−20
10	−0.006	−60	10	−0.002	−24
20	−0.012	−48	20	−0.005	−24
30	−0.016	−40	30	−0.007	−24
40	−0.020	−34	40	−0.010	−30
50	−0.023	−30	50	−0.013	−30
60	−0.026	00	60	−0.016	−24
70	−0.026	00	70	−0.018	−24
80	−0.027	00	80	−0.021	−30
90	−0.027	00	90	−0.024	−24
100	−0.026	13	100	−0.026	−20
110	−0.024	13	110	−0.028	−20
120	−0.023	15	120	−0.030	−20
130	−0.020	24	130	−0.032	−20
140	−0.018	20	140	−0.034	−20
150	−0.016	27	150	−0.036	−13
160	−0.012	34	160	−0.037	−10
170	−0.009	34	170	−0.038	−10
180	−0.005	40	180	−0.039	00
190	−0.001	40	190	−0.039	00
200	0.003	40	200	−0.040	00
210	0.007	40	210	−0.040	00
220	0.011	34	220	−0.040	00
230	0.014	34	230	−0.040	00
240	0.018	34	240	−0.040	00
250	0.021	30	250	−0.040	00
260	0.024	34	260	−0.040	00
270	0.028	27	270	−0.039	00
280	0.030	20	280	−0.039	00
290	0.032	20	290	−0.039	00
300	0.034	13	300	−0.039	00

Heat capacities and the corresponding temperatures published in the literature cannot always be considered strictly as raw values. The results of measurements are in some cases "smoothed" graphically or with a computer program before tabulation (listed temperatures are very often rounded values). There is usually little information on such data treatment in original data sources; it can be, however, assumed that during smoothing the heat capacity is already subject to shifts comparable or even larger than those corresponding to the scale changes; hence, real "raw values" are frequently not available. In these cases, adjustments in the heat capacities due to any temperature scale changes become highly questionable (see Tables 7 to 12 for examples) and application of the conversion formula by Douglas (69DOU) as presented by Goldberg and Weir (92GOL/WEI) is not possible.

It is obvious from the above considerations that conversion of heat capacities to the basis of ITS-90 is not always a straightforward process since the needed information is often missing. Also, since the changes in the heat capacities that would be produced are almost always less than the accuracy of the heat capacities, the conversion is very rarely called for.

d. *Magnitude of the changes in the heat capacity and in the uncertainty of recommended data*

The reliability of recommended data is expressed in this monograph by six uncertainty classes. The best uncertainty class I (error below 0.1 percent) is used only for three compounds (*n*-heptane, ethanol, benzothiazole) in a part of the temperature interval where experimental data measured by different authors were reported. The next class II (error below 0.3 percent) covers about 200 compounds. Considering the magnitude of changes in C_p due to the temperature scale shifts, it seems reasonable to correct only the data for "class I compounds". For the above three compounds with the class I uncertainty, the necessary auxiliary information required for conversion to ITS-90 is available only for two compounds: n-heptane (three data sets, see Table 13) and for ethanol (one data set 77HAI/SUG).

e. *Conclusions regarding temperature scale changes*

Upon consideration of the factors given above, the conversion of experimental data to ITS-90 turned out to be either questionable or unnecessary.

The following six tables illustrate the effect of temperature scale changes on heat capacities using as examples four compounds measured at the former Bureau of Mines at Bartlesville, Oklahoma; temperature increments were always available in the original literature. Tables 7 to 9 illustrate conversion from IPTS-48 to ITS-90, Tables 10 to 12 are examples of conversion from IPTS-68 to ITS-90. Temperature in the first column corresponds to IPTS-48 or IPTS-68 in the relationship to the heat capacities in the second column and to ITS-90 in the relationship to the third column. Differences (in percent) are listed in the last column of all tables.

Two different ways in converting the data are demonstrated:

1. In Tables 7,8,10 and 11 the temperatures and heat capacities reported in the original source were converted to ITS-90 (both the effects on T and ΔT were taken into consideration). Subsequently both original and converted data were

smoothed separately by third degree polynomials which served for generating the values at the listed rounded temperatures.

2. In Tables 9 and 12 the temperatures and heat capacities reported in the original source were converted to ITS-90 in the same way as under 1. Subsequently only the converted data were smoothed by a third degree polynomial which served for generating ITS-90 heat capacity values at the temperatures identical to those given in the original source.

Comparison of percent differences for the cases where heat capacities for an identical compound were treated differently (Tables 8 and 9 and Tables 11 and 12) shows that both approaches to the conversion give comparable results.

Examination of the magnitude of differences in heat capacities due to the temperature scale change reveals that they are at least two times smaller that the measurement error.

TABLE 7. Conversion of saturation heat capacities from IPTS-48 to ITS-90 for 2,2-dimethylpentane (61HUF/GRO, measurement precision and measurement accuracy 0.1 % and 0.2 %, respectively). Comparison of smoothed data.

T K	C_{48} $J \cdot K^{-1} \cdot mol^{-1}$	C_{90} $J \cdot K^{-1} \cdot mol^{-1}$	$100 \cdot (C_{48}-C_{90})/C_{90}$ %
160	167.07	166.91	0.10
170	170.12	169.99	0.08
180	173.33	173.22	0.06
190	176.68	176.61	0.04
200	180.16	180.14	0.01
210	183.79	183.81	−0.01
220	187.55	187.62	−0.04
230	191.45	191.55	−0.05
240	195.47	195.60	−0.07
250	199.61	199.77	−0.08
260	203.88	204.05	−0.08
270	208.26	208.43	−0.08
280	212.76	212.91	−0.07
290	217.36	217.48	−0.06
300	222.08	222.14	−0.03

TABLE 8. Conversion of saturation heat capacities from IPTS-48 to ITS-90 for 1-(ethylthio)propane (61MCC/FIN, measurement precision and measurement accuracy 0.1 % and 0.2 %, respectively). Comparison of smoothed data.

T K	C_{48} $J \cdot K^{-1} \cdot mol^{-1}$	C_{90} $J \cdot K^{-1} \cdot mol^{-1}$	$100 \cdot (C_{48}-C_{90})/C_{90}$ %
160	180.63	180.39	0.13
170	180.36	180.20	0.09
180	180.40	180.31	0.05
190	180.74	180.70	0.02
200	181.35	181.36	−0.01
210	182.22	182.27	−0.03
220	183.33	183.41	−0.04
230	184.66	184.77	−0.06
240	186.20	186.32	−0.06
250	187.93	188.06	−0.07
260	189.84	189.97	−0.07
270	191.89	192.03	−0.07
280	194.09	194.22	−0.07
290	196.41	196.53	−0.06
300	198.83	198.94	−0.06
310	201.34	201.43	−0.04
320	203.92	203.99	−0.03
330	206.56	206.61	−0.02
340	209.22	209.26	−0.02
350	211.91	211.93	−0.01
360	214.60	214.60	0.00
370	217.27	217.25	0.01

TABLE 9. Conversion of saturation heat capacities from IPTS-48 to ITS-90 for 1-(ethylthio)propane (61MCC/FIN, measurement precision and measurement accuracy 0.1 % and 0.2 %, respectively). Comparison of the smoothed data (ITS-90) with the raw values reported in the literature source (IPTS-48).

T K	C_{48} $J \cdot K^{-1} \cdot mol^{-1}$	C_{90} $J \cdot K^{-1} \cdot mol^{-1}$	$100 \cdot (C_{48}-C_{90})/C_{90}$ %
165.59	180.44	180.25	0.11
171.78	180.35	180.20	0.08
181.88	180.45	180.37	0.04
197.08	181.15	181.15	0.00
217.30	183.01	183.08	−0.04
237.22	185.76	185.87	−0.06
257.28	189.30	189.44	−0.07
278.36	193.73	193.85	−0.06
288.71	196.11	196.22	−0.06
298.39	198.44	198.54	−0.05
308.40	200.94	201.03	−0.04
328.08	206.05	206.10	−0.02
347.24	211.17	211.19	−0.01
356.67	213.71	213.71	0.00
366.02	216.22	216.20	0.01

TABLE 10. Conversion from IPTS-68 to ITS-90 for 1-methylpyrrole (88MES/TOD, measurement precision and measurement accuracy 0.05 % and 0.1 %, respectively). Comparison of smoothed data.

T K	C_{68} $J \cdot K^{-1} \cdot mol^{-1}$	C_{90} $J \cdot K^{-1} \cdot mol^{-1}$	$100 \cdot (C_{68}-C_{90})/C_{90}$ %
230	137.64	137.66	−0.01
240	139.03	139.05	−0.01
250	140.61	140.64	−0.02
260	142.35	142.38	−0.02
270	144.23	144.27	−0.03
280	146.22	146.26	−0.03
290	148.30	148.34	−0.03
300	150.46	150.50	−0.03
310	152.69	152.73	−0.03
320	154.97	155.02	−0.03
330	157.31	157.36	−0.03
340	159.69	159.73	−0.03
350	162.09	162.14	−0.03
360	164.52	164.56	−0.02

TABLE 11. Conversion from IPTS-68 to ITS-90 for (Z)-2-hexene (90MES/TOD, measurement precision and measurement accuracy 0.05 % and 0.1 %, respectively). Comparison of smoothed data.

T K	C_{68} $J \cdot K^{-1} \cdot mol^{-1}$	C_{90} $J \cdot K^{-1} \cdot mol^{-1}$	$100 \cdot (C_{68}-C_{90})/C_{90}$ %
130	154.07	154.06	0.01
140	153.39	153.38	0.01
150	153.03	153.03	0.00
160	152.98	152.98	0.00
170	153.24	153.24	0.00
180	153.78	153.78	0.00
190	154.60	154.61	−0.01
200	155.69	155.70	−0.01
210	157.03	157.05	−0.01
220	158.62	158.64	−0.01
230	160.44	160.47	−0.02
240	162.49	162.52	−0.02
250	164.75	164.78	−0.02
260	167.21	167.25	−0.02
270	169.86	169.90	−0.02
280	172.69	172.73	−0.02
290	175.69	175.74	−0.03
300	178.85	178.90	−0.03
310	182.15	182.20	−0.03
320	185.59	185.64	−0.03

TABLE 12. Conversion from IPTS-68 to ITS-90 for (Z)-2-hexene (90MES/TOD, measurement precision and measurement accuracy 0.05 % and 0.1 %, respectively). Comparison of the smoothed data (ITS-90) with the raw values reported in the literature source (IPTS-68).

T K	C_{68} $J \cdot K^{-1} \cdot mol^{-1}$	C_{90} $J \cdot K^{-1} \cdot mol^{-1}$	$100 \cdot (C_{68}-C_{90})/C_{90}$ %
129.062	154.15	154.14	0.01
137.174	153.55	153.54	0.01
162.026	153.01	153.01	0.00
180.930	153.85	153.85	0.00
200.557	155.76	155.77	–0.01
221.615	158.90	158.92	–0.01
242.141	162.96	162.99	–0.02
263.156	168.03	168.07	–0.02
284.157	173.92	173.96	–0.02
300.630	179.05	179.10	–0.03
314.445	183.66	183.72	–0.03

The only compound where the conversion was performed was n-heptane which is used as the primary calorimetric standard for heat capacity measurements of liquids. Accurate raw experimental data measured at the National Bureau of Standards (NBS), Washington D.C. (47OSB/GIN, 54DOU/FUR) and at the Bureau of Mines (BM), Bartlesville, Oklahoma (61MCC/MES) have been critically selected from all available sources and converted to ITS-90. The conversion procedure was rather complex and was described in 94ZAB/RUZ. Heat capacities converted to ITS-90 and differences between the original heat capacities that relate to IPTS-48 and those converted to ITS-90 are presented in Table 13. Both IPTS-48 and ITS-90 values were generated from a smoothing equation. Differences in heat capacities due to the conversion to ITS-90 are of the same magnitude as the measurement error of the NBS and the BM data. A joint correlation of converted data from the two laboratories provided recommended heat capacities of n-heptane that are given in the last column of Table 13.

Our recommendations differ to some extent from those published for n-heptane in the IUPAC publication on reference materials (88MAR). The previous recommendations are based exclusively on the NBS data (54DOU/FUR). The evaluators (A.J. Head and R. Sabbah) converted temperatures to IPTS-68; however no corrections were applied to the experimental heat capacities, although it was possible to do so by using information in the original paper. In producing our recommendations, we were converting both temperatures and heat capacity values to ITS-90 and also made a correction using the latest values of molecular masses.

TABLE 13. Heat capacity of n-heptane converted to ITS-90.

	NBS data		BM data		Recommended values
T_{90} K	C_{sat} $J \cdot K^{-1} \cdot mol^{-1}$	d[a] %	C_{sat} $J \cdot K^{-1} \cdot mol^{-1}$	d[a] %	C_{sat} $J \cdot K^{-1} \cdot mol^{-1}$
182.603	202.84[b]	0.15	202.77[b]	0.12	202.80[b]
190	201.79	0.08	201.89	0.03	201.84
200	201.32	–0.01	201.50	–0.02	201.38
210	201.67	0.01	201.90	–0.05	201.74
220	202.74	0.00	202.97	–0.04	202.81
230	204.41	–0.02	204.62	–0.05	204.48
240	206.60	–0.06	206.75	–0.07	206.66
250	209.18	–0.01	209.28	–0.08	209.24
260	212.07	–0.16	212.14	–0.08	212.12
270	215.17	–0.17	215.26	–0.11	215.21
280	218.45	–0.10	218.59	–0.11	218.48
290	221.92	–0.08	222.10	–0.08	221.94
298.15	224.88[c]	–0.06	225.08	–0.07	224.88
300	225.56	–0.05	225.76	–0.05	225.57
310	229.38	–0.05	229.55	–0.03	229.37
320	233.35	–0.04	233.47	–0.04	233.33
330	237.48	–0.04	237.50	–0.03	237.45
340	241.76	–0.04	241.69	–0.01	241.72
350	246.18	–0.04	246.04	–0.02	246.14
360[d]	250.74	–0.04	250.60	–0.01	250.69
370	255.43	–0.05	255.41[b]	–0.01	255.37
380	260.24	–0.05			260.18
390	265.16	–0.05			265.10
400[e]	270.19	–0.02			270.14
410	275.33	0.01			275.30
420	280.62	0.05			280.61
430	286.11	0.07			286.11
440	291.84	0.06			291.85
450	297.86	0.03			297.88
460	304.21	–0.02			304.24
470	310.94	–0.09			310.98
480	318.11	–0.12			318.13
490	325.74	0.00			325.75
500	334.91	0.22			334.89
510	350.69	–0.13			350.66
520	379.15	0.06			379.16

[a] $d = (C_{48} - C_{90}) \cdot 100/C_{90}$, heat capacities calculated at the same numerical value of temperature in IPTS–48 and in ITS–90
[b] extrapolated
[c] temperature in the original source 54DOU/FUR is 298.16 K
[d] data from NBS above this temperature were derived from enthalpy measurements by drop calorimetry; their accuracy is substantially lower compared to adiabatic measurements below this temperature. Only the temperature shift was considered at $T > 360$ K (no ΔT correction)
[e] n–heptane is not a calorimetric standard above this temperature

5.7. Units and Conversion Factors

All numerical data reported in the tables are given in SI units. Factors for converting the values from the original sources to SI units are as follows:

$1\ cal_{th} = 4.1840$ J
$1\ cal_{IT} = 4.1868$ J
$1\ Btu_{th}\ lb^{-1}(°F)^{-1} = 4.1840\ J \cdot K^{-1} \cdot g^{-1}$
$1\ Btu_{IT}\ lb^{-1}(°F)^{-1} = 4.1868\ J \cdot K^{-1} \cdot g^{-1}$
$T/K = t/°C + 273.15$
$T/K = T/°F \cdot 0.55555 + 255.37$
$R = 8.31451\ J \cdot K^{-1} \cdot mol^{-1}$

where subscript ''th'' and ''IT'' denote the thermochemical basis and the International Steam Tables basis, respectively. In many cases, it is not clear which basis was used in the original source; for data in calories it was usually the thermochemical basis and for those in $Btu_{IT}lb^{-1}(°F)^{-1}$ the IT basis. During the course of this project the recommended value of the universal gas constant R was modified by the CODATA Fundamental Constants Task Group (88COH/TAY) from 8.31441 to $8.31451\ J \cdot K^{-1} \cdot mol^{-1}$; the latter value was used throughout this compilation.

6. Guide to Tables

6.1. Division of Substances into Groups

To make the data presentation logical and easy to follow, the compounds are divided into *sections* according to the kinds of atoms they contain. These are further divided into *groups* according to their chemical structure. This arrangement makes it possible to illustrate specific features of individual families of substances and to show the connection between the magnitude of the quantities listed and the chemical structure of individual substances. The adopted ordering system can be seen in the following table. Each group is assigned a designated double digit.

Table 14. Division of Compounds into Groups

0. Inorganic substances	
	01. Elements
	02. Inorganic compounds
1. Compounds of Carbon and Hydrogen	
	11. Saturated aliphatic hydrocarbons
	12. Saturated cyclic hydrocarbons
	13. Unsaturated aliphatic hydrocarbons
	14. Aromatic and unsaturated cyclic hydrocarbons
2. Compounds of Carbon, Hydrogen and Halogen	
	21. Fluorine derivatives
	22. Chlorine derivatives
	23. Bromine derivatives
	24. Iodine derivatives
	25. Mixed halogen derivatives
3. Compounds of Carbon, Hydrogen and Nitrogen	
	31. Amines
	32. Nitriles
	33. Heterocyclic nitrogen compounds
	34. Miscellaneous nitrogen compounds
4. Compounds of Carbon, Hydrogen and Oxygen	
	41. Ethers
	42. Alcohols and phenols
	43. Carbonyl compounds
	44. Acids and anhydrides
	45. Esters
	46. Heterocyclic oxygen compounds
	47. Miscellaneous oxygen compounds
5. Compounds of Carbon, Hydrogen and Sulfur	
	51. Sulfides
	52. Thiols
	53. Heterocyclic sulfur compounds
6. Other Organic Compounds Containing Halogens, Nitrogen, Oxygen and Sulfur	
	61. Compounds of carbon, hydrogen, halogen and oxygen
	62. Compounds of carbon, hydrogen, nitrogen and oxygen
	63. Compounds of carbon, hydrogen, oxygen and sulfur
	64. Miscellaneous compounds
7. Organic Compounds Containing Other Elements than Halogens, Nitrogen, Oxygen and Sulfur	
	71. Organosilicon compounds
	72. Organic compounds containing phosphorus and boron
	73. Organometallic compounds
	74. Salts of organic acids

The substances of section 0 (elements and inorganic compounds) are ordered according to their formulas in alphabetical order. Organic compounds (sections 1 to 7) are ordered within each group according to the usual convention (Hill system); i.e., in order of increasing number of carbon and hydrogen atoms. Other elements are placed in alphabetical order. The substances having the same molecular formula are arranged in alphabetical order according to their names. Each substance is coded by two numbers separated by a dash. The first two-digit number indicates the group (family of substances), and the second three-digit number denotes the serial number of a substance inside the group. In naming the substances, we adopted the Chemical Abstracts rules for the names under which the substance was presented with exception of simple esters where we used shorter names, e.g., methyl acetate instead of methyl ester acetic acid. Codes for individual substances permitting their location inside the tables and the corresponding CAS registry numbers can be found in the formula and name indices at the end of the compilation.

6.2. Introductory Texts for Individual Groups

A short introduction precedes the tabular part for each family of compounds. The introduction is roughly divided into four parts. The division is not always strictly followed and information from different parts may be mixed together. The four parts contain the following information:

1. statistical data that give the total number of compounds in the family, number of compounds with data at a single temperature, most frequently 298.15 K;
2. a list and characteristics of main laboratories that specialized in the measurement of heat capacity for a particular family of compounds;
3. for multiple sources an explanation is sometimes given why particular data sets were selected for or rejected from the final correlations; this account is presented for most important compounds and in some additional cases where specifications are needed;
4. additional discussion of measurements from certain laboratories where reliability of data is difficult to establish and information in the original source is scarce or doubtful. Supplementary information is given for special cases.

No special attention has been paid in the introductory chapters to the measurements that have been performed at either a single temperature or in a narrow temperature interval; these data are often by-products of excess heat capacity measurements with mixtures (concentration extremes). In the latter investigations the authors did not often pay much attention to sample purity; impurities do not have much effect on excess heat capacities but impact upon the reliability of results for pure substances.

For the sake of clarity, we have sometimes used trivial names of compounds instead of systematic names. The list of all compound names and synonyms used in the compilation is given in the Name Index of Compounds.

We have used acronyms for referring to different laboratories whose names repeatedly appeared in introductory chapters. The list that follows gives an alphabetical list of acronyms and full names of the main laboratories that were engaged in the measurement of heat capacities.

6.2.1. Acronyms of Laboratories

A

ANAZ — Akademia Nauk (Academy of Sciences) Azerbaidzhanskoi SSR, Baku, Azerbaidzhan
ANSM — Akademia Nauk (Academy of Sciences) USSR, Moscow, USSR
APIB — Azerbaidzhanskii Politekh.-Institut, Baku, Azerbaidzhan (Politechnical Institute in Baku)
ASUB — Azerbaidzhan State University, Baku, Azerbaidzhan

B

BAF — Bergakademie Freiberg, Germany
BMB — Bureau of Mines (now see NIPER), Bartlesville, Oklahoma, USA
BSU — Byelorussian State University, Minsk, Byelorussia
BTIM — Byelorussian Technology Institute S.M.Kirova, Minsk, Byelorussia
BYUP — Brigham Young University, Provo, Utah, USA

C

CITP — California Institute of Technology, Pasadena, California, USA
CIUG — Chemistry Institute of Lobachevski State University, Gorki (now Nizhny Novgorod), USSR

D

DCM — Dow Chemical Co., Midland, Michigan, USA
DICP — Dalian Institute of Chemical Physics, Academia Sinica, China

E

EREL — Exxon Research and Equipment Co., Linden, NJ, USA

F

FSO — Faculté des Sciences Orsay, Seine-et-Oise, France
FUAN — Free University, Amsterdam, Netherlands

G

GPI — Groznenski Petroleum Institute, Grozny, Georgia (Oil Institute Academician M.D.Millionshchnikov, Grozny)
GITA — Georgia Institute of Technology, Atlanta, USA

I

ICG — Institute of Chemistry, AN USSR, Gor'kii (now Nizhny Novgorod), USSR
ICLO — The Inorganic Chemistry Laboratory, Oxford, England
ICTP — Institute of Chemical Technology, Prague, Czechoslovakia (now Czech Republic)
IICN — Institute of Inorganic Chemistry, AN USSR, Novosibirsk, USSR
IISB — Indian Institute of Science, Bangalore, India
IITD — Indian Institute of Technology, New Delhi, India
IKNR — Institut Khimii Nevodnykh Raztvorov, AN USSR, Ivanovo, USSR (Institute of Non-Aqueous Solution Chemistry)
IPCL — Institut für Physikalische Chemie, Leuna-Merseburg, Germany
ITN — Institut Teplofiziki, AN USSR, Novosibirsk, USSR
IVTM — Institut Vysokikh Temperatur, AN USSR, Moscow, USSR (Institute of High Temperatures)

J

JHUB — The John Hopkin's University, Baltimore, Maryland, USA

K

KCIW — Kinetic Chemicals Inc., Willmington, Delaware, USA
KGPI — Kurski Gosudarstvenyi Pedagogicheskii Institut, Kursk, USSR
KITH — Kaeltetechnischen Institut der Technische Hochschule Karlsruhe, Germany
KSUK — Kent State University, Kent, Ohio, USA

L

LCPP — Laboratoire de Chimie Physique, Paris, France
LCSP — Laboratoire de Chimie Structurale, Pau, France
LISI — Leningradskii Inzhenerno-Stroitel'nii Institut, Leningrad (now St.Petersburg), USSR (Leningrad Engineering-Architectural Institute)
LSU — Leningrad State University, Leningrad (now St.Petersburg), USSR

M

MCSL — Monsanto Company, St.Louis, Missouri, USA
MGUM — McGill University, Montreal, Canada
MITC — Massachusetts Institute of Technology, Cambridge, USA
MLUH — Martin-Luther University, Halle-Wittenberg, Germany
MRCE — Monsanto Research Corporation, Everett, Massachusetts, USA
MSUM — Moscow State University, Moscow, USSR (Lomonosov State University)

N

NBSB — National Bureau of Standards (from 1990 see NIST), Boulder, Colorado, USA
NBSW — National Bureau of Standards (from 1990 see NIST), Gaithersburg, Maryland, USA
NCLT — National Chemical Laboratory for Industry, Tokyo, Japan
NIPER — National Institute for Petroleum and Energy Research (formerly BMB, see), Bartlesville, Oklahoma, USA
NIST — National Institute of Standards and Technology, Boulder, Colorado and Gaithersburg, Maryland, USA (see formerly NBSB and NBSW)
NPLT — National Physical Laboratory, Teddington, England
NRCO — National Research Council, Ottawa, Canada
NRLW — Naval Research Laboratory, Washington, USA
NWUC — Northwestern University, Xi'an, People Republic China
NWUE — Northwestern University, Evanston, Illinois, USA

O

OCUO — Osaka City University, Osaka, Japan
OSU — Ohio State University, Columbus, USA

P

PASW — Polish Academy of Sciences, Warsaw, Poland
PIK — Polytechnical Institute in Khar'kov, Ukraine
PSC — Pennsylvania State College, Pennsylvania, USA
PTIL — Physical Technology Institute, Leningrad, (now St.Petersburg), USSR
PUP — Princeton University, Princeton, USA

R

RUH — (William Marsh) Rice University, Houston, Texas, USA
RAB — Rumanian Academy, Bucharest, Romania

S

SRIL — Soviet Research Institute of Petroleum Chemistry, Leningrad (now St.Petersburg), USSR
SRNO — Southern Regional Research Laboratory, New Orleans, Louisiana, USA
SUC — Stanford University, California, USA
SUU — State University of Utrecht, Utrecht, The Netherlands

T

THA — Technische Hochschule, Aachen, Germany
TUB — Technical University in Breslau, Germany (now Wroclaw, Poland)

U

UCB — University of California, Berkeley, USA
UCCC — Union Carbide Corporation, Oak Ridge, Tennessee, USA
UCCW — Union Carbide Corporation, South Charleston, West Virginia, USA
UCE — University of Cambridge, England
UGG — University of Goettingen, Germany
UHS — University of Halle/Saale, Germany
ULE — University of London, England
ULS — University of Lund, Sweden
UMAA — University of Michigan, Ann Arbor, USA
UMG — University of Munich, Germany
UME — University of Manchester, England
UMF — University of Miami, Florida, USA
UOE — University of Oxford, England
UOTO — University of Osaka, Toyonaka, Osaka, Japan
USC — University of Sherbrooke, Canada
USE — University of Sulford, England
UWA — University of Vienna, Austria
UWG — University of Wurzburg, Germany

W

WRUC — Western Reserve University, Cleveland, Ohio, USA
WUSL — Washington University, St.Louis, Missouri, USA

Y

YUNH — Yale University, New Haven, Connecticut, USA

6.3. Tables and Deviation Plot

Most of the information on the data and their processing is given for each substance in tables which are described in detail below. Methodology and procedure which led to this presentation of data are described in Chapter 5. The tables for each substance are preceded with name of the substance, its formula, code, and CAS registry number.

The characterization of raw data, their consistency, results of correlation and recommended data are summarized for each substance separately in three to five tables and a deviation plot. A table of *experimental heat capacities* (table type number 1) reviews all primary sources and characterizes briefly experimental conditions and the quality of data. The upper half of the table of *correlated heat capacities* (table type number 2) lists sources selected for establishing the recommended values; the statistical criteria defined in Section 5.2 indicate for both selected and discarded sources how the individual data sets compare with values generated from the correlation equation. The consistency of raw data is graphically demonstrated in the *deviation plot*. The table of correlated heat capacities is not present when the data are available at only one temperature or when there is only one literature source. The table listing *parameters of polynomial(s)* (table type number 3) obtained by fitting the selected data also gives information on the quality of the overall fit and reliability of the recommended data. The table of *recommended values* (table type number 4) contains discrete heat capacities generated from the listed parameters of polynomials. *Parameters of quasi-polynomial equation* (table type number 5) are presented for substances where application of this relationship is meaningful.

Each table is coded with three numbers separated by periods. The first two numbers are identical with the substance code (see Section 6.1) and the third relates to the type of table (1 to 5). Even when some tables are missing, the numbering related to the type of table is retained.

The symbol C without a subscript specification or the term heat capacity is used in tables whenever we refer both to C_p and C_{sat} at temperatures where the difference between their values is negligible compared with the expected experimental error.

Certain statistics and parameters are listed in the modified E notation. The first part of the number denotes the fractional part which is followed by the exponent to the base 10. The exponent is separated from the fractional part by a plus or a minus sign (e.g. −1.53−5 means $-1.53 \cdot 10^{-5}$).

6.3.1. Experimental Heat Capacities

Each line of the table contains information on one literature source of experimental data. When several distinct data sets were given for the same substance in one publication (for example, parameters of smoothing equations relating to different temperature subintervals or the data obtained from two different calorimeters), there are several lines for one data source, each relating to one data set.

First column: the abbreviated reference in the form YYAAA/BBBM where YY is the last two digits of the year of publication (for a year before 1900 the reference code is preceded by an asterisk), AAA and BBB are the first three letters of the last name of the first and second author (if present), respectively. M is a digit from 1 to 9 distinguishing papers published by the same author(s) within the same year. If the same data set appeared in two or more different publications the reference to the earliest source is given; an appropriate note is given below the table.

For some sources, a footnote is added below the table. In these cases, there is a capital N between the first and second column.

Second column: temperature range of the data set in kelvin.

Third column: number of experimental data points; symbol "eqn" is used in those cases where only parameters of a smoothing equation were presented in the original literature; symbol "S" denotes that the discrete values given in the original literature source were generated from a smoothing equation.

Fourth column: error of measurement, $\sigma_r C_{exp}$, in percent claimed by the author(s); abbreviation "nosp" is used when no specification is given in the original literature.

Fifth column: purity of the substance in percent (given with the same number of significant digits as in the original source) and analytical method used for its determination; the meaning of the abbreviations used is as follows: "anal"− analytical (used when the analytical method was not specified), "chrom"− gas or liquid chromatography, "estim"− the purity was estimated by the author(s), "melpt"− determination of impurities from the melting point depression.

Sixth column: type of heat capacity reported in the original literature; C_p—isobaric heat capacity, C_{sat}—saturation heat capacity, C_{avg}− average heat capacity determined over a temperature range usually greater than 10 K.

Seventh column: type of calorimetric method used for determining the data and reference to the publication where the instrument is described. The coding used for classification of experimental techniques is introduced in Table 2 and the individual types of calorimeters are summarized in Table 3 (both in Chapter 3). The criteria for classifying heat capacity calorimeters are discussed in Section 3.1. All references to calorimeters listed in tables of experimental heat capacities are summarized in Table 4.

More information on experimental data is given in Section 5.1.

6.3.2. Correlated Heat Capacities

This table contains information on the results of the correlation presented for the individual data sets. The statistics for the selected and rejected data sets are listed in the upper and lower half of the table, respectively. The meaning of the columns in the *upper part* of the table is as follows.

First column: the abbreviated reference (the same as in the table of experimental heat capacities).

Second column: temperature range (in kelvin) in which the data from a particular source were included in the correlation.

Third column: number of values used in the final determination of the correlation parameters.

The entries in the second and third columns are identical to those in the table of experimental heat capacities when the whole data set was included in the correlation; they differ if part of the data was discarded.

Fourth column: percentage error, $\sigma_r C$, used in Eq. (V-2) to estimate the variance of individual data points. This value is either $\sigma_r C_{exp}$ (column 4 in the table of experimental heat capacities) or is assigned by the evaluator in cases when no error is specified in the original source or the author's estimate does not seem to be realistic. When $\sigma_r C$ was assigned or modified by the evaluator, the value is followed by the sign #.

Fifth column: average weighted deviation d_w defined by Eq. (V-7).

Sixth column: average deviation d, defined by Eq. (V-8), divided by R (dimensionless).

Seventh column: average percentage deviation d_r defined by Eq. (V-9).

Eighth column: bias of the data set d_b, defined by Eq. (V-10), divided by R (dimensionless).

Ninth column: the difference between the number of experimental points with positive and negative deviation from the recommended values (denoted by +/−).

In the *lower part* of the table, the quantities d/R, d_r, d_b/R and +/– are presented in parenthesis for each reference rejected from the final correlation.

In some cases, the correlations for isobaric and saturation heat capacities were performed separately. We give, however, only one table of correlated heat capacities for these compounds as the selection of data sources did not differ and statistical values were practically identical in both correlations.

More information on the evaluation and selection processes can be found in Section 5.2.

6.3.3. Parameters of Polynomials

This table gives characteristics of the final correlation of the selected data in the upper part and, the corresponding parameters of the polynomial(s) in the lower part.

The *upper part* consists of two lines when both isobaric and saturation heat capacities are tabulated and of only one line when no distinction is made between the two heat capacities. The following items are listed.

First column: type of heat capacity listed — C_p and C_{sat} denote isobaric and saturation heat capacities, respectively. The C is used when no distinction was made between the two types of heat capacities (see Section 5.4.)

Second column: the total number of all experimental data points available.

Third column: the total number of experimental data points used in the correlation.

Fourth column: the standard weighted deviation s_w defined by Eqns. (V-1) and (V-3).

Fifth column: standard deviation s, defined by Eq. (V-4), divided by R (dimensionless).

Sixth column: standard percentage deviation s_r defined by Eq. (V-5).

Seventh column: bias s_b, defined by Eq. (V-6), divided by R (dimensionless).

Eighth column: the overall difference between the numbers of experimental points with positive and negative deviations from the recommended values (denoted by +/–).

The parameters of the cubic spline polynomials describing individual subintervals of the temperature range of the selected data are listed in the *lower part* of the table. When the isobaric and saturation heat capacities are distinguished, two sets of parameters are listed separately, the first set relating to the temperature correlation of C_p and the second to that of C_{sat}. When the whole temperature range of experimental data is described by only one equation the parameters were obtained by simple unconstrained fitting using a polynomial of third or lower degree. The meaning of the individual columns is as follows:

First column: temperature subinterval in kelvin to which the listed parameters relate.

Second to fifth columns: parameters of the polynomial defined by Eq. (V-11) valid in the given subinterval. Dimension of the parameters is K^{-j}.

Sixth column: level of uncertainty assigned by the evaluators to the data generated from the polynomial in the given temperature subinterval. This characteristic expresses the expected overall uncertainty of the recommended data and reflects both the uncertainty in the experimental values and possible error due to the fitting procedure. The following levels of uncertainty were assigned:

- I excellent data (uncertainty below 0.1 percent)
- II highly reliable data (uncertainty below 0.3 percent)
- III reliable data (uncertainty below 0.5 percent)
- IV medium quality data (uncertainty below 1 percent)
- V data of low reliability (uncertainty below 3 percent)
- VI very unreliable data with a possibility of gross systematic errors (uncertainty above 3 percent).

The level of uncertainty assigned to individual subintervals for cubic splines often differs due to the unequal reliability of experimental data in different parts of the temperature range. On the other hand in certain cases one set of parameters was used for describing two (or exceptionally three) experimental data sets of different accuracy, which related to different temperature ranges. Then the level of uncertainty assigned to recommended data relates always to the data set of lower quality; in reality the recommended data are more reliable than indicated at temperatures where the experimental data set of higher quality was available. The reader can obtain information regarding changes of reliability within the temperature interval of recommended data from the table of Correlated Heat Capacities.

More details on fitting data with polynomials can be found in Section 5.3.1.

6.3.4. Recommended Values

This table lists recommended specific and molar heat capacities generated from the parameters of the polynomials at typical temperatures 273.15 and 298.15 K (provided they lie inside the interval of experimental data) and over the whole temperature range of the selected experimental data. Isobaric and saturation heat capacities are tabulated when the parameters for both C_p and C_{sat} were determined, otherwise only one data set is presented. The number of significant digits is determined by the level of uncertainty of the recommended values.

For some compounds where most of experimental data are available above the normal boiling temperature the recommended isobaric and saturation heat capacities exhibit small differences in the temperature range where both quantities are in other cases equal. This artifact is exclusively due to the fitting procedure.

6.3.5. Parameters of the Quasi-Polynomial Equation

The quasi-polynomial equation (V-12) is useful as a simple expression permitting extrapolations to higher temperatures (see Section 5.3.2. for details). The quasi-polynomial fit is performed only for C_p data as its main application was in the calculation of enthalpy and entropy differences.

The table is organized analogously to the table presenting parameters of polynomials; the *upper part* of the table is identical. The items tabulated in the *lower part* have the following meaning.

First column: temperature interval in kelvin to which the listed parameters relate. This interval is not necessarily identical with the temperature range of the representation by polynomials.

Second column: critical temperature in kelvin

Third to eighth columns: four or six parameters of the quasi-polynomial equation ($m = 1$ or 3). The parameters are dimensionless.

Ninth column: level of uncertainty assigned by the evaluators to the data generated from the quasi-polynomial equation within the temperature range of the experimental data. When using the equation outside this range, the uncertainty decreases with the length of the extrapolation.

6.3.6. Deviation Plots

An overview of the deviations from the recommended values for all experimental data (both included in and rejected from the final correlation) measured by various authors is presented in graphical form in the deviation plots. The temperature is plotted along the x-axis and the relative percentage deviation for individual data points along the y-axis. Data from different sources are distinguished by different symbols. Points that lie outside the range of the ordinate in the plot are accompanied by the numerical value of the deviation. Some points that overlap each other or the accompanying figures are omitted. When too many data points were available some less important sets and values exhibiting large deviations from the recommended data were not included.

7. References

*45JOU Joule J. P., Mem. Proc. Manchester Lit. Phil. Soc. **2**, 559 (1845).
*62REG Regnault V., Mem. Acad. Sci. Inst. France **26**, 262 (1862).
*67HIR Hirn G. -A., Ann. Chim. Phys. [4] **10**, 32-92 (1867).
*77BER Berthelot, C. R. Hebd. Seances Acad. Sci. **85**, 648-51 (1877).
*79BER Berthelot, C. R. Hebd. Seances Acad. Sci. **89**, 119-21 (1879).
*80OGI Ogier J., Ann. Chim. Phys. [5] **20**, 5-66 (1880).
*81BER/OGI Berthelot, Ogier, Ann. Chim. Phys. [5] **23**, 199-274 (1881).
*81OGI Ogier J., C. R. Hebd. Seances Acad. Sci. **92**, 922-4 (1881).
*81VON von Reis M. A., Ann. Phys. (Leipzig) [3] **13**, 447-65 (1881).
*82OGI Ogier J., C. R. Hebd. Seances Acad. Sci. **94**, 82-6 (1882).
*84BAT Battelli, Atti R. Ist. Veneto Sci., Lett. Arti **3**, 1781- (1884), from: Washburn E. W. (editor), International Critical Tables, McGraw-Hill Co., New York **5**, 107-13 (1929).
*84WER Werner M. E., Ann. Chim. Phys. [6] **3**, 567-74 (1884).
*85STO/WIL Stohmann F., Wilsing H., J. Prakt. Chem. **32**, 80-93 (1885).
*86LUD Ludeking C., Ann. Phys. (Leipzig) [3] **27**, 72-90 (1886).
*86SCH Schiff R., Justus Liebigs Ann. Chem. **233**, 300-37 (1886).
*87SCH Schiff R., Z. Phys. Chem. Stoechiom. Verwandtschaftslehre **1**, 376-90 (1887).
*90PIC Pickering S. U., Proc. R. Soc. London **49**, 11-32 (1890).
*92STO Stortenbeker W., Z. Phys. Chem. Stoechiom. Verwandtschaftslehre **10**, 183-202 (1892).
*95MAS/GUI Massol G., Guillot A., C. R. Hebd. Seances Acad. Sci. **121**, 208-10 (1895).
*95PIC Pickering S. U., J. Chem. Soc. **67**, 664-84 (1895).
*95SCH Schlamp, Giessen. Oberhess. Ges. Ber. **31**, 100- (1895), from: Washburn E. W. (editor), International Critical Tables, McGraw-Hill Co., New York **5**, 107-13 (1929).
*97BER Berthelot M. P. E., Thermochimie, Vol. I and II, Gautier-Villars et Fils, Paris (1897).
*98LOU Louguinine W., Ann. Chim. Phys. [7] **13**, 289-377 (1898).
00LOU Louguinine W., Arch. Sci. Phys. Natur. [4] **9**, 5-26 (1900).
01DEF De Forcrand M., C. R. Hebd. Seances Acad. Sci. **132**, 569-71 (1901).
01KAH Kahlenberg L., J. Phys. Chem. **5**, 284-8 (1901).
02GAE Gaede W., Phys. Z. **4**, 105 (1902).
02LOU1 Louguinine W., Ann. Chim. Phys. [7] **26**, 228-47 (1902).
02LOU2 Louguinine W., Ann. Chim. Phys. [7] **27**, 105-44 (1902).
04LOU Louguinine W., J. Chim. Phys. Phys. -Chim. Biol. **2**, 1-6 (1904).
05DIE Dieterici C., Ann. Phys. (Leipzig) **16**, 593-620 (1905).
05LOU Louguinine W., J. Chim. Phys. Phys. -Chim. Biol. **3**, 640-7 (1905).
06BER Bernini A., Phys. Z. **7**, 168-72 (1906).
06WIG Wigand A., Marburger Sitzungsber. 186 (1906), from: Protz L., Ann. Phys. (Leipzig) [4] **31**, 127-48 (1910).
07BAT Battelli A., Atti Accad. Naz. Lincei, Cl. Sci. Fis. Mat. Natur., Rend. **16**, (1. sem.)243-57 (1907).
07GUI1 Guinchant, C. R. Hebd. Seances Acad. Sci. **145**, 68-70 (1907).
07GUI2 Guinchant, C. R. Hebd. Seances Acad. Sci. **145**, 320-2 (1907).
07WAL Walden P., Z. Phys. Chem. Stoechiom. Verwandtschaftslehre **58**, 479-511 (1907).
08BAT Battelli A., Phys. Z. **9**, 671-4 (1908).
08BOG/WIN Bogojawlensky A., Winogradow N., Z. Phys. Chem. Stoechiom. Verwandtschaftslehre **64**, 251-4 (1908).
08KAH/KOE Kahlenberg L., Koenig R., J. Phys. Chem. **12**, 290-2 (1908).
09EUC Eucken A., Phys. Z. **10**, 586 (1909).
09GOO/KAL Goodwin H. M., Kalmus H. T., Phys. Rev. **28**, 1-24 (1909).
09SCH Schwers F., Recl. Trav. Chim. Pays-Bas **28**, 42-67 (1909).
10NER Nernst W., Sitzungsber. Kgl. Preuss. Akad. Wiss. **12**, **13**, 261 (1910), from: Chem. Abstr. **4**, 2397 (1910).
10PRO Protz L., Ann. Phys. (Leipzig) [4] **31**, 127-48 (1910).
11LEW/RAN Lewis G. N., Randall M., J. Am. Chem. Soc. **33**, 476-88 (1911).
11NER Nernst W., Ann. Phys. (Leipzig) [4] **36**, 395 (1911).
11POM Poma G., Gazz. Chim. Ital. **41/2**, 518-38 (1911).
12BAU Baud M. E., Ann. Chim. Phys. [8] **27**, 89-116 (1912).
12LAN/BER Landolt-Börnstein Physikalisch-Chemische Tabellen, 4 Auflage, Börnstein R., Roth W. A., (editors), Springer-Verlag, Berlin (1912).
12LUS Lussana S., Nuovo Cimento [6] **4**, 207-32 (1912).
12SCH1 Schulze A., Ber. Deut. Phys. Ges. **14**, 189-210 (1912).
12SCH2 Schulze A., Ber. Deut. Phys. Ges. **14**, 258-74 (1912).
12SCH3 Schulze A., Ber. Deut. Phys. Ges. **14**, 344-55 (1912).
13CAM Campetti A., Atti R. Accad. Sci. Torino, Cl. Sci. Fis. Mat. Nat. **48**, 96 (1913), from: Washburn E. W. (editor), International Critical Tables, McGraw-Hill Co., New York **5**, 107-13 (1929).
13HER/RAT Herz W., Rathmann W., Chem. Ztg. **37**, 621-2 (1913).
13NAS/BRE Nasini R., Bresciani U., Atti. R. Accad. Naz. Lincei, Mem. Cl. Sci. Fis. Mat. Nat. **9**, 341-401 (1913), from: Washburn E. W. (editor), International Critical Tables, McGraw-Hill Co., New York **5**, 107-13 (1929).
13REN Rengade E., C. R. Hebd. Seances Acad. Sci. **156**, 1897-9 (1913).
14GRI Griffiths E., Proc. R. Soc. London, A **89**, 561-74 (1914).
14NER/SCH Nernst W., Schwers F., Sitzungsber. Kgl. Preuss. Akad. Wiss. 335 (1914), from: McCullough J. P., Scott D. W. (editors), Experimental Thermodynamics, Vol. I, Butterworth, London 133-211 (1968).

16BRA Bramley A., J. Chem. Soc. 496-519 (1916).

16EUC Eucken A., Ber. Deut. Phys. Ges. **18**, 4-17 (1916).

16KUR Kurbatov V. Ya., Zh. Russ. Fiz. -Khim. O-va. **48**, 192 (1916).

17HIL/DUS Hildebrand J. H., Duschak A. D., Foster A. H., Beebe C. W., J. Am. Chem. Soc. **39**, 2293-7 (1917).

17MAT/KRA Mathews J. H., Krause E. L., Van Bohnson L., J. Am. Chem. Soc. **39**, 398-413 (1917).

18NAR Narbutt J., Z. Elektrochem. **24**, 339-42 (1918).

19DEJ Dejardin M. G., Ann. Phys. (Paris) **11**, 253-91 (1919).

20COH/MOE Cohen E., Moesveld A. L. T., Z. Phys. Chem. Stoechiom. Verwandtschaftslehre **95**, 305-27 (1920).

20GIB/LAT Gibson G. E., Latimer W. M., Parks G. S., J. Am. Chem. Soc. **42**, 1533-42 (1920).

20GIB/PAR Gibson G. E., Parks G. S., Latimer W. M., J. Am. Chem. Soc. **42**, 1542-50 (1920).

20MAA/HAT Maass O., Hatcher W. H., J. Am. Chem. Soc. **42**, 2548-69 (1920).

21JAE/VON Jaeger W., Von Steinwehr H., Ann. Phys. (Leipzig) **64**, 305-66 (1921).

21TRE Trehin R., Ann. Phys. (Paris) **15**, 246-63 (1921).

22COH/MOE Cohen E., Moesveld A. L. T., Z. Phys. Chem. Stoechiom. Verwandtschaftslehre **100**, 151-8 (1922).

22HER/SCH Herz W., Schuftan P., Z. Phys. Chem. Stoechiom. Verwandtschaftslehre **101**, 269-85 (1922).

22LAT Latimer W. M., J. Am. Chem. Soc. **44**, 90-7 (1922).

22SIM Simon F., Ann. Phys. (Leipzig) [4] **68**, 241-80 (1922).

23GIB/GIA Gibson G. E., Giauque W. F., J. Am. Chem. Soc. **45**, 93-104 (1923).

23JEN/SHO Jenkin C. F., Shorthose D. N., G. B. Dep. Sci. Ind. Res., Food Invest. Board, Spec. Rep. No. 14, 1-35 (1923).

23LAN/BOR Landolt-Börnstein Physikalisch-Chemische Tabellen, 5 Auflage, Roth W. A., Scheel K. (editors), Springer-Verlag, Berlin (1923).

23SIM/LAN Simon F., Lange F., Z. Phys. **15**, 312-21 (1923).

24EUC/KAR Eucken A., Karwat E., Z. Phys. Chem. Stoechiom. Verwandtschaftslehre **112**, 467-85 (1924).

24GAR/RAN Garner W. E., Randall F. C., J. Chem. Soc. 881-96 (1924).

24HER/BLO Herz W., Bloch W., Z. Phys. Chem. Stoechiom. Verwandtschaftslehre **110**, 23-39 (1924).

24KEY/BEA Keyes F. G., Beattie J. A., J. Am. Chem. Soc. **46**, 1751-60 (1924).

24SHO Shorthose D. N., G. B. Dep. Sci. Ind. Res., Food Invest. Board, Spec. Rep. No. 19, 1-16 (1924).

24TIA Tian A., C. R. Hebd. Seances Acad. Sci. **178**, 705-7 (1924)

24WIL/DAN Williams J. W., Daniels F., J. Am. Chem. Soc. **46**, 903-17 (1924).

25DRU/WEI Drucker C., Weissbach H., Z. Phys. Chem. Stoechiom. Verwandtschaftslehre **117**, 209-41 (1925).

25LYN Lynn G., Thesis, University of Yale (1925), from: Washburn E. W. (editor), International Critical Tables, McGraw-Hill Co., New York **5**, 107-13 (1929).

25PAR Parks G. S., J. Am. Chem. Soc. **47**, 338-45 (1925).

25PAR/KEL Parks G. S., Kelley K. K., J. Am. Chem. Soc. **47**, 2089-97 (1925).

25WIL/DAN Williams J. W., Daniels F., J. Am. Chem. Soc. **47**, 1490-503 (1925).

26AND Andrews D. H., J. Am. Chem. Soc. **48**, 1287-98 (1926).

26AND/LYN Andrews D. H., Lynn G., Johnston J., J. Am. Chem. Soc. **48**, 1274-86 (1926).

26AUE Auerbach R., Z. Phys. Chem. Stoechiom. Verwandtschaftslehre **121**, 337-60 (1926).

26AWB/GRI Awbery J. H., Griffiths E., Proc. Phys. Soc., London **38**, 378-98 (1926).

26DAN/JEN Dana L. I., Jenkins A. C., Burdick J. N., Timm R. C., Refrig. Eng. **12**, 387-405 (1926).

26GAR/MAD Garner W. E., Madden F. C., Rushbrooke J. E., J. Chem. Soc. 2491-502 (1926).

26MAA/BAR Maass O., Barnes W. H., Proc. R. Soc. London, A **111**, 224-44 (1926).

26MON Mondain-Monval M. P., C. R. Hebd. Seances Acad. Sci. **182**, 1465-8 (1926).

26PAR/AND Parks G. S., Anderson C. T., J. Am. Chem. Soc. **48**, 1506-12 (1926).

26PAR/HUF Parks G. S., Huffman H. M., J. Am. Chem. Soc. **48**, 2788-93 (1926).

26SCH Schmidt G. C., Z. Phys. Chem. Stoechiom. Verwandtschaftslehre **121**, 221-53 (1926).

26UMI Umino S., Sci. Rep. Tohoku Imp. Univ., Ser. 1 **15**, 597-617 (1926).

27FER Ferguson J. B., J. Phys. Chem. **31**, 757-63 (1927).

27LAN/BER Landolt-Börnstein Physikalisch-Chemische Tabellen, 5 Auflage, Suppl. 1, Roth W. A., Scheel K. (editors), Springer-Verlag, Berlin (1927).

27PAR/HUF Parks G. S., Huffman H. M., J. Phys. Chem. **31**, 1842-55 (1927).

28EUC/HAU Eucken A., Hauck F., Z. Phys. Chem. Stoechiom. Verwandtschaftslehre **134**, 161-77 (1928).

28FIS/BIL Fischer W., Biltz W., Z. Anorg. Allg. Chem. **176**, 81-111 (1928).

28GIA/WIE1 Giauque W. F., Wiebe R., J. Am. Chem. Soc. **50**, 101-22 (1928).

28GIA/WIE2 Giauque W. F., Wiebe R., J. Am. Chem. Soc. **50**, 2193-202 (1928).

28LAN Lang H. R., Proc. R. Soc. London, A **118**, 138-56 (1928).

28LAT/GRE Latimer W. M., Greensfelder B. S., J. Am. Chem. Soc. **50**, 2202-13 (1928).

28PAR/KEL Parks G. S., Kelley K. K., J. Phys. Chem. **32**, 734-50 (1928).

28SKA/SAX Skau E. L., Saxton B., J. Am. Chem. Soc. **50**, 2693-701 (1928).

29AND Andrews D. H., J. Franklin Inst. **206**, 285 (1929).

29CLU1 Clusius K., Z. Phys. Chem., Abt. B **3**, 41-79 (1929).

29CLU2 Clusius K., Z. Phys. Chem., Abt. B **4**, 1-13 (1929).

29CLU/HIL Clusius K., Hiller K., Z. Phys. Chem., Abt. B **4**, 158-68 (1929).

29GIA/JOH Giauque W. F., Johnston H. L., J. Am. Chem. Soc. **51**, 2300-21 (1929).

29GIA/WIE Giauque W. F., Wiebe R., J. Am. Chem. Soc. **51**, 1441-9 (1929).

29HER/LOR Herz W., Lorentz E., Z. Phys. Chem., Abt. A **140**, 406-22 (1929).

29JOH/GIA Johnston H. L., Giauque W. F., J. Am. Chem. Soc. **51**, 3194-214 (1929).

29KEL1 Kelley K. K., J. Am. Chem. Soc. **51**, 180-7 (1929).

29KEL2 Kelley K. K., J. Am. Chem. Soc. **51**, 779-86 (1929).

29KEL3 Kelley K. K., J. Am. Chem. Soc. **51**, 1145-50 (1929).

29KEL4 Kelley K. K., J. Am. Chem. Soc. **51**, 1400-6 (1929).

29KEL5 Kelley K. K., J. Am. Chem. Soc. **51**, 2738-41 (1929).

29MIT/HAR1 Mitsukuri S., Hara K., Bull. Chem. Soc. Jpn. **4**, 77-81 (1929).

29MIT/HAR2 Mitsukuri S., Hara K., Proc. Imp. Acad. (Tokyo) **5**, 27-8 (1929).

29WAS Washburn E. W. (editor), International Critical Tables, McGraw-Hill Co., New York **5**, 107-13 (1929).

30BAR/MAA Barnes W. H., Maass O., Can. J. Res., Sect. B **3**, 70-79 (1930).

30BUT/MAA Butler K. H., Maass O., J. Am. Chem. Soc. **52**, 2184-98 (1930).

30CAR/STO Carpenter L. G., Stoodley L. G., Phil. Mag. J. Sci. [7] **10**, 249-65 (1930).

30HUF/PAR1 Huffman H. M., Parks G. S., Daniels A. C., J. Am. Chem. Soc. **52**, 1547-58 (1930).

30HUF/PAR2 Huffman H. M., Parks G. S., Thomas S. B., J. Am. Chem. Soc. **52**, 3241-51 (1930).

30PAR/HUF1 Parks G. S., Huffman H. M., J. Am. Chem. Soc. **52**, 4381-91 (1930).

30PAR/HUF2 Parks G. S., Huffman H. M., Thomas S. B., J. Am. Chem. Soc. **52**, 1032-41 (1930).

30SOU/AND Southard J. C., Andrews D. H., J. Franklin Inst. **209**, 349-60 (1930).

30WAS1	Wassermann A., Z. Phys. Chem., Abt. A **146**, 418-45 (1930).
30WAS2	Wassermann A., Z. Phys. Chem., Abt. A **146**, 446-61 (1930).
30WIE/BRE	Wiebe R., Brevoort M. J., J. Am. Chem. Soc. **52**, 622-33 (1930).
30WIE/HUB	Wiebe R., Hubbard K. H., Brevoort M. J., J. Am. Chem. Soc. **52**, 611-22 (1930).
31BLA/LEI	Blacet F. E., Leighton P. A., Bartlett E. P., J. Phys. Chem. **35**, 1935-43 (1931).
31BUF/FLE	Buffington R. M., Fleischer J., Ind. Eng. Chem. **23**, 1290-2 (1931).
31CLI/AND	Cline J. K., Andrews D. H., J. Am. Chem. Soc. **53**, 3668-73 (1931).
31DEE	Deese R. F., Jr., J. Am. Chem. Soc. **53**, 3673-83 (1931).
31FIO/GIN	Fiock E. F., Ginnings D. C., Holton W. B., Bur. Stand. J. Res. **6**, 881-900 (1931).
31FIS	Fischer W., Z. Anorg. Allg. Chem. **200**, 332-42 (1931).
31FOR/BRU	Forrest H. O., Brugmann E. W., Cummings L. W. T., Ind. Eng. Chem. **23**, 37-9 (1931).
31FRA	Frandsen M., Bur. Stand. J. Res. **7**, 477-83 (1931).
31GAR/VAN	Garner W. E., Van Bibber K., King A. M., J. Chem. Soc. 1533-41 (1931).
31HUF/PAR	Huffman H. M., Parks G. S., Barmore M., J. Am. Chem. Soc. **53**, 3876-88 (1931).
31LAN/BER	Landolt-Börnstein Physikalisch-Chemische Tabellen, 5 Auflage, Suppl. 2, Vol. II, Roth W. A., Scheel K. (editors), Springer-Verlag, Berlin (1931).
31NEW/KAU	Newton R. F., Kaura B. D., De Vries T., Ind. Eng. Chem. **23**, 35-7 (1931).
31SMI/AND1	Smith R. H., Andrews D. H., J. Am. Chem. Soc. **53**, 3644-60 (1931).
31SMI/AND2	Smith R. H., Andrews D. H., J. Am. Chem. Soc. **53**, 3661-7 (1931).
31SWI/RYB1	Swietoslawski W., Rybicka M., Solodkowska W., Bull. Int. Acad. Pol. Sci. Lett., Cl. Sci. Math. Nat. Ser. A 322-35 (1931).
31SWI/RYB2	Swietoslawski W., Rybicka S., Solodkowska W., Rocz. Chem. **11**, 65-77 (1931).
31THO/PAR	Thomas S. B., Parks G. S., J. Phys. Chem. **35**, 2091-102 (1931).
32CAR/HAR	Carpenter L. G., Harle T. F., Proc. R. Soc. London, A **136**, 243-50 (1932).
32CLA/GIA	Clayton J. O., Giauque W. F., J. Am. Chem. Soc. **54**, 2610-26 (1932).
32NEI	Neiman M. B., Zh. Obshch. Khim. **2**, 1-8 (1932).
32NEU	Neumann M. B., Z. Phys. Chem., Abt. A **158**, 258-64 (1932).
32NEU/KUR	Neumann M. B., Kurlyankin I. A., Zh. Obshch. Khim. **2**, 317-21 (1932).
32SOC	Socolik A. S., Z. Phys. Chem., Abt. A **158**, 305-12 (1932).
32SPA/THO	Spaght M. E., Thomas S. B., Parks G. S., J. Phys. Chem. **36**, 882-8 (1932).
32TRE	Trew V. C. G., Trans. Faraday Soc. **28**, 509-14 (1932).
33CLU	Clusius K., Z. Phys. Chem., Abt. B **23**, 213-25 (1933).
33DAH/JUN	Dahmlos J., Jung G., Z. Phys. Chem., Abt. B **21**, 317-22 (1933).
33FER/MIL	Ferguson A., Miller J. T., Proc. Phys. Soc., London **45**, 194-207 (1933).
33FER/THO	Ferry J. D., Thomas S. B., J. Phys. Chem. **37**, 253-5 (1933).
33GIA/CLA	Giauque W. F., Clayton J. O., J. Am. Chem. Soc. **55**, 4875-89 (1933).
33KOL/UDO	Kolosovskii N. A., Udovenko V. V., C. R. Hebd. Seances Acad. Sci. **197**, 519-20 (1933).
33LEB/MOE	LeBlanc M., Moebius E., Ber. Verh. Saechs. Akad. Wiss. Leipzig, Math. -Phys. Kl. **85**, 75-96 (1933).
33PAR/HUF	Parks G. S., Huffman H. M., Barmore M., J. Am. Chem. Soc. **55**, 2733-40 (1933).
33POH/MEH	Pohland E., Mehl W., Z. Phys. Chem., Abt. A **164**, 48-54 (1933).
33ROT/MEY1	Roth W. A., Meyer I., Z. Elektrochem. **39**, 35-7 (1933).
33ROT/MEY2	Roth W. A., Meyer I., Zeumer H., Z. Anorg. Allg. Chem. **214**, 309-20 (1933).
33SOU/BRI	Southard J. C., Brickwedde F. G., J. Am. Chem. Soc. **55**, 4378-84 (1933).
33STR/MAL	Straub J., Malotaux R. N. M. A., Recl. Trav. Chim. Pays-Bas **52**, 275-88 (1933).
33TRE/WAT	Trew V. C. G., Watkins G. M. C., Trans. Faraday Soc. **29**, 1310-8 (1933).
34JAC/PAR	Jacobs C. J., Parks G. S., J. Am. Chem. Soc. **56**, 1513-17 (1934).
34KOL/UDO1	Kolosovskii N. A., Udovenko V. V., C. R. Hebd. Seances Acad. Sci. **198**, 1394-5 (1934).
34KOL/UDO2	Kolosovskii N. A., Udovenko V. V., Zh. Obshch. Khim. **4**, 1027-33 (1934).
34MEH1	Mehl W., Chem. Fabr. **7**, 240-1 (1934).
34MEH2	Mehl W., Z. Gesamte Kaelte Ind. **41**, 152-3 (1934).
34MIL	Miller P., Ph. D. Thesis, Iowa State College (1934).
34PAR/LIG	Parks G. S., Light D. W., J. Am. Chem. Soc. **56**, 1511-3 (1934).
34PAR/TOD	Parks G. S., Todd S. S., J. Chem. Phys. **2**, 440-1 (1934).
34RAD/JUL	Radulescu D., Jula O., Z. Phys. Chem., Abt. B **26**, 390-4 (1934).
34ROT/MEY	Roth W. A., Meyer I., Zeumer H., Z. Anorg. Allg. Chem. **216**, 303-4 (1934).
35AOY/KAN	Aoyama S., Kanda E., Sci. Rep. Tohoku Imp. Univ., Ser. 1 **24**, 116-21 (1935).
35BAR/CLU	Bartholome E., Clusius K., Z. Phys. Chem., Abt. B **28**, 167-77 (1935).
35BLU/GIA	Blue R. W., Giauque W. F., J. Am. Chem. Soc. **57**, 991-7 (1935).
35BRO/BAR1	Brown R. S., Barnes W. H., Maass O., Can. J. Res., Sect. B **12**, 699-701 (1935).
35BRO/BAR2	Brown R. S., Barnes W. H., Maass O., Can. J. Res., Sect. B **13**, 167-9 (1935).
35CLU/BAR	Clusius K., Bartholome E., Z. Phys. Chem., Abt. B **30**, 237-57 (1935).
35HUF/ELL	Huffman H. M., Ellis E. L., J. Am. Chem. Soc. **57**, 46-8 (1935).
35JAC	Jacobs L., Trans. Faraday Soc. **31**, 813-21 (1935).
35KEE/KEE	Keesom W. H., Keesom A. P., Physica (Amsterdam) **2**, 557-69 (1935).
35SAG/LAC	Sage B. H., Lacey W. N., Ind. Eng. Chem. **27**, 1484-8 (1935).
35SKA	Skau E. L., J. Phys. Chem. **49**, 761-7 (1935).
36AST/MES	Aston J. G., Messerly G. H., J. Am. Chem. Soc. **58**, 2354-61 (1936).
36CLU	Clusius K., Z. Phys. Chem., Abt. B **31**, 459-74 (1936).
36CLU/FRA1	Clusius K., Frank A., Z. Phys. Chem., Abt. B **34**, 405-19 (1936).
36CLU/FRA2	Clusius K., Frank A., Z. Phys. Chem., Abt. B **34**, 420-31 (1936).
36CLU/GOL	Clusius K., Goldmann J., Z. Phys. Chem., Abt. B **31**, 256-62 (1936).
36ERN/WAT	Ernst R. C., Watkins C. H., Ruwe H. H., J. Phys. Chem. **40**, 627-35 (1936).
36EVA/EDL	Evans T. W., Edlund K. R., Ind. Eng. Chem. **28**, 1186-8 (1936).
36FER/COC	Ferguson A., Cockett A. H., Nature (London) **138**, 842-3 (1936).
36GIA/BLU	Giauque W. F., Blue R. W., J. Am. Chem. Soc. **58**, 831-7 (1936).
36GIA/STO	Giauque W. F., Stout J. W., J. Am. Chem. Soc. **58**, 1144-50 (1936).
36GLA/CHE	Glagoleva A. A., Cherbov S. I., Zh. Obshch. Khim. **6**, 685-90 (1936).
36KHO/KAL	Khokhlovkin M. A., Kalacheva A. V., Sint. Kauch. **5** (1), 25-7 (1936).
36KUR/VOS	Kurnakov N. S., Voskresenskaya N. K., Izv. Akad. Nauk SSSR, Ser. Khim. (3), 439-66 (1936).

36LAN/BOR Landolt-Börnstein Physikalisch-Chemische Tabellen, 5 Auflage, Suppl. 3, Vol. III, Roth W. A., Scheel K. (editors), Springer-Verlag, Berlin (1936).
36LON/KEM Long E. A., Kemp I. D., J. Am. Chem. Soc. **58**, 1829-34 (1936).
36LOR/BLA Lord R. C., Jr., Blanchard E. R., J. Chem. Phys. **4**, 707-10 (1936).
36NEG Negishi G. R., J. Am. Chem. Soc. **58**, 2293-6 (1936).
36PAR/THO Parks G. S., Thomas S. B., Light D. W., J. Chem. Phys. **4**, 64-9 (1936).
36PAR/TOD1 Parks G. S., Todd S. S., Moore W. A., J. Am. Chem. Soc. **58**, 398-401 (1936).
36PAR/TOD2 Parks G. S., Todd S. S., Shomate C. H., J. Am. Chem. Soc. **58**, 2505-8 (1936).
36PEA/BAK Pearce J. N., Bakke H. M., Proc. Iowa Acad. Sci. **43**, 171-4 (1936).
36TOD/PAR Todd S. S., Parks G. S., J. Am. Chem. Soc. **58**, 134-7 (1936).
37AST/SIL Aston J. G., Siller C. W., Messerly G. H., J. Am. Chem. Soc. **59**, 1743-51 (1937).
37BEK/WOO Bekkedahl N., Wood L. A., J. Res. Natl. Bur. Stand. (U.S.) **19**, 551-8 (1937).
37BRO/MAN Brown O. L. I., Manov G. G., J. Am. Chem. Soc. **59**, 500-2 (1937).
37CAR/HAR Carpenter L. G., Harle T. F., Phil. Mag. J. Sci. [7] **23**, 193-208 (1937).
37COH/BUI Cohen E., Buij J. S., Z. Phys. Chem., Abt. B **35**, 270-84 (1937).
37EGA/KEM Egan C. J., Kemp J. D., J. Am. Chem. Soc. **59**, 1264-8 (1937).
37ELL Ellyett C. D., Trans. Faraday Soc. **33**, 1218-24 (1937).
37GIA/EGA Giauque W. F., Egan C. J., J. Chem. Phys. **5**, 45-54 (1937).
37KAN Kanda E., Bull. Chem. Soc. Jpn. **12**, 511-20 (1937).
37KEM/GIA Kemp J. D., Giauque W. F., J. Am. Chem. Soc. **59**, 79-84 (1937).
37OSB/STI Osborne N. S., Stimson H. F., Ginnings D. C., J. Res. Natl. Bur. Stand. (U.S.) **18**, 389-447 (1937).
37OVE/GIA Overstreet R., Giauque W. F., J. Am. Chem. Soc. **59**, 254-9 (1937).
37PAR/SHO Parks G. S., Shomate C. H., Kennedy W. D., Crawford B. L., Jr., J. Chem. Phys. **5**, 359-63 (1937).
37PER1 Perlick A., Bull. Int. Inst. Refrig. **18** (1), A1-A9 (1937).
37PER2 Perlick A., Z. Gesamte Kaelte Ind. **44**, 201-6 (1937).
37STE/GIA Stephenson C. C., Giauque W. F., J. Chem. Phys. **5**, 149-58 (1937).
37STU Stull D. R., J. Am. Chem. Soc. **59**, 2726-33 (1937).
37VOL Vold R. D., J. Am. Chem. Soc. **59**, 1515-21 (1937).
37WIT/KEM Witt R. K., Kemp J. D., J. Am. Chem. Soc. **59**, 273-6 (1937).
38CLU/KRU Clusius K., Kruis A., Konnertz F., Ann. Phys. (Leipzig) [5] **33**, 642-50 (1938).
38CLU/RIC Clusius K., Riccoboni L., Z. Phys. Chem., Abt. B **38**, 81-95 (1938).
38EGA/KEM Egan C. J., Kemp J. D., J. Am. Chem. Soc. **60**, 2097-101 (1938).
38EUC/SCH Eucken A., Schroder E., Z. Phys. Chem., Abt. B **41**, 307-19 (1938).
38FRE/HIL1 Frederick K. J., Hildebrand J. H., J. Am. Chem. Soc. **60**, 1436-9 (1938).
38FRE/HIL2 Frederick K. J., Hildebrand J. H., J. Am. Chem. Soc. **60**, 2522-3 (1938).
38GIA/KEM Giauque W. F., Kemp J. D., J. Chem. Phys. **6**, 40-52 (1938).
38GIA/STE Giauque W. F., Stephenson C. C., J. Am. Chem. Soc. **60**, 1389-94 (1938).
38HIC Hicks J. F. G., Jr., J. Am. Chem. Soc. **60**, 1000-4 (1938).
38KAI Kaischew R., Z. Phys. Chem., Abt. B **40**, 273-80 (1938).
38KEM/EGA Kemp J. D., Egan C. J., J. Am. Chem. Soc. **60**, 1521-5 (1938).
38KEN/SHO Kennedy W. D., Shomate C. H., Parks G. S., J. Am. Chem. Soc. **60**, 1507-9 (1938).
38KRE Kreutzer C., Ann. Phys. (Leipzig) [5] **33**, 192-209 (1938).
38KRU/CLU Kruis A., Clusius K., Z. Phys. Chem., Abt. B **38**, 156-76 (1938).
38PAN/DUD Panov N., Dudnikov I., Sb. Tr. Opytnogo Zavoda im. Akad. S. V. Lebedeva 36-40 (1938), from: Chem. Abstr. **34**, 1514[3] (1940).
38RIE Riedel L., Z. Gesamte Kaelte Ind. **45**, 221-5 (1938).
38SAG/LAC Sage B. H., Lacey W. N., Ind. Eng. Chem. **30**, 673-81 (1938).
38VER Verdonk G., Chem. Weekbl. **35**, 741-3 (1938).
39AST/EID1 Aston J. G., Eidinoff M. L., J. Am. Chem. Soc. **61**, 1533-8 (1939).
39AST/EID2 Aston J. G., Eidinoff M. L., Forster W. S., J. Am. Chem. Soc. **61**, 1539-43 (1939).
39BYK Bykov V. T., Zh. Fiz. Khim. **13**, 1013-9 (1939).
39CAR/STE Carpenter L. G., Steward C. J., Phil. Mag. J. Sci. [7] **27**, 551-64 (1939).
39FRE/HIL Frederick K. J., Hildebrand J. H., J. Am. Chem. Soc. **61**, 1555-8 (1939).
39GIA/POW Giauque W. F., Powell T. M., J. Am. Chem. Soc. **61**, 1970-4 (1939).
39GIA/RUE Giauque W. F., Ruehrwein R. A., J. Am. Chem. Soc. **61**, 2626-33 (1939).
39MAZ1 Mazur J., Acta Phys. Pol. **7**, 290-304 (1939).
39MAZ2 Mazur J., Acta Phys. Pol. **7**, 318-26 (1939).
39MAZ3 Mazur J., Z. Phys. **113**, 710-20 (1939).
39OSB/STI Osborne N. S., Stimson H. F., Ginnings D. C., J. Res. Natl. Bur. Stand. (U.S.) **23**, 197-260 (1939).
39PHI Philip N. M., Proc. Indian Acad. Sci., Sect. A **9**, 109-20 (1939).
39POW/GIA Powell T. M., Giauque W. F., J. Am. Chem. Soc. **61**, 2366-70 (1939).
39RAI Railing W. E., J. Am. Chem. Soc. **61**, 3349-53 (1939).
39RIE1 Riedel L., Bull. Inst. Int. Froid **20** (3), B1-B10 (1939).
39RIE2 Riedel L., Bull. Inst. Int. Froid **20** (4), B1-B10 (1939).
39RIE3 Riedel L., Z. Gesamte Kaelte Ind. **46**, 105-6 (1939).
39RUE/GIA Ruehrwein R. A., Giauque W. F., J. Am. Chem. Soc. **61**, 2940-4 (1939).
39SAG/EVA Sage B. H., Evans H. D., Lacey W. N., Ind. Eng. Chem. **31**, 763-7 (1939).
40AST/KEN Aston J. G., Kennedy R. M., Schumann S. C., J. Am. Chem. Soc. **62**, 2059-63 (1940).
40AST/MES Aston J. G., Messerly G. H., J. Am. Chem. Soc. **62**, 1917-23 (1940).
40AWB/GRI Awbery J. H., Griffiths E., Proc. Phys. Soc., London **52**, 770-6 (1940).
40BEN/MCH Benning A. F., McHarness R. C., Markwood W. H., Jr., Smith W. J., Ind. Eng. Chem. **32**, 976-80 (1940).
40BUR Burlew J. S., J. Am. Chem. Soc. **62**, 681-9 (1940).
40CLU/POP Clusius K., Popp L., Z. Phys. Chem., Abt. B **46**, 63-81 (1940).
40COC/FER Cockett A. H., Ferguson A., Phil. Mag. J. Sci. [7] **29**, 185-99 (1940).
40KUB/SCH Kubaschewski O., Schrag G., Z. Elektrochem. Angew. Phys. Chem. **46**, 675-80 (1940).
40MES/AST Messerly G. H., Aston J. G., J. Am. Chem. Soc. **62**, 886-90 (1940).
40MES/KEN Messerly G. H., Kennedy R. M., J. Am. Chem. Soc. **62**, 2988-91 (1940).
40PIT1 Pitzer K. S., J. Am. Chem. Soc. **62**, 331-5 (1940).
40PIT2 Pitzer K. S., J. Am. Chem. Soc. **62**, 1224-7 (1940).
40RIE Riedel L., Z. Gesamte Kaelte Ind. **47**, 87 (1940).
40TUR/BAR Turovskii Ya. A., Bartenev G. M., Zh. Tekh. Fiz. **10**, 514-20 (1940).
41AST/KEN Aston J. G., Kennedy R. M., Messerly G. H., J. Am. Chem. Soc. **63**, 2343-8 (1941).
41GIA/MEA Giauque W. F., Meads P. F., J. Am. Chem. Soc. **63**, 1897-901 (1941).
41KEN/SAG Kennedy R. M., Sagenkahn M., Aston J. G., J. Am. Chem. Soc. **63**, 2267-72 (1941).

41NEL/NEW Nelson E. W., Newton R. F., J. Am. Chem. Soc. **63**, 2178-82 (1941).
41OSB/GAR Osborne D. W., Garner C. S., Doescher R. N., Yost D. M., J. Am. Chem. Soc. **63**, 3496-9 (1941).
41PAR/WES Parks G. S., West T. J., Moore G. E., J. Am. Chem. Soc. **63**, 1133-5 (1941).
41PIT/SCO Pitzer K. S., Scott D. W., J. Am. Chem. Soc. **63**, 2419-22 (1941).
41RIE1 Riedel L., Bull. Int. Inst. Refrig. **22** (4), A1-A3 (1941).
41RIE2 Riedel L., Z. Gesamte Kaelte Ind. **48**, 89-92 (1941).
41RIE3 Riedel L., Z. Gesamte Kaelte Ind. **48**, 105-7 (1941).
41RUS/RUN Russell H., Jr., Rundle R. E., Yost D. M., J. Am. Chem. Soc. **63**, 2825-8 (1941).
41SCH Schmidt W. R., Ph. D. Thesis, Washington University (St. Louis) (1941).
41STO/FIS Stout J. W., Fisher L. H., J. Chem. Phys. **9**, 163-8 (1941).
41YOS/GAR Yost D. M., Garner C. S., Osborne D. W., Rubin T. R., Russell H., Jr., J. Am. Chem. Soc. **63**, 3488-92 (1941).
41YOS/OSB Yost D. M., Osborne D. W., Garner C. S., J. Am. Chem. Soc. **63**, 3492-9 (1941).
41ZHD Zhdanov A. K., Zh. Obshch. Khim. **11**, 471-82 (1941).
41ZIE/MES Ziegler W. T., Messer C. E., J. Am. Chem. Soc. **63**, 2694-700 (1941).
42CLU/FAB Clusius K., Faber G., Z. Phys. Chem., Abt. B **51**, 352-70 (1942).
42FOR/GIA Forsythe W. R., Giauque W. F., J. Am. Chem. Soc. **64**, 48-61 (1942).
42OSB/DOE Osborne D. W., Doescher R. N., Yost D. M., J. Am. Chem. Soc. **64**, 169-72 (1942).
42RIE1 Riedel L., Bull. Int. Inst. Refrig. **23** (1), A1-A7 (1942).
42RIE2 Riedel L., Bull. Int. Inst. Refrig. **23** (2), A1-A5 (1942).
42RUS/OSB Russell H., Jr., Osborne D. W., Yost D. M., J. Am. Chem. Soc. **64**, 165-9 (1942).
42SCH/AST Schumann S. C., Aston J. G., Sagenkahn M., J. Am. Chem. Soc. **64**, 1039-43 (1942).
42TRI/ENG Trimble H. M., Engle C. J., Brown R. A., Schmuck R. F., J. Am. Chem. Soc. **64**, 679-81 (1942).
42YOU/HIL Young F. E., Hildebrand J. H., J. Am. Chem. Soc. **64**, 839-40 (1942).
42ZIE/AND Ziegler W. T., Andrews D. H., J. Am. Chem. Soc. **64**, 2482-5 (1942).
43AST/FIN Aston J. G., Finke H. L., Schumann S. C., J. Am. Chem. Soc. **65**, 341-6 (1943).
43AST/SZA Aston J. G., Szasz G. J., Fink H. L., J. Am. Chem. Soc. **65**, 1135-9 (1943).
43BAC/PER Backer H. J., Perdok W. G., Recl. Trav. Chim. Pays-Bas **62**, 533-49 (1943).
43GUT/HUF Guthrie G. B., Jr., Huffman H. M., J. Am. Chem. Soc. **65**, 1139-43 (1943).
43PIT/SCO Pitzer K. S., Scott D. W., J. Am. Chem. Soc. **65**, 803-11 (1943).
43RUE/HUF Ruehrwein R. A., Huffman H. M., J. Am. Chem. Soc. **65**, 1620-5 (1943).
44AST/SAG Aston J. G., Sagenkahn M. L., Szasz G. J., Moessen G. W., Zuhr H. F., J. Am. Chem. Soc. **66**, 1171-7 (1944).
44BAI/TOD Bailey A. E., Todd S. S., Singleton W. S., Oliver G. D., J. Am. Oil Chem. Soc. **21**, 293-7 (1944).
44EIB Eibert J., Thesis, Washington University (St. Louis) (1944).
44GUT/SPI Guthrie G. B., Jr., Spitzer R. W., Huffman H. M., J. Am. Chem. Soc. **66**, 2120-1 (1944).
44HIC/HOO Hicks J. F. G., Hooley J. G., Stephenson C. C., J. Am. Chem. Soc. **66**, 1064-7 (1944).
44RUB/LEV Rubin T. R., Levedahl B. H., Yost D. M., J. Am. Chem. Soc. **66**, 279-82 (1944).
44RUS/GOL Russell H., Jr., Golding D. R. V., Yost D. M., J. Am. Chem. Soc. **66**, 16-20 (1944).
44SCO/FER Scott R. B., Ferguson W. J., Brickwedde F. G., J. Res. Natl. Bur. Stand. (U.S.) **33**, 1-20 (1944).
45DAV/WIE Davis H. S., Wiedeman O. F., Ind. Eng. Chem. **37**, 482-5 (1945).
45FIS/NAY Fischl F. B., Naylor B. F., Ziemer C. W., Parks G. S., Aston J. G., J. Am. Chem. Soc. **67**, 2075-9 (1945).
45GUT/PIT Guttman L., Pitzer K. S., J. Am. Chem. Soc. **67**, 324-7 (1945).
45SCO/BRI Scott R. B., Brickwedde F. G., J. Res. Natl. Bur. Stand. (U.S.) **35**, 501-12 (1945).
45SCO/MEY Scott R. B., Meyers C. H., Rands R. D., Jr., Brickwedde F. G., Bekkedahl N., J. Res. Natl. Bur. Stand. (U.S.) **35**, 39-85 (1945).
45ZHD Zhdanov A. K., Zh. Obshch. Khim. **15**, 895-902 (1945).
46AST/FIN Aston J. G., Fink H. L., Bestul A. B., Pace E. L., Szasz G. J., J. Am. Chem. Soc. **68**, 52-7 (1946).
46DOU/HUF1 Douslin D. R., Huffman H. M., J. Am. Chem. Soc. **68**, 173-6 (1946).
46DOU/HUF2 Douslin D. R., Huffman H. M., J. Am. Chem. Soc. **68**, 1704-8 (1946).
46HOG Hoge H. J., J. Res. Natl. Bur. Stand. (U.S.) **36**, 111-8 (1946).
46KIL/PIT Kilpatrick J. E., Pitzer K. S., J. Am. Chem. Soc. **68**, 1066-72 (1946).
46PIT/GUT Pitzer K. S., Guttman L., Westrum E. F., Jr., J. Am. Chem. Soc. **68**, 2209-12 (1946).
46RUE/POW Ruehrwein R. A., Powell T. M., J. Am. Chem. Soc. **68**, 1063-72 (1946).
47AST/SZA Aston J. G., Szasz G. J., J. Am. Chem. Soc. **69**, 3108-14 (1947).
47BAR Bartenev G. M., Zh. Tekh. Fiz. **17**, 1321-4 (1947).
47CHA/SIN Charbonnet G. H., Singleton W. S., J. Am. Oil Chem. Soc. **24**, 140-2 (1947).
47CLU/WOL Clusius K., Wolf G., Z. Naturforsch. **2a**, 495-504 (1947).
47CON/ELV Conner A. Z., Elving P. J., Steingiser S., J. Am. Chem. Soc. **69**, 1532 (1947).
47COR/GIN Corruccini R. J., Ginnings D. C., J. Am. Chem. Soc. **69**, 2291-4 (1947).
47GIN/COR Ginnings D. C., Corruccini R. J., J. Res. Natl. Bur. Stand. (U.S.) **38**, 583-91 (1947).
47HUF Huffman H. M., Chem. Rev. **40**, 1-14 (1947).
47JON/GIA Jones W. M., Giauque W. F., J. Am. Chem. Soc. **69**, 983-7 (1947).
47KUR Kurbatov V. Ya., Zh. Obshch. Khim. **17**, 1999-2009 (1947).
47OSB/GIN Osborne N. S., Ginnings D. C., J. Res. Natl. Bur. Stand. (U.S.) **39**, 453-77 (1947).
47PUS/FED Pushin N. A., Fedjushkin A. V., Krgovic B., Glas. Hem. Drus. Beograd (Bull. Chem. Soc. Beograd) **11**, 12-24 (1947).
47ROS/PIT Rossini F. D., Pitzer K. S., Taylor W. J., Ebert J. P., Kilpatrick J. E., Beckett C. W., Williams M. D., Werner H. G., Selected Values of Hydrocarbons, NBS Circular C461, U.S. Gov't. Printing Office, Washington, D. C. (1947).
47SCH/ZOS Schildknecht C. E., Zoss A. O., McKinley C., Ind. Eng. Chem. **39**, 180-6 (1947).
47SKU Skuratov S. M., Kolloid. Zh. **9**, 133-40 (1947).
47SZA/MOR Szasz G. J., Morrison J. A., Pace E. L., Aston J. G., J. Chem. Phys. **15**, 562-4 (1947).
47TOD/OLI Todd S. S., Oliver G. D., Huffman H. M., J. Am. Chem. Soc. **69**, 1519-25 (1947).
48BRI/HOG Brickwedde F. G., Hoge H. J., Scott R. B., J. Chem. Phys. **16**, 429-36 (1948).
48EUC Eucken A., Z. Elektrochem. Angew. Phys. Chem. **52**, 255-68 (1948).
48GIA/JON Giauque W. F., Jones W. M., J. Am. Chem. Soc. **70**, 120-4 (1948).
48GIN/COR Ginnings D. C., Corruccini R. J., Ind. Eng. Chem. **40**, 1990-1 (1948).
48GOR/GIA Gordon J., Giauque W. F., J. Am. Chem. Soc. **70**, 1506-10 (1948).
48HUF/EAT Huffman H. M., Eaton M., Oliver G. D., J. Am. Chem. Soc. **70**, 2911-4 (1948).
48KUR Kurbatov V. Ya., Zh. Obshch. Khim. **18**, 372-87 (1948).

48MAZ Mazee W. M., Recl. Trav. Chim. Pays-Bas **67**, 197-213 (1948).

48OLI/EAT Oliver G. D., Eaton M., Huffman H. M., J. Am. Chem. Soc. **70**, 1502-5 (1948).

48PAC/AST Pace E. L., Aston J. G., J. Am. Chem. Soc. **70**, 566-70 (1948).

48TSC1 Tschamler H., Monatsh. Chem. **79**, 162-77 (1948).

48TSC2 Tschamler H., Monatsh. Chem. **79**, 223-32 (1948).

48TSC3 Tschamler H., Monatsh. Chem. **79**, 233-42 (1948).

48TSC4 Tschamler H., Monatsh. Chem. **79**, 243-7 (1948).

48TSC5 Tschamler H., Monatsh. Chem. **79**, 499-507 (1948).

48TSC6 Tschamler H., Monatsh. Chem. **79**, 508-20 (1948).

48WOO/SCO Woolley H. W., Scott R. B., Brickwedde F. G., J. Res. Natl. Bur. Stand. (U.S.) **41**, 379-475 (1948).

49DHO/JUN D'Hont M., Jungers J. C., Bull. Soc. Chim. Belg. **58**, 196-204 (1949).

49GIA/GOR Giauque W. F., Gordon J., J. Am. Chem. Soc. **71**, 2176-81 (1949).

49HOR/STO Horowitz K. H., Stow F. S., Jr., Anal. Chem. **21**, 1571-2 (1949).

49HUF/TOD Huffman H. M., Todd S. S., Oliver G. D., J. Am. Chem. Soc. **71**, 584-92 (1949).

49LEE Leech J. W., Proc. Phys. Soc., London Sect. B **62**, 390-8 (1949).

49PAR/HAT Parks G. S., Hatton J. A., J. Am. Chem. Soc. **71**, 2773-5 (1949).

49PAR/MOO Parks G. S., Moore G. E., Renquist M. L., Naylor B. F., McClaine L. A., Fujii P. S., Hatton J. A., J. Am. Chem. Soc. **71**, 3386-9 (1949).

49SCH/SAG Schlinger W. G., Sage B. H., Ind. Eng. Chem. **41**, 1779-82 (1949).

49SCO/GRO Scott D. W., Gross M. E., Oliver G. D., Huffman H. M., J. Am. Chem. Soc. **71**, 1634-6 (1949).

49SCO/OLI Scott D. W., Oliver G. D., Gross M. E., Hubbard W. N., Huffman H. M., J. Am. Chem. Soc. **71**, 2293-7 (1949).

49STA/GUP Staveley L. A. K., Gupta A. K., Trans. Faraday Soc. **45**, 50-61 (1949).

49TSC/RIC1 Tschamler H., Richter E., Monatsh. Chem. **80**, 510-6 (1949).

49TSC/RIC2 Tschamler H., Richter E., Wettig F., Monatsh. Chem. **80**, 749-58 (1949).

49TSC/RIC3 Tschamler H., Richter E., Wettig F., Monatsh. Chem. **80**, 856-63 (1949).

49TSC/WET Tschamler H., Wettig F., Richter E., Monatsh. Chem. **80**, 572-82 (1949).

49WAD/KNO Waddington G., Knowlton J. W., Scott D. W., Oliver G. D., Todd S. S., Hubbard W. N., Smith J. C., Huffman H. M, J. Am. Chem. Soc. **71**, 797-808 (1949).

49WEI Weissberger A. (editor), Physical Methods of Organic Chemistry, Interscience Publ. Co., New York Vol. **1/1**, 773 (1949).

49WUY/JUN Wuyts J., Jungers J. C., Bull. Soc. Chim. Belg. **58**, 80-6 (1949).

50AST/MAS Aston J. G., Mastrangelo V. R. S., Moessen G. W., J. Am. Chem. Soc. **72**, 5287-91 (1950).

50AUE/SAG Auerbach C. E., Sage B. H., Lacey W. N., Ind. Eng. Chem. **42**, 110-3 (1950).

50CRO/SMY1 Crowe R. W., Smyth C. P., J. Am. Chem. Soc. **72**, 1098-106 (1950).

50CRO/SMY2 Crowe R. W., Smyth C. P., J. Am. Chem. Soc. **72**, 4009-15 (1950).

50EGA/WAK Egan E. P., Jr., Wakefield Z. T., Elmore K. L., J. Am. Chem. Soc. **72**, 2418-21 (1950).

50GIN/DOU Ginnings D. C., Douglas T. B., Ball A. F., J. Res. Natl. Bur. Stand. (U.S.) **45**, 23-33 (1950).

50HOU/MAS1 Hough E. W., Mason D. M., Sage B. H., J. Am. Chem. Soc. **72**, 5774-5 (1950).

50HOU/MAS2 Hough E. W., Mason D. M., Sage B. H., J. Am. Chem. Soc. **72**, 5775-7 (1950).

50JOH/CLA Johnston H. L., Clarke J. T., Rifkin E. B., Kerr E. C., J. Am. Chem. Soc. **72**, 3933-8 (1950).

50KUR Kurbatov V. Ya., Zh. Obshch. Khim. **20**, 1139-44 (1950).

50KUS/CRO Kushner L. M., Crowe R. W., Smyth C. P., J. Am. Chem. Soc. **72**, 1091-8 (1950).

50SAG/HOU Sage B. H., Hough E. W., Anal. Chem. **22**, 1304-9 (1950).

50SCO/FIN1 Scott D. W., Finke H. L., Gross M. E., Guthrie G. B., Huffman H. M., J. Am. Chem. Soc. **72**, 2424-30 (1950).

50SCO/FIN2 Scott D. W., Finke H. L., Hubbard W. N., McCullough J. P., Gross M. E., Williamson K. D., Waddington G., Huffman H. M., J. Am. Chem. Soc. **72**, 4664-8 (1950).

50SIN/WAR Singleton W. S., Ward T. L., Dollear F. G., J. Am. Oil Chem. Soc. **27**, 143-6 (1950).

50TIM Timmermans J., Physicochemical Constants of Pure Organic Compounds, Vol. I, Elsevier, Amsterdam (1950).

50URA/SID Urazovskii S. S., Sidorov I. A., Dokl. Akad. Nauk. SSSR **70**, 859-62 (1950).

50YAG/UNT Yaggee F. L., Untermyer S., Argonne Natl. Lab. [Rep.] ANL-4458, 1-27 (1950).

51AST/FIN Aston J. G., Fink H. L., Janz G. J., Russell K. E., J. Am. Chem. Soc. **73**, 1939-43 (1951).

51AST/JAN Aston J. G., Janz G. J., Russell K. E., J. Am. Chem. Soc. **73**, 1943-5 (1951).

51BAT/SMI Bates A. G., Smith D. J., U.S. Atomic Energy Commission Report K729 (1951), from: Hultgren R., Orr R. L., Anderson P. D., Kelley K. K., Select. Values Thermodyn. Properties of Metals and Alloys, J. Wiley, New York (1963).

51CON/SAG Connolly T. J., Sage B. H., Lacey W. N., Ind. Eng. Chem. **43**, 946-50 (1951).

51DOL/HET Dole M., Hettinger W. P., Jr., Larson N., Wethington J. A., Worthington A. E, Rev. Sci. Instrum. **22**, 812-7 (1951).

51DOU/BAL Douglas T. B., Ball A. F., Ginnings D. C., J. Res. Natl. Bur. Stand. (U.S.) **46**, 334-48 (1951).

51EUC/EIG Eucken A., Eigen M., Z. Elektrochem. **55**, 343-54 (1951).

51FOL/GIG Foley W. T., Giguere P. A., Can. J. Chem. **29**, 895-903 (1951).

51FUR/GIN Furukawa G. T., Ginnings D. C., McCoskey R. E., Nelson R. A., J. Res. Natl. Bur. Stand. (U.S.) **46**, 195-206 (1951).

51FUR/MCC Furukawa G. T., McCoskey R. E., King G. J., J. Res. Natl. Bur. Stand. (U.S.) **47**, 256-61 (1951).

51GRI/BER Grisard J. W., Bernhardt H. A., Oliver G. D., J. Am. Chem. Soc. **73**, 5725-7 (1951).

51HUL/WIL Hull R. A., Wilkinson K. R., Wilks J., Proc. Phys. Soc., London Sect. A **64**, 379-88 (1951).

51KER/RIF Kerr E. C., Rifkin E. B., Johnston H. L., Clarke J. T., J. Am. Chem. Soc. **73**, 282-4 (1951).

51OLI/GRI1 Oliver G. D., Grisard J. W., J. Am. Chem. Soc. **73**, 1688-90 (1951).

51OLI/GRI2 Oliver G. D., Grisard J. W., Cunningham C. W., J. Am. Chem. Soc. **73**, 5719-22 (1951).

51PAR/SIM Parkinson D. H., Simon F. E., Spedding F. H., Proc. R. Soc. London, A **207**, 137-55 (1951).

51POP/GAL Popov M. M., Gal'chenko G. L., Zh. Obshch. Khim. **21**, 2220-35 (1951).

51SCO/FIN Scott D. W., Finke H. L., McCullough J. P., Gross M. E., Williamson K. D., Waddington G., Huffman H. M, J. Am. Chem. Soc. **73**, 261-5 (1951).

51SIE/CRU Sieg L., Crutzen J. L., Jost W., Z. Phys. Chem. (Leipzig) **198**, 263-9 (1951).

51TSC/KRI Tschamler H., Krischai H., Monatsh. Chem. **82**, 259-70 (1951).

52ADA/JOH Adams G. B., Jr., Johnston H. L., Kerr E. C., J. Am. Chem. Soc. **74**, 4784-7 (1952).

52CUR/JOH Curme G. O., Jr. (editor), Johnston F. (assoc. ed.), Glycols, Reinhold, New York 251, (1952).

52DOU/BAL Douglas T. B., Ball A. F., Ginnings D. C., Davis W. D., J. Am. Chem. Soc. **74**, 2472-8 (1952).

52ERD/JAG Erdös E., Jäger L., Pouchlý J., Chem. Listy **46**, 770 (1952).

52FIN/SCO Finke H. L., Scott D. W., Gross M. E., Waddington G., Huffman H. M., J. Am. Chem. Soc. **74**, 2804-6 (1952).

52GUT/SCO Guthrie G. B., Jr., Scott D. W., Hubbard W. N., Katz C., McCullough J. P., Gross M. E., Williamson K. D., Waddington G., J. Am. Chem. Soc. **74**, 4662-9 (1952).

52HOF Hoffman J. D., J. Am. Chem. Soc. **74**, 1696-700 (1952).

52HUB/FIN Hubbard W. N., Finke H. L., Scott D. W., McCullough J. P., Katz C., Gross M. E., Messerly J. F., Pennington R. E., Waddington G., J. Am. Chem. Soc. **74**, 6025-30 (1952).

52MCC/SCO McCullough J. P., Scott D. W., Finke H. L., Gross M. E., Williamson K. D., Pennington R. E., Waddington G., Huffman H. M, J. Am. Chem. Soc. **74**, 2801-4 (1952).

52OLI/GRI Oliver G. D., Grisard J. W., J. Am. Chem. Soc. **74**, 2705-7 (1952).

52RED/LON Redmond R. F., Lones J., U.S. Atomic Energy Commission Report ORNL-1342, 3-20 (1952).

52ROS/WAG Rossini F. D., Wagman D. D., Evans W. H., Levine S., Jaffe I., Selected Values of Chemical Thermodynamic Properties, National Bureau of Standards, Circular 500, U.S. Government Printing Office, Washington D. C. (1952).

52RUB/GIA Rubin T. R., Giauque W. F., J. Am. Chem. Soc. **74**, 800-4 (1952).

52SCH/SAG Schlinger W. G., Sage B. H., Ind. Eng. Chem. **44**, 2454-6 (1952).

52SCO/DOU Scott D. W., Douslin D. R., Gross M. E., Oliver G. D., Huffman H. M., J. Am. Chem. Soc. **74**, 883-7 (1952).

52SCO/FIN1 Scott D. W., Finke H. L., Hubbard W. N., McCulloughJ. P., Oliver G. D., Gross M. E., Katz C., Williamson K. D., Waddington G., Huffman H. M, J. Am. Chem. Soc. **74**, 4656-62 (1952).

52SCO/FIN2 Scott D. W., Finke H. L., McCullough J. P., Gross M. E., Pennington R. E., Waddington G., J. Am. Chem. Soc. **74**, 2478-83 (1952).

52STA/AMI Stallard R. D., Amis E. S., J. Am. Chem. Soc. **74**, 1781-90 (1952).

52WAR/SIN Ward T. L., Singleton W. S., J. Phys. Chem. **56**, 696-8 (1952).

53AST/WOO Aston J. G., Wood J. L., Zolki T. P., J. Am. Chem. Soc. **75**, 6202-4 (1953).

53BUS/GIA Busey R. H., Giauque W. F., J. Am. Chem. Soc. **75**, 806-9 (1953).

53CLA/RIF Clarke J. T., Rifkin E. B., Johnston H. L., J. Am. Chem. Soc. **75**, 781-5 (1953).

53FUR/MCC Furukawa G. T., McCoskey R. E., Reilly M. L., J. Res. Natl. Bur. Stand. (U.S.) **51**, 69-72 (1953).

53GIN/FUR Ginnings D. C., Furukawa G. T., J. Am. Chem. Soc. **75**, 522-7 (1953).

53GME Gmelin L. (founder): Schwefel-system number 9, Gmelins Handbuch Anorganischen Chemie, Verlag Chemie-GMBH, Weinheim, Vol. **B. 1,** 349 (1953).

53GRA/SMI Gray P., Smith P. L., J. Chem. Soc. 2380-5 (1953).

53GRO/OLI Gross M. E., Oliver G. D., Huffman H. M., J. Am. Chem. Soc. **75**, 2801-4 (1953).

53HIL/GIA Hildenbrand D. L., Giauque W. F., J. Am. Chem. Soc. **75**, 2811-8 (1953).

53HOF/DEC Hoffman J. D., Decker B. F., J. Phys. Chem. **57**, 520-9 (1953).

53HU /WHI1 Hu J. -H., White D., Johnston H. L., J. Am. Chem. Soc. **75**, 1232-6 (1953).

53HU /WHI2 Hu J. -H., White D., Johnston H. L., J. Am. Chem. Soc. **75**, 5642-5 (1953).

53LLE Llewellyn D. R., J. Chem. Soc. 28-36 (1953).

53MCC/SCO McCullough J. P., Scott D. W., Finke H. L., Hubbard W. N., Gross M. E., Katz C., Pennington R. E., Messerly J. F., Waddington G., J. Am. Chem. Soc. **75**, 1818-24 (1953).

53MCC/SUN McCullough J. P., Sunner S., Finke H. L., Hubbard W. N., Gross M. E., Pennington R. E., Messerly J. F., Good W. D., Waddington G., J. Am. Chem. Soc. **75**, 5075-81 (1953).

53RAT/GWI Rathjens G. W., Jr., Gwinn W. D., J. Am. Chem. Soc. **75**, 5629-33 (1953).

53RIF/KER Rifkin E. B., Kerr E. C., Johnston H. L., J. Am. Chem. Soc. **75**, 785-8 (1953).

53ROS/PIT Rossini F. D., Pitzer K. S., Arnett R. L., Brown R. M., Pimentel G. C., Selected Values of Physical and Thermodynamic Properties of Hydrocarbons and Related Compounds, Carnegie Press, Pittsburgh, PA (1953).

53SCO/FIN Scott D. W., Finke H. L., Hubbard W. N., McCullough J. P., Katz C., Gross M. E., Messerly J. F., Pennington R. E., Waddington G., J. Am. Chem. Soc. **75**, 2795-800 (1953).

53SEY Seyer W. F., J. Am. Chem. Soc. **75**, 616-21 (1953).

53WES/HAT Westrum E. F., Jr., Hatcher J. B., Osborne D. W., J. Chem. Phys. **21**, 419-23 (1953).

53WIL/DOL Wilhoit R. C., Dole M., J. Phys. Chem. **57**, 14-21 (1953).

54BRA/MOE Braune H., Moeller O., Z. Naturforsch. **9a**, 210-7 (1954).

54BUB Bubushyan M. B., Teploenergetika (Moscow) **1** (3), 49-51 (1954).

54DOU/FUR Douglas T. B., Furukawa G. T., McCoskey R. E., Ball A. F, J. Res. Natl. Bur. Stand. (U.S.) **53**, 139-53 (1954).

54FIN/GRO1 Finke H. L., Gross M. E., Messerly J. F., Waddington G., J. Am. Chem. Soc. **76**, 854-7 (1954).

54FIN/GRO2 Finke H. L., Gross M. E., Waddington G., Huffman H. M., J. Am. Chem. Soc. **76**, 333-41 (1954).

54FUR/MCC Furukawa G. T., McCoskey R. E., Reilly M. L., J. Res. Natl. Bur. Stand. (U.S.) **52**, 11-6 (1954).

54FUR/PAR Furukawa G. T., Park R. P., NBS Report No. 3649, 1-27 (1954).

54GRA/SMI Gray P., Smith P. L., J. Chem. Soc. 769-73 (1954).

54MCC/FIN1 McCullough J. P., Finke H. L., Hubbard W. N., Good W. D., Pennington R. E., Messerly J. F., Waddington W., J. Am. Chem. Soc. **76**, 2661-9 (1954).

54MCC/FIN2 McCullough J. P., Finke H. L., Scott D. W., Gross M. E., Messerly J. F., Pennington R. E., Waddington G., J. Am. Chem. Soc. **76**, 4796-802 (1954).

54OSB/ABR Osborne D. W., Abrham B. M., Weinstock B., Phys. Rev. **94**, 202-3 (1954).

54ROB/SYD Roberts T. R., Sydoriak S. G., Phys. Rev. **93**, 1418 (1954).

54SMI/HAL Smith A. L., Hallett N. C., Johnston H. L., J. Am. Chem. Soc. **76**, 1486-8 (1954).

54STA/WAR Staveley L. A. K., Warren J. B., Paget H. P., Dowrick D. J., J. Chem. Soc. 1992-2001 (1954).

54STR/ICK Strelkov P. G., Ickevich E. S., Kostryukov V. N., Mirskaya G. G., Samoilov V. N., Zh. Fiz. Khim. **28**, 459-72 (1954).

55AST/WIL Aston J. G., Wills P. E., Zolki T. P., J. Am. Chem. Soc. **77**, 3939-41 (1955).

55AST/ZOL Aston J. G., Zolki T. P., Wood J. L., J. Am. Chem. Soc. **77**, 281-4 (1955).

55DAU/MAR Dauphinee T. M., Martin D. L., Preston-Thomas H., Proc. R. Soc. London, A **233**, 214-22 (1955).

55DOU/EPS Douglas T. B., Epstein L. F., Dever J. L., Howland W. H., J. Am. Chem. Soc. **77**, 2144-50 (1955).

55FUR Furukawa G. T., NBS Report No. 3712, 1-16 (1955).

55FUR/MCC Furukawa G. T., McCoskey R. E., Reilly M. L., Harman A. W., J. Res. Natl. Bur. Stand. (U.S.) **55**, 201-7 (1955).

55FUR/PAR Furukawa G. T., Park R. P., J. Res. Natl. Bur. Stand. (U.S.) **55**, 255-60 (1955).

55GIG/MOR1 Giguere P. A., Morissette B. G., Olmos A. W., Can. J. Chem. **33**, 657-64 (1955).

55GIG/MOR2 Giguere P. A., Morissette B. G., Olmos A. W., Knop O., Can. J. Chem. **33**, 804-20 (1955).

55GUC/CHR Gucker F. T., Christens J. M., Proc. Indiana Acad. Sci. **64**, 97 (1955), from: Chem. Abstr. **50**, 4562g (1956).

55HEL/HEI Helfrey P. F., Heiser D. A., Sage B. H., Ind. Eng. Chem. **47**, 2385-8 (1955).

55HUT/MAN Hutchinson E., Manchester K. E., Rev. Sci. Instrum. **26**, 364-7 (1955).

55LYD Lydersen A. L., Estimation of Critical Properties of Organic Compounds, Univ. Wisconsin, Coll. Eng., Exp. Stn. Rep. 3, Madison, Wis. (1955).

55MAR/SMI Marx P., Smith C. W., Worthington A. E., Dole M., J. Am. Chem. Soc. **59**, 1015-9 (1955).

55MCC/FIN McCullough J. P., Finke H. L., Messerly J. F., Pennington R. E., Hossenlopp I. A., Waddington G., J. Am. Chem. Soc. **77**, 6119-25 (1955).

55PAC/PIE Pace E. L., Pierce L., Dennis K. S., Rev. Sci. Instrum. **26**, 20-2 (1955).

55PIE/PAC Pierce L., Pace E. L., J. Chem. Phys. **23**, 551-5 (1955).

55ROB/SYD Roberts T. R., Sydoriak S. G., Phys. Rev. **98**, 1672-8 (1955).

55RUI1 Ruiter L. H., Recl. Trav. Chim. Pays-Bas **74**, 1131-59 (1955).

55RUI2 Ruiter L. H., Recl. Trav. Chim. Pays-Bas **74**, 1467-81 (1955).

55SCO/FIN Scott D. W., Finke H. L., McCullough J. P., Gross M. E., Messerly J. F., Pennington R. E., Waddington G., J. Am. Chem. Soc. **77**, 4993-8 (1955).

55SEK/MOM Seki S., Momotani M., Nakatsu K., Oshima T., Bull. Chem. Soc. Jpn. **28**, 411-6 (1955).

55SHE/GIA Sherman R. H., Giauque W. F., J. Am. Chem. Soc. **77**, 2154-60 (1955).

55SPI/STA Spice J. E., Staveley L. A. K., Harrow G. A., J. Chem. Soc. 100-4 (1955).

55STA/TUP Staveley L. A. K., Tupman W. I., Hart K. R., Trans. Faraday Soc. **51**, 323-43 (1955).

55TAY/JOH Taylor R. D., Johnson B. H., Kilpatrick J. E., J. Chem. Phys. **23**, 1225-31 (1955).

55TAY/KIL Taylor R. D., Kilpatrick J. E., J. Chem. Phys. **23**, 1232-5 (1955).

55TOD Todd L. J., Ph. D. Thesis, John Hopkins University, Baltimore, Maryland (1955), from: McCullough J. P., Scott D. W. (editors), Experimental Thermodynamics, Vol. I, Butterworth, London 421-535 (1968).

55WAR/VIC Ward T. L., Vicknair E. J., Singleton W. S., Feuge R. O., J. Phys. Chem. **59**, 4-7 (1955).

56CAL/PRA Calvet E., Prat H., Microcalorimıtrie. Applications physico-chimiques et biologiques, Mason et Cie, Paris (1956).

56COO/BAL Coops J., Balk A. N., Tolk M. W., Recl. Trav. Chim. Pays-Bas **75**, 75-81 (1956).

56DUF/EVE Duff G. M., Everett D. H., Trans. Faraday Soc. **52**, 753-63 (1956).

56FIN/SCO Finke H. L., Scott D. W., Gross M. E., Messerly J. F., Waddington G., J. Am. Chem. Soc. **78**, 5469-76 (1956).

56FUR/DOU Furukawa G. T., Douglas T. B., McCoskey R. E., Ginnings D. C., J. Res. Natl. Bur. Std. (U.S.) **57**, 67-82 (1956).

56HAA Haase R., Thermodynamik der Mischphasen, Springer Verlag, Berlin-Göttingen-Heidelberg (1956).

56LI Li J. C. M., J. Am. Chem. Soc. **78**, 1081-3 (1956).

56LI /PIT Li J. C. M., Pitzer K. S., J. Am. Chem. Soc. **78**, 1077-80 (1956).

56MCE/MAL McEwen M., Malcolm, Monsanto Chemical Co. March 31 (1956), from: Walker B. E., Jr., Brouks M. S., Ewing C. T., Miller R. R., Ind. Eng. Chem., Chem. Eng. Data Ser. **3**, 280-2 (1958).

56MUR/VAN Murti P. S., Van Winkle M., J. Chem. Eng. Data **3**, 65-71 (1956).

56PAR/KEN Parks G. S., Kennedy W. D., Gates R. R., Mosley J. R., Moore G. E., Renquist M. L., J. Am. Chem. Soc. **78**, 56-9 (1956).

56PEN/FIN Pennington R. E., Finke H. L., Hubbard W. N., Messerly J. F., Frow F. R., Hossenlopp I. A., Waddington G., J. Am. Chem. Soc. **78**, 2055-60 (1956).

56PEN/SCO Pennington R. E., Scott D. W., Finke H. L., McCullough J. P., Messerly J. F., Hossenlopp I. A., Waddington G., J. Am. Chem. Soc. **78**, 3266-72 (1956).

56PLA Plank R., Kaeltetechnik **8**, 127-8 (1956).

56POP/KOL Popov M. M., Kolesov V. P., Zh. Obshch. Khim. **26**, 2385-93 (1956).

56SCH/GOT Schrock V. E., Gott R. E., Starkman E. S., WADC Technical Report 56-104, 1-144 (1956).

56SCH/HIL Schneider A., Hilmer O., Z. Anorg. Allg. Chem. **286**, 97-117 (1956).

56SCO/GOO Scott D. W., Good W. D., Waddington G., J. Phys. Chem. **60**, 1090-5 (1956).

56SCO/MCC1 Scott D. W., McCullough J. P., Good W. D., Messerly J. F., Pennington R. E., Kincheloe I. A., Hossenlopp I. A., Douslin D. R., Waddington G., J. Am. Chem. Soc. **78**, 5457-63 (1956).

56SCO/MCC2 Scott D. W., McCullough J. P., Hubbard W. N., Messerly J. F., Hossenlopp I. A., Frow F. R., Waddington G., J. Am. Chem. Soc. **78**, 5463-8 (1956).

56SMI/DOL Smith C. W., Dole M., J. Polym. Sci. **20**, 37-56 (1956).

56WAL/GRA Walker B. E., Grand J. A., Miller R. R., J. Phys. Chem. **60**, 231-3 (1956).

57CRU/JOS Crutzen J. L., Jost W., Sieg L., Ber. Bunsenges. Phys. Chem. **61**, 230-46 (1957).

57HAR/MOE Harrison D., Moelwyn-Hughes E. A., Proc. R. Soc. London, A **239**, 230-46 (1957).

57HIL/LOU Hill R. W., Lounasmaa O. V., Philos. Mag. [8] **2**, 143-8 (1957).

57KEN Kenttamaa J., Suom. Kemistil. **B30**, 15-8 (1957).

57KER Kerr E. C., Thesis, Ohio State University, Univ. Microfilms Publ. No. 21480, 1-80 (1957).

57MCC/DOU McCullough J. P., Douslin D. R., Messerly J. F., Hossenlopp I. A., Kincheloe T. C., Waddington G., J. Am. Chem. Soc. **79**, 4289-95 (1957).

57MCC/FIN1 McCullough J. P., Finke H. L., Gross M. E., Messerly J. F., Waddington G., J. Phys. Chem. **61**, 289-301 (1957).

57MCC/FIN2 McCullough J. P., Finke H. L., Messerly J. F., Todd S. S., Kincheloe T. C., Waddington G., J. Phys. Chem. **61**, 1105-16 (1957).

57NEI/WHI Neilson E. F., White D., J. Am. Chem. Soc. **79**, 5618-21 (1957).

57PIL Pilcher G., Anal. Chim. Acta **17**, 144-60 (1957).

57PUT/KIL Putnam W. E., Kilpatrick J. E, J. Chem. Phys. **27**, 1075-80 (1957).

57SCO/FIN Scott D. W., Finke H. L., McCullough J. P., Messerly J. F., Pennington R. E., Hossenlopp I. A., Waddington G., J. Am. Chem. Soc. **79**, 1062-8 (1957).

57YAR/KAY Yarrington R. M., Kay W. B., J. Phys. Chem. **61**, 1259-60 (1957).

58ANO Anon, Union Carbide Co., Physical Properties, Chemical Division, New York (1958), from: Wilhoit R. C., Chao J., Hall K. R., J. Phys. Chem. Ref. Data **14**, 1-175 (1985).

58BRA/WAT Brabets R. I., Waterman T. E., J. Chem. Phys. **28**, 1212-4 (1958).

58CHE Cherneeva L. I., Teploenergetika (Moscow) **5** (12), 71-8 (1958).

58DVO Dvořák J., Chem. Prum. **8**, 287-91 (1958).

58EGA/LUF Egan E. P., Jr., Luff B. B., Wakefield Z. T., J. Phys. Chem. **62**, 1091-5 (1958).

58HIL/KRA Hildenbrand D. L., Kramer W. R., McDonald R. A., Stull D. R., J. Am. Chem. Soc. **80**, 4129-32 (1958).

58HUB/DOU Hubbard W. N., Douslin D. R., McCullough J. P., Scott D. W., Todd S. S., Messerly J. F., Hossenlopp I. A., George A., Waddington G., J. Am. Chem. Soc. **80**, 3547-54 (1958).

58KOE/GIA Koehler J. K., Giauque W. F., J. Am. Chem. Soc. **80**, 2659-62 (1958).

58LIN/WAR Linevsky M. J., Wartik T., J. Phys. Chem. **62**, 1146-8 (1958).

58LUT/PAN Lutskii A. E., Panova A. N., Zh. Fiz. Khim. **32**, 2183-6 (1958).

58MCC/FIN McCullough J. P., Finke H. L., Scott D. W., Pennington R. E., Gross M. E., Messerly J. F., Waddington G., J. Am. Chem. Soc. **80**, 4786-93 (1958).

58PEP Peppel W. J., Ind. Eng. Chem. **50**, 767-70 (1958).

58SCO/MCC Scott D. W., McCullough J. P., Messerly J. F., Pennington R. E., Hossenlopp I. A., Finke H. L., Waddington G., J. Am. Chem. Soc. **80**, 55-9 (1958).

58SWI/ZIE1 Swietoslawski W., Zielenkiewicz A., Bull. Acad. Pol. Sci., Ser. Sci. Chim. Geol. Geogr. **6**, 365-6 (1958).

58SWI/ZIE2 Swietoslawski W., Zielenkiewicz A., Bull. Acad. Pol. Sci., Ser. Sci. Chim. Geol. Geogr. **6**, 367-9 (1958).

58WAL/BRO Walker B. E., Jr., Brouks M. S., Ewing C. T., Miller R. R., Ind. Eng. Chem., Chem. Eng. Data Ser. **3**, 280-2 (1958).

58WES/GIN West E. D., Ginnings D. C., J. Res. Natl. Bur. Stand. (U.S.) **60**, 309-16 (1958).

59ABA/MUS Abas-Zade A. K., Mustafaev R. A., Dokl. Akad. Nauk. Az. SSR **15**, 775-9 (1959).

59BAK Baker B. L., U.S. Atomic Energy Commission Report No. [38-1]-118, 1-50 (1959).

59BEN/THO Bengough W. I., Thomson R. A. M., Trans. Faraday Soc. **55**, 268-71 (1959).

59CLU/SPE Clusius K., Sperandio A., Piesbergen U., Z. Naturforsch. **14a**, 793-801 (1959).

59COX/SMI Cox R. A., Smith N. D., Proc. R. Soc. London, A **252**, 51-62 (1959).

59GOO/SCO Good W. D., Scott D. W., Lacina J. L., McCullough J. P., J. Phys. Chem. **63**, 1139-42 (1959).

59HIL/MCD Hildenbrand D. L., McDonald R. A., Kramer W. R., Stull D. R., J. Chem. Phys. **30**, 930-4 (1959).

59HIL/SIN Hildenbrand D. L., Sinke G. C., McDonald R. A., Kramer W. R., Stull D. R., J. Chem. Phys. **31**, 650-4 (1959).

59HUT/BAI Hutchinson E., Bailey L. G., Z. Phys. Chem. (Frankfurt am Main) **21**, 30-7 (1959).

59KEN/TOM Kenttämaa J., Tommila E., Martti M., Ann. Acad. Sci. Fenn., Ser. A2: (93), 1-20 (1959).

59KOL/PAU Kolesov V. P., Paukov I. E., Skuratov S. M., Seregin E. A., Dokl. Akad. Nauk. SSSR **128**, 130-2 (1959).

59LUT/PAN Lutskii A. E., Panova A. N., Zh. Fiz. Khim. **33**, 970-5 (1959).

59MCC/DOU McCullough J. P., Douslin D. R., Hubbard W. N., Todd S. S., Messerly J. F., Hossenlopp I. A., Frow F. R., Dawson J. P., Waddington G., J. Am. Chem. Soc. **81**, 5884-90 (1959).

59ONK Onken U., Z. Elektrochem. **63**, 321-7 (1959).

59PAU/KOL Paukov I. E., Kolesov V. P., Skuratov S. M., Dokl. Akad. Nauk. SSSR **126**, 325-6 (1959).

59RIB/EGO Ribkin S. L., Egorov B. N., At. Energ. **7**, 462-5 (1959).

59SCO/DOU Scott D. W., Douslin D. R., Messerly J. F., Todd S. S., Hossenlopp I. A., Kincheloe T. C., McCullough J. P., J. Am. Chem. Soc. **81**, 1015-20 (1959).

59STU/SIN Stull D. R., Sinke G. C., McDonald R. A., Hatton W. E., Hildenbrand D. L., Symp. Thermodyn. Fitzens-Wattens Tirol No. 48, 1-9 (1959).

59SUG/SEK Suga H., Seki S., Bull. Chem. Soc. Jpn. **32**, 1088-93 (1959).

59WES1 West E. D., J. Am. Chem. Soc. **81**, 29-37 (1959).

59WES2 Westrum E. F., Jr., Symp. Therm. Propert., Purdue University, Lafayette (1959).

60BAR/BOL Barger J. W., Bolze C. C., Hughes R. L., Medved T. M., Lubr. Eng. **16**, 105-7, 118 (1960).

60BOC/PET Bockhoff F. J., Petrella R. V., Pace E. L., J. Chem. Phys. **32**, 799-804 (1960).

60BRA/MYE Brady A. P., Myers O. E., Clauss J. K., J. Phys. Chem. **64**, 588-91 (1960).

60DIN/PAC Dininny R. E., Pace E. L., J. Chem. Phys. **32**, 805-9 (1960).

60GIA/OTT Giauque W. F., Ott J. B., J. Am. Chem. Soc. **82**, 2689-95 (1960).

60JOH Johnson P. R. (Du Pont de Nemours & Co), Prospectus (1960), from: Weast R. C. (editor), Kirk-Othmer: Encyclopedia of Chemical Technology, Third Edition, Vol. 5, Wiley-Interscience, 774 (1964).

60KAR/STR1 Karasharli K. A., Strelkov P. G., Dokl. Akad. Nauk. Az. SSR **16**, 249-53 (1960).

60KAR/STR2 Karasharli K. A., Strelkov P. G., Dokl. Akad. Nauk. Az. SSR **16**, 341-4 (1960).

60KAR/STR3 Karasharli K. A., Strelkov P. G., Dokl. Akad. Nauk. SSSR **131**, 568-9 (1960).

60KEN/LIN Kenttämaa J., Lindberg J. J., Suom. Kemistil. **B33**, 32-5 (1960).

60MAR Martin J. J., J. Chem. Eng. Data **5**, 334-6 (1960).

60OTT/GIA Ott J. B., Giauque W. F., J. Am. Chem. Soc. **82**, 1308-11 (1960).

60SWI/ZIE Swietoslawski W., Zielenkiewicz A., Bull. Acad. Pol. Sci., Ser. Sci. Chim. **8**, 651-3 (1960).

61BEA/CHI Beaumont R. H., Chihara H., Morrison J. A., Proc. Phys. Soc., London **78**, 1462-81 (1961).

61BER/SCO Berg W. T., Scott D. W., Hubbard W. N., Todd S. S., Messerly J. F., Hossenlopp I. A., Osborn A., Douslin D. R., McCullough J. P, J. Phys. Chem. **65**, 1425-30 (1961).

61BRO/FOC Brown I., Fock W., Aust. J. Chem. **14**, 387-96 (1961).

61EGA/LUF Egan E. P., Jr., Luff B. B., J. Phys. Chem. **65**, 523-6 (1961).

61FLU/LEA Flubacher P., Leadbetter A. J., Morrison J. A., Proc. Phys. Soc., London **78**, 1149-61 (1961).

61GEL Geller B. E., Zh. Fiz. Khim. **35**, 2210-6 (1961).

61GOO Goodwin R. D., J. Res. Natl. Bur. Stand. (U.S.) **65C**, 231-43 (1961).

61HES/WHI Hestermans P., White D., J. Phys. Chem. **65**, 362-5 (1961).

61HUF/GRO Huffman H. M., Gross M. E., Scott D. W., McCullough J. P., J. Phys. Chem. **65**, 495-503 (1961).

61LAN/BOR Landolt-Börnstein Physikalisch-Chemische Tabellen, 6 Auflage, Zahlenwert und Funktionen aus Naturwissenschaften und Technik, Vol. II, Part 4: Kalorische Zustandsgrossen, Springer-Verlag, Berlin (1961).

61MCC/FIN McCullough J. P., Finke H. L., Hubbard W. N., Todd S. S., Messerly J. F., Douslin D. R., Waddington G., J. Phys. Chem. **65**, 785-91 (1961).

61MCC/MES McCullough J. P., Messerly J. F., Bull. -U.S., Bur. Mines **596**, 1-15 (1961).

61PAC/BOB Pace E. L., Bobka R. J., J. Chem. Phys. **35**, 454-7 (1961).

61ROURousseau R. A., Ph. D. Thesis, John Hopkins University, Baltimore, Maryland (1961).

61SCO/MES Scott D. W., Messerly J. F., Todd S. S., Guthrie G. B., Hossenlopp I. A., Moore R. T., Osborn A., Berg W. T., McCullough J. P., J. Phys. Chem. **65**, 1320-6 (1961).

61STU/SIN Stull D. R., Sinke G. C., McDonald R. A., Hatton W. E., Hildenbrand D. L., Pure Appl. Chem. **2**, 315-22 (1961).

62DAI/EVA Dainton F. S., Evans D. M., Hoare F. E., Melia T. P., Polymer **3**, 271-6 (1962).

62EUB/SMI Eubank P. T., Smith J. M., J. Chem. Eng. Data **7**, 75-8 (1962).

62GIG/CAR Giguere P. A., Carmichael J. L., J. Chem. Eng. Data **7**, 526-7 (1962).

62GOL/BEL Gollis M. H., Belenyessy L. I., Gudzinowicz B. J., Koch S. D., Smith J. O., Wineman R. J., J. Chem. Eng. Data **7**, 311-6 (1962).

62HAT/HIL Hatton W. E., Hildenbrand D. L., Sinke G. C., Stull D. R., J. Chem. Eng. Data **7**, 229-31 (1962).

62KAT Katayama T., Kagaku Kogaku (Chemical Engineering, Japan) **26**, 361-72 (1962).

62KOL/PAU Kolesov V. P., Paukov I. E., Skuratov S. M., Zh. Fiz. Khim. **36**, 770-9 (1962).

62KOL/SER Kolesov V. P., Seregin E. A., Skuratov S. M., Zh. Fiz. Khim. **36**, 647-51 (1962).

62LEV Levinson L. S., Rev. Sci. Instrum. **33**, 639-42 (1962).

62LOW/MOE Low D. I. R., Moelwyn-Hughes E. A., Proc. R. Soc. London, A **267**, 384-94 (1962).

62MEL Melia T. P., Polymer **3**, 317-21 (1962).

62MUR/YAM Murakami Y., Yamada T., Kagaku Kogaku (Chemical Engineering, Japan) **26**, 867-72 (1962).

62OME1 Omel'chenko F. S., Izv. Vyssh. Ucheb. Zaved., Pishch. Tekhnol. (3), 97-8 (1962).

62OME2 Omel'chenko F. S., Izv. Vyssh. Ucheb. Zaved., Pishch. Tekhnol. (2), 151-2 (1962).

62PAC/PET Pace E. L., Petrella R. V., J. Chem. Phys. **36**, 2991-4 (1962).

62RAB/NIK Rabinovich I. B., Nikolaev P. N., Dokl. Akad. Nauk. SSSR **142**, 1335-8 (1962).

62SCO/DOU Scott D. W., Douslin D. R., Finke H. L., Hubbard W. N., Messerly J. F., Hossenlopp I. A., McCullough J. P., J. Phys. Chem. **66**, 1334-41 (1962).

62SCO/GOO Scott D. W., Good W. D., Todd S. S., Messerly J. F., Berg W. T., Hossenlopp I. A., Lacina J. L., Osborn A., McCullough J. P., J. Chem. Phys. **36**, 406-12 (1962).

62SCO/GUT Scott D. W., Guthrie G. B., Messerly J. F., Todd S. S., Berg W. T., Hossenlopp I. A., McCullough J. P., J. Phys. Chem. **66**, 911-4 (1962).

62SCO/MES Scott D. W., Messerly J. F., Todd S. S., Hossenlopp I. A., Douslin D. R., McCullough J. P., J. Chem. Phys. **37**, 867-73 (1962).

62SER/KOL Seregin E. A., Kolesov V. P., Belikova N. A., Skuratov S. M., Plate A. F., Dokl. Akad. Nauk. SSSR **145**, 580-3 (1962).

62STR/BAR Strepikheev Yu. A., Baranov Yu. I., Burmistrova O. A., Izv. Vyssh. Ucheb. Zaved., Khim. Khim. Tekhnol. **5**, 387-90 (1962).

62VAL/BRO Valentine R. H., Brodale G. E., Giauque W. F., J. Phys. Chem. **66**, 392-5 (1962).

62WEB/KIL Weber L. A., Kilpatric J. E., J. Chem. Phys. **36**, 829-34 (1962).

62YOU/DIL Younglove B. A., Diller D. E., Cryogenics **2**, 283-7 (1962).

63AND/COU1 Andon R. J. L., Counsell J. F., Herington E. F. G., Martin J. F., Trans. Faraday Soc. **59**, 830-5 (1963).

63AND/COU2 Andon R. J. L., Counsell J. F., Martin J. F., Trans. Faraday Soc. **59**, 1555-8 (1963).

63BEN Benjamin L., Can. J. Chem. **41**, 2210-8 (1963).

63COL/GIL Colwell J. H., Gill E. K., Morrison J. A., J. Chem. Phys. **39**, 635-53 (1963).

63FEH/SEY Feher F., Seyfried K., Z. Anorg. Allg. Chem. **322**, 155-61 (1963).

63GAT/KRE Gattow G., Krebs B., Z. Anorg. Allg. Chem. **322**, 113-28 (1963).

63GOO/TOD Good W. D., Todd S. S., Messerly J. F., Lacina J. L., Dawson J. P., Scott D. W., McCullough J. P., J. Phys. Chem. **67**, 1306-11 (1963).

63GUD/CAM Gudzinowicz B. J., Campbell R. H., Adamas J. S., J. Chem. Eng. Data **8**, 201-4 (1963).

63HUL/ORR Hultgren R., Orr R. L., Anderson P. D., Kelley K. K., Selected Values Thermodyn. Properties of Metals and Alloys, J. Wiley, New York (1963).

63KAZ/ORR Kaznoff A. I., Orr R. L., Hultgren R., Univ. Calif. Berkeley, Materials Res. Lab. Tech. Rept., Ser. 155 NONR-222, No. 2 (1963), from: Hultgren R., Orr R. L., Anderson P. D., Kelley K. K., Selected Values Thermodyn. Properties of Metals and Alloys, J. Wiley, New York (1963).

63MCC/MES McCullough J. P., Messerly J. F., Moore R. T., Todd S. S., J. Phys. Chem. **67**, 677-9 (1963).

63OET Oetting F. L., J. Phys. Chem. **67**, 2757-61 (1963).

63SCO/GOO Scott D. W., Good W. D., Guthrie D. B., Todd S. S., Hossenlopp I. A., Osborn A. G., McCullough J. P., J. Phys. Chem. **67**, 685-9 (1963).

63SCO/HUB Scott D. W., Hubbard W. N., Messerly J. F., Todd S. S., Hossenlopp I. A., Good W. D., Douslin D. R., McCullough J. P., J. Phys. Chem. **67**, 680-5 (1963).

63SCO/MES Scott D. W., Messerly J. F., Todd S. S., Hossenlopp I. A., Osborn A., McCullough J. P., J. Chem. Phys. **38**, 532-9 (1963).

63TRO/WES Trowbridge J. C., Westrum E. F., Jr., J. Phys. Chem. **67**, 2381-5 (1963).

63VAR/KOP Vargaftik N. B., Kopylov N. I., Lapushkin S. A., Pyatibratrov S. N., Sokolov S. N., Izv. Vyssh. Ucheb. Zaved., Neft Gaz **6** (3), 75-8 (1963).

63WUL/WES Wulff C. A., Westrum E. F., Jr., J. Phys. Chem. **67**, 2376-81 (1963).

63ZIE/MUL Ziegler W. T., Mullins J. C., Hwa S. C. P., Adv. Cryog. Eng. **8**, 268-77 (1963).

64ARN1 Arnold H., Z. Phys. Chem. (Leipzig) **225**, 45-56 (1964).

64ARN2 Arnold H., Z. Phys. Chem. (Leipzig) **226**, 146-56 (1964).

64CAM/NAG Campbell A. N., Nagarajan M. K., Can. J. Chem. **42**, 1616-26 (1964).

64CLE/WON Clever H. L., Wong W. -K., Wulff C. A., Westrum E. F., Jr., J. Phys. Chem. **68**, 1967-9 (1964).

64EDI/CUB Eding H., Cubicciotti D., J. Chem. Eng. Data **9**, 524-5 (1964).

64FUR/REI Furukawa G. T., Reilly M. L., Piccirelli J. H., Tanenbaum M., J. Res. Natl. Bur. Stand. (U.S.) **68A**, 367-79 (1964).

64GAT/HEI Gattow G., Heinrich G., Z. Anorg. Allg. Chem. **331**, 256-74 (1964).

64KRO/VEN Kroutil M., Vender M., Chem. Prum. **14**, 412-5 (1964).

64MCE/KIL McEachern D. M., Jr., Kilpatrick J. E., J. Chem. Phys. **41**, 3127-31 (1964).

64MOE/THO Moelwyn-Hughes E. A., Thorpe P. L., Proc. R. Soc. London, A **278**, 574-87 (1964).

64OET Oetting F. L., J. Chem. Phys. **41**, 149-53 (1964).

64RAS/BAS Rastogi R. P., Bassi P. S., J. Phys. Chem. **68**, 2398-406 (1964).

64SER/GOR Seregin E. A., Goroshko N. N., Kolesov V. P., Belikova N. A., Skuratov S. M., Plate A. F, Dokl. Akad. Nauk. SSSR **159**, 1381-4 (1964).

64SIN/OET Sinke G. C., Oetting F. L., J. Phys. Chem. **68**, 1354-8 (1964).

64TIM Timini B. A., Ph. D. Thesis, Bristol (1964), from: McCullough J. P., Scott D. W. (editors), Experimental Thermodynamics, Vol. I, Butterworth, London 421-535 (1968).

64TRO/WES1 Trowbridge J. C., Westrum E. F., Jr., J. Phys. Chem. **68**, 42-9 (1964).

64TRO/WES2 Trowbridge J. C., Westrum E. F., Jr., J. Phys. Chem. **68**, 255-60 (1964).

64VAS1 Vas'kov E. T., Inzh. -Fiz. Zh. **7** (6), 92-6 (1964).

64VAS2 Vas'kov E. T., Izv. Vyssh. Ucheb. Zaved., Priborostr. **7** (4), 155-7 (1964).

64VUK/RAS Vukalovich M. P., Rasskazov D. S., Popov V. N., Babikov Yu. M., Teploenergetika (Moscow) **11** (6), 56-8 (1964).

64WEA Weast R. C. (editor), Kirk-Othmer: Encyclopedia of Chemical Technology, Third Edition, Vol. 5, Wiley-Interscience, (1964).

64WUL/WES Wulff C. A., Westrum E. F., Jr., J. Phys. Chem. **68**, 430-1 (1964).

64ZIM/BAR Zimmer M. F., Baroody E. E., Schwartz M., McAllister M. P., J. Chem. Eng. Data **9**, 527-9 (1964).

65CAL/GIA Calder G. V., Giauque W. F., J. Phys. Chem. **69**, 2443-52 (1965).

65CAM/GIE Campbell A. N., Gieskes J. M. T. M., Can. J. Chem. **43**, 1004-11 (1965).

65CAR/WES1 Carlson H. G., Westrum E. F., Jr., J. Chem. Eng. Data **10**, 134-5 (1965).

65CAR/WES2 Carlson H. G., Westrum E. F., Jr., J. Phys. Chem. **69**, 1524-30 (1965).

65CLE/WON Clever H. L., Wong W. -K., Westrum E. F., Jr., J. Phys. Chem. **69**, 1209-13 (1965).

65CLE/WUL Clever H. L., Wulff C. A., Westrum E. F., Jr., J. Phys. Chem. **69**, 1983-8 (1965).

65COU/GRE Counsell J. F., Green J. H. S., Hales J. L., Martin J. F., Trans. Faraday Soc. **61**, 212-8 (1965).

65COU/HAL Counsell J. F., Hales J. L., Martin J. F., Trans. Faraday Soc. **61**, 1869-75 (1965).

65CUT/MOR Cutler A. J. B., Morrison J. A., Trans. Faraday Soc. **61**, 429-42 (1965).

65FIN/GRU Findenegg G. H., Gruber K., Pereira J. F., Kohler F., Monatsh. Chem. **96**, 669-78 (1965).

65FIN/MES Finke H. L., Messerly J. F., Todd S. S., J. Phys. Chem. **69**, 2094-100 (1965).

65GOD/BAR Godovskii Yu. K., Barskii Yu. P., Plast. Massy (7), 57-9 (1965).

65GUM/KOS1 Gumbatov D. O., Kostryukov V. N., Zh. Fiz. Khim. **39**, 116-22 (1965).

65GUM/KOS2 Gumbatov D. O., Kostryukov V. N., Shaulov Yu. Kh., Izv. Akad. Nauk Azerb. SSR, Ser. Fiz. - Tekn. Mat. Nauk (1), 53-8 (1965).

65GUM/KOS3 Gumbatov D. O., Kostryukov V. N., Shaulov Yu. Kh., Izv. Akad. Nauk Azerb. SSR, Ser. Fiz. - Tekn. Mat. Nauk (6), 89-93 (1965).

65KAU/BIT Kauer E., Bittrich H. J., Wiss. Z. Tech. Hochsch. Chem. "Carl Schorlemmer", Leuna-Merseburg **7**, 73-6 (1965).

65KRI/LOE Kriebel M., Loeffler H. J., Kaeltetechnik **17**, 266-71 (1965).

65MCD/KIL1 McDougall L. A., Kilpatrick J. E., J. Chem. Phys. **42**, 2307-10 (1965).

65MCD/KIL2 McDougall L. A., Kilpatrick J. E., J. Chem. Phys. **42**, 2311-21 (1965).

65MES/TOD1 Messerly J. F., Todd S. S., Finke H. L., J. Phys. Chem. **69**, 353-9 (1965).

65MES/TOD2 Messerly J. F., Todd S. S., Finke H. L., J. Phys. Chem. **69**, 4304-11 (1965).

65OET Oetting F. L., J. Chem. Eng. Data **10**, 122-5 (1965).

65PAC/TUR Pace E. L., Turnbull B. F., J. Chem. Phys. **43**, 1953-7 (1965).

65PUT/MCE Putnam W. E., McEachern D. M., Jr., Kilpatric J. E., J. Chem. Phys. **42**, 749-55 (1965).

65SOM/COO Somsen G., Coops J., Recl. Trav. Chim. Pays-Bas **84**, 985-1002 (1965).

65STE/BLA Sterrett K. F., Blackburn D. H., Bestul A. B., Chang S. S., Horman J., J. Res. Natl. Bur. Stand. (U.S.) **69C**, 19-26 (1965).

65SUG/SEK Suga H., Seki S., Bull. Chem. Soc. Jpn. **38**, 1000-6 (1965).

65TIM Timmermans J., Physicochemical Constants of Pure Organic Compounds, Vol. II, Elsevier, Amsterdam (1965).

65TUN/MIS Tungusov V. P., Mishchenko K. P., Zh. Fiz. Khim. **39**, 2968-72 (1965).

65ZAL/KOC Zalikin A. A., Kochetkov V. L., Strepikheev Yu. A., Khim. Prom-st. (Moscow) **41**, 338 (1965).

65ZIE Zielenkiewicz A., Przem. Chem. **44**, 664-7 (1965).

66BEA/CLE Beaumont R. H., Clegg B., Gee G., Herbert J. B. M., Marks D. J., Roberts R. C., Sims D., Polymer **7**, 401-17 (1966).

66DRA/LAN Drakin S. I., Lantukhova L. V., Karapet'yanc M. Kh., Zh. Fiz. Khim. **40**, 451-5 (1966).

66DWO/GUI Dworkin A., Guillamin M., J. Chim. Phys. Phys. -Chim. Biol. **63**, 53-8 (1966).

66EGA/LUF Egan E. P., Jr., Luff B. B., J. Chem. Eng. Data **11**, 194-6 (1966).

66ENI/PET Enig J. W., Petrone F. J., Phys. Fluids **9**, 398-408 (1966).

66FIE Fiegel H., NASA Contract. Rep. No. 5474, 1-184 (1966).

66GAT/DRA Gattow G., Draeger M., Z. Anorg. Allg. Chem. **343**, 232-43 (1966).

66GIL/MOR Gill E. K., Morrison J. A., J. Chem. Phys. **45**, 1585-90 (1966).

66HWA/ZIE Hwa S. C. P., Ziegler W. T., J. Phys. Chem. **70**, 2572-93 (1966).

66KLE Klesper I., Z. Phys. Chem. (Frankfurt am Main) **51**, 1-12 (1966).

66LIN/ZIE Lin K. F., Ziegler W. T., J. Chem. Eng. Data **11**, 187-9 (1966).

66NIK/LEB Nikolaev P. N., Lebedev B. V., Tr. Khim. Khim. Tekhnol. (2), 332-43 (1966).

66NIK/RAB Nikolaev P. N., Rabinovich I. B., Galperin V. A., Cvetkov V. G., Zh. Fiz. Khim. **40**, 1091-7 (1966).

66PER/COM Perkin-Elmer DSC-1, commercial calorimeter, Anal. Chem. **38**, 1331-6 (1966).

66SAV Savogina M. S., Izv. Vyssh. Ucheb. Zaved., Khim. Khim. Tekhnol. **9**, 379-81 (1966).

66SHI/ATA Shinoda T., Atake T., Chihara H., Mashiko Y., Seki S., Kogyo Kagaku Zasshi (J. Chem. Soc. Japan, Ind. Chem. Sect.) **69**, 1619-22 (1966).

66TEC Technical Brochure, Silanes, Union Carbide Corp., Silicones Division, New York (1966), from: Yaws C. L., Dickens L. L., Hopper J. R., Li K. -Y., Chu T. C. T., J. Chin. Inst. Chem. Eng. **14**, 205-14 (1983).

66VAS Vas'kov E. T., Inzh. -Fiz. Zh. **10**, 127-30 (1966).

66ZAL/STR Zalikin A. A., Strepikheev Yu. A., Zh. Prikl. Khim. (Leningrad) **39**, 2607 (1966).

67AND/COU Andon R. J. L., Counsell J. F., Lees E. B., Martin J. F., Mash C. J., Trans. Faraday Soc. **63**, 1115-21 (1967).

67BAR/POR Barrall E. M. II, Porter R. S., Johnson J. F., J. Phys. Chem. **71**, 895-900 (1967).

67CHA/HOR Chang S. S., Horman J. A., Bestul A. B., J. Res. Natl. Bur. Stand. (U.S.) **71A**, 293-305 (1967).

67GRA Grabner R. W., Thesis, Lehigh University, Univ. Microfilms Publ. No. 67-14935, 1-191 (1967).

67GRO Groenvold F., Acta Chem. Scand. **21**, 1695-713 (1967).

67KAR/MEK Karasharli K. A., Mekhtiev S. A., Azerb. Khim. Zh. (1), 76-9 (1967).

67LEB/RAB1 Lebedev B. V., Rabinovich I. B., Tr. Khim. Khim. Tekhnol. (2), 30-5 (1967).

67LEB/RAB2 Lebedev B. V., Rabinovich I. B., Budarina V. A., Vysokomol. Soedin., Ser. A **9**, 488-94 (1967).

67LEB/RAB3 Lebed'ev B. V., Rabinovich I. B., Martynenko L. Ya., Vysokomol. Soedin., Ser. A **9**, 1640-5 (1967).

67MAG Magill J. H., J. Chem. Phys. **47**, 2802-7 (1967).

67MES/GUT Messerly J. F., Guthrie G. B., Todd S. S., Finke H. L., J. Chem. Eng. Data **12**, 338-46 (1967).

67MES/TOD Messerly J. F., Todd S. S., Guthrie G. B., J. Chem. Eng. Data **12**, 426-9 (1967).

67NAG/KAR1 Nagiev M. F., Karasharli K. A., Mechtiev S. A., Dokl. Akad. Nauk. Az. SSR **23** (8), 8-10 (1967).

67NAG/KAR2 Nagiev M. F., Karasharli K. A., Mechtiev S. A., Dokl. Akad. Nauk. Az. SSR **23** (11), 23-5 (1967).

67NIK/RAB1 Nikolaev P. N., Rabinovich I. B., Zh. Obshch. Khim. **41**, 2191-4 (1967).

67NIK/RAB2 Nikolaev P. N., Rabinovich I. B., Lebedev B. V., Zh. Fiz. Khim. **41**, 1294-9 (1967).

67PAC Pacor P., Anal. Chim. Acta **37**, 200-8 (1967).

67PAC/PLA Pace E. L., Plaush A. C., J. Chem. Phys. **47**, 38-43 (1967).

67PLA/PAC Plaush A. C., Pace E. L., J. Chem. Phys. **47**, 44-8 (1967).

67RAB/LEB Rabinovich I. B., Lebedev B. V., Melent'eva T. I., Vysokomol. Soedin., Ser. A **9**, 1699-705 (1967).

67RAS/GAN Rastorguev Yu. L., Ganiev Yu. A., Izv. Vyssh. Ucheb. Zaved., Neft Gaz **10** (1), 79-82 (1967).

67RIB/WES Ribner A., Westrum E. F., Jr., J. Phys. Chem. **71**, 1208-15 (1967).

67SCO/BER Scott D. W., Berg W. T., Hossenlopp I. A., Hubbard W. N., Messerly J. F., Todd S. S., Douslin D. R., McCullough J. P., Waddington G., J. Phys. Chem. **71**, 2263-70 (1967).

67SMI/GOO1 Smith N. K., Good W. D., J. Chem. Eng. Data **12**, 570-2 (1967).

67SMI/GOO2 Smith N. K., Good W. D., J. Chem. Eng. Data **12**, 572-4 (1967).

67TAM/ZUZ Tamplin, W. S., Zuzic, D. A., Hydrocarbon Process. **46** (8), 145-146 (1967).

67WES/RIB Westrum E. F., Jr., Ribner A., J. Phys. Chem. **71**, 1216-24 (1967).

68ADA/SUG Adachi K., Suga H., Seki S., Bull. Chem. Soc. Jpn. **41**, 1073-87 (1968).

68AHL/YOU Ahlert R. C., Younts C. A., J. Chem. Eng. Data **13**, 402-5 (1968).

68AND/COU1 Andon R. J. L., Counsell J. F., Hales J. L., Lees E. B., Martin J. F., J. Chem. Soc. A 2357-61 (1968).

68AND/COU2 Andon R. J. L., Counsell J. F., Martin J. F., J. Chem. Soc. A 1894-7 (1968).

68BAG/KUC Bagatskii M. I., Kucheryavy V. A., Manzhelii V. G., Popov V. A., Phys. Status Solidi **26**, 453-60 (1968).

68CLE/MEL Clegg G. A., Melia T. P., Tyson A., Polymer **9**, 75-9 (1968).

68COU/HAL1 Counsell J. F., Hales J. L., Lees E. B., Martin J. F., J. Chem. Soc. A 2994-6 (1968).

68COU/HAL2 Counsell J. F., Hales J. L., Martin J. F., J. Chem. Soc. A 2042-4 (1968).

68COU/LEE Counsell J. F., Lees E. B., Martin J. F., J. Chem. Soc. A 1819-23 (1968).

68ELL/CHR Elliott J. H., Chris M. D., J. Chem. Eng. Data **13**, 475-9 (1968).

68GIR/WES Girdhar H. L., Westrum E. F., Jr, Wulff C. A., J. Chem. Eng. Data **13**, 239-42 (1968).

68GOU/GIR Goursot P., Girdhar H. L., Westrum E. F., Jr., C. R. Seances Acad. Sci., Ser. C **266**, 949-50 (1968).
68GOU/WES1 Goursot P., Westrum E. F., Jr., C. R. Seances Acad. Sci., Ser. C **266**, 667-8 (1968).
68GOU/WES2 Goursot P., Westrum E. F., Jr., C. R. Seances Acad. Sci., Ser. C **266**, 1267-8 (1968).
68GOU/WES3 Goursot P., Westrum E. F., Jr., J. Chem. Eng. Data **13**, 468-71 (1968).
68GOU/WES4 Goursot P., Westrum E. F., Jr., J. Chem. Eng. Data **13**, 471-5 (1968).
68GOU/WES5 Goursot P., Westrum E. F., Jr., Metzger J., C. R. Seances Acad. Sci., Ser. C **266**, 590-2 (1968).
68KOS/MOS Kostryukov V. N., Mosin A. M., Samorukov O. P., Shlyakova V. G., Zh. Fiz. Khim. **42**, 1836-40 (1968).
68LEA Leadbetter A. J., J. Phys. C: Solid State Phys. **1**, 1481-8 (1968).
68MCC/SCO McCullough J. P., Scott D. W. (editors), Experimental Thermodynamics, Vol. I, Calorimetry of Non-reacting Systems, Butterworths, London (1968).
68OST/DOB Ostrovskii V. E., Dobrovol'skii N. N., Karpovich I. R., Frolov F. Ya., Zh. Fiz. Khim. **42**, 550-5 (1968).
68PAC/REN Pace E. L., Reno M. A., J. Chem. Phys. **48**, 1231-5 (1968).
68PAZ/REC Paz Andrade M. I., Recacho E., Paz Fernandez J. M., Acta Cient. Compostelana **5** (2), 49-55 (1968).
68REC1 Recko W. M., Bull. Acad. Pol. Sci., Ser. Sci. Chim. **16**, 543-8 (1968).
68REC2 Recko W. M., Bull. Acad. Pol. Sci., Ser. Sci. Chim. **16**, 549-52 (1968).
68REC3 Recko W. M., Bull. Acad. Pol. Sci., Ser. Sci. Chim. **16**, 553-5 (1968).
68REE/SEE Reese B. O., Seely L. B., Shaw R., Tegg D., West. States Sect. Combust. Inst. Pap., WSS/CI 68-39, 1-7 (1968).
68SUG/ADA Sugisaki M., Adachi K., Suga H., Seki S., Bull. Chem. Soc. Jpn. **41**, 593-600 (1968).
68WAG/EVA Wagman D. D., Evans W. H., Parker V. B., Halow I., Bailey S. M., Schumm R. H., Selected Values of Chemical Thermodynamic Properties, NBS Technical Note 270-3, Washington D. C. (1968).
68WES/FUR Westrum E. F., Jr., Furukawa G. T., McCullough J. P., Experimental Thermodynamics, McCullough J. P., Scott D. W. (editors), Vol. I, Chap. 5, Butterworth, London 133-214 (1968).
68WES/WES West E. D., Westrum E. F., Jr., Experimental Thermodynamics, McCullough J. P., Scott D. W. (editors), Vol. I, Chap. 9, Butterworth, London 333-94 (1968).
69ATK/LAR Atkinson C. M. L., Larkin J. A., Richardson M. J., J. Chem. Thermodyn. **1**, 435-40 (1969).
69BER/WES Berman H. A., West E. D., J. Chem. Eng. Data **14**, 107-9 (1969).
69CLE/MEL1 Clegg G. A., Melia T. P., Makromol. Chem. **123**, 184-93 (1969).
69CLE/MEL2 Clegg G. A., Melia T. P., Makromol. Chem. **123**, 194-202 (1969).
69CLE/MEL3 Clegg G. A., Melia T. P., Makromol. Chem. **130**, 258-60 (1969).
69CLE/MEL4 Clegg G. A., Melia T. P., Polymer **10**, 912-22 (1969).
69DOU Douglas T. B., J. Res. Natl. Bur. Stand. (U.S.) **73A**, 451-70 (1969).
69ENO/SHI1 Enokido H., Shinoda T., Mashiko Y., Bull. Chem. Soc. Jpn. **42**, 84-91 (1969).
69ENO/SHI2 Enokido H., Shinoda T., Mashiko Y., Bull. Chem. Soc. Jpn. **42**, 3415-21 (1969).
69FEB/HER Feber R. C., Herrick C. C., Levinson L. S., J. Chem. Thermodyn. **1**, 169-75 (1969).
69GOO/WEB Goodwin R. D., Weber L. A., J. Res. Natl. Bur. Stand. (U.S.) **73A**, 1-13 (1969).
69GOU/WES Goursot P., Westrum E. F., Jr., J. Chem. Eng. Data **14**, 1-3 (1969).
69JOE/GJA Joens O., Gjaldbaek J. C., Dan. Tidsskr. Farm. **43**, 151-5 (1969).
69KAR/KOS Karasharli K. A., Kostryukov V. N., Dzhafarov O. I., Gejdarov Kh. I., Azerb. Khim. Zh. (1), 32-4 (1969).
69MEL/MER Melia T. P., Merrifield R., J. Appl. Chem. **19**, 79-82/3 (1969).
69NAG/DZH1 Nagiev M. F., Dzhafarov O. I., Karasharli K. A., Kostryukov V. N., Dokl. Akad. Nauk. SSSR **184**, 648-9 (1969).
69NAG/DZH2 Nagiev M. F., Dzhafarov O. I., Karasharli K. A., Kostryukov V. N., Gejdarov Kh. I., Dokl. Akad. Nauk. SSSR **188**, 370-1 (1969).
69PAU/GLU1 Paukov I. E., Glukhikh L. K., Zh. Fiz. Khim. **43**, 226-8 (1969).
69PAU/GLU2 Paukov I. E., Glukhikh L. K., Zh. Fiz. Khim. **43**, 1350-2 (1969).
69PAU/LAV1 Paukov I. E., Lavrent'eva M. N., Zh. Fiz. Khim. **43**, 2938-41 (1969).
69PAU/LAV2 Paukov I. E., Lavrent'eva M. N., Anisimov M. P., Zh. Fiz. Khim. **43**, 785-7 (1969).
69PAU/LAV3 Paukov I. E., Lavrent'eva M. N., Anisimov M. P., Zh. Fiz. Khim. **43**, 2941-3 (1969).
69PER/COM Perkin-Elmer DSC-2, commercial calorimeter, Prospectus (1969).
69RAB/MAR Rabinovich I. B., Martynenko L. Ya., Maslova V. A., Tr. Khim. Khim. Tekhnol. (2), 10-4 (1969).
69SAD/STE Sadowska K. W., Stepniewska G. B., Recko W. M., Przem. Chem. **48**, 282-5 (1969).
69SMI/MAT Smirnova E. P., Matveichuk N. L., Mikhailin A. D., Illarionov V. V., Dokl. Akad. Nauk. SSSR **189**, 344-6 (1969).
69SMI/PAC Smith J. H., Pace E. L., J. Phys. Chem. **73**, 4232-6 (1969).
69SOU/GOU1 Soulie M. A., Goursot P., Peneloux A., Metzger J., J. Chim. Phys. Phys. -Chim. Biol. **66**, 603-6 (1969).
69SOU/GOU2 Soulie M. A., Goursot P., Peneloux A., Metzger J., J. Chim. Phys. Phys. -Chim. Biol. **66**, 607-10 (1969).
69STO/MAR Stokes R. H., Marsh K. N., Tomlins R. P., J. Chem. Thermodyn. **1**, 211-21 (1969).
69TOM/LIN Tommila E., Lindell E., Virtalaine M. -L., Laakso R., Suom. Kemistil. **B42**, 95-104 (1969).
69WEB/JUS Webb D. U., Justice B. H., Prophet H., J. Chem. Thermodyn. **1**, 227-39 (1969).
69WIL/ROT Wilhelm E., Rott E., Kohler F., Proc. 1st Intern. Conf. Calorimetry and Thermodynamics, Warszaw 767-71 (1969).
69WIL/SCH Wilhelm E., Schano R., Becker G., Findenegg G. H., Kohler F., Trans. Faraday Soc. **65**, 1443-55 (1969).
70AKH Akhmedov A. G., Zh. Fiz. Khim. **44**, 2061-2 (1970).
70AND/COU Andon R. J. L., Counsell J. F., Lees E. B., Martin J. F., J. Chem. Soc. A 833-7 (1970).
70AND/WES Andrews J. T. S., Westrum E. F., Jr., J. Chem. Thermodyn. **2**, 245-53 (1970).
70ANO Anon, Product Information of Dimethylformamide, Du Pont de Nemours, Wilmington, USA (1970), from: Hradetzky G., Hammerl I., Kisan W., Wehner K., Bittrich H. -J., Data of Selective Solvents, VEB Deutscher Verlag der Wissenschaften, Berlin (1989).
70CHA/MCN Chang S., McNally D., Shary-Tehrany S., Hickey M. J., Boyd R. H., J. Am. Chem. Soc. **92**, 3109-18 (1970).
70CLE/WES Clever H. L., Westrum E. F., Jr., J. Phys. Chem. **74**, 1309-17 (1970).
70FIN/MCC Finke H. L., McCullough J. P., Messerly J. F., Guthrie G. B., Douslin D. R., J. Chem. Thermodyn. **2**, 27-41 (1970).
70FIN/TOD Finke H. L., Todd S. S., Messerly J. F., J. Chem. Thermodyn. **2**, 129-38 (1970).
70GOO/PRY Goodwin R. D., Prydz R., J. Res. Natl. Bur. Stand. (U.S.) **74A**, 499-505 (1970).
70GOU/GIR Goursot P., Girdhar H. L., Westrum E. F., Jr., J. Phys. Chem. **74**, 2538-41 (1970).
70HAR/HEA Harrop D., Head A. J., Lewis G. B., J. Chem. Thermodyn. **2**, 203-10 (1970).
70KRE Kresheck G. C., J. Chem. Phys. **52**, 5966-7 (1970).
70LKB/COM LKB 8721-1, commercial calorimeter, Prospectus (1970).

70MAR/RAB Martynenko L. Yu., Rabinovich I. B., Ovchinnikov Yu. B., Maslova V. A., Vysokomol. Soedin., Ser. A **12**, 841-8 (1970).

70MES/FIN Messerly J. F., Finke H. L., J. Chem. Thermodyn. **2**, 867-80 (1970).

70MES/TOD Messerly J. F., Todd S. S., Guthrie G. B., J. Chem. Eng. Data **15**, 227-32 (1970).

70PAC/JEP Pace E. L., Jepson B. E., J. Chem. Phys. **52**, 911-4 (1970).

70PAZ/PAZ Paz Andrade M. I., Paz J. M., Recacho E., An. Quim. **66**, 961-7 (1970).

70REC Recko W. M., Bull. Acad. Pol. Sci., Ser. Sci. Chim. **18**, 297-301 (1970).

70REE/SEE Reese B. O., Seely L. B., Shaw R., Tegg D., J. Chem. Eng. Data **15**, 140-2 (1970).

70TOU/MAK Touloukian Y. S., Makita T., Specific Heat of Nonmetallic Liquids and Gases, Vol. 6 in Thermophysical Properties of Matter, Plenum Press, New York (1970).

70VAL/KIL Valerga A. J., Kilpatrick J. E., J. Chem. Phys. **52**, 4545-9 (1970).

70WAD Wadsö I., Quart. Rev. Biophys. 3 (1970)383.

70WES/WON Westrum E. F., Jr., Wong W. -K., Morawetz E., J. Phys. Chem. **74**, 2542-7 (1970).

70WON/WES Wong W. -K., Westrum E. F., Jr., J. Phys. Chem. **74**, 1303-8 (1970).

71ADA/SUG Adachi K., Suga H., Seki S., Bull. Chem. Soc. Jpn. **44**, 78-89 (1971).

71AMI/ALI Amirkhanov Kh. I., Alibekov B. G., Vikhrov D. I., Mirskaya V. A., Levina L. N., Teplofiz. Vys. Temp. **9**, 1310-3 (1971).

71AND/CON Andon R. J. L., Connett J. E., Counsell J. F., Lees E. B., Martin J. F., J. Chem. Soc. A 661-4 (1971).

71BEL/HUL Bell H., Hultgren R., Met. Transition **2**, 3230-1 (1971).

71BOY/SAN Boyd R. H., Sanwal S. N., Shary-Tehrany S., McNally D., J. Phys. Chem. **75**, 1264-71 (1971).

71CAR/WES Carlson H. G., Westrum E. F., Jr., J. Chem. Phys. **54**, 1464-71 (1971).

71COU/LEE Counsell J. F., Lee D. A., Martin J. F., J. Chem. Soc. A 313-6 (1971).

71DES/BHA Deshpande D. D., Bhatgadde L. G., Aust. J. Chem. **24**, 1817-22 (1971).

71DU /COM Du Pont 990 DSC, commercial calorimeter, Prospectus (1971).

71GAM/VIS Gambhir R. D., Viswanathan B., Gopal E. S. R., Indian J. Pure Appl. Phys. **9**, 787-92 (1971).

71GOP/GAM Gopal E. S. R., Gambhir R. D., Govindarajan K., Viswanathan B., J. Indian Inst. Sci. **53**, 355-67 (1971).

71HAL/BAL Hall H. K., Jr., Baldt J. H., J. Am. Chem. Soc. **93**, 140-5 (1971).

71HOD Hodge P. E., Ph. D. Thesis, Case Western Reserve University (1971).

71KHA/SUB Khan V. H., Subrahmanyam S. V., Trans. Faraday Soc. **67**, 2282-91 (1971).

71KOL/VOR Kolesov V. P., Vorob'ev V. N., Zh. Fiz. Khim. **45**, 1293 (1971), data deposited in VINITI 1760-70.

71KON/SUU Konicek J., Suurkuusk J., Wadsö I., Chem. Scr. **1**, 217-20 (1971).

71KON/WAD Konicek J., Wadsö I., Acta Chem. Scand. **25**, 1541-51 (1971).

71LEB/RAB1 Lebedev B. V., Rabinovich I. B., Tr. Khim. Khim. Tekhnol. (1), 8-11 (1971).

71LEB/RAB2 Lebedev B. V., Rabinovich I. B., Tr. Khim. Khim. Tekhnol. (1), 12-5 (1971).

71MAR/CIO Marchidan D. I., Ciopec M., Rev. Roum. Chim. **16**, 1145-54 (1971).

71MAY/WAL Mayer J., Waluga T., Rap. -Inst. Fiz. Jad. (Krakow) 750/PL, 1-18 (1971).

71MES/FIN Messerly J. F., Finke H. L., J. Chem. Thermodyn. **3**, 675-87 (1971).

71MUS Mustafaev R. A., Izv. Vyssh. Ucheb. Zaved., Priborostr. **14** (7), 129-34 (1971).

71OSB/SCH Osborne D. W., Schreiner F., Selig H., J. Chem. Phys. **54**, 3790-8 (1971).

71PAU1 Paukov I. E., Zh. Fiz. Khim. **45**, 1293-4 (1971).

71PAU2 Paukov I. E., Zh. Fiz. Khim. **45**, 1297-8 (1971).

71PAU/RAK Paukov I. E., Rakhmenkulov F. S., Zh. Fiz. Khim. **45**, 1296-7 (1971).

71PIC/LED Picker P., Leduc P. A., Philip P. R., Desnoyers J. E., J. Chem. Thermodyn. **3**, 631-42 (1971).

71POP/MAN Popov V. A., Manzhelii V. G., Bagatskii M. I., J. Low. Temp. Phys. **5**, 427-33 (1971).

71PRI Privalko V. P., Zh. Fiz. Khim. **45**, 1585 (1971), data deposited in VINITI 2713-71.

71REC/SAD Recko W. M., Sadowska K. W., Woycicka M. K., Bull. Acad. Pol. Sci., Ser. Sci. Chim. **19**, 475-91 (1971).

71SHA/YAK Shaulov Yu. Kh., Yakubov E. S., Zh. Fiz. Khim. **45**, 188 (1971), data deposited in VINITI, from: Maslova V. A., Rabinovich I. B., Tr. Khim. Khim. Tekhnol. (1), 40-63 (1974).

71VAN Van Miltenburg J. C., Ph. D. Thesis, Rijksuniversiteit Utrecht (1971).

71VAS/PET Vasil'ev I. A., Petrov V. M., Ignat'ev V. M., Vvedenskii A. A., Zh. Fiz. Khim. **45**, 1316 (1971), data deposited in VINITI 2530-71.

71WON/WES Wong W. -K., Westrum E. F., Jr., J. Chem. Thermodyn. **3**, 105-24 (1971).

72ADA/SUG Adachi K., Suga H., Seki S., Bull. Chem. Soc. Jpn. **45**, 1960-72 (1972).

72ARE/VAN Arentsen J. G., Van Miltenburg J. C., J. Chem. Thermodyn. **4**, 789-91 (1972).

72CHA/BES Chang S. S., Bestul A. B., J. Chem. Phys. **56**, 503-16 (1972).

72COP/REA Cope A. F. G., Reamer H. H., Pings C. J., Ber. Bunsenges. Phys. Chem. **76**, 318-20 (1972).

72FIN/MCC Finke H. L., McCullough J. P., Messerly J. F., Osborne A., Douslin D. R., J. Chem. Thermodyn. **4**, 477-94 (1972).

72FIN/MES Finke H. L., Messerly J. F., Todd S. S., J. Chem. Thermodyn. **4**, 359-74 (1972).

72HOR Horvath A. L., Z. Phys. Chem. (Frankfurt am Main) **78**, 209-10 (1972).

72KAW/OTA Kawaizumi F., Otake T., Nomura H., Miyahara Y., Nippon Kagaku Kaishi 1773-6 (1972).

72KOL/VOR Kolesov V. P., Vorob'ev V. N., Dokl. Akad. Nauk. SSSR **203**, 116-9 (1972).

72MAN/TOL Manzhelii V. G., Tolkachev A. M., Krupskii I. N., Voitovich E. I., Popov V . A., Koloskova L. A., J. Low. Temp. Phys. **7**, 169-82 (1972).

72MAS/FAM Maslova V. A., Faminskaya L. A., Tr. Khim. Khim. Tekhnol. (2), 51-3 (1972).

72MAS/RAB Maslova V. A., Rabinovich I. B., Nistratov V. P., Faminskaya L. A., Tr. Khim. Khim. Tekhnol. (2), 44-50 (1972).

72PAC/HOD Pace E. L., Hodge P. E., J. Chem. Thermodyn. **4**, 441-7 (1972).

72REC/SAD Recko W. M., Sadowska K. W., Bull. Acad. Pol. Sci., Ser. Sci. Chim. **20**, 73-9 (1972).

72STR/NOV Strepikheev Yu. A., Novikova O. P., Chimishkyan A. L., Khim. Prom-st. (Moscow) **48**, 230-1 (1972).

72VAN Van Miltenburg J. C., J. Chem. Thermodyn. **4**, 773-82 (1972).

72VOR/KOL Vorob'ev V. N., Kolesov V. P., Kuznec V. M., Babushkina T. A., Zh. Obshch. Khim. **42**, 391-9 (1972).

72WAK/LUF Wakefield Z. T., Luff B. B., Reed R. B., J. Chem. Eng. Data **17**, 420-3 (1972).

73AKH/EKS Akhundov T. S., Eksaev R. A., Izv. Vyssh. Ucheb. Zaved., Neft Gaz **16** (2), 68-72 (1973).

73AND/COU Andon R. J. L., Counsell J. F., Lee D. A., Martin J. F., J. Chem. Soc., Faraday Trans. 1 **69**, 1721-6 (1973).

73AND/GOR Andrews J. T. S., Gordon J. E., J. Chem. Soc., Faraday Trans. 1 **69**, 546-54 (1973).

73AND/MAR1 Andon R. J. L., Martin J. F., J. Chem. Soc., Faraday Trans. 1 **69**, 761-70 (1973).

73AND/MAR2 Andon R. J. L., Martin J. F., J. Chem. Soc., Faraday Trans. 1 **69**, 871-5 (1973).
73DZH/GUS Dzhafarov O. I., Guseinov Z. A., Karasharli K. A., Zeinalov T. D., Zh. Fiz. Khim. **47**, 2448 (1973), data deposited in VINITI 5024-72.
73FIN/MES Finke H. L., Messerly J. F., J. Chem. Thermodyn. **5**, 247-57 (1973).
73GOO/FIN Good W. D., Finke H. L., Messerly J. F., Osborn A. G., Harrison R. H., Moore R. T., Douslin D. R., Air Force Off. Sci. Res. Rep. [Techn. Rep.] AFOSR-TR (U.S.) 73-0760, 1-6 (1973).
73GRO Groenvold F., J. Chem. Thermodyn. **5**, 525-31 (1973).
73GUS/DZH Guseinov Z. A., Dzhafarov O. I., Karasharli K. A., Zh. Fiz. Khim. **47**, 2716 (1973), data deposited in VINITI 6329-73.
73HAL/SMI Hall H. K., Jr., Smith C. D., Baldt J. H., J. Am. Chem. Soc. **95**, 3197-201 (1973).
73KAL/WOY Kalinowska B., Woycicka M., Bull. Acad. Pol. Sci., Ser. Sci. Chim. **21**, 845-8 (1973).
73KIS/SUG Kishimoto K., Suga H., Seki S., Bull. Chem. Soc. Jpn. **46**, 3020-31 (1973).
73KOL/VOR Kolesov V. P., Vorob'ev V. N., Sarzhina E. A., Third Int. Conf. Chem. Thermodyn., Baden **2**, 33-9 (1973).
73KOM/MIL Komarek K. L., Miller E., Schick G., Monatsh. Chem. **104**, 1139-53 (1973).
73KRI/LIC Krien G., Licht H. H., Zierath J., Thermochim. Acta **6**, 465-72 (1973).
73KUS/SUU Kusano K., Suurkuusk J., Wadsö I., J. Chem. Thermodyn. **5**, 757-67 (1973).
73LEB/TSV Lebedev B. V., Tsvetkova L. Ya., Rabinovich I. B., Tr. Khim. Khim. Tekhnol. (1), 17-20 (1973).
73MAR Martin J. F., Chem. Soc. Specialist Periodical Report, Chemical Thermodynamics, Vol. I, Chapt. 4, (1973).
73MAS/NOV Maslova V. A., Novoselova N. V., Moseeva E. M., Berezhnaya N. D., Rabinovich I. B., Tr. Khim. Khim. Tekhnol. (2), 51-2 (1973).
73MOS/NIK Moseeva E. M., Nikolaev P. N., Sechenova T. A., Rabinovich I. B., Tr. Khim. Khim. Tekhnol. (1), 153-5 (1973).
73PER/COM Perkin-Elmer DSC-1-B, commercial calorimeter, Prospectus (1973).
73SHI/ENO Shinoda T., Enokido H., Maeda Y., Tomita H., Mashiko Y., Bull. Chem. Soc. Jpn. **46**, 48-52 (1973).
73SOR/SEK Sorai M., Seki S., Mol. Cryst. Liq. Cryst. **23**, 299-327 (1973).
73SUB/RAJ Subrahmanyam S. V., Rajagopal E., Z. Phys. Chem. (Frankfurt am Main) **85**, 256-68 (1973).
73VAS/KOR Vasil'ev I. A., Korkhov A. D., Zh. Fiz. Khim. **47**, 2710 (1973), data deposited in VINITI 6326-73.
74AND/BAC Andrews J. T. S., Bacon W. E., J. Chem. Thermodyn. **6**, 515-23 (1974).
74AND/COU Andon R. J. L., Counsell J. F., Lee D. A., Martin J. F., J. Chem. Soc., Faraday Trans. 1 **70**, 1914-7 (1974).
74AND/MAR Andon R. J. L., Martin J. F., J. Chem. Soc., Faraday Trans. 1 **70**, 605-8 (1974).
74ATA/CHI Atake T., Chihara H., Bull. Chem. Soc. Jpn. **47**, 2126-36 (1974).
74COC/NOR Cochran M. A., North A. M., Pethrick R. A., J. Chem. Soc., Faraday Trans. 2 **70**, 1274-9 (1974).
74DEV/SOM De Visser C., Somsen G., Z. Phys. Chem. (Frankfurt am Main) **92**, 159-62 (1974).
74DIA/REN Diaz Pena M., Renuncio J. A. R., An. Quim. **70**, 113-20 (1974).
74DSM/COM DSM-2, comercial calorimeter (product of USSR), Vestn. Akad. Nauk SSSR **44** (7), 141 (1974).
74GEI/DZH1 Geidarov Kh. I., Dzhafarov O. I., Karasharli K. A., Kostryukov V. N., Zh. Fiz. Khim. **48**, 1058 (1974), data deposited in VINITI 7671-73.
74GEI/DZH2 Geidarov Kh. I., Dzhafarov O. I., Karasharli K. A., Kostryukov V. N., Zh. Fiz. Khim. **48**, 1147-9 (1974).
74GRU/SHE Gruzdev V. A., Sheludyakov E. P., Kiriyanenko A. A., Kolotov Ya. L., Lavrov V. A., Shilyakov A. A., Shestova A. I., Shumskaya A. I., Fluid Mech. -Sov. Res. **3**, 172-6/4 (1974).
74KOL/VOR Kolesov V. P., Vorob'ev V. N., Sarzhina E. A., Pentin Yu. A., Timoshenkova Yu. D., J. Chem. Thermodyn. **6**, 613-28 (1974).
74LEB/ARO Lebedev B. V., Aron B. M., Tr. Khim. Khim. Tekhnol. (1), 75-7 (1974).
74LEB/TSV Lebedev B. V., Tsvetkova L. Ya., Tr. Khim. Khim. Tekhnol. (1), 143-5 (1974).
74MAS/RAB Maslova V. A., Rabinovich I. B., Tr. Khim. Khim. Tekhnol. (1), 40-63 (1974).
74MES/FIN Messerly J. F., Finke H. L., Todd S. S., J. Chem. Thermodyn. **6**, 635-57 (1974).
74MIL/LEB Milov V. I., Lebedev B. V., Tsvetkova L. Ya., Sladkov A. M., Luneva L. K., Lebedev N. K., Evstropov A. A., Tr. Khim. Khim. Tekhnol. (1), 140-5 (1974).
74MOS/MOU Mosselman C., Mourik J., Dekker H., J. Chem. Thermodyn. **6**, 477-87 (1974).
74PET/TER Petit J. C., Ter Minassian L., J. Chem. Thermodyn. **6**, 1139-52 (1974).
74RAJ/SUB Rajagopal E., Subrahmanyam S. V., J. Chem. Thermodyn. **6**, 873-6 (1974).
74SHI/MAE Shinoda T., Maeda Y., Enokido H., J. Chem. Thermodyn. **6**, 921-34 (1974).
74SOR/NAK Sorai M., Nakamura T., Seki S., Bull. Chem. Soc. Jpn. **47**, 2192-6 (1974).
74SUU/WAD Suurkuusk J., Wadsö I., J. Chem. Thermodyn. **6**, 667-79 (1974).
74TOM/ADA Tomlins R. P., Adamson M., J. Chem. Thermodyn. **6**, 757-62 (1974).
74VAS/KOR Vasil'ev I. A., Korkhov A. D., Tr. Khim. Khim. Tekhnol. (1), 103-5 (1974).
74VOR/KOL Vorob'ev V. N., Kolesov V. P., Sarzhina E. A., Kuramshina G. M., Pentin Yu. A., Zh. Fiz. Khim. **48**, 239 (1974), data deposited in VINITI 6783-73.
74WIL/ZET Wilhelm E., Zettler M., Sackmann H, Ber. Bunsenges. Phys. Chem. **78**, 795-804 (1974).
74YOU Younglove B. A., J. Res. Natl. Bur. Stand. (U.S.) **78A**, 401-10 (1974).
75AMB/CON Ambrose D., Connett J. E., Green J. H. S., Hales J. L., Head A. J., Martin J. F., J. Chem. Thermodyn. **7**, 1143-57 (1975).
75AND/COU Andon R. J. L., Counsell J. F., Lee D. A., Martin J. F., J. Chem. Thermodyn. **7**, 587-92 (1975).
75AND/MAR Andon R. J. L., Martin J. F., J. Chem. Thermodyn. **7**, 593-606 (1975).
75CUC1 Cuccuru A., Thermochim. Acta **11**, 247-51 (1975).
75CUC2 Cuccuru A., Thermochim. Acta **13**, 96-9 (1975).
75ENG Engineering Science Data Unit Series, Heat Capacity and Enthalpy of Liquids. I: Aromatic Hydrocarbons, Vol. 4, (Item No. 75015), ESDU International Ltd., London (1975).
75FER/FRA Ferloni P., Franzosini P., Gazz. Chim. Ital. **105**, 391-401 (1975).
75FER/SAN Ferloni P., Sanesi M., Franzosini P., Z. Naturforsch. **30a**, 1447-54 (1975).
75GEI/DZH Geidarov Kh. I., Dzhafarov O. I., Karasharli K. A., Kostryukov V. N., Zh. Fiz. Khim. **49**, 493-4 (1975).
75GRI/RAS Grigor'ev B. A., Rastorguev Yu. L., Yanin G. S., Izv. Vyssh. Ucheb. Zaved., Neft Gaz **18** (10), 63-6 (1975).
75GRO/BEN Grolier J. -P. E., Benson G. C., Picker P., J. Chem. Eng. Data **20**, 243-6 (1975).
75GUS/DZH Guseinov Z. A., Dzhafarov O. I., Karasharli K. A., Zh. Fiz. Khim. **49**, 219-20 (1975).
75GUS/KAR Guseinov Z. A., Karasharli K. A., Dzhafarov O. I., Nurullaev G. G., Nametkin N. S., Gusel'nikov L. E., Volnina E. A., Burdasov E. N., Vdovin V. M., Dokl. Akad. Nauk. SSSR **222**, 1369-72 (1975).

75HOL/ZIE Holzhauer J. K., Ziegler W. T., J. Phys. Chem. **79**, 590-604 (1975).

75HOR Horvath A. L., Physical Properties of Inorganic Compounds, Edvard Arnold, London (1975).

75JOL/BOI Jolicoeur C., Boileau J., Bazinet S., Picker P., Can. J. Chem. **53**, 716-22 (1975).

75KOS/SAM Kostryukov V. N., Samorukov O. P., Samorukova N. Kh., Chaplygina E. K., Zh. Fiz. Khim. **49**, 1591 (1975), data deposited in VINITI 586-75.

75LEB/LEB Lebedev B. V., Lebedev N. K., Tsvetkova L. Ya., Rabinovich I. B., Tr. Khim. Khim. Tekhnol. (4), 41-5 (1975).

75LEB/MIL1 Lebedev B. V., Milov V. I., Rabinovich I. B., Sladkov A. M., Luneva L. K., Tr. Khim. Khim. Tekhnol. (4), 35-40 (1975).

75LEB/MIL2 Lebedev B. V., Milov V. I., Tsvetkova L. Ya., Lebedev N. K., Kiparisova E. G., Evstropov A. A., Zh. Fiz. Khim. **49**, 1592 (1975), data deposited in VINITI 605-75.

75LEB/RAB Lebedev B. V., Rabinovich I. B., Evstropov A. A., Rykhlevskaya M. S., Tr. Khim. Khim. Tekhnol. (4), 27-34 (1975).

75LEB/TSV Lebedev B. V., Tsvetkova L. Ya., Kiparisova E. G., Lebedev N. K., Zh. Fiz. Khim. **49**, 2152 (1975), data deposited in VINITI 1324-75.

75MAS/SCO Masi J. F., Scott R. B., J. Res. Natl. Bur. Stand. (U.S.) **79A**, 619-28 (1975).

75MEK/KAR1 Mekhtiev S. A., Karasharli K. A., Dzhafarov O. I., Zh. Fiz. Khim. **49**, 259 (1975), data deposited in VINITI 2585-74.

75MEK/KAR2 Mekhtiev S. A., Karasharli K. A., Dzhafarov O. I., Zh. Fiz. Khim. **49**, 259 (1975), data deposited in VINITI 2588-74.

75MEK/KAR3 Mekhtiev S. A., Karasharli K. A., Dzhafarov O. I., Zh. Fiz. Khim. **49**, 2578-9 (1975).

75MEK/KAR4 Mekhtiev S. A., Karasharli K. A., Dzhafarov O. I., Kuznetsova A. G., Zh. Fiz. Khim. **49**, 1914-6 (1975).

75MES/FIN Messerly J. F., Finke H. L., Osborn A. G., Douslin D. R., J. Chem. Thermodyn. **7**, 1029-46 (1975).

75MON Montgomery R. L., Thesis, Oklahoma State University, Stillwater, Univ. Microfilms Publ. No. 76-9730 (1975).

75NIC/WAD Nichols N., Wadsö I., J. Chem. Thermodyn. **7**, 329-36 (1975).

75PAU1 Paukov I. E., Zh. Fiz. Khim. **49**, 265 (1975), data deposited in VINITI 2593-74.

75PAU2 Paukov I. E., Zh. Fiz. Khim. **49**, 265 (1975), data deposited in VINITI 2594-74.

75PED/KAY Pedersen M. J., Kay W. B., Hershey H. C., J. Chem. Thermodyn. **7**, 1107-18 (1975).

75PHI/WAL Phillips J. C., Wall L. A., Aldridge M. H., Polym. Eng. Sci. **15**, 73-8 (1975).

75RAS/GRI Rastorguev Yu. L., Grigor'ev B. A., Yanin G. S., Izv. Vyssh. Ucheb. Zaved., Neft Gaz **18** (4), 73-7 (1975).

75SAF/GER Safir L. I., Gerasimov A. A., Grigor'ev B. A., Izv. Vyssh. Ucheb. Zaved., Neft Gaz **18** (11), 61-5 (1975).

75SAN San Jose J. L., Ph. D. Thesis, Massachussetts Institute of Technology (1975).

75STR/SUN Stridh G., Sunner S., J. Chem. Thermodyn. **7**, 161-8 (1975).

75VAS/VAS Vasil'ev I. A., Vasil'eva T. F., Korkhov A. D., Alkilenkarbonaty, VNII Neftekhim, Leningrad 152-60 (1975).

75WOY/KAL Woycicka M. K., Kalinowska B., Bull. Acad. Pol. Sci., Ser. Sci. Chim. **23**, 759-64 (1975).

76ADE/SIM Adeosun S. O., Sime S. J., Thermochim. Acta **17**, 351-9 (1976).

76AND/MAR Andon R. J. L., Martin J. F., J. Chem. Thermodyn. **8**, 1159-66 (1976).

76BON/CER Bonner O. D., Cerutti P. J., J. Chem. Thermodyn. **8**, 105-12 (1976).

76CAR/WES Carling R. W., Westrum E. F., Jr., J. Chem. Thermodyn. **8**, 269-76 (1976).

76CHI/INA1 Chihara H., Inaba A., J. Chem. Thermodyn. **8**, 915-34 (1976).

76CHI/INA2 Chihara H., Inaba A., J. Chem. Thermodyn. **8**, 935-58 (1976).

76CON/GIA Conti G., Gianni P., Matteoli E., Mengheri M., Chim. Ind. (Milan) **58**, 225 (1976).

76DES/DEV Desnoyers J. E., De Visser C., Perron G., Picker P. J. Solution Chem. **5**, 605-16 (1976).

76DWO/FIG Dworkin A., Figuiere P., Ghelfenstein M., Szwarc H., J. Chem. Thermodyn. **8**, 835-44 (1976).

76ENG Engineering Science Data Unit Series, Heat Capacity and Enthalpy of Liquids. II: Halogenated Methanes, Vol. 4, (Item No. 76010), ESDU International Ltd., London (1976).

76FIN/MES Finke H. L., Messerly J. F., Douslin D. R., J. Chem. Thermodyn. **8**, 965-83 (1976).

76FOR/BEN1 Fortier J. -L., Benson G. C., J. Chem. Thermodyn. **8**, 411-23 (1976).

76FOR/BEN2 Fortier J. -L., Benson G. C., Picker P., J. Chem. Thermodyn. **8**, 289-99 (1976).

76GEI/DZH Geidarov Kh. I., Dzhafarov O. I., Karasharli K. A., Kostryukov V. N., Zh. Fiz. Khim. **50**, 1112-4 (1976).

76GIE/WOL Giepel G., Wolf G., Z. Phys. Chem. (Leipzig) **257**, 587-93 (1976).

76GOO/LEE Good W. D., Lee S. H., J. Chem. Thermodyn. **8**, 643-50 (1976).

76KAR/GRO Karbalai Ghassemi M. H., Grolier J. -P. E., Int. Data Ser., Sel. Data Mixtures, Ser. A (2)95 (1976).

76KUL/DZH Kuliev A. M., Dzhafarov O. I., Karasharli K. A., Kuznetsova A. G., Zh. Fiz. Khim. **50**, 1903 (1976), data deposited in VINITI 1191-76.

76LEB/LIT Lebedev B. V., Lityagov V. Ya., Tr. Khim. Khim. Tekhnol. (5), 89-105 (1976).

76LEB/RYA Lebedeva N. D., Ryadnenko V. L., Termodinam. i Kinet. Khim. Protsessov, GIPKh Leningrad 79 (1976).

76MET/COM Mettler TA 2000, commercial calorimeter, Thermochim. Acta **17**, 351-9 (1976).

76MIL/SCH Miller J. W., Jr., Schorr G. R., Yaws C. L., Chem. Eng. (New York) **83** (23), 129-31 (1976).

76MUS Mustafaev R. A., Zh. Fiz. Khim. **50**, 1899 (1976), data deposited in VINITI 1224-76.

76PHI/MAT Phillips J. C., Mattamal M. M., J. Chem. Eng. Data **21**, 228-32 (1976).

76RAZ Raznjevic K., Handbook of Thermodynamic Tables and Charts, Hemisphere Publishing Corp., New York (1976).

76ROD1 Roder H. M., J. Chem. Phys. **65**, 1371-3 (1976).

76ROD2 Roder H. M., J. Res. Natl. Bur. Stand. (U.S.) **80A**, 739-59 (1976).

76SAN/MEL San Jose J. L., Mellinger G., Reid R. C., J. Chem. Eng. Data **21**, 414-7 (1976).

76SKO/SUU Skold R., Suurkuusk J., Wadsö I., J. Chem. Thermodyn. **8**, 1075-80 (1976).

76TOU/MAK Touloukian Y. S., Makita T., Thermophysical Properties of Matter, Suppl. to Vol. **6**, IFI/Plenum, New York (1976).

76VAS/KOR Vasil'ev I. A., Korkhov A. D., Minkin G. M., Rabinovich I. B., Sheiman M. S., Nistratov V. P, Termodin. Org. Soedin. (5), 21-5 (1976).

77BON/BED Bonner O. D., Bednarek J. M., Arisman R. K., J. Am. Chem. Soc. **99**, 2898-902 (1977).

77DEV/PER1 De Visser C., Perron G., Desnoyers J. E., Can. J. Chem. **55**, 856-62 (1977).

77DEV/PER2 De Visser C., Perron G., Desnoyers J. E., J. Am. Chem. Soc. **99**, 5894-900 (1977).

77DEV/PER3 De Visser C., Perron G., Desnoyers J. E., Heuvelsland W. J. M., Somsen G., J. Chem. Eng. Data **22**, 74-9 (1977).

77ENG Engineering Science Data Unit Series, Heat Capacity and Enthalpy of Liquids. III: Halogenated Ethanes and Ethylenes, Vol. 4, (Item No. 77007), ESDU International Ltd., London (1977).

77FIN/MES Finke H. L., Messerly J. F., Lee S. H., Osborn A. G., Douslin D. R., J. Chem. Thermodyn. **9**, 937-56 (1977).

77FOR/BEN Fortier J. -L., Benson G. C., J. Chem. Thermodyn. **9**, 1181-8 (1977).

77HAI/SUG1 Haida O., Suga H., Seki S., Bull. Chem. Soc. Jpn. **50**, 802-9 (1977).

77HAI/SUG2 Haida O., Suga H., Seki S., J. Chem. Thermodyn. **9**, 1133-48 (1977).

77HAR/ATA Harada M., Atake T., Chihara H., J. Chem. Thermodyn. **9**, 523-34 (1977).

77HOF/SAN Hoffman S. P., San Jose J. L., Reid R. C., J. Chem. Eng. Data **22**, 385-8 (1977).

77KOR/VAS Korkhov A. D., Vasil'ev I. A., Termodin. Org. Soedin. (6), 34-7 (1977).

77KOS/SAM Kostryukov V. N., Samorukov O. P., Samorukova N. Kh., Krasavin A. M., Petrunin A. B., 7th Vses. Konf. Kalorim. [Rosshir. Tezisy Dokl.] **2**, 387-90 (1977).

77KU /COM KU-300, commercial calorimeter VNIIFTRI (product of USSR), Termodin. Org. Soedin. (6), 89-92 (1977).

77KUL/DZH1 Kuliev A. M., Dzhafarov O. I., Karasharli K. A., Kuznetsova A. G., Zh. Fiz. Khim. **51**, 1564 (1977), data deposited in VINITI 986-77.

77KUL/DZH2 Kuliev A. M., Dzhafarov O. I., Karasharli K. A., Kuznetsova A. G., Zh. Fiz. Khim. **51**, 1564 (1977), data deposited in VINITI 987-77.

77LAG/PIE Lagnier R., Pierre J., Mortimer M. J., Cryogenics **17**, 349-53 (1977).

77LEB/EVS1 Lebedev B. V., Evstropov A. A., Belov V. I., Termodin. Org. Soedin. (6), 3-12 (1977).

77LEB/EVS2 Lebedev B. V., Evstropov A. A., Belov V. I., Sytov G. A., Fiz. -Khim. Osnovy Sinteza i Pererab. Polim. (2), 5-13 (1977).

77LEB/EVS3 Lebedev B. V., Evstropov A. A., Sadikov G. B., Larina L. F., Zh. Fiz. Khim. **51**, 1285 (1977), data deposited in VINITI 4328-76.

77LEB/RAB1 Lebedev B. V., Rabinovich I. B., Evstropov A. A., J. Chem. Thermodyn. **9**, 101-10 (1977).

77LEB/RAB2 Lebedev B. V., Rabinovich I. B., Lebedev N. K., Tsvetkov V. G., Durgar'yan S. G., Khotimskii V. S., Vysokomol. Soedin., Ser. A **19**, 513-8 (1977).

77MAK Makita T., Reito **52**, 26-33 (1977).

77MEI/BLO Meijer E. L., Blok J. G., Kroon J., Oonk H. A. J., Thermochim. Acta **20**, 325-34 (1977).

77MIL/LEB Milov V. I., Lebedev B. V., Kiparisova E. G., Bykova T. A., Sladkov A. M., Termodin. Org. Soedin. (6), 13-25 (1977).

77MUR/SUB Murthy N. M., Subrahmanyam S. V., Indian J. Pure Appl. Phys. **15**, 485-9 (1977).

77NAZ/MUS Naziev Ya. M., Mustafaev M. R., Abasov A. A., Proc. Symp. Thermophys. Prop. **7th**, 525-31 (1977).

77STE/WAT Steele W. V., Watt I., J. Chem. Thermodyn. **9**, 843-9 (1977).

77VOR/PRI Vorob'ev A. F., Privalova N. M., Rekharskii M. V., Zh. Fiz. Khim. **51**, 1843-6 (1977).

77VYU V'yugin A. I., Thesis, IKhTI Ivanovo (1977), from: Korolev V. P., Kukharenko V. A., Krestov G. A., Zh. Fiz. Khim. **60**, 1854-7 (1986).

77WIL/GRO Wilhelm E., Grolier J. -P. E., Karbalai Ghassemi M. H., Ber. Bunsenges. Phys. Chem. **81**, 925-30 (1977).

78ADE Adeosun S. O., J. Therm. Anal. **14**, 235-43 (1978).

78BUR/KAM Burchalova G. V., Kamelova G. P., Nistratov V. P., Sheiman M. S., Rabinovich I. B., Termodin. Org. Soedin. (7), 21-2 (1978).

78BYV/JAS Byval'tsev Yu. A., Jasiunas B. V., Perelygin V. M., Izv. Vyssh. Ucheb. Zaved., Pishch. Tekhnol. (5), 169-70 (1978).

78DEV/HEU De Visser C., Heuvelsland W. J. M., Dunn L. A., Somsen G., J. Chem. Soc., Faraday Trans. 1 **74**, 1159-69 (1978).

78EVS/BEL Evstropov A. A., Belov B. I., Lebedev B. V., Kiparisova E. G., Bykova T. A., Zh. Fiz. Khim. **52**, 242-3 (1978), data deposited in VINITI 2144-77.

78GEI/KAR1 Geidarov Kh. I., Karasharli K. A., Dzhafarov O. I., Zh. Fiz. Khim. **52**, 804 (1978), data deposited in VINITI 3882-77.

78GEI/KAR2 Geidarov Kh. I., Karasharli K. A., Dzhafarov O. I., Guseinov Z. A., Zh. Fiz. Khim. **52**, 805 (1978), data deposited in VINITI 3881-77.

78GOD/RAC Godlewska M., Rachwalska M., Phys. Status Solidi A **47**, 661-5 (1978).

78GOO Goodwin R. D., J. Res. Natl. Bur. Stand. (U.S.) **83**, 449-58 (1978).

78GRO/WIL Grolier J. -P. E., Wilhelm E., Hamedi M. H., Ber. Bunsenges. Phys. Chem. **82**, 1282-90 (1978).

78GUN/HEA Gundry H. A., Head A. J., J. Chem. Thermodyn. **10**, 195-200 (1978).

78INA/CHI Inaba A., Chihara H., J. Chem. Thermodyn. **10**, 45-64 (1978).

78KAR/RAB Karyakin N. V., Rabinovich I. B., Nistratov V. P., Termodin. Org. Soedin. (7), 12-6 (1978).

78KAR/SAP Karyakin N. V., Sapozhnikov V. N., Teplyakov M. M., Chebotarev V. P., Vysokomol. Soedin., Ser. A **20**, 1950-6 (1978).

78KIS/SUG Kishimoto K., Suga H., Seki S., Bull. Chem. Soc. Jpn. **51**, 1691-6 (1978).

78KON/RUF Konkoly-Thege I., Ruff I., Adeosun S. O., Sime S. J., Thermochim. Acta **24**, 89-96 (1978).

78KOS/KOL Kosarukina E. A., Kolesov V. P., Vorob'ev V. N., Fogel L., Zh. Fiz. Khim. **52**, 509 (1978), data deposited in VINITI 3512-77.

78LEB/EVS Lebedev B. V., Evstropov A. A., Kiparisova E. G., Belov V. I., Vysokomol. Soedin., Ser. A **20**, 20-37 (1978).

78LEB/LEB Lebedev B. V., Lebedev N. K., Lityagov V. Ya., Mukhina N. N., Termodin. Org. Soedin. (7), 23-37 (1978).

78LEB/RAB Lebedev B. V., Rabinovich I. B., Milov V. I., Lityagov V. Ya., J. Chem. Thermodyn. **10**, 321-9 (1978).

78LEB/TSV1 Lebedev B. V., Tsvetkova L. Ya., Rabinovich I. B., J. Chem. Thermodyn. **10**, 809-15 (1978).

78LEB/TSV2 Lebedev B. V., Tsvetkova L. Ya., Rabinovich I. B., Finkel'shtein E. Sh., Strel'chik B. S., Termodin. Org. Soedin. (7), 3-7 (1978).

78MAR/CIO1 Marchidan D. I., Ciopec M., J. Therm. Anal. **14**, 131-50 (1978).

78MAR/CIO2 Marchidan D. I., Ciopec M., Rev. Roum. Chim. **23**, 19-29 (1978).

78MCC/HAM McCormick D. G., Hamilton W. S., J. Chem. Thermodyn. **10**, 275-8 (1978).

78MIL Mills K. C., Thermochim. Acta **23**, 390-2 (1978).

78MOS/RAB Moseeva E. M., Rabinovich I. B., Busygina G. I., Safonov V. A., Ovchinnikov E. Yu., Termodin. Org. Soedin. (7), 8-11 (1978).

78ROU/PER1 Roux G., Perron G., Desnoyers J. E., Can. J. Chem. **56**, 2808-14 (1978).

78ROU/PER2 Roux G., Perron G., Desnoyers J. E., J. Solution Chem. **7**, 639-54 (1978).

78RYB/EME Rybalkin V. I., Emel'yanov V. M., Stupak P. M., Litovchenko N. R., Z'ola M . I., Neftepererab. Neftekhim. (Kiev) **16**, 48-50 (1978).

78SAF Safir L. I., Izv. Vyssh. Ucheb. Zaved., Neft Gaz **21** (12), 14, 82 (1978).

78SOL/SUK Solovev G. V., Sukhinin G. I., Stolyarov N. N., Chashkin Yu. R., Kholod. Tekh. (6), 30-3 (1978).

78SPI/AND Spinella G. M., Andrews J. T. S., Bacon W. E., Fort R. C., Jr., Sabo J., J. Chem. Thermodyn. **10**, 1023-32 (1978).

78WES/LAN Westrum E. F., Jr., Landee C. P., Takahashi Y., Chavret M., J. Chem. Thermodyn. **10**, 835-46 (1978).

78WOY/KAL Woycicka M., Kalinowska B., Bull. Acad. Pol. Sci., Ser. Sci. Chim. **26**, 371-5 (1978).

79AMI/LEB Amitin E. B., Lebedeva E. G., Paukov I. E., Zh. Fiz. Khim. **53**, 2666-70 (1979).

79AND/GRI Andolenko R. A., Grigor'ev B. A., Izv. Vyssh. Ucheb. Zaved., Neft Gaz **22** (11), 78, 90 (1979).

79BAB/NIS Babinkov A. G., Nistratov V. P., Larina V. N., Shvetsova K. G., Sapozhnikov V. N., Zhilitskaya O. M, Termodin. Org. Soedin. (8), 28-33 (1979).

79BAL/PET Balakina M. M., Petrova G. V., Vasil'ev I. A., Termodin. Org. Soedin. (8), 37-41 (1979).

79BRI/VAN Briels W. J., Van Miltenburg J. C., J. Chem. Phys. **70**, 1064-6 (1979).

79BRO/ZIE Brown G. N., Jr., Ziegler W. T., J. Chem. Eng. Data **24**, 319-30 (1979).

79CZA Czarnota I., Bull. Acad. Pol. Sci., Ser. Sci. Chim. **27**, 763-72 (1979).

79DEV/SOM De Visser C., Somsen G., J. Solution Chem. **8**, 593-600 (1979).

79DU /COM Du Pont 900 DTA, comercial calorimeter, Anal. Chim. Acta **27**, 763-72 (1979).

79DYK/REP Dykyj J., Repáš M., Tlak Nasýtených Par Organických Zlúčenin, Veda, Bratislava (1979).

79DZH/KAR Dzhafarov O. I., Karasharli K. A., Kuliev A. M., Azerb. Khim. Zh. (5), 112-4 (1979).

79ENG Engineering Science Data Unit Series, Heat Capacity and Enthalpy of Liquids: Aliphatic Alcohols, Vol. 4, (Item No. 79028), ESDU International Ltd., London (1979).

79EVS/LEB1 Evstropov A. A., Lebedev B. V., Kiparisova E. G., Prusakova I. L., Termodin. Org. Soedin. (8), 14-21 (1979).

79EVS/LEB2 Evstropov A. A., Lebedev B. V., Kulagina T. G., Lyudvig E. B., Belen'kaya B. G., Vysokomol. Soedin., Ser. A **21**, 2038-44 (1979).

79FOR/BEN Fortier J. -L., Benson G. C., J. Chem. Eng. Data **24**, 34-7 (1979).

79FOR/DAR Fortier J. -L., D'Arcy P. J., Benson G. C., Thermochim. Acta **28**, 37-43 (1979).

79FUC Fuchs R., J. Chem. Thermodyn. **11**, 959-61 (1979).

79FUC/PEA Fuchs R., Peacock L. A., Can. J. Chem. **57**, 2302-4 (1979).

79GME Gmelin E., Thermochim. Acta **29**, 1-39 (1979).

79GRI/YAN Grigor'ev B. A., Yanin G. S., Rastorguev Yu. L., Tr. GIAP **54**, 57-64 (1979).

79GRO/HAM Grolier J. -P. E., Hamedi M. H., Wilhelm E., Kehiaian H. V., Thermochim. Acta **31**, 79-84 (1979).

79HER Herington, E. F. G., Physical Constants of Some Constituents of Coal Tar, National Physical Laboratory, Teddington, NPL Report Chem. 102 (1979).

79KIY/DAR Kiyohara O., D'Arcy P. J., Benson G. C., Can. J. Chem. **57**, 1006-10 (1979).

79KOL Kollar-Hunek K., Ph. D. Thesis, Budapest Technical University, Budapest (1979).

79LEB/EVS Lebedev B. V., Evstropov A. A., Shurygina A. M., Termodin. Org. Soedin. (8), 5-13 (1979).

79LEB/LEB1 Lebedev B. V., Lebedev N. K., Kiparisova E. G., Khim. Elementoorg. Soedin. (7), 61-5 (1979).

79LEB/LEB2 Lebedev N. K., Lebedev B. V., Khotimskii V. S., Durgar'yan S. G., Vysokomol. Soedin., Ser. A **21**, 1031-6 (1979).

79LEB/LIT Lebedev B. V., Lityagov V. Ya., Krentsina T. I., Milov V. I., Zh. Fiz. Khim. **53**, 264-5 (1979), data deposited in VINITI 1077-78.

79LEE/HOS Lee-Bechtold S. H., Hossenlopp I. A., Scott D. W., Osborn A. G., Good W. D., J. Chem. Thermodyn. **11**, 469-82 (1979).

79MAR/BRA Martin Rubio F., Bravo R., Pintos M., Gomez-Gotor A., Acta Cient. Compostelana **16**, 79-105 (1979).

79MRA/NAA Mraw S. C., Naas D. F., J. Chem. Thermodyn. **11**, 567-84 (1979).

79MUR/SUB Murthy N. M., Subrahmanyam S. V., Indian J. Pure Appl. Phys. **17**, 620-2 (1979).

79NIS/BAB Nistratov V. P., Babinkov A. G., Shvetsova K. G., Lapteva S. A., Termodin. Org. Soedin. (8), 33-6 (1979).

79OTT/GOA Ott J. B., Goates J. R., Grigg R. B., J. Chem. Thermodyn. **11**, 1167-73 (1979).

79RIC/SAV Richardson M. J., Savill N. G., Thermochim. Acta **30**, 327-37 (1979).

79RIG/KLE Rigaudy J., Klesney S. P. (editors), Nomenclature of Organic Chemistry, IUPAC Publication, Pergamon Press, Oxford (1979).

79SAL/PEA Saluja P. P. S., Peacock L. A., Fuchs R., J. Am. Chem. Soc. **101**, 1958-62 (1979).

79SAM/GRI Samsonov D. P., Grinberg E. E., Efremov A. A., Zh. Prikl. Khim. (Leningrad) **52**, 1909-11 (1979).

79SCH/OFF Schaake R. C. F., Offringa J. C. A., Van der Berg G. J. K., Van Miltenburg J. C., Recl. Trav. Chim. Pays-Bas **98**, 408-12 (1979).

79STE Steele W. V., J. Chem. Thermodyn. **11**, 1185-8 (1979).

79STE/TAM Stephens M. A., Tamplin W. S., J. Chem. Eng. Data **24**, 81-2 (1979).

79SUN/SVE Sunner S., Svensson C., Zelepuga A. S., J. Chem. Thermodyn. **11**, 491-5 (1979).

79SVE Svensson C., J. Chem. Thermodyn. **11**, 593-6 (1979).

79TAK/YOK Takahashi Y., Yokokawa H., Kadokura H., Sekine Y., Mukaibo T., J. Chem. Thermodyn. **11**, 379-94 (1979).

79TSU/SOR Tsuji K., Sorai M., Suga H., Seki S., Mol. Cryst. Liq. Cryst. **55**, 71-88 (1979).

79VAN/ZEL1 Van Kasteren P. H. G., Zeldenrust H., Ind. Eng. Chem., Fundam. **18**, 333-9 (1979).

79VAN/ZEL2 Van Kasteren P. H. G., Zeldenrust H., Ind. Eng. Chem., Fundam. **18**, 339-45 (1979).

79VES/ZAB Veselý F., Zábranský M., Svoboda V., Pick J., Collect. Czech. Chem. Commun. **44**, 3529-32 (1979).

79WIL/FAR Wilhelm E., Faradjzadeh A., Grolier J. -P. E., J. Chem. Thermodyn. **11**, 979-84 (1979).

79WIL/GRO Wilhelm E., Grolier J. -P. E., Karbalai Ghassemi M. H., Thermochim. Acta **28**, 59-69 (1979).

79YUE/YOS Yuen H. K., Yosel C. J., Thermochim. Acta **33**, 281-91 (1979).

79ZHO/KOS Zhogin D. Yu., Kosarukina E. A., Kolesov V. P., Baburina I. I., Pentin Yu. A., Izmest'ev I. V., Danilov A. V., Zh. Obshch. Khim. **49**, 629-37 (1979).

80BAL/PET Balakina M. M., Petrova G. V., Vasil'ev I. A., CNIIT Eneftekhim. No50nch-80D (1980), from: Vasil'ev I. A., Petrov V. M., Termodin. Svoistva Kislorodsoderzhashchikh Organ. Soedineniy, Khimiya Leningrad (1984).

80BAL/VAS Balakina M. M., Vasil'ev I. A., data deposited in CNIIT Eneftekhim. No49nch-80D (1980), from: Vasil'ev I. A., Petrov V. M., Termodin. Svoistva Kislorodsoderzhashchikh Organ. Soedineniy, Khimiya Leningrad (1984).

80CZA Czarnota I., Bull. Acad. Pol. Sci., Ser. Sci. Chim. **28**, 651-9 (1980).

80FAL Falk B., J. Chem. Thermodyn. **12**, 967-74 (1980).

80FUC Fuchs R., Can. J. Chem. **58**, 2305-6 (1980).

80GUR/GAV Gurevich V. M., Gavrichev K. S., Gorbunov V. E., Probl. Kalorim. Khim. Thermodin., Dokl. Vses. Konf. **9th**, 231-3 (1980).

80KAL/JED Kalinowska B., Jedlinska J., Woycicki W., Stecki J., J. Chem. Thermodyn. **12**, 891-6 (1980).

80KUL/LEB Kulagina T. G., Lebedev B. V., Kiparisova E. G., Mukhina N. N., Fiz. -Khim. Osnovy Sinteza i Pererab. Polim. 104-8 (1980), from: Vasil'ev I. A., Petrov V. M., Termodin. Svoistva Kislorodsoderzhashchikh Organ. Soedineniy, Khimiya Leningrad (1984).

80LEB/DOB Lebedeva N. D., Dobychin S. L., Stepanov V. B., Katin Yu. A., Kozlova N. M, Zh. Prikl. Khim. (Leningrad) **53**, 803-6 (1980).

80LYA Lyashkov V. I., Izv. Vyssh. Ucheb. Zaved., Khim. Khim. Tekhnol. **23**, 1085-6 (1980).

80MRA/NAA Mraw S. C., Naas-O'Rourke D. F., J. Chem. Thermodyn. **12**, 691-704 (1980).

80NAK/SUG Nakamura N., Suga H., Seki S., Bull. Chem. Soc. Jpn. **53**, 2755-61 (1980).

80NEF/FIL Nefedov S. N., Filippov L. P., Izv. Vyssh. Ucheb. Zaved., Neft Gaz **23** (2), 51-4 (1980).

80ROU/ROB Roux G., Roberts D., Perron G., Desnoyers J. E., J. Solution Chem. **9**, 629-47 (1980).

80SHA/LYU Shakirov R. F., Lyubarskii M. V., deposited Doc. SPSTL 3Khp-D80, 1-19 (1980).

80SOR/TSU Sorai M., Tsuji K., Suga H., Seki S., Mol. Cryst. Liq. Cryst. **80**, 33-58 (1980).

80TUT/GAB Tutubalina V. P., Gabdrakhmanov F. G., Konyukhova T. M., Teplo- Massobmen Khim. Tekhnol. (8), 9-11 (1980).

80VAS/TRE Vasil'ev I. A., Treibsho E. I., Korkhov A. D., Petrov V. M., Orlova N. G., Balakina M. M., , Inzh. -Fiz. Zh. **39**, 1054-60 (1980).

80VOG/SCH Vogel L., Schuberth H., Chem. Tech. (Leipzig) **32**, 143-4 (1980).

80WON/WES Wong W. -K., Westrum E. F., Jr., Mol. Cryst. Liq. Cryst. **61**, 207-28 (1980).

80ZHO/KOS Zhogin D. Yu., Kosarukina E. A., Kolesov V. P., Zh. Obshch. Khim. **50**, 249-55 (1980).

81ARU Arutyunyan G. S., Izv. Akad. Nauk Azerb. SSR, Ser. Fiz. - Tekn. Mat. Nauk **2** (2), 97-9 (1981).

81ARU/BAG Arutyunyan G. S., Bagdasaryan S. S., Kerimov A. M., Izv. Akad. Nauk Azerb. SSR, Ser. Fiz. - Tekn. Mat. Nauk **2** (6), 94-7 (1981).

81ATA/ELS Atalla S. R., El-Sharkawy A. A., Gasser F. A., Int. J. Thermophys. **2**, 155-62 (1981).

81BYS Byström K., J. Chem. Thermodyn. **13**, 139-45 (1981).

81CAS/WIL Casanova C., Wilhelm E., Grolier J. -P. E., Kehiaian H. V., J. Chem. Thermodyn. **13**, 241-8 (1981).

81ENG Engineering Science Data Unit Series, Heat Capacity and Enthalpy of Liquids: Aliphatic Ketones, Vol. 4, (Item No. 81030), ESDU International Ltd., London (1981).

81EVS/LEB Evstropov A. A., Lebedev B. V., Kulagina T. G., Lebedev N. K., Termodin. Org. Soedin. 20-31 (1981).

81FIN/MES Finke H. L., Messerly J. F., Lee-Bechtold S. H., J. Chem. Thermodyn. **13**, 345-56 (1981).

81GRO/ING Grolier J. -P. E., Inglese A., Roux A. H., Wilhelm E., Ber. Bunsenges. Phys. Chem. **85**, 768-72 (1981).

81HOE Hoehne G. W. H., Polym. Bull. (Berlin) **6**, 41-6 (1981).

81KOL/KOS Kolesov V. P., Kosarukina E. A., Zhogin D. Yu., Poloznikova M. E., Pentin Yu. A., J. Chem. Thermodyn. **13**, 115-29 (1981).

81LAK/GOP Lakshmikumar S. T., Gopal E. S. R., J. Indian Inst. Sci. **63 (A)**, 277-329 (1981).

81LEB/EVS1 Lebedev B. V., Evstropov A. A., Kiparisova E. G., J. Chem. Thermodyn. **13**, 1185-204 (1981).

81LEB/EVS2 Lebedev B. V., Evstropov A. A., Kiparisova E. G., Termodin. Org. Soedin. 9-15 (1981).

81LEB/RYA Lebedeva N. D., Ryadnenko V. L., Kozlova N. M., Karpenko N. A., Stepanov V. B., Termodinam. i Kinet. Khim. Protsessov, GIPKh Leningrad 69-72 (1981).

81LEE/FIN Lee-Bechtold S. H., Finke H. L., Messerly J. F., Scott D. W., J. Chem. Thermodyn. **13**, 213-28 (1981).

81LUF Luff B. B., J. Chem. Eng. Data **26**, 70-4 (1981).

81MUS/GAN Mustafaev R. A., Ganiev D. K., Inzh. -Fiz. Zh. **40**, 898-900 (1981).

81ROG/PIT Rogers P. S. Z., Pitzer K. S., J. Phys. Chem. **85**, 2886-95 (1981).

81SHA/DZH Shakhtakhtinskii T. N., Dzhafarov O. I., Karasharli K. A., Kuliev A. M., Dokl. Akad. Nauk. Az. SSR **37** (9), 54-7 (1981).

81SMI/WOO Smith-Magowan D., Wood R. H., J. Chem. Thermodyn. **13**, 1047-73 (1981).

81SOR/SUG Sorai M., Suga H., Mol. Cryst. Liq. Cryst. **73**, 47-69 (1981).

81WEB Weber L. A., J. Chem. Thermodyn. **13**, 389-403 (1981).

81ZHO/KOS1 Zhogin D. Yu., Kosarukina E. A., Kolesov V. P., Zh. Fiz. Khim. **55**, 1955-60 (1981).

81ZHO/KOS2 Zhogin D. Yu., Kosarukina E. A., Kolesov V. P., Prokudin I. P., Mel'nichenko B. A., Zh. Fiz. Khim. **55**, 98-103 (1981).

82BEN/DAR Benson G. C., D'Arcy P. J., J. Chem. Eng. Data **27**, 439-42 (1982).

82CAM/REY Campos-Vallette M., Rey-Lafon M., Lagnier R., Chem. Phys. Lett. **89**, 189-92 (1982).

82DZH/KAR1 Dzhafarov O. I., Karasharli K. A., Kuliev A. M., Azerb. Khim. Zh. (2), 92-5 (1982).

82DZH/KAR2 Dzhafarov O. I., Karasharli K. A., Kuliev A. M., Azerb. Khim. Zh. (3), 111-3 (1982).

82EVS/LEB Evstropov A. A., Lebedev B. V., Kulagina T. G., Lebedev N. K., Vysokomol. Soedin., Ser. A **24**, 568-74 (1982).

82GAM/CAL Gammon B. E., Callanan J. E., Hossenlopp I. A., Osborn A. G., Good W. D., Proc. Symp. Thermophys. Prop. **8th**, 402-8/2 (1982).

82GOR/GRI Gorbunova N. I., Grigoriev V. A., Simonov V. M., Shipova V. A., Int. J. Thermophys. **3**, 1-15 (1982).

82GOR/SIM Gorbunova N. I., Simonov V. M., Shipova V. A., Dokl. Akad. Nauk. SSSR **266**, 850-3 (1982).

82GRO/ING Grolier J. -P. E., Inglese A., Wilhelm E., J. Chem. Thermodyn. **14**, 523-9 (1982).

82GUS/KAR Guseinov Z. A., Karasharli K. A., Azerb. Khim. Zh. (4), 107-9 (1982).

82KAR/IGA Karabaev M. K., Igamberdyev Kh. T., Abduzhaminov T. A., Turdiev N. Sh., Khabibullaev B. K., Izv. Akad. Nauk Uzb. SSR, Ser. Fiz. -Mat. Nauk (4), 46-9 (1982).

82KAR/SHV Karyakin N. V., Shvetsova K. G., Dzharimova E. S., Termodin. Org. Soedin. 33-7 (1982).

82KOS/KOL Kosarukina E. A., Kolesov V. P., Kuramshina G. M., Pentin Yu. A., Termodin. Org. Soedin. 11-7 (1982).

82KOS/ZHO Kosarukina E. A., Zhogin D. Yu., Kolesov V. P., Kuramshina G. M., Pentin Yu. A., Izmest'ev I. V., Danilov A. V., Zh. Fiz. Khim. **56**, 1892-6 (1982).

82KUL/DZH Kuliev A. M., Dzhafarov O. I., Karasharli K. A., Kostryukov V. N., Zh. Fiz. Khim. **56**, 310-3 (1982).

82KUL/LEB1 Kulagina T. G., Lebedev B. V., Fiz. -Khim. Osnovy Sinteza i Pererab. Polim. 54-8 (1982).

82KUL/LEB2 Kulagina T. G., Lebedev B. V., Termodin. Org. Soedin. 18-26 (1982).

82LAK/GOP Lakshmikumar S. T., Gopal E. S. R., Int. Rev. Phys. Chem. **2**, 197-235 (1982).

82LEB/BYK Lebedev B. V., Bykova T. A., Smirnova N. N., Kulagina T. G., Zh. Obshch. Khim. **52**, 2630-6 (1982).

82MAR/AND Martin J. F., Andon R. J. L., J. Chem. Thermodyn. **14**, 679-88 (1982).

82MIN/SAB Minadakis C., Sabbah R., Thermochim. Acta **55**, 147-59 (1982).

82NUZ Nuzzi M., Riv. Combust. **36**, 41-50 (1982).

82OGU/WAT Oguni M., Watanabe K., Matsuo T., Suga H., Seki S., Bull. Chem. Soc. Jpn. **55**, 77-84 (1982).

82ROW/SWI Rowlinson J. S., Swinton F. L., Liquid and Liquid Mixtures, Butterworths, London (1982).

82SCH/VAN1 Schaake R. C. F., Van Miltenburg J. C., De Kruif C. G., J. Chem. Thermodyn. **14**, 763-9 (1982).

82SCH/VAN2 Schaake R. C. F., Van Miltenburg J. C., De Kruif C. G., J. Chem. Thermodyn. **14**, 771-8 (1982).

82SOR/YOS Sorai M., Yoshioka H., Suga H., Mol. Cryst. Liq. Cryst. **84**, 39-54 (1982).

82TAN Tanaka R., J. Chem. Thermodyn. **14**, 259-68 (1982).

82TAN/ZHO Tan Z., Zhou L., Chen S., Yin A., Sun Y., Ye J., Diwen Wuli (Acta Phys. Temper. Hum. Sin.) **4**, 322-5 (1982).

82TSU/SOR1 Tsuji K., Sorai M., Suga H., Seki S., Mol. Cryst. Liq. Cryst. **87**, 293-304 (1982).

82TSU/SOR2 Tsuji K., Sorai M., Suga H., Seki S., Mol. Cryst. Liq. Cryst. **87**, 305-17 (1982).

82TSU/SOR3 Tsuji K., Sorai M., Suga H., Seki S., Mol. Cryst. Liq. Cryst. **90**, 85-96 (1982).

82TSU/SOR4 Tsuji K., Sorai M., Suga H., Seki S., Mol. Cryst. Liq. Cryst. **90**, 97-110 (1982).

82TUT/GAB Tutubalina V. P., Gabdrakhmanov F. G., Konyukhova T. M., Teplo- Massobmen Khim. Tekhnol. 27-8 (1982).

82VAS/PET Vasil'ev I. A., Petrov V. M., Treibsho E. I., Zh. Prikl. Khim. (Leningrad) **55**, 2116-7 (1982).

82VIL/CAS Villamanan M. A., Casanova C., Roux-Desgranges G., Grolier J. -P. E., Thermochim. Acta **52**, 279-83 (1982).

82VOR/YAK Vorob'ev A. F., Yakovlev P. N., Zh. Fiz. Khim. **56**, 1933-6 (1982).

82WAG/EVA Wagman D. D., Evans W. H., Parker V. B., Schumm R. H., Halow I., Bailey S. M., Churney K. L., Nuttall R. L., J. Phys. Chem. Ref. Data, Suppl. **11** (2), 1-394 (1982).

82WHI/WOO White D. E., Wood R. H. J. Solution Chem. **11**, 223-36 (1982).

82WIL/FAR Wilhelm E., Faradjzadeh A., Grolier J. -P. E., J. Chem. Thermodyn. **14**, 1199-200 (1982).

82WIL/ING Wilhelm E., Inglese A., Quint J. R., Grolier J. -P. E., J. Chem. Thermodyn. **14**, 303-8 (1982).

83AIC/KUM Aicart E., Kumaran M. K., Halpin C. J., Benson G. C., J. Chem. Thermodyn. **15**, 1189-97 (1983).

83ANI/VOR Anisimov M. A., Voronov V. P., Wiss. Beitr. Martin Luther Univ. Halle Wittenberg **41** (N12), 1-6 (1983).

83BAB/RAB Babinkov A. G., Rabinovich I. B., Zh. Obshch. Khim. **53**, 2744-7 (1983).

83BEN/DAR Benson G. C., D'Arcy P. J., Sugamori M. E., Thermochim. Acta **71**, 161-6 (1983).

83BER/STE Berezovskii G. A., Stenin Yu. G., Paukov I. E., Zh. Fiz. Khim. **57**, 983-5 (1983).

83BYK/LEB1 Bykova T. A., Lebedev B. V., Fiz. -Khim. Osnovy Sinteza i Pererab. Polim. 39-41 (1983).

83BYK/LEB2 Bykova T. A., Lebedev B. V., Tarasov E. N., Termodin. Org. Soedin. 48-52 (1983).

83CAM/DIA Campos-Vallette M., Diaz F. G., Semina (Londrina, Braz.) **4**, 405-9 (1983).

83CHA Chang S. S., J. Chem. Phys. **79**, 6229-36 (1983).

83DAP/DEL D'Aprano A., DeLisi R., Donato D. I., J. Solution Chem. **12**, 383-400 (1983).

83DEK/VAN De Kruif C. G., Van Miltenburg J. C., Blok J. G., J. Chem. Thermodyn. **15**, 129-36 (1983).

83DEW/DEK De Wit H. G. M., De Kruif C. G., Van Miltenburg J. C., J. Chem. Thermodyn. **15**, 891-902 (1983).

83DEW/OFF De Wit H. G. M., Offringa J. C. A., De Kruif C. G., Van Miltenburg J. C., Thermochim. Acta **65**, 43-51 (1983).

83DYA D'yakova G. N., Ph. D. Thesis, LGU Leningrad (1983), from: Vasil'ev I. A., Petrov V. M., Termodin. Svoistva Kislorodsoderzhashchikh Organ. Soedineniy, Khimiya Leningrad (1984).

83DZH/KUL Dzhafarov O. I., Kuliev A. M., deposited Doc. VINITI No. 3143-83, 1-9 (1983).

83FRA/WES Franzosini P., Westrum E. F., Jr., Plautz W. A., J. Chem. Thermodyn. **15**, 609-18 (1983).

83GEI/GUS Geidarov Kh. I., Guseinov Z. A., Kitachaev Sh. M., deposited Doc. VINITI No. 3788-83, 1-10 (1983).

83GEI/KAR Geidarov Kh. I., Karasharli K. I., Guseinov Z. A., Azerb. Khim. Zh. (5), 98-101 (1983).

83GOR/SIM Gorbunova N. I., Simonov V. M., Shipova V. A., Teplofiz. Vys. Temp. **21**, 270-5 (1983).

83GRO/FAR1 Grolier J. -P. E., Faradjzadeh A., Int. Data Ser., Sel. Data Mixtures, Ser. A (3)250 (1983).

83GRO/FAR2 Grolier J. -P. E., Faradjzadeh A., Int. Data Ser., Sel. Data Mixtures, Ser. A (3)251 (1983).

83GRO/FAR3 Grolier J. -P. E., Faradjzadeh A., Int. Data Ser., Sel. Data Mixtures, Ser. A (3)252 (1983).

83GUS/KLI Guseinov K. D., Klimova T. F., Ul'trazvuk. Termodin. Svoistva Veshchestva 97-102 (1983).

83JOA/PAL Joarder R. N., Palchaudhuri S., Rao R. V. G., Indian J. Pure Appl. Phys. **21**, 151-4 (1983).

83KAR/ABD1 Karabaev M. K., Abduzhaminov T. P., Igamberdyev Kh. T., Turdiev N. Sh., Izv. Akad. Nauk Uzb. SSR, Ser. Fiz. -Mat. Nauk (4), 72-4 (1983).

83KAR/ABD2 Karabaev M. K., Abduzhaminov T. P., Tursunov Sh. O., Igamberdyev Kh. T., Izv. Akad. Nauk Uzb. SSR, Ser. Fiz. -Mat. Nauk (6), 57-8 (1983).

83KIM/DAR Kimura F., D'Arcy P. J., Sugamori M. E., Benson G. C., Thermochim. Acta **64**, 149-54 (1983).

83KIM/TRE Kimura F., Treszczanowicz A. J., Helpin C. J., Benson G. C., J. Chem. Thermodyn. **15**, 503-10 (1983).

83KOR/DYA Korkhov A. D., D'yakova G. N., Vasil'ev I. A., Fiz. -Khim. Aspekty Razrab. Neftekhim Protsessov, M. 75-84 (1983), from: Vasil'ev I. A., Petrov V. M., Termodin. Svoistva Kislorodsoderzhashchikh Organ. Soedineniy, Khimiya Leningrad (1984).

83KUK/KOR Kukharenko V. A., Korolev V. P., Krestov G. A., Izv. Vyssh. Ucheb. Zaved., Khim. Khim. Tekhnol. **26**, 369-72 (1983).

83LEB/EVS Lebedev B. V., Evstropov A. A., J. Chem. Thermodyn. **15**, 115-28 (1983).

83MAC/STR Machado J. R. S., Streett W. B., J. Chem. Eng. Data **28**, 218-23 (1983).

83MCG McGarry J., Ind. Eng. Chem., Process Des. Develop. **2**, 313-22 (1983).

83ORO/MRA O'Rourke D. F., Mraw S. C., J. Chem. Thermodyn. **15**, 489-502 (1983).

83RAB/KHL Rabinovich I. B., Khlyustova T. B., Mochalov A. N., Kokurina N. Yu., Zh. Fiz. Khim. **57**, 2867-9 (1983).

83ROU/ROU Roux A. H., Roux-Desgranges G., Grolier J. -P. E., Viallard A., Calorim. Anal. Therm. **14**, 142-50 (1983).

83SID/SVE Siddiqi M. A., Svejda P., Kohler F., Ber. Bunsenges. Phys. Chem. **87**, 1176-81 (1983).

83SOR/TAN Sorai M., Tani K., Suga H., Mol. Cryst. Liq. Cryst. **97**, 365-86 (1983).

83TAK/KAD Takahashi Y., Kadokura H., Yokokawa H., J. Chem. Thermodyn. **15**, 65-81 (1983).

83TAN/ZHO Tan Z., Zhou L., Chen S., Yin A., Sun Y., Ye J., Wang X., Sci. Sin., Ser. B: Engl. Ed. **26**, 1014-26 (1983).

83YAW/DIC Yaws C. L., Dickens L. L., Hopper J. R., Li K. -Y., Chu T. C. T., J. Chin. Inst. Chem. Eng. **14**, 205-14 (1983).

83YOS/SOR1 Yoshikawa M., Sorai M., Suga H., Seki S., J. Phys. Chem. Solids **44**, 311-21 (1983).

83YOS/SOR2 Yoshioka H., Sorai M., Suga H., Mol. Cryst. Liq. Cryst. **95**, 11-30 (1983).

84AMI/MIN Amitin E. B., Minenkov Yu. F., Nabutovskaya O. A., Paukov I. E., Sokolova S. I, J. Chem. Thermodyn. **16**, 431-6 (1984).

84BAG/BAE Baglai A. K., Baev A. A., Belousov V. P., Beregovykh V. V., Grushenko M. M., Gurarii L. L., Konstantinov S. G., Kostyushko Yu. L. et al., Khim. -Farm. Zh. **18**, 1013-9 (1984).

84BEN/DAR Benson G. C., D'Arcy P. J., Kumaran M. K., Thermochim. Acta **75**, 353-60 (1984).

84BEN/KUM Benson G. C., Kumaran M. K., D'Arcy P. J., Thermochim. Acta **74**, 187-91 (1984).

84BER/STE Berezovskii G. A., Stenin Yu. G., Paukov I. E., Zh. Fiz. Khim. **58**, 511-3 (1984).

84BRA/PIN Bravo R., Pintos M., Baluja M. C., Paz Andrade M. I., Roux-Desgranges G., Grolier J. -P. E., J. Chem. Thermodyn. **16**, 73-9 (1984).

84DOM/EVA Domalski E. S., Evans W. H., Hearing E. D., J. Phys. Chem. Ref. Data, Suppl. **13** (1), 1-286 (1984).

84DZH/KUL Dzhafarov O. I., Kuliev A. M., Kuznetsova A. G., Zh. Fiz. Khim. **58**, 1630-3 (1984).

84FIL/LAU Filippov L. P., Laushkina L. A., Zh. Fiz. Khim. **58**, 1068-71 (1984).

84GEI/KAR Geidarov Kh. I., Karasharli K. A., Azerb. Khim. Zh. (1), 142-5 (1984).

84GRI/AND Grigor'ev B. A., Andolenko R. A., Izv. Vyssh. Ucheb. Zaved., Neft Gaz **27** (2), 60-2 (1984).

84GRO/BEN Grolier J. -P. E., Benson G. C., Can. J. Chem. **62**, 949-53 (1984).

84GRO/ING Grolier J. -P. E., Inglese A., Wilhelm E., J. Chem. Thermodyn. **16**, 67-71 (1984).

84GUS/MIR Guseinov S. O., Mirzaliev A. A., Izv. Vyssh. Ucheb. Zaved., Neft Gaz **27** (5), 41-5 (1984).

84GUS/SHU Guseinov K. D., Shubin V. V., Klimova T. F., Izv. Vyssh. Ucheb. Zaved., Neft Gaz **27** (8), 54-7 (1984).

84HAA/GAL Haar L., Gallager J. S., Kell G. S., NBS/NRC Steam Tables, Hemisphere Publishing Corp., New York, N. Y. (1984).

84HAN/HAW Handa Y. P., Hawkins R. E., Murray J. J., J. Chem. Thermodyn. **16**, 623-32 (1984).

84ING/GRO Inglese A., Grolier J. -P. E., Wilhelm E., Fluid Phase Equilib. **15**, 287-9 (1984).

84KAL/WOY Kalinowska B., Woycicki W., J. Chem. Thermodyn. **16**, 609-13 (1984).

84KUL/DZH Kuliev A. M., Dzhafarov O. I., deposited Doc. VINITI No. 1284-84, 1-8 (1984).

84KUM/BEN Kumaran M. K., Benson G. C., D'Arcy P. J., Halpin C. J., J. Chem. Thermodyn. **16**, 1181-9 (1984).

84LEB/BYK1 Lebedev B. V., Bykova T. a., Kiparisova E. G., Pankratov V. A., Korshak V. V., Laktionov V. M., Zh. Obshch. Khim. **54**, 417-24 (1984).

84LEB/BYK2 Lebedev B. V., Bykova T. A., Kiparisova E. G., Pankratov V. A., Mitina L. M., Korshak V. V., Zh. Obshch. Khim. **54**, 1352-8 (1984).

84LEB/GUT Lebedeva N. D., Gutner N. M., Katin Yu. A., Kozlova N. M., Kiseleva N. N., Makhinya E. F., Dobytchin S. L., Zh. Prikl. Khim. (Leningrad) **57**, 2297-301 (1984).

84LEB/KUL Lebedev B. V., Kulagina T. G., Svistunov V. S., Papkov V. S., Zhdanov A. A, Vysokomol. Soedin., Ser. A **26**, 2476-82 (1984).

84POD/RAC Podoski T., Rachwalska M., Mayer J., Physica B+C (Amsterdam) **125**, 215-8 (1984).

84RAB/KAR Rabinovich I. B., Karyakin N. V., Dzharimova E. S., Siling S. A., Ponomarev I. I., Vinogradova S. V., Izv. Akad. Nauk SSSR, Ser. Khim. 779-87 (1984).

84ROU/GRO Roux A. H., Grolier J. -P. E., Inglese A., Wilhelm E., Ber. Bunsenges. Phys. Chem. **88**, 986-92 (1984).

84SES Šesták J., Thermophysical Properties of Solids, Elsevier, Amsterdam (1984).

84SHE/NIS Sheiman M. S., Nistratov V. P., Kamelova G. P., Rabinovich I. B., Probl. Kalorim. Khim. Thermodin., Dokl. Vses. Konf. **10th**, 457-9/2 (1984).

84STE/OLS Stephens M., Olson J. D., Thermochim. Acta **76**, /79-85 (1984).

84STO/CHA Stolyarov N. N., Chashkin Yu. R., Kholod. Tekh. (2), 39-40 (1984).

84VAN/BOU Van Miltenburg J. C., Bouwstra J. A., J. Chem. Thermodyn. **16**, 61-5 (1984).

84VAS/PET Vasil'ev I. A., Petrov V. M., Termodin. Svoistva Kislorodsoderzhashchikh Organ. Soedineniy, Khimiya Leningrad (1984).

84ZEG/SOM1 Zegers H. C., Somsen G., Fluid Phase Equilib. **18**, 299-311 (1984).

84ZEG/SOM2 Zegers H. C., Somsen G., J. Chem. Thermodyn. **16**, 225-35 (1984).

85ALL/EVA Allen M., Evans D. F., Lumry R., J. Solution Chem. **14**, 549-60 (1985).

85BAL/BRA Baluja M. C., Bravo R., Pintos M., Paz Andrade M. I., Roux-Desgranges G., Grolier J. -P. E., Calorim. Anal. Therm. **16**, 138-44 (1985).

85BEN/DAR Benson G. C., D'Arcy P. J., Kumaran M. K., J. Chem. Thermodyn. **17**, 501-4 (1985).

85BOU/DEL Bouwstra J. A., De Leeuw V. V., Van Miltenburg J. C., J. Chem. Thermodyn. **17**, 685-95 (1985).

85CAM/THO Campbell S. W., Thodos G., J. Chem. Eng. Data **30**, 102-11 (1985).

85COS/PAT1 Costas M., Patterson D., Int. Data Ser., Sel. Data Mixtures, Ser. A (3)212 (1985).

85COS/PAT2 Costas M., Patterson D., Int. Data Ser., Sel. Data Mixtures, Ser. A (3)213 (1985).

85COS/PAT3 Costas M., Patterson D., Int. Data Ser., Sel. Data Mixtures, Ser. A (3)214 (1985).

85COS/PAT4 Costas M., Patterson D., Int. Data Ser., Sel. Data Mixtures, Ser. A (3)215 (1985).

85COS/PAT5 Costas M., Patterson D., Int. Data Ser., Sel. Data Mixtures, Ser. A (3)216 (1985).

85COS/PAT6 Costas M., Patterson D., Int. Data Ser., Sel. Data Mixtures, Ser. A (3)217 (1985).

85COS/PAT7 Costas M., Patterson D., Int. Data Ser., Sel. Data Mixtures, Ser. A (3)222 (1985).

85COS/PAT8 Costas M., Patterson D., J. Chem. Soc., Faraday Trans. 1 **81**, 635-54 (1985).

85COS/PAT9 Costas M., Patterson D., J. Chem. Soc., Faraday Trans. 1 **81**, 2381-98 (1985).

85CZA Czarnota I., High Temp. - High Pressures **17**, 543-6 (1985).

85DZH/KUL Dzhafarov O. I., Kuliev A. M., Zh. Fiz. Khim. **59**, 308-12 (1985).

85ENG Engineering Science Data Unit Series, Heat Capacity and Enthalpy of Liquids: Organic Sulphur Compounds (Thiols and Sulphides), Vol. 4 (Item No. 85024), ESDU International Ltd., London (1985).

85FIG/SZW Figuiere P., Szwarc H., Oguni M., Suga H., J. Chem. Thermodyn. **17**, 949-66 (1985).

85GUS/MIR Guseinov S. O., Mirzaliev A. A., Izv. Vyssh. Ucheb. Zaved., Neft Gaz **28** (5), 58-62 (1985).

85HAN Handa Y. P., Can. J. Chem. **63**, 68-70 (1985).

85HEP/KOO Hepler L. G., Kooner Z. S., Roux-Desgranges G., Grolier J. -P. E., J. Solution Chem. **14**, 579-94 (1985).

85KAL/WOY Kalinowska B., Woycicki W., J. Chem. Thermodyn. **17**, 829-34 (1985).

85KAR/ABD1 Karabaev M. K., Abduzhaminov T. P., Izv. Akad. Nauk Uzb. SSR, Ser. Fiz. -Mat. Nauk (2), 77-81 (1985).

85KAR/ABD2 Karabaev M. K., Abduzhaminov T. P., Saidov A. A., Igamberdyev Kh. T., Izv. Akad. Nauk Uzb. SSR, Ser. Fiz. -Mat. Nauk (4), 71-4 (1985).

85KAR/SAI Karabaev M. K., Saidov A. A., Abduzhaminov T. P., Kenisarin M. M., Izv. Akad. Nauk Uzb. SSR, Ser. Fiz. -Mat. Nauk (6), 51-4 (1985).

85LAI/GRO Lainez A., Grolier J. -P. E., Wilhelm E., Thermochim. Acta **91**, 243-8 (1985).

85LAI/ROD Lainez A., Rodrigo M., Roux A. H., Grolier J. -P. E., Wilhelm E., Calorim. Anal. Therm. **16**, 153-8 (1985).

85LAI/WIL Lainez A., Wilhelm E., Roux-Desgranges G., Grolier J. -P. E., J. Chem. Thermodyn. **17**, 1153-61 (1985).

85LEB/BYK1 Lebedev B. V., Bykova T. A., Kiparisova E. G., Chernomordik Yu. A., Kurapov A. S., Sergeev V. A., Izv. Akad. Nauk SSSR, Ser. Khim. 301-6 (1985).

85LEB/BYK2 Lebedev B. V., Bykova T. A., Kiparisova E. G., Pankratov V. A., Frenkel Ts. M., Korshak V. V., Vysokomol. Soedin., Ser. A **27**, 1411-6 (1985).

85NKI/CHA Nkinamubanzi P., Charlet G., Delmas G., Fluid Phase Equilib. **20**, 57-73 (1985).

85NOV/TSV Novoselova N. V., Tsvetkova L. Ya., Rabinovich I. B., Moseeva E. M., Faminskaya L. A., Zh. Fiz. Khim. **59**, 604-8 (1985).

85OGA Ogawa H. Murakami S., Thermochim. Acta **88**, 255-60 (1985).

85OVC/MOS Ovchinnikov E. Yu., Moseeva E. M., Rabinovich I. B., Zh. Fiz. Khim. **59**, 2836-7 (1985).

85RAB/NOV Rabinovich I. B., Novoselova N. V., Moseeva E. M., Babinkov A. G., Tsvetkova L. Ya., Zh. Fiz. Khim. **59**, 2127-30 (1985).

85RAB/SHE Rabinovich I. B., Sheiman M. S., Nistratov V. P., Kamelova G. P., Zorin A. D, Zh. Fiz. Khim. **59**, 2414-7 (1985).

85SHA/ZHU Shashkov A. G., Zhuk I. P., Karolik V. A., Inzh. -Fiz. Zh. **49**, 467-72 (1985).

85TAN Tanaka R., J. Chem. Eng. Data **30**, 267-9 (1985).

85TAN/NAK Tanaka R., Nakamichi T., Murakami S., J. Solution Chem. **14**, 795-803 (1985).

85VOG/SCH Vogel L., Schuberth H., Wiss. Z. Martin Luther Univ. Halle Wittenberg, Math. Naturwiss. Reihe **34** (3), 79-80 (1985).

85WAD Wadsö I., Thermochim. Acta **96**, 313-25 (1985).

85WIL/CHA Wilhoit R. C., Chao J., Hall K. R., J. Phys. Chem. Ref. Data **14**, 1-175 (1985).

86ACH/HAS Achorn P. J., Haseltine W. G., Miller J. K., J. Chem. Eng. Data **31**, 385-7 (1986).

86ALP/PES Al'per G. A., Peshekhodov P. B., Nikiforov M. Yu., Petrov A. N., Krestov G. A., Zh. Obshch. Khim. **56**, 1688-91 (1986).

86BEN/DAR1 Benson G. C., D'Arcy P. J., Can. J. Chem. **64**, 2139-41 (1986).

86BEN/DAR2 Benson G. C., D'Arcy P. J., J. Chem. Thermodyn. **18**, 493-8 (1986).

86BEN/DAR3 Benson G. C., D'Arcy P. J., Thermochim. Acta **102**, 75-81 (1986).

86BEN/KUM Benson G. C., Kumaran M. K., Halpin C. J., D'Arcy P. J., J. Chem. Thermodyn. **18**, 1007-14 (1986).

86BER/GUR Beregovykh V. V., Gurarii L. L., Belousov V. P., Podkovyrov A. I., Khim. -Farm. Zh. **20**, 1146-8 (1986).

86CDA/COM CDA, commercial calorimeter GSKB Leningrad, Khim. -Farm. Zh. **20**, 1146-8 (1986).

86CHI/NGU Chirico R. P., Nguyen A., Steele W. V., Strube M. M., Hossenlopp I. A., Gammon B. E., U.S.Dept. of Energy Report, NIPER, Bartlesville No. 135, 1-42 (1986).

86DEV/GUS Devyatykh G. G., Gusev A. V., Gibin A. M., Zhernenkov N. V., Kabanov A. V., Zh. Neorg. Khim. **31**, 2223-6 (1986).

86DZH/KUL Dzhafarov O. I., Kuliev A. M., Dokl. Akad. Nauk. Az. SSR **42** (11), 56-9 (1986).

86EMO/NAU Emons H. H., Naumann R., Jahn K., Flammersheim H. J., Thermochim. Acta **104**, 127-37 (1986).

86ENG1 Engineering Data Science Unit Series, Heat Capacity and Enthalpy of (Organic) Liquids: Alkanes, Vol. 4 (Item No. 86007), ESDU International Ltd., London (1986).

86ENG2 Engineering Science Data Unit Series, Heat Capacity and Enthalpy of Liquids: Alkenes and Alkadienes, Vol. 4, (Item No. 86023), ESDU International Ltd., London (1986).

86FAM/MAS Faminskaya L. A., Maslova V. A., Termodin. Org. Soedin. 6-7 (1986).

86GAT/WOO Gates J. A., Wood R. H., Cobos J. C., Casanova C., Roux A. H., Roux-Desgranges G., Grolier J. -P. E., Fluid Phase Equilib. **27**, 137-51 (1986).

86GUS/KLI Guseinov K. D., Klimova T. F., Izv. Vyssh. Ucheb. Zaved., Neft Gaz **29** (4), 60-2 (1986).

86JIM/ROM Jimenez E., Romani L., Paz Andrade M. I., Roux-Desgranges G., Grolier J. -P. E., J. Solution Chem. **15**, 879-90 (1986).

86KAL/JAC Kalinowska B., Jackowski A. W., J. Chem. Thermodyn. **18**, 715-7 (1986).

86KAU/HEI Kauschka G., Heinz D., Daebler L., Z. Phys. Chem. (Leipzig) **267**, 830-2 (1986).

86KOR/KUK Korolev V. P., Kukharenko V. A., Krestov G. A., Zh. Fiz. Khim. **60**, 1854-7 (1986).

86KRU/FED Kruchina T. I., Fedyainov N. V., Vasilev V. A., Nazimok V. F., Izv. Vyssh. Ucheb. Zaved., Khim. Khim. Tekhnol. **29** (1), 53-5 (1986).

86MAG/ELY Magee J. W., Ely J. F., Int. J. Thermophys. **7**, 1163-82 (1986).

86MER/BEN Mercier J., Benoist L., Le Parlouer P., Journees d'Etude sur les Capac. Calor. des Syst. Condenses, Cadarache, France 164-84 (1986).

86NAZ/BAS1 Naziev Ya. M., Bashirov M. M., Badalov Yu. A., Inzh. -Fiz. Zh. **51**, 789-95 (1986).

86NAZ/BAS2 Naziev Ya. M., Bashirov M. M., Badalov Yu. A., Inzh. -Fiz. Zh. **51**, 998-1004 (1986).

86NIL/WAD Nilsson S. -O., Wadsö I., J. Chem. Thermodyn. **18**, 673-81 (1986).

86NOV/RAB Novoselova N. V., Rabinovich I. B., Tsvetkova L. Ya., Moseeva E. M., Babinkov A. G, Zh. Fiz. Khim. **60**, 1627-30 (1986).

86OGA/MUR Ogawa H., Murakami S., Thermochim. Acta **109**, 145-54 (1986).

86ORT Ortega J., Rev. Latinoam. Ing. Quim. Quim. Apl. (Lat. Am. J. Chem. Eng. Appl. Chem.) **16**, 307-15 (1986).

86OVC/POD Ovchinnikov E. Yu., Podgornov V. A., Moseeva E. M., Zh. Fiz. Khim. **60**, 760-2 (1986).

86RAB/NOV Rabinovich I. B., Novoselova N. V., Moseeva E. M., Babinkov A. G., Tsvetkova L. Ya., Zh. Fiz. Khim. **60**, 545-8 (1986).

86RAB/PET Rabinovich I. B., Pet'kov V. I., Zarudaeva S. S., Ovchinnikov E. Yu., Zh. Fiz. Khim. **60**, 767-9 (1986).

86RAB/SHE Rabinovich I. B., Sheiman M. S., Kamelova G. P., Nistratov V. P., Termodin. Org. Soedin. 3-5 (1986).

86ROU/GRO Roux-Desgranges G., Grolier J. -P. E., Villamanan M. A., Casanova C., Fluid Phase Equilib. **25**, 209-30 (1986).

86ROU/ZIE Rouquerol J., Zielenkiewicz W., Thermochim. Acta **109**, 121-37 (1986).

86SMI/SRI Smith B. D., Srivastava R., Thermodynamic Data for Pure Compounds, Phys. Sci. Data 25, Part B, Elsevier, Amsterdam (1986).

86STE/CHI Steele W. V., Chirico R. D., Collier W. B., Hossenlopp I. A., Nguyen A., Strube M. M., U.S.Dept. of Energy Report, NIPER, Bartlesville, No. 188, 1-112 (1986).

86TAN/TOY Tanaka R., Toyama S., Murakami S., J. Chem. Thermodyn. **18**, 63-73 (1986).

86WIL/LAI Wilhelm E., Lainez A., Roux A. H., Grolier J. -P. E., Thermochim. Acta **105**, 101-10 (1986).

86ZEG/BOE Zegers H. C., Boegschoten R., Mels W., Somsen G., Can. J. Chem. **64**, 40-5 (1986).

86ZHA/BEN Zhang D., Benson G. C., Kumaran M. K., Lu B. C. Y., J. Chem. Thermodyn. **18**, 149-58 (1986).

86ZHO/CHE Zhou L., Chen S., Ye J., Yin A., Sun Y., Wuli Huaxue Xuebao (Acta Phys. Chim. Sinica) **2**, 229-33 (1986).

87COB/CAS Cobos J. C., Casanova C., Roux-Desgranges G., Grolier J. -P. E., J. Chem. Thermodyn. **19**, 791-6 (1987).

87CZA Czarnota I., Third CODATA Symp., Budapest C9. 4 (1987).

87DEV/GUS Devyatykh G. G., Gusev A. V., Gibin A. M., Polozkov S. A., Zh. Fiz. Khim. **61**, 1476-9 (1987).

87DZH/KUL1 Dzhafarov O. I., Kuliev A. M., Zh. Fiz. Khim. **61**, 602-5 (1987).

87DZH/KUL2 Dzhafarov O. I., Kuliev A. M., Zh. Fiz. Khim. **61**, 606-10 (1987).

87DZH/KUL3 Dzhafarov O. I., Kuliev A. M., Zh. Fiz. Khim. **61**, 1504-9 (1987).

87GIB/GRI Gibin A. M., Grishnova N. D., Gusev A. V., Moiseev A. N, Remeshkova I. I., Salganskii Yu. M., Vysokochist. Veshchestva (6), 28-32 (1987).

87GIB/GUS Gibin A. M., Gusev A. V., Remeshkova I. I., Vysokochist. Veshchestva (6), 33-5 (1987).

87GME Gmelin E., Thermochim. Acta **110**, 183-208 (1987).

87GUR/PET Gurarii L. L., Petrashkevich R. I., Baglai A. K., Kuleshov G. G., Khim. -Farm. Zh. **21**, 249-52 (1987).

87GUS/GIB Gusev A. V., Gibin A. M., Zhernenkov N. V., Kabanov A. V., Lazukin V. F., Vysokochist. Veshchestva (1), 83-7 (1987).

87KAL/KOH Kalali H., Kohler F., Svejda P., Monatsh. Chem. **118**, 1-18 (1987).

87KHO/BUG Khodzhaeva M. G., Bugakov Yu. V., Ismailov T. S., Khim. -Farm. Zh. **21**, 760-2 (1987).

87LAN/CRI Lankford J. I., Criss C. M., J. Solution Chem. **16**, 885-(1987).

87LES/LIC Lesbats C., Lichanot A., Thermochim. Acta **109**, 317-29 (1987).

87MES/TOD Messerly J. F., Todd S. S., Finke H. L., Gammon B. E., U.S. Dept. of Energy Report, NIPER, Bartlesville, No. 83, 1-37 (1987).

87MIL/FEN1 Mills P. L., Fenton R. L., J. Chem. Eng. Data **32**, 266-73 (1987).

87MIL/FEN2 Mills P. L., Fenton R. L., Schaefer G. F., J. Chem. Eng. Data **32**, 251-65 (1987).

87MIR/SHA Mirzaliev A. A., Shakhmuradov Sh. G., Guseinov S. O., Izv. Vyssh. Ucheb. Zaved., Neft Gaz **30** (4), 55-8 (1987).

87MOR/SLA Morachevskii A. G., Sladkov, I. B., Fiziko-khimicheskie Svoistva Molekulyarnykh Neorganicheskikh Soedinenii, Khimiya, Leningrad 1987.

87NAZ/BAD Naziev Ya. M., Badalov Yu. A., Bashirov M. M., Izv. Vyssh. Ucheb. Zaved., Neft Gaz **30** (5), 59-62 (1987).

87OKH/RAZ Okhotin V. S., Razumeichenko L. A., Skorodumov A. V., Teplofiz. Vys. Temp. **25**, 812-4 (1987).

87PER/COM Perkin-Elmer DSC-2C, commercial calorimeter, J. Chem. Eng. Data **32**, 251-65 (1987).

87PIE Piekarski H., Can. J. Chem. **65**, 2810-4 (1987).

87TAN Tanaka R., J. Chem. Eng. Data **32**, 176-7 (1987).

87TAN/YE Tan Z., Ye J., Yin A., Chen S., Wang W., Kexue Tongbao (Foreign. Lang. Ed.) **32**, 240-6 (1987), from: Tan Z., Sorai M., Suga H., Sci. China, Ser. B **32**, 1194-207 (1989).
87VAN/VAN Van Miltenburg J. C., Van den Berg J. K., Van Bommel M. J., J. Chem. Thermodyn. **19**, 1129-37 (1987).
87WIL/ING Wilhelm E., Inglese A., Roux A. H., Grolier J. -P. E., Fluid Phase Equilib. **34**, 49-67 (1987).
87WIL/LAI Wilhelm E., Lainez A., Berkane M., Roux-Desgranges G., Roux A. H., Grolier J. -P. E., Calorim. Anal. Therm. **18**, 95-9 (1987).
87ZAB/HYN Zábranský M., Hynek V., Finkeová-Haštabová J., Veselý F., Collect. Czech. Chem. Commun. **52**, 251-7 (1987).
88AND/PAT Andreoli-Ball L., Patterson D., Costas M., Caceres-Alonso M., J. Chem. Soc., Faraday Trans. 1 **84**, 3991-4012 (1988).
88BAG/GUR Baglai A. K., Gurarii L. L., Kuleshov G. G., J. Chem. Eng. Data **33**, 512-8 (1988).
88BAS/NIL Bastos M., Nilsson S. -O., Riberio da Silva M. D. M. C., Riberio da Silva M. A. V., Wadsö I., J. Chem. Thermodyn. **20**, 1353-9 (1988).
88BOB/KAM Bobylev V. A., Kamskaya O. I., Kossoi A. A., Koludarova E. Yu., Zh. Obshch. Khim. **58**, 2536-8 (1988).
88BOY/CRI Boyette S., Criss C. M., J. Chem. Eng. Data **33**, 426-9 (1988).
88BUN Bunyan P. F., Thermochim. Acta **130**, 335-44 (1988).
88CAC/COS Caceres-Alonso M., Costas M., Andreoli-Ball L., Patterson D., Can. J. Chem. **66**, 989-98 (1988).
88COH/TAY Cohen E. R., Taylor B. N., J. Phys. Chem. Ref. Data **17**, 1795-1803 (1988).
88COS/VAN Costas M., Van Tra H., Patterson D., Caceros-Alonso M., Tardajos G., Aicart E., J. Chem. Soc., Faraday Trans. 1 **84**, 1603-16 (1988).
88CZA1 Czarnota I., J. Chem. Thermodyn. **20**, 457-62 (1988).
88CZA2 Czarnota I., 10th IUPAC Conf. Chem. Thermodyn., Prague, A-32 (1988).
88GAM/BRO Gambino M., Bros J. P., Thermochim. Acta **127**, 223-36 (1988).
88GUS/MIR Guseinov S. O., Mirzaliev A. A., Shakhmuradov Sh. G., Izv. Vyssh. Ucheb. Zaved., Neft Gaz **31** (9), 13, 35 (1988).
88ING Inglese A., Int. Data Ser., Sel. Data Mixtures, Ser. A (3), 170-4 (1988).
88KAL/WOY Kalinowska B., Woycicki W., J. Chem. Thermodyn. **20**, 1131-5 (1988).
88KOZ/KRA Kozyro A. A., Krasulin A. P., Simirskii V. V., Markovnik V. S., Zh. Fiz. Khim. **62**, 1747-51 (1988).
88KUZ/KHA Kuznetsov M. A., Kharin V. E., Gerasimov A. A., Grigor'ev B. A., Izv. Vyssh. Ucheb. Zaved., Neft Gaz **31** (11), 49-52 (1988).
88LEB/KUL Lebedev B. V., Kulagina T. G., Smirnova N. N., J. Chem. Thermodyn. **20**, 1383-96 (1988).
88LEB/SMI Lebedev B. V., Smirnova N. N., Vasil'ev V. G., Karataev E. N., Dregich' E. N., Fishchenko I. A., Vysokochist. Veshchestva (5), 56-9 (1988).
88LEB/VAS Lebedev B. V., Vasil'ev V. G., Zh. Fiz. Khim. **62**, 3099-102 (1988).
88LIL/MAK Liley P. E., Makita T., Tanaka Y., Properties of Inorganic and Organic Fluids, CINDAS Vol. V-1, Hemisphere, New York (1988).
88LUS/RUB Lushnikov V. N., Rubtsov Yu. I., Eremenko L. T., Korolev A. M., Zh. Fiz. Khim. **62**, 1209-14 (1988).
88MAJ/GAT Majer V., Gates J. A., Inglese A., Wood R. H., J. Chem. Thermodyn. **20**, 949-68 (1988).
88MAR Marsh K. N. (editor), Recommended Reference Materials for the Realization of Physicochemical Properties, IUPAC Publication, Blackwell Sci. Publ., Oxford, 92 246-7 (1988).
88MEL/VER Mel'nikov G. A., Verveiko V. N., Otpushchennikov N. F., Zh. Fiz. Khim. **62**, 799-800 (1988).
88MES/FIN Messerly J. F., Finke H. L., Good W. D., Gammon B. E., J. Chem. Thermodyn. **20**, 485-501 (1988).
88MES/TOD Messerly J. F., Todd S. S., Finke H. L., Good W. D., Gammon B. E., J. Chem. Thermodyn. **20**, 209-24 (1988).
88NAZ/BAS Naziev Ya. M., Bashirov M. M., Teplofiz. Vys. Temp. **26**, 58-62 (1988).
88NIS/SHE Nistratov V. P., Sheiman M. S., Rabinovich I. B., Vasil'ev V. G., Karataev E. N., Feshchenko I. A., Zh. Fiz. Khim. **62**, 2219-21 (1988).
88OKA/OGA Okano T., Ogawa H., Murakami S., Can. J. Chem. **66**, 713-7 (1988).
88PER/AIC Perez-Casas S., Aicart E., Trejo L. M., Costas M., Int. Data Ser., Sel. Data Mixtures, Ser. A (2)114 (1988).
88PIE/SOM1 Piekarski H., Somsen G., J. Chem. Soc., Faraday Trans. 1 **84**, 529-37 (1988).
88PIE/SOM2 Piekarski H., Somsen G., 10th IUPAC Conf. Chem. Thermodyn., Prague I-54 (1988).
88PIN/BRA Pintos M., Bravo R., Baluja M. C., Paz Andrade M. I., Roux-Desgranges G., Grolier J. -P. E., Can. J. Chem. **66**, 1179-86 (1988).
88RAB/NIS Rabinovich I. B., Nistratov V. P., Fedoseev V. B., Sheiman M. S., Kamelova G. P., Karataev E. N., Feshchenko I. A., Zh. Fiz. Khim. **62**, 1349-52 (1988).
88ROD/MAR Rodante F., Marrosu G., Thermochim. Acta **136**, 209-18 (1988).
88SAI/TAN Saito A., Tanaka R., J. Chem. Thermodyn. **20**, 859-65 (1988).
88SHI/OGA1 Shiohama Y., Ogawa H., Murakami S., Fujihara I., J. Chem. Thermodyn. **20**, 1183-9 (1988).
88SHI/OGA2 Shiohama Y., Ogawa H., Murakami S., Fujihara I., J. Chem. Thermodyn. **20**, 1307-14 (1988).
88STE/ARC Steele W. V., Archer D. G., Chirico R. D., Collier W. B., Hossenlopp I. A., Nguyen A., Smith N. K., Gammon B. E., J. Chem. Thermodyn. **20**, 1233-64 (1988).
88SVO/VES Svoboda V., Veselý F., Hynek V., Chvátal Z., J. Chem. Thermodyn. **20**, 639-40 (1988).
88TAN/LUO Tanaka R., Luo B., Benson G. C., Lu B. C. -Y., Thermochim. Acta **127**, 15-23 (1988).
88VAN/VAN Van Bommel M. J., Van Miltenburg J. C., Schuijff A., J. Chem. Thermodyn. **20**, 397-403 (1988).
88VES/ZAB Veselý F., Zábranský M., Barcal P., Svoboda V., J. Chem. Thermodyn. **20**, 1283-5 (1988).
88WHI/WOO White D. E., Wood R. H., Biggerstaff D. R. J. Chem. Thermodyn. **20**, 159-68 (1988).
88ZHA/ZOU Zhang Z., Zou H., Yang M., Gaodeng Xuexiao Huaxue Xuebao **9**, 1085-7 (1988).
89BAR/KOO1 Barta L., Kooner Z. S., Hepler L. G., Roux-Desgranges G., Grolier J. P. -E., J. Solution Chem. **18**, 663-73 (1989).
89BAR/KOO2 Barta L., Kooner Z. S., Hepler L. G., Roux-Desgranges G., Grolier J. P. -E., Can. J. Chem. **67**, 1225-9 (1989).
89BRE/LIC Bret-Dibat P., Lichanot A., Thermochim. Acta **147**, 261-71 (1989).
89CHI/HOS Chirico R. D., Hossenlopp I. A., Nguyen A., Steele W. V., Gammon B. E., J. Chem. Thermodyn. **21**, 179-201 (1989).
89CHI/KNI Chirico R. D., Knipmeyer S. E., Nguyen A., Steele W. V., J. Chem. Thermodyn. **21**, 1307-31 (1989).
89COB/GAR Cobos J. C., Garcia I., Casanova C., Roux-Desgranges G., Grolier J. -P. E., Int. Data Ser., Sel. Data Mixtures, Ser. A (1)59-63 (1989).
89COS/YAO Costas M., Yao Z., Patterson D., J. Chem. Soc., Faraday Trans. 1 **85**, 2211-27 (1989).
89DAU/DAN Daubert T. E., Danner R. P., Physical and Thermodynamic Properties of Pure Chemicals: Data Compilation, DIPPR Database, 4 volumes, Hemisphere Publishing Corp., New York (1989).
89DOU/KHA Douheret G., Khadir A., Pal A., Thermochim. Acta **142**, 219-43 (1989).
89GUS/BAI Guseinov K. D., Bairamov N. M., Aliev A. E., Izv. Vyssh. Ucheb. Zaved., Neft Gaz **32** (1), 50-2 (1989).
89HRA/HAM Hradetzky G., Hammerl I., Kisan W., Wehner K., Bittrich H. -J., Data of Selective Solvents, VEB Deutscher Verlag der Wissenschaften, Berlin (1989).

89HUI/VAN Huinink J., Van Miltenburg K., Oonk H. A. J., Schuijff A., Groen P., J. Chem. Eng. Data **34**, 99-100 (1989).

89ING Inglese A., Int. Data Ser., Sel. Data Mixtures, Ser. A (2)145-8 (1989).

89JAD/FRA Jadot R., Fraiha M., Int. J. Refrig. **12**, 198-203 (1989).

89JIM/ROU Jimenez E., Roux-Desgranges G., Grolier J. -P. E., Wilhelm E., Thermochim. Acta **151**, 99-107 (1989).

89KHO/PUL Khodzhayeva M. G., Pulatova M. A., Nadirova D. A., Gabzalilova N. R., Ismailov T. S., Khim. -Farm. Zh. **23**, 378-80 (1989).

89KNI/ARC Knipmeyer S. E., Archer D. G., Chirico R. D., Gammon B. E., Hossenlopp I. A, Nguyen A., Smith N. K., Steele W. V., Strube M. M., Fluid Phase Equilib. **52**, 185-92 (1989).

89KON/STR Kondrat'ev S. N., Strizhov B. V., Zh. Fiz. Khim. **63**, 1181-92 (1989).

89KOZ/MAR Kozyro A. A., Marachuk L. I., Krasulin A. P., Yursha I. A., Kabo G. Ya., Zh. Prikl. Khim. (Leningrad) **62**, 595-6 (1989).

89KUL/KRE Kulikov M. V., Krestov A. G., Safonova L. P., Kolker A. M., Termodinam. Rastvorov Neelektrolitov, Ivanovo 27-32 (1989).

89LAI/ROD Lainez A., Rodrigo M. -M., Wilhelm E., Grolier J. -P. E., J. Chem. Eng. Data **34**, 332-5 (1989).

89LEB/SMI Lebedev B. V., Smirnova N. N., Zh. Fiz. Khim. **63**, 1444-50 (1989).

89MAJ/SVO Majer V., Svoboda V., Pick J., Heats of Vaporization of Fluids, Elsevier, Amsterdam (1989).

89NIS/RAB Nistratov V. P., Rabinovich I. B., Sheiman M. S., Smirnova N. N., Zelyaev I. A., Feshchenko I. A., Zh. Fiz. Khim. **63**, 1779-83 (1989).

89OHN/FUJ Ohnishi K., Fujihara I., Murakami S., Fluid Phase Equilib. **46**, 59-72 (1989).

89PET/PES Petrov A. N., Peshekhodov P. B., Al'per G. A., Termodinam. Rastvorov Neelektrolitov, Ivanovo 21-6 (1989).

89PRA/RAJ Prasad T. E. V., Rajiah A., Prasad D. H. L., Swamy V. N., Phys. Chem. Liq. **20**, 157-65 (1989).

89RAB/NIS1 Rabinovich I. B., Nistratov V. P., Sheiman M. S., Karataev E. N., Kamelova G. P., Feshchenko I. A., Zh. Fiz. Khim. **63**, 2520-4 (1989).

89RAB/NIS2 Rabinovich I. B., Nistratov V. P., Sheiman M. S., Klimov K. N., Kamelov G. P., Zorin A. D., Zh. Fiz. Khim. **63**, 522-5 (1989).

89SHE/RAB Sheiman M. S., Rabinovich I. B., Nistratov V. P., Kamelova G. P., Karataev E. N., Feshchenko I. A., Zh. Fiz. Khim. **63**, 836-8 (1989).

89STE/CHI1 Steele W. V., Chirico R. D., Hossenlopp I. A., Nguyen A., Smith N. K., Gammon B. E, J. Chem. Thermodyn. **21**, 81-107 (1989).

89STE/CHI2 Steele W. V., Chirico R. D., Hossenlopp I. A., Nguyen A., Smith N. K., Gammon B. E., J. Chem. Thermodyn. **21**, 1121-49 (1989).

89STE/CHI3 Steele W. V., Chirico R. D., Nguyen A., Hossenlopp I. A., Smith N. K., AIChE Symp. Ser. 271 **85**, 140-62 (1989).

89TAN/SOR Tan Z., Sorai M., Suga H., Sci. China, Ser. B **32**, 1194-207 (1989).

89VAS/LEB Vasil'ev V. G., Lebedev B. V., Zh. Obshch. Khim. **59**, 2451-20 (1989).

89VAS/NOV Vasilev V. A., Novikov A. N., Izv. Vyssh. Ucheb. Zaved., Khim. Khim. Tekhnol. **2** (12), 53-6 (1989).

89VES/BAR Veselý F., Barcal P., Zábranský M., Svoboda V., Collect. Czech. Chem. Commun. **54**, 602-7 (1989).

89VOG/SCH Vogel L., Schuberth H., Wiss. Z. Martin Luther Univ. Halle Wittenberg, Math. Naturwiss. Reihe **38**, 153-6 (1989).

89WIL/LAI Wilhelm E., Lainez A., Grolier J. -P. E., Fluid Phase Equilib. **49**, 233-50 (1989).

89ZHA/YAN1 Zhang Z., Yang M., Int. Conf. Chem. Thermodyn. Calorim., Beijing, China C28 (1989).

89ZHA/YAN2 Zhang Z. Y., Yang M. L., Thermochim. Acta **156**, 157-61 (1989).

90ALP/PES Al'per G. A., Peshekhodov P. B., Nikiforov M. Yu., Krestov G. A., Zh. Fiz. Khim. **64**, 652-5 (1990).

90BRI/WAD Briggner L. -E., Wadsö I., J. Chem. Thermodyn. **22**, 143-8 (1990).

90CHA/GAD Chao J., Gadella N. A. M., Gammon B. E., Marsh K. N., Rodgers A. S., Somayajulu G. R., Wilhoit R. C., J. Phys. Chem. Ref. Data **19**, 1547-1615 (1990).

90CHI/ARC Chirico R. D., Archer D. G., Hossenlopp I. A., Nguyen A., Steele W. V., Gammon B. E., J. Chem. Thermodyn. **22**, 665-82 (1990).

90CHI/GAM Chirico R. D., Gammon B. E., Knipmeyer S. E., Nguyen A., Strube M. M., Tsonopoulos C., Steele W. V., J. Chem. Thermodyn. **22**, 1075-96 (1990).

90DOM/HEA Domalski E. S., Hearing E. D., J. Phys. Chem. Ref. Data **19**, 881-1047 (1990).

90FIN/PAY Finch A., Payne J., Thermochim. Acta **164**, 55-63 (1990).

90JAL/ROB Jalabert D., Robert J. B., Roux-Buisson H., Calorim. Anal. Therm. **20-21**, 441-7 (1990).

90JIN/WUN Jin Y., Wunderlich B., J. Therm. Anal. **36**, 765-89 (1990).

90KOZ/SIM Kozyro A. A., Simirkii V. V., Kabo G. Ya., Frenkel M. L., Krasulin A. P., Sebruk V. M., Sokolov N. A., Zh. Fiz. Khim. **64**, 2360-5 (1990).

90LEB/KUL Lebedev B. V., Kulagina T. G., J. Chem. Thermodyn. **22**, 21-34 (1990).

90MES/TOD Messerly J. F., Todd S. S., Finke H. L., Lee-Bechtold S. H., Guthrie G. B., Steele W. V., Chirico R. D., J. Chem. Thermodyn. **22**, 1107-28 (1990).

90MEV/LIC Meva'a L. M., Lichanot A., Thermochim. Acta **158**, 335-45 (1990).

90RAO/RAJ Rao H. U., Rajiah A., Prasad D. H. L., Indian Chem. Eng. **32**, 50-2 (1990).

90STE/CHI Steele W. V., Chirico R. D., Nguyen A., Hossenlopp I. A., Smith N. K., AIChE Symp. Ser. 279 **86**, 138-54 (1990).

90YAM/OGA Yamanaka K., Ogawa H., Murakami S., Thermochim. Acta **169**, 193-201 (1990).

90ZAB/RUZ Zábranský M., Růžička V., Jr., Majer V., J. Phys. Chem. Ref. Data **19**, 719-62 (1990).

90ZHA/YAN Zhang Z. Y., Yang M. L., J. Chem. Thermodyn. **22**, 617-22 (1990).

91BAN/GAR Banipal T. S., Garg S. K., Ahluwalia J. C., J. Chem. Thermodyn. **23**, 923-31 (1991).

91CAR/WOO Carter R. W., Wood R. H., J. Chem. Thermodyn. **23**, 1037-56 (1991).

91CDA CDATA, A Databank of Physicochemical Properties, Prague Institute of Chemical Technology, Prague (1991).

91CHI/KNI1 Chirico R. D., Knipmeyer S. E., Nguyen A., Steele W. V., J. Chem. Thermodyn. **23**, 431-50 (1991).

91CHI/KNI2 Chirico R. D., Knipmeyer S. E., Nguyen A., Steele W. V., J. Chem. Thermodyn. **23**, 759-79 (1991).

91COB/GAR Cobos J. C., Garcia I., Casanova C., Roux A. H., Roux-Desgranges G., Grolier J. -P. E., Fluid Phase Equilib. **69**, 223-33 (1991).

91COM IUPAC Commission on Atomic Weights and Isotopic Abundances, Pure Appl. Chem. **63**, 975-90 (1991).

91COX/QUI Coxam J. -Y., Quint J. R., Grolier J. -P. E. J. Chem. Thermodyn. **23**, 1075-83 (1991).

91CZA Czarnota I., J. Chem. Thermodyn. **23**, 25-30 (1991).

91DES/PAT Deshpande D. D., Patterson D., Andreoli-Ball L., Costas M., Trejo L. M., J. Chem. Soc., Faraday Trans. **87**, 1133-9 (1991).

91DOU/PAL Douheret G., Pal A., Hoiland H., Anowi O., Davis M. I., J. Chem. Thermodyn. **23**, 569-80 (1991).

91GAR/PRE1 Gardner P. J., Preston S. R., Thermochim. Acta **180**, 281-7 (1991).

91GAR/PRE2 Gardner P. J., Preston S. R., Thermochim. Acta **185**, 219-26 (1991).

91GRO/ROU Grolier J. -P. E., Roux-Desgranges G., Berkane M., Wilhelm E., J. Chem. Thermodyn. **23**, 421-9 (1991).

91ISK/ISM Iskhakov B. G., Ismailov T. S., Khodzhaeva M. G., Igamberdyev Kh. T., Mamadalimov A. T., Inzh. -Fiz. Zh. **61**, 427-30 (1991).

91JIN/WUN Jin Y., Wunderlich B., J. Phys. Chem. **95**, 9000-7 (1991).

91KAL/KOH Kalali H., Kohler F., Svejda P., J. Chem. Eng. Data **36**, 326-9 (1991).

91KOZ/SHE Kozyro A. A., Sheiman M. S., Kabo G. Ya., Yursha I. A., Kruk V. S., Zh. Prikl. Khim. (Leningrad) **64**, 1704-9 (1991).

91LAB/WES Labban A. K., Westrum E. F., Jr., Rodriguez Caheda J. A., Can. J. Chem. **69**, 1796-803 (1991).

91LEB/KUL Lebedev B. V., Kulagina T. G., J. Chem. Thermodyn. **23**, 1097-106 (1991).

91LIC Lichanot A., Thermochim. Acta **177**, 265-72 (1991).

91MAG Magee J. W., J. Res. Natl. Bur. Stand. (U.S.) **96**, 725-40 (1991).

91MAR Marsh K. N. et al., Selected Values of Thermodynamic Properties of Organic Compounds, TRC Data Project (extant data sheets), Thermodynamics Research Center, Texas A&M University, College Station TX (1991).

91NAK/SOE Nakai H., Soejima H., Tamura K., Ogawa H., Murakami S., Toshiyasu Y., Thermochim. Acta **183**, 15-27 (1991).

91OGA/MIT Ogawa H., Mitsuo H., Inoue H., Tamura K., Murakami S., Netsu Sokutei **18**, 150-6 (1991).

91RAB/SHE Rabinovich I. B., Sheiman M. S., Kamelova G. P., Nistratov V. P., Karataev E. N., Zh. Fiz. Khim. **65**, 2071-6 (1991).

91ROZ/SAF Rozhnov A. M., Safronov V. V., Verevkin S. P., Sharonov K. G., Alenin V. I., J. Chem. Thermodyn. **23**, 629-35 (1991).

91RUZ/ZAB Růžička V., Jr., Zábranský M., Majer V., J. Phys. Chem. Ref. Data **20**, 405-44 (1991)

91SOE/NAK Soejima H., Nakai H., Tamura K., Ogawa H., Murakami S., Thermochim. Acta **177**, 305-15 (1991).

91STE/CHI Steele W. V., Chirico R. D., Knipmeyer S. E., Nguyen A., J. Chem. Thermodyn. **23**, 957-77 (1991).

91SVO/ZAB1 Svoboda V., Zábranský M., Collect. Czech. Chem. Commun. **56**, 2786-90 (1991).

91SVO/ZAB2 Svoboda V., Zábranský M., Bárta M., J. Chem. Thermodyn. **23**, 711-2 (1991).

91TAN/ADA Tanaka R., Adachi M., J. Chem. Thermodyn. **23**, 1023-7 (1991).

91TRE/COS Trejo L. M., Costas M., Patterson D., Int. Data Ser., Sel. Data Mixtures, Ser. A (1)30-50 (1991).

91VAN/VER Van Miltenburg J. C., Verdonk M. L., J. Chem. Thermodyn. **23**, 273-9 (1991).

91VAS/BYK Vasil'ev V. G., Bykova T. A., Lebedev B. V., Zh. Fiz. Khim. **65**, 51-4 (1991).

91WIL/JIM Wilhelm E., Jimenez E., Roux-Desgranges G., Grolier J. -P. E., J. Solution Chem. **20**, 17-28 (1991).

92BAO/CAC Baonza V. G., Caceres M., Nunez J., J. Phys. Chem. **96**, 1932-8 (1992).

92GOL/WEI Goldberg R. N., Weir R. D., Pure Appl. Chem. **64**, 1545-62 (1992).

92KAB/KOZ Kabo G. J., Kozyro A. A., Krouk V. S., Sevruk V. M., Yursha I. A., Simirsky V. V., Gogolinsky V. I., J. Chem. Thermodyn. **24**, 1-13 (1992).

92KAL/KOH Kalali H., Kohler F., Svejda P., J. Chem. Eng. Data **37**, 133-6 (1992).

92KOZ/KAB Kozyro A. A., Kabo G. J., Kronk V. S., Sheiman M. S., Yurska I. A., Simirsky V. V., Krasulin A. P., Sevruk V. M., Gogolinsky V. I., J. Chem. Thermodyn. **24**, 883-95 (1992).

92KUL/LEB Kulagina T. G., Lebedev B. V., Karataev E. N., Amosov Yu. I., Zh. Fiz. Khim. **66**, 1694-7 (1992).

92LAI/ROD Lainez A., Rodrigo M. M., Wilhelm E., Grolier J. -P. E., J. Solution Chem. **21**, 49-65 (1992).

92LEB/KUL1 Lebedev B. V., Kulagina T. G., J. Chem. Thermodyn. **24**, 751-60 (1992).

92LEB/KUL2 Lebedev B. V., Kulagina T. G., Sheiman M. S., Nistratov V. P., J. Chem. Thermodyn. **24**, 673-83 (1992).

92MIY/TAM Miyanaga S., Tamura K., Murakami S., J. Chem. Thermodyn. **24**, 291-6 (1992).

92PIE/SOM Piekarski H., Somsen G., J. Solution Chem. **21**, 557-67 (1992).

92STE/CHI1 Steele W. V., Chirico R. D., Knipmeyer S. E., Nguyen A., J. Chem. Thermodyn. **24**, 245-71 (1992).

92STE/CHI2 Steele W. V., Chirico R. D., Knipmeyer S. E., Nguyen A., J. Chem. Thermodyn. **24**, 499-529 (1992).

92VAN/COR Van der Vis M. G. M., Cordfunke E. H. P., Konings R. J. M., Van den Berg G. J. K., Van Miltenburg J. C., J. Chem. Thermodyn. **24**, 1103-8 (1992).

93DUR/AOU Durupt N., Aoulmi A., Rogalski M., private communication (1993).

93GAR/BAN Garg S. K., Banipal T. S., Ahluwalia J. C., J. Chem. Thermodyn. **25**, 57-62 (1993).

93HAL Hallen D., J. Chem. Thermodyn. **25**, 519-24 (1993).

93PAG/HUO Page M., Huot J. -Y., Jolicoeur C., J. Chem. Thermodyn. **25**, 139-54 (1993).

93ZAB/BUR Zábranský M., Bureš M., Růžička V., Jr., Thermochim. Acta **216**, 25-45 (1993).

94SVO/CEJ Svoboda V., Čejka D., Zábranský M., J. Chem. Thermodyn. **26**, 147-50 (1994).

94ZAB/RUZ Zábranský M., Růžička V., Jr., J. Phys. Chem. Ref. Data **23**, 55-61 (1994)

Formula Index of Compounds

The formula index of compounds lists all compounds for which data were compiled and evaluated. Compounds are sorted in the order of empirical formulas. The empirical formula is listed on the first line. The second line introduced by the sign • gives the name of a compound and the Chemical Abstracts Service Registry Number (CAS RN) enclosed in brackets. The next line(s), if present, gives synonyms, common names or commercial names. The last line for each compound lists the group-member number. The system of dividing compounds into groups is described in detail in the Section 6.1. The first two-digit number indicates the group, and the second three-digit number denotes the serial number of a compound inside the group. Tables in the compilation are coded with three numbers separated by full stops. The first two numbers are identical with the compound group and member number.

If there are more compounds of the same empirical formula, they are listed under a common formula and each different compound is introduced by the sign •.

For geometric isomers which are common among alkenes the symbols (Z) and (E) are used to identify similar groups either on the same side of the double bond or across from the double bond. For simple systems (Z) and (E) are equivalent to *cis* and *trans*.

$AgNO_3$
• Silver(1+) salt nitric acid [7761-88-8]
Silver nitrate
Group-member nr.: 02-001

$AlBr_3$
• Aluminium bromide [7727-15-3]
Group-member nr.: 02-002

$AlCl_3$
• Aluminium chloride [7446-70-0]
Group-member nr.: 02-003

AlI_3
• Aluminium iodide [7784-23-8]
Group-member nr.: 02-004

Ar
• Argon [7440-37-1]
Group-member nr.: 01-001

$AsCl_3$
• Arsenous trichloride [7784-34-1]
Arsenic trichloride
Arsenic chloride
Group-member nr.: 02-005

AsF_3
• Arsenous trifluoride [7784-35-2]
Arsenic trifluoride
Group-member nr.: 02-006

AsH_3
• Arsine [7784-42-1]
Group-member nr.: 02-007

BF_3
• Trifluoroborane [7637-07-2]
Boron fluoride (BF_3)
Boron trifluoride
Group-member nr.: 02-008

B_2Cl_4
• Boron chloride (B_2Cl_4) [13701-67-2]
Diboron tetrachloride
Group-member nr.: 02-009

B_2H_6
• Diborane(6) [19287-45-7]
Group-member nr.: 02-010

$B_3H_6N_3$
• Borazine [6569-51-3]
Borazole
Group-member nr.: 02-011

B_5H_9
• Pentaborane(9) [19624-22-7]
Group-member nr.: 02-012

$B_{10}H_{14}$
• Decaborane(14) [17702-41-9]
Group-member nr.: 02-013

Bi
• Bismuth [7440-69-9]
Group-member nr.: 01-002

BrD
• Hydrobromic acid-*d* [13536-59-9]
Deuterium bromide
Group-member nr.: 02-014

BrF_3
• Bromine fluoride (BrF_3) [7787-71-5]
Bromine trifluoride
Group-member nr.: 02-015

BrH
• Hydrobromic acid [10035-10-6]
Hydrogen bromide
Group-member nr.: 02-016

BrIn
- Indium bromide (InBr) [14280-53-6]
  Group-member nr.: 02-017

Br_2
- Bromine [7726-95-6]
  Group-member nr.: 01-003

Br_2Sn
- Tin bromide ($SnBr_2$) [10031-24-0]
  Group-member nr.: 02-018

Br_4Sn
- Tetrabromostannane [7789-67-5]
  Tin(IV) bromide
  Tin tetrabromide
  Group-member nr.: 02-019

$CBrCl_3$
- Bromotrichloromethane [75-62-7]
  Group-member nr.: 25-001

$CBrF_3$
- Bromotrifluoromethane [75-63-8]
  R13B1
  Group-member nr.: 25-002

CBr_4
- Tetrabromomethane [558-13-4]
  Carbon tetrabromide
  Perbromomethane
  Group-member nr.: 23-001

$CClF_3$
- Chlorotrifluoromethane [75-72-9]
  R13
  Group-member nr.: 25-003

$CClN_3O_6$
- Chlorotrinitromethane [1943-16-4]
  Group-member nr.: 64-001

CCl_2F_2
- Dichlorodifluoromethane [75-71-8]
  R12
  Group-member nr.: 25-004

CCl_2O
- Carbonic dichloride [75-44-5]
  Phosgene
  Group-member nr.: 02-020

CCl_3F
- Trichlorofluoromethane [75-69-4]
  R11
  Group-member nr.: 25-005

CCl_4
- Tetrachloromethane [56-23-5]
  Carbon tetrachloride
  Perchloromethane
  Group-member nr.: 22-001

CD_4
- Methane-d_4 [558-20-3]
  Deuteromethane
  Group-member nr.: 11-001

CF_2O
- Carbonic difluoride [353-50-4]
  Carbonyl fluoride
  Group-member nr.: 02-021

CF_4
- Tetrafluoromethane [75-73-0]
  Carbon tetrafluoride
  Perfluoromethane
  R14
  Group-member nr.: 21-001

$CHBr_3$
- Tribromomethane [75-25-2]
  Bromoform
  Methenyl tribromide
  Group-member nr.: 23-002

$CHClF_2$
- Chlorodifluoromethane [75-45-6]
  R22
  Group-member nr.: 25-006

$CHCl_2F$
- Dichlorofluoromethane [75-43-4]
  R21
  Group-member nr.: 25-007

$CHCl_3$
- Trichloromethane [67-66-3]
  Chloroform
  Methylidyne trichloride
  Group-member nr.: 22-002

CHF_3
- Trifluoromethane [75-46-7]
  Fluoroform
  R23
  Group-member nr.: 21-002

CHF_3S
- Trifluoromethanethiol [1493-15-8]
  Group-member nr.: 64-002

CHN
- Hydrocyanic acid [74-90-8]
  Hydrogen cyanide
  Group-member nr.: 02-022

$CHNaO_2$
- Sodium salt formic acid [141-53-7]
  Sodium formate
  Natrium formate
  Group-member nr.: 74-001

CH_2Br_2
- Dibromomethane [74-95-3]
  Methylene dibromide
  Group-member nr.: 23-003

CH_2Cl_2
- Dichloromethane [75-09-2]
  Methylene dichloride
  Group-member nr.: 22-003

CH_2I_2
- Diiodomethane [75-11-6]
  Methylene diiodide
  Group-member nr.: 24-001

CH_2N_2
- Cyanamide [420-04-2]
  Group-member nr.: 34-001

CH_2O_2
- Formic acid [64-18-6]
  Methanoic acid
  Hydrogencarboxylic acid
  Group-member nr.: 44-001

CH_2S_3
- Trithiocarbonic acid [594-08-1]
  Carbon sulfide dihydrosulfide
  Group-member nr.: 02-023

CH_3Br
- Bromomethane [74-83-9]
  Methyl bromide
  Group-member nr.: 23-004

CH_3Cl
- Chloromethane [74-87-3]
  Methyl chloride
  Group-member nr.: 22-004

CH_3ClFOP
- Methylphosphonic chloride fluoride [753-71-9]
  Methylphosphonic chlorofluoride
  Group-member nr.: 72-001

CH_3Cl_2OP
- Methylphosphonic dichloride [676-97-1]
  Group-member nr.: 72-002

CH_3D
- Methane-*d* [676-49-3]
  Monodeuteromethane
  Group-member nr.: 11-002

CH_3DO
- Methanol-*d* [1455-13-6]
  Monodeuteromethanol
  Group-member nr.: 42-001

CH_3F_2OP
- Methylphosphonic difluoride [676-99-3]
  Group-member nr.: 72-003

CH_3I
- Iodomethane [74-88-4]
  Methyl iodide
  Group-member nr.: 24-002

CH_3NO
- Formamide [75-12-7]
  Methanamide
  Group-member nr.: 62-001

CH_3NO_2
- Nitromethane [75-52-5]
  Group-member nr.: 62-002

CH_3NO_3
- Methyl ester nitric acid [598-58-3]
  Methyl nitrate
  Group-member nr.: 62-003

CH_4
- Methane [74-82-8]
  Group-member nr.: 11-003

CH_4N_2O
- Urea [57-13-6]
  Carbamide
  Group-member nr.: 62-004

CH_4O
- Methanol [67-56-1]
  Methyl alcohol
  Carbinol
  Wood alcohol
  Group-member nr.: 42-002

CH_4S
- Methanethiol [74-93-1]
  Methyl mercaptan
  Group-member nr.: 52-001

CH_5N
- Methanamine [74-89-5]
  Methylamine
  Aminomethane
  Group-member nr.: 31-001

CH_6N_2
- Methylhydrazine [60-34-4]
  Monomethylhydrazine
  Group-member nr.: 34-002

CO
- Carbon monoxide [630-08-0]
  Group-member nr.: 02-024

COS
- Carbon oxide sulfide [463-58-1]
  Carbonyl sulfide
  Group-member nr.: 02-025

CO_2
- Carbon dioxide [124-38-9]
  Group-member nr.: 02-026

CS_2
- Carbon disulfide [75-15-0]
  Group-member nr.: 02-027

CSe_2
- Carbon selenide (CSe_2) [506-80-9]
  Group-member nr.: 02-028

$C_2Br_2D_4$
- 1,2-Dibromoethane-1,1,2,2-d_4 [22581-63-1]
  1,2-Dibromotetradeuteroethane
  Group-member nr.: 23-005

$C_2Br_2F_4$
- 1,2-Dibromo-1,1,2,2-tetrafluoroethane [124-73-2]
  R114B2
  Group-member nr.: 25-008

$C_2Br_3D_3$
- 1,1,2-Tribromoethane-1,2,2-d_3 [unknown]
  Group-member nr.: 23-006

C_2ClF_3
- Chlorotrifluoroethene [79-38-9]
  Chlorotrifluoroethylene
  Group-member nr.: 25-009

C_2ClF_5
- Chloropentafluoroethane [76-15-3]
  R115
  Group-member nr.: 25-010

$C_2Cl_2F_4$
- 1,2-Dichloro-1,1,2,2-tetrafluoroethane [76-14-2]
  sym-Dichlorotetrafluoroethane
  Fluorocarbon-114
  R114
  Group-member nr.: 25-012
- Dichlorotetrafluoroethane (unspecified isomer) [1320-37-2]
  Group-member nr.: 25-011

$C_2Cl_3F_3$
- 1,1,1-Trichloro-2,2,2-trifluoroethane [354-58-5]
  Group-member nr.: 25-013
- 1,1,2-Trichloro-1,2,2-trifluoroethane [76-13-1]
  R113
  Group-member nr.: 25-014

C_2Cl_4
- Tetrachloroethene [127-18-4]
  Perchloroethylene
  Group-member nr.: 22-005

$C_2Cl_4F_2$
- 1,1,2,2-Tetrachloro-1,2-difluoroethane [76-12-0]
  sym-Tetrachlorodifluoroethane
  R112
  Group-member nr.: 25-015

$C_2F_2O_2$
- Ethanedioyl difluoride [359-40-0]
  Oxalyl difluoride
  Group-member nr.: 61-001

C_2F_3N
- Trifluoroacetonitrile [353-85-5]
  Trifluoroethanenitrile
  Trifluoromethyl cyanide
  Group-member nr.: 64-003

C_2F_4
- Tetrafluoroethene [116-14-3]
  Perfluoroethylene
  Group-member nr.: 21-003

C_2F_4O
- Trifluoroacetyl fluoride [354-34-7]
  Group-member nr.: 61-002

C_2F_6
- Hexafluoroethane [76-16-4]
  Perfluoroethane
  Group-member nr.: 21-004

$C_2HBrClF_3$
- 1-Bromo-2-chloro-1,1,2-trifluoroethane [354-06-3]
  Group-member nr.: 25-016
- 2-Bromo-2-chloro-1,1,1-trifluoroethane [151-67-7]
  Halothane
  Group-member nr.: 25-017

$C_2HBr_2D_3$
- 1,2-Dibromoethane-1,1,2-d_3 [117164-17-7]
  Group-member nr.: 23-007

$C_2HBr_3D_2$
- 1,1,2-Tribromoethane-2,2-d_2 [unknown]
  Group-member nr.: 23-008

C_2HCl_3
- Trichloroethene [79-01-6]
  Trichloroethylene
  Group-member nr.: 22-006

$C_2HCl_3F_2$
- 1,1,2-Trichloro-1,2-difluoroethane [354-15-4]
  R122
  Group-member nr.: 25-018

C_2HCl_3O
- Trichloroacetaldehyde [75-87-6]
  Trichloroethanal
  Chloral
  Group-member nr.: 61-003

$C_2HCl_3O_2$
- Trichloroacetic acid [76-03-9]
  Trichloroethanoic acid
  Group-member nr.: 61-004

C_2HCl_5
- Pentachloroethane [76-01-7]
  Pentalin
  Group-member nr.: 22-007

$C_2H_2Br_2D_2$
- 1,2-Dibromoethane-1,1-d_2 [unknown]
  Group-member nr.: 23-009
- 1,2-Dibromoethane-1,2-d_2 [unknown]
  Group-member nr.: 23-010

$C_2H_2Br_3D$
- 1,1,2-Tribromoethane-1-*d* [unknown]
  Group-member nr.: 23-011
- 1,1,2-Tribromoethane-2-*d* [unknown]
  Group-member nr.: 23-012

$C_2H_2Br_4$
- 1,1,2,2-Tetrabromoethane [79-27-6]
  sym-Tetrabromoethane
  Acetylene tetrabromide
  Group-member nr.: 23-013

$C_2H_2Cl_2$
- 1,1-Dichloroethene [75-35-4]
  1,1-Dichloroethylene
  Vinylidene chloride
  Group-member nr.: 22-008
- 1,2-Dichloroethene [540-59-0]
  1,2-Dichloroethylene (unspecified stereoisomer)
  Group-member nr.: 22-009
- (*E*)-1,2-Dichloroethene [156-60-5]
  trans-1,2-Dichloroethene
  trans-1,2-Dichloroethylene
  Group-member nr.: 22-010
- (*Z*)-1,2-Dichloroethene [156-59-2]
  cis-1,2-Dichloroethene
  cis-1,2-Dichloroethylene
  Group-member nr.: 22-011

$C_2H_2Cl_2F_2$
- 1,1-Dichloro-1,2-difluoroethane [1842-05-3]
  R132
  Group-member nr.: 25-019

$C_2H_2Cl_2O_2$
- Dichloroacetic acid [79-43-6]
  Dichloroethanoic acid
  Group-member nr.: 61-005

$C_2H_2Cl_4$
- 1,1,1,2-Tetrachloroethane [630-20-6]
  Group-member nr.: 22-012
- 1,1,2,2-Tetrachloroethane [79-34-5]
  sym-Tetrachloroethane
  Acetylene tetrachloride
  Group-member nr.: 22-013

C_2H_3Br
- Bromoethene [593-60-2]
  Bromoethylene
  Vinyl bromide
  Group-member nr.: 23-014

$C_2H_3Br_2D$
- 1,2-Dibromoethane-*d* [unknown]
  Group-member nr.: 23-015

$C_2H_3Br_3$
- 1,1,2-Tribromoethane [78-74-0]
  Group-member nr.: 23-016

C_2H_3Cl
- Chloroethene [75-01-4]
  Vinyl chloride
  Group-member nr.: 22-014

$C_2H_3ClF_2$
- 1-Chloro-1,1-difluoroethane [75-68-3]
  R142
  Group-member nr.: 25-020

C_2H_3ClO
- Acetyl chloride [75-36-5]
  Ethanoyl chloride
  Group-member nr.: 61-006

$C_2H_3ClO_2$
- Chloroacetic acid [79-11-8]
  Chloroethanoic acid
  Group-member nr.: 61-007

$C_2H_3Cl_3$
- 1,1,1-Trichloroethane [71-55-6]
  Methylchloroform
  Group-member nr.: 22-015
- 1,1,2-Trichloroethane [79-00-5]
  β-Trichloroethane
  Group-member nr.: 22-016

$C_2H_3Cl_3Si$
- Trichloroethenylsilane [75-94-5]
  Trichlorovinylsilane
  Group-member nr.: 71-001

$C_2H_3F_3$
- 1,1,1-Trifluoroethane [420-46-2]
  Group-member nr.: 21-005

$C_2H_3F_3O$
- 2,2,2-Trifluoroethanol [75-89-8]
  Group-member nr.: 61-008

C_2H_3N
- Acetonitrile [75-05-8]
  Ethanenitrile
  Group-member nr.: 32-002

C_2H_4
- Ethene [74-85-1]
  Ethylene
  Group-member nr.: 13-001

C_2H_4BrCl
- 1-Bromo-2-chloroethane [107-04-0]
  Ethylene chlorobromide
  Group-member nr.: 25-021

$C_2H_4Br_2$
- 1,2-Dibromoethane [106-93-4]
  Ethylene dibromide
  Group-member nr.: 23-017

$C_2H_4Cl_2$
- 1,1-Dichloroethane [75-34-3]
  Ethylidene dichloride
  Group-member nr.: 22-017
- 1,2-Dichloroethane [107-06-2]
  Ethylene dichloride
  Group-member nr.: 22-018

$C_2H_4D_2O_2$
- 1,2-Ethanediol-d_2 [2219-52-5]
  Dideutero-1,2-ethanediol
  Dideuteroethylene glycol
  Group-member nr.: 42-003

C_2H_4O
- Acetaldehyde [75-07-0]
  Ethanal
  Group-member nr.: 43-001
- Oxirane [75-21-8]
  Epoxyethane
  Ethylene oxide
  Group-member nr.: 46-001

$C_2H_4O_2$
- Acetic acid [64-19-7]
  Ethanoic acid
  Methanecarboxylic acid
  Vinegar acid
  Group-member nr.: 44-002
- Methyl formate [107-31-3]
  Methyl methanoate
  Group-member nr.: 45-001

C_2H_5Br
- Bromoethane [74-96-4]
  Ethyl bromide
  Group-member nr.: 23-018

C_2H_5Cl
- Chloroethane [75-00-3]
  Ethyl chloride
  Group-member nr.: 22-019

$C_2H_5Cl_3Si$
- Trichloroethylsilane [115-21-9]
  Group-member nr.: 71-002

C_2H_5DO
- Ethanol-*d* [925-93-9]
  Monodeuteroethanol
  Group-member nr.: 42-004

C_2H_5I
- Iodoethane [75-03-6]
  Ethyl iodide
  Group-member nr.: 24-003

C_2H_5N
- Aziridine [151-56-4]
  Ethylene imine
  Group-member nr.: 33-001

C_2H_5NO
- Acetamide [60-35-5]
  Ethanamide
  Group-member nr.: 62-005
- *N*-Methylformamide [123-39-7]
  Group-member nr.: 62-006

$C_2H_5NO_2$
- Nitroethane [79-24-3]
  Group-member nr.: 62-007

$C_2H_5NO_3$
- Ethyl nitrate [625-58-1]
  Ethyl ester nitric acid
  Group-member nr.: 62-008

C_2H_6
- Ethane [74-84-0]
  Group-member nr.: 11-004

C_2H_6Cd
- Dimethylcadmium [506-82-1]
  Group-member nr.: 73-001

$C_2H_6Cl_2Si$
- Dichloroethylsilane [1789-58-8]
  Group-member nr.: 71-003

C_2H_6O
- Ethanol [64-17-5]
  Ethyl alcohol
  Grain alcohol
  Group-member nr.: 42-005
- Oxydimethane [115-10-6]
  Dimethyl ether
  Methyl ether
  Methyl oxide
  Wood ether
  Group-member nr.: 41-001

C_2H_6OS
- Sulfinyldimethane [67-68-5]
  Dimethyl sulfoxide
  DMSO
  Group-member nr.: 63-001

$C_2H_6O_2$
- 1,2-Ethanediol [107-21-1]
  Ethylene glycol
  Glycol
  Group-member nr.: 42-006

$C_2H_6O_2S$
- Sulfonyldimethane [67-71-0]
  Dimethyl sulfone
  Group-member nr.: 63-002

C_2H_6S
- Ethanethiol [75-08-1]
  Ethyl mercaptan
  Group-member nr.: 52-002
- Thiodimethane [75-18-3]
  Dimethyl sulfide
  Group-member nr.: 51-001

$C_2H_6S_2$
- 2,3-Dithiabutane [624-92-0]
  Dimethyl disulfide
  Methyl disulfide
  Group-member nr.: 51-002

C_2H_6Se
- Selenodimethane [593-79-3]
  Dimethyl selenide
  Methyl selenide
  Group-member nr.: 73-002

$C_2H_6Se_2$
- Dimethyl diselenide [7101-31-7]
  Methyl diselenide
  Group-member nr.: 73-003

C_2H_6Zn
- Dimethylzinc [544-97-8]
  Group-member nr.: 73-004

C_2H_7N
- Ethanamine [75-04-7]
  Ethylamine
  Group-member nr.: 31-002
- *N*-Methylmethanamine [124-40-3]
  Dimethylamine
  Group-member nr.: 31-003

$C_2H_8N_2$
- 1,1-Dimethylhydrazine [57-14-7]
  unsym-Dimethylhydrazine
  Group-member nr.: 34-003
- 1,2-Dimethylhydrazine [540-73-8]
  sym-Dimethylhydrazine
  Group-member nr.: 34-004
- 1,2-Ethanediamine [107-15-3]
  Ethylenediamine
  1,2-Diaminoethane
  Group-member nr.: 31-004

$C_2H_8N_2O_3$
- Nitrate ethylamine [22113-86-6]
  Ethylammonium nitrate
  Group-member nr.: 62-009

$C_2H_{11}B_2N$
- (Dimethylamino)diborane(6) [22580-01-4]
  Group-member nr.: 72-004

C_2N_2
- Ethanedinitrile [460-19-5]
  Cyanogen
  Group-member nr.: 32-001

C_3F_6O
- 1,1,1,3,3,3-Hexafluoro-2-propanone [684-16-2]
  Hexafluoroacetone
  Perfluoroacetone
  Group-member nr.: 61-009

C_3F_8
- Octafluoropropane [76-19-7]
  Perfluoropropane
  Group-member nr.: 21-006

$C_3H_2ClF_5$
- 1-Chloro-1,1,3,3,3-pentafluoropropane [460-92-4]
  Group-member nr.: 25-022

$C_3H_2Cl_3F_3$
- 1,1,1-Trichloro-3,3,3-trifluoropropane [7125-84-0]
  Group-member nr.: 25-023

$C_3H_2N_2$
- Propanedinitrile [109-77-3]
  Malononitrile
  Dicyanomethane
  Group-member nr.: 32-003

$C_3H_3Cl_2F_3$
- 3,3-Dichloro-1,1,1-trifluoropropane [460-69-5]
  Group-member nr.: 25-024

$C_3H_3Cl_3O_2$
- Methyl trichloroacetate [598-99-2]
  Methyl ester trichloroacetic acid
  Group-member nr.: 61-010

C_3H_3N
- 2-Propenenitrile [107-13-1]
  Acrylonitrile
  Vinyl cyanide
  Group-member nr.: 32-004

C_3H_3NO
- Isoxazole [288-14-2]
  Group-member nr.: 62-010
- Oxazole [288-42-6]
  Group-member nr.: 62-011

C_3H_3NS
- Thiazole [288-47-1]
  Group-member nr.: 64-004

$C_3H_3N_3$
- 1,3,5-Triazine [290-87-9]
  sym-Triazine
  Group-member nr.: 33-002

$C_3H_4ClF_3$
- 3-Chloro-1,1,1-trifluoropropane [460-35-5]
  Group-member nr.: 25-025

$C_3H_4Cl_2O_2$
- Methyl dichloroacetate [116-54-1]
  Methyl ester dichloroacetic acid
  Group-member nr.: 61-011

$C_3H_4Cl_3NSi$
- 3-Trichlorosilylpropanenitrile [1071-22-3]
  Group-member nr.: 71-004

$C_3H_4Cl_4$
- 1,1,1,3-Tetrachloropropane [1070-78-6]
  Group-member nr.: 22-020

$C_3H_4N_2$
- 1*H*-Imidazole [288-32-4]
  1,3-Diazole
  Glyoxaline
  Group-member nr.: 33-003
- 1*H*-Pyrazole [288-13-1]
  1,2-Diazole
  Group-member nr.: 33-004

$C_3H_4N_2O$
- 2-Cyanoacetamide [107-91-5]
  Group-member nr.: 62-012

$C_3H_4O_2$
- 2-Oxetanone [57-57-8]
  β-Propiolactone
  Hydracrylolactone
  USAN
  Group-member nr.: 47-001
- 2-Propenoic acid [79-10-7]
  Acrylic acid
  Acroleic acid
  Group-member nr.: 44-003

$C_3H_4O_3$
- 1,3-Dioxolan-2-one [96-49-1]
  Cyclic ethylene ester carbonic acid
  Ethylene carbonate
  Group-member nr.: 47-002

C_3H_5Br
- 3-Bromo-1-propene [106-95-6]
  Allyl bromide
  Group-member nr.: 23-019

$C_3H_5Br_3$
- 1,2,3-Tribromopropane [96-11-7]
  Group-member nr.: 23-020

C_3H_5Cl
- 3-Chloro-1-propene [107-05-1]
  3-Chloropropene
  Allyl chloride
  Group-member nr.: 22-021

C_3H_5ClO
- Propanoyl chloride [79-03-8]
  Propionyl chloride
  Group-member nr.: 61-012

$C_3H_5ClO_2$
- Methyl chloroacetate [96-34-4]
  Methyl ester chloroacetic acid
  Group-member nr.: 61-013

$C_3H_5Cl_3$
- 1,2,3-Trichloropropane [96-18-4]
  Group-member nr.: 22-022

$C_3H_5D_3O_3$
- 1,2,3-Propanetriol-*O*,*O*,*O*-d_3 [7325-16-8]
  Glycerol-*O*,*O*,*O*-d_3
  Trideuteroglycerol
  Group-member nr.: 42-007

C_3H_5N
- Propanenitrile [107-12-0]
  Propionitrile
  Ethyl cyanide
  Group-member nr.: 32-005

C_3H_5NO
- 2-Propenamide [79-06-1]
  Acrylamide
  Group-member nr.: 62-013

$C_3H_5NO_4$
- Methyl ester nitroacetic acid [2483-57-0]
  Methyl nitroacetate
  Group-member nr.: 62-014

C_3H_5NS
- Isothiocyanatoethane [542-85-8]
  Ethyl isothiocyanate
  Group-member nr.: 64-005

$C_3H_5NaO_2$
- Sodium salt propanoic acid [137-40-6]
  Sodium propanoate
  Natrium propionate
  Group-member nr.: 74-002

C_3H_6
- Cyclopropane [75-19-4]
  Trimethylene
  Group-member nr.: 12-001
- 1-Propene [115-07-1]
  Propylene
  Group-member nr.: 13-002

$C_3H_6Br_2$
- 1,2-Dibromopropane [78-75-1]
  Propylene bromide
  Group-member nr.: 23-021
- 1,3-Dibromopropane [109-64-8]
  Trimethylene bromide
  Group-member nr.: 23-022

$C_3H_6Cl_2$
- 1,2-Dichloropropane [78-87-5]
  Propylene chloride
  Group-member nr.: 22-023
- 1,3-Dichloropropane [142-28-9]
  Trimethylene dichloride
  Group-member nr.: 22-024
- 2,2-Dichloropropane [594-20-7]
  Isopropylidene chloride
  Group-member nr.: 22-025

$C_3H_6F_4N_2$
- *N,N,N',N'*-Tetrafluoro-1,2-propanediamine [15403-25-5]
  Group-member nr.: 64-006

$C_3H_6N_2O_4$
- 2,2-Dinitropropane [595-49-3]
  Group-member nr.: 62-015

C_3H_6O
- Methyloxirane [75-56-9]
  1,2-Epoxypropane
  Propene oxide
  Group-member nr.: 46-002
- Oxetane [503-30-0]
  Trimethylene oxide
  Group-member nr.: 46-003
- Propanal [123-38-6]
  Propionaldehyde
  Group-member nr.: 43-003
- 2-Propanone [67-64-1]
  Dimethyl ketone
  Acetone
  Group-member nr.: 43-004
- 2-Propen-1-ol [107-18-6]
  Allyl alcohol
  Group-member nr.: 42-008

$C_3H_6O_2$
- 1,3-Dioxolane [646-06-0]
  Group-member nr.: 46-004
- Ethyl formate [109-94-4]
  Ethyl methanoate
  Group-member nr.: 45-002
- Methyl acetate [79-20-9]
  Methyl ethanoate
  Group-member nr.: 45-003
- Propanoic acid [79-09-4]
  Propionic acid
  Methylacetic acid
  Group-member nr.: 44-004

$C_3H_6O_2S$
- 3-Mercaptopropanoic acid [107-96-0]
  3-Sulfanylpropionic acid
  Group-member nr.: 63-003

$C_3H_6O_3$
- Dimethyl carbonate [616-38-6]
  Methyl carbonate
  Group-member nr.: 45-004
- (*R,S*)-2-Hydroxypropanoic acid [598-82-3]
  Lactic acid (racemic)
  Group-member nr.: 47-003
- 1,3,5-Trioxane [110-88-3]
  sym-Trioxane
  Group-member nr.: 46-005

C_3H_6S
- Thietane [287-27-4]
  Thiacyclobutane
  Trimethylene sulfide
  Group-member nr.: 53-001

C_3H_7Br
- 1-Bromopropane [106-94-5]
  Propyl bromide
  Group-member nr.: 23-023
- 2-Bromopropane [75-26-3]
  Isopropyl bromide
  Group-member nr.: 23-024

C_3H_7Cl
- 1-Chloropropane [540-54-5]
  Propyl chloride
  Group-member nr.: 22-026

C_3H_7I
- 1-Iodopropane [107-08-4]
  Propyl iodide
  Group-member nr.: 24-004

$C_3H_7N_2$
- Cyclopropanamine [765-30-0]
  Cyclopropylamine
  Group-member nr.: 31-005

C_3H_7NO
- *N*, *N*-Dimethylformamide [68-12-2]
  N, *N*-Dimethylmethanamide
  Group-member nr.: 62-016

$C_3H_7NO_2$
- Ethyl ester carbamic acid [51-79-6]
  Ethyl carbamate
  Ethyl urethane
  Group-member nr.: 62-017

$C_3H_7NO_3$
- 1-Methylethyl ester nitric acid [1712-64-7]
  Isopropyl nitrate
  Group-member nr.: 62-018

C_3H_8
- Propane [74-98-6]
  Group-member nr.: 11-005

C_3H_8O
- 1-Propanol [71-23-8]
  Propyl alcohol
  Group-member nr.: 42-009
- 2-Propanol [67-63-0]
  Isopropyl alcohol
  Group-member nr.: 42-010

$C_3H_8O_2$
- Dimethoxymethane [109-87-5]
  2,4-Dioxapentane
  Formaldehyde dimethyl acetal
  Methylal
  Group-member nr.: 41-002
- 2-Methoxyethanol [109-86-4]
  Methyl cellosolve
  Glycol monomethyl ether
  Group-member nr.: 47-004
- 1,2-Propanediol [57-55-6]
  Propylene glycol
  Group-member nr.: 42-011

$C_3H_8O_3$
- 1,2,3-Propanetriol [56-81-5]
  Glycerol
  Glycyl alcohol
  Glycerine
  Group-member nr.: 42-012

C_3H_8S
- (Methylthio)ethane [624-89-5]
  Ethyl methyl sulfide
  Group-member nr.: 51-003
- 1-Propanethiol [107-03-9]
  1-Propyl mercaptan
  Group-member nr.: 52-003
- 2-Propanethiol [75-33-2]
  Isopropyl mercaptan
  Group-member nr.: 52-004

C_3H_9Al
- Trimethylaluminium [75-24-1]
  Group-member nr.: 73-005

C_3H_9As
- Trimethylarsine [593-88-4]
  Group-member nr.: 73-006

C_3H_9B
- Trimethylborane [593-90-8]
  Group-member nr.: 72-005

C_3H_9Ga
- Trimethylgallium [1445-79-0]
  Group-member nr.: 73-007

C_3H_9N
- *N*, *N*-Dimethylmethanamine [75-50-3]
  Trimethylamine
  Group-member nr.: 31-006
- 1-Propanamine [107-10-8]
  n-Propylamine
  Group-member nr.: 31-007
- 2-Propanamine [75-31-0]
  Isopropylamine
  Group-member nr.: 31-008

$C_3H_{10}N_2$
- (*R*,*S*)-1,2-Propanediamine [10424-38-1]
  1,2-Diaminopropane (racemic)
  Group-member nr.: 31-009
- Trimethylhydrazine [1741-01-1]
  Group-member nr.: 34-005

$C_3H_{12}BN$
- Trimethylamineborane [75-22-9]
  N, *N*-Dimethylmethanamine compd. with borane (1:1)
  Group-member nr.: 72-006

C_3O_2
- 1,2-Propadiene-1,3-dione [504-64-3]
  Carbon suboxide
  Group-member nr.: 43-002

$C_4Br_2Cl_2F_6$
- 1,4-Dibromo-2,3-dichloro-1,1,2,3,4,4-hexafluorobutane [375-42-8]
  Group-member nr.: 25-026

C_4F_8
- Octafluorocyclobutane [115-25-3]
  Perfluorocyclobutane
  Group-member nr.: 21-007

$C_4H_2O_3$
- 2,5-Furandione [108-31-6]
  (*Z*)-Butenedioic acid anhydride
  Maleic anhydride
  Group-member nr.: 44-005

$C_4H_3Cl_3OS$
- *O*-Methyl ester 2,3,3-trichloro-2-propenethioic acid [76619-91-5]
  O-Methyl ester trichlorothioacrylic acid
  Methyl trichlorothioacrylate
  Group-member nr.: 64-007

$C_4H_4N_2$
- Butanedinitrile [110-61-2]
  Succinonitrile
  Ethylene dicyanide
  Group-member nr.: 32-006

C_4H_4O
- Furan [110-00-9]
  Furfuran
  Tetrol
  Group-member nr.: 46-006

$C_4H_4O_4$
- 1,4-Dioxane-2,5-dione [502-97-6]
  p-Dioxane-2,5-dione
  Glycolide
  Diglycolide
  Group-member nr.: 47-005

C_4H_4S
- Thiophene [110-02-1]
  Thiofuran
  Group-member nr.: 53-002

C_4H_5Cl
- 2-Chloro-1,3-butadiene [126-99-8]
  Chloroprene
  Group-member nr.: 22-027

$C_4H_5ClO_2$
- (*E*)-3-Chloro-2-butenoic acid [6214-28-4]
  trans-3-Chloro-2-butenoic acid
  3-Chlorocrotonic acid
  Group-member nr.: 61-014
- (*Z*)-3-Chloro-2-butenoic acid [6213-90-7]
  cis-3-Chloro-2-butenoic acid
  3-Chloroisocrotonic acid
  Group-member nr.: 61-015

$C_4H_5Cl_3O$
- 2,2,3-Trichlorobutanal [76-36-8]
  2,2,3-Trichlorobutyraldehyde
  Butylchloral
  Group-member nr.: 61-016

$C_4H_5Cl_3O_2$
- Ethyl trichloroacetate [515-84-4]
  Ethyl ester trichloroacetic acid
  Group-member nr.: 61-017

C_4H_5N
- Cyclopropanecarbonitrile [5500-21-0]
  Cyclopropyl cyanide
  Cyanocyclopropane
  Group-member nr.: 32-007
- 2-Methyl-2-propenenitrile [126-98-7]
  Methacrylonitrile
  Group-member nr.: 32-008
- 1*H*-Pyrrole [109-97-7]
  Azole
  Group-member nr.: 33-005

C_4H_5NO
- 4-Oxobutanenitrile [3515-93-3]
  β-Cyanopropionaldehyde
  Group-member nr.: 62-019

$C_4H_5NO_2$
- 2,5-Pyrrolidinedione [123-56-8]
  Succinimide
  Group-member nr.: 62-020

C_4H_5NS
- 3-Isothiocyanato-1-propene [57-06-7]
  Allyl isothiocyanate
  Group-member nr.: 64-008
- 2-Methylthiazole [3581-87-1]
  Group-member nr.: 64-009

C_4H_6
- 1,2-Butadiene [590-19-2]
  Group-member nr.: 13-003
- 1,3-Butadiene [106-99-0]
  Vinylethylene
  Divinyl
  Erythrene
  Group-member nr.: 13-004
- 1-Butyne [107-00-6]
  Ethylacetylene
  Group-member nr.: 13-005
- 2-Butyne [503-17-3]
  Dimethylacetylene
  Crotonylene
  Group-member nr.: 13-006

$C_4H_6Cl_2O_2$
- Ethyl dichloroacetate [535-15-9]
  Ethyl ester dichloroacetic acid
  Group-member nr.: 61-018

C_4H_6O
- 2-Butenal [4170-30-3]
  Crotonoaldehyde
  Group-member nr.: 43-005
- 2,3-Dihydrofuran [1191-99-7]
  Group-member nr.: 46-007

$C_4H_6O_2$
- (*E*)-2-Butenoic acid [107-93-7]
  trans-2-Butenoic acid
  Crotonic acid
  Group-member nr.: 44-006
- 2(3*H*)-Dihydrofuranone [96-48-0]
  β-Butyrolactone
  4-Butanolide
  Group-member nr.: 47-006
- Ethenyl acetate [108-05-4]
  Vinyl acetate
  Ethenyl ethanoate
  Group-member nr.: 45-005
- Methyl propenoate [96-33-3]
  Methyl acrylate
  Group-member nr.: 45-006
- 2-Methyl-2-propenoic acid [79-41-4]
  Methacrylic acid
  α-Methacrylic acid
  Group-member nr.: 44-007

$C_4H_6O_3$
- Acetic acid anhydride [108-24-7]
  Acetic anhydride
  Ethanoic acid anhydride
  Group-member nr.: 44-008
- 4-Methyl-1,3-dioxolan-2-one [108-32-7]
  Cyclic propylene ester carbonic acid
  Propylene carbonate
  Group-member nr.: 47-007

C_4H_7ClO
- Butanoyl chloride [141-75-3]
  Butyryl chloride
  Group-member nr.: 61-019
- 2-Methylpropanoyl chloride [79-30-1]
  2-Methylpropionyl chloride
  Isobutyryl chloride
  Group-member nr.: 61-020

$C_4H_7ClO_2$
- Ethyl chloroacetate [105-39-5]
  Ethyl ester chloroacetic acid
  Group-member nr.: 61-021

$C_4H_7Cl_2NSi$
- 3-(Dichloromethylsilyl)propanenitrile [1071-21-2]
  3-(Dichloromethylsilyl)propionitrile
  Dichloro-(2-cyanethyl)methylsilane
  Group-member nr.: 71-005

C_4H_7N
- Butanenitrile [109-74-0]
  Butyronitrile
  Propyl cyanide
  Group-member nr.: 32-009
- 2-Methylpropanenitrile [78-82-0]
  2-Methylpropionitrile
  Isobutyronitrile
  Isopropyl cyanide
  Group-member nr.: 32-010

C_4H_7NO
- 3-Methoxypropionitrile [110-67-8]
  Group-member nr.: 62-021
- 2-Pyrrolidinone [616-45-5]
  2-Pyrrolidone
  β-Butyrolactam
  Group-member nr.: 62-022

$C_4H_7NaO_2$
- Sodium salt butanoic acid [156-54-7]
  Sodium butanoate
  Natrium butyrate
  Group-member nr.: 74-003

C_4H_8
- 1-Butene [106-98-9]
  Group-member nr.: 13-007
- (*E*)-2-Butene [624-64-6]
  trans-2-Butene
  trans-Dimethylethylene
  Group-member nr.: 13-008
- (*Z*)-2-Butene [590-18-1]
  cis-2-Butene
  cis-Dimethylethylene
  Group-member nr.: 13-009
- Cyclobutane [287-23-0]
  Tetramethylene
  Group-member nr.: 12-002
- 2-Methyl-1-propene [115-11-7]
  Isobutene
  Group-member nr.: 13-010

$C_4H_8Cl_2$
- 1,4-Dichlorobutane [110-56-5]
  Tetramethylene dichloride
  Group-member nr.: 22-028

$C_4H_8Cl_2O$
- 1,1'-Oxybis(2-chloroethane) [111-44-4]
  Bis(2-chloroethyl) ether
  β,β'-Dichlorodiethyl ether
  Chlorex
  Group-member nr.: 61-022

C_4H_8O
- Butanal [123-72-8]
  Butyraldehyde
  Group-member nr.: 43-006
- 2-Butanone [78-93-3]
  Ethyl methyl ketone
  Group-member nr.: 43-007
- Ethyloxirane [106-88-7]
  1,2-Epoxybutane
  1,2-Butylene oxide
  Group-member nr.: 46-008
- 2-Methoxy-1-propene [116-11-0]
  Isopropenyl methyl ether
  Group-member nr.: 41-003
- Tetrahydrofuran [109-99-9]
  Oxolane
  Tetramethylene oxide
  Group-member nr.: 46-009

$C_4H_8O_2$
- Butanoic acid [107-92-6]
  Butyric acid
  Ethylacetic acid
  Propylformic acid
  Group-member nr.: 44-009
- 1,3-Dioxane [505-22-6]
  m-Dioxane
  Group-member nr.: 46-010
- 1,4-Dioxane [123-91-1]
  p-Dioxane
  1,4-Diethylene dioxide
  Dioxyethylene ether
  Diethylene ether
  Group-member nr.: 46-011
- Ethyl acetate [141-78-6]
  Ethyl ethanoate
  Group-member nr.: 45-007
- Methyl propanoate [554-12-1]
  Methyl propionate
  Group-member nr.: 45-008
- 2-Methylpropanoic acid [79-31-2]
  Isobutyric acid
  Group-member nr.: 44-010
- Propyl formate [110-74-7]
  Propyl methanoate
  Group-member nr.: 45-009

$C_4H_8O_2S$
- Tetrahydrothiophene 1,1-dioxide [126-33-0]
  Tetramethylene sulfone
  Sulfolane
  Group-member nr.: 63-004

$C_4H_8O_4$
- 1,3,5,7-Tetroxocane [293-30-1]
  1,3,5,7-Tetraoxacyclooctane
  Group-member nr.: 46-012

C_4H_8S
- Tetrahydrothiophene [110-01-0]
  Thiolane
  Tetramethylene sulfide
  Group-member nr.: 53-003

$C_4H_8S_2$
- 1,3-Dithiane [505-23-7]
  m-Dithiane
  Group-member nr.: 53-004
- 1,4-Dithiane [505-29-3]
  p-Dithiane
  Group-member nr.: 53-005

C_4H_9Br
- 1-Bromobutane [109-65-9]
  n-Butyl bromide
  Group-member nr.: 23-025
- 1-Bromo-2-methylpropane [78-77-3]
  Isobutyl bromide
  Group-member nr.: 23-026
- 2-Bromo-2-methylpropane [507-19-7]
  tert-Butyl bromide
  Group-member nr.: 23-027

C_4H_9Cl
- 1-Chlorobutane [109-69-3]
  n-Butyl chloride
  Group-member nr.: 22-029
- 1-Chloro-2-methylpropane [513-36-0]
  Isobutyl chloride
  Group-member nr.: 22-030
- 2-Chloro-2-methylpropane [507-20-0]
  tert-Butyl chloride
  Group-member nr.: 22-031

C_4H_9I
- 1-Iodo-2-methylpropane [513-38-2]
  Isobutyl iodide
  Group-member nr.: 24-005

C_4H_9N
- Pyrrolidine [123-75-1]
  Tetrahydropyrrole
  Azolidine
  Group-member nr.: 33-006

C_4H_9NO
- 2-Butanone oxime [96-29-7]
  Ethyl methyl ketoxime (unspecified stereoisomer)
  Group-member nr.: 62-023
- *N*,*N*-Dimethylacetamide [127-19-5]
  Group-member nr.: 62-024
- *N*-Ethylacetamide [625-50-3]
  N-Ethylethanamide
  Group-member nr.: 62-025

• *N*-Methylpropanamide [1187-58-2]
N-Methylpropionamide
Group-member nr.: 62-026
• Morpholine [110-91-8]
Tetrahydro-4*H*-1,4-oxazine
Diethylenimide oxide
Group-member nr.: 62-027

C_4H_{10}
• Butane [106-97-8]
Group-member nr.: 11-006
• 2-Methylpropane [75-28-5]
Isobutane
Trimethylmethane
Group-member nr.: 11-007

$C_4H_{10}Cl_2Si$
• Dichlordiethylsilane [1719-53-5]
Group-member nr.: 71-006

$C_4H_{10}Hg$
• Diethylmercury [627-44-1]
Group-member nr.: 73-008

$C_4H_{10}N_2$
• Piperazine [110-85-0]
Hexahydropyrazine
Perhydro-1,4-diazine
Diethylenediamine
Group-member nr.: 33-007

$C_4H_{10}O$
• 1-Butanol [71-36-3]
Butyl alcohol
Group-member nr.: 42-013
• 2-Butanol [78-92-2]
sec-Butyl alcohol
Methyl ethyl carbinol (unspecified chirality)
Group-member nr.: 42-014
• (*R*,*S*)-2-Butanol [15892-23-6]
sec-Butyl alcohol
Methyl ethyl carbinol (racemic)
Group-member nr.: 42-015
• (*S*)-2-Butanol [4221-99-2]
D-*sec*-Butyl alcohol
D-Methyl ethyl carbinol
Group-member nr.: 42-016
• 1-Methoxypropane [557-17-5]
Methyl propyl ether
Group-member nr.: 41-004
• 2-Methoxypropane [598-53-8]
Methyl 1-methylethyl ether
Isopropyl methyl ether
Group-member nr.: 41-005
• 2-Methyl-1-propanol [78-83-1]
Isobutyl alcohol
Group-member nr.: 42-017
• 2-Methyl-2-propanol [75-65-0]
tert-Butyl alcohol
Trimethyl carbinol
Group-member nr.: 42-018
• 1,1'-Oxybisethane [60-29-7]
Diethyl ether
Ethyl ether
Ethyl oxide
Diethyl oxide
Group-member nr.: 41-006

$C_4H_{10}O_2$
• 1,3-Butanediol [107-88-0]
1,3-Butylene glycol
Group-member nr.: 42-019
• 1,4-Butanediol [110-63-4]
1,4-Butylene glycol
Tetramethylene glycol
Group-member nr.: 42-020
• 2,3-Butanediol [513-85-9]
2,3-Butylene glycol
Group-member nr.: 42-021
• 1,2-Dimethoxyethane [110-71-4]
2,5-Dioxahexane
Ethylene glycol dimethyl ether
Monoglyme
GDME
Group-member nr.: 41-007
• 2-Ethoxyethanol [110-80-5]
3-Oxa-1-pentanol
Ethyl cellosolve
Ethylene glycol monoethyl ether
Group-member nr.: 47-008

$C_4H_{10}O_2Se$
• 2,2'-Selenodiethanol [27974-49-8]
β-Selenium diglycol
Group-member nr.: 73-009

$C_4H_{10}O_3$
• 2,2'-Oxybisethanol [111-46-6]
1,5-Dihydroxy-3-oxapentane
2,2'-Dihydroxydiethyl ether
Diethylene glycol
Group-member nr.: 47-009

$C_4H_{10}O_4$
• (*R**, *S**)-1,2,3,4-Butanetetrol [149-32-6]
Erythro-1,2,3,4-butanetetrol
Erythritol
Group-member nr.: 42-022

$C_4H_{10}S$
• 1-Butanethiol [109-79-5]
1-Butyl mercaptan
Group-member nr.: 52-005
• 2-Butanethiol [513-53-1]
2-Butyl mercaptan
sec-Butyl mercaptan
Group-member nr.: 52-006

- 2-Methyl-1-propanethiol [513-44-0]
  Isobutyl mercaptan
  Group-member nr.: 52-007
- 2-Methyl-2-propanethiol [75-66-1]
  tert-Butyl mercaptan
  Group-member nr.: 52-008
- 1-(Methylthio)propane [3877-15-4]
  Methyl propyl sulfide
  Group-member nr.: 51-004
- 2-(Methylthio)propane [1551-21-9]
  Methyl 1-methylethyl sulfide
  Isopropyl methyl sulfide
  Group-member nr.: 51-005
- 1,1'-Thiobisethane [352-93-2]
  Diethyl sulfide
  Group-member nr.: 51-006

$C_4H_{10}S_2$
- 3,4-Dithiahexane [110-81-6]
  Diethyl disulfide
  Ethyl disulfide
  Group-member nr.: 51-007

$C_4H_{10}Zn$
- Diethylzinc [557-20-0]
  Group-member nr.: 73-010

$C_4H_{11}N$
- 1-Butanamine [109-73-9]
  n-Butylamine
  Group-member nr.: 31-010
- *N*-Ethylethanamine [109-89-7]
  Diethylamine
  Group-member nr.: 31-011
- 2-Methyl-1-propanamine [78-81-9]
  Isobutylamine
  Group-member nr.: 31-012
- 2-Methyl-2-propanamine [75-64-9]
  tert-Butylamine
  Group-member nr.: 31-013

$C_4H_{11}NO$
- 2-Amino-2-methyl-1-propanol [124-68-5]
  Isobutanolamine
  Group-member nr.: 62-028
- *N*, *N*-Diethylhydroxylamine [3710-84-7]
  Group-member nr.: 62-029
- 3-Methoxy-1-propanamine [5332-73-0]
  3-Methoxypropylamine
  4-Oxapentanamine
  Group-member nr.: 62-030

$C_4H_{11}NO_2$
- 2-Amino-2-methyl-1,3-propanediol [115-69-5]
  Group-member nr.: 62-031

$C_4H_{11}NO_3$
- 2-Amino-2-(hydroxymethyl)-1,3-propanediol [77-86-1]
  Trometamol
  Group-member nr.: 62-032

$C_4H_{12}CdSe$
- Dimethyl[selenobis(methane)]cadmium [143481-65-6]
  Group-member nr.: 73-011

$C_4H_{12}CdTe$
- Dimethyl[tellurobis(methane)]cadmium [143481-66-7]
  Group-member nr.: 73-012

$C_4H_{12}Ge$
- Tetramethylgermane [865-52-1]
  Group-member nr.: 73-013

$C_4H_{12}N_2$
- 2-Methyl-1,2-propanediamine [811-93-8]
  1,2-Diamino-2-methylpropane
  Group-member nr.: 31-014

$C_4H_{12}O_4Si$
- Tetramethyl ester silicic acid [681-84-5]
  Tetramethyl orthosilicate
  Tetramethoxysilane
  Group-member nr.: 71-007

$C_4H_{12}Pb$
- Tetramethylplumbane [75-74-1]
  Tetramethyllead
  Group-member nr.: 73-014

$C_4H_{12}SeZn$
- Dimethyl[selenobis(methane)]zinc [108430-95-1]
  Complex dimethylzinc with dimethylselenium
  Group-member nr.: 73-015

$C_4H_{12}Si$
- Tetramethylsilane [75-76-3]
  Group-member nr.: 71-008

$C_4H_{12}Sn$
- Tetramethylstannane [594-27-4]
  Tetramethyltin
  Group-member nr.: 73-016

$C_4H_{12}TeZn$
- Dimethyl[tellurobis(methane)]zinc [127283-03-8]
  Complex dimethylzinc with dimethyltellurium
  Group-member nr.: 73-017

$C_4H_{13}N_3$
- *N*-(2-Aminoethyl)-1,2-ethanediamine [111-40-0]
  Diethylenetriamine
  Group-member nr.: 31-015

C_4NiO_4
- Nickel carbonyl [13463-39-3]
  Group-member nr.: 02-029

$C_5F_{11}N$
- Undecafluoropiperidine [836-77-1]
  Perfluoropiperidine
  Group-member nr.: 64-010

C_5F_{12}
- Dodecafluoropentane [678-26-2]
  Perfluoropentane
  Group-member nr.: 21-008

$C_5F_{13}N$
- 1,1,2,2,2-Pentafluoro-*N*-(pentafluoroethyl)-*N*-(trifluoromethyl)-ethanamine [758-48-5]
  Group-member nr.: 64-011

$C_5H_3F_7O_2$
- Methyl heptafluorobutanoate [356-24-1]
  Methyl ester heptafluorobutanoic acid
  Methyl perfluorobutyrate
  Group-member nr.: 61-023

$C_5H_4O_2$
- 2-Furancarboxaldehyde [98-01-1]
  Furfural
  Pyromucic aldehyde
  Furfuraldehyde
  Group-member nr.: 47-010

$C_5H_5Cl_3O_2$
- 2-Propenyl trichloroacetate [6304-34-3]
  2-Propenyl ester trichloroacetic acid
  Allyl trichloroacetate
  Group-member nr.: 61-024

C_5H_5N
- Bicyclo[1.1.0]butane-1-carbonitrile [16955-35-4]
  1-Bicyclobutyl cyanide
  1-Cyanobicyclobutane
  Group-member nr.: 32-011
- Pyridine [110-86-1]
  Azine
  Group-member nr.: 33-008

$C_5H_6Cl_2O_2$
- 2-Propenyl dichloroacetate [30895-77-3]
  2-Propenyl ester dichloroacetic acid
  Allyl dichloroacetate
  Group-member nr.: 61-025

$C_5H_6N_2$
- Dimethylpropanedinitrile [7321-55-3]
  2,2-Dicyanopropane
  Dimethylmalononitrile
  Group-member nr.: 32-012
- Pentanedinitrile [544-13-8]
  1,3-Dicyanopropane
  Glutaronitrile
  Trimethylenedicyanide
  Group-member nr.: 32-013

C_5H_6O
- 2-Methylfuran [534-22-5]
  Sylvan
  Group-member nr.: 46-013

$C_5H_6O_2$
- 2-Furanmethanol [98-00-0]
  Furfuryl alcohol
  Furyl carbinol
  Group-member nr.: 47-011

C_5H_6S
- 2-Methylthiophene [554-14-3]
  Group-member nr.: 53-006
- 3-Methylthiophene [616-44-4]
  Group-member nr.: 53-007

$C_5H_7ClO_2$
- 2-Propenyl chloroacetate [2916-14-5]
  2-Propenyl ester chloroacetic acid
  Allyl chloroacetate
  Group-member nr.: 61-026

$C_5H_7Cl_3O_2$
- Propyl trichloroacetate [13313-91-2]
  Propyl ester trichloroacetic acid
  Group-member nr.: 61-027

C_5H_7N
- Cyclobutanecarbonitrile [4426-11-3]
  Cyclobutyl cyanide
  Cyanocyclobutane
  Group-member nr.: 32-014
- 1-Methyl-1*H*-pyrrole [96-54-8]
  Group-member nr.: 33-009

$C_5H_7NO_2$
- Ethyl ester cyanoacetic acid [105-56-6]
  Ethyl cyanoacetate
  Group-member nr.: 62-033

C_5H_8
- Cyclopentene [142-29-0]
  Group-member nr.: 14-001
- 2-Methyl-1,3-butadiene [78-79-5]
  Isoprene
  Group-member nr.: 13-011
- 3-Methyl-1,2-butadiene [598-25-4]
  Group-member nr.: 13-012
- Methylenecyclobutane [1120-56-5]
  Group-member nr.: 12-003
- 1,2-Pentadiene [591-95-7]
  Group-member nr.: 13-013
- (*E*)-1,3-Pentadiene [2004-70-8]
  trans-1,3-Pentadiene
  Group-member nr.: 13-014
- (*Z*)-1,3-Pentadiene [1574-41-0]
  cis-1,3-Pentadiene
  Group-member nr.: 13-015
- 1,4-Pentadiene [591-93-5]
  Group-member nr.: 13-016
- 2,3-Pentadiene [591-96-8]
  Group-member nr.: 13-017
- Spiropentane [157-40-4]
  Spirocyclane
  Cyclopropanespirocyclopropane
  Group-member nr.: 12-004

$C_5H_8Br_4$
- 1,3-Dibromo-2,2-bis(bromomethyl)propane [3229-00-3]
  2,2-Bis(bromomethyl)-1,3-dibromopropane
  Pentaerythrityl tetrabromide
  Group-member nr.: 23-028

$C_5H_8Cl_2O$
- 3,3-Bis(chloromethyl)oxetane [78-71-7]
  3,3-Bis(chloromethyl)oxacyclobutane
  Group-member nr.: 61-028

$C_5H_8Cl_2O_2$
- Ethyl 2,3-dichloropropanoate [6628-21-3]
  Ethyl ester 2,3-dichloropropanoic acid
  Group-member nr.: 61-029
- Propyl dichloroacetate [37587-81-8]
  Propyl ester dichloroacetic acid
  Group-member nr.: 61-030

$C_5H_8Cl_4$
- 1,3-Dichloro-2,2-bis(chloromethyl)propane [3228-99-7]
  2,2-Bis(chloromethyl)-1,3-dichloropropane
  Pentaerythrityl tetrachloride
  Group-member nr.: 22-032

$C_5H_8F_4$
- 1,3-Difluoro-2,2-bis(fluoromethyl)propane [338-23-8]
  2,2-Bis(fluoromethyl)-1,3-difluoropropane
  Pentaerythrityl tetrafluoride
  Group-member nr.: 21-009

C_5H_8O
- Cyclopentanone [120-92-3]
  Group-member nr.: 43-008

$C_5H_8O_2$
- Methyl 2-methyl-2-propenoate [80-62-6]
  Methyl methacrylate
  Group-member nr.: 45-010
- 2,4-Pentanedione [123-54-6]
  Acetylacetone
  Diacetylmethane
  Group-member nr.: 43-009
- 2-Propenyl acetate [591-87-7]
  2-Propenyl ethanoate
  Allyl acetate
  Group-member nr.: 45-011
- Tetrahydro-2*H*-pyran-2-one [542-28-9]
  δ-Valerolactone
  5-Pentanolide
  Group-member nr.: 47-012

C_5H_9ClO
- Pentanoyl chloride [638-29-9]
  Valeryl chloride
  Group-member nr.: 61-031

$C_5H_9ClO_2$
- Ethyl 2-chloropropanoate [535-13-7]
  Ethyl ester 2-chloropropanoic acid
  Ethyl α-chloropropionate
  Group-member nr.: 61-032
- Propyl chloroacetate [5396-24-7]
  Propyl ester chloroacetic acid
  Group-member nr.: 61-033

C_5H_9N
- 2,2-Dimethylpropanenitrile [630-18-2]
  2-Cyano-2-methylpropane
  Pivalonitrile
  Trimethylacetonitrile
  Group-member nr.: 32-015
- Pentanenitrile [110-59-8]
  Valeronitrile
  Butyl cyanide
  Group-member nr.: 32-016

C_5H_9NO
- 1-Methyl-2-pyrrolidinone [872-50-4]
  1-Methyl-2-pyrrolidone
  Group-member nr.: 62-034
- 2-Piperidinone [675-20-7]
  δ-Valerolactam
  2-Piperidone
  Group-member nr.: 62-035

C_5H_{10}
- Cyclopentane [287-92-3]
  Pentamethylene
  Group-member nr.: 12-005
- 2-Methyl-1-butene [563-46-2]
  Group-member nr.: 13-018
- 2-Methyl-2-butene [513-35-9]
  Trimethylethylene
  β-Isoamylene
  Group-member nr.: 13-019
- 3-Methyl-1-butene [563-45-1]
  Isopropylethylene
  α-Isoamylene
  Group-member nr.: 13-020
- 1-Pentene [109-67-1]
  Group-member nr.: 13-021
- 2-Pentene [109-68-2]
  β-Amylene
  Methylethylethylene (unspecified stereoisomer)
  Group-member nr.: 13-022
- (*E*)-2-Pentene [646-04-8]
  trans-2-Pentene
  trans-β-Amylene
  trans-Methylethylethylene
  Group-member nr.: 13-023
- (*Z*)-2-Pentene [627-20-3]
  cis-2-Pentene
  cis-β-Amylene
  cis-Methylethylethylene
  Group-member nr.: 13-024

$C_5H_{10}Cl_2$
- 1,5-Dichloropentane [628-76-2]
  Pentamethylene dichloride
  Group-member nr.: 22-033

$C_5H_{10}N_2$
- 3-(Dimethylamino)propanenitrile [1738-25-6]
  3-(Dimethylamino)propionitrile
  Group-member nr.: 34-006

$C_5H_{10}O$
- Cyclopentanol [96-41-3]
  Cyclopentyl alcohol
  Group-member nr.: 42-023

• 2,2-Dimethylpropanal [630-19-3]
Pivalaldehyde
Trimethylacetaldehyde
Group-member nr.: 43-010
• 3-Methyl-2-butanone [563-80-4]
Isopropyl methyl ketone
Group-member nr.: 43-011
• 2-Methyl-3-buten-2-ol [115-18-4]
Group-member nr.: 42-024
• Pentanal [110-62-3]
Valeraldehyde
Pentyl aldehyde
Amyl aldehyde
Group-member nr.: 43-012
• 2-Pentanone [107-87-9]
Methyl propyl ketone
Group-member nr.: 43-013
• 3-Pentanone [96-22-0]
Diethyl ketone
Metacetone
Ethyl propionyl
Propione
Group-member nr.: 43-014
• Tetrahydropyran [142-68-7]
Oxane
Pentamethylene oxide
Group-member nr.: 46-014

$C_5H_{10}O_2$
• Butyl formate [592-84-7]
Butyl methanoate
Group-member nr.: 45-012
• 1,3-Dioxepane [505-65-7]
m-Dioxepane
Group-member nr.: 46-015
• Ethyl propanoate [105-37-3]
Ethyl propionate
Group-member nr.: 45-013
• Methyl butanoate [623-42-7]
Methyl butyrate
Group-member nr.: 45-014
• 3-Methylbutanoic acid [503-74-2]
Isovaleric acid
Group-member nr.: 44-011
• 1-Methylethyl acetate [108-21-4]
Isopropyl acetate
1-Methylethyl ethanoate
Group-member nr.: 45-016
• Methyl 2-methylpropanoate [547-63-7]
Methyl isobutyrate
Group-member nr.: 45-015
• 2-Methylpropyl formate [542-55-2]
2-Methylpropyl methanoate
Isobutyl formate
Group-member nr.: 45-017
• Pentanoic acid [109-52-4]
Valeric acid
Group-member nr.: 44-012
• Propyl acetate [109-60-4]
Propyl ethanoate
Group-member nr.: 45-018
• Tetrahydro-2-furanmethanol [97-99-4]
Tetrahydrofurfuryl alcohol
Group-member nr.: 47-013

$C_5H_{10}O_3$
• Diethyl carbonate [105-58-8]
Ethyl carbonate
Group-member nr.: 45-019
• 2-Methoxyethanol acetate [110-49-6]
2-Methoxyethyl acetate
Group-member nr.: 47-014

$C_5H_{10}O_5$
• 1,3,5,7,9-Pentoxecane [16528-92-0]
Pentoxane
Group-member nr.: 46-016

$C_5H_{10}S$
• Cyclopentanethiol [1679-07-8]
Cyclopentyl mercaptan
Group-member nr.: 52-009
• 2-Methyltetrahydrothiophene [1795-09-1]
2-Methylthiolane
2-Methylcyclothiapentane
Group-member nr.: 53-008
• 3-Methyltetrahydrothiophene [4740-00-5]
3-Methylthiolane
3-Methylcyclothiapentane
Group-member nr.: 53-009
• Tetrahydro-2*H*-thiopyran [1613-51-0]
Thiacyclohexane
Pentamethylene sulfide
Group-member nr.: 53-010

$C_5H_{11}Br$
• 1-Bromo-3-methylbutane [107-82-4]
Isoamyl bromide
Group-member nr.: 23-029
• 1-Bromopentane [110-53-2]
n-Amyl bromide
Group-member nr.: 23-030

$C_5H_{11}Cl$
• 1-Chloro-3-methylbutane [107-84-6]
Isoamyl chloride
Group-member nr.: 22-034

$C_5H_{11}I$
• 1-Iodo-3-methylbutane [541-28-6]
Isoamyl iodide
Group-member nr.: 24-006

$C_5H_{11}N$
• Cyclopentanamine [1003-03-8]
Cyclopentylamine
Group-member nr.: 31-016

- 1-Methylpyrrolidine [120-94-5]
  N-Methylpyrrolidine
  Group-member nr.: 33-010
- 3-Methylpyrrolidine [34375-89-8]
  Group-member nr.: 33-011
- Piperidine [110-89-4]
  Pentamethylenimine
  Hexahydropyridine
  Perhydroazine
  Group-member nr.: 33-012

$C_5H_{11}NO$
- *N*,2-Dimethylpropanamide [2675-88-9]
  N-Methylisobutyramide
  Group-member nr.: 62-036
- *N*-Methylbutanamide [17794-44-4]
  N-Methylbutyramide
  Group-member nr.: 62-037
- *N*-(1-Methylethyl)acetamide [1118-69-0]
  N-(1-Methylethyl)ethanamide
  N-Isopropylacetamide
  Group-member nr.: 62-038
- *N*-Propylacetamide [5331-48-6]
  N-Propylethanamide
  Group-member nr.: 62-039

C_5H_{12}
- 2,2-Dimethylpropane [463-82-1]
  Neopentane
  Tetramethylmethane
  Group-member nr.: 11-008
- 2-Methylbutane [78-78-4]
  Isopentane
  Group-member nr.: 11-009
- Pentane [109-66-0]
  Group-member nr.: 11-010

$C_5H_{12}N_2O$
- Tetramethylurea [632-22-4]
  Group-member nr.: 62-040

$C_5H_{12}O$
- 1-Ethoxypropane [628-32-0]
  Ethyl propyl ether
  Group-member nr.: 41-008
- 1-Methoxybutane [628-28-4]
  Butyl methyl ether
  Group-member nr.: 41-009
- 2-Methoxy-2-methylpropane [1634-04-4]
  1,1-Dimethylethyl methyl ether
  tert-Butyl methyl ether
  Group-member nr.: 41-010
- 2-Methyl-1-butanol [137-32-6]
  sec-Butyl carbinol
  Group-member nr.: 42-025
- 2-Methyl-2-butanol [75-85-4]
  tert-Amyl alcohol
  tert-Pentyl alcohol
  Dimethyl ethyl carbinol
  Group-member nr.: 42-026
- 3-Methyl-1-butanol [123-51-3]
  Isopentyl alcohol
  Isoamyl alcohol
  Group-member nr.: 42-027
- 3-Methyl-2-butanol [598-75-4]
  sec-Isoamyl alcohol
  Group-member nr.: 42-028
- 1-Pentanol [71-41-0]
  n-Amyl alcohol
  Group-member nr.: 42-029
- 2-Pentanol [6032-29-7]
  sec-Amyl alcohol
  Methyl propyl carbinol
  Group-member nr.: 42-030
- 3-Pentanol [584-02-1]
  1-Ethyl-1-propanol
  Diethyl carbinol
  Group-member nr.: 42-031

$C_5H_{12}O_2$
- 2,2-Dimethoxypropane [77-76-9]
  Acetone dimethyl acetal
  Group-member nr.: 41-011
- 2,2-Dimethyl-1,3-propanediol [126-30-7]
  Group-member nr.: 42-032
- 1-Ethoxy-2-methoxyethane [5137-45-1]
  2,5-Dioxaheptane
  Ethylene glycol ethyl ether methyl ether
  Group-member nr.: 41-012
- 2-(1-Methylethoxy)ethanol [109-59-1]
  4-Methyl-3-oxa-1-pentanol
  2-Isopropoxyethanol
  Ethylene glycol monoisopropyl ether
  Group-member nr.: 47-015
- 1,5-Pentanediol [111-29-5]
  Group-member nr.: 42-033
- 2-Propoxyethanol [2807-30-9]
  3-Oxa-1-hexanol
  Ethylene glycol monopropyl ether
  Propyl cellosolve
  Group-member nr.: 47-016

$C_5H_{12}O_3$
- 2-(2-Methoxyethoxy)ethanol [111-77-3]
  Diethylene glycol monomethyl ether
  Group-member nr.: 47-017

$C_5H_{12}S$
- 1-(Ethylthio)propane [4110-50-3]
  Ethyl propyl sulfide
  Group-member nr.: 51-008
- 2-Methyl-2-butanethiol [1679-09-0]
  2-Methyl-2-butyl mercaptan
  tert-Amyl mercaptan
  Group-member nr.: 52-010
- 3-Methyl-1-butanethiol [541-31-1]
  3-Methylbutyl mercaptan
  Isopentyl mercaptan
  Isoamyl mercaptan
  Group-member nr.: 52-011

- 3-Methyl-2-butanethiol [2084-18-6]
  3-Methyl-2-butyl mercaptan
  sec-Isoamyl mercaptan
  Group-member nr.: 52-012
- 2-Methyl-2-(methylthio)propane [6163-64-0]
  1,1-Dimethylethyl methyl sulfide
  tert-Butyl methyl sulfide
  Group-member nr.: 51-009
- 1-(Methylthio)butane [628-29-5]
  Butyl methyl sulfide
  Group-member nr.: 51-010
- 2-(Methylthio)butane [10359-64-5]
  3-Methyl-2-thiapentane
  Methyl 1-methylpropyl sulfide
  sec-Butyl methyl sulfide
  Group-member nr.: 51-011
- 1-Pentanethiol [110-66-7]
  1-Pentyl mercaptan
  Group-member nr.: 52-013
- Tetrakis(methylthia)methane [6156-25-8]
  Tetramethyl ester tetrathiacarbonic acid
  Group-member nr.: 51-012

$C_5H_{12}Si$
- 1,1-Dimethylsilacyclobutane [2295-12-7]
  Group-member nr.: 71-009
- Ethenyltrimethylsilane [754-05-2]
  Trimethylvinylsilane
  Group-member nr.: 71-010

$C_5H_{13}N$
- *N*-Ethyl-*N*-methylethanamine [616-39-7]
  N-Methyldiethylamine
  N, *N*-Diethylmethylamine
  Group-member nr.: 31-017
- 1-Pentanamine [110-58-7]
  Pentylamine
  n-Amylamine
  Group-member nr.: 31-018

$C_5H_{13}NO$
- 2-(Ethylmethylamino)ethanol [2893-43-8]
  N, *N*-Ethylmethylethanolamine
  Group-member nr.: 62-041

$C_5H_{14}N_2$
- *N*, *N*-Dimethyl-1,3-propanediamine [109-55-7]
  N, *N*-Dimethylpropylenediamine
  Group-member nr.: 31-019

C_6BrF_5
- Bromopentafluorobenzene [344-04-7]
  Group-member nr.: 25-027

$C_6Br_2Cl_3F_9$
- 1,6-Dibromo-2,3,5-trichloro-1,1,2,3,4,4,5,6,6-nonafluorohexane [85131-86-8]
  Group-member nr.: 25-028

C_6ClF_5
- Chloropentafluorobenzene [344-07-0]
  Group-member nr.: 25-029

$C_6Cl_3F_3$
- 1,3,5-Trichloro-2,4,6-trifluorobenzene [319-88-0]
  Group-member nr.: 25-030

C_6D_6
- Benzene-d_6 [1076-43-3]
  Hexadeuterobenzene
  Group-member nr.: 14-002

C_6D_{12}
- Cyclohexane-d_{12} [1735-17-7]
  Dodecadeuterocyclohexane
  Group-member nr.: 12-006

$C_6F_5NO_2$
- Pentafluoronitrobenzene [880-78-4]
  Group-member nr.: 64-012

C_6F_6
- Hexafluorobenzene [392-56-3]
  Perfluorobenzene
  Group-member nr.: 21-010

C_6F_{14}
- Tetradecafluorohexane [355-42-0]
  Perfluorohexane
  Group-member nr.: 21-011

$C_6F_{15}N$
- 1,1,2,2,2-Pentafluoro-*N*, *N*-bis(pentafluoroethyl)-ethanamine [359-70-6]
  Group-member nr.: 64-013

C_6HF_5
- Pentafluorobenzene [363-72-4]
  Group-member nr.: 21-012

C_6HF_5O
- Pentafluorophenol [771-61-9]
  Group-member nr.: 61-034

$C_6H_2F_4$
- 1,2,3,4-Tetrafluorobenzene [551-62-2]
  Group-member nr.: 21-013
- 1,2,3,5-Tetrafluorobenzene [2367-82-0]
  Group-member nr.: 21-014
- 1,2,4,5-Tetrafluorobenzene [327-54-8]
  Group-member nr.: 21-015

$C_6H_2F_5N$
- 2,3,4,5,6-Pentafluorobenzenamine [771-60-8]
  Pentafluoroaniline
  Group-member nr.: 64-014

$C_6H_3Cl_3$
- 1,2,4-Trichlorobenzene [120-82-1]
  Group-member nr.: 22-035

$C_6H_3Cl_4N$
- 2-Chloro-6-(trichloromethyl)pyridine [1929-82-4]
  Group-member nr.: 64-015

C_6H_4BrCl
- 1-Bromo-2-chlorobenzene [694-80-4]
  o-Bromochlorobenzene
  Group-member nr.: 25-031
- 1-Bromo-3-chlorobenzene [108-37-2]
  m-Bromochlorobenzene
  Group-member nr.: 25-032
- 1-Bromo-4-chlorobenzene [106-39-8]
  p-Bromochlorobenzene
  Group-member nr.: 25-033

C_6H_4BrI
- 1-Bromo-2-iodobenzene [583-55-1]
  o-Bromoiodobenzene
  Group-member nr.: 25-034
- 1-Bromo-3-iodobenzene [591-18-4]
  m-Bromoiodobenzene
  Group-member nr.: 25-035

$C_6H_4BrNO_2$
- 1-Bromo-3-nitrobenzene [585-79-5]
  m-Bromonitrobenzene
  Group-member nr.: 64-016

$C_6H_4Br_2$
- 1,2-Dibromobenzene [583-53-9]
  o-Dibromobenzene
  Group-member nr.: 23-031
- 1,3-Dibromobenzene [108-36-1]
  m-Dibromobenzene
  Group-member nr.: 23-032

$C_6H_4Br_2O$
- 2,4-Dibromophenol [615-58-7]
  Group-member nr.: 61-035

$C_6H_4ClNO_2$
- 1-Chloro-3-nitrobenzene [121-73-3]
  m-Chloronitrobenzene
  Group-member nr.: 64-017
- 1-Chloro-4-nitrobenzene [100-00-5]
  p-Chloronitrobenzene
  Group-member nr.: 64-018

$C_6H_4Cl_2$
- 1,2-Dichlorobenzene [95-50-1]
  o-Dichlorobenzene
  Group-member nr.: 22-036
- 1,3-Dichlorobenzene [541-73-1]
  m-Dichlorobenzene
  Group-member nr.: 22-037
- 1,4-Dichlorobenzene [106-46-7]
  p-Dichlorobenzene
  Group-member nr.: 22-038

$C_6H_4F_2$
- 1,2-Difluorobenzene [367-11-3]
  o-Difluorobenzene
  Group-member nr.: 21-016
- 1,3-Difluorobenzene [372-18-9]
  m-Difluorobenzene
  Group-member nr.: 21-017
- 1,4-Difluorobenzene [540-36-3]
  p-Difluorobenzene
  Group-member nr.: 21-018

$C_6H_4I_2$
- 1,2-Diiodobenzene [615-42-9]
  o-Diiodobenzene
  Group-member nr.: 24-007
- 1,3-Diiodobenzene [626-00-6]
  m-Diiodobenzene
  Group-member nr.: 24-008

$C_6H_4N_2O_4$
- 1,2-Dinitrobenzene [528-29-0]
  o-Dinitrobenzene
  Group-member nr.: 62-042
- 1,3-Dinitrobenzene [99-65-0]
  m-Dinitrobenzene
  Group-member nr.: 62-043
- 1,4-Dinitrobenzene [100-25-4]
  p-Dinitrobenzene
  Group-member nr.: 62-044

$C_6H_4O_2$
- 2,5-Cyclohexadiene-1,4-dione [106-51-4]
  p-Benzoquinone
  p-Quinone
  Group-member nr.: 43-015

C_6H_5Br
- Bromobenzene [108-86-1]
  Phenyl bromide
  Group-member nr.: 23-033

C_6H_5BrO
- 4-Bromophenol [106-41-2]
  p-Bromophenol
  Group-member nr.: 61-036

C_6H_5Cl
- Chlorobenzene [108-90-7]
  Phenyl chloride
  Group-member nr.: 22-039

C_6H_5ClO
- 2-Chlorophenol [95-57-8]
  o-Chlorophenol
  Group-member nr.: 61-037

$C_6H_5Cl_3Si$
- Trichlorophenylsilane [98-13-5]
  Group-member nr.: 71-011

C_6H_5F
- Fluorobenzene [462-06-6]
  Group-member nr.: 21-019

C_6H_5FO
- 4-Fluorophenol [371-41-5]
  p-Fluorophenol
  Group-member nr.: 61-038

C_6H_5I
- Iodobenzene [591-50-4]
  Group-member nr.: 24-009

$C_6H_5NO_2$
- Nitrobenzene [98-95-3]
  Group-member nr.: 62-045

$C_6H_5NO_3$
- 2-Nitrophenol [88-75-5]
  o-Nitrophenol
  Group-member nr.: 62-046
- 3-Nitrophenol [554-84-7]
  m-Nitrophenol
  Group-member nr.: 62-047
- 4-Nitrophenol [100-02-7]
  p-Nitrophenol
  Group-member nr.: 62-048

C_6H_6
- Benzene [71-43-2]
  Group-member nr.: 14-003

C_6H_6ClN
- 3-Chlorobenzenamine [108-42-9]
  m-Chloroaniline
  Group-member nr.: 64-019

C_6H_6FN
- 4-Fluorobenzenamine [371-40-4]
  p-Fluoroaniline
  Group-member nr.: 64-020

$C_6H_6N_2O_2$
- 2-Nitrobenzenamine [88-74-4]
  o-Nitroaniline
  Group-member nr.: 62-049
- 3-Nitrobenzenamine [99-09-2]
  m-Nitroaniline
  Group-member nr.: 62-050
- 4-Nitrobenzenamine [100-01-6]
  p-Nitroaniline
  Group-member nr.: 62-051

C_6H_6O
- Phenol [108-95-2]
  Group-member nr.: 42-034

$C_6H_6O_2$
- 1,2-Benzenediol [120-80-9]
  1,2-Dihydroxybenzene
  Pyrocatechol
  Catechol
  Group-member nr.: 42-035
- 1,3-Benzenediol [108-46-3]
  1,3-Dihydroxybenzene
  Resorcinol
  Group-member nr.: 42-036
- 1,4-Benzenediol [123-31-9]
  1,4-Dihydroxybenzene
  Hydroquinone
  Group-member nr.: 42-037

C_6H_6S
- Benzenethiol [108-98-5]
  Thiophenol
  Phenyl mercaptan
  Group-member nr.: 52-014

C_6H_7N
- Benzenamine [62-53-3]
  Aniline
  Aminobenzene
  Phenylamine
  Group-member nr.: 31-020
- Bicyclo[2.1.0]pentane-1-carbonitrile [31357-71-8]
  1-Bicyclo[2.1.0]pentyl cyanide
  1-Cyanobicyclo[2.1.0]pentane
  Group-member nr.: 32-017
- 3-Methylenecyclobutanecarbonitrile [15760-35-7]
  3-Methylenecyclobutyl cyanide
  1-Cyano-3-methylenecyclobutane
  Group-member nr.: 32-018
- 2-Methylpyridine [109-06-8]
  α-Picoline
  Group-member nr.: 33-013
- 3-Methylpyridine [108-99-6]
  β-Picoline
  Group-member nr.: 33-014
- 4-Methylpyridine [108-89-4]
  γ-Picoline
  Group-member nr.: 33-015

C_6H_8
- 1,3-Cyclohexadiene [592-57-4]
  Group-member nr.: 14-004
- 1,4-Cyclohexadiene [628-41-1]
  Group-member nr.: 14-005

$C_6H_8N_2$

- 1,3-Benzenediamine [108-45-2]
  1,3-Phenylenediamine
  Group-member nr.: 31-021
- Phenylhydrazine [100-63-0]
  Group-member nr.: 34-007

$C_6H_8O_2$

- 1,4-Cyclohexanedione [637-88-7]
  Tetrahydroquinone
  Group-member nr.: 43-016
- Methyl bicyclo[1.1.0]butane-1-carboxylate [4935-01-7]
  Methyl ester bicyclo[1.1.0]butane-1-carboxylic acid
  Group-member nr.: 45-020

$C_6H_8O_4$

- Dimethyl (*Z*)-2-butenedioate [624-48-6]
  Dimethyl ester (*Z*)-2-butenedioic acid
  Dimethyl *cis*-2-butenedioate
  Dimethyl maleate
  Group-member nr.: 45-021
- 3,6-Dimethyl-1,4-dioxane-2,5-dione [95-96-5]
  Group-member nr.: 47-018

C_6H_8S

- 2,5-Dimethylthiophene [638-02-8]
  Group-member nr.: 53-011

C_6H_9N

- Cyclopentanecarbonitrile [4254-02-8]
  Cyclopentyl cyanide
  Cyanocyclopentane
  Group-member nr.: 32-019
- 2,4-Dimethyl-1*H*-pyrrole [625-82-1]
  Group-member nr.: 33-016
- 2,5-Dimethyl-1*H*-pyrrole [625-84-3]
  Group-member nr.: 33-017

C_6H_{10}

- Cyclohexene [110-83-8]
  1,2,3,4-Tetrahydrobenzene
  Group-member nr.: 14-006
- 1,5-Hexadiene [592-42-7]
  Diallyl
  Group-member nr.: 13-025
- 1-Methylcyclopentene [693-89-0]
  Group-member nr.: 14-007
- 3-Methylcyclopentene [1120-62-3]
  Group-member nr.: 14-008

$C_6H_{10}O$

- Cyclohexanone [108-94-1]
  Pimelic ketone
  Ketohexamethylene
  Group-member nr.: 43-017
- 4-Methyl-3-penten-2-one [141-79-7]
  Mesityl oxide
  Group-member nr.: 43-018
- 7-Oxabicyclo[4.1.0]heptane [286-20-4]
  1,2-Epoxycyclohexane
  Cyclohexene oxide
  Group-member nr.: 46-017

$C_6H_{10}O_2$

- Ethyl cyclopropanecarboxylate [4606-07-9]
  Ethyl ester cyclopropanecarboxylic acid
  Group-member nr.: 45-022
- Methyl cyclobutanecarboxylate [765-85-5]
  Methyl ester cyclobutanecarboxylic acid
  Group-member nr.: 45-023
- 2-Oxepanone [502-44-3]
  ϵ-Caprolactone
  6-Hexanolide
  Group-member nr.: 47-019
- 2-Propenyl propanoate [2408-20-0]
  Allyl propionate
  Group-member nr.: 45-024

$C_6H_{10}O_3$

- Ethyl ester 3-oxobutanoic acid [141-97-9]
  Ethyl 3-oxobutanoate
  Ethyl acetoacetate
  Ethyl ester acetoacetic acid
  Group-member nr.: 47-020

$C_6H_{10}O_4$

- Diethyl ethanedioate [95-92-1]
  Diethyl ester ethanedioic acid
  Diethyl oxalate
  Group-member nr.: 45-025
- 1,2-Ethanediyl diacetate [111-55-7]
  1,2-Ethanediyl ester acetic acid
  Ethylene glycol diacetate
  Group-member nr.: 45-026
- Hexanedioic acid [124-04-9]
  Adipic acid
  1,4-Butanedicarboxylic acid
  Group-member nr.: 44-013

$C_6H_{11}N$

- Hexanenitrile [628-73-9]
  Capronitrile
  Amyl cyanide
  Group-member nr.: 32-020

$C_6H_{11}NO$

- Cyclohexanone oxime [100-64-1]
  Group-member nr.: 62-052
- Hexahydro-2*H*-azepin-2-one [105-60-2]
  ϵ-Caprolactam
  6-Hexanelactam
  6-Aminohexanoic lactam
  Group-member nr.: 62-053

$C_6H_{11}NO_2$

- 4,4-Dimethoxybutanenitrile [14618-78-1]
  β-Cyanopropionaldehyde dimethyl acetal
  Group-member nr.: 62-054

C_6H_{12}

- Cyclohexane [110-82-7]
  Hexamethylene
  Group-member nr.: 12-007
- 2,3-Dimethyl-2-butene [563-79-1]
  Tetramethylethylene
  Group-member nr.: 13-026
- 3,3-Dimethyl-1-butene [558-37-2]
  tert-Butylethylene
  Neohexene
  Group-member nr.: 13-027
- 1-Hexene [592-41-6]
  Hexylene
  Group-member nr.: 13-028
- (*Z*)-2-Hexene [7688-21-3]
  cis-2-Hexene
  Group-member nr.: 13-029
- Methylcyclopentane [96-37-7]
  Group-member nr.: 12-008

$C_6H_{12}BNO_3$

- 2,8,9-Trioxa-5-aza-1-borabicyclo[3.3.3] - undecane [283-56-7]
  2,2',2''-Nitrilotriethanol cyclic ester with boric acid (1:1)
  Triethanolamine borate
  Group-member nr.: 72-007

$C_6H_{12}Cl_2$

- 1,6-Dichlorohexane [2163-00-0]
  Hexamethylene chloride
  Group-member nr.: 22-040

$C_6H_{12}Cl_3O_4P$

- 2-Chloroethanol phosphate (3:1) [115-96-8]
  Tri(2-chloroethyl)phosphate
  Group-member nr.: 72-008

$C_6H_{12}N_2$

- 1,4-Diazobicyclo[2.2.2]octane [280-57-9]
  Triethylenediamine
  Group-member nr.: 33-018
- (1-Methylethylidene)hydrazone 2-propanone [627-70-3]
  Acetone azine
  Dimethyl ketazine
  Group-member nr.: 34-008

$C_6H_{12}O$

- Cyclohexanol [108-93-0]
  Cyclohexyl alcohol
  Hexahydrophenol
  Group-member nr.: 42-038
- 3,3-Dimethyl-2-butanone [75-97-8]
  tert-Butyl methyl ketone
  Pinacolone
  Group-member nr.: 43-019
- 1-(Ethenyloxy)butane [111-34-2]
  Butyl vinyl ether
  Butoxyethylene
  Group-member nr.: 41-013
- 1-(Ethenyloxy)-2-methylpropane [109-53-5]
  Isobutyl vinyl ether
  Isobutoxyethylene
  Group-member nr.: 41-014
- Hexanal [66-25-1]
  Hexaldehyde
  Caproaldehyde
  Group-member nr.: 43-020
- 2-Hexanone [591-78-6]
  Butyl methyl ketone
  Group-member nr.: 43-021
- 3-Hexanone [589-38-8]
  Ethyl propyl ketone
  Group-member nr.: 43-022
- 4-Methyl-2-pentanone [108-10-1]
  Isobutyl methyl ketone
  Group-member nr.: 43-023

$C_6H_{12}O_2$

- Butyl acetate [123-86-4]
  Butyl ethanoate
  Group-member nr.: 45-027
- 1,1-Dimethylethyl acetate [540-88-5]
  1,1-Dimethylethyl ethanoate
  tert-Butyl acetate
  Group-member nr.: 45-028
- Ethyl butanoate [105-54-4]
  Ethyl butyrate
  Group-member nr.: 45-029
- Ethyl 2-methylpropanoate [97-62-1]
  Ethyl isobutyrate
  Group-member nr.: 45-030
- Hexanoic acid [142-62-1]
  Caproic acid
  Group-member nr.: 44-014
- 3-Methylbutyl formate [110-45-2]
  3-Methylbutyl methanoate
  Isoamyl formate
  Isopentyl formate
  Group-member nr.: 45-033
- Methyl 2,2-dimethylpropanoate [598-98-1]
  Methyl ester 2,2-dimethylpropanoic acid
  Methyl pivalate
  Methyl trimethylacetate
  Group-member nr.: 45-031
- Methyl pentanoate [624-24-8]
  Methyl valerate
  Group-member nr.: 45-032
- 2-Methylpropyl acetate [110-19-0]
  2-Methylpropyl ethanoate
  Isobutyl acetate
  Group-member nr.: 45-034
- Propyl propanoate [106-36-5]
  Propyl propionate
  Group-member nr.: 45-035
- Tetrahydro-2*H*-pyran-2-methanol [100-72-1]
  Group-member nr.: 47-021

$C_6H_{12}O_3$
- 2-Ethoxyethanol acetate [111-15-9]
  2-Ethoxyethyl acetate
  Cellosolve acetate
  Group-member nr.: 47-022
- 2,4,6-Trimethyl-1,3,5-trioxane [123-63-7]
  Paraldehyde
  Group-member nr.: 46-018

$C_6H_{12}S$
- Cyclohexanethiol [1569-69-3]
  Cyclohexyl mercaptan
  Group-member nr.: 52-015
- Methylthiocyclopentane [7133-36-0]
  Cyclopentyl methyl sulfide
  Group-member nr.: 51-013

$C_6H_{13}Br$
- 1-Bromohexane [111-25-1]
  n-Hexyl bromide
  Group-member nr.: 23-034
- 3-Bromohexane [3377-87-5]
  Group-member nr.: 23-035

$C_6H_{13}N$
- Cyclohexanamine [108-91-8]
  Cyclohexylamine
  Group-member nr.: 31-022
- Hexahydro-1*H*-azepine [111-49-9]
  Perhydroazepine
  Hexamethylenimine
  Azacycloheptane
  Group-member nr.: 33-019
- 1-Methylpiperidine [626-67-5]
  N-Methylpiperidine
  Group-member nr.: 33-020
- 2-Methylpiperidine [109-05-7]
  2-Pipecoline
  α-Pipecoline
  Group-member nr.: 33-021
- 4-Methylpiperidine [626-58-4]
  4-Pipecoline
  β-Pipecoline
  Group-member nr.: 33-022

$C_6H_{13}NO$
- *N*-Butylacetamide [1119-49-9]
  N-Butylethanamide
  Group-member nr.: 62-055
- *N*, *N*-Diethylacetamide [685-91-6]
  N, *N*-Diethylethanamide
  Group-member nr.: 62-056
- *N*-Methylpentanamide [6225-10-1]
  N-Methylvaleramide
  Group-member nr.: 62-057

C_6H_{14}
- 2,2-Dimethylbutane [75-83-2]
  Neohexane
  Group-member nr.: 11-011
- 2,3-Dimethylbutane [79-29-8]
  Diisopropyl
  Group-member nr.: 11-012
- Hexane [110-54-3]
  Group-member nr.: 11-013
- 2-Methylpentane [107-83-5]
  Dimethylpropylmethane
  Group-member nr.: 11-014
- 3-Methylpentane [96-14-0]
  Diethylmethylmethane
  Group-member nr.: 11-015

$C_6H_{14}N_2O$
- Dipropyldiazene 1-oxide [17697-55-1]
  Group-member nr.: 62-058

$C_6H_{14}O$
- 3,3-Dimethyl-1-butanol [624-95-3]
  Group-member nr.: 42-039
- 2-Ethoxy-2-methylpropane [637-92-3]
  tert-Butyl ethyl ether
  Group-member nr.: 41-015
- 2-Ethyl-1-butanol [97-95-0]
  2-Ethylbutyl alcohol
  Group-member nr.: 42-040
- 1-Hexanol [111-27-3]
  Hexyl alcohol
  Group-member nr.: 42-041
- 2-Hexanol [626-93-7]
  Group-member nr.: 42-042
- 3-Hexanol [623-37-0]
  Ethyl propyl carbinol
  Group-member nr.: 42-043
- 2-Methoxy-2-methylbutane [994-05-8]
  Methyl *tert*-pentyl ether
  3,3-Dimethyl-2-oxapentane
  tert-Amyl methyl ether
  Group-member nr.: 41-016
- 2-Methyl-1-pentanol [105-30-6]
  Group-member nr.: 42-044
- 2-Methyl-2-pentanol [590-36-3]
  Group-member nr.: 42-045
- 3-Methyl-2-pentanol [565-60-6]
  Group-member nr.: 42-046
- 3-Methyl-3-pentanol [77-74-7]
  Group-member nr.: 42-047
- 4-Methyl-2-pentanol [108-11-2]
  Group-member nr.: 42-048
- 1,1'-Oxybispropane [111-43-3]
  Dipropyl ether
  Group-member nr.: 41-017
- 2,2'-Oxybispropane [108-20-3]
  1,1'-Bis(methylethyl) ether
  Diisopropyl ether
  Group-member nr.: 41-018

$C_6H_{14}O_2$
- 2-Butoxyethanol [111-76-2]
  3-Oxa-1-heptanol
  Ethylene glycol monobutyl ether
  Butyl cellosolve
  Group-member nr.: 47-023
- 1,1-Diethoxyethane [105-57-7]
  Acetaldehyde diethyl acetal
  Acetal
  Group-member nr.: 41-019
- 1,2-Diethoxyethane [629-14-1]
  3,6-Dioxaoctane
  Ethylene glycol diethyl ether
  Diethyl cellosolve
  Group-member nr.: 41-020
- 1-Methoxy-2-propoxyethane [17081-22-0]
  2,5-Dioxaoctane
  Ethylene glycol methyl ether propyl ether
  Group-member nr.: 41-021

$C_6H_{14}O_3$
- 2-(2-Ethoxyethoxy)ethanol [111-90-0]
  Diethylene glycol monoethyl ether
  Group-member nr.: 47-024
- 2-Ethyl-2-(hydroxymethyl)-1,3-propanediol [77-99-6]
  1,1,1-Tris(hydroxymethyl)propane
  Trimethylolpropane
  Group-member nr.: 42-049
- 1,1'-Oxybis(2-methoxyethane) [111-96-6]
  2,5,8-Trioxanonane
  Diethylene glycol dimethyl ether
  Diglyme
  Group-member nr.: 41-022

$C_6H_{14}O_4$
- 2,2'-[1,2-Ethanediylbis(oxy)]bisethanol [112-27-6]
  1,8-Dihydroxy-3,6-dioxaoctane
  Triethylene glycol
  Group-member nr.: 47-025

$C_6H_{14}O_6$
- *D*-Mannitol [69-65-8]
  Group-member nr.: 42-050

$C_6H_{14}S$
- 1-Hexanethiol [111-31-9]
  1-Hexyl mercaptan
  Group-member nr.: 52-016
- 2-(Methylthio)pentane [13286-91-4]
  3-Methyl-2-thiahexane
  Methyl 1-methylbutyl sulfide
  Group-member nr.: 51-014
- 1,1'-Thiobispropane [111-47-7]
  Dipropyl sulfide
  Group-member nr.: 51-015
- 2,2'-Thiobispropane [625-80-9]
  Bis(1-methylethyl) sulfide
  Diisopropyl sulfide
  Group-member nr.: 51-016

$C_6H_{14}S_2$
- Dipropyl disulfide [629-19-6]
  4,5-Dithiaoctane
  Group-member nr.: 51-017

$C_6H_{15}Al$
- Triethylaluminium [97-93-8]
  Group-member nr.: 73-018

$C_6H_{15}As$
- Triethylarsine [617-75-4]
  Group-member nr.: 73-019

$C_6H_{15}B$
- Triethylborane [97-94-9]
  Group-member nr.: 72-009

$C_6H_{15}Bi$
- Triethylbismuthine [617-77-6]
  Group-member nr.: 73-020

$C_6H_{15}ClSi$
- Chlorotriethylsilane [994-30-9]
  Group-member nr.: 71-012

$C_6H_{15}Ga$
- Triethylgallium [1115-99-7]
  Group-member nr.: 73-021

$C_6H_{15}In$
- Triethylindium [923-34-2]
  Group-member nr.: 73-022

$C_6H_{15}N$
- *N*,*N*-Diethylethanamine [121-44-8]
  Triethylamine
  Group-member nr.: 31-023
- 1-Hexanamine [111-26-2]
  Hexylamine
  Group-member nr.: 31-024
- *N*-Propyl-1-propanamine [142-84-7]
  Dipropylamine
  Group-member nr.: 31-025

$C_6H_{15}NO_3$
- 2,2',2''-Nitrilotrisethanol [102-71-6]
  Tri(2-hydroxyethyl)amine
  Triethanolamine
  Group-member nr.: 62-059

$C_6H_{15}N_3$
- 1-Piperazineethanamine [140-31-8]
  1-(2-Aminoethyl)piperazine
  Group-member nr.: 33-023

$C_6H_{15}Sb$
- Triethylstibine [617-85-6]
  Triethylantimony
  Group-member nr.: 73-023

$C_6H_{16}Si_2$
- 1,1,3,3-Tetramethyl-1,3-disilacyclobutane [1627-98-1]
  Group-member nr.: 71-013

$C_6H_{18}BN$
- Triethylamineborane [1722-26-5]
  N, *N*-Dimethylethanamine compd. with borane (1:1)
  Group-member nr.: 72-010

$C_6H_{18}N_3OP$
- Hexamethyl phosphoric triamide [680-31-9]
  Hexamethylphosphoramide
  Group-member nr.: 72-011

$C_6H_{18}N_4$
- *N*, *N'*-Bis(2-aminoethyl)-1,2-ethanediamine [112-24-3]
  Triethylenetetramine
  Group-member nr.: 31-026

$C_6H_{18}OSi_2$
- Hexamethyldisiloxane [107-46-0]
  Group-member nr.: 71-014

$C_6H_{18}O_3Si_3$
- Hexamethylcyclotrisiloxane [541-05-9]
  Group-member nr.: 71-015

$C_6H_{18}Si_2$
- Hexamethyldisilane [1450-14-2]
  Group-member nr.: 71-016

$C_6H_{21}N_3Si_3$
- 2,2,4,4,6,6-Hexamethylcyclotrisilazane [1009-93-4]
  Dimethylaminosilane trimer
  Group-member nr.: 71-017

C_7F_8
- Pentafluoro(trifluoromethyl)benzene [434-64-0]
  Octafluorotoluene
  Perfluorotoluene
  Group-member nr.: 21-020

C_7F_{14}
- (Trifluoromethyl)undecafluorocyclohexane [355-02-2]
  Perfluoromethylcyclohexane
  Group-member nr.: 21-021

C_7F_{16}
- Hexadecafluoroheptane [335-57-9]
  Perfluoroheptane
  Group-member nr.: 21-022

$C_7H_3F_5$
- Pentafluoromethylbenzene [771-56-2]
  2,3,4,5,6-Pentafluorotoluene
  Group-member nr.: 21-023

C_7H_4ClNO
- 1-Chloro-3-isocyanatobenzene [2909-38-8]
  m-Chlorophenylisocyanate
  Group-member nr.: 64-021

$C_7H_4F_3NO_2$
- 1-Nitro-3-(trifluoromethyl)benzene [98-46-4]
  m-Trifluoromethylnitrobenzene
  Group-member nr.: 64-022

C_7H_5ClO
- Benzoyl chloride [98-88-4]
  Group-member nr.: 61-039

$C_7H_5ClO_2$
- 2-Chlorobenzoic acid [118-91-2]
  o-Chlorobenzoic acid
  Group-member nr.: 61-040
- 3-Chlorobenzoic acid [535-80-8]
  m-Chlorobenzoic acid
  Group-member nr.: 61-041
- 4-Chlorobenzoic acid [74-11-3]
  p-Chlorobenzoic acid
  Group-member nr.: 61-042

$C_7H_5F_3$
- (Trifluoromethyl)benzene [98-08-8]
  α,α,α-Trifluorotoluene
  Group-member nr.: 21-024

C_7H_5N
- Benzonitrile [100-47-0]
  Phenyl cyanide
  Group-member nr.: 32-021

C_7H_5NO
- Benzoxazole [273-53-0]
  Group-member nr.: 62-060

$C_7H_5NO_4$
- 2-Nitrobenzoic acid [552-16-9]
  o-Nitrobenzoic acid
  o-Nitrodracylic acid
  Group-member nr.: 62-061
- 3-Nitrobenzoic acid [121-92-6]
  m-Nitrobenzoic acid
  m-Nitrodracylic acid
  Group-member nr.: 62-062
- 4-Nitrobenzoic acid [62-23-7]
  p-Nitrobenzoic acid
  p-Nitrodracylic acid
  Group-member nr.: 62-063

C_7H_5NS
- Benzothiazole [95-16-9]
  Group-member nr.: 64-023
- Isothiocyanatobenzene [103-72-0]
  Phenyl ester isothiocyanic acid
  Group-member nr.: 64-024

$C_7H_5N_3O_6$
- 2-Methyl-1,3,5-trinitrobenzene [118-96-7]
  2,4,6-Trinitrotoluene
  Tritol
  TNT
  Group-member nr.: 62-064

$C_7H_5N_5O_8$
- *N*-Methyl-*N*,2,4,6-tetranitrobenzenamine [479-45-8]
  Trinitrophenylmethylnitroamine
  Tetryl
  Group-member nr.: 62-065

$C_7H_6F_8O_3$
- Bis(2,2,3,3-tetrafluoropropyl) ester carbonic acid [1422-70-4]
  Bis(2,2,3,3-tetrafluoropropyl) carbonate
  Group-member nr.: 61-043

$C_7H_6N_2O_4$
- 2-Methyl-1,3-dinitrobenzene [606-20-2]
  2,6-Dinitrotoluene
  Group-member nr.: 62-066

C_7H_6O
- Benzaldehyde [100-52-7]
  Benzoic aldehyde
  Group-member nr.: 43-024

$C_7H_6O_2$
- Benzoic acid [65-85-0]
  Benzenecarboxylic acid
  Carboxybenzene
  Phenylformic acid
  Group-member nr.: 44-015
- 2-Hydroxybenzaldehyde [90-02-8]
  Salicyl aldehyde
  Group-member nr.: 47-026

$C_7H_6O_3$
- 2-Hydroxybenzoic acid [69-72-7]
  Salicylic acid
  Group-member nr.: 47-027

C_7H_7Cl
- (Chloromethyl)benzene [100-44-7]
  Benzyl chloride
  Group-member nr.: 22-041
- 1-Chloro-2-methylbenzene [95-49-8]
  2-Chlorotoluene
  o-Chlorotoluene
  o-Tolyl chloride
  Group-member nr.: 22-042

C_7H_7F
- 1-Fluoro-2-methylbenzene [95-52-3]
  2-Fluorotoluene
  o-Tolyl fluoride
  Group-member nr.: 21-025
- 1-Fluoro-3-methylbenzene [352-70-5]
  3-Fluorotoluene
  m-Tolyl fluoride
  Group-member nr.: 21-026
- 1-Fluoro-4-methylbenzene [352-32-9]
  p-Fluorotoluene
  p-Tolyl fluoride
  Group-member nr.: 21-027

C_7H_7NO
- Benzamide [55-21-0]
  Group-member nr.: 62-067

$C_7H_7NO_2$
- 1-Methyl-3-nitrobenzene [99-08-1]
  m-Nitrotoluene
  Group-member nr.: 62-068
- 1-Methyl-4-nitrobenzene [99-99-0]
  p-Nitrotoluene
  Group-member nr.: 62-069

$C_7H_7NO_3$
- 1-Methoxy-2-nitrobenzene [91-23-6]
  o-Nitroanisole
  Group-member nr.: 62-070
- 1-Methoxy-3-nitrobenzene [555-03-3]
  m-Nitroanisole
  Group-member nr.: 62-071
- 1-Methoxy-4-nitrobenzene [100-17-4]
  p-Nitroanisole
  Group-member nr.: 62-072

C_7H_8
- Bicyclo[2.2.1]hepta-2,5-diene [121-46-0]
  2,5-Norbornadiene
  Group-member nr.: 14-009
- 1,3,5-Cycloheptatriene [544-25-2]
  Group-member nr.: 14-010
- Methylbenzene [108-88-3]
  Toluene
  Phenyl methane
  Group-member nr.: 14-011
- Tetracyclo[3.2.0.0^{2,7}.0^{4,6}]heptane [278-06-8]
  Quadricyclane
  Group-member nr.: 12-009

$C_7H_8Cl_2Si$
- Dichloromethylphenylsilane [149-74-6]
  Group-member nr.: 71-018

C_7H_8O

- Benzenemethanol [100-51-6]
  Benzyl alcohol
  Group-member nr.: 42-051
- Methoxybenzene [100-66-3]
  Methyl phenyl ether
  Anisole
  Group-member nr.: 41-023
- 2-Methylphenol [95-48-7]
  o-Hydroxytoluene
  o-Cresol
  Group-member nr.: 42-052
- 3-Methylphenol [108-39-4]
  m-Hydroxytoluene
  m-Cresol
  Group-member nr.: 42-053
- 4-Methylphenol [106-44-5]
  p-Hydroxytoluene
  p-Cresol
  Group-member nr.: 42-054

$C_7H_8O_2$

- 2,6-Dimethyl-4*H*-pyran-4-one [1004-36-0]
  2,6-Dimethyl-β-pyrone
  Group-member nr.: 47-028

C_7H_8S

- Methylthiobenzene [100-68-5]
  Methyl phenyl sulfide
  Group-member nr.: 51-018

C_7H_9N

- Benzenemethanamine [100-46-9]
  Benzylamine
  Phenylmethylamine
  Aminotoluene
  Group-member nr.: 31-027
- Bicyclo[3.1.0]hexane-1-carbonitrile [31357-72-9]
  1-Bicyclo[3.1.0]hexyl cyanide
  1-Cyanobicyclo[3.1.0]hexane
  Group-member nr.: 32-022
- 2,3-Dimethylpyridine [583-61-9]
  2,3-Lutidine
  Group-member nr.: 33-024
- 2,4-Dimethylpyridine [108-47-4]
  2,4-Lutidine
  Group-member nr.: 33-025
- 2,5-Dimethylpyridine [589-93-5]
  2,5-Lutidine
  Group-member nr.: 33-026
- 2,6-Dimethylpyridine [108-48-5]
  2,6-Lutidine
  Group-member nr.: 33-027
- 3,4-Dimethylpyridine [583-58-4]
  3,4-Lutidine
  Group-member nr.: 33-028
- 3,5-Dimethylpyridine [591-22-0]
  3,5-Lutidine
  Group-member nr.: 33-029
- *N*-Methylbenzenamine [100-61-8]
  N-Methylaniline
  Methylphenylamine
  Group-member nr.: 31-028
- 2-Methylbenzenamine [95-53-4]
  2-Methylaniline
  o-Toluidine
  Group-member nr.: 31-029
- 3-Methylbenzenamine [108-44-1]
  3-Methylaniline
  m-Toluidine
  Group-member nr.: 31-030
- 4-Methylbenzenamine [106-49-0]
  4-Methylaniline
  p-Toluidine
  Group-member nr.: 31-031

$C_7H_{11}N$

- Cyclohexanecarbonitrile [766-05-2]
  Cyclohexyl cyanide
  Cyanocyclohexane
  Group-member nr.: 32-023

C_7H_{12}

- Bicyclo[4.1.0]heptane [286-08-8]
  Norcarane
  Group-member nr.: 12-010
- (*Z*)-Cycloheptene [45510-00-7]
  cis-Cycloheptene
  Group-member nr.: 14-012
- 1-Ethylcyclopentene [2146-38-5]
  Group-member nr.: 14-013
- Ethylidenecyclopentane [2146-37-4]
  Group-member nr.: 12-011
- 4-Methylcyclohexene [591-47-9]
  Group-member nr.: 14-014
- Methylenecyclohexane [1192-37-6]
  Group-member nr.: 12-012

$C_7H_{12}O$

- Cycloheptanone [502-42-1]
  Suberone
  Group-member nr.: 43-025
- 2-Methylcyclohexanone [583-60-8]
  o-Methylcyclohexanone
  Group-member nr.: 43-026
- 3-Methylcyclohexanone [591-24-2]
  m-Methylcyclohexanone
  Group-member nr.: 43-027
- 4-Methylcyclohexanone [589-92-4]
  p-Methylcyclohexanone
  Group-member nr.: 43-028

$C_7H_{12}O_2$

- Butyl 2-propenoate [141-32-2]
  Butyl ester 2-propenoic acid
  Butyl acrylate
  Group-member nr.: 45-036

- 2-Propenyl butanoate [2051-78-7]
  Allyl butyrate
  Group-member nr.: 45-037
- 2-Propenyl 2-methylpropanoate [15727-77-2]
  2-Propenyl ester 2-methylpropanoic acid
  Allyl isobutyrate
  Group-member nr.: 45-038

$C_7H_{12}O_4$
- Diethyl propanedioate [105-53-3]
  Diethyl ester propanedioic acid
  Diethyl malonate
  Group-member nr.: 45-039

$C_7H_{13}NO$
- Hexahydro-2(1*H*)-azocinone [673-66-5]
  β-Enantholactam
  Group-member nr.: 62-073

C_7H_{14}
- Cycloheptane [291-64-5]
  Group-member nr.: 12-013
- 1,1-Dimethylcyclopentane [1638-26-2]
  Group-member nr.: 12-014
- *cis*-1,2-Dimethylcyclopentane [1192-18-3]
  Group-member nr.: 12-015
- *trans*-1,2-Dimethylcyclopentane [822-50-4]
  Group-member nr.: 12-016
- *trans*-1,3-Dimethylcyclopentane [1759-58-6]
  Group-member nr.: 12-017
- Ethylcyclopentane [1640-89-7]
  Group-member nr.: 12-018
- 1-Heptene [592-76-7]
  1-Heptylene
  Group-member nr.: 13-030
- Methylcyclohexane [108-87-2]
  Hexahydrotoluene
  Group-member nr.: 12-019

$C_7H_{14}ClNO$
- Dipropylcarbamic chloride [27086-19-7]
  N, *N*-Dipropylcarbamoyl chloride
  Group-member nr.: 64-025

$C_7H_{14}O$
- Cycloheptanol [502-41-0]
  Group-member nr.: 42-055
- 3,4-Dimethylpentanal [19353-21-0]
  3,4-Dimethylvaleraldehyde
  Group-member nr.: 43-029
- 2,4-Dimethyl-3-pentanone [565-80-0]
  Diisopropyl ketone
  Group-member nr.: 43-030
- Heptanal [111-71-7]
  Heptyl aldehyde
  Enanthaldehyde
  Group-member nr.: 43-031
- 2-Heptanone [110-43-0]
  Methyl pentyl ketone
  Group-member nr.: 43-032
- 4-Heptanone [123-19-3]
  Dipropyl ketone
  Group-member nr.: 43-033
- 1-Methylcyclohexanol [590-67-0]
  Group-member nr.: 42-056
- 2-Methylcyclohexanol [583-59-5]
  o-Methylcyclohexanol (unspecified stereoisomer and chirality)
  Group-member nr.: 42-057
- *cis*-2-Methylcyclohexanol [7443-70-1]
  Group-member nr.: 42-058
- *trans*-2-Methylcyclohexanol [7443-52-9]
  Group-member nr.: 42-059
- 3-Methylcyclohexanol [591-23-1]
  m-Methylcyclohexanol (unspecified stereoisomer and chirality)
  Group-member nr.: 42-060
- 4-Methylcyclohexanol [589-91-3]
  p-Methylcyclohexanol (unspecified stereoisomer)
  Group-member nr.: 42-061
- 3-Methylhexanal [19269-28-4]
  Group-member nr.: 43-034
- 2-Methyl-3-hexanone [7379-12-6]
  Group-member nr.: 43-035

$C_7H_{14}O_2$
- Ethyl 2,2-dimethylpropanoate [3938-95-2]
  Ethyl ester 2,2-dimethylpropanoic acid
  Ethyl pivalate
  Group-member nr.: 45-040
- Ethyl 3-methylbutanoate [108-64-5]
  Ethyl isovalerate
  Group-member nr.: 45-041
- Ethyl pentanoate [539-82-2]
  Ethyl valerate
  Group-member nr.: 45-042
- Heptanoic acid [111-14-8]
  Enanthic acid
  Group-member nr.: 44-016
- 3-Methylbutyl acetate [123-92-2]
  3-Methylbutyl ethanoate
  Isoamyl acetate
  Isopentyl acetate
  Group-member nr.: 45-043
- 2-Methylpropyl propanoate [540-42-1]
  Isobutyl propionate
  Group-member nr.: 45-044
- Pentyl acetate [628-63-7]
  Pentyl ethanoate
  Amyl acetate
  Group-member nr.: 45-045
- Propyl butanoate [105-66-8]
  Propyl butyrate
  Group-member nr.: 45-046
- Propyl 2-methylpropanoate [644-49-5]
  Propyl isobutyrate
  Group-member nr.: 45-047

$C_7H_{15}Br$
- 1-Bromoheptane [629-04-9]
  n-Heptyl bromide
  Group-member nr.: 23-036

$C_7H_{15}Cl$
- 3-Chloroheptane [999-52-0]
  Group-member nr.: 22-043

$C_7H_{15}N$
- Octahydroazocine [1121-92-2]
  Perhydroazocine
  Heptamethylenimine
  Group-member nr.: 33-030

C_7H_{16}
- 2,2-Dimethylpentane [590-35-2]
  Group-member nr.: 11-016
- 2,3-Dimethylpentane [565-59-3]
  Group-member nr.: 11-017
- 2,4-Dimethylpentane [108-08-7]
  Group-member nr.: 11-018
- 3,3-Dimethylpentane [562-49-2]
  Group-member nr.: 11-019
- 3-Ethylpentane [617-78-7]
  Triethylmethane
  Group-member nr.: 11-020
- Heptane [142-82-5]
  Dipropylmethane
  Group-member nr.: 11-021
- 2-Methylhexane [591-76-4]
  Isoheptane
  Ethylisobutylmethane
  Group-member nr.: 11-022
- 3-Methylhexane [589-34-4]
  Group-member nr.: 11-023
- 2,2,3-Trimethylbutane [464-06-2]
  Isopropyltrimethylmethane
  Triptane
  Group-member nr.: 11-024

$C_7H_{16}O$
- 2-Ethoxy-2-methylbutane [919-94-8]
  Ethyl *tert*-pentyl ether
  4,4-Dimethyl-3-oxahexane
  tert-Amyl ethyl ether
  Group-member nr.: 41-024
- 3-Ethyl-3-pentanol [597-49-9]
  Group-member nr.: 42-062
- 1-Heptanol [111-70-6]
  Heptyl alcohol
  Group-member nr.: 42-063
- 2-Heptanol [543-49-7]
  Group-member nr.: 42-064
- 4-Heptanol [589-55-9]
  Dipropyl carbinol
  Group-member nr.: 42-065

$C_7H_{16}O_2$
- 1-Butoxy-2-methoxyethane [13343-98-1]
  2,5-Dioxanonane
  Ethylene glycol butyl ether methyl ether
  Group-member nr.: 41-025
- 1,3-Diethoxypropane [3459-83-4]
  3,7-Dioxanonane
  Propylene glycol diethyl ether
  Group-member nr.: 41-026
- 1,5-Dimethoxypentane [111-89-7]
  2,8-Dioxanonane
  Pentylene glycol dimethyl ether
  Group-member nr.: 41-027
- 1-Ethoxy-4-methoxybutane [36865-47-1]
  2,7-Dioxanonane
  Butylene glycol ethyl ether methyl ether
  Group-member nr.: 41-028
- 1-Ethoxy-2-propoxyethane [18854-31-4]
  3,6-Dioxanonane
  Ethylene glycol ethyl ether propyl ether
  Group-member nr.: 41-029
- 1-Methoxy-3-propoxypropane [89851-49-0]
  2,6-Dioxanonane
  Propylene glycol methyl ether propyl ether
  Group-member nr.: 41-030

$C_7H_{16}O_3$
- 1,1',1''-[Methylidynetris(oxy)]trisethane [122-51-0]
  Triethoxymethane
  Triethyl ester orthoformic acid
  Triethyl orthoformate
  Group-member nr.: 41-031

$C_7H_{16}S$
- 1-Heptanethiol [1639-09-4]
  1-Heptyl mercaptan
  Group-member nr.: 52-017
- 2-(Methylthio)hexane [76858-84-9]
  3-Methyl-2-thiaheptane
  Methyl 1-methylpentyl sulfide
  Group-member nr.: 51-019

$C_7H_{20}Si_2$
- Methylenebis[trimethylsilane] [2117-28-4]
  2,2,4,4-Tetramethyl-2,4-disilapentane
  Hexamethyldisilamethane
  Bis(trimethylsilyl)methane
  Group-member nr.: 71-019

C_8F_{16}
- Decafluorobis(trifluoromethyl)cyclohexane [26637-68-3]
  Perfluorodimethylcyclohexane
  Hexadecafluorodimethyl-cyclohexane
  (unspecified isomer)
  Group-member nr.: 21-028

$C_8F_{16}O$
- 2,2,3,3,4,5,5-Heptafluorotetrahydro-4-(nonafluorobutyl)furan [646-85-5]
  Perfluoro-3-butyltetrahydrofuran
  Group-member nr.: 61-044

C_8F_{18}
- Octadecafluorooctane [307-34-6]
  Perfluorooctane
  Group-member nr.: 21-029

$C_8H_4Cl_2O_2$
- 1,2-Benzenedicarbonyl dichloride [88-95-9]
  Phthaloyl chloride
  Phthaloyl dichloride
  Group-member nr.: 61-045

$C_8H_4N_2$
- 1,2-Benzenedicarbonitrile [91-15-6]
  1,2-Dicyanobenzene
  Phthalonitrile
  Group-member nr.: 32-024

C_8H_6
- Ethynylbenzene [536-74-3]
  Phenylacetylene
  Group-member nr.: 14-015

C_8H_6O
- Benzofuran [271-89-6]
  Benzo[*b*]furan
  Cumaron
  Group-member nr.: 46-019

C_8H_6S
- Benzo[*b*]thiophene [95-15-8]
  Group-member nr.: 53-012

$C_8H_7N_5O_8$
- *N*-Ethyl-*N*,2,4,6-tetranitrobenzenamine [6052-13-7]
  Trinitrophenylethylnitroamine
  Ethyltetryl
  Group-member nr.: 62-074
- 3-Methyl-2,4,6-trinitro-*N*-(nitromethyl)benzenamine [43072-20-4]
  2,4,6-Trinitro-*N*-(methylnitro)-*m*-toluidine
  Methyltetryl
  Group-member nr.: 62-075

C_8H_8
- 1,3,5,7-Cyclooctatetraene [629-20-9]
  Group-member nr.: 14-016
- Ethenylbenzene [100-42-5]
  Vinylbenzene
  Styrene
  Group-member nr.: 14-017

C_8H_8O
- 2,3-Dihydrobenzofuran [496-16-2]
  Coumaran
  Group-member nr.: 46-020
- 1-Phenylethanone [98-86-2]
  Methyl phenyl ketone
  Acetophenone
  Acetyl benzene
  Hypnone
  Group-member nr.: 43-036

$C_8H_8O_2$
- Methyl benzoate [93-58-3]
  Methyl ester benzoic acid
  Group-member nr.: 45-048
- 2-Methylbenzoic acid [118-90-1]
  o-Toluic acid
  o-Toluylic acid
  Group-member nr.: 44-017
- 3-Methylbenzoic acid [99-04-7]
  m-Toluic acid
  Group-member nr.: 44-018
- 4-Methylbenzoic acid [99-94-5]
  p-Toluic acid
  Group-member nr.: 44-019

$C_8H_8O_3$
- Methyl 2-hydroxybenzoate [119-36-8]
  Methyl salicylate
  Group-member nr.: 47-029
- 3a,4,7,7a-Tetrahydro-1,3-isobenzofurandione [85-43-8]
  4-Cyclohexene-1,2-dicarboxylic acid anhydride
  1,2,3,6-Tetrahydrophthalic anhydride
  Group-member nr.: 44-020

$C_8H_9NO_2$
- *N*-(2-Hydroxyphenyl)acetamide [614-80-2]
  o-Hydroxyacetanilide
  Group-member nr.: 62-076
- Methyl ester phenylcarbamic acid [2603-10-3]
  Methyl phenylcarbamate
  Methyl ester carbanilic acid
  Group-member nr.: 62-077

C_8H_{10}
- 1,2-Dimethylbenzene [95-47-6]
  o-Xylene
  Group-member nr.: 14-018
- 1,3-Dimethylbenzene [108-38-3]
  m-Xylene
  Group-member nr.: 14-019
- 1,4-Dimethylbenzene [106-42-3]
  p-Xylene
  Group-member nr.: 14-020
- Ethylbenzene [100-41-4]
  Phenylethane
  Group-member nr.: 14-021

$C_8H_{10}N_2$
- Phenylhydrazone acetaldehyde [935-07-9]
  Group-member nr.: 34-009

$C_8H_{10}O$
- Benzeneethanol [60-12-8]
  Phenethyl alcohol
  2-Phenylethanol
  Group-member nr.: 42-066
- Ethoxybenzene [103-73-1]
  Ethyl phenyl ether
  Phenetole
  Group-member nr.: 41-032
- 1-Methoxy-4-methylbenzene [104-93-8]
  Methyl 4-methylphenyl ether
  p-Methylanisole
  Methyl *p*-tolyl ether
  p-Cresol methyl ether
  Group-member nr.: 41-033

$C_8H_{10}O_2$
- 2-Phenoxyethanol [122-99-6]
  Ethylene glycol monophenyl ether
  Group-member nr.: 47-030

$C_8H_{10}O_4$
- Di-2-propenyl ethanedioate [615-99-6]
  Di-2-propenyl ester ethanedioic acid
  Diallyl oxalate
  Group-member nr.: 45-049

$C_8H_{11}N$
- Benzeneethanamine [64-04-0]
  2-Phenylethylamine
  Phenethylamine
  Group-member nr.: 31-032
- *endo*-Bicyclo[2.2.1]heptane-2-carbonitrile [3211-87-8]
  endo-2-Cyanobicyclo[2.2.1]heptane
  endo-2-Norbornanecarbonitrile
  Group-member nr.: 32-025
- *exo*-Bicyclo[2.2.1]heptane-2-carbonitrile [3211-90-3]
  exo-2-Cyanobicyclo[2.2.1]heptane
  exo-2-Norbornanecarbonitrile
  Group-member nr.: 32-026
- *N*, *N*-Dimethylbenzenamine [121-69-7]
  N, *N*-Dimethylaniline
  Group-member nr.: 31-033
- 2,6-Dimethylbenzenamine [87-62-7]
  2,6-Dimethylaniline
  2,6-Xylidine
  Group-member nr.: 31-034
- 2,3,6-Trimethylpyridine [1462-84-6]
  2,3,6-Collidine
  Group-member nr.: 33-031
- 2,4,6-Trimethylpyridine [108-75-8]
  2,4,6-Collidine
  Group-member nr.: 33-032

C_8H_{12}
- Bicyclo[2.2.2]oct-2-ene [931-64-6]
  Group-member nr.: 14-022
- 1,5-Cyclooctadiene (unspecified stereoisomer) [111-78-4]
  Group-member nr.: 14-023

$C_8H_{12}N_2$
- 1,6-Diisocyanohexane [929-57-7]
  1,6-Hexamethylene isocyanide
  1,6-Hexamethylene diisocyanide
  Group-member nr.: 34-010

$C_8H_{12}N_2O_2$
- 1,6-Diisocyanatohexane [822-06-0]
  Hexamethylene diisocyanate
  Hexamethylene ester isocyanic acid
  Group-member nr.: 62-078

C_8H_{14}
- Bicyclo[2.2.2]octane [280-33-1]
  Group-member nr.: 12-020
- *cis*-Bicyclo[4.2.0]octane [28282-35-1]
  Group-member nr.: 12-021
- Cyclooctene (unspecified stereoisomer) [931-88-4]
  Group-member nr.: 14-024
- Ethylidenecyclohexane [1003-64-1]
  Group-member nr.: 12-022
- *endo*-2-Methylbicyclo[2.2.1]heptane [765-90-2]
  endo-2-Methylnorbornane
  Group-member nr.: 12-023
- *exo*-2-Methylbicyclo[2.2.1]heptane [872-78-6]
  exo-2-Methylnorbornane
  Group-member nr.: 12-024
- *cis*-Octahydropentalene [1755-05-1]
  cis-Bicyclo[3.3.0]octane
  Group-member nr.: 12-025
- *trans*-Octahydropentalene [5597-89-7]
  trans-Bicyclo[3.3.0]octane
  Group-member nr.: 12-026
- (2-Propenyl)cyclopentane [3524-75-2]
  Allylcyclopentane
  Group-member nr.: 12-027

$C_8H_{14}O$
- 6-Methyl-5-hepten-2-one [110-93-0]
  Group-member nr.: 43-037
- 3-Oxabicyclo[3.2.2]nonane [283-27-2]
  Group-member nr.: 46-021

$C_8H_{14}O_2$
- Butyl 2-methyl-2-propenoate [97-88-1]
  Butyl methacrylate
  Group-member nr.: 45-050
- 2-Propenyl pentanoate [6321-45-5]
  Allyl valerate
  Group-member nr.: 45-051

$C_8H_{14}O_4$
- Diethyl butanedioate [123-25-1]
  Diethyl ester butanedioic acid
  Diethyl succinate
  Group-member nr.: 45-052
- Dipropyl ethanedioate [615-98-5]
  Dipropyl ester ethanedioic acid
  Dipropyl oxalate
  Group-member nr.: 45-053
- 1,2-Ethanediyl dipropanoate [123-80-8]
  1,2-Ethanediyl ester propanoic acid
  Ethylene
  glycol dipropionate
  Group-member nr.: 45-054

$C_8H_{14}O_6$
- Diethyl ester 2,3-dihydroxy-[*R*-(*R**, *R**)]-butanedioic acid [87-91-2]
  Diethyl *L*(+)-tartarate
  Group-member nr.: 47-031

$C_8H_{15}N$
- 3-Azabicyclo[3.2.2]nonane [283-24-9]
  Group-member nr.: 33-033

C_8H_{16}
- Cyclooctane [292-64-8]
  Group-member nr.: 12-028
- 1,1-Dimethylcyclohexane [590-66-9]
  Group-member nr.: 12-029
- *cis*-1,2-Dimethylcyclohexane [2207-01-4]
  Group-member nr.: 12-030
- *trans*-1,2-Dimethylcyclohexane [6876-23-9]
  Group-member nr.: 12-031
- *cis*-1,3-Dimethylcyclohexane [638-04-0]
  Group-member nr.: 12-032
- *trans*-1,3-Dimethylcyclohexane [2207-03-6]
  Group-member nr.: 12-033
- *cis*-1,4-Dimethylcyclohexane [624-29-3]
  Group-member nr.: 12-034
- *trans*-1,4-Dimethylcyclohexane [2207-04-7]
  Group-member nr.: 12-035
- Ethylcyclohexane [1678-91-7]
  Group-member nr.: 12-036
- 1-Octene [111-66-0]
  1-Octylene
  1-Caprylene
  Group-member nr.: 13-031
- Propylcyclopentane [2040-96-2]
  Group-member nr.: 12-037
- 2,4,4-Trimethyl-1-pentene [107-39-1]
  Group-member nr.: 13-032
- 2,4,4-Trimethyl-2-pentene [107-40-4]
  Group-member nr.: 13-033

$C_8H_{16}O$
- Octanal [124-13-0]
  Octyl aldehyde
  Caprilic aldehyde
  Group-member nr.: 43-038
- 2-Octanone [111-13-7]
  Hexyl methyl ketone
  Group-member nr.: 43-039

$C_8H_{16}O_2$
- Butyl butanoate [109-21-7]
  Butyl butyrate
  Group-member nr.: 45-055
- Ethyl hexanoate [123-66-0]
  Ethyl caproate
  Group-member nr.: 45-056
- Hexyl acetate [142-92-7]
  Hexyl ethanoate
  Group-member nr.: 45-057
- 3-Methylbutyl propanoate [105-68-0]
  Isoamyl propionate
  Isopentyl propionate
  Group-member nr.: 45-059
- Methyl heptanoate [106-73-0]
  Methyl enanthoate
  Group-member nr.: 45-058
- 2-Methylpropyl butanoate [539-90-2]
  Isobutyl butyrate
  Group-member nr.: 45-060
- 2-Methylpropyl 2-methylpropanoate [97-85-8]
  Isobutyl isobutyrate
  Group-member nr.: 45-061
- Octanoic acid [124-07-2]
  Caprylic acid
  Group-member nr.: 44-021
- Pentyl propanoate [624-54-4]
  Amyl propionate
  Group-member nr.: 45-062
- Propyl pentanoate [141-06-0]
  Propyl valerate
  Group-member nr.: 45-063

$C_8H_{16}O_4$
- 2-(2-Ethoxyethoxy)ethanol acetate [112-15-2]
  2-(2-Ethoxyethoxy)ethyl acetate
  Diethylene glycol ethyl ether acetate
  Group-member nr.: 47-032
- 1,4,7,10-Tetraoxacyclododecane [294-93-9]
  Group-member nr.: 46-022

$C_8H_{17}Br$
- 1-Bromooctane [111-83-1]
  n-Octyl bromide
  Group-member nr.: 23-037

$C_8H_{17}NO$
- *N*, *N*-Dipropylacetamide [1116-24-1]
  Group-member nr.: 62-079

C_8H_{18}
- 2,3-Dimethylhexane [584-94-1]
  Group-member nr.: 11-025
- 2,5-Dimethylhexane [592-13-2]
  Group-member nr.: 11-026

- 3,3-Dimethylhexane [563-16-6]
  Group-member nr.: 11-027
- 3-Ethyl-2-methylpentane [609-26-7]
  Group-member nr.: 11-028
- 2-Methylheptane [592-27-8]
  Group-member nr.: 11-029
- 3-Methylheptane [589-81-1]
  Group-member nr.: 11-030
- 4-Methylheptane [589-53-7]
  Methyldipropylmethane
  Group-member nr.: 11-031
- Octane [111-65-9]
  Group-member nr.: 11-032
- 2,2,3,3-Tetramethylbutane [594-82-1]
  Group-member nr.: 11-033
- 2,2,4-Trimethylpentane [540-84-1]
  Isooctane
  Group-member nr.: 11-034
- 2,3,3-Trimethylpentane [560-21-4]
  Group-member nr.: 11-035
- 2,3,4-Trimethylpentane [565-75-3]
  Group-member nr.: 11-036

$C_8H_{18}N_2O$
- Bis(1,1-dimethylethyl)diazene 1-oxide [16649-52-8]
  Di-*tert*-butyldiazene *N*-oxide
  Group-member nr.: 62-080

$C_8H_{18}O$
- 2-Ethyl-1-hexanol [104-76-7]
  Group-member nr.: 42-067
- 2-Methyl-1-heptanol [106-67-2]
  Group-member nr.: 42-068
- 2-Methyl-2-heptanol [625-25-2]
  Group-member nr.: 42-069
- 2-Methyl-4-heptanol [21570-35-4]
  Group-member nr.: 42-070
- 3-Methyl-2-heptanol [31367-46-1]
  Group-member nr.: 42-071
- 4-Methyl-2-heptanol [56298-90-9]
  Group-member nr.: 42-072
- 4-Methyl-3-heptanol [14979-39-6]
  Group-member nr.: 42-073
- 4-Methyl-4-heptanol [598-01-6]
  Group-member nr.: 42-074
- 5-Methyl-1-heptanol [7212-53-5]
  Group-member nr.: 42-075
- 5-Methyl-2-heptanol [54630-50-1]
  Group-member nr.: 42-076
- 6-Methyl-2-heptanol [4730-22-7]
  Group-member nr.: 42-077
- 6-Methyl-3-heptanol [18720-66-6]
  Group-member nr.: 42-078
- 1-Octanol [111-87-5]
  Octyl alcohol
  Group-member nr.: 42-079
- 2-Octanol [123-96-6]
  Group-member nr.: 42-080
- 3-Octanol [589-98-0]
  Group-member nr.: 42-081
- 4-Octanol [589-62-8]
  Group-member nr.: 42-082
- 1,1'-Oxybisbutane [142-96-1]
  Dibutyl ether
  Group-member nr.: 41-034

$C_8H_{18}O_2$
- 1,2-Dipropoxyethane [18854-56-3]
  4,7-Dioxadecane
  Ethylene glycol dipropyl ether
  Group-member nr.: 41-035

$C_8H_{18}O_3$
- 2-(2-Butoxyethoxy)ethanol [112-34-5]
  Diethylene glycol monobutyl ether
  Group-member nr.: 47-033
- 1,1'-Oxybis(2-ethoxyethane) [112-36-7]
  3,6,9-Trioxaundecane
  Diethylene glycol diethyl ether
  Group-member nr.: 41-036

$C_8H_{18}O_4$
- 2,5,8,11-Tetraoxadodecane [112-49-2]
  Triethylene glycol dimethyl ether
  Triglyme
  Group-member nr.: 41-037

$C_8H_{18}O_5$
- 2,2'-[Oxybis(2,1-ethanediyloxy)]bisethanol [112-60-7]
  Tetraethylene glycol
  Group-member nr.: 47-034

$C_8H_{18}S$
- 2-(Methylthio)heptane [54063-12-6]
  3-Methyl-2-thiaoctane
  Methyl 1-methylhexyl sulfide
  Group-member nr.: 51-020
- 1,1'-Thiobisbutane [544-40-1]
  Dibutyl sulfide
  Group-member nr.: 51-021

$C_8H_{19}N$
- 2-Methyl-*N*-(2-methylpropyl)-1-propanamine [110-96-3]
  Diisobutylamine
  Group-member nr.: 31-035

$C_8H_{19}NSi$
- 1-[2-(Trimethylsilyl)ethyl]azetidine [42525-64-4]
  N-(β-Trimethylsilylethyl)trimethylenimine
  Group-member nr.: 71-020

$C_8H_{20}Ge$
- Tetraethylgermane [597-63-7]
  Group-member nr.: 73-024

$C_8H_{20}N_4$
- 1,4-Piperazinediethanamine [6531-38-0]
  1,4-Bis(2-aminoethyl)piperazine
  Group-member nr.: 33-034
- *N*-[2-(1-Piperazinyl)ethyl]-1,2-ethanediamine [24028-46-4]
  1-[2-[(2-Aminoethyl)amino]ethyl]piperazine
  Group-member nr.: 33-035

$C_8H_{20}O_4Si$
- Tetraethyl ester silicic acid [78-10-4]
  Tetraethyl orthosilicate
  Tetraethoxysilane
  Group-member nr.: 71-021

$C_8H_{20}Pb$
- Tetraethylplumbane [78-00-2]
  Tetraethyllead
  TEL
  Group-member nr.: 73-025

$C_8H_{20}Si$
- Tetraethylsilane [631-36-7]
  Group-member nr.: 71-022

$C_8H_{20}Sn$
- Tetraethylstannane [597-64-8]
  Tetraethyltin
  Group-member nr.: 73-026

$C_8H_{23}N_5$
- *N*-(2-Aminoethyl)-*N*'-[2-[(2-aminoethyl)amino]ethyl]-1,2-ethanediamine [112-57-2]
  Tetraethylenepentamine
  Group-member nr.: 31-036

$C_8H_{24}O_4Si_4$
- Octamethylcyclotetrasiloxane [556-67-2]
  Group-member nr.: 71-023

$C_8H_{28}N_4Si_4$
- 2,2,4,4,6,6,8,8-Octamethylcyclotetrasilazane [1020-84-4]
  Dimethylaminosilane tetramer
  Group-member nr.: 71-024

$C_9H_4O_5$
- 1,3-Dihydro-1,3-dioxo-5-isobenzofurancarboxylic acid [552-30-7]
  1,2,4-Benzenetricarboxylic 1:2 anhydride
  Trimellitic anhydride
  Group-member nr.: 47-035

$C_9H_6N_2O_2$
- 2,4-Diisocyanato-1-methylbenzene [584-84-9]
  4-Methyl-*m*-phenyl diisocyanate
  4-Methyl-1,3-phenylene ester isocyanic acid
  2,4-Diisocyanate tolylene
  Group-member nr.: 62-081

C_9H_7ClO
- 2-Chloro-3-phenyl-2-propenal [18365-42-9]
  α-Chlorocinnamaldehyde
  Group-member nr.: 61-046

C_9H_7N
- Isoquinoline [119-65-3]
  Benzo[c]pyridine
  Group-member nr.: 33-036
- Quinoline [91-22-5]
  Benzo[*b*]pyridine
  Group-member nr.: 33-037

C_9H_8
- 1*H*-Indene [95-13-6]
  Group-member nr.: 14-025

C_9H_8O
- 3-Phenyl-2-propenal [104-55-2]
  Cinnamaldehyde
  Group-member nr.: 43-040

C_9H_{10}
- 2,3-Dihydroindene [496-11-7]
  Indan
  Group-member nr.: 14-026
- (1-Methylethenyl)benzene [98-83-9]
  Isopropenylbenzene
  α-Methylstyrene
  Group-member nr.: 14-027

$C_9H_{10}O$
- 3,4-Dihydro-1*H*-2-benzopyran [493-05-0]
  Isochroman
  Group-member nr.: 46-023
- 3,4-Dihydro-2*H*-1-benzopyran [493-08-3]
  Chroman
  Group-member nr.: 46-024
- 1-Phenyl-1-propanone [93-55-0]
  Ethyl phenyl ketone
  Propiophenone
  Group-member nr.: 43-041

$C_9H_{10}O_2$
- Ethyl benzoate [93-89-0]
  Ethyl ester benzoic acid
  Group-member nr.: 45-064
- (Phenoxymethyl)oxirane [122-60-1]
  Phenyl glycidyl ether
  Group-member nr.: 47-036
- Phenylmethyl acetate [140-11-4]
  Phenylmethyl ester acetic acid
  Benzyl acetate
  Group-member nr.: 45-065

$C_9H_{10}O_3$
- 3a,4,7,7a-Tetrahydro-4-methyl-1,3-isobenzofurandione [5333-84-6]
  3-Methyl-4-cyclohexene-1,2-dicarboxylic acid anhydride
  3-Methyl-1,2,3,6-tetrahydrophthalic anhydride
  Group-member nr.: 44-022

$C_9H_{11}N$
- 1,2,3,4-Tetrahydroquinoline [635-46-1]
  Group-member nr.: 33-038
- 5,6,7,8-Tetrahydroquinoline [10500-57-9]
  2,3-Cyclohexenopyridine
  Group-member nr.: 33-039

$C_9H_{11}NO_2$
- Ethyl ester phenylcarbamic acid [101-99-5]
  Ethyl phenylcarbamate
  Ethyl ester carbanilic acid
  Group-member nr.: 62-082

C_9H_{12}
- (1-Methylethyl)benzene [98-82-8]
  Isopropylbenzene
  Cumene
  Group-member nr.: 14-028
- Propylbenzene [103-65-1]
  Group-member nr.: 14-029
- 1,2,3-Trimethylbenzene [526-73-8]
  Hemimellitene
  Group-member nr.: 14-030
- 1,2,4-Trimethylbenzene [95-63-6]
  Pseudocumene
  Group-member nr.: 14-031
- 1,3,5-Trimethylbenzene [108-67-8]
  Mesitylene
  Group-member nr.: 14-032

$C_9H_{12}O$
- Benzenepropanol [122-97-4]
  3-Phenyl-1-propanol
  Phenylethyl carbinol
  Hydrocinnamyl alcohol
  Group-member nr.: 42-083
- 1-Ethoxy-4-methylbenzene [622-60-6]
  Ethyl-4-methylphenyl ether
  p-Methylphenetole
  Ethyl *p*-tolyl ether
  Group-member nr.: 41-038
- 1-Methoxy-2,4-dimethylbenzene [6738-23-4]
  Methyl 2,4-dimethylphenyl ether
  2,4-Dimethylanisole
  Group-member nr.: 41-039
- Propoxybenzene [622-85-5]
  Phenyl propyl ether
  Group-member nr.: 41-040

$C_9H_{13}N$
- Benzenepropanamine [2038-57-5]
  3-Phenylpropylamine
  Group-member nr.: 31-037
- *N*, *N*, 2-Trimethylbenzenamine [609-72-3]
  N, *N*-Dimethyl-*o*-toluidine
  Group-member nr.: 31-038

$C_9H_{14}O$
- 2,5,6-Trimethyl-2-cyclohexen-1-one [20030-30-2]
  Group-member nr.: 43-042

$C_9H_{14}O_6$
- 1,2,3-Propanetriol triacetate [102-76-1]
  1,2,3-Propanetriyl ester acetic acid
  Triacetin
  Group-member nr.: 45-066

C_9H_{16}
- *cis*-Bicyclo[6.1.0]nonane [13757-43-2]
  Group-member nr.: 12-038
- Octahydro-1*H*-indene [496-10-6]
  Hexahydroindan
  Hydrindan (unspecified stereoisomer)
  Group-member nr.: 12-039
- *cis*-Octahydro-1*H*-indene [4551-51-3]
  cis-Hexahydroindan
  cis-Hydrindan
  Group-member nr.: 12-040
- *trans*-Octahydro-1*H*-indene [3296-50-2]
  trans-Hexahydroindan
  trans-Hydrindan
  Group-member nr.: 12-041
- (2-Propenyl)cyclohexane [2114-42-3]
  Allylcyclohexane
  Group-member nr.: 12-042

$C_9H_{16}O_2$
- Ethyl cyclohexanecarboxylate [3289-28-9]
  Ethyl ester cyclohexanecarboxylic acid
  Group-member nr.: 45-067

$C_9H_{16}O_4$
- Dipropyl propanedioate [1117-19-7]
  Dipropyl ester propanedioic acid
  Dipropyl malonate
  Group-member nr.: 45-068

C_9H_{18}
- Butylcyclopentane [2040-95-1]
  Group-member nr.: 12-043
- 1-Nonene [124-11-8]
  Group-member nr.: 13-034
- Propylcyclohexane [1678-92-8]
  Group-member nr.: 12-044

$C_9H_{18}O$
- 2,6-Dimethyl-4-heptanone [108-83-8]
  Diisobutyl ketone
  Group-member nr.: 43-043
- Nonanal [124-19-6]
  Nonyl aldehyde
  Pelargonic aldehyde
  Group-member nr.: 43-044
- 5-Nonanone [502-56-7]
  Dibutyl ketone
  Group-member nr.: 43-045

$C_9H_{18}O_2$
- Butyl pentanoate [591-68-4]
  Butyl valerate
  Group-member nr.: 45-069
- 3-Methylbutyl butanoate [106-27-4]
  Isoamyl butyrate
  Isopentyl butyrate
  Group-member nr.: 45-071

- 3-Methylbutyl 2-methylpropanoate [2050-01-3]
  Isoamyl isobutyrate
  Isopentyl isobutyrate
  Group-member nr.: 45-072
- Methyl octanoate [111-11-5]
  Methyl caprylate
  Group-member nr.: 45-070
- 2-Methylpropyl pentanoate [10588-10-0]
  Isobutyl valerate
  Group-member nr.: 45-073
- Nonanoic acid [112-05-0]
  Pelargonic acid
  Group-member nr.: 44-023
- Pentyl butanoate [540-18-1]
  Pentyl butyrate
  Amyl butyrate
  Group-member nr.: 45-074

$C_9H_{19}Br$
- 1-Bromononane [693-58-3]
  n-Nonyl bromide
  Group-member nr.: 23-038

$C_9H_{19}NOS$
- *S*-Ethyl ester dipropylcarbamothionic acid [759-94-4]
  S-Ethyl dipropylthiocarbamate
  Group-member nr.: 64-026

C_9H_{20}
- 3,3-Diethylpentane [1067-20-5]
  Group-member nr.: 11-037
- Nonane [111-84-2]
  Group-member nr.: 11-038
- 2,2,3,3-Tetramethylpentane [7154-79-2]
  Group-member nr.: 11-039
- 2,2,4,4-Tetramethylpentane [1070-87-7]
  Group-member nr.: 11-040

$C_9H_{20}N_2O$
- Tetraethyl urea [1187-03-7]
  Tetraethylcarbamide
  Group-member nr.: 62-083

$C_9H_{20}O$
- 1-Nonanol [143-08-8]
  Nonyl alcohol
  Group-member nr.: 42-084

$C_9H_{21}Al$
- Tripropylaluminium [102-67-0]
  Group-member nr.: 73-027

$C_9H_{21}ClO_3Si$
- (3-Chloropropyl)triethoxysilane [5089-70-3]
  Group-member nr.: 71-025

$C_9H_{23}NO_3Si$
- 3-(Triethoxysilyl)-1-propanamine [919-30-2]
  3-(Triethoxysilyl)propylamine
  3-Aminopropyltriethoxysilane
  Group-member nr.: 71-026

$C_9H_{24}Si_2$
- 1,3-Propanediylbis[trimethylsilane] [2295-05-8]
  2,2,6,6-Tetramethyl-2,6-disilaheptane
  Hexamethyldisilylpropane
  Group-member nr.: 71-027

$C_9H_{24}Si_3$
- 1,1,3,3,5,5-Hexamethyl-1,3,5-trisilacyclohexane [1627-99-2]
  Group-member nr.: 71-028

$C_{10}F_{16}$
- Hexadecafluoro-1,2,3,4,5,6,7,8-octahydronaphthalene [54939-04-7]
  Group-member nr.: 21-030

$C_{10}F_{18}$
- *cis*-Octadecafluorodecahydronaphthalene [60433-11-6]
  cis-Perfluorodecaline
  Group-member nr.: 21-031
- *trans*-Octadecafluorodecahydronaphthalene [60433-12-7]
  trans-Perfluorodecaline
  Group-member nr.: 21-032

$C_{10}H_2O_6$
- 1*H*,3*H*-Benzo[1,2-*c*:4,5-*c*']difuran-1,3,5,7-tetrone [89-32-7]
  1,2,4,5-Benzenetetracarboxylic 1,2:4,5 dianhydride
  Pyromellitic dianhydride
  Group-member nr.: 44-024

$C_{10}H_6O_2S$
- 1-Oxide naphtho[1,8-*cd*]-1,2-dithiole [49833-12-7]
  Naphthalene 1,8-disulfide-*S*-oxide
  Group-member nr.: 63-005

$C_{10}H_7Cl$
- 1-Chloronaphthalene [90-13-1]
  Group-member nr.: 22-044
- 2-Chloronaphthalene [91-58-7]
  Group-member nr.: 22-045

$C_{10}H_7NO_2$
- 1-Nitronaphthalene [86-57-7]
  α-Nitronaphthalene
  Group-member nr.: 62-084

$C_{10}H_8$
- Naphthalene [91-20-3]
  Group-member nr.: 14-033

$C_{10}H_8O$
- 1-Naphthol [90-15-3]
  α-Naphthol
  Group-member nr.: 42-085
- 2-Naphthol [135-19-3]
  β-Naphthol
  Group-member nr.: 42-086

$C_{10}H_9N$
- 2-Naphthalenamine [91-59-8]
  β-Naphthalenamine
  2-Aminonaphthalene
  Group-member nr.: 31-039

$C_{10}H_{10}$
- Tricyclo[3.3.2.0^{2,8}]deca-3,6,9-triene [1005-51-2]
  Group-member nr.: 14-034

$C_{10}H_{10}Fe$
- Ferrocene [102-54-5]
  Dicyclopentadienyliron
  Group-member nr.: 73-028

$C_{10}H_{10}O_2$
- 2-Propenyl benzoate [583-04-0]
  2-Propenyl ester benzoic acid
  Allyl benzoate
  Group-member nr.: 45-075

$C_{10}H_{10}O_4$
- Dimethyl 1,2-benzenedicarboxylate [131-11-3]
  Dimethyl phthalate
  Dimethyl ester 1,2-benzenedicarboxylic acid
  Group-member nr.: 45-076
- Dimethyl 1,4-benzenedicarboxylate [120-61-6]
  Dimethyl terephthalate
  Dimethyl ester 1,4-benzenedicarboxylic acid
  Group-member nr.: 45-077

$C_{10}H_{12}$
- 1,2,3,4-Tetrahydronaphthalene [119-64-2]
  Tetralin
  Group-member nr.: 14-035

$C_{10}H_{12}O$
- 1-(4-Ethylphenyl)ethanone [937-30-4]
  4'-Ethylacetophenone
  Group-member nr.: 43-046
- 1-Methoxy-4-(1-propenyl)benzene [104-46-1]
  p-Propenylanisole
  Anethole
  Group-member nr.: 41-041

$C_{10}H_{12}O_2$
- 2-Methoxy-4-(2-propenyl)phenol [97-53-0]
  4-Allyl-2-methoxyphenol
  Eugenol
  Group-member nr.: 47-037
- Propyl benzoate [2315-68-6]
  Propyl ester benzoic acid
  Group-member nr.: 45-078

$C_{10}H_{13}NO_2$
- Propyl ester phenylcarbamic acid [5532-90-1]
  Propyl phenylcarbamate
  Propyl ester carbanilic acid
  Group-member nr.: 62-085

$C_{10}H_{14}$
- Butylbenzene [104-51-8]
  1-Phenylbutane
  Group-member nr.: 14-036
- (1,1-Dimethylethyl)benzene [98-06-6]
  tert-Butylbenzene
  2-Methyl-2-phenylpropane
  Group-member nr.: 14-037
- 1-Methyl-4-(1-methylethyl)benzene [99-87-6]
  1-Isopropyl-4-methylbenzene
  p-Cymene
  Group-member nr.: 14-038
- (1-Methylpropyl)benzene [135-98-8]
  sec-Butylbenzene
  2-Phenylbutane
  Group-member nr.: 14-039
- 1,2,3,4-Tetramethylbenzene [488-23-3]
  Prehnitene
  Group-member nr.: 14-040
- 1,2,3,5-Tetramethylbenzene [527-53-7]
  Isodurene
  Group-member nr.: 14-041
- 1,2,4,5-Tetramethylbenzene [95-93-2]
  Durene
  Group-member nr.: 14-042

$C_{10}H_{14}O$
- 2-Methyl-5-(1-methylethyl)phenol [499-75-2]
  5-Isopropyl-2-methylphenol
  Carvacrol
  Group-member nr.: 42-087

$C_{10}H_{14}O_4$
- Di-2-propenyl butanedioate [925-16-6]
  Di-2-propenyl ester butanedioic acid
  Diallyl succinate
  Group-member nr.: 45-079

$C_{10}H_{14}Si$
- Ethenyldimethylphenylsilane [1125-26-4]
  Group-member nr.: 71-029

$C_{10}H_{15}N$
- *N*, *N*-Diethylbenzenamine [91-66-7]
  N, *N*-Diethylaniline
  Group-member nr.: 31-040

$C_{10}H_{15}NO$
- (*E*)-(±)-2-Methyl-5-(1-methylethenyl)-2-cyclohexen-1-one oxime [55658-55-4]
  dl-Carvoxime
  Group-member nr.: 62-086

- [*R*-(*E*)]-2-Methyl-5-(1-methylethenyl)-2-cyclohexen-1-one oxime [60827-56-7]
  l-Carvoxime
  Group-member nr.: 62-087

$C_{10}H_{16}$
- 6,6-Dimethyl-2-methylenebicyclo[3.1.1]heptane [127-91-3]
  β-Pinene
  Nopinene
  Terebenthene
  Group-member nr.: 12-045
- 4-Methylene-1-(1-methylethyl)bicyclo[3.1.0] -hexane [3387-41-5]
  Sabinene
  Group-member nr.: 12-046
- 1-Methyl-4-(1-methylethenyl)cyclohexene [138-86-3]
  Limonene
  Group-member nr.: 14-043
- Octahydro-4,7-methano-1*H*-indene [6004-38-2]
  Hexahydro-4,7-methanoindan
  Tricyclo[5.2.1.0^{2,6}]decane
  Tetrahydrodicyclopentadiene
  Group-member nr.: 12-047

$C_{10}H_{16}O$
- Camphor [76-22-2]
  1,7,7-Trimethylbicyclo[2.2.1]heptan-2-one
  Group-member nr.: 43-047
- 3,7-Dimethyl-2,6-octadienal [5392-40-5]
  Citral
  Group-member nr.: 43-048
- 3,7-Dimethyl-6-octen-1-yn-3-ol [29171-20-8]
  Dehydrolynalol
  Group-member nr.: 42-088
- 5-Methyl-2-(1-methylethylidene)cyclohexanone [15932-80-6]
  p-Menth-4(8)-en-3-one
  1-Isopropylidene-4-methylcyclo-hexan-2-on
  Pulegone
  Group-member nr.: 43-049

$C_{10}H_{18}$
- 1,1'-Bicyclopentyl [1636-39-1]
  Cyclopentylcyclopentane
  Group-member nr.: 12-048
- *cis*-Decahydroazulene [16189-46-1]
  cis-Perhydroazulene
  cis-Bicyclo[5.3.0]decane
  Group-member nr.: 12-049
- Decahydronaphthalene [91-17-8]
  Decalin (unspecified stereoisomer)
  Group-member nr.: 12-050
- *cis*-Decahydronaphthalene [493-01-6]
  cis-Decalin
  Group-member nr.: 12-051
- *trans*-Decahydronaphthalene [493-02-7]
  trans-Decalin
  Group-member nr.: 12-052
- Octahydromethyl-1*H*-indene [unknown]
  Hexahydromethylindan
  Methylhydrindan (unspecified isomer)
  Group-member nr.: 12-053
- 2,6,6-Trimethylbicyclo[3.1.1]heptane [473-55-2]
  Pinane
  Group-member nr.: 12-054

$C_{10}H_{18}O$
- 3,7-Dimethyl-1,6-octadien-3-ol [78-70-6]
  Linalol
  Group-member nr.: 42-089

$C_{10}H_{18}O_4$
- Bis(2-methylpropyl) ethanedioate [2050-61-5]
  Bis(2-methylpropyl) ester ethanedioic acid
  Diisobutyl oxalate
  Group-member nr.: 45-080
- Dipropyl butanedioate [925-15-5]
  Dipropyl ester butanedioic acid
  Dipropyl succinate
  Group-member nr.: 45-081
- 1,2-Ethanediyl dibutanoate [105-72-6]
  1,2-Ethanediyl ester butanoic acid
  1,2-Ethanediol dibutyrate
  Ethylene glycol dibutyrate
  Group-member nr.: 45-082

$C_{10}H_{20}$
- Butylcyclohexane [1678-93-9]
  Group-member nr.: 12-055
- 1-Decene [872-05-9]
  1-Decylene
  α-Decylene
  Group-member nr.: 13-035
- Diethylcyclohexane (unspecified isomer) [1331-43-7]
  Group-member nr.: 12-056
- 1,4-Diethylcyclohexane [1679-00-1]
  Group-member nr.: 12-057
- (1,1-Dimethylethyl)cyclohexane [3178-22-1]
  tert-Butylcyclohexane
  Group-member nr.: 12-058

$C_{10}H_{20}O$
- Decanal [112-31-2]
  Decyl aldehyde
  Capric aldehyde
  Group-member nr.: 43-050

$C_{10}H_{20}O_2$
- Decanoic acid [334-48-5]
  Capric acid
  Group-member nr.: 44-025
- Hexyl butanoate [2639-63-6]
  Hexyl butyrate
  Group-member nr.: 45-083

- 3-Methylbutyl pentanoate [2050-09-1]
  Isoamyl valerate
  Isopentyl valerate
  Group-member nr.: 45-084
- Octyl acetate [112-14-1]
  Group-member nr.: 45-085

$C_{10}H_{20}O_4$
- 2-(2-Butoxyethoxy)ethanol acetate [124-17-4]
  2-(2-Butoxyethoxy)ethyl acetate
  Diethylene glycol butyl ether acetate
  Group-member nr.: 47-038

$C_{10}H_{20}O_5$
- 1,4,7,10,13-Pentaoxacyclopentadecane [33100-27-5]
  Group-member nr.: 46-025

$C_{10}H_{21}NO$
- *N*, *N*-Dibutylacetamide [1563-90-2]
  Group-member nr.: 62-088

$C_{10}H_{22}$
- Decane [124-18-5]
  Decyl hydride
  Group-member nr.: 11-041
- 2,7-Dimethyloctane [1072-16-8]
  Group-member nr.: 11-042
- 2-Methylnonane [871-83-0]
  Isodecane
  Group-member nr.: 11-043
- 3-Methylnonane [5911-04-6]
  Group-member nr.: 11-044
- 4-Methylnonane [17301-94-9]
  Group-member nr.: 11-045
- 5-Methylnonane [15869-85-9]
  Group-member nr.: 11-046

$C_{10}H_{22}N_4$
- 1,1'-(1,2-Ethanediyl)bispiperazine [19479-83-5]
  Dipiperazinylethane
  Group-member nr.: 33-040

$C_{10}H_{22}O$
- 1-Decanol [112-30-1]
  Decyl alcohol
  Group-member nr.: 42-090
- 5-Decanol [5205-34-5]
  Group-member nr.: 42-091
- 3,7-Dimethyl-1-octanol [106-21-8]
  Group-member nr.: 42-092
- 4-Propyl-4-heptanol [2198-72-3]
  Group-member nr.: 42-093

$C_{10}H_{22}O_2$
- 1,1'-[1,2-Ethanediylbis(oxy)]bisbutane [112-48-1]
  1,2-Dibutoxyethane
  Ethylene glycol dibutyl ether
  Group-member nr.: 41-042
- 1,1'-[Ethylidenebis(oxy)]bisbutane [871-22-7]
  1,1-Dibutoxyethane
  Acetaldehyde dibutylacetal
  6-Methyl-5,7-dioxaundecane
  Group-member nr.: 41-043

$C_{10}H_{22}O_5$
- 2,5,8,11,14-Pentaoxapentadecane [143-24-8]
  Bis(methoxyethoxyethyl) ether
  Tetraethylene glycol dimethyl ether
  Group-member nr.: 41-044

$C_{10}H_{22}O_6$
- 3,6,9,12-Tetraoxatetradecane-1,14-diol [4792-15-8]
  Pentaethylene glycol
  Group-member nr.: 47-039

$C_{10}H_{22}S$
- 1-Decanethiol [143-10-2]
  1-Decyl mercaptan
  Group-member nr.: 52-018
- 1,1'-Thiobispentane [872-10-6]
  6-Thiaundecane
  Dipentyl sulfide
  Group-member nr.: 51-022

$C_{10}H_{25}N_5$
- *N*-(2-Aminoethyl)-1,4-piperazinediethanamine [31295-54-2]
  1-(2-Aminoethyl)-4-[2-[(2-aminoethyl)-amino]ethyl]piperazine
  Group-member nr.: 33-041

$C_{10}H_{26}O_3Si_3$
- 2,2,4,4-Tetraethyl-6,6-dimethylcyclotrisiloxane [110505-51-6]
  Group-member nr.: 71-030

$C_{11}H_{10}$
- 1-Methylnaphthalene [90-12-0]
  α-Methylnaphthalene
  Group-member nr.: 14-044
- 2-Methylnaphthalene [91-57-6]
  β-Methylnaphthalene
  Group-member nr.: 14-045

$C_{11}H_{12}O_2$
- Ethyl 3-phenyl-2-propenoate [103-36-6]
  Ethyl ester 3-phenyl-2-propenoic acid
  Ethyl cinnamate (unspecified stereoisomer)
  Group-member nr.: 45-086
- Phenylmethyl 2-methyl-2-propenoate [2495-37-6]
  Phenylmethyl ester 2-methyl-2-propenoic acid
  Benzyl methacrylate
  Group-member nr.: 45-087

$C_{11}H_{14}$
- 1,1-Dimethylindan [4912-92-9]
  Group-member nr.: 14-046

- 4,6-Dimethylindan [1685-82-1]
  Group-member nr.: 14-047
- 4,7-Dimethylindan [6682-71-9]
  Group-member nr.: 14-048

$C_{11}H_{14}O_2$
- Ethyl benzenepropanoate [2021-28-5]
  Ethyl ester benzenepropanoic acid
  Ethyl 3-phenylpropanoate
  Ethyl hydrocinnamate
  Group-member nr.: 45-088

$C_{11}H_{16}$
- Pentamethylbenzene [700-12-9]
  Group-member nr.: 14-049

$C_{11}H_{16}Si$
- Ethenyldimethyl(phenylmethyl)silane [18001-46-2]
  Benzyldimethylvinylsilane
  Group-member nr.: 71-031

$C_{11}H_{20}$
- Decahydro-1-methylnaphthalene [2958-75-0]
  α-Methyldecalin
  Group-member nr.: 12-059
- Decahydro-2-methylnaphthalene [2958-76-1]
  β-Methyldecalin
  Group-member nr.: 12-060
- Ethyloctahydro-1*H*-indene [95098-51-4]
  Ethylhexahydroindan
  Ethylhydrindan (unspecified isomer)
  Group-member nr.: 12-061

$C_{11}H_{20}O_2$
- Oxacyclododecan-2-one [1725-03-7]
  Undecanolide
  Undecanolactone
  Group-member nr.: 47-040

$C_{11}H_{22}$
- 1-Undecene [821-95-4]
  α-Undecene
  Group-member nr.: 13-036

$C_{11}H_{22}O_2$
- Methyl decanoate [110-42-9]
  Methyl caprate
  Group-member nr.: 45-089
- Undecanoic acid [112-37-8]
  Undecylic acid
  Group-member nr.: 44-026

$C_{11}H_{24}$
- 2-Methyldecane [6975-98-0]
  Group-member nr.: 11-047
- Undecane [1120-21-4]
  Hendecane
  Group-member nr.: 11-048

$C_{11}H_{24}O$
- 1-Methoxydecane [7289-52-3]
  Decyl methyl ether
  Group-member nr.: 41-045
- 1-Undecanol [112-42-5]
  Undecyl alcohol
  Decyl carbinol
  1-Hendecanol
  Group-member nr.: 42-094

$C_{12}F_{10}$
- 2,2',3,3',4,4',5,5',6,6'-Decafluoro-1,1'-biphenyl [434-90-2]
  Decafluorobiphenyl
  Perfluorobiphenyl
  Group-member nr.: 21-033

$C_{12}H_8$
- Acenaphthylene [208-96-8]
  Group-member nr.: 14-050

$C_{12}H_8Cl_2O_2S$
- 1,1'-Sulfonylbis(4-chlorobenzene) [80-07-9]
  Bis(*p*-chlorophenyl)sulfone
  4,4'-Dichlorodiphenyl sulfone
  Group-member nr.: 64-027

$C_{12}H_8N_2O_5$
- 1,1'-Oxybis(4-nitrobenzene) [101-63-3]
  Bis(*p*-nitrophenyl) ether
  4,4'-Dinitrodiphenyl ether
  Group-member nr.: 62-089

$C_{12}H_8O$
- Dibenzofuran [132-64-9]
  Diphenylene oxide
  Group-member nr.: 46-026

$C_{12}H_8OS_2$
- Dibenzo[*c,e*][1,2]dithiin-5-oxide [49833-13-8]
  Diphenylene 2,2'-disulfide-*S*-oxide
  Group-member nr.: 63-006

$C_{12}H_8S$
- Dibenzothiophene [132-65-0]
  Group-member nr.: 53-013

$C_{12}H_9Cl$
- 2-Chloro-1,1'-biphenyl [2051-60-7]
  o-Chlorobiphenyl
  o-Chlorodiphenyl
  Group-member nr.: 22-046
- 4-Chloro-1,1'-biphenyl [2051-62-9]
  p-Chlorobiphenyl
  p-Chlorodiphenyl
  Group-member nr.: 22-047

$C_{12}H_9Cl_3Si$
- [1,1'-Biphenyl]-2-yltrichlorosilane [18030-62-1]
  o-(Trichlorosilyl)biphenyl
  o-Trichlorosilanediphenyl
  Group-member nr.: 71-032
- [1,1'-Biphenyl]-4-yltrichlorosilane [18030-61-0]
  p-(Trichlorosilyl)biphenyl
  p-Trichlorosilanediphenyl
  Group-member nr.: 71-033

$C_{12}H_{10}$
- Acenaphthene [83-32-9]
  1,8-Dihydroacenaphthalene
  Ethylenenaphthalene
  Group-member nr.: 14-051
- 1,1'-Biphenyl [92-52-4]
  Diphenyl
  Group-member nr.: 14-052

$C_{12}H_{10}N_2$
- Diphenyldiazene [103-33-3]
  Azobenzene
  Diphenyldiimide (unspecified stereoisomer)
  Group-member nr.: 34-011
- (*E*)-Diphenyldiazene [17082-12-1]
  trans-Azobenzene
  trans-Diphenyldiimide
  Group-member nr.: 34-012

$C_{12}H_{10}O$
- 1,1'-Oxybisbenzene [101-84-8]
  Diphenyl ether
  Diphenyl oxide
  Group-member nr.: 41-046

$C_{12}H_{10}S$
- 1,1'-Thiobisbenzene [139-66-2]
  Diphenyl sulfide
  Phenyl sulfide
  Group-member nr.: 51-023

$C_{12}H_{11}N$
- [1,1'-Biphenyl]-2-amine [90-41-5]
  2-Aminobiphenyl
  Group-member nr.: 31-041
- *N*-Phenylbenzenamine [122-39-4]
  Diphenylamine
  Group-member nr.: 31-042

$C_{12}H_{12}$
- 1,8-Dimethylnaphthalene [569-41-5]
  Group-member nr.: 14-053
- 2,3-Dimethylnaphthalene [581-40-8]
  Group-member nr.: 14-054
- 2,6-Dimethylnaphthalene [581-42-0]
  Group-member nr.: 14-055
- 2,7-Dimethylnaphthalene [582-16-1]
  Group-member nr.: 14-056

$C_{12}H_{12}Ge$
- Diphenylgermane [1675-58-7]
  Group-member nr.: 73-029

$C_{12}H_{12}N_2O$
- 4,4'-Oxybisbenzenamine [101-80-4]
  4,4'-Oxydianiline
  4,4'-Diaminodiphenyl ether
  Group-member nr.: 62-090

$C_{12}H_{12}N_2O_2S$
- 4,4'-Sulfonylbisbenzenamine [80-08-0]
  4,4'-Sulfonyldianiline
  4-Aminophenyl sulfone
  Dapsone
  Group-member nr.: 64-028

$C_{12}H_{14}O_4$
- Diethyl 1,2-benzenedicarboxylate [84-66-2]
  Diethyl phthalate
  Diethyl ester
  1,2-benzenedicarboxylic acid
  Group-member nr.: 45-090
- Diethyl 1,4-benzenedicarboxylate [636-09-9]
  Diethyl terephthalate
  Diethyl ester 1,4-benzenedicarboxylic acid
  Group-member nr.: 45-091

$C_{12}H_{16}$
- Cyclohexylbenzene [827-52-1]
  Phenylcyclohexane
  Group-member nr.: 14-057

$C_{12}H_{18}$
- (*E*, *E*, *Z*)-1,5,9-Cyclododecatriene [706-31-0]
  trans,trans,cis-1,5,9-Cyclododecatriene
  Group-member nr.: 14-058
- Decahydrodimethanonaphthalene [unknown]
  Dimethanodecalin (unspecified isomer)
  Group-member nr.: 12-062
- Hexamethylbenzene [87-85-4]
  Mellitene
  Group-member nr.: 14-059
- Hexylbenzene [1077-16-3]
  1-Phenylhexane
  Group-member nr.: 14-060

$C_{12}H_{20}$
- 1,3-Dimethyltricyclo[3.3.1.1^{3,7}]decane [702-79-4]
  1,3-Dimethyladamantane
  Group-member nr.: 12-063
- Tricyclo[6.2.1.1^{3,6}]dodecane [281-84-5]
  Group-member nr.: 12-064

$C_{12}H_{20}O_6$
- 1,2,3-Propanetriyl tripropanoate [139-45-7]
  1,2,3-Propanetriyl ester propanoic acid
  Tripropionin
  Group-member nr.: 45-092

$C_{12}H_{22}$
- 1,1'-Bicyclohexyl [92-51-3]
  Cyclohexylcyclohexane
  Dodecahydrobiphenol
  Group-member nr.: 12-065
- Decahydrodimethylnaphthalene [28777-88-0]
  Dimethyldecalin (unspecified isomer)
  Group-member nr.: 12-066
- Ethyldecahydronaphthalene [25551-49-9]
  Ethyldecalin (unspecified isomer)
  Group-member nr.: 12-067
- 1-Ethyldecahydronaphthalene [1008-17-9]
  1-Ethyldecalin
  α-Ethyldecalin (unspecified stereoisomer)
  Group-member nr.: 12-068
- 2-Ethyldecahydronaphthalene [1618-23-1]
  2-Ethyldecalin
  β-Ethyldecalin (unspecified stereoisomer)
  Group-member nr.: 12-069
- Octahydro(1-methylethyl)-1*H*-indene [88889-26-3]
  Hexahydroisopropylindan
  Isopropylhydrindan (unspecified isomer)
  Group-member nr.: 12-070

$C_{12}H_{22}O_2$
- Nonyl 2-propenoate [2664-55-3]
  Nonyl ester 2-propenoic acid
  Nonyl acrylate
  Group-member nr.: 45-093
- Octyl 2-methyl-2-propenoate [2157-01-9]
  Octyl ester 2-methyl-2-propenoic acid
  Octyl methacrylate
  Group-member nr.: 45-094

$C_{12}H_{22}O_4$
- Bis(3-methylbutyl) ethanedioate [2051-00-5]
  Bis(3-methylbutyl) ester ethanedioic acid
  Diisoamyl oxalate
  Group-member nr.: 45-095
- Bis(2-methylpropyl) butanedioate [925-06-4]
  Bis(2-methylpropyl) ester butanedioic acid
  Diisobutyl succinate
  Group-member nr.: 45-096
- Dibutyl butanedioate [141-03-7]
  Dibutyl ester butanedioic acid
  Dibutyl succinate
  Group-member nr.: 45-097

$C_{12}H_{22}O_4Pb$
- Lead(2+) salt hexanoic acid [15773-53-2]
  Lead(II) hexanoate
  Lead(II) caproate
  Group-member nr.: 74-004

$C_{12}H_{24}$
- 1-Dodecene [112-41-4]
  α-Dodecylene
  Group-member nr.: 13-037

$C_{12}H_{24}N_2O_2$
- *N*, *N*'-Dipropylhexanediamine [10263-96-4]
  N, *N*'-Dipropyladipamide
  Group-member nr.: 62-091

$C_{12}H_{24}O_2$
- Decyl acetate [112-17-4]
  Group-member nr.: 45-098
- Dodecanoic acid [143-07-7]
  Lauric acid
  Group-member nr.: 44-027

$C_{12}H_{25}NO$
- *N*, *N*-Dipentylacetamide [16238-16-7]
  Group-member nr.: 62-092

$C_{12}H_{26}$
- Dodecane [112-40-3]
  Dihexyl
  Group-member nr.: 11-049
- 2,2,4,6,6-Pentamethylheptane [13475-82-6]
  Group-member nr.: 11-050

$C_{12}H_{26}O$
- 1-Dodecanol [112-53-8]
  Dodecyl alcohol
  Lauryl alcohol
  Group-member nr.: 42-095

$C_{12}H_{26}O_3$
- 1,1'-[Oxybis(2,1-ethanediyloxy)]bisbutane [112-73-2]
  Bis(2-butoxyethyl)ether
  Diethylene glycol dibutyl ether
  Group-member nr.: 41-047

$C_{12}H_{26}S$
- 1,1'-Thiobishexane [6294-31-1]
  7-Thiatridecane
  Dihexyl sulfide
  Group-member nr.: 51-024

$C_{12}H_{28}O_4Si$
- Tetrapropyl ester silicic acid [682-01-9]
  Tetrapropyl orthosilicate
  Tetrapropoxysilane
  Group-member nr.: 71-034

$C_{12}H_{30}OSi_2$
- Hexaethyldisiloxane [994-49-0]
  Group-member nr.: 71-035

$C_{12}H_{30}O_3Si_3$
- Hexaethylcyclotrisiloxane [2031-79-0]
  Group-member nr.: 71-036

$C_{13}H_9N$
- Acridine [260-94-6]
  Benzo[*b*]quinoline
  2,3,5,6-Dibenzopyridine
  Group-member nr.: 33-042
- Benzo[*h*]quinoline [230-27-3]
  7,8-Benzoquinoline
  Group-member nr.: 33-043
- Phenanthridine [229-87-8]
  Benzo[c]quinoline
  Group-member nr.: 33-044

$C_{13}H_{10}$
- 9*H*-Fluorene [86-73-7]
  α-Diphenylenemethane
  Group-member nr.: 14-061

$C_{13}H_{10}N_2$
- *N*, *N*'-Methanetetraylbisbenzenamine [622-16-2]
  Diphenylcarbodiimide
  Group-member nr.: 34-013

$C_{13}H_{10}O$
- Diphenyl methanone [119-61-9]
  Diphenyl ketone
  Benzophenone
  Group-member nr.: 43-051

$C_{13}H_{10}O_3$
- Phenyl ester 2-hydroxybenzoic acid [118-55-8]
  Phenyl salicylate
  Salol
  Group-member nr.: 47-041

$C_{13}H_{11}Cl$
- 1,1'-(Chloromethylene)bisbenzene [90-99-3]
  Diphenylchloromethane
  Group-member nr.: 22-048

$C_{13}H_{11}N$
- 9-Methyl-9*H*-carbazole [1484-12-4]
  Group-member nr.: 33-045

$C_{13}H_{12}$
- 1,1'-Methylenebisbenzene [101-81-5]
  Diphenylmethane
  Benzylbenzene
  Group-member nr.: 14-062

$C_{13}H_{13}N$
- *N*-Methyl-*N*-phenylbenzenamine [552-82-9]
  N-Methyldiphenylamine
  Group-member nr.: 31-043

$C_{13}H_{14}N_2$
- 4,4'-Methylenebisbenzenamine [101-77-9]
  4,4'-Methylenedianiline
  Bis-(4-aminophenyl)methane
  Group-member nr.: 31-044

$C_{13}H_{15}N$
- 2,3,4,9-Tetrahydro-9-methyl-1*H*-carbazole [6303-88-4]
  1,2,3,4-Tetrahydro-9-methylcarbazole
  Group-member nr.: 33-046

$C_{13}H_{15}NO$
- 1-(1-Isocyanato-1-methylethyl)-3-(1-methylethenyl) benzene [2094-99-7]
  m-Isopropenyl-α,α-dimethylbenzyl ester isocyanic acid
  Group-member nr.: 62-093
- 1-(1-Isocyanato-1-methylethyl)-4-(1-methylethenyl) benzene [2889-58-9]
  p-Isopropenyl-α,α-dimethylbenzyl ester isocyanic acid
  Group-member nr.: 62-094

$C_{13}H_{18}$
- 1,1,4,6-Tetramethylindan [941-60-6]
  Group-member nr.: 14-063
- 1,1,4,7-Tetramethylindan [1078-04-2]
  Group-member nr.: 14-064

$C_{13}H_{19}NO_2$
- Hexyl ester phenylcarbamic acid [7461-26-9]
  Hexyl phenylcarbamate
  Hexyl ester carbanilic acid
  Group-member nr.: 62-095

$C_{13}H_{20}$
- Heptylbenzene [1078-71-3]
  1-Phenylheptane
  Group-member nr.: 14-065

$C_{13}H_{20}O$
- 6,10-Dimethyl-3,5,9-undecatriene-2-one [141-10-6]
  Pseudoionone
  ψ-Ionone
  Group-member nr.: 43-052
- 6,10-Dimethyl-4,5,9-undecatriene-2-one [16647-05-5]
  Group-member nr.: 43-053
- 4-(2,6,6-Trimethyl-1-cyclohexen-1-yl)-3-buten-2-one [14901-07-6]
  β-Ionone
  Group-member nr.: 43-054

$C_{13}H_{22}$
- 1,3,5-Trimethyltricyclo[3.3.1.1^{3,7}]decane [707-35-7]
  1,3,5-Trimethyladamantane
  Group-member nr.: 12-071

$C_{13}H_{24}$
- Decahydro(1-methylethyl)naphthalene [27193-29-9]
  Isopropyldecalin (unspecified isomer)
  Group-member nr.: 12-072
- Decahydro-1-(1-methylethyl)naphthalene [1010-74-8]
  Decahydro-1-isopropylnaphthalene
  α-Isopropyldecalin
  Group-member nr.: 12-073

- Decahydro-1-propylnaphthalene [91972-45-1]
  1-Propyldecalin
  α-Propyldecalin
  Group-member nr.: 12-074
- 2-Methyl-1,1'-bicyclohexyl [unknown]
  1-Methylcyclohexylcyclohexane
  Group-member nr.: 12-075
- 1,1'-Methylenebiscyclohexane [3178-23-2]
  Dicyclohexylmethane
  Group-member nr.: 12-076

$C_{13}H_{24}O_2$
- Nonyl 2-methyl-2-propenoate [2696-43-7]
  Nonyl ester 2-methyl-2-propenoic acid
  Nonyl methacrylate
  Group-member nr.: 45-099
- Oxacyclotetradecan-2-one [1725-04-8]
  Tridecanolide
  Tridecanolactone
  Group-member nr.: 47-042

$C_{13}H_{26}$
- Heptylcyclohexane [5617-41-4]
  Group-member nr.: 12-077

$C_{13}H_{26}O$
- 6,10-Dimethyl-2-undecanone [1604-34-8]
  Hexahydropseudoionone
  Group-member nr.: 43-055

$C_{13}H_{26}O_2$
- Tridecanoic acid [638-53-9]
  Tridecylic acid
  Group-member nr.: 44-028

$C_{13}H_{26}O_2Si_3$
- 1,1,1,3,5,5,5-Heptamethyl-3-phenyltrisiloxane [546-44-1]
  Group-member nr.: 71-037

$C_{13}H_{28}$
- Tridecane [629-50-5]
  Group-member nr.: 11-051

$C_{13}H_{28}O$
- 5-Butyl-5-nonanol [597-93-3]
  Group-member nr.: 42-096
- 1-Tridecanol [112-70-9]
  n-Tridecyl alcohol
  Group-member nr.: 42-097

$C_{14}H_8O_2$
- 9,10-Anthracenedione [84-65-1]
  9,10-Antraquinone
  Group-member nr.: 43-056

$C_{14}H_{10}$
- Anthracene [120-12-7]
  Group-member nr.: 14-066
- 1,1'-(1,2-Ethynediyl)bisbenzene [501-65-5]
  1,2-Diphenylacetylene
  Tolan
  Group-member nr.: 14-067
- Phenanthrene [85-01-8]
  Group-member nr.: 14-068

$C_{14}H_{12}$
- (*E*)-1,1'-(1,2-Ethenediyl)bisbenzene [103-30-0]
  trans-1,2-Diphenylethylene
  Stilbene
  Toluylene
  Group-member nr.: 14-070
- 9,10-Dihydrophenanthrene [776-35-2]
  Group-member nr.: 14-069
- 1,1'-Ethenylidenebisbenzene [530-48-3]
  1,1-Diphenylethylene
  Group-member nr.: 14-071

$C_{14}H_{12}O$
- (3-Methylphenyl)phenylmethanone [643-65-2]
  3-Methylbenzophenone
  Phenyl *m*-tolyl ketone
  Group-member nr.: 43-057

$C_{14}H_{14}$
- 1,1'-(1,2-Ethanediyl)bisbenzene [103-29-7]
  Bibenzyl
  1,2-Diphenylethane
  sym-Diphenylethane
  Dibenzyl
  Group-member nr.: 14-072
- 1,1'-Ethylidenebisbenzene [612-00-0]
  1,1-Diphenylethane
  unsym-Diphenylethane
  Group-member nr.: 14-073

$C_{14}H_{14}N_2O_3$
- Bis(4-methoxyphenyl)diazene-1-oxide [1562-94-3]
  4,4'-Dimethoxyazoxybenzene
  p-Azoxyanisole
  Group-member nr.: 62-096

$C_{14}H_{16}N_2O_2$
- 1,3-Bis(1-isocyanato-1-methylethyl)benzene [2778-42-9]
  α,α,α',α'-Tetramethyl-*m*-xylylene ester isocyanic acid
  Group-member nr.: 62-097
- 1,4-Bis(1-isocyanato-1-methylethyl)benzene [2778-41-8]
  α,α,α',α'-Tetramethyl-*p*-xylylene ester isocyanic acid
  Group-member nr.: 62-098

$C_{14}H_{18}$
- 1,2,3,4,5,6,7,8-Octahydroanthracene [1079-71-6]
  Group-member nr.: 14-074

$C_{14}H_{20}$
- Decahydro-3,5,1,7-[1,2,3,4]butanetetraylnaphthalene [2292-79-7]
  Diamantane
  Group-member nr.: 12-078

$C_{14}H_{20}O_3$
- 1-[4-(1,1-Diethoxyethyl)phenyl]ethanone [64533-95-5]
  p-Diacetylbenzene diethyl ketal
  Group-member nr.: 47-043

$C_{14}H_{22}$
- Octylbenzene [2189-60-8]
  1-Phenyloctane
  Group-member nr.: 14-075

$C_{14}H_{24}$
- Dodecahydro-9-methylfluorene [92431-75-9]
  9-Methylperhydrofluorene
  Group-member nr.: 12-079
- Tetradecahydrophenanthrene [5743-97-5]
  Perhydrophenanthrene (unspecified stereoisomer)
  Group-member nr.: 12-080
- (4aα,4bα,8aα,10aβ)-Tetradecahydrophenanthrene [27425-35-0]
  cis-syn-trans-Perhydrophenantrene
  Group-member nr.: 12-081
- (4aα,4bβ,8aα,10aα)-Tetradecahydrophenanthrene [27389-73-7]
  trans-anti-cis-Perhydrophenanthrene
  Group-member nr.: 12-082
- (4aα,4bβ,8aα,10aβ)-Tetradecahydrophenanthrene [2108-89-6]
  trans-anti-trans-Perhydrophenanthrene
  Group-member nr.: 12-083

$C_{14}H_{26}$
- 1-Butyldecahydronaphthalene [92369-80-7]
  1-Butyldecalin
  α-Butyldecalin
  Group-member nr.: 12-084
- 1-(Cyclohexylmethyl)-2-methylcyclohexane [66826-96-8]
  Cyclohexyl(2-methylcyclohexyl)methane
  Group-member nr.: 12-085
- Decahydro-1-(1-methylpropyl)naphthalene [92369-82-9]
  1-*sec*-Butyldecahydronaphthalene
  α-*sec*-Butyldecalin
  Group-member nr.: 12-086
- Decahydro-1-(2-methylpropyl)naphthalene [92369-83-0]
  Decahydro-1-isobutylnaphthalene
  α-Isobutyldecalin
  Group-member nr.: 12-087
- (1,1-Dimethylethyl)decahydronaphthalene [27193-30-2]
  tert-Butyldecalin (unspecified isomer)
  Group-member nr.: 12-088
- 1,1'-(1,2-Ethanediyl)biscyclohexane [3321-50-4]
  1,2-Dicyclohexylethane
  Group-member nr.: 12-089
- 2-Ethyl-1,1'-bicyclohexyl [66826-94-6]
  2-Ethylcyclohexylcyclohexane
  Group-member nr.: 12-090
- 1,1'-Ethylidenebiscyclohexane [2319-61-1]
  1,1-Dicyclohexylethane
  Group-member nr.: 12-091

$C_{14}H_{26}O$
- 2-(1,2-Dimethylpropyl)-5,6-dimethyl-2-heptenal [99914-84-8]
  Group-member nr.: 43-058
- 2-Pentyl-2-nonenal [3021-89-4]
  Group-member nr.: 43-059

$C_{14}H_{26}O_2$
- Decyl 2-methyl-2-propenoate [3179-47-3]
  Decyl ester 2-methyl-2-propenoic acid
  Decyl methacrylate
  Group-member nr.: 45-100

$C_{14}H_{26}O_4$
- Bis(3-methylbutyl) butanedioate [818-04-2]
  Bis(3-methylbutyl) ester butanedioic acid
  Diisoamyl succinate
  Group-member nr.: 45-101

$C_{14}H_{28}O$
- 2-Tetradecanone [2345-27-9]
  Dodecyl methyl ketone
  Group-member nr.: 43-060

$C_{14}H_{28}O_2$
- Tetradecanoic acid [544-63-8]
  Myristic acid
  Group-member nr.: 44-029

$C_{14}H_{30}$
- Tetradecane [629-59-4]
  Group-member nr.: 11-052

$C_{14}H_{30}O$
- 1-Tetradecanol [112-72-1]
  n-Tetradecyl alcohol
  Myristyl alcohol
  Group-member nr.: 42-098

$C_{14}H_{30}S$
- 1,1'-Thiobisheptane [629-65-2]
  8-Thiapentadecane
  Diheptyl sulfide
  Group-member nr.: 51-025

$C_{15}H_{10}N_2O_2$
- 1,1'-Methylenebis[4-isocyanatobenzene] [101-68-8]
  Methylenedi-*p*-phenylene ester isocyanic acid
  Diphenylmethane-4,4'-diisocyanate
  Group-member nr.: 62-099

$C_{15}H_{12}$
- 4-Methylphenanthrene [832-64-4]
  Group-member nr.: 14-076

$C_{15}H_{16}$
- (1-Methylethyl)-1,1'-biphenyl [25640-78-2]
  Isopropylbiphenyl
  Isopropyldiphenyl (unspecified isomer)
  Group-member nr.: 14-077

$C_{15}H_{16}N_2O_3$
- (4-Ethoxyphenyl)(4-methoxyphenyl)diazene-*N*-oxide [56095-14-8]
  4-Ethoxy-4'-methoxyazoxybenzene
  p-Azoxyanisoyl-phenetole
  Group-member nr.: 62-100

$C_{15}H_{16}O_2$
- 4,4'-(1-Methylethylidene)bisphenol [80-05-7]
  4,4'-Isopropylidenediphenol
  4,4'-Dihydroxydiphenyl-2,2-propane
  Group-member nr.: 42-099

$C_{15}H_{26}$
- 1,1':3',1''-Tercyclopentane [6051-40-7]
  1,3-Dicyclopentylcyclopentane
  Group-member nr.: 12-092

$C_{15}H_{26}O_6$
- 1,2,3-Propanetriyl tributanoate [60-01-5]
  1,2,3-Propanetriyl ester butanoic acid
  Tributyrin
  Group-member nr.: 45-102

$C_{15}H_{28}$
- Cyclohexyl(ethylcyclohexyl)methane (unspecified isomer) [97239-02-6]
  Group-member nr.: 12-093
- 1-(Cyclohexylmethyl)-2-ethylcyclohexane [66374-71-8]
  Cyclohexyl(2-ethylcyclohexyl)methane
  Group-member nr.: 12-094
- 1,2-Dicyclohexylpropane [41851-34-7]
  Group-member nr.: 12-095
- (1-Methylethyl)-1,1'-bicyclohexyl [31624-59-6]
  Isopropylbicyclohexyl
  Isopropylcyclohexylcyclohexane (unspecified isomer and stereoisomer)
  Group-member nr.: 12-096

$C_{15}H_{28}O$
- 3,7,11-Trimethyl-1-dodecyn-3-ol [1604-35-9]
  Group-member nr.: 42-100

$C_{15}H_{28}O_2$
- Oxacyclohexadecan-2-one [106-02-5]
  Pentadecanolide
  Pentadecanolactone
  15-Hydroxypentadecanoic acid lactone
  Group-member nr.: 47-044

$C_{15}H_{30}$
- Decylcyclopentane [1795-21-7]
  Cyclopentyldecane
  Group-member nr.: 12-097

$C_{15}H_{30}O$
- 2-Pentadecanone [2345-28-0]
  Methyl tridecyl ketone
  Group-member nr.: 43-061

$C_{15}H_{30}O_2$
- Methyl tetradecanoate [124-10-7]
  Methyl myristate
  Group-member nr.: 45-103
- Pentadecanoic acid [1002-84-2]
  Pentadecylic acid
  Group-member nr.: 44-030

$C_{15}H_{32}$
- Pentadecane [629-62-9]
  Group-member nr.: 11-053

$C_{15}H_{32}O$
- 1-Pentadecanol [629-76-5]
  n-Pentadecyl alcohol
  Group-member nr.: 42-101

$C_{16}H_{10}$
- Fluoranthene [206-44-0]
  Group-member nr.: 14-078
- Pyrene [129-00-0]
  Group-member nr.: 14-079

$C_{16}H_{12}Ge$
- Diethynyldiphenylgermane [1675-59-8]
  Group-member nr.: 73-030

$C_{16}H_{12}Si$
- Diethynyldiphenylsilane [1675-57-6]
  Group-member nr.: 71-038

$C_{16}H_{14}O_2$
- 1,4-Diphenyl-1,4-butanedione [495-71-6]
  1,2-Dibenzoylethane
  Group-member nr.: 43-062

$C_{16}H_{16}N_2O_2$
- [(4-Methoxyphenyl)methylene]hydrazone-4-methoxybenzaldehyde [2299-73-2]
  p-Anisaldehydeazine
  Anisaldazine
  Group-member nr.: 62-101

$C_{16}H_{18}N_2O_3$
- Bis(4-ethoxyphenyl)diazene-1-oxide [4792-83-0]
  4,4'-Diethoxyazoxybenzene
  p-Azoxyphenetole
  Group-member nr.: 62-102

$C_{16}H_{22}OSi_2$
- 1,1,3,3-Tetramethyl-1,3-diphenyldisiloxane [56-33-7]
  Group-member nr.: 71-039

$C_{16}H_{22}O_3Si_3$
- 2,2,4,4-Tetramethyl-6,6-diphenylcyclotrisiloxane [1693-51-2]
  Group-member nr.: 71-040

$C_{16}H_{22}O_4$
- Dibutyl 1,2-benzenedicarboxylate [84-74-2]
  Dibutyl ester 1,2-benzenedicarboxylic acid
  Dibutyl phthalate
  Group-member nr.: 45-104

$C_{16}H_{25}NO_2$
- Nonyl phenylcarbamate [33689-71-3]
  Nonyl ester phenylcarbamic acid
  Nonyl ester carbanilic acid
  Group-member nr.: 62-103

$C_{16}H_{28}$
- 1-Cyclohexyloctahydro-3-methyl-1*H*-indene [unknown]
  1-Cyclohexyl-3-methylhydrindan
  Group-member nr.: 12-098
- 2-Ethyltetradecahydrophenanthrene [90591-84-7]
  2-Ethylperhydrophenanthrene
  Group-member nr.: 12-099

$C_{16}H_{30}$
- Cyclohexyl[(1-methylethyl)cyclohexyl]methane [97676-41-0]
  Cyclohexyl(isopropylcyclohexyl)methane (unspecified isomer)
  Group-member nr.: 12-101
- 1,1'-(1-Methyl-1,3-propanediyl)biscyclohexane [41851-35-8]
  1,3-Dicyclohexylbutane
  Group-member nr.: 12-100

$C_{16}H_{30}HgO_4$
- Mercury(2+) salt octanoic acid [28043-54-1]
  Mercury(II) octanoate
  Mercury(II) caprylate
  Group-member nr.: 74-005

$C_{16}H_{30}O_4Pb$
- Lead(2+) salt octanoic acid [7319-86-0]
  Lead(II) octanoate
  Lead(II) caprylate
  Group-member nr.: 74-006

$C_{16}H_{32}$
- Decylcyclohexane [1795-16-0]
  Cyclohexyldecane
  Group-member nr.: 12-102
- 1-Hexadecene [629-73-2]
  α-Hexadecylene
  Cetene
  Group-member nr.: 13-038

$C_{16}H_{32}O_2$
- Hexadecanoic acid [57-10-3]
  Palmitic acid
  Group-member nr.: 44-031

$C_{16}H_{34}$
- 2,2,4,4,6,8,8-Heptamethylnonane [4390-04-9]
  Group-member nr.: 11-054
- Hexadecane [544-76-3]
  Cetane
  Group-member nr.: 11-055

$C_{16}H_{34}O$
- 1-Hexadecanol [36653-82-4]
  Hexadecyl alcohol
  Cetyl alcohol
  Group-member nr.: 42-102

$C_{16}H_{34}S$
- 1,1'-Thiobisoctane [2690-08-6]
  9-Thiaheptadecane
  Dioctyl sulfide
  Group-member nr.: 51-026

$C_{16}H_{35}N$
- *N*, *N*-Dimethyl-2-pentyl-1-nonanamine [99916-30-0]
  N, *N*-Dimethyl-2-pentylnonylamine
  Group-member nr.: 31-045

$C_{16}H_{36}Ge$
- Tetrabutylgermane [1067-42-1]
  Group-member nr.: 73-031

$C_{16}H_{36}O_4Si$
- Tetrabutyl ester silicic acid [4766-57-8]
  Tetrabutyl orthosilicate
  Tetrabutoxysilane
  Group-member nr.: 71-041

$C_{16}H_{36}O_4Ti$
- Titanium(4+) salt 1-butanol [5593-70-4]
  Tetrabutoxytitanium
  Group-member nr.: 74-007

$C_{16}H_{40}O_4Si_4$
- Octaethylcyclotetrasiloxane [1451-99-6]
  Group-member nr.: 71-042

$C_{17}H_{17}NSi$
- 1-[2-(Trimethylsilyl)ethyl]aziridine [18387-12-7]
  N-[(β-Trimethylsilyl)ethyl]ethylenimine
  Group-member nr.: 71-043

$C_{17}H_{30}$
- Cyclopentylbicyclohexyl (unspecified isomer) [26447-22-3]
  Group-member nr.: 12-103

$C_{17}H_{32}$
- Bis(ethylcyclohexyl)methane (unspecified isomer) [98028-64-9]
  Group-member nr.: 12-104
- 1-Cyclohexyl-1-(1-methylethyl)cyclohexylethane [26637-18-3]
  1-Cyclohexyl-1-(isopropylcyclohexyl)ethane (unspecified isomer)
  Group-member nr.: 12-105

$C_{17}H_{34}O_2$
- Heptadecanoic acid [506-12-7]
  Margaric acid
  Group-member nr.: 44-032

$C_{17}H_{36}$
- Heptadecane [629-78-7]
  Group-member nr.: 11-056

$C_{18}H_{12}$
- Triphenylene [217-59-4]
  9,10-Benzophenanthrene
  Group-member nr.: 14-080

$C_{18}H_{14}$
- 1,1':2',1''-Terphenyl [84-15-1]
  o-Terphenyl
  Group-member nr.: 14-081
- 1,1':3',1''-Terphenyl [92-06-8]
  m-Terphenyl
  Group-member nr.: 14-082
- 1,1':4',1''-Terphenyl [92-94-4]
  p-Terphenyl
  Group-member nr.: 14-083

$C_{18}H_{15}ClSi$
- Chlorotriphenylsilane [76-86-8]
  Group-member nr.: 71-044

$C_{18}H_{15}OP$
- Triphenylphosphine oxide [791-28-6]
  Oxotriphenylphosphorane
  Oxotriphenylphosphorus
  Group-member nr.: 72-012

$C_{18}H_{15}O_4P$
- Triphenyl ester phosphoric acid [115-86-6]
  Triphenyl phosphate
  Group-member nr.: 72-013

$C_{18}H_{18}$
- 1-Methyl-7-(1-methylethyl)phenanthrene [483-65-8]
  7-Isopropyl-1-methylphenanthrene
  Retene
  Group-member nr.: 14-084

$C_{18}H_{18}N_2O_5$
- Diethyl ester 4,4'-azoxybisbenzoic acid [6421-04-1]
  Diethyl 4,4'-azoxybisbenzoate
  Ethyl-*p,p*-azoxybenzoate
  Group-member nr.: 62-104

$C_{18}H_{21}NO$
- 4-Butyl-*N*-[(4-methoxyphenyl)methylene]-benzenamine [26227-73-6]
  N-(*p*-Methoxybenzylidene)-*p*-butylaniline
  MBBA
  Group-member nr.: 62-105

$C_{18}H_{21}NO_2$
- 2-[[(4-Butylphenyl)imino]methyl]-4-methoxyphenol [52218-22-1]
  N-(*o*-Hydroxy-*p*-methoxybenzylidene)-*p*-butylaniline
  2,4-OHMBBA
  Group-member nr.: 62-106

$C_{18}H_{22}$
- *ar,ar*'-Bis(1-methylethyl)-1,1'-biphenyl [36876-13-8]
  Diisopropylbiphenyl (unspecified isomer)
  Group-member nr.: 14-085

$C_{18}H_{22}N_2O$
- (*E*)-(4-Butylphenyl)(4-ethoxyphenyl)diazene [98644-12-3]
  Group-member nr.: 62-107

$C_{18}H_{22}N_2O_3$
- Bis(4-propoxyphenyl)diazene-1-oxide [23315-55-1]
  4,4'-Dipropoxyazoxybenzene
  Group-member nr.: 62-108

$C_{18}H_{28}O_2Si_3$
- 1,1,1,5,5,5-Hexamethyl-3,3-diphenyltrisiloxane [797-77-3]
  Group-member nr.: 71-045

$C_{18}H_{28}O_4Si_4$
- 2,2,4,4,6,6-Hexamethyl-8,8-diphenylcyclotetra-siloxane [30026-85-8]
  Group-member nr.: 71-046

$C_{18}H_{30}$
- 3-Ethylhexadecahydropyrene [94262-24-5]
  3-Ethylperhydropyrene
  Group-member nr.: 12-106

$C_{18}H_{30}O_4$
- 1,4-Bis(1,1-diethoxyethyl)benzene [47189-08-2]
  p-Diacetylbenzene tetraethyl ketal
  Ethylacetal *p*-diacetyl benzene
  Group-member nr.: 41-048

$C_{18}H_{32}$
- 1-Cyclohexyloctahydro-1,3,3-trimethyl-1*H*-indene [22236-61-9]
  1-Cyclohexyl-1,3,3-trimethylhydrindan
  Group-member nr.: 12-107
- 1,1':2',1''-Tercyclohexane [2456-43-1]
  o-Tercyclohexane
  Group-member nr.: 12-108
- 1,1':3',1''-Tercyclohexane [1706-50-9]
  m-Tercyclohexane
  Group-member nr.: 12-109
- 1,1':4',1''-Tercyclohexane [1795-19-3]
  p-Tercyclohexane
  Group-member nr.: 12-110

$C_{18}H_{32}O$
- 6,10,14-Trimethyl-3,5-pentadecadien-2-one [1604-32-6]
  Group-member nr.: 43-063

$C_{18}H_{34}$
- 1,1,-Bis(dimethylcyclohexyl)ethane (unspecified isomer) [98803-06-6]
  Group-member nr.: 12-112

- 1,1-Bis(ethylcyclohexyl)ethane (unspecified isomer) [98803-07-7]
  Group-member nr.: 12-111

$C_{18}H_{34}O_2$
- (*Z*)-9-Octadecenoic acid [112-80-1]
  Oleic acid
  Group-member nr.: 44-033

$C_{18}H_{34}O_4$
- Dibutyl decanedioate [109-43-3]
  Dibutyl ester decanedioic acid
  Dibutyl sebacate
  Group-member nr.: 45-105

$C_{18}H_{36}$
- Dodecylcyclohexane [1795-17-1]
  Cyclohexyldodecane
  Group-member nr.: 12-113
- Hexaethylcyclohexane (unspecified isomer) [98803-61-3]
  Group-member nr.: 12-114

$C_{18}H_{36}N_2O_2$
- *N*, *N*'-Dihexylhexanediamide [21150-82-3]
  N, *N*'-Dihexyladipamide
  Group-member nr.: 62-109

$C_{18}H_{36}O$
- 6,10,14-Trimethyl-2-pentadecanone [502-69-2]
  Phytone
  Group-member nr.: 43-064

$C_{18}H_{36}O_2$
- Octadecanoic acid [57-11-4]
  Stearic acid
  Group-member nr.: 44-034

$C_{18}H_{37}Cl$
- 1-Chlorooctadecane [3386-33-2]
  Group-member nr.: 22-049

$C_{18}H_{38}$
- Octadecane [593-45-3]
  Group-member nr.: 11-057

$C_{18}H_{38}O$
- 1-Octadecanol [112-92-5]
  Octadecyl alcohol
  Stearyl alcohol
  Group-member nr.: 42-103

$C_{18}H_{38}S$
- 1,1'-Thiobisnonane [929-98-6]
  10-Thianonadecane
  Dinonyl sulfide
  Group-member nr.: 51-027

$C_{19}H_{16}$
- 1,1',1''-Methylidynetrisbenzene [519-73-3]
  Triphenylmethane
  Tritane
  Group-member nr.: 14-086

$C_{19}H_{22}ClNO$
- 4-Chloro-*N*-[[4-(hexyloxy)phenyl]methylene] benzenamine [5219-48-7]
  p-*n*-Hexyloxybenzylideneamino-*p*'-chlorobenzene
  Group-member nr.: 64-029

$C_{19}H_{22}FNO$
- 4-Fluoro-*N*-[[4-(hexyloxy)]methylene] benzenamine [56544-26-4]
  p-*n*-Hexyloxybenzylideneamino-*p*'-fluorobenzene
  Group-member nr.: 64-030

$C_{19}H_{23}NO$
- 4-Butyl-*N*-[(4-ethoxyphenyl)methylene] benzenamine [29743-08-6]
  p-Butyl-*N*-(*p*-ethoxybenzylidene)aniline
  Group-member nr.: 62-110
- *N*-[[4-(Hexyloxy)phenyl]methylene] benzenamine [5219-49-8]
  p-*n*-Hexyloxybenzylideneaniline
  Group-member nr.: 62-111

$C_{19}H_{36}$
- Bis(2,4,6-trimethylcyclohexyl)methane [94380-80-0]
  Group-member nr.: 12-115
- 4-Heptyl-1,1'-bicyclohexyl [96667-88-8]
  4-Heptylcyclohexylcyclohexane
  Group-member nr.: 12-116
- 1,1'-Heptylidenebiscyclohexane [2090-15-5]
  1,1-Dicyclohexylheptane
  Group-member nr.: 12-117

$C_{19}H_{38}O_2$
- Nonadecanoic acid [646-30-0]
  Nonadecylic acid
  Group-member nr.: 44-035

$C_{19}H_{40}$
- Nonadecane [629-92-5]
  Group-member nr.: 11-058
- 2,6,10,14-Tetramethylpentadecane [1921-70-6]
  Pristane
  Group-member nr.: 11-059

$C_{20}H_{12}$
- Perylene [198-55-0]
  Group-member nr.: 14-087

$C_{20}H_{14}$
- 9,10-Dihydro-9,10[1',2']-benzenoanthracene [477-75-8]
  Triptycene
  Group-member nr.: 14-088

$C_{20}H_{22}N_2O$
• 4-[[[4-(Hexyloxy)phenyl]methylene]amino] - benzonitrile [35280-78-5]
p-n-Hexyloxybenzylideneamino-*p*'-benzonitrile
Group-member nr.: 62-112

$C_{20}H_{24}N_2O_3$
• (*E*)-4-[(4-Methoxyphenyl)azo]phenyl ester heptanoic acid [97402-83-0]
Group-member nr.: 62-113

$C_{20}H_{25}NO$
• *N*-[[4-(Hexyloxy)phenyl]methylene]-4-methyl-benzenamine [25959-51-7]
p-n-Hexyloxybenzylidene-*p*'-toluidine
Group-member nr.: 62-114

$C_{20}H_{26}N_2O_3$
• Bis(4-butoxyphenyl)diazene-1-oxide [17051-01-3]
4,4'-Dibutoxyazoxybenzene
Group-member nr.: 62-115

$C_{20}H_{30}O_3Si_3$
• 2,2,4,4-Tetraethyl-6,6-diphenylcyclotrisiloxane [108543-32-4]
Group-member nr.: 71-047

$C_{20}H_{34}$
• Diethylhexadecahydropyrene [26446-93-5]
Diethylperhydropyrene
Group-member nr.: 12-118

$C_{20}H_{34}O_5Si_5$
• Octamethyldiphenylcyclopentasiloxane (unspecified isomer) [51134-26-0]
Group-member nr.: 71-048

$C_{20}H_{38}HgO_4$
• Mercury(2+) salt decanoic acid [27394-49-6]
Mercury(II) decanoate;
Mercury(II) caprate
Group-member nr.: 74-008

$C_{20}H_{38}O$
• 3,7,11,15-Tetramethyl-1-hexadecyn-3-ol [29171-23-1]
Group-member nr.: 42-104

$C_{20}H_{38}PbO_4$
• Lead(2+) salt decanoic acid [15773-52-1]
Lead(II) decanoate;
Lead(II) caprate
Group-member nr.: 74-009

$C_{20}H_{40}O$
• 3,7,11,15-Tetramethyl-1-hexadecen-3-ol [505-32-8]
Isophytol
Group-member nr.: 42-105

$C_{20}H_{40}O_2$
• Eicosanoic acid [506-30-9]
Icosanic acid;
Arachidic acid
Group-member nr.: 44-036

$C_{20}H_{42}$
• Eicosane [112-95-8]
Icosane
Group-member nr.: 11-060

$C_{20}H_{42}S$
• 1,1'-Thiobisdecane [693-83-4]
11-Thiaheneicosane;
11-Thiahenicosane;
Didecyl sulfide
Group-member nr.: 51-028

$C_{20}H_{44}Sn$
• Tetrapentylstannane [3765-65-9]
Tetrapentyltin
Group-member nr.: 73-032

$C_{21}H_{14}N_2O_4$
• 1,1'-(Methylenedi-4,1-phenylene)bis-1*H*-pyrrole-2,5-dione [13676-54-5]
Bis-(4-maleic acidimidphenyl)methane
Group-member nr.: 62-116

$C_{21}H_{16}$
• 1-(2-Naphthalenylmethyl)naphthalene [611-48-3]
1,2'-Dinaphthylmethane
Group-member nr.: 14-089

$C_{21}H_{21}O_4P$
• Tris(methylphenyl) ester phosphoric acid [1330-78-5]
Tricresyl phosphate;
Tritolyl phosphate (unspecified isomer)
Group-member nr.: 72-014

$C_{21}H_{24}O_3Si_3$
• (2α,4α,6α)-2,4,6-Trimethyl-2,4,6-triphenylcyclo-trisiloxane [3424-57-5]
cis-2,4,6-Trimethyl-2,4,6-triphenylcyclotrisiloxane
Group-member nr.: 71-049
• (2α,4α,6β)-2,4,6-Trimethyl-2,4,6-triphenylcyclotrisil-oxane [6138-53-0]
trans-2,4,6-Trimethyl-2,4,6-triphenylcyclotrisiloxane
Group-member nr.: 71-050

$C_{21}H_{24}O_4$
• 2,2'-[(1-Methylethylidene)bis(4,1-phenyleneoxy-methylene)]bisoxirane [1675-54-3]
Group-member nr.: 47-045

$C_{21}H_{38}$
• 9-(2-Ethylhexyl)dodecahydrofluorene [95135-48-1]
9-(2-Ethylhexyl)perhydrofluorene
Group-member nr.: 12-119

- 1,1',1''-(1-Propanyl-2-ylidene)triscyclohexane [55682-89-8]
  1,1,3-Tricyclohexylpropane
  Group-member nr.: 12-120

$C_{21}H_{38}O_6$
- 1,2,3-Propanetriyl ester hexanoic acid [621-70-5]
  1,2,3-Propanetriyl trihexanoate
  Trihexanoin
  Tricaproin
  Group-member nr.: 45-106

$C_{21}H_{40}$
- 4-Nonyl-1,1'-bicyclohexyl [95135-87-8]
  4-Nonylcyclohexylcyclohexane
  Group-member nr.: 12-121

$C_{21}H_{42}O_4$
- 2,3-Dihydroxypropyl ester octadecanoic acid [123-94-4]
  1-Monostearin
  Group-member nr.: 47-046

$C_{21}H_{44}$
- Heneicosane [629-94-7]
  Henicosane
  Group-member nr.: 11-061

$C_{22}H_{14}$
- Pentacene [135-48-8]
  2,3,6,7-Dibenzoanthracene
  Group-member nr.: 14-090

$C_{22}H_{24}O_6$
- *trans*-Bis(4-methoxyphenyl) ester 1,4-cyclohexanedicarboxylic acid [26379-55-5]
  Di(*p*-methoxyphenyl)-*trans*-cyclohexane-1,4-dicarboxylate
  Group-member nr.: 47-047

$C_{22}H_{29}NO$
- 4-Butyl-*N*-[[4-(pentyloxy)phenyl]methylene]-benzenamine [29743-10-0]
  N-*p*-*n*-Pentyloxybenzylidene-*p*'-*n*-butylaniline
  Group-member nr.: 62-117

$C_{22}H_{30}N_2O_3$
- Bis(4-pentyloxyphenyl)diazene-1-oxide [19482-05-4]
  4,4'-Bis(pentyloxy)azoxybenzene
  Group-member nr.: 62-118

$C_{22}H_{42}O_4$
- Bis(2-ethylhexyl) hexanedioate [103-23-1]
  Bis(2-ethylhexyl) adipate
  Bis(2-ethylhexyl) ester hexanedioic acid
  Group-member nr.: 45-107
- Dihexyl decanedioate [2449-10-7]
  Dihexyl ester decanedioic acid
  Dihexyl sebacate
  Group-member nr.: 45-108

$C_{22}H_{44}N_2O_2$
- *N*, *N*'-Dihexyldecanediamine [31827-03-9]
  N, *N*'-Dihexylsebacamide
  Group-member nr.: 62-119

$C_{22}H_{44}O_2$
- Butyl octadecanoate [123-95-5]
  Butyl ester octadecanoic acid
  Butyl stearate
  Group-member nr.: 45-109

$C_{22}H_{45}Br$
- 1-Bromodocosane [6938-66-5]
  n-Docosyl bromide
  Group-member nr.: 23-039

$C_{22}H_{46}$
- Docosane [629-97-0]
  Group-member nr.: 11-062

$C_{23}H_{31}NO$
- 4-Butyl-*N*-[[4-(hexyloxy)phenyl]methylene]-benzenamine [29743-11-1]
  N-*p*-*n*-Hexyloxybenzylidene-*p*'-*n*-butylaniline
  Group-member nr.: 62-120

$C_{23}H_{44}O_5$
- 3-(Acetyloxy)-2-hydroxypropyl ester octadecanoic acid [820-17-7]
  3-Aceto-1-stearin
  Group-member nr.: 47-048

$C_{23}H_{48}$
- Tricosane [638-67-5]
  Group-member nr.: 11-063

$C_{24}H_{18}$
- 5'-Phenyl-1,1':3',1''-terphenyl [612-71-5]
  1,3,5-Triphenylbenzene
  Group-member nr.: 14-091
- 1,1':3',1'':3'',1'''-Quaterphenyl [1166-18-3]
  m-Quaterphenyl
  Group-member nr.: 14-092

$C_{24}H_{32}O_3$
- 4-(Heptyloxy)phenyl ester 4-butylbenzoic acid [38454-35-2]
  p-(Heptyloxy)phenyl *p*-butylbenzoate
  Group-member nr.: 47-049

$C_{24}H_{34}$
- 1,1-Diphenyldodecane [1603-53-8]
  Group-member nr.: 14-093

$C_{24}H_{34}N_2O_3$
- Bis(4-hexyloxyphenyl)diazene-1-oxide [2587-42-0]
  4,4'-Bis(hexyloxy)azoxybenzene
  Group-member nr.: 62-121

$C_{24}H_{38}O_4$
- Bis(2-ethylhexyl)-1,2-benzenedicarboxylate [117-81-7]
  Bis(2-ethylhexyl) ester phthalic acid
  Group-member nr.: 45-110

$C_{24}H_{40}$
- (1-Cyclohexyldodecyl)benzene [62155-50-4]
  1-Cyclohexyl-1-phenyldodecane
  Group-member nr.: 14-094

$C_{24}H_{46}CdO_4$
- Cadmium salt dodecanoic acid [2605-44-9]
  Cadmium dodecanoate
  Cadmium laurate
  Group-member nr.: 74-010

$C_{24}H_{46}HgO_4$
- Mercury(2+) salt dodecanoic acid [23186-25-6]
  Mercury(II) dodecanoate
  Mercury(II) laurate
  Group-member nr.: 74-011

$C_{24}H_{46}O_4Pb$
- Lead(2+) salt dodecanoic acid [15773-55-4]
  Lead(II) dodecanoate
  Lead(II) laurate
  Group-member nr.: 74-012

$C_{24}H_{50}$
- Tetracosane [646-31-1]
  Group-member nr.: 11-064

$C_{24}H_{52}ClNO_4$
- *N*, *N*, *N*-Trihexyl-1-hexanaminium perchlorate [4656-81-9]
  Tetrahexylammonium perchlorate
  Group-member nr.: 74-013

$C_{24}H_{52}O_4Si$
- Tetrakis(2-ethylbutyl) ester silicic acid [78-13-7]
  Tetrakis(2-ethylbutyl) orthosilicate
  Tetrakis(2-ethylbutoxy)silane
  Group-member nr.: 71-051

$C_{25}H_{46}$
- 4'-Heptyl-1,1':3',1''-tercyclohexane [unknown]
  4-Heptyl-*m*-tercyclohexyl
  Group-member nr.: 12-122

$C_{25}H_{46}O_6$
- 2,3-Bis(acetyloxy)propyl ester octadecanoic acid [33599-07-4]
  1,2-Diacetostearin
  Group-member nr.: 45-111

$C_{25}H_{48}O_4$
- Bis(2-ethylhexyl) nonanedioate [103-24-2]
  Bis(2-ethylhexyl) azelate
  Bis(2-ethylhexyl) ester nonanedioic acid
  Group-member nr.: 45-112

$C_{25}H_{52}$
- Pentacosane [629-99-2]
  Group-member nr.: 11-065

$C_{26}H_{18}$
- 9,10-Diphenylanthracene [1499-10-1]
  Group-member nr.: 14-095

$C_{26}H_{26}OSi_2$
- 1,3-Dimethyl-1,1,3,3-tetraphenyldisiloxane [807-28-3]
  Group-member nr.: 71-052

$C_{26}H_{36}O_3Si_3$
- 2,2-Dimethyl-4,4,6,6-tetraphenylcyclotrisiloxane [1438-86-4]
  Group-member nr.: 71-053

$C_{26}H_{38}N_2O_3$
- Bis(4-heptyloxyphenyl)diazene-1-oxide [2635-26-9]
  4,4'-Bis(heptyloxy)azoxybenzene
  Group-member nr.: 62-122

$C_{26}H_{50}O_4$
- Dioctyl decanedioate [2432-87-3]
  Dioctyl ester decanedioic acid
  Dioctyl sebacate
  Group-member nr.: 45-113

$C_{26}H_{54}$
- Hexacosane [630-01-3]
  Group-member nr.: 11-066

$C_{27}H_{48}$
- 11-Phenylheneicosane [6703-80-6]
  (1-Decylundecyl)benzene
  11-Phenylhenicosane
  Group-member nr.: 14-096

$C_{27}H_{50}$
- 4'-Nonyl-1,1':3',1''-tercyclohexane [unknown]
  4-Nonyl-*m*-tercyclohexyl
  Group-member nr.: 12-123

$C_{27}H_{50}O_6$
- 1,2,3-Propanetriyl ester octanoic acid [538-23-8]
  1,2,3-Propanetriyl trioctanoate
  Trioctanoin
  Group-member nr.: 45-114

$C_{27}H_{54}$
- (1-Decylundecyl)cyclohexane [6703-99-7]
  11-Cyclohexylheneicosane
  11-Cyclohexylhenicosane
  Group-member nr.: 12-124

$C_{27}H_{56}$
- Heptacosane [593-49-7]
  Group-member nr.: 11-067

$C_{28}H_{28}P$
- 1,4-Butanediylbis(diphenylphosphine) [7688-25-7]
  1,4-Bis(diphenylphosphino)butane
  Tetramethylenebis(diphenylphosphine)
  Group-member nr.: 72-015

$C_{28}H_{32}O_4Si_4$
- 2,2,4,4-Tetramethyl-6,6,8,8-tetraphenylcyclotetrasiloxane [1693-47-6]
  Group-member nr.: 71-054
- 2,4,6,8-Tetramethyl-2,4,6,8-tetraphenylcyclotetrasiloxane [77-63-4]
  Group-member nr.: 71-055

$C_{28}H_{42}N_2O_3$
- Bis(4-octyloxyphenyl)diazene-1-oxide [25729-12-8]
  4,4'-Bis(octyloxy)azoxybenzene
  Group-member nr.: 62-123

$C_{28}H_{54}CdO_4$
- Cadmium salt tetradecanoic acid [10196-67-5]
  Cadmium tetradecanoate
  Cadmium myristate
  Group-member nr.: 74-014

$C_{28}H_{54}HgO_4$
- Mercury(2+) salt tetradecanoic acid [36215-49-3]
  Mercury(II) tetradecanoate
  Mercury(II) myristate
  Group-member nr.: 74-015

$C_{28}H_{54}O_4Pb$
- Lead(2+) salt tetradecanoic acid [32112-52-0]
  Lead(II) tetradecanoate
  Lead(II) myristate
  Group-member nr.: 74-016

$C_{28}H_{58}$
- Octacosane [630-02-4]
  Group-member nr.: 11-068

$C_{29}H_{42}O_4$
- 4-(Hexyloxy)phenyl ester 4-(decyloxy)benzoic acid [68162-09-4]
  4-Hexyloxyphenyl-4'-*n*-decyloxybenzoate
  Group-member nr.: 47-050

$C_{30}H_{46}N_2O_3$
- Bis(4-nonyloxyphenyl)diazene-1-oxide [25729-13-9]
  4,4'-Bis(nonyloxy)azoxybenzene
  Group-member nr.: 62-124

$C_{30}H_{58}O_4$
- Didecyl decanedioate [2432-89-5]
  Didecyl ester decanedioic acid
  Didecyl sebacate
  Group-member nr.: 45-115

$C_{30}H_{61}Br$
- 1-Bromotriacontane [4209-22-7]
  n-Triacontyl bromide
  Group-member nr.: 23-040

$C_{30}H_{62}$
- 2,6,10,15,19,23-Hexamethyltetracosane [111-01-3]
  Squalane
  Group-member nr.: 11-069
- Triacontane [638-68-6]
  Group-member nr.: 11-070

$C_{31}H_{52}O_3$
- 3,4-Dihydro-2,5,7,8-tetramethyl-2-(4,8,12-trimethyltridecyl)-2*H*-1-benzopyran-6-ol acetate [7695-91-2]
  α-Tocopherol acetate
  Group-member nr.: 47-051

$C_{31}H_{64}$
- 11-Decylheneicosane [55320-06-4]
  11-Decylhenicosane
  Group-member nr.: 11-071
- Hentriacontane [630-04-6]
  Unatriacontane
  Group-member nr.: 11-072

$C_{32}H_{50}N_2O_3$
- Bis(4-decyloxyphenyl)diazene-1-oxide [2312-12-1]
  4,4'-Bis(decyloxy)azoxybenzene
  Group-member nr.: 62-125

$C_{32}H_{62}CdO_4$
- Cadmium salt hexadecanoic acid [6427-86-7]
  Cadmium hexadecanoate
  Cadmium palmitate
  Group-member nr.: 74-017

$C_{32}H_{62}HgO_4$
- Mercury(2+) salt hexadecanoic acid [16888-60-1]
  Mercury(II) hexadecanoate
  Mercury(II) palmitate
  Group-member nr.: 74-018

$C_{32}H_{62}O_4Pb$
- Lead(2+) salt hexadecanoic acid [15773-56-5]
  Lead(II) hexadecanoate
  Lead(II) palmitate
  Group-member nr.: 74-019

$C_{32}H_{66}$
- Dotriacontane [544-85-4]
  Group-member nr.: 11-073

$C_{33}H_{62}O_6$
- 1,2,3-Propanetriyl ester decanoic acid [621-71-6]
  1,2,3-Propanetriyl tridecanoate
  Tridecanoin
  Group-member nr.: 45-116

$C_{33}H_{68}$
- Tritriacontane [630-05-7]
  Group-member nr.: 11-074

$C_{34}H_{54}N_2O_3$
- Bis[4-(undecyloxy)phenyl]diazene-1-oxide [2312-13-2]
  4,4'-Bis(undecyloxy)azoxybenzene
  Group-member nr.: 62-126

$C_{34}H_{66}O_4$
- Didodecyl decanedioate [2432-88-4]
  Didodecyl ester decanedioic acid
  Didodecyl sebacate
  Group-member nr.: 45-117

$C_{34}H_{70}$
- Tetratriacontane [14167-59-0]
  Group-member nr.: 11-075

$C_{35}H_{72}$
- Pentatriacontane [630-07-9]
  Group-member nr.: 11-076

$C_{36}H_{24}$
- 1,1',1''-(1,3,5-Benzenetriyl)trisnaphthalene [7059-70-3]
  1,3,5-Tri-α-naphthylbenzene
  Group-member nr.: 14-097

$C_{36}H_{58}N_2O_3$
- Bis[4-(dodecyloxy)phenyl]diazene-1-oxide [2312-14-3]
  4,4'-Bis(dodecyloxy)azoxybenzene
  Group-member nr.: 62-127

$C_{36}H_{70}CdO_4$
- Cadmium salt octadecanoic acid [2223-93-0]
  Cadmium octadecanoate
  Cadmium stearate
  Group-member nr.: 74-020

$C_{36}H_{70}HgO_4$
- Mercury(2+) salt octadecanoic acid [645-99-8]
  Mercury(II) octadecanoate
  Mercury(II) stearate
  Group-member nr.: 74-021

$C_{36}H_{70}O_4Pb$
- Lead(2+) salt octadecanoic acid [1072-35-1]
  Lead(II) octadecanoate
  Lead(II) stearate
  Group-member nr.: 74-022

$C_{36}H_{74}$
- Hexatriacontane [630-06-8]
  Group-member nr.: 11-077

$C_{37}H_{54}Si$
- Tribenzylhexadecylsilane [4033-52-7]
  Group-member nr.: 71-056

$C_{38}H_{70}O_8$
- 1,6-Hexanediyldihexyl ester decanedioic acid [55205-81-7]
  Dihexyl hexamethylenesebacate
  Group-member nr.: 45-118

$C_{38}H_{74}O_4$
- Ditetradecyl decanedioate [26719-47-1]
  Ditetradecyl ester decanedioic acid
  Ditetradecyl sebacate
  Group-member nr.: 45-119

$C_{39}H_{74}O_6$
- 1,2,3-Propanetriyl ester dodecanoic acid [538-24-9]
  1,2,3-Propanetriyl tridodecanoate
  Trilaurin
  Group-member nr.: 45-120

$C_{40}H_{82}$
- Tetracontane [4181-95-7]
  Group-member nr.: 11-078

$C_{41}H_{72}O_2$
- Cholest-5-en-3-ol (3β) tetradecanoate [1989-52-2]
  Cholesterol myristate
  Group-member nr.: 45-121

$C_{42}H_{66}O_{12}$
- 1,2,3,4,5,6-Benzenehexayl ester hexanoic acid [65201-69-6]
  Benzene hexa-*n*-hexanoate
  Group-member nr.: 45-122

$C_{42}H_{82}O_4$
- Dihexadecyl decanedioate [26719-48-2]
  Dihexadecyl ester decanedioic acid
  Dihexadecyl sebacate
  Group-member nr.: 45-123

$C_{42}H_{86}$
- Dotetracontane [7098-20-6]
  Group-member nr.: 11-079

$C_{43}H_{88}$
- Tritetracontane [7098-21-7]
  Group-member nr.: 11-080

$C_{44}H_{90}$
- Tetratetracontane [7098-22-8]
  Group-member nr.: 11-081

$C_{45}H_{86}O_6$
- 1,2,3-Propanetriyl ester tetradecanoic acid [555-45-3]
  1,2,3-Propanetriyl tritetradecanoate
  Trimyristin
  Group-member nr.: 45-124

$C_{46}H_{90}O_4$
- Dioctadecyl decanedioate [3072-03-5]
  Dioctadecyl ester decanedioic acid
  Dioctadecyl sebacate
  Group-member nr.: 45-125

$C_{48}H_{78}O_{12}$
• 1,2,3,4,5,6-Benzenehexayl ester heptanoic acid [65201-70-9]
Benzene hexa-*n*-heptanoate
Group-member nr.: 45-126

$C_{48}H_{98}$
• Octatetracontane [7098-26-2]
Group-member nr.: 11-082

$C_{50}H_{102}$
• Pentacontane [6596-40-3]
Group-member nr.: 11-083

$C_{51}H_{98}O_6$
• 1,2,3-Propanetriyl ester hexadecanoic acid [555-44-2]
1,2,3-Propanetriyl trihexadecanoate
Tripalmitin
Group-member nr.: 45-127

$C_{54}H_{90}O_{12}$
• 1,2,3,4,5,6-Benzenehexayl ester octanoic acid [65201-71-0]
Benzene hexa-*n*-octanoate
Group-member nr.: 45-128

$C_{54}H_{98}O_{12}$
• 10,19,28,37-Tetraoxodihexyl ester 11,18,29,36-tetraoxa-hexatetracosanedioic acid [55205-82-8]
Dihexyl bis(hexamethylenesebacate)
Group-member nr.: 45-129

$C_{57}H_{110}O_6$
• 1,2,3-Propanetriyl ester octadecanoic acid [555-43-1]
1,2,3-Propanetriyl trioctadecanoate
Tristearin
Group-member nr.: 45-130

ClD
• Hydrochlorid acid-*d* [7698-05-7]
Deuterium chloride
Group-member nr.: 02-030

$ClFO_3$
• Perchloryl fluoride ($(ClO_3)F$) [7616-94-6]
Group-member nr.: 02-031

ClF_3
• Chlorine fluoride (ClF_3) [7790-91-2]
Chlorine trifluoride
Group-member nr.: 02-032

ClH
• Hydrochlorid acid [7647-01-0]
Hydrogen chloride
Group-member nr.: 02-033

$ClHO_4$
• Perchloric acid [7601-90-3]
Group-member nr.: 02-034

ClI
• Iodine chloride [7790-99-0]
Iodine monochloride
Group-member nr.: 02-035

$ClNaO_3$
• Sodium salt chloric acid [7775-09-9]
Sodium chlorate
Group-member nr.: 02-036

Cl_2
• Chlorine [7782-50-5]
Group-member nr.: 01-004

Cl_2OS
• Thionyl chloride [7719-09-7]
Group-member nr.: 02-037

Cl_2O_2S
• Sulfuryl chloride [7791-25-5]
Group-member nr.: 02-038

$Cl_2O_5S_2$
• Disulfuryl chloride [7791-27-7]
Pyrosulfuryl chloride
Group-member nr.: 02-039

Cl_2S_2
• Sulfur chloride (S_2Cl_2) [10025-67-9]
Sulfur monochloride
Group-member nr.: 02-040

Cl_2Sn
• Tin chloride ($SnCl_2$) [7772-99-8]
Group-member nr.: 02-041

Cl_3HSi
• Trichlorosilane [10025-78-2]
Group-member nr.: 02-042

Cl_3OP
• Phosphoryl chloride [10025-87-3]
Group-member nr.: 02-043

Cl_3P
• Phosphorous trichloride [7719-12-2]
Phosphorus chloride (PCl_3)
Group-member nr.: 02-044

Cl_4Ge
• Tetrachlorogermane [10038-98-9]
Germanium tetrachloride
Group-member nr.: 02-045

Cl_4Si
- Tetrachlorosilane [10026-04-7]
  Silicon chloride
    Group-member nr.: 02-046

Cl_4Sn
- Tetrachlorostannane [7646-78-8]
  Tin(IV) chloride
  Tin tetrachloride
    Group-member nr.: 02-047

Cl_4Te
- Tellurium chloride ($TeCl_4$) [10026-07-0]
  Tellurium tetrachloride
    Group-member nr.: 02-048

Cl_4Ti
- Titanium chloride ($TiCl_4$) [7550-45-0]
  Titanium tetrachloride
  Titanic chloride
    Group-member nr.: 02-049

Cs
- Cesium [7440-46-2]
    Group-member nr.: 01-005

CsF_2H
- Cesium fluoride ($Cs(HF_2)$) [12280-52-3]
  Cesium hydrogen difluoride
    Group-member nr.: 02-050

DH
- Deuterium molecule with hydrogen [13983-20-5]
  Hydrogen deuteride
    Group-member nr.: 02-051

DI
- Hydroiodic acid-*d* [14104-45-1]
  Deuterium iodide
    Group-member nr.: 02-052

D_2
- Deuterium (D_2) [7782-39-0]
    Group-member nr.: 01-006
- *ortho*-Deuterium (D_2) [7782-39-0]
    Group-member nr.: 01-007

D_2O
- Water-d_2 [7789-20-0]
  Heavy water
  Deuterium oxide
    Group-member nr.: 02-053

D_2O_2
- Hydrogen-d_2 peroxide [6909-54-2]
  Deuterium peroxide
    Group-member nr.: 02-054

D_2S
- Hydrogen sulfide-d_2 [13536-94-2]
  Deuterium sulfide
    Group-member nr.: 02-055

D_2Se
- Hydrogen selenide-d_2 [13536-95-3]
  Deuterium selenide
    Group-member nr.: 02-056

D_3N
- Ammonia-d_3 [13550-49-7]
  Deuterium ammonia
  Trideuterated ammonia
    Group-member nr.: 02-057

FH
- Hydrofluoric acid [7664-39-3]
  Hydrogen fluoride
    Group-member nr.: 02-058

F_2
- Fluorine [7782-41-4]
    Group-member nr.: 01-008

F_2HRb
- Rubidium fluoride ($Rb(HF_2)$) [12280-64-7]
  Rubidium hydrogen difluoride
    Group-member nr.: 02-059

F_2H_5N
- Ammonium fluoride ($(NH_4)(HF_2)$) [1341-49-7]
  Ammonium hydrogen difluoride
    Group-member nr.: 02-060

F_2O
- Oxygen fluoride [7783-41-7]
    Group-member nr.: 02-061

F_2OS
- Thionyl fluoride [7783-42-8]
    Group-member nr.: 02-062

F_2O_2S
- Sulfuryl fluoride [2699-79-8]
    Group-member nr.: 02-063

F_3N
- Nitrogen fluoride (NF_3) [7783-54-2]
  Nitrogen trifluoride
    Group-member nr.: 02-064

F_3OP
- Phosphoryl fluoride [13478-20-1]
    Group-member nr.: 02-065

F_3P
- Phosphorous trifluoride [7783-55-3]
  Phosphorus trifluoride
  Phosphorous fluoride (PF_3)
    Group-member nr.: 02-066

F_5I
- Iodine fluoride (IF_5) [7783-66-6]
  Iodine pentafluoride
    Group-member nr.: 02-067

F_5Nb
• Niobium fluoride (NbF_5) [7783-68-8]
Niobium pentafluoride
Group-member nr.: 02-068

F_6Mo
• Molybdenum fluoride (MoF_6) [7783-77-9]
Molybdenum hexafluoride
Group-member nr.: 02-069

F_6S
• Sulfur fluoride (SF_6) [2551-62-4]
Sulfur hexafluoride
Group-member nr.: 02-070

F_6U
• Uranium(VI) fluoride [7783-81-5]
Uranium hexafluoride
Group-member nr.: 02-071

Ga
• Gallium [7440-55-3]
Group-member nr.: 01-009

GeH_4
• Germane [7782-65-2]
Group-member nr.: 02-072

HI
• Hydroiodic acid [10034-85-2]
Hydrogen iodide
Group-member nr.: 02-073

HNO_3
• Nitric acid [7697-37-2]
Group-member nr.: 02-074

H_2
• Hydrogen [1333-74-0]
Group-member nr.: 01-010
• *para*-Hydrogen [1333-74-0]
Group-member nr.: 01-011

H_2O
• Water [7732-18-5]
Group-member nr.: 02-075

H_2O_2
• Hydrogen peroxide [7722-84-1]
Group-member nr.: 02-076

H_2O_4S
• Sulfuric acid [7664-93-9]
Group-member nr.: 02-077

$H_2O_7S_2$
• Disulfuric acid [7783-05-3]
Pyrosulfuric acid
Group-member nr.: 02-078

H_2S
• Hydrogen sulfide [7783-06-4]
Group-member nr.: 02-079

H_2S_2
• Hydrogen sulfide (H_2S_2) [13465-07-1]
Dihydrogen disulfide
Group-member nr.: 02-080

H_2S_3
• Hydrogen sulfide (H_2S_3) [13845-23-3]
Dihydrogen trisulfide
Group-member nr.: 02-081

H_2S_4
• Hydrogen sulfide (H_2S_4) [13845-25-5]
Dihydrogen tetrasulfide
Group-member nr.: 02-082

H_2S_5
• Hydrogen sulfide (H_2S_5) [13845-24-4]
Dihydrogen pentasulfide
Group-member nr.: 02-083

H_2S_6
• Hydrogen sulfide (H_2S_6) [13845-51-7]
Dihydrogen hexasulfide
Group-member nr.: 02-084

H_2Se
• Hydrogen selenide [7783-07-5]
Group-member nr.: 02-085

H_3N
• Ammonia [7664-41-7]
Group-member nr.: 02-086

H_3O_4P
• Phosphoric acid [7664-38-2]
Group-member nr.: 02-087

H_3P
• Phosphine [7803-51-2]
Group-member nr.: 02-088

H_4N_2
• Hydrazine [302-01-2]
Group-member nr.: 02-089

H_4Si
• Silane [7803-62-5]
Group-member nr.: 02-090

H_5NO
• Ammonium hydroxide [1336-21-6]
Group-member nr.: 02-091

$H_5N_3O_3$
• Hydrazine mononitrate [13464-97-6]
Hydrazine nitrate
Group-member nr.: 02-092

H_8N_2O
- Ammonium oxide [12161-77-2]
  Group-member nr.: 02-093

He
- Helium [7440-59-7]
  Group-member nr.: 01-012
- Helium (isotope of mass 3) [14762-55-1]
  Group-member nr.: 01-013

Hg
- Mercury [7439-97-6]
  Group-member nr.: 01-014

HgI_2
- Mercury iodide (HgI_2) [7774-29-0]
  Group-member nr.: 02-094

I_2
- Iodine [7553-56-2]
  Group-member nr.: 01-015

I_4Sn
- Tetraiodostannane [7790-47-8]
  Tin iodide
  Stannic iodide
  Group-member nr.: 02-095

In
- Indium [7440-74-6]
  Group-member nr.: 01-016

K
- Potassium [7440-09-7]
  Kalium
  Group-member nr.: 01-017

Kr
- Krypton [7439-90-9]
  Group-member nr.: 01-018

Li
- Lithium [7439-93-2]
  Group-member nr.: 01-019

$LiNO_3$
- Lithium salt nitric acid [7790-69-4]
  Lithium nitrate
  Group-member nr.: 02-096

NO
- Nitrogen oxide (NO) [10102-43-9]
  Nitrogen monoxide
  Nitric oxide
  Group-member nr.: 02-097

NO_2
- Nitrogen oxide (NO_2) [10102-44-0]
  Nitrogen dioxide
  Group-member nr.: 02-098

N_2
- Nitrogen [7727-37-9]
  Group-member nr.: 01-020

N_2O
- Nitrogen oxide (N_2O) [10024-97-2]
  Dinitrogen monoxide
  Nitrous oxide
  Group-member nr.: 02-099

N_2O_4
- Nitrogen oxide (N_2O_4) [10544-72-6]
  Dinitrogen tetroxide
  Nitrogen peroxide
  Nitrogen tetroxide
  Group-member nr.: 02-100

Na
- Sodium [7440-23-5]
  Natrium
  Group-member nr.: 01-021

Ne
- Neon [7440-01-9]
  Group-member nr.: 01-022

O_2
- Oxygen [7782-44-7]
  Group-member nr.: 01-023

O_2S
- Sulfur dioxide [7446-09-5]
  Group-member nr.: 02-101

O_3
- Ozone [10028-15-6]
  Group-member nr.: 01-024

O_3S
- Sulfur trioxide [7446-11-9]
  Group-member nr.: 02-102

O_6P_4
- Phosphorus oxide (P_4O_6) [12440-00-5]
  Group-member nr.: 02-103

P_4
- Phosphorus mol.(P_4) [12185-10-3]
  Group-member nr.: 01-025

Rb
- Rubidium [7440-17-7]
  Group-member nr.: 01-026

S
- Sulfur [7704-34-9]
  Group-member nr.: 01-027

Se
- Selenium [7782-49-2]
  Group-member nr.: 01-028

Sn
- Tin [7440-31-5]
  Stannane
  Group-member nr.: 01-029

Xe
- Xenon [7440-63-3]
  Group-member nr.: 01-030

Name Index of Compounds

0. Inorganic Substances

01. Elements

The elements family contains 30 members that have melting temperatures below 573 K which is the upper temperature limit for including a compound in this monograph. All measurements of the 30 elements were performed over a range of temperatures.

For the sake of clarity, this introduction is subdivided into several parts, each covering one group of elements depending on their physical state at ambient temperature (gases, liquids and solids). Measurements on these elements were carried out in diverse laboratories by various experimental techniques which differed largely because of their individual properties and unique characteristics.

Gaseous elements, hydrogen, deuterium, nitrogen, oxygen, and rare gases, were measured mainly before World War II in German laboratories at TUB, UMG, UWG (16EUC, 28EUC/HAU, 29CLU2, 29CLU/HIL, 35CLU/BAR, 36CLU, 38CLU/KRU, 38CLU/RIC) and in USA at UCB (29GIA/JOH, 33GIA/CLA). Ortho-deuterium and para-hydrogen were measured after World War II at OSU (50JOH/CLA, 51KER/RIF). Some of the elements have never been reinvestigated since, and therefore, the above measurements served in most cases as a basis for the final correlation. This was particularly the case for all of the rare gases (except helium, see below). Regarding more recent sources, extensive data sets from NBSB were published for para-hydrogen (62YOU/DIL), nitrogen (81WEB, 91MAG), and oxygen (69GOO/WEB). These data were measured in an adiabatic spherical calorimeter (61GOO) up to the critical region with an estimated uncertainty of 0.3 to 0.7 % (the higher value applies to the vicinity of the critical point) and were preferred in the correlation to the older sources. The latter data have been, however, considered in certain cases when they were consistent with the NBSB data.

Interpreting the experimental data for two isotopes of helium (^{3}He and ^{4}He) was outside the scope of this monograph. Liquid helium exists only over a very narrow temperature range and a sharp λ-transition requires special treatment as part of the correlation which would not be consistent with the approach used for other compounds. We have, therefore, presented only a survey of the main data sources for the heat capacity of helium in this compilation.

The four halogen elements exist at ambient temperature in three different phases (fluorine and chlorine are gases, bromine is a liquid, and iodine is a solid). Very extensive and reliable measurements of fluorine are available from NBSB (70GOO/PRY) with data covering the entire range of temperatures up to the critical temperature. Data for chlorine from UCB (39GIA/POW) have been preferred to older and less accurate data from TUB (24EUC/KAR). Only one data set, measured at DCM (58HIL/KRA), is available for bromine. Two data sets were available for iodine; the older but apparently more reliable source (37CAR/HAR) has been selected. All halogens exhibit a minimum on the heat capacity curve just below the normal boiling temperature. This phenomenon is frequently observed for chemical elements and is discussed in a more general manner in the literature (93ZAB/BUR).

Liquid heat capacity data are available in addition for two non-metallic elements, phosphorus and sulfur, which are solids at ambient temperature. Only smoothed data presented as parameters of a quadratic equation are available for liquid phosphorus from UCB (42YOU/HIL) in the temperature range from 298 to 370 K. This equation partly describes the undercooled liquid for which the lower temperature limit is almost 20 K below the melting temperature. Certain properties of sulfur, such as the enthalpy of fusion, melting temperature and heat capacity, are very sensitive to impurities in the sample. Five available sets of experimental data differ significantly from each other and we have selected the values originating from the publication presenting the combined measurements in three Austrian laboratories and one American laboratory (73KOM/MIL) using a single sample of highest purity. The heat capacity curve $C_p = f(T)$ exhibits a sharp maximum around 435 K that is due to the decay of the eight-member S_8 ring into a linear chain. In order to avoid the anomalous heat capacity values caused by this sharp transition, we have divided the entire temperature interval into two subintervals. The lower temperature subinterval covers the range from the melting temperature to 432.15 K and the other subinterval from 438.15 K to 598.15 K; data in each subinterval were correlated by a separate polynomial.

The last part of this introduction deals with solid metallic elements. Most metals have melting temperatures high above ambient temperature and the interval between the melting and normal boiling temperature is generally large. Heat capacity of liquid metals usually decreases slowly with temperature or data remain constant within the experimental error over a wide temperature interval. For that reason, most data for many metals were calculated from measurements which were carried out in high temperature drop calorimeters under extreme temperature conditions. In those cases, when the data reached into the temperature region of our interest (melting temperature below 573 K), we have frequently taken the data recommended in the monograph by Hultgren (63HUL/ORR) as well as data supplemented with more recent references.

Mercury, gallium, and alkali metals have melting temperatures substantially lower than most other metals, however, their $C_p = f(T)$ curve is similar to that for other metals. Three reliable and consistent data sets have been selected for correlating heat capacity of mercury: an extensive data set from NBSW (51DOU/BAL), additional data sets from UCB (53BUS/GIA), and data from IICN (79AMI/LEB). In the case of gallium, data by (83TAK/KAD) measured jointly at two Japanese laboratories and measurements from ICCN (84AMI/MIN) have been selected as the most reliable. Alkali metals were studied systematically at NBSW; data have been selected for lithium (55DOU/EPS) (a joint research effort with Knolls Atomic Laboratory, Schenectady, New York), sodium (50GIN/DOU) and potassium (52DOU/BAL) with uncertainties between 0.3 and 0.4 %. Parameters of a linear smoothing equation are available for rubidium (13REN), however, for cesium only two data points were measured (55DAU/MAR).

Name: Argon
Formula: Ar

CAS–RN: 7440–37–1
Group No.: 1–001
Molar Mass: 39.95

TABLE 1.1.1. Experimental heat capacities

Reference	Temp. range K	No. pnts	Error %	Purity %	Purity method	Type capacity	Calorimeter Type	Calorimeter Reference
16EUC	84.9–93.9	5	nosp	not specified		C_{sat}	BSIO	16EUC
28EUC/HAU	90.0–140.0	6S	nosp	not specified		C_{sat}	BSIO	28EUC/HAU
36CLU	86.6–88.9	3	nosp	not specified		C_{sat}	BSIO	36CLU/GOL
61FLU/LEA	84.6–86.3	3	0.50	99.9997	melpt	C_p	BSAO	61FLU/LEA

TABLE 1.1.2. Correlated heat capacities

Reference	Temp. range K	No. pnts used	$\sigma_r C$ %	d_w	d/R	d_r %	d_b/R	+/–
				Selected data				
16EUC	84.9–93.9	5	3.00#	1.086	1.74–1	3.26	–1.50–1	–3
28EUC/HAU	90.0–140.0	6	2.00#	1.254	1.54–1	2.51	–2.73–2	0
61FLU/LEA	84.6–86.3	3	0.50	2.860	7.78–2	1.43	1.42–2	–1
				Rejected data				
36CLU	(3.51–1, 6.91, –3.49–1, –3)							

TABLE 1.1.3. Parameters of cubic spline polynomials

Heat capacity type	No. data points total	No. data points used	s_w	s/R	s_r %	s_b/R	+/–
C_p	17	12	0.744	8.14–2	1.26	–5.32–3	–1
C_{sat}	17	12	0.492	3.96–2	0.69	–6.60–3	–2

Temp. range K	A_1	A_2	A_3	A_4	Level of uncertainty
84.6–105.0	5.21208+1	–1.50242+2	1.57885+2	–5.41279+1	V
105.0–140.0	–1.28816+2	3.66722+2	–3.34461+2	1.02172+2	V
84.6–110.0	2.05119+1	–4.98369+1	5.23003+1	–1.74277+1	V
110.0–140.0	–9.06124+1	2.53229+2	–2.23214+2	6.60616+1	V

TABLE 1.1.4. Recommended values of heat capacities

Temp. (K)	90	100	110	120	130	140
c_p (J K^{-1}g^{-1})	1.11	1.17	1.22	1.29	1.49	1.96
C_p (J K^{-1}mol^{-1})	44.3	46.9	48.8	51.4	59.5	78.2
c_{sat} (J K^{-1}g^{-1})	1.11	1.15	1.20	1.25	1.35	1.60
C_{sat} (J K^{-1}mol^{-1})	44.2	46.1	48.0	49.8	54.0	63.9

TABLE 1.1.5. Parameters of quasi-polynomial equation

Heat capacity type	No. data points total	No. data points used	s_w	s/R	s_r %	s_b/R	+/–
C_p	17	12	0.871	6.54–2	1.12	1.15–2	–2

Temp. range K	T_c K	A_1	A_2	A_3	A_4	Level of uncertainty
84.6–140.0	150.80	1.03253–1	3.75540–1	4.50207	7.09719–3	V

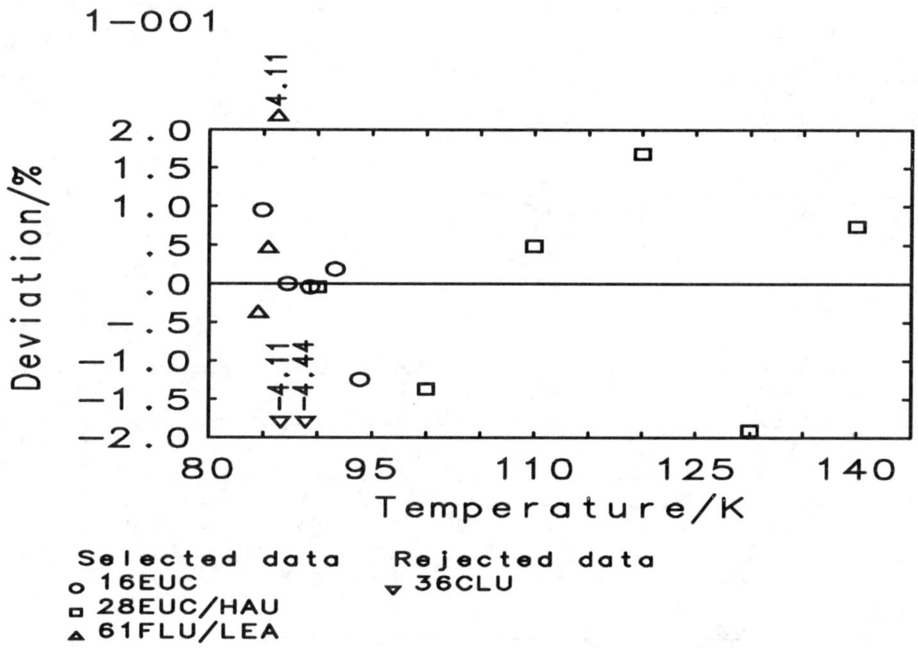

Name: Bismuth
Formula: Bi

CAS-RN: 7440-69-9
Group No.: 1-002
Molar Mass: 208.98

TABLE 1.2.1. Experimental heat capacities

Reference		Temp. range K	No. pnts	Error %	Purity %	Purity method	Type capacity	Calorimeter Type	Calorimeter Reference
26AWB/GRI	N	543.2–673.2	2	nosp	not specified		C_p	DSIO	26AWB/GRI
26UMI		573.2–873.2	5S	nosp	99.875	anal	C_{avg}	not specified	
32CAR/HAR		545.6–644.1	10	0.50	99.986	anal	C_p	BSIO	30CAR/STO
40KUB/SCH	N	548.2–798.2	2	nosp	not specified		C_{avg}	DSIO	40KUB/SCH
71BEL/HUL		544.8–801.8	21	nosp	99.9999	anal	C_p	DSIO	71BEL/HUL
83JOA/PAL	N	544.5	1	nosp	not specified		C_p	not specified	

26AWB/GRI constant value calculated from temperature dependence of enthalpy by the authors; suspect value
40KUB/SCH constant value calculated from temperature dependence of enthalpy by the authors; suspect value
83JOA/PAL the origin of data unclear

TABLE 1.2.2. Correlated heat capacities

Reference	Temp. range K	No. pnts used	$\sigma_r C$ %	d_w	d/R	d_r %	d_b/R	+/–
				Selected data				
32CAR/HAR	545.6–644.1	10	0.50	1.202	2.16–2	0.60	–1.48–3	0
71BEL/HUL	544.8–801.8	21	0.50#	1.263	2.23–2	0.63	–1.11–3	–2
				Rejected data				
26AWB/GRI	(1.46, 73.25, –1.46, –1)			26UMI	(3.29–1, 8.20, 2.24–1, 2)			
40KUB/SCH	(4.75–1, 11.93, 4.55–1, 2)							

TABLE 1.2.3. Parameters of regression polynomial

Heat capacity type	No. data points total	used	s_w	s/R	s_r %	s_b/R	+/–
C_p	41	31	1.308	2.32–2	0.65	2.77–4	–2

Temp. range K	A_1	A_2	A_3	Level of uncertainty
544.8–801.8	5.87111	–6.09214–1	3.73303–2	IV

TABLE 1.2.4. Recommended values of heat capacities

Temp. (K)	550	570	590	610	630	650	670
c_p (J K^{-1}g^{-1})	0.145	0.144	0.142	0.141	0.140	0.139	0.138
C_p (J K^{-1}mol^{-1})	30.35	30.03	29.73	29.47	29.22	29.00	28.81
Temp. (K)	690	710	730	750	770	790	810
c_p (J K^{-1}g^{-1})	0.137	0.136	0.136	0.135	0.135	0.135	0.135
C_p (J K^{-1}mol^{-1})	28.64	28.50	28.38	28.28	28.22	28.17	28.15

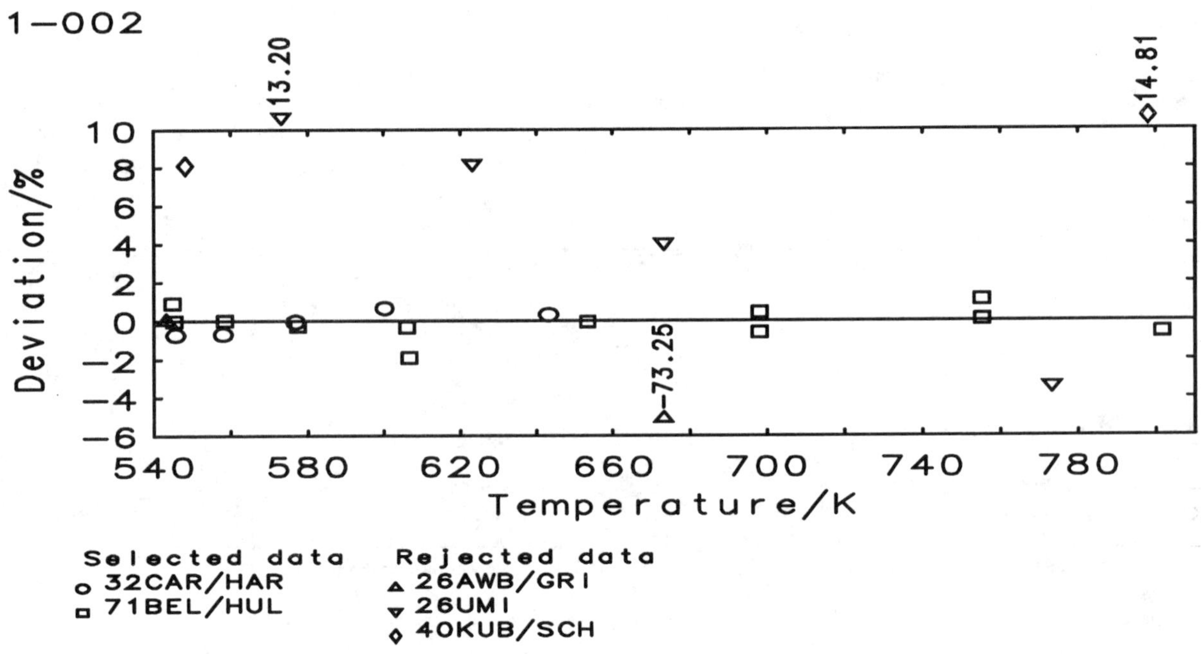

Name: Bromine
Formula: Br_2

CAS–RN: 7726–95–6
Group No.: 1–003
Molar Mass: 159.81

TABLE 1.3.1. Experimental heat capacities

Reference	Temp. range K	No. pnts	Error %	Purity %	Purity method	Type capacity	Calorimeter Type	Calorimeter Reference
58HIL/KRA	265.9–300.0	6	nosp	99.999	melpt	C_p	BSAO	58HIL/KRA

TABLE 1.3.3. Parameters of regression polynomial

Heat capacity type	No. data points total	used	s_w	s/R	s_r %	s_b/R	+/–
C_p	6	6	0.009	2.36–4	0.00	4.77–7	0

Temp. range K	A_1	A_2	A_3	Level of uncertainty
265.9–300.0	2.14218+1	–7.90841	1.26668	III

TABLE 1.3.4. Recommended values of heat capacities

Temp. (K)	270	273.15	280	290	298.15	300
c_p (J K^{-1}g^{-1})	0.484	0.482	0.479	0.476	0.474	0.473
C_p (J K^{-1}mol^{-1})	77.35	77.08	76.57	76.00	75.69	75.63

Name: Chlorine
Formula: Cl_2

CAS–RN: 7782–50–5
Group No.: 1–004
Molar Mass: 70.91

TABLE 1.4.1. Experimental heat capacities

Reference	Temp. range K	No. pnts	Error %	Purity %	method	Type capacity	Calorimeter Type	Reference
24EUC/KAR	187.7–197.0	6	3.00	not specified		C_{sat}	BSIO	24EUC/KAR
39GIA/POW	179.0–236.8	11	nosp	99.997	melpt	C_p	BSIO	37GIA/EGA

TABLE 1.4.2. Correlated heat capacities

Reference	Temp. range K	No. pnts used	$\sigma_r C$ %	d_w	d/R	d_r %	d_b/R	+/–
				Selected data				
39GIA/POW	179.0–236.8	11	0.40#	0.199	6.37–3	0.08	1.01–5	1
				Rejected data				
24EUC/KAR	(1.57–1, 1.90, 1.32–1, 6)							

TABLE 1.4.3. Parameters of regression polynomial

Heat capacity type	No. data points total	used	s_w	s/R	s_r %	s_b/R	+/–
C_p	17	11	0.233	7.46–3	0.09	1.01–5	1

Temp. range K	A_1	A_2	A_3	Level of uncertainty
179.0–236.8	7.69170	5.55367–1	–1.94679–1	III

TABLE 1.4.4. Recommended values of heat capacities

Temp. (K)	180	190	200	210	220	230
c_p (J K^{-1}g^{-1})	0.945	0.943	0.941	0.938	0.935	0.931
C_p (J K^{-1}mol^{-1})	67.02	66.88	66.71	66.51	66.28	66.01

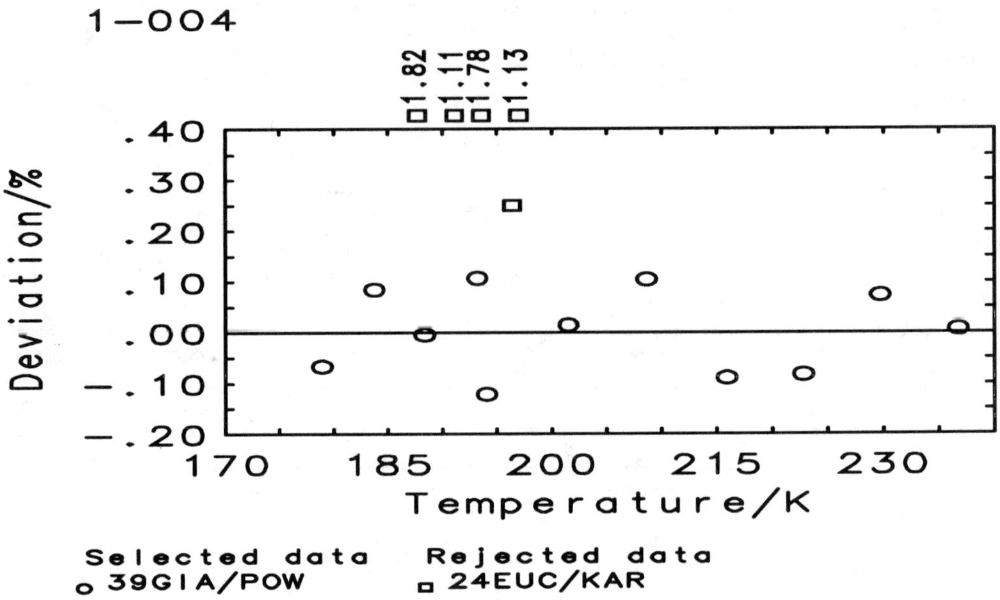

Name: Cesium
Formula: Cs

CAS–RN: 7440–46–2
Group No.: 1–005
Molar Mass: 132.91

TABLE 1.5.1. Experimental heat capacities

Reference	Temp. range K	No. pnts	Error %	Purity %	Purity method	Type capacity	Calorimeter Type	Calorimeter Reference
13REN	302.0–373.0	eqn	nosp	not specified		C_p	DSIO	13REN
55DAU/MAR	310.0–320.0	2S	0.20	not specified		C_p	BSAO	55DAU/MAR

TABLE 1.5.2. Correlated heat capacities

Reference	Temp. range K	No. pnts used	$\sigma_r C$ %	d_w	d/R	d_r %	d_b/R	+/–
				Selected data				
55DAU/MAR	310.0–320.0	2	0.20	0.000	0.00	0.00	0.00	0
				Rejected data				
13REN	(1.17–1, 2.95, 1.17–1, 1)							

TABLE 1.5.3. Parameters of regression polynomial

Heat capacity type	No. data points total	No. data points used	s_w	s/R	s_r %	s_b/R	+/–
C_p	10	2	0.000	0.00	0.00	0.00	0

Temp. range K	A_1	Level of uncertainty
310.0–320.0	3.83451	III

TABLE 1.5.4. Recommended values of heat capacities

Temp. (K)	310	320
c_p (J K^{-1}g^{-1})	0.240	0.240
C_p (J K^{-1}mol^{-1})	31.88	31.88

Name: Deuterium (D_2)
Formula: D_2

CAS–RN: 7782–39–0
Group No.: 1–006
Molar Mass: 4.03

TABLE 1.6.1. Experimental heat capacities

Reference		Temp. range K	No. pnts	Error %	Purity %	Purity method	Type capacity	Calorimeter Type	Calorimeter Reference
35CLU/BAR	N	19.4–21.7	8	nosp	99.7	estim	C_{sat}	BSIO	36CLU/GOL

35CLU/BAR content of HD 0.3 %

TABLE 1.6.3. Parameters of regression polynomial

Heat capacity type	No. data points total	No. data points used	s_w	s/R	s_r %	s_b/R	+/–
C_{sat}	8	8	0.611	1.56–2	0.61	1.41–4	1

Temp. range K	A_1	A_2	Level of uncertainty
19.4–21.7	3.23134–1	1.11682+1	IV

TABLE 1.6.4. Recommended values of heat capacities

Temp. (K)	19.50	20	20.50	21	21.50
c_{sat} (J K^{-1}g^{-1})	5.162	5.278	5.393	5.508	5.623
C_{sat} (J K^{-1}mol^{-1})	20.79	21.26	21.72	22.19	22.65

Name: *ortho*–Deuterium (D_2)
Formula: D_2

CAS–RN: 7782–39–0
Group No.: 1–007
Molar Mass: 4.03

TABLE 1.7.1. Experimental heat capacities

Reference		Temp. range K	No. pnts	Error %	Purity %	Purity method	Type capacity	Calorimeter Type	Calorimeter Reference
51KER/RIF	N	19.9–22.7	8	nosp	99.2	anal	C_p	BSIO	50JOH/CLA

51KER/RIF sample contained 97.8 % *ortho*–Deuterium; C_p values were corrected for presence of HD (1.1 %)

TABLE 1.7.3. Parameters of regression polynomial

Heat capacity type	No. data points total	No. data points used	s_w	s/R	s_r %	s_b/R	+/–
C_p	8	8	0.528	7.47–3	0.26	2.98–5	0

Temp. range K	A_1	A_2	Level of uncertainty
19.9–22.7	7.29466–1	1.01918+1	III

TABLE 1.7.4. Recommended values of heat capacities

Temp. (K)	20	20.50	21	21.50	22	22.50
c_p ($J K^{-1} g^{-1}$)	5.713	5.818	5.924	6.029	6.134	6.239
C_p ($J K^{-1} mol^{-1}$)	23.01	23.44	23.86	24.28	24.71	25.13

Name: Fluorine
Formula: F_2

CAS–RN: 7782–41–4
Group No.: 1–008
Molar Mass: 38.00

TABLE 1.8.1. Experimental heat capacities

Reference	Temp. range K	No. pnts	Error %	Purity %	Purity method	Type capacity	Calorimeter Type	Calorimeter Reference
37KAN	57.5–83.4	5	nosp	not specified		C_p	BSIO	37KAN
53HU/WHI2	58.1–81.3	6	nosp	99.97	melpt	C_p	BSIO	50JOH/CLA
70GOO/PRY	55.2–143.4	36	nosp	99.99	anal	C_{sat}	BSAO	61GOO

TABLE 1.8.2. Correlated heat capacities

Reference	Temp. range K	No. pnts used	$\sigma_r C$ %	d_w	d/R	d_r %	d_b/R	+/–
				Selected data				
70GOO/PRY	55.2–140.7	34	0.30#	1.687	5.94–2	0.51	8.54–4	–2
				Rejected data				
37KAN	(1.21, 21.89, –1.21, –5)			53HU/WHI2	(1.46–1, 2.11, 1.20–1, 6)			

TABLE 1.8.3. Parameters of cubic spline polynomials

Heat capacity type	No. data points total	No. data points used	s_w	s/R	s_r %	s_b/R	+/–
C_p	47	34	1.894	6.67–2	0.57	8.54–4	–2
C_{sat}	47	34	1.330	4.05–2	0.40	3.40–4	–3

Temp. range K	A_1	A_2	A_3	A_4	Level of uncertainty
55.2–85.0	–2.66035	3.71743+1	–4.94806+1	2.23401+1	III
85.0–105. 0	1.48979+1	–2.47958+1	2.34254+1	–6.25044	III
105.0–132. 0	–1.30711+2	3.91230+2	–3.72789+2	1.19532+2	IV
132.0–140. 7	–2.00495+4	4.56611+4	–3.46682+4	8.77998+3	V
55.2–85.0	–1.67553	3.28266+1	–4.31301+1	1.92659+1	III
85.0–105.0	1.30737+1	–1.92295+1	1.81125+1	–4.75079	III
105.0–132.0	–8.11488+1	2.49978+2	–2.38275+2	7.66421+1	IV
132.0–140.7	–1.16463+4	2.65345+4	–2.01508+4	5.10505+3	V

TABLE 1.8.4. Recommended values of heat capacities

Temp. (K)	60	70	80	90	100	110	120
c_p (J $K^{-1}g^{-1}$)	1.457	1.483	1.499	1.532	1.592	1.677	1.860
C_p (J $K^{-1}mol^{-1}$)	55.35	56.36	56.95	58.20	60.50	63.72	70.67
c_{sat} (J $K^{-1}g^{-1}$)	1.456	1.483	1.498	1.526	1.577	1.646	1.782
C_{sat} (J $K^{-1}mol^{-1}$)	55.33	56.35	56.93	57.99	59.91	62.56	67.73
Temp. (K)	130	140					
c_p (J $K^{-1}g^{-1}$)	2.294	4.11					
C_p (J $K^{-1}mol^{-1}$)	87.18	156					
c_{sat} (J $K^{-1}g^{-1}$)	2.083	3.21					
C_{sat} (J $K^{-1}mol^{-1}$)	79.15	122					

TABLE 1.8.5. Parameters of quasi-polynomial equation

Heat capacity type	No. data points total	used	s_w	s/R	s_r %	s_b/R	+/–
C_p	47	34	2.094	5.88–2	0.63	–1.47–4	4

Temp. range K	T_c K	A_1	A_2	A_3	A_4	A_5	A_6	Level of uncertainty
55.2–140.7	144.30	–1.41175	2.83069–1	6.31565	–1.11994	–3.37656	2.28286	IV

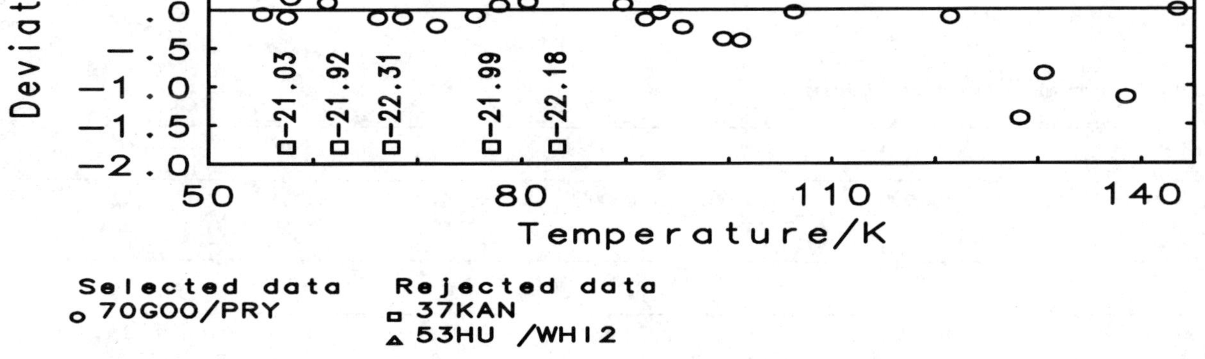

Name: Gallium
Formula: Ga

CAS–RN: 7440–55–3
Group No.: 1–009
Molar Mass: 69.72

TABLE 1.9.1. Experimental heat capacities

Reference		Temp. range K	No. pnts	Error %	Purity %	Purity method	Type capacity	Calorimeter Type	Calorimeter Reference
33ROT/MEY2	N	298.1–313.1	2	nosp	not specified		C_p	DSIO	33ROT/MEY2
34ROT/MEY	N	298.1–373.1	2S	nosp	not specified		C_{avg}	DSIO	33ROT/MEY2
52ADA/JOH		306.9–322.8	4	nosp	99.98	melpt	C_p	BSAO	52ADA/JOH
83TAK/KAD		298.3–613.4	87	1.00	99.9999	anal	C_p	BDHO	79TAK/YOK
84AMI/MIN		304.4–319.2	27	nosp	99.9999	anal	C_p	BSAO	79AMI/LEB

33ROT/MEY2 constant value calculated from temperature dependence of enthalpy by the authors; correct data in 34ROT/MEY
34ROT/MEY corrected data from 33ROT/MEY2; a constant value calculated from temperature dependence of enthalpy by the authors

TABLE 1.9.2. Correlated heat capacities

Reference	Temp. range K	No. pnts used	$\sigma_r C$ %	d_w	d/R	d_r %	d_b/R	+/–
				Selected data				
83TAK/KAD	298.3–613.4	87	1.00	0.301	1.01–2	0.30	–3.71–3	–27
84AMI/MIN	304.4–319.2	27	0.30#	0.215	2.20–3	0.06	1.10–3	11
				Rejected data				
33ROT/MEY2	(4.34–2, 1.26, 4.34–2, 1)			34ROT/MEY	(1.13–1, 3.28, 1.13–1, 1)			
52ADA/JOH	(3.89–2, 1.16, –3.88–2, –4)							

TABLE 1.9.3. Parameters of regression polynomial

Heat capacity type	No. data points total	No. data points used	s_w	s/R	s_r %	s_b/R	+/–
C_p	122	114	0.288	9.03–3	0.27	–2.57–3	–16

Temp. range K	A_1	A_2	A_3	A_4	Level of uncertainty
298.3–613.4	4.50213	–6.03882–1	9.73838–2	–5.57025–3	IV

TABLE 1.9.4. Recommended values of heat capacities

Temp. (K)	300	320	340	360	380	400	420
c_p (J K^{-1}g^{-1})	0.407	0.404	0.400	0.397	0.394	0.392	0.390
C_p (J K^{-1}mol^{-1})	28.41	28.14	27.90	27.69	27.50	27.34	27.20
Temp. (K)	440	460	480	500	520	540	560
c_p (J K^{-1}g^{-1})	0.388	0.387	0.385	0.384	0.383	0.382	0.381
C_p (J K^{-1}mol^{-1})	27.07	26.96	26.87	26.78	26.71	26.64	26.57
Temp. (K)	580	600					
c_p (J K^{-1}g^{-1})	0.380	0.379					
C_p (J K^{-1}mol^{-1})	26.51	26.45					

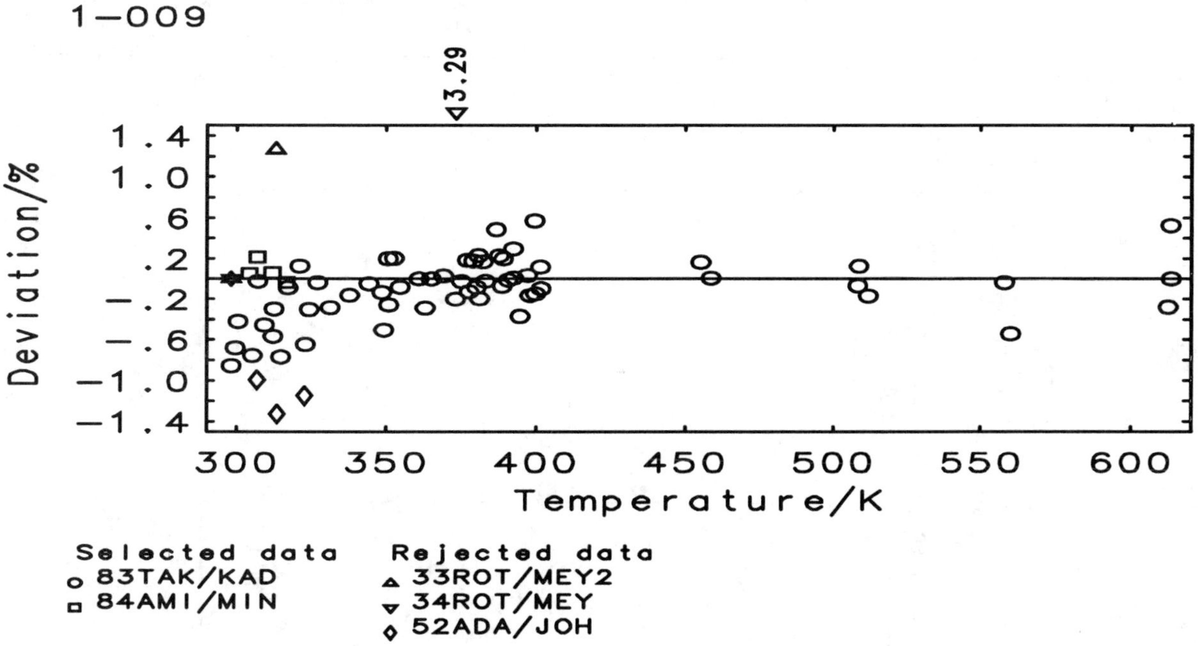

Name: Hydrogen
Formula: H_2

CAS–RN: 1333–74–0
Group No.: 1–010
Molar Mass: 2.02

TABLE 1.10.1. Experimental heat capacities

Reference	Temp. range K	No. pnts	Error %	Purity %	Purity method	Type capacity	Calorimeter Type	Calorimeter Reference
16EUC	17.4–21.3	5	nosp	not specified		C_{sat}	BSIO	16EUC
23SIM/LAN	15.4–18.6	8	nosp	not specified		C_{sat}	BSIO	23SIM/LAN

TABLE 1.10.2. Correlated heat capacities

Reference	Temp. range K	No. pnts used	$\sigma_r C$ %	d_w	d/R	d_r %	d_b/R	+/–
				Selected data				
16EUC	17.4–21.3	5	2.00#	0.937	4.11–2	1.87	2.00–2	3
23SIM/LAN	15.4–18.6	8	1.00#	0.777	1.55–2	0.78	–2.60–3	0

TABLE 1.10.3. Parameters of regression polynomial

Heat capacity type	No. data points total	No. data points used	s_w	s/R	s_r %	s_b/R	+/–
C_p	13	13	0.916	3.07–2	1.43	6.08–3	3
C_{sat}	13	13	0.786	2.83–2	1.33	3.78–3	2

Temp. range K	A_1	A_2	Level of uncertainty
15.4–21.3	1.00111–1	1.07833+1	V
15.4–21.3	2.05678–1	1.01391+1	V

TABLE 1.10.4. Recommended values of heat capacities

Temp. (K)	15	17	19	21
c_p (J $K^{-1}g^{-1}$)	7.08	7.97	8.86	9.75
C_p (J $K^{-1}mol^{-1}$)	14.3	16.1	17.9	19.7
c_{sat} (J $K^{-1}g^{-1}$)	7.12	7.96	8.79	9.63
C_{sat} (J $K^{-1}mol^{-1}$)	14.4	16.0	17.7	19.4

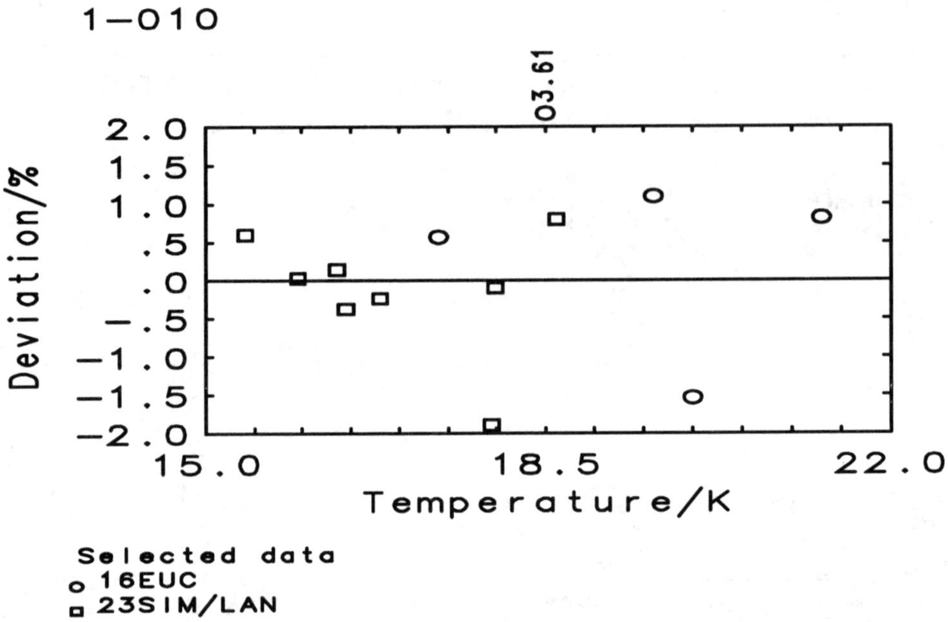

Name: *para*–Hydrogen
Formula: H_2

CAS–RN: 1333–74–0
Group No.: 1–011
Molar Mass: 2.02

TABLE 1.11.1. Experimental heat capacities

Reference		Temp. range K	No. pnts	Error %	Purity %	Purity method	Type capacity	Calorimeter Type	Calorimeter Reference
29CLU/HIL	N	15.1–18.0	6	1.50	not specified		C_p	BSIO	29CLU1
50JOH/CLA	N	15.1–19.0	10	0.50	not specified		C_p	BSIO	50JOH/CLA
54SMI/HAL	N	18.3–31.5	8	1.00	99.79	anal	C_{sat}	BSIO	53RIF/KER
62YOU/DIL	N	14.8–31.5	32	0.30	99.99	estim	C_{sat}	BSAO	61GOO

29CLU/HIL sample contained 94.0 % *para*–Hydrogen
50JOH/CLA sample contained 99.8 % *para*–Hydrogen
54SMI/HAL sample contained 0.21 % *ortho*–Hydrogen
62YOU/DIL error increased near critical temperature

TABLE 1.11.2. Correlated heat capacities

Reference	Temp. range K	No. pnts used	$\sigma_r C$ %	d_w	d/R	d_r %	d_b/R	+/–
				Selected data				
50JOH/CLA	16.3–19.0	7	0.50	1.635	1.64–2	0.82	–5.29–3	–3
62YOU/DIL	14.8–31.5	32	0.30	1.543	2.92–2	0.46	8.48–4	4
				Rejected data				
29CLU/HIL	(5.22–2, 2.83, 1.96–2, 2)			54SMI/HAL	(2.16–1, 3.06, 1.50–2, 0)			

TABLE 1.11.3. Parameters of cubic spline polynomials

Heat capacity type	No. data points total	used	s_w	s/R	s_r %	s_b/R	+/–
C_p	56	39	1.695	2.98–2	0.59	–2.54–4	1
C_{sat}	56	39	1.356	1.77–2	0.52	–5.30–4	–1

Temp. range K	A_1	A_2	A_3	A_4	Level of uncertainty
14.8–21.0	1.70966	–1.43623+1	1.15069+2	–1.39395+2	III
21.0–28.0	1.88860+1	2.79861+2	–1.28599+3	2.08452+3	IV
28.0–31.5	–1.03753+3	1.11939+4	–4.02648+4	4.84879+4	V
14.8–21.0	1.84338–1	9.74303	–6.81781	5.38039+1	III
21.0–28.0	–1.15732+1	1.77709+2	–8.06654+2	1.32339+3	IV
28.0–31.5	–5.77711+2	6.24347+3	–2.24701+4	2.71132+4	V

TABLE 1.11.4. Recommended values of heat capacities

Temp. (K)	15	17	19	21	23	25	27
c_p (J K^{-1}g^{-1})	6.904	7.872	8.986	10.22	11.61	13.51	16.32
C_p (J K^{-1}mol^{-1})	13.92	15.87	18.11	20.60	23.41	27.23	32.90
c_{sat} (J K^{-1}g^{-1})	6.904	7.869	8.902	10.01	11.26	12.85	15.06
C_{sat} (J K^{-1}mol^{-1})	13.92	15.86	17.95	20.19	22.69	25.91	30.36
Temp. (K)	29	31					
c_p (J K^{-1}g^{-1})	20.7	31.5					
C_p (J K^{-1}mol^{-1})	41.6	63.5					
c_{sat} (J K^{-1}g^{-1})	18.2	25.3					
C_{sat} (J K^{-1}mol^{-1})	36.8	50.9					

TABLE 1.11.5. Parameters of quasi-polynomial equation

Heat capacity type	No. data points total	used	s_w	s/R	s_r %	s_b/R	+/–
C_p	56	39	2.067	4.03–2	0.66	3.88–4	–3

Temp. range K	T_c K	A_1	A_2	A_3	A_4	Level of uncertainty
14.8–31.5	33.00	–9.76684–1	2.18035–1	1.85514–1	1.09376	V

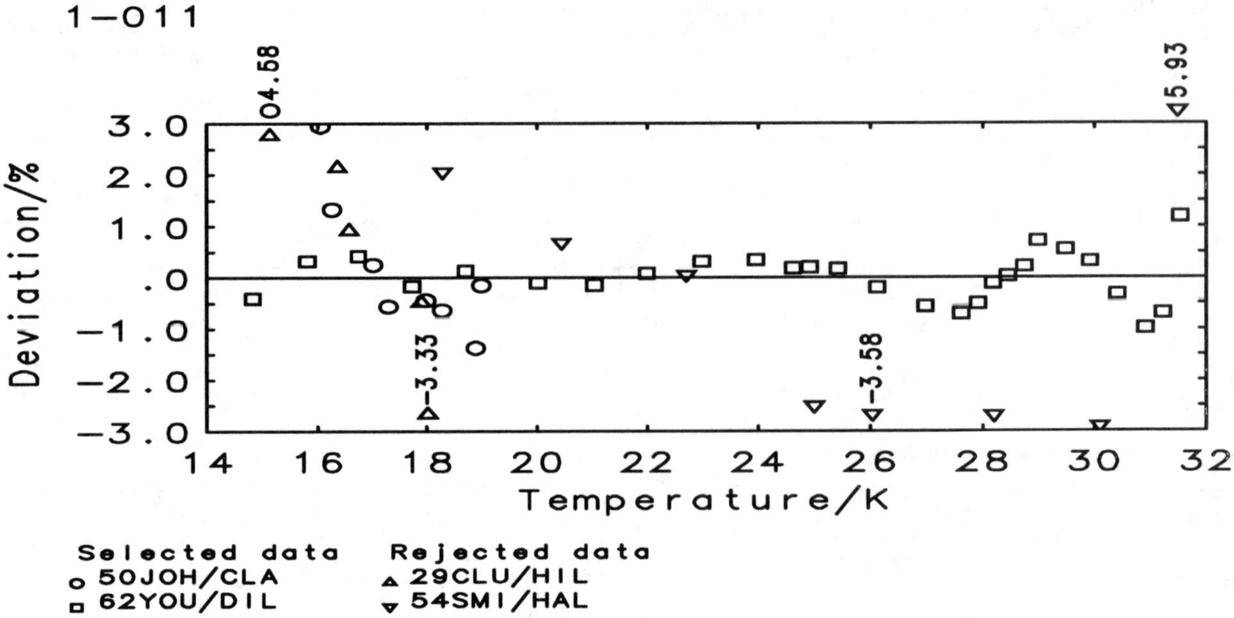

Name: Helium
Formula: He

CAS–RN: 7440–59–7
Group No.: 1–012
Molar Mass: 4.00

TABLE 1.12.1. Experimental heat capacities

Reference	Temp. range K	No. pnts	Error %	Purity %	Purity method	Type capacity	Calorimeter Type	Calorimeter Reference
35KEE/KEE	1.2–2.3	30	nosp	not	specified	C_{sat}	BSIO	35KEE/KEE
51HUL/WIL	0.6–1.6	11S	5.00	not	specified	C_{sat}	BSIO	51HUL/WIL
57HIL/LOU	1.8–5.1	46S	1.00	not	specified	C_{sat}	BSIO	57HIL/LOU

Name: Helium (isotope of mass 3)
Formula: He

CAS–RN: 14762–55–1
Group No.: 1–013
Molar Mass: 3.02

TABLE 1.13.1. Experimental heat capacities

Reference		Temp. range K	No. pnts	Error %	Purity %	Purity method	Type capacity	Calorimeter Type	Calorimeter Reference
54OSB/ABR		.0–1.0	eqn	5.00	not	specified	C_{sat}	BSIO	54OSB/ABR
54ROB/SYD	N	.5–1.7	14	nosp	not	specified	C_{sat}	BSIO	54ROB/SYD
55ROB/SYD	N	.4–2.4	21	2.00	not	specified	C_{sat}	BSIO	54ROB/SYD

54ROB/SYD content of He(4) isotope below 0.1 %
55ROB/SYD content of He(4) isotope 0.12 %

Name: Mercury
Formula: Hg

CAS–RN: 7439–97–6
Group No.: 1–014
Molar Mass: 200.59

TABLE 1.14.1. Experimental heat capacities

Reference		Temp. range K	No. pnts	Error %	Purity %	Purity method	Type capacity	Calorimeter Type	Calorimeter Reference
12LUS	N	335.6–363.1	2	nosp	not specified		C_{avg}	DSIO	12LUS
30CAR/STO		236.5–285.1	2	nosp	not specified		C_p	BSIO	30CAR/STO
51DOU/BAL		273.1–733.2	26S	0.10	99.999	anal	C_p	DSTO	50GIN/DOU
53BUS/GIA		239.0–325.9	20	0.10	99.9994	anal	C_p	BSIO	41GIA/MEA
79AMI/LEB		237.9–299.0	14	nosp	99.9998	melpt	C_p	BSAO	79AMI/LEB
88BUN		349.6	1	nosp	not specified		C_p	DDCT	85WAD

TABLE 1.14.2. Correlated heat capacities

Reference	Temp. range K	No. pnts used	$\sigma_r C$ %	d_w	d/R	d_r %	d_b/R	+/–
				Selected data				
51DOU/BAL	273.1–733.2	26	0.10	1.301	4.37–3	0.13	–1.56–3	–3
53BUS/GIA	239.0–325.9	20	0.10	1.148	3.89–3	0.11	2.22–3	10
79AMI/LEB	237.9–299.0	14	0.20#	0.371	2.50–3	0.07	–9.32–4	–6
				Rejected data				
12LUS	(1.16–2, 0.35, –1.16–2, –2)			30CAR/STO	(1.87–3, 0.06, –1.87–3, –1)			
88BUN	(6.73–2, 2.07, –6.73–2, –1)							

TABLE 1.14.3. Parameters of cubic spline polynomials

Heat capacity type	No. data points total	No. data points used	s_w	s/R	s_r %	s_b/R	+/–
C_p	65	60	1.146	4.01–3	0.12–1.	55–4	1

Temp. range K	A_1	A_2	A_3	A_4	Level of uncertainty
237.9–550.0	3.94048	–3.22219–1	5.08152–2	–2.64935–3	II
550.0–733.2	3.18634	8.91258–2	–2.39748–2	1.88337–3	II

TABLE 1.14.4. Recommended values of heat capacities

Temp. (K)	240	260	273.15	280	298.15	300	320
c_p (J K^{-1}g^{-1})	0.142	0.141	0.140	0.140	0.139	0.139	0.139
C_p (J K^{-1}mol^{-1})	28.46	28.27	28.15	28.09	27.95	27.93	27.79
Temp. (K)	340	360	380	400	420	440	460
c_p (J K^{-1}g^{-1})	0.138	0.137	0.137	0.137	0.136	0.136	0.136
C_p (J K^{-1}mol^{-1})	27.67	27.57	27.47	27.40	27.33	27.28	27.24
Temp. (K)	480	500	520	540	560	580	600
c_p (J K^{-1}g^{-1})	0.136	0.135	0.135	0.135	0.135	0.135	0.135
C_p (J K^{-1}mol^{-1})	27.20	27.18	27.16	27.15	27.14	27.14	27.15
Temp. (K)	620	640	660	680	700	720	740
c_p (J K^{-1}g^{-1})	0.135	0.135	0.136	0.136	0.136	0.136	0.137
C_p (J K^{-1}mol^{-1})	27.16	27.18	27.20	27.24	27.28	27.34	27.40

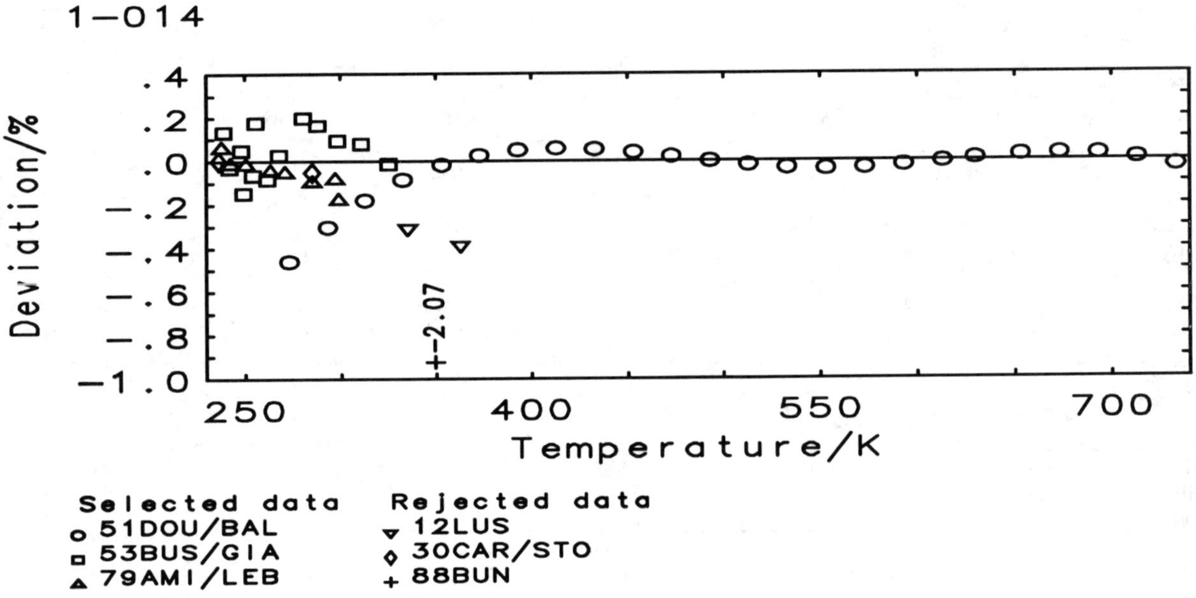

Name: Iodine
Formula: I_2

CAS–RN: 7553–56–2
Group No.: 1–015
Molar Mass: 253.81

TABLE 1.15.1. Experimental heat capacities

Reference		Temp. range K	No. pnts	Error %	Purity %	Purity method	Type capacity	Calorimeter Type	Calorimeter Reference
37CAR/HAR		389.4–432.7	5	nosp	99.81	anal	C_p	BSIO	30CAR/STO
38FRE/HIL1	N	386.8–433.1	2	1.50	not specified		C_{avg}	DSIO	38FRE/HIL1

38FRE/HIL1 constant value calculated from temperature dependence of enthalpy by the authors

TABLE 1.15.2. Correlated heat capacities

Reference	Temp. range K	No. pnts used	$\sigma_r C$ %	d_w	d/R	d_r %	d_b/R	+/–
				Selected data				
37CAR/HAR	389.4–432.7	5	1.50#	0.800	1.12–1	1.20	2.69–3	–1

TABLE 1.15.3. Parameters of regression polynomial

Heat capacity type	No. data points total	No. data points used	s_w	s/R	s_r %	s_b/R	+/–
C_p	7	5	1.033	1.45–1	1.55	2.69–3	–1

Temp. range K	A_1	A_2	Level of uncertainty
389.4–432.7	1.69634+1	–1.78919	V

TABLE 1.15.4. Recommended values of heat capacities

Temp. (K)	390	400	410	420	430
c_p (J K^{-1}g^{-1})	0.33	0.32	0.32	0.31	0.30
C_p (J K^{-1}mol^{-1})	83.0	81.5	80.1	78.6	77.1

Name: Indium
Formula: In

CAS–RN: 7440–74–6
Group No.: 1–016
Molar Mass: 114.82

TABLE 1.16.1. Experimental heat capacities

Reference		Temp. range K	No. pnts	Error %	Purity %	Purity method	Type capacity	Calorimeter Type	Calorimeter Reference
33ROT/MEY2		429.1–457.1	2	nosp	not specified		C_p	DSIO	33ROT/MEY2
56SCH/HIL	N	470.5	1	nosp	99.9	anal	C_{avg}	DSIO	56SCH/HIL
63KAZ/ORR		429.3–800.0	5S	nosp	not specified		C_p	not specified	

56SCH/HIL average value in temperature range 433-508 K

TABLE 1.16.2. Correlated heat capacities

Reference	Temp. range K	No. pnts used	$\sigma_r C$ %	d_w	d/R	d_r %	d_b/R	+/–
				Selected data				
33ROT/MEY2	429.1–457.1	2	0.70#	0.131	3.24–3	0.09	–2.33–3	–1
63KAZ/ORR	429.3–800.0	5	0.50#	0.083	1.47–3	0.04	4.79–4	1
				Rejected data				
56SCH/HIL	(4.24–2, 1.21, –4.24–2, –1)							

TABLE 1.16.3. Parameters of regression polynomial

Heat capacity type	No. data points total	No. data points used	s_w	s/R	s_r %	s_b/R	+/–
C_p	8	7	0.117	2.52–3	0.07	–3.25–4	0

Temp. range K	A_1	A_2	Level of uncertainty
429.1–800.0	3.61555	–1.60987–2	III

TABLE 1.16.4. Recommended values of heat capacities

Temp. (K)	430	480	530	580	630	680	730
c_p (J K^{-1}g^{-1})	0.257	0.256	0.256	0.255	0.254	0.254	0.253
C_p (J K^{-1}mol^{-1})	29.49	29.42	29.35	29.29	29.22	29.15	29.08
Temp. (K)	780	830					
c_p (J K^{-1}g^{-1})	0.253	0.252					
C_p (J K^{-1}mol^{-1})	29.02	28.95					

Name: Potassium
Formula: K

CAS–RN: 7440–09–7
Group No.: 1–017
Molar Mass: 39.10

TABLE 1.17.1. Experimental heat capacities

Reference	Temp. range K	No. pnts	Error %	Purity %	Purity method	Type capacity	Calorimeter Type	Calorimeter Reference
06BER	343.1–401.6	3	nosp	not specified		C_{avg}	DSTO	06BER
06WIG	340.8–363.1	2	nosp	not specified		C_p	not specified	
13REN	337.0–373.0	eqn	nosp	not specified		C_p	DSIO	13REN
39CAR/STE	336.9–609.5	31	nosp	99.92	melpt	C_p	BSIO	30CAR/STO
52DOU/BAL	348.1-1073.2	9S	0.30	99.85	melpt	C_{sat}	DSTO	50GIN/DOU

TABLE 1.17.2. Correlated heat capacities

Reference	Temp. range K	No. pnts used	$\sigma_r C$ %	d_w	d/R	d_r %	d_b/R	+/–
				Selected data				
52DOU/BAL	348.1	9	0.30	0.016	1.73–4	0.00	–2.65–7	0
				Rejected data				
06BER	(5.43–1, 12.42, 5.34–1, 2)			06WIG	(3.40–1, 8.15, 3.40–1, 1)			
13REN	(2.91–1, 7.07, 2.91–1, 1)			39CAR/STE	(2.53–1, 6.35, 2.43–1, 23)			

TABLE 1.17.3. Parameters of regression polynomial

Heat capacity type	No. data points total	used	s_w	s/R	s_r %	s_b/R	+/–
C_{sat}	47	9	0.019	2.11–4	0.01	–2.65–7	0

Temp. range K	A_1	A_2	A_3	Level of uncertainty
348.1–1073.2	4.47054	–2.29700–1	1.47774–2	II

TABLE 1.17.4. Recommended values of heat capacities

Temp. (K)	350	400	450	500	550	600	650
c_{sat} (J K^{-1}g^{-1})	0.818	0.806	0.795	0.785	0.777	0.771	0.766
C_{sat} (J K^{-1}mol^{-1})	31.99	31.50	31.06	30.69	30.38	30.13	29.95
Temp. (K)	700	750	800	850	900	950	1000
c_{sat} (J K^{-1}g^{-1})	0.763	0.761	0.761	0.763	0.766	0.770	0.776
C_{sat} (J K^{-1}mol^{-1})	29.82	29.76	29.76	29.81	29.93	30.12	30.36
Temp. (K)	1050						
c_{sat} (J K^{-1}g^{-1})	0.784						
C_{sat} (J K^{-1}mol^{-1})	30.66						

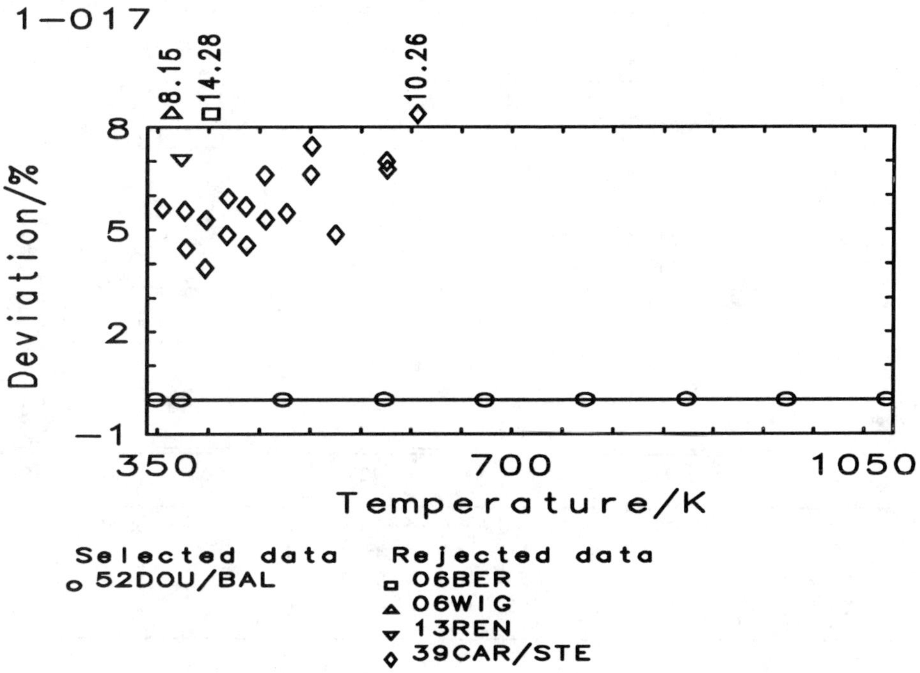

Name: Krypton
Formula: Kr

CAS–RN: 7439–90–9
Group No.: 1–018
Molar Mass: 83.80

TABLE 1.18.1. Experimental heat capacities

Reference	Temp. range K	No. pnts	Error %	Purity %	Purity method	Type capacity	Calorimeter Type	Calorimeter Reference
36CLU	118.3–120.5	4	nosp	not specified		C_{sat}	BSIO	36CLU/GOL
38CLU/KRU	117.1–123.1	6	nosp	not specified		C_{sat}	BSIO	36CLU/GOL
61BEA/CHI	116.5–118.4	5	0.50	99.9995	melpt	C_p	BSAO	61FLU/LEA

TABLE 1.18.2. Correlated heat capacities

Reference	Temp. range K	No. pnts used	$\sigma_r C$ %	d_w	d/R	d_r %	d_b/R	+/–
				Selected data				
38CLU/KRU	117.1–123.1	6	1.00#	1.034	5.54–2	1.03	2.67–2	2
61BEA/CHI	116.5–118.4	5	0.50	0.817	2.14–2	0.41	–7.49–3	–3
				Rejected data				
36CLU	(1.15–1, 2.12, 1.14–1, 4)							

TABLE 1.18.3. Parameters of regression polynomial

Heat capacity type	No. data points total	used	s_w	s/R	s_r %	s_b/R	+/–
C_p	15	11	1.041	4.79–2	0.90	1.11–2	–1
C_{sat}	15	11	1.045	4.80–2	0.90	1.12–2	–1

Temp. range K	A_1	A_2	Level of uncertainty
116.5–123.1	1.48874	3.20386	IV
116.5–123.1	1.59216	3.10704	IV

TABLE 1.18.4. Recommended values of heat capacities

Temp. (K)	117	119	121	123
c_p (J K^{-1}g^{-1})	0.520	0.526	0.532	0.539
C_p (J K^{-1}mol^{-1})	43.55	44.08	44.61	45.14
c_{sat} (J K^{-1}g^{-1})	0.519	0.525	0.531	0.537
C_{sat} (J K^{-1}mol^{-1})	43.46	43.98	44.50	45.01

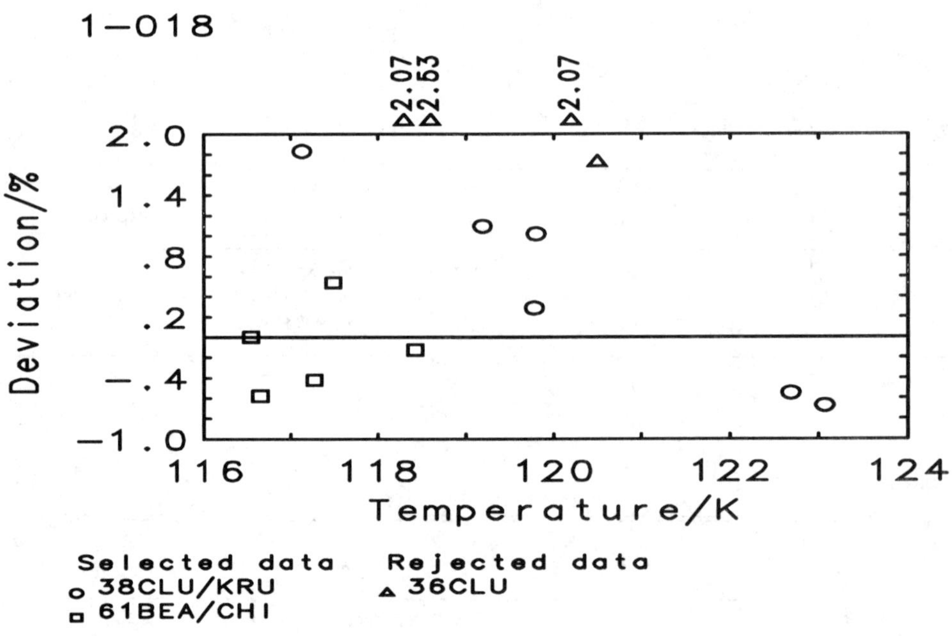

Name: Lithium
Formula: Li

CAS–RN: 7439–93–2
Group No.: 1–019
Molar Mass: 6.94

TABLE 1.19.1. Experimental heat capacities

Reference		Temp. range K	No. pnts	Error %	Purity %	Purity method	Type capacity	Calorimeter Type	Calorimeter Reference
50YAG/UNT		473.1–773.2	4	10.00	99.7	chrom	C_p	BDHO	50YAG/UNT
51BAT/SMI	N	673.0-1273.0	2	nosp	not specified		C_p	not specified	
52RED/LON	N	523.0-1373.0	eqn	5.00	99.9	anal	C_p	DSTO	52RED/LON
55DOU/EPS		453.7-1200.0	24S	0.30	99.989	melpt	C_p	DSTO	50GIN/DOU

51BAT/SMI authors give a constant C_p over the entire temperature range
52RED/LON calculated from temperature dependence of enthalpy by the authors

TABLE 1.19.2. Correlated heat capacities

Reference	Temp. range K	No. pnts used	$\sigma_r C$ %	d_w	d/R	d_r %	d_b/R	+/–
				Selected data				
55DOU/EPS	453.7–1200.0	24	0.30	0.648	6.92–3	0.19	2.61–5	4
				Rejected data				
50YAG/UNT	(5.22–1, 16.88, 1.48–2, 2)			51BAT/SMI	(1.50–1, 4.47, –1.50–1, –1)			
52RED/LON	(1.01–1, 2.81, 8.12–2, 10)							

TABLE 1.19.3. Parameters of regression polynomial

Heat capacity type	No. data points total	No. data points used	s_w	s/R	s_r %	s_b/R	+/–
C_p	48	24	0.710	7.59–3	0.21	2.61–5	4

Temp. range K	A_1	A_2	A_3	A_4	Level of uncertainty
453.7–1200.0	4.68445	–3.64142–1	3.63908–2	–1.21532–3	III

TABLE 1.19.4. Recommended values of heat capacities

Temp. (K)	460	510	560	610	660	710	760
c_p (J K^{-1}g^{-1})	4.386	4.328	4.280	4.242	4.213	4.191	4.175
C_p (J K^{-1}mol^{-1})	30.44	30.04	29.71	29.45	29.24	29.09	28.98
Temp. (K)	810	860	910	960	1010	1060	1110
c_p (J K^{-1}g^{-1})	4.165	4.158	4.155	4.153	4.153	4.152	4.150
C_p (J K^{-1}mol^{-1})	28.91	28.86	28.84	28.83	28.82	28.82	28.80
Temp. (K)	1160	1210					
c_p (J K^{-1}g^{-1})	4.145	4.137					
C_p (J K^{-1}mol^{-1})	28.77	28.71					

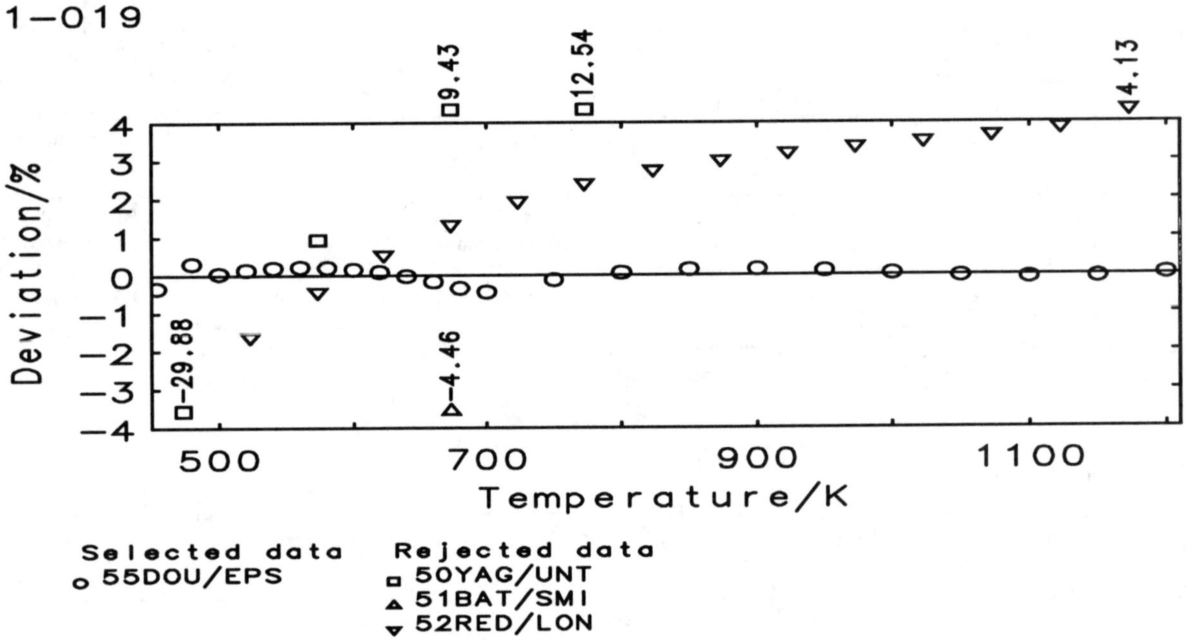

Name: Nitrogen
Formula: N_2

CAS–RN: 7727–37–9
Group No.: 1–020
Molar Mass: 28.01

TABLE 1.20.1. Experimental heat capacities

Reference	Temp. range K	No. pnts	Error %	Purity %	Purity method	Type capacity	Calorimeter Type	Calorimeter Reference
16EUC	64.6–72.7	5	nosp	not specified		C_{sat}	BSIO	16EUC
29CLU1	66.9–73.5	5	0.70	not specified		C_p	BSIO	29CLU1
30WIE/BRE	79.2–117.0	14	2.00	not specified		C_{sat}	BSIO	30WIE/HUB
33GIA/CLA	65.0–77.7	7	0.80	99.997	melpt	C_p	BSIO	28GIA/WIE1
59CLU/SPE	65.9–68.2	4	nosp	not specified		C_p	not specified	
81WEB	78.5–124.8	43	0.60	99.99	estim	C_{sat}	BSAO	61GOO
91MAG	64.8–120.7	102	0.50	99.9998	anal	C_{sat}	BSAO	61GOO

TABLE 1.20.2. Correlated heat capacities

Reference	Temp. range K	No. pnts used	$\sigma_r C$ %	d_w	d/R	d_r %	d_b/R	+/–
				Selected data				
33GIA/CLA	65.0–77.7	7	0.80	0.323	1.75–2	0.26	4.24–3	1
81WEB	78.5–124.8	43	0.60	1.887	2.39–1	1.13	1.06–3	1
91MAG	64.8–120.7	102	0.50	0.471	2.23–2	0.24	1.81–3	18
				Rejected data				
16EUC	(6.91–2, 1.03, –6.32–2, –4)			29CLU1	(8.44–2, 1.23, 8.23–2, 5)			
30WIE/BRE	(2.22–1, 2.14, –7.46–2, –2)			59CLU/SPE	(6.00–2, 0.88, 4.84–2, 3)			

TABLE 1.20.3. Parameters of cubic spline polynomials

Heat capacity type	No. data points total	used	s_w	s/R	s_r %	s_b/R	+/–
C_p	180	152	1.100	1.31–1	0.65	1.71–3	20
C_{sat}	180	152	1.279	1.31–1	0.76	2.06–3	15

Temp. range K	A_1	A_2	A_3	A_4	Level of uncertainty
64.8–105.0	–1.14490+1	7.04695+1	–9.20497+1	4.08251+1	III
105.0–118.0	–7.74321+2	2.25011+3	–2.16789+3	6.99823+2	IV
118.0–124.8	–4.28969+4	1.09341+5	–9.29231+4	2.63369+4	VI
64.8–105.0	–5.70805	4.80807+1	–6.28551+1	2.80389+1	III
105.0–118.0	–4.62439+2	1.35303+3	–1.30566+3	4.22581+2	IV
118.0–124.8	–3.73354+4	9.50979+4	–8.07505+4	2.28646+4	VI

TABLE 1.20.4. Recommended values of heat capacities

Temp. (K)	65	70	75	80	85	90	95
c_p (J K^{-1}g^{-1})	1.982	2.012	2.033	2.053	2.082	2.130	2.204
C_p (J K^{-1}mol^{-1})	55.51	56.36	56.94	57.51	58.33	59.66	61.73
c_{sat} (J K^{-1}g^{-1})	1.985	2.008	2.026	2.044	2.068	2.105	2.161
C_{sat} (J K^{-1}mol^{-1})	55.61	56.26	56.75	57.25	57.93	58.97	60.54

Temp. (K)	100	105	110	115	120	125
c_p (J K^{-1}g^{-1})	2.314	2.469	2.703	3.147	4.02	7.90
C_p (J K^{-1}mol^{-1})	64.82	69.17	75.73	88.17	113	221
c_{sat} (J K^{-1}g^{-1})	2.243	2.356	2.521	2.819	3.40	6.47
C_{sat} (J K^{-1}mol^{-1})	62.83	66.00	70.63	78.96	95.1	181

TABLE 1.20.5. Parameters of quasi-polynomial equation

Heat capacity type	No. data points total	used	s_w	s/R	s_r %	s_b/R	+/–
C_p	180	152	1.724	1.17–1	0.95	–1.39–3	15

Temp. range K	T_c K	A_1	A_2	A_3	A_4	A_5	A_6	Level of uncertainty
64.8–124.8	126.20	–2.64196	1.38507–1	1.50586	2.58845+1	–5.31133+1	2.86788+1	V

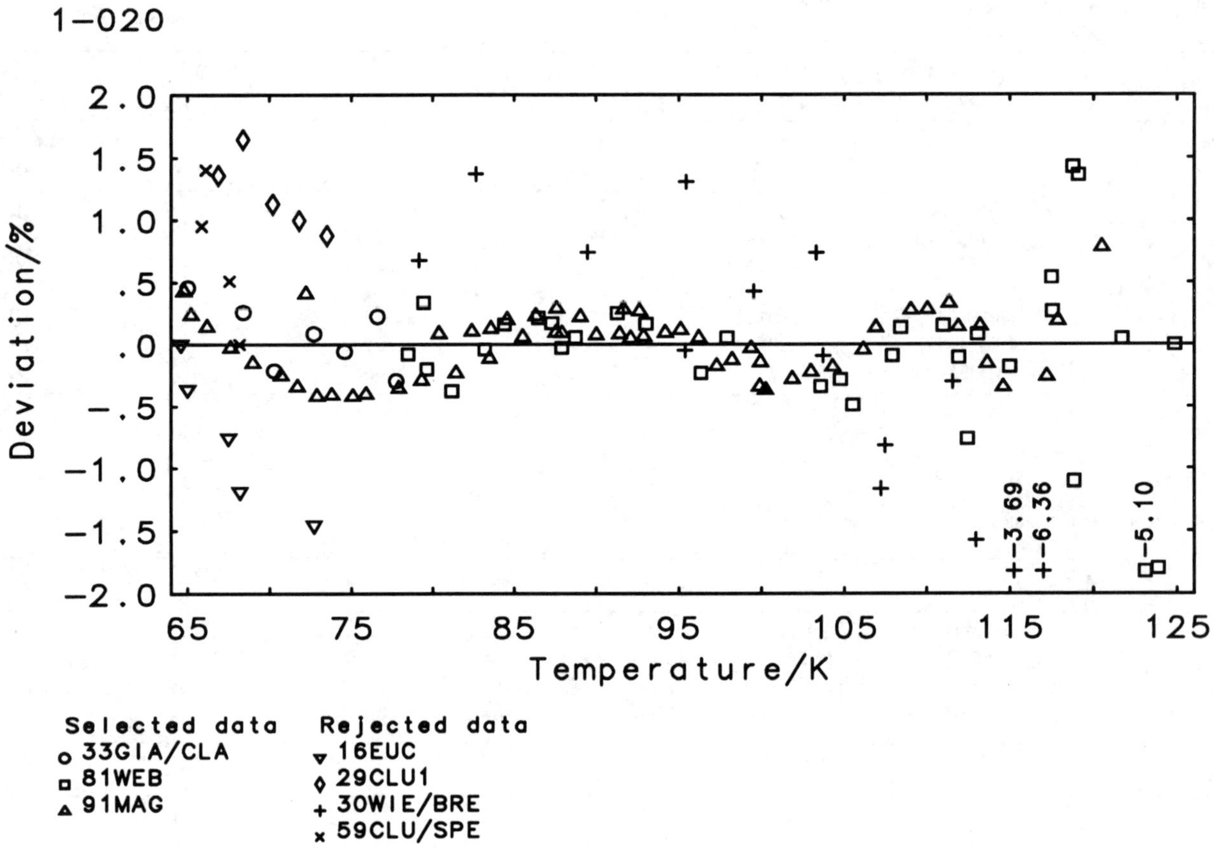

Name: Sodium
Formula: Na

CAS–RN: 7440–23–5
Group No.: 1–021
Molar Mass: 22.99

TABLE 1.21.1. Experimental heat capacities

Reference		Temp. range K	No. pnts	Error %	Purity %	Purity method	Type capacity	Calorimeter Type	Calorimeter Reference
06BER	N	401.6	1	nosp	not specified		C_{avg}	DSTO	06BER
13REN		371.1	1	nosp	not specified		C_p	DSIO	13REN
14GRI		374.0–411.6	5	nosp	not specified		C_p	BSIO	14GRI
50GIN/DOU		371.0-1173.0	10S	0.40	99.9984	melpt	C_{sat}	DSTO	50GIN/DOU
56SCH/HIL	N	483.0	1	nosp	not specified		C_{avg}	DSIO	56SCH/HIL

06BER average value in temperature range 373–430 K
56SCH/HIL average value in temperature range 463–503 K

TABLE 1.21.2. Correlated heat capacities

Reference	Temp. range K	No. pnts used	$\sigma_r C$ %	d_w	d/R	d_r %	d_b/R	+/–
				Selected data				
50GIN/DOU	371.0–1173.0	10	0.40	0.017	2.53–4	0.01	1.19–7	–1
				Rejected data				
06BER	(6.48–2, 1.68, 6.48–2, 1)			13REN	(1.00–2, 0.26, –1.00–2, –1)			
14GRI	(8.24–2, 2.21, –8.24–2, –5)			56SCH/HIL	(1.19–1, 3.34, –1.19–1, –1)			

TABLE 1.21.3. Parameters of regression polynomial

Heat capacity type	No. data points total	used	s_w	s/R	s_r %	s_b/R	+/–
C_{sat}	18	10	0.021	3.02–4	0.01	1.19–7	–1

Temp. range K	A_1	A_2	A_3	Level of uncertainty
371.0–1173.0	4.50689	–2.30446–1	1.27872–2	III

TABLE 1.21.4. Recommended values of heat capacities

Temp. (K)	380	430	480	530	580	630	680
c_{sat} (J K^{-1}g^{-1})	1.380	1.357	1.336	1.318	1.302	1.288	1.277
C_{sat} (J K^{-1}mol^{-1})	31.73	31.20	30.73	30.30	29.94	29.62	29.36
Temp. (K)	730	780	830	880	930	980	1030
c_{sat} (J K^{-1}g^{-1})	1.268	1.261	1.257	1.255	1.255	1.257	1.262
C_{sat} (J K^{-1}mol^{-1})	29.15	29.00	28.89	28.84	28.85	28.91	29.02
Temp. (K)	1080	1130	1180				
c_{sat} (J K^{-1}g^{-1})	1.269	1.279	1.290				
C_{sat} (J K^{-1}mol^{-1})	29.18	29.40	29.67				

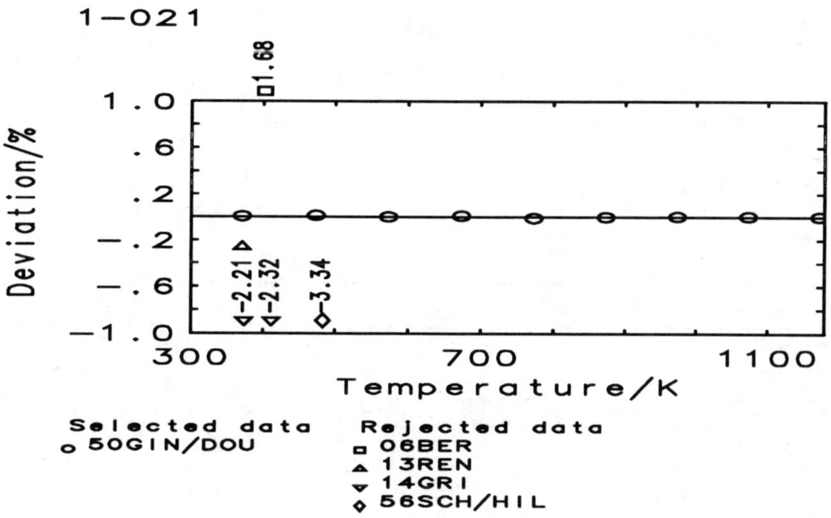

Name: Neon
Formula: Ne

CAS–RN: 7440–01–9
Group No.: 1–022
Molar Mass: 20.18

TABLE 1.22.1. Experimental heat capacities

Reference	Temp. range K	No. pnts	Error %	Purity %	Purity method	Type capacity	Calorimeter Type	Calorimeter Reference
29CLU2	27.8–41.8	6	1.50	not specified		C_{sat}	BSIO	29CLU1
36CLU	25.7–26.8	4	nosp	not specified		C_{sat}	BSIO	36CLU/GOL

TABLE 1.22.2. Correlated heat capacities

Reference	Temp. range K	No. pnts used	$\sigma_r C$ %	d_w	d/R	d_r %	d_b/R	+/–
				Selected data				
29CLU2	37.9–41.8	4	1.50	1.135	1.12–1	1.70	5.24–3	2
36CLU	25.7–26.8	4	1.50#	0.161	1.06–2	0.24	–7.46–4	0

TABLE 1.22.3. Parameters of regression polynomial

Heat capacity type	No. data points total	No. data points used	s_w	s/R	s_r %	s_b/R	+/–
C_p	10	8	1.147	1.12–1	1.72	2.25–3	2
C_{sat}	10	8	1.370	1.28–1	2.06	2.52–3	2

Temp. range K	A_1	A_2	A_3	A_4	Level of uncertainty
25.7–41.8	–6.26990+1	6.65671+2	–2.20106+3	2.43611+3	V
25.7–41.8	–2.64403+1	2.98063+2	–9.68020+2	1.06744+3	V

TABLE 1.22.4. Recommended values of heat capacities

Temp. (K)	25	30	35	40
c_p (J K^{-1}g^{-1})	1.74	1.93	2.10	3.01
C_p (J K^{-1}mol^{-1})	35.1	38.9	42.4	60.8
c_{sat} (J K^{-1}g^{-1})	1.75	1.93	2.09	2.56
C_{sat} (J K^{-1}mol^{-1})	35.4	38.9	42.1	51.7

TABLE 1.22.5. Parameters of quasi-polynomial equation

Heat capacity type	No. data points total	No. data points used	s_w	s/R	s_r %	s_b/R	+/–
C_p	10	8	1.331	1.42–1	2.00	6.09–3	0

Temp. range K	T_c K	A_1	A_2	A_3	A_4	Level of uncertainty
25.7–41.8	44.40	–7.46473–1	1.94151–1	2.83415	7.17511–1	V

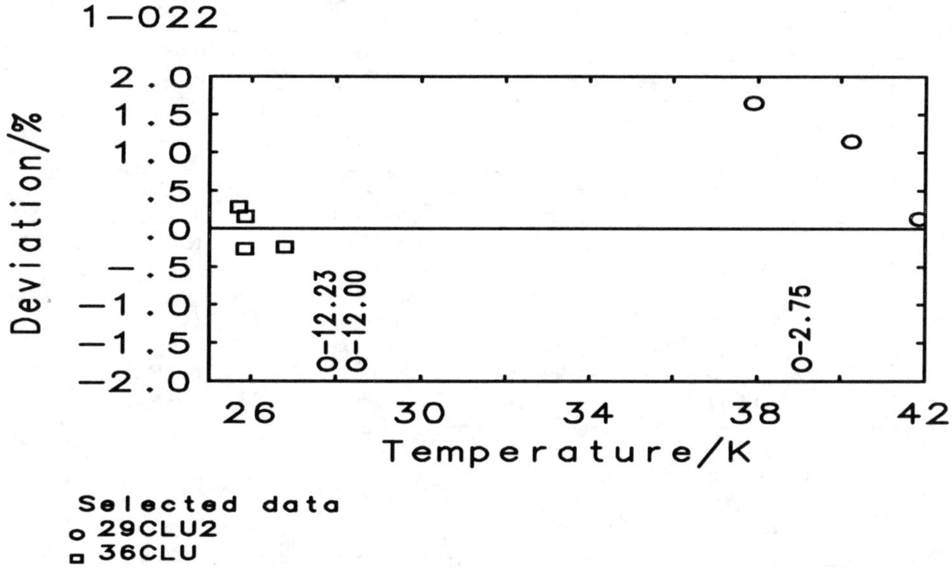

Name: Oxygen
Formula: O_2

CAS–RN: 7782–44–7
Group No.: 1–023
Molar Mass: 32.00

TABLE 1.23.1. Experimental heat capacities

Reference		Temp. range K	No. pnts	Error %	Purity %	Purity method	Type capacity	Calorimeter Type	Calorimeter Reference
16EUC		57.4–73.0	7	nosp	not specified		C_{sat}	BSIO	16EUC
29CLU1		56.6–72.8	13	0.70	not specified		C_p	BSIO	29CLU1
29GIA/JOH	N	57.0–90.3	24	nosp	not specified		C_p	BSIO	28GIA/WIE1
69GOO/WEB		56.4–152.4	88	nosp	99.99	anal	C_{sat}	BSAO	61GOO

29GIA/JOH high sample purity

TABLE 1.23.2. Correlated heat capacities

Reference	Temp. range K	No. pnts used	$\sigma_r C$ %	d_w	d/R	d_r %	d_b/R	+/–
				Selected data				
29GIA/JOH	57.0–90.3	24	0.50#	0.568	1.84–2	0.28	1.15–2	11
69GOO/WEB	56.4–152.4	88	0.30#	1.166	3.20–2	0.35	–7.98–4	–6
				Rejected data				
16EUC	(3.42–2, 0.54, 3.68–3, 1)			29CLU1	(9.37–2, 1.48, –5.18–2, –6)			

TABLE 1.23.3. Parameters of cubic spline polynomials

Heat capacity type	No. data points total	used	s_w	s/R	s_r %	s_b/R	+/–
C_p	132	112	1.102	3.05–2	0.35	1.84–3	5
C_{sat}	132	112	1.045	2.94–2	0.33	1.79–3	5

Temp. range K	A_1	A_2	A_3	A_4	Level of uncertainty
56.4–90.0	6.11269	1.89527	–3.84283	2.49240	II
90.0–120.0	1.01011	1.89039+1	–2.27413+1	9.49182	III
120.0–144.0	–2.54928+2	6.58750+2	–5.55946+2	1.57604+2	III
144.0–152.4	–3.67917+4	7.67771+4	–5.34159+4	1.23937+4	V
56.4–90.0	6.40740	5.63030–1	–1.84056	1.49122	II
90.0–120.0	2.39592	1.39347+1	–1.66979+1	6.99395	III
120.0–144.0	–1.60486+2	4.21139+2	–3.56035+2	1.01254+2	III
144.0–152.4	–3.17377+4	6.62069+4	–4.60406+4	1.06764+4	V

TABLE 1.23.4. Recommended values of heat capacities

Temp. (K)	60	70	80	90	100	110	120
c_p (J K^{-1}g^{-1})	1.664	1.666	1.675	1.695	1.732	1.798	1.910
C_p (J K^{-1}mol^{-1})	53.25	53.31	53.59	54.23	55.41	57.55	61.10
c_{sat} (J K^{-1}g^{-1})	1.664	1.666	1.674	1.692	1.722	1.774	1.860
C_{sat} (J K^{-1}mol^{-1})	53.25	53.31	53.57	54.13	55.10	56.78	59.52
Temp. (K)	**130**	**140**	**150**				
c_p (J K^{-1}g^{-1})	2.119	2.633	4.38				
C_p (J K^{-1}mol^{-1})	67.80	84.25	140				
c_{sat} (J K^{-1}g^{-1})	2.014	2.370	3.68				
C_{sat} (J K^{-1}mol^{-1})	64.45	75.84	118				

TABLE 1.23.5. Parameters of quasi-polynomial equation

Heat capacity type	No. data points total	used	s_w	s/R	s_r %	s_b/R	+/–
C_p	132	112	2.314	9.11–2	0.72	–4.45–3	–15

Temp. range K	T_c K	A_1	A_2	A_3	A_4	A_5	A_6	Level of uncertainty
56.4–152.4	154.58	–2.28256	1.49832–1	4.54619	8.83694	–2.50571+1	1.49298+1	V

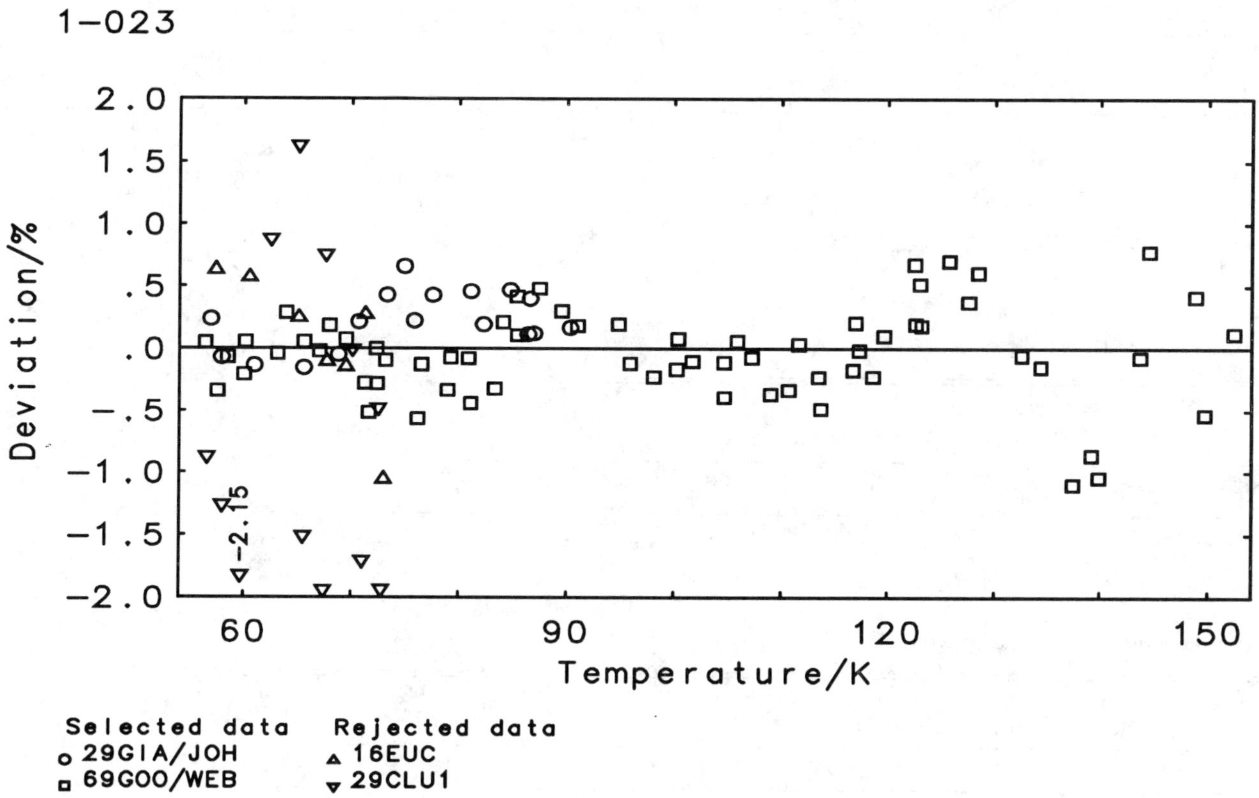

Name: Ozone
Formula: O_3

CAS–RN: 10028–15–6
Group No.: 1–024
Molar Mass: 48.00

TABLE 1.24.1. Experimental heat capacities

Reference		Temp. range K	No. pnts	Error %	Purity %	Purity method	Type capacity	Calorimeter Type	Calorimeter Reference
58BRA/WAT	N	96.5–122.9	6	nosp	not specified		C_p	BDHT	58BRA/WAT

58BRA/WAT error increasing from 2 to 8 % between lower and upper temperature limits

TABLE 1.24.3. Parameters of regression polynomial

Heat capacity type	No. data points total	No. data points used	s_w	s/R	s_r %	s_b/R	+/–
C_p	6	6	1.191	6.70–1	5.95	5.28–2	2

Temp. range K	A_1	A_2	Level of uncertainty
96.6–122.9	8.51603	2.15555	V

TABLE 1.24.4. Recommended values of heat capacities

Temp. (K)	100	110	120
c_p (J K^{-1}g^{-1})	1.85	1.89	1.92
C_p (J K^{-1}mol^{-1})	88.7	90.5	92.3

Name: Phosphorus mol.(P_4)
Formula: P_4

CAS–RN: 12185–10–3
Group No.: 1–025
Molar Mass: 123.90

TABLE 1.25.1. Experimental heat capacities

Reference		Temp. range K	No. pnts	Error %	Purity % method	Type capacity	Calorimeter Type	Calorimeter Reference
42YOU/HIL	N	298.0–370.0	eqn	nosp	not specified	C_p	DSIO	42YOU/HIL

42YOU/HIL low temperature limit is below n.m.t.; undercooled liquid

TABLE 1.25.3. Parameters of regression polynomial

Heat capacity type	No. data points total	used	s_w	s/R	s_r %	s_b/R	+/–
C_p	10	10	0.000	7.21–7	0.00	–3.82–7	0

Temp. range K	A_1	A_2	A_3	Level of uncertainty
298.0–370.0	1.21480+1	6.00450–1	–1.97613–1	IV

TABLE 1.25.4. Recommended values of heat capacities

Temp. (K)	300	310	320	330	340	350	360
c_p (J K^{-1}g^{-1})	0.817	0.813	0.808	0.804	0.799	0.794	0.788
C_p (J K^{-1}mol^{-1})	101.2	100.7	100.2	99.59	98.99	98.35	97.68

Temp. (K)	370
c_p (J K^{-1}g^{-1})	0.783
C_p (J K^{-1}mol^{-1})	96.98

Name: Rubidium
Formula: Rb

CAS–RN: 7440–17–7
Group No.: 1–026
Molar Mass: 85.47

TABLE 1.26.1. Experimental heat capacities

Reference	Temp. range K	No. pnts	Error %	Purity % method	Type capacity	Calorimeter Type	Calorimeter Reference
13REN	312.0–373.0	eqn	nosp	not specified	C_p	DSIO	13REN

TABLE 1.26.3. Parameters of regression polynomial

Heat capacity type	No. data points total	used	s_w	s/R	s_r %	s_b/R	+/–
C_p	7	7	0.000	1.51–7	0.00	6.81–8	0

Temp. range K	A_1	A_2	Level of uncertainty
312.0–373.2	4.26656	–1.11823–1	VI

TABLE 1.26.4. Recommended values of heat capacities

Temp. (K)	320	330	340	350	360	370
c_p (J K^{-1}g^{-1})	0.38	0.38	0.38	0.38	0.38	0.37
C_p (J K^{-1}mol^{-1})	32.5	32.4	32.3	32.2	32.1	32.0

Name: Sulfur
Formula: S

CAS–RN: 7704–34–9
Group No.: 1–027
Molar Mass: 121.75

TABLE 1.27.1. Experimental heat capacities

Reference	Temp. range K	No. pnts	Error %	Purity %	Purity method	Type capacity	Calorimeter Type	Calorimeter Reference
54BRA/MOE	393.1–693.2	31S	0.50	not specified		C_p	BSAO	54BRA/MOE
59WES1	388.4–717.8	43S	0.20	99.999	melpt	C_p	BSAO	68WES/WES
67PAC	418.1	1	nosp	not specified		C_p	BDHT	79DU/COM
73KOM/MIL	393.1–598.2	43S	nosp	99.999	melpt	C_p	BDHO	73KOM/MIL
75MON	405.8–433.3	14	nosp	99.999	melpt	C_{sat}	BSAO	75MON

TABLE 1.27.2a. Correlated heat capacities

Reference	Temp. range K	No. pnts used	$\sigma_r C$ %	d_w	d/R	d_r %	d_b/R	+/–
				Selected data				
73KOM/MIL	393.1–432.1	9	0.50#	1.189	2.33–2	0.59	1.34–4	1
				Rejected data				
54BRA/MOE	(9.25–2, 2.38, 8.72–2, 4)			59WES1	(8.37–1, 15.59, 6.49–1, 8)			
67PAC	(5.84–1, 13.41, 5.84–1, 1)			75MON	(1.08, 17.93, 7.59–1, 11)			

TABLE 1.27.2b. Correlated heat capacities

Reference	Temp. range K	No. pnts used	$\sigma_r C$ %	d_w	d/R	d_r %	d_b/R	+/–
				Selected data				
73KOM/MIL	438.1–598.2	33	0.50#	1.089	2.87–2	0.54	3.93–4	1
				Rejected data				
54BRA/MOE	(1.59–1, 3.50, 2.96–2, 0)			59WES1	(2.16–1, 4.65, 1.04–1, 4)			

TABLE 1.27.3a. Parameters of regression polynomial

Heat capacity type	No. data points total	used	s_w	s/R	s_r %	s_b/R	+/–
C	132	9	1.596	3.13–2	0.80	1.34–4	1
Temp. range K		A_1	A_2	A_3	A_4		Level of uncertainty
393.1–432.1		–1.85689+3	1.37277+3	–3.37560+2	2.76644+1		IV

TABLE 1.27.3b. Parameters of cubic spline polynomials

Heat capacity type	No. data points total	used	s_w	s/R	s_r %	s_b/R	+/–
C_p	132	33	1.182	3.12–2	0.59 3.	93–4	1
Temp. range K		A_1	A_2	A_3	A_4		Level of uncertainty
438.1–510.0		5.90051+2	–3.46559+2	6.83281+1	–4.48944		IV
510.0–598.2		–6.70282+1	3.99583+1	–7.45952	4.64001–1		IV

TABLE 1.27.4. Recommended values of heat capacities

Temp. (K)	390	400	410	420	430		
c_p (J K^{-1}g^{-1})	0.948	0.975	0.972	0.984	1.052		
C_p (J K^{-1}mol^{-1})	30.40	31.25	31.17	31.54	33.74		
Temp. (K)	440	450	460	470	480	490	500
c_p (J K^{-1}g^{-1})	1.451	1.317	1.224	1.163	1.129	1.113	1.110
C_p (J K^{-1}mol^{-1})	46.53	42.25	39.24	37.30	36.19	35.70	35.59
Temp. (K)	510	520	530	540	550	560	570
c_p (J K^{-1}g^{-1})	1.112	1.113	1.113	1.112	1.112	1.113	1.116
C_p (J K^{-1}mol^{-1})	35.65	35.68	35.69	35.67	35.67	35.70	35.79
Temp. (K)	580	590	600				
c_p (J K^{-1}g^{-1})	1.121	1.129	1.142				
C_p (J K^{-1}mol^{-1})	35.95	36.22	36.61				

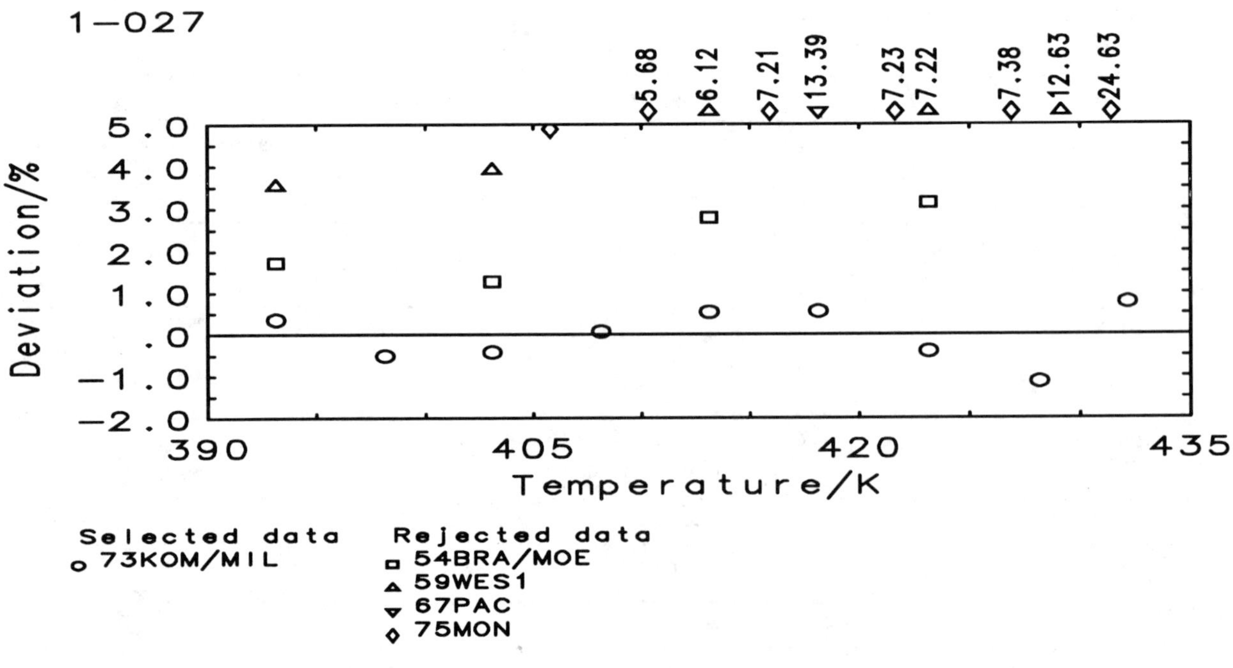

1–027

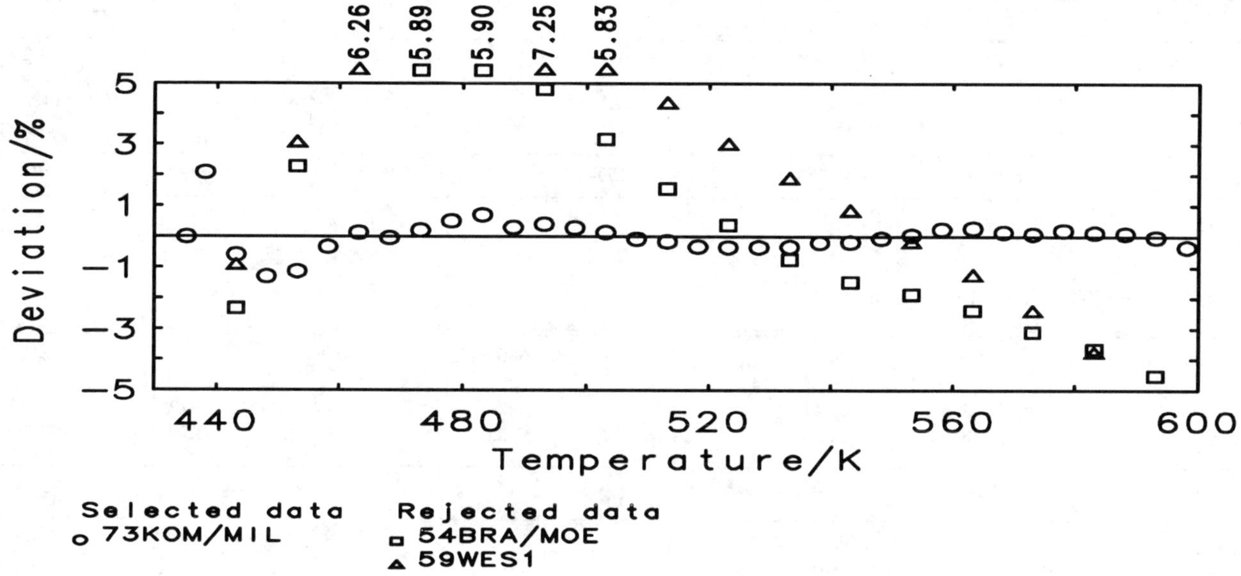

Name: Selenium
Formula: Se

CAS–RN: 7782–49–2
Group No.: 1–028
Molar Mass: 78.96

TABLE 1.28.1. Experimental heat capacities

Reference		Temp. range K	No. pnts	Error %	Purity %	Purity method	Type capacity	Calorimeter Type	Calorimeter Reference
26MON		490.1	1	nosp	not specified		C_p	DSIO	26MON
64GAT/HEI	N	503.1–533.2	4S	nosp	not specified		C_p	DSIO	63GAT/KRE
73GRO		502.9–994.1	48	nosp	99.9992	melpt	C_p	BSAO	67GRO
90JIN/WUN		523.2–548.2	2	nosp	99.999	anal	C_p	BDHT	90JIN/WUN

64GAT/HEI high purity of substance with traces of Al and Cu (0.0001 %)

TABLE 1.28.2. Correlated heat capacities

Reference	Temp. range K	No. pnts used	$\sigma_r C$ %	d_w	d/R	d_r %	d_b/R	+/–
				Selected data				
73GRO	509.3–994.1	47	0.50#	0.921	1.91–2	0.46	1.76–4	–3
				Rejected data				
64GAT/HEI	(2.60, 37.68, 2.59, 3)		90JIN/WUN	(7.38–2, 1.77, –7.34–2, –2)				

TABLE 1.28.3. Parameters of regression polynomial

Heat capacity type	No. data points total	used	s_w	s/R	s_r %	s_b/R	+/–
C_p	55	47	0.963	2.00–2	0.48	1.76–4	–3
Temp. range K	A_1		A_2	A_3	A_4		**Level of uncertainty**
509.3–994.1	5.18010		–4.35497–2	–4.66559–2	4.21206–3		III

TABLE 1.28.4. Recommended values of heat capacities

Temp. (K)	510	560	610	660	710	760	810
c_p (J K^{-1}g^{-1})	0.453	0.444	0.435	0.429	0.424	0.422	0.422
C_p (J K^{-1}mol^{-1})	35.78	35.03	34.38	33.85	33.48	33.29	33.30
Temp. (K)	860	910	960				
c_p (J K^{-1}g^{-1})	0.425	0.431	0.441				
C_p (J K^{-1}mol^{-1})	33.54	34.04	34.83				

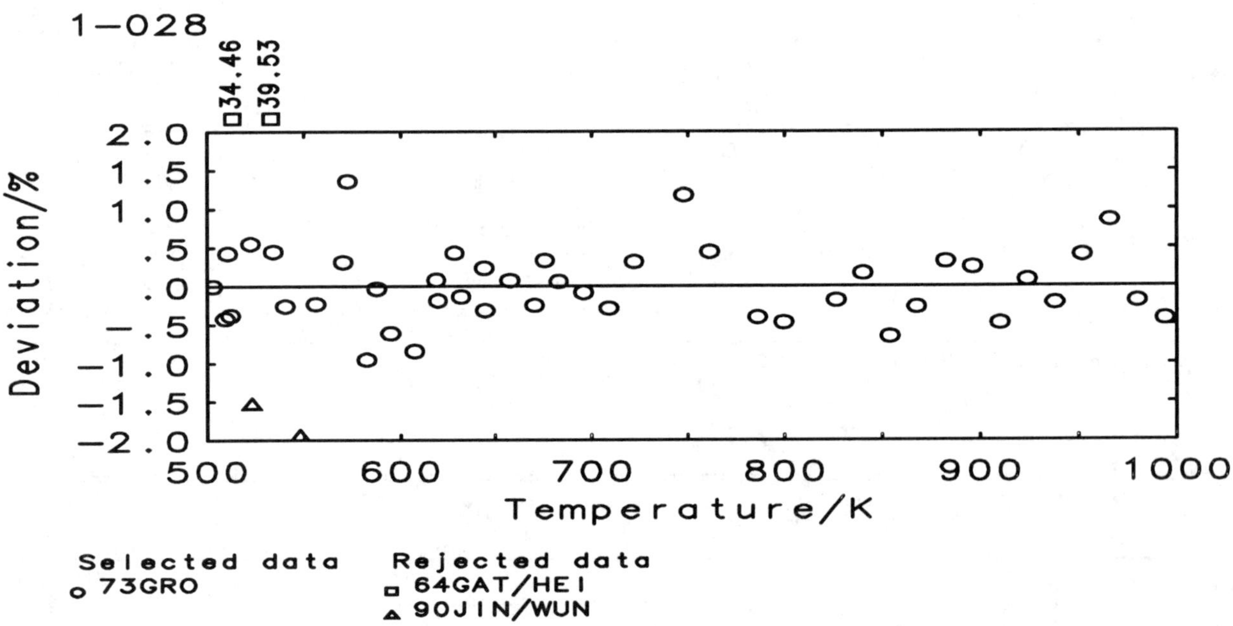

HEAT CAPACITY OF LIQUIDS

Name: Tin
Formula: Sn

CAS–RN: 7440–31–5
Group No.: 1–029
Molar Mass: 118.71

TABLE 1.29.1. Experimental heat capacities

Reference		Temp. range K	No. pnts	Error %	Purity %	Purity method	Type capacity	Calorimeter Type	Calorimeter Reference
26AWB/GRI	N	503.1–573.2	2	nosp	99.98	anal	C_p	DSIO	26AWB/GRI
26UMI		523.2–873.2	8S	nosp	99.989	anal	C_{avg}	not specified	
47BAR		523.2–623.2	3S	1.00	99.84	anal	C_p	BDHO	40TUR/BAR
63KAZ/ORR		505.0–800.0	5S	nosp	not specified		C_p	not specified	
69FEB/HER	N	1222.0–1745.0	2	0.70	not specified		C_p	DSAO	62LEV

26AWB/GRI constant value calculated from temperature dependence of enthalpy by the authors
69FEB/HER constant value calculated from temperature dependence of enthalpy by the authors

TABLE 1.29.2. Correlated heat capacities

Reference	Temp. range K	No. pnts used	$\sigma_r C$ %	d_w	d/R	d_r %	d_b/R	+/–
				Selected data				
63KAZ/ORR	505.0–800.0	5	0.80#	1.295	3.66–2	1.04	–4.74–3	–3
69FEB/HER	1222.0–1745.0	2	0.70	0.916	2.23–2	0.64	1.08–2	0
				Rejected data				
26AWB/GRI	(1.49–1, 4.47, –1.49–1, –1)			26UMI	(7.34–1, 28.79, –6.31–1, –6)			
47BAR	(5.25–2, 1.47, 4.15–2, 3)							

TABLE 1.29.3. Parameters of regression polynomial

Heat capacity type	No. data points total	No. data points used	s_w	s/R	s_r %	s_b/R	+/–
C_p	20	7	1.586	4.38–2	1.24	–2.99–4	–3

Temp. range K	A_1	A_2	A_3	Level of uncertainty
505.0–1745.0	3.63461	–3.20295–2	1.33065–3	V

TABLE 1.29.4. Recommended values of heat capacities

Temp. (K)	510	610	710	810	910	1010	1110
c_p (J K^{-1}g^{-1})	0.25	0.24	0.24	0.24	0.24	0.24	0.24
C_p (J K^{-1}mol^{-1})	29.1	29.0	28.9	28.8	28.7	28.7	28.6

Temp. (K)	1210	1310	1410	1510	1610	1710
c_p (J K^{-1}g^{-1})	0.24	0.24	0.24	0.24	0.24	0.24
C_p (J K^{-1}mol^{-1})	28.6	28.6	28.7	28.7	28.8	28.9

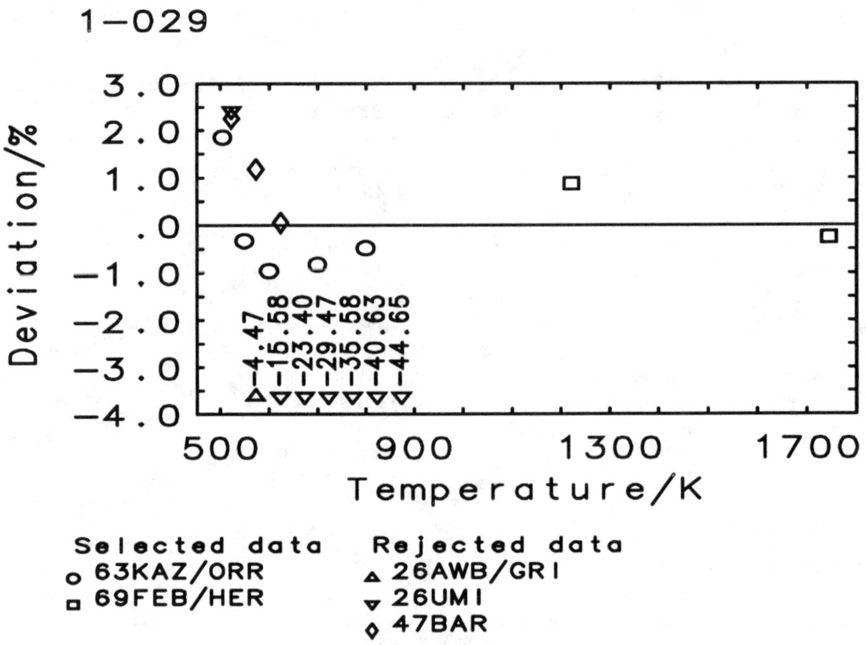

Name: Xenon
Formula: Xe

CAS–RN: 7440–63–3
Group No.: 1–030
Molar Mass: 131.29

TABLE 1.30.1. Experimental heat capacities

Reference	Temp. range K	No. pnts	Error %	Purity %	Purity method	Type capacity	Calorimeter Type	Calorimeter Reference
38CLU/RIC	163.2–165.9	7	nosp	not specified		C_{sat}	BSIO	36CLU/GOL

TABLE 1.30.3. Parameters of regression polynomial

Heat capacity type	No. data points total	No. data points used	s_w	s/R	s_r %	s_b/R	+/–
C_{sat}	7	7	0.710	2.67–2	0.50	1.90–4	–1

Temp. range K	A_1	A_2	Level of uncertainty
163.2 –165.9	5.18191	1.17549–1	IV

TABLE 1.30.4. Recommended values of heat capacities

Temp. (K)	163	164	165	166
c_{sat} (J K^{-1}g^{-1})	0.340	0.340	0.340	0.341
C_{sat} (J K^{-1}mol^{-1})	44.68	44.69	44.70	44.71

02. Inorganic Compounds

The family of inorganic compounds contains 103 compounds, 12 of which were measured at one temperature only.

Similar to the family containing elements, the family of inorganic compounds has a significantly different character from most of the organic compounds in this compilation. In general, the only unifying feature may be that their melting temperature is below 573 K. Measurements were performed in a large number of diverse laboratories. Special attention is paid to some industrially important compounds.

The largest number of measurements, 22, were carried out at UCB by the group of Professor W.F. Giauque (1949 Nobel Laureate in chemistry) over a period of 30 years. All measurements from this laboratory were carried out in an isoperibol calorimeter and were included in the correlation with a claimed uncertainty of 0.3 % (with the exception of nitric oxide, NO, where the claimed error was 1 %). The group of compounds measured at UCB is very diverse and some compounds were measured only at this laboratory. Hydroiodic acid and hydrogen sulfide were investigated in other laboratories, but only the data from UCB (29GIA/WIE, 36GIA/BLU) have been considered in the final correlation. For sulfuric acid, ammonia, and nitric oxide, NO, the data measured at UCB (52RUB/GIA, 37OVE/GIA, 29JOH/GIA) have been merged with less accurate data from other laboratories (24EUC/KAR, 26DAN/JEN, 32SOC, 57KER, 71POP/MAN) in order to extend the temperature interval of the recommended data. Reliable data from UCB for hydrobromic acid (28GIA/WIE2), hydrochloric acid (28GIA/WIE1) and nitrous oxide, N_2O, (35BLU/GIA) have been supplemented with other good quality data from OCUO (76CHI/INA2, 76CHI/INA1, 74ATA/CHI) in the final correlation.

A large number of measurements (16 data sets) were performed in German laboratories (TUB, UMG, and UWG) before the World War II. The uncertainty of their data is 0.5 % or larger, especially for older measurements. These laboratories were particularly involved in investigations of deuterated compounds (38KRU/CLU, 47CLU/WOL), silane (33CLU), germane (42CLU/FAB), and hydrogen selenide (38KRU/CLU). The data for phosphine (36CLU/FRA1) have been supplemented by data from UCB (37STE/GIA). For a correlation of data for carbon monoxide, we have supplemented experimental data from the above German laboratories (16EUC, 29CLU1, 38KAI) with data from UCB (32CLA/GIA).

Very accurate and extensive measurements of four different fluorine compounds were carried out at WRUC (55PIE/PAC, 60BOC/PET, 62PAC/PET, 65PAC/TUR, 70PAC/JEP). Data for five chlorides were measured at ICG (86DEV/GUS, 87DEV/GUS, 87GIB/GUS, 87GUS/GIB, 87GUS/GUB); we have concluded that the real error in these data is about 0.5 %; this is an uncertainty about twice as large as that claimed by the authors.

The inorganic family includes several industrially important compounds. Heat capacity of carbon dioxide was measured up to the critical temperature at NBSB (86MAG/ELY); the reported uncertainty of the data is 0.5 % and seems to be realistic. The vapor pressure of carbon dioxide at the triple point temperature is about five times larger than the atmospheric pressure. Thus, all measurements in the liquid phase were carried out at an elevated pressure. Experimental data for nitrogen fluoride were also measured at NBSB (81WEB) as a function of temperature up to the vicinity of the critical temperature. The published data were the saturation heat capacities; as no pVT data were available for conversion to the isobaric heat capacity, all the presented recommended values relate to C_{sat}. These data have been supplemented for the correlation by the results from WRUC (55PIE/PAC); the average deviation between the two data sets is about 1 %.

For sulfur dioxide, we have jointly correlated data from two independent sources, UCB (38GIA/STE) and KITH (39RIE2), despite the fact that the heat capacity varied differently with temperature in these two sources. Due to the lack of information in the two original papers, it was impossible to decide which temperature dependence was closer to the reality. For that reason, we have assigned the level of uncertainty IV to the recommended data. For sulfur trioxide, we had only one primary set of calorimetric values in a very narrow temperature range (12 K) from The Kazan Institute of Chemical Technology (89KON/STR). We have opted, however, for secondary data on SO_3 published as parameters of a cubic equation in a review paper by (76MIL/SCH). As the primary data (obtained in US industrial laboratories) were unavailable to us, an exception has been made in the policy of not using secondary sources. The recommended data for sulfur trioxide have been assigned the level of uncertainty VI because information on the experimental procedure was not available.

Experimental data for carbon disulfide are very abundant. From 11 sets of data we have selected for the final correlation the data from UCB (37BRO/MAN), ICLO (55STA/TUP) and data generated from parameters of a quadratic equation published by Indian authors from IISB (70GOP/GAM).

Experimental data for hydrazine measured at BMB (49SCO/OLI) represent the only set of data from BMB for this family of compounds. We have supplemented these data by results from CITP (50HOU/MAS) for extending the BMB measurements towards higher temperatures, and with values reported by American Aviation, Inc., Canoga Park, California (68AHL/YOU).

Heat capacities of phosphoric acid have been calculated by us from heat capacity data on water solutions of phosphorus pentoxide, P_2O_5, published by Tennessee Valley Authority, Alabama (58EGA/LUF, 81LUF).

For the correlation of heat capacities of heavy water, the data from Queen Mary College, London (40COC/FER), UGG (51EUC/EIG) and the most extensive measurements published by U.S. Atomic Energy Commission (59BAK) have been selected from 9 available data sources.

The heat capacity of water was measured repeatedly (very often at the single temperature, 298.15 K) for verifying the reliability of a calorimeter. For this reason, we have not included all available sets in the table of experimental data. We have based our selection on two sets of data from NBSW (37OSB/STI, 39OSB/STI). The former set contains data measured up to the critical temperature with uncertainties between 1 to 1.5 %; the latter set contains highly accurate data (uncertainty 0.1 %) from the melting temperature up to 368 K. Three data points in the proximity of the critical temperature were obviously subject to a large error and have been rejected from the final correlation. The above two sets from NBSW were supplemented with high precision data measured in the interval from 290 to 321 K at Queen Mary College, London (40COC/FER). For the conversion of C_{sat} to C_p, we have used the recent equation of state for water (84HAA/GAL).

Due to a frequent occurrence of a minimum on the $C_p=f(T)$ curve for many inorganic compounds, parameters of a quasi-polynomial equation either have not been calculated or were obtained for a limited temperature interval from the minimum on the curve to the upper temperature limit of the experimental data.

Name: Silver(1+) salt nitric acid
Formula: $AgNO_3$

CAS–RN: 7761–88–8
Group No.: 2–001
Molar Mass: 169.87

TABLE 2.1.1. Experimental heat capacities

Reference		Temp. range K	No. pnts	Error %	Purity % method	Type capacity	Calorimeter Type	Calorimeter Reference
07GUI2	N	502.1	1	nosp	not specified	C_{avg}	DSIO	07GUI2
09GOO/KAL	N	491.1–541.2	2	nosp	not specified	C_{avg}	DSIO	09GOO/KAL

07GUI2 average value in temperature range 481–523 K
09GOO/KAL constant value calculated from temperature dependence of enthalpy by the authors

TABLE 2.1.2. Correlated heat capacities

Reference	Temp. range K	No. pnts used	$\sigma_r C$ %	d_w	d/R	d_r %	d_b/R	+/–
				Selected data				
09GOO/KAL	491.1–541.2	2	3.00#	0.000	0.00	0.00	0.00	0
				Rejected data				
07GUI2	(6.84–1, 4.28, –6.84–1, –1)							

TABLE 2.1.3. Parameters of regression polynomial

Heat capacity type	No. data points total	No. data points used	s_w	s/R	s_r %	s_b/R	+/–
C	3	2	0.000	0.00	0.00	0.00	0

Temp. range K	A_1	Level of uncertainty
491.1–541.2	1.66692+1	V

TABLE 2.1.4. Recommended values of heat capacities

Temp. (K)	500	540
c (J $K^{-1}g^{-1}$)	0.82	0.82
C (J $K^{-1}mol^{-1}$)	139	139

Name: Aluminium bromide
Formula: $AlBr_3$

CAS–RN: 7727–15–3
Group No.: 2–002
Molar Mass: 266.69

TABLE 2.2.1. Experimental heat capacities

Reference		Temp. range K	No. pnts	Error %	Purity % method	Type capacity	Calorimeter Type	Calorimeter Reference
31FIS	N	370.6–407.1	2	nosp	not specified	C_p	DSTO	28FIS/BIL
69WEB/JUS	N	370.6–529.0	3	nosp	not specified	C_p	not specified	

31FIS constant value calculated from a graph (enthalpy versus temperature) by the compilers
69WEB/JUS data calculated by the authors from 31FIS

TABLE 2.2.2. Correlated heat capacities

Reference	Temp. range K	No. pnts used	$\sigma_r C$ %	d_w	d/R	d_r %	d_b/R	+/–
				Selected data				
69WEB/JUS	370.6–529.0	3	3.00#	0.000	0.00	0.00	0.00	0
				Rejected data				
31FIS	(8.05–1, 5.08, 8.05–1, 2)							

TABLE 2.2.3. Parameters of regression polynomial

Heat capacity type	No. data points total	used	s_w	s/R	s_r %	s_b/R	+/–
C_p	5	3	0.000	0.00	0.00	0.00	0

Temp. range K	A_1	Level of uncertainty
370.6–529.0	1.50311+1	VI

TABLE 2.2.4. Recommended values of heat capacities

Temp. (K)	370	450	530
c_p (J K^{-1}g^{-1})	0.47	0.47	0.47
C_p (J K^{-1}mol^{-1})	125	125	125

Name: Aluminium chloride
Formula: $AlCl_3$

CAS–RN: 7446–70–0
Group No.: 2–003
Molar Mass: 133.34

TABLE 2.3.1. Experimental heat capacities

Reference		Temp. range K	No. pnts	Error %	Purity %	Purity method	Type capacity	Calorimeter Type	Calorimeter Reference
31FIS	N	469.1–505.1	2	nosp	not specified		C_p	DSTO	28FIS/BIL

31FIS constant value calculated from a graph (enthalpy versus temperature) by the compilers

TABLE 2.3.3. Parameters of regression polynomial

Heat capacity type	No. data points total	used	s_w	s/R	s_r %	s_b/R	+/–
C_p	2	2	0.000	0.00	0.00	0.00	0

Temp. range K	A_1	Level of uncertainty
469.1–505.1	1.74415+1	VI

TABLE 2.3.4. Recommended values of heat capacities

Temp. (K)	470	510
c_p (J K^{-1}g^{-1})	1.09	1.09
C_p (J K^{-1}mol^{-1})	145	145

Name: Aluminium iodide
Formula: AlI_3

CAS–RN: 7784–23–8
Group No.: 2–004
Molar Mass: 407.69

TABLE 2.4.1. Experimental heat capacities

Reference		Temp. range K	No. pnts	Error %	Purity %	Purity method	Type capacity	Calorimeter Type	Calorimeter Reference
31FIS	N	464.1–480.1	2	nosp	not specified		C_p	DSTO	28FIS/BIL

31FIS constant value calculated from a graph (enthalpy versus temperature) by the compilers

TABLE 2.4.3. Parameters of regression polynomial

Heat capacity type	No. data points total	No. data points used	s_w	s/R	s_r %	s_b/R	+/–
C_p	2	2	0.000	0.00	0.00	0.00	0

Temp. range K	A_1	Level of uncertainty
464.1–480.1	1.49908+1	V

TABLE 2.4.4. Recommended values of heat capacities

Temp. (K)	470	480
c_p (J K^{-1}g^{-1})	0.31	0.31
C_p (J K^{-1}mol^{-1})	125	125

Name: Arsenous trichloride
Formula: $AsCl_3$

CAS–RN: 7784–34–1
Group No.: 2–005
Molar Mass: 181.28

TABLE 2.5.1. Experimental heat capacities

Reference	Temp. range K	No. pnts	Error %	Purity %	Purity method	Type capacity	Calorimeter Type	Calorimeter Reference
87GIB/GUS	255.3–272.7	10	0.20	99.95	melpt	C_p	BSAO	86DEV/GUS

TABLE 2.5.3. Parameters of regression polynomial

Heat capacity type	No. data points total	No. data points used	s_w	s/R	s_r %	s_b/R	+/–
C_p	10	10	0.213	6.97–3	0.04	5.34–6	1

Temp. range K	A_1	A_2	Level of uncertainty
255.3–272.7	1.67743+1	–1.55815–1	II

TABLE 2.5.4. Recommended values of heat capacities

Temp. (K)	260	270
c_p (J K^{-1}g^{-1})	0.751	0.750
C_p (J K^{-1}mol^{-1})	136.1	136.0

Name: Arsenous trifluoride
Formula: AsF_3

CAS–RN: 7784–35–2
Group No.: 2–006
Molar Mass: 131.92

TABLE 2.6.1. Experimental heat capacities

Reference	Temp. range K	No. pnts	Error %	Purity %	Purity method	Type capacity	Calorimeter Type	Calorimeter Reference
41RUS/RUN	276.2–287.8	4	0.10	99.98	melpt	C_p	BSAO	41YOS/GAR

TABLE 2.6.3. Parameters of regression polynomial

Heat capacity type	No. data points total	No. data points used	s_w	s/R	s_r %	s_b/R	+/–
C_p	4	4	0.733	1.11–2	0.07	7.63–6	0

Temp. range K	A_1	A_2	Level of uncertainty
276.2–287.8	1.37157+1	4.94961–1	II

TABLE 2.6.4. Recommended values of heat capacities

Temp. (K)	280	290
c_p (J K^{-1}g^{-1})	0.952	0.955
C_p (J K^{-1}mol^{-1})	125.6	126.0

Name: Arsine
Formula: AsH_3

CAS–RN: 7784–42–1
Group No.: 2–007
Molar Mass: 77.95

TABLE 2.7.1. Experimental heat capacities

Reference	Temp. range K	No. pnts	Error %	Purity %	Purity method	Type capacity	Calorimeter Type	Calorimeter Reference
55SHE/GIA	160.9–207.6	8	nosp	99.96	melpt	C_p	BSIO	28GIA/WIE1

TABLE 2.7.3. Parameters of regression polynomial

Heat capacity type	No. data points total	No. data points used	s_w	s/R	s_r %	s_b/R	+/–
C_p	8	8	0.262	5.72–3	0.08	5.25–6	0

Temp. range K	A_1	A_2	A_3	Level of uncertainty
160.9–207.6	8.58301	–1.47190	4.08470–1	II

TABLE 2.7.4. Recommended values of heat capacities

Temp. (K)	170	180	190	200	210
c_p (J $K^{-1}g^{-1}$)	0.775	0.774	0.775	0.776	0.778
C_p (J $K^{-1}mol^{-1}$)	60.37	60.34	60.37	60.47	60.64

Name: Trifluoroborane
Formula: BF_3

CAS–RN: 7637–07–2
Group No.: 2–008
Molar Mass: 67.81

TABLE 2.8.1. Experimental heat capacities

Reference	Temp. range K	No. pnts	Error %	Purity %	Purity method	Type capacity	Calorimeter Type	Calorimeter Reference
38EUC/SCH	145.0–173.0	5S	nosp	99.3	melpt	C_p	BSAO	38EUC/SCH

TABLE 2.8.3. Parameters of regression polynomial

Heat capacity type	No. data points total	No. data points used	s_w	s/R	s_r %	s_b/R	+/–
C_p	5	5	0.032	2.83–3	0.02	7.63–7	0

Temp. range K	A_1	A_2	Level of uncertainty
145.0–173.0	1.08556+1	1.02776	IV

TABLE 2.8.4. Recommended values of heat capacities

Temp. (K)	150	160	170
c_p (J $K^{-1}g^{-1}$)	1.520	1.533	1.545
C_p (J $K^{-1}mol^{-1}$)	103.1	103.9	104.8

Name: Boron chloride (B_2Cl_4)
Formula: B_2Cl_4

CAS–RN: 13701–67–2
Group No.: 2–009
Molar Mass: 163.43

TABLE 2.9.1. Experimental heat capacities

Reference	Temp. range K	No. pnts	Error %	Purity %	Purity method	Type capacity	Calorimeter Type	Calorimeter Reference
58LIN/WAR	183.3–214.5	7	0.50	98.5	melpt	C_p	BSIO	36AST/MES

TABLE 2.9.3. Parameters of regression polynomial

Heat capacity type	No. data points total	No. data points used	s_w	s/R	s_r %	s_b/R	+/–
C_p	7	7	0.440	3.65–2	0.22	1.14–4	–1

Temp. range K	A_1	A_2	Level of uncertainty
183.3–214.5	1.68135+1	–1.34233–1	IV

TABLE 2.9.4. Recommended values of heat capacities

Temp. (K)	190	200	210
c_p (J K^{-1}g^{-1})	0.842	0.842	0.841
C_p (J K^{-1}mol^{-1})	137.7	137.6	137.5

Name: Diborane (6)
Formula: B_2H_6

CAS-RN: 19287–45–7
Group No.: 2–010
Molar Mass: 27.67

TABLE 2.10.1. Experimental heat capacities

Reference	Temp. range K	No. pnts	Error %	Purity %	Purity method	Type capacity	Calorimeter Type	Calorimeter Reference
53CLA/RIF	112.9–176.7	16	0.10	99.992	melpt	C_p	BSIO	50JOH/CLA
53RIF/KER	170.1–282.3	14	1.00	99.992	melpt	C_{sat}	BSIO	53RIF/KER

TABLE 2.10.2. Correlated heat capacities

Reference	Temp. range K	No. pnts used	$\sigma_r C$ %	d_w	d/R	d_r %	d_b/R	+/–
				Selected data				
53CLA/RIF	112.9–176.7	16	0.10	2.372	2.16–2	0.24	8.56–5	–2
53RIF/KER	170.–282.3	14	1.00	1.769	2.22–1	1.77	1.09–2	–2

TABLE 2.10.3. Parameters of cubic spline polynomials

Heat capacity type	No. data points total	No. data points used	s_w	s/R	s_r %	s_b/R	+/–
C	30	30	2.314	1.67–1	1.34	5.14–3	–4

Temp. range K	A_1	A_2	A_3	A_4	Level of uncertainty
112.9–250.0	9.50739	6.57368–1	–1.63625	6.70162–1	IV
250.0–282.3	–3.26859+3	3.93437+3	–1.57512+3	2.10468+2	V

TABLE 2.10.4. Recommended values of heat capacities

Temp. (K)	110	120	130	140	150	160	170
c (J K^{-1}g^{-1})	2.747	2.734	2.725	2.722	2.727	2.739	2.761
C (J K^{-1}mol^{-1})	76.02	75.65	75.40	75.33	75.44	75.79	76.40
Temp. (K)	180	190	200	210	220	230	240
c (J K^{-1}g^{-1})	2.794	2.839	2.896	2.968	3.056	3.160	3.283
C (J K^{-1}mol^{-1})	77.30	78.54	80.14	82.13	84.56	87.45	90.83
Temp. (K)	250	260	270	273.15	280		
c (J K^{-1}g^{-1})	3.424	3.65	4.27	4.61	5.68		
C (J K^{-1}mol^{-1})	94.75	101	118	128	157		

TABLE 2.10.5. Parameters of quasi-polynomial equation

Heat capacity type	No. data points total	used	s_w	s/R	s_r %	s_b/R	+/–
C	30	30	3.378	1.87–1	1.59	3.20–2	8

Temp. range K	T_c K	A_1	A_2	A_3	A_4	A_5	A_6	Level of uncertainty
112.9–282.3	289.80	–2.22120	1.76504–1	7.70639	5.76824	–1.94083+1	1.16871+1	V

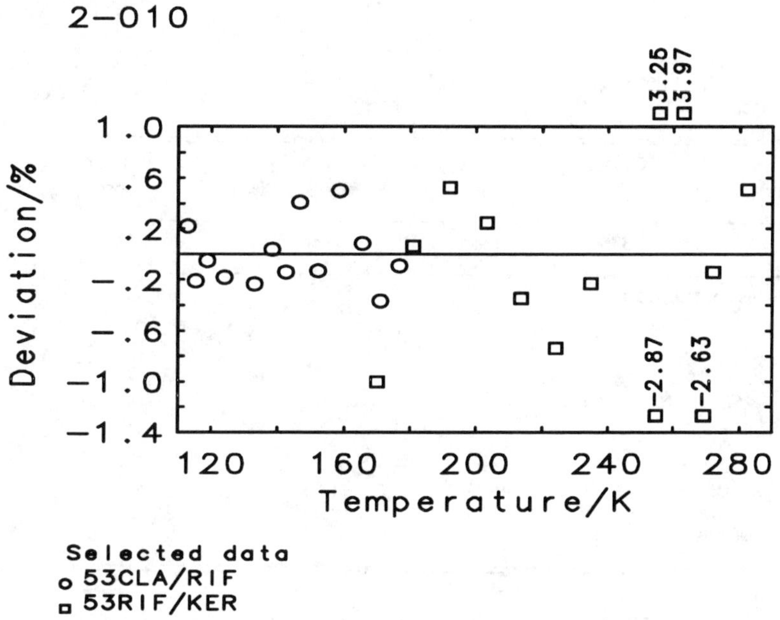

Name: Borazine
Formula: $B_3H_6N_3$

CAS–RN: 6569–51–3
Group No.: 2–011
Molar Mass: 80.50

TABLE 2.11.1. Experimental heat capacities

Reference		Temp. range K	No. pnts	Error %	Purity %	Purity method	Type capacity	Calorimeter Type	Calorimeter Reference
91LEB/KUL	N	225.9–310.4	29	0.20	99.94	melpt	C_p	BSAO	76LEB/LIT

TABLE 2.11.3. Parameters of regression polynomial

Heat capacity type	No. data points total	used	s_w	s/R	s_r %	s_b/R	+/–
C_p	29	29	1.092	7.31–2	0.44	5.76–4	2

Temp. range K	A_1	A_2	A_3	Level of uncertainty
225.9–310.4	1.16116+1	1.32871–1	6.36502–1	III

TABLE 2.11.4. Recommended values of heat capacities

Temp. (K)	230	240	250	260	270	273.15	280
c_p (J $K^{-1}g^{-1}$)	1.579	1.611	1.644	1.679	1.716	1.727	1.753
C_p (J $K^{-1}mol^{-1}$)	127.1	129.7	132.4	135.2	138.1	139.0	141.1
Temp. (K)	290	298.15	300	310			
c_p (J $K^{-1}g^{-1}$)	1.792	1.825	1.832	1.874			
C_p (J $K^{-1}mol^{-1}$)	144.3	146.9	147.5	150.8			

Name: Pentaborane(9)
Formula: B_5H_9

CAS–RN: 19624–22–7
Group No.: 2–012
Molar Mass: 63.13

TABLE 2.12.1. Experimental heat capacities

Reference		Temp. K	Capac. J/(K.g)	Error %	Purity %	Purity Method	Type capacity	Calorimeter Type	Calorimeter Reference
59WES2	N	298.15	2.373	nosp	not specified		C_p	not specified	

59WES2 unpublished data from Johnston H.L. et al.

Name: Decaborane (14)
Formula: $B_{10}H_{14}$

CAS–RN: 17702–41–9
Group No.: 2–013
Molar Mass: 122.22

TABLE 2.13.1. Experimental heat capacities

Reference	Temp. range K	No. pnts	Error %	Purity %	Purity method	Type capacity	Calorimeter Type	Calorimeter Reference
55FUR/PAR	373.0–380.0	6	0.30	99.984	melpt	C_{sat}	BSAO	45SCO/MEY

TABLE 2.13.3. Parameters of regression polynomial

Heat capacity type	No. data points total	No. data points used	s_w	s/R	s_r %	s_b/R	+/–
C_{sat}	6	6	1.272	1.44–1	0.38	1.81–4	0

Temp. range K	A_1	A_2	A_3	Level of uncertainty
373.0–380.0	–5.97490+3	3.17983+3	–4.20378+2	III

TABLE 2.13.4. Recommended values of heat capacities

Temp. (K)	372	374	376	378	380
c_{sat} (J $K^{-1}g^{-1}$)	2.498	2.558	2.594	2.608	2.599
C_{sat} (J $K^{-1}mol^{-1}$)	305.3	312.6	317.1	318.8	317.7

Name: Hydrobromic acid–*d*
Formula: BrD

CAS–RN: 13536–59–9
Group No.: 2–014
Molar Mass: 81.92

TABLE 2.14.1. Experimental heat capacities

Reference	Temp. range K	No. pnts	Error %	Purity %	Purity method	Type capacity	Calorimeter Type	Calorimeter Reference
47CLU/WOL	190.5–205.8	5	nosp	not specified		C_p	BSIO	36CLU/GOL
76CHI/INA2	187.6–188.2	2	0.20	99.927	melpt	C_p	BSAO	74ATA/CHI

TABLE 2.14.2. Correlated heat capacities

Reference	Temp. range K	No. pnts used	$\sigma_r C$ %	d_w	d/R	d_r %	d_b/R	+/–
				Selected data				
47CLU/WOL	190.5–205.8	5	0.50#	1.054	3.98–2	0.53	5.66–3	3
76CHI/INA2	187.6–188.2	2	0.20	1.942	2.91–2	0.39	–1.87–3	0

TABLE 2.14.3. Parameters of regression polynomial

Heat capacity type	No. data points total	No. data points used	s_w	s/R	s_r %	s_b/R	+/–
C_p	7	7	1.809	4.90–2	0.65	3.51–3	3

Temp. range K	A_1	A_2	A_3	Level of uncertainty
187.6–205.8	–2.10881	9.49143	–2.32871	IV

TABLE 2.14.4. Recommended values of heat capacities

Temp. (K)	190	200
c_p (J K^{-1}g^{-1})	0.763	0.767
C_p (J K^{-1}mol^{-1})	62.51	62.85

Name: Bromine fluoride (BrF_3)
Formula: BrF_3

CAS–RN: 7787–71–5
Group No.: 2–015
Molar Mass: 136.90

TABLE 2.15.1. Experimental heat capacities

Reference	Temp. range K	No. pnts	Error %	Purity %	Purity method	Type capacity	Calorimeter Type	Calorimeter Reference
52OLI/GRI	285.5–308.1	8	0.20	99.91	melpt	C_{sat}	BSAO	43RUE/HUF

TABLE 2.15.3. Parameters of regression polynomial

Heat capacity type	No. data points total	used	s_w	s/R	s_r %	s_b/R	+/–
C_{sat}	8	8	0.525	1.57–2	0.10	2.50–5	0

Temp. range K	A_1	A_2	Level of uncertainty
285.5–308.1	1.30998+1	6.33129–1	II

TABLE 2.15.4. Recommended values of heat capacities

Temp. (K)	290	298.15	300	310
c_{sat} (J $K^{-1}g^{-1}$)	0.907	0.910	0.911	0.915
C_{sat} (J $K^{-1}mol^{-1}$)	124.2	124.6	124.7	125.2

Name: Hydrobromic acid
Formula: BrH

CAS–RN: 10035–10–6
Group No.: 2–016
Molar Mass: 80.91

TABLE 2.16.1. Experimental heat capacities

Reference	Temp. range K	No. pnts	Error %	Purity %	method	Type capacity	Calorimeter Type	Reference
24EUC/KAR	190.7–198.3	4	3.00	not specified		C_{sat}	BSIO	24EUC/KAR
28GIA/WIE2	189.9–205.1	6	0.30	not specified		C_{sat}	BSIO	28GIA/WIE1
76CHI/INA2	188.3–190.3	3	0.20	99.987	melpt	C_p	BSAO	74ATA/CHI

TABLE 2.16.2. Correlated heat capacities

Reference	Temp. range K	No. pnts used	$\sigma_r C$ %	d_w	d/R	d_r %	d_b/R	+/–
				Selected data				
28GIA/WIE2	189.9–205.1	6	0.30	1.089	2.35–2	0.33	2.58–3	0
76CHI/INA2	188.3–190.3	3	0.20	4.853	6.89–2	0.97	–8.12–4	1
				Rejected data				
24EUC/KAR	(1.46–1, 1.99, 1.38–1, 4)							

TABLE 2.16.3. Parameters of regression polynomial

Heat capacity type	No. data points total	used	s_w	s/R	s_r %	s_b/R	+/–
C	13	9	3.334	5.01–2	0.70	1.45–3	1

Temp. range K	A_1	A_2	Level of uncertainty
188.3–205.1	6.11645	5.39060–1	IV

TABLE 2.16.4. Recommended values of heat capacities

Temp. (K)	190	195	200	205
c (J K^{-1}g^{-1})	0.734	0.737	0.739	0.742
C (J K^{-1}mol^{-1})	59.37	59.60	59.82	60.04

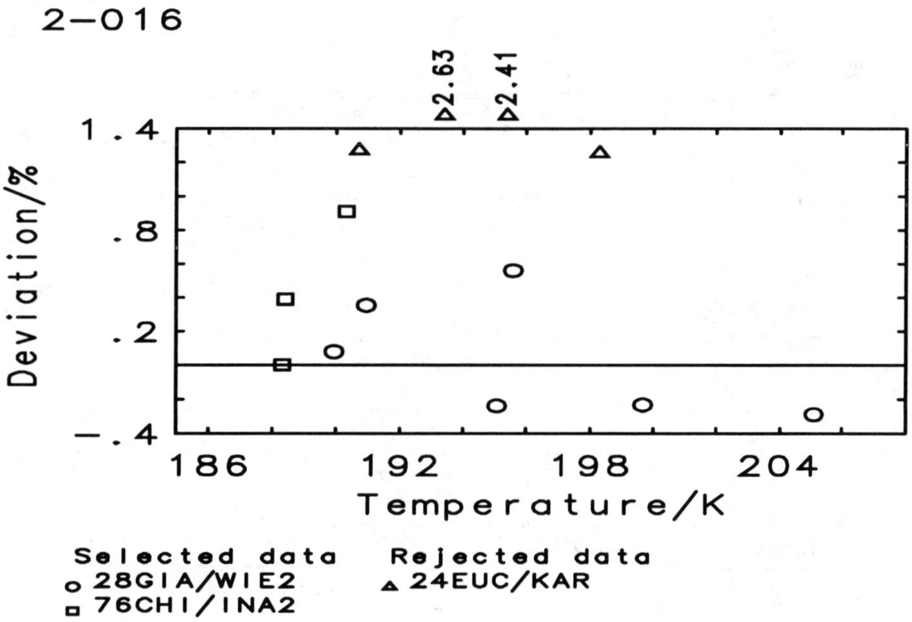

Name: Indium bromide (InBr)
Formula: BrIn

CAS–RN: 14280–53–6
Group No.: 2–017
Molar Mass: 194.72

TABLE 2.17.1. Experimental heat capacities

Reference	Temp. range K	No. pnts	Error %	Purity %	Purity method	Type capacity	Calorimeter Type	Calorimeter Reference
91GAR/PRE1	580.0–700.0	eqn	nosp	99.999	anal	C_p	BDHT	69PER/COM

TABLE 2.17.3. Parameters of regression polynomial

Heat capacity type	No. data points total	No. data points used	s_w	s/R	s_r %	s_b/R	+/–
C_p	12	12	0.000	3.99–7	0.00	1.19–7	0

Temp. range K	A_1	A_2	Level of uncertainty
580.0–699.9	5.76462	2.36214–1	V

TABLE 2.17.4. Recommended values of heat capacities

Temp. (K)	580	600	620	640	660	680	700
c_p ($J K^{-1}g^{-1}$)	0.30	0.31	0.31	0.31	0.31	0.31	0.32
C_p ($J K^{-1}mol^{-1}$)	59.3	59.7	60.1	60.5	60.9	61.3	61.7

Name: Tin bromide ($SnBr_2$)
Formula: Br_2Sn

CAS–RN: 10031–24–0
Group No.: 2–018
Molar Mass: 278.52

TABLE 2.18.1. Experimental heat capacities

Reference	Temp. range K	No. pnts	Error %	Purity %	Purity method	Type capacity	Calorimeter Type	Calorimeter Reference
91GAR/PRE2	520.0–800.0	eqn	nosp	not specified		C_p	BDHT	69PER/COM

TABLE 2.18.3. Parameters of regression polynomial

Heat capacity type	No. data points total	No. data points used	s_w	s/R	s_r %	s_b/R	+/–
C_p	15	15	0.000	6.48–7	0.00	–2.54–7	0

Temp. range K	A_1	A_2	Level of uncertainty
520.0–800.0	8.44067	6.01839–1	V

TABLE 2.18.4. Recommended values of heat capacities

Temp. (K)	520	540	560	580	600	620	640
c_p ($J K^{-1}g^{-1}$)	0.35	0.35	0.35	0.36	0.36	0.36	0.37
C_p ($J K^{-1}mol^{-1}$)	96.2	97.2	98.2	99.2	100	101	102
Temp. (K)	660	680	700	720	740	760	780
c_p ($J K^{-1}g^{-1}$)	0.37	0.37	0.38	0.38	0.38	0.39	0.39
C_p ($J K^{-1}mol^{-1}$)	103	104	105	106	107	108	109
Temp. (K)	800						
c_p ($J K^{-1}g^{-1}$)	0.40						
C_p ($J K^{-1}mol^{-1}$)	110						

Name: Tetrabromostannane
Formula: Br_4Sn

CAS–RN: 7789–67–5
Group No.: 2–019
Molar Mass: 438.33

TABLE 2.19.1. Experimental heat capacities

Reference		Temp. range K	No. pnts	Error %	Purity %	Purity method	Type capacity	Calorimeter Type	Calorimeter Reference
36KUR/VOS	N	311.6	1	nosp	not specified		C_{avg}	DSIO	36KUR/VOS
83BER/STE		306.9–336.2	8	nosp	99.98	melpt	C_p	BSAO	79AMI/LEB

36KUR/VOS average value in temperature range 290–333 K

TABLE 2.19.2. Correlated heat capacities

Reference	Temp. range K	No. pnts used	$\sigma_r C$ %	d_w	d/R	d_r %	d_b/R	+/–
				Selected data				
83BER/STE	306.9–336.2	8	0.30#	0.142	8.35–3	0.04	7.15–6	–2
				Rejected data				
36KUR/VOS	(8.61, 30.50, 8.61, 1)							

TABLE 2.19.3. Parameters of regression polynomial

Heat capacity type	No. data points total	used	s_w	s/R	s_r %	s_b/R	+/–
C_p	9	8	0.164	9.65–3	0.05	7.15–6	–2

Temp. range K	A_1	A_2	Level of uncertainty
306.9–336.2	2.17989+1	–6.98827–1	III

TABLE 2.19.4. Recommended values of heat capacities

Temp. (K)	310	320	330	340
c_p (J K^{-1}g^{-1})	0.372	0.371	0.370	0.368
C_p (J K^{-1}mol^{-1})	163.2	162.7	162.1	161.5

Name: Carbonic dichloride
Formula: CCl_2O

CAS–RN: 75–44–5
Group No.: 2–020
Molar Mass: 98.92

TABLE 2.20.1. Experimental heat capacities

Reference	Temp. range K	No. pnts	Error %	Purity %	Purity method	Type capacity	Calorimeter Type	Calorimeter Reference
48GIA/JON	148.4–279.4	24	nosp	99.993	melpt	C_p	BSIO	37GIA/EGA
60GIA/OTT	145.6–160.6	3	0.20	99.963	melpt	C_p	BSIO	37GIA/EGA

TABLE 2.20.2. Correlated heat capacities

Reference	Temp. range K	No. pnts used	$\sigma_r C$ %	d_w	d/R	d_r %	d_b/R	+/–
				Selected data				
48GIA/JON	148.4–79.4	24	0.30#	0.525	1.90–2	0.16	–1.26–3	–4
60GIA/OTT	145.6–160.6	3	0.20	0.977	2.44–2	0.20	4.76–3	1

TABLE 2.20.3. Parameters of regression polynomial

Heat capacity type	No. data points total	used	s_w	s/R	s_r %	s_b/R	+/–
C_p	27	27	0.642	2.13–2	0.18	–5.87–4	–3

Temp. range K	A_1	A_2	A_3	A_4	Level of uncertainty
145.6–279.4	2.18191+1	–1.17651+1	4.58699	–5.79124–1	II

TABLE 2.20.4. Recommended values of heat capacities

Temp. (K)	150	160	170	180	190	200	210
c_p (J K^{-1}g^{-1})	1.054	1.039	1.028	1.019	1.013	1.009	1.007
C_p (J K^{-1}mol^{-1})	104.2	102.8	101.7	100.8	100.2	99.81	99.59
Temp. (K)	220	230	240	250	260	270	273.15
c_p (J K^{-1}g^{-1})	1.006	1.007	1.009	1.011	1.014	1.017	1.017
C_p (J K^{-1}mol^{-1})	99.53	99.59	99.76	99.99	100.3	100.6	100.6
Temp. (K)	280						
c_p (J K^{-1}g^{-1})	1.019						
C_p (J K^{-1}mol^{-1})	100.8						

Name: Carbonic difluoride
Formula: CF_2O

CAS–RN: 353–50–4
Group No.: 2–021
Molar Mass: 66.01

TABLE 2.21.1. Experimental heat capacities

Reference	Temp. range K	No. pnts	Error %	Purity %	Purity method	Type capacity	Calorimeter Type	Calorimeter Reference
68PAC/REN	165.9–187.9	10	0.10	99.85	melpt	C_{sat}	BSAO	55PAC/PIE

TABLE 2.21.3. Parameters of regression polynomial

Heat capacity type	No. data points total	used	s_w	s/R	s_r %	s_b/R	+/–
C_{sat}	10	10	1.815	2.41–2	0.18	5.96–5	–2

Temp. range K	A_1	A_2	A_3	Level of uncertainty
165.9–187.9	–1.06008	1.81301+1	–5.64513	II

TABLE 2.21.4. Recommended values of heat capacities

Temp. (K)	170	180	190
c_{sat} (J K^{-1}g^{-1})	1.694	1.673	1.639
C_{sat} (J K^{-1}mol^{-1})	111.8	110.4	108.2

Name: Hydrocyanic acid
Formula: CHN

CAS–RN: 74–90–8
Group No.: 2–022
Molar Mass: 27.03

TABLE 2.22.1. Experimental heat capacities

Reference	Temp. range K	No. pnts	Error %	Purity %	Purity method	Type capacity	Calorimeter Type	Calorimeter Reference
39GIA/RUE	266.6–298.3	12	0.20	99.999	melpt	C_p	BSIO	37GIA/EGA

TABLE 2.22.3. Parameters of regression polynomial

Heat capacity type	No. data points total	used	s_w	s/R	s_r %	s_b/R	+/–
C_p	12	12	1.528	2.60–2	0.31	1.32–4	2

Temp. range K	A_1	A_2	Level of uncertainty
266.6–298.3	8.12927	1.33473–1	III

TABLE 2.24.4. Recommended values of heat capacities

Temp. (K)	70	75	80	85
c_p (J K^{-1}g^{-1})	2.138	2.152	2.167	2.182
C_p (J K^{-1}mol^{-1})	59.87	60.28	60.70	61.11
c_{sat} (J K^{-1}g^{-1})	2.138	2.149	2.161	2.172
C_{sat} (J K^{-1}mol^{-1})	59.89	60.20	60.52	60.83

Name: Trithiocarbonic acid
Formula: CH_2S_3

CAS–RN: 594–08–1
Group No.: 2–023
Molar Mass: 110.22

TABLE 2.23.1. Experimental heat capacities

Reference		Temp. K	Capac. J/(K.g)	Error %	Purity %	Purity Method	Type capacity	Calorimeter Type	Calorimeter Reference
63GAT/KRE	N	272.15	1.347	nosp	not specified		C_{avg}	DSIO	63GAT/KRE

63GAT/KRE average value in temperature range 246–293 K

Name: Carbon monoxide
Formula: CO

CAS–RN: 630–08–0
Group No.: 2–024
Molar Mass: 28.01

TABLE 2.24.1. Experimental heat capacities

Reference	Temp. range K	No. pnts	Error %	Purity %	Purity method	Type capacity	Calorimeter Type	Calorimeter Reference
16EUC	69.3–77.9	4	nosp	not specified		C_{sat}	BSIO	16EUC
29CLU1	70.1–81.3	6	0.70	not specified		C_p	BSIO	29CLU1
32CLA/GIA	70.0–84.7	9	nosp	99.999	melpt	C_{sat}	BSIO	28GIA/WIE1
38KAI	70.9–75.8	10	0.40	not specified		C_p	BSIO	38KAI
66GIL/MOR	68.9–75.7	5	2.00	97.4	anal	C_p	BSAO	61FLU/LEA

TABLE 2.24.2. Correlated heat capacities

Reference	Temp. range K	No. pnts used	$\sigma_r C$ %	d_w	d/R	d_r %	d_b/R	+/–
				Selected data				
16EUC	69.3–77.9	4	1.00#	0.678	4.96–2	0.68	3.49–2	2
29CLU1	70.1–81.3	6	0.70	1.220	6.30–2	0.85	5.35–2	6
32CLA/GIA	70.0–84.7	9	0.60#	0.841	3.67–2	0.50	4.20–3	1
38KAI	70.9–75.8	10	0.40	1.108	3.18–2	0.44	–1.37–2	–4
				Rejected data				
66GIL/MOR	(1.27–1, 1.78, –1.17–1, –4)							

TABLE 2.24.3. Parameters of regression polynomial

Heat capacity type	No. data points total	used	s_w	s/R	s_r %	s_b/R	+/–
C_p	34	29	1.044	4.55–2	0.62	1.25–2	5
C_{sat}	34	29	1.042	4.58–2	0.63	1.25–2	4

Temp. range K	A_1	A_2	Level of uncertainty
69.3–84.7	6.50679	9.91676–1	IV
69.3–84.7	6.67275	7.57508–1	IV

TABLE 2.24.4. Recommended values of heat capacities

Temp. (K)	70	75	80	85
c_p (J K^{-1}g^{-1})	2.138	2.152	2.167	2.182
C_p (J K^{-1}mol^{-1})	59.87	60.28	60.70	61.11
c_{sat} (J K^{-1}g^{-1})	2.138	2.149	2.161	2.172
C_{sat} (J K^{-1}mol^{-1})	59.89	60.20	60.52	60.83

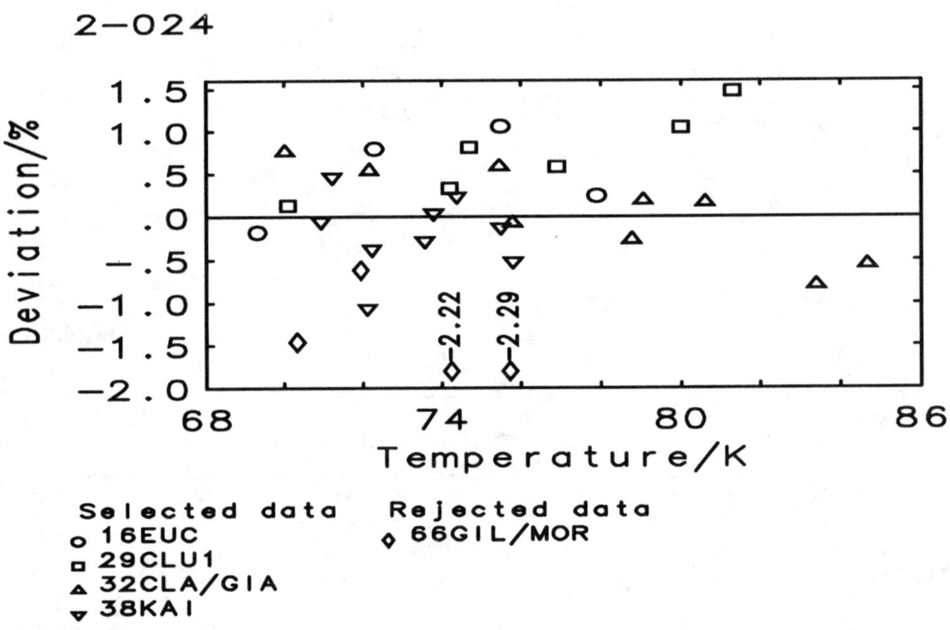

Name: Carbon oxide sulfide
Formula: COS

CAS-RN: 463–58–1
Group No.: 2–025
Molar Mass: 60.08

TABLE 2.25.1. Experimental heat capacities

Reference	Temp. range K	No. pnts	Error %	Purity %	Purity method	Type capacity	Calorimeter Type	Calorimeter Reference
37KEM/GIA	137.3–220.9	14	nosp	99.995	melpt	C_p	BSIO	28GIA/WIE1

TABLE 2.25.3. Parameters of regression polynomial

Heat capacity type	No. data points total	No. data points used	s_w	s/R	s_r %	s_b/R	+/–
C_p	14	14	0.294	7.62–3	0.09	9.54–6	–2

Temp. range K	A_1	A_2	A_3	A_4	Level of uncertainty
137.3–220.9	1.39741+1	–7.04145	2.89638	–3.68937–1	II

TABLE 2.25.4. Recommended values of heat capacities

Temp. (K)	140	150	160	170	180	190	200
c_p (J K^{-1}g^{-1})	1.215	1.202	1.192	1.185	1.181	1.179	1.180
C_p (J K^{-1}mol^{-1})	73.01	72.20	71.60	71.19	70.94	70.85	70.88
Temp. (K)	210	220					
c_p (J K^{-1}g^{-1})	1.182	1.186					
C_p (J K^{-1}mol^{-1})	71.03	71.28					

Name: Carbon dioxide
Formula: CO_2

CAS-RN: 124–38–9
Group No.: 2–026
Molar Mass: 44.01

TABLE 2.26.1. Experimental heat capacities

Reference		Temp. range K	No. pnts	Error %	Purity % method	Type capacity	Calorimeter Type	Calorimeter Reference
26MAA/BAR	N	257.5	1	nosp	not specified	C_{avg}	DSIO	20MAA/HAT
28EUC/HAU		220.0–290.0	8S	nosp	not specified	C_{sat}	BSIO	28EUC/HAU
86MAG/ELY		220.4–303.0	77	0.50	99.9946 chrom	C_{sat}	BSAO	61GOO

26MAA/BAR average value in temperature range 217–298 K

TABLE 2.26.2. Correlated heat capacities

Reference	Temp. range K	No. pnts used	$\sigma_r C$ %	d_w	d/R	d_r %	d_b/R	+/–
				Selected data				
86MAG/ELY	220.4–303.0	77	0.70#	0.883	1.70–1	0.62	2.25–3	–6
				Rejected data				
26MAA/BAR	(4.57–1, 4.05, –4.57–1, –1)			28EUC/HAU	(4.56–1, 4.18, –4.15–1, –7)			

TABLE 2.26.3. Parameters of cubic spline polynomials

Heat capacity type	No. data points total	used	s_w	s/R	s_r %	s_b/R	+/–
C_p	86	77	0.927	1.78–1	0.65	2.25–3	–6
C_{sat}	86	77	0.821	1.15–1	0.57	1.29–3	0

Temp. range K	A_1	A_2	A_3	A_4	Level of uncertainty
220.4–270.0	−1.78222+2	2.47707+2	−1.09285+2	1.62093+1	IV
270.0–290.0	−5.21131+3	5.84002+3	−2.18051+3	2.71916+2	IV
290.0–300.0	−2.20867+5	2.28932+5	−7.91088+4	9.11424+3	V
300.0–303.0	−2.68798+7	2.68879+7	−8.96543+6	9.96484+5	V
220.4–270.0	−1.03008+2	1.50085+2	−6.68718+1	1.00298+1	IV
270.0–290.0	−3.25145+3	3.64836+3	−1.36253+3	1.69987+2	IV
290.0–300.0	−1.21857+5	1.26343+5	−4.36711+4	5.03304+3	V
300.0–303.0	−1.40300+7	1.40345+7	−4.67972+6	5.20149+5	V

TABLE 2.26.4. Recommended values of heat capacities

Temp. (K)	220	230	240	250	260	270	273.15
c_p (J K^{-1}g^{-1})	1.963	2.004	2.054	2.132	2.257	2.446	2.524
C_p (J K^{-1}mol^{-1})	86.41	88.19	90.40	93.84	99.32	107.7	111.1
c_{sat} (J K^{-1}g^{-1})	1.949	1.978	2.015	2.073	2.162	2.294	2.347
C_{sat} (J K^{-1}mol^{-1})	85.78	87.04	88.69	91.22	95.15	101.0	103.3

Temp. (K)	280	290	298.15	300
c_p (J K^{-1}g^{-1})	2.767	3.480	5.47	6.56
C_p (J K^{-1}mol^{-1})	121.8	153.1	241	289
c_{sat} (J K^{-1}g^{-1})	2.511	2.974	4.17	4.80
C_{sat} (J K^{-1}mol^{-1})	110.5	130.9	183	211

TABLE 2.26.5. Parameters of quasi-polynomial equation

Heat capacity type	No. data points total	used	s_w	s/R	s_r %	s_b/R	+/–
C_p	86	77	1.384	3.99–1	0.97	–8.34–3	1

Temp. range K	T_c K	A_1	A_2	A_3	A_4	A_5	A_6	Level of uncertainty
220.4–303.0	304.17	−3.85134	2.07142–1	−1.90613+1	1.25641+2	−1.90385+2	8.61326+1	V

2–026

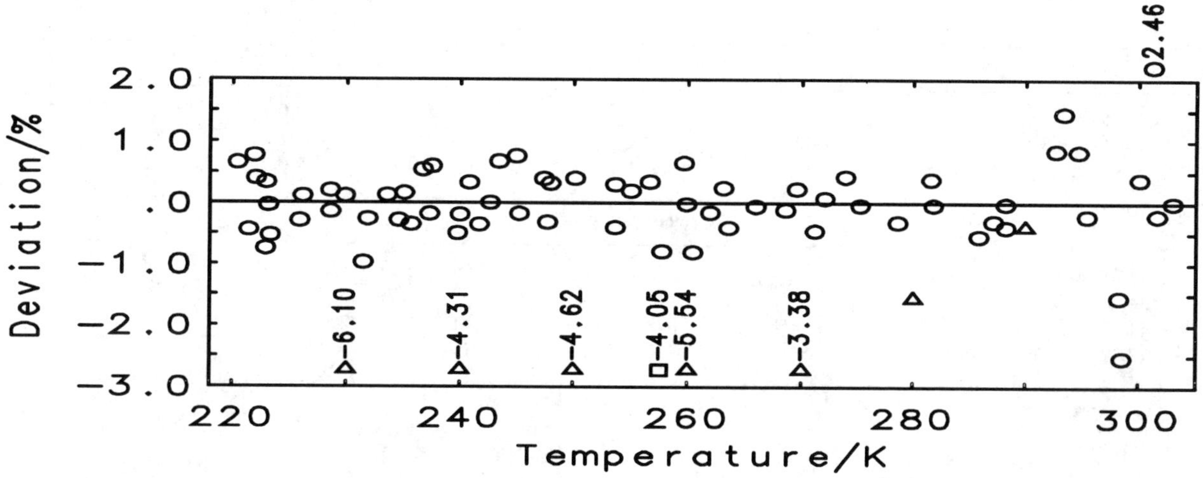

Name: Carbon disulfide
Formula: CS_2

CAS–RN: 75–15–0
Group No.: 2–027
Molar Mass: 76.14

TABLE 2.27.1. Experimental heat capacities

Reference		Temp. range K	No. pnts	Error %	Purity %	Purity method	Type capacity	Calorimeter Type	Calorimeter Reference
07BAT	N	177.4–273.2	7	nosp	not specified		C_p	BSIO	07BAT
12SCH3		243.1–293.1	3	nosp	not specified		C_p	BSIO	12SCH1
25DRU/WEI		293.1	1	nosp	not specified		C_p	BSIO	25DRU/WEI
26SCH		288.1	1	nosp	not specified		C_p	not specified	
37BRO/MAN		163.9–297.4	18	nosp	99.999	melpt	C_p	BSIO	28LAT/GRE
39MAZ3	N	172.1–292.4	16	nosp	not specified		C_p	BSIO	39MAZ3
39PHI		301.1	1	nosp	not specified		C_p	BSIO	49WEI
45ZHD		280.4–304.3	3	nosp	not specified		C_p	BSIT	34KOL/UDO2
55STA/TUP		286.0–316.8	7	1.00	not specified		C_p	BSAO	55STA/TUP
71GAM/VIS	N	303.1–313.1	2	nosp	not specified		C_p	BSAO	71GOP/GAM
71GOP/GAM		303.0–320.0	eqn	1.00	not specified		C_p	BSAO	71GOP/GAM

07BAT same data in 08BAT
39MAZ3 data from a graph only
71GAM/VIS data from a graph only

TABLE 2.27.2. Correlated heat capacities

Reference	Temp. range K	No. pnts used	$\sigma_r C$ %	d_w	d/R	d_r %	d_b/R	+/−
				Selected data				
37BRO/MAN	163.9–297.4	18	0.70#	1.009	6.44–2	0.71	−1.80–2	0
55STA/TUP	286.0–316.8	7	1.00	1.247	1.18–1	1.25	8.47–2	3
71GOP/GAM	303.0–320.2	5	1.00	0.670	6.41–2	0.67	2.52–2	1
				Rejected data				
07BAT	(1.22, 16.07, −1.10, −7)			12SCH3	(1.09–1, 1.21, −1.08–1, −3)			
25DRU/WEI	(4.48–2, 0.48, −4.48–2, −1)			26SCH	(5.89–1, 6.79, −5.89–1, −1)			
39MAZ3	(1.17, 15.51, −9.94–1, −14)			39PHI	(2.51–1, 2.74, −2.51–1, −1)			
45ZHD	(3.37–1, 3.76, −3.36–1, −3)			71GAM/VIS	(8.13–2, 0.85, 7.45–2, 2)			

TABLE 2.27.3. Parameters of regression polynomial

Heat capacity type	No. data points total	used	s_w	s/R	s_r %	s_b/R	+/−
C_p	64	30	1.100	8.61–2	0.92	1.32–2	4

Temp. range K	A_1	A_2	A_3	A_4	Level of uncertainty
163.9–320.2	6.38046	4.30422	−2.26739	3.89181–1	IV

TABLE 2.27.4. Recommended values of heat capacities

Temp. (K)	170	180	190	200	210	220	230
c_p (J K^{-1}g^{-1})	0.989	0.988	0.987	0.986	0.985	0.985	0.985
C_p (J K^{-1}mol^{-1})	75.30	75.26	75.18	75.10	75.03	74.99	75.00
Temp. (K)	240	250	260	270	273.15	280	290
c_p (J K^{-1}g^{-1})	0.986	0.988	0.992	0.997	0.999	1.005	1.014
C_p (J K^{-1}mol^{-1})	75.08	75.25	75.53	75.93	76.09	76.49	77.21
Temp. (K)	298.15	300	310	320			
c_p (J K^{-1}g^{-1})	1.023	1.026	1.040	1.058			
C_p (J K^{-1}mol^{-1})	77.93	78.11	79.22	80.56			

TABLE 2.27.5. Parameters of quasi-polynomial equation

Heat capacity type	No. data points total	used	s_w	s/R	s_r %	s_b/R	+/−
C_p	64	30	1.127	8.51–2	0.91	1.76–3	−2

Temp. range K	T_c K	A_1	A_2	A_3	A_4	Level of uncertainty
163.9–320.2	552.00	4.16496+1	1.38732+1	−5.31148	3.12597+1	V

2-027

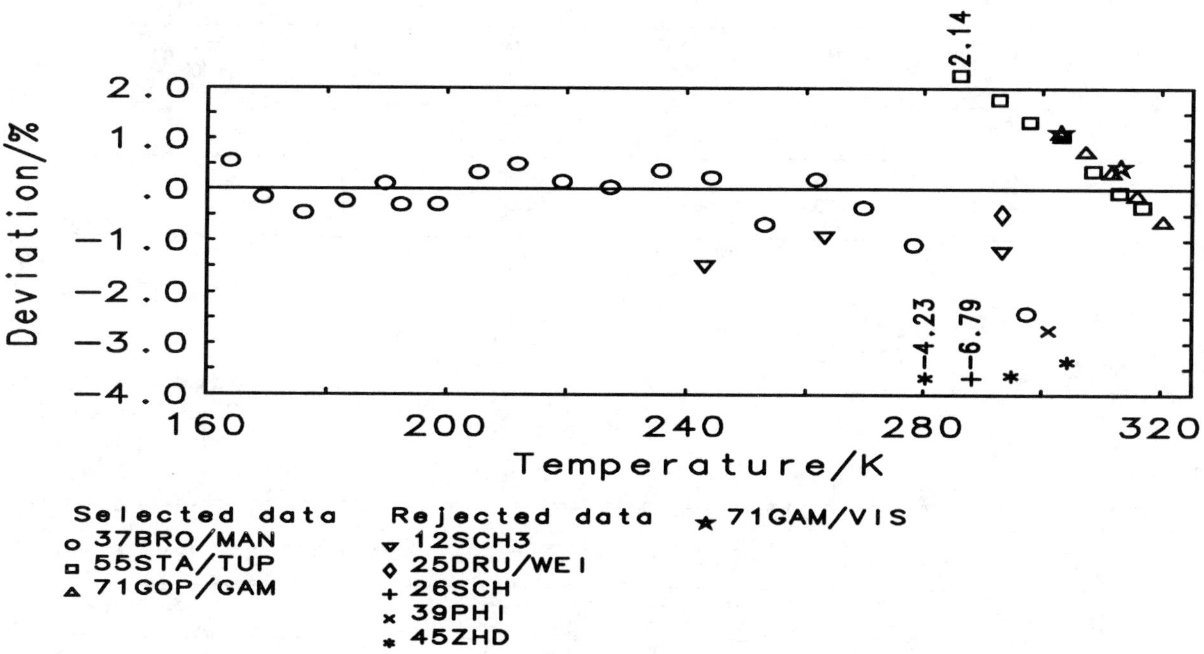

Name: Carbon selenide (CSe_2)
Formula: CSe_2

CAS-RN: 506-80-9
Group No.: 2-028
Molar Mass: 169.93

TABLE 2.28.1. Experimental heat capacities

Reference	Temp. range K	No. pnts	Error %	Purity %	Purity method	Type capacity	Calorimeter Type	Calorimeter Reference
66GAT/DRA	233.1-323.1	10S	nosp	not specified		C_p	DSIO	63GAT/KRE

TABLE 2.28.3. Parameters of regression polynomial

Heat capacity type	No. data points total	No. data points used	s_w	s/R	s_r %	s_b/R	+/-
C_p	10	10	1.108	9.37-2	0.89	1.35-3	0

Temp. range K	A_1	A_2	Level of uncertainty
233.1-323.1	7.82043	9.69064-1	IV

TABLE 2.28.4. Recommended values of heat capacities

Temp. (K)	230	240	250	260	270	273.15	280
c_p ($J K^{-1} g^{-1}$)	0.492	0.496	0.501	0.506	0.511	0.512	0.515
C_p ($J K^{-1} mol^{-1}$)	83.55	84.36	85.17	85.97	86.78	87.03	87.58
Temp. (K)	**290**	**298.15**	**300**	**310**	**320**		
c_p ($J K^{-1} g^{-1}$)	0.520	0.524	0.525	0.530	0.534		
C_p ($J K^{-1} mol^{-1}$)	88.39	89.05	89.19	90.00	90.81		

Name: Nickel carbonyl
Formula: C_4NiO_4

CAS–RN: 13463–39–3
Group No.: 2–029
Molar Mass: 170.73

TABLE 2.29.1. Experimental heat capacities

Reference	Temp. range K	No. pnts	Error %	Purity %	Purity method	Type capacity	Calorimeter Type	Calorimeter Reference
55SPI/STA	255.0–300.0	10S	0.50	99.99	melpt	C_p	BSAO	49STA/GUP

TABLE 2.29.3. Parameters of regression polynomial

Heat capacity type	No. data points total	No. data points used	s_w	s/R	s_r %	s_b/R	+/–
C_p	10	10	0.020	2.42–3	0.01	1.14–6	0

Temp. range K	A_1	A_2	A_3	Level of uncertainty
255.0–300.0	3.03711+1	–5.36978	1.15206	IV

TABLE 2.29.4. Recommended values of heat capacities

Temp. (K)	260	270	273.15	280	290	298.15	300
c_p ($J K^{-1} g^{-1}$)	1.178	1.182	1.183	1.187	1.193	1.198	1.199
C_p ($J K^{-1} mol^{-1}$)	201.2	201.8	202.0	202.6	203.6	204.6	204.8

Name: Hydrochlorid acid–*d*
Formula: ClD

CAS–RN: 7698–05–7
Group No.: 2–030
Molar Mass: 37.47

TABLE 2.30.1. Experimental heat capacities

Reference	Temp. range K	No. pnts	Error %	Purity %	Purity method	Type capacity	Calorimeter Type	Calorimeter Reference
47CLU/WOL	162.1–176.2	4	nosp	not specified		C_p	BSIO	36CLU/GOL
76CHI/INA1	160.7–162.9	2	0.20	99.976	melpt	C_p	BSAO	74ATA/CHI

TABLE 2.30.2. Correlated heat capacities

Reference	Temp. range K	No. pnts used	$\sigma_r C$ %	d_w	d/R	d_r %	d_b/R	+/–
				Selected data				
47CLU/WOL	162.1–176.2	4	0.80#	0.983	5.86–2	0.79	4.46–2	2
76CHI/INA1	160.7–162.9	2	0.20	0.759	1.11–2	0.15	–5.42–3	0

TABLE 2.30.3. Parameters of regression polynomial

Heat capacity type	No. data points total	used	s_w	s/R	s_r %	s_b/R	+/–
C_p	6	6	1.120	5.91–2	0.79	2.79–2	2

Temp. range K	A_1	A_2	Level of uncertainty
160.7–176.2	6.46866	5.45731–1	IV

TABLE 2.30.4. Recommended values of heat capacities

Temp. (K)	160	170	180
c_p (J K^{-1}g^{-1})	1.629	1.641	1.653
C_p (J K^{-1}mol^{-1})	61.04	61.50	61.95

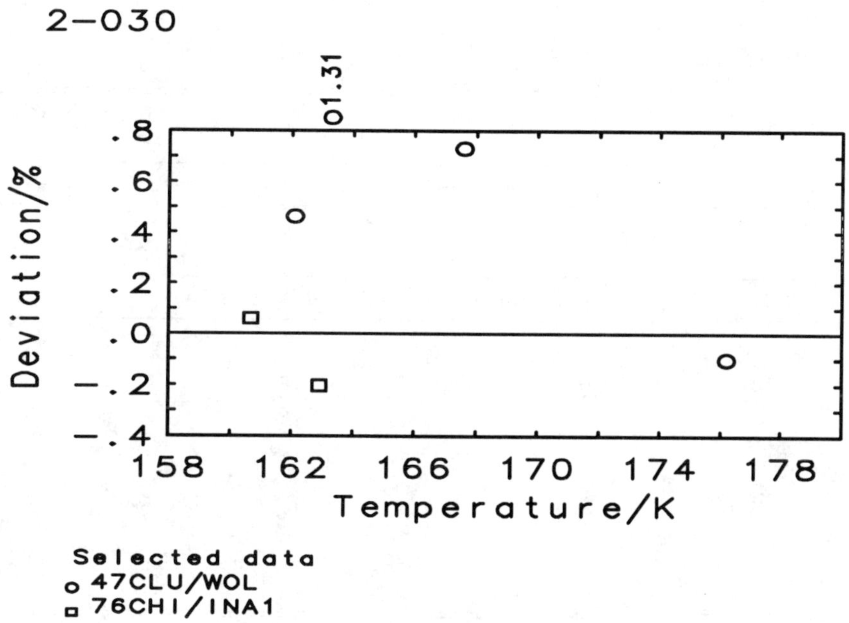

Name: Perchloryl fluoride ((ClO_3)F)
Formula: $ClFO_3$

CAS–RN: 7616–94–6
Group No.: 2–031
Molar Mass: 102.45

TABLE 2.31.1. Experimental heat capacities

Reference	Temp. range K	No. pnts	Error %	Purity %	Purity method	Type capacity	Calorimeter Type	Calorimeter Reference
58KOE/GIA	130.5–223.3	15	nosp	99.999	melpt	C_{sat}	BSIO	28GIA/WIE1

TABLE 2.31.3. Parameters of regression polynomial

Heat capacity type	No. data points total	No. data points used	s_w	s/R	s_r %	s_b/R	+/–
C_{sat}	15	15	0.426	9.21–3	0.09	1.14–5	–3

Temp. range K	A_1	A_2	A_3	Level of uncertainty
130.5–223.3	1.15736+1	–1.83073	8.10623–1	II

TABLE 2.31.4. Recommended values of heat capacities

Temp. (K)	130	140	150	160	170	180	190
c_{sat} (J K^{-1}g^{-1})	0.857	0.860	0.864	0.870	0.877	0.885	0.894
C_{sat} (J K^{-1}mol^{-1})	87.83	88.13	88.56	89.13	89.83	90.67	91.64
Temp. (K)	**200**	**210**	**220**				
c_{sat} (J K^{-1}g^{-1})	0.905	0.917	0.931				
C_{sat} (J K^{-1}mol^{-1})	92.75	93.99	95.36				

TABLE 2.31.5. Parameters of quasi-polynomial equation

Heat capacity type	No. data points total	No. data points used	s_w	s/R	s_r %	s_b/R	+/–
C_{sat}	15	15	0.734	1.59–2	0.15	3.64–5	1

Temp. range K	T_c K	A_1	A_2	A_3	A_4	Level of uncertainty
130.5–223.3	368.00	2.29888	1.97504	8.25697	6.68955–1	II

Name: Chlorine fluoride (ClF_3)
Formula: ClF_3

CAS–RN: 7790-91-2
Group No.: 2–032
Molar Mass: 92.45

TABLE 2.32.1. Experimental heat capacities

Reference	Temp. range K	No. pnts	Error %	Purity %	Purity method	Type capacity	Calorimeter Type	Calorimeter Reference
51GRI/BER	196.8–278.3	12	0.20	99.96	melpt	C_p	BSAO	43RUE/HUF

TABLE 2.32.3. Parameters of regression polynomial

Heat capacity type	No. data points total	used	s_w	s/R	s_r %	s_b/R	+/−
C_p	12	12	0.268	7.30–3	0.05	5.80–6	1

Temp. range K	A_1	A_2	A_3	Level of uncertainty
196.8–278.3	1.41865+1	−1.22917	4.30487−1	II

TABLE 2.32.4. Recommended values of heat capacities

Temp. (K)	200	210	220	230	240	250	260
c_p (J K^{-1}g^{-1})	1.210	1.214	1.220	1.226	1.234	1.242	1.250
C_p (J K^{-1}mol^{-1})	111.8	112.3	112.8	113.4	114.0	114.8	115.6
Temp. (K)	270	273.15	280				
c_p (J K^{-1}g^{-1})	1.260	1.263	1.270				
C_p (J K^{-1}mol^{-1})	116.5	116.7	117.4				

Name: Hydrochlorid acid
Formula: ClH

CAS–RN: 7647–01–0
Group No.: 2–033
Molar Mass: 36.46

TABLE 2.33.1. Experimental heat capacities

Reference	Temp. range K	No. pnts	Error %	Purity %	method	Type capacity	Calorimeter Type	Reference
24EUC/KAR	164.3–188.2	9	3.00	not specified		C_{sat}	BSIO	24EUC/KAR
28GIA/WIE1	163.7–185.2	5	nosp	not specified		C_p	BSIO	28GIA/WIE1
29CLU1	163.0–173.4	6	0.70	not specified		C_p	BSIO	29CLU1
76CHI/INA1	161.8–164.1	2	0.20	99.98	melpt	C_p	BSAO	74ATA/CHI

TABLE 2.33.2. Correlated heat capacities

Reference	Temp. range K	No. pnts used	$\sigma_r C$ %	d_w	d/R	d_r %	d_b/R	+/−
				Selected data				
28GIA/WIE1	163.7–185.2	5	0.50#	0.069	2.42–3	0.03	−1.17–3	−3
76CHI/INA1	161.8–164.1	2	0.20	0.065	9.13–4	0.01	4.71–4	1
				Rejected data				
24EUC/KAR	(5.27–1, 6.97, 5.25–1, 7)			29CLU1	(2.43–1, 3.35, 2.42–1, 6)			

TABLE 2.33.3. Parameters of regression polynomial

Heat capacity type	No. data points total	used	s_w	s/R	s_r %	s_b/R	+/−
C_p	22	7	0.080	2.48–3	0.04	−7.04–4	−2

Temp. range K	A_1	A_2	Level of uncertainty
161.8–185.2	6.33282	4.02127−1	III

TABLE 2.33.4. Recommended values of heat capacities

Temp. (K)	160	165	170	175	180	185
c_p (J $K^{-1}g^{-1}$)	1.591	1.595	1.600	1.605	1.609	1.614
C_p (J $K^{-1}mol^{-1}$)	58.00	58.17	58.34	58.51	58.67	58.84

TABLE 2.33.5. Parameters of quasi-polynomial equation

Heat capacity type	No. data points total	used	s_w	s/R	s_r %	s_b/R	+/−
C_p	22	7	0.078	1.27–3	0.02	−6.34–5	1

Temp. range K	T_c K	A_1	A_2	A_3	A_4	Level of uncertainty
161.8–185.2	324.70	5.96279–2	3.15811–1	6.39398	2.81457–3	III

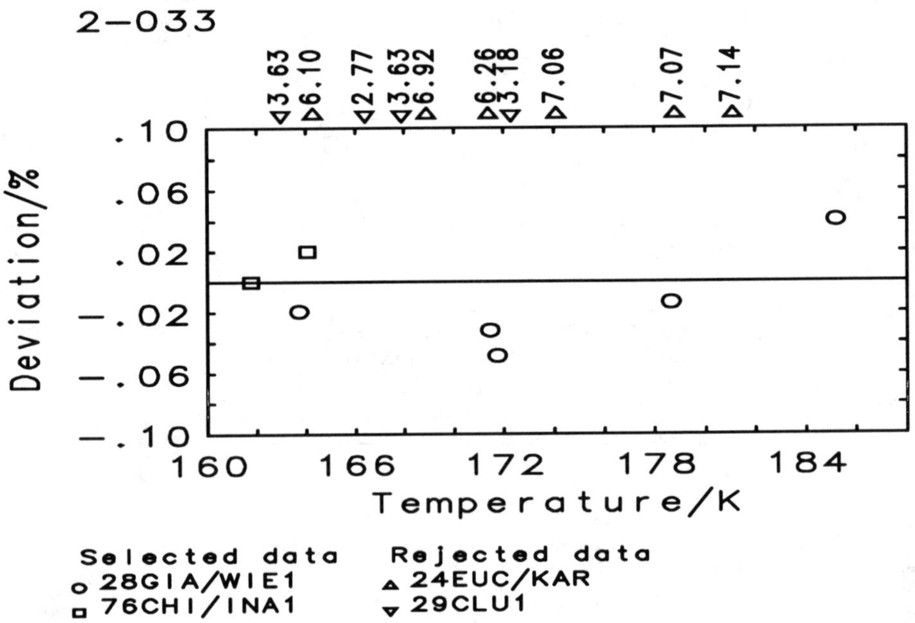

Name: Perchloric acid
Formula: $ClHO_4$

CAS–RN: 7601–90–3
Group No.: 2–034
Molar Mass: 100.46

TABLE 2.34.1. Experimental heat capacities

Reference	Temp. range K	No. pnts	Error %	Purity %	Purity method	Type capacity	Calorimeter Type	Calorimeter Reference
64TRO/WES1	173.3–295.1	20	0.10	97.5	melpt	C_{sat}	BSAO	68WES/FUR

TABLE 2.34.3. Parameters of regression polynomial

Heat capacity type	No. data points total	used	s_w	s/R	s_r %	s_b/R	+/–
C_{sat}	20	20	1.544	2.10–2	0.15	5.58–5	–4

Temp. range K	A_1	A_2	A_3	Level of uncertainty
173.3–295.1	1.31461+1	–7.18856–1	3.97487–1	II

TABLE 2.34.4. Recommended values of heat capacities

Temp. (K)	180	190	200	210	220	230	240
c_{sat} ($J\ K^{-1}g^{-1}$)	1.088	1.094	1.101	1.108	1.116	1.125	1.135
C_{sat} ($J\ K^{-1}mol^{-1}$)	109.3	109.9	110.6	111.3	112.1	113.0	114.0
Temp. (K)	250	260	270	273.15	280	290	298.15
c_{sat} ($J\ K^{-1}g^{-1}$)	1.145	1.156	1.167	1.171	1.179	1.192	1.203
C_{sat} ($J\ K^{-1}mol^{-1}$)	115.0	116.1	117.3	117.6	118.5	119.8	120.9
Temp. (K)	300						
c_{sat} ($J\ K^{-1}g^{-1}$)	1.206						
C_{sat} ($J\ K^{-1}mol^{-1}$)	121.1						

Name: Iodine chloride
Formula: ClI

CAS–RN: 7790–99–0
Group No.: 2–035
Molar Mass: 162.36

TABLE 2.35.1. Experimental heat capacities

Reference		Temp. range K	No. pnts	Error %	Purity %	method	Type capacity	Calorimeter Type	Reference
*92STO	N	318.8	1	nosp	not specified		C_{avg}	DSIO	*92STO
65CAL/GIA		303.7–317.8	3	nosp	99.99	melpt	C_p	BSIO	37GIA/EGA

*92STO average value in temperature range 288–350 K

TABLE 2.35.2. Correlated heat capacities

Reference	Temp. range K	No. pnts used	$\sigma_r C$ %	d_w	d/R	d_r %	d_b/R	+/–
				Selected data				
65CAL/GIA	303.7–317.8	3	0.30#	0.443	1.65–2	0.13	4.42–5	–1

TABLE 2.35.3. Parameters of regression polynomial

Heat capacity type	No. data points total	used	s_w	s/R	s_r %	s_b/R	+/–
C_p	4	3	0.768	2.86–2	0.23	4.42–5	–1

Temp. range K	A_1	A_2	Level of uncertainty
303.7–317.8	1.26430+1	–7.79397–2	III

TABLE 2.35.4. Recommended values of heat capacities

Temp. (K)	305	310	315
c_p (J $K^{-1}g^{-1}$)	0.635	0.635	0.635
C_p (J $K^{-1}mol^{-1}$)	103.1	103.1	103.1

Name: Sodium salt chloric acid
Formula: $ClNaO_3$

CAS–RN: 7775–09–9
Group No.: 2–036
Molar Mass: 106.44

TABLE 2.36.1. Experimental heat capacities

Reference		Temp. range K	No. pnts	Error %	Purity %	Purity method	Type capacity	Calorimeter Type	Calorimeter Reference
09GOO/KAL	N	528.2–572.2	2	nosp	not specified		C_{avg}	DSIO	09GOO/KAL

09GOO/KAL constant value calculated from temperature dependence of enthalpy by the authors

TABLE 2.36.3. Parameters of regression polynomial

Heat capacity type	No. data points total	No. data points used	s_w	s/R	s_r %	s_b/R	+/–
C	2	2	0.000	0.00	0.00	0.00	0

Temp. range K	A_1	Level of uncertainty
528.2–572.2	1.74079+1	V

TABLE 2.36.4. Recommended values of heat capacities

Temp. (K)	530	570
c (J $K^{-1}g^{-1}$)	1.36	1.36
C (J $K^{-1}mol^{-1}$)	145	145

Name: Thionyl chloride
Formula: Cl_2OS

CAS–RN: 7719–09–7
Group No.: 2–037
Molar Mass: 118.97

TABLE 2.37.1. Experimental heat capacities

Reference		Temp. K	Capac. J/(K.g)	Error %	Purity %	Purity Method	Type capacity	Calorimeter Type	Calorimeter Reference
*82OGI	N	311.65	1.015	nosp	not specified		C_{avg}	DSIO	*79BER

*82OGI average value in temperature range 290–333 K

Name: Sulfuryl chloride
Formula: Cl_2O_2S

CAS–RN: 7791–25–5
Group No.: 2–038
Molar Mass: 134.97

TABLE 2.38.1. Experimental heat capacities

Reference		Temp. K	Capac. J/(K.g)	Error %	Purity %	Purity Method	Type capacity	Calorimeter Type	Calorimeter Reference
*82OGI	N	312.15	0.9749	nosp	not specified		C_{avg}	DSIO	*79BER

*82OGI average value in temperature range 288–336 K

Name: Disulfuryl chloride
Formula: $Cl_2O_5S_2$

CAS–RN: 7791–27–7
Group No.: 2–039
Molar Mass: 215.03

TABLE 2.39.1. Experimental heat capacities

Reference		Temp. K	Capac. J/(K.g)	Error %	Purity %	Purity Method	Type capacity	Calorimeter Type	Calorimeter Reference
*82OGI	N	345.65	1.079	nosp	not specified		C_{avg}	DSIO	*79BER

*82OGI average value in temperature range 288–403 K

Name: Sulfur chloride (S_2Cl_2)
Formula: Cl_2S_2

CAS–RN: 10025–67–9
Group No.: 2–040
Molar Mass: 135.04

TABLE 2.40.1. Experimental heat capacities

Reference		Temp. K	Capac. J/(K.g)	Error %	Purity %	Purity Method	Type capacity	Calorimeter Type	Calorimeter Reference
*81OGI	N	314.15	0.9205	nosp	not specified		C_{avg}	DSIO	*79BER

*81OGI average value in temperature range 285–343 K

Name: Tin chloride ($SnCl_2$)
Formula: Cl_2Sn

CAS–RN: 7772–99–8
Group No.: 2–041
Molar Mass: 189.62

TABLE 2.41.1. Experimental heat capacities

Reference	Temp. range K	No. pnts	Error %	Purity %	Purity method	Type capacity	Calorimeter Type	Calorimeter Reference
91GAR/PRE2	520.0–800.0	eqn	nosp	not specified		C_p	BDHT	69PER/COM

TABLE 2.41.3. Parameters of regression polynomial

Heat capacity type	No. data points total	used	s_w	s/R	s_r %	s_b/R	+/–
C_p	15	15	0.000	5.29–7	0.00	2.54–7	0

Temp. range K	A_1	A_2	Level of uncertainty
520.0–800.0	8.86042	4.87220–1	V

TABLE 2.41.4. Recommended values of heat capacities

Temp. (K)	520	560	600	640	680	720	760
c_p (J $K^{-1}g^{-1}$)	0.50	0.51	0.52	0.53	0.53	0.54	0.55
C_p (J $K^{-1}mol^{-1}$)	94.7	96.4	98.0	99.6	101	103	104

Temp. (K)	800
c_p (J $K^{-1}g^{-1}$)	0.56
C_p (J $K^{-1}mol^{-1}$)	106

Name: Trichlorosilane
Formula: Cl_3HSi

CAS–RN: 10025–78–2
Group No.: 2–042
Molar Mass: 135.45

TABLE 2.42.1. Experimental heat capacities

Reference		Temp. K	Capac. J/(K.g)	Error %	Purity %	Purity Method	Type capacity	Calorimeter Type	Calorimeter Reference
66TEC	N	315.65	0.9623	nosp	not specified		C_{avg}	not specified	

66TEC average value in temperature range 298–333K

Name: Phosphoryl chloride
Formula: Cl_3OP

CAS–RN: 10025–87–3
Group No.: 2–043
Molar Mass: 153.33

TABLE 2.43.1. Experimental heat capacities

Reference	Temp. range K	No. pnts	Error %	Purity %	Purity method	Type capacity	Calorimeter Type	Calorimeter Reference
60OTT/GIA	280.9–300.2	6	nosp	99.98	melpt	C_p	BSIO	28GIA/WIE1

TABLE 2.43.3. Parameters of regression polynomial

Heat capacity type	No. data points total	used	s_w	s/R	s_r %	s_b/R	+/–
C_p	6	6	0.484	1.61–2	0.10	2.07–5	–1

Temp. range K	A_1	A_2	Level of uncertainty
280.9–300.2	1.51391+1	5.18557–1	II

TABLE 2.43.4. Recommended values of heat capacities

Temp. (K)	280	290	298.15	300
c_p (J K^{-1}g^{-1})	0.900	0.902	0.905	0.905
C_p (J K^{-1}mol^{-1})	137.9	138.4	138.7	138.8

Name: Phosphorous trichloride
Formula: Cl_3P

CAS–RN: 7719–12–2
Group No.: 2–044
Molar Mass: 137.33

TABLE 2.44.1. Experimental heat capacities

Reference	Temp. range K	No. pnts	Error %	Purity %	Purity method	Type capacity	Calorimeter Type	Calorimeter Reference
87DEV/GUS	186.2–272.5	25	0.20	99.999	chrom	C_p	BSAO	86DEV/GUS

TABLE 2.44.3. Parameters of regression polynomial

Heat capacity type	No. data points total	No. data points used	s_w	s/R	s_r %	s_b/R	+/–
C_p	25	25	0.386	1.14–2	0.08	1.64–5	–2

Temp. range K	A_1	A_2	A_3	A_4	Level of uncertainty
186.2–272.5	2.22974+1	–1.00511+1	4.27091	–5.81474–1	II

TABLE 2.44.4. Recommended values of heat capacities

Temp. (K)	170	180	190	200	210	220	230
c_p (J K^{-1}g^{-1})	0.890	0.887	0.886	0.886	0.886	0.888	0.890
C_p (J K^{-1}mol^{-1})	122.2	121.8	121.6	121.6	121.7	121.9	122.2
Temp. (K)	240	250	260	270			
c_p (J K^{-1}g^{-1})	0.892	0.895	0.897	0.899			
C_p (J K^{-1}mol^{-1})	122.5	122.9	123.2	123.5			

TABLE 2.44.5. Parameters of quasi-polynomial equation

Heat capacity type	No. data points total	No. data points used	s_w	s/R	s_r %	s_b/R	+/–
C_p	25	25	0.517	1.53–2	0.10	2.84–5	–1

Temp. range K	T_c K	A_1	A_2	A_3	A_4	Level of uncertainty
186.2–272.5	563.00	1.04426+1	4.68709	9.88347	5.81639	II

Name: Tetrachlorogermane
Formula: Cl_4Ge

CAS–RN: 10038–98–9
Group No.: 2–045
Molar Mass: 214.42

TABLE 2.45.1. Experimental heat capacities

Reference	Temp. range K	No. pnts	Error %	Purity %	method	Type capacity	Calorimeter Type	Reference
86DEV/GUS	221.7–273.1	7S	0.20	99.991	melpt	C_{sat}	BSAO	86DEV/GUS

TABLE 2.45.3. Parameters of regression polynomial

Heat capacity type	No. data points total	used	s_w	s/R	s_r %	s_b/R	+/–
C_{sat}	7	7	0.082	2.83–3	0.02	2.72–7	–1

Temp. range K	A_1	A_2	A_3	Level of uncertainty
221.7–273.1	1.46572+1	1.95136	–3.52805–1	II

TABLE 2.45.4. Recommended values of heat capacities

Temp. (K)	230	240	250	260	270
c_{sat} (J $K^{-1}g^{-1}$)	0.670	0.671	0.672	0.673	0.673
C_{sat} (J $K^{-1}mol^{-1}$)	143.7	143.9	144.1	144.2	144.3

Name: Tetrachlorosilane
Formula: Cl_4Si

CAS–RN: 10026–04–7
Group No.: 2–046
Molar Mass: 169.90

TABLE 2.46.1. Experimental heat capacities

Reference		Temp. range K	No. pnts	Error %	Purity %	method	Type capacity	Calorimeter Type	Reference
08KAH/KOE	N	303.1	1	nosp	not specified		C_{avg}	DSIO	01KAH
22LAT		208.8–294.3	2	2.00	not specified		C_p	BSIO	20GIB/LAT
37VOL		298.1	1	nosp	not specified		C_p	BSIO	37VOL
66SAV		293.0–323.0	eqn	1.20	not specified		C_p	BSIO	66SAV
74WIL/ZET		273.1–323.1	6S	nosp	not specified		C_p	BSAO	64ARN1
87GUS/GIB		205.4–272.1	16	0.20	99.995	chrom	C_p	BSAO	86DEV/GUS

08KAH/KOE average value in temperature range 293–313 K

TABLE 2.46.2. Correlated heat capacities

Reference	Temp. range K	No. pnts used	$\sigma_r C$ %	d_w	d/R	d_r %	d_b/R	+/–
Selected data								
74WIL/ZET	273.1–323.1	6	0.80#	2.699	3.61–1	2.16	2.46–1	4
87GUS/GIB	205.4–272.1	16	0.40#	1.306	8.27–2	0.52	–2.07–2	0

Rejected data

Reference		Reference	
08KAH/KOE	(3.45–1, 2.12, –3.45–1, –1)	22LAT	(1.07, 6.19, 1.06, 2)
37VOL	(6.59–1, 3.84, 6.59–1, 1)	66SAV	(1.60, 8.48, 1.46, 6)

TABLE 2.46.3. Parameters of regression polynomial

Heat capacity type	No. data points total	used	s_w	s/R	s_r %	s_b/R	+/–
C_p	32	22	1.933	2.16–1	1.30	5.20–2	4

Temp. range K	A_1	A_2	A_3	Level of uncertainty
205.4–323.1	2.48455+1	–7.81485	1.68321	V

TABLE 2.46.4. Recommended values of heat capacities

Temp. (K)	210	220	230	240	250	260	270
c_p (J K^{-1}g^{-1})	0.78	0.77	0.77	0.77	0.77	0.78	0.78
C_p (J K^{-1}mol^{-1})	132	131	131	131	132	132	133
Temp. (K)	273.15	280	290	298.15	300	310	320
c_p (J K^{-1}g^{-1})	0.79	0.79	0.80	0.81	0.81	0.82	0.84
C_p (J K^{-1}mol^{-1})	134	134	136	137	138	140	142

TABLE 2.46.5. Parameters of quasi-polynomial equation

Heat capacity type	No. data points total	used	s_w	s/R	s_r %	s_b/R	+/–
C_p	32	22	2.729	2.21–1	1.36	2.91–2	4

Temp. range K	T_c K	A_1	A_2	A_3	A_4	Level of uncertainty
205.4–323.1	506.70	9.16339	4.65359	1.07622+1	4.51090	V

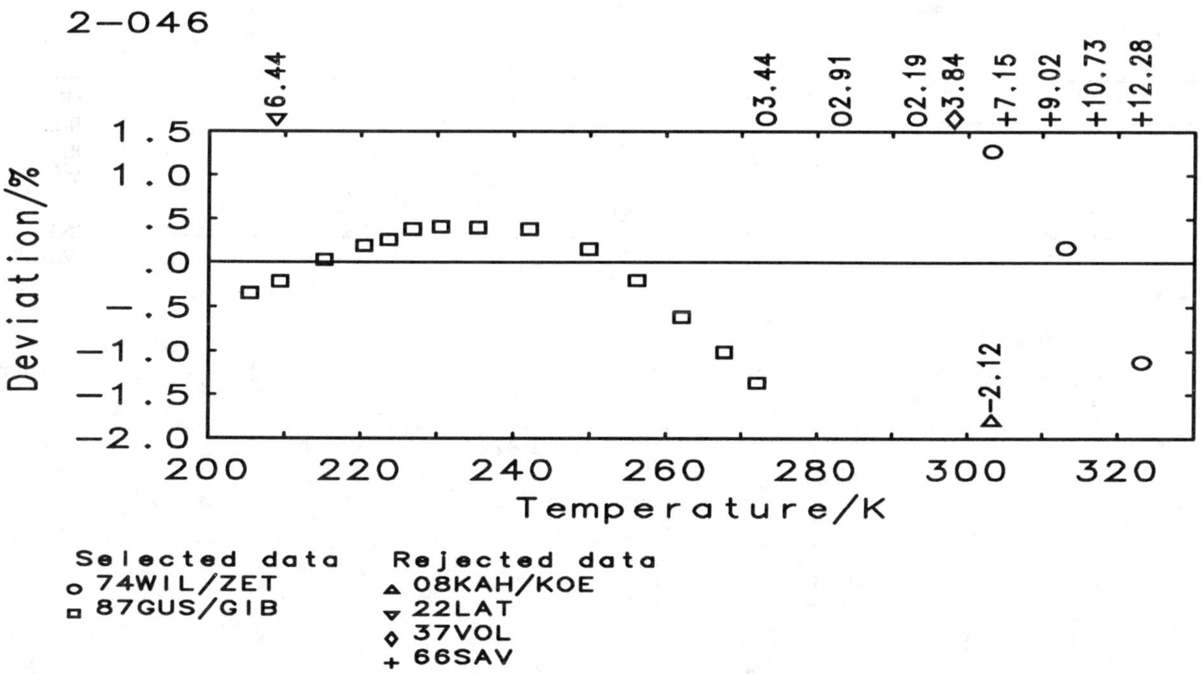

Name: Tetrachlorostannane
Formula: Cl_4Sn

CAS–RN: 7646–78–8
Group No.: 2–047
Molar Mass: 260.52

TABLE 2.47.1. Experimental heat capacities

Reference		Temp. range K	No. pnts	Error %	Purity %	Purity method	Type capacity	Calorimeter Type	Calorimeter Reference
22LAT		266.1–294.0	2	2.00	not specified		C_p	BSIO	20GIB/LAT
36KUR/VOS	N	311.6	1	nosp	not specified		C_{avg}	DSIO	36KUR/VOS
37VOL		298.1	1	nosp	not specified		C_p	BSIO	37VOL
74WIL/ZET		273.1–323.1	6	nosp	not specified		C_p	BSAO	64ARN1
84BER/STE		243.3–327.3	18	nosp	99.95	mclpt	C_p	BSAO	79AMI/LEB
87GUS/GIB		242.4–272.1	13	0.20	99.995	chrom	C_p	BSAO	86DEV/GUS

36KUR/VOS average value in temperature range 290–333 K

TABLE 2.47.2. Correlated heat capacities

Reference	Temp. range K	No. pnts used	$\sigma_r C$ %	d_w	d/R	d_r %	d_b/R	+/–
				Selected data				
37VOL	298.1	1	0.80#	0.495	7.54–2	0.40	7.54–2	1
74WIL/ZET	273.1–323.1	6	0.80#	0.837	1.28–1	0.67	1.25–1	6
84BER/STE	243.3–327.3	18	0.30#	0.425	2.42–2	0.13	–1.77–2	–12
87GUS/GIB	242.4–272.1	13	0.20	0.294	1.13–2	0.06	7.05–3	6
				Rejected data				
22LAT	(5.72–1, 2.89, 3.71–1, 0)			36KUR/VOS	(5.40–1, 2.78, 5.40–1, 1)			

TABLE 2.47.3. Parameters of regression polynomial

Heat capacity type	No. data points total	No. data points used	s_w	s/R	s_r %	s_b/R	+/–
C_p	41	38	0.502	5.76–2	0.30	1.57–2	1

Temp. range K	A_1	A_2	A_3	Level of uncertainty
242.4–327.3	1.68997+1	1.94005	–4.20380–1	III

TABLE 2.47.4. Recommended values of heat capacities

Temp. (K)	240	250	260	270	273.15	280	290
c_p ($J K^{-1} g^{-1}$)	0.611	0.610	0.610	0.609	0.608	0.608	0.606
C_p ($J K^{-1} mol^{-1}$)	159.1	159.0	158.8	158.6	158.5	158.3	157.9

Temp. (K)	298.15	300	310	320	330
c_p ($J K^{-1} g^{-1}$)	0.605	0.604	0.602	0.600	0.598
C_p ($J K^{-1} mol^{-1}$)	157.5	157.4	156.9	156.3	155.7

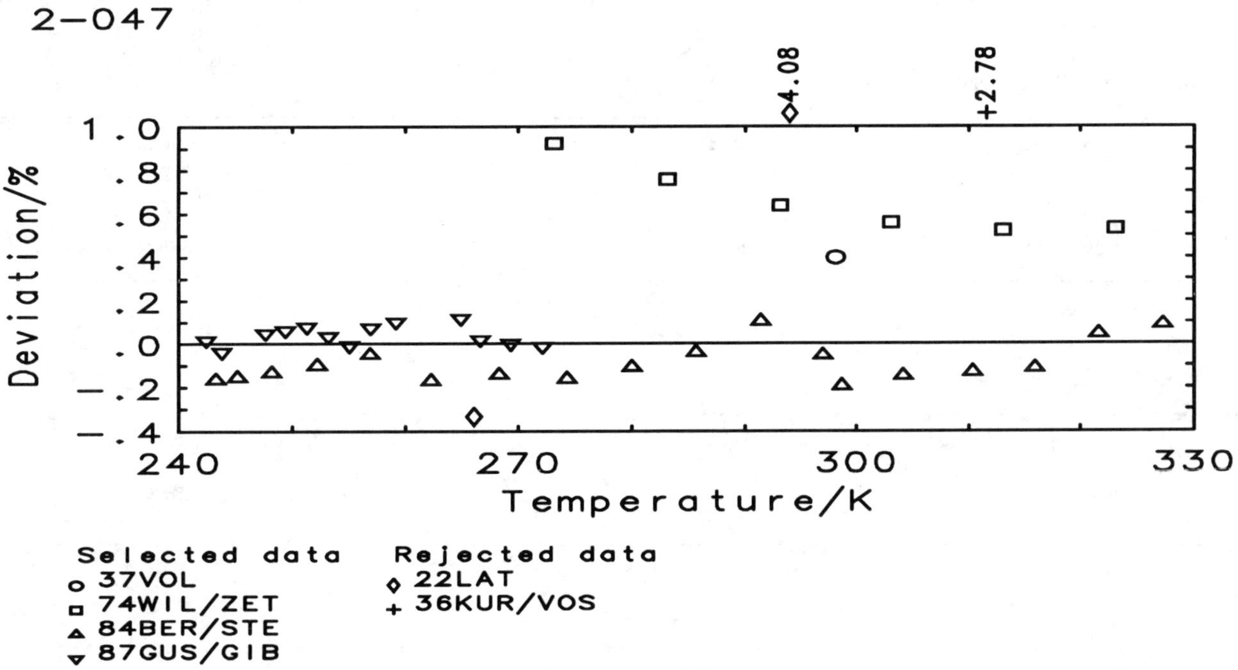

Name: Tellurium chloride ($TeCl_4$)
Formula: Cl_4Te

CAS–RN: 10026–07–0
Group No.: 2–048
Molar Mass: 269.41

TABLE 2.48.1. Experimental heat capacities

Reference		Temp. range K	No. pnts	Error %	Purity %	Purity method	Type capacity	Calorimeter Type	Calorimeter Reference
38FRE/HIL2	N	497.3–538.2	2	1.00	not specified		C_{avg}	DSIO	38FRE/HIL1

38FRE/HIL2 constant value calculated from temperature dependence of enthalpy by the authors

TABLE 2.48.3. Parameters of regression polynomial

Heat capacity type	No. data points total	No. data points used	s_w	s/R	s_r %	s_b/R	+/–
C	2	2	0.000	0.00	0.00	0.00	0

Temp. range K	A_1	Level of uncertainty
497.3–538.2	2.76769+1	V

TABLE 2.48.4. Recommended values of heat capacities

Temp. (K)	500	540
c ($J K^{-1} g^{-1}$)	0.85	0.85
C ($J K^{-1} mol^{-1}$)	230	230

Name: Titanium chloride ($TiCl_4$)
Formula: Cl_4Ti

CAS–RN: 7550–45–0
Group No.: 2–049
Molar Mass: 189.69

TABLE 2.49.1. Experimental heat capacities

Reference	Temp. range K	No. pnts	Error %	Purity %	Purity method	Type capacity	Calorimeter Type	Calorimeter Reference
22LAT	251.6–294.3	2	2.00	not specified		C_p	BSIO	20GIB/LAT
66SAV	293.5–331.4	7	1.20	not specified		C_p	BSIO	66SAV

TABLE 2.49.2. Correlated heat capacities

Reference	Temp. range K	No. pnts used	$\sigma_r C$ %	d_w	d/R	d_r %	d_b/R	+/–
				Selected data				
66SAV	293.5–331.4	7	1.20	0.195	4.49–2	0.23	2.65–4	1
				Rejected data				
22LAT	(4.31–1, 2.34, 4.31–1, 1)							

TABLE 2.49.3. Parameters of regression polynomial

Heat capacity type	No. data points total	No. data points used	s_w	s/R	s_r %	s_b/R	+/–
C_p	9	7	0.230	5.31–2	0.28	2.65–4	1

Temp. range K	A_1	A_2	Level of uncertainty
293.5–331.4	–4.63803	7.68756	V

TABLE 2.49.4. Recommended values of heat capacities

Temp. (K)	290	298.15	300	310	320	330
c_p ($J\,K^{-1}g^{-1}$)	0.77	0.80	0.81	0.84	0.87	0.91
C_p ($J\,K^{-1}mol^{-1}$)	147	152	153	160	166	172

Name: Cesium fluoride ($Cs(HF_2)$)
Formula: CsF_2H

CAS–RN: 12280–52–3
Group No.: 2–050
Molar Mass: 171.91

TABLE 2.50.1. Experimental heat capacities

Reference	Temp. range K	No. pnts	Error %	Purity %	Purity method	Type capacity	Calorimeter Type	Calorimeter Reference
78WES/LAN	460.9–528.4	10	nosp	99.92	melpt	C_p	BSAO	68WES/WES

TABLE 2.50.3. Parameters of regression polynomial

Heat capacity type	No. data points total	used	s_w	s/R	s_r %	s_b/R	+/–
C_p	10	10	0.427	2.40–2	0.17	6.54–5	0

Temp. range K	A_1	A_2	Level of uncertainty
460.9–528.4	1.16474+1	5.03795–1	III

TABLE 2.50.4. Recommended values of heat capacities

Temp. (K)	460	470	480	490	500	510	520
c_p (J $K^{-1}g^{-1}$)	0.675	0.678	0.680	0.683	0.685	0.688	0.690
C_p (J $K^{-1}mol^{-1}$)	116.1	116.5	116.9	117.4	117.8	118.2	118.6

Temp. (K)	530
c_p (J $K^{-1}g^{-1}$)	0.692
C_p (J $K^{-1}mol^{-1}$)	119.0

Name: Deuterium molecule with hydrogen
Formula: DH

CAS–RN: 13983–20–5
Group No.: 2–051
Molar Mass: 3.02

TABLE 2.51.1. Experimental heat capacities

Reference		Temp. range K	No. pnts	Error %	Purity %	method	Type capacity	Calorimeter Type	Reference
48WOO/SCO	N	16.6–22.0	7S	nosp	not specified		C_{sat}	not specified	

48WOO/SCO unpublished data measured by F.G. Brickwedde and R.B. Scott

TABLE 2.51.3. Parameters of regression polynomial

Heat capacity type	No. data points total	used	s_w	s/R	s_r %	s_b/R	+/–
C_{sat}	7	7	0.398	5.32–3	0.20	1.03–5	1

Temp. range K	A_1	A_2	A_3	Level of uncertainty
16.6–22.0	–1.54338	2.78828+1	–3.16203+1	IV

TABLE 2.51.4. Recommended values of heat capacities

Temp. (K)	17	18	19	20	21	22
c_{sat} (J $K^{-1}g^{-1}$)	6.281	6.744	7.189	7.617	8.027	8.420
C_{sat} (J $K^{-1}mol^{-1}$)	18.98	20.38	21.72	23.02	24.26	25.45

Name: Hydroiodic acid–*d*
Formula: DI

CAS–RN: 14104–45–1
Group No.: 2–052
Molar Mass: 128.92

TABLE 2.52.1. Experimental heat capacities

Reference	Temp. range K	No. pnts	Error %	Purity %	Purity method	Type capacity	Calorimeter Type	Calorimeter Reference
47CLU/WOL	224.8–227.7	3	nosp	not specified		C_p	BSIO	36CLU/GOL
78INA/CHI	223.3	1	0.20	99.994	melpt	C_p	BSAO	74ATA/CHI

TABLE 2.52.2. Correlated heat capacities

Reference	Temp. range K	No. pnts used	$\sigma_r C$ %	d_w	d/R	d_r %	d_b/R	+/–
				Selected data				
47CLU/WOL	224.8–227.7	3	0.50#	2.331	9.09–2	1.16	3.54–2	1
78INA/CHI	223.3	1	0.40#	2.071	6.30–2	0.83	–6.30–2	–1

TABLE 2.52.3. Parameters of regression polynomial

Heat capacity type	No. data points total	No. data points used	s_w	s/R	s_r %	s_b/R	+/–
C_p	4	4	3.208	1.20–1	1.54	1.08–2	0

Temp. range K	A_1	A_2	Level of uncertainty
223.3–227.7	3.59995–1	3.26943	V

TABLE 2.52.4. Recommended values of heat capacities

Temp. (K)	223	225	227
c_p (J K^{-1}g^{-1})	0.49	0.50	0.50
C_p (J K^{-1}mol^{-1})	63.6	64.2	64.7

Name: Water–d_2
Formula: D_2O

CAS-RN: 7789–20–0
Group No.: 2–053
Molar Mass: 20.03

TABLE 2.53.1. Experimental heat capacities

Reference		Temp. range K	No. pnts	Error %	Purity %	Purity method	Type capacity	Calorimeter Type	Calorimeter Reference
35BAR/CLU	N	284.1	1	1.00	99.7	anal	C_{avg}	DSTO	35BAR/CLU
35BRO/BAR1	N	287.6	1	1.00	98.	anal	C_{avg}	DSAO	30BAR/MAA
35BRO/BAR2	N	288.1–318.6	3	1.00	98.	anal	C_{avg}	DSAO	30BAR/MAA
35JAC	N	276.9	1	nosp	97.7	anal	C_{sat}	BSIO	35JAC
36LON/KEM		279.4–295.2	4	0.20	99.997	melpt	C_p	BSIO	36GIA/STO
40COC/FER	N	285.1–325.8	31	nosp	99.2	anal	C_{sat}	BDHO	40COC/FER
51EUC/EIG	N	293.0–398.0	12S	nosp	96.	estim	C_{sat}	BSAO	51EUC/EIG
59BAK	N	304.0–473.1	37	nosp	99.77	anal	C_{sat}	BSIO	59BAK
70KRE		298.1	1	nosp	99.7	anal	C_p	BSAO	63BEN

35BAR/CLU average value in temperature range 277–291 K
35BRO/BAR1 average value in temperature range 277–298 K
35BRO/BAR2 average values in temperature ranges 277–299 K, 299–318 K and 299–338 K
35JAC value extrapolated to 100 % D2O
40COC/FER preliminary data published in 36FER/COC
51EUC/EIG purity was estimated from density by the authors
59BAK data in temperature range 373–473 K measured in an isoperibol calorimeter (but not described)

TABLE 2.53.2. Correlated heat capacities

Reference	Temp. range K	No. pnts used	$\sigma_r C$ %	d_w	d/R	d_r %	d_b/R	+/–
				Selected data				
40COC/FER	285.1–325.8	31	0.50#	0.276	1.40–2	0.14	1.99–3	–3
51EUC/EIG	293.0–398.0	11	0.30#	0.490	1.48–2	0.15	–9.00–3	–7
59BAK	304.0–473.1	37	0.70#	1.068	7.55–2	0.75	1.28–2	2
70KRE	298.1	1	0.30#	0.050	1.51–3	0.01	–1.51–3	–1
				Rejected data				
35BRO/BAR1	(2.03–1, 1.96, 2.03–1, 1)			35BRO/BAR2	(7.08–2, 0.69, 5.34–2, 1)			
36LON/KEM	(1.53–1, 1.54, –1.47–1, –2)							

TABLE 2.53.3. Parameters of regression polynomial

Heat capacity type	No. data points total	No. data points used	s_w	s/R	s_r %	s_b/R	+/–
C_p	91	80	0.788	5.37–2	0.53	5.42–3	–9
C_{sat}	91	80	0.788	5.37–2	0.53	5.48–3	–11

Temp. range K	A_1	A_2	A_3	A_4	Level of uncertainty
285.1–473.1	1.15134	8.71318	–2.74578	2.79911–1	IV
285.1–473.1	1.69083	8.21175	–2.59048	2.63876–1	IV

TABLE 2.53.4. Recommended values of heat capacities

Temp. (K)	280	290	298.15	300	310	320	330
c_p (J K^{-1}g^{-1})	4.220	4.216	4.210	4.208	4.199	4.188	4.177
C_p (J K^{-1}mol^{-1})	84.52	84.43	84.31	84.28	84.09	83.88	83.66
c_{sat} (J K^{-1}g^{-1})	4.221	4.216	4.210	4.208	4.199	4.188	4.177
C_{sat} (J K^{-1}mol^{-1})	84.53	84.43	84.31	84.28	84.09	83.88	83.66
Temp. (K)	340	350	360	370	380	390	400
c_p (J K^{-1}g^{-1})	4.167	4.157	4.149	4.143	4.140	4.141	4.146
C_p (J K^{-1}mol^{-1})	83.45	83.25	83.09	82.97	82.91	82.92	83.03
c_{sat} (J K^{-1}g^{-1})	4.167	4.157	4.148	4.142	4.138	4.138	4.142
C_{sat} (J K^{-1}mol^{-1})	83.45	83.25	83.08	82.95	82.88	82.88	82.96
Temp. (K)	410	420	430	440	450	460	470
c_p (J K^{-1}g^{-1})	4.156	4.172	4.194	4.224	4.262	4.308	4.363
C_p (J K^{-1}mol^{-1})	83.23	83.55	84.00	84.60	85.35	86.28	87.39
c_{sat} (J K^{-1}g^{-1})	4.151	4.166	4.186	4.213	4.248	4.291	4.342
C_{sat} (J K^{-1}mol^{-1})	83.14	83.43	83.84	84.38	85.08	85.93	86.96

TABLE 2.53.5. Parameters of quasi-polynomial equation

Heat capacity type	No. data points total	used	s_w	s/R	s_r %	s_b/R	+/−
C_p	91	14	0.722	4.02–2	0.40	−1.08–3	0

Temp. range K	T_c K	A_1	A_2	A_3	A_4	Level of uncertainty
373.0–463.1	644.65	1.94231+1	4.52475	3.97269	2.08442+1	IV

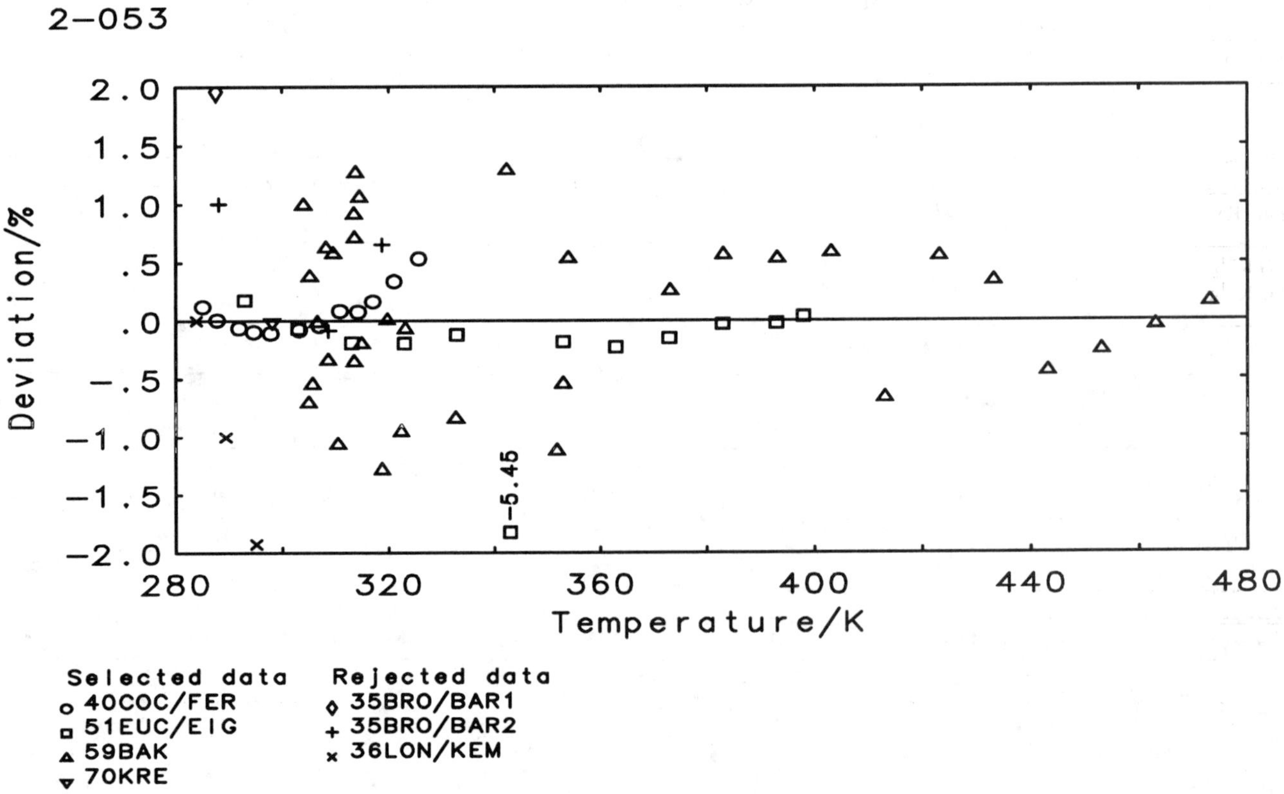

Name: Hydrogen–d_2 peroxide
Formula: D_2O_2

CAS–RN: 6909–54–2
Group No.: 2–054
Molar Mass: 36.03

TABLE 2.54.1. Experimental heat capacities

Reference	Temp. K	Capac. J/(K.g)	Error %	Purity %	Purity Method	Type capacity	Calorimeter Type	Calorimeter Reference
55GIG/MOR2	298.15	2.662	nosp	not specified		C_p	DSTO	55GIG/MOR1

Name: Hydrogen sulfide–d_2
Formula: D_2S

CAS–RN: 13536–94–2
Group No.: 2–055
Molar Mass: 36.09

TABLE 2.55.1. Experimental heat capacities

Reference		Temp. range K	No. pnts	Error %	Purity %	Purity method	Type capacity	Calorimeter Type	Calorimeter Reference
38KRU/CLU	N	188.8–202.5	9	nosp	not specified		C_p	BSIO	36CLU/GOL

38KRU/CLU sample contained 0.8 % HDS

TABLE 2.55.3. Parameters of regression polynomial

Heat capacity type	No. data points total	No. data points used	s_w	s/R	s_r %	s_b/R	+/–
C_p	9	9	0.688	4.11–2	0.48	3.08–4	–3

Temp. range K	A_1	A_2	Level of uncertainty
188.8–202.5	8.82456	–1.66007–1	IV

TABLE 2.55.4. Recommended values of heat capacities

Temp. (K)	190	195	200
c_p (J $K^{-1}g^{-1}$)	1.960	1.958	1.956
C_p (J $K^{-1}mol^{-1}$)	70.75	70.68	70.61

Name: Hydrogen selenide–d_2
Formula: D_2Se

CAS–RN: 13536–95–3
Group No.: 2–056
Molar Mass: 82.99

TABLE 2.56.1. Experimental heat capacities

Reference		Temp. range K	No. pnts	Error %	Purity %	Purity method	Type capacity	Calorimeter Type	Calorimeter Reference
38KRU/CLU	N	210.9–232.7	12	nosp	not specified		C_p	BSIO	36CLU/GOL

38KRU/CLU sample contained 0.8 % HDSe

TABLE 2.56.3. Parameters of regression polynomial

Heat capacity type	No. data points total	used	s_w	s/R	s_r %	s_b/R	+/–
C_p	12	12	0.748	4.48–2	0.52	3.92–4	2

Temp. range K	A_1	A_2	Level of uncertainty
210.9–232.7	8.90311	–1.52128–1	IV

TABLE 2.56.4. Recommended values of heat capacities

Temp. (K)	210	220	230
c_p (J K^{-1}g^{-1})	0.860	0.858	0.857
C_p (J K^{-1}mol^{-1})	71.37	71.24	71.12

Name: Ammonia–d_3
Formula: D_3N

CAS–RN: 13550–49–7
Group No.: 2–057
Molar Mass: 20.05

TABLE 2.57.1. Experimental heat capacities

Reference		Temp. range K	No. pnts	Error %	Purity %	method	Type capacity	Calorimeter Type	Reference
72MAN/TOL	N	202.4–216.7	7	1.00	99.99	anal	C_p	BSAO	68BAG/KUC

72MAN/TOL content of heavy isotope 98.2 %

TABLE 2.57.3. Parameters of regression polynomial

Heat capacity type	No. data points total	used	s_w	s/R	s_r %	s_b/R	+/–
C_p	7	7	0.194	1.90–2	0.19	5.24–5	–2

Temp. range K	A_1	A_2	Level of uncertainty
202.4–216.7	6.87015	1.38543	V

TABLE 2.57.4. Recommended values of heat capacities

Temp. (K)	200	205	210	215
c_p (J K^{-1}g^{-1})	4.00	4.03	4.06	4.08
C_p (J K^{-1}mol^{-1})	80.2	80.7	81.3	81.9

Name: Hydrofluoric acid
Formula: FH

CAS–RN: 7664–39–3
Group No.: 2–058
Molar Mass: 20.01

TABLE 2.58.1. Experimental heat capacities

Reference		Temp. range K	No. pnts	Error %	Purity %	Purity method	Type capacity	Calorimeter Type	Calorimeter Reference
33DAH/JUN	N	200.0–270.0	8S	nosp	not specified		C_{sat}	BSIO	24EUC/KAR
53HU/WHI1		197.9–289.1	13	nosp	99.8	anal	C_p	BSIO	50JOH/CLA

33DAH/JUN probably the best data for HF (information in 72HOR)

TABLE 2.58.2. Correlated heat capacities

Reference	Temp. range K	No. pnts used	$\sigma_r C$ %	d_w	d/R	d_r %	d_b/R	+/–
				Selected data				
53HU/WHI1	197.9–289.1	13	0.50#	0.292	8.15–3	0.15	2.26–5	1
				Rejected data				
33DAH/JUN	(2.12, 27.67, 2.09, 8)							

TABLE 2.58.3. Parameters of regression polynomial

Heat capacity type	No. data points total	No. data points used	s_w	s/R	s_r %	s_b/R	+/–
C_p	21	13	0.333	9.29–3	0.17	2.26–5	1

Temp. range K	A_1	A_2	A_3	Level of uncertainty
197.9–289.1	7.53087	–2.69185	7.59270–1	III

TABLE 2.58.4. Recommended values of heat capacities

Temp. (K)	200	210	220	230	240	250	260
c_p (J K^{-1}g^{-1})	2.155	2.172	2.196	2.226	2.262	2.305	2.354
C_p (J K^{-1}mol^{-1})	43.10	43.45	43.93	44.53	45.26	46.12	47.10

Temp. (K)	270	273.15	280	290	298.15
c_p (J K^{-1}g^{-1})	2.410	2.428	2.471	2.539	2.599
C_p (J K^{-1}mol^{-1})	48.21	48.58	49.44	50.80	52.00

TABLE 2.58.5. Parameters of quasi-polynomial equation

Heat capacity type	No. data points total	No. data points used	s_w	s/R	s_r %	s_b/R	+/–
C_p	21	13	0.487	1.38–2	0.24	4.78–5	0

Temp. range K	T_c K	A_1	A_2	A_3	A_4	Level of uncertainty
197.9–289.1	461.00	4.13160	2.54785	2.29853	1.67495	III

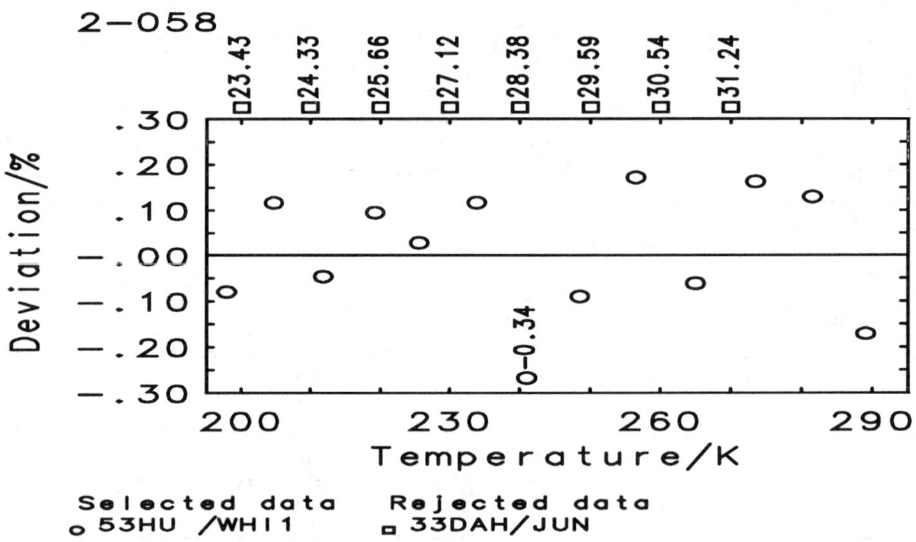

Name: Rubidium fluoride ($Rb(HF_2)$)
Formula: F_2HRb

CAS–RN: 12280–64–7
Group No.: 2–059
Molar Mass: 124.47

TABLE 2.59.1. Experimental heat capacities

Reference	Temp. range K	No. pnts	Error %	Purity %	Purity method	Type capacity	Calorimeter Type	Calorimeter Reference
78WES/LAN	492.4–530.2	7	nosp	99.92	melpt	C_p	BSAO	68WES/WES

TABLE 2.59.3. Parameters of regression polynomial

Heat capacity type	No. data points total	used	s_w	s/R	s_r %	s_b/R	+/–
C_p	7	7	0.582	3.33–2	0.23	1.11–4	–1

Temp. range K	A_1	A_2	Level of uncertainty
492.4–530.2	1.06165+1	7.28154–1	III

TABLE 2.59.4. Recommended values of heat capacities

Temp. (K)	490	500	510	520	530
c_p (J $K^{-1}g^{-1}$)	0.947	0.952	0.957	0.962	0.967
C_p (J $K^{-1}mol^{-1}$)	117.9	118.5	119.1	119.8	120.4

Name: Ammonium fluoride (($(NH_4)(HF_2)$))
Formula: F_2H_5N

CAS–RN: 1341–49–7
Group No.: 2–060
Molar Mass: 57.04

TABLE 2.60.1. Experimental heat capacities

Reference	Temp. range K	No. pnts	Error %	Purity %	Purity method	Type capacity	Calorimeter Type	Calorimeter Reference
76CAR/WES	406.9–443.5	8	nosp	99.88	melpt	C_p	BSAO	68WES/WES

TABLE 2.60.3. Parameters of regression polynomial

Heat capacity type	No. data points total	No. data points used	s_w	s/R	s_r %	s_b/R	+/–
C_p	8	8	0.251	1.73–2	0.10	2.62–5	–1

Temp. range K	A_1	A_2	Level of uncertainty
406.9–443.5	1.17567+1	1.30981	III

TABLE 2.60.4. Recommended values of heat capacities

Temp. (K)	410	420	430	440
c_p ($J\,K^{-1}g^{-1}$)	2.496	2.515	2.535	2.554
C_p ($J\,K^{-1}mol^{-1}$)	142.4	143.5	144.6	145.7

Name: Oxygen fluoride
Formula: F_2O

CAS–RN: 7783–41–7
Group No.: 2–061
Molar Mass: 54.00

TABLE 2.61.1. Experimental heat capacities

Reference	Temp. range K	No. pnts	Error %	Purity %	Purity method	Type capacity	Calorimeter Type	Calorimeter Reference
66FIE	82.0–197.8	26	nosp	99.5	anal	C_{sat}	BSAO	66FIE

TABLE 2.61.3. Parameters of regression polynomial

Heat capacity type	No. data points total	No. data points used	s_w	s/R	s_r %	s_b/R	+/–
C_{sat}	26	26	0.971	6.38–2	0.68	8.16–4	–1

Temp. range K	A_1	A_2	Level of uncertainty
82.1–197.9	8.03871	1.01577	IV

TABLE 2.61.4. Recommended values of heat capacities

Temp. (K)	90	100	110	120	130	140	150
c_{sat} (J K^{-1}g^{-1})	1.379	1.394	1.410	1.426	1.441	1.457	1.472
C_{sat} (J K^{-1}mol^{-1})	74.44	75.28	76.13	76.97	77.82	78.66	79.51
Temp. (K)	160	170	180	190	200		
c_{sat} (J K^{-1}g^{-1})	1.488	1.504	1.519	1.535	1.551		
C_{sat} (J K^{-1}mol^{-1})	80.35	81.20	82.04	82.88	83.73		

TABLE 2.61.5. Parameters of quasi-polynomial equation

Heat capacity type	No. data points total	used	s_w	s/R	s_r %	s_b/R	+/–
C_{sat}	26	26	0.978	6.43–2	0.68	7.77–4	–1

Temp. range K	T_c K	A_1	A_2	A_3	A_4	Level of uncertainty
82.1–197.9	215.10	–5.56381–2	3.90601–4	8.10077	1.98130	IV

Name: Thionyl fluoride
Formula: F_2OS

CAS–RN: 7783–42–8
Group No.: 2–062
Molar Mass: 86.06

TABLE 2.62.1. Experimental heat capacities

Reference	Temp. range K	No. pnts	Error %	Purity %	Purity method	Type capacity	Calorimeter Type	Calorimeter Reference
65PAC/TUR	147.0–229.7	39	nosp	99.974	melpt	C_{sat}	BSAO	55PAC/PIE

TABLE 2.62.3. Parameters of regression polynomial

Heat capacity type	No. data points total	used	s_w	s/R	s_r %	s_b/R	+/–
C_{sat}	39	39	1.104	4.09–2	0.33	2.51–4	5

Temp. range K	A_1	A_2	A_3	Level of uncertainty
147.0–229.7	1.44510+1	–2.31294	6.26343–1	III

TABLE 2.62.4. Recommended values of heat capacities

Temp. (K)	150	160	170	180	190	200	210
c_{sat} (J K^{-1}g^{-1})	1.197	1.194	1.191	1.190	1.190	1.191	1.194
C_{sat} (J K^{-1}mol^{-1})	103.0	102.7	102.5	102.4	102.4	102.5	102.7
Temp. (K)	220	230					
c_{sat} (J K^{-1}g^{-1})	1.197	1.202					
C_{sat} (J K^{-1}mol^{-1})	103.1	103.5					

Name: Sulfuryl fluoride
Formula: F_2O_2S

CAS–RN: 2699–79–8
Group No.: 2–063
Molar Mass: 102.06

TABLE 2.63.1. Experimental heat capacities

Reference	Temp. range K	No. pnts	Error %	Purity %	Purity method	Type capacity	Calorimeter Type	Calorimeter Reference
60BOC/PET	142.3–209.6	20	0.10	99.71	melpt	C_{sat}	BSAO	55PAC/PIE

TABLE 2.63.3. Parameters of regression polynomial

Heat capacity type	No. data points total	used	s_w	s/R	s_r %	s_b/R	+/–
C_{sat}	20	20	0.676	8.60–3	0.07	9.92–6	–4

Temp. range K	A_1	A_2	A_3	A_4	Level of uncertainty
142.3–209.6	–1.17359+1	4.51180+1	–2.68893+1	5.19590	II

TABLE 2.63.4. Recommended values of heat capacities

Temp. (K)	140	150	160	170	180	190	200
c_{sat} (J K^{-1}g^{-1})	1.058	1.057	1.051	1.041	1.031	1.023	1.019
C_{sat} (J K^{-1}mol^{-1})	108.0	107.9	107.2	106.3	105.2	104.4	104.0

Temp. (K)	210
c_{sat} (J K^{-1}g^{-1})	1.022
C_{sat} (J K^{-1}mol^{-1})	104.3

Name: Nitrogen fluoride (NF_3)
Formula: F_3N

CAS–RN: 7783–54–2
Group No.: 2–064
Molar Mass: 71.00

TABLE 2.64.1. Experimental heat capacities

Reference	Temp. range K	No. pnts	Error %	Purity %	Purity method	Type capacity	Calorimeter Type	Calorimeter Reference
55PIE/PAC	70.1–143.8	44	0.10	99.999	melpt	C_{sat}	BSAO	55PAC/PIE
81WEB	67.9–228.9	100	0.60	99.88	melpt	C_{sat}	BSAO	61GOO

TABLE 2.64.2. Correlated heat capacities

Reference	Temp. range K	No. pnts used	$\sigma_r C$ %	d_w	d/R	d_r %	d_b/R	+/–
				Selected data				
55PIE/PAC	70.1–143.8	44	0.10	2.998	2.58–2	0.30	5.01–3	4
81WEB	67.9–228.9	100	0.60	1.725	9.68–2	1.03	–7.49–2	–88

TABLE 2.64.3. Parameters of cubic spline polynomials

Heat capacity type	No. data points total	used	s_w	s/R	s_r %	s_b/R	+/–
C_{sat}	144	144	2.241	8.37–2	0.90	–5.05–2	–84

Temp. range K	A_1	A_2	A_3	A_4	Level of uncertainty
67.9–150.0	1.12335+1	–5.62316	3.12811	–3.10740–1	III
150.0–210.0	–9.04221	3.49283+1	–2.39062+1	5.69688	IV
210.0–228.9	–7.08988+3	1.01504+4	–4.84080+3	7.70284+2	V

TABLE 2.64.4. Recommended values of heat capacities

Temp. (K)	70	80	90	100	110	120	130
c_{sat} (J K^{-1}g^{-1})	1.022	1.004	0.993	0.987	0.986	0.990	0.999
C_{sat} (J K^{-1}mol^{-1})	72.53	71.32	70.51	70.07	70.00	70.28	70.90
Temp. (K)	140	150	160	170	180	190	200
c_{sat} (J K^{-1}g^{-1})	1.012	1.029	1.051	1.082	1.124	1.182	1.261
C_{sat} (J K^{-1}mol^{-1})	71.83	73.07	74.64	76.79	79.79	83.94	89.50
Temp. (K)	210	220	230				
c_{sat} (J K^{-1}g^{-1})	1.36	1.58	2.37				
C_{sat} (J K^{-1}mol^{-1})	96.8	112	168				

TABLE 2.64.5. Parameters of quasi-polynomial equation

Heat capacity type	No. data points total	used	s_w	s/R	s_r %	s_b/R	+/–
C_{sat}	144	73	4.742	5.08–2	0.56	9.92–3	5

Temp. range K	T_c K	A_1	A_2	A_3	A_4	A_5	A_6	Level of uncertainty
111.8–228.9	234.00	–1.86809	1.23546–1	7.88781	8.10292–1	–9.48440	7.30941	IV

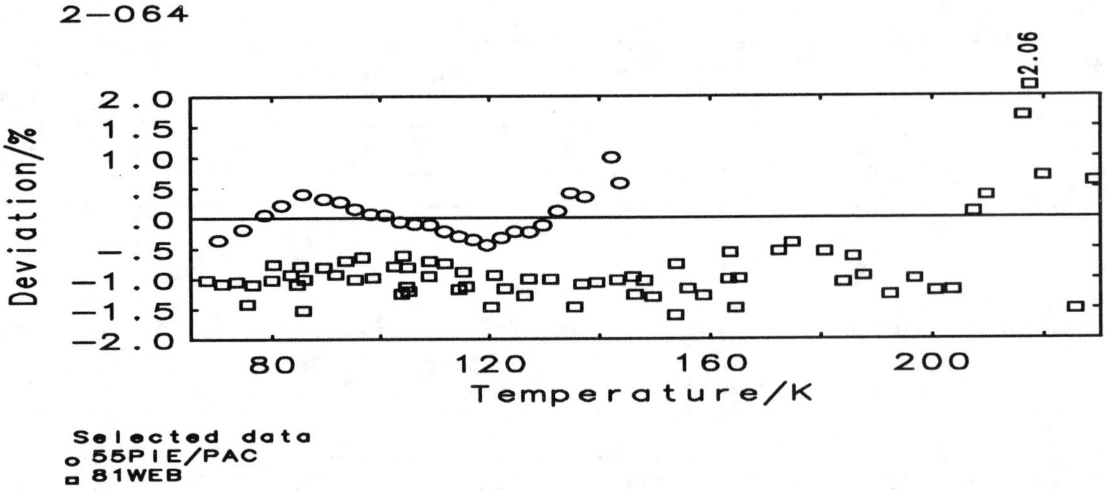

Name: Phosphoryl fluoride
Formula: F_3OP

CAS–RN: 13478–20–1
Group No.: 2–065
Molar Mass: 103.97

TABLE 2.65.1. Experimental heat capacities

Reference	Temp. range K	No. pnts	Error %	Purity %	Purity method	Type capacity	Calorimeter Type	Calorimeter Reference
70PAC/JEP	235.3–241.2	4	nosp	99.926	melpt	C_{sat}	BSAO	55PAC/PIE

TABLE 2.65.3. Parameters of regression polynomial

Heat capacity type	No. data points total	No. data points used	s_w	s/R	s_r %	s_b/R	+/–
C_{sat}	4	4	0.661	2.09–2	0.13	2.77–5	–2

Temp. range K	A_1	A_2	Level of uncertainty
235.3–241.2	1.61491+1	–1.62846–1	II

TABLE 2.65.4. Recommended values of heat capacities

Temp. (K)	235	237	239	241
c_{sat} (J $K^{-1}g^{-1}$)	1.261	1.261	1.260	1.260
C_{sat} (J $K^{-1}mol^{-1}$)	131.1	131.1	131.0	131.0

Name: Phosphorous trifluoride
Formula: F_3P

CAS–RN: 7783–55–3
Group No.: 2–066
Molar Mass: 87.97

TABLE 2.66.1. Experimental heat capacities

Reference	Temp. range K	No. pnts	Error %	Purity %	Purity method	Type capacity	Calorimeter Type	Calorimeter Reference
62PAC/PET	123.4–168.7	18	nosp	99.69	melpt	C_{sat}	BSAO	55PAC/PIE

TABLE 2.66.3. Parameters of regression polynomial

Heat capacity type	No. data points total	No. data points used	s_w	s/R	s_r %	s_b/R	+/–
C_{sat}	18	18	0.841	1.93–2	0.17	5.47–5	0

Temp. range K	A_1	A_2	A_3	Level of uncertainty
123.4–168.7	1.48344+1	–4.51807	1.50975	II

TABLE 2.66.4. Recommended values of heat capacities

Temp. (K)	130	140	150	160	170
c_{sat} (J $K^{-1}g^{-1}$)	1.088	1.084	1.083	1.084	1.089
C_{sat} (J $K^{-1}mol^{-1}$)	95.72	95.35	95.24	95.37	95.76

Name: Iodine fluoride (IF_5)
Formula: F_5I

CAS–RN: 7783–66–6
Group No.: 2–067
Molar Mass: 221.90

TABLE 2.67.1. Experimental heat capacities

Reference	Temp. range K	No. pnts	Error %	Purity %	Purity method	Type capacity	Calorimeter Type	Calorimeter Reference
71OSB/SCH	288.2–345.5	7	0.10	99.985	melpt	C_{sat}	BSAO	43RUE/HUF

TABLE 2.67.3. Parameters of regression polynomial

Heat capacity type	No. data points total	No. data points used	s_w	s/R	s_r %	s_b/R	+/–
C_{sat}	7	7	0.135	2.83–3	0.01	2.72–7	0

Temp. range K	A_1	A_2	A_3	Level of uncertainty
288.2–345.5	2.74430+1	–3.83116	5.60825–1	II

TABLE 2.67.4. Recommended values of heat capacities

Temp. (K)	290	298.15	300	310	320	330	340
c_{sat} ($J K^{-1}g^{-1}$)	0.789	0.787	0.787	0.785	0.784	0.783	0.783
C_{sat} ($J K^{-1}mol^{-1}$)	175.0	174.7	174.6	174.2	174.0	173.8	173.8

Temp. (K)	350
c_{sat} ($J K^{-1}g^{-1}$)	0.783
C_{sat} ($J K^{-1}mol^{-1}$)	173.8

Name: Niobium fluoride (NbF_5)
Formula: F_5Nb

CAS–RN: 7783–68–8
Group No.: 2–068
Molar Mass: 187.90

TABLE 2.68.1. Experimental heat capacities

Reference		Temp. range K	No. pnts	Error %	Purity %	Purity method	Type capacity	Calorimeter Type	Calorimeter Reference
60BRA/MYE	N	375.0–475.0	3S	nosp	99.1	anal	C_{avg}	DSIO	62BRA/MYE

60BRA/MYE values calculated from temperature dependence of enthalpy by the compilers

TABLE 2.68.3. Parameters of regression polynomial

Heat capacity type	No. data points total	No. data points used	s_w	s/R	s_r %	s_b/R	+/–
C	3	3	0.000	0.00	0.00	0.00	0

Temp. range K	A_1	A_2	A_3	Level of uncertainty
375.0–475.0	–5.16983	8.62509	–7.34692–1	V

TABLE 2.68.4. Recommended values of heat capacities

Temp. (K)	380	390	400	410	420	430	440
c (J K^{-1}g^{-1})	0.75	0.77	0.78	0.79	0.80	0.81	0.82
C (J K^{-1}mol^{-1})	141	144	146	148	150	152	154
Temp. (K)	450	460	470				
c (J K^{-1}g^{-1})	0.83	0.84	0.85				
C (J K^{-1}mol^{-1})	156	158	159				

Name: Molybdenum fluoride (MoF_6)
Formula: F_6Mo

CAS–RN: 7783–77–9
Group No.: 2–069
Molar Mass: 209.93

TABLE 2.69.1. Experimental heat capacities

Reference	Temp. K	Capac. J/(K.g)	Error %	Purity %	Purity Method	Type capacity	Calorimeter Type	Calorimeter Reference
60BRA/MYE	298.15	0.7894	nosp	99.98	melpt	C_p	BSIO	36PEA/BAK

Name: Sulfur fluoride (SF_6)
Formula: F_6S

CAS–RN: 2551–62–4
Group No.: 2–070
Molar Mass: 146.06

TABLE 2.70.1. Experimental heat capacities

Reference	Temp. range K	No. pnts	Error %	Purity %	Purity method	Type capacity	Calorimeter Type	Calorimeter Reference
38EUC/SCH	225.0–230.0	2	nosp	99.6	melpt	C_p	BSAO	38EUC/SCH

TABLE 2.70.3. Parameters of regression polynomial

Heat capacity type	No. data points total	No. data points used	s_w	s/R	s_r %	s_b/R	+/–
C_p	2	2	0.000	0.00	0.00	1.91–6	0

Temp. range K	A_1	A_2	Level of uncertainty
225.0–230.0	–3.33130+1	2.07326+1	IV

TABLE 2.70.4. Recommended values of heat capacities

Temp. (K)	225	230
c_p (J K^{-1}g^{-1})	0.759	0.818
C_p (J K^{-1}mol^{-1})	110.9	119.5

Name: Uranium(VI) fluoride
Formula: F_6U

CAS–RN: 7783–81–5
Group No.: 2–071
Molar Mass: 352.02

TABLE 2.71.1. Experimental heat capacities

Reference	Temp. range K	No. pnts	Error %	Purity %	Purity method	Type capacity	Calorimeter Type	Calorimeter Reference
48BRI/HOG	337.2–370.0	8S	1.00	99.98	melpt	C_{sat}	BSAO	45SCO/MEY
53LLE	343.1–373.1	4S	1.00	not specified		C_{sat}	BSIO	53LLE

TABLE 2.71.2. Correlated heat capacities

Reference	Temp. range K	No. pnts used	$\sigma_r C$ %	d_w	d/R	d_r %	d_b/R	+/–
				Selected data				
48BRI/HOG	337.2–370.0	8	1.00	0.033	7.58–3	0.03	5.01–6	–1
				Rejected data				
53LLE	(1.02, 4.19, 1.01, 3)							

TABLE 2.71.3. Parameters of regression polynomial

Heat capacity type	No. data points total	No. data points used	s_w	s/R	s_r %	s_b/R	+/–
C_{sat}	12	8	0.038	8.75–3	0.04	5.01–6	–1

Temp. range K	A_1	A_2	Level of uncertainty
337.2–370.0	1.75393+1	1.60265	IV

TABLE 2.71.4. Recommended values of heat capacities

Temp. (K)	340	350	360	370
c_{sat} (J $K^{-1}g^{-1}$)	0.543	0.547	0.551	0.554
C_{sat} (J $K^{-1}mol^{-1}$)	191.1	192.5	193.8	195.1

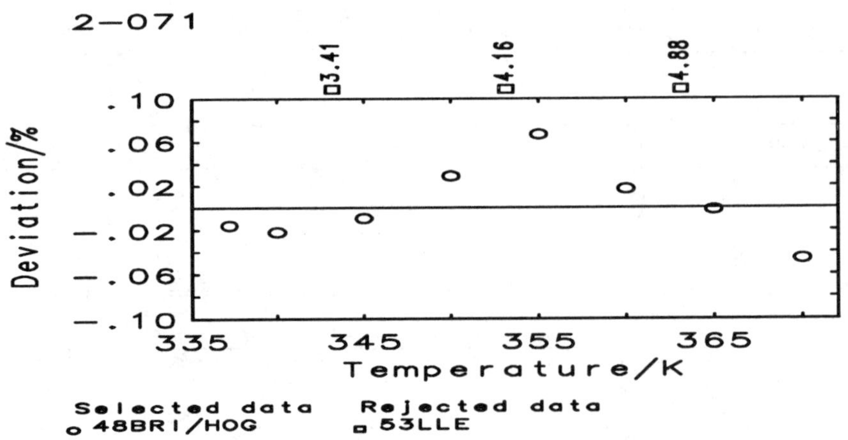

Name: Germane
Formula: GeH_4

CAS-RN: 7782–65–2
Group No.: 2–072
Molar Mass: 76.64

TABLE 2.72.1. Experimental heat capacities

Reference	Temp. range K	No. pnts	Error %	Purity %	Purity method	Type capacity	Calorimeter Type	Calorimeter Reference
42CLU/FAB	112.7–165.4	8	nosp	not specified		C_p	BSIO	36CLU/GOL

TABLE 2.72.3. Parameters of regression polynomial

Heat capacity type	No. data points total	No. data points used	s_w	s/R	s_r %	s_b/R	+/–
C_p	8	8	1.089	6.43–2	0.87	6.97–4	–2

Temp. range K	A_1	A_2	A_3	Level of uncertainty
112.7–165.4	1.21719+1	–7.34650	2.75449	IV

TABLE 2.72.4. Recommended values of heat capacities

Temp. (K)	110	120	130	140	150	160	170
c_p ($J\,K^{-1}g^{-1}$)	0.805	0.794	0.789	0.790	0.797	0.810	0.829
C_p ($J\,K^{-1}mol^{-1}$)	61.72	60.88	60.50	60.58	61.11	62.10	63.55

Name: Hydroiodic acid
Formula: HI

CAS-RN: 10034–85–2
Group No.: 2–073
Molar Mass: 127.91

TABLE 2.73.1. Experimental heat capacities

Reference	Temp. range K	No. pnts	Error %	Purity %	Purity method	Type capacity	Calorimeter Type	Calorimeter Reference
24EUC/KAR	224.0–238.0	3	3.00	not specified		C_{sat}	BSIO	24EUC/KAR
29GIA/WIE	227.1–236.1	3	nosp	not specified		C_{sat}	BSIO	28GIA/WIE1

TABLE 2.73.2. Correlated heat capacities

Reference	Temp. range K	No. pnts used	$\sigma_r C$ %	d_w	d/R	d_r %	d_b/R	+/–
				Selected data				
29GIA/WIE	227.1–236.1	3	0.50#	0.109	3.90–3	0.05	4.45–6	–1
				Rejected data				
24EUC/KAR	(9.20–1, 11.39, 9.20–1, 1)							

TABLE 2.73.3. Parameters of regression polynomial

Heat capacity type	No. data points total	used	s_w	s/R	s_r %	s_b/R	+/–
C_{sat}	6	3	0.189	6.76–3	0.09	4.45–6	–1

Temp. range K	A_1	A_2	Level of uncertainty
227.1–236.1	9.60416	–1.05055	IV

TABLE 2.73.4. Recommended values of heat capacities

Temp. (K)	230	232	234	236
c_{sat} ($J K^{-1}g^{-1}$)	0.467	0.466	0.464	0.463
C_{sat} ($J K^{-1}mol^{-1}$)	59.76	59.59	59.41	59.24

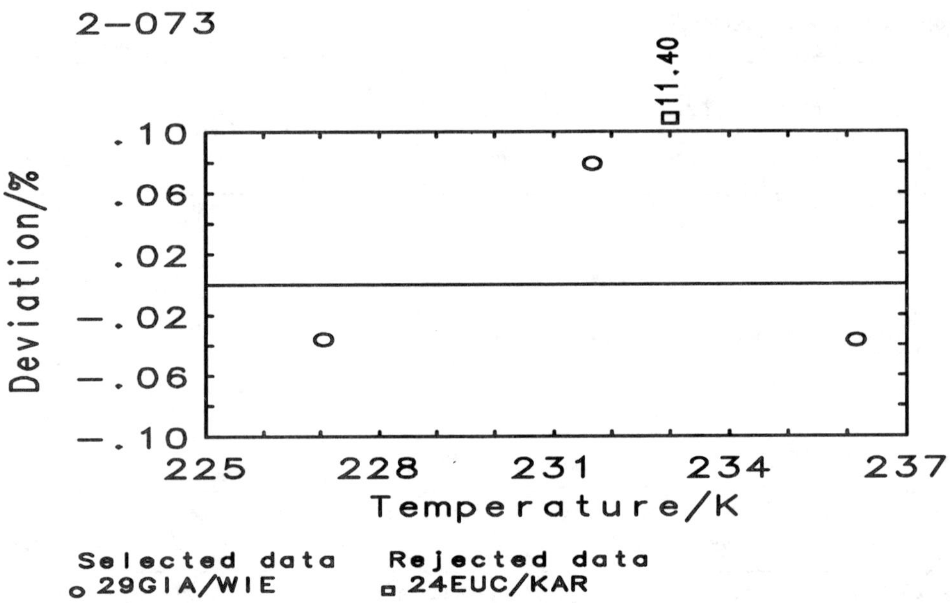

Name: Nitric acid
Formula: HNO_3

CAS–RN: 7697–37–2
Group No.: 2–074
Molar Mass: 63.01

TABLE 2.74.1. Experimental heat capacities

Reference	Temp. range K	No. pnts	Error %	Purity %	Purity method	Type capacity	Calorimeter Type	Calorimeter Reference
42FOR/GIA	236.6–302.9	10	0.10	99.99	melpt	C_p	BSIO	37GIA/EGA

TABLE 2.74.3. Parameters of regression polynomial

Heat capacity type	No. data points total	used	s_w	s/R	s_r %	s_b/R	+/–
C_p	10	10	0.576	3.06–2	0.23	1.12–4	2

Temp. range K	A_1	A_2	Level of uncertainty
236.6–302.9	1.42869+1	–3.55116–1	III

TABLE 2.74.4. Recommended values of heat capacities

Temp. (K)	240	250	260	270	273.15	280	290
c_p ($J K^{-1} g^{-1}$)	1.773	1.768	1.763	1.759	1.757	1.754	1.749
C_p ($J K^{-1} mol^{-1}$)	111.7	111.4	111.1	110.8	110.7	110.5	110.2
Temp. (K)	298.15	300					
c_p ($J K^{-1} g^{-1}$)	1.745	1.745					
C_p ($J K^{-1} mol^{-1}$)	110.0	109.9					

Name: Water
Formula: H_2O

CAS–RN: 7732–18–5
Group No.: 2–075
Molar Mass: 18.02

TABLE 2.75.1. Experimental heat capacities

Reference		Temp. range K	No. pnts	Error %	Purity % method	Type capacity	Calorimeter Type	Calorimeter Reference
05DIE	N	287.7–576.4	20	nosp	not specified	C_{avg}	DSTO	05DIE
21JAE/VON		277.9–322.8	67	nosp	not specified	C_p	BSIO	21JAE/VON
37OSB/STI	N	375.6–647.2	63S	nosp	not specified	C_{sat}	BSAO	37OSB/STI
39OSB/STI	N	273.6–368.1	20S	nosp	not specified	C_p	BSAO	39OSB/STI
40COC/FER	N	290.0–321.2	25	nosp	not specified	C_{sat}	BDHO	40COC/FER
56SCH/GOT		277.0–293.1	19	2.00	not specified	C_p	BSAO	33SOU/BRI
59COX/SMI		276.9–295.3	3	nosp	not specified	C_p	BSAO	59COX/SMI
70KRE		298.1	1	nosp	not specified	C_p	BSAO	63BEN
78BYV/JAS		293.1–333.1	3	2.00	not specified	C_p	BDCT	78BYV/JAS
86NAZ/BAS1		318.6–364.1	3	2.00	not specified	C_p	BDHO	86NAZ/BAS1

05DIE average values between the given temperatures and 273.15 K
37OSB/STI values calculated from temperature dependence of enthalpy by the compiler
39OSB/STI same data in 53GIN/FUR
40COC/FER preliminary data published in 36FER/COC

TABLE 2.75.2. Correlated heat capacities

Reference	Temp. range K	No. pnts used	$\sigma_r C$ %	d_w	d/R	d_r %	d_b/R	+/–
				Selected data				
37OSB/STI	375.6–644.7	60	1.00#	2.609	1.98	2.61	1.32–1	13
39OSB/STI	273.6–368.1	20	0.10#	0.893	8.10–3	0.09	–2.51–3	–2
40COC/FER	290.0–321.2	25	0.10#	0.250	2.27–3	0.03	1.76–3	18
59COX/SMI	277.0–295.3	3	0.20#	0.353	6.40–3	0.07	–5.00–3	–2
70KRE	298.1	1	0.20#	0.095	1.73–3	0.02	–1.73–3	–1

TABLE 2.75.3. Parameters of cubic spline polynomials

Heat capacity type	No. data points total	No. data points used	s_w	s/R	s_r %	s_b/R	+/–
C_p	112	109	2.045	1.52	2.00	7.23–2	26
C_{sat}	224	109	1.054	2.48–1	0.96	1.16–2	21

Temp. range K	A_1	A_2	A_3	A_4	Level of uncertainty
273.6–380.0	2.09634+1	–1.01344+1	2.82530	–2.56738–1	II
380.0–590.0	–2.20666+1	2.38366+1	–6.11445	5.27450–1	IV
590.0–635.0	–4.01518+4	2.04288+4	–3.46458+3	1.95921+2	V
635.0–644.6	–2.38034+7	1.12472+7	–1.77145+6	9.30037+4	V
273.6–380.0	1.91935+1	–8.46779	2.30596	–2.03214–1	II
380.0–590.0	–1.20682+1	1.62126+1	–4.18887	3.66508–1	IV
590.0–635.0	–1.75899+4	8.95411+3	–1.51909+3	8.59539+1	V
635.0–644.6	–5.92160+6	2.79825+6	–4.40779+5	2.31442+4	V

TABLE 2.75.4. Recommended values of heat capacities

Temp. (K)	280	290	298.15	300	310	320	330
c_p (J K^{-1}g^{-1})	4.201	4.187	4.181	4.180	4.177	4.178	4.182
C_p (J K^{-1}mol^{-1})	75.67	75.44	75.31	75.30	75.24	75.26	75.34
c_{sat} (J K^{-1}g^{-1})	4.201	4.188	4.181	4.180	4.177	4.177	4.181
C_{sat} (J K^{-1}mol^{-1})	75.67	75.45	75.32	75.30	75.24	75.25	75.32

Temp. (K)	340	350	360	370	380	390	400
c_p (J K^{-1}g^{-1})	4.189	4.198	4.208	4.218	4.229	4.238	4.249
C_p (J K^{-1}mol^{-1})	75.46	75.62	75.80	75.99	76.18	76.36	76.54
c_{sat} (J K^{-1}g^{-1})	4.187	4.196	4.206	4.217	4.229	4.241	4.254
C_{sat} (J K^{-1}mol^{-1})	75.44	75.59	75.78	75.98	76.19	76.40	76.63

Temp. (K)	410	420	430	440	450	460	470
c_p (J K^{-1}g^{-1})	4.261	4.277	4.297	4.324	4.359	4.403	4.458
C_p (J K^{-1}mol^{-1})	76.76	77.04	77.41	77.90	78.53	79.32	80.31
c_{sat} (J K^{-1}g^{-1})	4.268	4.286	4.308	4.334	4.367	4.406	4.454
C_{sat} (J K^{-1}mol^{-1})	76.90	77.21	77.60	78.08	78.67	79.38	80.24

Temp. (K)	480	490	500	510	520	530	540
c_p (J K^{-1}g^{-1})	4.525	4.606	4.701	4.814	4.944	5.094	5.265
C_p (J K^{-1}mol^{-1})	81.52	82.97	84.70	86.72	89.07	91.78	94.86
c_{sat} (J K^{-1}g^{-1})	4.511	4.577	4.655	4.745	4.848	4.965	5.097
C_{sat} (J K^{-1}mol^{-1})	81.26	82.46	83.86	85.48	87.34	89.44	91.83

Temp. (K)	550	560	570	580	590	600	610
c_p (J K^{-1}g^{-1})	5.459	5.676	5.919	6.188	6.486	6.90	7.89
C_p (J K^{-1}mol^{-1})	98.34	102.3	106.6	111.5	116.8	124	142
c_{sat} (J K^{-1}g^{-1})	5.245	5.411	5.594	5.797	6.020	6.30	6.85
C_{sat} (J K^{-1}mol^{-1})	94.50	97.48	100.8	104.4	108.5	114	123

Temp. (K)	620	630	640
c_p (J K^{-1}g^{-1})	10.0	13.8	25.1
C_p (J K^{-1}mol^{-1})	180	248	452
c_{sat} (J K^{-1}g^{-1})	7.89	9.66	13.7
C_{sat} (J K^{-1}mol^{-1})	142	174	248

2–075

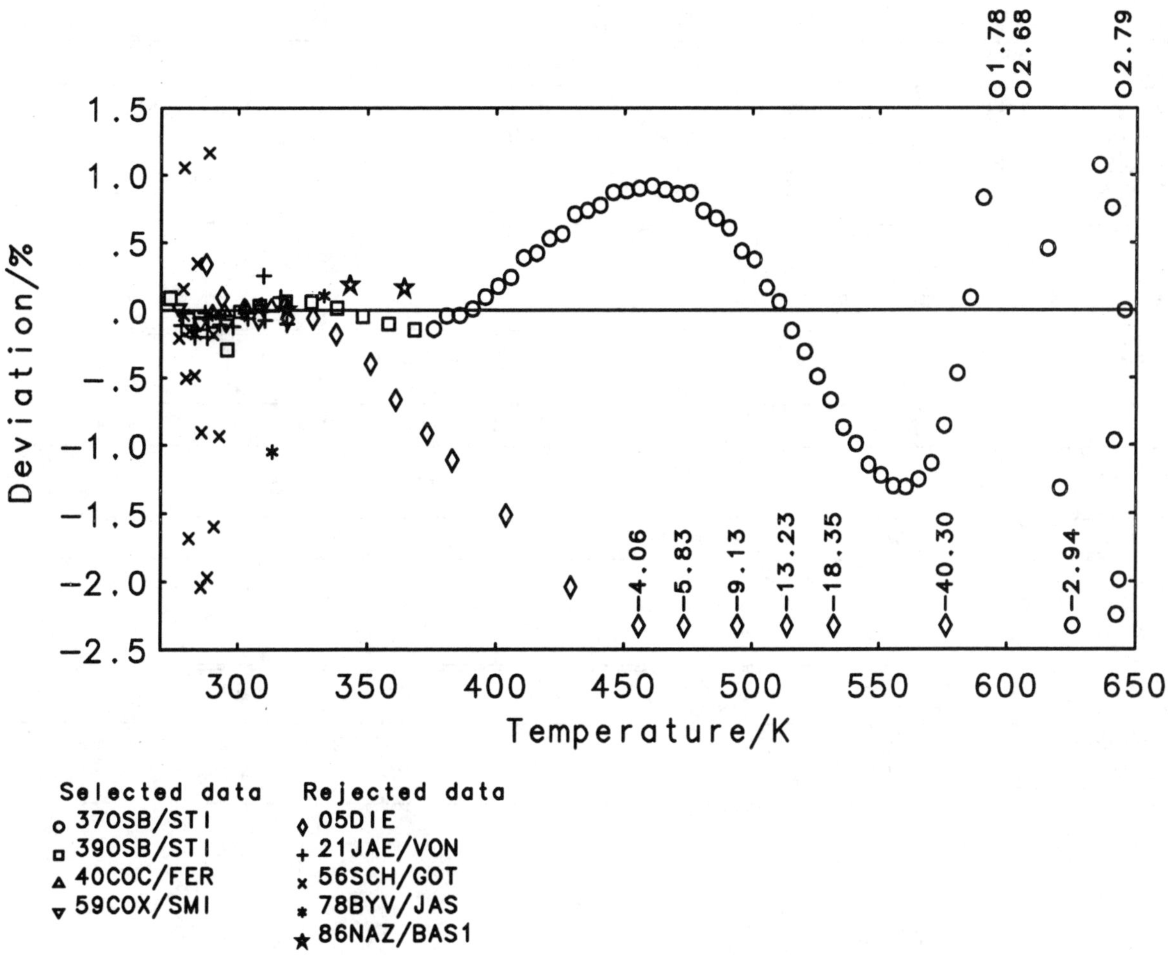

Name: Hydrogen peroxide
Formula: H_2O_2

CAS–RN: 7722–84–1
Group No.: 2–076
Molar Mass: 34.01

TABLE 2.76.1. Experimental heat capacities

Reference		Temp. range K	No. pnts	Error %	Purity %	Purity method	Type capacity	Calorimeter Type	Calorimeter Reference
20MAA/HAT	N	282.4	1	2.00	not specified		C_{avg}	DSIO	20MAA/HAT
51FOL/GIG	N	285.6	1	0.25	99.8	anal	C_{avg}	DSTO	51FOL/GIG
55GIG/MOR2		298.1	1	nosp	not specified		C_p	DSTO	55GIG/MOR1
62GIG/CAR	N	309.1–316.6	2S	nosp	not specified		C_{avg}	DSTO	55GIG/MOR1
64KRO/VEN	N	293.1–333.1	3S	nosp	not specified		C_p	BSIO	64KRO/VEN

20MAA/HAT average value in temperature range 273–292 K
51FOL/GIG average value in temperature range 273–298 K
62GIG/CAR extrapolated data obtained from measurement with aqueous solutions (max. concentration 90 mass %)
64KRO/VEN extrapolated data obtained from measurement with aqueous solutions (max. concentration 80 mass %)

TABLE 2.76.2. Correlated heat capacities

Reference	Temp. range K	No. pnts used	$\sigma_r C$ %	d_w	d/R	d_r %	d_b/R	+/–
				Selected data				
51FOL/GIG	285.6	1	0.25	0.146	3.93–3	0.04	3.93–3	1
55GIG/MOR2	298.1	1	0.40#	0.016	6.83–4	0.01	–6.83–4	0
62GIG/CAR	309.1–316.6	2	0.40#	0.218	9.35–3	0.09	–7.63–3	–2
64KRO/VEN	293.1–333.1	3	0.60#	0.567	3.67–2	0.34	4.69–3	–1

TABLE 2.76.3. Parameters of regression polynomial

Heat capacity type	No. data points total	used	s_w	s/R	s_r %	s_b/R	+/–
C_p	8	7	0.520	3.25–2	0.30	2.94–4	–2

Temp. range K	A_1	A_2	A_3	Level of uncertainty
285.6–333.1	3.00483+1	–1.26522+1	2.07198	IV

TABLE 2.76.4. Recommended values of heat capacities

Temp. (K)	280	290	298.15	300	310	320	330
c_p (J K^{-1}g^{-1})	2.656	2.636	2.626	2.625	2.625	2.635	2.655
C_p (J K^{-1}mol^{-1})	90.35	89.65	89.33	89.29	89.28	89.62	90.30

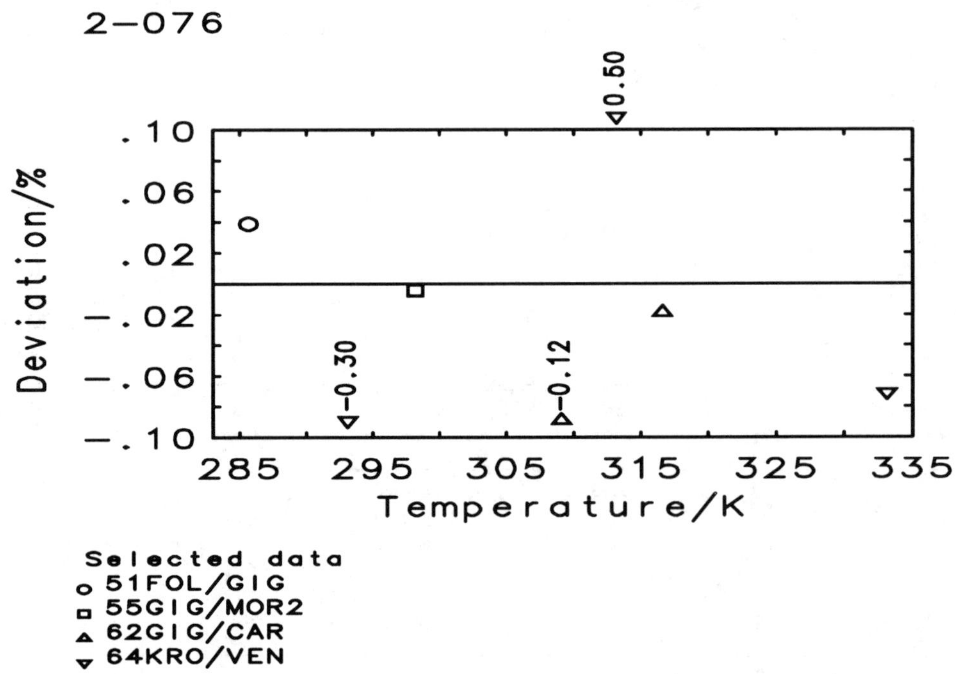

Name: Sulfuric acid
Formula: H_2O_4S

CAS–RN: 7664–93–9
Group No.: 2–077
Molar Mass: 98.08

TABLE 2.77.1. Experimental heat capacities

Reference		Temp. range K	No. pnts	Error %	Purity %	Purity method	Type capacity	Calorimeter Type	Calorimeter Reference
*90PIC		284.0–334.0	eqn	4.00	not specified		C_p	BDHO	*90PIC
32SOC	N	295.6–353.1	4S	nosp	not specified		C_p	not specified	
52RUB/GIA		287.7–305.7	4	0.20	99.94	melpt	C_p	BSIO	37GIA/EGA

32SOC values at 295.65 K and 313.15 K are unpublished data by Savarizky

TABLE 2.77.2. Correlated heat capacities

Reference	Temp. range K	No. pnts used	$\sigma_r C$ %	d_w	d/R	d_r %	d_b/R	+/–
				Selected data				
32SOC	295.6–353.1	4	1.50#	0.872	2.26–1	1.31	–2.13–1	–4
52RUB/GIA	287.7–305.7	4	0.20	1.321	4.38–2	0.26	4.12–3	2
				Rejected data				
*90PIC	(4.26–1, 2.44, 4.24–1, 4)							

TABLE 2.77.3. Parameters of regression polynomial

Heat capacity type	No. data points total	No. data points used	s_w	s/R	s_r %	s_b/R	+/–
C_p	13	8	1.293	1.88–1	1.09	–1.05–1	–2

Temp. range K	A_1	A_2	Level of uncertainty
287.7–353.1	9.22774	2.49400	IV

TABLE 2.77.4. Recommended values of heat capacities

Temp. (K)	290	298.15	300	310	320	330	340
c_p (J $K^{-1}g^{-1}$)	1.395	1.413	1.417	1.438	1.459	1.480	1.501
C_p (J $K^{-1}mol^{-1}$)	136.9	138.5	138.9	141.0	143.1	145.2	147.2

Temp. (K)	350
c_p (J $K^{-1}g^{-1}$)	1.522
C_p (J $K^{-1}mol^{-1}$)	149.3

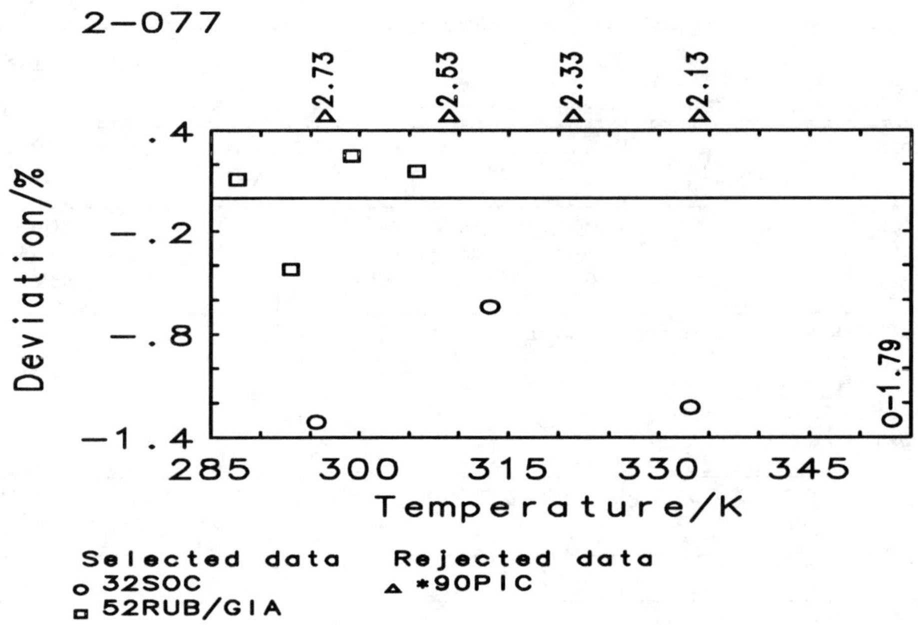

Name: Disulfuric acid
Formula: $H_2O_7S_2$

CAS–RN: 7783–05–3
Group No.: 2–078
Molar Mass: 178.14

TABLE 2.78.1. Experimental heat capacities

Reference	Temp. K	Capac. J/(K.g)	Error %	Purity %	Purity Method	Type capacity	Calorimeter Type	Calorimeter Reference
26AUE	308.15	1.351	nosp	not specified		C_{avg}	DSIO	26AUE

Name: Hydrogen sulfide
Formula: H_2S

CAS–RN: 7783–06–4
Group No.: 2–079
Molar Mass: 34.08

TABLE 2.79.1. Experimental heat capacities

Reference	Temp. range K	No. pnts	Error %	Purity %	Purity method	Type capacity	Calorimeter Type	Calorimeter Reference
36CLU/FRA2	193.6–208.9	4	nosp	not specified		C_p	BSIO	36CLU/GOL
36GIA/BLU	189.9–211.0	6	nosp	99.999	melpt	C_p	BSIO	28GIA/WIE1

TABLE 2.79.2. Correlated heat capacities

Reference	Temp. range K	No. pnts used	$\sigma_r C$ %	d_w	d/R	d_r %	d_b/R	+/–
				Selected data				
36GIA/BLU	189.9–211.0	6	0.30#	0.231	5.67–3	0.07	8.11–6	1
				Rejected data				
36CLU/FRA2	(7.87–2, 0.95, 5.34–2, 2)							

TABLE 2.79.3. Parameters of regression polynomial

Heat capacity type	No. data points total	used	s_w	s/R	s_r %	s_b/R	+/–
C_p	10	6	0.283	6.94–3	0.08	8.11–6	1

Temp. range K	A_1	A_2	Level of uncertainty
189.9–211.0	7.64899	2.64168–1	II

TABLE 2.79.4. Recommended values of heat capacities

Temp. (K)	190	195	200	205	210
c_p (J K^{-1}g^{-1})	1.988	1.992	1.995	1.998	2.001
C_p (J K^{-1}mol^{-1})	67.77	67.88	67.99	68.10	68.21

2–079

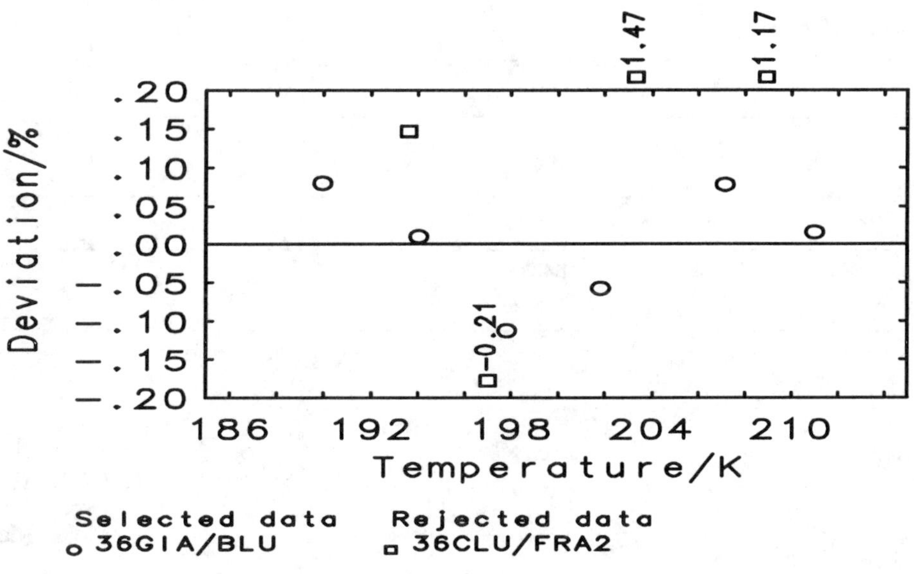

Name: Hydrogen sulfide (H_2S_2)
Formula: H_2S_2

CAS–RN: 13465–07–1
Group No.: 2–080
Molar Mass: 66.15

TABLE 2.80.1. Experimental heat capacities

Reference		Temp. range K	No. pnts	Error %	Purity %	Purity method	Type capacity	Calorimeter Type	Calorimeter Reference
30BUT/MAA	N	184.1–298.1	2	5.00	not specified		C_{avg}	DSIO	20MAA/HAT
63FEH/SEY		239.8–313.1	5	2.00	not specified		C_p	BSAO	63FEH/SEY

30BUT/MAA constant value calculated from temperature dependence of enthalpy by the authors

TABLE 2.80.2. Correlated heat capacities

Reference	Temp. range K	No. pnts used	$\sigma_r C$ %	d_w	d/R	d_r %	d_b/R	+/–
				Selected data				
63FEH/SEY	239.8–313.1	5	2.00	0.252	5.63–2	0.50	5.73–4	–1
				Rejected data				
30BUT/MAA	(1.16–1, 1.05, –1.16–1, –1)							

TABLE 2.80.3. Parameters of regression polynomial

Heat capacity type	No. data points total	No. data points used	s_w	s/R	s_r %	s_b/R	+/–
C_p	7	5	0.325	7.27–2	0.65	5.73–4	–1

Temp. range K	A_1	A_2	Level of uncertainty
239.8–313.1	9.74820	4.87109–1	V

TABLE 2.80.4. Recommended values of heat capacities

Temp. (K)	240	250	260	270	273.15	280	290
c_p ($J\,K^{-1}g^{-1}$)	1.37	1.38	1.38	1.39	1.39	1.40	1.40
C_p ($J\,K^{-1}mol^{-1}$)	90.8	91.2	91.6	92.0	92.1	92.4	92.8

Temp. (K)	298.15	300	310
c_p ($J\,K^{-1}g^{-1}$)	1.41	1.41	1.42
C_p ($J\,K^{-1}mol^{-1}$)	93.1	93.2	93.6

Name: Hydrogen sulfide (H_2S_3)
Formula: H_2S_3

CAS–RN: 13845–23–3
Group No.: 2–081
Molar Mass: 98.21

TABLE 2.81.1. Experimental heat capacities

Reference	Temp. range K	No. pnts	Error %	Purity %	Purity method	Type capacity	Calorimeter Type	Calorimeter Reference
63FEH/SEY	236.9–295.9	5	2.00	not specified		C_p	BSAO	63FEH/SEY

TABLE 2.81.3. Parameters of regression polynomial

Heat capacity type	No. data points total	used	s_w	s/R	s_r %	s_b/R	+/–
C_p	5	5	0.312	9.19–2	0.62	6.84–4	–1
Temp. range K	A_1	A_2					Level of uncertainty
236.9–296.0	1.32735+1	5.23184–1					V

TABLE 2.81.4. Recommended values of heat capacities

Temp. (K)	240	250	260	270	273.15	280	290
c_p $(J\,K^{-1}g^{-1})$	1.23	1.23	1.24	1.24	1.24	1.25	1.25
C_p $(J\,K^{-1}mol^{-1})$	121	121	122	122	122	123	123
Temp. (K)	298.15	300					
c_p $(J\,K^{-1}g^{-1})$	1.26	1.26					
C_p $(J\,K^{-1}mol^{-1})$	123	123					

Name: Hydrogen sulfide (H_2S_4)
Formula: H_2S_4

CAS–RN: 13845–25–5
Group No.: 2–082
Molar Mass: 130.28

TABLE 2.82.1. Experimental heat capacities

Reference	Temp. range K	No. pnts	Error %	Purity %	Purity method	Type capacity	Calorimeter Type	Calorimeter Reference
63FEH/SEY	245.8–293.1	5	2.00	not specified		C_p	BSAO	63FEH/SEY

TABLE 2.82.3. Parameters of regression polynomial

Heat capacity type	No. data points total	used	s_w	s/R	s_r %	s_b/R	+/–
C_p	5	5	0.171	6.28–2	0.34	2.59–4	1
Temp. range K	A_1	A_2					Level of uncertainty
245.8–293.1	1.64246+1	7.33620–1					V

TABLE 2.82.4. Recommended values of heat capacities

Temp. (K)	250	260	270	273.15	280	290	298.15
c_p $(J\,K^{-1}g^{-1})$	1.17	1.17	1.17	1.18	1.18	1.18	1.19
C_p $(J\,K^{-1}mol^{-1})$	152	152	153	153	154	154	155

Name: Hydrogen sulfide (H_2S_5)
Formula: H_2S_5

CAS–RN: 13845–24–4
Group No.: 2–083
Molar Mass: 162.35

TABLE 2.83.1. Experimental heat capacities

Reference	Temp. range K	No. pnts	Error %	Purity %	Purity method	Type capacity	Calorimeter Type	Calorimeter Reference
63FEH/SEY	244.1–293.1	5	2.00	not specified		C_p	BSAO	63FEH/SEY

TABLE 2.83.3. Parameters of regression polynomial

Heat capacity type	No. data points total	No. data points used	s_w	s/R	s_r %	s_b/R	+/–
C_p	5	5	0.405	1.79–1	0.81	1.75–3	–1

Temp. range K	A_1	A_2	Level of uncertainty
244.1–293.1	1.86800+1	1.26655	V

TABLE 2.83.4. Recommended values of heat capacities

Temp. (K)	250	260	270	273.15	280	290	298.15
c_p ($J\ K^{-1}g^{-1}$)	1.12	1.13	1.13	1.13	1.14	1.14	1.15
C_p ($J\ K^{-1}mol^{-1}$)	182	183	184	184	185	186	187

Name: Hydrogen sulfide (H_2S_6)
Formula: H_2S_6

CAS–RN: 13845–51–7
Group No.: 2–084
Molar Mass: 194.41

TABLE 2.84.1. Experimental heat capacities

Reference	Temp. range K	No. pnts	Error %	Purity %	Purity method	Type capacity	Calorimeter Type	Calorimeter Reference
63FEH/SEY	246.8–312.9	5	2.00	not specified		C_p	BSAO	63FEH/SEY

TABLE 2.84.3. Parameters of regression polynomial

Heat capacity type	No. data points total	No. data points used	s_w	s/R	s_r %	s_b/R	+/–
C_p	5	5	0.807	4.16–1	1.62	8.05–3	0

Temp. range K	A_1	A_2	Level of uncertainty
246.8–313.0	2.05033+1	1.84926	V

TABLE 2.84.4. Recommended values of heat capacities

Temp. (K)	250	260	270	273.15	280	290	298.15
c_p (J K^{-1}g^{-1})	1.07	1.08	1.09	1.09	1.10	1.11	1.11
C_p (J K^{-1}mol^{-1})	209	210	212	212	214	215	216

Temp. (K)	300	310
c_p (J K^{-1}g^{-1})	1.11	1.12
C_p (J K^{-1}mol^{-1})	217	218

Name: Hydrogen selenide
Formula: H_2Se

CAS–RN: 7783–07–5
Group No.: 2–085
Molar Mass: 80.98

TABLE 2.85.1. Experimental heat capacities

Reference	Temp. range K	No. pnts	Error %	Purity %	Purity method	Type capacity	Calorimeter Type	Calorimeter Reference
38KRU/CLU	210.4–229.7	10	nosp	not specified		C_p	BSIO	36CLU/GOL

TABLE 2.85.3. Parameters of regression polynomial

Heat capacity type	No. data points total	No. data points used	s_w	s/R	s_r %	s_b/R	+/–
C_p	10	10	0.411	2.70–2	0.33	1.42–4	–2

Temp. range K	A_1	A_2	Level of uncertainty
210.4–229.7	8.93420	–3.44971–1	IV

TABLE 2.85.4. Recommended values of heat capacities

Temp. (K)	210	220	230
c_p (J K^{-1}g^{-1})	0.843	0.839	0.836
C_p (J K^{-1}mol^{-1})	68.26	67.97	67.69

Name: Ammonia
Formula: H_3N

CAS–RN: 7664–41–7
Group No.: 2–086
Molar Mass: 17.03

TABLE 2.86.1. Experimental heat capacities

Reference	Temp. range K	No. pnts	Error %	Purity %	Purity method	Type capacity	Calorimeter Type	Calorimeter Reference
24EUC/KAR	199.5–221.1	7	3.00	not specified		C_{sat}	BSIO	24EUC/KAR
26DAN/JEN	246.8–290.4	10	2.00	not specified		C_{sat}	BSIO	26DAN/JEN
37OVE/GIA	197.8–238.3	13	nosp	99.999	melpt	C_{sat}	BSIO	37GIA/EGA
71POP/MAN	199.2–200.8	2	1.00	99.99	anal	C_p	BSAO	68BAG/KUC

TABLE 2.86.2. Correlated heat capacities

Reference	Temp. range K	No. pnts used	$\sigma_r C$ %	d_w	d/R	d_r %	d_b/R	+/–
				Selected data				
26DAN/JEN	246.9–290.4	10	2.00	1.025	1.99–1	2.05	9.62–2	4
37OVE/GIA	197.8–238.3	13	0.30#	0.750	2.03–2	0.22	–1.46–3	–1
71POP/MAN	199.2–200.8	2	1.00	0.095	8.43–3	0.10	2.20–3	0
				Rejected data				
24EUC/KAR	(3.34–1, 3.61, 3.13–1, 7)							

TABLE 2.86.3. Parameters of regression polynomial

Heat capacity type	No. data points total	used	s_w	s/R	s_r %	s_b/R	+/–
C_p	32	25	0.900	1.35–1	1.39	3.79–2	3
C_{sat}	32	25	0.894	1.35–1	1.40	3.78–2	3

Temp. range K	A_1	A_2	A_3	Level of uncertainty
197.8–290.4	1.01781+1	–1.87331	6.06344–1	V
197.8–290.4	9.89140	–1.59465	5.38618–1	V

TABLE 2.86.4. Recommended values of heat capacities

Temp. (K)	200	210	220	230	240	250	260
c_p (J K^{-1}g^{-1})	4.32	4.35	4.39	4.43	4.48	4.53	4.59
C_p (J K^{-1}mol^{-1})	73.6	74.1	74.8	75.5	76.3	77.2	78.2
c_{sat} (J K^{-1}g^{-1})	4.32	4.35	4.39	4.43	4.48	4.53	4.58
C_{sat} (J K^{-1}mol^{-1})	73.6	74.1	74.7	75.4	76.2	77.1	78.0

Temp. (K)	270	273.15	280	290	298.15
c_p (J K^{-1}g^{-1})	4.66	4.68	4.73	4.81	4.87
C_p (J K^{-1}mol^{-1})	79.3	79.7	80.5	81.9	83.0
c_{sat} (J K^{-1}g^{-1})	4.64	4.66	4.71	4.78	4.85
C_{sat} (J K^{-1}mol^{-1})	79.1	79.4	80.2	81.5	82.5

TABLE 2.86.5. Parameters of quasi-polynomial equation

Heat capacity type	No. data points total	used	s_w	s/R	s_r %	s_b/R	+/–
C_p	32	23	5.286	1.85–1	2.02	6.89–2	5

Temp. range K	T_c K	A_1	A_2	A_3	A_4	Level of uncertainty
197.8–290.4	405.60	5.03149	3.84393–1	3.05052	1.64648+1	V

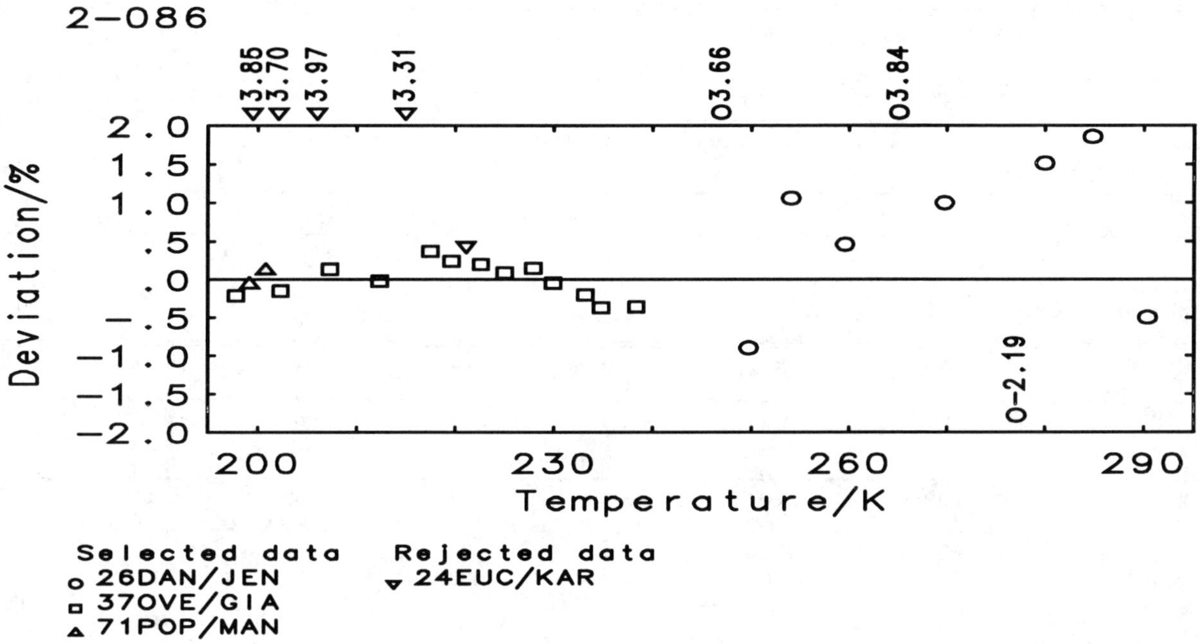

Name: Phosphoric acid
Formula: H_3O_4P

CAS–RN: 7664–38–2
Group No.: 2–087
Molar Mass: 98.00

TABLE 2.87.1. Experimental heat capacities

Reference		Temp. range K	No. pnts	Error %	Purity % method	Type capacity	Calorimeter Type	Calorimeter Reference
58EGA/LUF	N	288.1–353.1	6S	nosp	not specified	C_p	BSAO	45SCO/MEY
69SMI/MAT	N	335.6	1	nosp	not specified	C_{avg}	DSIO	69SMI/MAT
72WAK/LUF	N	323.1–473.1	16S	nosp	not specified	C_p	DSIO	50EGA/WAK
81LUF	N	300.0–530.0	24S	nosp	not specified	C_{sat}	BDHT	69PER/COM

58EGA/LUF extrapolated data obtained from measurement in aqueous solutions (max. concentration 85 mass %)
69SMI/MAT average value in temperature range 298–373 K; interpolated from data in aqueous solution of P_2O_5
72WAK/LUF interpolated by the compilers from C_p of solutions of P_2O_5 in H_3PO_4 (concentration range of P_2O_5: 72.3 to 84.8 mass %)
81LUF interpolated from measurements in aqueous solutions of P_2O_5

TABLE 2.87.2. Correlated heat capacities

Reference	Temp. range K	No. pnts used	$\sigma_r C$ %	d_w	d/R	d_r %	d_b/R	+/–
				Selected data				
58EGA/LUF	288.1–353.1	6	0.50#	1.578	1.56–1	0.79	–5.30–2	–1
69SMI/MAT	335.6	1	1.00#	0.160	3.16–2	0.16	3.16–2	1
72WAK/LUF	323.1–473.1	16	0.70#	0.641	9.19–2	0.45	3.74–2	2
81LUF	300.0–530.0	24	2.00#	0.919	3.89–1	1.84	3.14–2	–2

TABLE 2.87.3. Parameters of regression polynomial

Heat capacity type	No. data points total	used	s_w	s/R	s_r %	s_b/R	+/–
C	47	47	0.986	3.01–1	1.43	2.27–2	0

Temp. range K	A_1	A_2	A_3	A_4	Level of uncertainty
288.1–530.0	3.53509+1	–1.58700+1	4.68941	–4.01211–1	V

TABLE 2.87.4. Recommended values of heat capacities

Temp. (K)	290	298.15	300	310	320	330	340
c ($J\,K^{-1}g^{-1}$)	1.61	1.62	1.62	1.63	1.65	1.67	1.68
C ($J\,K^{-1}mol^{-1}$)	158	159	159	160	162	163	165
Temp. (K)	350	360	370	380	390	400	410
c ($J\,K^{-1}g^{-1}$)	1.70	1.72	1.74	1.76	1.78	1.80	1.82
C ($J\,K^{-1}mol^{-1}$)	167	169	171	172	174	176	178
Temp. (K)	420	430	440	450	460	470	480
c ($J\,K^{-1}g^{-1}$)	1.84	1.86	1.88	1.90	1.91	1.93	1.94
C ($J\,K^{-1}mol^{-1}$)	180	182	184	186	187	189	190
Temp. (K)	490	500	510	520	530		
c ($J\,K^{-1}g^{-1}$)	1.95	1.96	1.97	1.97	1.97		
C ($J\,K^{-1}mol^{-1}$)	191	192	193	193	193		

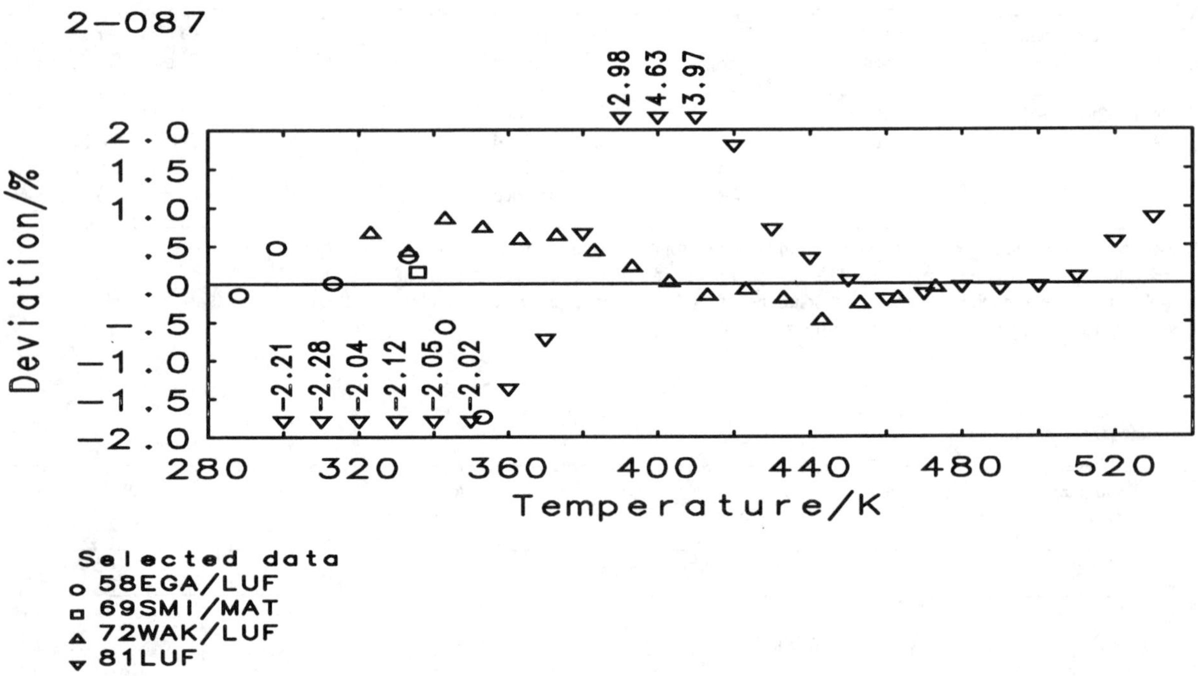

Name: Phosphine
Formula: H_3P

CAS–RN: 7803–51–2
Group No.: 2–088
Molar Mass: 34.00

TABLE 2.88.1. Experimental heat capacities

Reference	Temp. range K	No. pnts	Error %	Purity %	Purity method	Type capacity	Calorimeter Type	Calorimeter Reference
36CLU/FRA1	143.2–185.8	9	nosp	not specified		C_p	BSIO	36CLU/GOL
37STE/GIA	146.1–182.7	8	nosp	99.996	melpt	C_p	BSIO	28GIA/WIE1

TABLE 2.88.2. Correlated heat capacities

Reference	Temp. range K	No. pnts used	$\sigma_r C$ %	d_w	d/R	d_r %	d_b/R	+/–
				Selected data				
36CLU/FRA1	143.2–185.8	9	0.70#	0.775	3.99–2	0.54	1.96–2	5
37STE/GIA	146.1–182.7	8	0.30#	0.414	9.05–3	0.12	–3.95–3	–4

TABLE 2.88.3. Parameters of regression polynomial

Heat capacity type	No. data points total	No. data points used	s_w	s/R	s_r %	s_b/R	+/–
C_p	17	17	0.696	3.27–2	0.45	8.54–3	1

Temp. range K	A_1	A_2	A_3	Level of uncertainty
143.2–185.8	1.06796+1	–4.08758	1.22669	II

TABLE 2.88.4. Recommended values of heat capacities

Temp. (K)	140	150	160	170	180
c_p (J $K^{-1}g^{-1}$)	1.800	1.787	1.780	1.779	1.784
C_p (J $K^{-1}mol^{-1}$)	61.21	60.76	60.53	60.50	60.67

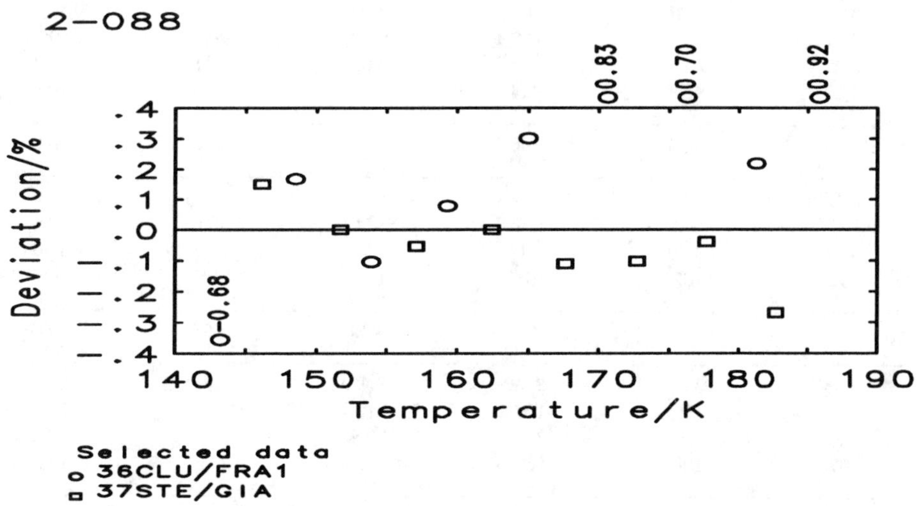

Name: Hydrazine
Formula: H_4N_2

CAS–RN: 302–01–2
Group No.: 2–089
Molar Mass: 32.05

TABLE 2.89.1. Experimental heat capacities

Reference	Temp. range K	No. pnts	Error %	Purity %	Purity method	Type capacity	Calorimeter Type	Calorimeter Reference
49SCO/OLI	274.7–340.0	9S	0.30	99.75	melpt	C_{sat}	BSAO	43RUE/HUF
50HOU/MAS1	313.1–353.1	5S	1.00	99.9	anal	C_p	BSAO	50SAG/HOU
68AHL/YOU	331.1–464.1	6	2.00	97.	estim	C_p	FSIO	68AHL/YOU

TABLE 2.89.2. Correlated heat capacities

Reference	Temp. range K	No. pnts used	$\sigma_r C$ %	d_w	d/R	d_r %	d_b/R	+/–
				Selected data				
49SCO/OLI	274.7–340.0	9	0.30	0.154	5.60–3	0.05	1.77–3	1
50HOU/MAS1	313.1–353.1	5	1.00	0.604	7.21–2	0.60	–5.55–2	–3
68AHL/YOU	331.1–464.1	6	2.00	0.588	1.52–1	1.18	7.34–2	4

TABLE 2.89.3. Parameters of regression polynomial

Heat capacity type	No. data points total	No. data points used	s_w	s/R	s_r %	s_b/R	+/–
C_p	20	20	0.492	9.82–2	0.77	8.96–3	2
C_{sat}	20	20	0.500	1.00–1	0.79	9.15–3	1

Temp. range K	A_1	A_2	A_3	Level of uncertainty
274.7–464.1	1.10165+1	–1.79459–1	1.58186–1	V
274.7–464.1	1.08647+1	–8.00113–2	1.41975–1	V

TABLE 2.89.4. Recommended values of heat capacities

Temp. (K)	280	290	298.15	300	310	320	330
c_p (J K^{-1}g^{-1})	3.05	3.07	3.08	3.09	3.11	3.13	3.15
C_p (J K^{-1}mol^{-1})	97.7	98.3	98.8	99.0	99.6	100	101
c_{sat} (J K^{-1}g^{-1})	3.05	3.07	3.08	3.09	3.11	3.13	3.15
C_{sat} (J K^{-1}mol^{-1})	97.7	98.3	98.8	99.0	99.6	100	101

Temp. (K)	340	350	360	370	380	390	400
c_p (J K^{-1}g^{-1})	3.17	3.20	3.22	3.25	3.27	3.30	3.33
C_p (J K^{-1}mol^{-1})	102	102	103	104	105	106	107
c_{sat} (J K^{-1}g^{-1})	3.17	3.20	3.22	3.25	3.27	3.30	3.33
C_{sat} (J K^{-1}mol^{-1})	102	102	103	104	105	106	107

Temp. (K)	410	420	430	440	450	460
c_p (J K^{-1}g^{-1})	3.36	3.39	3.42	3.45	3.48	3.51
C_p (J K^{-1}mol^{-1})	108	109	109	110	112	113
c_{sat} (J K^{-1}g^{-1})	3.35	3.38	3.41	3.44	3.47	3.50
C_{sat} (J K^{-1}mol^{-1})	107	108	109	110	111	112

TABLE 2.89.5. Parameters of quasi-polynomial equation

Heat capacity type	No. data points total	used	s_w	s/R	s_r %	s_b/R	+/−
C_p	20	20	0.836	1.31–1	1.02	4.63–2	1 1

Temp. range K	T_c K	A_1	A_2	A_3	A_4	Level of uncertainty
274.7–464.1	653.00	−1.03550	6.07044–1	9.92212	4.41589–1	V

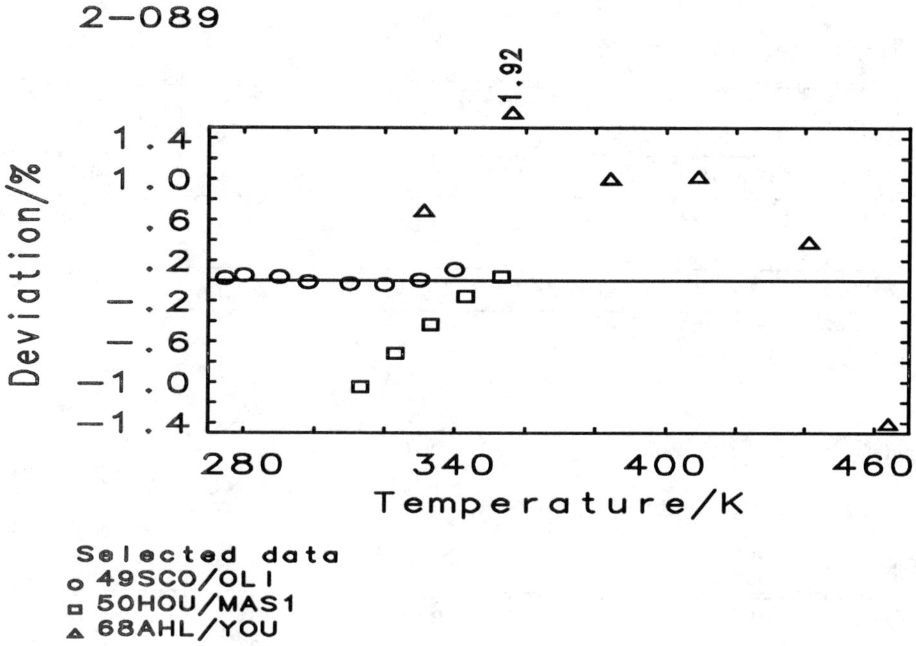

Name: Silane
Formula: H_4Si

CAS-RN: 7803–62–5
Group No.: 2–090
Molar Mass: 32.12

TABLE 2.90.1. Experimental heat capacities

Reference	Temp. range K	No. pnts	Error %	Purity %	Purity method	Type capacity	Calorimeter Type	Calorimeter Reference
33CLU	90.3–159.2	15	nosp	not specified		C_{sat}	BSIO	29CLU1

TABLE 2.90.3. Parameters of regression polynomial

Heat capacity type	No. data points total	used	s_w	s/R	s_r %	s_b/R	+/−
C_{sat}	15	15	0.852	4.35–2	0.60	4.50–4	3

Temp. range K	A_1	A_2	Level of uncertainty
90.3–159.2	6.96847	3.07401–1	IV

TABLE 2.90.4. Recommended values of heat capacities

Temp. (K)	90	100	110	120	130	140	150
c_{sat} (J K^{-1}g^{-1})	1.876	1.884	1.892	1.899	1.907	1.915	1.923
C_{sat} (J K^{-1}mol^{-1})	60.24	60.50	60.75	61.01	61.26	61.52	61.77
Temp. (K)	160						
c_{sat} (J K^{-1}g^{-1})	1.931						
C_{sat} (J K^{-1}mol^{-1})	62.03						

TABLE 2.90.5. Parameters of quasi-polynomial equation

Heat capacity type	No. data points total	used	s_w	s/R	s_r %	s_b/R	+/–
C_{sat}	15	15	0.871	4.46–2	0.61	4.35–4	1

Temp. range K	T_c K	A_1	A_2	A_3	A_4	Level of uncertainty
90.3–159.2	508.10	2.61284+1	7.32745	–7.06326–1	2.32922+1	IV

Name: Ammonium hydroxide
Formula: H_5NO

CAS–RN: 1336–21–6
Group No.: 2–091
Molar Mass: 35.05

TABLE 2.91.1. Experimental heat capacities

Reference		Temp. range K	No. pnts	Error %	Purity %	Purity method	Type capacity	Calorimeter Type	Calorimeter Reference
53HIL/GIA	N	197.1–290.2	18	nosp	99.62	anal	C_p	BSIO	28GIA/WIE1

53HIL/GIA sample contained 0.33 mol.% Ammonium oxide; results corrected to pure Ammonium hydroxide

TABLE 2.91.3. Parameters of regression polynomial

Heat capacity type	No. data points total	used	s_w	s/R	s_r %	s_b/R	+/–
C_p	18	18	0.116	5.77–3	0.03	3.60–6	–2

Temp. range K	A_1	A_2	A_3	A_4	Level of uncertainty
197.1–290.2	1.69315+1	–1.06230+1	6.56821	–9.45264–1	II

TABLE 2.91.4. Recommended values of heat capacities

Temp. (K)	200	210	220	230	240	250	260
c_p (J K^{-1}g^{-1})	3.415	3.520	3.627	3.735	3.844	3.952	4.057
C_p (J K^{-1}mol^{-1})	119.7	123.3	127.1	130.9	134.7	138.5	142.2
Temp. (K)	270	273.15	280	290	298.15		
c_p (J K^{-1}g^{-1})	4.158	4.189	4.254	4.344	4.411		
C_p (J K^{-1}mol^{-1})	145.7	146.8	149.1	152.2	154.6		

Name: Hydrazine mononitrate
Formula: $H_5N_3O_3$

CAS–RN: 13464–97–6
Group No.: 2–092
Molar Mass: 95.06

TABLE 2.92.1. Experimental heat capacities

Reference	Temp. range K	No. pnts	Error %	Purity %	Purity method	Type capacity	Calorimeter Type	Calorimeter Reference
86ZHO/CHE	350.0–370.0	3S	0.20	99.86	melpt	C_p	BSAO	83TAN/ZHO

TABLE 2.92.3. Parameters of regression polynomial

Heat capacity type	No. data points total	No. data points used	s_w	s/R	s_r %	s_b/R	+/–
C_p	3	3	0.021	9.82–4	0.00	–6.36–7	0

Temp. range K	A_1	A_2	Level of uncertainty
350.0–370.0	2.29531+1	1.68377–1	IV

TABLE 2.92.4. Recommended values of heat capacities

Temp. (K)	350	360	370
c_p ($J\,K^{-1}g^{-1}$)	2.059	2.061	2.062
C_p ($J\,K^{-1}mol^{-1}$)	195.7	195.9	196.0

Name: Ammonium oxide
Formula: H_8N_2O

CAS–RN: 12161–77–2
Group No.: 2–093
Molar Mass: 52.08

TABLE 2.93.1. Experimental heat capacities

Reference		Temp. range K	No. pnts	Error %	Purity %	Purity method	Type capacity	Calorimeter Type	Calorimeter Reference
53HIL/GIA	N	196.8–270.2	14	nosp	99.795	anal	C_p	BSIO	28GIA/WIE1

53HIL/GIA sample contained 0.155 mol.% Ammonium hydroxide; results corrected to pure Ammonium oxide

TABLE 2.93.3. Parameters of regression polynomial

Heat capacity type	No. data points total	No. data points used	s_w	s/R	s_r %	s_b/R	+/–
C_p	14	14	0.155	1.15–2	0.05	5.72–6	–1

Temp. range K	A_1	A_2	A_3	A_4	Level of uncertainty
196.8–270.2	2.15748+1	–7.08088	6.17904	–1.02491	II

TABLE 2.93.4. Recommended values of heat capacities

Temp. (K)	200	210	220	230	240	250	260
c_p (J K^{-1}g^{-1})	3.821	3.906	3.990	4.072	4.152	4.227	4.298
C_p (J K^{-1}mol^{-1})	199.0	203.4	207.8	212.1	216.2	220.1	223.8

Temp. (K)	270
c_p (J K^{-1}g^{-1})	4.363
C_p (J K^{-1}mol^{-1})	227.2

Name: Mercury iodide (HgI_2)
Formula: HgI_2

CAS–RN: 7774–29–0
Group No.: 2–094
Molar Mass: 454.40

TABLE 2.94.1. Experimental heat capacities

Reference		Temp. range K	No. pnts	Error %	Purity % method	Type capacity	Calorimeter Type	Calorimeter Reference
07GUI1	N	523.2–603.0	2	4.00	not specified	C_p	DSIO	07GUI1

07GUI1 values calculated from temperature dependence of enthalpy by the author

TABLE 2.94.3. Parameters of regression polynomial

Heat capacity type	No. data points total	used	s_w	s/R	s_r %	s_b/R	+/–
C_p	2	2	0.000	0.00	0.00	0.00	0

Temp. range K	A_1	Level of uncertainty
523.2–603.0	1.26678+1	VI

TABLE 2.94.4. Recommended values of heat capacities

Temp. (K)	530	600
c_p (J K^{-1}g^{-1})	0.23	0.23
C_p (J K^{-1}mol^{-1})	105	105

Name: Tetraiodostannane
Formula: I_4Sn

CAS–RN: 7790–47–8
Group No.: 2–095
Molar Mass: 626.33

TABLE 2.95.1. Experimental heat capacities

Reference		Temp. K	Capac. J/(K.g)	Error %	Purity % method	Type capacity	Calorimeter Type	Calorimeter Reference
36NEG	N	430.40	0.2679	nosp	not specified	C_{avg}	DSIO	36NEG

36NEG average value in temperature range 418–443 K

Name: Lithium salt nitric acid
Formula: $LiNO_3$

CAS–RN: 7790–69–4
Group No.: 2–096
Molar Mass: 68.95

TABLE 2.96.1. Experimental heat capacities

Reference		Temp. range K	No. pnts	Error %	Purity %	Purity method	Type capacity	Calorimeter Type	Calorimeter Reference
09GOO/KAL	N	523.2–576.2	2	nosp	not specified		C_{avg}	DSIO	09GOO/KAL

09GOO/KAL constant value calculated from temperature dependence of enthalpy by the authors

TABLE 2.96.3. Parameters of regression polynomial

Heat capacity type	No. data points total	No. data points used	s_w	s/R	s_r %	s_b/R	+/–
C	2	2	0.000	0.00	0.00	0.00	0

Temp. range K	A_1	Level of uncertainty
523.2–576.2	1.35309+1	V

TABLE 2.96.4. Recommended values of heat capacities

Temp. (K)	525	575
c (J K^{-1}g^{-1})	1.63	1.63
C (J K^{-1}mol^{-1})	113	113

Name: Nitrogen oxide (NO)
Formula: NO

CAS–RN: 10102–43–9
Group No.: 2–097
Molar Mass: 30.01

TABLE 2.97.1. Experimental heat capacities

Reference	Temp. range K	No. pnts	Error %	Purity %	Purity method	Type capacity	Calorimeter Type	Calorimeter Reference
24EUC/KAR	115.4–117.0	3	3.00	not specified		C_{sat}	BSIO	24EUC/KAR
29JOH/GIA	112.8–120.6	6	1.00	99.999	melpt	C_{sat}	BSIO	28GIA/WIE1
57KER	113.4–155.7	11	nosp	99.999	estim	C_{sat}	BSIO	30WIE/HUB

TABLE 2.97.2. Correlated heat capacities

Reference	Temp. range K	No. pnts used	$\sigma_r C$ %	d_w	d/R	d_r %	d_b/R	+/–
				Selected data				
24EUC/KAR	115.4–117.0	3	3.00	0.262	6.83–2	0.78	5.82–2	3
29JOH/GIA	112.8–120.6	6	1.00	0.941	8.18–2	0.94	–4.30–2	–4
57KER	113.4–155.7	11	0.60#	1.095	9.14–2	0.66	9.44–3	1

TABLE 2.97.3. Parameters of cubic spline polynomials

Heat capacity type	No. data points total	used	s_w	s/R	s_r %	s_b/R	+/–
C_p	20	20	1.117	9.87–2	0.89	1.00–3	0
C_{sat}	20	20	1.119	9.86–2	0.89	1.05–3	0

Temp. range K	A_1	A_2	A_3	A_4	Level of uncertainty
112.8–140.0	–4.97348+2	1.25489+3	–1.05137+3	2.97920+2	IV
140.0–155.7	4.50967+3	–9.47444+3	6.61243+3	–1.52680+3	IV
112.8–140.0	–4.90222+2	1.23646+3	–1.03541+3	2.93282+2	IV
140.0–155.7	4.58450+3	–9.63795+3	6.73203+3	–1.55611+3	IV

TABLE 2.97.4. Recommended values of heat capacities

Temp. (K)	110	115	120	125	130	135	140
c_p (J K^{-1}g^{-1})	2.051	2.337	2.590	2.873	3.247	3.774	4.516
C_p (J K^{-1}mol^{-1})	61.53	70.11	77.72	86.21	97.43	113.2	135.5
c_{sat} (J K^{-1}g^{-1})	2.050	2.334	2.586	2.866	3.236	3.755	4.486
C_{sat} (J K^{-1}mol^{-1})	61.51	70.05	77.60	86.01	97.09	112.7	134.6

Temp. (K)	145	150	155
c_p (J K^{-1}g^{-1})	5.472	6.388	6.945
C_p (J K^{-1}mol^{-1})	164.2	191.7	208.4
c_{sat} (J K^{-1}g^{-1})	5.424	6.311	6.824
C_{sat} (J K^{-1}mol^{-1})	162.8	189.4	204.8

TABLE 2.97.5. Parameters of quasi-polynomial equation

Heat capacity type	No. data points total	used	s_w	s/R	s_r %	s_b/R	+/–
C_p	20	20	3.334	2.57–1	2.38	–6.03–2	–6

Temp. range K	T_c K	A_1	A_2	A_3	A_4	A_5	A_6	Level of uncertainty
112.8–155.7	180.00	1.84380+2	6.08075	–2.73952+2	1.67806+3	–2.67048+3	1.80793+3	V

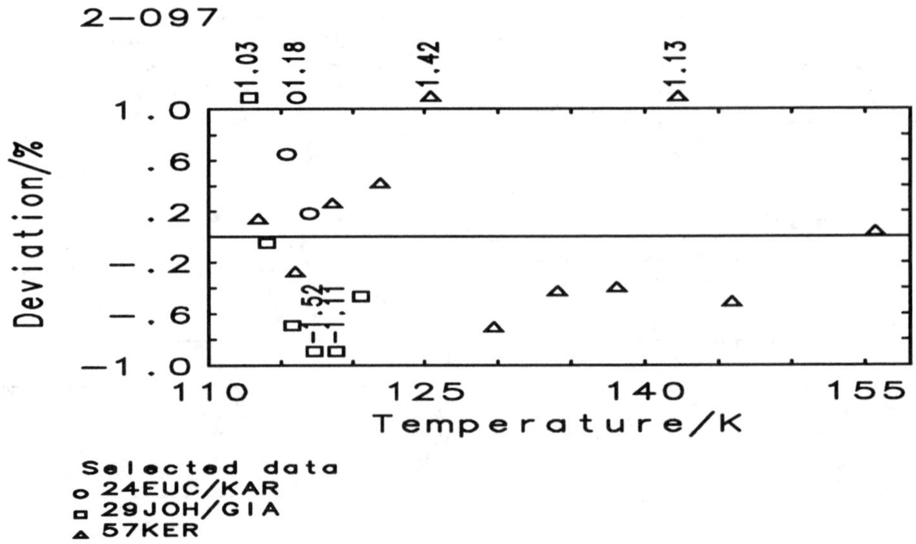

Name: Nitrogen oxide (NO_2)
Formula: NO_2

CAS–RN: 10102–44–0
Group No.: 2–098
Molar Mass: 46.01

TABLE 2.98.1. Experimental heat capacities

Reference	Temp. range K	No. pnts	Error %	Purity %	Purity method	Type capacity	Calorimeter Type	Calorimeter Reference
50SAG/HOU	294.3–377.6	16	2.00	not specified		C_p	BSAO	50SAG/HOU

TABLE 2.98.3. Parameters of regression polynomial

Heat capacity type	No. data points total	No. data points used	s_w	s/R	s_r %	s_b/R	+/–
C_p	16	16	0.193	3.61–2	0.39	3.22–4	4

Temp. range K	A_1	A_2	A_3	Level of uncertainty
294.3–377.6	6.86227+1	–3.92134+1	6.45157	V

TABLE 2.98.4. Recommended values of heat capacities

Temp. (K)	290	298.15	300	310	320	330	340
c_p (J $K^{-1}g^{-1}$)	1.66	1.64	1.63	1.64	1.66	1.71	1.79
C_p (J $K^{-1}mol^{-1}$)	76.2	75.3	75.2	75.3	76.5	78.8	82.1

Temp. (K)	350	360	370	380
c_p (J $K^{-1}g^{-1}$)	1.88	2.00	2.14	2.31
C_p (J $K^{-1}mol^{-1}$)	86.5	92.0	98.6	106

Name: Nitrogen oxide (N_2O)
Formula: N_2O

CAS–RN: 10024–97–2
Group No.: 2–099
Molar Mass: 44.01

TABLE 2.99.1. Experimental heat capacities

Reference	Temp. range K	No. pnts	Error %	Purity %	Purity method	Type capacity	Calorimeter Type	Calorimeter Reference
35BLU/GIA	183.6–187.1	5	nosp	99.999	estim	C_{sat}	BSIO	28GIA/WIE1
74ATA/CHI	183.1–184.4	6	nosp	99.999	melpt	C_{sat}	BSAO	74ATA/CHI

TABLE 2.99.2. Correlated heat capacities

Reference	Temp. range K	No. pnts used	$\sigma_r C$ %	d_w	d/R	d_r %	d_b/R	+/–
				Selected data				
35BLU/GIA	183.6–187.1	5	0.50#	1.227	5.73–2	0.61	2.14–2	2
74ATA/CHI	183.1–184.4	6	0.40#	1.573	5.83–2	0.63	–1.03–2	–2

TABLE 2.99.3. Parameters of regression polynomial

Heat capacity type	No. data points total	used	s_w	s/R	s_r %	s_b/R	+/–
C_{sat}	11	11	1.577	6.39–2	0.69	4.10–3	0

Temp. range K	A_1	A_2	Level of uncertainty
183.1–187.1	9.17947–1	4.54050	IV

TABLE 2.99.4. Recommended values of heat capacities

Temp. (K)	183	184	185	186	187
c_{sat} (J K^{-1}g^{-1})	1.743	1.752	1.760	1.769	1.777
C_{sat} (J K^{-1}mol^{-1})	76.72	77.10	77.47	77.85	78.23

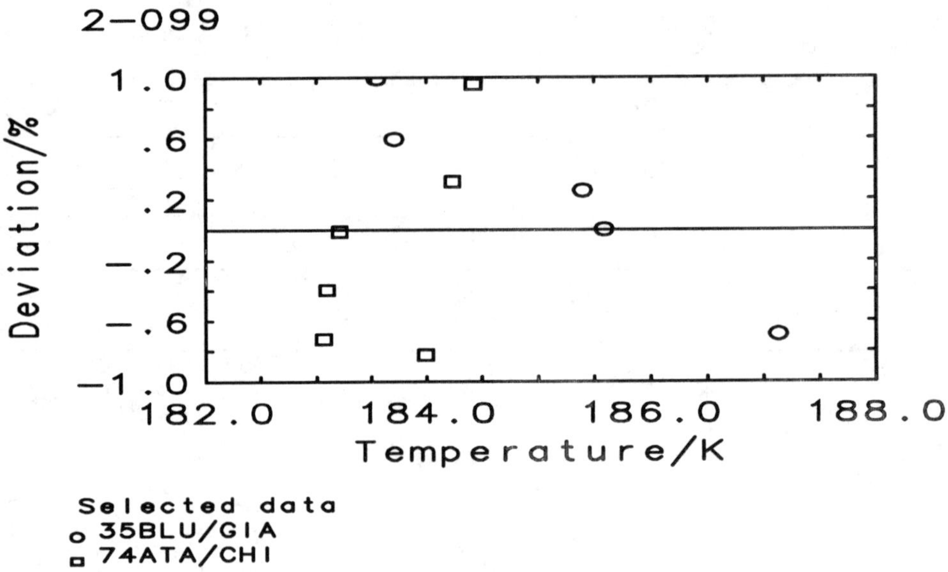

Name: Nitrogen oxide (N_2O_4)
Formula: N_2O_4

CAS–RN: 10544–72–6
Group No.: 2–100
Molar Mass: 92.01

TABLE 2.100.1. Experimental heat capacities

Reference	Temp. range K	No. pnts	Error %	Purity %	Purity method	Type capacity	Calorimeter Type	Calorimeter Reference
38GIA/KEM	265.4–291.3	6	nosp	99.99	melpt	C_p	BSIO	28GIA/WIE1

TABLE 2.100.3. Parameters of regression polynomial

Heat capacity type	No. data points total	used	s_w	s/R	s_r %	s_b/R	+/-
C_p	6	6	0.176	1.47-2	0.09	1.84-5	-1

Temp. range K	A_1	A_2	Level of uncertainty
265.4-291.3	1.14261+1	1.90591	IV

TABLE 2.100.4. Recommended values of heat capacities

Temp. (K)	270	273.15	280	290	298.15
c_p (J $K^{-1}g^{-1}$)	1.498	1.503	1.515	1.532	1.546
C_p (J $K^{-1}mol^{-1}$)	137.8	138.3	139.4	141.0	142.2

Name: Sulfur dioxide
Formula: O_2S

CAS-RN: 7446-09-5
Group No.: 2-101
Molar Mass: 64.06

TABLE 2.101.1. Experimental heat capacities

Reference		Temp. range K	No. pnts	Error %	Purity %	Purity method	Type capacity	Calorimeter Type	Calorimeter Reference
37PER2	N	202.3-291.4	23	nosp	not specified		C_{sat}	BSIO	37PER1
38GIA/STE		201.7-260.9	12	0.20	99.999	melpt	C_p	BSIO	37GIA/EGA
39RIE2		233.6-272.9	10	nosp	not specified		C_{sat}	BSIO	39RIE3

37PER2 sample contained traces of water (information in 39RIE2)

TABLE 2.101.2. Correlated heat capacities

Reference	Temp. range K	No. pnts used	$\sigma_r C$ %	d_w	d/R	d_r %	d_b/R	+/-
				Selected data				
38GIA/STE	201.7-260.9	12	0.20	0.522	1.09-2	0.10	-3.00-3	-1
39RIE2	233.6-273.0	10	0.70#	0.694	5.11-2	0.49	4.50-2	10
				Rejected data				
37PER2	(2.61-1, 2.43, 2.51-1, 19)							

TABLE 2.101.3. Parameters of regression polynomial

Heat capacity type	No. data points total	used	s_w	s/R	s_r %	s_b/R	+/-
C_p	45	22	0.670	3.91-2	0.37	1.88-2	9
C_{sat}	45	22	0.671	3.91-2	0.37	1.89-2	9

Temp. range K	A_1	A_2	A_3	A_4	Level of uncertainty
201.7-273.0	-1.03956+1	2.82843+1	-1.26047+1	1.84895	IV
201.7-273.0	-1.05434+1	2.83858+1	-1.26030+1	1.84108	IV

TABLE 2.101.4. Recommended values of heat capacities

Temp. (K)	210	220	230	240	250	260	270
c_p (J K^{-1}g^{-1})	1.368	1.364	1.360	1.355	1.353	1.354	1.360
C_p (J K^{-1}mol^{-1})	87.62	87.39	87.10	86.83	86.68	86.74	87.11
c_{sat} (J K^{-1}g^{-1})	1.368	1.364	1.359	1.355	1.352	1.353	1.358
C_{sat} (J K^{-1}mol^{-1})	87.62	87.39	87.09	86.81	86.63	86.66	86.98

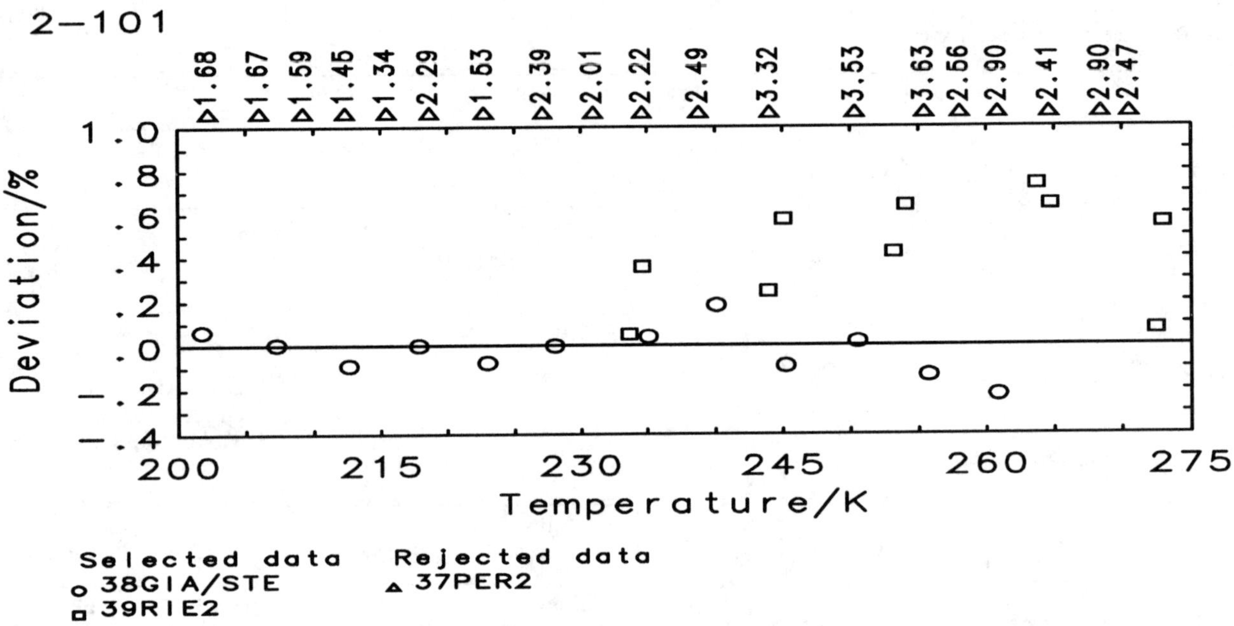

Name: Sulfur trioxide
Formula: O_3S

CAS–RN: 7446–11–9
Group No.: 2–102
Molar Mass: 80.06

TABLE 2.102.1. Experimental heat capacities

Reference		Temp. range K	No. pnts	Error %	Purity %	Purity method	Type capacity	Calorimeter Type	Calorimeter Reference
53GME	N	303.1	1	nosp	not specified		C_{avg}	not specified	
76MIL/SCH	N	290.0–473.0	eqn	nosp	not specified		C_p	not specified	
89KON/STR	N	290.1–302.0	7S	0.30	not specified		C_{sat}	BSAO	89KON/STR

53GME gamma form; average value in temperature range 298–308 K
76MIL/SCH unspecified origin of experimental data
89KON/STR SO_2 content 0.01 mass %

TABLE 2.102.2. Correlated heat capacities

Reference	Temp. range K	No. pnts used	$\sigma_r C$ %	d_w	d/R	d_r %	d_b/R	+/–
				Selected data				
76MIL/SCH	311.6–473.6	16	3.00#	1.093	1.34	3.28	–3.07–1	–2
89KON/STR	290.1–302.0	7	0.30	1.305	1.07–1	0.39	8.85–3	–1
				Rejected data				
53GME	(2.12, 6.84, 2.12, 1)							

TABLE 2.102.3. Parameters of cubic spline polynomials

Heat capacity type	No. data points total	used	s_w	s/R	s_r %	s_b/R	+/–
C_p	26	23	1.313	1.26	3.10	–2.11–1	–3
C_{sat}	26	23	1.394	1.35	3.30	–2.11–1	–3

Temp. range K	A_1	A_2	A_3	A_4	Level of uncertainty
290.1–305.0	–3.72738+3	3.52845+3	–1.11164+3	1.17580+2	IV
305.0–473.6	–8.91729+2	7.39284+2	–1.97155+2	1.76369+1	V
290.1–305.0	–4.31398+3	4.11091+3	–1.30428+3	1.38804+2	IV
305.0–473.6	–8.34788+2	6.88754+2	–1.82265+2	1.61787+1	V

TABLE 2.102.4. Recommended values of heat capacities

Temp. (K)	290	298.15	300	310	320	330	340
c_p ($J K^{-1} g^{-1}$)	2.485	2.833	2.898	3.20	3.43	3.60	3.73
C_p ($J K^{-1} mol^{-1}$)	199.0	226.8	232.1	256	275	289	299
c_{sat} ($J K^{-1} g^{-1}$)	2.484	2.832	2.897	3.19	3.42	3.60	3.73
C_{sat} ($J K^{-1} mol^{-1}$)	198.9	226.7	232.0	256	274	288	298
Temp. (K)	**350**	**360**	**370**	**380**	**390**	**400**	**410**
c_p ($J K^{-1} g^{-1}$)	3.82	3.89	3.94	3.99	4.05	4.12	4.23
C_p ($J K^{-1} mol^{-1}$)	306	311	315	319	324	330	338
c_{sat} ($J K^{-1} g^{-1}$)	3.82	3.88	3.94	3.98	4.03	4.09	4.18
C_{sat} ($J K^{-1} mol^{-1}$)	306	311	315	319	323	328	335
Temp. (K)	**420**	**430**	**440**	**450**	**460**	**470**	
c_p ($J K^{-1} g^{-1}$)	4.38	4.57	4.84	5.17	5.60	6.11	
C_p ($J K^{-1} mol^{-1}$)	350	366	387	414	448	489	
c_{sat} ($J K^{-1} g^{-1}$)	4.31	4.48	4.70	4.99	5.35	5.80	
C_{sat} ($J K^{-1} mol^{-1}$)	345	358	376	399	428	464	

TABLE 2.102.5. Parameters of quasi-polynomial equation

Heat capacity type	No. data points total	used	s_w	s/R	s_r %	s_b/R	+/–
C_p	26	23	12.610	1.80	5.94	5.09–1	5

Temp. range K	T_c K	A_1	A_2	A_3	A_4	A_5	A_6	Level of uncertainty
290.1–473.6	491.00	–3.41225	2.63366–2	2.06294+1	–2.28477	–1.45276+1	4.60172+1	VI

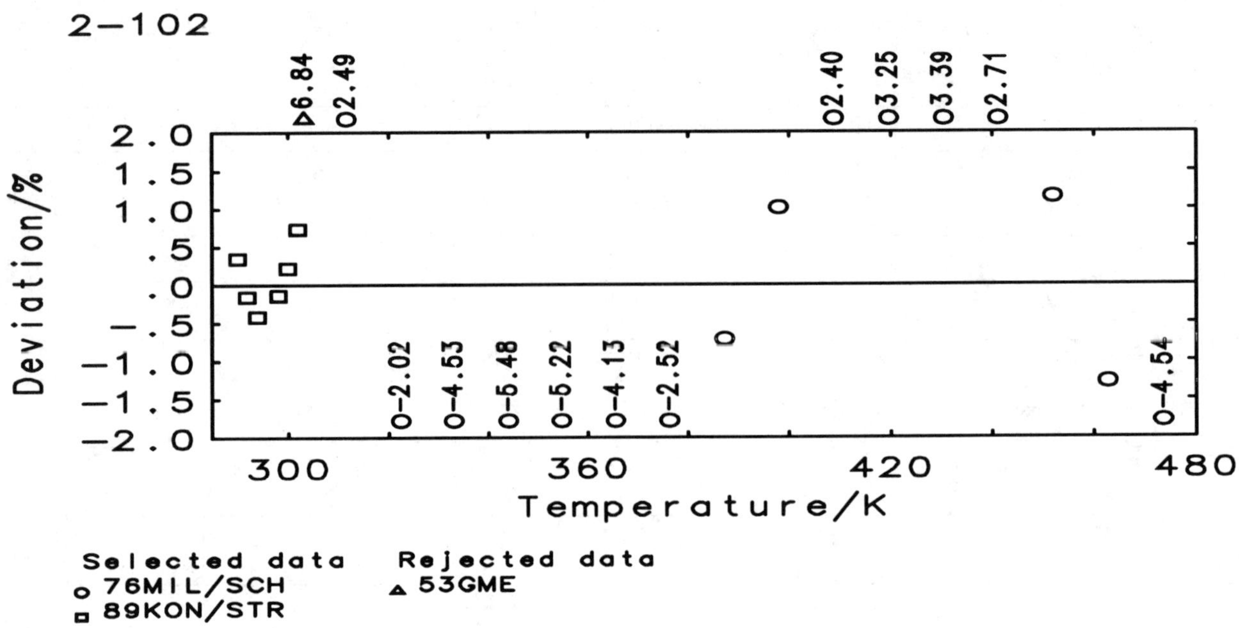

Name: Phosphorus oxide P_4O_6
Formula: O_6P_4

CAS–RN: 12440–00–5
Group No.: 2–103
Molar Mass: 219.89

TABLE 2.103.1. Experimental heat capacities

Reference	Temp. K	Capac. J/(K.g)	Error %	Purity %	Purity method	Type capacity	Calorimeter Type	Calorimeter Reference
86KAU/HEI	298.00	0.9509	0.65	not specified		C_p	BSIO	86KAU/HEI

1. Compounds of Carbon and Hydrogen

11. Saturated Aliphatic Hydrocarbons

This family consists of 83 saturated aliphatic hydrocarbons; for 7 compounds data are available at a single temperature only. A survey of experimental sources and the recommended data for the C_1 to C_{18} n-alkanes have been recently published (91RUZ/ZAB). For these compounds, some newer measurements have been considered when establishing recommendations for this monograph.

Most of the measurements in this family were performed at BMB in connection with investigations of properties of compounds contained in petroleum and coal. A total of 33 hydrocarbons containing five or more carbon atoms were measured in an adiabatic calorimeter (43RUE/HUF, stated measurement error of 0.2 %) using high purity samples (43GUT/HUF, 46DOU/HUF2, 52SCO/DOU, 54FIN/GRO2, 61HUF/GRO, 61MCC/MES, 67MES/GUT, 71MES/FIN, 73FIN/MES, 76FIN/MES). In view of the high reliability of these sources, all the data have been selected for the generation of recommended values.

NBSW is another laboratory that was systematically involved in the measurement of the heat capacities of alkanes. Osborne and Ginnings (47OSB/GIN) published data for 11 compounds determined over a narrow temperature interval (20 K) with a claimed error of 0.1 %. All the experimental values have been considered in the final selection for producing the recommended data. More recent measurements were carried out over a wide temperature range at NBSB using a spherical adiabatic calorimeter (61GOO). The measurements were carried out for C_1 to C_3 n-alkane (74YOU, 76ROD2, 78GOO) hydrocarbons from the melting temperature to the critical temperature with the stated measurement error around 0.5 %. These results have been the basis for recommendations and have been supplemented by accurate data from NRCO (65CUT/MOR) and from UCB (37WIT/KEM, 38KEM/EGA).

The low-temperature data for some C_4 hydrocarbons and 2,2-dimethylpropane obtained at the PSC (36AST/MES, 40AST/KEN, 40AST/MES) have been the basis for recommendations and have been supplemented by the superambient data from CITP (35SAG/LAC, 38SAG/LAC). The latter data are of lower accuracy as they were obtained by conversion from the high-pressure measurements. The data from the CITP have been also used for extending the temperature interval of the correlation for some other compounds (50AUE/SAG, 51CON/SAG).

The data from GPI for higher n-alkanes at superambient temperatures (75GRI/RAS, 82KUZ/KHA, 84GRI/AND, 88KUZ/KHA) have been mainly used for extending the low-temperature data from other sources to higher temperatures. The measurement error of 0.9 % reported by the authors is reasonable, and the data are fully compatible with other selected sources. This is not, however, the case for the data from APIB (76MUS, 77NAZ/MUS), where the reported error was much lower than the differences from other reliable data sources. The APIB data have been generally discarded with the exception of high-temperature values for dodecane where no other data were available.

Results for n-hexane from PASW (80KAL/JED) with a reported error of 0.2 % have been selected for the final correlation. Similarly, the data published for C_9, C_{11} and C_{16} n-alkanes from IITD (91BAN/GAR) having a reported error of 0.4 % have been considered as reliable and included in the final selection.

In the 1930's and the beginning of the 1940's, a laboratory at SUC (31HUF/PAR) was involved in the systematic measurement of heat capacities for a variety of organic compounds. Results gave reported errors of around 1 %. The data from this source have been considered when no other values were available for the following: 2,7-dimethyloctane (30PAR/HUF), C_{25} and C_{33} n-alkanes (32SPA/THO), and four isomers of methylnonanes (41PAR/WES). In the case of n-hexane, the values from SUC have been also included in the final correlation as they exhibited good agreement with other selected sources.

Measurements of heat capacities of n-heptane are very frequent as n-heptane is the calorimetric standard (solid and liquid) in the temperature range from 0 to 400 K. Very often this substance was used for calibration purposes or for testing the performance of a calorimeter. Data of this type are numerous in literature and have not been compiled by us. The recommended data for n-heptane as a calorimetric standard were originally established in the temperature range from 182.6 K to 400 K by Ginnings and Furukawa at NBSW (53GIN/FUR). In the liquid phase, the data were based on their extensive measurements performed in an adiabatic calorimeter from 185 K up to 370 K and on data from a drop calorimeter at higher temperatures up to 520 K (54DOU/FUR). These data have been supplemented in the temperature range between 280 and 360 K by additional data from another adiabatic calorimeter (47OSB/GIN). New recommendations for liquid n-heptane in the temperature range from the triple point temperature 182.603 K up to 480 K have been recently published by Zábranský and Růžička (94ZAB/RUZ). The new recommendations also considered the accurate data from BMB (61MCC/MES) summarized in the reference (61HUF/GRO) in addition to data from the above two sources. In all three sources of experimental data, detailed information is available regarding the performance of experiments and raw results; this made it possible to convert the original data to the actually valid temperature scale ITS-90. For more general discussion of conversion between different temperature scales see chapter 5.6. The parameters of cubic splines describing the recommended data for n-heptane in this monograph are identical with those published in the reference (94ZAB/RUZ). In addition, the parameters of a quasi-polynomial equation are also given for a larger temperature range up to the vicinity of the critical temperature.

No experimental data are available for branched alkanes higher than C_{11} with the exception of 11-decyleicosane. This compound was measured independently at SUC and at PSC (45FIS/MAY) with data from the two laboratories in good agreement.

There is also a lack of reliable data for n-alkanes higher than C_{19} with the single exception of n-hexacosane $C_{26}H_{54}$. This compound was measured over a narrow temperature interval of 26 K at NPLT (76AND/MAR). All remaining studies present usually experimental data on several higher molar mass n-alkanes and the reliability of results is low (48MAZ, 69ATK/LAR, 81HOE). This also holds true for data from Laboratoire de Thermodynamique, INPL-ENSIC, Nancy, France (93DUR/AOU), where a DSC calorimeter (SETARAM) was used for measurements of n-alkanes from C_{18} to C_{50}, with two branched alkanes C_{19} and squalane C_{30}; here, the uncertainty in the data is up to 5 %.

Name: Methane–d_4
Formula: CD_4

CAS–RN: 558–20–3
Group No.: 11–001
Molar Mass: 20.07

TABLE 11.1.1. Experimental heat capacities

Reference	Temp. range K	No. pnts	Error %	Purity %	Purity method	Type capacity	Calorimeter Type	Calorimeter Reference
40CLU/POP	92.0–108.9	9	nosp	99.2	melpt	C_p	BSIO	40CLU/POP
63COL/GIL	91.0–95.0	3	nosp	99.995	chrom	C_p	BSAO	61FLU/LEA

TABLE 11.1.2. Correlated heat capacities

Reference	Temp. range K	No. pnts used	$\sigma_r C$ %	d_w	d/R	d_r %	d_b/R	+/–
				Selected data				
40CLU/POP	92.0–108.9	9	3.00#	1.033	2.12–1	3.10	2.00–1	9
63COL/GIL	91.0–95.0	3	0.60#	1.227	4.80–2	0.74	–2.18–2	–1

TABLE 11.1.3. Parameters of regression polynomial

Heat capacity type	No. data points total	No. data points used	s_w	s/R	s_r %	s_b/R	+/–
C_p	12	12	1.189	2.02–1	2.97	1.45–1	8

Temp. range K	A_1	A_2	Level of uncertainty
91.0–108.9	4.51684	2.17371	V

TABLE 11.1.4. Recommended values of heat capacities

Temp. (K)	90	95	100	105	110
c_p (J K^{-1}g^{-1})	2.68	2.73	2.77	2.82	2.86
C_p (J K^{-1}mol^{-1})	53.8	54.7	55.6	56.5	57.4

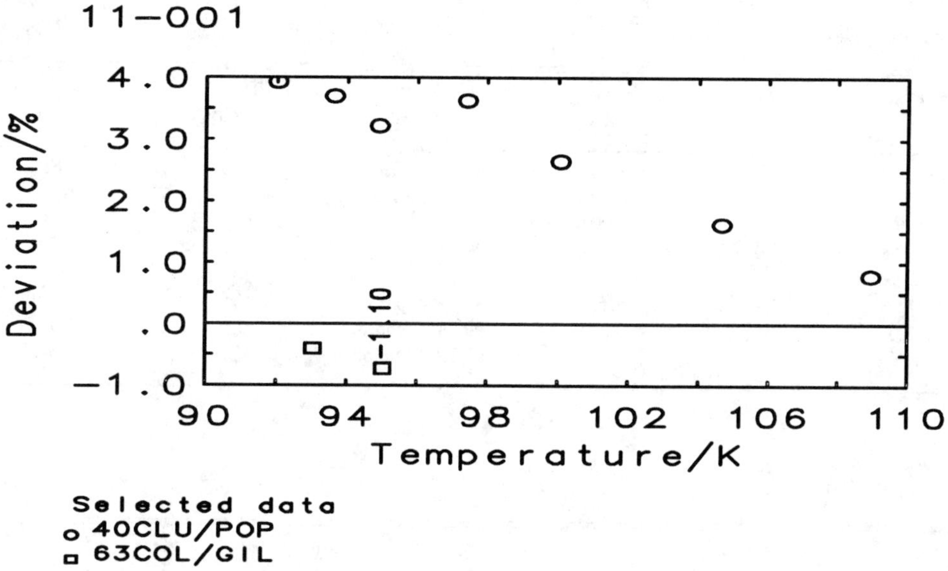

Name: Methane–*d*
Formula: CH_3D

CAS–RN: 676–49–3
Group No.: 11–002
Molar Mass: 17.05

TABLE 11.2.1. Experimental heat capacities

Reference	Temp. range K	No. pnts	Error %	Purity %	Purity method	Type capacity	Calorimeter Type	Calorimeter Reference
40CLU/POP	92.5–100.3	13	nosp	99.3	anal	C_p	BSIO	40CLU/POP

TABLE 11.2.3. Parameters of regression polynomial

Heat capacity type	No. data points total	No. data points used	s_w	s/R	s_r %	s_b/R	+/–
C_p	13	13	1.034	6.86–2	1.03	1.20–3	–1

Temp. range K	A_1	A_2	Level of uncertainty
92.5–100.3	4.88467	1.79811	V

TABLE 11.2.4. Recommended values of heat capacities

Temp. (K)	92.50	95	97.50	100
c_p (J $K^{-1}g^{-1}$)	3.19	3.22	3.24	3.26
C_p (J $K^{-1}mol^{-1}$)	54.4	54.8	55.2	55.6

Name: Methane
Formula: CH_4

CAS–RN: 74–82–8
Group No.: 11–003
Molar Mass: 16.04

TABLE 11.3.1. Experimental heat capacities

Reference	Temp. range K	No. pnts	Error %	Purity %	Purity method	Type capacity	Calorimeter Type	Calorimeter Reference
24EUC/KAR	96.3–108.7	7	3.00	not specified		C_{sat}	BSIO	24EUC/KAR
29CLU1	95.4–105.3	6	0.70	not specified		C_p	BSIO	29CLU1
30WIE/BRE	110.9–188.2	18	2.00	not specified		C_{sat}	BSIO	30WIE/HUB
61HES/WHI	114.5–187.5	9	nosp	99.96	melpt	C_{sat}	BSIO	53RIF/KER
63COL/GIL	93.4	1	nosp	99.995	chrom	C_p	BSAO	61FLU/LEA
65CUT/MOR	93.9–107.0	5	0.20	99.98	chrom	C_{sat}	BSAO	61FLU/LEA
74YOU	95.4–187.6	66	0.50	99.99	melpt	C_{sat}	BSAO	61GOO

TABLE 11.3.2. Correlated heat capacities

Reference	Temp. range K	No. pnts used	$\sigma_r C$ %	d_w	d/R	d_r %	d_b/R	+/–
				Selected data				
29CLU1	95.4–105.3	5	0.70	0.390	1.78–2	0.27	–5.19–3	–1
63COL/GIL	93.4	1	0.70#	0.597	2.66–2	0.42	–2.66–2	–1
65CUT/MOR	93.9–107.0	4	0.20	2.718	3.54–2	0.54	–2.93–2	–4
74YOU	95.4–184.2	61	0.50	2.294	1.24–1	1.15	1.62–2	12
				Rejected data				
24EUC/KAR	(2.79–1, 4.09, 2.67–1, 7)			30WIE/BRE	(1.11–1, 1.21, –9.50–2, –14)			
61HES/WHI	(4.32–1, 3.38, 2.76–1, 6)							

TABLE 11.3.3. Parameters of cubic spline polynomials

Heat capacity type	No. data points total	No. data points used	s_w	s/R	s_r %	s_b/R	+/–
C_p	112	71	2.344	1.21–1	1.13	1.15–2	6
C_{sat}	112	71	1.888	8.00–2	0.90	9.98–3	7

Temp. range K	A_1	A_2	A_3	A_4	Level of uncertainty
93.4–120.0	–7.80294	3.79280+1	–3.44502+1	1.08305+1	III
120.0–150.0	1.43599+1	–1.74792+1	1.17225+1	–1.99521	III
150.0–180.0	–4.62540+2	9.36320+2	–6.24144+2	1.39308+2	IV
180.0–184.2	–1.68405+5	2.80841+5	–1.56127+5	2.89361+4	V
93.4–120.0	2.20300–1	1.55051+1	–1.36198+1	4.39165	III
120.0–150.0	6.34746	1.87213–1	–8.54884–1	8.45838–1	III
150.0–180.0	–2.79355+2	5.71593+2	–3.81792+2	8.54985+1	IV
180.0–184.2	–9.66777+4	1.61236+5	–8.96395+4	1.66147+4	V

TABLE 11.3.4. Recommended values of heat capacities

Temp. (K)	100	110	120	130	140	150	160
c_p (J K^{-1}g^{-1})	3.372	3.446	3.533	3.661	3.830	4.034	4.339
C_p (J K^{-1}mol^{-1})	54.09	55.28	56.68	58.74	61.45	64.71	69.61
c_{sat} (J K^{-1}g^{-1})	3.367	3.442	3.526	3.630	3.760	3.918	4.150
C_{sat} (J K^{-1}mol^{-1})	54.02	55.22	56.56	58.24	60.32	62.85	66.58
Temp. (K)	170	180					
c_p (J K^{-1}g^{-1})	5.106	6.767					
C_p (J K^{-1}mol^{-1})	81.91	108.6					
c_{sat} (J K^{-1}g^{-1})	4.679	5.770					
C_{sat} (J K^{-1}mol^{-1})	75.06	92.57					

TABLE 11.3.5. Parameters of quasi-polynomial equation

Heat capacity type	No. data points total	used	s_w	s/R	s_r %	s_b/R	+/−
C_p	112	71	1.895	5.98–2	0.71	−1.95–3	−1

Temp. range K	T_c K	A_1	A_2	A_3	A_4	A_5	A_6	Level of uncertainty
93.4–184.2	190.55	−1.14331	3.00241–1	5.99170	−1.42092	−1.21632	1.01112	IV

11−003

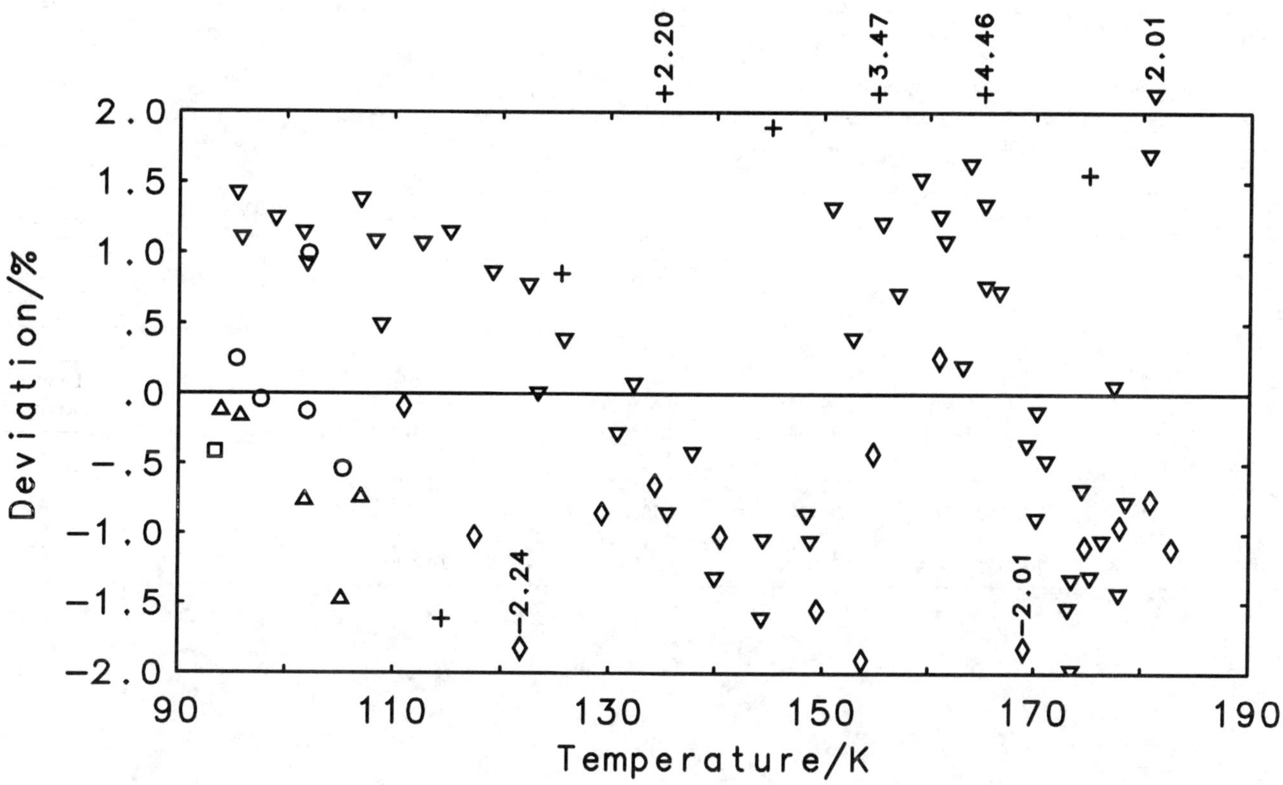

Selected data
○ 29CLU1
□ 63COL/GIL
△ 65CUT/MOR
▽ 74YOU

Rejected data
◇ 30WIE/BRE
+ 61HES/WHI

Name: Ethane
Formula: C_2H_6

CAS–RN: 74–84–0
Group No.: 11–004
Molar Mass: 30.07

TABLE 11.4.1. Experimental heat capacities

Reference		Temp. range K	No. pnts	Error %	Purity %	Purity method	Type capacity	Calorimeter Type	Calorimeter Reference
28EUC/HAU		100.0–270.0	18S	nosp	not specified		C_{sat}	BSIO	28EUC/HAU
30WIE/HUB	N	96.8–138.2	15	1.00	99.0	estim	C_{sat}	BSIO	30WIE/HUB
37WIT/KEM		91.6–180.9	29	0.40	99.5	melpt	C_p	BSIO	28GIA/WIE1
76ROD1		92.2	1	nosp	not specified		C_p	BSAO	61GOO
76ROD2		93.7–301.5	106	0.50	99.98	chrom	C_{sat}	BSAO	61GOO

30WIE/HUB error below and above n.b.t. 0.5 % and 1.0 %, respectively

TABLE 11.4.2. Correlated heat capacities

Reference	Temp. range K	No. pnts used	$\sigma_r C$ %	d_w	d/R	d_r %	d_b/R	+/–
				Selected data				
30WIE/HUB	96.8–138.2	15	1.00	0.423	3.52–2	0.42	4.33–3	2
37WIT/KEM	91.6–180.9	29	0.40	0.943	3.21–2	0.38	4.88–3	12
76ROD1	92.2	1	0.30#	1.511	3.73–2	0.45	3.73–2	1
76ROD2	93.7–300.5	105	0.50	1.524	1.21–1	0.76	–8.65–4	–14
				Rejected data				
28EUC/HAU	(1.70, 15.55, 1.64, 18)							

TABLE 11.4.3. Parameters of cubic spline polynomials

Heat capacity type	No. data points total	No. data points used	s_w	s/R	s_r %	s_b/R	+/–
C_p	169	150	1.386	1.05–1	0.69	1.02–3	1
C_{sat}	169	150	1.017	6.46–2	0.50	1.59–4	–5

Temp. range K	A_1	A_2	A_3	A_4	Level of uncertainty
91.6–150.0	5.15482	7.37389	–5.93045	1.64512	III
150.0–220.0	1.16122+1	–5.54078	2.67933	–2.68162–1	III
220.0–285.0	–1.45955+2	2.09323+2	–9.49860+1	1.45296+1	IV
285.0–300.6	–6.63205+4	6.98668+4	–2.45362+4	2.87315+3	V
91.6–150.0	5.78529	5.82763	–4.69303	1.32210	III
150.0–220.0	1.08828+1	–4.36730	2.10359	–1.88259–1	III
220.0–285.0	–9.30994+1	1.37427+2	–6.23482+1	9.57716	IV
285.0–300.6	–3.81156+4	4.01611+4	–1.41057+4	1.65208+3	V

TABLE 11.4.4. Recommended values of heat capacities

Temp. (K)	100	110	120	130	140	150	160
c_p (J K^{-1}g^{-1})	2.279	2.289	2.297	2.304	2.314	2.329	2.352
C_p (J K^{-1}mol^{-1})	68.54	68.84	69.06	69.28	69.58	70.05	70.74
c_{sat} (J K^{-1}g^{-1})	2.279	2.289	2.296	2.305	2.315	2.331	2.353
C_{sat} (J K^{-1}mol^{-1})	68.53	68.82	69.05	69.30	69.62	70.09	70.75
Temp. (K)	170	180	190	200	210	220	230
c_p (J K^{-1}g^{-1})	2.383	2.421	2.466	2.517	2.574	2.637	2.708
C_p (J K^{-1}mol^{-1})	71.66	72.80	74.15	75.68	77.40	79.28	81.43
c_{sat} (J K^{-1}g^{-1})	2.382	2.417	2.457	2.504	2.556	2.613	2.678
C_{sat} (J K^{-1}mol^{-1})	71.61	72.66	73.90	75.30	76.87	78.58	80.53
Temp. (K)	240	250	260	270	273.15	280	290
c_p (J K^{-1}g^{-1})	2.809	2.963	3.194	3.527	3.657	3.985	4.69
C_p (J K^{-1}mol^{-1})	84.46	89.09	96.05	106.1	110.0	119.8	141
c_{sat} (J K^{-1}g^{-1})	2.763	2.885	3.059	3.302	3.395	3.629	4.11
C_{sat} (J K^{-1}mol^{-1})	83.09	86.75	91.99	99.29	102.1	109.1	124
Temp. (K)	298.15	300					
c_p (J K^{-1}g^{-1})	7.01	8.04					
C_p (J K^{-1}mol^{-1})	211	242					
c_{sat} (J K^{-1}g^{-1})	5.52	6.13					
C_{sat} (J K^{-1}mol^{-1})	166	184					

TABLE 11.4.5. Parameters of quasi-polynomial equation

Heat capacity type	No. data points total	used	s_w	s/R	s_r %	s_b/R	+/–
C_p	169	150	0.817	8.00–2	0.40	–1.34–3	6

Temp. range K	T_c K	A_1	A_2	A_3	A_4	A_5	A_6	Level of uncertainty
91.6–300.5	305.33	–2.17287	2.58198–1	7.94568	–1.95607	–3.99364	3.70838	IV

11–004

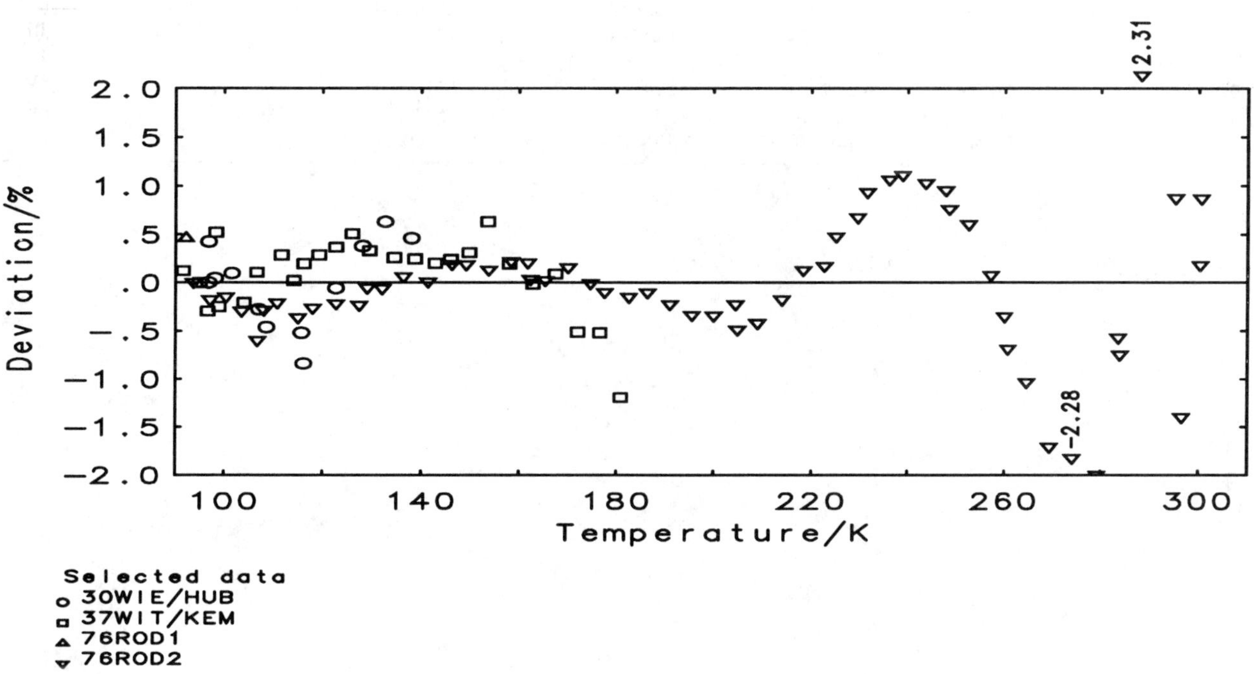

Name: Propane
Formula: C_3H_8

CAS–RN: 74–98–6
Group No.: 11–005
Molar Mass: 44.10

TABLE 11.5.1. Experimental heat capacities

Reference	Temp. range K	No. pnts	Error %	Purity %	Purity method	Type capacity	Calorimeter Type	Calorimeter Reference
26DAN/JEN	241.8–291.6	12	2.00	not specified		C_{sat}	BSIO	26DAN/JEN
38KEM/EGA	89.7–229.8	22	nosp	99.9	melpt	C_p	BSIO	28GIA/WIE1
65CUT/MOR	91.1–104.8	7	0.20	99.99	chrom	C_{sat}	BSAO	61FLU/LEA
78GOO	81.1–288.8	78	2.00	99.99	melpt	C_{sat}	BSAO	61GOO

TABLE 11.5.2. Correlated heat capacities

Reference	Temp. range K	No. pnts used	$\sigma_r C$ %	d_w	d/R	d_r %	d_b/R	+/–
				Selected data				
38KEM/EGA	89.7–229.8	22	0.50#	0.512	2.76–2	0.26	8.07–4	1
65CUT/MOR	91.1–104.8	7	0.20	1.678	3.40–2	0.34	–1.75–2	–3
78GOO	81.1–288.8	78	0.50#	0.501	2.57–2	0.25	9.87–3	34
				Rejected data				
26DAN/JEN	(4.88–1, 4.02, –3.47–1, –9)							

TABLE 11.5.3. Parameters of cubic spline polynomials

Heat capacity type	No. data points total	No. data points used	s_w	s/R	s_r %	s_b/R	+/–
C_p	119	107	0.668	2.75–2	0.27	6.22–3	32
C_{sat}	119	107	0.668	2.75–2	0.27	6.22–3	34

Temp. range K	A_1	A_2	A_3	A_4	Level of uncertainty
81.1–115.0	1.06718+1	–2.47874	2.76584	–7.79883–1	III
115.0–200.0	8.91893	2.09400	–1.21047	3.72671–1	III
200.0–288.8	8.45191	2.79454	–1.56074	4.31049–1	III
81.1–115.0	1.06179+1	–2.32023	2.61241	–7.31099–1	III
115.0–200.0	8.96953	1.97996	–1.12687	3.52752–1	III
200.0–288.8	9.77871	7.66188–1	–5.19990–1	2.51605–1	III

TABLE 11.5.4. Recommended values of heat capacities

Temp. (K)	90	100	110	120	130	140	150
c_p (J K^{-1}g^{-1})	1.907	1.919	1.933	1.948	1.964	1.980	1.998
C_p (J K^{-1}mol^{-1})	84.08	84.63	85.26	85.91	86.59	87.31	88.09
c_{sat} (J K^{-1}g^{-1})	1.907	1.919	1.933	1.948	1.964	1.980	1.998
C_{sat} (J K^{-1}mol^{-1})	84.08	84.63	85.25	85.91	86.59	87.31	88.09
Temp. (K)	160	170	180	190	200	210	220
c_p (J K^{-1}g^{-1})	2.017	2.039	2.063	2.090	2.121	2.155	2.194
C_p (J K^{-1}mol^{-1})	88.94	89.89	90.96	92.16	93.51	95.03	96.75
c_{sat} (J K^{-1}g^{-1})	2.017	2.039	2.063	2.090	2.120	2.154	2.192
C_{sat} (J K^{-1}mol^{-1})	88.94	89.90	90.96	92.15	93.49	94.99	96.67
Temp. (K)	230	240	250	260	270	273.15	280
c_p (J K^{-1}g^{-1})	2.238	2.287	2.342	2.403	2.471	2.494	2.546
C_p (J K^{-1}mol^{-1})	98.67	100.8	103.3	106.0	109.0	110.0	112.3
c_{sat} (J K^{-1}g^{-1})	2.235	2.282	2.333	2.390	2.453	2.474	2.521
C_{sat} (J K^{-1}mol^{-1})	98.54	100.6	102.9	105.4	108.2	109.1	111.2
Temp. (K)	290	298.15	300				
c_p (J K^{-1}g^{-1})	2.629	2.703	2.720				
C_p (J K^{-1}mol^{-1})	115.9	119.2	120.0				
c_{sat} (J K^{-1}g^{-1})	2.595	2.660	2.676				
C_{sat} (J K^{-1}mol^{-1})	114.4	117.3	118.0				

TABLE 11.5.5. Parameters of quasi-polynomial equation

Heat capacity type	No. data points total	No. data points used	s_w	s/R	s_r %	s_b/R	+/–
C_p	119	107	0.856	3.75–2	0.33	–3.53–3	–16

Temp. range K	T_c K	A_1	A_2	A_3	A_4	Level of uncertainty
81.1–288.8	369.85	–4.67531–1	1.02374	8.63274	5.33790–2	III

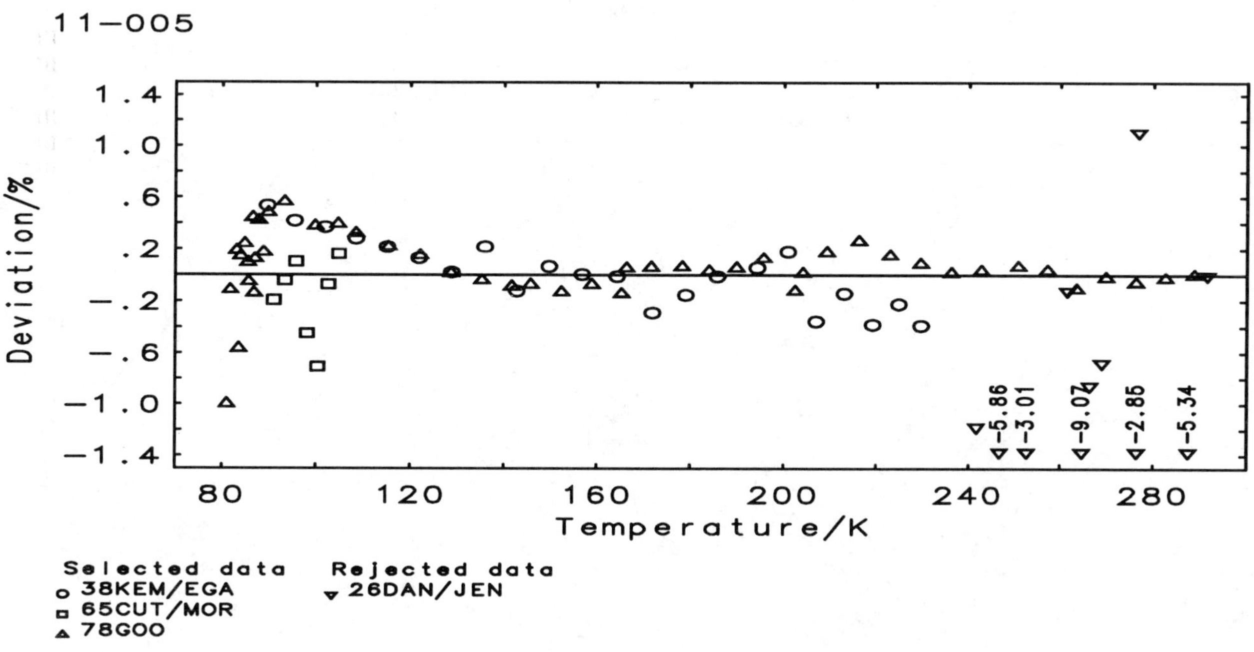

Name: Butane
Formula: C_4H_{10}

CAS–RN: 106–97–8
Group No.: 11–006
Molar Mass: 58.12

TABLE 11.6.1. Experimental heat capacities

Reference		Temp. range K	No. pnts	Error %	Purity %	Purity method	Type capacity	Calorimeter Type	Calorimeter Reference
26DAN/JEN		256.9–294.4	10	2.00	not specified		C_{sat}	BSIO	26DAN/JEN
31HUF/PAR		139.7–261.8	8	1.00	not specified		C_p	BSIO	25PAR
35SAG/LAC	N	295.4–365.0	18	nosp	99.21	anal	C_{sat}	BSAO	35SAG/LAC
40AST/MES		139.9–268.1	21	nosp	99.999	melpt	C_p	BSIO	36AST/MES

35SAG/LAC data from a graph only

TABLE 11.6.2. Correlated heat capacities

Reference	Temp. range K	No. pnts used	$\sigma_r C$ %	d_w	d/R	d_r %	d_b/R	+/–
				Selected data				
35SAG/LAC	295.4–365.0	18	1.50#	0.811	2.21–1	1.22	–3.53–2	–4
40AST/MES	139.9–268.1	21	0.30#	1.070	4.78–2	0.32	1.70–3	–3
				Rejected data				
26DAN/JEN	(3.72–1, 2.22, 1.58–1, 4)			31HUF/PAR	(1.04–1, 0.73, –9.38–2, –7)			

TABLE 11.6.3. Parameters of cubic spline polynomials

Heat capacity type	No. data points total	No. data points used	s_w	s/R	s_r %	s_b/R	+/–
C_p	57	39	1.043	1.67–1	0.93	–1.54–2	–7
C_{sat}	57	39	1.041	1.67–1	0.94	–1.56–2	–7

Temp. range K	A_1	A_2	A_3	A_4	Level of uncertainty
139.9–210.0	4.83762	1.35958+1	–7.10150	1.34439	III
210.0–300.0	1.67875+1	–3.47554	1.02771	5.40420–2	IV
300.0–365.0	–9.55521+1	1.08864+2	–3.64188+1	4.21477	V
139.9–210.0	4.97370	1.33646+1	–6.97262	1.32080	III
210.0–300.0	1.72549+1	–4.17994	1.38194	–5.31594–3	IV
300.0–365.0	–7.32884+1	8.63633+1	–2.87991+1	3.34814	V

TABLE 11.6.4. Recommended values of heat capacities

Temp. (K)	140	150	160	170	180	190	200
c_p (J K^{-1}g^{-1})	1.951	1.973	1.991	2.007	2.023	2.039	2.057
C_p (J K^{-1}mol^{-1})	113.4	114.7	115.7	116.7	117.6	118.5	119.5
c_{sat} (J K^{-1}g^{-1})	1.951	1.973	1.991	2.007	2.023	2.039	2.057
C_{sat} (J K^{-1}mol^{-1})	113.4	114.7	115.7	116.7	117.6	118.5	119.6
Temp. (K)	210	220	230	240	250	260	270
c_p (J K^{-1}g^{-1})	2.077	2.102	2.130	2.162	2.198	2.238	2.283
C_p (J K^{-1}mol^{-1})	120.7	122.1	123.8	125.7	127.8	130.1	132.7
c_{sat} (J K^{-1}g^{-1})	2.077	2.102	2.130	2.161	2.197	2.237	2.280
C_{sat} (J K^{-1}mol^{-1})	120.7	122.1	123.8	125.6	127.7	130.0	132.5
Temp. (K)	273.15	280	290	298.15	300	310	320
c_p (J K^{-1}g^{-1})	2.298	2.332	2.385	2.431	2.442	2.50	2.57
C_p (J K^{-1}mol^{-1})	133.6	135.5	138.6	141.3	141.9	146	150
c_{sat} (J K^{-1}g^{-1})	2.294	2.327	2.378	2.423	2.433	2.49	2.56
C_{sat} (J K^{-1}mol^{-1})	133.4	135.3	138.2	140.8	141.4	145	149
Temp. (K)	330	340	350	360			
c_p (J K^{-1}g^{-1})	2.66	2.75	2.87	3.01			
C_p (J K^{-1}mol^{-1})	154	160	167	175			
c_{sat} (J K^{-1}g^{-1})	2.63	2.72	2.82	2.95			
C_{sat} (J K^{-1}mol^{-1})	153	158	164	171			

TABLE 11.6.5. Parameters of quasi-polynomial equation

Heat capacity type	No. data points total	used	s_w	s/R	s_r %	s_b/R	+/–
C_p	57	39	4.651	2.49–1	1.62	9.50–2	15
Temp. range K	T_c K	A_1	A_2	A_3	A_4		Level of uncertainty
139.9–365.0	425.25	–2.13769	7.79723–1	1.07168+1	1.46518		V

11–006

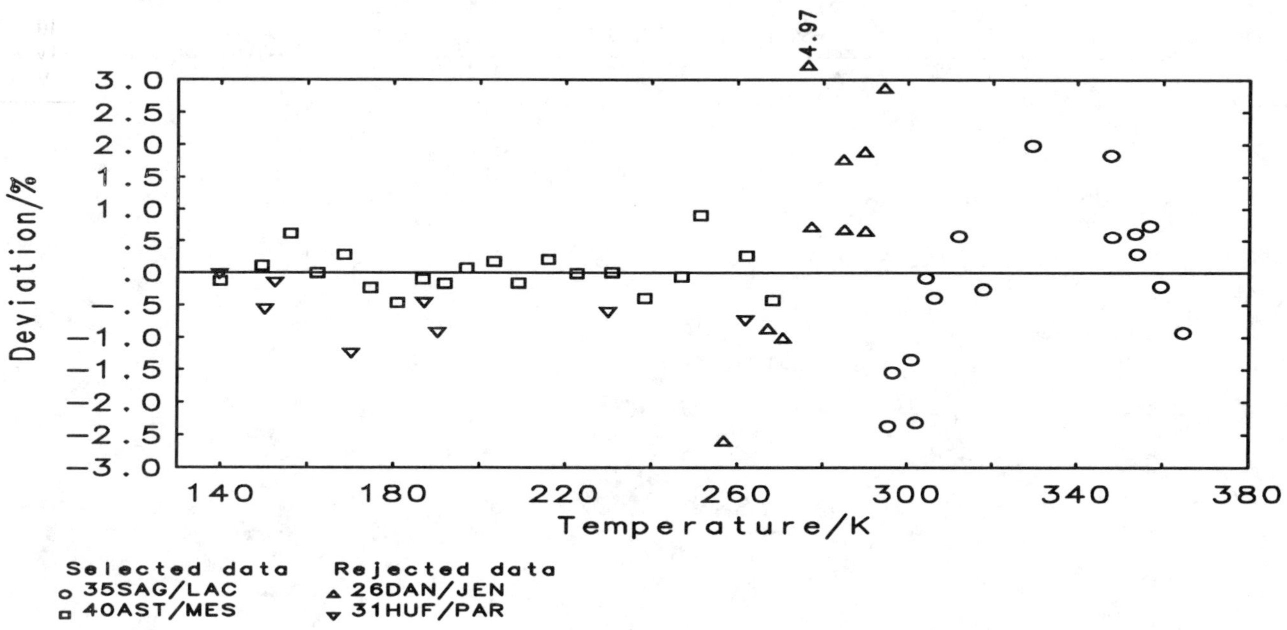

Name: 2–Methylpropane
Formula: C_4H_{10}

CAS–RN: 75–28–5
Group No.: 11–007
Molar Mass: 58.12

TABLE 11.7.1. Experimental heat capacities

Reference	Temp. range K	No. pnts	Error %	Purity %	Purity method	Type capacity	Calorimeter Type	Calorimeter Reference
26DAN/JEN	259.4–290.6	11	2.00	not specified		C_{sat}	BSIO	26DAN/JEN
37PAR/SHO	115.4–258.3	19	0.70	99.7	melpt	C_p	BSIO	25PAR
38SAG/LAC	294.3–377.6	6S	1.00	99.97	anal	C_p	BSAO	35SAG/LAC
40AST/KEN	116.9–257.0	25	nosp	99.9985	melpt	C_p	BSIO	36AST/MES

TABLE 11.7.2. Correlated heat capacities

Reference	Temp. range K	No. pnts used	$\sigma_r C$ %	d_w	d/R	d_r %	d_b/R	+/–
				Selected data				
37PAR/SHO	115.4–258.3	19	0.70	0.472	4.67–2	0.33	–2.58–2	–5
38SAG/LAC	294.3–377.6	6	1.00	0.712	1.24–1	0.71	–6.06–2	–4
40AST/KEN	116.9–257.0	25	0.40#	0.844	4.81–2	0.34	9.24–3	5
				Rejected data				
26DAN/JEN	(1.18, 6.65, 5.76–1, 5)							

TABLE 11.7.3. Parameters of cubic spline polynomials

Heat capacity type	No. data points total	No. data points used	s_w	s/R	s_r %	s_b/R	+/–
C_p	61	50	0.747	6.52–2	0.42	–1.25–2	–4
C_{sat}	61	50	0.724	6.54–2	0.43	–1.21–2	–2

Temp. range K	A_1	A_2	A_3	A_4	Level of uncertainty
115.4–250.0	1.11419+1	–3.29263–1	1.03263	–1.01518–1	III
250.0–377.6	–2.41559+1	4.20281+1	–1.59103+1	2.15754	V
115.4–250.0	1.08896+1	1.00531–1	7.99188–1	–6.11147–2	III
250.0–377.6	–1.31097+1	2.88997+1	–1.07205+1	1.47484	V

TABLE 11.7.4. Recommended values of heat capacities

Temp. (K)	120	130	140	150	160	170	180
c_p (J K^{-1}g^{-1})	1.725	1.750	1.778	1.807	1.837	1.869	1.903
C_p (J K^{-1}mol^{-1})	100.3	101.7	103.3	105.0	106.8	108.7	110.6
c_{sat} (J K^{-1}g^{-1})	1.725	1.750	1.778	1.807	1.838	1.870	1.903
C_{sat} (J K^{-1}mol^{-1})	100.2	101.7	103.3	105.0	106.8	108.7	110.6
Temp. (K)	190	200	210	220	230	240	250
c_p (J K^{-1}g^{-1})	1.938	1.974	2.012	2.051	2.090	2.131	2.172
C_p (J K^{-1}mol^{-1})	112.6	114.8	116.9	119.2	121.5	123.9	126.3
c_{sat} (J K^{-1}g^{-1})	1.938	1.974	2.011	2.050	2.089	2.130	2.172
C_{sat} (J K^{-1}mol^{-1})	112.6	114.7	116.9	119.1	121.4	123.8	126.2
Temp. (K)	260	270	273.15	280	290	298.15	300
c_p (J K^{-1}g^{-1})	2.22	2.26	2.28	2.31	2.37	2.42	2.43
C_p (J K^{-1}mol^{-1})	129	131	132	134	138	141	141
c_{sat} (J K^{-1}g^{-1})	2.21	2.26	2.27	2.31	2.36	2.41	2.42
C_{sat} (J K^{-1}mol^{-1})	129	131	132	134	137	140	141
Temp. (K)	310	320	330	340	350	360	370
c_p (J K^{-1}g^{-1})	2.50	2.59	2.69	2.81	2.94	3.09	3.26
C_p (J K^{-1}mol^{-1})	146	151	156	163	171	180	190
c_{sat} (J K^{-1}g^{-1})	2.49	2.56	2.65	2.74	2.85	2.98	3.11
C_{sat} (J K^{-1}mol^{-1})	145	149	154	160	166	173	181
Temp. (K)	380						
c_p (J K^{-1}g^{-1})	3.46						
C_p (J K^{-1}mol^{-1})	201						
c_{sat} (J K^{-1}g^{-1})	3.27						
C_{sat} (J K^{-1}mol^{-1})	190						

TABLE 11.7.5. Parameters of quasi-polynomial equation

Heat capacity type	No. data points total	used	s_w	s/R	s_r %	s_b/R	+/–
C_p	61	50	0.913	9.97–2	0.57	1.21–2	15

Temp. range K	T_c K	A_1	A_2	A_3	A_4	Level of uncertainty
115.4–377.6	408.20	–2.42548	3.44901–1	9.38882	4.26424	V

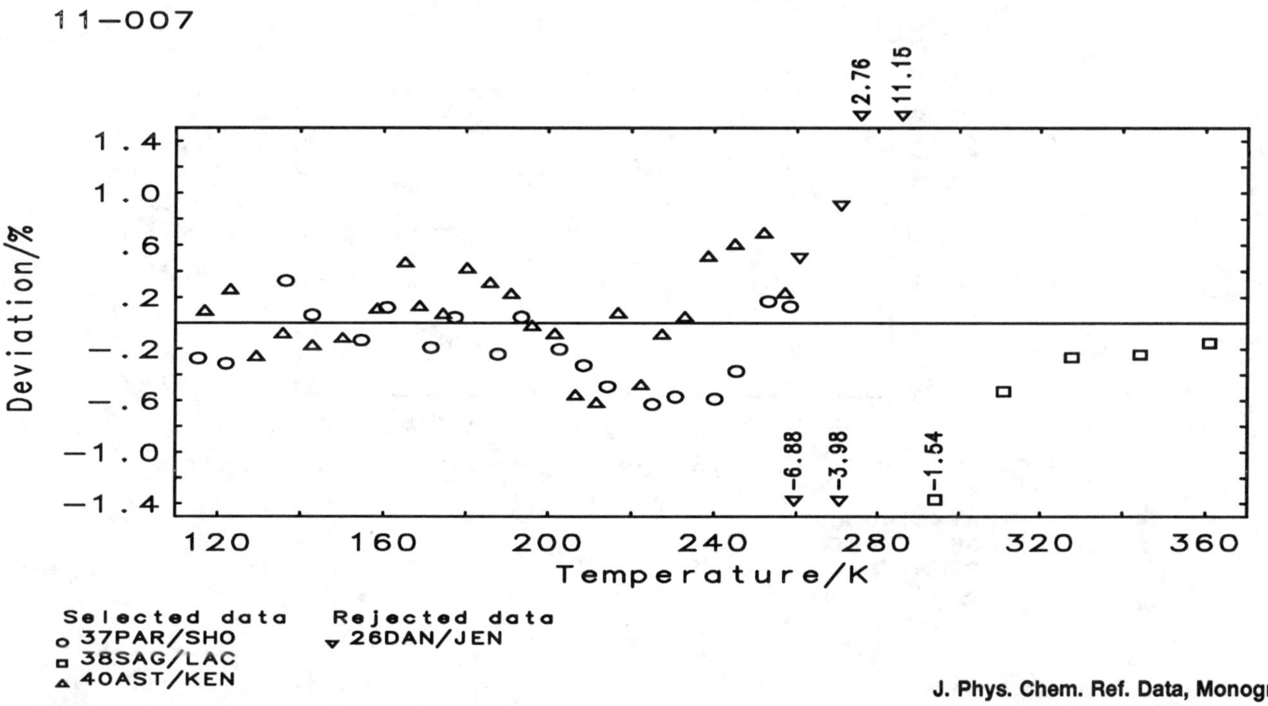

Name: 2,2–Dimethylpropane
Formula: C_5H_{12}

CAS–RN: 463–82–1
Group No.: 11–008
Molar Mass: 72.15

TABLE 11.8.1. Experimental heat capacities

Reference	Temp. range K	No. pnts	Error %	Purity %	Purity method	Type capacity	Calorimeter Type	Calorimeter Reference
36AST/MES	259.0–278.9	4	0.50	99.27	melpt	C_p	BSIO	36AST/MES
69ENO/SHI1	258.1–259.9	3	0.10	99.997	melpt	C_p	BSAO	66SHI/ATA

TABLE 11.8.2. Correlated heat capacities

Reference	Temp. range K	No. pnts used	$\sigma_r C$ %	d_w	d/R	d_r %	d_b/R	+/–
				Selected data				
36AST/MES	259.0–278.9	4	1.00#	0.984	1.86–1	0.98	–8.81–2	–2
69ENO/SHI1	258.0–259.9	3	0.30#	0.844	4.63–2	0.25	1.13–2	1

TABLE 11.8.3. Parameters of regression polynomial

Heat capacity type	No. data points total	No. data points used	s_w	s/R	s_r %	s_b/R	+/–
C_p	7	7	1.096	1.70–1	0.90	–4.55–2	–1

Temp. range K	A_1	A_2	Level of uncertainty
258.0–278.9	2.84682	5.97751	IV

TABLE 11.8.4. Recommended values of heat capacities

Temp. (K)	260	265	270	273.15	275	280
c_p (J K^{-1}g^{-1})	2.119	2.153	2.188	2.210	2.222	2.257
C_p (J K^{-1}mol^{-1})	152.9	155.4	157.9	159.4	160.3	162.8

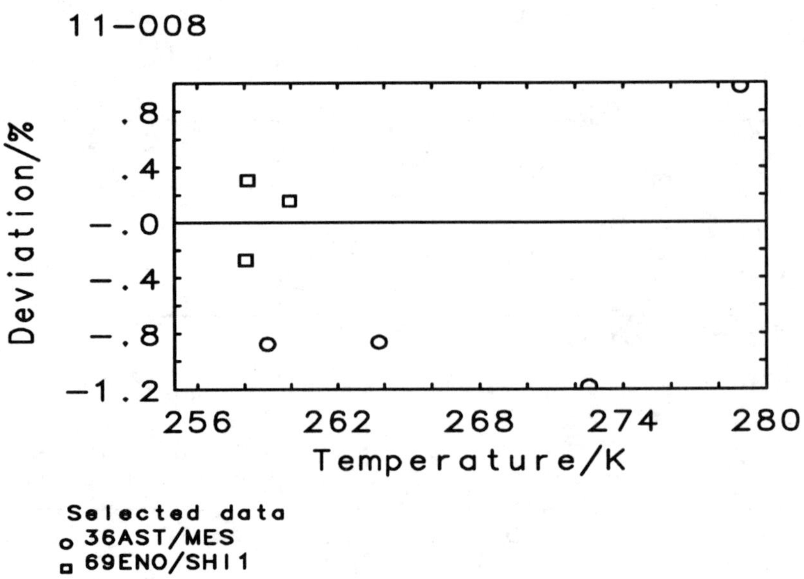

Name: 2–Methylbutane
Formula: C_5H_{12}

CAS–RN: 78–78–4
Group No.: 11–009
Molar Mass: 72.15

TABLE 11.9.1. Experimental heat capacities

Reference		Temp. range K	No. pnts	Error %	Purity %	Purity method	Type capacity	Calorimeter Type	Calorimeter Reference
30PAR/HUF2		120.5–275.8	14	1.00	not specified		C_p	BSIO	25PAR
42SCH/AST		120.0–290.0	13S	0.20	99.995	melpt	C_p	BSAO	39AST/EID1
43GUT/HUF	N	115.7–297.8	83	0.20	99.987	melpt	C_p	BSAO	43RUE/HUF
68SUG/ADA		113.4–298.3	88	nosp	99.986	melpt	C_{sat}	BSAO	65SUG/SEK
79VAN/ZEL2		120.0–270.0	16	nosp	99.9	anal	C_p	FSIO	79VAN/ZEL1
87CZA		289.0–298.3	3	1.00	not specified		C_p	BSIO	79CZA

43GUT/HUF smoothed values in 71MES/FIN

TABLE 11.9.2. Correlated heat capacities

Reference	Temp. range K	No. pnts used	$\sigma_r C$ %	d_w	d/R	d_r %	d_b/R	+/–
				Selected data				
43GUT/HUF	115.7–297.8	82	0.20	0.858	3.08–2	0.17	–2.42–2	–75
68SUG/ADA	113.4–298.3	87	0.20#	0.879	3.10–2	0.18	2.30–2	67
				Rejected data				
30PAR/HUF2	(8.71–2, 0.49, –6.76–2, –12)			42SCH/AST	(2.39–1, 1.21, 2.49–2, –7)			
79VAN/ZEL2	(1.45–1, 0.87, –1.40–1, –16)			87CZA	(1.02–1, 0.52, –1.01–1, –3)			

TABLE 11.9.3. Parameters of cubic spline polynomials

Heat capacity type	No. data points total	No. data points used	s_w	s/R	s_r %	s_b/R	+/–
C	217	169	0.882	3.14–2	0.18	1.07–4	–8

Temp. range K	A_1	A_2	A_3	A_4	Level of uncertainty
113.4–205.0	1.03413+1	6.50157	–3.03180	6.72000–1	II
205.0–298.3	1.77248+1	–4.30352	2.23898	–1.85037–1	II

TABLE 11.9.4. Recommended values of heat capacities

Temp. (K)	120	130	140	150	160	170	180
c ($J K^{-1} g^{-1}$)	1.722	1.745	1.768	1.791	1.813	1.836	1.860
C ($J K^{-1} mol^{-1}$)	124.2	125.9	127.6	129.2	130.8	132.5	134.2
Temp. (K)	190	200	210	220	230	240	250
c ($J K^{-1} g^{-1}$)	1.885	1.912	1.942	1.973	2.007	2.044	2.082
C ($J K^{-1} mol^{-1}$)	136.0	138.0	140.1	142.4	144.8	147.5	150.2
Temp. (K)	260	270	273.15	280	290	298.15	300
c ($J K^{-1} g^{-1}$)	2.123	2.165	2.178	2.209	2.254	2.292	2.301
C ($J K^{-1} mol^{-1}$)	153.1	156.2	157.2	159.4	162.6	165.4	166.0

TABLE 11.9.5. Parameters of quasi-polynomial equation

Heat capacity type	No. data points total	used	s_w	s/R	s_r %	s_b/R	+/–
C	217	169	1.081	3.85–2	0.22	1.55–4	–7

Temp. range K	T_c K	A_1	A_2	A_3	A_4	Level of uncertainty
113.4–298.3	460.40	–2.88905	1.59142	1.15863+1	1.31119	III

11–009

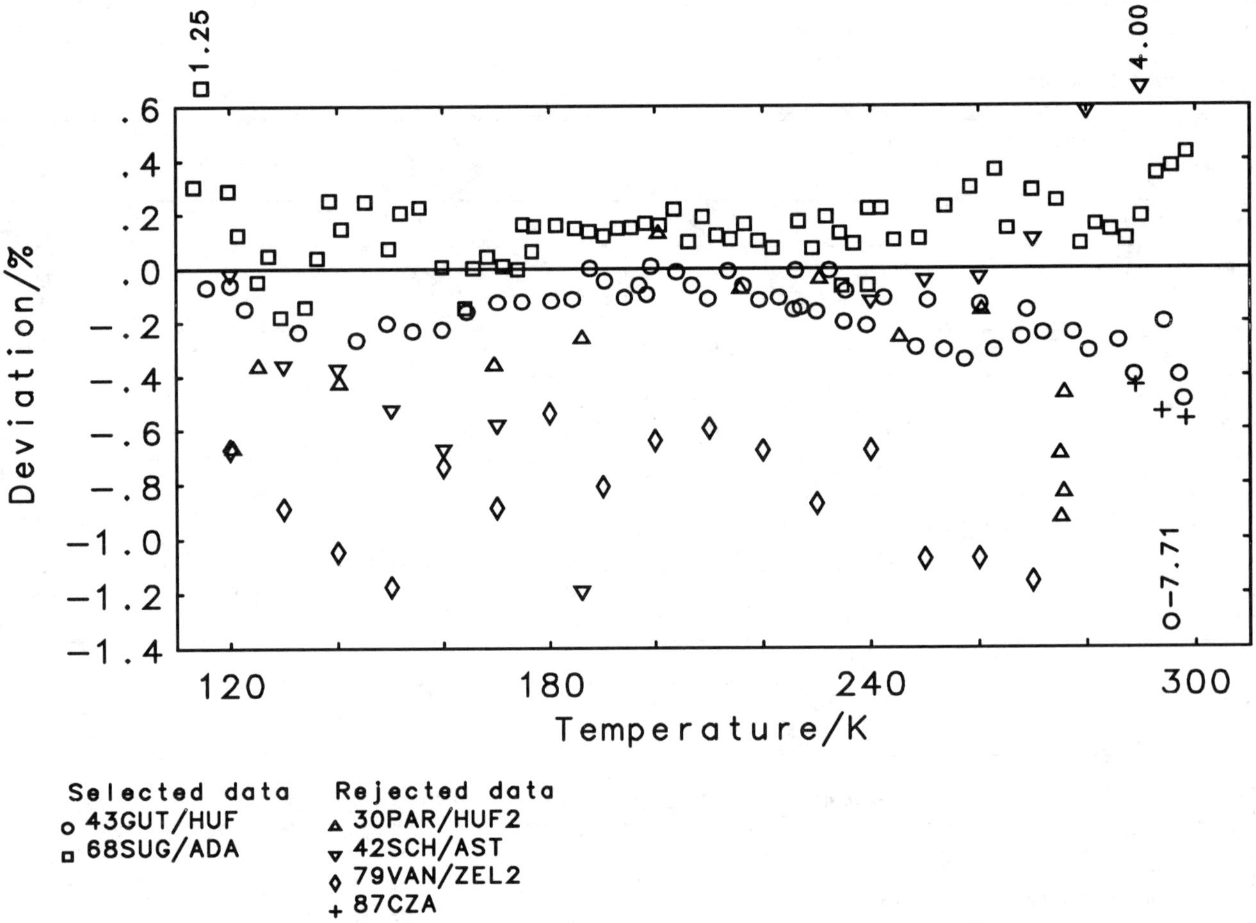

Name: Pentane
Formula: C_5H_{12}

CAS–RN: 109–66–0
Group No.: 11–010
Molar Mass: 72.15

TABLE 11.10.1. Experimental heat capacities

Reference		Temp. range K	No. pnts	Error %	Purity %	Purity method	Type capacity	Calorimeter Type	Calorimeter Reference
30PAR/HUF1		149.9–290.0	14	1.00	not specified		C_p	BSIO	25PAR
40MES/KEN		151.3–286.4	19	nosp	99.961	melpt	C_p	BSAO	39AST/EID1
67MES/GUT		148.6–302.9	25	0.20	99.86	melpt	C_{sat}	BSAO	47HUF
71AMI/ALI	N	313.1–468.1	19S	2.00	not specified		C_p	not specified	
75GRI/RAS	N	299.9–383.6	9	1.00	not specified		C_p	BDAO	75RAS/GRI
85CZA		299.0	1	nosp	not specified		C_p	BSIO	79CZA
88MEL/VER		153.1–433.1	10	2.50	not specified		C_{sat}	not specified	

71AMI/ALI calculated from Cv measured at the saturation line
75GRI/RAS all values (except the first one) at pressures above the vapour pressure

TABLE 11.10.2. Correlated heat capacities

Reference	Temp. range K	No. pnts used	$\sigma_r C$ %	d_w	d/R	d_r %	d_b/R	+/–
				Selected data				
67MES/GUT	148.6–302.9	25	0.10	0.235	4.39–3	0.02	2.35–4	2
75GRI/RAS	299.9–383.6	9	1.00	0.527	1.16–1	0.53	–6.76–2	–5
88MEL/VER	153.1–393.1	9	2.50	0.200	1.10–1	0.50	2.76–2	1
				Rejected data				
30PAR/HUF1	(1.11–1, 0.60, –8.97–2, –12)			40MES/KEN	(1.28–1, 0.66, 2.25–2, –5)			
71AMI/ALI	(3.91–1, 1.78, 3.37–1, 9)			85CZA	(2.86–1, 1.44, –2.86–1, –1)			

TABLE 11.10.3. Parameters of cubic spline polynomials

Heat capacity type	No. data points total	No. data points used	s_w	s/R	s_r %	s_b/R	+/–
C_p	97	43	0.338	7.90–2	0.36	–8.23–3	–2
C_{sat}	97	43	0.311	7.50–2	0.34	–7.60–3	–3

Temp. range K	A_1	A_2	A_3	A_4	Level of uncertainty
148.6–210.0	2.06484+1	–4.67090	1.47800	1.78866–2	II
210.0–290.0	2.35162+1	–8.76772	3.42886	–2.91773–1	II
290.0–393.1	1.33314+1	1.76823	–2.04216–1	1.25822–1	IV
148.6–210.0	2.04884+1	–4.40249	1.32945	4.49989–2	II
210.0–290.0	2.41030+1	–9.56614	3.78833	–3.45299–1	II
290.0–393.1	1.54904+1	–6.56541–1	7.16052–1	7.83648–3	IV

TABLE 11.10.4. Recommended values of heat capacities

Temp. (K)	150	160	170	180	190	200	210
c_p (J K^{-1}g^{-1})	1.962	1.963	1.967	1.974	1.986	2.001	2.019
C_p (J K^{-1}mol^{-1})	141.6	141.6	141.9	142.5	143.3	144.4	145.7
c_{sat} (J K^{-1}g^{-1})	1.962	1.963	1.967	1.974	1.986	2.001	2.019
C_{sat} (J K^{-1}mol^{-1})	141.6	141.6	141.9	142.5	143.3	144.4	145.7
Temp. (K)	220	230	240	250	260	270	273.15
c_p (J K^{-1}g^{-1})	2.042	2.067	2.096	2.128	2.163	2.201	2.213
C_p (J K^{-1}mol^{-1})	147.3	149.2	151.2	153.6	156.1	158.8	159.7
c_{sat} (J K^{-1}g^{-1})	2.042	2.067	2.096	2.128	2.163	2.200	2.213
C_{sat} (J K^{-1}mol^{-1})	147.3	149.2	151.3	153.6	156.1	158.8	159.6
Temp. (K)	280	290	298.15	300	310	320	330
c_p (J K^{-1}g^{-1})	2.241	2.283	2.319	2.327	2.374	2.422	2.474
C_p (J K^{-1}mol^{-1})	161.7	164.7	167.3	167.9	171.3	174.8	178.5
c_{sat} (J K^{-1}g^{-1})	2.240	2.282	2.317	2.325	2.370	2.418	2.466
C_{sat} (J K^{-1}mol^{-1})	161.6	164.6	167.2	167.8	171.0	174.4	178.0
Temp. (K)	340	350	360	370	380	390	
c_p (J K^{-1}g^{-1})	2.527	2.583	2.641	2.703	2.766	2.833	
C_p (J K^{-1}mol^{-1})	182.3	186.4	190.6	195.0	199.6	204.4	
c_{sat} (J K^{-1}g^{-1})	2.517	2.570	2.624	2.681	2.739	2.799	
C_{sat} (J K^{-1}mol^{-1})	181.6	185.4	189.3	193.4	197.6	201.9	

TABLE 11.10.5. Parameters of quasi-polynomial equation

Heat capacity type	No. data points total	used	s_w	s/R	s_r %	s_b/R	+/−
C_p	97	43	2.099	2.56–1	1.09	−2.45–3	−14

Temp. range K	T_c K	A_1	A_2	A_3	A_4	A_5	A_6	Level of uncertainty
148.6–393.1	469.80	5.13461+1	2.77849	1.31715+1	6.14601+1	−3.33298+1	1.00059+2	V

11–010

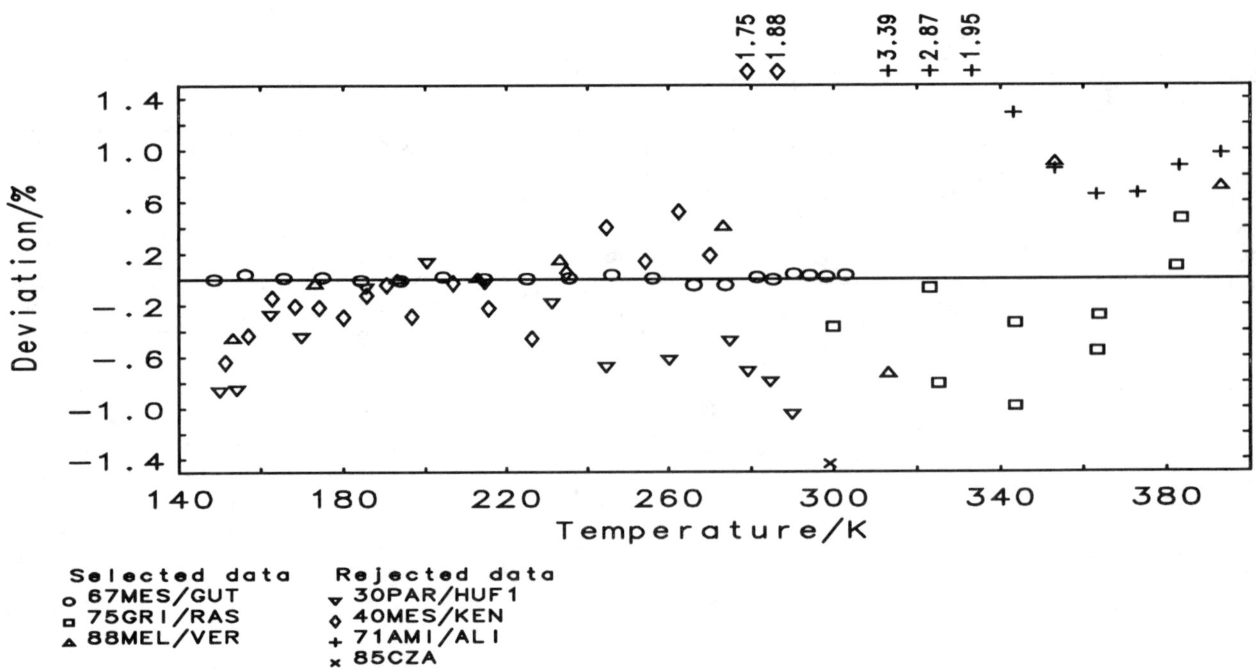

Name: 2,2–Dimethylbutane
Formula: C_6H_{14}

CAS–RN: 75–83–2
Group No.: 11–011
Molar Mass: 86.18

TABLE 11.11.1. Experimental heat capacities

Reference	Temp. range K	No. pnts	Error %	Purity %	Purity method	Type capacity	Calorimeter Type	Calorimeter Reference
37STU	180.0–320.0	15S	nosp	not specified		C_p	BDHO	37STU
46DOU/HUF2	177.3–296.1	17	0.20	99.985	melpt	C_p	BSAO	43RUE/HUF
46KIL/PIT	178.7–279.6	12	0.20	99.995	melpt	C_p	BSAO	45GUT/PIT
50AUE/SAG	299.8–366.5	13	nosp	99.0	melpt	C_p	BSAO	39SAG/EVA
83BEN/DAR	298.1	1	nosp	not specified		C_p	FSIT	71PIC/LED
84BEN/DAR	298.1	1	nosp	99.88	estim	C_p	FSIT	71PIC/LED
86BEN/DAR1	298.1	1	nosp	99.88	anal	C_p	FSIT	71PIC/LED
88COS/VAN	298.1	1	nosp	99.	anal	C_p	FSIT	71PIC/LED
89OHN/FUJ	298.1	1	nosp	99.9	chrom	C_p	FSIO	85OGA

TABLE 11.11.2. Correlated heat capacities

Reference	Temp. range K	No. pnts used	$\sigma_r C$ %	d_w	d/R	d_r %	d_b/R	+/–
				Selected data				
46DOU/HUF2	177.3–296.1	17	0.10	0.815	1.68–2	0.08	–4.75–3	–3
46KIL/PIT	178.7–279.6	12	0.20	0.922	3.93–2	0.18	2.02–2	4
50AUE/SAG	299.8–366.5	13	1.00#	0.619	1.47–1	0.62	9.69–2	5
83BEN/DAR	298.1	1	0.50#	0.000	3.81–5	0.00	–3.81–5	0
84BEN/DAR	298.1	1	0.50#	0.021	2.37–3	0.01	2.37–3	1
86BEN/DAR1	298.1	1	0.50#	0.316	3.60–2	0.16	3.60–2	1
89OHN/FUJ	298.1	1	0.30#	0.931	6.37–2	0.28	6.37–2	1
				Rejected data				
37STU	(7.00–1, 3.50, –6.22–1, –13)			88COS/VAN	(3.29–1, 1.43, 3.29–1, 1)			

TABLE 11.11.3. Parameters of cubic spline polynomials

Heat capacity type	No. data points total	No. data points used	s_w	s/R	s_r %	s_b/R	+/–
C_p	62	46	0.818	8.67–2	0.37	3.31–2	9
C_{sat}	62	46	0.936	8.63–2	0.37	3.32–2	11

Temp. range K	A_1	A_2	A_3	A_4	Level of uncertainty
177.3–270.0	1.54212+1	1.56234–2	8.74132–1	–1.84652–2	II
270.0–366.5	1.59731+1	–5.97585–1	1.10125	–4.65041–2	IV
177.3–270.0	1.73539+1	–2.65154	2.08685	–2.00112–1	II
270.0–366.5	1.12115+1	4.17332	–4.40878–1	1.11954–1	IV

TABLE 11.11.4. Recommended values of heat capacities

Temp. (K)	180	190	200	210	220	230	240
c_p (J K^{-1}g^{-1})	1.753	1.783	1.814	1.846	1.880	1.916	1.953
C_p (J K^{-1}mol^{-1})	151.1	153.7	156.3	159.1	162.0	165.1	168.3
c_{sat} (J K^{-1}g^{-1})	1.754	1.783	1.814	1.846	1.880	1.916	1.953
C_{sat} (J K^{-1}mol^{-1})	151.1	153.6	156.3	159.1	162.1	165.1	168.3
Temp. (K)	**250**	**260**	**270**	**273.15**	**280**	**290**	**298.15**
c_p (J K^{-1}g^{-1})	1.991	2.031	2.072	2.085	2.114	2.158	2.195
C_p (J K^{-1}mol^{-1})	171.6	175.0	178.5	179.7	182.2	186.0	189.1
c_{sat} (J K^{-1}g^{-1})	1.991	2.031	2.071	2.084	2.113	2.155	2.190
C_{sat} (J K^{-1}mol^{-1})	171.6	175.0	178.5	179.6	182.1	185.7	188.8
Temp. (K)	**300**	**310**	**320**	**330**	**340**	**350**	**360**
c_p (J K^{-1}g^{-1})	2.203	2.250	2.298	2.347	2.397	2.449	2.501
C_p (J K^{-1}mol^{-1})	189.9	193.9	198.0	202.2	206.6	211.0	215.5
c_{sat} (J K^{-1}g^{-1})	2.198	2.243	2.289	2.335	2.384	2.433	2.484
C_{sat} (J K^{-1}mol^{-1})	189.5	193.3	197.2	201.3	205.4	209.7	214.1

TABLE 11.11.5. Parameters of quasi-polynomial equation

Heat capacity type	No. data points total	used	s_w	s/R	s_r %	s_b/R	+/–
C_p	62	46	1.403	1.32–1	0.56	5.52–2	12

Temp. range K	T_c K	A_1	A_2	A_3	A_4	Level of uncertainty
177.3–366.5	488.80	–4.16296	5.24633–1	1.23638+1	8.25827	IV

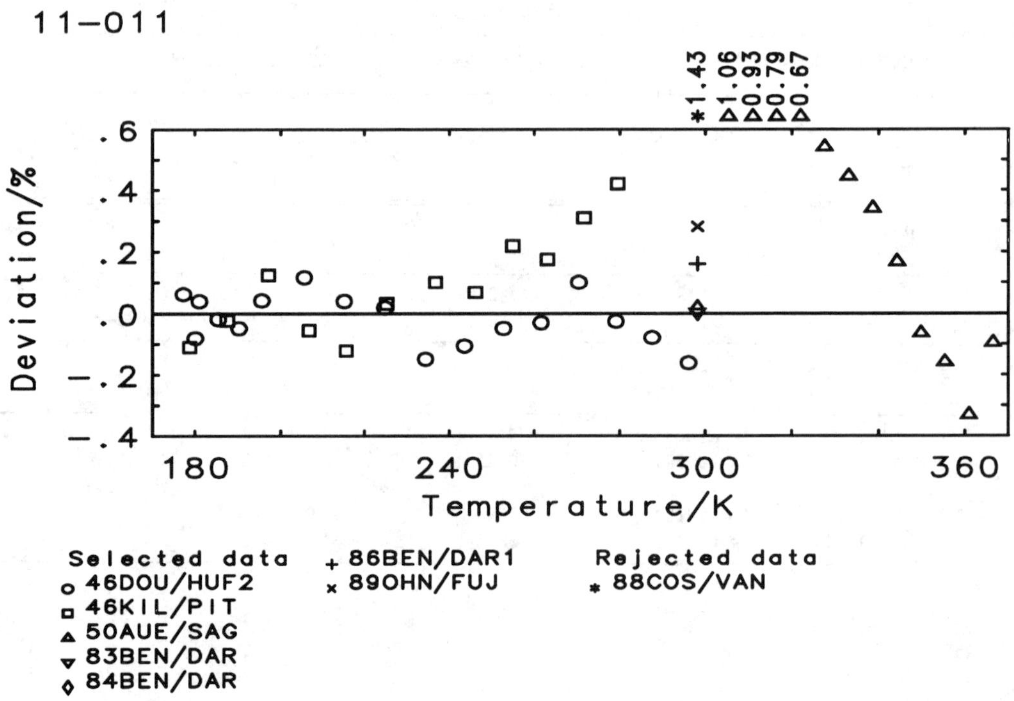

Name: 2,3–Dimethylbutane
Formula: C_6H_{14}

CAS–RN: 79–29–8
Group No.: 11–012
Molar Mass: 86.18

TABLE 11.12.1. Experimental heat capacities

Reference	Temp. range K	No. pnts	Error %	Purity %	Purity method	Type capacity	Calorimeter Type	Calorimeter Reference
37STU	140.0–320.0	19S	nosp	not specified		C_p	BDHO	37STU
46DOU/HUF2	149.4–306.5	38	0.20	99.985	melpt	C_p	BSAO	43RUE/HUF
71ADA/SUG	146.9–302.9	48	nosp	99.87	melpt	C_p	BSAO	65SUG/SEK
82WIL/FAR	298.1	1	nosp	99.	estim	C_p	FSIT	71PIC/LED
83BEN/DAR	298.1	1	nosp	not specified		C_p	FSIT	71PIC/LED
84BEN/DAR	298.1	1	nosp	99.98	melpt	C_p	FSIT	71PIC/LED
84FIL/LAU	293.0	1	2.00	98.	anal	C_p	BDHO	84FIL/LAU
86BEN/DAR1	298.1	1	nosp	99.88	anal	C_p	FSIT	71PIC/LED
89OHN/FUJ	298.1	1	nosp	99.9	chrom	C_p	FSIO	85OGA

TABLE 11.12.2. Correlated heat capacities

Reference	Temp. range K	No. pnts used	$\sigma_r C$ %	d_w	d/R	d_r %	d_b/R	+/–
				Selected data				
46DOU/HUF2	149.4–306.5	38	0.20#	1.045	4.23–2	0.21	–3.18–2	–29
71ADA/SUG	146.9–302.9	48	0.20#	1.020	4.41–2	0.20	2.59–2	41
82WIL/FAR	298.1	1	0.50#	0.283	3.21–2	0.14	–3.21–2	–1
83BEN/DAR	298.1	1	0.50#	0.315	3.57–2	0.16	–3.57–2	–1
84BEN/DAR	298.1	1	0.50#	0.421	4.78–2	0.21	–4.78–2	–1
86BEN/DAR1	298.1	1	0.50#	0.029	3.26–3	0.01	–3.26–3	–1
89OHN/FUJ	298.1	1	0.30#	0.083	5.67–3	0.02	–5.67–3	–1
				Rejected data				
37STU	(8.09–1, 4.07, 5.42–1, 8)			84FIL/LAU	(1.14–1, 0.51, –1.14–1, –1)			

TABLE 11.12.3. Parameters of regression polynomial

Heat capacity type	No. data points total	No. data points used	s_w	s/R	s_r %	s_b/R	+/–
C_p	111	91	1.027	4.37–2	0.21	–9.64–4	7

Temp. range K	A_1	A_2	A_3	A_4	Level of uncertainty
146.9–306.5	1.46648+1	1.50058	1.57352–1	8.30773–2	II

TABLE 11.12.4. Recommended values of heat capacities

Temp. (K)	150	160	170	180	190	200	210
c_p (J K^{-1}g^{-1})	1.693	1.718	1.744	1.771	1.800	1.829	1.860
C_p (J K^{-1}mol^{-1})	145.9	148.1	150.3	152.7	155.1	157.6	160.3
Temp. (K)	220	230	240	250	260	270	273.15
c_p (J K^{-1}g^{-1})	1.892	1.926	1.961	1.997	2.035	2.074	2.087
C_p (J K^{-1}mol^{-1})	163.1	166.0	169.0	172.1	175.4	178.8	179.8
Temp. (K)	280	290	298.15	300	310		
c_p (J K^{-1}g^{-1})	2.115	2.158	2.194	2.202	2.248		
C_p (J K^{-1}mol^{-1})	182.3	186.0	189.1	189.8	193.8		

TABLE 11.12.5. Parameters of quasi-polynomial equation

Heat capacity type	No. data points total	used	s_w	s/R	s_r %	s_b/R	+/–
C_p	111	91	1.075	4.64–2	0.22	–1.91–3	0

Temp. range K	T_c K	A_1	A_2	A_3	A_4	Level of uncertainty
146.9–306.5	500.00	–4.66092	1.52954	1.26220+1	3.55076	II

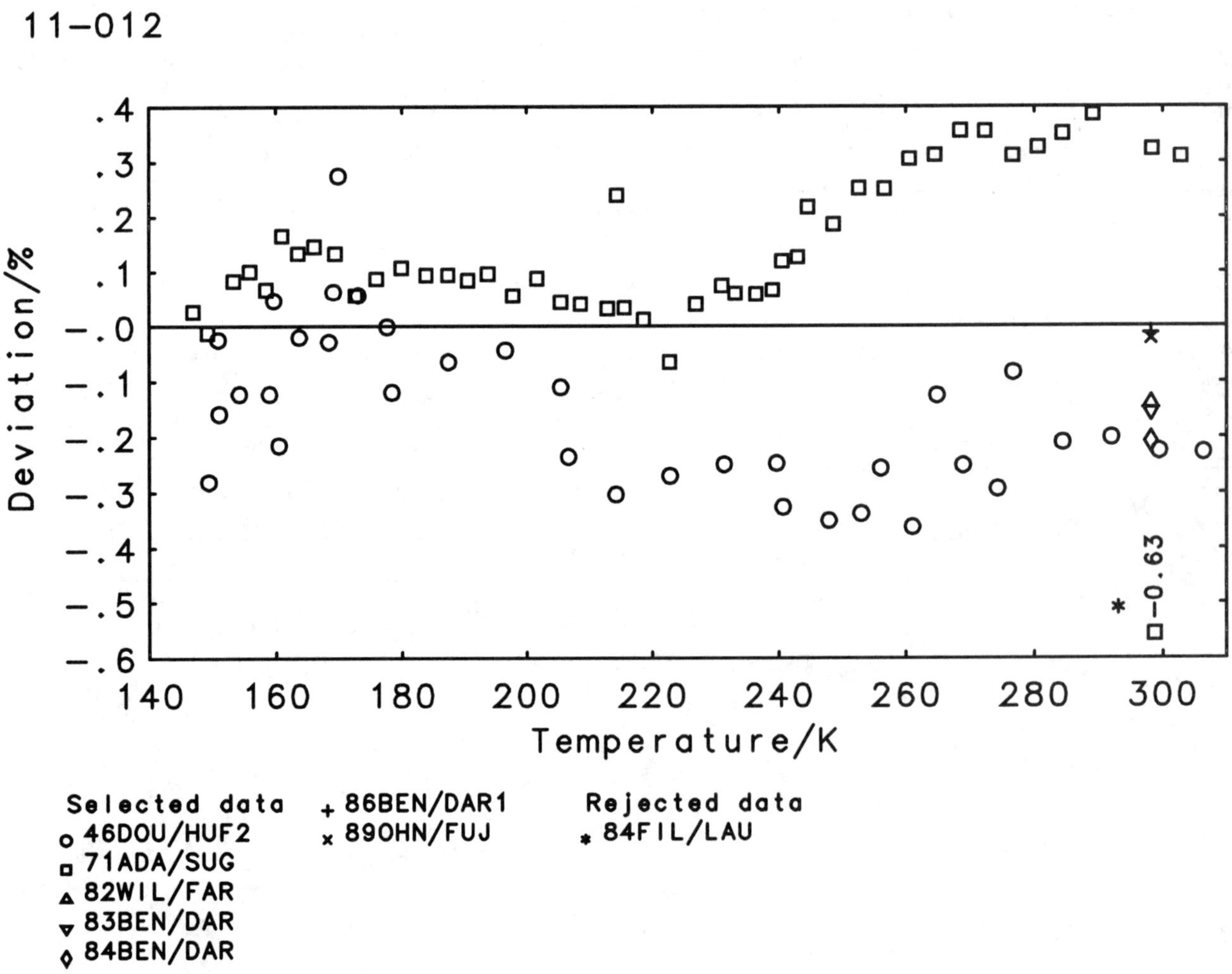

Name: Hexane
Formula: C_6H_{14}

CAS–RN: 110–54–3
Group No.: 11–013
Molar Mass: 86.18

TABLE 11.13.1. Experimental heat capacities

Reference		Temp. range K	No. pnts	Error %	Purity %	Purity method	Type capacity	Calorimeter Type	Calorimeter Reference
30PAR/HUF2		183.5–295.1	8	nosp	not specified		C_p	BSIO	25PAR
31HUF/PAR		188.8–293.5	5	1.00	not specified		C_p	BSIO	25PAR
37STU		180.0–320.0	15S	nosp	not specified		C_p	BDHO	37STU
39PHI		300.6	1	nosp	not specified		C_p	BSIO	49WEI
46DOU/HUF2	N	180.4–301.0	24	0.20	99.992	melpt	C_p	BSAO	43RUE/HUF
51CON/SAG		299.8–366.5	13S	0.70	not specified		C_p	BSAO	39SAG/EVA
69WIL/ROT		293.1	1	0.40	not specified		C_p	BDAO	65FIN/GRU
71AMI/ALI	N	343.1–506.1	19S	2.00	not specified		C_p	not specified	
71REC/SAD		303.1	1	0.30	not specified		C_p	BSIO	70REC
74DIA/REN		298.2–324.7	14	0.30	not specified		C_p	BSAO	74DIA/REN
75GRI/RAS	N	304.5–463.1	9	1.00	not specified		C_p	BDAO	75RAS/GRI
76KAR/GRO		298.2	1	nosp	not specified		C_p	FSIT	71PIC/LED
80KAL/JED		180.6–301.4	33	0.20	99.95	chrom	C_p	BSAO	80KAL/JED
81GRO/ING		298.1	1	nosp	99.5	chrom	C_p	FSIT	71PIC/LED
82WIL/ING		298.1	1	nosp	99.5	estim	C_p	FSIT	71PIC/LED
84BEN/DAR		298.1	1	nosp	99.98	melpt	C_p	FSIT	71PIC/LED
84BRA/PIN		298.1	1	nosp	99.0	melpt	C_p	FSIT	71PIC/LED
85COS/PAT1		298.2	1	nosp	99.	anal	C_p	FSIT	71PIC/LED
85COS/PAT8		298.1–313.1	2	nosp	99.	estim	C_p	FSIT	71PIC/LED
85CZA		299.9	1	nosp	not specified		C_p	BSIO	79CZA
86BEN/DAR1		298.1	1	nosp	99.88	anal	C_p	FSIT	71PIC/LED
86NAZ/BAS2		308.4–333.1	2	2.00	99.8	estim	C_p	BDHO	86NAZ/BAS1
88MEL/VER		183.1–473.1	11	2.50	not specified		C_{sat}	not specified	
88SAI/TAN		298.1	1	nosp	99.95	anal	C_p	FSIT	71PIC/LED
89OHN/FUJ		298.1	1	nosp	99.9	chrom	C_p	FSIO	85OGA
89VOG/SCH		333.1	1	nosp	not specified		C_p	BDHT	69PER/COM
91OGA/MIT		298.1	1	nosp	not specified		C_p	FSIO	85OGA
91TRE/COS		298.1	1	nosp	99.	anal	C_p	FSIT	71PIC/LED

46DOU/HUF2 smoothed data in 67MES/GUT
71AMI/ALI calculated from Cv measured at the saturation line
75GRI/RAS all values (except the first one) at pressures above the vapour pressure

TABLE 11.13.2. Correlated heat capacities

Reference	Temp. range K	No. pnts used	$\sigma_r C$ %	d_w	d/R	d_r %	d_b/R	+/–
				Selected data				
31HUF/PAR	188.8–293.5	5	1.00	0.242	5.46–2	0.24	–3.68–2	–1
46DOU/HUF2	180.4–301.0	24	0.20#	0.589	2.48–2	0.12	–2.99–3	–1
51CON/SAG	299.8–366.5	13	0.70	0.382	6.90–2	0.27	5.70–2	9
69WIL/ROT	293.1	1	0.40	1.053	9.78–2	0.42	–9.78–2	–1
75GRI/RAS	304.5–463.1	9	1.00	0.631	1.88–1	0.63	–3.49–2	–1
76KAR/GRO	298.2	1	0.50#	0.457	5.39–2	0.23	5.39–2	1
80KAL/JED	180.6–301.4	33	0.20	1.125	4.81–2	0.23	2.74–3	–1
81GRO/ING	298.1	1	0.50#	0.005	6.26–4	0.00	6.26–4	0
82WIL/ING	298.1	1	0.50#	0.343	4.03–2	0.17	–4.03–2	–1
84BEN/DAR	298.1	1	0.50#	0.138	1.63–2	0.07	1.63–2	1
84BRA/PIN	298.1	1	0.50#	0.343	4.03–2	0.17	–4.03–2	–1
86BEN/DAR1	298.1	1	0.50#	0.179	2.11–2	0.09	2.11–2	1
88MEL/VER	183.1–413.1	10	2.50	0.137	8.13–2	0.34	–3.73–2	–4
88SAI/TAN	298.1	1	0.50#	0.025	2.98–3	0.01	–2.98–3	–1
89OHN/FUJ	298.1	1	0.30#	0.760	5.35–2	0.23	–5.35–2	–1

Rejected data

30PAR/HUF2	(2.39–1, 1.08, –2.19–1, –8)	37STU	(1.24, 5.21, 6.35–1, 4)
39PHI	(1.26, 5.64, –1.26, –1)	71AMI/ALI	(5.45–1, 1.89, –3.33–1, –5)
71REC/SAD	(3.04–1, 1.30, –3.04–1, –1)	74DIA/REN	(6.32–1, 2.50, 5.70–1, 14)
85COS/PAT1	(2.38–1, 1.00, 2.38–1, 1)	85COS/PAT8	(2.88–1, 1.17, 1.57–1, 0)
85CZA	(1.80–1, 0.77, –1.80–1, –1)	86NAZ/BAS2	(5.19–1, 2.06, 5.15–1, 2)
89VOG/SCH	(8.22–1, 3.16, 8.22–1, 1)	91OGA/MIT	(1.46–1, 0.62, 1.46–1, 1)
91TRE/COS	(2.40–1, 1.01, 2.40–1, 1)		

TABLE 11.13.3. Parameters of cubic spline polynomials

Heat capacity type	No. data points total	No. data points used	s_w	s/R	s_r %	s_b/R	+/–
C_p	171	103	0.775	7.67–2	0.29	–2.47–3	–1
C_{sat}	171	103	0.764	6.78–2	0.27	–2.69–3	3

Temp. range K	A_1	A_2	A_3	A_4	Level of uncertainty
180.4–250.0	4.62532+1	–3.38510+1	1.40593+1	–1.77381	II
250.0–330.0	1.25487+1	6.59447	–2.11887	3.83276–1	III
330.0–463.1	5.66582+1	–3.35051+1	1.00325+1	–8.44136–1	IV
180.4–250.0	4.63534+1	–3.39839+1	1.41172+1	–1.78209	II
250.0–330.0	1.28287+1	6.24568	–1.97461	3.63490–1	III
330.0–463.1	6.93905+1	–4.51742+1	1.36072+1	–1.21043	IV

TABLE 11.13.4. Recommended values of heat capacities

Temp. (K)	180	190	200	210	220	230	240
c_p (J K^{-1}g^{-1})	1.981	1.980	1.987	2.001	2.020	2.044	2.072
C_p (J K^{-1}mol^{-1})	170.7	170.6	171.3	172.4	174.1	176.2	178.5
c_{sat} (J K^{-1}g^{-1})	1.981	1.980	1.987	2.001	2.020	2.044	2.072
C_{sat} (J K^{-1}mol^{-1})	170.7	170.6	171.3	172.4	174.1	176.2	178.5
Temp. (K)	250	260	270	273.15	280	290	298.15
c_p (J K^{-1}g^{-1})	2.101	2.133	2.166	2.177	2.201	2.238	2.270
C_p (J K^{-1}mol^{-1})	181.1	183.8	186.7	187.6	189.7	192.9	195.7
c_{sat} (J K^{-1}g^{-1})	2.101	2.133	2.166	2.177	2.201	2.238	2.270
C_{sat} (J K^{-1}mol^{-1})	181.1	183.8	186.7	187.6	189.7	192.9	195.6
Temp. (K)	300	310	320	330	340	350	360
c_p (J K^{-1}g^{-1})	2.278	2.320	2.365	2.413	2.464	2.518	2.574
C_p (J K^{-1}mol^{-1})	196.3	199.9	203.8	207.9	212.3	217.0	221.8
c_{sat} (J K^{-1}g^{-1})	2.278	2.320	2.364	2.412	2.462	2.515	2.570
C_{sat} (J K^{-1}mol^{-1})	196.3	199.9	203.8	207.9	212.2	216.8	221.5
Temp. (K)	370	380	390	400	410	420	430
c_p (J K^{-1}g^{-1})	2.632	2.691	2.751	2.811	2.871	2.930	2.988
C_p (J K^{-1}mol^{-1})	226.8	231.9	237.0	242.2	247.4	252.5	257.5
c_{sat} (J K^{-1}g^{-1})	2.626	2.682	2.738	2.792	2.845	2.896	2.943
C_{sat} (J K^{-1}mol^{-1})	226.3	231.1	235.9	240.6	245.2	249.5	253.6
Temp. (K)	440	450	460				
c_p (J K^{-1}g^{-1})	3.045	3.099	3.151				
C_p (J K^{-1}mol^{-1})	262.4	267.1	271.5				
c_{sat} (J K^{-1}g^{-1})	2.986	3.025	3.058				
C_{sat} (J K^{-1}mol^{-1})	257.3	260.7	263.6				

TABLE 11.13.5. Parameters of quasi-polynomial equation

Heat capacity type	No. data points total	used	s_w	s/R	s_r %	s_b/R	+/–
C_p	171	102	3.699	3.17–1	1.27	6.79–2	–8

Temp. range K	T_c K	A_1	A_2	A_3	A_4	Level of uncertainty
180.4–443.1	507.90	–3.44495	4.67981–1	1.56153+1	6.33982	V

11–013

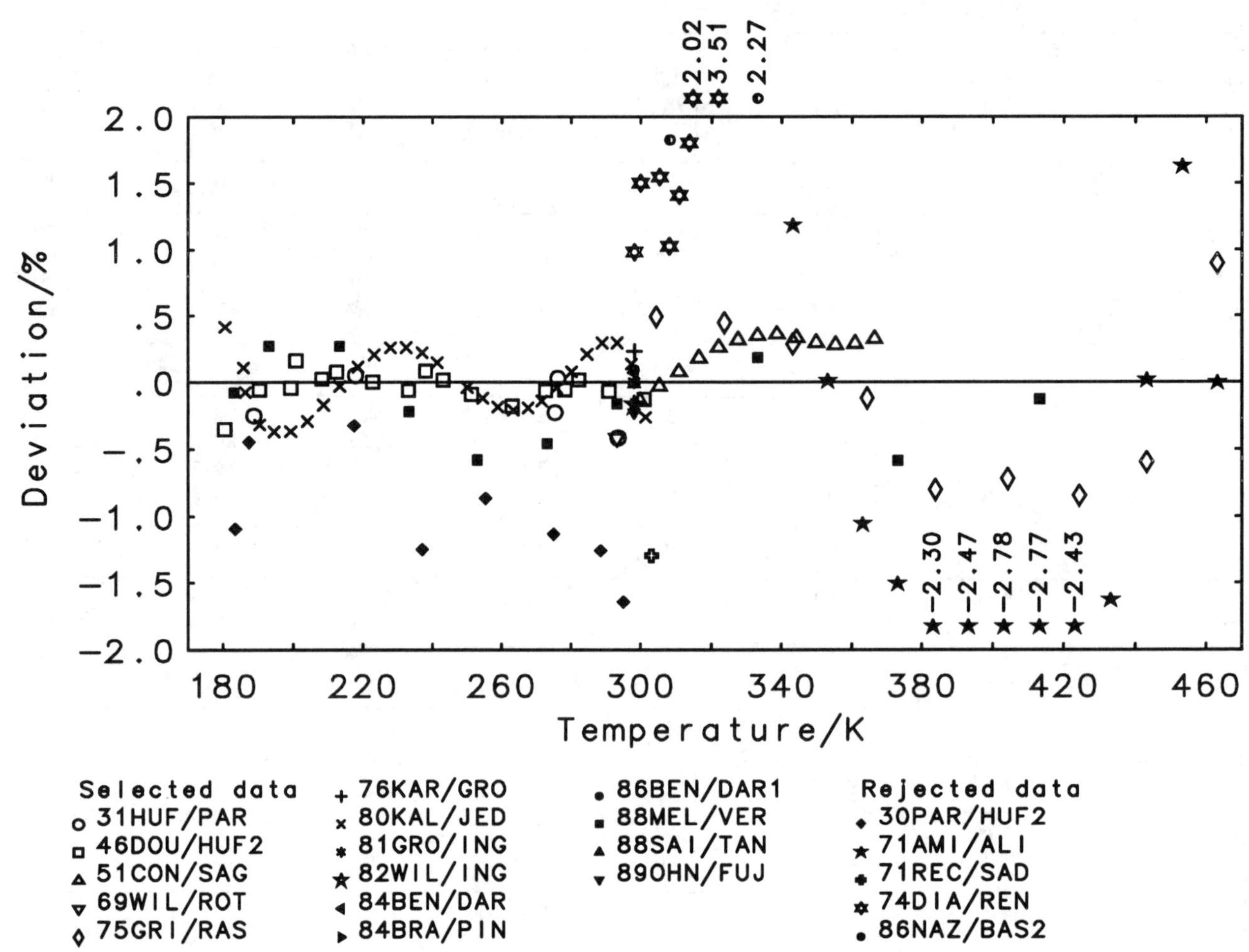

Name: 2–Methylpentane
Formula: C_6H_{14}

CAS–RN: 107–83–5
Group No.: 11–014
Molar Mass: 86.18

TABLE 11.14.1. Experimental heat capacities

Reference		Temp. range K	No. pnts	Error %	Purity %	Purity method	Type capacity	Calorimeter Type	Calorimeter Reference
37STU		120.0–320.0	21S	nosp	not specified		C_p	BDHO	37STU
46DOU/HUF2	N	117.4–303.3	28	0.20	99.971	melpt	C_p	BSAO	43RUE/HUF
83BEN/DAR		298.1	1	nosp	not specified		C_p	FSIT	71PIC/LED
84BEN/DAR		298.1	1	nosp	99.98	melpt	C_p	FSIT	71PIC/LED
84FIL/LAU		293.0	1	2.00	97.	anal	C_p	BDHO	84FIL/LAU
86BEN/DAR1		298.1	1	nosp	99.88	anal	C_p	FSIT	71PIC/LED
89OHN/FUJ		298.1	1	nosp	99.9	chrom	C_p	FSIO	85OGA

46DOU/HUF2 smoothed values in 71MES/FIN

TABLE 11.14.2. Correlated heat capacities

Reference	Temp. range K	No. pnts used	$\sigma_r C$ %	d_w	d/R	d_r %	d_b/R	+/−
				Selected data				
46DOU/HUF2	117.4–303.3	28	0.20	0.411	1.57–2	0.08	−3.61–4	2
83BEN/DAR	298.1	1	0.50#	0.132	1.53–2	0.07	1.53–2	1
84BEN/DAR	298.1	1	0.50#	0.214	2.50–2	0.11	2.50–2	1
86BEN/DAR1	298.1	1	0.50#	0.255	2.98–2	0.13	2.98–2	1
				Rejected data				
37STU	(7.01–1, 3.28, 5.78–1, 13)			84FIL/LAU	(8.35–1, 3.76, −8.35–1, −1)			
89OHN/FUJ	(5.74–2, 0.25, 5.74–2, 1)							

TABLE 11.14.3. Parameters of regression polynomial

Heat capacity type	No. data points total	used	s_w	s/R	s_r %	s_b/R	+/−
C_p	54	31	0.417	1.76–2	0.09	1.93–3	5

Temp. range K	A_1	A_2	A_3	Level of uncertainty
117.4–303.3	1.71510+1	−6.17987–1	8.98789–1	II

TABLE 11.14.4. Recommended values of heat capacities

Temp. (K)	120	130	140	150	160	170	180
c_p (J K^{-1}g^{-1})	1.708	1.724	1.741	1.760	1.781	1.804	1.828
C_p (J K^{-1}mol^{-1})	147.2	148.6	150.1	151.7	153.5	155.5	157.6
Temp. (K)	190	200	210	220	230	240	250
c_p (J K^{-1}g^{-1})	1.855	1.882	1.912	1.943	1.976	2.011	2.048
C_p (J K^{-1}mol^{-1})	159.8	162.2	164.8	167.5	170.3	173.3	176.5
Temp. (K)	260	270	273.15	280	290	298.15	300
c_p (J K^{-1}g^{-1})	2.086	2.126	2.139	2.168	2.211	2.248	2.256
C_p (J K^{-1}mol^{-1})	179.8	183.2	184.3	186.8	190.5	193.7	194.4

TABLE 11.14.5. Parameters of quasi-polynomial equation

Heat capacity type	No. data points total	used	s_w	s/R	s_r %	s_b/R	+/−
C_p	54	31	0.968	4.46–2	0.21	−6.52–3	−3

Temp. range K	T_c K	A_1	A_2	A_3	A_4	Level of uncertainty
117.4–303.3	497.50	−2.53223	3.33295	1.24549+1	4.80969–1	II

11-014

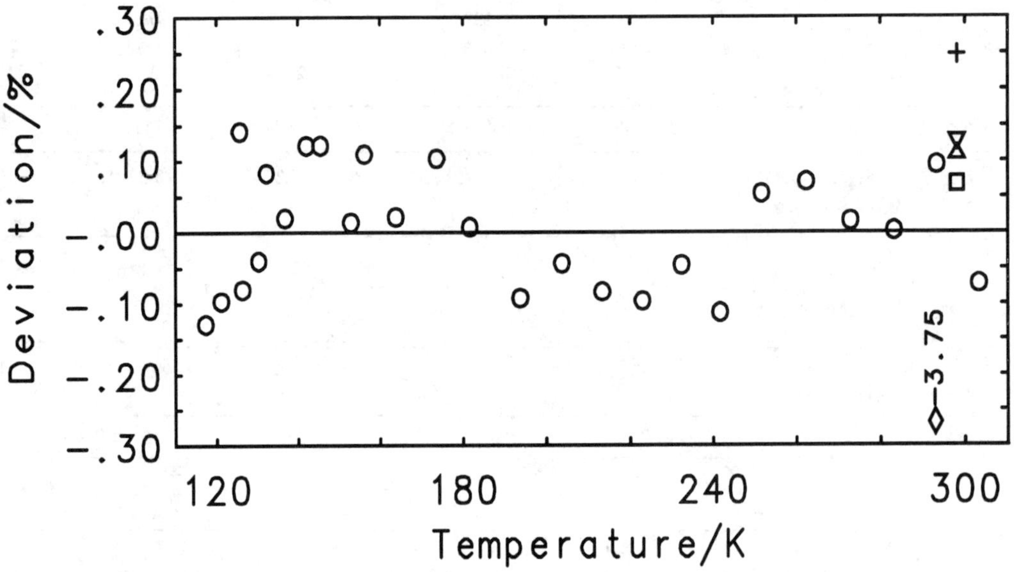

Selected data
○ 46DOU/HUF2
□ 83BEN/DAR
△ 84BEN/DAR
▽ 86BEN/DAR1

Rejected data
◊ 84FIL/LAU
+ 89OHN/FUJ

Name: 3–Methylpentane
Formula: C_6H_{14}

CAS–RN: 96–14–0
Group No.: 11–015
Molar Mass: 86.18

TABLE 11.15.1. Experimental heat capacities

Reference	Temp. range K	No. pnts	Error %	Purity %	Purity method	Type capacity	Calorimeter Type	Calorimeter Reference
37STU	120.0–320.0	21S	3.00	not specified		C_p	BDHO	37STU
46DOU/HUF2	125.3–302.1	19	0.20	not specified		C_p	BSAO	43RUE/HUF
73FIN/MES	119.0–327.3	24	0.20	99.994	melpt	C_{sat}	BSAO	43RUE/HUF
80CZA	298.9	1	nosp	not specified		C_p	BSIO	79CZA
82OGU/WAT	119.5–288.2	64	1.00	not specified		C_p	BSAO	82OGU/WAT
83BEN/DAR	298.1	1	nosp	not specified		C_p	FSIT	71PIC/LED
84BEN/DAR	298.1	1	0.20	99.98	chrom	C_p	FSIT	71PIC/LED
84FIL/LAU	293.0	1	2.00	99.	anal	C_p	BDHO	84FIL/LAU
86BEN/DAR1	298.1	1	nosp	99.88	anal	C_p	FSIT	71PIC/LED
89OHN/FUJ	298.1	1	nosp	99.9	chrom	C_p	FSIO	85OGA

TABLE 11.15.2. Correlated heat capacities

Reference	Temp. range K	No. pnts used	$\sigma_r C$ %	d_w	d/R	d_r %	d_b/R	+/–
				Selected data				
46DOU/HUF2	125.3–302.1	19	0.20	0.541	2.09–2	0.11	–1.04–3	0
73FIN/MES	119.1–327.3	24	0.20	0.246	1.03–2	0.05	–1.61–3	–1
83BEN/DAR	298.1	1	0.50#	0.118	1.36–2	0.06	1.36–2	1
84BEN/DAR	298.1	1	0.20	0.532	2.44–2	0.11	2.44–2	1
86BEN/DAR1	298.1	1	0.50#	0.349	4.00–2	0.17	4.00–2	1
89OHN/FUJ	298.1	1	0.30#	0.877	6.05–2	0.26	6.05–2	1
				Rejected data				
37STU	(7.11–1, 3.56, 5.15–1, 13)			80CZA	(5.03–1, 2.24, –5.03–1, –1)			
82OGU/WAT	(9.09–2, 0.46, 4.86–2, 34)			84FIL/LAU	(9.01–2, 0.40, 9.01–2, 1)			

TABLE 11.15.3. Parameters of regression polynomial

Heat capacity type	No. data points total	used	s_w	s/R	s_r %	s_b/R	+/–
C	134	47	0.432	1.96–2	0.09	1.71–3	3

Temp. range K	A_1	A_2	A_3	Level of uncertainty
119.1–327.3	1.69688+1	–4.86827–1	8.33953–1	II

TABLE 11.15.4. Recommended values of heat capacities

Temp. (K)	120	130	140	150	160	170	180
c (J K^{-1}g^{-1})	1.697	1.712	1.729	1.748	1.768	1.790	1.813
C (J K^{-1}mol^{-1})	146.2	147.5	149.0	150.6	152.4	154.2	156.3
Temp. (K)	190	200	210	220	230	240	250
c (J K^{-1}g^{-1})	1.838	1.865	1.893	1.923	1.955	1.988	2.023
C (J K^{-1}mol^{-1})	158.4	160.7	163.2	165.7	168.5	171.3	174.3
Temp. (K)	260	270	273.15	280	290	298.15	300
c (J K^{-1}g^{-1})	2.059	2.097	2.109	2.136	2.178	2.212	2.220
C (J K^{-1}mol^{-1})	177.4	180.7	181.8	184.1	187.7	190.7	191.3
Temp. (K)	310	320	330				
c (J K^{-1}g^{-1})	2.265	2.311	2.358				
C (J K^{-1}mol^{-1})	195.2	199.1	203.2				

TABLE 11.15.5. Parameters of quasi-polynomial equation

Heat capacity type	No. data points total	used	s_w	s/R	s_r %	s_b/R	+/–
C	134	47	1.343	6.08–2	0.27	–7.69–3	0

Temp. range K	T_c K	A_1	A_2	A_3	A_4	Level of uncertainty
119.1–327.3	504.50	–3.36575	2.60533	1.29200+1	1.08703	II

11–015

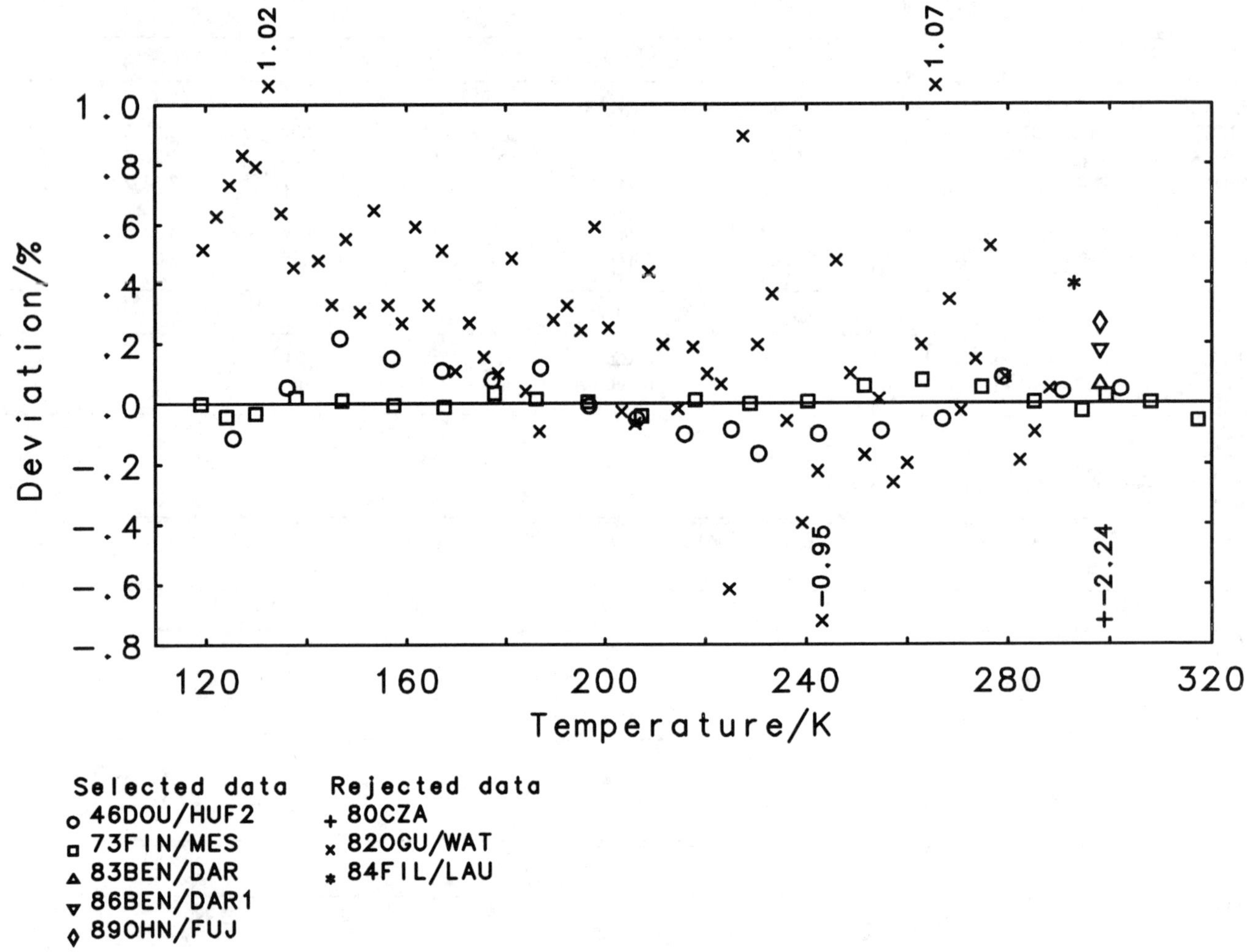

Name: 2,2–Dimethylpentane
Formula: C_7H_{16}

CAS–RN: 590–35–2
Group No.: 11–016
Molar Mass: 100.20

TABLE 11.16.1. Experimental heat capacities

Reference	Temp. range K	No. pnts	Error %	Purity %	Purity method	Type capacity	Calorimeter Type	Calorimeter Reference
30HUF/PAR2	153.2–294.1	14	1.00	not specified		C_p	BSIO	25PAR
61HUF/GRO	154.7–298.4	20	0.20	99.83	melpt	C_{sat}	BSAO	47HUF

TABLE 11.16.2. Correlated heat capacities

Reference	Temp. range K	No. pnts used	$\sigma_r C$ %	d_w	d/R	d_r %	d_b/R	+/–
				Selected data				
61HUF/GRO	154.7298.4	20	0.20#	0.141	6.82–3	0.03	–1.91–7	0
				Rejected data				
30HUF/PAR2	(2.04–1, 0.88, –1.94–1, –12)							

TABLE 11.16.3. Parameters of regression polynomial

Heat capacity type	No. data points total	used	s_w	s/R	s_r %	s_b/R	+/–
C_{sat}	34	20	0.153	7.39–3	0.03	–1.91–7	0
Temp. range K		A_1	A_2	A_3			Level of uncertainty
154.7–298.4		1.62607+1	1.14508	7.80583–1			II

TABLE 11.16.4. Recommended values of heat capacities

Temp. (K)	160	170	180	190	200	210	220
c_{sat} (J K^{-1}g^{-1})	1.667	1.698	1.730	1.764	1.798	1.834	1.872
C_{sat} (J K^{-1}mol^{-1})	167.0	170.1	173.4	176.7	180.2	183.8	187.6
Temp. (K)	230	240	250	260	270	273.15	280
c_{sat} (J K^{-1}g^{-1})	1.910	1.950	1.992	2.034	2.078	2.092	2.123
C_{sat} (J K^{-1}mol^{-1})	191.4	195.4	199.6	203.8	208.2	209.6	212.7
Temp. (K)	290	298.15	300				
c_{sat} (J K^{-1}g^{-1})	2.169	2.208	2.217				
C_{sat} (J K^{-1}mol^{-1})	217.4	221.3	222.2				

TABLE 11.16.5. Parameters of quasi-polynomial equation

Heat capacity type	No. data points total	used	s_w	s/R	s_r %	s_b/R	+/–
C_{sat}	34	20	0.458	2.15–2	0.09	3.34–5	1

Temp. range K	T_c K	A_1	A_2	A_3	A_4	Level of uncertainty
154.7–298.4	520.50	–6.75415	1.76171	1.30664+1	6.47362	II

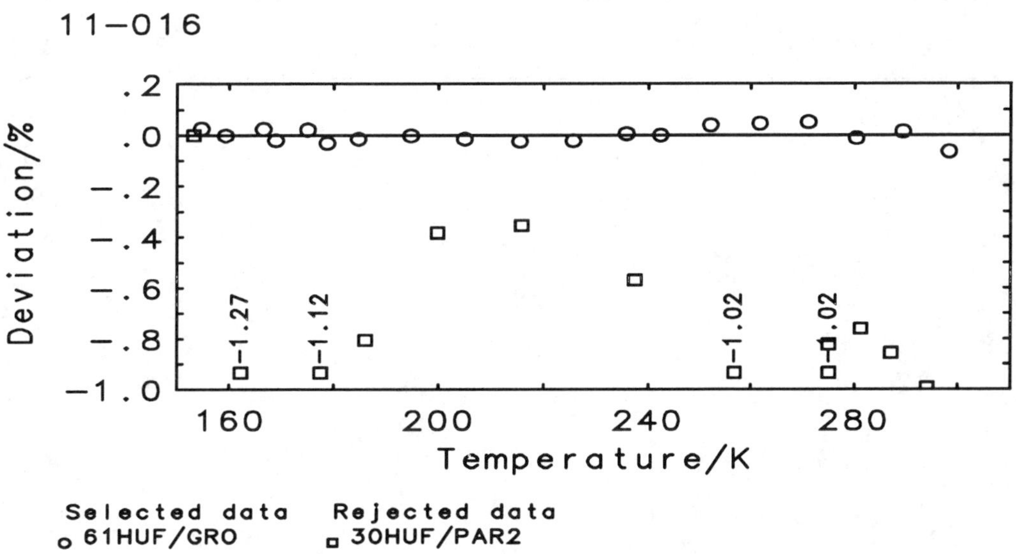

Name: 2,3–Dimethylpentane
Formula: C_7H_{16}

CAS–RN: 565–59–3
Group No.: 11–017
Molar Mass: 100.20

TABLE 11.17.1. Experimental heat capacities

Reference	Temp. range K	No. pnts	Error %	Purity %	Purity method	Type capacity	Calorimeter Type	Calorimeter Reference
30HUF/PAR2	90.4–291.3	15	1.00	not specified		C_p	BSIO	25PAR
76FIN/MES	84.0–385.3	69	0.20	99.8	estim	C_{sat}	BSAO	43RUE/HUF
84FIL/LAU	293.0	1	2.00	95.	anal	C_p	BDHO	84FIL/LAU

TABLE 11.17.2. Correlated heat capacities

Reference	Temp. range K	No. pnts used	$\sigma_r C$ %	d_w	d/R	d_r %	d_b/R	+/–
				Selected data				
76FIN/MES	85.0–385.3	68	0.20	1.009	4.39–2	0.20	1.46–4	4
				Rejected data				
30HUF/PAR2	(7.25–2, 0.31, –5.06–2, –6)			84FIL/LAU	(6.09–1, 2.41, –6.09–1, –1)			

TABLE 11.17.3. Parameters of regression polynomial

Heat capacity type	No. data points total	No. data points used	s_w	s/R	s_r %	s_b/R	+/–
C_p	85	68	1.032	4.49–2	0.21	1.46–4	4
C_{sat}	85	68	1.069	4.78–2	0.21	1.43–4	4

Temp. range K	A_1	A_2	A_3	Level of uncertainty
85.0–385.3	1.76512+1	6.68788–1	7.36103–1	II
85.0–385.3	1.76323+1	6.95008–1	7.28393–1	II

TABLE 11.17.4. Recommended values of heat capacities

Temp. (K)	90	100	110	120	130	140	150
c_p (J K^{-1}g^{-1})	1.564	1.581	1.600	1.619	1.640	1.662	1.685
C_p (J K^{-1}mol^{-1})	156.7	158.4	160.3	162.2	164.3	166.5	168.9
c_{sat} (J K^{-1}g^{-1})	1.564	1.581	1.600	1.619	1.640	1.662	1.686
C_{sat} (J K^{-1}mol^{-1})	156.7	158.4	160.3	162.3	164.4	166.6	168.9
Temp. (K)	160	170	180	190	200	210	220
c_p (J K^{-1}g^{-1})	1.710	1.735	1.762	1.791	1.820	1.851	1.882
C_p (J K^{-1}mol^{-1})	171.3	173.9	176.6	179.4	182.4	185.4	188.6
c_{sat} (J K^{-1}g^{-1})	1.710	1.736	1.763	1.791	1.820	1.851	1.882
C_{sat} (J K^{-1}mol^{-1})	171.4	173.9	176.6	179.4	182.4	185.4	188.6
Temp. (K)	230	240	250	260	270	273.15	280
c_p (J K^{-1}g^{-1})	1.915	1.950	1.985	2.022	2.060	2.072	2.099
C_p (J K^{-1}mol^{-1})	191.9	195.4	198.9	202.6	206.4	207.6	210.3
c_{sat} (J K^{-1}g^{-1})	1.915	1.950	1.985	2.022	2.059	2.072	2.098
C_{sat} (J K^{-1}mol^{-1})	191.9	195.4	198.9	202.6	206.4	207.6	210.3
Temp. (K)	290	298.15	300	310	320	330	340
c_p (J K^{-1}g^{-1})	2.139	2.173	2.181	2.224	2.268	2.313	2.359
C_p (J K^{-1}mol^{-1})	214.4	217.7	218.5	222.8	227.2	231.8	236.4
c_{sat} (J K^{-1}g^{-1})	2.139	2.172	2.180	2.223	2.266	2.312	2.358
C_{sat} (J K^{-1}mol^{-1})	214.3	217.7	218.4	222.7	227.1	231.6	236.3
Temp. (K)	350	360	370	380	390		
c_p (J K^{-1}g^{-1})	2.407	2.456	2.506	2.557	2.610		
C_p (J K^{-1}mol^{-1})	241.2	246.1	251.1	256.3	261.5		
c_{sat} (J K^{-1}g^{-1})	2.405	2.454	2.504	2.555	2.607		
C_{sat} (J K^{-1}mol^{-1})	241.0	245.9	250.9	256.0	261.3		

TABLE 11.17.5. Parameters of quasi-polynomial equation

Heat capacity type	No. data points total	used	s_w	s/R	s_r %	s_b/R	+/–
C_p	85	68	2.730	1.35–1	0.55	1.35–3	0

Temp. range K	T_c K	A_1	A_2	A_3	A_4	Level of uncertainty
85.0–385.3	537.40	–5.71710	1.77122	1.48250+1	4.61336	IV

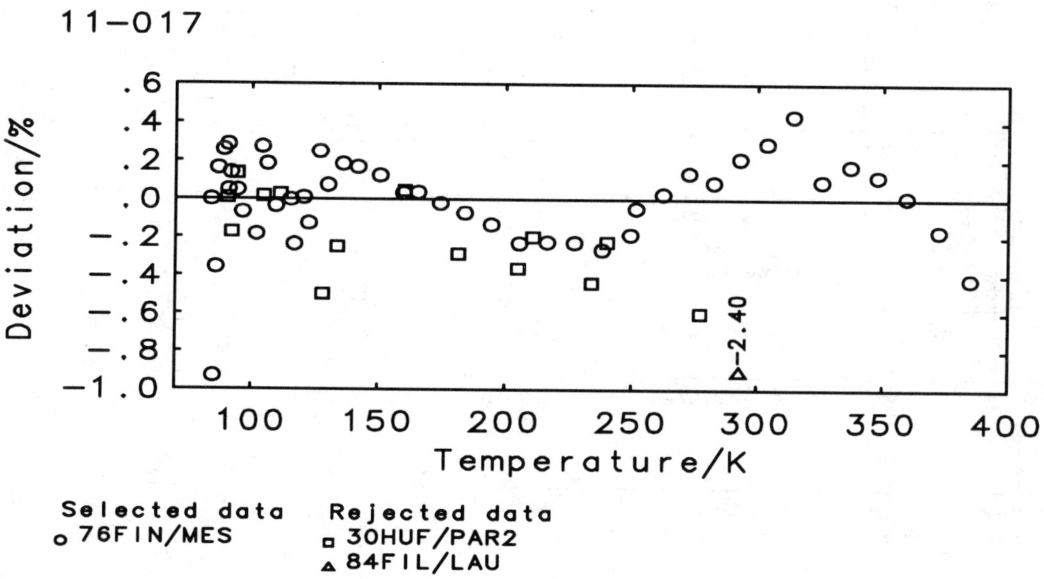

Name: 2,4–Dimethylpentane
Formula: C_7H_{16}

CAS–RN: 108–08–7
Group No.: 11–018
Molar Mass: 100.20

TABLE 11.18.1. Experimental heat capacities

Reference	Temp. range K	No. pnts	Error %	Purity %	Purity method	Type capacity	Calorimeter Type	Calorimeter Reference
30HUF/PAR2	160.2–294.4	15	1.00	not specified		C_p	BSIO	25PAR
61HUF/GRO	160.8–307.1	40	0.20	99.91	melpt	C_{sat}	BSAO	47HUF

TABLE 11.18.2. Correlated heat capacities

Reference	Temp. range K	No. pnts used	$\sigma_r C$ %	d_w	d/R	d_r %	d_b/R	+/–
				Selected data				
61HUF/GRO	160.8–307.1	40	0.20#	0.175	8.52–3	0.03	6.68–6	–1
				Rejected data				
30HUF/PAR2	(1.54–1, 0.64, –1.35–1, –14)							

TABLE 11.18.3. Parameters of regression polynomial

Heat capacity type	No. data points total	No. data points used	s_w	s/R	s_r %	s_b/R	+/–
C_{sat}	55	40	0.184	8.98–3	0.04	6.68–6	–1

Temp. range K	A_1	A_2	A_3	A_4	Level of uncertainty
160.8–307.1	1.86799+1	–1.84428	2.13249	–1.94744–1	II

TABLE 11.18.4. Recommended values of heat capacities

Temp. (K)	160	170	180	190	200	210	220
c_{sat} (J K^{-1}g^{-1})	1.692	1.722	1.754	1.787	1.822	1.859	1.898
C_{sat} (J K^{-1}mol^{-1})	169.5	172.5	175.7	179.1	182.6	186.3	190.2
Temp. (K)	230	240	250	260	270	273.15	280
c_{sat} (J K^{-1}g^{-1})	1.937	1.979	2.021	2.064	2.109	2.123	2.154
C_{sat} (J K^{-1}mol^{-1})	194.1	198.3	202.5	206.8	211.3	212.7	215.8
Temp. (K)	290	298.15	300	310			
c_{sat} (J K^{-1}g^{-1})	2.200	2.238	2.247	2.295			
C_{sat} (J K^{-1}mol^{-1})	220.5	224.3	225.2	229.9			

TABLE 11.18.5. Parameters of quasi-polynomial equation

Heat capacity type	No. data points total	No. data points used	s_w	s/R	s_r %	s_b/R	+/–
C_{sat}	55	40	0.609	3.00–2	0.12	6.93–5	2

Temp. range K	T_c K	A_1	A_2	A_3	A_4	Level of uncertainty
160.8–307.1	519.80	–6.75355	1.46590	1.33425+1	7.77858	II

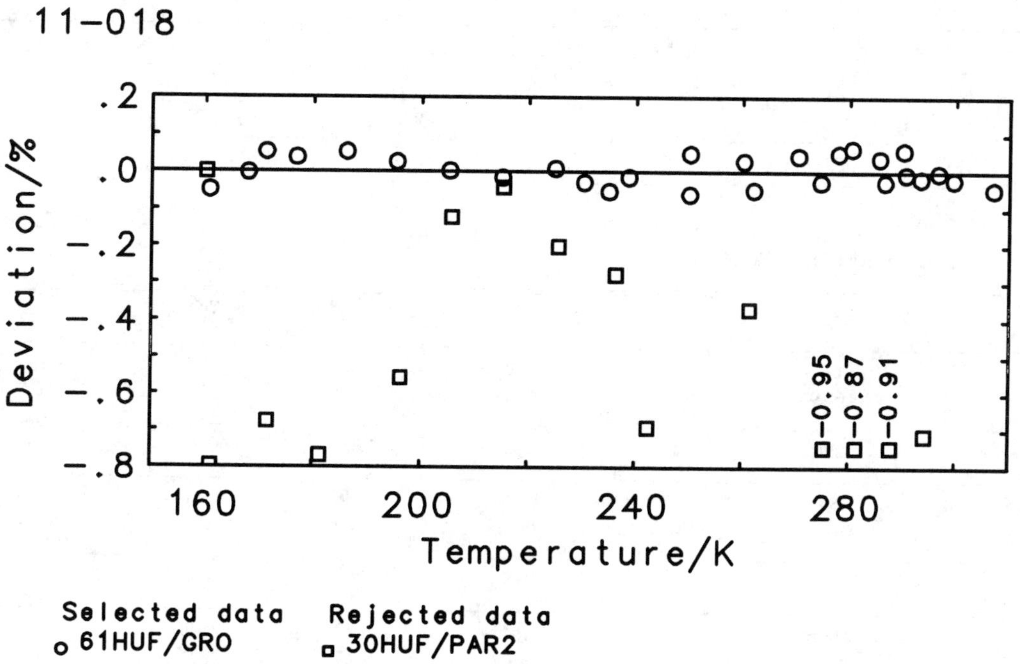

Name: 3,3-Dimethylpentane
Formula: C_7H_{16}

CAS-RN: 562-49-2
Group No.: 11-019
Molar Mass: 100.20

TABLE 11.19.1. Experimental heat capacities

Reference	Temp. range K	No. pnts	Error %	Purity % method	Type capacity	Calorimeter Type Reference
30HUF/PAR2	145.7–292.9	14	1.00	not specified	C_p	BSIO 25PAR
76FIN/MES	146.9–350.8	28	0.20	99.982 melpt	C_{sat}	BSAO 43RUE/HUF

TABLE 11.19.2. Correlated heat capacities

Reference	Temp. range K	No. pnts used	$\sigma_r C$ %	d_w	d/R	d_r %	d_b/R	+/–
				Selected data				
76FIN/MES	146.9–350.8	28	0.20	0.276	1.47–2	0.06	1.59–5	0
				Rejected data				
30HUF/PAR2	(1.19–1, 0.55, –1.07–1, –13)							

TABLE 11.19.3. Parameters of regression polynomial

Heat capacity type	No. data points total	used	s_w	s/R	s_r %	s_b/R	+/−
C_{sat}	42	28	0.298	1.59–2	0.06	1.59–5	0

Temp. range K	A_1	A_2	A_3	A_4	Level of uncertainty
146.9–350.8	2.06869+1	−3.23326	2.10782	−1.48624–1	II

TABLE 11.19.4. Recommended values of heat capacities

Temp. (K)	150	160	170	180	190	200	210
c_{sat} (J K^{-1}g^{-1})	1.666	1.684	1.705	1.728	1.754	1.781	1.810
C_{sat} (J K^{-1}mol^{-1})	166.9	168.8	170.9	173.2	175.7	178.5	181.4
Temp. (K)	220	230	240	250	260	270	273.15
c_{sat} (J K^{-1}g^{-1})	1.841	1.875	1.910	1.946	1.985	2.024	2.037
C_{sat} (J K^{-1}mol^{-1})	184.5	187.8	191.3	195.0	198.9	202.9	204.1
Temp. (K)	280	290	298.15	300	310	320	330
c_{sat} (J K^{-1}g^{-1})	2.066	2.109	2.145	2.153	2.198	2.245	2.293
C_{sat} (J K^{-1}mol^{-1})	207.0	211.3	214.9	215.7	220.3	224.9	229.7
Temp. (K)	340	350					
c_{sat} (J K^{-1}g^{-1})	2.341	2.391					
C_{sat} (J K^{-1}mol^{-1})	234.6	239.6					

TABLE 11.19.5. Parameters of quasi-polynomial equation

Heat capacity type	No. data points total	used	s_w	s/R	s_r %	s_b/R	+/−
C_{sat}	42	28	1.883	8.97–2	0.38	5.61–4	0

Temp. range K	T_c K	A_1	A_2	A_3	A_4	Level of uncertainty
146.9–350.8	536.40	−5.50206	2.66040	1.36968+1	2.84474	III

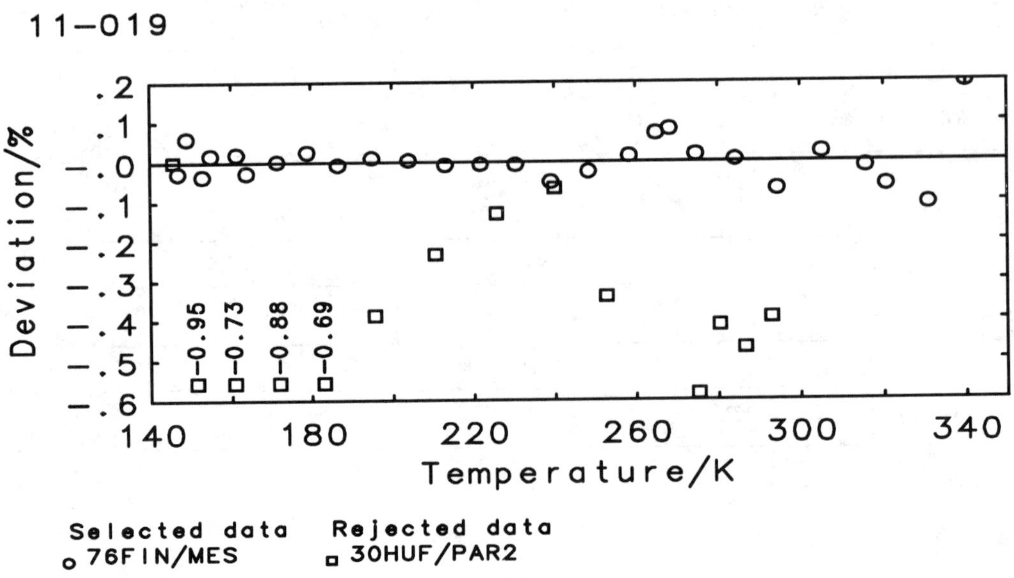

Name: 3–Ethylpentane
Formula: C_7H_{16}

CAS–RN: 617–78–7
Group No.: 11–020
Molar Mass: 100.20

TABLE 11.20.1. Experimental heat capacities

Reference	Temp. range K	No. pnts	Error %	Purity %	Purity method	Type capacity	Calorimeter Type	Calorimeter Reference
30HUF/PAR2	161.4–294.8	11	1.00	not specified		C_p	BSIO	25PAR
61HUF/GRO	161.2–298.8	44	nosp	99.995	melpt	C_{sat}	BSAO	47HUF
84FIL/LAU	293.0	1	2.00	98.	anal	C_p	BDHO	84FIL/LAU

TABLE 11.20.2. Correlated heat capacities

Reference	Temp. range K	No. pnts used	$\sigma_r C$ %	d_w	d/R	d_r %	d_b/R	+/–
				Selected data				
61HUF/GRO	161.2–298.8	44	0.20#	0.189	8.84–3	0.04	7.67–6	–1
				Rejected data				
30HUF/PAR2	(9.38–2, 0.38, –3.04–2, –3)			84FIL/LAU	(2.82, 9.73, 2.82, 1)			

TABLE 11.20.3. Parameters of regression polynomial

Heat capacity type	No. data points total	No. data points used	s_w	s/R	s_r %	s_b/R	+/–
C_{sat}	56	44	0.196	9.16–3	0.04	7.67–6	–1

Temp. range K	A_1	A_2	A_3	Level of uncertainty
161.2–298.8	1.83550+1	3.45606–1	7.90238–1	II

TABLE 11.20.4. Recommended values of heat capacities

Temp. (K)	160	170	180	190	200	210	220
c_{sat} (J K^{-1}g^{-1})	1.737	1.761	1.787	1.814	1.843	1.872	1.903
C_{sat} (J K^{-1}mol^{-1})	174.0	176.5	179.1	181.8	184.6	187.6	190.7
Temp. (K)	230	240	250	260	270	273.15	280
c_{sat} (J K^{-1}g^{-1})	1.936	1.970	2.005	2.041	2.078	2.091	2.117
C_{sat} (J K^{-1}mol^{-1})	194.0	197.4	200.9	204.5	208.3	209.5	212.2
Temp. (K)	290	298.15	300				
c_{sat} (J K^{-1}g^{-1})	2.158	2.191	2.199				
C_{sat} (J K^{-1}mol^{-1})	216.2	219.6	220.4				

TABLE 11.20.5. Parameters of quasi-polynomial equation

Heat capacity type	No. data points total	used	s_w	s/R	s_r %	s_b/R	+/–
C_{sat}	56	44	0.300	1.44–2	0.06	1.56–5	–2
Temp. range K	T_c K	A_1	A_2	A_3	A_4		Level of uncertainty
161.2–298.8	540.60	–5.87119	2.44074	1.43331+1	3.53078		II

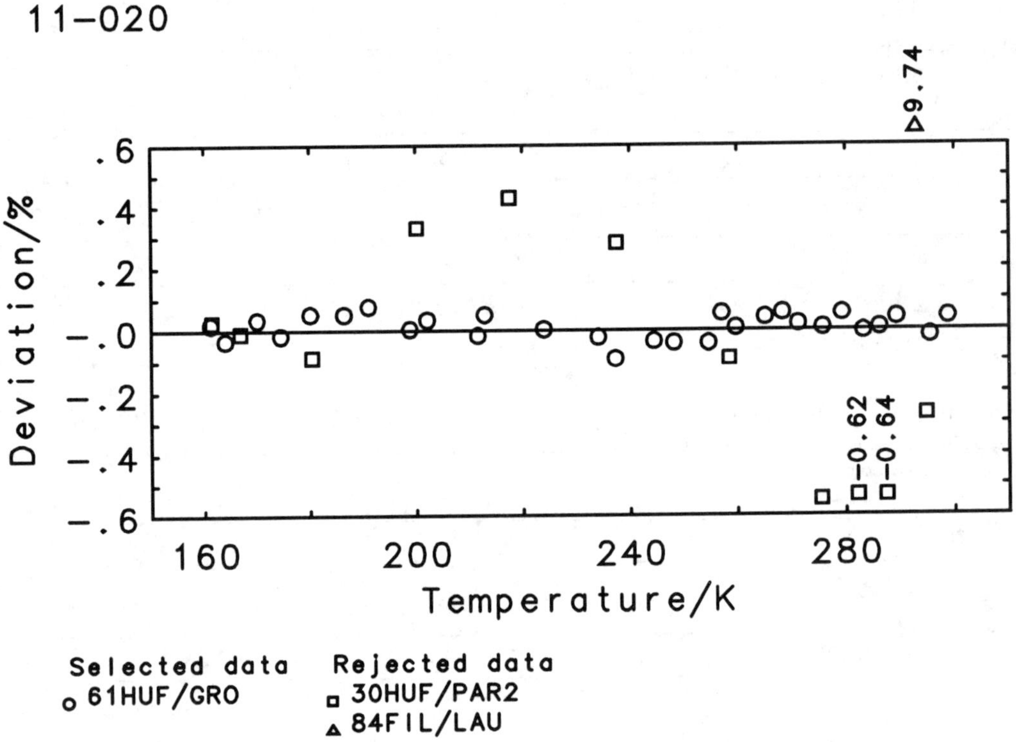

Name: Heptane
Formula: C_7H_{16}

CAS–RN: 142–82–5
Group No.: 11–021
Molar Mass: 100.20

TABLE 11.21.1. Experimental heat capacities

Reference		Temp. range K	No. pnts	Error %	Purity %	Purity method	Type capacity	Calorimeter Type	Calorimeter Reference
24WIL/DAN		309.0–380.0	eqn	nosp	not specified		C_p	BSAO	24WIL/DAN
30PAR/HUF2		188.0–299.2	14	nosp	not specified		C_p	BSIO	25PAR
37VOL		298.1	1	nosp	not specified		C_p	BSIO	37VOL
39BYK		298.1–298.1	2	nosp	not specified		C_p	BSIT	39BYK
39PHI		300.8	1	nosp	not specified		C_p	BSIO	49WEI
40PIT2		194.6–317.6	6	0.50	99.85	melpt	C_p	BSIO	28LAT/GRE
47OSB/GIN	N	280.6–358.1	9S	0.10	99.96	estim	C_{sat}	BSAO	47OSB/GIN
47OSB/GIN		285.6–305.6	5S	0.10	99.96	estim	C_{sat}	BSAO	47OSB/GIN
54DOU/FUR	N	185.0–520.0	47S	0.10	99.999	melpt	C_{sat}	BSAO	45SCO/MEY
58SWI/ZIE1	N	332.6	1	nosp	not specified		avg	DSIO	58SWI/ZIE1
61HUF/GRO	N	182.6–370.0	22S	0.20	not specified		C_{sat}	BSAO	47HUF
61MCC/MES	N	188.5–307.8	20	nosp	99.97	melpt	C_{sat}	BSAO	43RUE/HUF
61MCC/MES	N	183.0–320.8	25	nosp	99.996	melpt	C_{sat}	BSAO	43RUE/HUF
61MCC/MES	N	188.9–320.4	23	nosp	99.996	melpt	C_{sat}	BSAO	43RUE/HUF
61MCC/MES	N	192.9–340.8	17	nosp	99.996	melpt	C_{sat}	BSAO	43RUE/HUF
61MCC/MES	N	190.9–367.0	19	nosp	99.94	melpt	C_{sat}	BSAO	43RUE/HUF
66KLE		293.1–343.1	11S	0.10	not specified		C_p	BSAO	66KLE
70AKH		293.1	1	nosp	not specified		C_p	BDHO	59ABA/MUS
71AMI/ALI	N	373.1–538.2	21S	2.00	not specified		C_p	not specified	
72VAN		185.0–250.0	14S	0.50	99.968	melpt	C_p	BSAO	72VAN
75GRI/RAS	N	303.5–342.8	3	nosp	not specified		C_p	BDAO	75RAS/GRI
75HOL/ZIE		182.0–312.0	eqn	0.20	99.73	anal	C_p	BSAO	45SCO/MEY
75SAN	N	413.1–513.2	8S	1.00	not specified		C_p	FSIO	75SAN
77MEI/BLO		187.0–354.6	35	nosp	not specified		C_{sat}	BSAO	72VAN
79BRO/ZIE		183.0–302.0	eqn	nosp	99.7	melpt	C_p	BSAO	45SCO/MEY
79SCH/OFF		190.0–285.0	20S	nosp	99.88	melpt	C_{sat}	BSAO	79SCH/OFF
80KAL/JED		185.0–301.0	96	0.20	99.95	chrom	C_p	BSAO	80KAL/JED
83TAN/ZHO		214.2–365.2	40	0.20	99.99	chrom	C_{sat}	BSAO	83TAN/ZHO
84GRI/AND		293.0–323.6	14	0.50	not specified		C_p	BSAO	67RAS/GAN
84GUS/MIR		303.1–363.1	4S	1.60	not specified		C_p	BDHT	84GUS/MIR
85CZA		299.3	1	nosp	not specified		C_p	BSIO	79CZA
87VAN/VAN		184.2–345.7	37	0.20	99.75	melpt	C_p	BSAO	87VAN/VAN
87WIL/ING		298.1	1	nosp	99.5	anal	C_p	FSIT	71PIC/LED
88MEL/VER		193.1–523.2	11	2.50	not specified		C_{sat}	not specified	

47OSB/GIN calorimeter with a vessel of large volume
54DOU/FUR same data in 53GIN/FUR
58SWI/ZIE1 average value in temperature range 294–369 K
61HUF/GRO same data in 67MES/GUT
61MCC/MES measured in 1947
61MCC/MES measured in 1949
61MCC/MES measured in 1951(A)
61MCC/MES measured in 1951(B); results less accurate near and above room temperature
61MCC/MES measured in 1954
71AMI/ALI calculated from Cv measured at the saturation line
75GRI/RAS grade: pure
75SAN same data in 76SAN/MEL; C_p at the saturation pressure extrapolated from high pressure measurement

TABLE 11.21.2a. Correlated heat capacities

Reference	Temp. range K	No. pnts used	$\sigma_r C$ %	d_w	d/R	d_r %	d_b/R	+/–
				Selected data				
54DOU/FUR	185.0–370.0	32	0.10	0.092	2.35–3	0.01	–1.78–3	–9
54DOU/FUR	380.0–480.0	11	0.30	0.053	5.78–3	0.02	1.44–3	3
61HUF/GRO	182.6–370.0	22	0.20	0.304	1.58–2	0.06	1.00–2	14

TABLE 11.21.2b. Correlated heat capacities

Reference	Temp. range K	No. pnts used	$\sigma_r C$ %	d_w	d/R	d_r %	d_b/R	+/–
				Selected data				
54DOU/FUR	200.0–520.0	44	0.10	7.832	2.57–1	0.78	–1.68–2	–4
61HUF/GRO	200.0–370.0	20	0.10	6.914	1.76–1	0.69	–1.46–2	–4
				Rejected data				
24WIL/DAN	(6.03–1, 2.19, –2.07–1, –1)			30PAR/HUF2	(3.71–1, 1.44, –2.67–1, –9)			
37VOL	(3.72–1, 1.39, –3.72–1, –1)			39BYK	(1.69, 6.65, –1.69, –2)			
39PHI	(1.90, 7.49, –1.90, –1)			40PIT2	(6.55–2, 0.26, 4.29–3, 1)			
47OSB/GIN	(1.09–1, 0.40, –2.61–2, –1)			47OSB/GIN	(1.39–1, 0.52, –1.37–1, –5)			
58SWI/ZIE1	(1.10, 3.68, 1.10, 1)			61MCC/MES	(1.88–1, 0.74, –8.78–2, –10)			
61MCC/MES	(2.04–1, 0.83, 3.54–2, 1)			61MCC/MES	(1.76–1, 0.71, –2.64–2, –8)			
61MCC/MES	(1.87–1, 0.74, –5.73–2, –10)			61MCC/MES	(1.06, 3.62, 2.62–1, –2)			
66KLE	(9.09–2, 0.33, –4.96–2, –4)			70AKH	(2.29–1, 0.86, –2.29–1, –1)			
71AMI/ALI	(7.68–1, 2.12, –5.76–1, –11)			72VAN	(2.56–1, 1.05, 7.64–2, 1)			
75GRI/RAS	(1.84–1, 0.66, –1.73–2, 1)			75HOL/ZIE	(1.72–1, 0.67, –1.18–1, –2)			
75SAN	(9.64–1, 2.57, –7.24–1, –6)			77MEI/BLO	(1.53–1, 0.62, –4.93–2, –15)			
79BRO/ZIE	(1.00, 3.95, –9.95–1, –2)			79SCH/OFF	(2.08–1, 0.83, –6.74–2, –8)			
80KAL/JED	(1.94–1, 0.77, –8.04–2, –40)			83TAN/ZHO	(1.88–1, 0.73, –1.35–1, –24)			
84GRI/AND	(3.18–1, 1.14, 3.10–1, 14)			84GUS/MIR	(3.76–1, 1.33, –3.64–1, –4)			
85CZA	(8.58–2, 0.32, –8.58–2, –1)			87VAN/VAN	(1.67–1, 0.65, –1.02–1, –18)			
87WIL/ING	(1.31–1, 0.48, –1.31–1, –1)			88MEL/VER	(2.09–1, 0.70, 4.56–2, –1)			

TABLE 11.21.3. Parameters of cubic spline polynomials

Heat capacity type	No. data points total	used	s_w	s/R	s_r %	s_b/R	+/–
C_p	65	65	0.199	1.01–2	0.04	2.77–3	8
C_{sat}	65	65	0.197	9.80–3	0.04	2.71–3	2

Temp. range K	A_1	A_2	A_3	A_4	Level of uncertainty
182.6–260.0	6.11411+1	–4.53403+1	1.76764+1	–2.11880	I
260.0–400.0	2.54134+1	–4.11613	1.82098	–8.60461–2	I
400.0–480.0	–8.76637+1	8.06917+1	–1.93810+1	1.68078	I
182.6–260.0	6.09258+1	–4.50536+1	1.75504+1	–2.10051	I
260.0–400.0	2.62792+1	–5.07669	2.17466	–1.29263–1	I
400.0–480.0	–4.11778+1	4.55161+1	–1.04735+1	9.24753–1	I

TABLE 11.21.4. Recommended values of heat capacities

Temp. (K)	190	200	210	220	230	240	250
c_p (J K^{-1}g^{-1})	2.0142	2.0094	2.0128	2.0235	2.0402	2.0620	2.0878
C_p (J K^{-1}mol^{-1})	201.82	201.34	201.69	202.75	204.43	206.62	209.20
c_{sat} (J K^{-1}g^{-1})	2.0142	2.0094	2.0128	2.0235	2.0402	2.0620	2.0878
C_{sat} (J K^{-1}mol^{-1})	201.82	201.35	201.68	202.75	204.43	206.62	209.20
Temp. (K)	260	270	273.15	280	290	298.15	300
c_p (J K^{-1}g^{-1})	2.1167	2.1476	2.1577	2.1803	2.2149	2.2444	2.2512
C_p (J K^{-1}mol^{-1})	212.09	215.19	216.21	218.47	221.93	224.89	225.58
c_{sat} (J K^{-1}g^{-1})	2.1167	2.1476	2.1577	2.1803	2.2149	2.2444	2.2513
C_{sat} (J K^{-1}mol^{-1})	212.09	215.19	216.21	218.47	221.93	224.89	225.59
Temp. (K)	310	320	330	340	350	360	370
c_p (J K^{-1}g^{-1})	2.2892	2.3291	2.3706	2.4136	2.4582	2.5043	2.5519
C_p (J K^{-1}mol^{-1})	229.39	233.38	237.53	241.85	246.32	250.94	255.71
c_{sat} (J K^{-1}g^{-1})	2.2892	2.3289	2.3701	2.4128	2.4568	2.5023	2.5490
C_{sat} (J K^{-1}mol^{-1})	229.38	233.36	237.49	241.76	246.18	250.73	255.41
Temp. (K)	380	390	400	410	420	430	440
c_p (J K^{-1}g^{-1})	2.6010	2.6514	2.7032	2.756	2.812	2.870	2.932
C_p (J K^{-1}mol^{-1})	260.62	265.68	270.87	276.2	281.8	287.6	293.8
c_{sat} (J K^{-1}g^{-1})	2.5969	2.6461	2.6963	2.748	2.800	2.855	2.912
C_{sat} (J K^{-1}mol^{-1})	260.22	265.14	270.17	275.3	280.6	286.1	291.8
Temp. (K)	450	460	470	480			
c_p (J K^{-1}g^{-1})	2.999	3.072	3.151	3.237			
C_p (J K^{-1}mol^{-1})	300.5	307.8	315.7	324.3			
c_{sat} (J K^{-1}g^{-1})	2.973	3.036	3.103	3.175			
C_{sat} (J K^{-1}mol^{-1})	297.9	304.2	310.9	318.1			

TABLE 11.21.5. Parameters of quasi-polynomial equation

Heat capacity type	No. data points total	used	s_w	s/R	s_r %	s_b/R	+/–
C_p	542	64	7.805	2.43–1	0.78	–1.61–2	–8

Temp. range K	T_c K	A_1	A_2	A_3	A_4	A_5	A_6	Level of uncertainty
200.0–520.0	540.15	–4.08402	1.74311–1	1.93025+1	3.65193	5.97785	2.07661	IV

11–021

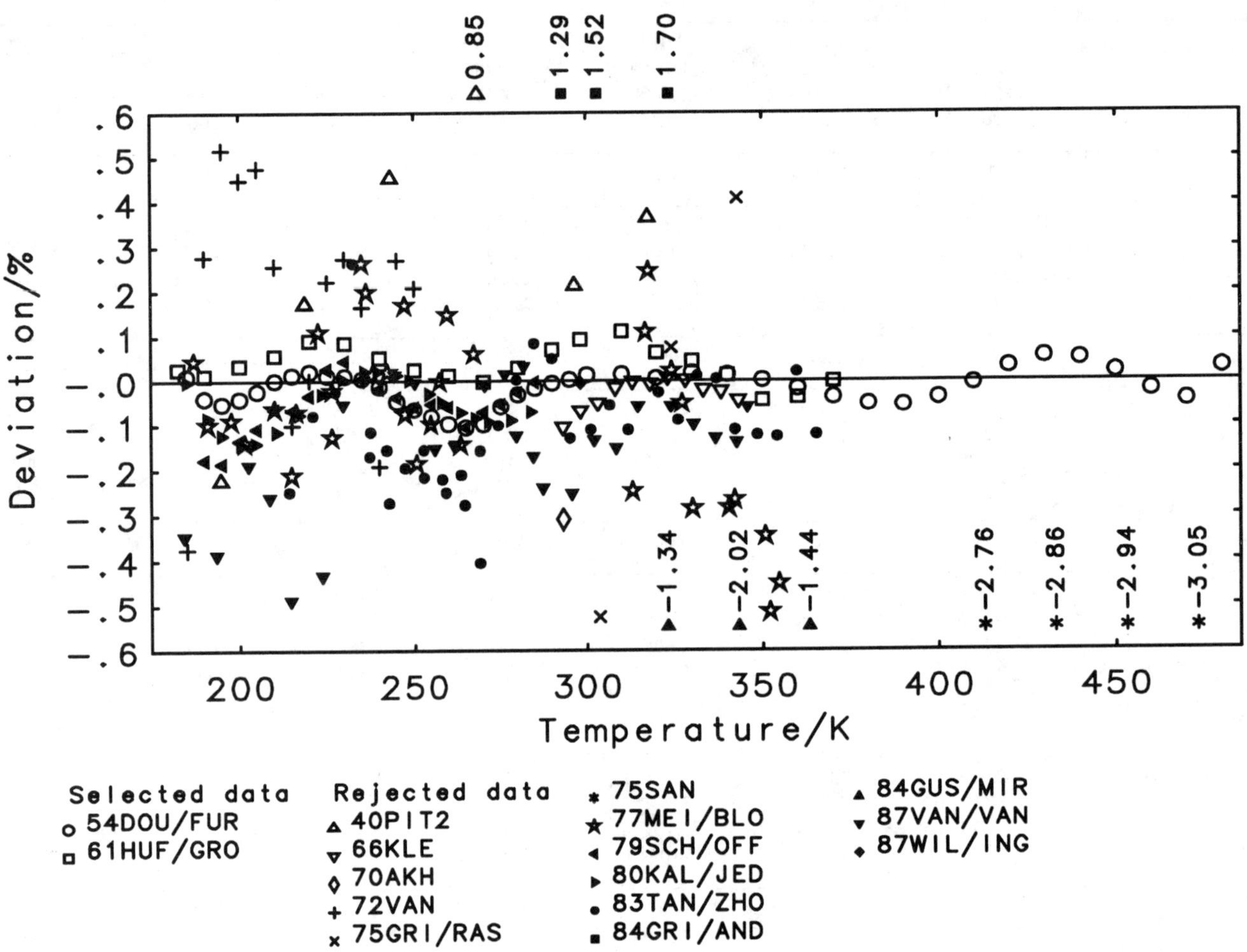

Name: 2–Methylhexane
Formula: C_7H_{16}

CAS–RN: 591–76–4
Group No.: 11–022
Molar Mass: 100.20

TABLE 11.22.1. Experimental heat capacities

Reference		Temp. range K	No. pnts	Error %	Purity %	Purity method	Type capacity	Calorimeter Type	Calorimeter Reference
30PAR/HUF2		160.0–292.4	12	1.00	not specified		C_p	BSIO	25PAR
61HUF/GRO	N	160.4–301.2	19	0.20	99.84	melpt	C_{sat}	BSAO	47HUF
84FIL/LAU		293.0	1	2.00	99.	anal	C_p	BDHO	84FIL/LAU

61HUF/GRO smoothed values in 71MES/FIN

TABLE 11.22.2. Correlated heat capacities

Reference	Temp. range K	No. pnts used	$\sigma_r C$ %	d_w	d/R	d_r %	d_b/R	+/–
				Selected data				
61HUF/GRO	160.4–301.2	19	0.20	0.159	8.11–3	0.03	2.91–6	1
				Rejected data				
30PAR/HUF2	(9.54–2, 0.39, –1.02–2, –1)			84FIL/LAU	(5.52–1, 2.13, –5.52–1, –1)			

TABLE 11.22.3. Parameters of regression polynomial

Heat capacity type	No. data points total	used	s_w	s/R	s_r %	s_b/R	+/–
C_{sat}	32	19	0.174	8.84–3	0.03	2.91–6	1

Temp. range K	A_1	A_2	A_3	Level of uncertainty
160.4–301.2	2.09916+1	–1.30228	1.09026	II

TABLE 11.22.4. Recommended values of heat capacities

Temp. (K)	160	170	180	190	200	210	220
c_{sat} (J K^{-1}g^{-1})	1.800	1.820	1.840	1.863	1.888	1.914	1.942
C_{sat} (J K^{-1}mol^{-1})	180.4	182.3	184.4	186.7	189.1	191.8	194.6
Temp. (K)	230	240	250	260	270	273.15	280
c_{sat} (J K^{-1}g^{-1})	1.972	2.004	2.037	2.072	2.110	2.122	2.148
C_{sat} (J K^{-1}mol^{-1})	197.6	200.8	204.1	207.7	211.4	212.6	215.3
Temp. (K)	290	298.15	300				
c_{sat} (J K^{-1}g^{-1})	2.189	2.224	2.232				
C_{sat} (J K^{-1}mol^{-1})	219.4	222.8	223.6				

TABLE 11.22.5. Parameters of quasi-polynomial equation

Heat capacity type	No. data points total	used	s_w	s/R	s_r %	s_b/R	+/–
C_{sat}	32	19	0.580	2.83–2	0.12	4.93–5	2

Temp. range K	T_c K	A_1	A_2	A_3	A_4	Level of uncertainty
160.4–301.2	530.40	–2.87205	4.35021	1.42552+1	4.74040–1	II

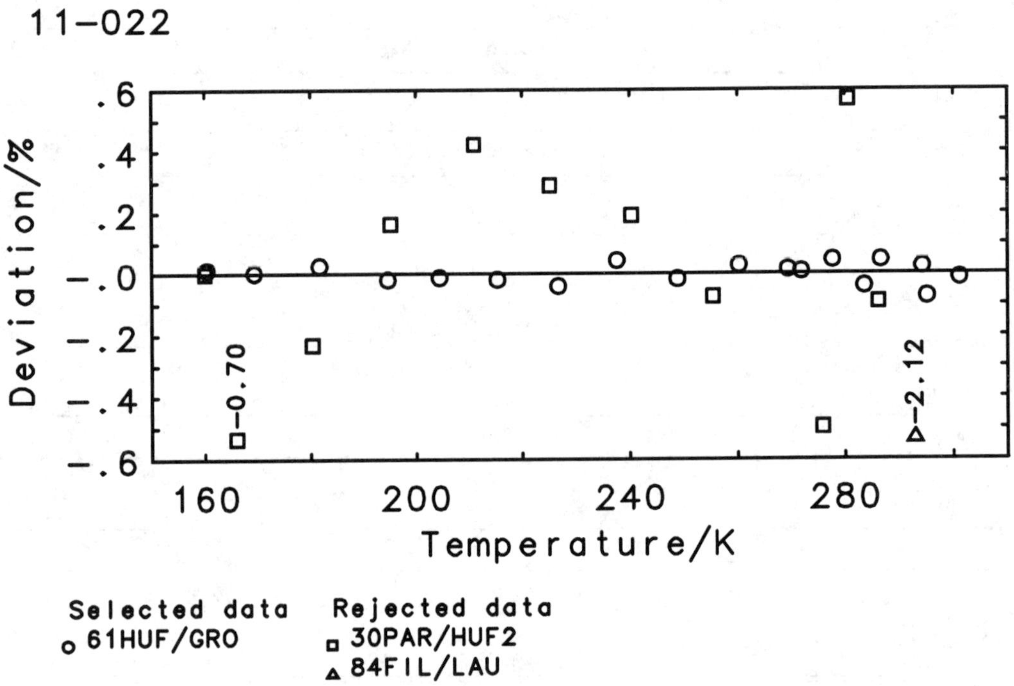

Name: 3–Methylhexane
Formula: C_7H_{16}

CAS–RN: 589–34–4
Group No.: 11–023
Molar Mass: 100.20

TABLE 11.23.1. Experimental heat capacities

Reference	Temp. range K	No. pnts	Error %	Purity %	Purity method	Type capacity	Calorimeter Type	Calorimeter Reference
30HUF/PAR2	99.1–289.2	11	1.00	not specified		C_p	BSIO	25PAR
36PAR/THO	91.0–289.2	26	1.00	not specified		C_p	BSIO	25PAR

TABLE 11.23.2. Correlated heat capacities

Reference	Temp. range K	No. pnts used	$\sigma_r C$ %	d_w	d/R	d_r %	d_b/R	+/–
				Selected data				
30HUF/PAR2	99.1–289.2	11	1.00	0.347	8.61–2	0.35	–3.16–2	–3
36PAR/THO	91.0–289.2	26	1.00	0.222	5.22–2	0.22	1.38–2	5

TABLE 11.23.3. Parameters of regression polynomial

Heat capacity type	No. data points total	used	s_w	s/R	s_r %	s_b/R	+/–
C_p	37	37	0.277	6.69–2	0.28	3.19–4	2

Temp. range K	A_1	A_2	A_3	Level of uncertainty
91.0–289.2	1.98577+1	–7.93355–1	1.00239	IV

TABLE 11.23.4. Recommended values of heat capacities

Temp. (K)	90	100	110	120	130	140	150
c_p (J K^{-1}g^{-1})	1.656	1.665	1.676	1.688	1.703	1.719	1.736
C_p (J K^{-1}mol^{-1})	165.9	166.8	167.9	169.2	170.6	172.2	174.0
Temp. (K)	160	170	180	190	200	210	220
c_p (J K^{-1}g^{-1})	1.755	1.776	1.799	1.823	1.849	1.876	1.905
C_p (J K^{-1}mol^{-1})	175.9	178.0	180.2	182.7	185.3	188.0	190.9
Temp. (K)	230	240	250	260	270	273.15	280
c_p (J K^{-1}g^{-1})	1.936	1.969	2.003	2.039	2.076	2.088	2.115
C_p (J K^{-1}mol^{-1})	194.0	197.3	200.7	204.3	208.1	209.3	212.0
Temp. (K)	290	298.15					
c_p (J K^{-1}g^{-1})	2.156	2.191					
C_p (J K^{-1}mol^{-1})	216.1	219.5					

TABLE 11.23.5. Parameters of quasi-polynomial equation

Heat capacity type	No. data points total	used	s_w	s/R	s_r %	s_b/R	+/–
C_p	37	37	0.391	8.96–2	0.39	2.22–3	5

Temp. range K	T_c K	A_1	A_2	A_3	A_4	Level of uncertainty
91.0–289.2	535.30	2.83394	7.94241	1.08092+1	2.52795–1	IV

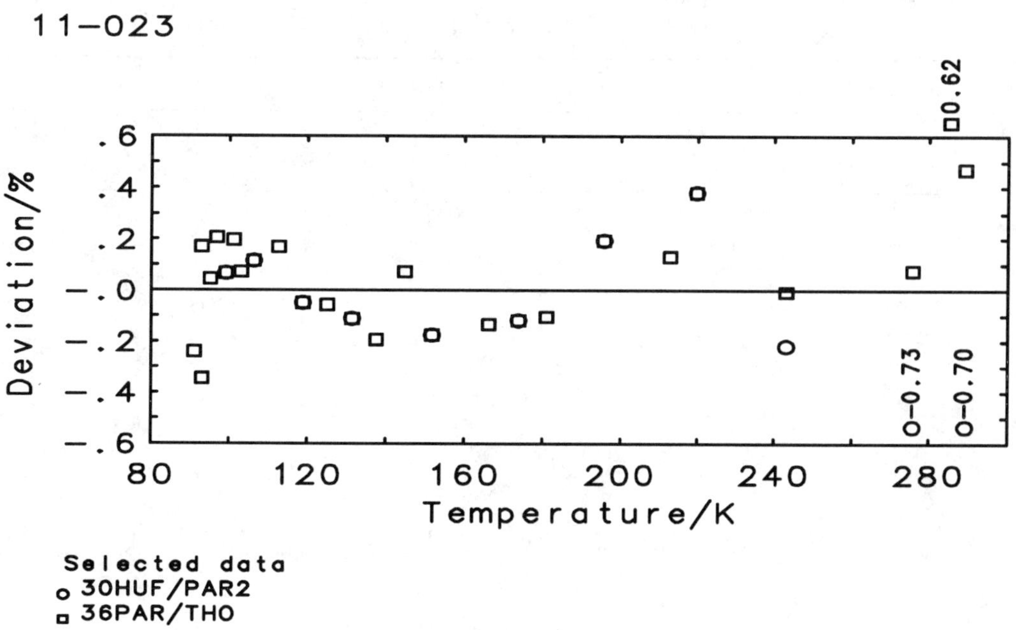

Name: 2,2,3–Trimethylbutane
Formula: C_7H_{16}

CAS–RN: 464–06–2
Group No.: 11–024
Molar Mass: 100.20

TABLE 11.24.1. Experimental heat capacities

Reference	Temp. range K	No. pnts	Error %	Purity %	Purity method	Type capacity	Calorimeter Type	Calorimeter Reference
30HUF/PAR2	253.2–293.9	6	1.00	not specified		C_p	BSIO	25PAR
61HUF/GRO	253.0–313.3	10	0.20	99.95	melpt	C_{sat}	BSAO	47HUF

TABLE 11.24.2. Correlated heat capacities

Reference	Temp. range K	No. pnts used	$\sigma_r C$ %	d_w	d/R	d_r %	d_b/R	+/–
				Selected data				
61HUF/GRO	253.0–313.3	10	0.20	0.220	1.10–2	0.04	8.39–6	–1
				Rejected data				
30HUF/PAR2	(3.03–1, 1.25, –2.93–1, –6)							

TABLE 11.24.3. Parameters of regression polynomial

Heat capacity type	No. data points total	No. data points used	s_w	s/R	s_r %	s_b/R	+/–
C_{sat}	16	10	0.246	1.22–2	0.05	8.39–6	–1

Temp. range K	A_1	A_2	Level of uncertainty
253.0–313.3	1.00856+1	5.22893	II

TABLE 11.24.4. Recommended values of heat capacities

Temp. (K)	250	260	270	273.15	280	290	298.15
c_{sat} (J K^{-1}g^{-1})	1.922	1.965	2.008	2.022	2.052	2.095	2.130
C_{sat} (J K^{-1}mol^{-1})	192.5	196.9	201.2	202.6	205.6	209.9	213.5

Temp. (K)	300	310
c_{sat} (J K^{-1}g^{-1})	2.138	2.182
C_{sat} (J K^{-1}mol^{-1})	214.3	218.6

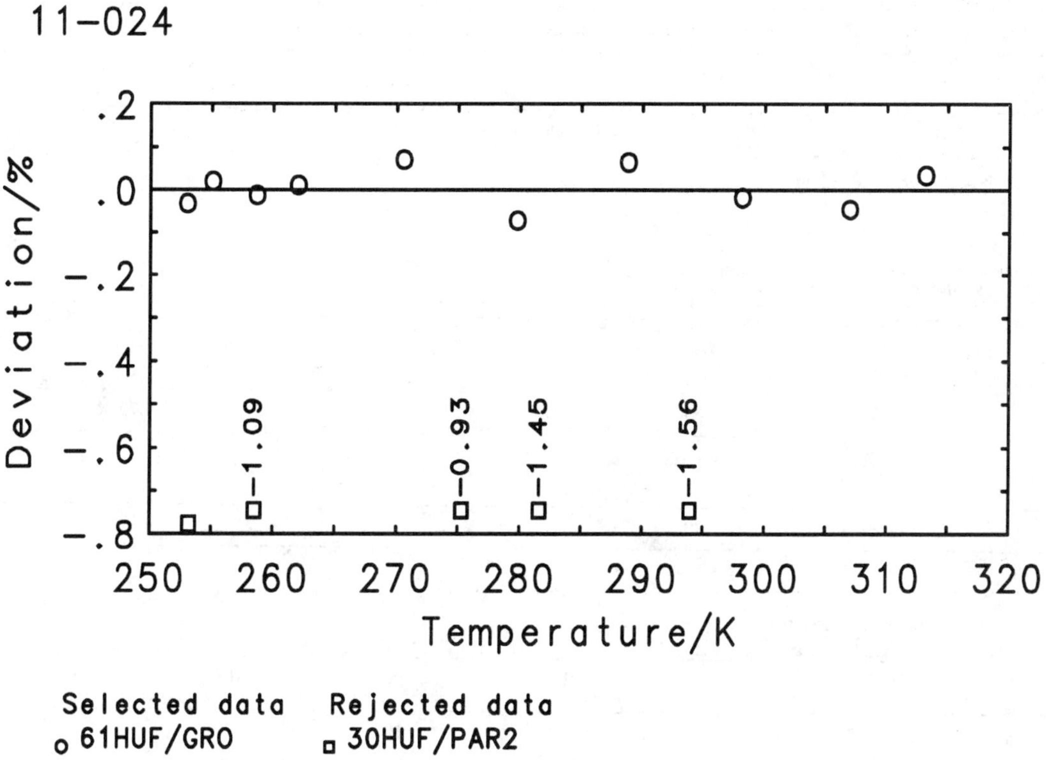

Name: 2,3–Dimethylhexane
Formula: C_8H_{18}

CAS–RN: 584–94–1
Group No.: 11–025
Molar Mass: 114.23

TABLE 11.25.1. Experimental heat capacities

Reference	Temp. K	Capac. J/(K.g)	Error %	Purity %	Purity method	Type capacity	Calorimeter Type	Calorimeter Reference
84FIL/LAU	293.00	2.190	2.00	99.0	estim	C_p	BDHO	84FIL/LAU

Name: 2,5–Dimethylhexane
Formula: C_8H_{18}

CAS–RN: 592–13–2
Group No.: 11–026
Molar Mass: 114.23

TABLE 11.26.1. Experimental heat capacities

Reference	Temp. range K	No. pnts	Error %	Purity %	Purity method	Type capacity	Calorimeter Type	Calorimeter Reference
47OSB/GIN	285.6–313.1	4S	0.10	99.96	melpt	C_{sat}	BSAO	47OSB/GIN
84FIL/LAU	293.0	1	2.00	95.0	estim	C_p	BDHO	84FIL/LAU

TABLE 11.26.2. Correlated heat capacities

Reference	Temp. range K	No. pnts used	$\sigma_r C$ %	d_w	d/R	d_r %	d_b/R	+/–
				Selected data				
47OSB/GIN	285.6–313.1	4	0.30#	0.964	8.74–2	0.29	4.72–4	0
				Rejected data				
84FIL/LAU	(6.58–1, 2.17, 6.58–1, 1)							

TABLE 11.26.3. Parameters of regression polynomial

Heat capacity type	No. data points total	used	s_w	s/R	s_r %	s_b/R	+/–
C_{sat}	5	4	1.363	1.24–1	0.41	4.72–4	0

Temp. range K	A_1	A_2	Level of uncertainty
285.6–313.1	1.02999+1	6.59931	III

TABLE 11.26.4. Recommended values of heat capacities

Temp. (K)	280	290	298.15	300	310
c_{sat} (J K^{-1}g^{-1})	2.095	2.143	2.182	2.191	2.239
C_{sat} (J K^{-1}mol^{-1})	239.3	244.8	249.2	250.2	255.7

Name: 3,3–Dimethylhexane
Formula: C_8H_{18}

CAS–RN: 563–16–6
Group No.: 11–027
Molar Mass: 114.23

TABLE 11.27.1. Experimental heat capacities

Reference	Temp. range K	No. pnts	Error %	Purity %	Purity method	Type capacity	Calorimeter Type	Calorimeter Reference
47OSB/GIN	290.6–305.6	4S	0.10	99.96	melpt	C_{sat}	BSAO	47OSB/GIN

TABLE 11.27.3. Parameters of regression polynomial

Heat capacity type	No. data points total	used	s_w	s/R	s_r %	s_b/R	+/–
C_{sat}	4	4	0.660	1.98–2	0.07	1.76–5	1

Temp. range K	A_1	A_2	Level of uncertainty
290.6–305.6	–8.57274	1.28264+1	II

TABLE 11.27.4. Recommended values of heat capacities

Temp. (K)	290	295	298.15	300	305
c_{sat} (J K^{-1}g^{-1})	2.083	2.130	2.160	2.177	2.223
C_{sat} (J K^{-1}mol^{-1})	238.0	243.3	246.7	248.7	254.0

Name: 3–Ethyl–2–methylpentane
Formula: C_8H_{18}

CAS–RN: 609–26–7
Group No.: 11–028
Molar Mass: 114.23

TABLE 11.28.1. Experimental heat capacities

Reference	Temp. K	Capac. J/(K.g)	Error %	Purity %	Purity method	Type capacity	Calorimeter Type	Calorimeter Reference
84FIL/LAU	293.00	2.248	2.00	99.0	estim	C_p	BDHO	84FIL/LAU

Name: 2–Methylheptane
Formula: C_8H_{18}

CAS–RN: 592–27–8
Group No.: 11–029
Molar Mass: 114.23

TABLE 11.29.1. Experimental heat capacities

Reference	Temp. range K	No. pnts	Error %	Purity %	Purity method	Type capacity	Calorimeter Type	Calorimeter Reference
47OSB/GIN	285.6–305.6	5S	0.10	99.96	melpt	C_{sat}	BSAO	47OSB/GIN
71MES/FIN	177.6–369.6	23	0.20	99.969	melpt	C_{sat}	BSAO	43RUE/HUF
84FIL/LAU	293.0	1	2.00	99.	anal	C_p	BDHO	84FIL/LAU

TABLE 11.29.2. Correlated heat capacities

Reference	Temp. range K	No. pnts used	$\sigma_r C$ %	d_w	d/R	d_r %	d_b/R	+/–
				Selected data				
47OSB/GIN	285.6–305.6	5	0.10	0.609	1.86–2	0.06	–1.32–2	–3
71MES/FIN	177.6–369.6	23	0.20#	0.486	2.76–2	0.10	1.15–2	11
				Rejected data				
84FIL/LAU	(1.32–1, 0.44, 1.32–1, 1)							

TABLE 11.29.3. Parameters of regression polynomial

Heat capacity type	No. data points total	No. data points used	s_w	s/R	s_r %	s_b/R	+/–
C_{sat}	29	28	0.550	2.83–2	0.10	7.11–3	8

Temp. range K	A_1	A_2	A_3	A_4	Level of uncertainty
177.6–369.6	3.06623+1	–8.37686	3.62516	–2.88494–1	II

TABLE 11.29.4. Recommended values of heat capacities

Temp. (K)	180	190	200	210	220	230	240
c_{sat} ($J\,K^{-1}g^{-1}$)	1.867	1.882	1.900	1.921	1.944	1.970	1.998
C_{sat} ($J\,K^{-1}mol^{-1}$)	213.2	215.0	217.0	219.4	222.1	225.0	228.2
Temp. (K)	250	260	270	273.15	280	290	298.15
c_{sat} ($J\,K^{-1}g^{-1}$)	2.029	2.061	2.096	2.107	2.132	2.171	2.203
C_{sat} ($J\,K^{-1}mol^{-1}$)	231.7	235.4	239.4	240.7	243.6	247.9	251.6
Temp. (K)	300	310	320	330	340	350	360
c_{sat} ($J\,K^{-1}g^{-1}$)	2.210	2.252	2.295	2.339	2.384	2.430	2.477
C_{sat} ($J\,K^{-1}mol^{-1}$)	252.5	257.2	262.1	267.1	272.3	277.6	282.9
Temp. (K)	370						
c_{sat} ($J\,K^{-1}g^{-1}$)	2.524						
C_{sat} ($J\,K^{-1}mol^{-1}$)	288.4						

TABLE 11.29.5. Parameters of quasi-polynomial equation

Heat capacity type	No. data points total	used	s_w	s/R	s_r %	s_b/R	+/−
C_{sat}	29	28	2.238	1.24–1	0.42	−1.10–2	2

Temp. range K	T_c K	A_1	A_2	A_3	A_4	Level of uncertainty
177.6–369.6	559.60	−4.68388	3.83163	1.76597+1	1.43142	III

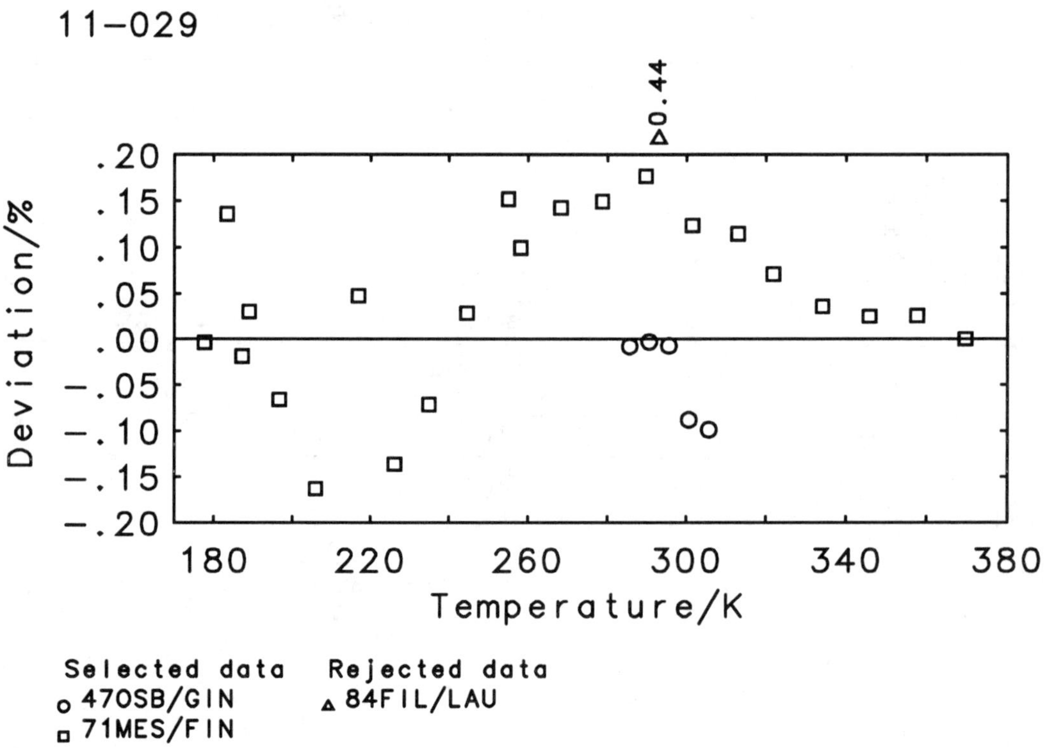

Name: 3–Methylheptane
Formula: C_8H_{18}

CAS–RN: 589–81–1
Group No.: 11–030
Molar Mass: 114.23

TABLE 11.30.1. Experimental heat capacities

Reference	Temp. range K	No. pnts	Error %	Purity %	Purity method	Type capacity	Calorimeter Type	Calorimeter Reference
47OSB/GIN	285.6–305.6	5S	0.10	99.96	melpt	C_{sat}	BSAO	47OSB/GIN
73FIN/MES	157.2–376.0	28	0.20	99.973	melpt	C_{sat}	BSAO	43RUE/HUF
84FIL/LAU	293.0	1	2.00	97.	anal	C_p	BDHO	84FIL/LAU

TABLE 11.30.2. Correlated heat capacities

Reference	Temp. range K	No. pnts used	$\sigma_r C$ %	d_w	d/R	d_r %	d_b/R	+/–
				Selected data				
47OSB/GIN	285.605.6	5	0.20#	0.745	4.45–2	0.15	–4.42–2	–5
73FIN/MES	157.2–376.0	28	0.20#	0.438	2.62–2	0.09	7.97–3	5
				Rejected data				
84FIL/LAU	(1.73–1, 0.58, 1.73–1, 1)							

TABLE 11.30.3. Parameters of regression polynomial

Heat capacity type	No. data points total	No. data points used	s_w	s/R	s_r %	s_b/R	+/–
C_{sat}	34	33	0.530	3.17–2	0.11	5.97–5	0

Temp. range K	A_1	A_2	A_3	A_4	Level of uncertainty
157.2–376.0	2.55387+1	–3.63138	2.14998	–1.41712–1	II

TABLE 11.30.4. Recommended values of heat capacities

Temp. (K)	160	170	180	190	200	210	220
c_{sat} (J K^{-1}g^{-1})	1.794	1.811	1.830	1.851	1.874	1.898	1.925
C_{sat} (J K^{-1}mol^{-1})	205.0	206.9	209.0	211.4	214.0	216.9	219.9
Temp. (K)	230	240	250	260	270	273.15	280
c_{sat} (J K^{-1}g^{-1})	1.953	1.983	2.015	2.048	2.083	2.094	2.119
C_{sat} (J K^{-1}mol^{-1})	223.1	226.6	230.2	234.0	237.9	239.2	242.1
Temp. (K)	290	298.15	300	310	320	330	340
c_{sat} (J K^{-1}g^{-1})	2.157	2.189	2.196	2.236	2.278	2.320	2.364
C_{sat} (J K^{-1}mol^{-1})	246.4	250.0	250.8	255.4	260.2	265.0	270.0
Temp. (K)	350	360	370	380			
c_{sat} (J K^{-1}g^{-1})	2.409	2.454	2.501	2.548			
C_{sat} (J K^{-1}mol^{-1})	275.1	280.3	285.7	291.1			

TABLE 11.30.5. Parameters of quasi-polynomial equation

Heat capacity type	No. data points total	used	s_w	s/R	s_r %	s_b/R	+/–
C_{sat}	34	33	1.867	1.08–1	0.37	7.45–3	3

Temp. range K	T_c K	A_1	A_2	A_3	A_4	Level of uncertainty
157.2–376.0	563.70	–6.47015	2.46348	1.76550+1	4.24835	III

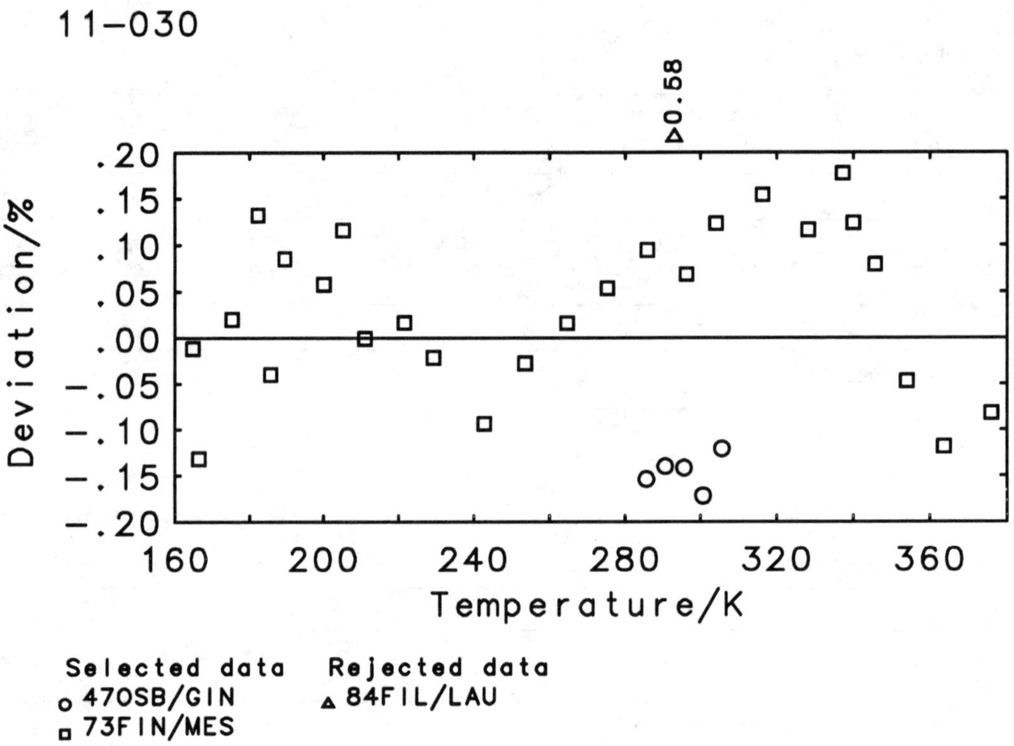

Name: 4–Methylheptane
Formula: C_8H_{18}

CAS–RN: 589–53–7
Group No.: 11–031
Molar Mass: 114.23

TABLE 11.31.1. Experimental heat capacities

Reference	Temp. range K	No. pnts	Error %	Purity %	Purity method	Type capacity	Calorimeter Type	Calorimeter Reference
47OSB/GIN	285.6–313.1	4S	0.10	99.96	melpt	C_{sat}	BSAO	47OSB/GIN

TABLE 11.31.3. Parameters of regression polynomial

Heat capacity type	No. data points total	used	s_w	s/R	s_r %	s_b/R	+/–
C_{sat}	4	4	1.339	1.22–1	0.40	4.56–4	0
Temp. range K	A_1	A_2					Level of uncertainty
285.6–313.1	1.05644+1	6.58446					III

TABLE 11.31.4. Recommended values of heat capacities

Temp. (K)	280	290	298.15	300	310
c_{sat} (J $K^{-1}g^{-1}$)	2.111	2.159	2.198	2.207	2.255
C_{sat} (J $K^{-1}mol^{-1}$)	241.1	246.6	251.1	252.1	257.6

Name: Octane
Formula: C_8H_{18}

CAS–RN: 111–65–9
Group No.: 11–032
Molar Mass: 114.23

TABLE 11.32.1. Experimental heat capacities

Reference		Temp. range K	No. pnts	Error %	Purity %	Purity method	Type capacity	Calorimeter Type	Calorimeter Reference
*98LOU	N	344.9	1	nosp	not specified		avg	DSIO	*98LOU
30PAR/HUF2		223.0–293.7	8	nosp	not specified		C_p	BSIO	25PAR
31HUF/PAR		227.0–298.3	5	nosp	not specified		C_p	BSIO	25PAR
47OSB/GIN		285.6–305.6	5S	0.10	99.96	estim	C_{sat}	BSAO	47OSB/GIN
49TSC/WET		298.1	1	nosp	not specified		C_p	BSIO	49TSC/RIC1
51CON/SAG		299.8–366.5	13S	nosp	not specified		C_p	BSAO	39SAG/EVA
54FIN/GRO2	N	222.6–297.6	18	0.20	99.94	melpt	C_{sat}	BSAO	43RUE/HUF
61ROU		283.1–312.3	15	nosp	99.83	anal	C_p	BSAO	61ROU
70AKH		293.1	1	nosp	not specified		C_p	BDHO	59ABA/MUS
71AMI/ALI	N	403.1–567.2	21S	2.00	not specified		C_p	not specified	
75GRI/RAS	N	303.9–462.1	9	1.00	not specified		C_p	BDAO	75RAS/GRI
77NAZ/MUS		303.2–383.2	5	1.50	not specified		C_p	BSAO	77NAZ/MUS
80SHA/LYU		216.4–300.0	12S	0.50	99.57	melpt	C_p	BSAO	80SHA/LYU
81GRO/ING		298.1	1	nosp	99.0	chrom	C_p	FSIT	71PIC/LED
84GRI/AND	N	297.5–386.1	7	0.50	not specified		C_p	BSAO	67RAS/GAN
84ROU/GRO		298.1	1	0.30	99.	estim	C_p	FSIT	71PIC/LED
85COS/PAT2		298.2	1	nosp	99.	anal	C_p	FSIT	71PIC/LED
85LAI/GRO		298.1	1	nosp	not specified		C_p	FSIT	71PIC/LED
85LAI/ROD		298.1	1	nosp	not specified		C_p	FSIT	71PIC/LED
86BEN/DAR1		298.1	1	nosp	99.	anal	C_p	FSIT	71PIC/LED
91BAN/GAR		318.1–373.1	12	0.40	98.8	chrom	C_p	BDCT	91BAN/GAR
91OGA/MIT		298.1	1	nosp	not specified		C_p	FSIO	85OGA
91SOE/NAK		298.1	1	nosp	99.99	chrom	C_p	FSIO	85OGA
91TRE/COS		298.1	1	nosp	99.	anal	C_p	FSIT	71PIC/LED
92LAI/ROD		298.1	1	nosp	99.5	anal	C_p	FSIT	71PIC/LED

*98LOU average value in temperature range 294–396 K
54FIN/GRO2 smoothed data in 67MES/GUT
71AMI/ALI calculated from C_v measured at the saturation line
75GRI/RAS grade: pure; the last five values at pressures higher than vapour pressure
84GRI/AND grade: pure

TABLE 11.32.2. Correlated heat capacities

Reference	Temp. range K	No. pnts used	$\sigma_r C$ %	d_w	d/R	d_r %	d_b/R	+/–
				Selected data				
47OSB/GIN	285.6–305.6	5	0.30#	1.444	1.33–1	0.43	1.32–1	5
49TSC/WET	298.1	1	0.50#	0.820	1.26–1	0.41	1.26–1	1
54FIN/GRO2	222.6–297.6	18	0.20	0.706	4.26–2	0.14	–1.90–2	–3
75GRI/RAS	304.0–462.1	9	1.00	0.510	1.73–1	0.51	–6.58–2	–3
77NAZ/MUS	303.2–383.2	5	1.50	0.080	3.94–2	0.12	–2.42–2	–1
85COS/PAT2	298.2	1	0.50#	0.415	6.39–2	0.21	6.39–2	1
85LAI/GRO	298.1	1	0.50#	0.257	3.93–2	0.13	–3.93–2	–1
86BEN/DAR1	298.1	1	0.50#	0.013	2.06–3	0.01	–2.06–3	0
91OGA/MIT	298.1	1	0.30#	1.098	1.01–1	0.33	1.01–1	1
91TRE/COS	298.1	1	0.50#	0.433	6.65–2	0.22	6.65–2	1

Rejected data

*98LOU	(1.40–3, 0.00, –1.40–3, 0)	30PAR/HUF2	(4.84–1, 1.67, –4.68–1, –8)
31HUF/PAR	(3.74–1, 1.28, –3.69–1, –5)	51CON/SAG	(5.69–1, 1.75, –5.45–1, –13)
61ROU	(3.43–1, 1.14, –2.18–1, –9)	70AKH	(2.13–1, 0.69, 2.13–1, 1)
71AMI/ALI	(1.36, 3.63, –1.33, –6)	80SHA/LYU	(1.45–1, 0.48, –5.94–2, 0)
81GRO/ING	(1.27–1, 0.42, –1.27–1, –1)	84GRI/AND	(3.29–1, 0.99, –3.22–1, –7)
84ROU/GRO	(1.33–1, 0.44, –1.33–1, –1)	85LAI/ROD	(1.69–1, 0.55, –1.69–1, –1)
91BAN/GAR	(1.61–1, 0.49, –1.47–1, –12)	91SOE/NAK	(1.12–1, 0.36, –1.12–1, –1)
92LAI/ROD	(1.89–1, 0.62, –1.89–1, –1)		

TABLE 11.32.3. Parameters of cubic spline polynomials

Heat capacity type	No. data points total	used	s_w	s/R	s_r %	s_b/R	+/–
C_p	143	43	0.796	1.07–1	0.33	–1.79–3	1
C_{sat}	143	43	0.795	1.07–1	0.33	–2.03–3	1

Temp. range K	A_1	A_2	A_3	A_4	Level of uncertainty
222.6–300.0	7.59487+1	–5.42060+1	1.93384+1	–2.09614	II
300.0–462.1	1.52779+1	6.46485	–8.85159–1	1.50930–1	IV
222.6–300.0	7.57257+1	–5.39481+1	1.92395+1	–2.08356	II
300.0–462.1	1.70184+1	4.75936	–3.29600–1	9.07863–2	IV

TABLE 11.32.4. Recommended values of heat capacities

Temp. (K)	220	230	240	250	260	270	273.15
c_p (J K^{-1}g^{-1})	2.036	2.043	2.057	2.078	2.103	2.134	2.144
C_p (J K^{-1}mol^{-1})	232.6	233.4	235.0	237.4	240.3	243.7	244.9
c_{sat} (J K^{-1}g^{-1})	2.036	2.043	2.057	2.078	2.103	2.133	2.144
C_{sat} (J K^{-1}mol^{-1})	232.6	233.4	235.0	237.4	240.3	243.7	244.9
Temp. (K)	280	290	298.15	300	310	320	330
c_p (J K^{-1}g^{-1})	2.167	2.203	2.233	2.240	2.279	2.318	2.358
C_p (J K^{-1}mol^{-1})	247.5	251.6	255.1	255.9	260.3	264.8	269.4
c_{sat} (J K^{-1}g^{-1})	2.167	2.203	2.233	2.240	2.279	2.318	2.358
C_{sat} (J K^{-1}mol^{-1})	247.5	251.6	255.1	255.9	260.3	264.8	269.4
Temp. (K)	340	350	360	370	380	390	400
c_p (J K^{-1}g^{-1})	2.399	2.441	2.484	2.528	2.573	2.619	2.666
C_p (J K^{-1}mol^{-1})	274.0	278.8	283.7	288.7	293.9	299.2	304.6
c_{sat} (J K^{-1}g^{-1})	2.399	2.441	2.483	2.527	2.571	2.617	2.663
C_{sat} (J K^{-1}mol^{-1})	274.0	278.8	283.7	288.6	293.7	298.9	304.2
Temp. (K)	410	420	430	440	450	460	
c_p (J K^{-1}g^{-1})	2.715	2.766	2.818	2.871	2.926	2.983	
C_p (J K^{-1}mol^{-1})	310.2	315.9	321.9	328.0	334.2	340.7	
c_{sat} (J K^{-1}g^{-1})	2.711	2.760	2.810	2.861	2.914	2.968	
C_{sat} (J K^{-1}mol^{-1})	309.7	315.3	321.0	326.9	332.9	339.0	

TABLE 11.32.5. Parameters of quasi-polynomial equation

Heat capacity type	No. data points total	used	s_w	s/R	s_r %	s_b/R	+/–
C_p	143	43	2.106	2.31–1	0.70	9.14–2	14

Temp. range K	T_c K	A_1	A_2	A_3	A_4	Level of uncertainty
222.6–462.1	568.95	–5.81193	9.16902–1	1.95769+1	9.20996	IV

11–032

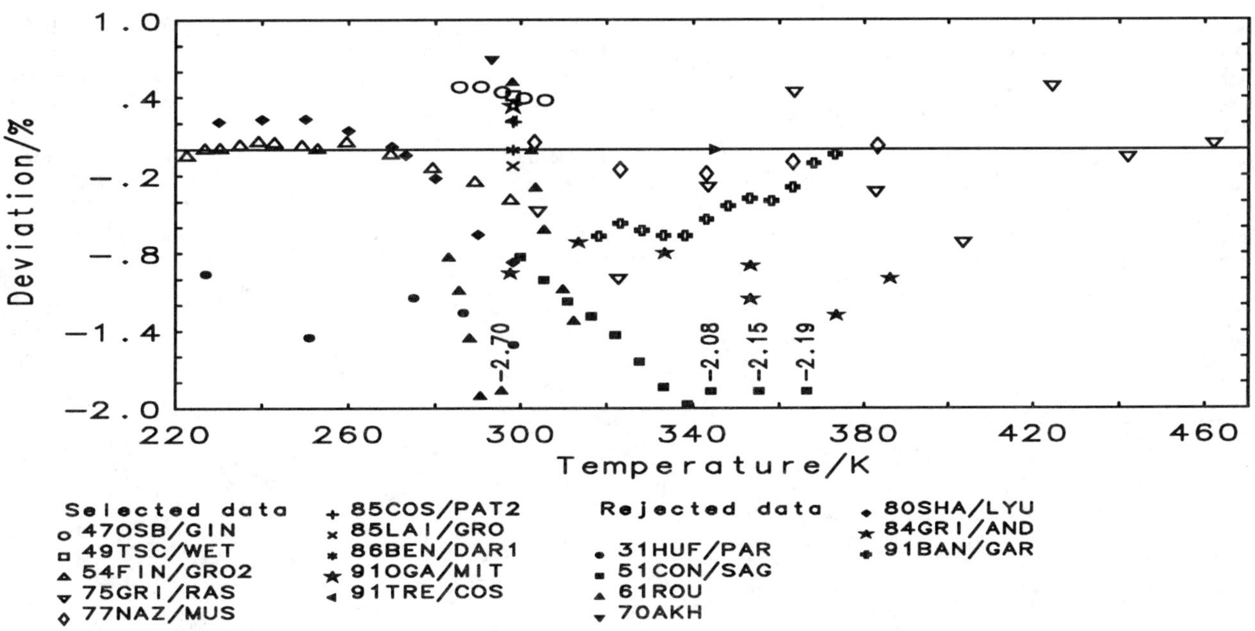

Name: 2,2,3,3–Tetramethylbutane
Formula: C_8H_{18}

CAS–RN: 594–82–1
Group No.: 11–033
Molar Mass: 114.23

TABLE 11.33.1. Experimental heat capacities

Reference	Temp. range K	No. pnts	Error %	Purity %	Purity method	Type capacity	Calorimeter Type	Calorimeter Reference
52SCO/DOU	375.4–378.0	3	0.20	99.976	melpt	C_{sat}	BSAO	43RUE/HUF

TABLE 11.33.3. Parameters of regression polynomial

Heat capacity type	No. data points total	No. data points used	s_w	s/R	s_r %	s_b/R	+/–
C_{sat}	3	3	0.353	2.39–2	0.07	1.14–5	–1

Temp. range K	A_1	A_2	Level of uncertainty
375.4–378.0	5.74080	7.44551	II

TABLE 11.33.4. Recommended values of heat capacities

Temp. (K)	375	378
c_{sat} (J K^{-1}g^{-1})	2.450	2.466
C_{sat} (J K^{-1}mol^{-1})	279.9	281.7

Name: 2,2,4–Trimethylpentane
Formula: C_8H_{18}

CAS–RN: 540–84–1
Group No.: 11–034
Molar Mass: 114.23

TABLE 11.34.1. Experimental heat capacities

Reference	Temp. range K	No. pnts	Error %	Purity %	Purity method	Type capacity	Calorimeter Type	Calorimeter Reference
30PAR/HUF2	169.6–295.2	15	1.00	not specified		C_p	BSIO	25PAR
40PIT2	171.1–317.3	8	0.20	99.99	melpt	C_p	BSIO	28LAT/GRE
47OSB/GIN	285.6–305.6	5S	0.10	99.96	melpt	C_{sat}	BSAO	47OSB/GIN
49TSC/WET	298.1	1	nosp	not specified		C_p	BSIO	49TSC/RIC1
50AUE/SAG	299.8–366.5	13	1.00	99.9	melpt	C_p	BSAO	39SAG/EVA
61ROU	300.4–315.4	12	nosp	99.99	anal	C_p	BSAO	61ROU
73SUB/RAJN	298.1–323.1	3	0.30	not specified		C_p	BSIO	64MOE/THO
76FOR/BEN1	298.1	1	nosp	not specified		C_p	FSIT	71PIC/LED
84FIL/LAU	293.0	1	2.00	not specified		C_p	BDHO	84FIL/LAU
87KAL/KOH	293.1–313.1	2	nosp	99.98	chrom	C_p	FSIT	71PIC/LED
88COS/VAN	298.1	1	nosp	99.	anal	C_p	FSIT	71PIC/LED
88SHI/OGA1	298.1	1	nosp	99.9	chrom	C_p	FSIO	85OGA

TABLE 11.34.2. Correlated heat capacities

Reference	Temp. range K	No. pnts used	$\sigma_r C$ %	d_w	d/R	d_r %	d_b/R	+/–
				Selected data				
40PIT2	171.1–317.3	8	0.20	0.771	4.21–2	0.15	2.54–2	4
47OSB/GIN	285.6–305.6	5	0.10	0.545	1.54–2	0.05	–8.22–3	–2
50AUE/SAG	299.8–366.5	13	1.00	0.310	9.25–2	0.31	–8.23–2	–11
76FOR/BEN1	298.1	1	0.50#	0.265	3.81–2	0.13	3.81–2	1
84FIL/LAU	293.0	1	2.00	0.014	8.12–3	0.03	–8.12–3	–1
				Rejected data				
30PAR/HUF2	(2.62–1, 0.98, –2.35–1, –14)			49TSC/WET	(8.14–1, 2.92, –8.14–1, –1)			
61ROU	(1.35–1, 0.46, 7.33–2, 4)			73SUB/RAJ	(3.39–1, 1.15, –2.93–1, –3)			
87KAL/KOH	(5.71–1, 1.89, 4.93–1, 2)			88COS/VAN	(4.71–1, 1.62, 4.71–1, 1)			
88SHI/OGA1	(1.37–1, 0.47, 1.37–1, 1)							

TABLE 11.34.3. Parameters of regression polynomial

Heat capacity type	No. data points total	used	s_w	s/R	s_r %	s_b/R	+/–
C	63	28	0.561	7.31–2	0.25	–3.14–2	–9

Temp. range K	A_1	A_2	A_3	A_4	Level of uncertainty
171.1–366.5	2.00353+1	–1.43004	1.91141	–1.53543–1	III

TABLE 11.34.4. Recommended values of heat capacities

Temp. (K)	170	180	190	200	210	220	230
c (J K^{-1}g^{-1})	1.629	1.657	1.686	1.717	1.750	1.784	1.819
C (J K^{-1}mol^{-1})	186.0	189.2	192.6	196.2	199.9	203.8	207.8
Temp. (K)	240	250	260	270	273.15	280	290
c (J K^{-1}g^{-1})	1.855	1.893	1.932	1.972	1.984	2.012	2.054
C (J K^{-1}mol^{-1})	211.9	216.2	220.7	225.2	226.7	229.9	234.6
Temp. (K)	298.15	300	310	320	330	340	350
c (J K^{-1}g^{-1})	2.089	2.096	2.140	2.184	2.228	2.273	2.319
C (J K^{-1}mol^{-1})	238.6	239.5	244.4	249.4	254.5	259.7	264.9
Temp. (K)	360	370					
c (J K^{-1}g^{-1})	2.365	2.412					
C (J K^{-1}mol^{-1})	270.2	275.5					

TABLE 11.34.5. Parameters of quasi-polynomial equation

Heat capacity type	No. data points total	used	s_w	s/R	s_r %	s_b/R	+/–
C	63	28	0.928	8.12–2	0.28	–3.23–2	–8

Temp. range K	T_c K	A_1	A_2	A_3	A_4	Level of uncertainty
171.1–366.5	544.00	–6.41602	7.86779–1	1.46601+1	1.30804+1	III

11–034

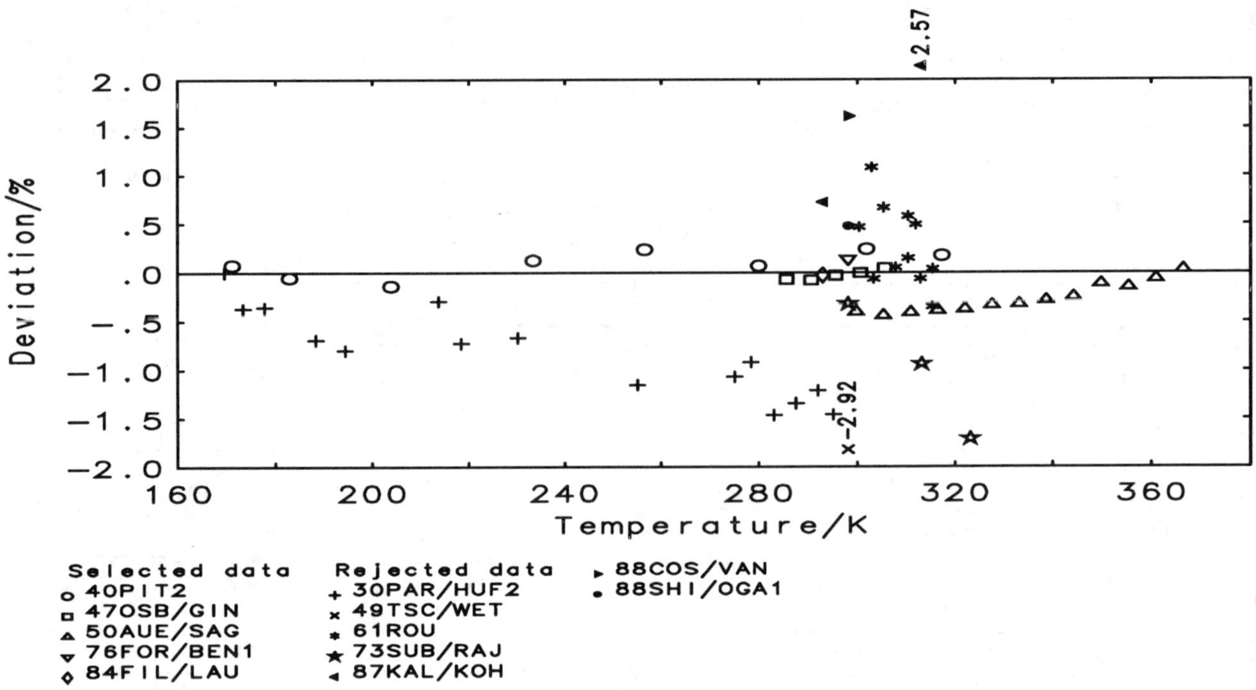

Name: 2,3,3–Trimethylpentane
Formula: C_8H_{18}

CAS–RN: 560–21–4
Group No.: 11–035
Molar Mass: 114.23

TABLE 11.35.1. Experimental heat capacities

Reference	Temp. range K	No. pnts	Error %	Purity %	Purity method	Type capacity	Calorimeter Type	Calorimeter Reference
47OSB/GIN	290.6–305.6	4	0.10	99.96	melpt	C_{sat}	BSAO	47OSB/GIN

TABLE 11.35.3. Parameters of regression polynomial

Heat capacity type	No. data points total	No. data points used	s_w	s/R	s_r %	s_b/R	+/–
C_{sat}	4	4	0.593	1.75–2	0.06	1.62–5	0

Temp. range K	A_1	A_2	A_3	Level of uncertainty
290.6–305.6	1.33968+2	−8.15968+1	1.56188+1	II

TABLE 11.35.4. Recommended values of heat capacities

Temp. (K)	290	298.15	300	310
c_{sat} (J K^{-1}g^{-1})	2.088	2.149	2.165	2.265
C_{sat} (J K^{-1}mol^{-1})	238.6	245.5	247.3	258.7

Name: 2,3,4–Trimethylpentane
Formula: C_8H_{18}

CAS–RN: 565–75–3
Group No.: 11–036
Molar Mass: 114.23

TABLE 11.36.1. Experimental heat capacities

Reference	Temp. range K	No. pnts	Error %	Purity %	Purity method	Type capacity	Calorimeter Type	Calorimeter Reference
41PIT/SCO	172.8–323.6	14	0.20	99.75	melpt	C_p	BSIO	28LAT/GRE
47OSB/GIN	290.6–305.6	4S	0.10	99.96	melpt	C_{sat}	BSAO	47OSB/GIN

TABLE 11.36.2. Correlated heat capacities

Reference	Temp. range K	No. pnts used	$\sigma_r C$ %	d_w	d/R	d_r %	d_b/R	+/–
				Selected data				
41PIT/SCO	172.8–323.6	14	0.70#	1.534	3.20–1	1.07	1.67–2	2
47OSB/GIN	290.6–305.6	4	0.20#	3.399	2.03–1	0.68	–7.15–5	0

TABLE 11.36.3. Parameters of regression polynomial

Heat capacity type	No. data points total	No. data points used	s_w	s/R	s_r %	s_b/R	+/–
C	18	18	2.378	3.38–1	1.13	1.30–2	2

Temp. range K	A_1	A_2	A_3	A_4	Level of uncertainty
172.8–323.6	–1.39272+1	4.42905+1	–1.79226+1	2.67701	IV

TABLE 11.36.4. Recommended values of heat capacities

Temp. (K)	170	180	190	200	210	220	230
c (J K^{-1}g^{-1})	1.654	1.699	1.739	1.775	1.808	1.839	1.871
C (J K^{-1}mol^{-1})	188.9	194.1	198.6	202.7	206.5	210.1	213.7
Temp. (K)	240	250	260	270	273.15	280	290
c (J K^{-1}g^{-1})	1.903	1.937	1.974	2.016	2.030	2.063	2.116
C (J K^{-1}mol^{-1})	217.4	221.3	225.5	230.3	231.9	235.6	241.8
Temp. (K)	298.15	300	310	320			
c (J K^{-1}g^{-1})	2.166	2.178	2.248	2.329			
C (J K^{-1}mol^{-1})	247.4	248.8	256.8	266.0			

TABLE 11.36.5. Parameters of quasi-polynomial equation

Heat capacity type	No. data points total	No. data points used	s_w	s/R	s_r %	s_b/R	+/–
C	18	18	3.456	2.15–1	0.75	–7.21–2	–1 0

Temp. range K	T_c K	A_1	A_2	A_3	A_4	Level of uncertainty
172.8–323.6	566.30	–7.64507	9.15039–1	1.39061+1	1.59685+1	IV

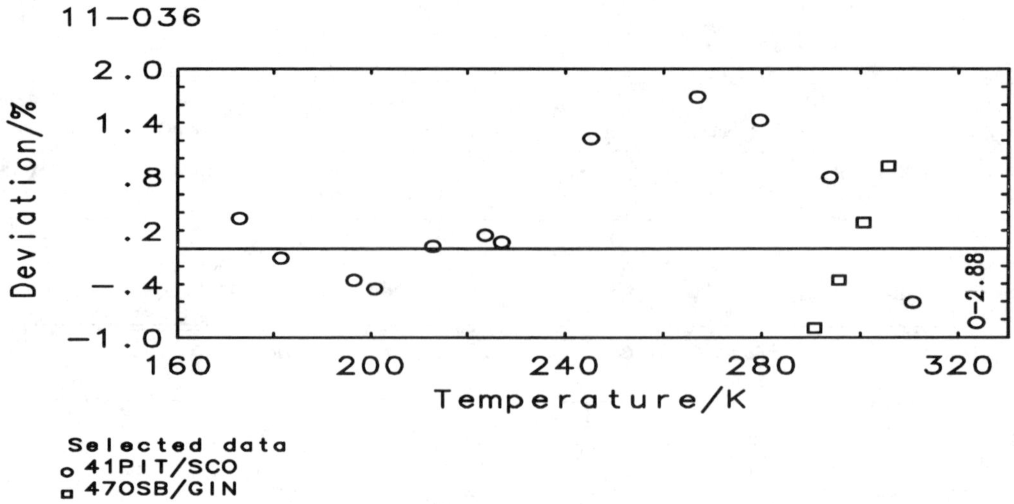

Name: 3,3–Diethylpentane
Formula: C_9H_{20}

CAS–RN: 1067–20–5
Group No.: 11–037
Molar Mass: 128.26

TABLE 11.37.1. Experimental heat capacities

Reference	Temp. range K	No. pnts	Error %	Purity %	Purity method	Type capacity	Calorimeter Type	Calorimeter Reference
54STA/WAR	250.0–260.0	2S	0.50	99.50	melpt	C_p	BSAO	49STA/GUP
76FIN/MES	243.2–365.4	16	0.20	99.985	melpt	C_{sat}	BSAO	43RUE/HUF
79FUC/PEA	298.1	1	nosp	99.	chrom	C_p	BSIO	80FUC

TABLE 11.37.2. Correlated heat capacities

Reference	Temp. range K	No. pnts used	$\sigma_r C$ %	d_w	d/R	d_r %	d_b/R	+/–
				Selected data				
76FIN/MES	243.2–365.4	16	0.20#	0.384	2.57–2	0.08	5.18–4	–1
79FUC/PEA	298.1	1	0.50#	0.287	4.80–2	0.14	–4.80–2	–1
				Rejected data				
54STA/WAR	(5.71–1, 1.84, 5.71–1, 2)							

TABLE 11.37.3. Parameters of regression polynomial

Heat capacity type	No. data points total	No. data points used	s_w	s/R	s_r %	s_b/R	+/–
C	19	17	0.418	3.04–2	0.09	–2.34–3	–2

Temp. range K	A_1	A_2	A_3	Level of uncertainty
243.2–365.4	1.45855+1	5.47668	2.92227–1	II

TABLE 11.37.4. Recommended values of heat capacities

Temp. (K)	250	260	270	273.15	280	290	298.15
c (J K^{-1}g^{-1})	1.952	1.997	2.042	2.057	2.088	2.134	2.172
C (J K^{-1}mol^{-1})	250.3	256.1	261.9	263.8	267.8	273.8	278.6
Temp. (K)	300	310	320	330	340	350	360
c (J K^{-1}g^{-1})	2.181	2.228	2.276	2.323	2.372	2.420	2.469
C (J K^{-1}mol^{-1})	279.7	285.8	291.9	298.0	304.2	310.4	316.7
Temp. (K)	370						
c (J K^{-1}g^{-1})	2.518						
C (J K^{-1}mol^{-1})	323.0						

TABLE 11.37.5. Parameters of quasi-polynomial equation

Heat capacity type	No. data points total	used	s_w	s/R	s_r %	s_b/R	+/–
C	19	17	0.479	3.47–2	0.10	–3.33–3	–2

Temp. range K	T_c K	A_1	A_2	A_3	A_4	Level of uncertainty
243.2–365.4	616.00	–5.83313	2.64844–1	1.35863+1	3.21184+1	II

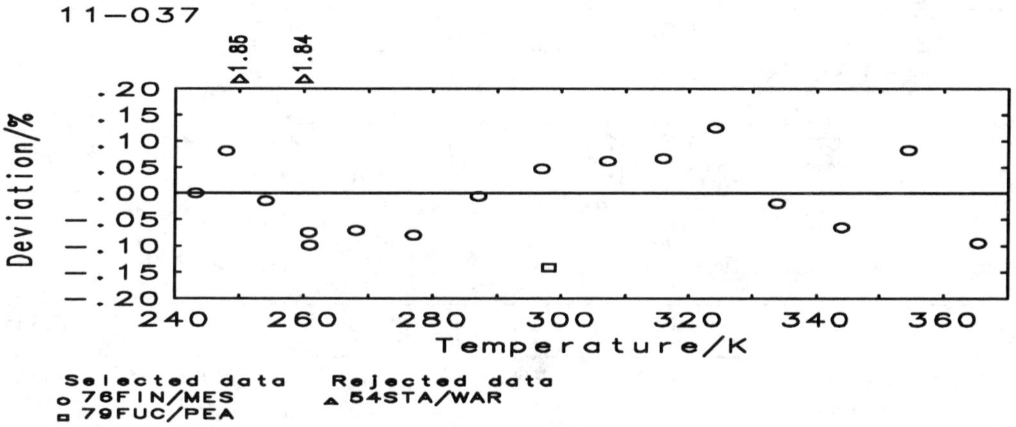

Name: Nonane
Formula: C_9H_{20}

CAS–RN: 111–84–2
Group No.: 11–038
Molar Mass: 128.26

TABLE 11.38.1. Experimental heat capacities

Reference		Temp. range K	No. pnts	Error %	Purity %	method	Type capacity	Calorimeter Type	Reference
30PAR/HUF2		224.5–299.1	8	1.00	not specified		C_p	BSIO	25PAR
31HUF/PAR		228.3–297.9	8	1.00	not specified		C_p	BSIO	25PAR
47OSB/GIN		283.1–313.1	4S	0.10	99.96	estim	C_{sat}	BSAO	47OSB/GIN
54FIN/GRO2	N	225.0–313.9	22	0.20	99.88	melpt	C_{sat}	BSAO	43RUE/HUF
58SWI/ZIE1	N	348.8	1	nosp	not specified		avg	DSIO	58SWI/ZIE1
70AKH		293.1	1	nosp	not specified		C_p	BDHO	59ABA/MUS
76MUS		307.8–417.8	10	2.50	not specified		C_p	BDAO	71MUS
79GRO/HAM		298.1	1	nosp	not specified		C_p	FSIT	71PIC/LED
82WIL/ING		298.1	1	nosp	99.0	estim	C_p	FSIT	71PIC/LED
88AND/PAT		298.1	1	nosp	98.	anal	C_p	FSIT	71PIC/LED
91BAN/GAR		318.1–373.1	12	0.40	99.8	chrom	C_p	BDCT	91BAN/GAR
91OGA/MIT		298.1	1	nosp	not specified		C_p	FSIO	85OGA
91TRE/COS		298.1	1	nosp	99.	anal	C_p	FSIT	71PIC/LED

54FIN/GRO2 smoothed data in 67MES/GUT
58SWI/ZIE1 average value in temperature range 295–402 K

TABLE 11.38.2. Correlated heat capacities

Reference	Temp. range K	No. pnts used	$\sigma_r C$ %	d_w	d/R	d_r %	d_b/R	+/−
				Selected data				
47OSB/GIN	283.1–313.1	4	0.10	1.657	5.66–2	0.17	2.37–3	0
54FIN/GRO2	225.0–313.9	22	0.20	0.305	2.04–2	0.06	5.40–3	2
79GRO/HAM	298.1	1	0.50#	0.282	4.82–2	0.14	−4.82–2	−1
82WIL/ING	298.1	1	0.50#	0.494	8.43–2	0.25	−8.43–2	−1
88AND/PAT	298.1	1	0.50#	0.224	3.84–2	0.11	3.84–2	1
91BAN/GAR	318.1–373.1	12	0.40#	0.564	8.16–2	0.23	−4.50–2	−7
91TRE/COS	298.1	1	0.50#	0.071	1.22–2	0.04	−1.22–2	−1
				Rejected data				
30PAR/HUF2	(4.25–1, 1.29, −4.24–1, −7)			31HUF/PAR	(3.05–1, 0.93, −2.99–1, −8)			
58SWI/ZIE1	(1.62, 4.18, 1.62, 1)			70AKH	(5.79–1, 1.68, 5.79–1, 1)			
76MUS	(9.12–1, 2.47, −8.09–1, −6)			91OGA/MIT	(2.04–1, 0.59, 2.04–1, 1)			

TABLE 11.38.3. Parameters of cubic spline polynomials

Heat capacity type	No. data points total	used	s_w	s/R	s_r %	s_b/R	+/−
C_p	71	42	0.683	5.53–2	0.16	−1.23–2	−7

Temp. range K	A_1	A_2	A_3	A_4	Level of uncertainty
225.0–300.0	1.23077+2	−1.00722+2	3.58380+1	−4.04253	II
300.0–373.1	−8.26076+1	1.04963+2	−3.27236+1	3.57543	III

TABLE 11.38.4. Recommended values of heat capacities

Temp. (K)	230	240	250	260	270	273.15	280
c_p (J K^{-1}g^{-1})	2.062	2.067	2.081	2.101	2.128	2.137	2.158
C_p (J K^{-1}mol^{-1})	264.5	265.1	266.9	269.5	272.9	274.0	276.7
Temp. (K)	290	298.15	300	310	320	330	340
c_p (J K^{-1}g^{-1})	2.190	2.218	2.224	2.257	2.291	2.327	2.367
C_p (J K^{-1}mol^{-1})	280.9	284.4	285.2	289.5	293.9	298.5	303.6
Temp. (K)	350	360	370				
c_p (J K^{-1}g^{-1})	2.411	2.462	2.520				
C_p (J K^{-1}mol^{-1})	309.3	315.8	323.2				

TABLE 11.38.5. Parameters of quasi-polynomial equation

Heat capacity type	No. data points total	used	s_w	s/R	s_r %	s_b/R	+/−
C_p	71	41	1.618	9.62–2	0.28	−7.53–3	−8

Temp. range K	T_c K	A_1	A_2	A_3	A_4	Level of uncertainty
230.1–373.1	594.90	1.14187+1	1.20016+1	1.66310+1	2.71605	III

11–038

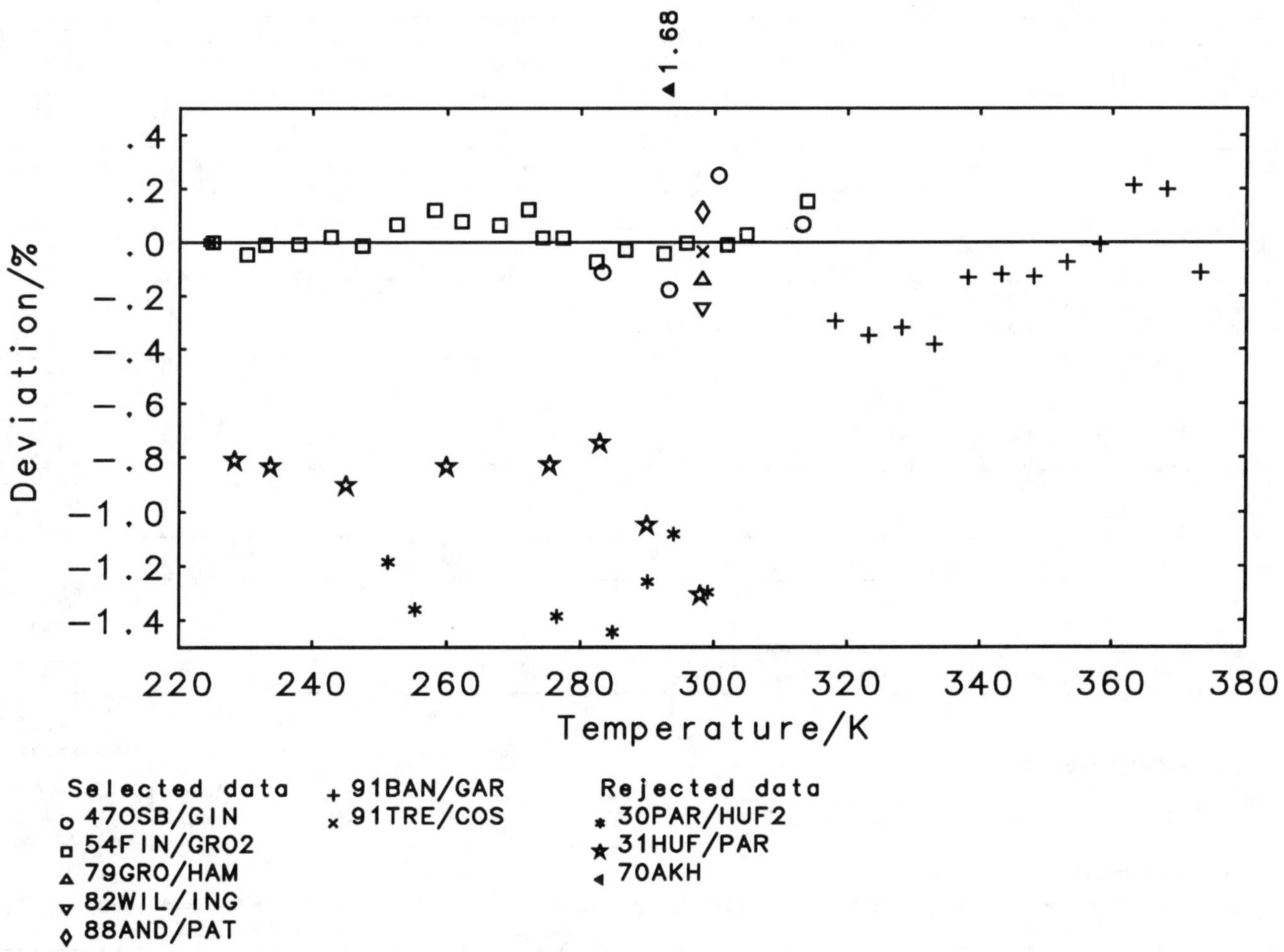

Name: 2,2,3,3–Tetramethylpentane
Formula: C_9H_{20}

CAS–RN: 7154–79–2
Group No.: 11–039
Molar Mass: 128.26

TABLE 11.39.1. Experimental heat capacities

Reference	Temp. range K	No. pnts	Error %	Purity %	Purity method	Type capacity	Calorimeter Type	Calorimeter Reference
76FIN/MES	269.2–392.7	19	0.20	99.993	melpt	C_{sat}	BSAO	43RUE/HUF

TABLE 11.39.3. Parameters of regression polynomial

Heat capacity type	No. data points total	No. data points used	s_w	s/R	s_r %	s_b/R	+/–
C_{sat}	19	19	0.225	1.57–2	0.04	1.28–5	–1

Temp. range K	A_1	A_2	A_3	Level of uncertainty
269.2–392.7	1.26743+1	6.40507	9.95780–2	II

TABLE 11.39.4. Recommended values of heat capacities

Temp. (K)	270	273.15	280	290	298.15	300	310
c_{sat} (J K^{-1}g^{-1})	1.990	2.004	2.035	2.080	2.117	2.125	2.171
C_{sat} (J K^{-1}mol^{-1})	255.2	257.0	261.0	266.8	271.5	272.6	278.4
Temp. (K)	320	330	340	350	360	370	380
c_{sat} (J K^{-1}g^{-1})	2.216	2.262	2.308	2.354	2.400	2.446	2.493
C_{sat} (J K^{-1}mol^{-1})	284.3	290.1	296.0	301.9	307.8	313.8	319.7
Temp. (K)	390						
c_{sat} (J K^{-1}g^{-1})	2.539						
C_{sat} (J K^{-1}mol^{-1})	325.7						

TABLE 11.39.5. Parameters of quasi-polynomial equation

Heat capacity type	No. data points total	used	s_w	s/R	s_r %	s_b/R	+/–
C_{sat}	19	19	0.222	1.56–2	0.04	1.15–5	0

Temp. range K	T_c K	A_1	A_2	A_3	A_4	Level of uncertainty
269.2–392.7	607.60	–1.47040	1.36537–2	1.22125+1	3.95878+1	II

Name: 2,2,4,4–Tetramethylpentane
Formula: C_9H_{20}

CAS–RN: 1070–87–7
Group No.: 11–040
Molar Mass: 128.26

TABLE 11.40.1. Experimental heat capacities

Reference	Temp. range K	No. pnts	Error %	Purity %	Purity method	Type capacity	Calorimeter Type	Calorimeter Reference
76FIN/MES	212.7–379.5	19	0.20	99.985	melpt	C_{sat}	BSAO	43RUE/HUF

TABLE 11.40.3. Parameters of regression polynomial

Heat capacity type	No. data points total	used	s_w	s/R	s_r %	s_b/R	+/–
C_{sat}	19	19	0.253	1.65–2	0.05	1.31–5	3

Temp. range K	A_1	A_2	A_3	A_4	Level of uncertainty
212.7–379.5	2.11552+1	–1.50750	2.34209	–2.05882–1	II

TABLE 11.40.4. Recommended values of heat capacities

Temp. (K)	210	220	230	240	250	260	270
c_{sat} (J K^{-1}g^{-1})	1.712	1.749	1.787	1.827	1.867	1.909	1.952
C_{sat} (J K^{-1}mol^{-1})	219.6	224.3	229.3	234.3	239.5	244.9	250.3
Temp. (K)	273.15	280	290	298.15	300	310	320
c_{sat} (J K^{-1}g^{-1})	1.965	1.995	2.039	2.076	2.084	2.130	2.176
C_{sat} (J K^{-1}mol^{-1})	252.1	255.9	261.6	266.3	267.3	273.2	279.1
Temp. (K)	330	340	350	360	370	380	
c_{sat} (J K^{-1}g^{-1})	2.223	2.270	2.317	2.365	2.412	2.460	
C_{sat} (J K^{-1}mol^{-1})	285.1	291.1	297.2	303.3	309.4	315.5	

TABLE 11.40.5. Parameters of quasi-polynomial equation

Heat capacity type	No. data points total	used	s_w	s/R	s_r %	s_b/R	+/−
C_{sat}	19	19	0.638	4.08–2	0.13	1.09–4	0

Temp. range K	T_c K	A_1	A_2	A_3	A_4	Level of uncertainty
212.7–379.5	574.70	−6.23952	3.98970–1	1.39685+1	2.43951+1	II

Name: Decane
Formula: $C_{10}H_{22}$

CAS–RN: 124–18–5
Group No.: 11–041
Molar Mass: 142.28

TABLE 11.41.1. Experimental heat capacities

Reference		Temp. range K	No. pnts	Error %	Purity %	Purity method	Type capacity	Calorimeter Type	Calorimeter Reference
*98LOU	N	360.9	1	nosp	not specified		avg	DSIO	*98LOU
30PAR/HUF2		242.3–295.5	6	1.00	not specified		C_p	BSIO	25PAR
31HUF/PAR		251.2–297.7	6	1.00	not specified		C_p	BSIO	25PAR
47OSB/GIN		290.6–305.6	4S	0.10	99.96	estim	C_{sat}	BSAO	47OSB/GIN
52SCH/SAG		299.8–366.5	13S	1.00	99.7	estim	C_{sat}	BSAO	39SAG/EVA
54FIN/GRO2	N	247.0–318.6	17	0.20	99.91	melpt	C_{sat}	BSAO	43RUE/HUF
70AKH		293.1	1	nosp	not specified		C_p	BDHO	59ABA/MUS
75GRI/RAS	N	303.3–462.4	9	1.00	not specified		C_p	BDAO	75RAS/GRI
79GRO/HAM		298.1	1	nosp	not specified		C_p	FSIT	71PIC/LED
82WIL/ING		298.1	1	nosp	99.0	estim	C_p	FSIT	71PIC/LED
83SID/SVE		293.1	1	nosp	99.9	melpt	C_p	FSIT	71PIC/LED
84GRO/ING		298.1	1	nosp	99.0	melpt	C_p	FSIT	71PIC/LED
84ROU/GRO		298.1	1	nosp	98.0	estim	C_p	FSIT	71PIC/LED
85BAL/BRA		298.1	1	nosp	99.	anal	C_p	FSIT	71PIC/LED
85COS/PAT3		298.2	1	nosp	99.	anal	C_p	FSIT	71PIC/LED
85COS/PAT8		283.1–313.1	3	nosp	99.	estim	C_p	FSIT	71PIC/LED
85LAI/ROD		298.1	1	nosp	not specified		C_p	FSIT	71PIC/LED
85LAI/WIL		298.1	1	0.30	99.	estim	C_p	FSIT	71PIC/LED
86GAT/WOO		298.1–368.1	4	nosp	99.	anal	C_p	BDCT	83ROU/ROU
87WIL/ING		298.1	1	nosp	99.	anal	C_p	FSIT	71PIC/LED
88COS/VAN		298.1	1	nosp	99.	anal	C_p	FSIT	71PIC/LED
88KUZ/KHA		293.8–335.2	15	nosp	99.8	chrom	C_p	FSIO	75SAF/GER
88KUZ/KHA		310.1–421.9	16	nosp	not specified		C_p	BSAO	67RAS/GAN
88PIN/BRA		298.1	1	nosp	99.	anal	C_p	FSIT	71PIC/LED
91BAN/GAR		318.1–373.1	12	0.40	98.9	chrom	C_p	BDCT	91BAN/GAR
91TRE/COS		298.1	1	nosp	99.	anal	C_p	FSIT	71PIC/LED
92LAI/ROD		298.1	1	nosp	99.	anal	C_p	FSIT	71PIC/LED

*98LOU average value in temperature range 295–427 K
54FIN/GRO2 smoothed data in 67MES/GUT
75GRI/RAS last two values at pressures above the vapour pressure

TABLE 11.41.2. Correlated heat capacities

Reference	Temp. range K	No. pnts used	$\sigma_r C$ %	d_w	d/R	d_r %	d_b/R	+/–
				Selected data				
54FIN/GRO2	247.0–318.6	17	0.20	0.158	1.17–2	0.03	–1.09–4	–2
75GRI/RAS	303.3–462.4	9	1.00	0.667	2.77–1	0.67	–7.66–2	–1
84GRO/ING	298.1	1	0.50#	0.310	5.84–2	0.15	–5.84–2	–1
84ROU/GRO	298.1	1	0.50#	0.131	2.47–2	0.07	–2.47–2	–1
86GAT/WOO	298.1–368.1	4	0.30#	1.126	1.36–1	0.34	–1.35–1	–4
88KUZ/KHA	293.8–335.2	15	0.50#	0.419	8.04–2	0.21	7.32–2	15
88KUZ/KHA	310.1–421.9	16	0.60#	0.330	8.54–2	0.20	3.50–2	4
91BAN/GAR	318.1–373.1	12	0.40	0.327	5.50–2	0.13	1.64–2	1

Rejected data

*98LOU	(3.97–1, 0.94, 3.97–1, 1)	30PAR/HUF2	(6.60–1, 1.85, –6.22–1, –5)
31HUF/PAR	(3.10–1, 0.85, –3.06–1, –6)	47OSB/GIN	(3.28–1, 0.87, –2.70–2, 0)
52SCH/SAG	(4.05–1, 1.04, –3.67–1, –11)	70AKH	(4.51–1, 1.22, –4.51–1, –1)
79GRO/HAM	(1.34–1, 0.36, –1.34–1, –1)	82WIL/ING	(2.28–1, 0.61, –2.28–1, –1)
83SID/SVE	(2.26–1, 0.61, –2.26–1, –1)	85BAL/BRA	(1.41–1, 0.38, –1.41–1, –1)
85COS/PAT3	(1.23–1, 0.32, 1.23–1, 1)	85COS/PAT8	(1.55–1, 0.40, 4.26–2, 1)
85LAI/ROD	(1.71–1, 0.46, –1.71–1, –1)	85LAI/WIL	(1.60–1, 0.42, –1.60–1, –1)
87WIL/ING	(1.80–1, 0.48, –1.80–1, –1)	88COS/VAN	(1.09–1, 0.29, 1.09–1, 1)
88PIN/BRA	(1.55–1, 0.41, –1.55–1, –1)	91TRE/COS	(1.26–1, 0.33, 1.26–1, 1)
92LAI/ROD	(2.18–1, 0.58, –2.18–1, –1)		

TABLE 11.41.3. Parameters of cubic spline polynomials

Heat capacity type	No. data points total	used	s_w	s/R	s_r %	s_b/R	+/–
C_p	121	75	0.467	1.21–1	0.29	7.22–3	11
C_{sat}	121	75	0.468	1.21–1	0.29	0.25–3	10

Temp. range K	A_1	A_2	A_3	A_4	Level of uncertainty
247.0–314.0	8.33283+1	–5.10491+1	1.70127+1	–1.68060	II
314.0–462.4	3.55264+1	–5.37839	2.46788	–1.36562–1	III
247.0–314.0	8.29151+1	–5.06122+1	1.68594+1	–1.66276	II
314.0–462.4	3.67462+1	–6.50171	2.81150	–1.71470–1	III

TABLE 11.41.4. Recommended values of heat capacities

Temp. (K)	250	260	270	273.15	280	290	298.15
c_p (J K^{-1}g^{-1})	2.091	2.108	2.129	2.137	2.155	2.184	2.210
C_p (J K^{-1}mol^{-1})	297.5	299.9	303.0	304.1	306.6	310.8	314.4
c_{sat} (J K^{-1}g^{-1})	2.091	2.108	2.129	2.137	2.155	2.184	2.210
C_{sat} (J K^{-1}mol^{-1})	297.5	299.9	303.0	304.1	306.6	310.8	314.4

Temp. (K)	300	310	320	330	340	350	360
c_p (J K^{-1}g^{-1})	2.216	2.250	2.286	2.323	2.361	2.400	2.441
C_p (J K^{-1}mol^{-1})	315.3	320.1	325.2	330.5	335.9	341.5	347.4
c_{sat} (J K^{-1}g^{-1})	2.216	2.250	2.286	2.323	2.361	2.400	2.441
C_{sat} (J K^{-1}mol^{-1})	315.3	320.1	325.2	330.5	335.9	341.6	347.4

TABLE 11.41.4. Recommended values of heat capacities—Continued

Temp. (K)	370	380	390	400	410	420	430
c_p (J K^{-1}g^{-1})	2.483	2.526	2.570	2.616	2.662	2.709	2.757
C_p (J K^{-1}mol^{-1})	353.3	359.4	365.7	372.1	378.7	385.4	392.2
c_{sat} (J K^{-1}g^{-1})	2.483	2.526	2.570	2.615	2.661	2.707	2.755
C_{sat} (J K^{-1}mol^{-1})	353.3	359.4	365.7	372.1	378.6	385.2	391.9
Temp. (K)	440	450	460				
c_p (J K^{-1}g^{-1})	2.805	2.855	2.905				
C_p (J K^{-1}mol^{-1})	399.2	406.2	413.3				
c_{sat} (J K^{-1}g^{-1})	2.803	2.851	2.901				
C_{sat} (J K^{-1}mol^{-1})	398.8	405.7	412.7				

TABLE 11.41.5. Parameters of quasi-polynomial equation

Heat capacity type	No. data points total	used	s_w	s/R	s_r %	s_b/R	+/–
C_p	121	75	1.223	2.15–1	0.51	3.02–3	2

Temp. range K	T_c K	A_1	A_2	A_3	A_4	Level of uncertainty
247.0–462.4	617.65	–8.49622	1.93863	2.41305+1	9.30887	III

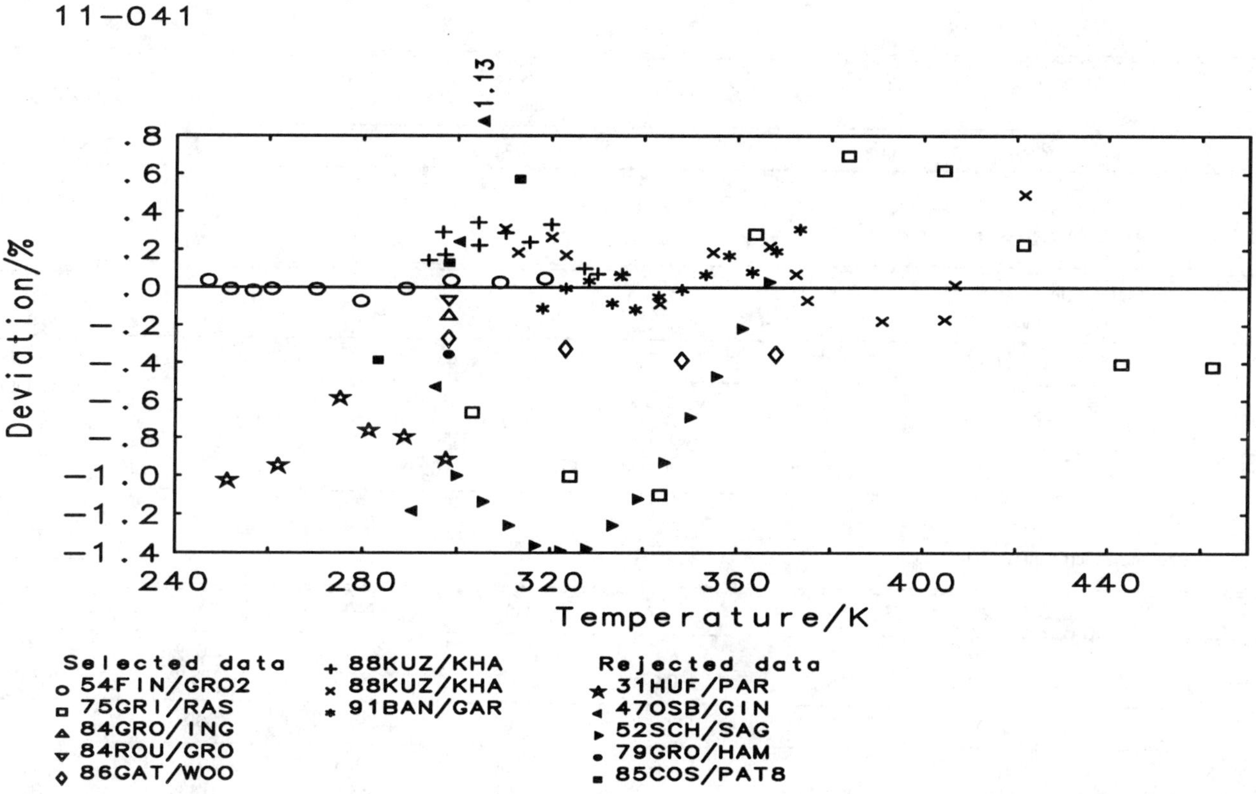

Name: 2,7–Dimethyloctane
Formula: $C_{10}H_{22}$

CAS–RN: 1072–16–8
Group No.: 11–042
Molar Mass: 142.28

TABLE 11.42.1. Experimental heat capacities

Reference	Temp. range K	No. pnts	Error %	Purity %	Purity method	Type capacity	Calorimeter Type	Calorimeter Reference
30PAR/HUF2	223.2–295.0	8	1.00	not specified		C_p	BSIO	25PAR

TABLE 11.42.3. Parameters of regression polynomial

Heat capacity type	No. data points total	used	s_w	s/R	s_r %	s_b/R	+/–
C_p	8	8	0.194	6.70–2	0.19	2.19–4	0

Temp. range K	A_1	A_2	Level of uncertainty
223.2–295.0	2.03020+1	5.40560	IV

TABLE 11.42.4. Recommended values of heat capacities

Temp. (K)	230	240	250	260	270	273.15	280
c_p (J K^{-1}g^{-1})	1.913	1.944	1.976	2.008	2.039	2.049	2.071
C_p (J K^{-1}mol^{-1})	272.2	276.7	281.2	285.7	290.2	291.6	294.6
Temp. (K)	290	298.15	300				
c_p (J K^{-1}g^{-1})	2.102	2.128	2.134				
C_p (J K^{-1}mol^{-1})	299.1	302.8	303.6				

Name: 2–Methylnonane
Formula: $C_{10}H_{22}$

CAS–RN: 871–83–0
Group No.: 11–043
Molar Mass: 142.28

TABLE 11.43.1. Experimental heat capacities

Reference	Temp. range K	No. pnts	Error %	Purity %	Purity method	Type capacity	Calorimeter Type	Calorimeter Reference
41PAR/WES	200.0–298.0	11S	0.25	99.1	melpt	C_p	BSIO	25PAR

TABLE 11.43.3. Parameters of regression polynomial

Heat capacity type	No. data points total	used	s_w	s/R	s_r %	s_b/R	+/–
C_p	11	11	0.258	2.22–2	0.06	2.32–5	1

Temp. range K	A_1	A_2	A_3	A_4	Level of uncertainty
200.0–298.0	7.94595+1	–5.37085+1	1.95595+1	–2.09386	III

TABLE 11.43.4. Recommended values of heat capacities

Temp. (K)	200	210	220	230	240	250	260
c_p (J K^{-1}g^{-1})	1.959	1.960	1.968	1.982	2.003	2.029	2.059
C_p (J K^{-1}mol^{-1})	278.8	278.9	280.0	282.1	285.0	288.7	293.0
Temp. (K)	270	273.15	280	290	298.15	300	
c_p (J K^{-1}g^{-1})	2.093	2.105	2.130	2.170	2.203	2.211	
C_p (J K^{-1}mol^{-1})	297.8	299.5	303.1	308.7	313.5	314.6	

TABLE 11.43.5. Parameters of quasi-polynomial equation

Heat capacity type	No. data points total	used	s_w	s/R	s_r %	s_b/R	+/–
C_p	11	11	1.205	1.06–1	0.30	4.30–4	1

Temp. range K	T_c K	A_1	A_2	A_3	A_4	Level of uncertainty
200.0–298.0	609.30	2.05696+2	8.86112+1	–5.58771+1	1.19373+2	III

Name: 3–Methylnonane
Formula: $C_{10}H_{22}$

CAS–RN: 5911–04–6
Group No.: 11–044
Molar Mass: 142.28

TABLE 11.44.1. Experimental heat capacities

Reference		Temp. range K	No. pnts	Error %	Purity %	Purity method	Type capacity	Calorimeter Type	Calorimeter Reference
41PAR/WES	N	190.0–298.0	12S	0.25	99.7	melpt	C_p	BSIO	25PAR

41PAR/WES racemic mixture

TABLE 11.44.3. Parameters of regression polynomial

Heat capacity type	No. data points total	No. data points used	s_w	s/R	s_r %	s_b/R	+/–
C_p	12	12	0.548	4.67–2	0.14	9.46–5	1

Temp. range K	A_1	A_2	A_3	Level of uncertainty
190.0–298.0	3.38710+1	–5.08070	2.08079	III

TABLE 11.44.4. Recommended values of heat capacities

Temp. (K)	190	200	210	220	230	240	250
c_p ($J\,K^{-1}g^{-1}$)	1.854	1.872	1.892	1.915	1.940	1.967	1.997
C_p ($J\,K^{-1}mol^{-1}$)	263.8	266.3	269.2	272.4	276.0	279.9	284.1

Temp. (K)	260	270	273.15	280	290	298.15	300
c_p ($J\,K^{-1}g^{-1}$)	2.029	2.064	2.076	2.101	2.141	2.175	2.183
C_p ($J\,K^{-1}mol^{-1}$)	288.7	293.7	295.3	299.0	304.6	309.5	310.6

TABLE 11.44.5. Parameters of quasi-polynomial equation

Heat capacity type	No. data points total	No. data points used	s_w	s/R	s_r %	s_b/R	+/–
C_p	12	12	0.624	5.47–2	0.16	1.19–4	2

Temp. range K	T_c K	A_1	A_2	A_3	A_4	Level of uncertainty
190.0–298.0	613.70	2.77878+1	2.49312+1	3.48471	7.74292	III

Name: 4–Methylnonane
Formula: $C_{10}H_{22}$

CAS–RN: 17301–94–9
Group No.: 11–045
Molar Mass: 142.28

TABLE 11.45.1. Experimental heat capacities

Reference		Temp. range K	No. pnts	Error %	Purity %	Purity method	Type capacity	Calorimeter Type	Calorimeter Reference
41PAR/WES	N	180.0–298.0	13S	0.25	99.3	melpt	C_p	BSIO	25PAR

41PAR/WES racemic mixture

TABLE 11.45.3. Parameters of regression polynomial

Heat capacity type	No. data points total	used	s_w	s/R	s_r %	s_b/R	+/–
C_p	13	13	0.666	5.39–2	0.17	1.31–4	0

Temp. range K	A_1	A_2	A_3	Level of uncertainty
180.0–298.0	3.55303+1	–6.71178	2.54999	II

TABLE 11.45.4. Recommended values of heat capacities

Temp. (K)	180	190	200	210	220	230	240
c_p ($J K^{-1} g^{-1}$)	1.853	1.869	1.888	1.910	1.935	1.962	1.993
C_p ($J K^{-1} mol^{-1}$)	263.7	265.9	268.6	271.7	275.3	279.2	283.6
Temp. (K)	250	260	270	273.15	280	290	298.15
c_p ($J K^{-1} g^{-1}$)	2.027	2.064	2.104	2.117	2.146	2.192	2.231
C_p ($J K^{-1} mol^{-1}$)	288.4	293.6	299.3	301.2	305.4	311.9	317.5
Temp. (K)	300						
c_p ($J K^{-1} g^{-1}$)	2.241						
C_p ($J K^{-1} mol^{-1}$)	318.8						

TABLE 11.45.5. Parameters of quasi-polynomial equation

Heat capacity type	No. data points total	used	s_w	s/R	s_r %	s_b/R	+/–
C_p	13	13	0.834	6.84–2	0.21	2.07–4	2

Temp. range K	T_c K	A_1	A_2	A_3	A_4	Level of uncertainty
180.0–298.0	613.70	5.02490+1	3.62593+1	–7.31356	1.74091+1	III

Name: 5–Methylnonane
Formula: $C_{10}H_{22}$

CAS–RN: 15869–85–9
Group No.: 11–046
Molar Mass: 142.28

TABLE 11.46.1. Experimental heat capacities

Reference	Temp. range K	No. pnts	Error %	Purity %	Purity method	Type capacity	Calorimeter Type	Calorimeter Reference
41PAR/WES	190.0–298.0	12S	0.25	99.7	melpt	C_p	BSIO	25PAR

TABLE 11.46.3. Parameters of regression polynomial

Heat capacity type	No. data points total	used	s_w	s/R	s_r %	s_b/R	+/–
C_p	12	12	0.221	1.97–2	0.06	1.19–5	1

Temp. range K	A_1	A_2	A_3	Level of uncertainty
190.0–298.0	3.47624+1	–6.04279	2.37084	III

TABLE 11.46.4. Recommended values of heat capacities

Temp. (K)	190	200	210	220	230	240	250
c_p (J K^{-1}g^{-1})	1.861	1.879	1.901	1.925	1.952	1.982	2.014
C_p (J K^{-1}mol^{-1})	264.7	267.4	270.5	273.9	277.8	282.0	286.6
Temp. (K)	260	270	273.15	280	290	298.15	300
c_p (J K^{-1}g^{-1})	2.050	2.088	2.101	2.129	2.172	2.210	2.219
C_p (J K^{-1}mol^{-1})	291.7	297.1	298.9	302.9	309.1	314.5	315.7

TABLE 11.46.5. Parameters of quasi-polynomial equation

Heat capacity type	No. data points total	used	s_w	s/R	s_r %	s_b/R	+/–
C_p	12	12	0.426	3.67–2	0.11	5.26–5	0
Temp. range K	T_c K	A_1	A_2	A_3	A_4		Level of uncertainty
190.0–298.0	613.70	3.59812+1	2.96818+1	−1.23427	1.09044+1		III

Name: 2–Methyldecane
Formula: $C_{11}H_{24}$

CAS–RN: 6975–98–0
Group No.: 11–047
Molar Mass: 156.31

TABLE 11.47.1. Experimental heat capacities

Reference	Temp. range K	No. pnts	Error %	Purity %	Purity method	Type capacity	Calorimeter Type	Calorimeter Reference
71MES/FIN	235.4–384.8	22	0.20	99.979	melpt	C_{sat}	BSAO	43RUE/HUF

TABLE 11.47.3. Parameters of regression polynomial

Heat capacity type	No. data points total	used	s_w	s/R	s_r %	s_b/R	+/–
C_{sat}	22	22	0.342	2.91–2	0.07	2.51–5	3
Temp. range K		A_1	A_2	A_3	A_4		Level of uncertainty
235.4–384.8		5.68879+1	−2.46238+1	8.88342	−8.08008–1		II

TABLE 11.47.4. Recommended values of heat capacities

Temp. (K)	230	240	250	260	270	273.15	280
c_{sat} (J K^{-1}g^{-1})	1.990	2.010	2.033	2.059	2.088	2.098	2.120
C_{sat} (J K^{-1}mol^{-1})	311.1	314.2	317.8	321.9	326.4	327.9	331.3
Temp. (K)	290	298.15	300	310	320	330	340
c_{sat} (J K^{-1}g^{-1})	2.153	2.182	2.189	2.226	2.265	2.305	2.346
C_{sat} (J K^{-1}mol^{-1})	336.6	341.1	342.1	348.0	354.0	360.3	366.7
Temp. (K)	350	360	370	380			
c_{sat} (J K^{-1}g^{-1})	2.387	2.429	2.472	2.514			
C_{sat} (J K^{-1}mol^{-1})	373.2	379.7	386.3	392.9			

TABLE 11.47.5. Parameters of quasi-polynomial equation

Heat capacity type	No. data points total	used	s_w	s/R	s_r %	s_b/R	+/–
C_{sat}	22	22	1.158	9.43–2	0.23	3.66–4	1

Temp. range K	T_c K	A_1	A_2	A_3	A_4	Level of uncertainty
235.4–384.8	623.00	–9.83062	4.11511	2.39812+1	5.87111	II

Name: Undecane
Formula: $C_{11}H_{24}$

CAS–RN: 1120–21–4
Group No.: 11–048
Molar Mass: 156.31

TABLE 11.48.1. Experimental heat capacities

Reference		Temp. range K	No. pnts	Error %	Purity %	Purity method	Type capacity	Calorimeter Type	Calorimeter Reference
31HUF/PAR		258.5–298.0	4	1.00	not specified		C_p	BSIO	25PAR
54FIN/GRO2	N	251.7–298.9	12	0.20	99.98	melpt	C_{sat}	BSAO	43RUE/HUF
70AKH		293.1	1	nosp	not specified		C_p	BDHO	59ABA/MUS
76MUS		307.8–458.6	14	2.50	not specified		C_p	BDAO	71MUS
84GRI/AND		292.3–433.4	8	0.80	not specified		C_p	BSAO	67RAS/GAN

54FIN/GRO2 smoothed data in 67MES/GUT

TABLE 11.48.2. Correlated heat capacities

Reference	Temp. range K	No. pnts used	$\sigma_r C$ %	d_w	d/R	d_r %	d_b/R	+/–
				Selected data				
54FIN/GRO2	251.7–298.9	12	0.20	0.127	1.04–2	0.03	1.24–3	1
84GRI/AND	292.3–433.4	8	0.80	0.257	8.81–2	0.21	–2.92–2	0
				Rejected data				
31HUF/PAR	(1.96–1, 0.48, –1.87–1, –4)			70AKH	(4.84–1, 1.16, 4.84–1, 1)			
76MUS	(3.05, 6.87, –2.94, –11)							

TABLE 11.48.3. Parameters of cubic spline polynomials

Heat capacity type	No. data points total	used	s_w	s/R	s_r %	s_b/R	+/–
C_p	39	20	0.220	6.50–2	0.15	–1.09–2	1

Temp. range K	A_1	A_2	A_3	A_4	Level of uncertainty
251.7–300.0	1.64553+2	–1.29722+2	4.35318+1	–4.65086	II
300.0–433.4	5.51273+1	–2.02967+1	7.05661	–5.98064–1	IV

TABLE 11.48.4. Recommended values of heat capacities

Temp. (K)	250	260	270	273.15	280	290	298.15
c_p (J K^{-1}g^{-1})	2.109	2.117	2.133	2.140	2.156	2.183	2.207
C_p (J K^{-1}mol^{-1})	329.7	331.0	333.5	334.5	336.9	341.2	345.0
Temp. (K)	300	310	320	330	340	350	360
c_p (J K^{-1}g^{-1})	2.213	2.245	2.279	2.314	2.350	2.388	2.426
C_p (J K^{-1}mol^{-1})	345.9	350.9	356.2	361.7	367.4	373.2	379.2
Temp. (K)	370	380	390	400	410	420	430
c_p (J K^{-1}g^{-1})	2.465	2.504	2.544	2.584	2.623	2.662	2.701
C_p (J K^{-1}mol^{-1})	385.3	391.5	397.6	403.8	410.0	416.1	422.2

TABLE 11.48.5. Parameters of quasi-polynomial equation

Heat capacity type	No. data points total	used	s_w	s/R	s_r %	s_b/R	+/–
C_p	39	20	1.284	3.59–1	0.77	8.58–2	2

Temp. range K	T_c K	A_1	A_2	A_3	A_4	Level of uncertainty
251.7–433.4	638.85	–7.00457–1	8.14112	2.56974+1	1.50667–2	V

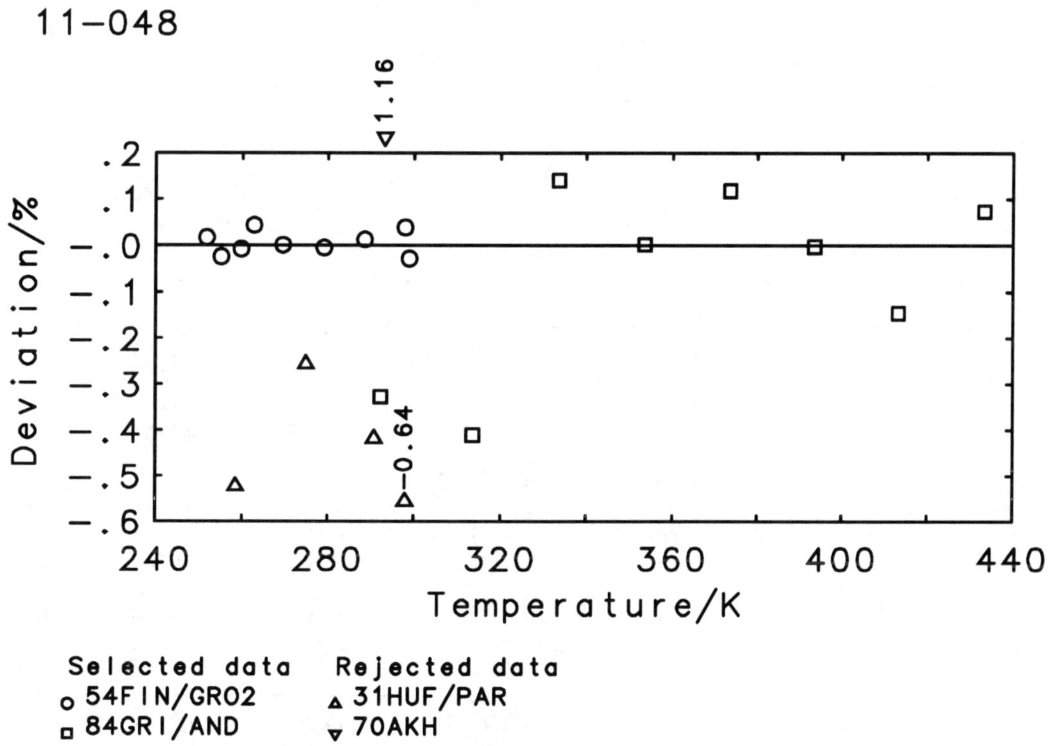

Name: Dodecane
Formula: $C_{12}H_{26}$

CAS–RN: 112–40–3
Group No.: 11–049
Molar Mass: 170.34

TABLE 11.49.1. Experimental heat capacities

Reference		Temp. range K	No. pnts	Error %	Purity %	Purity method	Type capacity	Calorimeter Type	Calorimeter Reference
31HUF/PAR		275.1–297.7	4	1.00	not specified		C_p	BSIO	25PAR
54FIN/GRO2	N	266.7–317.4	11	0.20	99.93	melpt	C_{sat}	BSAO	43RUE/HUF
71REC/SAD		303.1	1	0.30	not specified		C_p	BSIO	70REC
73KAL/WOY		303.1	1	0.20	99.5	chrom	C_p	BSIO	70REC
77NAZ/MUS		303.2–483.2	10	1.50	not specified		C_p	BSAO	77NAZ/MUS
81GRO/ING		298.1	1	nosp	99.0	chrom	C_p	FSIT	71PIC/LED
84GRO/BEN		298.1	1	nosp	not specified		C_p	FSIT	71PIC/LED
84KUM/BEN		298.1	1	nosp	not specified		C_p	FSIT	71PIC/LED
84ROU/GRO		298.1	1	nosp	98.	estim	C_p	FSIT	71PIC/LED
85COS/PAT4		298.2	1	nosp	99.	anal	C_p	FSIT	71PIC/LED
85COS/PAT8		283.1–313.1	3	nosp	99.0	estim	C_p	FSIT	71PIC/LED
85LAI/ROD		298.1	1	nosp	not specified		C_p	FSIT	71PIC/LED
86BEN/DAR3		298.1	1	nosp	99.	anal	C_p	FSIT	71PIC/LED
86WIL/LAI		298.1	1	nosp	99.	anal	C_p	FSIT	71PIC/LED
88AND/PAT		298.1	1	nosp	98.	anal	C_p	FSIT	71PIC/LED
88COS/VAN		298.1	1	nosp	99.	anal	C_p	FSIT	71PIC/LED
89LAI/ROD		298.1	1	nosp	99.	anal	C_p	FSIT	71PIC/LED
89VOG/SCH		333.1	1	nosp	not specified		C_p	BDHT	69PER/COM
91TRE/COS		298.1	1	nosp	99.	anal	C_p	FSIT	71PIC/LED
92LAI/ROD		298.1	1	nosp	99.	anal	C_p	FSIT	71PIC/LED

54FIN/GRO2 smoothed data in 67MES/GUT

TABLE 11.49.2. Correlated heat capacities

Reference	Temp. range K	No. pnts used	$\sigma_r C$ %	d_w	d/R	d_r %	d_b/R	+/–
				Selected data				
54FIN/GRO2	266.7–317.4	11	0.20	0.169	1.49–2	0.03	1.40–3	3
77NAZ/MUS	303.2–483.2	10	1.50	0.108	8.03–2	0.16	9.02–3	2
84GRO/BEN	298.1	1	0.50#	0.343	7.74–2	0.17	–7.74–2	–1
84KUM/BEN	298.1	1	0.50#	0.322	7.26–2	0.16	–7.26–2	–1
85COS/PAT4	298.2	1	0.50#	0.038	8.65–3	0.02	8.65–3	1
86BEN/DAR3	298.1	1	0.50#	0.284	6.41–2	0.14	–6.41–2	–1
88AND/PAT	298.1	1	0.50#	0.296	6.70–2	0.15	6.70–2	1
88COS/VAN	298.1	1	0.50#	0.099	2.25–2	0.05	2.25–2	1
91TRE/COS	298.1	1	0.50#	0.051	1.16–2	0.03	1.16–2	1
				Rejected data				
31HUF/PAR	(3.87–1, 0.87, –3.77–1, –4)			71REC/SAD	(4.39–1, 0.97, –4.39–1, –1)			
73KAL/WOY	(9.29–1, 2.08, –9.29–1, –1)			81GRO/ING	(1.30–1, 0.29, –1.30–1, –1)			
84ROU/GRO	(1.17–1, 0.26, –1.17–1, –1)			85COS/PAT8	(2.21–1, 0.48, 7.63–2, 1)			
85LAI/ROD	(3.13–1, 0.70, –3.13–1, –1)			86WIL/LAI	(3.13–1, 0.70, –3.13–1, –1)			
89LAI/ROD	(4.55–1, 1.02, –4.55–1, –1)			89VOG/SCH	(1.27, 2.60, 1.27, 1)			
92LAI/ROD	(3.55–1, 0.79, –3.55–1, –1)							

TABLE 11.49.3. Parameters of cubic spline polynomials

Heat capacity type	No. data points total	used	s_w	s/R	s_r %	s_b/R	+/–
C_p	44	28	0.179	5.46–2	0.12	8.99–4	6

Temp. range K	A_1	A_2	A_3	A_4	Level of uncertainty
266.7–310.0	1.93881+2	−1.53376+2	5.06932+1	−5.35794	II
310.0–483.2	2.35104+1	1.14984+1	−2.49216	3.60918–1	IV

TABLE 11.49.4. Recommended values of heat capacities

Temp. (K)	270	273.15	280	290	298.15	300	310
c_p (J K^{-1}g^{-1})	2.141	2.146	2.160	2.184	2.207	2.212	2.243
C_p (J K^{-1}mol^{-1})	364.7	365.6	367.9	372.0	375.9	376.9	382.1
Temp. (K)	320	330	340	350	360	370	380
c_p (J K^{-1}g^{-1})	2.275	2.308	2.342	2.377	2.414	2.451	2.490
C_p (J K^{-1}mol^{-1})	387.6	393.2	398.9	404.9	411.1	417.5	424.2
Temp. (K)	390	400	410	420	430	440	450
c_p (J K^{-1}g^{-1})	2.531	2.574	2.618	2.664	2.712	2.763	2.815
C_p (J K^{-1}mol^{-1})	431.2	438.4	446.0	453.8	462.0	470.6	479.5
Temp. (K)	460	470	480				
c_p (J K^{-1}g^{-1})	2.870	2.927	2.987				
C_p (J K^{-1}mol^{-1})	488.9	498.6	508.8				

TABLE 11.49.5. Parameters of quasi-polynomial equation

Heat capacity type	No. data points total	used	s_w	s/R	s_r %	s_b/R	+/–
C_p	44	28	0.708	1.56–1	0.30	2.46–2	–1

Temp. range K	T_c K	A_1	A_2	A_3	A_4	Level of uncertainty
266.7–483.2	658.65	−1.01500+1	3.88605	2.90559+1	6.62768	IV

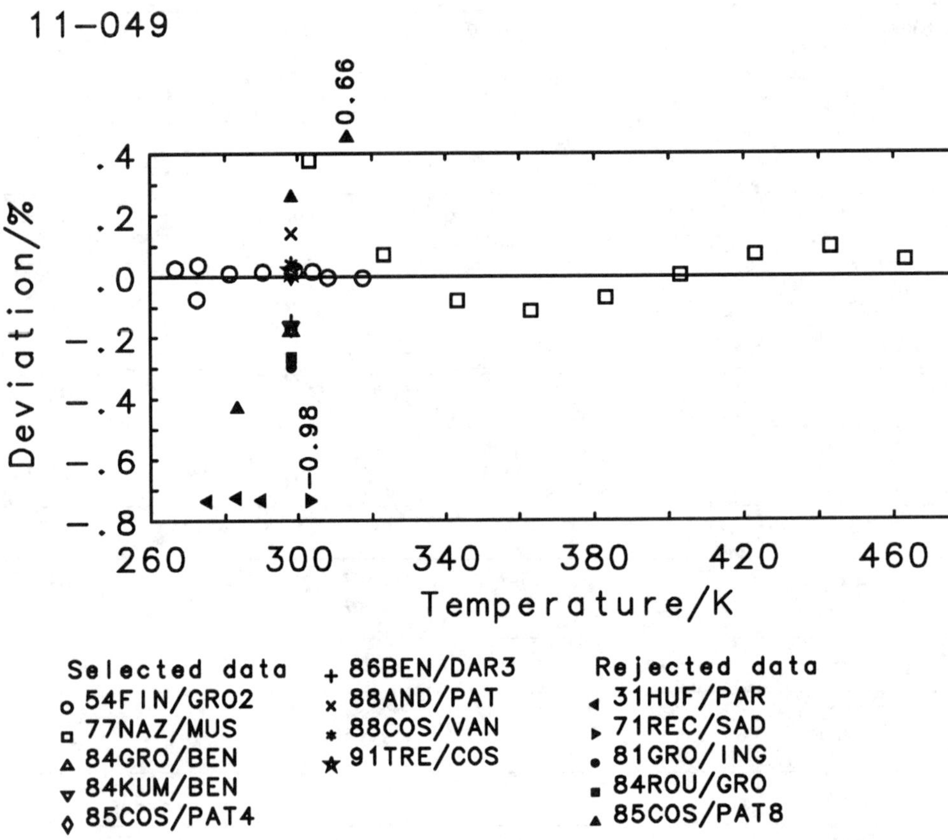

Name: 2,2,4,6,6–Pentamethylheptane
Formula: $C_{12}H_{26}$

CAS–RN: 13475–82–6
Group No.: 11–050
Molar Mass: 170.34

TABLE 11.50.1. Experimental heat capacities

Reference	Temp. K	Capac. J/(K.g)	Error %	Purity %	Purity method	Type capacity	Calorimeter Type	Calorimeter Reference
88COS/VAN	298.15	2.060	nosp	99.	anal	C_p	FSIT	71PIC/LED

Name: Tridecane
Formula: $C_{13}H_{28}$

CAS–RN: 629–50–5
Group No.: 11–051
Molar Mass: 184.37

TABLE 11.51.1. Experimental heat capacities

Reference		Temp. range K	No. pnts	Error %	Purity %	Purity method	Type capacity	Calorimeter Type	Calorimeter Reference
54FIN/GRO2	N	271.7–306.4	8	0.20	99.95	melpt	C_{sat}	BSAO	43RUE/HUF
75WOY/KAL		303.1	1	nosp	99.97	chrom	C_p	BSIO	70REC
76MUS		307.8–478.6	16	2.50	not specified		C_p	BDAO	71MUS

54FIN/GRO2 smoothed data in 67MES/GUT

TABLE 11.51.2. Correlated heat capacities

Reference	Temp. range K	No. pnts used	$\sigma_r C$ %	d_w	d/R	d_r %	d_b/R	+/–
				Selected data				
54FIN/GRO2	271.7–306.4	8	0.20	0.200	1.92–2	0.04	6.63–5	–2
75WOY/KAL	303.1	1	0.50#	0.011	2.59–3	0.01	–2.59–3	0

TABLE 11.51.3. Parameters of regression polynomial

Heat capacity type	No. data points total	used	s_w	s/R	s_r %	s_b/R	+/–
C	25	9	0.231	2.23–2	0.05	–2.29–4	–2

Temp. range K	A_1	A_2	A_3	Level of uncertainty
271.7–306.4	9.31514+1	–3.59261+1	7.07296	II

TABLE 11.51.4. Recommended values of heat capacities

Temp. (K)	270	273.15	280	290	298.15	300	310
c ($J K^{-1} g^{-1}$)	2.152	2.155	2.165	2.185	2.206	2.211	2.244
C ($J K^{-1} mol^{-1}$)	396.7	397.4	399.2	402.8	406.7	407.7	413.7

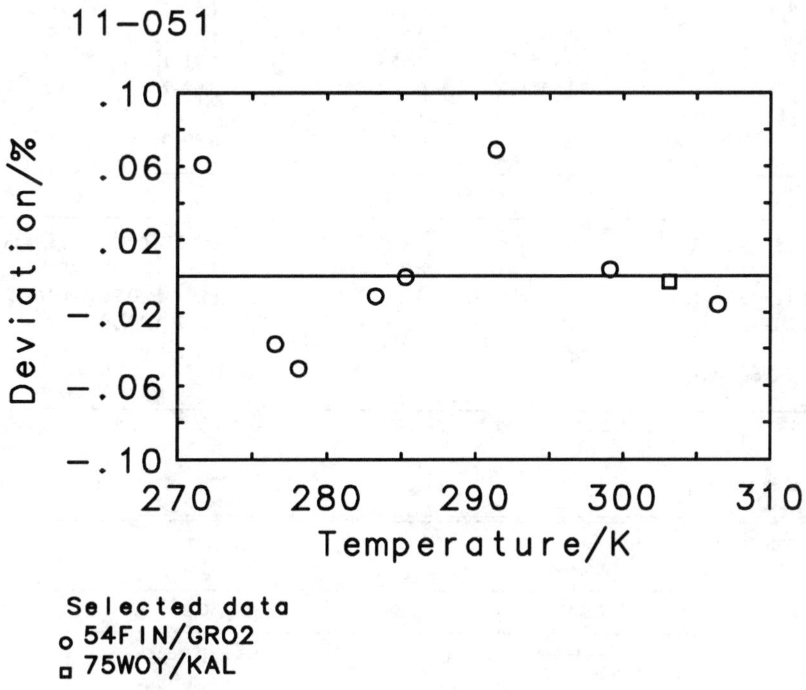

Name: Tetradecane
Formula: $C_{14}H_{30}$

CAS–RN: 629–59–4
Group No.: 11–052
Molar Mass: 198.39

TABLE 11.52.1. Experimental heat capacities

Reference		Temp. range K	No. pnts	Error %	Purity %	Purity method	Type capacity	Calorimeter Type	Calorimeter Reference
34PAR/LIG		280.6–290.6	4	1.00	not specified		C_p	BSIO	25PAR
54FIN/GRO2	N	282.7–302.8	7	0.20	99.93	melpt	C_{sat}	BSAO	43RUE/HUF
76MUS		307.8–501.5	18	2.50	not specified		C_p	BDAO	71MUS
84GRI/AND		296.2–433.3	8	0.80	not specified		C_p	BSAO	67RAS/GAN
84GRO/BEN		298.1	1	nosp	99.5	estim	C_p	FSIT	71PIC/LED
84GRO/ING		298.1	1	nosp	99.	estim	C_p	FSIT	71PIC/LED
84ROU/GRO		298.1	1	nosp	99.	estim	C_p	FSIT	71PIC/LED
85BAL/BRA		298.1	1	nosp	99.	anal	C_p	FSIT	71PIC/LED
85LAI/WIL		298.1	1	0.30	99.5	estim	C_p	FSIT	71PIC/LED
86WIL/LAI		298.1	1	nosp	99.	anal	C_p	FSIT	71PIC/LED
87WIL/ING		298.1	1	nosp	99.	anal	C_p	FSIT	71PIC/LED
88COS/VAN		298.1	1	nosp	99.	anal	C_p	FSIT	71PIC/LED
91TRE/COS		298.1	1	nosp	99.	anal	C_p	FSIT	71PIC/LED
92LAI/ROD		298.1	1	nosp	99.	anal	C_p	FSIT	71PIC/LED

54FIN/GRO2 smoothed data in 67MES/GUT

TABLE 11.52.2. Correlated heat capacities

Reference	Temp. range K	No. pnts used	$\sigma_r C$ %	d_w	d/R	d_r %	d_b/R	+/–
				Selected data				
54FIN/GRO2	282.7–302.8	7	0.20	0.158	1.67–2	0.03	8.95–3	2
84GRI/AND	296.2–433.3	8	0.80	0.344	1.53–1	0.28	–5.44–2	0
84ROU/GRO	298.1	1	0.50#	0.653	1.71–1	0.33	–1.71–1	–1
88COS/VAN	298.1	1	0.50#	0.149	3.92–2	0.07	–3.92–2	–1
91TRE/COS	298.1	1	0.50#	0.025	6.69–3	0.01	–6.69–3	–1
				Rejected data				
34PAR/LIG	(9.61–2, 0.18, –9.59–2, –3)			76MUS	(5.82, 10.99, –5.74, –11)			
84GRO/BEN	(2.33–1, 0.44, –2.33–1, –1)			84GRO/ING	(2.18–1, 0.42, –2.18–1, –1)			
85BAL/BRA	(5.78–1, 1.11, –5.78–1, –1)			85LAI/WIL	(4.97–1, 0.95, –4.97–1, –1)			
86WIL/LAI	(5.24–1, 1.00, –5.24–1, –1)			87WIL/ING	(5.02–1, 0.96, –5.02–1, –1)			
92LAI/ROD	(6.04–1, 1.16, –6.04–1, –1)							

TABLE 11.52.3. Parameters of cubic spline polynomials

Heat capacity type	No. data points total	No. data points used	s_w	s/R	s_r %	s_b/R	+/–
C_p	47	18	0.347	1.30–1	0.24	–3.28–2	–1

Temp. range K	A_1	A_2	A_3	A_4	Level of uncertainty
282.7–310.0	1.43242+1	3.48980+1	–1.22889+1	1.64457	II
310.0–433.3	9.76193+1	–4.57102+1	1.37138+1	–1.15142	IV

TABLE 11.52.4. Recommended values of heat capacities

Temp. (K)	280	290	298.15	300	310	320	330
c_p (J K^{-1}g^{-1})	2.171	2.191	2.209	2.214	2.238	2.265	2.294
C_p (J K^{-1}mol^{-1})	430.7	434.8	438.3	439.2	444.0	449.4	455.1
Temp. (K)	340	350	360	370	380	390	400
c_p (J K^{-1}g^{-1})	2.325	2.358	2.392	2.427	2.463	2.499	2.536
C_p (J K^{-1}mol^{-1})	461.3	467.8	474.5	481.5	488.6	495.8	503.1
Temp. (K)	410	420	430				
c_p (J K^{-1}g^{-1})	2.572	2.608	2.644				
C_p (J K^{-1}mol^{-1})	510.3	517.5	524.5				

TABLE 11.52.5. Parameters of quasi-polynomial equation

Heat capacity type	No. data points total	used	s_w	s/R	s_r %	s_b/R	+/-
C_p	47	18	0.605	1.92–1	0.33	3.04–2	6

Temp. range K	T_c K	A_1	A_2	A_3	A_4	Level of uncertainty
282.7–433.3	693.00	-8.99323	7.25409	3.37132+1	2.78733	IV

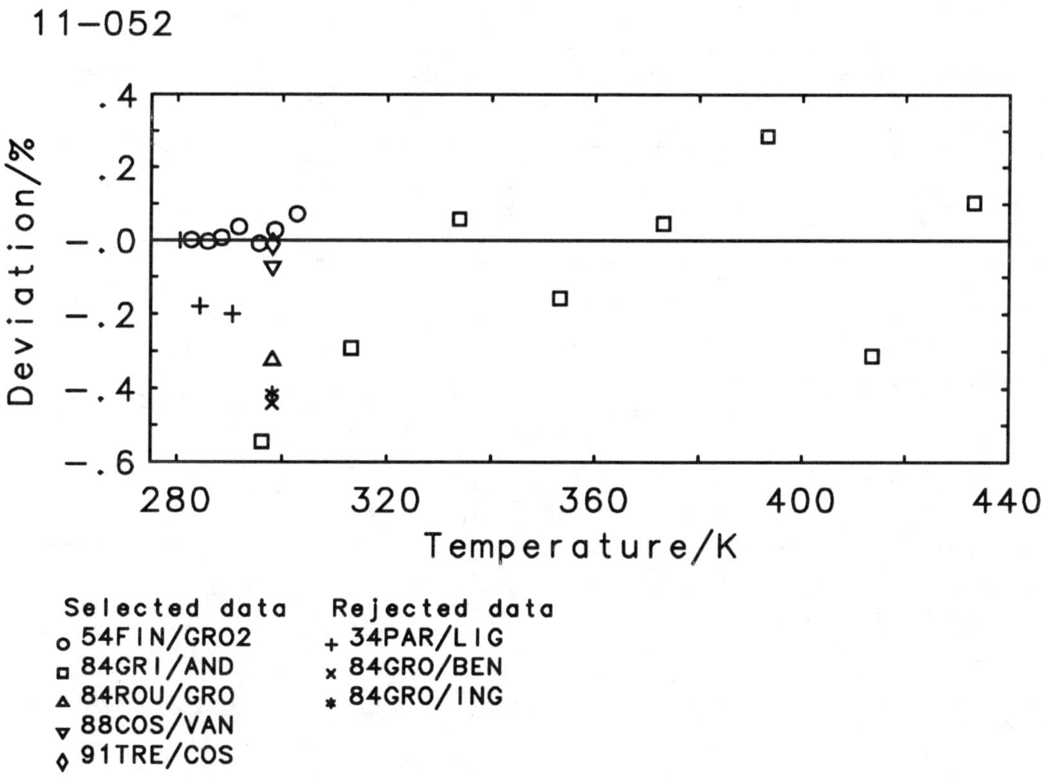

Name: Pentadecane
Formula: $C_{15}H_{32}$

CAS–RN: 629–62–9
Group No.: 11–053
Molar Mass: 212.42

TABLE 11.53.1. Experimental heat capacities

Reference		Temp. range K	No. pnts	Error %	Purity %	Purity method	Type capacity	Calorimeter Type	Calorimeter Reference
54FIN/GRO2	N	285.5–312.8	7	0.20	99.95	melpt	C_{sat}	BSAO	43RUE/HUF
70AKH		293.1	1	nosp	not specified		C_p	BDHO	59ABA/MUS
81GRO/ING		298.1	1	nosp	not specified		C_p	FSIT	71PIC/LED
88COS/VAN		298.1	1	nosp	99.	anal	C_p	FSIT	71PIC/LED
91TRE/COS		298.1	1	nosp	99.	anal	C_p	FSIT	71PIC/LED

54FIN/GRO2 smoothed data in 67MES/GUT

TABLE 11.53.2. Correlated heat capacities

Reference	Temp. range K	No. pnts used	$\sigma_r C$ %	d_w	d/R	d_r %	d_b/R	+/–
				Selected data				
54FIN/GRO2	285.5–312.8	7	0.20	0.078	8.81–3	0.02	–1.37–3	–1
91TRE/COS	298.1	1	0.50#	0.214	6.05–2	0.11	6.05–2	1
				Rejected data				
70AKH	(7.09–1, 1.25, 7.09–1, 1)			81GRO/ING	(2.61–1, 0.46, –2.61–1, –1)			
88COS/VAN	(1.40–1, 0.25, –1.40–1, –1)							

TABLE 11.53.3. Parameters of regression polynomial

Heat capacity type	No. data points total	No. data points used	s_w	s/R	s_r %	s_b/R	+/–
C	11	8	0.133	2.90–2	0.05	6.36–3	0

Temp. range K	A_1	A_2	A_3	Level of uncertainty
285.5–312.8	1.00964+2	–3.53605+1	6.86082	II

TABLE 11.53.4. Recommended values of heat capacities

Temp. (K)	280	290	298.15	300	310
c (J $K^{-1}g^{-1}$)	2.182	2.197	2.213	2.217	2.242
C (J $K^{-1}mol^{-1}$)	463.5	466.6	470.0	470.9	476.2

Name: 2,2,4,4,6,8,8–Heptamethylnonane
Formula: $C_{16}H_{34}$

CAS–RN: 4390–04–9
Group No.: 11–054
Molar Mass: 226.45

TABLE 11.54.1. Experimental heat capacities

Reference	Temp. K	Capac. J/(K.g)	Error %	Purity %	Purity method	Type capacity	Calorimeter Type	Calorimeter Reference
88COS/VAN	298.15	2.026	nosp	99.	anal	C_p	FSIT	71PIC/LED

Name: Hexadecane
Formula: $C_{16}H_{34}$

CAS–RN: 544–76–3
Group No.: 11–055
Molar Mass: 226.45

TABLE 11.55.1. Experimental heat capacities

Reference		Temp. range K	No. pnts	Error %	Purity %	Purity method	Type capacity	Calorimeter Type	Calorimeter Reference
49PAR/MOO		290.0–300.0	2S	1.00	95.	estim	C_p	BSIO	25PAR
54FIN/GRO2	N	295.4–320.3	9	0.20	99.88	melpt	C_{sat}	BSAO	43RUE/HUF
56SCH/GOT		292.1–294.6	4	2.00	not specified		C_p	BSAO	33SOU/BRI
62GOL/BEL		310.9–422.0	3	nosp	96.9	chrom	C_p	BDHT	63GUD/CAM
68REC1	N	298.0–313.0	eqn	nosp	not specified		C_p	BSAO	68REC1
70AKH		293.1	1	nosp	not specified		C_p	BDHO	59ABA/MUS
71REC/SAD		303.1	1	0.30	not specified		C_p	BSIO	70REC
72REC/SAD		298.1	1	0.30	not specified		C_p	BSIO	70REC
73KAL/WOY		303.1	1	0.20	99.5	chrom	C_p	BSIO	70REC
74DIA/REN		300.1–323.5	9	0.30	not specified		C_p	BSAO	74DIA/REN
74PET/TER		297.8–453.5	14	1.00	98.	melpt	C_p	BDCT	74PET/TER
81GRO/ING		298.1	1	nosp	not specified		C_p	FSIT	71PIC/LED
85COS/PAT5		298.2	1	nosp	99.	anal	C_p	FSIT	71PIC/LED
86WIL/LAI		298.1	1	nosp	99.	anal	C_p	FSIT	71PIC/LED
88COS/VAN		298.1	1	nosp	99.	anal	C_p	FSIT	71PIC/LED
89LAI/ROD		298.1	1	nosp	99.	anal	C_p	FSIT	71PIC/LED
91BAN/GAR		318.1–373.1	12	0.40	99.4	chrom	C_p	BDCT	91BAN/GAR
91TRE/COS		298.1	1	nosp	99.	anal	C_p	FSIT	71PIC/LED
92LAI/ROD		298.1	1	nosp	99.	anal	C_p	FSIT	71PIC/LED

54FIN/GRO2 smoothed data in 67MES/GUT
68REC1 same data in 68REC3

TABLE 11.55.2. Correlated heat capacities

Reference	Temp. range K	No. pnts used	$\sigma_r C$ %	d_w	d/R	d_r %	d_b/R	+/–
				Selected data				
54FIN/GRO2	295.4–320.3	9	0.20	0.320	3.93–2	0.06	6.54–3	–1
73KAL/WOY	303.1	1	0.20	0.361	4.38–2	0.07	4.38–2	1
74DIA/REN	300.1–323.5	9	0.30	0.599	1.11–1	0.18	2.45–2	5
74PET/TER	297.8–395.8	11	1.00	0.853	5.35–1	0.85	–1.07–1	–5
88COS/VAN	298.1	1	0.50#	0.555	1.67–1	0.28	–1.67–1	–1
91BAN/GAR	318.1–373.1	12	0.40	0.265	6.81–2	0.11	–4.00–2	–8

Rejected data

49PAR/MOO	(2.94–1, 0.48, 2.94–1, 1)	62GOL/BEL	(2.42, 4.05, –2.38, –2)
68REC1	(2.09–1, 0.34, –1.97–1, –4)	71REC/SAD	(6.87–1, 1.15, –6.87–1, –1)
72REC/SAD	(6.01–1, 1.01, –6.01–1, –1)	81GRO/ING	(1.96–1, 0.33, –1.96–1, –1)
85COS/PAT5	(2.41–1, 0.40, –2.41–1, –1)	86WIL/LAI	(7.06–1, 1.18, –7.06–1, –1)
89LAI/ROD	(5.34–1, 0.89, –5.34–1, –1)	91TRE/COS	(2.26–1, 0.38, –2.26–1, –1)
92LAI/ROD	(6.56–1, 1.10, –6.56–1, –1)		

TABLE 11.55.3. Parameters of regression polynomial

Heat capacity type	No. data points total	used	s_w	s/R	s_r %	s_b/R	+/–
C	68	43	0.587	2.94–1	0.47	–3.49–2	–9

Temp. range K	A_1	A_2	A_3	A_4	Level of uncertainty
295.4–395.8	1.12967+2	–5.42802+1	1.65941+1	–1.44561	IV

TABLE 11.55.4. Recommended values of heat capacities

Temp. (K)	300	310	320	330	340	350	360
c (J K^{-1}g^{-1})	2.219	2.244	2.270	2.299	2.329	2.360	2.393
C (J K^{-1}mol^{-1})	502.5	508.0	514.0	520.5	527.3	534.5	541.9
Temp. (K)	370	380	390	400			
c (J K^{-1}g^{-1})	2.426	2.460	2.494	2.527			
C (J K^{-1}mol^{-1})	549.4	557.1	564.7	572.3			

TABLE 11.55.5. Parameters of quasi-polynomial equation

Heat capacity type	No. data points total	used	s_w	s/R	s_r %	s_b/R	+/–
C	68	43	0.655	2.94–1	0.47	–2.01–2	–1

Temp. range K	T_c K	A_1	A_2	A_3	A_4	Level of uncertainty
295.4–395.8	722.00	–1.32513+1	8.18227	3.70095+1	5.36517	IV

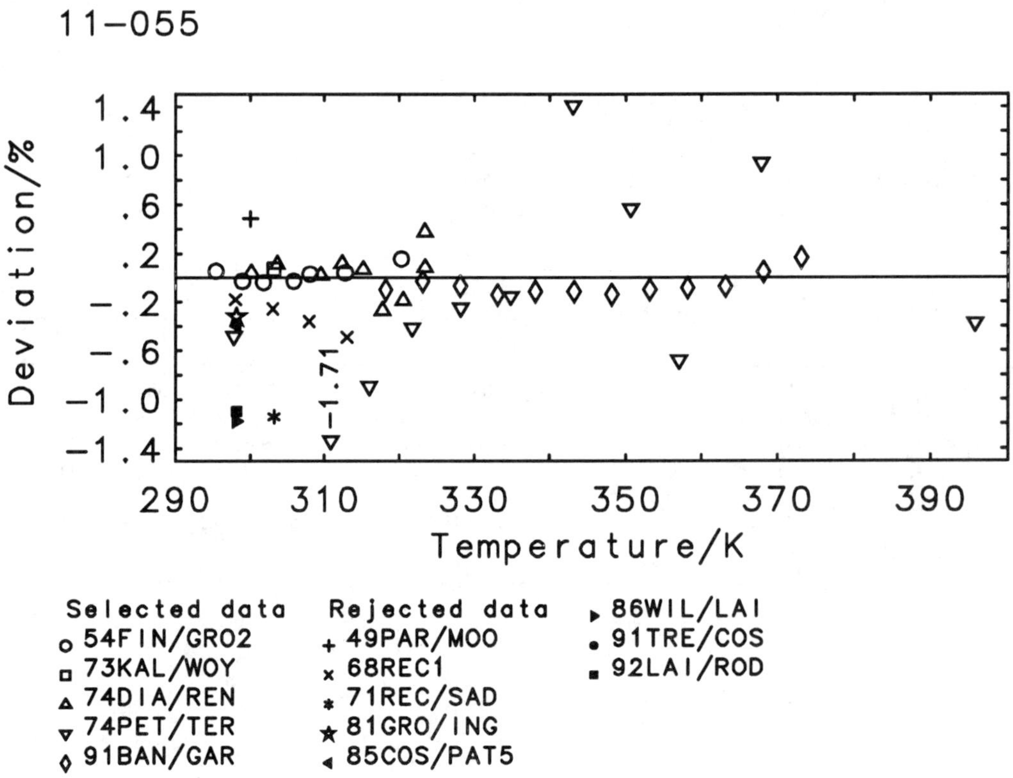

Name: Heptadecane
Formula: $C_{17}H_{36}$

CAS–RN: 629–78–7
Group No.: 11–056
Molar Mass: 240.47

TABLE 11.56.1. Experimental heat capacities

Reference	Temp. range K	No. pnts	Error %	Purity %	Purity method	Type capacity	Calorimeter Type	Calorimeter Reference
67MES/GUT	301.9–384.4	11	0.20	99.89	melpt	C_{sat}	BSAO	47HUF
70AKH	293.1	1	nosp	not specified		C_p	BDHO	59ABA/MUS

TABLE 11.56.2. Correlated heat capacities

Reference	Temp. range K	No. pnts used	$\sigma_r C$ %	d_w	d/R	d_r %	d_b/R	+/–
				Selected data				
67MES/GUT	301.9–384.4	11	0.20#	0.305	4.07–2	0.06	4.51–5	1

TABLE 11.56.3. Parameters of regression polynomial

Heat capacity type	No. data points total	No. data points used	s_w	s/R	s_r %	s_b/R	+/–
C_{sat}	12	11	0.357	4.78–2	0.07	4.51–5	1

Temp. range K	A_1	A_2	A_3	Level of uncertainty
301.9–384.4	5.77157+1	–3.13378	1.76927	II

TABLE 11.56.4. Recommended values of heat capacities

Temp. (K)	300	310	320	330	340	350	360
c_{sat} ($J\ K^{-1}g^{-1}$)	2.221	2.248	2.275	2.304	2.334	2.366	2.398
C_{sat} ($J\ K^{-1}mol^{-1}$)	534.1	540.5	547.1	554.1	561.3	568.9	576.7
Temp. (K)	370	380					
c_{sat} ($J\ K^{-1}g^{-1}$)	2.432	2.467					
C_{sat} ($J\ K^{-1}mol^{-1}$)	584.9	593.3					

Name: Octadecane
Formula: $C_{18}H_{38}$

CAS–RN: 593–45–3
Group No.: 11–057
Molar Mass: 254.50

TABLE 11.57.1. Experimental heat capacities

Reference	Temp. range K	No. pnts	Error %	Purity %	Purity method	Type capacity	Calorimeter Type	Calorimeter Reference
49PAR/MOO	300.0	1	1.00	96.	estim	C_p	BSIO	25PAR
67MES/GUT	304.4–378.7	11	0.20	99.98	melpt	C_{sat}	BSAO	47HUF
81HOE	325.0–375.0	3S	5.00	not specified		C_p	BDHT	69PER/COM
93DUR/AOU	373.0–473.0	51	nosp	not specified		C_p	BDCT	86MER/BEN

TABLE 11.57.2. Correlated heat capacities

Reference	Temp. range K	No. pnts used	$\sigma_r C$ %	d_w	d/R	d_r %	d_b/R	+/−
				Selected data				
49PAR/MOO	300.0	1	1.00	0.140	9.51−2	0.14	−9.51−2	−1
67MES/GUT	304.4–378.7	11	0.20#	0.378	5.29−2	0.08	−1.35−2	−4
93DUR/AOU	373.0–473.0	51	3.00#	0.628	1.50	1.88	7.31−1	19
				Rejected data				
81HOE	(2.38, 3.42, −2.36, −3)							

TABLE 11.57.3. Parameters of regression polynomial

Heat capacity type	No. data points total	used	s_w	s/R	s_r %	s_b/R	+/−
C	66	63	0.601	1.38	1.74	5.88−1	14

Temp. range K	A_1	A_2	A_3	Level of uncertainty
300.0–473.0	5.77808+1	−1.26076	1.55254	V

TABLE 11.57.4. Recommended values of heat capacities

Temp. (K)	300	310	320	330	340	350	360
c ($J\,K^{-1}g^{-1}$)	2.221	2.247	2.275	2.304	2.334	2.365	2.397
C ($J\,K^{-1}mol^{-1}$)	565.1	572.0	579.1	586.4	594.0	601.9	610.0
Temp. (K)	370	380	390	400	410	420	430
c ($J\,K^{-1}g^{-1}$)	2.430	2.464	2.499	2.534	2.571	2.609	2.648
C ($J\,K^{-1}mol^{-1}$)	618.4	627.0	635.9	645.0	654.4	664.1	674.0
Temp. (K)	440	450	460	470			
c ($J\,K^{-1}g^{-1}$)	2.688	2.729	2.771	2.815			
C ($J\,K^{-1}mol^{-1}$)	684.2	694.6	705.3	716.3			

TABLE 11.57.5. Parameters of quasi-polynomial equation

Heat capacity type	No. data points total	used	s_w	s/R	s_r %	s_b/R	+/−
C	66	63	1.375	1.32	1.62	−2.16−1	−7

Temp. range K	T_c K	A_1	A_2	A_3	A_4	Level of uncertainty
300.0–473.0	746.00	−1.66944+1	1.59290	3.87559+1	4.37413+1	V

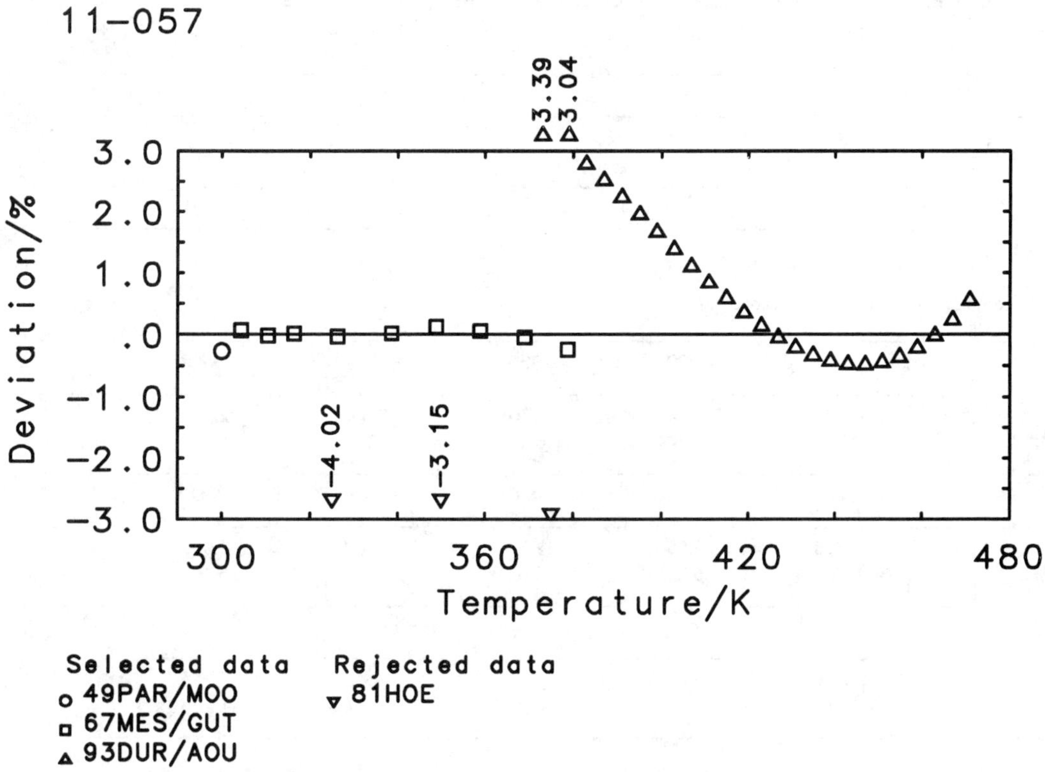

Name: Nonadecane
Formula: $C_{19}H_{40}$

CAS–RN: 629–92–5
Group No.: 11–058
Molar Mass: 268.53

TABLE 11.58.1. Experimental heat capacities

Reference	Temp. range K	No. pnts	Error %	Purity %	Purity method	Type capacity	Calorimeter Type	Calorimeter Reference
69ATK/LAR	305.0–453.0	eqn	nosp	99.9	chrom	C_p	BSAO	58WES/GIN
84GRI/AND	313.1–433.2	10	0.70	not specified		C_p	BSAO	67RAS/GAN
93DUR/AOU	373.0–473.0	51	nosp	not specified		C_p	BDCT	86MER/BEN

TABLE 11.58.2. Correlated heat capacities

Reference	Temp. range K	No. pnts used	$\sigma_r C$ %	d_w	d/R	d_r %	d_b/R	+/–
				Selected data				
69ATK/LAR	305.0–453.0	9	0.50#	0.251	1.01–1	0.13	1.00–1	9
84GRI/AND	313.1–433.2	10	0.70	0.484	2.67–1	0.34	–1.74–1	–4
				Rejected data				
93DUR/AOU	(1.88, 2.25, 6.01–1, 8)							

TABLE 11.58.3. Parameters of regression polynomial

Heat capacity type	No. data points total	used	s_w	s/R	s_r %	s_b/R	+/–
C_p	70	19	0.414	2.17–1	0.28	–4.42–2	5

Temp. range K	A_1	A_2	Level of uncertainty
305.0–453.0	3.83583+1	1.08986+1	IV

TABLE 11.58.4. Recommended values of heat capacities

Temp. (K)	310	320	330	340	350	360	370
c_p ($J\,K^{-1}g^{-1}$)	2.234	2.268	2.301	2.335	2.369	2.403	2.436
C_p ($J\,K^{-1}mol^{-1}$)	599.8	608.9	618.0	627.0	636.1	645.1	654.2
Temp. (K)	380	390	400	410	420	430	440
c_p ($J\,K^{-1}g^{-1}$)	2.470	2.504	2.538	2.571	2.605	2.639	2.673
C_p ($J\,K^{-1}mol^{-1}$)	663.3	672.3	681.4	690.5	699.5	708.6	717.6
Temp. (K)	450						
c_p ($J\,K^{-1}g^{-1}$)	2.706						
C_p ($J\,K^{-1}mol^{-1}$)	726.7						

TABLE 11.58.5. Parameters of quasi-polynomial equation

Heat capacity type	No. data points total	used	s_w	s/R	s_r %	s_b/R	+/–
C_p	70	19	0.435	2.11–1	0.26	4.93–3	3

Temp. range K	T_c K	A_1	A_2	A_3	A_4	Level of uncertainty
305.0–453.0	756.20	–1.01421+1	4.26039–1	4.13677+1	6.03592+1	IV

11–058

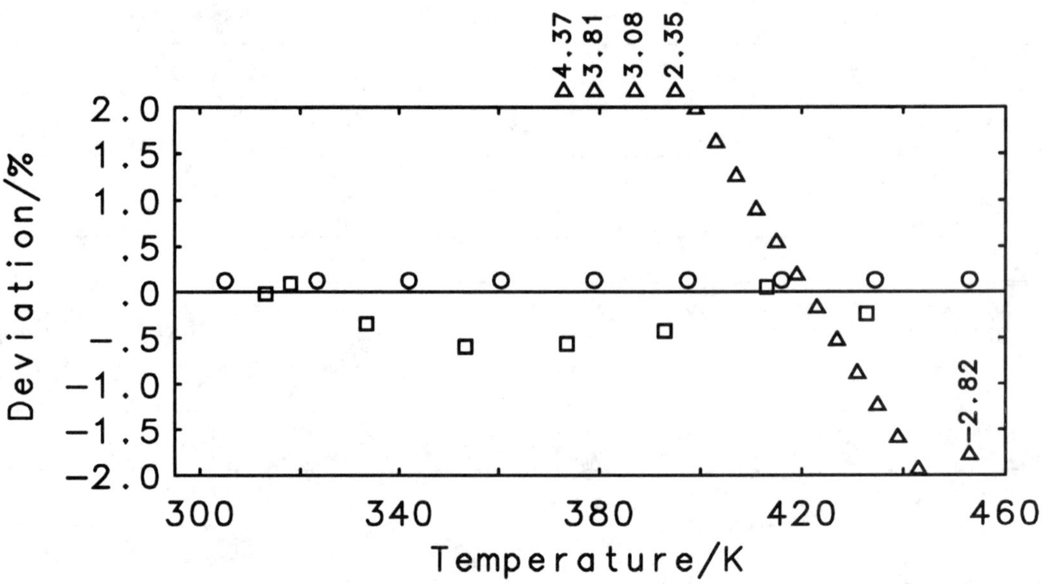

Selected data ○ 69ATK/LAR □ 84GRI/AND

Rejected data Δ 93DUR/AOU

Name: 2,6,10,14–Tetramethylpentadecane
Formula: $C_{19}H_{40}$

CAS–RN: 1921–70–6
Group No.: 11–059
Molar Mass: 268.53

TABLE 11.59.1. Experimental heat capacities

Reference	Temp. range K	No. pnts	Error %	Purity %	Purity method	Type capacity	Calorimeter Type	Calorimeter Reference
91TRE/COS	298.1	1	nosp	99.	anal	C_p	FSIT	71PIC/LED
93DUR/AOU	304.0–404.0	51	nosp	not specified		C_p	BDCT	86MER/BEN

TABLE 11.59.2. Correlated heat capacities

Reference	Temp. range K	No. pnts used	$\sigma_r C$ %	d_w	d/R	d_r %	d_b/R	+/–
Selected data								
91TRE/COS	298.1	1	0.50#	0.417	1.43–1	0.21	–1.43–1	–1
93DUR/AOU	304.0–404.0	51	5.00#	0.187	6.83–1	0.93	2.92–1	14

TABLE 11.59.3. Parameters of regression polynomial

Heat capacity type	No. data points total	No. data points used	s_w	s/R	s_r %	s_b/R	+/–
C_p	52	52	0.198	6.90–1	0.94	2.84–1	13

Temp. range K	A_1	A_2	Level of uncertainty
298.1–404.0	3.43208+1	1.15204+1	VI

TABLE 11.59.4. Recommended values of heat capacities

Temp. (K)	300	310	320	330	340	350	360
c_p (J K^{-1}g^{-1})	2.13	2.17	2.20	2.24	2.28	2.31	2.35
C_p (J K^{-1}mol^{-1})	573	582	592	601	611	621	630

Temp. (K)	370	380	390	400
c_p (J K^{-1}g^{-1})	2.38	2.42	2.45	2.49
C_p (J K^{-1}mol^{-1})	640	649	659	669

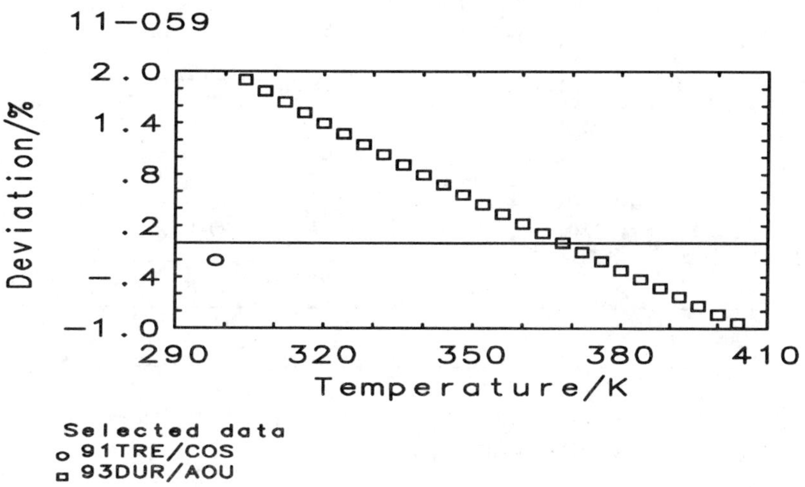

Name: Eicosane
Formula: $C_{20}H_{42}$

CAS–RN: 112–95–8
Group No.: 11–060
Molar Mass: 282.55

TABLE 11.60.1. Experimental heat capacities

Reference	Temp. range K	No. pnts	Error %	Purity %	Purity method	Type capacity	Calorimeter Type	Calorimeter Reference
81HOE	325.0–400.0	4S	5.00	not specified		C_p	BDHT	69PER/COM

TABLE 11.60.3. Parameters of regression polynomial

Heat capacity type	No. data points total	No. data points used	s_w	s/R	s_r %	s_b/R	+/–
C_p	4	4	0.143	6.05–1	0.71	5.18–3	0

Temp. range K	A_1	A_2	Level of uncertainty
325.0–400.0	2.89612+1	1.52350+1	VI

TABLE 11.60.4. Recommended values of heat capacities

Temp. (K)	320	330	340	350	360	370	380
c_p ($J K^{-1} g^{-1}$)	2.29	2.33	2.38	2.42	2.47	2.51	2.56
C_p ($J K^{-1} mol^{-1}$)	646	659	671	684	697	709	722

Temp. (K)	390	400
c_p ($J K^{-1} g^{-1}$)	2.60	2.65
C_p ($J K^{-1} mol^{-1}$)	735	747

Name: Heneicosane
Formula: $C_{21}H_{44}$

CAS–RN: 629–94–7
Group No.: 11–061
Molar Mass: 296.58

TABLE 11.61.1. Experimental heat capacities

Reference	Temp. range K	No. pnts	Error %	Purity %	Purity method	Type capacity	Calorimeter Type	Calorimeter Reference
48MAZ	318.1–328.1	2	nosp	not specified		C_p	BDHO	33STR/MAL
84GRI/AND	315.9–433.1	8	0.80	not specified		C_p	BSAO	67RAS/GAN

TABLE 11.61.2. Correlated heat capacities

Reference	Temp. range K	No. pnts used	$\sigma_r C$ %	d_w	d/R	d_r %	d_b/R	+/–
				Selected data				
84GRI/AND	315.9–433.1	8	0.80	0.260	1.90–1	0.21	7.71–4	–2
				Rejected data				
48MAZ	(4.30, 5.05, 4.27, 2)							

TABLE 11.61.3. Parameters of regression polynomial

Heat capacity type	No. data points total	used	s_w	s/R	s_r %	s_b/R	+/−
C_p	10	8	0.329	2.40−1	0.26	7.71−4	−2

Temp. range K	A_1	A_2	A_3	Level of uncertainty
315.9–433.1	6.54526+1	−2.32741−1	1.54098	IV

TABLE 11.61.4. Recommended values of heat capacities

Temp. (K)	320	330	340	350	360	370	380
c_p (J K^{-1}g^{-1})	2.256	2.284	2.312	2.341	2.371	2.402	2.434
C_p (J K^{-1}mol^{-1})	669.2	677.3	685.7	694.4	703.3	712.4	721.9
Temp. (K)	390	400	410	420	430		
c_p (J K^{-1}g^{-1})	2.467	2.500	2.534	2.570	2.606		
C_p (J K^{-1}mol^{-1})	731.5	741.5	751.7	762.1	772.8		

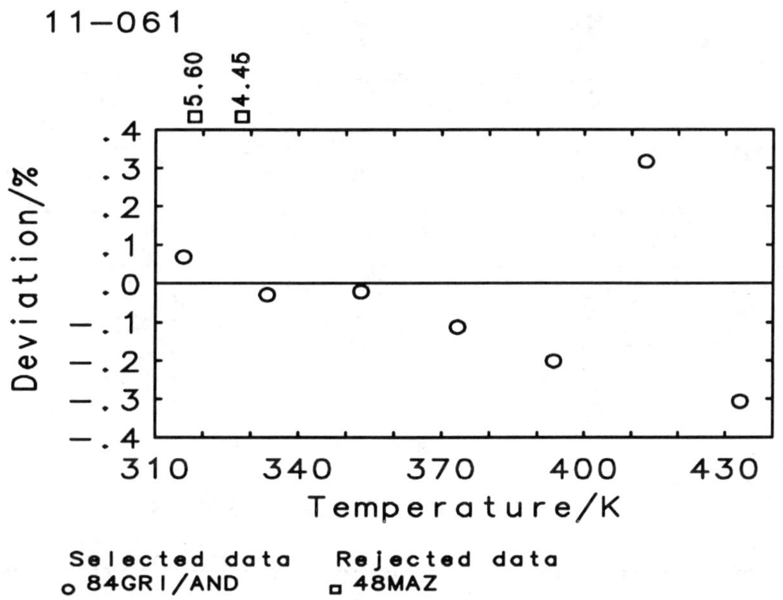

Name: Docosane
Formula: $C_{22}H_{46}$

CAS–RN: 629–97–0
Group No.: 11–062
Molar Mass: 310.61

TABLE 11.62.1. Experimental heat capacities

Reference	Temp. range K	No. pnts	Error %	Purity %	Purity method	Type capacity	Calorimeter Type	Calorimeter Reference
31GAR/VAN	317.5	1	nosp	not specified		C_p	DSIO	24GAR/RAN
69ATK/LAR	318.0–453.0	eqn	nosp	99.7	chrom	C_p	BSAO	58WES/GIN
81HOE	325.0–500.0	8S	5.00	not specified		C_p	BDHT	69PER/COM
93DUR/AOU	373.0–473.0	51	nosp	not specified		C_p	BDCT	86MER/BEN

TABLE 11.62.2. Correlated heat capacities

Reference	Temp. range K	No. pnts used	$\sigma_r C$ %	d_w	d/R	d_r %	d_b/R	+/–
				Selected data				
69ATK/LAR	318.0–453.0	10	0.50#	0.000	4.18–6	0.00	–7.63–7	0
				Rejected data				
81HOE	(1.92+1, 25.97, –1.91+1, –6)			93DUR/AOU	(5.50, 5.29, 5.28, 41)			

TABLE 11.62.3. Parameters of regression polynomial

Heat capacity type	No. data points total	used	s_w	s/R	s_r %	s_b/R	+/–
C_p	70	10	0.000	4.67–6	0.00	–7.63–7	0

Temp. range K	A_1	A_2	Level of uncertainty
318.0–453.0	4.38783+1	1.27388+1	V

TABLE 11.62.4. Recommended values of heat capacities

Temp. (K)	320	330	340	350	360	370	380
c_p (J K^{-1}g^{-1})	2.27	2.30	2.33	2.37	2.40	2.44	2.47
C_p (J K^{-1}mol^{-1})	704	714	725	736	746	757	767
Temp. (K)	390	400	410	420	430	440	450
c_p (J K^{-1}g^{-1})	2.50	2.54	2.57	2.61	2.64	2.67	2.71
C_p (J K^{-1}mol^{-1})	778	788	799	810	820	831	841

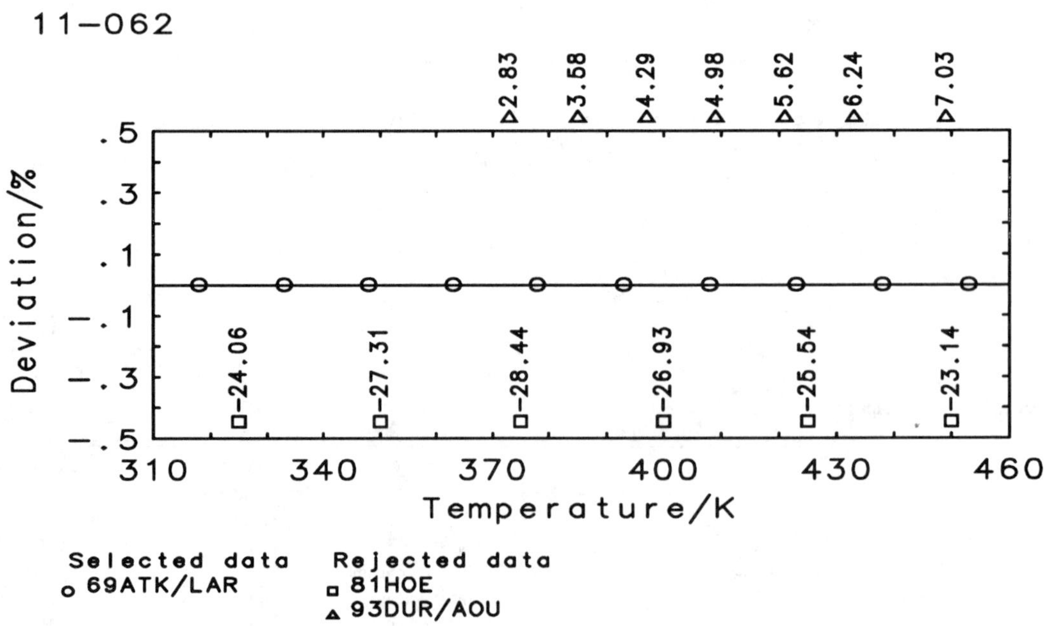

Name: Tricosane
Formula: $C_{23}H_{48}$

CAS–RN: 638–67–5
Group No.: 11–063
Molar Mass: 324.63

TABLE 11.63.1. Experimental heat capacities

Reference	Temp. range K	No. pnts	Error %	Purity %	Purity method	Type capacity	Calorimeter Type	Calorimeter Reference
69ATK/LAR	321.0–453.0	eqn	nosp	99.9	chrom	C_p	BSAO	58WES/GIN

TABLE 11.63.3. Parameters of regression polynomial

Heat capacity type	No. data points total	No. data points used	s_w	s/R	s_r %	s_b/R	+/–
C_p	10	10	0.000	5.39–6	0.00	3.05–6	0

Temp. range K	A_1	A_2	Level of uncertainty
321.0–453.3	4.44029+1	1.37045+1	V

TABLE 11.63.4. Recommended values of heat capacities

Temp. (K)	320	330	340	350	360	370	380
c_p (J K^{-1}g^{-1})	2.26	2.30	2.33	2.37	2.40	2.44	2.47
C_p (J K^{-1}mol^{-1})	734	745	757	768	779	791	802
Temp. (K)	390	400	410	420	430	440	450
c_p (J K^{-1}g^{-1})	2.51	2.54	2.58	2.61	2.65	2.68	2.72
C_p (J K^{-1}mol^{-1})	814	825	836	848	859	871	882

Name: Tetracosane
Formula: $C_{24}H_{50}$

CAS–RN: 646–31–1
Group No.: 11–064
Molar Mass: 338.66

TABLE 11.64.1. Experimental heat capacities

Reference	Temp. range K	No. pnts	Error %	Purity %	Purity method	Type capacity	Calorimeter Type	Calorimeter Reference
48MAZ	325.1–334.1	2	nosp	not specified		C_p	BDHO	33STR/MAL
69ATK/LAR	324.0–453.0	eqn	nosp	99.9	chrom	C_p	BSAO	58WES/GIN
81HOE	350.0–425.0	4S	5.00	not specified		C_p	BDHT	69PER/COM
84GRI/AND	330.6–433.4	8	0.80	not specified		C_p	BSAO	67RAS/GAN

TABLE 11.64.2. Correlated heat capacities

Reference	Temp. range K	No. pnts used	$\sigma_r C$ %	d_w	d/R	d_r %	d_b/R	+/–
				Selected data				
69ATK/LAR	324.0–452.8	8	0.80#	0.861	6.87–1	0.69	6.81–1	8
84GRI/AND	330.6–433.4	8	0.80	0.900	7.16–1	0.72	–6.60–1	–8
				Rejected data				
48MAZ	(5.81, 5.88, 5.78, 2)	81HOE		(9.38–1, 0.90, 2.17–1, 0)				

TABLE 11.64.3. Parameters of regression polynomial

Heat capacity type	No. data points total	used	s_w	s/R	s_r %	s_b/R	+/–
C_p	22	16	0.942	7.50–1	0.75	1.03–2	0

Temp. range K	A_1	A_2	Level of uncertainty
324.0–452.8	4.84202+1	1.35427+1	IV

TABLE 11.64.4. Recommended values of heat capacities

Temp. (K)	330	340	350	360	370	380	390
c_p (J K^{-1}g^{-1})	2.286	2.319	2.352	2.386	2.419	2.452	2.485
C_p (J K^{-1}mol^{-1})	774.2	785.4	796.7	808.0	819.2	830.5	841.7
Temp. (K)	400	410	420	430	440	450	
c_p (J K^{-1}g^{-1})	2.519	2.552	2.585	2.618	2.652	2.685	
C_p (J K^{-1}mol^{-1})	853.0	864.3	875.5	886.8	898.0	909.3	

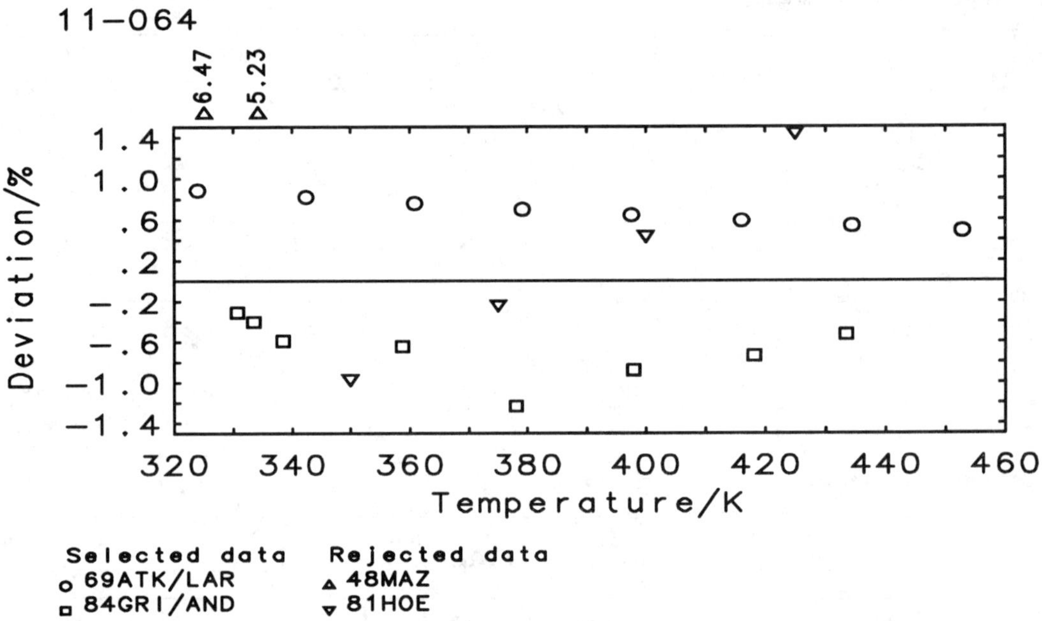

Name: Pentacosane
Formula: $C_{25}H_{52}$

CAS–RN: 629–99–2
Group No.: 11–065
Molar Mass: 352.69

TABLE 11.65.1. Experimental heat capacities

Reference	Temp. range K	No. pnts	Error %	Purity %	Purity method	Type capacity	Calorimeter Type	Calorimeter Reference
32SPA/THO	333.1–373.1	5S	1.00	not specified		C_p	BDHO	31THO/PAR

TABLE 11.65.3. Parameters of regression polynomial

Heat capacity type	No. data points total	used	s_w	s/R	s_r %	s_b/R	+/−
C_p	5	5	0.056	5.61−2	0.06	4.42−5	0

Temp. range K	A_1	A_2	Level of uncertainty
333.1–373.1	4.90533+1	1.47254+1	V

TABLE 11.65.4. Recommended values of heat capacities

Temp. (K)	330	340	350	360	370
c_p ($J\,K^{-1}g^{-1}$)	2.30	2.34	2.37	2.41	2.44
C_p ($J\,K^{-1}mol^{-1}$)	812	824	836	849	861

Name: Hexacosane
Formula: $C_{26}H_{54}$

CAS–RN: 630–01–3
Group No.: 11–066
Molar Mass: 366.71

TABLE 11.66.1. Experimental heat capacities

Reference	Temp. range K	No. pnts	Error %	Purity %	method	Type capacity	Calorimeter Type	Reference
31GAR/VAN	329.3	1	nosp	not specified		C_p	DSIO	24GAR/RAN
69ATK/LAR	329.0–453.0	eqn	nosp	99.8	chrom	C_p	BSAO	58WES/GIN
76AND/MAR	332.2–358.6	10	0.10	99.77	melpt	C_p	BSAO	68WES/FUR

TABLE 11.66.2. Correlated heat capacities

Reference	Temp. range K	No. pnts used	$\sigma_r C$ %	d_w	d/R	d_r %	d_b/R	+/−
				Selected data				
69ATK/LAR	329.0–453.6	15	0.80#	0.523	4.34−1	0.42	−1.24−1	0
76AND/MAR	332.2–358.6	10	0.10	0.295	3.06−2	0.03	3.02−3	4
				Rejected data				
31GAR/VAN	(1.38, 1.33, 1.38, 1)							

TABLE 11.66.3. Parameters of regression polynomial

Heat capacity type	No. data points total	used	s_w	s/R	s_r %	s_b/R	+/−
C_p	26	25	0.475	3.59−1	0.35	−7.35−2	4

Temp. range K	A_1	A_2	A_3	Level of uncertainty
329.0–453.6	9.12413+1	−5.44480	2.64279	III

TABLE 11.66.4. Recommended values of heat capacities

Temp. (K)	330	340	350	360	370	380	390
c_p (J K^{-1}g^{-1})	2.314	2.342	2.371	2.401	2.432	2.465	2.499
C_p (J K^{-1}mol^{-1})	848.5	858.7	869.4	880.4	891.9	903.9	916.3
Temp. (K)	400	410	420	430	440	450	
c_p (J K^{-1}g^{-1})	2.534	2.570	2.607	2.646	2.686	2.727	
C_p (J K^{-1}mol^{-1})	929.1	942.4	956.1	970.3	984.8	999.9	

TABLE 11.66.5. Parameters of quasi-polynomial equation

Heat capacity type	No. data points total	used	s_w	s/R	s_r %	s_b/R	+/–
C_p	25	24	3.255	3.47–1	0.34	1.47–1	8

Temp. range K	T_c K	A_1	A_2	A_3	A_4	Level of uncertainty
329.0–452.5	817.00	–1.31216+1	4.26776–1	5.29851+1	1.00859+2	IV

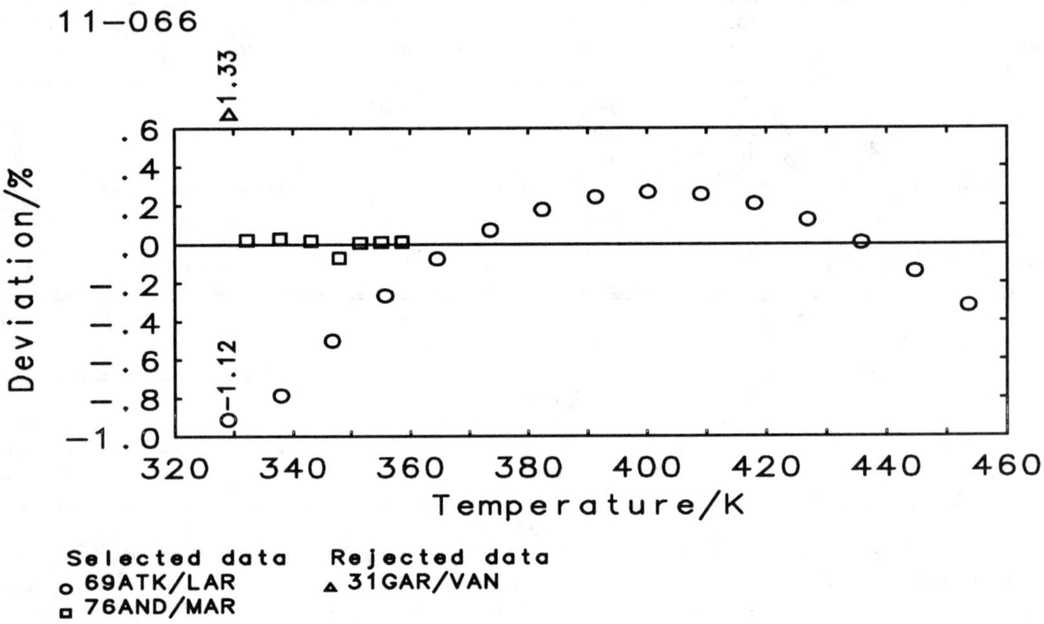

Name: Heptacosane
Formula: $C_{27}H_{56}$

CAS–RN: 593–49–7
Group No.: 11–067
Molar Mass: 380.74

TABLE 11.67.1. Experimental heat capacities

Reference		Temp. K	Capac. J/(K.g)	Error %	Purity %	Purity method	Type capacity	Calorimeter Type	Calorimeter Reference
38VER	N	338.15	2.176	nosp	not specified		avg	BDHO	33STR/MAL

38VER average value in temperature range 332–342 K

Name: Octacosane
Formula: $C_{28}H_{58}$

CAS–RN: 630–02–4
Group No.: 11–068
Molar Mass: 394.77

TABLE 11.68.1. Experimental heat capacities

Reference	Temp. range K	No. pnts	Error %	Purity %	Purity method	Type capacity	Calorimeter Type	Calorimeter Reference
53HOF/DEC	343.1–353.1	2	4.00	90.	estim	C_p	BDHO	52HOF
69ATK/LAR	334.0–453.0	eqn	nosp	99.9	chrom	C_p	BSAO	58WES/GIN
93DUR/AOU	373.0–473.0	51	nosp	not specified		C_p	BDCT	86MER/BEN

TABLE 11.68.2. Correlated heat capacities

Reference	Temp. range K	No. pnts used	$\sigma_r C$ %	d_w	d/R	d_r %	d_b/R	+/–
				Selected data				
69ATK/LAR	334.0–452.8	13	0.80#	0.000	5.60–6	0.00	–4.11–6	0
				Rejected data				
53HOF/DEC	(1.70, 1.55, –1.69, –2)			93DUR/AOU	(1.71, 1.47, –1.42, –34)			

TABLE 11.68.3. Parameters of regression polynomial

Heat capacity type	No. data points total	No. data points used	s_w	s/R	s_r %	s_b/R	+/–
C_p	66	13	0.000	6.09–6	0.00	–4.11–6	0

Temp. range K	A_1	A_2	Level of uncertainty
334.0–452.8	5.55775+1	1.61905+1	IV

TABLE 11.68.4. Recommended values of heat capacities

Temp. (K)	340	350	360	370	380	390	400
c_p (J $K^{-1}g^{-1}$)	2.330	2.364	2.398	2.432	2.466	2.500	2.535
C_p (J $K^{-1}mol^{-1}$)	919.8	933.3	946.7	960.2	973.6	987.1	1001
Temp. (K)	410	420	430	440	450		
c_p (J $K^{-1}g^{-1}$)	2.569	2.603	2.637	2.671	2.705		
C_p (J $K^{-1}mol^{-1}$)	1014	1027	1041	1054	1068		

Name: 2,6,10,15,19,23–Hexamethyltetracosane
Formula: $C_{30}H_{62}$

CAS–RN: 111–01–3
Group No.: 11–069
Molar Mass: 422.82

TABLE 11.69.1. Experimental heat capacities

Reference	Temp. range K	No. pnts	Error %	Purity %	Purity method	Type capacity	Calorimeter Type	Calorimeter Reference
91TRE/COS	298.1	1	nosp	99.	anal	C_p	FSIT	71PIC/LED
93DUR/AOU	304.0–404.0	51	nosp	not specified		C_p	BDCT	86MER/BEN

TABLE 11.69.2. Correlated heat capacities

Reference	Temp. range K	No. pnts used	$\sigma_r C$ %	d_w	d/R	d_r %	d_b/R	+/–
				Selected data				
91TRE/COS	298.1	1	0.50#	0.882	4.70–1	0.44	–4.70–1	–1
93DUR/AOU	304.0–404.0	51	3.00#	0.229	7.51–1	0.69	3.45–1	15

TABLE 11.69.3. Parameters of regression polynomial

Heat capacity type	No. data points total	used	s_w	s/R	s_r %	s_b/R	+/–
C_p	52	52	0.262	7.61–1	0.70	3.29–1	14

Temp. range K	A_1	A_2	Level of uncertainty
298.1–404.0	9.25703+1	4.86460	V

TABLE 11.69.4. Recommended values of heat capacities

Temp. (K)	300	310	320	330	340	350	360
c_p (J K^{-1}g^{-1})	2.11	2.12	2.13	2.14	2.15	2.16	2.16
C_p (J K^{-1}mol^{-1})	891	895	899	903	907	911	915

Temp. (K)	370	380	390	400
c_p (J K^{-1}g^{-1})	2.17	2.18	2.19	2.20
C_p (J K^{-1}mol^{-1})	919	923	927	931

Name: Triacontane
Formula: $C_{30}H_{62}$

CAS–RN: 638–68–6
Group No.: 11–070
Molar Mass: 422.82

TABLE 11.70.1. Experimental heat capacities

Reference	Temp. range K	No. pnts	Error %	Purity %	Purity method	Type capacity	Calorimeter Type	Calorimeter Reference
31GAR/VAN	338.5	1	nosp	not specified		C_p	DSIO	24GAR/RAN
48MAZ	340.1–350.1	2	nosp	not specified		C_p	BDHO	33STR/MAL
81HOE	350.0–450.0	5S	5.00	not specified		C_p	BDHT	69PER/COM
93DUR/AOU	373.0–473.0	51	nosp	not specified		C_p	BDCT	86MER/BEN

TABLE 11.70.2. Correlated heat capacities

Reference	Temp. range K	No. pnts used	$\sigma_r C$ %	d_w	d/R	d_r %	d_b/R	+/–
				Selected data				
93DUR/AOU	373.0–473.0	51	3.00#	0.000	3.32–4	0.00	2.99–6	0
				Rejected data				
81HOE	(2.08+1, 19.82, –2.07+1, –4)							

TABLE 11.70.3. Parameters of regression polynomial

Heat capacity type	No. data points total	used	s_w	s/R	s_r %	s_b/R	+/–
C_p	59	51	0.000	3.39–4	0.00	2.99–6	0

Temp. range K	A_1	A_2	Level of uncertainty
373.0–473.0	7.74907+1	1.17846+1	VI

TABLE 11.70.4. Recommended values of heat capacities

Temp. (K)	370	380	390	400	410	420	430
c_p (J K^{-1}g^{-1})	2.38	2.40	2.43	2.45	2.47	2.50	2.52
C_p (J K^{-1}mol^{-1})	1010	1020	1030	1040	1050	1060	1070
Temp. (K)	**440**	**450**	**460**	**470**			
c_p (J K^{-1}g^{-1})	2.54	2.57	2.59	2.61			
C_p (J K^{-1}mol^{-1})	1080	1090	1100	1100			

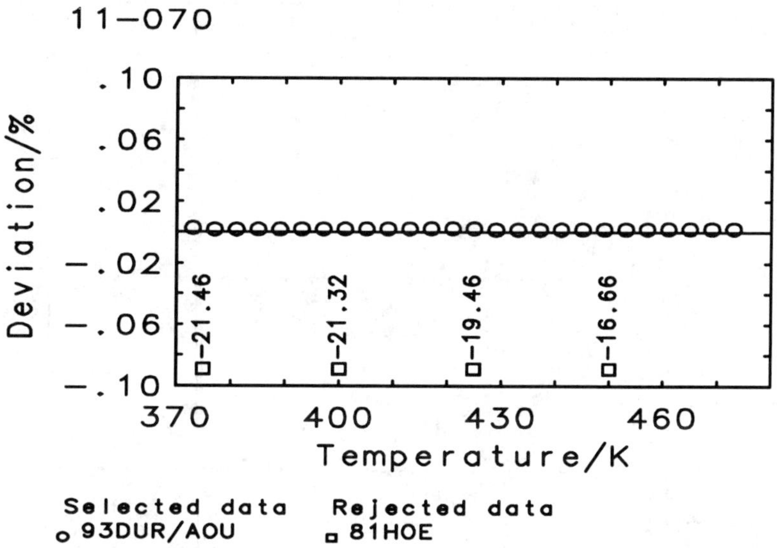

Name: 11–Decylheneicosane
Formula: $C_{31}H_{64}$

CAS–RN: 55320–06–4
Group No.: 11–071
Molar Mass: 436.85

TABLE 11.71.1. Experimental heat capacities

Reference		Temp. range K	No. pnts	Error %	Purity %	Purity method	Type capacity	Calorimeter Type	Calorimeter Reference
45FIS/NAY	N	282.0–297.4	5	0.70	96.6	melpt	C_p	BSIO	25PAR
45FIS/NAY	N	283.8–297.8	23	0.20	96.48	melpt	C_p	BSAO	47AST/SZA

45FIS/NAY the same sample measured at Stanford University, Stanford, CA
45FIS/NAY the same sample measured at Pennsylvania State University, University Park, PA

TABLE 11.71.2. Correlated heat capacities

Reference	Temp. range K	No. pnts used	$\sigma_r C$ %	d_w	d/R	d_r %	d_b/R	+/–
				Selected data				
45FIS/NAY	282.0–297.4	5	0.70	1.489	1.19	1.04	1.18	5
45FIS/NAY	283.8–297.8	23	0.20	0.851	1.92–1	0.17	–1.98–2	–1

TABLE 11.71.3. Parameters of regression polynomial

Heat capacity type	No. data points total	used	s_w	s/R	s_r %	s_b/R	+/–
C_p	28	28	1.033	5.52–1	0.48	1.94–1	4

Temp. range K	A_1	A_2	Level of uncertainty
282.0–297.8	6.68867+1	1.58226+1	III

TABLE 11.71.4. Recommended values of heat capacities

Temp. (K)	290	295	298.15	300
c_p (J K^{-1}g^{-1})	2.146	2.161	2.171	2.176
C_p (J K^{-1}mol^{-1})	937.6	944.2	948.4	950.8

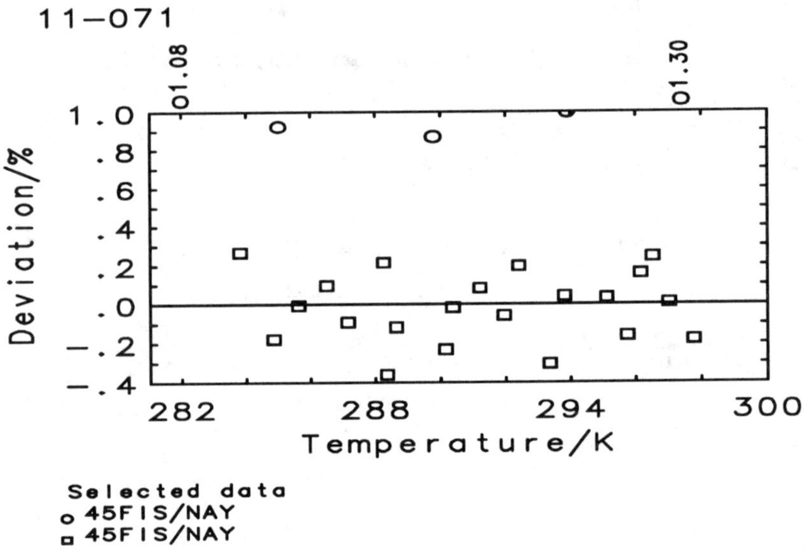

Name: Hentriacontane
Formula: $C_{31}H_{64}$

CAS–RN: 630–04–6
Group No.: 11–072
Molar Mass: 436.85

TABLE 11.72.1. Experimental heat capacities

Reference		Temp. K	Capac. J/(K.g)	Error %	Purity % method	Type capacity	Calorimeter Type	Reference
38VER	N	346.65	2.510	nosp	not specified	avg	BDHO	33STR/MAL

38VER average value in temperature range 341–351 K

Name: Dotriacontane
Formula: $C_{32}H_{66}$

CAS–RN: 544–85–4
Group No.: 11–073
Molar Mass: 450.88

TABLE 11.73.1. Experimental heat capacities

Reference		Temp. range K	No. pnts	Error %	Purity %	Purity method	Type capacity	Calorimeter Type	Calorimeter Reference
38VER	N	348.1	1	nosp	not specified		avg	BDHO	33STR/MAL
81HOE		350.0–475.0	6S	5.00	not specified		C_p	BDHT	69PER/COM

38VER average value in temperature range 343–353 K

TABLE 11.73.2. Correlated heat capacities

Reference	Temp. range K	No. pnts used	$\sigma_r C$ %	d_w	d/R	d_r %	d_b/R	+/–
				Selected data				
81HOE	350.0–475.0	6	5.00	0.195	1.23	0.98	2.57–2	2

TABLE 11.73.3. Parameters of regression polynomial

Heat capacity type	No. data points total	No. data points used	s_w	s/R	s_r %	s_b/R	+/–
C_p	7	6	0.239	1.51	1.20	2.57–2	2

Temp. range K	A_1	A_2	Level of uncertainty
350.0–475.0	7.27042+1	1.30208+1	VI

TABLE 11.73.4. Recommended values of heat capacities

Temp. (K)	350	360	370	380	390	400	410
c_p (J K^{-1}g^{-1})	2.18	2.21	2.23	2.25	2.28	2.30	2.33
C_p (J K^{-1}mol^{-1})	983	994	1010	1020	1030	1040	1050
Temp. (K)	420	430	440	450	460	470	480
c_p (J K^{-1}g^{-1})	2.35	2.37	2.40	2.42	2.45	2.47	2.49
C_p (J K^{-1}mol^{-1})	1060	1070	1080	1090	1100	1110	1120

Name: Tritriacontane
Formula: $C_{33}H_{68}$

CAS–RN: 630–05–7
Group No.: 11–074
Molar Mass: 464.90

TABLE 11.74.1. Experimental heat capacities

Reference	Temp. range K	No. pnts	Error %	Purity %	Purity method	Type capacity	Calorimeter Type	Calorimeter Reference
32SPA/THO	353.1–383.1	4S	1.00	not specified		C_p	BDHO	31THO/PAR

TABLE 11.74.3. Parameters of regression polynomial

Heat capacity type	No. data points total	used	s_w	s/R	s_r %	s_b/R	+/–
C_p	4	4	0.066	9.06–2	0.07	4.96–5	0

Temp. range K	A_1	A_2	Level of uncertainty
353.1–383.1	7.84717+1	1.56849+1	V

TABLE 11.74.4. Recommended values of heat capacities

Temp. (K)	350	360	370	380
c_p (J K^{-1}g^{-1})	2.39	2.41	2.44	2.47
C_p (J K^{-1}mol^{-1})	1110	1120	1130	1150

Name: Tetratriacontane
Formula: $C_{34}H_{70}$

CAS–RN: 14167–59–0
Group No.: 11–075
Molar Mass: 478.93

TABLE 11.75.1. Experimental heat capacities

Reference	Temp. range K	No. pnts	Error %	Purity %	Purity method	Type capacity	Calorimeter Type	Calorimeter Reference
31GAR/VAN	346.3	1	nosp	not specified		C_p	DSIO	24GAR/RAN
69ATK/LAR	346.0–453.0	eqn	nosp	99.8	chrom	C_p	BSAO	58WES/GIN

TABLE 11.75.2. Correlated heat capacities

Reference	Temp. range K	No. pnts used	$\sigma_r C$ %	d_w	d/R	d_r %	d_b/R	+/–
				Selected data				
69ATK/LAR	346.0–452.7	12	0.50#	0.000	7.63–6	0.00	–3.82–6	0
				Rejected data				
31GAR/VAN	(4.80–1, 0.35, 4.80–1, 1)							

TABLE 11.75.3. Parameters of regression polynomial

Heat capacity type	No. data points total	used	s_w	s/R	s_r %	s_b/R	+/–
C_p	13	12	0.000	8.36–6	0.00	–3.82–6	0

Temp. range K	A_1	A_2	Level of uncertainty
346.0–452.7	6.80640+1	1.98726+1	IV

TABLE 11.75.4. Recommended values of heat capacities

Temp. (K)	350	360	370	380	390	400	410
c_p (J K^{-1}g^{-1})	2.389	2.424	2.458	2.493	2.527	2.562	2.596
C_p (J K^{-1}mol^{-1})	1144	1161	1177	1194	1210	1227	1243
Temp. (K)	420	430	440	450			
c_p (J K^{-1}g^{-1})	2.631	2.665	2.700	2.734			
C_p (J K^{-1}mol^{-1})	1260	1276	1293	1309			

Name: Pentatriacontane
Formula: $C_{35}H_{72}$

CAS–RN: 630–07–9
Group No.: 11–076
Molar Mass: 492.96

TABLE 11.76.1. Experimental heat capacities

Reference	Temp. K	Capac. J/(K.g)	Error %	Purity %	Purity method	Type capacity	Calorimeter Type	Calorimeter Reference
31GAR/VAN	347.15	2.368	nosp	not specified		C_p	DSIO	24GAR/RAN

Name: Hexatriacontane
Formula: $C_{36}H_{74}$

CAS–RN: 630–06–8
Group No.: 11–077
Molar Mass: 506.98

TABLE 11.77.1. Experimental heat capacities

Reference	Temp. range K	No. pnts	Error %	Purity %	Purity method	Type capacity	Calorimeter Type	Calorimeter Reference
48MAZ	350.1–359.1	2	nosp	not specified		C_p	BDHO	33STR/MAL
69ATK/LAR	349.0–453.0	eqn	nosp	not specified		C_p	BSAO	58WES/GIN
81HOE	375.0–500.0	6S	5.00	not specified		C_p	BDHT	69PER/COM
91JIN/WUN	360.0–430.0	8	1.00	98.	anal	C_p	BDHT	90JIN/WUN
93DUR/AOU	373.0–473.0	51	nosp	not specified		C_p	BDCT	86MER/BEN

TABLE 11.77.2. Correlated heat capacities

Reference	Temp. range K	No. pnts used	$\sigma_r C$ %	d_w	d/R	d_r %	d_b/R	+/–
				Selected data				
69ATK/LAR	349.0–453.4	10	0.80#	0.428	5.42–1	0.34	5.29–1	10
91JIN/WUN	360.0–430.0	8	1.00	0.725	1.13	0.73	–1.01	–8
				Rejected data				
48MAZ	(1.06+1, 6.83, 1.06+1, 2)			81HOE	(9.95, 6.72, –9.91, –4)			
93DUR/AOU	(4.59, 2.95, –4.17, –41)							

TABLE 11.77.3. Parameters of regression polynomial

Heat capacity type	No. data points total	used	s_w	s/R	s_r %	s_b/R	+/–
C_p	77	18	0.615	9.07–1	0.58	–1.54–1	2

Temp. range K	A_1	A_2	Level of uncertainty
349.0–453.4	6.97449+1	2.12293+1	IV

TABLE 11.77.4. Recommended values of heat capacities

Temp. (K)	350	360	370	380	390	400	410
c_p (J K^{-1}g^{-1})	2.362	2.397	2.432	2.467	2.502	2.536	2.571
C_p (J K^{-1}mol^{-1})	1198	1215	1233	1251	1268	1286	1304
Temp. (K)	420	430	440	450			
c_p (J K^{-1}g^{-1})	2.606	2.641	2.676	2.711			
C_p (J K^{-1}mol^{-1})	1321	1339	1357	1374			

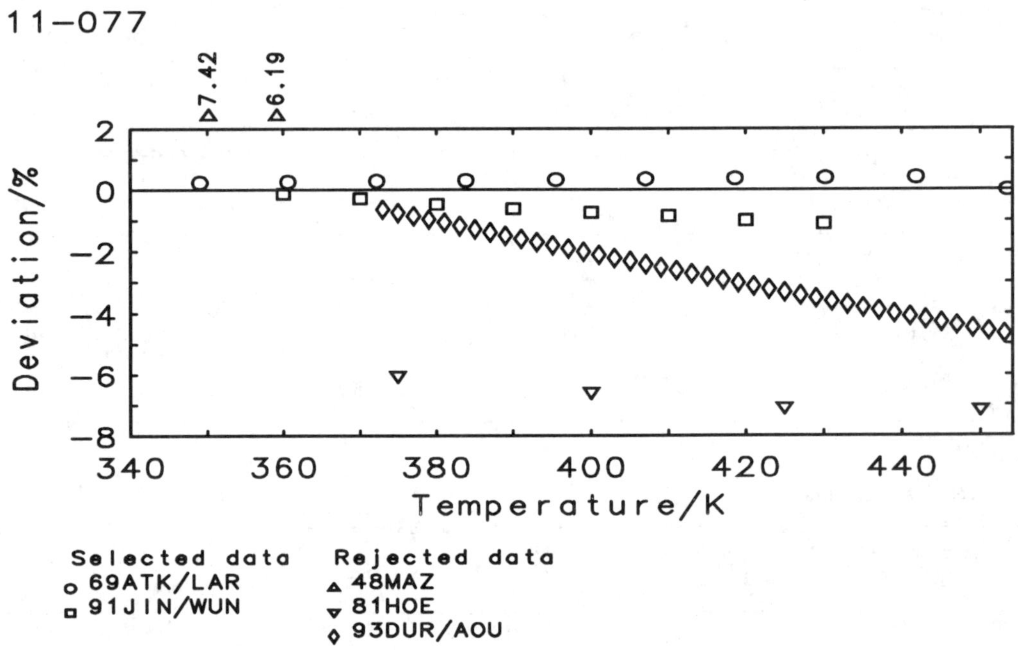

Name: Tetracontane
Formula: $C_{40}H_{82}$

CAS–RN: 4181–95–7
Group No.: 11–078
Molar Mass: 563.09

TABLE 11.78.1. Experimental heat capacities

Reference	Temp. range K	No. pnts	Error %	Purity %	Purity method	Type capacity	Calorimeter Type	Calorimeter Reference
48MAZ	355.1–363.1	2	nosp	not specified		C_p	BDHO	33STR/MAL

TABLE 11.78.3. Parameters of regression polynomial

Heat capacity type	No. data points total	used	s_w	s/R	s_r %	s_b/R	+/–
C_p	2	2	0.000	0.00	0.00	0.00	0

Temp. range K	A_1	Level of uncertainty
355.1–363.1	1.70014+2	VI

TABLE 11.78.4. Recommended values of heat capacities

Temp. (K)	355	365
c_p (J $K^{-1}g^{-1}$)	2.51	2.51
C_p (J $K^{-1}mol^{-1}$)	1410	1410

Name: Dotetracontane
Formula: $C_{42}H_{86}$

CAS–RN: 7098–20–6
Group No.: 11–079
Molar Mass: 591.14

TABLE 11.79.1. Experimental heat capacities

Reference	Temp. range K	No. pnts	Error %	Purity %	Purity method	Type capacity	Calorimeter Type	Calorimeter Reference
69ATK/LAR	357.0–453.0	eqn	nosp	99.8	chrom	C_p	BSAO	58WES/GIN

TABLE 11.79.3. Parameters of regression polynomial

Heat capacity type	No. data points total	used	s_w	s/R	s_r %	s_b/R	+/–
C_p	10	10	0.000	7.63–6	0.00	0.00	0

Temp. range K	A_1	A_2	Level of uncertainty
357.0–453.3	7.44141+1	2.74438+1	IV

TABLE 11.79.4. Recommended values of heat capacities

Temp. (K)	360	370	380	390	400	410	420
c_p (J $K^{-1}g^{-1}$)	2.436	2.475	2.513	2.552	2.591	2.629	2.668
C_p (J $K^{-1}mol^{-1}$)	1440	1463	1486	1509	1531	1554	1577
Temp. (K)	430	440	450				
c_p (J $K^{-1}g^{-1}$)	2.706	2.745	2.784				
C_p (J $K^{-1}mol^{-1}$)	1600	1623	1646				

Name: Tritetracontane
Formula: $C_{43}H_{88}$

CAS–RN: 7098–21–7
Group No.: 11–080
Molar Mass: 605.17

TABLE 11.80.1. Experimental heat capacities

Reference	Temp. range K	No. pnts	Error %	Purity %	Purity method	Type capacity	Calorimeter Type	Calorimeter Reference
48MAZ	359.1–364.1	2	nosp	not specified		C_p	BDHO	33STR/MAL

TABLE 11.80.3. Parameters of regression polynomial

Heat capacity type	No. data points total	used	s_w	s/R	s_r %	s_b/R	+/–
C_p	2	2	0.000	0.00	0.00	0.00	0

Temp. range K	A_1	A_2	Level of uncertainty
359.1–364.1	1.91855+2		VI

TABLE 11.80.4. Recommended values of heat capacities

Temp. (K)	360
c_p (J K^{-1}g^{-1})	2.64
C_p (J K^{-1}mol^{-1})	1600

Name: Tetratetracontane
Formula: $C_{44}H_{90}$

CAS–RN: 7098–22–8
Group No.: 11–081
Molar Mass: 619.20

TABLE 11.81.1. Experimental heat capacities

Reference	Temp. range K	No. pnts	Error %	Purity %	Purity method	Type capacity	Calorimeter Type	Calorimeter Reference
91JIN/WUN	370.0–430.0	7	1.00	99.	anal	C_p	BDHT	90JIN/WUN

TABLE 11.81.3. Parameters of regression polynomial

Heat capacity type	No. data points total	used	s_w	s/R	s_r %	s_b/R	+/–
C_p	7	7	0.044	8.24–2	0.04	6.97–5	0

Temp. range K	A_1	A_2	Level of uncertainty
370.0–430.0	1.05179+2	2.08976+1	IV

TABLE 11.81.4. Recommended values of heat capacities

Temp. (K)	370	380	390	400	410	420	430
c_p (J K^{-1}g^{-1})	2.451	2.479	2.507	2.535	2.563	2.591	2.619
C_p (J K^{-1}mol^{-1})	1517	1535	1552	1570	1587	1604	1622

Name: Octatetracontane
Formula: $C_{48}H_{98}$

CAS–RN: 7098–26–2
Group No.: 11–082
Molar Mass: 675.31

TABLE 11.82.1. Experimental heat capacities

Reference	Temp. range K	No. pnts	Error %	Purity %	Purity method	Type capacity	Calorimeter Type	Calorimeter Reference
69ATK/LAR	363.0–453.0	eqn	nosp	95.	estim	C_p	BSAO	58WES/GIN

TABLE 11.82.3. Parameters of regression polynomial

Heat capacity type	No. data points total	used	s_w	s/R	s_r %	s_b/R	+/–
C_p	10	10	0.000	9.34–6	0.00	4.58–6	0

Temp. range K	A_1	A_2	Level of uncertainty
363.0–453.0	9.63604+1	2.70463+1	IV

TABLE 11.82.4. Recommended values of heat capacities

Temp. (K)	370	380	390	400	410	420	430
c_p (J K^{-1}g^{-1})	2.419	2.452	2.485	2.518	2.552	2.585	2.618
C_p (J K^{-1}mol^{-1})	1633	1656	1678	1701	1723	1746	1768
Temp. (K)	440	450					
c_p (J K^{-1}g^{-1})	2.652	2.685					
C_p (J K^{-1}mol^{-1})	1791	1813					

Name: Pentacontane
Formula: $C_{50}H_{102}$

CAS–RN: 6596–40–3
Group No.: 11–083
Molar Mass: 703.36

TABLE 11.83.1. Experimental heat capacities

Reference	Temp. range K	No. pnts	Error %	Purity %	Purity method	Type capacity	Calorimeter Type	Calorimeter Reference
91JIN/WUN	380.0–430.0	6	1.00	99.	anal	C_p	BDHT	90JIN/WUN
93DUR/AOU	373.0–473.0	102	nosp	not specified		C_p	BDCT	86MER/BEN

TABLE 11.83.2. Correlated heat capacities

Reference	Temp. range K	No. pnts used	$\sigma_r C$ %	d_w	d/R	d_r %	d_b/R	+/–
				Selected data				
91JIN/WUN	380.0–430.0	6	1.00	0.161	3.50–1	0.16	3.27–1	6
93DUR/AOU	373.0–473.0	102	3.00#	0.269	1.74	0.81	–1.45–1	0

TABLE 11.83.3. Parameters of regression polynomial

Heat capacity type	No. data points total	used	s_w	s/R	s_r %	s_b/R	+/–
C_p	108	108	0.266	1.71	0.79	–1.19–1	6

Temp. range K	A_1	A_2	Level of uncertainty
373.0–473.0	1.21050+2	2.31803+1	V

TABLE 11.83.4. Recommended values of heat capacities

Temp. (K)	370	380	390	400	410	420	430
c_p (J $K^{-1}g^{-1}$)	2.445	2.472	2.500	2.527	2.554	2.582	2.609
C_p (J $K^{-1}mol^{-1}$)	1720	1739	1758	1777	1797	1816	1835
Temp. (K)	440	450	460	470			
c_p (J $K^{-1}g^{-1}$)	2.637	2.664	2.691	2.719			
C_p (J $K^{-1}mol^{-1}$)	1855	1874	1893	1912			

11–083

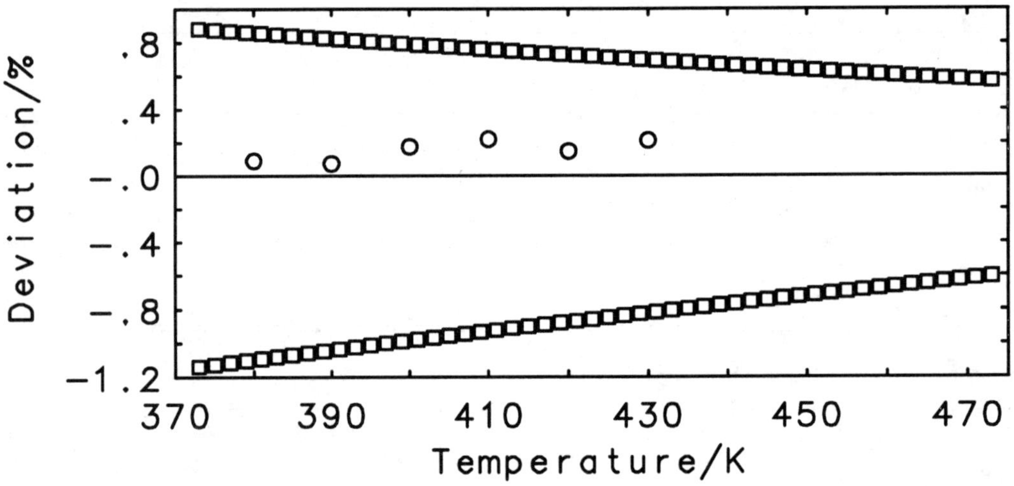

Selected data
o 91JIN/WUN
□ 93DUR/AOU

12. Saturated Cyclic Hydrocarbons

This family contains data on 124 cyclic hydrocarbons; for 14 compounds data are available at a single temperature only. By definition, this family includes hydrocarbons without an unsaturated bond in the ring; however, an unsaturated bond may appear in a side chain.

Measurements for more than one half of the compounds were carried out at MRCE (62GOL/BEL, 63GUD/CAM) which was interested in this group of compounds because of their potential use as jet fuels. However, the purity of the samples was low and they were not sufficiently identified with respect to the specification of isomers and stereoisomers. For each compound, the heat capacity was usually measured at four temperatures in the range from 313 to 483 K using a differential heat conduction calorimeter developed by the authors; the data were obtained with the reported error of around 1 %, which we considered as realistic.

A great deal of effort was devoted to the measurement of the heat capacity of cycloalkanes and substituted cycloalkanes at BMB. Altogether, high-purity samples for 27 compounds were studied. The investigations gave results with a reported error of 0.2 %. All the experimental data (43RUE/HUF, 46DOU/HUF1, 49HUF/TOD, 50SCO/FIN2, 53GRO/OLI, 56FIN/SCO, 57MCC/FIN2, 65FIN/MES, 65MES/TOD1, 72FIN/MCC, 81FIN/MES) have been used as a basis for the final correlation of this monograph.

The data for the first two members of the homologous series, cyclopropane and cyclobutane, were measured only at UCB (46RUE/POW, 53RAT/GWI) using an isoperibol calorimeter (37GIA/EGA) where results showed a reported measurement error of 0.2 %.

Reliable data for cyclopentane were obtained at BMB (46DOU/HUF1). Data for the same compound were also measured at PSC (43AST/FIN) and repeatedly investigated by the same laboratory (47SZA/MOR). The authors of the former data reported an error of about 1 %, and, therefore, we have included only the values from the latter source in the final correlation.

The data on cyclohexane are the most abundant and this abundance corresponds to the importance of this compound to the chemical industry; however, data from different sources often differ by several percent. Altogether, 10 data sets containing heat capacity as a function of temperature have been selected for the final correlation. The data measured at BMB (43RUE/HUF) have been considered to be the most reliable source in the literature, unfortunately, measurements are available only over a narrow temperature interval (20 K). These values have been supplemented by the data obtained at GPI (78SAF) which were measured in two different calorimeters (a flow isoperibol calorimeter and an adiabatic calorimeter with continuous heating) and were mutually consistent. At the same time, these values agreed reasonably with the Bartlesville data (43RUE/HUF). Further, several additional data sets have been used in the correlation, such as: measurements from UGG (66KLE) in the temperature range from 293 to 343 K, high-temperature data up to 523 K from MITC (75SAN), and data from CITP (50AUE/SAG) with a reported error about 1.5 %. The data from the last mentioned laboratory (51CON/SAG) have been selected for the correlation of methylcyclopentane together with the BMB data (46DOU/HUF1).

The data on methylcyclohexane are the second most abundant in the family. We have selected three data sets which show excellent consistency and agreement: they are the BMB values (46DOU/HUF1), the direct experimental data from GITA (66HWA/ZIE) and later data available as a smoothing equation from the same laboratory (75HOL/ZIE).

Accurate data measured at the BMB (57MCC/FIN2) for two stereoisomers of the decahydronaphthalenes have been supplemented for the *trans* isomer by the less accurate data from the University of California at Los Angeles (53SEY) which were obtained by graphically smoothing the experimental values. The latter data have been included for extending the temperature range of recommendations. Seyer (53SEY) reported a sharp lambda peak around 323 K for *cis*-decahydronaphthalene. This anomaly was not confirmed by the accurate measurements at Bartlesville and for that reason the Seyer's data have not been not considered.

Bicyclic and condensed cycloalkanes were investigated primarily because of their interesting molecular structure. The main contributions originate from: UMAA (70WON/WES), MSUM (64SER/GOR) and KSUK (78SPI/AND).

Name: Cyclopropane
Formula: C_3H_6

CAS-RN: 75–19–4
Group No.: 12–001
Molar Mass: 42.08

TABLE 12.1.1. Experimental heat capacities

Reference	Temp. range K	No. pnts	Error %	Purity %	Purity method	Type capacity	Calorimeter Type	Calorimeter Reference
46RUE/POW	153.6–242.8	14	0.20	99.97	melpt	C_p	BSIO	28GIA/WIE1

TABLE 12.1.3. Parameters of regression polynomial

Heat capacity type	No. data points total	No. data points used	s_w	s/R	s_r %	s_b/R	+/–
C_p	14	14	1.115	2.15–2	0.22	7.28–5	–2
C_{sat}	14	14	1.133	2.18–2	0.23	7.51–5	–2

Temp. range K	A_1	A_2	A_3	Level of uncertainty
153.6–242.8	1.28210+1	–4.50403	1.34254	II
153.6–242.8	1.27017+1	–4.36964	1.30505	II

TABLE 12.1.4. Recommended values of heat capacities

Temp. (K)	160	170	180	190	200	210	220
c_p (J K^{-1}g^{-1})	1.788	1.787	1.791	1.800	1.814	1.834	1.859
C_p (J K^{-1}mol^{-1})	75.26	75.20	75.36	75.74	76.35	77.18	78.24
c_{sat} (J K^{-1}g^{-1})	1.788	1.787	1.791	1.800	1.814	1.834	1.858
C_{sat} (J K^{-1}mol^{-1})	75.26	75.20	75.37	75.75	76.35	77.16	78.20

Temp. (K)	230	240
c_p (J K^{-1}g^{-1})	1.890	1.925
C_p (J K^{-1}mol^{-1})	79.52	81.02
c_{sat} (J K^{-1}g^{-1})	1.888	1.923
C_{sat} (J K^{-1}mol^{-1})	79.45	80.91

TABLE 12.1.5. Parameters of quasi-polynomial equation

Heat capacity type	No. data points total	No. data points used	s_w	s/R	s_r %	s_b/R	+/–
C_p	14	13	1.055	2.04–2	0.21	4.42–5	–3

Temp. range K	T_c K	A_1	A_2	A_3	A_4	A_5	A_6	Level of uncertainty
160.4–242.8	398.30	4.32889+2	4.04788+1	–4.66083+1	5.18080+2	–1.88430+2	5.03569+2	II

Name: Cyclobutane
Formula: C_4H_8

CAS-RN: 287–23–0
Group No.: 12–002
Molar Mass: 56.11

TABLE 12.2.1. Experimental heat capacities

Reference	Temp. range K	No. pnts	Error %	Purity %	Purity method	Type capacity	Calorimeter Type	Calorimeter Reference
53RAT/GWI	184.8–284.6	19	0.20	99.81	melpt	C_p	BSIO	28GIA/WIE1

TABLE 12.2.3. Parameters of regression polynomial

Heat capacity type	No. data points total	used	s_w	s/R	s_r %	s_b/R	+/–
C_p	19	19	1.626	3.90–2	0.33	2.06–4	5
C_{sat}	19	19	1.643	3.94–2	0.33	2.10–4	5

Temp. range K	A_1	A_2	A_3	Level of uncertainty
184.8–284.6	1.18155+1	–2.22040	8.95024–1	III
184.8–284.6	1.16660+1	–2.07935	8.62119–1	III

TABLE 12.2.4. Recommended values of heat capacities

Temp. (K)	190	200	210	220	230	240	250
c_p (J K^{-1}g^{-1})	1.605	1.623	1.645	1.669	1.696	1.725	1.757
C_p (J K^{-1}mol^{-1})	90.03	91.08	92.29	93.64	95.15	96.80	98.60
c_{sat} (J K^{-1}g^{-1})	1.605	1.624	1.645	1.669	1.696	1.725	1.757
C_{sat} (J K^{-1}mol^{-1})	90.03	91.09	92.30	93.66	95.15	96.79	98.58

Temp. (K)	260	270	273.15	280	290	298.15
c_p (J K^{-1}g^{-1})	1.792	1.829	1.842	1.869	1.912	1.949
C_p (J K^{-1}mol^{-1})	100.5	102.6	103.3	104.9	107.3	109.3
c_{sat} (J K^{-1}g^{-1})	1.791	1.828	1.840	1.868	1.910	1.946
C_{sat} (J K^{-1}mol^{-1})	100.5	102.6	103.3	104.8	107.1	109.2

TABLE 12.2.5. Parameters of quasi-polynomial equation

Heat capacity type	No. data points total	used	s_w	s/R	s_r %	s_b/R	+/–
C_p	19	19	1.820	4.35–2	0.36	2.63–4	4

Temp. range K	T_c K	A_1	A_2	A_3	A_4	Level of uncertainty
184.8–284.6	460.00	–1.82709–1	2.01495	7.29198	4.14188–3	III

Name: Methylenecyclobutane
Formula: C_5H_8

CAS-RN: 1120–56–5
Group No.: 12–003
Molar Mass: 68.12

TABLE 12.3.1. Experimental heat capacities

Reference		Temp. range K	No. pnts	Error %	Purity %	Purity method	Type capacity	Calorimeter Type	Calorimeter Reference
78LEB/TSV1	N	139.8–301.0	35	0.20	99.50	melpt	C_p	BSAO	76LEB/LIT
81FIN/MES		146.4–300.8	17	0.10	99.65	melpt	C_{sat}	BSAO	43RUE/HUF

78LEB/TSV1 same data in 78LEB/TSV2

TABLE 12.3.2. Correlated heat capacities

Reference	Temp. range K	No. pnts used	$\sigma_r C$ %	d_w	d/R	d_r %	d_b/R	+/–
				Selected data				
78LEB/TSV1	139.8–196.8	17	0.40#	0.238	1.24–2	0.10	3.27–3	2
81FIN/MES	146.4–300.8	17	0.20#	0.215	6.26–3	0.04	–8.05–4	–1

TABLE 12.3.3. Parameters of regression polynomial

Heat capacity type	No. data points total	No. data points used	s_w	s/R	s_r %	s_b/R	+/–
C	52	34	0.241	1.05–2	0.08	1.23–3	1

Temp. range K	A_1	A_2	A_3	A_4	Level of uncertainty
139.8–300.8	1.91024+1	–8.78714	3.60681	–3.47169–1	II

TABLE 12.3.4. Recommended values of heat capacities

Temp. (K)	140	150	160	170	180	190	200
c (J K^{-1}g^{-1})	1.577	1.570	1.569	1.572	1.580	1.592	1.609
C (J K^{-1}mol^{-1})	107.4	107.0	106.9	107.1	107.6	108.5	109.6
Temp. (K)	210	220	230	240	250	260	270
c (J K^{-1}g^{-1})	1.628	1.652	1.678	1.708	1.740	1.774	1.811
C (J K^{-1}mol^{-1})	110.9	112.5	114.3	116.3	118.5	120.9	123.4
Temp. (K)	273.15	280	290	298.15	300		
c (J K^{-1}g^{-1})	1.823	1.850	1.890	1.924	1.932		
C (J K^{-1}mol^{-1})	124.2	126.0	128.8	131.1	131.6		

TABLE 12.3.5. Parameters of quasi-polynomial equation

Heat capacity type	No. data points total	used	s_w	s/R	s_r %	s_b/R	+/−
C	52	23	0.371	1.27–2	0.09	5.65–4	3

Temp. range K	T_c K	A_1	A_2	A_3	A_4	A_5	A_6	Level of uncertainty
161.5–300.8	500.00	1.75463+2	1.46617+1	−3.07278	1.76862+2	−4.12338	1.77725+2	II

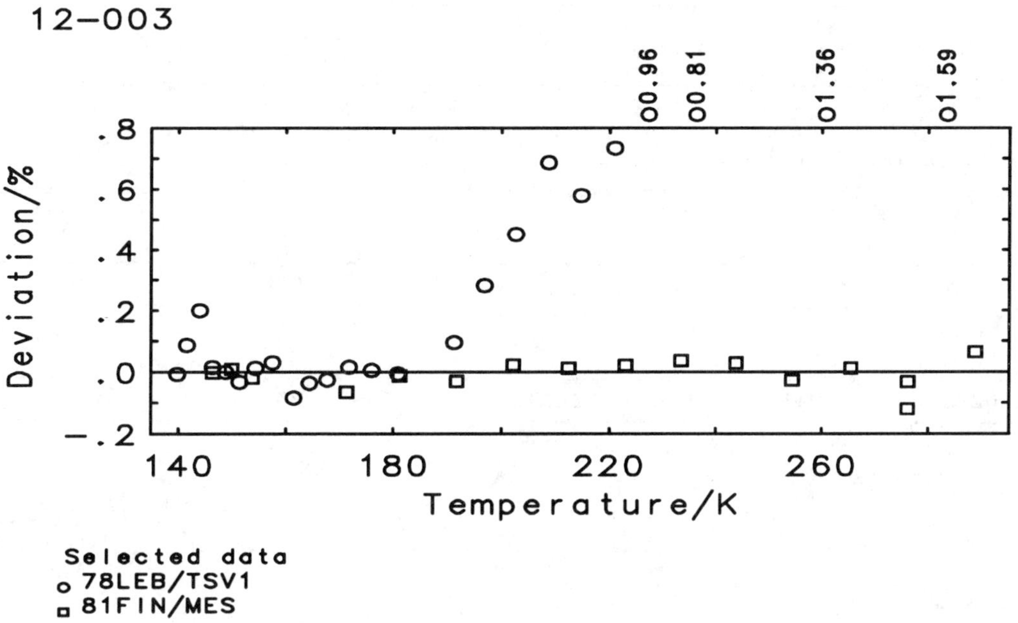

Name: Spiropentane
Formula: C_5H_8

CAS-RN: 157–40–4
Group No.: 12–004
Molar Mass: 68.12

TABLE 12.4.1. Experimental heat capacities

Reference	Temp. range K	No. pnts	Error %	Purity %	Purity method	Type capacity	Calorimeter Type	Calorimeter Reference
50SCO/FIN2	176.4–312.1	18	0.20	99.87	melpt	C_{sat}	BSAO	43RUE/HUF

TABLE 12.4.3. Parameters of regression polynomial

Heat capacity type	No. data points total	used	s_w	s/R	s_r %	s_b/R	+/−
C_{sat}	18	18	0.133	3.90–3	0.03	1.59–7	0

Temp. range K	A_1	A_2	A_3	A_4	Level of uncertainty
176.4–312.1	1.84093+1	−7.98160	3.35441	−3.11378–1	II

TABLE 12.4.4. Recommended values of heat capacities

Temp. (K)	180	190	200	210	220	230	240
c_{sat} (J $K^{-1}g^{-1}$)	1.598	1.613	1.632	1.655	1.681	1.710	1.742
C_{sat} (J $K^{-1}mol^{-1}$)	108.9	109.9	111.2	112.7	114.5	116.5	118.7
Temp. (K)	250	260	270	273.15	280	290	298.15
c_{sat} (J $K^{-1}g^{-1}$)	1.777	1.814	1.853	1.866	1.895	1.938	1.975
C_{sat} (J $K^{-1}mol^{-1}$)	121.0	123.6	126.2	127.1	129.1	132.0	134.5
Temp. (K)	300	310					
c_{sat} (J $K^{-1}g^{-1}$)	1.983	2.029					
C_{sat} (J $K^{-1}mol^{-1}$)	135.1	138.2					

TABLE 12.4.5. Parameters of quasi-polynomial equation

Heat capacity type	No. data points total	used	s_w	s/R	s_r %	s_b/R	+/–
C_{sat}	18	18	0.191	5.62–3	0.04	6.52–6	1

Temp. range K	T_c K	A_1	A_2	A_3	A_4	A_5	A_6	Level of uncertainty
176.4–312.1	512.00	1.45703+2	1.17743+1	4.33588–1	1.42795+2	9.42793	1.43898+2	II

Name: Cyclopentane
Formula: C_5H_{10}

CAS–RN: 287–92–3
Group No.: 12–005
Molar Mass: 70.13

TABLE 12.5.1. Experimental heat capacities

Reference	Temp. range K	No. pnts	Error %	Purity %	method	Type capacity	Calorimeter Type	Reference
34JAC/PAR	186.0–293.7	15	nosp	not specified		C_p	BSIO	25PAR
43AST/FIN	184.1–291.4	19	nosp	99.999	melpt	C_p	BSAO	39AST/EID1
46DOU/HUF1	185.8–300.1	16	0.20	99.995	melpt	C_p	BSAO	43RUE/HUF
47SZA/MOR	180.0–300.0	21S	0.10	not specified		C_p	BSAO	47AST/SZA
75JOL/BOI	298.1	1	0.30	99.9	melpt	C_p	FSIT	71PIC/LED
79FOR/DAR	298.1	1	0.30	not specified		C_p	FSIT	71PIC/LED
83SID/SVE	293.1	1	nosp	99.9	chrom	C_p	FSIT	71PIC/LED
85TAN	298.1	1	0.30	not specified		C_p	FSIT	71PIC/LED

TABLE 12.5.2. Correlated heat capacities

Reference	Temp. range K	No. pnts used	$\sigma_r C$ %	d_w	d/R	d_r %	d_b/R	+/–
				Selected data				
46DOU/HUF1	185.8–300.1	16	0.20#	0.606	1.55–2	0.12	9.98–3	8
47SZA/MOR	180.0–300.0	21	0.10	0.716	9.22–3	0.07	–1.89–3	–2
79FOR/DAR	298.1	1	0.30#	0.182	8.33–3	0.05	8.33–3	1
85TAN	298.1	1	0.30	0.160	7.30–3	0.05	–7.30–3	–1
				Rejected data				
34JAC/PAR	(9.27–2, 0.71, –4.29–2, –7)			43AST/FIN	(1.01–1, 0.72, 7.70–2, 11)			
75JOL/BOI	(5.76–2, 0.38, 5.76–2, 1)			83SID/SVE	(1.21–1, 0.80, 1.21–1, 1)			

TABLE 12.5.3. Parameters of regression polynomial

Heat capacity type	No. data points total	used	s_w	s/R	s_r %	s_b/R	+/–
C_p	75	39	0.691	1.28–2	0.10	3.11–3	6

Temp. range K	A_1	A_2	A_3	A_4	Level of uncertainty
180.0–300.1	1.79538+1	–8.74268	3.62016	–3.32700–1	II

TABLE 12.5.4. Recommended values of heat capacities

Temp. (K)	180	190	200	210	220	230	240
c_p (J K^{-1}g^{-1})	1.423	1.438	1.457	1.479	1.505	1.535	1.568
C_p (J K^{-1}mol^{-1})	99.82	100.9	102.2	103.7	105.6	107.7	110.0
Temp. (K)	250	260	270	273.15	280	290	298.15
c_p (J K^{-1}g^{-1})	1.603	1.642	1.682	1.696	1.725	1.770	1.808
C_p (J K^{-1}mol^{-1})	112.5	115.1	118.0	118.9	121.0	124.1	126.8
Temp. (K)	300						
c_p (J K^{-1}g^{-1})	1.817						
C_p (J K^{-1}mol^{-1})	127.4						

TABLE 12.5.5. Parameters of quasi-polynomial equation

Heat capacity type	No. data points total	used	s_w	s/R	s_r %	s_b/R	+/–
C_p	75	39	0.837	1.27–2	0.10	–3.30–4	0

Temp. range K	T_c K	A_1	A_2	A_3	A_4	A_5	A_6	Level of uncertainty
180.0–300.1	511.70	1.51777+2	1.21864+1	–1.13315	1.49227+2	8.25525	1.51974+2	II

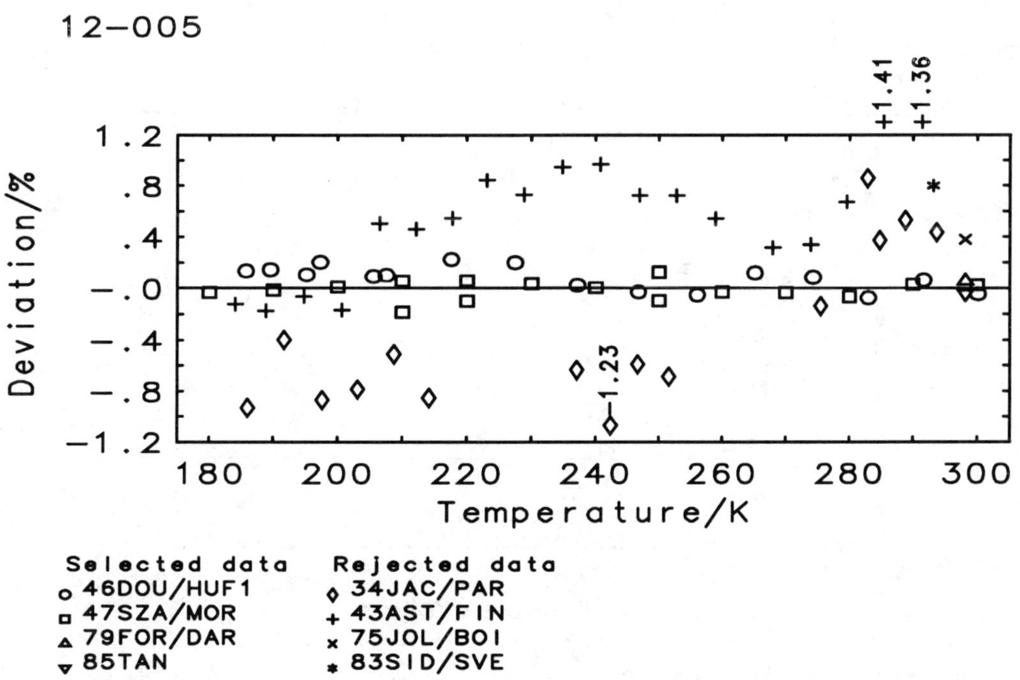

Name: Cyclohexane–d_{12}
Formula: C_6D_{12}

CAS–RN: 1735–17–7
Group No.: 12–006
Molar Mass: 96.23

TABLE 12.6.1. Experimental heat capacities

Reference		Temp. range K	No. pnts	Error %	Purity %	Purity method	Type capacity	Calorimeter Type	Calorimeter Reference
66NIK/RAB	N	283.1–333.1	11S	0.30	not specified		C_p	BSAO	47SKU
80MRA/NAA	N	285.0–320.0	9	1.00	not specified		C_p	BDHT	69PER/COM

66NIK/RAB deuteration: 97 mol.%
80MRA/NAA deuterated to 99.5 mol.%, content of impurities below 0.3 mol.%

TABLE 12.6.2. Correlated heat capacities

Reference	Temp. range K	No. pnts used	$\sigma_r C$ %	d_w	d/R	d_r %	d_b/R	+/–
				Selected data				
66NIK/RAB	283.1–333.1	11	0.30	0.223	1.52–2	0.07	1.92–5	1
				Rejected data				
80MRA/NAA	(4.95–1, 2.21, 3.58–1, 5)							

TABLE 12.6.3. Parameters of regression polynomial

Heat capacity type	No. data points total	No. data points used	s_w	s/R	s_r %	s_b/R	+/–
C_p	20	11	0.261	1.79–2	0.08	1.92–5	1

Temp. range K	A_1	A_2	A_3	Level of uncertainty
283.1–333.1	5.13584+1	–2.79697+1	6.08981	IV

TABLE 12.6.4. Recommended values of heat capacities

Temp. (K)	280	290	298.15	300	310	320	330
c_p (J K^{-1}g^{-1})	1.796	1.854	1.910	1.923	2.002	2.092	2.192
C_p (J K^{-1}mol^{-1})	172.8	178.4	183.8	185.1	192.7	201.3	211.0

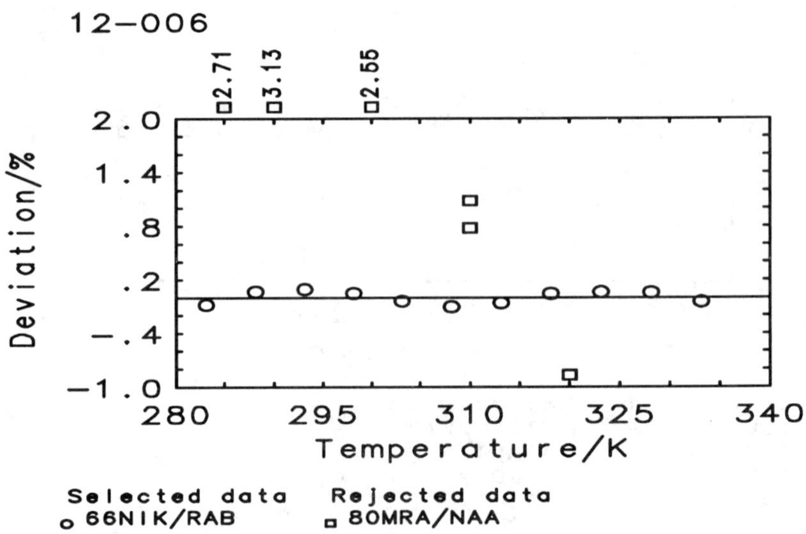

Name: Cyclohexane
Formula: C_6H_{12}

CAS–RN: 110–82–7
Group No.: 12–007
Molar Mass: 84.16

TABLE 12.7.1. Experimental heat capacities

Reference		Temp. range K	No. pnts	Error %	Purity %	Purity method	Type capacity	Calorimeter Type	Calorimeter Reference
19DEJ		295.6–323.3	10	nosp	not specified		C_p	BSIO	19DEJ
30PAR/HUF2		283.1–298.9	5	0.30	not specified		C_p	BSIO	25PAR
39PHI		304.1	1	nosp	not specified		C_p	BSIO	49WEI
43AST/SZA		281.6–292.5	4	nosp	99.905	melpt	C_p	BSAO	39AST/EID1
43RUE/HUF		282.3–301.3	7	0.20	99.985	melpt	C_p	BSAO	43RUE/HUF
48TSC2		292.6	1	nosp	not specified		C_p	BSIO	48TSC1
48TSC6		293.6	1	nosp	not specified		C_p	BSIO	48TSC1
50AUE/SAG		299.8–366.5	13	nosp	99.9	melpt	C_p	BSAO	39SAG/EVA
60SWI/ZIE	N	311.6–322.2	2	nosp	not specified		C_{avg}	DSIO	58SWI/ZIE1
61ROU		300.4–315.8	7	nosp	99.94	anal	C_p	BSAO	61ROU
64MOE/THO		298.0–327.7	4	0.50	not specified		C_p	BSIO	64MOE/THO
66KLE		293.1–343.1	11S	0.10	not specified		C_p	BSAO	66KLE
66NIK/RAB		283.1–333.1	11S	0.30	not specified		C_p	BSAO	47SKU
68REC1	N	298.0–313.0	eqn	nosp	not specified		C_p	BSAO	68REC1
69WIL/SCH		293.1–313.1	3	0.40	not specified		C_p	BDAO	65FIN/GRU
73SUB/RAJ		298.1–323.1	3	0.30	not specified		C_p	BSIO	64MOE/THO
74WIL/ZET		283.1–323.1	5	nosp	not specified		C_p	BSAO	64ARN1
75JOL/BOI		298.1	1	0.30	99.9	melpt	C_p	FSIT	71PIC/LED
75SAN	N	398.1–523.2	9S	1.00	not specified		C_p	FSIO	75SAN
78GRO/WIL		298.1	1	0.30	not specified		C_p	FSIT	71PIC/LED
78SAF		299.1–311.7	7	nosp	not specified		C_p	BSAO	67RAS/GAN
78SAF		299.1–311.7	7	nosp	not specified		C_p	FSIO	75SAF/GER
78SAF		299.1–311.7	7	nosp	not specified		C_p	BSAO	67RAS/GAN
79FOR/DAR		298.1	1	0.30	not specified		C_p	FSIT	71PIC/LED
79MAR/BRA		298.1	1	1.00	not specified		C_p	BDCT	70PAZ/PAZ
79WIL/GRO		298.1	1	0.30	not specified		C_p	FSIT	71PIC/LED
82GRO/ING		298.1	1	0.30	99.5	melpt	C_p	FSIT	71PIC/LED
82TAN		293.1–303.1	3S	0.30	99.99	chrom	C_p	FSIT	71PIC/LED
83SID/SVE		293.1	1	0.30	99.9	chrom	C_p	FSIT	71PIC/LED
85NKI/CHA		298.0	1	0.30	not specified		C_p	FSIT	71PIC/LED
85TAN/NAK		283.1–318.1	3	0.30	100.0	chrom	C_p	FSIT	71PIC/LED
86JIM/ROM		298.1	1	nosp	99.5	anal	C_p	FSIT	71PIC/LED
86ORT		298.1	1	1.00	99.5	anal	C_p	BDCT	70PAZ/PAZ
87KAL/KOH		293.1–313.1	2	nosp	not specified		C_p	FSIT	71PIC/LED
88SHI/OGA2		298.1	1	nosp	99.9	chrom	C_p	FSIO	85OGA
89JIM/ROU		298.1	1	nosp	not specified		C_p	FSIT	71PIC/LED
89LAI/ROD		298.1	1	nosp	not specified		C_p	FSIT	71PIC/LED
90JAL/ROB		293.0	1	5.00	99.	anal	C_p	BDHT	87PER/COM
91OGA/MIT		298.1	1	nosp	not specified		C_p	FSIO	85OGA
91TRE/COS		298.1	1	nosp	99.98	anal	C_p	FSIT	71PIC/LED

60SWI/ZIE average values in temperature ranges 294–329 K and 294–350 K
68REC1 same data in 68REC2
75SAN same data in 76SAN/MEL; C_p at the saturation pressure extrapolated from high pressure measurement

TABLE 12.7.2. Correlated heat capacities

Reference	Temp. range K	No. pnts used	$\sigma_r C$ %	d_w	d/R	d_r %	d_b/R	+/–
				Selected data				
43RUE/HUF	282.3–301.3	7	0.20#	0.224	8.35–3	0.04	5.52–3	4
50AUE/SAG	322.0–366.5	9	1.00#	1.154	2.41–1	1.15	–2.40–1	–9
64MOE/THO	298.0–327.7	4	0.50	0.604	5.67–2	0.30	–3.28–2	0
66KLE	293.1–343.1	11	0.10	0.502	9.91–3	0.05	–3.39–3	–3
74WIL/ZET	283.1–323.1	5	0.35#	0.930	6.25–2	0.33	6.20–2	5
75JOL/BOI	298.1	1	0.30	0.068	3.82–3	0.02	–3.82–3	–1
75SAN	398.1–523.2	9	1.00	1.266	4.50–1	1.27	1.17–2	–1
78GRO/WIL	298.1	1	0.30	0.252	1.42–2	0.08	1.42–2	1
78SAF	299.1–311.7	7	0.30	0.959	5.51–2	0.29	5.50–2	7
78SAF	299.1–311.7	7	0.70#	0.434	5.82–2	0.30	5.78–2	7
79FOR/DAR	298.1	1	0.30	0.175	9.84–3	0.05	–9.84–3	–1
79WIL/GRO	298.1	1	0.30	0.380	2.14–2	0.11	2.14–2	1
82GRO/ING	298.1	1	0.30	0.465	2.62–2	0.14	2.62–2	1
82TAN	293.1–303.1	3	0.30	0.494	2.77–2	0.15	–2.68–2	–3
85TAN/NAK	283.1–318.1	3	0.30	0.695	3.80–2	0.21	–3.27–2	–3
91OGA/MIT	298.1	1	0.30#	0.474	2.67–2	0.14	–2.67–2	–1
91TRE/COS	298.1	1	0.30#	0.495	2.79–2	0.15	–2.79–2	–1

Rejected data

19DEJ	(2.48, 11.37, 2.47, 10)	30PAR/HUF2	(2.42–1, 1.34, –2.39–1, –5)
39PHI	(7.00, 57.96, –7.00, –1)	43AST/SZA	(1.76–1, 0.95, 1.75–1, 3)
48TSC2	(7.26–2, 0.39, –7.26–2, –1)	48TSC6	(2.03–1, 1.11, –2.03–1, –1)
60SWI/ZIE	(7.29–1, 3.85, –7.29–1, –2)	61ROU	(2.05–1, 1.05, 2.05–1, 7)
66NIK/RAB	(3.33–1, 1.82, –2.05–1, –5)	68REC1	(1.53–1, 0.80, –1.46–1, –4)
69WIL/SCH	(1.42–1, 0.75, –1.39–1, –3)	73SUB/RAJ	(2.59–1, 1.34, 2.03–1, 3)
78SAF	(9.00–2, 0.47, 8.97–2, 7)	79MAR/BRA	(7.43–2, 0.40, –7.43–2, –1)
83SID/SVE	(5.85–2, 0.31, 5.85–2, 1)	85NKI/CHA	(4.39–2, 0.23, 4.39–2, 1)
86JIM/ROM	(9.96–2, 0.53, 9.96–2, 1)	86ORT	(4.59–2, 0.25, –4.59–2, –1)
87KAL/KOH	(1.99–1, 1.01, 1.81–1, 2)	88SHI/OGA2	(2.30–1, 1.24, –2.30–1, –1)
89JIM/ROU	(1.01–1, 0.53, 1.01–1, 1)	89LAI/ROD	(8.04–2, 0.43, 8.04–2, 1)
90JAL/ROB	(2.33–1, 1.27, –2.33–1, –1)		

TABLE 12.7.3. Parameters of cubic spline polynomials

Heat capacity type	No. data points total	used	s_w	s/R	s_r %	s_b/R	+/–
C_p	146	72	0.837	1.92–1	0.66	–1.76–2	3
C_{sat}	146	72	0.821	1.78–1	0.64	–1.76–2	2

Temp. range K	A_1	A_2	A_3	A_4	Level of uncertainty
282.3–360.0	–2.05560+1	3.05528+1	–8.75552	9.84165–1	II
360.0–440.0	1.01668+2	–7.13008+1	1.95371+1	–1.63553	IV
440.0–523.2	–1.26591+3	8.61138+2	–1.92381+2	1.44189+1	V
282.3–360.0	–2.09531+1	3.07362+1	–8.74351	9.74511–1	II
360.0–440.0	1.03511+2	–7.29843+1	2.00677+1	–1.69320	IV
440.0–523.2	–1.06245+3	7.21990+2	–1.60608+2	1.19944+1	V

TABLE 12.7.4. Recommended values of heat capacities

Temp. (K)	280	290	298.15	300	310	320	330
c_p (J K^{-1}g^{-1})	1.774	1.819	1.856	1.865	1.910	1.957	2.004
C_p (J K^{-1}mol^{-1})	149.3	153.1	156.2	156.9	160.8	164.7	168.7
c_{sat} (J K^{-1}g^{-1})	1.773	1.819	1.856	1.865	1.910	1.956	2.004
C_{sat} (J K^{-1}mol^{-1})	149.3	153.1	156.2	156.9	160.8	164.6	168.6
Temp. (K)	340	350	360	370	380	390	400
c_p (J K^{-1}g^{-1})	2.054	2.106	2.162	2.220	2.282	2.345	2.409
C_p (J K^{-1}mol^{-1})	172.9	177.3	181.9	186.9	192.0	197.4	202.8
c_{sat} (J K^{-1}g^{-1})	2.053	2.104	2.158	2.216	2.276	2.338	2.400
C_{sat} (J K^{-1}mol^{-1})	172.8	177.1	181.7	186.5	191.6	196.8	202.0
Temp. (K)	410	420	430	440	450	460	470
c_p (J K^{-1}g^{-1})	2.473	2.536	2.596	2.654	2.71	2.77	2.84
C_p (J K^{-1}mol^{-1})	208.1	213.4	218.5	223.4	228	233	239
c_{sat} (J K^{-1}g^{-1})	2.462	2.522	2.579	2.634	2.68	2.74	2.80
C_{sat} (J K^{-1}mol^{-1})	207.2	212.2	217.1	221.7	226	230	236
Temp. (K)	480	490	500	510	520		
c_p (J K^{-1}g^{-1})	2.94	3.06	3.22	3.43	3.70		
C_p (J K^{-1}mol^{-1})	247	258	271	289	311		
c_{sat} (J K^{-1}g^{-1})	2.88	2.99	3.12	3.29	3.51		
C_{sat} (J K^{-1}mol^{-1})	243	251	263	277	296		

TABLE 12.7.5. Parameters of quasi-polynomial equation

Heat capacity type	No. data points total	used	s_w	s/R	s_r %	s_b/R	+/–
C_p	146	72	1.386	1.94–1	0.65	–9.47–3	–7

Temp. range K	T_c K	A_1	A_2	A_3	A_4	Level of uncertainty
282.3–523.2	553.50	–4.19042	3.31133–1	7.71652	1.32572+1	V

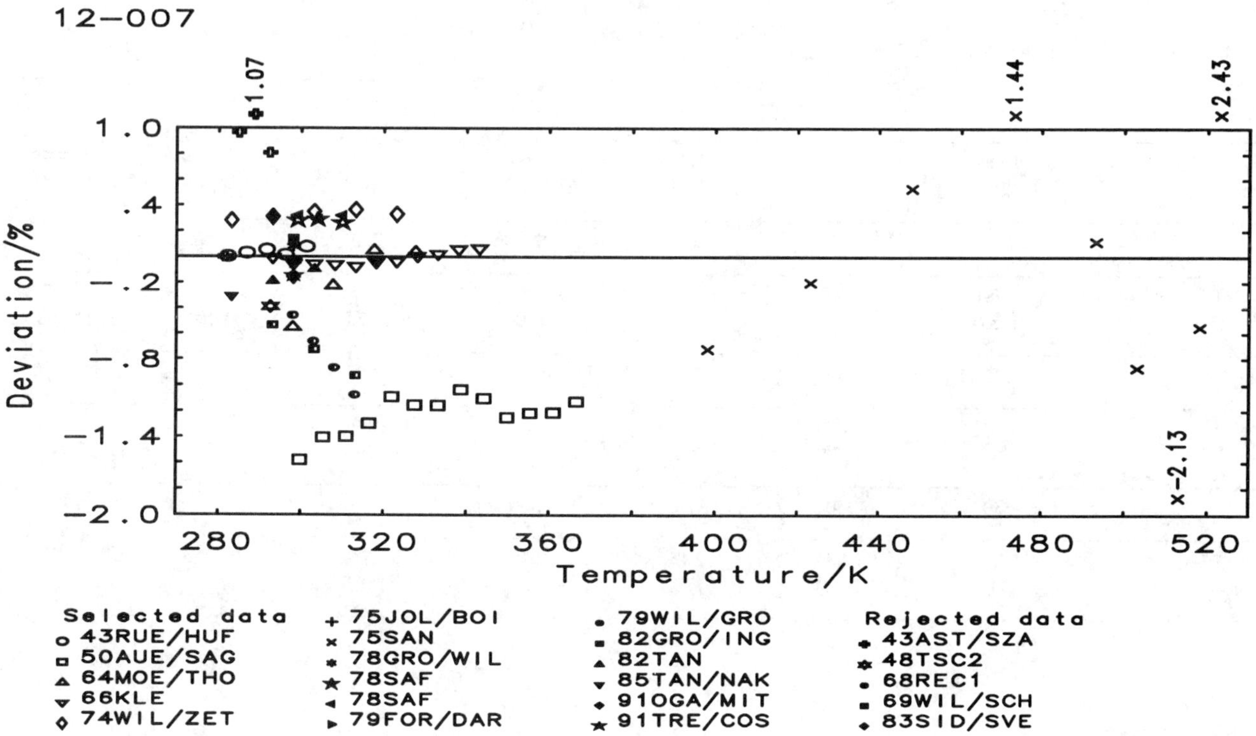

Name: Methylcyclopentane
Formula: C_6H_{12}

CAS–RN: 96–37–7
Group No.: 12–008
Molar Mass: 84.16

TABLE 12.8.1. Experimental heat capacities

Reference	Temp. range K	No. pnts	Error %	Purity %	Purity method	Type capacity	Calorimeter Type	Calorimeter Reference
31HUF/PAR	139.0–293.7	8	1.00	not specified		C_p	BSIO	25PAR
46DOU/HUF1	136.1–307.5	31	0.20	99.907	melpt	C_p	BSAO	43RUE/HUF
51CON/SAG	299.8–366.5	13S	0.70	99.9	estim	C_p	BSAO	39SAG/EVA
83SID/SVE	293.1	1	0.30	99.9	melpt	C_p	FSIT	71PIC/LED

TABLE 12.8.2. Correlated heat capacities

Reference	Temp. range K	No. pnts used	$\sigma_r C$ %	d_w	d/R	d_r %	d_b/R	+/–
Selected data								
46DOU/HUF1	136.1–307.5	31	0.20#	0.416	1.44–2	0.08	9.70–5	–4
51CON/SAG	299.8–366.5	13	0.70	0.187	2.83–2	0.13	–2.06–3	–10
Rejected data								
31HUF/PAR	(1.14–1, 0.71, –9.83–2, –6)			83SID/SVE	(9.69–2, 0.51, 9.69–2, 1)			

TABLE 12.8.3. Parameters of regression polynomial

Heat capacity type	No. data points total	No. data points used	s_w	s/R	s_r %	s_b/R	+/–
C_p	53	44	0.381	2.06–2	0.10	–5.40–4	–14

Temp. range K	A_1	A_2	A_3	A_4	Level of uncertainty
136.1–366.5	1.86504+1	–5.70883	2.48334	–1.74291–1	III

TABLE 12.8.4. Recommended values of heat capacities

Temp. (K)	140	150	160	170	180	190	200
c_p (J K^{-1}g^{-1})	1.487	1.490	1.498	1.508	1.522	1.538	1.558
C_p (J K^{-1}mol^{-1})	125.1	125.4	126.0	126.9	128.1	129.5	131.1
Temp. (K)	210	220	230	240	250	260	270
c_p (J K^{-1}g^{-1})	1.581	1.606	1.634	1.664	1.697	1.732	1.769
C_p (J K^{-1}mol^{-1})	133.0	135.1	137.5	140.0	142.8	145.8	148.9
Temp. (K)	273.15	280	290	298.15	300	310	320
c_p (J K^{-1}g^{-1})	1.782	1.809	1.850	1.885	1.894	1.939	1.986
C_p (J K^{-1}mol^{-1})	149.9	152.2	155.7	158.7	159.4	163.2	167.1
Temp. (K)	330	340	350	360	370		
c_p (J K^{-1}g^{-1})	2.034	2.084	2.136	2.188	2.242		
C_p (J K^{-1}mol^{-1})	171.2	175.4	179.7	184.2	188.7		

TABLE 12.8.5. Parameters of quasi-polynomial equation

Heat capacity type	No. data points total	used	s_w	s/R	s_r %	s_b/R	+/–
C_p	53	44	0.681	4.50–2	0.22	2.29–3	–8

Temp. range K	T_c K	A_1	A_2	A_3	A_4	A_5	A_6	Level of uncertainty
136.1–366.5	532.73	1.07170+2	7.53370	7.13094	1.04425+2	1.05510+1	1.19907+2	III

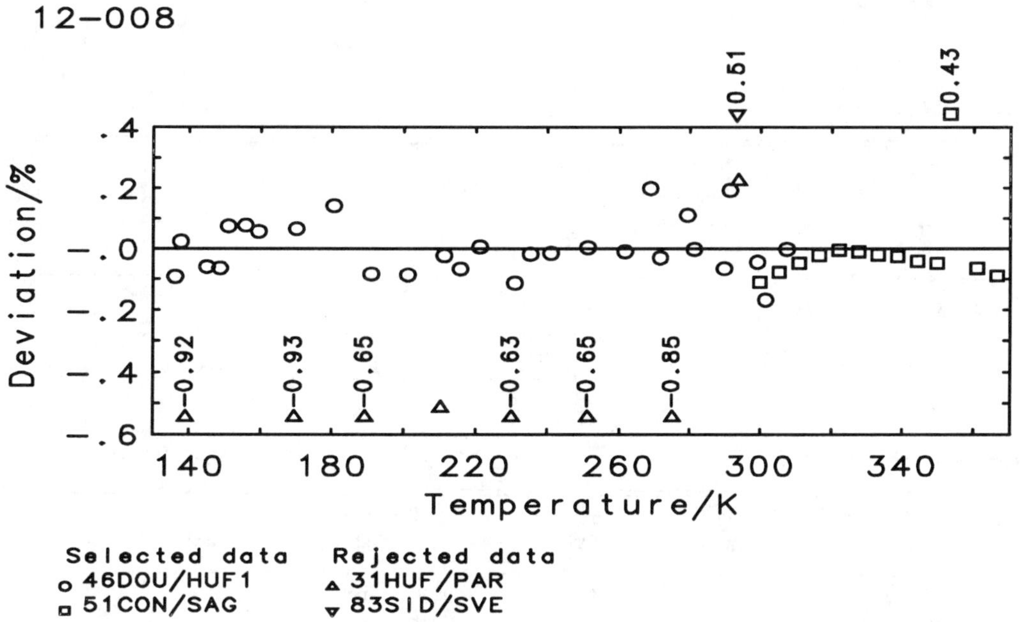

Name: Tetracyclo[3.2.0.0^{2,7}.0^{4,6}]heptane
Formula: C_7H_8

CAS-RN: 278–06–8
Group No.: 12–009
Molar Mass: 92.14

TABLE 12.9.1. Experimental heat capacities

Reference		Temp. K	Capac. J/(K.g)	Error %	Purity %	method	Type capacity	Calorimeter Type	Reference
73HAL/SMI	N	297.15	1.510	nosp	99.9	chrom	C_p	BDHT	71DU/COM

73HAL/SMI same data in 88CZA1 and 88CZA2

Name: Bicyclo[4.1.0]heptane
Formula: C_7H_{12}

CAS–RN: 286–08–8
Group No.: 12–010
Molar Mass: 96.17

TABLE 12.10.1. Experimental heat capacities

Reference	Temp. K	Capac. J/(K.g)	Error %	Purity %	method	Type capacity	Calorimeter Type	Reference
70CHA/MCN	315.00	1.953	2.00	not specified		C_p	BDHT	73PER/COM

Name: Ethylidenecyclopentane
Formula: C_7H_{12}

CAS–RN: 2146–37–4
Group No.: 12–011
Molar Mass: 96.17

TABLE 12.11.1. Experimental heat capacities

Reference	Temp. K	Capac. J/(K.g)	Error %	Purity %	Purity method	Type capacity	Calorimeter Type	Calorimeter Reference
79FUC/PEA	298.15	1.884	1.50	99.	chrom	C_p	BSIO	80FUC

Name: Methylenecyclohexane
Formula: C_7H_{12}

CAS–RN: 1192–37–6
Group No.: 12–012
Molar Mass: 96.17

TABLE 12.12.1. Experimental heat capacities

Reference	Temp. K	Capac. J/(K.g)	Error %	Purity %	Purity method	Type capacity	Calorimeter Type	Calorimeter Reference
79FUC/PEA	298.15	1.845	1.50	99.	chrom	C_p	BSIO	80FUC

Name: Cycloheptane
Formula: C_7H_{14}

CAS–RN: 291–64–5
Group No.: 12–013
Molar Mass: 98.19

TABLE 12.13.1. Experimental heat capacities

Reference	Temp. range K	No. pnts	Error %	Purity %	Purity method	Type capacity	Calorimeter Type	Calorimeter Reference
56FIN/SCO	269.3–300.4	7	0.20	99.951	melpt	C_{sat}	BSAO	43RUE/HUF
75JOL/BOI	298.1	1	0.30	99.0	melpt	C_p	FSIT	71PIC/LED
79FOR/DAR	298.1	1	0.30	not specified		C_p	FSIT	71PIC/LED

TABLE 12.13.2. Correlated heat capacities

Reference	Temp. range K	No. pnts used	$\sigma_r C$ %	d_w	d/R	d_r %	d_b/R	+/–
				Selected data				
56FIN/SCO	269.3–300.4	7	0.20#	0.105	4.43–3	0.02	6.97–4	1
79FOR/DAR	298.1	1	0.30	0.168	1.09–2	0.05	–1.09–2	–1
				Rejected data				
75JOL/BOI	(2.78–2, 0.13, –2.78–2, –1)							

TABLE 12.13.3. Parameters of regression polynomial

Heat capacity type	No. data points total	No. data points used	s_w	s/R	s_r %	s_b/R	+/–
C	9	8	0.133	6.54–3	0.03	–7.55–4	0

Temp. range K	A_1	A_2	Level of uncertainty
269.3–300.4	8.12315	4.56483	II

TABLE 12.13.4. Recommended values of heat capacities

Temp. (K)	270	273.15	280	290	298.15	300
c (J $K^{-1}g^{-1}$)	1.732	1.744	1.770	1.809	1.840	1.848
C (J $K^{-1}mol^{-1}$)	170.0	171.2	173.8	177.6	180.7	181.4

Name: 1,1–Dimethylcyclopentane
Formula: C_7H_{14}

CAS–RN: 1638–26–2
Group No.: 12–014
Molar Mass: 98.19

TABLE 12.14.1. Experimental heat capacities

Reference	Temp. range K	No. pnts	Error %	Purity %	Purity method	Type capacity	Calorimeter Type	Calorimeter Reference
53GRO/OLI	205.9–299.8	14	0.20	99.998	melpt	C_{sat}	BSAO	43RUE/HUF

TABLE 12.14.3. Parameters of regression polynomial

Heat capacity type	No. data points total	used	s_w	s/R	s_r %	s_b/R	+/–
C_{sat}	14	14	0.307	1.26–2	0.06	7.49–6	–2

Temp. range K	A_1	A_2	A_3	A_4	Level of uncertainty
205.9–299.8	2.95924+1	–1.80455+1	8.11618	–9.61125–1	II

TABLE 12.14.4. Recommended values of heat capacities

Temp. (K)	210	220	230	240	250	260	270
c_{sat} (J $K^{-1}g^{-1}$)	1.574	1.604	1.637	1.672	1.709	1.748	1.788
C_{sat} (J $K^{-1}mol^{-1}$)	154.6	157.5	160.7	164.2	167.8	171.7	175.6
Temp. (K)	273.15	280	290	298.15	300		
c_{sat} (J $K^{-1}g^{-1}$)	1.801	1.829	1.869	1.902	1.910		
C_{sat} (J $K^{-1}mol^{-1}$)	176.8	179.6	183.6	186.8	187.5		

TABLE 12.14.5. Parameters of quasi–polynomial equation

Heat capacity type	No. data points total	used	s_w	s/R	s_r %	s_b/R	+/–
C_{sat}	14	14	0.705	2.94–2	0.14	6.53–5	0

Temp. range K	T_c K	A_1	A_2	A_3	A_4	Level of uncertainty
205.9–299.8	547.00	–6.41924	1.96454	1.02632+1	5.24382	II

Name: *cis*–1,2–Dimethylcyclopentane
Formula: C_7H_{14}

CAS–RN: 1192–18–3
Group No.: 12–015
Molar Mass: 98.19

TABLE 12.15.1. Experimental heat capacities

Reference	Temp. range K	No. pnts	Error %	Purity %	Purity method	Type capacity	Calorimeter Type	Calorimeter Reference
53GRO/OLI	223.1–302.8	14	0.20	99.992	melpt	C_{sat}	BSAO	43RUE/HUF

TABLE 12.15.3. Parameters of regression polynomial

Heat capacity type	No. data points total	No. data points used	s_w	s/R	s_r %	s_b/R	+/–
C_{sat}	14	14	0.220	9.25–3	0.04	5.59–6	–4

Temp. range K	A_1	A_2	A_3	Level of uncertainty
223.1–302.8	1.51102+1	2.57996–1	7.68422–1	II

TABLE 12.15.4. Recommended values of heat capacities

Temp. (K)	230	240	250	260	270	273.15	280
c_{sat} (J K^{-1}g^{-1})	1.674	1.707	1.741	1.776	1.813	1.825	1.851
C_{sat} (J K^{-1}mol^{-1})	164.4	167.6	170.9	174.4	178.0	179.2	181.7
Temp. (K)	290	298.15	300				
c_{sat} (J K^{-1}g^{-1})	1.890	1.923	1.931				
C_{sat} (J K^{-1}mol^{-1})	185.6	188.8	189.6				

TABLE 12.15.5. Parameters of quasi–polynomial equation

Heat capacity type	No. data points total	No. data points used	s_w	s/R	s_r %	s_b/R	+/–
C_{sat}	14	14	0.276	1.16–2	0.06	9.94–6	–3

Temp. range K	T_c K	A_1	A_2	A_3	A_4	Level of uncertainty
223.1–302.8	565.00	–6.42470	1.97572	1.09516+1	5.22301	II

Name: *trans*–1,2–Dimethylcyclopentane
Formula: C_7H_{14}

CAS–RN: 822–50–4
Group No.: 12–016
Molar Mass: 98.19

TABLE 12.16.1. Experimental heat capacities

Reference		Temp. range K	No. pnts	Error %	Purity %	Purity method	Type capacity	Calorimeter Type	Calorimeter Reference
31HUF/PAR	N	161.5–294.2	8	1.00	not specified		C_p	BSIO	25PAR

31HUF/PAR unspecified stereoisomer or mixture; authors supposed *trans*–isomer (according to n.m.t. and n.b.t.)

TABLE 12.16.3. Parameters of regression polynomial

Heat capacity type	No. data points total	used	s_w	s/R	s_r %	s_b/R	+/–
C_p	8	8	0.349	6.98–2	0.35	3.28–4	–2
Temp. range K		A_1	A_2	A_3			**Level of uncertainty**
161.5–294.2		1.54912+1	5.64126–1	6.14681–1			V

TABLE 12.16.4. Recommended values of heat capacities

Temp. (K)	170	180	190	200	210	220	230
c_p (J K^{-1}g^{-1})	1.54	1.57	1.59	1.62	1.64	1.67	1.70
C_p (J K^{-1}mol^{-1})	152	154	156	159	161	164	167
Temp. (K)	240	250	260	270	273.15	280	290
c_p (J K^{-1}g^{-1})	1.73	1.76	1.79	1.82	1.83	1.85	1.89
C_p (J K^{-1}mol^{-1})	169	172	176	179	180	182	185
Temp. (K)	298.15	300					
c_p (J K^{-1}g^{-1})	1.92	1.92					
C_p (J K^{-1}mol^{-1})	188	189					

TABLE 12.16.5. Parameters of quasi–polynomial equation

Heat capacity type	No. data points total	used	s_w	s/R	s_r %	s_b/R	+/–
C_p	8	8	0.305	6.00–2	0.31	2.31–4	–2
Temp. range K	T_c K	A_1	A_2	A_3	A_4		**Level of uncertainty**
161.5–294.2	553.00	–5.05442	2.44579	1.20389+1	2.61135		V

Name: *trans*–1,3–Dimethylcyclopentane
Formula: C_7H_{14}

CAS–RN: 1759–58–6
Group No.: 12–017
Molar Mass: 98.19

TABLE 12.17.1. Experimental heat capacities

Reference	Temp. range K	No. pnts	Error %	Purity %	Purity method	Type capacity	Calorimeter Type	Calorimeter Reference
53GRO/OLI	143.7–304.0	22	0.20	99.98	melpt	C_{sat}	BSAO	43RUE/HUF

TABLE 12.17.3. Parameters of regression polynomial

Heat capacity type	No. data points total	used	s_w	s/R	s_r %	s_b/R	+/–
C_{sat}	22	22	0.185	7.42–3	0.04	4.16–6	4
Temp. range K		A_1	A_2	A_3	A_4		**Level of uncertainty**
143.7–304.0		1.78458+1	–2.42108	1.56831	–7.17062–2		II

TABLE 12.17.4. Recommended values of heat capacities

Temp. (K)	150	160	170	180	190	200	210
c_{sat} ($J K^{-1} g^{-1}$)	1.482	1.498	1.517	1.537	1.559	1.584	1.610
C_{sat} ($J K^{-1} mol^{-1}$)	145.5	147.1	148.9	150.9	153.1	155.5	158.1
Temp. (K)	220	230	240	250	260	270	273.15
c_{sat} ($J K^{-1} g^{-1}$)	1.638	1.668	1.700	1.734	1.769	1.806	1.818
C_{sat} ($J K^{-1} mol^{-1}$)	160.9	163.8	166.9	170.2	173.7	177.4	178.5
Temp. (K)	280	290	298.15	300			
c_{sat} ($J K^{-1} g^{-1}$)	1.845	1.885	1.920	1.927			
C_{sat} ($J K^{-1} mol^{-1}$)	181.2	185.1	188.5	189.2			

TABLE 12.17.5. Parameters of quasi-polynomial equation

Heat capacity type	No. data points total	used	s_w	s/R	s_r %	s_b/R	+/–
C_{sat}	22	22	1.029	4.14–2	0.21	1.25–4	0

Temp. range K	T_c K	A_1	A_2	A_3	A_4	Level of uncertainty
143.7–304.0	553.00	2.40978–1	6.69964	8.35847	2.16693–3	II

Name: Ethylcyclopentane
Formula: C_7H_{14}

CAS–RN: 1640–89–7
Group No.: 12–018
Molar Mass: 98.19

TABLE 12.18.1. Experimental heat capacities

Reference	Temp. range K	No. pnts	Error %	Purity %	Purity method	Type capacity	Calorimeter Type	Calorimeter Reference
53GRO/OLI	140.2–301.8	21	0.20	99.97	melpt	C_{sat}	BSAO	43RUE/HUF

TABLE 12.18.3. Parameters of regression polynomial

Heat capacity type	No. data points total	used	s_w	s/R	s_r %	s_b/R	+/–
C_{sat}	21	21	0.171	6.66–3	0.03	1.54–6	–1

Temp. range K	A_1	A_2	A_3	A_4	Level of uncertainty
140.2–301.8	2.12846+1	–5.96739	2.67319	–1.83805–1	II

TABLE 12.18.4. Recommended values of heat capacities

Temp. (K)	140	150	160	170	180	190	200
c_{sat} (J K^{-1}g^{-1})	1.496	1.501	1.510	1.521	1.535	1.553	1.573
C_{sat} (J K^{-1}mol^{-1})	146.9	147.4	148.2	149.3	150.8	152.5	154.4
Temp. (K)	210	220	230	240	250	260	270
c_{sat} (J K^{-1}g^{-1})	1.595	1.621	1.648	1.678	1.711	1.745	1.782
C_{sat} (J K^{-1}mol^{-1})	156.6	159.1	161.8	164.8	168.0	171.4	175.0
Temp. (K)	273.15	280	290	298.15	300		
c_{sat} (J K^{-1}g^{-1})	1.794	1.821	1.861	1.895	1.903		
C_{sat} (J K^{-1}mol^{-1})	176.1	178.8	182.7	186.1	186.9		

TABLE 12.18.5. Parameters of quasi-polynomial equation

Heat capacity type	No. data points total	used	s_w	s/R	s_r %	s_b/R	+/-
C_{sat}	21	21	0.192	7.66–3	0.04	2.00–6	1

Temp. range K	T_c K	A_1	A_2	A_3	A_4	A_5	A_6	Level of uncertainty
140.2–301.8	569.50	4.42530+2	4.37635+1	−2.67447+1	4.09232+2	1.01638+2	3.01333+2	II

Name: Methylcyclohexane
Formula: C_7H_{14}

CAS-RN: 108–87–2
Group No.: 12–019
Molar Mass: 98.19

TABLE 12.19.1. Experimental heat capacities

Reference	Temp. range K	No. pnts	Error %	Purity %	Purity method	Type capacity	Calorimeter Type	Calorimeter Reference
30PAR/HUF1	151.4–294.2	12	1.00	not specified		C_p	BSIO	25PAR
46DOU/HUF1	155.1–285.8	15	0.20	99.989	melpt	C_p	BSAO	43RUE/HUF
48TSC2	292.6	1	nosp	not specified		C_p	BSIO	48TSC1
48TSC6	293.6	1	nosp	not specified		C_p	BSIO	48TSC1
66HWA/ZIE	174.8–308.1	32	nosp	99.0	melpt	C_p	BSAO	45SCO/MEY
75HOL/ZIE	144.0–312.0	eqn	0.20	99.88	anal	C_p	BSAO	45SCO/MEY
79WIL/GRO	298.1	1	0.30	not specified		C_p	FSIT	71PIC/LED
82GRO/ING	298.1	1	0.30	99.5	estim	C_p	FSIT	71PIC/LED
83SID/SVE	293.1	1	0.30	99.9	chrom	C_p	FSIT	71PIC/LED
85TAN	298.1	1	0.30	not specified		C_p	FSIT	71PIC/LED
88SHI/OGA2	298.1	1	nosp	99.9	chrom	C_p	FSIO	85OGA

TABLE 12.19.2. Correlated heat capacities

Reference	Temp. range K	No. pnts used	$\sigma_r C$ %	d_w	d/R	d_r %	d_b/R	+/–
				Selected data				
46DOU/HUF1	155.1–285.8	15	0.20#	0.583	2.09–2	0.12	5.56–3	2
66HWA/ZIE	174.8–308.1	32	0.20#	0.583	2.23–2	0.12	1.26–2	19
75HOL/ZIE	144.0–309.0	34	0.20	0.410	1.49–2	0.08	–1.42–2	–34
85TAN	298.1	1	0.30	0.021	1.37–3	0.01	1.37–3	0
				Rejected data				
30PAR/HUF1	(7.51–2, 0.38, –3.55–2, –2)			48TSC2	(7.81–2, 0.36, –7.81–2, –1)			
48TSC6	(7.87–2, 0.36, –7.87–2, –1)			79WIL/GRO	(2.39–2, 0.11, –2.39–2, –1)			
82GRO/ING	(4.11–2, 0.18, 4.11–2, 1)			83SID/SVE	(1.24–1, 0.56, 1.24–1, 1)			
88SHI/OGA2	(6.84–2, 0.31, –6.84–2, –1)							

TABLE 12.19.3. Parameters of regression polynomial

Heat capacity type	No. data points total	used	s_w	s/R	s_r %	s_b/R	+/–
C_p	100	82	0.524	1.95–2	0.10	5.87–5	–1 3

Temp. range K	A_1	A_2	A_3	Level of uncertainty
144.0–309.0	1.57165+1	–7.17916–1	9.75107–1	II

TABLE 12.19.4. Recommended values of heat capacities

Temp. (K)	150	160	170	180	190	200	210
c_p (J K^{-1}g^{-1})	1.425	1.445	1.466	1.489	1.513	1.540	1.567
C_p (J K^{-1}mol^{-1})	140.0	141.9	144.0	146.2	148.6	151.2	153.9
Temp. (K)	220	230	240	250	260	270	273.15
c_p (J K^{-1}g^{-1})	1.597	1.628	1.661	1.695	1.731	1.769	1.781
C_p (J K^{-1}mol^{-1})	156.8	159.8	163.0	166.4	170.0	173.7	174.9
Temp. (K)	280	290	298.15	300	310		
c_p (J K^{-1}g^{-1})	1.808	1.849	1.884	1.892	1.936		
C_p (J K^{-1}mol^{-1})	177.5	181.5	184.9	185.7	190.1		

TABLE 12.19.5. Parameters of quasi-polynomial equation

Heat capacity type	No. data points total	used	s_w	s/R	s_r %	s_b/R	+/–
C_p	100	82	0.674	2.61–2	0.14	–2.39–3	–5

Temp. range K	T_c K	A_1	A_2	A_3	A_4	Level of uncertainty
144.0–309.0	572.20	–2.64050	5.81048	8.03223	2.99985–1	II

12–019

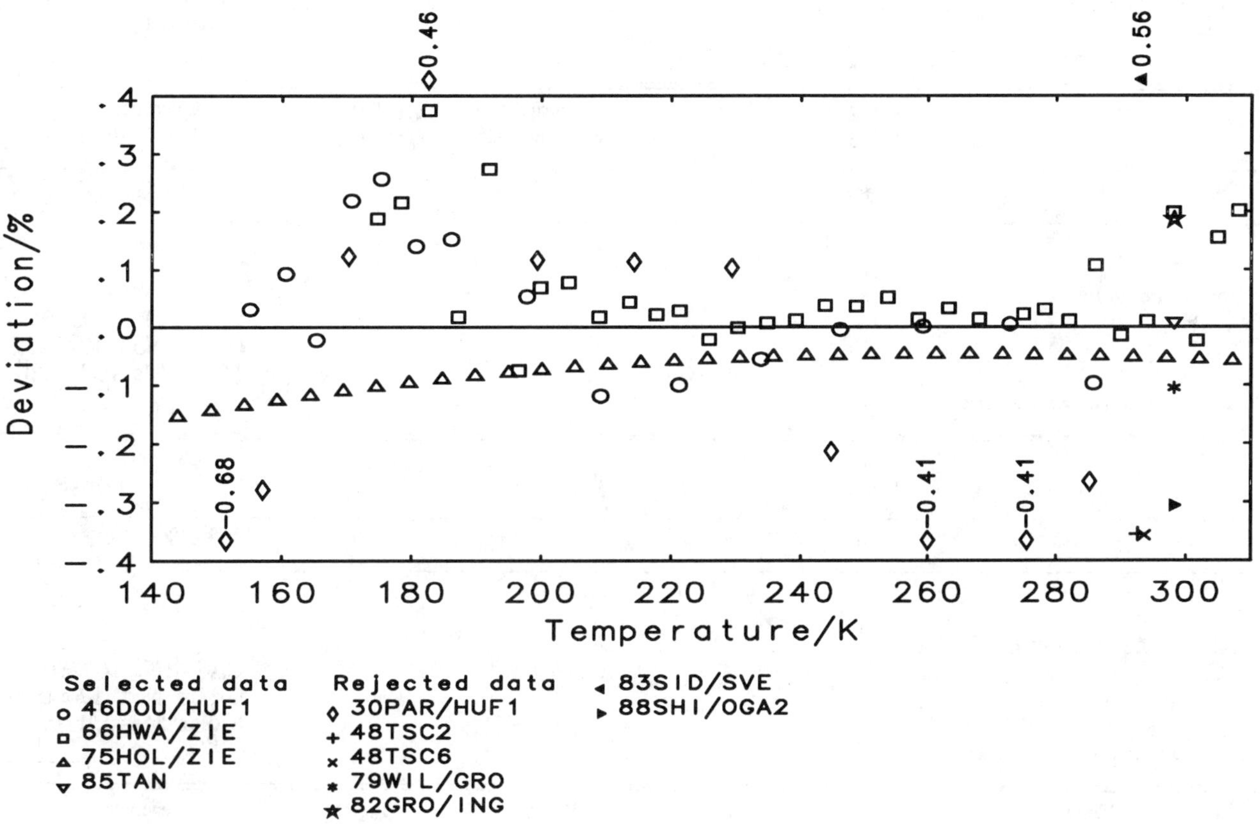

Name: Bicyclo[2.2.2]octane
Formula: C_8H_{14}

CAS–RN: 280–33–1
Group No.: 12–020
Molar Mass: 110.20

TABLE 12.20.1. Experimental heat capacities

Reference	Temp. range K	No. pnts	Error %	Purity %	Purity method	Type capacity	Calorimeter Type	Calorimeter Reference
70WON/WES	450.6–482.1	10	nosp	not specified		C_{sat}	BSAO	68WES/FUR

TABLE 12.20.3. Parameters of regression polynomial

Heat capacity type	No. data points total	No. data points used	s_w	s/R	s_r %	s_b/R	+/–
C_{sat}	10	10	0.856	8.29–2	0.26	3.49–4	2

Temp. range K	A_1	A_2	Level of uncertainty
450.6–482.1	5.46925	5.76028	III

TABLE 12.20.4. Recommended values of heat capacities

Temp. (K)	450	460	470	480
c_{sat} (J K^{-1}g^{-1})	2.368	2.412	2.455	2.499
C_{sat} (J K^{-1}mol^{-1})	261.0	265.8	270.6	275.4

Name: *cis*–Bicyclo[4.2.0]octane
Formula: C_8H_{14}

CAS–RN: 28282–35–1
Group No.: 12–021
Molar Mass: 110.20

TABLE 12.21.1. Experimental heat capacities

Reference	Temp. K	Capac. J/(K.g)	Error %	Purity %	Purity method	Type capacity	Calorimeter Type	Calorimeter Reference
70CHA/MCN	345.00	2.346	2.00	not specified		C_p	BDHT	73PER/COM

Name: Ethylidenecyclohexane
Formula: C_8H_{14}

CAS–RN: 1003–64–1
Group No.: 12–022
Molar Mass: 110.20

TABLE 12.22.1. Experimental heat capacities

Reference	Temp. K	Capac. J/(K.g)	Error %	Purity %	Purity method	Type capacity	Calorimeter Type	Calorimeter Reference
79FUC/PEA	298.15	1.849	1.50	99.	chrom	C_p	BSIO	80FUC

Name: *endo*–2–Methylbicyclo[2.2.1]heptane
Formula: C_8H_{14}

CAS–RN: 765–90–2
Group No.: 12–023
Molar Mass: 110.20

TABLE 12.23.1. Experimental heat capacities

Reference	Temp. range K	No. pnts	Error %	Purity %	Purity method	Type capacity	Calorimeter Type	Calorimeter Reference
64SER/GOR	280.0–310.0	5S	nosp	99.74	melpt	C_p	BSAO	62KOL/SER

TABLE 12.23.3. Parameters of regression polynomial

Heat capacity type	No. data points total	No. data points used	s_w	s/R	s_r %	s_b/R	+/–
C_p	5	5	0.008	9.01–4	0.00	–3.82–7	0

Temp. range K	A_1	A_2	Level of uncertainty
280.0–310.0	5.65347	5.53641	IV

TABLE 12.23.4. Recommended values of heat capacities

Temp. (K)	280	290	298.15	300	310
c_p (J $K^{-1}g^{-1}$)	1.596	1.638	1.672	1.680	1.721
C_p (J $K^{-1}mol^{-1}$)	175.9	180.5	184.3	185.1	189.7

Name: *exo*–2–Methylbicyclo[2.2.1]heptane
Formula: C_8H_{14}

CAS–RN: 872–78–6
Group No.: 12–024
Molar Mass: 110.20

TABLE 12.24.1. Experimental heat capacities

Reference		Temp. range K	No. pnts	Error %	Purity %	Purity method	Type capacity	Calorimeter Type	Calorimeter Reference
64SER/GOR	N	170.0–310.0	16S	nosp	not specified		C_p	BSAO	62KOL/SER

64SER/GOR authors expect impurities of 0.2–0.3 mol.%

TABLE 12.24.3. Parameters of regression polynomial

Heat capacity type	No. data points total	No. data points used	s_w	s/R	s_r %	s_b/R	+/–
C_p	16	16	0.100	9.26–3	0.05	8.11–6	1

Temp. range K	A_1	A_2	A_3	A_4	Level of uncertainty
170.0–310.0	2.03398+1	–7.56553	3.69354	–3.11857–1	IV

TABLE 12.24.4. Recommended values of heat capacities

Temp. (K)	170	180	190	200	210	220	230
c_p (J K^{-1}g^{-1})	1.254	1.273	1.295	1.319	1.347	1.377	1.410
C_p (J K^{-1}mol^{-1})	138.2	140.3	142.7	145.4	148.4	151.8	155.3
Temp. (K)	240	250	260	270	273.15	280	290
c_p (J K^{-1}g^{-1})	1.445	1.482	1.521	1.562	1.575	1.605	1.649
C_p (J K^{-1}mol^{-1})	159.2	163.3	167.6	172.1	173.6	176.8	181.7
Temp. (K)	298.15	300	310				
c_p (J K^{-1}g^{-1})	1.686	1.695	1.742				
C_p (J K^{-1}mol^{-1})	185.8	186.8	192.0				

Name: *cis*–Octahydropentalene
Formula: C_8H_{14}

CAS–RN: 1755–05–1
Group No.: 12–025
Molar Mass: 110.20

TABLE 12.25.1. Experimental heat capacities

Reference	Temp. range K	No. pnts	Error %	Purity %	Purity method	Type capacity	Calorimeter Type	Calorimeter Reference
70CHA/MCN	308.0–334.0	2	2.00	not specified		C_p	BDHT	73PER/COM

TABLE 12.25.3. Parameters of regression polynomial

Heat capacity type	No. data points total	No. data points used	s_w	s/R	s_r %	s_b/R	+/–
C_p	2	2	0.000	0.00	0.00	0.00	0

Temp. range K	A_1	A_2	Level of uncertainty
308.0–334.0	5.39602	6.58053	V

TABLE 12.25.4. Recommended values of heat capacities

Temp. (K)	310	320	330
c_p (J $K^{-1}g^{-1}$)	1.946	1.996	2.046
C_p (J $K^{-1}mol^{-1}$)	214.5	219.9	225.4

Name: *trans*–Octahydropentalene
Formula: C_8H_{14}

CAS–RN: 5597–89–7
Group No.: 12–026
Molar Mass: 110.20

TABLE 12.26.1. Experimental heat capacities

Reference	Temp. range K	No. pnts	Error %	Purity %	Purity method	Type capacity	Calorimeter Type	Calorimeter Reference
70CHA/MCN	308.0–334.0	2	2.00	not specified		C_p	BDHT	73PER/COM

TABLE 12.26.3. Parameters of regression polynomial

Heat capacity type	No. data points total	No. data points used	s_w	s/R	s_r %	s_b/R	+/–
C_p	2	2	0.000	0.00	0.00	0.00	0

Temp. range K	A_1	A_2	Level of uncertainty
308.0–334.0	−1.34823+1	1.14192+1	V

TABLE 12.26.4. Recommended values of heat capacities

Temp. (K)	310	320	330
c_p (J $K^{-1}g^{-1}$)	1.65	1.74	1.83
C_p (J $K^{-1}mol^{-1}$)	182	192	201

Name: (2–Propenyl)cyclopentane
Formula: C_8H_{14}

CAS–RN: 3524–75–2
Group No.: 12–027
Molar Mass: 110.20

TABLE 12.27.1. Experimental heat capacities

Reference	Temp. K	Capac. J/(K.g)	Error %	Purity %	Purity method	Type capacity	Calorimeter Type	Calorimeter Reference
79FUC/PEA	298.15	1.841	1.50	99.	chrom	C_p	BSIO	80FUC

Name: Cyclooctane
Formula: C_8H_{16}

CAS–RN: 292–64–8
Group No.: 12–028
Molar Mass: 112.22

TABLE 12.28.1. Experimental heat capacities

Reference	Temp. range K	No. pnts	Error %	Purity %	Purity method	Type capacity	Calorimeter Type	Calorimeter Reference
56FIN/SCO	294.6–321.6	6	0.20	99.82	melpt	C_{sat}	BSAO	43RUE/HUF
75JOL/BOI	298.1	1	0.30	99.7	melpt	C_p	FSIT	71PIC/LED
79FOR/DAR	298.1	1	0.30	not specified		C_p	FSIT	71PIC/LED
79WIL/FAR	298.1	1	0.30	99.0	melpt	C_p	FSIT	71PIC/LED
85TAN	298.1	1	0.30	not specified		C_p	FSIT	71PIC/LED
88SHI/OGA2	298.1	1	nosp	99.9	chrom	C_p	FSIO	85OGA

TABLE 12.28.2. Correlated heat capacities

Reference	Temp. range K	No. pnts used	$\sigma_r C$ %	d_w	d/R	d_r %	d_b/R	+/–
				Selected data				
56FIN/SCO	294.6–321.6	6	0.20#	0.186	9.76–3	0.04	–5.19–4	2
79FOR/DAR	298.1	1	0.30	0.008	6.41–4	0.00	–6.41–4	0
85TAN	298.1	1	0.30	0.100	7.78–3	0.03	7.78–3	1
				Rejected data				
75JOL/BOI	(1.47–1, 0.57, –1.47–1, –1)			79WIL/FAR	(1.12–1, 0.44, –1.12–1, –1)			
88SHI/OGA2	(3.78–2, 0.15, 3.78–2, 1)							

TABLE 12.28.3. Parameters of regression polynomial

Heat capacity type	No. data points total	No. data points used	s_w	s/R	s_r %	s_b/R	+/–
C	11	8	0.191	1.03–2	0.04	5.02–4	3

Temp. range K	A_1	A_2	Level of uncertainty
294.6–321.6	1.02950+1	5.23878	II

TABLE 12.28.4. Recommended values of heat capacities

Temp. (K)	290	298.15	300	310	320
c ($J\,K^{-1}g^{-1}$)	1.888	1.920	1.927	1.966	2.005
C ($J\,K^{-1}mol^{-1}$)	211.9	215.5	216.3	220.6	225.0

Name: 1,1–Dimethylcyclohexane
Formula: C_8H_{16}

CAS–RN: 590–66–9
Group No.: 12–029
Molar Mass: 112.22

TABLE 12.29.1. Experimental heat capacities

Reference	Temp. range K	No. pnts	Error %	Purity %	Purity method	Type capacity	Calorimeter Type	Calorimeter Reference
49HUF/TOD	242.7–303.5	13	0.20	99.97	melpt	C_p	BSAO	47HUF

TABLE 12.29.3. Parameters of regression polynomial

Heat capacity type	No. data points total	No. data points used	s_w	s/R	s_r %	s_b/R	+/–
C_p	13	13	0.202	9.46–3	0.04	4.84–6	0

Temp. range K	A_1	A_2	A_3	Level of uncertainty
242.7–303.5	1.61860+1	9.75223–2	9.77710–1	II

TABLE 12.29.4. Recommended values of heat capacities

Temp. (K)	250	260	270	273.15	280	290	298.15
c_p (J $K^{-1}g^{-1}$)	1.670	1.708	1.747	1.760	1.787	1.829	1.865
C_p (J $K^{-1}mol^{-1}$)	187.4	191.6	196.0	197.4	200.6	205.3	209.3

Temp. (K)	300
c_p (J $K^{-1}g^{-1}$)	1.873
C_p (J $K^{-1}mol^{-1}$)	210.2

TABLE 12.29.5. Parameters of quasi–polynomial equation

Heat capacity type	No. data points total	No. data points used	s_w	s/R	s_r %	s_b/R	+/–
C_p	13	13	0.205	9.56–3	0.04	5.43–6	0

Temp. range K	T_c K	A_1	A_2	A_3	A_4	Level of uncertainty
242.7–303.5	585.80	–8.35188	2.83840	1.03202+1	6.14379	II

Name: *cis*–1,2–Dimethylcyclohexane
Formula: C_8H_{16}

CAS–RN: 2207–01–4
Group No.: 12–030
Molar Mass: 112.22

TABLE 12.30.1. Experimental heat capacities

Reference	Temp. range K	No. pnts	Error %	Purity %	Purity method	Type capacity	Calorimeter Type	Calorimeter Reference
49HUF/TOD	228.3–298.8	16	0.20	99.9979	melpt	C_p	BSAO	47HUF

TABLE 12.30.3. Parameters of regression polynomial

Heat capacity type	No. data points total	used	s_w	s/R	s_r %	s_b/R	+/–
C_p	16	16	0.328	1.49–2	0.07	1.60–5	0

Temp. range K	A_1	A_2	A_3	Level of uncertainty
228.3–298.8	1.80627+1	–7.56275–1	1.06595	II

TABLE 12.30.4. Recommended values of heat capacities

Temp. (K)	230	240	250	260	270	273.15	280
c_p (J K^{-1}g^{-1})	1.627	1.659	1.692	1.727	1.763	1.775	1.801
C_p (J K^{-1}mol^{-1})	182.6	186.1	189.9	193.7	197.8	199.1	202.1
Temp. (K)	290	298.15	300				
c_p (J K^{-1}g^{-1})	1.840	1.873	1.881				
C_p (J K^{-1}mol^{-1})	206.5	210.2	211.1				

TABLE 12.30.5. Parameters of quasi–polynomial equation

Heat capacity type	No. data points total	used	s_w	s/R	s_r %	s_b/R	+/–
C_p	16	16	0.325	1.47–2	0.06	1.43–5	1

Temp. range K	T_c K	A_1	A_2	A_3	A_4	Level of uncertainty
228.3–298.8	601.00	–6.35113	4.95764	1.00864+1	2.03408	II

Name: *trans*–1,2–Dimethylcyclohexane
Formula: C_8H_{16}

CAS–RN: 6876–23–9
Group No.: 12–031
Molar Mass: 112.22

TABLE 12.31.1. Experimental heat capacities

Reference	Temp. range K	No. pnts	Error %	Purity %	Purity method	Type capacity	Calorimeter Type	Calorimeter Reference
49HUF/TOD	183.7–301.4	27	0.20	99.995	melpt	C_p	BSAO	47HUF

TABLE 12.31.3. Parameters of regression polynomial

Heat capacity type	No. data points total	used	s_w	s/R	s_r %	s_b/R	+/–
C_p	27	27	0.170	7.60–3	0.03	3.67–6	0

Temp. range K	A_1	A_2	A_3	Level of uncertainty
183.7–301.4	1.86498+1	–1.70433	1.30736	II

TABLE 12.31.4. Recommended values of heat capacities

Temp. (K)	190	200	210	220	230	240	250
c_p (J K^{-1}g^{-1})	1.492	1.517	1.544	1.573	1.604	1.637	1.672
C_p (J K^{-1}mol^{-1})	167.4	170.2	173.2	176.5	180.0	183.7	187.6
Temp. (K)	260	270	273.15	280	290	298.15	300
c_p (J K^{-1}g^{-1})	1.708	1.747	1.760	1.788	1.830	1.866	1.875
C_p (J K^{-1}mol^{-1})	191.7	196.0	197.5	200.6	205.4	209.4	210.4

TABLE 12.31.5. Parameters of quasi-polynomial equation

Heat capacity type	No. data points total	used	s_w	s/R	s_r %	s_b/R	+/–
C_p	27	27	0.380	1.73–2	0.08	1.86–5	–3

Temp. range K	T_c K	A_1	A_2	A_3	A_4	Level of uncertainty
183.7–301.4	591.60	–8.60285–1	8.86336	6.72416	2.08750–2	II

Name: *cis*–1,3–Dimethylcyclohexane
Formula: C_8H_{16}

CAS–RN: 638–04–0
Group No.: 12–032
Molar Mass: 112.22

TABLE 12.32.1. Experimental heat capacities

Reference	Temp. range K	No. pnts	Error %	Purity %	method	Type capacity	Calorimeter Type	Reference
49HUF/TOD	204.2–299.5	18	0.20	99.995	melpt	C_p	BSAO	47HUF

TABLE 12.32.3. Parameters of regression polynomial

Heat capacity type	No. data points total	used	s_w	s/R	s_r %	s_b/R	+/–
C_p	18	18	0.489	2.21–2	0.10	2.93–5	2

Temp. range K	A_1	A_2	A_3	Level of uncertainty
204.2–299.5	1.75729+1	–7.94236–1	1.12319	II

TABLE 12.32.4. Recommended values of heat capacities

Temp. (K)	210	220	230	240	250	260	270
c_p (J K^{-1}g^{-1})	1.545	1.575	1.607	1.640	1.675	1.712	1.750
C_p (J K^{-1}mol^{-1})	173.4	176.8	180.3	184.1	188.0	192.1	196.4
Temp. (K)	273.15	280	290	298.15	300		
c_p (J K^{-1}g^{-1})	1.762	1.790	1.831	1.866	1.875		
C_p (J K^{-1}mol^{-1})	197.7	200.8	205.5	209.4	210.3		

TABLE 12.32.5. Parameters of quasi-polynomial equation

Heat capacity type	No. data points total	used	s_w	s/R	s_r %	s_b/R	+/–
C_p	18	18	0.599	2.69–2	0.12	5.02–5	2

Temp. range K	T_c K	A_1	A_2	A_3	A_4	Level of uncertainty
204.2–299.5	586.60	–6.04943	5.16520	9.49600	1.77126	II

Name: *trans*–1,3–Dimethylcyclohexane
Formula: C_8H_{16}

CAS-RN: 2207–03–6
Group No.: 12–033
Molar Mass: 112.22

TABLE 12.33.1. Experimental heat capacities

Reference	Temp. range K	No. pnts	Error %	Purity %	Purity method	Type capacity	Calorimeter Type	Calorimeter Reference
49HUF/TOD	187.6–299.0	30	0.20	99.99	melpt	C_p	BSAO	47HUF

TABLE 12.33.3. Parameters of regression polynomial

Heat capacity type	No. data points total	used	s_w	s/R	s_r %	s_b/R	+/–
C_p	30	30	0.254	1.13–2	0.05	1.13–5	3

Temp. range K	A_1	A_2	A_3	Level of uncertainty
187.6–299.0	1.90500+1	–1.38442	1.20050	II

TABLE 12.33.4. Recommended values of heat capacities

Temp. (K)	190	200	210	220	230	240	250
c_p (J K^{-1}g^{-1})	1.538	1.562	1.588	1.616	1.646	1.678	1.711
C_p (J K^{-1}mol^{-1})	172.6	175.3	178.2	181.4	184.7	188.3	192.0
Temp. (K)	260	270	273.15	280	290	298.15	300
c_p (J K^{-1}g^{-1})	1.746	1.783	1.795	1.822	1.862	1.896	1.904
C_p (J K^{-1}mol^{-1})	195.9	200.1	201.4	204.4	209.0	212.8	213.7

TABLE 12.33.5. Parameters of quasi-polynomial equation

Heat capacity type	No. data points total	used	s_w	s/R	s_r %	s_b/R	+/–
C_p	30	30	0.304	1.37–2	0.06	1.35–5	–1 1

Temp. range K	T_c K	A_1	A_2	A_3	A_4	Level of uncertainty
187.6–299.0	593.20	–1.54221	8.10594	8.20165	7.33545–2	II

Name: *cis*–1,4–Dimethylcyclohexane
Formula: C_8H_{16}

CAS–RN: 624–29–3
Group No.: 12–034
Molar Mass: 112.22

TABLE 12.34.1. Experimental heat capacities

Reference	Temp. range K	No. pnts	Error %	Purity %	Purity method	Type capacity	Calorimeter Type	Calorimeter Reference
49HUF/TOD	189.5–303.2	22	0.20	99.995	melpt	C_p	BSAO	47HUF

TABLE 12.34.3. Parameters of regression polynomial

Heat capacity type	No. data points total	used	s_w	s/R	s_r %	s_b/R	+/–
C_p	22	22	0.266	1.25–2	0.05	6.85–6	–1

Temp. range K	A_1	A_2	A_3	Level of uncertainty
189.5–303.2	2.11664+1	–2.84248	1.44287	II

TABLE 12.34.4. Recommended values of heat capacities

Temp. (K)	190	200	210	220	230	240	250
c_p (J K^{-1}g^{-1})	1.554	1.575	1.597	1.622	1.649	1.679	1.710
C_p (J K^{-1}mol^{-1})	174.4	176.7	179.3	182.1	185.1	188.4	191.9
Temp. (K)	260	270	273.15	280	290	298.15	300
c_p (J K^{-1}g^{-1})	1.743	1.779	1.791	1.817	1.857	1.891	1.899
C_p (J K^{-1}mol^{-1})	195.6	199.6	200.9	203.9	208.3	212.2	213.1

TABLE 12.34.5. Parameters of quasi–polynomial equation

Heat capacity type	No. data points total	used	s_w	s/R	s_r %	s_b/R	+/–
C_p	22	22	0.530	2.48–2	0.11	4.02–5	–1

Temp. range K	T_c K	A_1	A_2	A_3	A_4	Level of uncertainty
189.5–303.2	593.00	6.14959	1.17733+1	5.74797	8.03033–1	II

Name: *trans*–1,4–Dimethylcyclohexane
Formula: C_8H_{16}

CAS–RN: 2207–04–7
Group No.: 12–035
Molar Mass: 112.22

TABLE 12.35.1. Experimental heat capacities

Reference	Temp. range K	No. pnts	Error %	Purity %	Purity method	Type capacity	Calorimeter Type	Calorimeter Reference
49HUF/TOD	242.2–299.3	9	0.20	99.995	melpt	C_p	BSAO	47HUF

TABLE 12.35.3. Parameters of regression polynomial

Heat capacity type	No. data points total	used	s_w	s/R	s_r %	s_b/R	+/–
C_p	9	9	0.309	1.42–2	0.06	1.23–5	1

Temp. range K	A_1	A_2	A_3	Level of uncertainty
242.2–299.3	2.13562+1	–3.12542	1.49133	II

TABLE 12.35.4. Recommended values of heat capacities

Temp. (K)	240	250	260	270	273.15	280	290
c_p (J K^{-1}g^{-1})	1.663	1.694	1.727	1.763	1.774	1.800	1.840
C_p (J K^{-1}mol^{-1})	186.6	190.1	193.8	197.8	199.1	202.0	206.5
Temp. (K)	298.15	300					
c_p (J K^{-1}g^{-1})	1.874	1.882					
C_p (J K^{-1}mol^{-1})	210.3	211.2					

TABLE 12.35.5. Parameters of quasi-polynomial equation

Heat capacity type	No. data points total	used	s_w	s/R	s_r %	s_b/R	+/–
C_p	9	9	0.320	1.47–2	0.06	1.36–5	0

Temp. range K	T_c K	A_1	A_2	A_3	A_4	Level of uncertainty
242.2–299.3	587.70	–1.43304	7.61948	8.78229	6.73799–2	II

Name: Ethylcyclohexane
Formula: C_8H_{16}

CAS–RN: 1678–91–7
Group No.: 12–036
Molar Mass: 112.22

TABLE 12.36.1. Experimental heat capacities

Reference	Temp. range K	No. pnts	Error %	Purity %	Purity method	Type capacity	Calorimeter Type	Calorimeter Reference
49HUF/TOD	167.4–299.2	21	0.20	99.996	melpt	C_p	BSAO	47HUF
49PAR/MOO	160.0–300.0	15S	1.00	99.1	melpt	C_p	BSIO	25PAR

TABLE 12.36.2. Correlated heat capacities

Reference	Temp. range K	No. pnts used	$\sigma_r C$ %	d_w	d/R	d_r %	d_b/R	+/–
				Selected data				
49HUF/TOD	167.4–299.2	21	0.20#	0.194	8.32–3	0.04	1.54–6	–3
				Rejected data				
49PAR/MOO	(1.06–1, 0.45, 3.59–2, 4)							

TABLE 12.36.3. Parameters of regression polynomial

Heat capacity type	No. data points total	used	s_w	s/R	s_r %	s_b/R	+/–
C_p	36	21	0.210	8.99–3	0.04	1.54–6	–3
Temp. range K		A_1	A_2	A_3			Level of uncertainty
167.4–299.2		1.82969+1	–1.18127	1.20478			II

TABLE 12.36.4. Recommended values of heat capacities

Temp. (K)	170	180	190	200	210	220	230
c_p (J K^{-1}g^{-1})	1.465	1.487	1.512	1.538	1.566	1.595	1.627
C_p (J K^{-1}mol^{-1})	164.4	166.9	169.6	172.6	175.7	179.0	182.5
Temp. (K)	240	250	260	270	273.15	280	290
c_p (J K^{-1}g^{-1})	1.660	1.695	1.732	1.770	1.783	1.810	1.853
C_p (J K^{-1}mol^{-1})	186.3	190.2	194.3	198.6	200.0	203.2	207.9
Temp. (K)	298.15	300					
c_p (J K^{-1}g^{-1})	1.888	1.897					
C_p (J K^{-1}mol^{-1})	211.9	212.8					

TABLE 12.36.5. Parameters of quasi–polynomial equation

Heat capacity type	No. data points total	used	s_w	s/R	s_r %	s_b/R	+/–
C_p	36	21	0.524	2.35–2	0.10	3.41–5	0
Temp. range K	T_c K	A_1	A_2	A_3	A_4		Level of uncertainty
167.4–299.2	602.40	–2.84865–1	9.64722	6.21621	2.10289–3		II

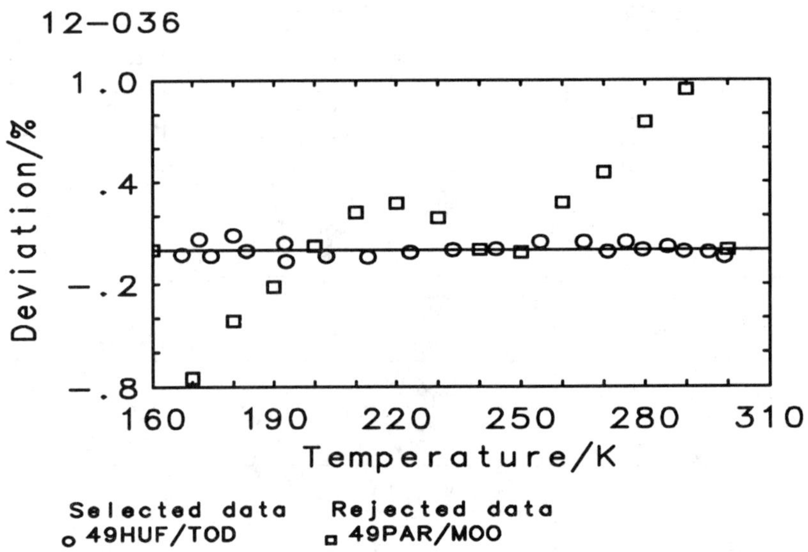

Name: Propylcyclopentane
Formula: C_8H_{16}

CAS–RN: 2040–96–2
Group No.: 12–037
Molar Mass: 112.22

TABLE 12.37.1. Experimental heat capacities

Reference	Temp. range K	No. pnts	Error %	Purity %	Purity method	Type capacity	Calorimeter Type	Calorimeter Reference
65MES/TOD1	162.1–364.6	25	0.20	99.95	melpt	C_{sat}	BSAO	47HUF

TABLE 12.37.3. Parameters of regression polynomial

Heat capacity type	No. data points total	No. data points used	s_w	s/R	s_r %	s_b/R	+/–
C_{sat}	25	25	0.593	2.73–2	0.12	5.52–5	–4

Temp. range K	A_1	A_2	A_3	A_4	Level of uncertainty
162.1–364.6	2.61195+1	–8.71538	3.97711	–3.57738–1	II

TABLE 12.37.4. Recommended values of heat capacities

Temp. (K)	160	170	180	190	200	210	220
c_{sat} (J K^{-1}g^{-1})	1.548	1.559	1.573	1.590	1.610	1.633	1.659
C_{sat} (J K^{-1}mol^{-1})	173.7	174.9	176.5	178.5	180.7	183.3	186.1
Temp. (K)	230	240	250	260	270	273.15	280
c_{sat} (J K^{-1}g^{-1})	1.686	1.716	1.749	1.783	1.818	1.830	1.856
C_{sat} (J K^{-1}mol^{-1})	189.2	192.6	196.2	200.0	204.0	205.3	208.2
Temp. (K)	290	298.15	300	310	320	330	340
c_{sat} (J K^{-1}g^{-1})	1.894	1.927	1.934	1.976	2.018	2.061	2.104
C_{sat} (J K^{-1}mol^{-1})	212.6	216.2	217.1	221.7	226.4	231.3	236.1
Temp. (K)	350	360	370				
c_{sat} (J K^{-1}g^{-1})	2.149	2.193	2.238				
C_{sat} (J K^{-1}mol^{-1})	241.1	246.1	251.1				

TABLE 12.37.5. Parameters of quasi–polynomial equation

Heat capacity type	No. data points total	No. data points used	s_w	s/R	s_r %	s_b/R	+/–
C_{sat}	25	25	0.699	3.24–2	0.14	5.61–5	–6

Temp. range K	T_c K	A_1	A_2	A_3	A_4	A_5	A_6	Level of uncertainty
162.1–364.6	596.20	2.55862+2	2.21400+1	–2.00627	2.36140+2	7.57297+1	1.92612+2	II

Name: *cis*–Bicyclo[6.1.0]nonane
Formula: C_9H_{16}

CAS–RN: 13757–43–2
Group No.: 12–038
Molar Mass: 124.23

TABLE 12.38.1. Experimental heat capacities

Reference	Temp. K	Capac. J/(K.g)	Error %	Purity %	Purity method	Type capacity	Calorimeter Type	Calorimeter Reference
70CHA/MCN	315.00	1.893	2.00	not specified		C_p	BDHT	73PER/COM

Name: Octahydro–1*H*–indene
Formula: C_9H_{16}

CAS–RN: 496–10–6
Group No.: 12–039
Molar Mass: 124.23

TABLE 12.39.1. Experimental heat capacities

Reference	Temp. range K	No. pnts	Error %	Purity %	Purity method	Type capacity	Calorimeter Type	Calorimeter Reference
62GOL/BEL	310.9–422.0	3	nosp	99.8	chrom	C_p	BDHT	63GUD/CAM
63GUD/CAM	313.1–423.1	3	nosp	99.8	chrom	C_p	BDHT	63GUD/CAM

TABLE 12.39.2. Correlated heat capacities

Reference	Temp. range K	No. pnts used	$\sigma_r C$ %	d_w	d/R	d_r %	d_b/R	+/–
				Selected data				
62GOL/BEL	310.9–422.0	3	1.00#	0.095	2.72–2	0.10	5.18–3	1
63GUD/CAM	313.1–423.1	3	1.00#	0.101	2.88–2	0.10	–5.03–3	–1

TABLE 12.39.3. Parameters of regression polynomial

Heat capacity type	No. data points total	No. data points used	s_w	s/R	s_r %	s_b/R	+/–
C_p	6	6	0.139	3.96–2	0.14	7.18–5	0

Temp. range K	A_1	A_2	A_3	Level of uncertainty
310.9–423.1	6.64917+1	–2.83722+1	4.98570	IV

TABLE 12.39.4. Recommended values of heat capacities

Temp. (K)	310	320	330	340	350	360	370
c_p (J K^{-1}g^{-1})	1.770	1.791	1.818	1.851	1.892	1.939	1.992
C_p (J K^{-1}mol^{-1})	219.9	222.4	225.8	230.0	235.0	240.8	247.5

Temp. (K)	380	390	400	410	420
c_p (J K^{-1}g^{-1})	2.053	2.120	2.194	2.274	2.361
C_p (J K^{-1}mol^{-1})	255.0	263.3	272.5	282.5	293.3

Name: *cis*–Octahydro–1*H*–indene
Formula: C_9H_{16}

CAS–RN: 4551–51–3
Group No.: 12–040
Molar Mass: 124.23

TABLE 12.40.1. Experimental heat capacities

Reference	Temp. range K	No. pnts	Error %	Purity %	Purity method	Type capacity	Calorimeter Type	Calorimeter Reference
72FIN/MCC	239.4–366.7	17	0.20	99.989	melpt	C_{sat}	BSAO	47HUF

TABLE 12.40.3. Parameters of regression polynomial

Heat capacity type	No. data points total	No. data points used	s_w	s/R	s_r %	s_b/R	+/–
C_{sat}	17	17	0.143	7.49–3	0.03	1.23–6	0

Temp. range K	A_1	A_2	A_3	A_4	Level of uncertainty
239.4–366.7	2.53888+1	–9.93412	4.61579	–4.16499–1	II

TABLE 12.40.4. Recommended values of heat capacities

Temp. (K)	240	250	260	270	273.15	280	290
c_{sat} (J K^{-1}g^{-1})	1.498	1.532	1.569	1.608	1.620	1.648	1.689
C_{sat} (J K^{-1}mol^{-1})	186.0	190.4	194.9	199.7	201.2	204.7	209.9

Temp. (K)	298.15	300	310	320	330	340	350
c_{sat} (J K^{-1}g^{-1})	1.724	1.732	1.777	1.822	1.868	1.914	1.961
C_{sat} (J K^{-1}mol^{-1})	214.2	215.2	220.7	226.3	232.0	237.8	243.7

Temp. (K)	360
c_{sat} (J K^{-1}g^{-1})	2.009
C_{sat} (J K^{-1}mol^{-1})	249.6

TABLE 12.40.5. Parameters of quasi–polynomial equation

Heat capacity type	No. data points total	No. data points used	s_w	s/R	s_r %	s_b/R	+/–
C_{sat}	17	17	0.576	3.02–2	0.12	5.73–5	0

Temp. range K	T_c K	A_1	A_2	A_3	A_4	Level of uncertainty
239.4–366.7	659.00	–1.13270+1	2.68298	8.63717	1.19551+1	II

Name: *trans*–Octahydro–1*H*–indene
Formula: C_9H_{16}

CAS–RN: 3296–50–2
Group No.: 12–041
Molar Mass: 124.23

TABLE 12.41.1. Experimental heat capacities

Reference	Temp. range K	No. pnts	Error %	Purity %	Purity method	Type capacity	Calorimeter Type	Calorimeter Reference
72FIN/MCC	211.9–391.0	25	0.20	99.978	melpt	C_{sat}	BSAO	47HUF

TABLE 12.41.3. Parameters of regression polynomial

Heat capacity type	No. data points total	used	s_w	s/R	s_r %	s_b/R	+/−
C_{sat}	25	25	0.167	8.15–3	0.03	3.13–6	−1

Temp. range K	A_1	A_2	A_3	A_4	Level of uncertainty
211.9–391.0	2.65175+1	−1.00744+1	4.29136	−3.55052–1	II

TABLE 12.41.4. Recommended values of heat capacities

Temp. (K)	220	230	240	250	260	270	273.15
c_{sat} (J K^{-1}g^{-1})	1.429	1.454	1.482	1.513	1.546	1.580	1.592
C_{sat} (J K^{-1}mol^{-1})	177.5	180.7	184.2	187.9	192.0	196.3	197.7
Temp. (K)	280	290	298.15	300	310	320	330
c_{sat} (J K^{-1}g^{-1})	1.617	1.655	1.688	1.695	1.737	1.780	1.824
C_{sat} (J K^{-1}mol^{-1})	200.9	205.6	209.7	210.6	215.8	221.1	226.5
Temp. (K)	340	350	360	370	380	390	
c_{sat} (J K^{-1}g^{-1})	1.869	1.914	1.961	2.008	2.056	2.104	
C_{sat} (J K^{-1}mol^{-1})	232.1	237.8	243.6	249.5	255.4	261.4	

TABLE 12.41.5. Parameters of quasi-polynomial equation

Heat capacity type	No. data points total	used	s_w	s/R	s_r %	s_b/R	+/−
C_{sat}	25	25	0.283	1.46–2	0.06	1.73–5	−2

Temp. range K	T_c K	A_1	A_2	A_3	A_4	A_5	A_6	Level of uncertainty
211.9–391.0	649.00	2.21586+2	1.77407+1	8.68620–1	2.04168+2	8.25614+1	1.71172+2	II

Name: (2–Propenyl)cyclohexane
Formula: C_9H_{16}

CAS–RN: 2114–42–3
Group No.: 12–042
Molar Mass: 124.23

TABLE 12.42.1. Experimental heat capacities

Reference	Temp. K	Capac. J/(K.g)	Error %	Purity %	Purity method	Type capacity	Calorimeter Type	Calorimeter Reference
79FUC/PEA	298.15	1.879	1.50	99.	chrom	C_p	BSIO	80FUC

Name: Butylcyclopentane
Formula: C_9H_{18}

CAS–RN: 2040–95–1
Group No.: 12–043
Molar Mass: 126.24

TABLE 12.43.1. Experimental heat capacities

Reference	Temp. range K	No. pnts	Error %	Purity %	Purity method	Type capacity	Calorimeter Type	Calorimeter Reference
65MES/TOD1	167.9–367.6	24	0.20	99.98	melpt	C_{sat}	BSAO	47HUF

TABLE 12.43.3. Parameters of regression polynomial

Heat capacity type	No. data points total	No. data points used	s_w	s/R	s_r %	s_b/R	+/–
C_{sat}	24	24	0.170	9.33–3	0.03	5.25–6	1

Temp. range K	A_1	A_2	A_3	A_4	Level of uncertainty
167.9–367.6	3.23222+1	–1.21006+1	5.18240	–4.83250–1	II

TABLE 12.43.4. Recommended values of heat capacities

Temp. (K)	170	180	190	200	210	220	230
c_{sat} (J K^{-1}g^{-1})	1.604	1.615	1.628	1.646	1.666	1.689	1.714
C_{sat} (J K^{-1}mol^{-1})	202.5	203.8	205.6	207.7	210.3	213.2	216.4
Temp. (K)	240	250	260	270	273.15	280	290
c_{sat} (J K^{-1}g^{-1})	1.742	1.772	1.805	1.839	1.850	1.875	1.912
C_{sat} (J K^{-1}mol^{-1})	219.9	223.7	227.8	232.1	233.5	236.7	241.4
Temp. (K)	298.15	300	310	320	330	340	350
c_{sat} (J K^{-1}g^{-1})	1.943	1.950	1.990	2.031	2.072	2.114	2.156
C_{sat} (J K^{-1}mol^{-1})	245.3	246.2	251.2	256.4	261.6	266.9	272.2
Temp. (K)	360	370					
c_{sat} (J K^{-1}g^{-1})	2.198	2.241					
C_{sat} (J K^{-1}mol^{-1})	277.5	282.9					

TABLE 12.43.5. Parameters of quasi–polynomial equation

Heat capacity type	No. data points total	No. data points used	s_w	s/R	s_r %	s_b/R	+/–
C_{sat}	24	24	0.323	1.81–2	0.06	4.85–6	1

Temp. range K	T_c K	A_1	A_2	A_3	A_4	A_5	A_6	Level of uncertainty
167.9–367.6	621.20	5.23910+2	5.08852+1	–2.74838+1	4.81266+2	1.38362+2	3.51213+2	II

Name: Propylcyclohexane
Formula: C_9H_{18}

CAS–RN: 1678–92–8
Group No.: 12–044
Molar Mass: 126.24

TABLE 12.44.1. Experimental heat capacities

Reference	Temp. range K	No. pnts	Error %	Purity %	Purity method	Type capacity	Calorimeter Type	Calorimeter Reference
65FIN/MES	185.5–373.2	24	0.20	99.96	melpt	C_{sat}	BSAO	47HUF

TABLE 12.44.3. Parameters of regression polynomial

Heat capacity type	No. data points total	No. data points used	s_w	s/R	s_r %	s_b/R	+/–
C_{sat}	24	24	0.094	5.68–3	0.02	2.46–6	–2

Temp. range K	A_1	A_2	A_3	A_4	Level of uncertainty
185.5–373.2	2.60337+1	–7.47579	3.93610	–3.62869–1	II

TABLE 12.44.4. Recommended values of heat capacities

Temp. (K)	190	200	210	220	230	240	250
c_{sat} (J K^{-1}g^{-1})	1.551	1.576	1.603	1.632	1.663	1.696	1.731
C_{sat} (J K^{-1}mol^{-1})	195.8	198.9	202.3	206.0	209.9	214.1	218.5
Temp. (K)	260	270	273.15	280	290	298.15	300
c_{sat} (J K^{-1}g^{-1})	1.767	1.805	1.817	1.844	1.884	1.918	1.925
C_{sat} (J K^{-1}mol^{-1})	223.1	227.8	229.4	232.8	237.8	242.1	243.1
Temp. (K)	310	320	330	340	350	360	370
c_{sat} (J K^{-1}g^{-1})	1.968	2.011	2.054	2.098	2.142	2.187	2.231
C_{sat} (J K^{-1}mol^{-1})	248.4	253.8	259.3	264.9	270.5	276.1	281.7

TABLE 12.44.5. Parameters of quasi–polynomial equation

Heat capacity type	No. data points total	No. data points used	s_w	s/R	s_r %	s_b/R	+/–
C_{sat}	24	24	0.153	8.58–3	0.03	–8.74–7	1

Temp. range K	T_c K	A_1	A_2	A_3	A_4	A_5	A_6	Level of uncertainty
185.5–373.2	625.60	2.08062+2	1.75359+1	3.18405	1.90809+2	9.49718+1	1.34979+2	II

Name: 6,6–Dimethyl–2–methylenebicyclo[3.1.1]heptane
Formula: $C_{10}H_{16}$

CAS–RN: 127–91–3
Group No.: 12–045
Molar Mass: 136.24

TABLE 12.45.1. Experimental heat capacities

Reference	Temp. range K	No. pnts	Error %	Purity %	Purity method	Type capacity	Calorimeter Type	Calorimeter Reference
*81BER/OGI	243.1–303.1	3S	nosp	not specified		C_{avg}	DSIO	*79BER

TABLE 12.45.3. Parameters of regression polynomial

Heat capacity type	No. data points total	used	s_w	s/R	s_r %	s_b/R	+/–
C	3	3	0.002	2.80–3	0.01	1.27–6	0

Temp. range K	A_1	A_2	Level of uncertainty
243.1–303.1	2.51898+1	4.05602	VI

TABLE 12.45.4. Recommended values of heat capacities

Temp. (K)	240	250	260	270	273.15	280	290
c (J K^{-1}g^{-1})	2.13	2.16	2.18	2.21	2.21	2.23	2.26
C (J K^{-1}mol^{-1})	290	294	297	300	302	304	307
Temp. (K)	298.15	300					
c (J K^{-1}g^{-1})	2.28	2.28					
C (J K^{-1}mol^{-1})	310	311					

Name: 4–Methylene–1–(1–methylethyl)bicyclo[3.1.0]hexane
Formula: $C_{10}H_{16}$

CAS-RN: 3387–41–5
Group No.: 12–046
Molar Mass: 136.24

TABLE 12.46.1. Experimental heat capacities

Reference		Temp. K	Capac. J/(K.g)	Error %	Purity %	Purity method	Type capacity	Calorimeter Type	Calorimeter Reference
33KOL/UDO	N	296.99	1.852	nosp	not specified		C_p	BSIT	34KOL/UDO2

33KOL/UDO same datum in 34KOL/UDO2

Name: Octahydro–4,7–methano–1*H*–indene
Formula: $C_{10}H_{16}$

CAS-RN: 6004–38–2
Group No.: 12–047
Molar Mass: 136.24

TABLE 12.47.1. Experimental heat capacities

Reference	Temp. range K	No. pnts	Error %	Purity %	Purity method	Type capacity	Calorimeter Type	Calorimeter Reference
71BOY/SAN	372.0–390.0	2	nosp	99.9	melpt	C_p	BDHT	73PER/COM

TABLE 12.47.3. Parameters of regression polynomial

Heat capacity type	No. data points total	used	s_w	s/R	s_r %	s_b/R	+/–
C_p	2	2	0.000	0.00	0.00	0.00	0

Temp. range K	A_1	A_2	Level of uncertainty
372.0–390.0	5.58572	7.54824	V

TABLE 12.47.4. Recommended values of heat capacities

Temp. (K)	370	380	390
c_p (J $K^{-1}g^{-1}$)	2.05	2.09	2.14
C_p (J $K^{-1}mol^{-1}$)	279	285	291

Name: 1,1'–Bicyclopentyl
Formula: $C_{10}H_{18}$

CAS-RN: 1636–39–1
Group No.: 12–048
Molar Mass: 138.25

TABLE 12.48.1. Experimental heat capacities

Reference		Temp. range K	No. pnts	Error %	Purity %	Purity method	Type capacity	Calorimeter Type	Calorimeter Reference
62GOL/BEL		310.9–422.0	3	nosp	98.9	chrom	C_p	BDHT	63GUD/CAM
63GUD/CAM	N	313.1–423.1	3	nosp	not specified		C_p	BDHT	63GUD/CAM
63GUD/CAM	N	313.1–423.1	3	nosp	not specified		C_p	BDHT	63GUD/CAM
76GOO/LEE		298.1	1	nosp	99.96	melpt	C_p	not specified	

63GUD/CAM substance contained 0.01 % Cyclopentyl bromide
63GUD/CAM substance contained 0.10 % Cyclopentyl bromide

TABLE 12.48.2. Correlated heat capacities

Reference	Temp. range K	No. pnts used	$\sigma_r C$ %	d_w	d/R	d_r %	d_b/R	+/–
				Selected data				
62GOL/BEL	310.9–422.0	3	1.00#	0.172	5.51–2	0.17	–1.03–2	0
63GUD/CAM	313.1–423.1	3	1.00#	0.609	1.70–1	0.61	8.87–2	0
63GUD/CAM	313.1–423.1	3	1.00#	0.720	1.98–1	0.72	–7.33–2	1

TABLE 12.48.3. Parameters of regression polynomial

Heat capacity type	No. data points total	No. data points used	s_w	s/R	s_r %	s_b/R	+/–
C_p	10	9	0.678	1.89–1	0.68	1.70–3	1

Temp. range K	A_1	A_2	A_3	Level of uncertainty
310.9–423.1	6.44337+1	–2.52925+1	4.32120	IV

TABLE 12.48.4. Recommended values of heat capacities

Temp. (K)	310	320	330	340	350	360	370
c_p (J $K^{-1}g^{-1}$)	1.657	1.669	1.686	1.708	1.735	1.767	1.805
C_p (J $K^{-1}mol^{-1}$)	229.1	230.7	233.0	236.1	239.8	244.3	249.5
Temp. (K)	380	390	400	410	420		
c_p (J $K^{-1}g^{-1}$)	1.848	1.896	1.949	2.007	2.071		
C_p (J $K^{-1}mol^{-1}$)	255.4	262.1	269.4	277.5	286.3		

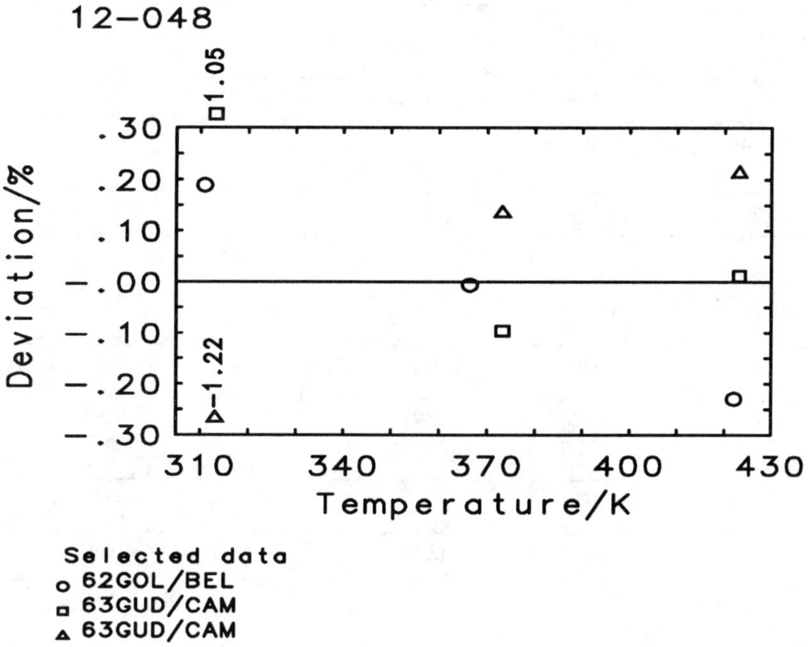

Name: *cis*–Decahydroazulene
Formula: $C_{10}H_{18}$

CAS-RN: 16189–46–1
Group No.: 12–049
Molar Mass: 138.25

TABLE 12.49.1. Experimental heat capacities

Reference	Temp. K	Capac. J/(K.g)	Error %	Purity %	Purity method	Type capacity	Calorimeter Type	Calorimeter Reference
70CHA/MCN	377.00	2.252	2.00	not specified		C_p	BDHT	73PER/COM

Name: Decahydronaphthalene
Formula: $C_{10}H_{18}$

CAS-RN: 91–17–8
Group No.: 12–050
Molar Mass: 138.25

TABLE 12.50.1. Experimental heat capacities

Reference		Temp. range K	No. pnts	Error %	Purity %	Purity method	Type capacity	Calorimeter Type	Calorimeter Reference
22HER/SCH		289.6	1	nosp	not specified		C_p	DSIO	22HER/SCH
62GOL/BEL	N	310.9–422.0	3	nosp	94.9	chrom	C_p	BDHT	63GUD/CAM
63GUD/CAM	N	313.1–423.1	3	nosp	not specified		C_p	BDHT	63GUD/CAM

62GOL/BEL mixture of isomers
63GUD/CAM mixture of isomers

TABLE 12.50.2. Correlated heat capacities

Reference	Temp. range K	No. pnts used	$\sigma_r C$ %	d_w	d/R	d_r %	d_b/R	+/–
				Selected data				
62GOL/BEL	310.9–422.0	3	1.00#	0.112	3.20–2	0.11	–5.51–3	–1
63GUD/CAM	313.1–423.1	3	1.00#	0.664	2.03–1	0.66	9.79–3	1

TABLE 12.50.3. Parameters of regression polynomial

Heat capacity type	No. data points total	used	s_w	s/R	s_r %	s_b/R	+/–
C_p	7	6	0.583	1.78–1	0.58	2.14–3	0

Temp. range K	A_1	A_2	Level of uncertainty
310.9–423.1	4.04425	7.19381	IV

TABLE 12.50.4. Recommended values of heat capacities

Temp. (K)	310	320	330	340	350	360	370
c_p (J K^{-1}g^{-1})	1.584	1.628	1.671	1.714	1.757	1.801	1.844
C_p (J K^{-1}mol^{-1})	219.0	225.0	231.0	237.0	243.0	249.0	254.9

Temp. (K)	380	390	400	410	420
c_p (J K^{-1}g^{-1})	1.887	1.930	1.974	2.017	2.060
C_p (J K^{-1}mol^{-1})	260.9	266.9	272.9	278.9	284.8

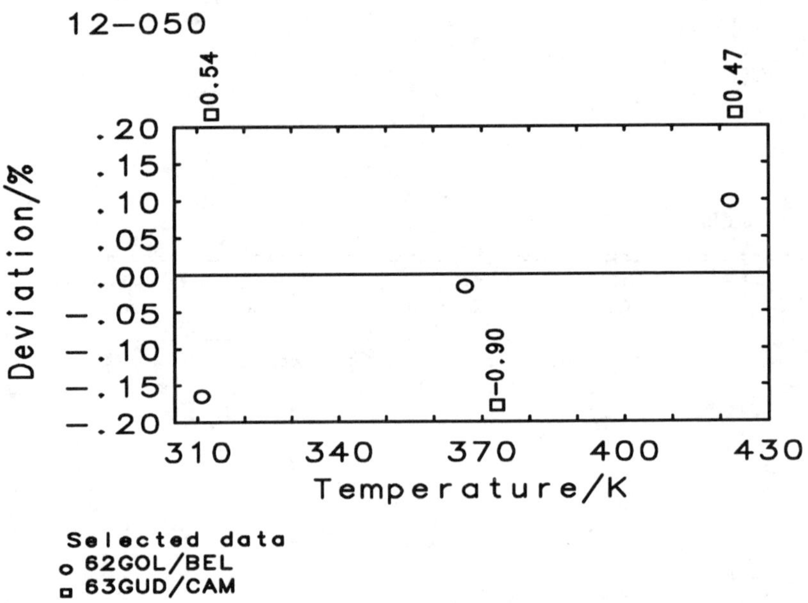

Name: *cis*–Decahydronaphthalene
Formula: $C_{10}H_{18}$

CAS-RN: 493–01–6
Group No.: 12–051
Molar Mass: 138.25

TABLE 12.51.1. Experimental heat capacities

Reference		Temp. range K	No. pnts	Error %	Purity %	Purity method	Type capacity	Calorimeter Type	Calorimeter Reference
49PAR/HAT		230.0–298.2	8S	0.70	99.6	melpt	C_p	BSIO	25PAR
53SEY	N	293.1–343.1	6S	nosp	not specified		C_p	BSAO	24WIL/DAN
57MCC/FIN2		234.2–344.5	37	0.20	99.88	melpt	C_{sat}	BSAO	47HUF
63GUD/CAM		313.1–423.1	3S	nosp	99.4	melpt	C_p	BDHT	63GUD/CAM
88SHI/OGA1		298.1	1	nosp	99.9	chrom	C_p	FSIO	85OGA
89OHN/FUJ		298.1	1	nosp	99.9	chrom	C_p	FSIO	85OGA

53SEY lambda transition observed in the temperature range 323.2–323.5 K

TABLE 12.51.2. Correlated heat capacities

Reference	Temp. range K	No. pnts used	$\sigma_r C$ %	d_w	d/R	d_r %	d_b/R	+/–
				Selected data				
57MCC/FIN2	234.2–344.5	37	0.20	0.461	2.58–2	0.09	−1.20–3	–1
63GUD/CAM	313.1–423.1	3	1.00#	2.381	7.20–1	2.38	3.63–1	–1
88SHI/OGA1	298.1	1	0.50#	0.188	2.62–2	0.09	2.62–2	1
89OHN/FUJ	298.1	1	0.50#	0.085	1.18–2	0.04	1.18–2	1
				Rejected data				
49PAR/HAT	(1.81–1, 0.70, 1.03–2, 1)			53SEY	(4.35–1, 1.41, 3.89–1, 6)			

TABLE 12.51.3. Parameters of regression polynomial

Heat capacity type	No. data points total	used	s_w	s/R	s_r %	s_b/R	+/–
C	56	42	0.799	2.01–1	0.67	2.58–2	0
Temp. range K	A_1		A_2	A_3			Level of uncertainty
234.2–423.1	1.67823+1		6.02504–1	1.04870			III

TABLE 12.51.4. Recommended values of heat capacities

Temp. (K)	240	250	260	270	273.15	280	290
c (J K^{-1}g^{-1})	1.460	1.494	1.530	1.567	1.579	1.605	1.645
C (J K^{-1}mol^{-1})	201.8	206.6	211.5	216.6	218.3	221.9	227.4
Temp. (K)	298.15	300	310	320	330	340	350
c (J K^{-1}g^{-1})	1.678	1.686	1.728	1.771	1.816	1.862	1.909
C (J K^{-1}mol^{-1})	232.0	233.0	238.9	244.9	251.0	257.4	263.9
Temp. (K)	360	370	380	390	400	410	420
c (J K^{-1}g^{-1})	1.957	2.007	2.058	2.110	2.163	2.218	2.274
C (J K^{-1}mol^{-1})	270.6	277.4	284.5	291.7	299.1	306.6	314.4

TABLE 12.51.5. Parameters of quasi-polynomial equation

Heat capacity type	No. data points total	used	s_w	s/R	s_r %	s_b/R	+/–
C	56	42	1.106	1.98–1	0.66	−4.71–3	–29
Temp. range K	T_c K	A_1	A_2	A_3	A_4		Level of uncertainty
234.2–423.1	702.30	−1.36540+1	3.29021	8.68558	1.41656+1		IV

12–051

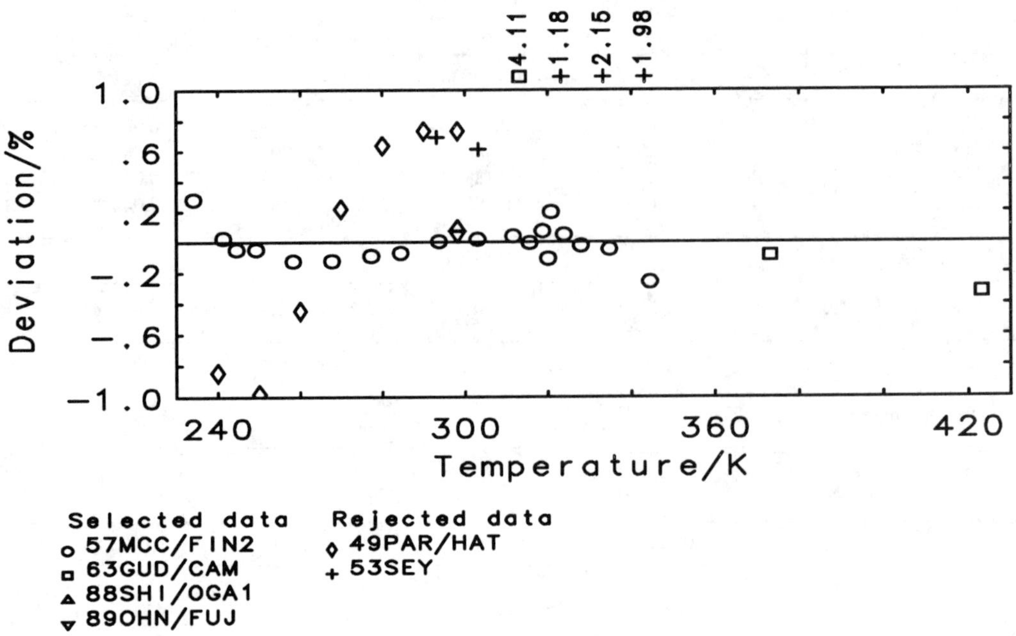

Name: *trans*–Decahydronaphthalene
Formula: $C_{10}H_{18}$

CAS-RN: 493–02–7
Group No.: 12–052
Molar Mass: 138.25

TABLE 12.52.1. Experimental heat capacities

Reference	Temp. range K	No. pnts	Error %	Purity %	Purity method	Type capacity	Calorimeter Type	Calorimeter Reference
49PAR/HAT	240.0–298.2	7S	0.70	98.9	melpt	C_p	BSIO	25PAR
53SEY	293.1–413.1	13S	nosp	not specified		C_p	BSAO	24WIL/DAN
57MCC/FIN2	248.8–344.3	13	0.20	99.98	melpt	C_{sat}	BSAO	47HUF
63GUD/CAM	313.1–423.1	3S	nosp	99.8	melpt	C_p	BDHT	63GUD/CAM
88SHI/OGA1	298.1	1	nosp	99.9	chrom	C_p	FSIO	85OGA
89OHN/FUJ	298.1	1	nosp	99.9	chrom	C_p	FSIO	85OGA

TABLE 12.52.2. Correlated heat capacities

Reference	Temp. range K	No. pnts used	$\sigma_r C$ %	d_w	d/R	d_r %	d_b/R	+/–
				Selected data				
53SEY	293.1–413.1	13	1.00#	0.918	2.97–1	0.92	1.86–1	6
57MCC/FIN2	248.8–344.3	13	0.20	0.626	3.76–2	0.13	–7.11–3	0
				Rejected data				
49PAR/HAT	(2.29–1, 0.87, 3.16–2, 0)			63GUD/CAM	(1.13, 3.74, 9.33–1, 2)			
88SHI/OGA1	(1.02–1, 0.37, 1.02–1, 1)			89OHN/FUJ	(1.02–1, 0.37, 1.02–1, 1)			

TABLE 12.52.3. Parameters of regression polynomial

Heat capacity type	No. data points total	used	s_w	s/R	s_r %	s_b/R	+/–
C	38	26	0.836	2.25–1	0.70	8.95–2	6
Temp. range K		A_1	A_2	A_3			Level of uncertainty
248.8–413.1		1.68061+1	4.81994–2	1.18237			IV

TABLE 12.52.4. Recommended values of heat capacities

Temp. (K)	250	260	270	273.15	280	290	298.15
c (J K^{-1}g^{-1})	1.462	1.499	1.537	1.549	1.576	1.617	1.651
C (J K^{-1}mol^{-1})	202.2	207.2	212.5	214.2	217.9	223.6	228.3
Temp. (K)	300	310	320	330	340	350	360
c (J K^{-1}g^{-1})	1.659	1.703	1.748	1.795	1.843	1.892	1.943
C (J K^{-1}mol^{-1})	229.4	235.5	241.7	248.1	254.7	261.6	268.6
Temp. (K)	370	380	390	400	410		
c (J K^{-1}g^{-1})	1.995	2.049	2.104	2.160	2.218		
C (J K^{-1}mol^{-1})	275.8	283.2	290.8	298.6	306.6		

TABLE 12.52.5. Parameters of quasi-polynomial equation

Heat capacity type	No. data points total	used	s_w	s/R		s_r %	s_b/R	+/–
C	38	26	2.017	1.40–1		0.47	1.75–2	3
Temp. range K	T_c K	A_1	A_2	A_3	A_4	A_5	A_6	Level of uncertainty
248.8–413.1	687.10	1.90456+3	2.02414+2	–1.84462+2	1.77417+3	3.51300+2	1.24819+3	IV

12–052

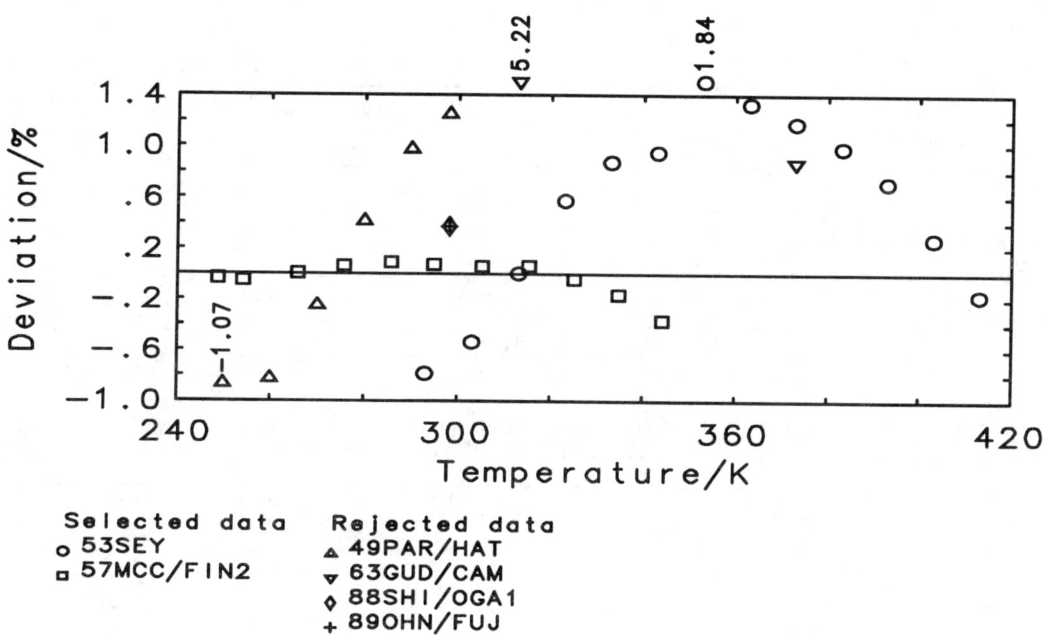

Name: Octahydromethyl–1*H*–indene
Formula: $C_{10}H_{18}$

CAS–RN: unknown
Group No.: 12–053
Molar Mass: 138.25

TABLE 12.53.1. Experimental heat capacities

Reference	Temp. range K	No. pnts	Error %	Purity %	Purity method	Type capacity	Calorimeter Type	Calorimeter Reference
63GUD/CAM	313.1–423.1	3	nosp	95.9	chrom	C_p	BDHT	63GUD/CAM

TABLE 12.53.3. Parameters of regression polynomial

Heat capacity type	No. data points total	No. data points used	s_w	s/R	s_r %	s_b/R	+/–
C_p	3	3	0.000	0.00	0.00	3.82–6	0

Temp. range K	A_1	A_2	A_3	Level of uncertainty
313.1–423.1	5.33793+1	–1.68620+1	3.18347	V

TABLE 12.53.4. Recommended values of heat capacities

Temp. (K)	310	320	330	340	350	360	370
c_p (J K^{-1}g^{-1})	1.91	1.93	1.95	1.98	2.01	2.04	2.08
C_p (J K^{-1}mol^{-1})	264	266	269	273	277	282	287

Temp. (K)	380	390	400	410	420
c_p (J K^{-1}g^{-1})	2.12	2.17	2.22	2.27	2.33
C_p (J K^{-1}mol^{-1})	293	300	307	314	322

Name: 2,6,6–Trimethylbicyclo[3.1.1]heptane
Formula: $C_{10}H_{18}$

CAS–RN: 473–55–2
Group No.: 12–054
Molar Mass: 138.25

TABLE 12.54.1. Experimental heat capacities

Reference	Temp. range K	No. pnts	Error %	Purity %	Purity method	Type capacity	Calorimeter Type	Calorimeter Reference
62GOL/BEL	310.9–422.0	3	nosp	98.	chrom	C_p	BDHT	63GUD/CAM
63GUD/CAM	313.1–423.1	3	nosp	not specified		C_p	BDHT	63GUD/CAM

TABLE 12.54.2. Correlated heat capacities

Reference	Temp. range K	No. pnts used	$\sigma_r C$ %	d_w	d/R	d_r %	d_b/R	+/–
				Selected data				
62GOL/BEL	310.9–422.0	3	1.00#	0.060	1.86–2	0.06	1.28–2	2
63GUD/CAM	313.1–423.1	3	1.00#	0.062	1.91–2	0.06	–1.27–2	–1

TABLE 12.54.3. Parameters of regression polynomial

Heat capacity type	No. data points total	used	s_w	s/R	s_r %	s_b/R	+/–
C_p	6	6	0.086	2.66–2	0.09	2.67–5	1

Temp. range K	A_1	A_2	A_3	Level of uncertainty
310.9–423.1	4.70112+1	–1.65904+1	3.35032	IV

TABLE 12.54.4. Recommended values of heat capacities

Temp. (K)	310	320	330	340	350	360	370
c_p (J K^{-1}g^{-1})	1.671	1.698	1.729	1.764	1.803	1.847	1.894
C_p (J K^{-1}mol^{-1})	231.0	234.7	239.0	243.9	249.3	255.3	261.8
Temp. (K)	380	390	400	410	420		
c_p (J K^{-1}g^{-1})	1.945	2.001	2.060	2.123	2.191		
C_p (J K^{-1}mol^{-1})	268.9	276.6	284.8	293.6	302.9		

Name: Butylcyclohexane
Formula: $C_{10}H_{20}$

CAS-RN: 1678–93–9
Group No.: 12–055
Molar Mass: 140.27

TABLE 12.55.1. Experimental heat capacities

Reference	Temp. range K	No. pnts	Error %	Purity %	Purity method	Type capacity	Calorimeter Type	Calorimeter Reference
65FIN/MES	207.5–365.4	21	0.20	99.97	melpt	C_{sat}	BSAO	47HUF

TABLE 12.55.3. Parameters of regression polynomial

Heat capacity type	No. data points total	used	s_w	s/R	s_r %	s_b/R	+/–
C_{sat}	21	21	0.096	5.99–3	0.02	9.08–7	1

Temp. range K	A_1	A_2	A_3	A_4	Level of uncertainty
207.5–365.4	3.18295+1	–1.06364+1	5.08397	–4.79809–1	II

TABLE 12.55.4. Recommended values of heat capacities

Temp. (K)	210	220	230	240	250	260	270
c_{sat} (J K^{-1}g^{-1})	1.628	1.655	1.685	1.716	1.750	1.785	1.821
C_{sat} (J K^{-1}mol^{-1})	228.4	232.2	236.3	240.7	245.4	250.3	255.5
Temp. (K)	273.15	280	290	298.15	300	310	320
c_{sat} (J K^{-1}g^{-1})	1.833	1.860	1.899	1.932	1.940	1.981	2.023
C_{sat} (J K^{-1}mol^{-1})	257.2	260.9	266.4	271.0	272.1	277.9	283.8
Temp. (K)	330	340	350	360	370		
c_{sat} (J K^{-1}g^{-1})	2.066	2.109	2.152	2.196	2.239		
C_{sat} (J K^{-1}mol^{-1})	289.8	295.8	301.9	308.0	314.0		

TABLE 12.55.5. Parameters of quasi-polynomial equation

Heat capacity type	No. data points total	used	s_w	s/R	s_r %	s_b/R	+/–
C_{sat}	21	21	1.004	6.44–2	0.20	2.16–4	0

Temp. range K	T_c K	A_1	A_2	A_3	A_4	Level of uncertainty
207.5–365.4	648.40	–1.14068+1	3.89369	1.44810+1	8.35427	II

Name: Diethylcyclohexane (unspecified isomer)
Formula: $C_{10}H_{20}$

CAS-RN: 1331–43–7
Group No.: 12–056
Molar Mass: 140.27

TABLE 12.56.1. Experimental heat capacities

Reference		Temp. range K	No. pnts	Error %	Purity %	method	Type capacity	Calorimeter Type	Reference
62GOL/BEL	N	310.9–422.0	3	nosp	99.8	chrom	C_p	BDHT	63GUD/CAM
63GUD/CAM	N	313.1–423.1	3	nosp	not specified		C_p	BDHT	63GUD/CAM

62GOL/BEL mixture of isomers
63GUD/CAM sample contained 95 mol.% of major component (not specified)

TABLE 12.56.2. Correlated heat capacities

Reference	Temp. range K	No. pnts used	$\sigma_r C$ %	d_w	d/R	d_r %	d_b/R	+/–
				Selected data				
62GOL/BEL	310.9–422.0	3	1.00#	1.753	6.60–1	1.75	1.46–1	1
63GUD/CAM	313.1–423.1	3	1.00#	0.452	1.77–1	0.45	–1.17–1	–1

TABLE 12.56.3. Parameters of regression polynomial

Heat capacity type	No. data points total	used	s_w	s/R	s_r %	s_b/R	+/–
C_p	6	6	1.568	5.91–1	1.57	1.43–2	0

Temp. range K	A_1	A_2	Level of uncertainty
310.9–423.1	6.80683	7.86799	V

TABLE 12.56.4. Recommended values of heat capacities

Temp. (K)	310	320	330	340	350	360	370
c_p (J K^{-1}g^{-1})	1.85	1.90	1.94	1.99	2.04	2.08	2.13
C_p (J K^{-1}mol^{-1})	259	266	272	279	286	292	299
Temp. (K)	380	390	400	410	420		
c_p (J K^{-1}g^{-1})	2.18	2.22	2.27	2.32	2.36		
C_p (J K^{-1}mol^{-1})	305	312	318	325	331		

Name: 1,4–Diethylcyclohexane
Formula: $C_{10}H_{20}$

CAS-RN: 1679–00–1
Group No.: 12–057
Molar Mass: 140.27

TABLE 12.57.1. Experimental heat capacities

Reference		Temp. range K	No. pnts	Error %	Purity %	Purity method	Type capacity	Calorimeter Type	Calorimeter Reference
63GUD/CAM	N	313.1–423.1	3	nosp	99.95	chrom	C_p	BDHT	63GUD/CAM
63GUD/CAM	N	313.1–423.1	3	nosp	99.97	chrom	C_p	BDHT	63GUD/CAM

63GUD/CAM mixture of isomers (30 % *cis* and 70 % *trans*)
63GUD/CAM mixture of isomers (80 % *cis*)

TABLE 12.57.2. Correlated heat capacities

Reference	Temp. range K	No. pnts used	$\sigma_r C$ %	d_w	d/R	d_r %	d_b/R	+/–
				Selected data				
63GUD/CAM	313.1–423.1	3	1.00#	1.138	3.83–1	1.14	1.03–1	1
63GUD/CAM	313.1–423.1	3	1.00#	1.138	3.82–1	1.14	–8.54–2	–1

TABLE 12.57.3. Parameters of regression polynomial

Heat capacity type	No. data points total	No. data points used	s_w	s/R	s_r %	s_b/R	+/–
C_p	6	6	1.610	5.41–1	1.61	8.69–3	0

Temp. range K	A_1	A_2	A_3	Level of uncertainty
313.1–423.1	6.28796+1	–2.12034+1	3.60397	IV

TABLE 12.57.4. Recommended values of heat capacities

Temp. (K)	310	320	330	340	350	360	370
c_p (J K^{-1}g^{-1})	1.884	1.893	1.906	1.923	1.945	1.971	2.001
C_p (J K^{-1}mol^{-1})	264.3	265.5	267.4	269.8	272.9	276.5	280.7

Temp. (K)	380	390	400	410	420
c_p (J K^{-1}g^{-1})	2.036	2.075	2.118	2.165	2.217
C_p (J K^{-1}mol^{-1})	285.6	291.0	297.1	303.7	311.0

Name: (1,1–Dimethylethyl)cyclohexane
Formula: $C_{10}H_{20}$

CAS-RN: 3178–22–1
Group No.: 12–058
Molar Mass: 140.27

TABLE 12.58.1. Experimental heat capacities

Reference	Temp. range K	No. pnts	Error %	Purity %	Purity method	Type capacity	Calorimeter Type	Calorimeter Reference
63GUD/CAM	313.1–423.1	3	nosp	99.5	chrom	C_p	BDHT	63GUD/CAM

TABLE 12.58.3. Parameters of regression polynomial

Heat capacity type	No. data points total	used	s_w	s/R	s_r %	s_b/R	+/−
C_p	3	3	0.595	2.28−1	0.60	2.11−3	1

Temp. range K	A_1	A_2	Level of uncertainty
313.1–423.1	1.42431	9.69204	IV

TABLE 12.58.4. Recommended values of heat capacities

Temp. (K)	320	330	340	350	360	370	380
c_p (J K^{-1}g^{-1})	1.923	1.980	2.038	2.095	2.153	2.210	2.268
C_p (J K^{-1}mol^{-1})	269.7	277.8	285.8	293.9	301.9	310.0	318.1
Temp. (K)	390	400	410	420			
c_p (J K^{-1}g^{-1})	2.325	2.382	2.440	2.497			
C_p (J K^{-1}mol^{-1})	326.1	334.2	342.2	350.3			

Name: Decahydro–1–methylnaphthalene
Formula: $C_{11}H_{20}$

CAS-RN: 2958–75–0
Group No.: 12–059
Molar Mass: 152.28

TABLE 12.59.1. Experimental heat capacities

Reference		Temp. range K	No. pnts	Error %	Purity %	Purity method	Type capacity	Calorimeter Type	Calorimeter Reference
62GOL/BEL		310.9–422.0	3	nosp	98.5	chrom	C_p	BDHT	63GUD/CAM
63GUD/CAM	N	313.1–423.1	3	nosp	98.5	chrom	C_p	BDHT	63GUD/CAM

63GUD/CAM sample contained 27.2 % *cis*– and 71.3 % *trans*–isomers

TABLE 12.59.2. Correlated heat capacities

Reference	Temp. range K	No. pnts used	$\sigma_r C$ %	d_w	d/R	d_r %	d_b/R	+/−
				Selected data				
62GOL/BEL	310.9–422.0	3	1.00#	0.298	1.01−1	0.30	9.24−2	3
63GUD/CAM	313.1–423.1	3	1.00#	1.192	4.44−1	1.19	−8.43−2	−1

TABLE 12.59.3. Parameters of regression polynomial

Heat capacity type	No. data points total	used	s_w	s/R	s_r %	s_b/R	+/−
C_p	6	6	1.064	3.95−1	1.06	4.08−3	2

Temp. range K	A_1	A_2	Level of uncertainty
310.9–423.1	6.18535	8.36547	IV

TABLE 12.59.4. Recommended values of heat capacities

Temp. (K)	310	320	330	340	350	360	370
c_p (J K^{-1}g^{-1})	1.754	1.799	1.845	1.891	1.936	1.982	2.028
C_p (J K^{-1}mol^{-1})	267.0	274.0	281.0	287.9	294.9	301.8	308.8
Temp. (K)	380	390	400	410	420		
c_p (J K^{-1}g^{-1})	2.073	2.119	2.165	2.210	2.256		
C_p (J K^{-1}mol^{-1})	315.7	322.7	329.6	336.6	343.6		

Name: Decahydro–2–methylnaphthalene
Formula: $C_{11}H_{20}$

CAS-RN: 2958–76–1
Group No.: 12–060
Molar Mass: 152.28

TABLE 12.60.1. Experimental heat capacities

Reference		Temp. range K	No. pnts	Error %	Purity %	Purity method	Type capacity	Calorimeter Type	Calorimeter Reference
62GOL/BEL		310.9–422.0	3	nosp	99.5	chrom	C_p	BDHT	63GUD/CAM
63GUD/CAM	N	313.1–423.1	3	nosp	99.5	chrom	C_p	BDHT	63GUD/CAM

63GUD/CAM sample contained 61.6 % *cis*– and 37.9 % *trans*–isomers

TABLE 12.60.2. Correlated heat capacities

Reference	Temp. range K	No. pnts used	$\sigma_r C$ %	d_w	d/R	d_r %	d_b/R	+/–
				Selected data				
62GOL/BEL	310.9–422.0	3	1.00#	0.288	9.35–2	0.29	–8.33–2	–3
63GUD/CAM	313.1–423.1	3	1.00#	1.324	4.81–1	1.32	1.00–1	1

TABLE 12.60.3. Parameters of regression polynomial

Heat capacity type	No. data points total	No. data points used	s_w	s/R	s_r %	s_b/R	+/–
C_p	6	6	1.174	4.24–1	1.17	8.48–3	–2

Temp. range K	A_1	A_2	Level of uncertainty
310.9–423.1	2.40572	9.05443	V

TABLE 12.60.4. Recommended values of heat capacities

Temp. (K)	310	320	330	340	350	360	370
c_p (J K^{-1}g^{-1})	1.66	1.71	1.76	1.81	1.86	1.91	1.96
C_p (J K^{-1}mol^{-1})	253	261	268	276	283	291	299
Temp. (K)	380	390	400	410	420		
c_p (J K^{-1}g^{-1})	2.01	2.06	2.11	2.16	2.21		
C_p (J K^{-1}mol^{-1})	306	314	321	329	336		

Name: Ethyloctahydro–1*H*–indene
Formula: $C_{11}H_{20}$

CAS-RN: 95098–51–4
Group No.: 12–061
Molar Mass: 152.28

TABLE 12.61.1. Experimental heat capacities

Reference	Temp. range K	No. pnts	Error %	Purity %	Purity method	Type capacity	Calorimeter Type	Calorimeter Reference
62GOL/BEL	310.9–422.0	3	nosp	99.4	chrom	C_p	BDHT	63GUD/CAM
63GUD/CAM	313.1–423.1	3	nosp	99.4	chrom	C_p	BDHT	63GUD/CAM

TABLE 12.61.2. Correlated heat capacities

Reference	Temp. range K	No. pnts used	$\sigma_r C$ %	d_w	d/R	d_r %	d_b/R	+/–
				Selected data				
62GOL/BEL	310.9–422.0	3	1.00#	0.151	5.60–2	0.15	–1.32–2	–1
63GUD/CAM	313.1–423.1	3	1.00#	0.153	6.06–2	0.15	1.30–2	0

TABLE 12.61.3. Parameters of regression polynomial

Heat capacity type	No. data points total	No. data points used	s_w	s/R	s_r %	s_b/R	+/–
C_p	6	6	0.186	7.15–2	0.19	–1.37–4	–1

Temp. range K	A_1	A_2	Level of uncertainty
310.9–423.1	1.01602+1	7.64425	IV

TABLE 12.61.4. Recommended values of heat capacities

Temp. (K)	310	320	330	340	350	360	370
c_p (J K^{-1}g^{-1})	1.849	1.890	1.932	1.974	2.016	2.057	2.099
C_p (J K^{-1}mol^{-1})	281.5	287.9	294.2	300.6	306.9	313.3	319.6

Temp. (K)	380	390	400	410	420
c_p (J K^{-1}g^{-1})	2.141	2.183	2.224	2.266	2.308
C_p (J K^{-1}mol^{-1})	326.0	332.4	338.7	345.1	351.4

Name: Decahydrodimethanonaphthalene
Formula: $C_{12}H_{18}$

CAS–RN: unknown
Group No.: 12–062
Molar Mass: 162.27

TABLE 12.62.1. Experimental heat capacities

Reference	Temp. range K	No. pnts	Error %	Purity %	Purity method	Type capacity	Calorimeter Type	Calorimeter Reference
63GUD/CAM	313.1–423.1	3	nosp	not specified		C_p	BDHT	63GUD/CAM

TABLE 12.62.3. Parameters of regression polynomial

Heat capacity type	No. data points total	used	s_w	s/R	s_r %	s_b/R	+/–
C_p	3	3	0.000	0.00	0.00	2.54–6	0

Temp. range K	A_1	A_2	A_3	Level of uncertainty
313.1–423.1	–4.78160+1	3.94169+1	–4.35526	V

TABLE 12.62.4. Recommended values of heat capacities

Temp. (K)	310	320	330	340	350	360	370
c_p (J K^{-1}g^{-1})	1.67	1.73	1.78	1.84	1.89	1.93	1.97
C_p (J K^{-1}mol^{-1})	270	280	290	298	306	313	319
Temp. (K)	380	390	400	410	420		
c_p (J K^{-1}g^{-1})	2.00	2.03	2.06	2.08	2.10		
C_p (J K^{-1}mol^{-1})	325	330	334	337	340		

Name: 1,3–Dimethyltricyclo[3.3.1.1^{3,7}]decane
Formula: $C_{12}H_{20}$

CAS-RN: 702–79–4
Group No.: 12–063
Molar Mass: 164.29

TABLE 12.63.1. Experimental heat capacities

Reference		Temp. K	Capac. J/(K.g)	Error %	Purity %	Purity method	Type capacity	Calorimeter Type	Calorimeter Reference
77STE/WAT	N	298.15	2.160	nosp	not specified		C_p	not specified	

77STE/WAT unspecified DSC calorimeter was used

Name: Tricyclo[6.2.1.1^{3,6}]dodecane
Formula: $C_{12}H_{20}$

CAS-RN: 281–84–5
Group No.: 12–064
Molar Mass: 164.29

TABLE 12.64.1. Experimental heat capacities

Reference		Temp. range K	No. pnts	Error %	Purity %	Purity method	Type capacity	Calorimeter Type	Calorimeter Reference
62GOL/BEL	N	310.9–422.0	3	nosp	98.5	chrom	C_p	BDHT	63GUD/CAM

62GOL/BEL probably an error in the name of substance; identified as Tetracyclo[6.2.1.1^{3,6}]dodecane by the authors

TABLE 12.64.3. Parameters of regression polynomial

Heat capacity type	No. data points total	used	s_w	s/R	s_r %	s_b/R	+/–
C_p	3	3	0.000	0.00	0.00	0.00	0

Temp. range K	A_1	A_2	Level of uncertainty
310.9–422.0	1.00505+1	7.58954	V

TABLE 12.64.4. Recommended values of heat capacities

Temp. (K)	310	320	330	340	350	360	370
c_p (J $K^{-1}g^{-1}$)	1.70	1.74	1.78	1.81	1.85	1.89	1.93
C_p (J $K^{-1}mol^{-1}$)	279	285	292	298	304	311	317
Temp. (K)	380	390	400	410	420		
c_p (J $K^{-1}g^{-1}$)	1.97	2.01	2.05	2.08	2.12		
C_p (J $K^{-1}mol^{-1}$)	323	330	336	342	349		

Name: 1,1'–Bicyclohexyl
Formula: $C_{12}H_{22}$

CAS-RN: 92–51–3
Group No.: 12–065
Molar Mass: 166.31

TABLE 12.65.1. Experimental heat capacities

Reference	Temp. range K	No. pnts	Error %	Purity %	Purity method	Type capacity	Calorimeter Type	Calorimeter Reference
57PIL	276.9	1	nosp	99.9985	melpt	C_p	BSAO	57PIL
62GOL/BEL	310.9–422.0	3S	nosp	99.2	chrom	C_p	BDHT	63GUD/CAM
63GUD/CAM	313.1–483.1	4	nosp	99.2	chrom	C_p	BDHT	63GUD/CAM
83ORO/MRA	279.1–470.0	47S	1.00	99.9	chrom	C_p	BDHT	69PER/COM

TABLE 12.65.2. Correlated heat capacities

Reference	Temp. range K	No. pnts used	$\sigma_r C$ %	d_w	d/R	d_r %	d_b/R	+/–
				Selected data				
57PIL	276.9	1	0.50#	0.433	6.96–2	0.22	6.96–2	1
83ORO/MRA	279.1–470.0	47	1.00	0.774	2.84–1	0.77	–7.60–4	–7
				Rejected data				
62GOL/BEL	(1.01, 2.63, –3.17–1, –1)			63GUD/CAM	(9.54–1, 2.45, –3.72–1, –1)			

TABLE 12.65.3. Parameters of regression polynomial

Heat capacity type	No. data points total	No. data points used	s_w	s/R	s_r %	s_b/R	+/–
C_p	55	48	0.785	2.87–1	0.78	7.07–4	–6

Temp. range K	A_1	A_2	Level of uncertainty
276.9–470.0	6.73499	9.15668	IV

TABLE 12.65.4. Recommended values of heat capacities

Temp. (K)	280	290	298.15	300	310	320	330
c_p (J K^{-1}g^{-1})	1.619	1.664	1.702	1.710	1.756	1.802	1.847
C_p (J K^{-1}mol^{-1})	269.2	276.8	283.0	284.4	292.0	299.6	307.2
Temp. (K)	340	350	360	370	380	390	400
c_p (J K^{-1}g^{-1})	1.893	1.939	1.985	2.031	2.076	2.122	2.168
C_p (J K^{-1}mol^{-1})	314.9	322.5	330.1	337.7	345.3	352.9	360.5
Temp. (K)	410	420	430	440	450	460	470
c_p (J K^{-1}g^{-1})	2.214	2.259	2.305	2.351	2.397	2.443	2.488
C_p (J K^{-1}mol^{-1})	368.1	375.8	383.4	391.0	398.6	406.2	413.8

TABLE 12.65.5. Parameters of quasi-polynomial equation

Heat capacity type	No. data points total	used	s_w	s/R	s_r %	s_b/R	+/–
C_p	55	48	0.784	2.87–1	0.78	2.67–3	–12

Temp. range K	T_c K	A_1	A_2	A_3	A_4	Level of uncertainty
276.9–470.0	899.00	-8.87719	2.95407–1	7.89791	6.66915+1	IV

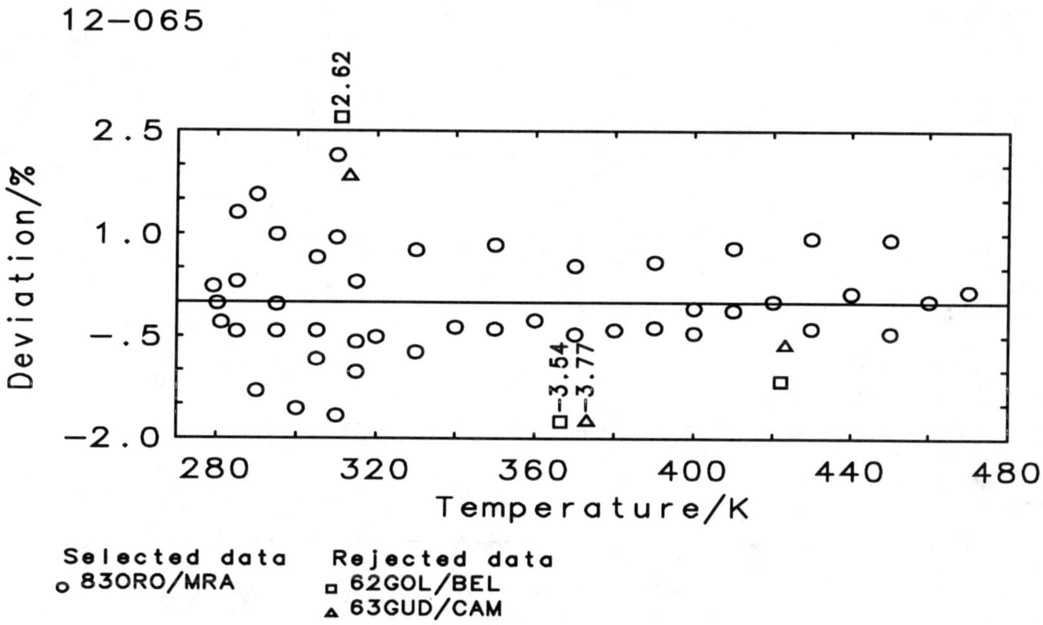

Name: Decahydrodimethylnaphthalene
Formula: $C_{12}H_{22}$

CAS-RN: 28777–88–0
Group No.: 12–066
Molar Mass: 166.31

TABLE 12.66.1. Experimental heat capacities

Reference		Temp. range K	No. pnts	Error %	Purity %	method	Type capacity	Calorimeter Type	Reference
63GUD/CAM	N	313.1–483.1	4	nosp	99.9	chrom	C_p	BDHT	63GUD/CAM

63GUD/CAM sample contained 37.3 mol.% of major component (not specified)

TABLE 12.66.3. Parameters of regression polynomial

Heat capacity type	No. data points total	used	s_w	s/R	s_r %	s_b/R	+/–
C_p	4	4	4.737	2.07	4.74	5.53–2	0

Temp. range K	A_1	A_2	A_3	Level of uncertainty
313.1–483.1	7.61407+1	-2.77146+1	4.83646	V

TABLE 12.66.4. Recommended values of heat capacities

Temp. (K)	310	320	330	340	350	360	370
c_p ($J K^{-1} g^{-1}$)	1.84	1.85	1.87	1.89	1.92	1.95	1.99
C_p ($J K^{-1} mol^{-1}$)	305	307	311	314	319	325	331
Temp. (K)	380	390	400	410	420	430	440
c_p ($J K^{-1} g^{-1}$)	2.03	2.08	2.13	2.19	2.25	2.32	2.39
C_p ($J K^{-1} mol^{-1}$)	338	346	355	364	375	386	398
Temp. (K)	450	460	470	480			
c_p ($J K^{-1} g^{-1}$)	2.47	2.55	2.64	2.73			
C_p ($J K^{-1} mol^{-1}$)	410	424	438	453			

Name: Ethyldecahydronaphthalene
Formula: $C_{12}H_{22}$

CAS-RN: 25551–49–9
Group No.: 12–067
Molar Mass: 166.31

TABLE 12.67.1. Experimental heat capacities

Reference		Temp. range K	No. pnts	Error %	Purity %	Purity method	Type capacity	Calorimeter Type	Calorimeter Reference
63GUD/CAM	N	313.1–483.1	4	nosp	98.7	chrom	C_p	BDHT	63GUD/CAM

63GUD/CAM sample contained 40.2 mol.% of major component (not specified)

TABLE 12.67.3. Parameters of regression polynomial

Heat capacity type	No. data points total	used	s_w	s/R	s_r %	s_b/R	+/–
C_p	4	4	3.342	1.48	3.34	2.78–2	0

Temp. range K	A_1	A_2	A_3	Level of uncertainty
313.1–483.1	5.03128+1	-1.48325+1	3.27849	VI

TABLE 12.67.4. Recommended values of heat capacities

Temp. (K)	310	320	330	340	350	360	370
c_p (J K^{-1}g^{-1})	1.79	1.82	1.85	1.89	1.93	1.97	2.02
C_p (J K^{-1}mol^{-1})	298	303	308	314	321	328	335
Temp. (K)	380	390	400	410	420	430	440
c_p (J K^{-1}g^{-1})	2.06	2.12	2.17	2.23	2.29	2.36	2.43
C_p (J K^{-1}mol^{-1})	343	352	361	371	381	392	403
Temp. (K)	450	460	470	480			
c_p (J K^{-1}g^{-1})	2.50	2.57	2.65	2.73			
C_p (J K^{-1}mol^{-1})	415	428	441	454			

Name: 1–Ethyldecahydronaphthalene
Formula: $C_{12}H_{22}$

CAS-RN: 1008–17–9
Group No.: 12–068
Molar Mass: 166.31

TABLE 12.68.1. Experimental heat capacities

Reference		Temp. range K	No. pnts	Error %	Purity %	Purity method	Type capacity	Calorimeter Type	Calorimeter Reference
62GOL/BEL		310.9–422.0	3	nosp	99.2	chrom	C_p	BDHT	63GUD/CAM
63GUD/CAM	N	313.1–423.1	3	nosp	99.2	chrom	C_p	BDHT	63GUD/CAM

63GUD/CAM mixture of isomers (31.4 % *cis* and 67.8 % *trans*)

TABLE 12.68.2. Correlated heat capacities

Reference	Temp. range K	No. pnts used	$\sigma_r C$ %	d_w	d/R	d_r %	d_b/R	+/–
				Selected data				
62GOL/BEL	310.9–422.0	3	1.00#	0.058	2.56–2	0.06	9.30–3	1
63GUD/CAM	313.1–423.1	3	1.00#	0.363	1.51–1	0.36	–7.28–3	1

TABLE 12.68.3. Parameters of regression polynomial

Heat capacity type	No. data points total	No. data points used	s_w	s/R	s_r %	s_b/R	+/–
C_p	6	6	0.319	1.32–1	0.32	1.01–3	2

Temp. range K	A_1	A_2	Level of uncertainty
310.9–423.1	1.15152+1	8.02962	IV

TABLE 12.68.4. Recommended values of heat capacities

Temp. (K)	310	320	330	340	350	360	370
c_p (J K^{-1}g^{-1})	1.820	1.860	1.900	1.941	1.981	2.021	2.061
C_p (J K^{-1}mol^{-1})	302.7	309.4	316.1	322.7	329.4	336.1	342.8
Temp. (K)	380	390	400	410	420		
c_p (J K^{-1}g^{-1})	2.101	2.141	2.181	2.222	2.262		
C_p (J K^{-1}mol^{-1})	349.4	356.1	362.8	369.5	376.1		

Name: 2–Ethyldecahydronaphthalene
Formula: $C_{12}H_{22}$

CAS-RN: 1618–23–1
Group No.: 12–069
Molar Mass: 166.31

TABLE 12.69.1. Experimental heat capacities

Reference		Temp. range K	No. pnts	Error %	Purity %	Purity method	Type capacity	Calorimeter Type	Calorimeter Reference
62GOL/BEL		310.9–422.0	3	nosp	99.8	chrom	C_p	BDHT	63GUD/CAM
63GUD/CAM	N	313.1–423.1	3	nosp	99.8	chrom	C_p	BDHT	63GUD/CAM

63GUD/CAM sample contained 31.7 % *cis*– and 68.1 % *trans*–isomers

TABLE 12.69.2. Correlated heat capacities

Reference	Temp. range K	No. pnts used	$\sigma_r C$ %	d_w	d/R	d_r %	d_b/R	+/–
				Selected data				
62GOL/BEL	310.9–422.0	3	1.00#	0.056	2.25–2	0.06	–8.76–3	0
63GUD/CAM	313.1–423.1	3	1.00#	0.214	8.68–2	0.21	9.84–3	1

TABLE 12.69.3. Parameters of regression polynomial

Heat capacity type	No. data points total	No. data points used	s_w	s/R	s_r %	s_b/R	+/–
C_p	6	6	0.192	7.77–2	0.19	5.42–4	1

Temp. range K	A_1	A_2	Level of uncertainty
310.9–423.1	8.01315	8.64761	IV

TABLE 12.69.4. Recommended values of heat capacities

Temp. (K)	310	320	330	340	350	360	370
c_p (J K^{-1}g^{-1})	1.741	1.784	1.827	1.871	1.914	1.957	2.000
C_p (J K^{-1}mol^{-1})	289.5	296.7	303.9	311.1	318.3	325.5	332.7
Temp. (K)	380	390	400	410	420		
c_p (J K^{-1}g^{-1})	2.044	2.087	2.130	2.173	2.216		
C_p (J K^{-1}mol^{-1})	339.8	347.0	354.2	361.4	368.6		

Name: Octahydro(1–methylethyl)–1*H*–indene
Formula: $C_{12}H_{22}$

CAS-RN: 88889–26–3
Group No.: 12–070
Molar Mass: 166.31

TABLE 12.70.1. Experimental heat capacities

Reference	Temp. range K	No. pnts	Error %	Purity %	Purity method	Type capacity	Calorimeter Type	Calorimeter Reference
62GOL/BEL	310.9–422.0	3	nosp	93.5	chrom	C_p	BDHT	63GUD/CAM
63GUD/CAM	313.1–423.1	3	nosp	93.5	chrom	C_p	BDHT	63GUD/CAM

TABLE 12.70.2. Correlated heat capacities

Reference	Temp. range K	No. pnts used	$\sigma_r C$ %	d_w	d/R	d_r %	d_b/R	+/–
				Selected data				
62GOL/BEL	310.9–422.0	3	1.00#	0.644	2.76–1	0.64	4.84–3	–1
63GUD/CAM	313.1–423.1	3	1.00#	0.637	2.78–1	0.64	1.12–3	–1

TABLE 12.70.3. Parameters of regression polynomial

Heat capacity type	No. data points total	used	s_w	s/R	s_r %	s_b/R	+/–
C_p	6	6	0.785	3.39–1	0.78	2.98–3	–2

Temp. range K	A_1	A_2	Level of uncertainty
310.9–423.1	2.38940+1	5.13264	V

TABLE 12.70.4. Recommended values of heat capacities

Temp. (K)	310	320	330	340	350	360	370
c_p (J K^{-1}g^{-1})	1.99	2.02	2.04	2.07	2.09	2.12	2.14
C_p (J K^{-1}mol^{-1})	331	335	339	344	348	352	357

Temp. (K)	380	390	400	410	420
c_p (J K^{-1}g^{-1})	2.17	2.20	2.22	2.25	2.27
C_p (J K^{-1}mol^{-1})	361	365	369	374	378

Name: 1,3,5–Trimethyltricyclo[3.3.1.1^{3,7}]decane
Formula: $C_{13}H_{22}$

CAS-RN: 707–35–7
Group No.: 12–071
Molar Mass: 178.32

TABLE 12.71.1. Experimental heat capacities

Reference		Temp. K	Capac. J/(K.g)	Error %	Purity %	Purity method	Type capacity	Calorimeter Type	Calorimeter Reference
77STE/WAT	N	298.15	1.920	nosp	99.95	melpt	C_p	not specified	

77STE/WAT unspecified DSC calorimeter was used

Name: Decahydro(1–methylethyl)naphthalene
Formula: $C_{13}H_{24}$

CAS-RN: 27193–29–9
Group No.: 12–072
Molar Mass: 180.33

TABLE 12.72.1. Experimental heat capacities

Reference		Temp. range K	No. pnts	Error %	Purity %	Purity method	Type capacity	Calorimeter Type	Calorimeter Reference
63GUD/CAM	N	373.1–483.1	3	nosp	99.9	chrom	C_p	BDHT	63GUD/CAM

63GUD/CAM sample contained 52.2 mol.% of major component (not specified)

TABLE 12.72.3. Parameters of regression polynomial

Heat capacity type	No. data points total	used	s_w	s/R	s_r %	s_b/R	+/–
C_p	3	3	0.000	0.00	0.00	7.63–6	0

Temp. range K	A_1	A_2	A_3	Level of uncertainty
373.1–483.1	1.50315+2	–5.92345+1	8.30490	V

TABLE 12.72.4. Recommended values of heat capacities

Temp. (K)	370	380	390	400	410	420	430
c_p ($J K^{-1} g^{-1}$)	2.07	2.08	2.10	2.13	2.17	2.21	2.27
C_p ($J K^{-1} mol^{-1}$)	373	375	379	385	391	399	409
Temp. (K)	440	450	460	470	480		
c_p ($J K^{-1} g^{-1}$)	2.33	2.39	2.47	2.55	2.64		
C_p ($J K^{-1} mol^{-1}$)	420	432	445	460	477		

Name: Decahydro–1–(1–methylethyl)naphthalene
Formula: $C_{13}H_{24}$

CAS-RN: 1010–74–8
Group No.: 12–073
Molar Mass: 180.33

TABLE 12.73.1. Experimental heat capacities

Reference		Temp. range K	No. pnts	Error %	Purity %	method	Type capacity	Calorimeter Type	Reference
62GOL/BEL		310.9–422.0	3	nosp	98.9	chrom	C_p	BDHT	63GUD/CAM
63GUD/CAM	N	312.0–483.1	4	nosp	98.9	chrom	C_p	BDHT	63GUD/CAM

63GUD/CAM mixture of isomers (22.3 % *cis* and 76.6 % *trans*)

TABLE 12.73.2. Correlated heat capacities

Reference	Temp. range K	No. pnts used	$\sigma_r C$ %	d_w	d/R	d_r %	d_b/R	+/–
				Selected data				
62GOL/BEL	310.9–422.0	3	1.00#	0.087	3.74–2	0.09	6.70–4	0
63GUD/CAM	312.0–483.1	4	1.00#	0.310	1.35–1	0.31	1.11–3	0

TABLE 12.73.3. Parameters of regression polynomial

Heat capacity type	No. data points total	used	s_w	s/R	s_r %	s_b/R	+/–
C_p	7	7	0.286	1.24–1	0.29	9.24–4	0

Temp. range K	A_1	A_2	Level of uncertainty
310.9–483.1	6.83533	1.00451+1	IV

TABLE 12.73.4. Recommended values of heat capacities

Temp. (K)	310	320	330	340	350	360	370
c_p (J K^{-1}g^{-1})	1.751	1.797	1.844	1.890	1.936	1.982	2.029
C_p (J K^{-1}mol^{-1})	315.7	324.1	332.4	340.8	349.2	357.5	365.9
Temp. (K)	380	390	400	410	420	430	440
c_p (J K^{-1}g^{-1})	2.075	2.121	2.168	2.214	2.260	2.307	2.353
C_p (J K^{-1}mol^{-1})	374.2	382.6	390.9	399.3	407.6	416.0	424.3
Temp. (K)	450	460	470	480			
c_p (J K^{-1}g^{-1})	2.399	2.446	2.492	2.538			
C_p (J K^{-1}mol^{-1})	432.7	441.0	449.4	457.7			

Name: Decahydro–1–propylnaphthalene
Formula: $C_{13}H_{24}$

CAS-RN: 91972–45–1
Group No.: 12–074
Molar Mass: 180.33

TABLE 12.74.1. Experimental heat capacities

Reference		Temp. range K	No. pnts	Error %	Purity %	Purity method	Type capacity	Calorimeter Type	Calorimeter Reference
62GOL/BEL		310.9–422.0	3S	nosp	98.2	chrom	C_p	BDHT	63GUD/CAM
63GUD/CAM	N	313.1–483.1	4	nosp	98.2	chrom	C_p	BDHT	63GUD/CAM

63GUD/CAM mixture of isomers (*cis* 32.0 % and *trans* 66.2 %)

TABLE 12.74.2. Correlated heat capacities

Reference	Temp. range K	No. pnts used	$\sigma_r C$ %	d_w	d/R	d_r %	d_b/R	+/–
				Selected data				
62GOL/BEL	310.9–422.0	3	1.00#	0.543	2.50–1	0.54	3.58–2	–1
63GUD/CAM	313.1–483.1	3	1.00#	0.332	1.75–1	0.33	–3.11–2	–1

TABLE 12.74.3. Parameters of regression polynomial

Heat capacity type	No. data points total	No. data points used	s_w	s/R	s_r %	s_b/R	+/–
C_p	7	6	0.636	3.05–1	0.64	2.33–3	–2

Temp. range K	A_1	A_2	A_3	Level of uncertainty
310.9–483.1	3.60317+1	–4.18632	1.80306	V

TABLE 12.74.4. Recommended values of heat capacities

Temp. (K)	310	320	330	340	350	360	370
c_p (J K^{-1}g^{-1})	1.86	1.89	1.93	1.97	2.00	2.04	2.09
C_p (J K^{-1}mol^{-1})	336	342	348	355	361	369	376
Temp. (K)	380	390	400	410	420	430	440
c_p (J K^{-1}g^{-1})	2.13	2.17	2.22	2.27	2.32	2.37	2.42
C_p (J K^{-1}mol^{-1})	384	392	400	409	418	427	437
Temp. (K)	450	460	470	480			
c_p (J K^{-1}g^{-1})	2.48	2.53	2.59	2.65			
C_p (J K^{-1}mol^{-1})	447	457	467	478			

Name: 2–Methyl–1,1'–bicyclohexyl
Formula: $C_{13}H_{24}$

CAS–RN: unknown
Group No.: 12–075
Molar Mass: 180.33

TABLE 12.75.1. Experimental heat capacities

Reference	Temp. range K	No. pnts	Error %	Purity %	Purity method	Type capacity	Calorimeter Type	Calorimeter Reference
63GUD/CAM	313.1–483.1	4	nosp	98.8	chrom	C_p	BDHT	63GUD/CAM

TABLE 12.75.3. Parameters of regression polynomial

Heat capacity type	No. data points total	No. data points used	s_w	s/R	s_r %	s_b/R	+/–
C_p	4	4	1.604	7.38–1	1.60	1.90–2	2

Temp. range K	A_1	A_2	Level of uncertainty
313.1–483.1	2.16555	1.19129+1	V

TABLE 12.75.4. Recommended values of heat capacities

Temp. (K)	310	320	330	340	350	360	370
c_p ($J\,K^{-1}g^{-1}$)	1.80	1.86	1.91	1.97	2.02	2.08	2.13
C_p ($J\,K^{-1}mol^{-1}$)	325	335	345	355	365	375	384
Temp. (K)	380	390	400	410	420	430	440
c_p ($J\,K^{-1}g^{-1}$)	2.19	2.24	2.30	2.35	2.41	2.46	2.52
C_p ($J\,K^{-1}mol^{-1}$)	394	404	414	424	434	444	454
Temp. (K)	450	460	470	480			
c_p ($J\,K^{-1}g^{-1}$)	2.57	2.63	2.68	2.74			
C_p ($J\,K^{-1}mol^{-1}$)	464	474	484	493			

Name: 1,1'–Methylenebiscyclohexane
Formula: $C_{13}H_{24}$

CAS–RN: 3178–23–2
Group No.: 12–076
Molar Mass: 180.33

TABLE 12.76.1. Experimental heat capacities

Reference	Temp. range K	No. pnts	Error %	Purity %	Purity method	Type capacity	Calorimeter Type	Calorimeter Reference
62GOL/BEL	310.9–422.0	3S	nosp	99.9	chrom	C_p	BDHT	63GUD/CAM
63GUD/CAM	313.1–483.1	4	nosp	100.0	chrom	C_p	BDHT	63GUD/CAM

TABLE 12.76.2. Correlated heat capacities

Reference	Temp. range K	No. pnts used	$\sigma_r C$ %	d_w	d/R	d_r %	d_b/R	+/–
				Selected data				
62GOL/BEL	310.9–422.0	3	1.00#	0.607	2.90–1	0.61	1.92–1	1
63GUD/CAM	373.1–483.1	3	1.00#	0.707	3.63–1	0.71	–1.82–1	–1

TABLE 12.76.3. Parameters of regression polynomial

Heat capacity type	No. data points total	used	s_w	s/R	s_r %	s_b/R	+/–
C_p	7	6	0.932	4.65–1	0.93	4.73–3	0

Temp. range K	A_1	A_2	A_3	Level of uncertainty
310.9–483.1	4.89274+1	–9.19220	2.32660	IV

TABLE 12.76.4. Recommended values of heat capacities

Temp. (K)	310	320	330	340	350	360	370
c_p (J K^{-1}g^{-1})	1.973	1.998	2.025	2.055	2.087	2.120	2.156
C_p (J K^{-1}mol^{-1})	355.8	360.3	365.3	370.6	376.3	382.4	388.8
Temp. (K)	380	390	400	410	420	430	440
c_p (J K^{-1}g^{-1})	2.194	2.235	2.277	2.321	2.368	2.417	2.468
C_p (J K^{-1}mol^{-1})	395.7	403.0	410.6	418.6	427.0	435.8	445.0
Temp. (K)	450	460	470	480			
c_p (J K^{-1}g^{-1})	2.521	2.576	2.634	2.693			
C_p (J K^{-1}mol^{-1})	454.6	464.6	474.9	485.6			

Name: Heptylcyclohexane
Formula: $C_{13}H_{26}$

CAS-RN: 5617–41–4
Group No.: 12–077
Molar Mass: 182.35

TABLE 12.77.1. Experimental heat capacities

Reference	Temp. range K	No. pnts	Error %	Purity %	Purity method	Type capacity	Calorimeter Type	Calorimeter Reference
49PAR/MOO	240.0–300.0	7S	1.00	97.1	melpt	C_p	BSIO	25PAR

TABLE 12.77.3. Parameters of regression polynomial

Heat capacity type	No. data points total	used	s_w	s/R	s_r %	s_b/R	+/–
C_p	7	7	0.192	7.86–2	0.19	2.93–4	0

Temp. range K	A_1	A_2	Level of uncertainty
240.0–300.0	2.02178+1	7.85827	V

TABLE 12.77.4. Recommended values of heat capacities

Temp. (K)	240	250	260	270	273.15	280	290
c_p (J K^{-1}g^{-1})	1.78	1.82	1.85	1.89	1.90	1.93	1.96
C_p (J K^{-1}mol^{-1})	325	331	338	345	347	351	358
Temp. (K)	298.15	300					
c_p (J K^{-1}g^{-1})	1.99	2.00					
C_p (J K^{-1}mol^{-1})	363	364					

Name: Decahydro–3,5,1,7–[1,2,3,4]butanetetraylnaphthalene
Formula: $C_{14}H_{20}$

CAS-RN: 2292–79–7
Group No.: 12–078
Molar Mass: 188.31

TABLE 12.78.1. Experimental heat capacities

Reference	Temp. range K	No. pnts	Error %	Purity %	Purity method	Type capacity	Calorimeter Type	Calorimeter Reference
78SPI/AND	519.2–537.1	7	nosp	99.91	melpt	C_p	BSAO	68WES/WES

TABLE 12.78.3. Parameters of regression polynomial

Heat capacity type	No. data points total	No. data points used	s_w	s/R	s_r %	s_b/R	+/–
C_p	7	7	0.080	1.72–2	0.03	8.72–6	0

Temp. range K	A_1	A_2	Level of uncertainty
519.2–537.1	6.49683	8.92994	III

TABLE 12.78.4. Recommended values of heat capacities

Temp. (K)	520	530	540
c_p (J K^{-1}g^{-1})	2.337	2.377	2.416
C_p (J K^{-1}mol^{-1})	440.1	447.5	455.0

Name: Dodecahydro–9–methylfluorene
Formula: $C_{14}H_{24}$

CAS-RN: 92431–75–9
Group No.: 12–079
Molar Mass: 192.34

TABLE 12.79.1. Experimental heat capacities

Reference	Temp. range K	No. pnts	Error %	Purity %	Purity method	Type capacity	Calorimeter Type	Calorimeter Reference
63GUD/CAM	313.1–483.1	4	nosp	99.9	chrom	C_p	BDHT	63GUD/CAM

TABLE 12.79.3. Parameters of regression polynomial

Heat capacity type	No. data points total	No. data points used	s_w	s/R	s_r %	s_b/R	+/–
C_p	4	4	1.319	6.12–1	1.32	3.78–3	–2

Temp. range K	A_1	A_2	Level of uncertainty
313.1–483.1	4.49872	1.10730+1	V

TABLE 12.79.4. Recommended values of heat capacities

Temp. (K)	310	320	330	340	350	360	370
c_p (J K^{-1}g^{-1})	1.68	1.73	1.77	1.82	1.87	1.92	1.97
C_p (J K^{-1}mol^{-1})	323	332	341	350	360	369	378
Temp. (K)	380	390	400	410	420	430	440
c_p (J K^{-1}g^{-1})	2.01	2.06	2.11	2.16	2.20	2.25	2.30
C_p (J K^{-1}mol^{-1})	387	396	406	415	424	433	442
Temp. (K)	450	460	470	480			
c_p (J K^{-1}g^{-1})	2.35	2.40	2.44	2.49			
C_p (J K^{-1}mol^{-1})	452	461	470	479			

Name: Tetradecahydrophenanthrene
Formula: $C_{14}H_{24}$

CAS-RN: 5743–97–5
Group No.: 12–080
Molar Mass: 192.34

TABLE 12.80.1. Experimental heat capacities

Reference	Temp. range K	No. pnts	Error %	Purity %	Purity method	Type capacity	Calorimeter Type	Calorimeter Reference
63GUD/CAM	313.1–483.1	4	nosp	not specified		C_p	BDHT	63GUD/CAM

TABLE 12.80.3. Parameters of regression polynomial

Heat capacity type	No. data points total	No. data points used	s_w	s/R	s_r %	s_b/R	+/–
C_p	4	4	1.056	5.44–1	1.06	1.50–3	0

Temp. range K	A_1	A_2	Level of uncertainty
313.1–483.1	9.60103	9.63400	V

TABLE 12.80.4. Recommended values of heat capacities

Temp. (K)	310	320	330	340	350	360	370
c_p (J K^{-1}g^{-1})	1.71	1.75	1.79	1.83	1.87	1.91	1.96
C_p (J K^{-1}mol^{-1})	328	336	344	352	360	368	376
Temp. (K)	380	390	400	410	420	430	440
c_p (J K^{-1}g^{-1})	2.00	2.04	2.08	2.12	2.16	2.21	2.25
C_p (J K^{-1}mol^{-1})	384	392	400	408	416	424	432
Temp. (K)	450	460	470	480			
c_p (J K^{-1}g^{-1})	2.29	2.33	2.37	2.41			
C_p (J K^{-1}mol^{-1})	440	448	456	464			

Name: (4aα,4bα,8aα,10aβ)–Tetradecahydrophenanthrene
Formula: $C_{14}H_{24}$

CAS-RN: 27425–35–0
Group No.: 12–081
Molar Mass: 192.34

TABLE 12.81.1. Experimental heat capacities

Reference	Temp. range K	No. pnts	Error %	Purity %	Purity method	Type capacity	Calorimeter Type	Calorimeter Reference
82NUZ	283.0–403.0	14S	nosp	not specified		C_p	BDHT	73PER/COM

TABLE 12.81.3. Parameters of regression polynomial

Heat capacity type	No. data points total	No. data points used	s_w	s/R	s_r %	s_b/R	+/–
C_p	14	14	0.920	6.20–1	1.38	1.36–2	0

Temp. range K	A_1	A_2	A_3	Level of uncertainty
283.0–403.0	5.19497+1	–1.44099+1	3.64063	V

TABLE 12.81.4. Recommended values of heat capacities

Temp. (K)	290	298.15	300	310	320	330	340
c_p ($J K^{-1}g^{-1}$)	1.76	1.79	1.79	1.83	1.86	1.90	1.95
C_p ($J K^{-1}mol^{-1}$)	339	344	345	351	359	366	375
Temp. (K)	350	360	370	380	390	400	
c_p ($J K^{-1}g^{-1}$)	1.99	2.04	2.10	2.15	2.21	2.27	
C_p ($J K^{-1}mol^{-1}$)	383	393	403	414	425	437	

Name: (4aα,4bβ,8aα,10aα)–Tetradecahydrophenanthrene
Formula: $C_{14}H_{24}$

CAS-RN: 27389–73–7
Group No.: 12–082
Molar Mass: 192.34

TABLE 12.82.1. Experimental heat capacities

Reference	Temp. range K	No. pnts	Error %	Purity %	Purity method	Type capacity	Calorimeter Type	Calorimeter Reference
82NUZ	323.0–403.0	9S	nosp	not specified		C_p	BDHT	73PER/COM

TABLE 12.82.3. Parameters of regression polynomial

Heat capacity type	No. data points total	No. data points used	s_w	s/R	s_r %	s_b/R	+/–
C_p	9	9	1.822	1.04	2.73	4.67–2	5

Temp. range K	A_1	A_2	Level of uncertainty
323.0–403.0	–4.44900+1	2.43424+1	V

TABLE 12.82.4. Recommended values of heat capacities

Temp. (K)	320	330	340	350	360	370	380
c_p (J K^{-1}g^{-1})	1.44	1.55	1.65	1.76	1.86	1.97	2.08
C_p (J K^{-1}mol^{-1})	278	298	318	338	359	379	399
Temp. (K)	390	400					
c_p (J K^{-1}g^{-1})	2.18	2.29					
C_p (J K^{-1}mol^{-1})	419	440					

Name: (4aα,4bβ,8aα,10aβ)–Tetradecahydrophenanthrene
Formula: $C_{14}H_{24}$

CAS-RN: 2108–89–6
Group No.: 12–083
Molar Mass: 192.34

TABLE 12.83.1. Experimental heat capacities

Reference	Temp. range K	No. pnts	Error %	Purity %	Purity method	Type capacity	Calorimeter Type	Calorimeter Reference
82NUZ	293.0–403.0	13S	nosp	not specified		C_p	BDHT	73PER/COM

TABLE 12.83.3. Parameters of regression polynomial

Heat capacity type	No. data points total	No. data points used	s_w	s/R	s_r %	s_b/R	+/–
C_p	13	13	2.151	1.38	3.23	6.81–2	–1

Temp. range K	A_1	A_2	Level of uncertainty
293.0–403.0	–2.35193+1	1.93147+1	VI

TABLE 12.83.4. Recommended values of heat capacities

Temp. (K)	290	298.15	300	310	320	330	340
c_p (J K^{-1}g^{-1})	1.40	1.47	1.49	1.57	1.66	1.74	1.82
C_p (J K^{-1}mol^{-1})	270	283	286	302	318	334	350
Temp. (K)	350	360	370	380	390	400	
c_p (J K^{-1}g^{-1})	1.91	1.99	2.07	2.16	2.24	2.32	
C_p (J K^{-1}mol^{-1})	367	383	399	415	431	447	

Name: 1–Butyldecahydronaphthalene
Formula: $C_{14}H_{26}$

CAS-RN: 92369–80–7
Group No.: 12–084
Molar Mass: 194.36

TABLE 12.84.1. Experimental heat capacities

Reference		Temp. range K	No. pnts	Error %	Purity %	Purity method	Type capacity	Calorimeter Type	Calorimeter Reference
62GOL/BEL		310.9–422.0	3	nosp	99.5	chrom	C_p	BDHT	63GUD/CAM
63GUD/CAM	N	313.1–483.1	4	nosp	99.5	chrom	C_p	BDHT	63GUD/CAM

63GUD/CAM mixture of isomers (25.2 % *cis* and 74.3 % *trans*)

TABLE 12.84.2. Correlated heat capacities

Reference	Temp. range K	No. pnts used	$\sigma_r C$ %	d_w	d/R	d_r %	d_b/R	+/–
				Selected data				
62GOL/BEL	310.9–422.0	3	1.00#	3.258	1.63	3.26	–1.20	–1
63GUD/CAM	313.1–483.1	4	1.00#	2.000	1.13	2.00	1.04	4

TABLE 12.84.3. Parameters of regression polynomial

Heat capacity type	No. data points total	used	s_w	s/R	s_r %	s_b/R	+/–
C_p	7	7	3.094	1.61	3.09	8.01–2	3

Temp. range K	A_1	A_2	Level of uncertainty
310.9–483.1	1.16844+1	9.96577	VI

TABLE 12.84.4. Recommended values of heat capacities

Temp. (K)	310	320	330	340	350	360	370
c_p (J K^{-1}g^{-1})	1.82	1.86	1.91	1.95	1.99	2.03	2.08
C_p (J K^{-1}mol^{-1})	354	362	371	379	387	395	404
Temp. (K)	380	390	400	410	420	430	440
c_p (J K^{-1}g^{-1})	2.12	2.16	2.21	2.25	2.29	2.33	2.38
C_p (J K^{-1}mol^{-1})	412	420	429	437	445	453	462
Temp. (K)	450	460	470	480			
c_p (J K^{-1}g^{-1})	2.42	2.46	2.50	2.55			
C_p (J K^{-1}mol^{-1})	470	478	487	495			

Name: 1–(Cyclohexylmethyl)–2–methylcyclohexane
Formula: $C_{14}H_{26}$

CAS-RN: 66826–96–8
Group No.: 12–085
Molar Mass: 194.36

TABLE 12.85.1. Experimental heat capacities

Reference	Temp. range K	No. pnts	Error %	Purity %	Purity method	Type capacity	Calorimeter Type	Calorimeter Reference
63GUD/CAM	313.1–483.1	4	nosp	not specified		C_p	BDHT	63GUD/CAM

TABLE 12.85.3. Parameters of regression polynomial

Heat capacity type	No. data points total	used	s_w	s/R	s_r %	s_b/R	+/–
C_p	4	4	2.491	1.25	2.49	1.89–2	0

Temp. range K	A_1	A_2	A_3	Level of uncertainty
313.1–483.1	9.18962+1	–3.24988+1	5.45448	V

TABLE 12.85.4. Recommended values of heat capacities

Temp. (K)	310	320	330	340	350	360	370
c_p (J $K^{-1}g^{-1}$)	1.86	1.87	1.88	1.90	1.92	1.95	1.98
C_p (J $K^{-1}mol^{-1}$)	362	364	366	370	374	379	385
Temp. (K)	380	390	400	410	420	430	440
c_p (J $K^{-1}g^{-1}$)	2.02	2.06	2.10	2.15	2.21	2.27	2.33
C_p (J $K^{-1}mol^{-1}$)	392	400	409	419	429	441	453
Temp. (K)	450	460	470	480			
c_p (J $K^{-1}g^{-1}$)	2.40	2.47	2.55	2.63			
C_p (J $K^{-1}mol^{-1}$)	466	481	496	512			

Name: Decahydro–1–(1–methylpropyl)naphthalene
Formula: $C_{14}H_{26}$

CAS-RN: 92369–82–9
Group No.: 12–086
Molar Mass: 194.36

TABLE 12.86.1. Experimental heat capacities

Reference		Temp. range K	No. pnts	Error %	Purity %	Purity method	Type capacity	Calorimeter Type	Calorimeter Reference
62GOL/BEL		310.9–422.0	3	nosp	98.3	chrom	C_p	BDHT	63GUD/CAM
63GUD/CAM	N	313.1–483.1	4	nosp	98.3	chrom	C_p	BDHT	63GUD/CAM

63GUD/CAM mixture of isomers (37.4 % *cis* and 60.9 % *trans*)

TABLE 12.86.2. Correlated heat capacities

Reference	Temp. range K	No. pnts used	$\sigma_r C$ %	d_w	d/R	d_r %	d_b/R	+/–
				Selected data				
62GOL/BEL	310.9–422.0	3	1.00#	0.741	3.84–1	0.74	2.66–1	1
63GUD/CAM	313.1–483.1	4	1.00#	1.480	7.73–1	1.48	–1.71–1	0

TABLE 12.86.3. Parameters of regression polynomial

Heat capacity type	No. data points total	No. data points used	s_w	s/R	s_r %	s_b/R	+/–
C_p	7	7	1.613	8.42–1	1.61	1.63–2	1

Temp. range K	A_1	A_2	A_3	Level of uncertainty
310.9–483.1	3.41286+1	–2.86511	1.76221	V

TABLE 12.86.4. Recommended values of heat capacities

Temp. (K)	310	320	330	340	350	360	370
c_p (J $K^{-1}g^{-1}$)	1.80	1.84	1.88	1.91	1.95	2.00	2.04
C_p (J $K^{-1}mol^{-1}$)	351	358	365	372	380	388	396
Temp. (K)	380	390	400	410	420	430	440
c_p (J $K^{-1}g^{-1}$)	2.08	2.13	2.18	2.22	2.27	2.33	2.38
C_p (J $K^{-1}mol^{-1}$)	405	414	423	432	442	452	463
Temp. (K)	450	460	470	480			
c_p (J $K^{-1}g^{-1}$)	2.43	2.49	2.55	2.61			
C_p (J $K^{-1}mol^{-1}$)	473	484	495	507			

Name: Decahydro–1–(2–methylpropyl)naphthalene
Formula: $C_{14}H_{26}$

CAS-RN: 92369–83–0
Group No.: 12–087
Molar Mass: 194.36

TABLE 12.87.1. Experimental heat capacities

Reference		Temp. range K	No. pnts	Error %	Purity %	Purity method	Type capacity	Calorimeter Type	Calorimeter Reference
63GUD/CAM	N	313.1–483.1	4	nosp	98.7	chrom	C_p	BDHT	63GUD/CAM

63GUD/CAM mixture of isomers (44.6 % *cis* and 54.1 % *trans*)

TABLE 12.87.3. Parameters of regression polynomial

Heat capacity type	No. data points total	No. data points used	s_w	s/R	s_r %	s_b/R	+/–
C_p	4	4	1.966	9.85–1	1.97	1.16–2	0

Temp. range K	A_1	A_2	A_3	Level of uncertainty
313.1–483.1	7.49411+1	–2.34307+1	4.25354	V

TABLE 12.87.4. Recommended values of heat capacities

Temp. (K)	310	320	330	340	350	360	370
c_p (J $K^{-1}g^{-1}$)	1.85	1.86	1.88	1.90	1.93	1.96	1.99
C_p (J $K^{-1}mol^{-1}$)	359	362	365	370	374	380	386
Temp. (K)	380	390	400	410	420	430	440
c_p (J $K^{-1}g^{-1}$)	2.02	2.06	2.11	2.16	2.21	2.26	2.32
C_p (J $K^{-1}mol^{-1}$)	393	401	410	419	429	439	451
Temp. (K)	450	460	470	480			
c_p (J $K^{-1}g^{-1}$)	2.38	2.45	2.51	2.59			
C_p (J $K^{-1}mol^{-1}$)	463	475	489	503			

Name: (1,1–Dimethylethyl)decahydronaphthalene
Formula: $C_{14}H_{26}$

CAS-RN: 27193–30–2
Group No.: 12–088
Molar Mass: 194.36

TABLE 12.88.1. Experimental heat capacities

Reference	Temp. range K	No. pnts	Error %	Purity %	Purity method	Type capacity	Calorimeter Type	Calorimeter Reference
63GUD/CAM	313.1–423.1	3	nosp	99.4	chrom	C_p	BDHT	63GUD/CAM

TABLE 12.88.3. Parameters of regression polynomial

Heat capacity type	No. data points total	No. data points used	s_w	s/R	s_r %	s_b/R	+/–
C_p	3	3	0.402	1.97–1	0.40	–1.33–4	–1

Temp. range K	A_1	A_2	Level of uncertainty
313.1–423.1	1.13587+1	9.91416	V

TABLE 12.88.4. Recommended values of heat capacities

Temp. (K)	310	320	330	340	350	360	370
c_p (J K^{-1}g^{-1})	1.80	1.84	1.89	1.93	1.97	2.01	2.06
C_p (J K^{-1}mol^{-1})	350	358	366	375	383	391	399

Temp. (K)	380	390	400	410	420
c_p (J K^{-1}g^{-1})	2.10	2.14	2.18	2.22	2.27
C_p (J K^{-1}mol^{-1})	408	416	424	432	441

Name: 1,1'–(1,2–Ethanediyl)biscyclohexane
Formula: $C_{14}H_{26}$

CAS-RN: 3321–50–4
Group No.: 12–089
Molar Mass: 194.36

TABLE 12.89.1. Experimental heat capacities

Reference	Temp. range K	No. pnts	Error %	Purity %	Purity method	Type capacity	Calorimeter Type	Calorimeter Reference
63GUD/CAM	313.1–483.1	4	nosp	99.7	chrom	C_p	BDHT	63GUD/CAM

TABLE 12.89.3. Parameters of regression polynomial

Heat capacity type	No. data points total	No. data points used	s_w	s/R	s_r %	s_b/R	+/–
C_p	4	4	0.527	2.76–1	0.53	1.18–3	0

Temp. range K	A_1	A_2	Level of uncertainty
313.1–483.1	1.33815+1	9.86702	IV

TABLE 12.89.4. Recommended values of heat capacities

Temp. (K)	310	320	330	340	350	360	370
c_p (J K^{-1}g^{-1})	1.881	1.923	1.965	2.008	2.050	2.092	2.134
C_p (J K^{-1}mol^{-1})	365.6	373.8	382.0	390.2	398.4	406.6	414.8
Temp. (K)	380	390	400	410	420	430	440
c_p (J K^{-1}g^{-1})	2.176	2.219	2.261	2.303	2.345	2.387	2.430
C_p (J K^{-1}mol^{-1})	423.0	431.2	439.4	447.6	455.8	464.0	472.2
Temp. (K)	450	460	470	480			
c_p (J K^{-1}g^{-1})	2.472	2.514	2.556	2.599			
C_p (J K^{-1}mol^{-1})	480.4	488.6	496.8	505.0			

Name: 2–Ethyl–1,1'–bicyclohexyl
Formula: $C_{14}H_{26}$

CAS-RN: 66826–94–6
Group No.: 12–090
Molar Mass: 194.36

TABLE 12.90.1. Experimental heat capacities

Reference	Temp. range K	No. pnts	Error %	Purity %	Purity method	Type capacity	Calorimeter Type	Calorimeter Reference
63GUD/CAM	313.1–483.1	4	nosp	99.9	chrom	C_p	BDHT	63GUD/CAM

TABLE 12.90.3. Parameters of regression polynomial

Heat capacity type	No. data points total	No. data points used	s_w	s/R	s_r %	s_b/R	+/–
C_p	4	4	0.066	3.55–2	0.07	3.81–5	0

Temp. range K	A_1	A_2	A_3	Level of uncertainty
313.1–483.1	3.64918+1	–3.24819	1.84457	IV

TABLE 12.90.4. Recommended values of heat capacities

Temp. (K)	310	320	330	340	350	360	370
c_p (J K^{-1}g^{-1})	1.889	1.924	1.962	2.001	2.041	2.084	2.127
C_p (J K^{-1}mol^{-1})	367.1	374.0	381.3	388.9	396.8	405.0	413.4
Temp. (K)	380	390	400	410	420	430	440
c_p (J K^{-1}g^{-1})	2.172	2.219	2.268	2.318	2.369	2.423	2.477
C_p (J K^{-1}mol^{-1})	422.2	431.4	440.8	450.5	460.5	470.9	481.5
Temp. (K)	450	460	470	480			
c_p (J K^{-1}g^{-1})	2.534	2.592	2.651	2.712			
C_p (J K^{-1}mol^{-1})	492.4	503.7	515.3	527.1			

Name: 1,1'–Ethylidenebiscyclohexane
Formula: $C_{14}H_{26}$

CAS-RN: 2319–61–1
Group No.: 12–091
Molar Mass: 194.36

TABLE 12.91.1. Experimental heat capacities

Reference	Temp. range K	No. pnts	Error %	Purity %	Purity method	Type capacity	Calorimeter Type	Calorimeter Reference
63GUD/CAM	313.1–483.1	4	nosp	99.	chrom	C_p	BDHT	63GUD/CAM

TABLE 12.91.3. Parameters of regression polynomial

Heat capacity type	No. data points total	No. data points used	s_w	s/R	s_r %	s_b/R	+/–
C_p	4	4	1.839	9.19–1	1.84	9.13–3	–2

Temp. range K	A_1	A_2	Level of uncertainty
313.1–483.1	5.56176	1.17308+1	V

TABLE 12.91.4. Recommended values of heat capacities

Temp. (K)	310	320	330	340	350	360	370
c_p ($J K^{-1} g^{-1}$)	1.79	1.84	1.89	1.94	1.99	2.04	2.09
C_p ($J K^{-1} mol^{-1}$)	349	358	368	378	388	397	407
Temp. (K)	380	390	400	410	420	430	440
c_p ($J K^{-1} g^{-1}$)	2.14	2.20	2.25	2.30	2.35	2.40	2.45
C_p ($J K^{-1} mol^{-1}$)	417	427	436	446	456	466	475
Temp. (K)	450	460	470	480			
c_p ($J K^{-1} g^{-1}$)	2.50	2.55	2.60	2.65			
C_p ($J K^{-1} mol^{-1}$)	485	495	505	514			

Name: 1,1':3',1''–Tercyclopentane
Formula: $C_{15}H_{26}$

CAS-RN: 6051–40–7
Group No.: 12–092
Molar Mass: 206.37

TABLE 12.92.1. Experimental heat capacities

Reference	Temp. range K	No. pnts	Error %	Purity %	Purity method	Type capacity	Calorimeter Type	Calorimeter Reference
63GUD/CAM	313.1–483.1	4	nosp	not specified		C_p	BDHT	63GUD/CAM

TABLE 12.92.3. Parameters of regression polynomial

Heat capacity type	No. data points total	No. data points used	s_w	s/R	s_r %	s_b/R	+/–
C_p	4	4	1.167	6.14–1	1.17	7.20–3	0

Temp. range K	A_1	A_2	Level of uncertainty
313.1–483.1	1.31200+1	9.80976	V

TABLE 12.92.4. Recommended values of heat capacities

Temp. (K)	310	320	330	340	350	360	370
c_p (J K^{-1}g^{-1})	1.75	1.79	1.83	1.87	1.91	1.95	1.99
C_p (J K^{-1}mol^{-1})	362	370	378	386	395	403	411
Temp. (K)	380	390	400	410	420	430	440
c_p (J K^{-1}g^{-1})	2.03	2.07	2.11	2.15	2.19	2.23	2.27
C_p (J K^{-1}mol^{-1})	419	427	435	443	452	460	468
Temp. (K)	450	460	470	480			
c_p (J K^{-1}g^{-1})	2.31	2.35	2.39	2.43			
C_p (J K^{-1}mol^{-1})	476	484	492	501			

Name: Cyclohexyl(ethylcyclohexyl)methane (unspecified isomer)
Formula: $C_{15}H_{28}$

CAS-RN: 97239–02–6
Group No.: 12–093
Molar Mass: 208.39

TABLE 12.93.1. Experimental heat capacities

Reference		Temp. range K	No. pnts	Error %	Purity %	Purity method	Type capacity	Calorimeter Type	Calorimeter Reference
62GOL/BEL		422.0	1	nosp	100.0	chrom	C_p	BDHT	63GUD/CAM
63GUD/CAM	N	313.1–483.1	4	nosp	100.0	chrom	C_p	BDHT	63GUD/CAM

63GUD/CAM sample contained 49.2 % of major component (not specified)

TABLE 12.93.2. Correlated heat capacities

Reference	Temp. range K	No. pnts used	$\sigma_r C$ %	d_w	d/R	d_r %	d_b/R	+/–
				Selected data				
62GOL/BEL	422.0	1	1.00#	0.150	8.88–2	0.15	8.88–2	1
63GUD/CAM	313.1–483.1	4	1.00#	0.054	3.03–2	0.05	–2.22–2	–2

TABLE 12.93.3. Parameters of regression polynomial

Heat capacity type	No. data points total	No. data points used	s_w	s/R	s_r %	s_b/R	+/–
C_p	5	5	0.131	7.61–2	0.13	4.88–5	–1

Temp. range K	A_1	A_2	A_3	Level of uncertainty
313.1–483.1	3.81731+1	–1.52507	1.53372	V

TABLE 12.93.4. Recommended values of heat capacities

Temp. (K)	310	320	330	340	350	360	370
c_p (J K^{-1}g^{-1})	1.92	1.95	1.99	2.02	2.06	2.10	2.14
C_p (J K^{-1}mol^{-1})	401	407	414	422	429	437	445
Temp. (K)	380	390	400	410	420	430	440
c_p (J K^{-1}g^{-1})	2.18	2.22	2.26	2.30	2.35	2.39	2.44
C_p (J K^{-1}mol^{-1})	453	462	471	480	489	499	508
Temp. (K)	450	460	470	480			
c_p (J K^{-1}g^{-1})	2.49	2.54	2.59	2.64			
C_p (J K^{-1}mol^{-1})	519	529	539	550			

Name: 1–(Cyclohexylmethyl)–2–ethylcyclohexane
Formula: $C_{15}H_{28}$

CAS-RN: 66374–71–8
Group No.: 12–094
Molar Mass: 208.39

TABLE 12.94.1. Experimental heat capacities

Reference	Temp. range K	No. pnts	Error %	Purity %	Purity method	Type capacity	Calorimeter Type	Calorimeter Reference
62GOL/BEL	310.9–422.0	3	nosp	99.8	chrom	C_p	BDHT	63GUD/CAM
63GUD/CAM	313.1–483.1	4	nosp	99.8	chrom	C_p	BDHT	63GUD/CAM

TABLE 12.94.2. Correlated heat capacities

Reference	Temp. range K	No. pnts used	$\sigma_r C$ %	d_w	d/R	d_r %	d_b/R	+/–
				Selected data				
62GOL/BEL	310.9–422.0	3	1.00#	0.296	1.65–1	0.30	–3.39–3	–1
63GUD/CAM	313.1–483.1	4	1.00#	0.256	1.54–1	0.26	4.79–3	0

TABLE 12.94.3. Parameters of regression polynomial

Heat capacity type	No. data points total	No. data points used	s_w	s/R	s_r %	s_b/R	+/–
C_p	7	7	0.362	2.10–1	0.36	1.28–3	–1

Temp. range K	A_1	A_2	A_3	Level of uncertainty
310.9–483.1	6.46970+1	–1.50303+1	3.26580	IV

TABLE 12.94.4. Recommended values of heat capacities

Temp. (K)	310	320	330	340	350	360	370
c_p ($J\,K^{-1}g^{-1}$)	1.975	1.997	2.021	2.049	2.079	2.111	2.146
C_p ($J\,K^{-1}mol^{-1}$)	411.5	416.1	421.2	426.9	433.2	439.9	447.3
Temp. (K)	380	390	400	410	420	430	440
c_p ($J\,K^{-1}g^{-1}$)	2.184	2.224	2.267	2.313	2.361	2.412	2.465
C_p ($J\,K^{-1}mol^{-1}$)	455.1	463.5	472.5	482.0	492.0	502.6	513.8
Temp. (K)	450	460	470	480			
c_p ($J\,K^{-1}g^{-1}$)	2.521	2.580	2.641	2.705			
C_p ($J\,K^{-1}mol^{-1}$)	525.4	537.6	550.4	563.7			

Name: 1,2–Dicyclohexylpropane
Formula: $C_{15}H_{28}$

CAS-RN: 41851–34–7
Group No.: 12–095
Molar Mass: 208.39

TABLE 12.95.1. Experimental heat capacities

Reference	Temp. range K	No. pnts	Error %	Purity %	Purity method	Type capacity	Calorimeter Type	Calorimeter Reference
62GOL/BEL	422.0	1	nosp	100.0	chrom	C_p	BDHT	63GUD/CAM
63GUD/CAM	313.1–483.1	4	nosp	100.0	chrom	C_p	BDHT	63GUD/CAM

TABLE 12.95.2. Correlated heat capacities

Reference	Temp. range K	No. pnts used	$\sigma_r C$ %	d_w	d/R	d_r %	d_b/R	+/–
				Selected data				
62GOL/BEL	422.0	1	1.00#	0.041	2.47–2	0.04	–2.47–2	–1
63GUD/CAM	313.1–483.1	4	1.00#	0.604	3.43–1	0.60	1.10–2	0

TABLE 12.95.3. Parameters of regression polynomial

Heat capacity type	No. data points total	No. data points used	s_w	s/R	s_r %	s_b/R	+/–
C_p	5	5	0.698	3.97–1	0.70	3.85–3	–1

Temp. range K	A_1	A_2	Level of uncertainty
313.1–483.1	1.29566+1	1.11236+1	V

TABLE 12.95.4. Recommended values of heat capacities

Temp. (K)	310	320	330	340	350	360	370
c_p (J K^{-1}g^{-1})	1.89	1.94	1.98	2.03	2.07	2.11	2.16
C_p (J K^{-1}mol^{-1})	394	404	413	422	431	441	450
Temp. (K)	380	390	400	410	420	430	440
c_p (J K^{-1}g^{-1})	2.20	2.25	2.29	2.34	2.38	2.43	2.47
C_p (J K^{-1}mol^{-1})	459	468	478	487	496	505	515
Temp. (K)	450	460	470	480			
c_p (J K^{-1}g^{-1})	2.51	2.56	2.60	2.65			
C_p (J K^{-1}mol^{-1})	524	533	542	552			

Name: (1–Methylethyl)–1,1'–bicyclohexyl (unspecified isomer)
Formula: $C_{15}H_{28}$

CAS-RN: 31624–59–6
Group No.: 12–096
Molar Mass: 208.39

TABLE 12.96.1. Experimental heat capacities

Reference		Temp. range K	No. pnts	Error %	Purity %	Purity method	Type capacity	Calorimeter Type	Calorimeter Reference
62GOL/BEL	N	310.9–422.0	3	nosp	99.6	chrom	C_p	BDHT	63GUD/CAM
63GUD/CAM	N	313.1–483.1	4	nosp	99.	chrom	C_p	BDHT	63GUD/CAM

62GOL/BEL isomer with n.m.t. 239.3 K
63GUD/CAM low boiling isomer

TABLE 12.96.2. Correlated heat capacities

Reference	Temp. range K	No. pnts used	$\sigma_r C$ %	d_w	d/R	d_r %	d_b/R	+/–
				Selected data				
62GOL/BEL	310.9–422.0	3	1.00#	0.125	7.33–2	0.13	4.67–2	1
63GUD/CAM	313.1–483.1	4	1.00#	0.280	1.58–1	0.28	–3.48–2	–2

TABLE 12.96.3. Parameters of regression polynomial

Heat capacity type	No. data points total	No. data points used	s_w	s/R	s_r %	s_b/R	+/–
C_p	7	7	0.268	1.53–1	0.27	1.41–4	–1

Temp. range K	A_1	A_2	Level of uncertainty
310.9–483.1	2.42297+1	8.36946	V

TABLE 12.96.4. Recommended values of heat capacities

Temp. (K)	310	320	330	340	350	360	370
c_p (J $K^{-1}g^{-1}$)	2.00	2.04	2.07	2.10	2.14	2.17	2.20
C_p (J $K^{-1}mol^{-1}$)	417	424	431	438	445	452	459
Temp. (K)	380	390	400	410	420	430	440
c_p (J $K^{-1}g^{-1}$)	2.24	2.27	2.30	2.34	2.37	2.40	2.44
C_p (J $K^{-1}mol^{-1}$)	466	473	480	487	494	501	508
Temp. (K)	450	460	470	480			
c_p (J $K^{-1}g^{-1}$)	2.47	2.50	2.54	2.57			
C_p (J $K^{-1}mol^{-1}$)	515	522	529	535			

Name: Decylcyclopentane
Formula: $C_{15}H_{30}$

CAS-RN: 1795–21–7
Group No.: 12–097
Molar Mass: 210.40

TABLE 12.97.1. Experimental heat capacities

Reference	Temp. range K	No. pnts	Error %	Purity %	Purity method	Type capacity	Calorimeter Type	Calorimeter Reference
65MES/TOD1	258.2–321.4	12	0.20	99.9	melpt	C_{sat}	BSAO	47HUF

TABLE 12.97.3. Parameters of regression polynomial

Heat capacity type	No. data points total	used	s_w	s/R	s_r %	s_b/R	+/–
C_{sat}	12	12	0.425	4.27–2	0.08	5.28–5	0

Temp. range K	A_1	A_2	A_3	Level of uncertainty
258.2–321.4	6.86653+1	–1.97027+1	4.64983	II

TABLE 12.97.4. Recommended values of heat capacities

Temp. (K)	260	270	273.15	280	290	298.15	300
c_{sat} (J $K^{-1}g^{-1}$)	1.931	1.951	1.958	1.974	2.001	2.025	2.031
C_{sat} (J $K^{-1}mol^{-1}$)	406.3	410.4	411.9	415.3	421.0	426.2	427.4
Temp. (K)	310	320					
c_{sat} (J $K^{-1}g^{-1}$)	2.066	2.104					
C_{sat} (J $K^{-1}mol^{-1}$)	434.6	442.6					

TABLE 12.97.5. Parameters of quasi–polynomial equation

Heat capacity type	No. data points total	used	s_w	s/R	s_r %	s_b/R	+/–
C_{sat}	12	12	0.539	5.41–2	0.11	8.36–5	0

Temp. range K	T_c K	A_1	A_2	A_3	A_4	Level of uncertainty
258.2–321.4	729.10	3.10354+2	1.45945+2	–9.99464+1	1.64994+2	II

Name: 1–Cyclohexyloctahydro–3–methyl–1*H*–indene
Formula: $C_{16}H_{28}$

CAS–RN: unknown
Group No.: 12–098
Molar Mass: 220.40

TABLE 12.98.1. Experimental heat capacities

Reference	Temp. range K	No. pnts	Error %	Purity %	Purity method	Type capacity	Calorimeter Type	Calorimeter Reference
63GUD/CAM	313.1–483.1	4	nosp	99.3	chrom	C_p	BDHT	63GUD/CAM

TABLE 12.98.3. Parameters of regression polynomial

Heat capacity type	No. data points total	No. data points used	s_w	s/R	s_r %	s_b/R	+/–
C_p	4	4	2.305	1.38	2.31	3.55–2	0

Temp. range K	A_1	A_2	Level of uncertainty
313.1–483.1	1.43593+1	1.09394+1	V

TABLE 12.98.4. Recommended values of heat capacities

Temp. (K)	310	320	330	340	350	360	370
c_p (J K^{-1}g^{-1})	1.82	1.86	1.90	1.94	1.99	2.03	2.07
C_p (J K^{-1}mol^{-1})	401	410	420	429	438	447	456
Temp. (K)	380	390	400	410	420	430	440
c_p (J K^{-1}g^{-1})	2.11	2.15	2.19	2.23	2.27	2.32	2.36
C_p (J K^{-1}mol^{-1})	465	474	483	492	501	510	520
Temp. (K)	450	460	470	480			
c_p (J K^{-1}g^{-1})	2.40	2.44	2.48	2.52			
C_p (J K^{-1}mol^{-1})	529	538	547	556			

Name: 2–Ethyltetradecahydrophenanthrene
Formula: $C_{16}H_{28}$

CAS-RN: 90591–84–7
Group No.: 12–099
Molar Mass: 220.40

TABLE 12.99.1. Experimental heat capacities

Reference	Temp. range K	No. pnts	Error %	Purity %	Purity method	Type capacity	Calorimeter Type	Calorimeter Reference
63GUD/CAM	313.1–483.1	4	nosp	not specified		C_p	BDHT	63GUD/CAM

TABLE 12.99.3. Parameters of regression polynomial

Heat capacity type	No. data points total	No. data points used	s_w	s/R	s_r %	s_b/R	+/–
C_p	4	4	2.278	1.30	2.28	1.35–2	0

Temp. range K	A_1	A_2	A_3	Level of uncertainty
313.1–483.1	5.14849+1	–8.78911	2.47961	V

TABLE 12.99.4. Recommended values of heat capacities

Temp. (K)	310	320	330	340	350	360	370
c_p (J $K^{-1}g^{-1}$)	1.81	1.84	1.87	1.90	1.93	1.96	2.00
C_p (J $K^{-1}mol^{-1}$)	400	405	411	418	425	432	440
Temp. (K)	380	390	400	410	420	430	440
c_p (J $K^{-1}g^{-1}$)	2.03	2.07	2.11	2.16	2.20	2.25	2.29
C_p (J $K^{-1}mol^{-1}$)	448	457	466	475	485	495	506
Temp. (K)	450	460	470	480			
c_p (J $K^{-1}g^{-1}$)	2.34	2.40	2.45	2.51			
C_p (J $K^{-1}mol^{-1}$)	517	528	540	552			

Name: 1,1'–(1–Methyl–1,3–propanediyl)biscyclohexane
Formula: $C_{16}H_{30}$

CAS-RN: 41851–35–8
Group No.: 12–100
Molar Mass: 222.41

TABLE 12.100.1. Experimental heat capacities

Reference	Temp. range K	No. pnts	Error %	Purity %	Purity method	Type capacity	Calorimeter Type	Calorimeter Reference
63GUD/CAM	313.1–483.1	4	nosp	96.0	chrom	C_p	BDHT	63GUD/CAM

TABLE 12.100.3. Parameters of regression polynomial

Heat capacity type	No. data points total	No. data points used	s_w	s/R	s_r %	s_b/R	+/–
C_p	4	4	0.387	2.44–1	0.39	2.27–3	2

Temp. range K	A_1	A_2	Level of uncertainty
313.1–483.1	1.03655+1	1.20838+1	IV

TABLE 12.100.4. Recommended values of heat capacities

Temp. (K)	310	320	330	340	350	360	370
c_p (J $K^{-1}g^{-1}$)	1.788	1.833	1.878	1.923	1.969	2.014	2.059
C_p (J $K^{-1}mol^{-1}$)	397.6	407.7	417.7	427.8	437.8	447.9	457.9
Temp. (K)	380	390	400	410	420	430	440
c_p (J $K^{-1}g^{-1}$)	2.104	2.149	2.194	2.240	2.285	2.330	2.375
C_p (J $K^{-1}mol^{-1}$)	468.0	478.0	488.1	498.1	508.2	518.2	528.3
Temp. (K)	450	460	470	480			
c_p (J $K^{-1}g^{-1}$)	2.420	2.465	2.511	2.556			
C_p (J $K^{-1}mol^{-1}$)	538.3	548.4	558.4	568.4			

Name: Cyclohexyl[(1–methylethyl)cyclohexyl]methane
Formula: $C_{16}H_{30}$

CAS-RN: 97676–41–0
Group No.: 12–101
Molar Mass: 222.41

TABLE 12.101.1. Experimental heat capacities

Reference		Temp. range K	No. pnts	Error %	Purity %	Purity method	Type capacity	Calorimeter Type	Calorimeter Reference
63GUD/CAM	N	313.1–483.1	4	nosp	99.6	chrom	C_p	BDHT	63GUD/CAM

63GUD/CAM sample contained 66.7 % of major component (not specified)

TABLE 12.101.3. Parameters of regression polynomial

Heat capacity type	No. data points total	No. data points used	s_w	s/R	s_r %	s_b/R	+/–
C_p	4	4	0.820	5.27–1	0.82	6.05–3	2

Temp. range K	A_1	A_2	Level of uncertainty
313.1–483.1	1.69477+1	1.10552+1	V

TABLE 12.101.4. Recommended values of heat capacities

Temp. (K)	310	320	330	340	350	360	370
c_p (J K^{-1}g^{-1})	1.91	1.96	2.00	2.04	2.08	2.12	2.16
C_p (J K^{-1}mol^{-1})	426	435	444	453	463	472	481
Temp. (K)	380	390	400	410	420	430	440
c_p (J K^{-1}g^{-1})	2.20	2.25	2.29	2.33	2.37	2.41	2.45
C_p (J K^{-1}mol^{-1})	490	499	509	518	527	536	545
Temp. (K)	450	460	470	480			
c_p (J K^{-1}g^{-1})	2.49	2.53	2.58	2.62			
C_p (J K^{-1}mol^{-1})	555	564	573	582			

Name: Decylcyclohexane
Formula: $C_{16}H_{32}$

CAS-RN: 1795–16–0
Group No.: 12–102
Molar Mass: 224.43

TABLE 12.102.1. Experimental heat capacities

Reference	Temp. range K	No. pnts	Error %	Purity %	Purity method	Type capacity	Calorimeter Type	Calorimeter Reference
65FIN/MES	274.3–300.5	9	0.20	99.91	melpt	C_{sat}	BSAO	47HUF

TABLE 12.102.3. Parameters of regression polynomial

Heat capacity type	No. data points total	used	s_w	s/R	s_r %	s_b/R	+/–
C_{sat}	9	9	0.192	2.04–2	0.04	1.40–5	0

Temp. range K	A_1	A_2	Level of uncertainty
274.3–300.5	2.86344+1	8.63524	II

TABLE 12.102.4. Recommended values of heat capacities

Temp. (K)	280	290	298.15	300
c_{sat} (J K^{-1}g^{-1})	1.957	1.989	2.015	2.021
C_{sat} (J K^{-1}mol^{-1})	439.1	446.3	452.1	453.5

Name: Cyclopentylbicyclohexyl (unspecified isomer)
Formula: $C_{17}H_{30}$

CAS-RN: 26447–22–3
Group No.: 12–103
Molar Mass: 234.43

TABLE 12.103.1. Experimental heat capacities

Reference	Temp. range K	No. pnts	Error %	Purity %	method	Type capacity	Calorimeter Type	Reference
62GOL/BEL	310.9–422.0	3	nosp	99.7	chrom	C_p	BDHT	63GUD/CAM
63GUD/CAM	313.1–483.1	4	nosp	not specified		C_p	BDHT	63GUD/CAM

TABLE 12.103.2. Correlated heat capacities

Reference	Temp. range K	No. pnts used	$\sigma_r C$ %	d_w	d/R	d_r %	d_b/R	+/–
				Selected data				
62GOL/BEL	310.9–422.0	3	1.00#	1.781	1.11	1.78	–4.02–1	–1
63GUD/CAM	313.1–483.1	4	1.00#	1.206	8.10–1	1.21	3.54–1	0

TABLE 12.103.3. Parameters of regression polynomial

Heat capacity type	No. data points total	used	s_w	s/R	s_r %	s_b/R	+/–
C_p	7	7	1.751	1.13	1.75	3.00–2	–1

Temp. range K	A_1	A_2	Level of uncertainty
310.9–483.1	1.54668+1	1.17371+1	V

TABLE 12.103.4. Recommended values of heat capacities

Temp. (K)	310	320	330	340	350	360	370
c_p (J K^{-1}g^{-1})	1.84	1.88	1.92	1.96	2.01	2.05	2.09
C_p (J K^{-1}mol^{-1})	431	441	451	460	470	480	490
Temp. (K)	380	390	400	410	420	430	440
c_p (J K^{-1}g^{-1})	2.13	2.17	2.21	2.26	2.30	2.34	2.38
C_p (J K^{-1}mol^{-1})	499	509	519	529	538	548	558
Temp. (K)	450	460	470	480			
c_p (J K^{-1}g^{-1})	2.42	2.46	2.51	2.55			
C_p (J K^{-1}mol^{-1})	568	578	587	597			

Name: Bis(ethylcyclohexyl)methane (unspecified isomer)
Formula: $C_{17}H_{32}$

CAS-RN: 98028–64–9
Group No.: 12–104
Molar Mass: 236.44

TABLE 12.104.1. Experimental heat capacities

Reference		Temp. range K	No. pnts	Error %	Purity %	Purity method	Type capacity	Calorimeter Type	Calorimeter Reference
63GUD/CAM	N	313.1–483.1	4	nosp	99.9	chrom	C_p	BDHT	63GUD/CAM

63GUD/CAM sample contained 58.8 % of major component (not specified)

TABLE 12.104.3. Parameters of regression polynomial

Heat capacity type	No. data points total	No. data points used	s_w	s/R	s_r %	s_b/R	+/–
C_p	4	4	0.719	4.97–1	0.72	1.81–3	–2

Temp. range K	A_1	A_2	Level of uncertainty
313.1–483.1	2.38015+1	1.03230+1	V

TABLE 12.104.4. Recommended values of heat capacities

Temp. (K)	310	320	330	340	350	360	370
c_p (J K^{-1}g^{-1})	1.96	2.00	2.03	2.07	2.11	2.14	2.18
C_p (J K^{-1}mol^{-1})	464	473	481	490	498	507	515
Temp. (K)	380	390	400	410	420	430	440
c_p (J K^{-1}g^{-1})	2.22	2.25	2.29	2.33	2.36	2.40	2.43
C_p (J K^{-1}mol^{-1})	524	533	541	550	558	567	576
Temp. (K)	450	460	470	480			
c_p (J K^{-1}g^{-1})	2.47	2.51	2.54	2.58			
C_p (J K^{-1}mol^{-1})	584	593	601	610			

Name: 1–Cyclohexyl–1–(1–methylethyl)cyclohexylethane (unspecified isomer)
Formula: $C_{17}H_{32}$

CAS-RN: 26637–18–3
Group No.: 12–105
Molar Mass: 236.44

TABLE 12.105.1. Experimental heat capacities

Reference		Temp. range K	No. pnts	Error %	Purity %	Purity method	Type capacity	Calorimeter Type	Calorimeter Reference
63GUD/CAM	N	313.1–483.1	4	nosp	99.5	chrom	C_p	BDHT	63GUD/CAM

63GUD/CAM sample contained 65.1 mol.% of major component (not specified)

TABLE 12.105.3. Parameters of regression polynomial

Heat capacity type	No. data points total	No. data points used	s_w	s/R	s_r %	s_b/R	+/–
C_p	4	4	0.213	1.41–1	0.21	1.66–4	0

Temp. range K	A_1	A_2	Level of uncertainty
313.1–483.1	1.79794+1	1.22042+1	V

TABLE 12.105.4. Recommended values of heat capacities

Temp. (K)	310	320	330	340	350	360	370
c_p (J K^{-1}g^{-1})	1.96	2.01	2.05	2.09	2.13	2.18	2.22
C_p (J K^{-1}mol^{-1})	464	474	484	494	505	515	525
Temp. (K)	380	390	400	410	420	430	440
c_p (J K^{-1}g^{-1})	2.26	2.31	2.35	2.39	2.43	2.48	2.52
C_p (J K^{-1}mol^{-1})	535	545	555	566	576	586	596
Temp. (K)	450	460	470	480			
c_p (J K^{-1}g^{-1})	2.56	2.61	2.65	2.69			
C_p (J K^{-1}mol^{-1})	606	616	626	637			

Name: 3–Ethylhexadecahydropyrene
Formula: $C_{18}H_{30}$

CAS-RN: 94262–24–5
Group No.: 12–106
Molar Mass: 246.44

TABLE 12.106.1. Experimental heat capacities

Reference	Temp. range K	No. pnts	Error %	Purity %	Purity method	Type capacity	Calorimeter Type	Calorimeter Reference
62GOL/BEL	310.9–422.0	3	nosp	99.9	chrom	C_p	BDHT	63GUD/CAM

TABLE 12.106.3. Parameters of regression polynomial

Heat capacity type	No. data points total	No. data points used	s_w	s/R	s_r %	s_b/R	+/–
C_p	3	3	1.024	6.10–1	1.02	5.09–3	1

Temp. range K	A_1	A_2	Level of uncertainty
310.9–422.0	3.03179+1	7.96830	IV

TABLE 12.106.4. Recommended values of heat capacities

Temp. (K)	310	320	330	340	350	360	370
c_p (J K^{-1}g^{-1})	1.856	1.883	1.910	1.937	1.964	1.991	2.018
C_p (J K^{-1}mol^{-1})	457.5	464.1	470.7	477.3	484.0	490.6	497.2
Temp. (K)	380	390	400	410	420		
c_p (J K^{-1}g^{-1})	2.044	2.071	2.098	2.125	2.152		
C_p (J K^{-1}mol^{-1})	503.8	510.5	517.1	523.7	530.3		

Name: 1–Cyclohexyloctahydro–1,3,3–trimethyl–1*H*–indene
Formula: $C_{18}H_{32}$

CAS-RN: 22236–61–9
Group No.: 12–107
Molar Mass: 248.45

TABLE 12.107.1. Experimental heat capacities

Reference	Temp. range K	No. pnts	Error %	Purity %	Purity method	Type capacity	Calorimeter Type	Calorimeter Reference
62GOL/BEL	310.9–422.0	3	nosp	99.3	chrom	C_p	BDHT	63GUD/CAM
63GUD/CAM	313.1–483.1	4	nosp	97.3	chrom	C_p	BDHT	63GUD/CAM

TABLE 12.107.2. Correlated heat capacities

Reference	Temp. range K	No. pnts used	$\sigma_r C$ %	d_w	d/R	d_r %	d_b/R	+/–
				Selected data				
62GOL/BEL	310.9–422.0	3	1.00#	0.337	2.24–1	0.34	–4.54–2	–1
63GUD/CAM	313.1–483.1	4	1.00#	0.264	1.78–1	0.26	3.51–2	0

TABLE 12.107.3. Parameters of regression polynomial

Heat capacity type	No. data points total	No. data points used	s_w	s/R	s_r %	s_b/R	+/–
C_p	7	7	0.352	2.36–1	0.35	6.03–4	–1

Temp. range K	A_1	A_2	Level of uncertainty
310.9–483.1	9.46173	1.46815+1	IV

TABLE 12.107.4. Recommended values of heat capacities

Temp. (K)	310	320	330	340	350	360	370
c_p (J K^{-1}g^{-1})	1.840	1.889	1.938	1.987	2.036	2.085	2.135
C_p (J K^{-1}mol^{-1})	457.1	469.3	481.5	493.7	505.9	518.1	530.3
Temp. (K)	380	390	400	410	420	430	440
c_p (J K^{-1}g^{-1})	2.184	2.233	2.282	2.331	2.380	2.429	2.478
C_p (J K^{-1}mol^{-1})	542.5	554.7	566.9	579.2	591.4	603.6	615.8
Temp. (K)	450	460	470	480			
c_p (J K^{-1}g^{-1})	2.528	2.577	2.626	2.675			
C_p (J K^{-1}mol^{-1})	628.0	640.2	652.4	664.6			

Name: 1,1':2',1''–Tercyclohexane
Formula: $C_{18}H_{32}$

CAS-RN: 2456–43–1
Group No.: 12–108
Molar Mass: 248.45

TABLE 12.108.1. Experimental heat capacities

Reference	Temp. range K	No. pnts	Error %	Purity %	Purity method	Type capacity	Calorimeter Type	Calorimeter Reference
62GOL/BEL	310.9–422.0	3	nosp	97.8	chrom	C_p	BDHT	63GUD/CAM
63GUD/CAM	313.1–483.1	4	nosp	95.6	chrom	C_p	BDHT	63GUD/CAM

TABLE 12.108.2. Correlated heat capacities

Reference	Temp. range K	No. pnts used	$\sigma_r C$ %	d_w	d/R	d_r %	d_b/R	+/–
				Selected data				
62GOL/BEL	310.9–422.0	3	1.00#	0.261	1.55–1	0.26	1.51–1	3
63GUD/CAM	313.1–483.1	4	1.00#	0.877	5.20–1	0.88	–9.75–2	2

TABLE 12.108.3. Parameters of regression polynomial

Heat capacity type	No. data points total	No. data points used	s_w	s/R	s_r %	s_b/R	+/–
C_p	7	7	0.810	4.80–1	0.81	9.07–3	5

Temp. range K	A_1	A_2	Level of uncertainty
310.9–483.1	5.86940	1.44709+1	V

TABLE 12.108.4. Recommended values of heat capacities

Temp. (K)	310	320	330	340	350	360	370
c_p (J K^{-1}g^{-1})	1.70	1.75	1.79	1.84	1.89	1.94	1.99
C_p (J K^{-1}mol^{-1})	422	434	446	458	470	482	494
Temp. (K)	380	390	400	410	420	430	440
c_p (J K^{-1}g^{-1})	2.04	2.09	2.13	2.18	2.23	2.28	2.33
C_p (J K^{-1}mol^{-1})	506	518	530	542	554	566	578
Temp. (K)	450	460	470	480			
c_p (J K^{-1}g^{-1})	2.38	2.42	2.47	2.52			
C_p (J K^{-1}mol^{-1})	590	602	614	626			

Name: 1,1':3',1''–Tercyclohexane
Formula: $C_{18}H_{32}$

CAS-RN: 1706–50–9
Group No.: 12–109
Molar Mass: 248.45

TABLE 12.109.1. Experimental heat capacities

Reference	Temp. range K	No. pnts	Error %	Purity %	Purity method	Type capacity	Calorimeter Type	Calorimeter Reference
62GOL/BEL	310.9–422.0	3	nosp	99.	chrom	C_p	BDHT	63GUD/CAM
63GUD/CAM	373.1–483.1	3	nosp	97.7	chrom	C_p	BDHT	63GUD/CAM

TABLE 12.109.2. Correlated heat capacities

Reference	Temp. range K	No. pnts used	$\sigma_r C$ %	d_w	d/R	d_r %	d_b/R	+/–
				Selected data				
62GOL/BEL	310.9–422.0	3	1.00#	0.940	5.79–1	0.94	–1.01–1	–1
63GUD/CAM	373.1–483.1	3	1.00#	0.906	6.11–1	0.91	1.28–1	1

TABLE 12.109.3. Parameters of regression polynomial

Heat capacity type	No. data points total	used	s_w	s/R	s_r %	s_b/R	+/–
C_p	6	6	1.306	8.41–1	1.31	1.34–2	0

Temp. range K	A_1	A_2	A_3	Level of uncertainty
310.9–483.1	9.06303+1	–2.69194+1	5.00584	V

TABLE 12.109.4. Recommended values of heat capacities

Temp. (K)	310	320	330	340	350	360	370
c_p (J K^{-1}g^{-1})	1.85	1.87	1.88	1.91	1.93	1.96	1.99
C_p (J K^{-1}mol^{-1})	460	464	468	474	480	487	495
Temp. (K)	380	390	400	410	420	430	440
c_p (J K^{-1}g^{-1})	2.03	2.07	2.11	2.16	2.20	2.26	2.31
C_p (J K^{-1}mol^{-1})	504	514	524	536	548	561	575
Temp. (K)	450	460	470	480			
c_p (J K^{-1}g^{-1})	2.37	2.43	2.50	2.57			
C_p (J K^{-1}mol^{-1})	589	605	621	638			

Name: 1,1':4',1''–Tercyclohexane
Formula: $C_{18}H_{32}$

CAS-RN: 1795–19–3
Group No.: 12–110
Molar Mass: 248.45

TABLE 12.110.1. Experimental heat capacities

Reference	Temp. K	Capac. J/(K.g)	Error %	Purity %	Purity method	Type capacity	Calorimeter Type	Calorimeter Reference
63GUD/CAM	423.15	2.272	nosp	91.7	chrom	C_p	BDHT	63GUD/CAM

Name: 1,1–Bis(ethylcyclohexyl)ethane (unspecified isomer)
Formula: $C_{18}H_{34}$

CAS-RN: 98803–07–7
Group No.: 12–111
Molar Mass: 250.47

TABLE 12.111.1. Experimental heat capacities

Reference		Temp. range K	No. pnts	Error %	Purity %	Purity method	Type capacity	Calorimeter Type	Calorimeter Reference
63GUD/CAM	N	313.1–483.1	4	nosp	97.9	chrom	C_p	BDHT	63GUD/CAM

63GUD/CAM sample contained 78.8 % of major component (not specified)

TABLE 12.111.3. Parameters of regression polynomial

Heat capacity type	No. data points total	used	s_w	s/R	s_r %	s_b/R	+/–
C_p	4	4	1.303	9.36–1	1.30	5.28–3	0

Temp. range K	A_1	A_2	Level of uncertainty
313.1–483.1	1.82936+1	1.26410+1	V

TABLE 12.111.4. Recommended values of heat capacities

Temp. (K)	310	320	330	340	350	360	370
c_p (J K^{-1}g^{-1})	1.91	1.95	1.99	2.03	2.08	2.12	2.16
C_p (J K^{-1}mol^{-1})	478	488	499	509	520	530	541
Temp. (K)	380	390	400	410	420	430	440
c_p (J K^{-1}g^{-1})	2.20	2.24	2.29	2.33	2.37	2.41	2.45
C_p (J K^{-1}mol^{-1})	551	562	573	583	594	604	615
Temp. (K)	450	460	470	480			
c_p (J K^{-1}g^{-1})	2.50	2.54	2.58	2.62			
C_p (J K^{-1}mol^{-1})	625	636	646	657			

Name: 1,1,–Bis(dimethylcyclohexyl)ethane (unspecified isomer)
Formula: $C_{18}H_{34}$

CAS-RN: 98803–06–6
Group No.: 12–112
Molar Mass: 250.47

TABLE 12.112.1. Experimental heat capacities

Reference		Temp. range K	No. pnts	Error %	Purity %	Purity method	Type capacity	Calorimeter Type	Calorimeter Reference
63GUD/CAM	N	313.1–483.1	4	nosp	99.0	chrom	C_p	BDHT	63GUD/CAM

63GUD/CAM sample contained 57.3 % of major component (not specified)

TABLE 12.112.3. Parameters of regression polynomial

Heat capacity type	No. data points total	used	s_w	s/R	s_r %	s_b/R	+/–
C_p	4	4	0.611	4.32–1	0.61	9.60–4	0

Temp. range K	A_1	A_2	A_3	Level of uncertainty
313.1–483.1	–2.90192+1	3.47508+1	–2.47710	V

TABLE 12.112.4. Recommended values of heat capacities

Temp. (K)	310	320	330	340	350	360	370
c_p (J K^{-1}g^{-1})	1.82	1.89	1.95	2.01	2.07	2.12	2.18
C_p (J K^{-1}mol^{-1})	456	472	488	503	518	532	546
Temp. (K)	380	390	400	410	420	430	440
c_p (J K^{-1}g^{-1})	2.23	2.28	2.34	2.38	2.43	2.48	2.52
C_p (J K^{-1}mol^{-1})	559	572	585	597	609	620	631
Temp. (K)	450	460	470	480			
c_p (J K^{-1}g^{-1})	2.56	2.60	2.64	2.68			
C_p (J K^{-1}mol^{-1})	642	652	662	671			

Name: Dodecylcyclohexane
Formula: $C_{18}H_{36}$

CAS-RN: 1795–17–1
Group No.: 12–113
Molar Mass: 252.48

TABLE 12.113.1. Experimental heat capacities

Reference	Temp. range K	No. pnts	Error %	Purity %	Purity method	Type capacity	Calorimeter Type	Calorimeter Reference
49PAR/MOO	290.0–300.0	2S	1.00	98.8	melpt	C_p	BSIO	25PAR

TABLE 12.113.3. Parameters of regression polynomial

Heat capacity type	No. data points total	used	s_w	s/R	s_r %	s_b/R	+/–
C_p	2	2	0.000	0.00	0.00	1.91–6	0

Temp. range K	A_1	A_2	Level of uncertainty
290.0–300.0	3.25258+1	1.01643+1	IV

TABLE 12.113.4. Recommended values of heat capacities

Temp. (K)	290	298.15	300
c_p (J K^{-1}g^{-1})	2.042	2.069	2.075
C_p (J K^{-1}mol^{-1})	515.5	522.4	524.0

Name: Hexaethylcyclohexane (unspecified isomer)
Formula: $C_{18}H_{36}$

CAS-RN: 98803–61–3
Group No.: 12–114
Molar Mass: 252.48

TABLE 12.114.1. Experimental heat capacities

Reference		Temp. range K	No. pnts	Error %	Purity %	Purity method	Type capacity	Calorimeter Type	Calorimeter Reference
63GUD/CAM	N	313.1–483.1	4	nosp	99.0	chrom	C_p	BDHT	63GUD/CAM

63GUD/CAM sample contained 47.7 % of 1,2,3,4,5,6–Hexaethylcyclohexane

TABLE 12.114.3. Parameters of regression polynomial

Heat capacity type	No. data points total	used	s_w	s/R	s_r %	s_b/R	+/–
C_p	4	4	2.149	1.57	2.15	2.57–2	0

Temp. range K	A_1	A_2	Level of uncertainty
313.1–483.1	2.98133+1	1.11418+1	V

TABLE 12.114.4. Recommended values of heat capacities

Temp. (K)	310	320	330	340	350	360	370
c_p ($J K^{-1}g^{-1}$)	2.12	2.16	2.19	2.23	2.27	2.30	2.34
C_p ($J K^{-1}mol^{-1}$)	535	544	554	563	572	581	591
Temp. (K)	380	390	400	410	420	430	440
c_p ($J K^{-1}g^{-1}$)	2.38	2.41	2.45	2.49	2.52	2.56	2.60
C_p ($J K^{-1}mol^{-1}$)	600	609	618	628	637	646	655
Temp. (K)	450	460	470	480			
c_p ($J K^{-1}g^{-1}$)	2.63	2.67	2.71	2.74			
C_p ($J K^{-1}mol^{-1}$)	665	674	683	693			

Name: Bis(2,4,6–trimethylcyclohexyl)methane
Formula: $C_{19}H_{36}$

CAS-RN: 94380–80–0
Group No.: 12–115
Molar Mass: 264.49

TABLE 12.115.1. Experimental heat capacities

Reference	Temp. range K	No. pnts	Error %	Purity %	Purity method	Type capacity	Calorimeter Type	Calorimeter Reference
63GUD/CAM	373.1–483.1	3	nosp	99.0	chrom	C_p	BDHT	63GUD/CAM

TABLE 12.115.3. Parameters of regression polynomial

Heat capacity type	No. data points total	used	s_w	s/R	s_r %	s_b/R	+/–
C_p	3	3	1.183	9.02–1	1.18	8.72–3	1

Temp. range K	A_1	A_2	Level of uncertainty
373.1–483.1	2.89109+1	1.12055+1	V

TABLE 12.115.4. Recommended values of heat capacities

Temp. (K)	370	380	390	400	410	420	430
c_p (J K^{-1}g^{-1})	2.21	2.25	2.28	2.32	2.35	2.39	2.42
C_p (J K^{-1}mol^{-1})	585	594	604	613	622	632	641
Temp. (K)	440	450	460	470	480		
c_p (J K^{-1}g^{-1})	2.46	2.49	2.53	2.56	2.60		
C_p (J K^{-1}mol^{-1})	650	660	669	678	688		

Name: 4–Heptyl–1,1'–bicyclohexyl
Formula: $C_{19}H_{36}$

CAS-RN: 96667–88–8
Group No.: 12–116
Molar Mass: 264.49

TABLE 12.116.1. Experimental heat capacities

Reference	Temp. range K	No. pnts	Error %	Purity %	Purity method	Type capacity	Calorimeter Type	Calorimeter Reference
63GUD/CAM	313.1–483.1	4	nosp	100.0	chrom	C_p	BDHT	63GUD/CAM

TABLE 12.116.3. Parameters of regression polynomial

Heat capacity type	No. data points total	No. data points used	s_w	s/R	s_r %	s_b/R	+/–
C_p	4	4	0.551	3.75–1	0.55	1.33–3	0

Temp. range K	A_1	A_2	Level of uncertainty
313.1–483.1	1.80496+1	1.28913+1	IV

TABLE 12.116.4. Recommended values of heat capacities

Temp. (K)	310	320	330	340	350	360	370
c_p (J K^{-1}g^{-1})	1.824	1.864	1.905	1.945	1.986	2.026	2.067
C_p (J K^{-1}mol^{-1})	482.3	493.1	503.8	514.5	525.2	535.9	546.7
Temp. (K)	380	390	400	410	420	430	440
c_p (J K^{-1}g^{-1})	2.107	2.148	2.188	2.229	2.269	2.310	2.350
C_p (J K^{-1}mol^{-1})	557.4	568.1	578.8	589.5	600.2	611.0	621.7
Temp. (K)	450	460	470	480			
c_p (J K^{-1}g^{-1})	2.391	2.432	2.472	2.513			
C_p (J K^{-1}mol^{-1})	632.4	643.1	653.8	664.6			

Name: 1,1'–Heptylidenebiscyclohexane
Formula: $C_{19}H_{36}$

CAS-RN: 2090–15–5
Group No.: 12–117
Molar Mass: 264.49

TABLE 12.117.1. Experimental heat capacities

Reference	Temp. range K	No. pnts	Error %	Purity %	Purity method	Type capacity	Calorimeter Type	Calorimeter Reference
62GOL/BEL	310.9–422.0	3	nosp	95.7	chrom	C_p	BDHT	63GUD/CAM
63GUD/CAM	313.1–483.1	4	nosp	95.7	chrom	C_p	BDHT	63GUD/CAM

TABLE 12.117.2. Correlated heat capacities

Reference	Temp. range K	No. pnts used	$\sigma_r C$ %	d_w	d/R	d_r %	d_b/R	+/–
				Selected data				
62GOL/BEL	310.9–422.0	3	1.00#	0.884	6.71–1	0.88	–2.85–1	–1
63GUD/CAM	313.1–483.1	4	1.00#	0.435	3.40–1	0.43	2.26–1	0

TABLE 12.117.3. Parameters of regression polynomial

Heat capacity type	No. data points total	used	s_w	s/R	s_r %	s_b/R	+/–
C_p	7	7	0.787	6.02–1	0.79	7.16–3	–1

Temp. range K	A_1	A_2	Level of uncertainty
310.9–483.1	2.52775+1	1.24652+1	V

TABLE 12.117.4. Recommended values of heat capacities

Temp. (K)	310	320	330	340	350	360	370
c_p (J K^{-1}g^{-1})	2.01	2.05	2.09	2.13	2.17	2.21	2.24
C_p (J K^{-1}mol^{-1})	531	542	552	563	573	583	594
Temp. (K)	380	390	400	410	420	430	440
c_p (J K^{-1}g^{-1})	2.28	2.32	2.36	2.40	2.44	2.48	2.52
C_p (J K^{-1}mol^{-1})	604	614	625	635	645	656	666
Temp. (K)	450	460	470	480			
c_p (J K^{-1}g^{-1})	2.56	2.60	2.64	2.68			
C_p (J K^{-1}mol^{-1})	677	687	697	708			

Name: Diethylhexadecahydropyrene
Formula: $C_{20}H_{34}$

CAS-RN: 26446–93–5
Group No.: 12–118
Molar Mass: 274.49

TABLE 12.118.1. Experimental heat capacities

Reference	Temp. range K	No. pnts	Error %	Purity %	Purity method	Type capacity	Calorimeter Type	Calorimeter Reference
63GUD/CAM	313.1–483.1	4	nosp	not specified		C_p	BDHT	63GUD/CAM

TABLE 12.118.3. Parameters of regression polynomial

Heat capacity type	No. data points total	used	s_w	s/R	s_r %	s_b/R	+/–
C_p	4	4	0.783	5.50–1	0.78	3.16–3	0

Temp. range K	A_1	A_2	A_3	Level of uncertainty
313.1–483.1	1.12014+2	–3.49886+1	6.06955	V

TABLE 12.118.4. Recommended values of heat capacities

Temp. (K)	310	320	330	340	350	360	370
c_p (J K^{-1}g^{-1})	1.87	1.88	1.90	1.91	1.94	1.96	1.99
C_p (J K^{-1}mol^{-1})	514	517	521	526	531	538	546
Temp. (K)	380	390	400	410	420	430	440
c_p (J K^{-1}g^{-1})	2.02	2.06	2.10	2.14	2.18	2.24	2.29
C_p (J K^{-1}mol^{-1})	555	564	575	587	600	614	628
Temp. (K)	450	460	470	480			
c_p (J K^{-1}g^{-1})	2.35	2.41	2.47	2.54			
C_p (J K^{-1}mol^{-1})	644	661	679	698			

Name: 9–(2–Ethylhexyl)dodecahydrofluorene
Formula: $C_{21}H_{38}$

CAS-RN: 95135–48–1
Group No.: 12–119
Molar Mass: 290.53

TABLE 12.119.1. Experimental heat capacities

Reference	Temp. range K	No. pnts	Error %	Purity %	Purity method	Type capacity	Calorimeter Type	Calorimeter Reference
63GUD/CAM	313.1–483.1	4	nosp	not specified		C_p	BDHT	63GUD/CAM

TABLE 12.119.3. Parameters of regression polynomial

Heat capacity type	No. data points total	No. data points used	s_w	s/R	s_r %	s_b/R	+/–
C_p	4	4	1.673	1.32	1.67	2.68–2	2

Temp. range K	A_1	A_2	Level of uncertainty
313.1–483.1	3.31552+1	1.26149+1	V

TABLE 12.119.4. Recommended values of heat capacities

Temp. (K)	310	320	330	340	350	360	370
c_p (J K^{-1}g^{-1})	2.07	2.10	2.14	2.18	2.21	2.25	2.28
C_p (J K^{-1}mol^{-1})	601	611	622	632	643	653	664
Temp. (K)	380	390	400	410	420	430	440
c_p (J K^{-1}g^{-1})	2.32	2.36	2.39	2.43	2.47	2.50	2.54
C_p (J K^{-1}mol^{-1})	674	685	695	706	716	727	737
Temp. (K)	450	460	470	480			
c_p (J K^{-1}g^{-1})	2.57	2.61	2.65	2.68			
C_p (J K^{-1}mol^{-1})	748	758	769	779			

Name: 1,1',1''–(1–Propanyl–2–ylidene)triscyclohexane
Formula: $C_{21}H_{38}$

CAS-RN: 55682–89–8
Group No.: 12–120
Molar Mass: 290.53

TABLE 12.120.1. Experimental heat capacities

Reference	Temp. range K	No. pnts	Error %	Purity %	Purity method	Type capacity	Calorimeter Type	Calorimeter Reference
63GUD/CAM	373.1–483.1	3	nosp	99.1	chrom	C_p	BDHT	63GUD/CAM

TABLE 12.120.3. Parameters of regression polynomial

Heat capacity type	No. data points total	No. data points used	s_w	s/R	s_r %	s_b/R	+/–
C_p	3	3	0.671	5.67–1	0.67	1.28–3	–1

Temp. range K	A_1	A_2	Level of uncertainty
373.1–483.1	2.52463+1	1.38599+1	IV

TABLE 12.120.4. Recommended values of heat capacities

Temp. (K)	370	380	390	400	410	420	430
c_p (J K^{-1}g^{-1})	2.190	2.230	2.269	2.309	2.349	2.388	2.428
C_p (J K^{-1}mol^{-1})	636.3	647.8	659.3	670.9	682.4	693.9	705.4
Temp. (K)	440	450	460	470	480		
c_p (J K^{-1}g^{-1})	2.468	2.507	2.547	2.587	2.626		
C_p (J K^{-1}mol^{-1})	717.0	728.5	740.0	751.5	763.1		

Name: 4–Nonyl–1,1'–bicyclohexyl
Formula: $C_{21}H_{40}$

CAS-RN: 95135–87–8
Group No.: 12–121
Molar Mass: 292.55

TABLE 12.121.1. Experimental heat capacities

Reference	Temp. range K	No. pnts	Error %	Purity %	Purity method	Type capacity	Calorimeter Type	Calorimeter Reference
63GUD/CAM	313.1–483.1	4	nosp	99.3	chrom	C_p	BDHT	63GUD/CAM

TABLE 12.121.3. Parameters of regression polynomial

Heat capacity type	No. data points total	No. data points used	s_w	s/R	s_r %	s_b/R	+/–
C_p	4	4	1.154	9.05–1	1.15	4.80–3	0

Temp. range K	A_1	A_2	A_3	Level of uncertainty
313.1–483.1	6.22338+1	–4.34298	2.05202	V

TABLE 12.121.4. Recommended values of heat capacities

Temp. (K)	310	320	330	340	350	360	370
c_p (J $K^{-1}g^{-1}$)	1.95	1.97	2.00	2.02	2.05	2.08	2.11
C_p (J $K^{-1}mol^{-1}$)	569	577	584	592	600	609	617
Temp. (K)	380	390	400	410	420	430	440
c_p (J $K^{-1}g^{-1}$)	2.14	2.17	2.21	2.24	2.28	2.32	2.35
C_p (J $K^{-1}mol^{-1}$)	627	636	646	656	667	678	689
Temp. (K)	450	460	470	480			
c_p (J $K^{-1}g^{-1}$)	2.39	2.44	2.48	2.52			
C_p (J $K^{-1}mol^{-1}$)	700	712	725	737			

Name: 4'–Heptyl–1,1':3',1''–tercyclohexane
Formula: $C_{25}H_{46}$

CAS–RN: unknown
Group No.: 12–122
Molar Mass: 346.64

TABLE 12.122.1. Experimental heat capacities

Reference	Temp. range K	No. pnts	Error %	Purity %	Purity method	Type capacity	Calorimeter Type	Calorimeter Reference
62GOL/BEL	310.9–422.0	3	nosp	99.0	chrom	C_p	BDHT	63GUD/CAM
63GUD/CAM	373.1–483.1	3	nosp	99.0	chrom	C_p	BDHT	63GUD/CAM

TABLE 12.122.2. Correlated heat capacities

Reference	Temp. range K	No. pnts used	$\sigma_r C$ %	d_w	d/R	d_r %	d_b/R	+/–
				Selected data				
62GOL/BEL	310.9–422.0	3	1.00#	0.531	4.96–1	0.53	3.64–1	1
63GUD/CAM	373.1–483.1	3	1.00#	1.545	1.50	1.54	–3.09–1	1

TABLE 12.122.3. Parameters of regression polynomial

Heat capacity type	No. data points total	No. data points used	s_w	s/R	s_r %	s_b/R	+/–
C_p	6	6	1.415	1.37	1.41	2.78–2	2

Temp. range K	A_1	A_2	Level of uncertainty
310.9–483.1	3.27457+1	1.53670+1	V

TABLE 12.122.4. Recommended values of heat capacities

Temp. (K)	310	320	330	340	350	360	370
c_p (J $K^{-1}g^{-1}$)	1.93	1.96	2.00	2.04	2.08	2.11	2.15
C_p (J $K^{-1}mol^{-1}$)	668	681	694	707	719	732	745
Temp. (K)	380	390	400	410	420	430	440
c_p (J $K^{-1}g^{-1}$)	2.19	2.22	2.26	2.30	2.33	2.37	2.41
C_p (J $K^{-1}mol^{-1}$)	758	771	783	796	809	822	834
Temp. (K)	450	460	470	480			
c_p (J $K^{-1}g^{-1}$)	2.44	2.48	2.52	2.55			
C_p (J $K^{-1}mol^{-1}$)	847	860	873	886			

Name: 4'–Nonyl–1,1':3',1''–tercyclohexane
Formula: $C_{27}H_{50}$

CAS–RN: unknown
Group No.: 12–123
Molar Mass: 374.69

TABLE 12.123.1. Experimental heat capacities

Reference	Temp. range K	No. pnts	Error %	Purity %	Purity method	Type capacity	Calorimeter Type	Calorimeter Reference
63GUD/CAM	373.1–483.1	3	nosp	96.8	chrom	C_p	BDHT	63GUD/CAM

TABLE 12.123.3. Parameters of regression polynomial

Heat capacity type	No. data points total	No. data points used	s_w	s/R	s_r %	s_b/R	+/–
C_p	3	3	0.737	7.87–1	0.74	5.42–3	1

Temp. range K	A_1	A_2	Level of uncertainty
373.1–483.1	3.64436+1	1.65801+1	V

able 12.123.4. Recommended values of heat capacities

Temp. (K)	370	380	390	400	410	420	430
c_p (J $K^{-1}g^{-1}$)	2.17	2.21	2.24	2.28	2.32	2.35	2.39
C_p (J $K^{-1}mol^{-1}$)	813	827	841	854	868	882	896
Temp. (K)	440	450	460	470	480		
c_p (J $K^{-1}g^{-1}$)	2.43	2.46	2.50	2.54	2.57		
C_p (J $K^{-1}mol^{-1}$)	910	923	937	951	965		

Name: (1–Decylundecyl)cyclohexane
Formula: $C_{27}H_{54}$

CAS-RN: 6703–99–7
Group No.: 12–124
Molar Mass: 378.73

TABLE 12.124.1. Experimental heat capacities

Reference	Temp. range K	No. pnts	Error %	Purity %	Purity method	Type capacity	Calorimeter Type	Calorimeter Reference
49PAR/MOO	270.0–300.0	4S	nosp	not specified		C_p	BSIO	25PAR

TABLE 12.124.3. Parameters of regression polynomial

Heat capacity type	No. data points total	No. data points used	s_w	s/R	s_r %	s_b/R	+/–
C_p	4	4	0.000	1.53–5	0.00	–7.63–6	0

Temp. range K	A_1	A_2	A_3	Level of uncertainty
270.0–300.0	–1.10918+2	1.25783+2	–1.90581+1	IV

TABLE 12.124.4. Recommended values of heat capacities

Temp. (K)	270	273.15	280	290	298.15	300
c_p ($J\,K^{-1}g^{-1}$)	1.971	1.986	2.017	2.054	2.079	2.084
C_p ($J\,K^{-1}mol^{-1}$)	746.3	752.2	763.8	778.0	787.3	789.1

13. Unsaturated Aliphatic Hydrocarbons

In this family the temperature dependent data are available for 36 alkenes (including 10 dienes) and 2 alkynes.

Systematic measurements were carried out for 21 C_5 to C_{16} substances at BMB with the reported error of 0.2 % (47TOD/OLI, 55SCO/FIN, 57MCC/FIN1, 70MES/TOD). Later new data were published by the former BMB (now called NIPER) (90MES/TOD) for two new alkenes and the earlier measurements were revised for additional two compounds having a claimed error of 0.1 %. The Bartlesville results have been considered as highly reliable and have been used alone for producing the recommended values with exception of 1-pentene where also the less accurate high pressure data from CIPT (49SCH/SAG) have been included for increasing the upper temperature limit of the recommended values. For the industrially important isoprene the extensive measurements at NBSW (37BEK/WOO) have been discarded as the data differed considerably from the Bartlesville values. The latter laboratory contributed this discrepancy to possible dissolved impurities (up to 1.7 %) in the NBSW sample; the contents was estimated from the shift in the triple point temperature of the sample (70MES/TOD).

Several low temperature isoperibol calorimeters were developed at UCB (37EGA/KEM, 39POW/GIA, 45GUT/PIT) which were used for supplying data for lowest alkenes C_2H_4 to C_4H_8 with the stated error of 0.2 %. Lower molar mass isomers of alkenes, dienes and 1-butyne were also measured at SUC (31HUF/PAR, 36TOD/PAR), NBSW (44SCO/FER, 45SCO/MAY) and PSC (46AST/FIN, 47AST/SZA, 50AST/MAS); the data from these sources are all in agreement to about 1 % and this is reasonable considering the time period of the measurements. The above mentioned data have been supplemented by less accurate high pressure data measured at CITP (49SCH/SAG, 50AUE/SAG, 52SCH/SAG) having an uncertainty which we have estimated to be about 1.5 %. For 2-butyne the low temperature measurements from CITP (41YOS/OSB) have been till recently the only available source. Later data obtained at the University of Complutense, Madrid, Spain (92BAO/CAC) have not been considered in the final selection as they seemed to be less reliable than the previous source.

Measurements for 3,3-dimethyl-1-butene (38KEM/SHO) and for two isomers of diisobutylene (denoted in the original paper as low and high boiling isomers, 36PAR/TOD2) were performed at SUC. The two isomers were not, however, sufficiently identified in the latter paper; according to the published melting and normal boiling temperatures, we guessed that the low boiling isomer was 2,4,4-trimethyl-1-pentene and 2,4,4-trimethyl-2-pentene corresponded to the high boiling one.

The only diene above C_4, 1,5-hexadiene, was measured in the last century at THA using a drop calorimeter (*81VON).

Name: Ethene
Formula: C_2H_4

CAS-RN: 74–85–1
Group No.: 13–001
Molar Mass: 28.05

TABLE 13.1.1. Experimental heat capacities

Reference	Temp. range K	No. pnts	Error %	Purity %	Purity method	Type capacity	Calorimeter Type	Calorimeter Reference
28EUC/HAU	110.0–170.0	7S	nosp	not specified		C_{sat}	BSIO	28EUC/HAU
37EGA/KEM	106.7–168.7	12	nosp	99.999	melpt	C_p	BSIO	37GIA/EGA

TABLE 13.1.2. Correlated heat capacities

Reference	Temp. range K	No. pnts used	$\sigma_r C$ %	d_w	d/R	d_r %	d_b/R	+/–
				Selected data				
37EGA/KEM	106.7–168.7	12	0.30#	0.374	9.11–3	0.11	2.05–5	–2
				Rejected data				
28EUC/HAU	(9.55–1, 10.32, 8.70–1, 6)							

TABLE 13.1.3. Parameters of regression polynomial

Heat capacity type	No. data points total	used	s_w	s/R	s_r %	s_b/R	+/–
C_p	19	12	0.431	1.05–2	0.13	2.05–5	–2
C_{sat}	19	12	0.425	1.04–2	0.13	1.95–5	–1

Temp. range K	A_1	A_2	A_3	Level of uncertainty
106.7–168.7	1.00884+1	–2.48003	7.65137–1	III
106.7–168.7	9.95951	–2.27249	6.82698–1	III

TABLE 13.1.4. Recommended values of heat capacities

Temp. (K)	110	120	130	140	150	160	170
c_p (J K^{-1}g^{-1})	2.456	2.434	2.418	2.405	2.398	2.394	2.396
C_p (J K^{-1}mol^{-1})	68.90	68.30	67.83	67.48	67.26	67.17	67.21
c_{sat} (J K^{-1}g^{-1})	2.456	2.435	2.418	2.405	2.397	2.392	2.392
C_{sat} (J K^{-1}mol^{-1})	68.89	68.31	67.84	67.48	67.24	67.11	67.09

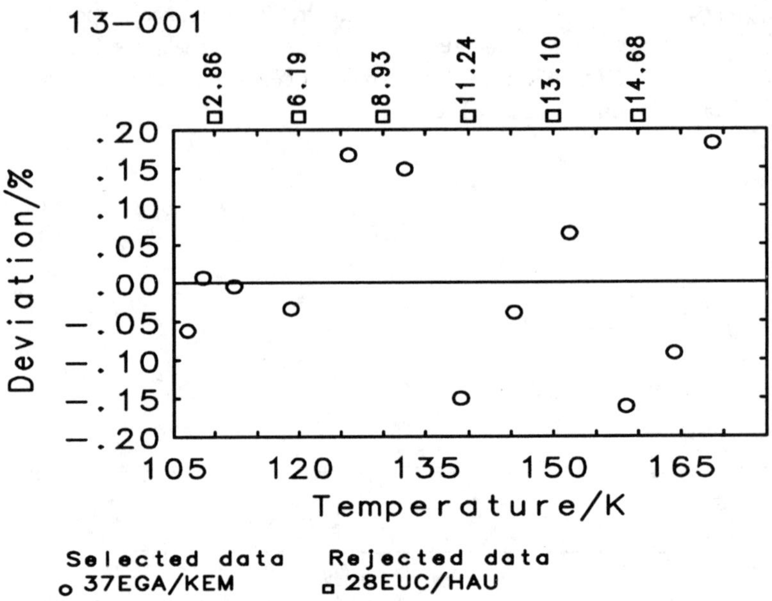

Name: 1–Propene
Formula: C_3H_6

CAS-RN: 115–07–1
Group No.: 13–002
Molar Mass: 42.08

TABLE 13.2.1. Experimental heat capacities

Reference	Temp. range K	No. pnts	Error %	Purity %	Purity method	Type capacity	Calorimeter Type	Calorimeter Reference
31HUF/PAR	93.1–210.3	10	nosp	99.9	estim	C_p	BSIO	25PAR
39POW/GIA	93.9–223.4	18	0.20	99.978	melpt	C_p	BSIO	28GIA/WIE1
50AUE/SAG	299.8–344.3	9	3.00	not specified		C_p	BSAO	39SAG/EVA

TABLE 13.2.2. Correlated heat capacities

Reference	Temp. range K	No. pnts used	$\sigma_r C$ %	d_w	d/R	d_r %	d_b/R	+/–
				Selected data				
39POW/GIA	93.9–223.4	18	0.20	0.991	2.13–2	0.20	9.48–5	–1
50AUE/SAG	299.8–344.3	9	3.00	0.552	2.10–1	1.66	3.33–3	1
				Rejected data				
31HUF/PAR	(7.37–2, 0.70, –6.12–2, –6)							

TABLE 13.2.3. Parameters of cubic spline polynomials

Heat capacity type	No. data points total	used	s_w	s/R	s_r %	s_b/R	+/–
C_p	37	27	0.986	1.39–1	1.10	1.17–3	0

Temp. range K	A_1	A_2	A_3	A_4	Level of uncertainty
93.9–200.0	1.60047+1	–8.83788	4.41175	–6.55201–1	III
200.0–310.0	8.60837	2.25666	–1.13552	2.69344–1	IV
310.0–344.3	–3.19330+3	3.10087+3	–1.00069+3	1.07748+2	V

TABLE 13.2.3. Parameters of cubic spline polynomials

Heat capacity type	No. data points total	used	s_w	s/R	s_r %	s_b/R	+/–
C_p	37	27	0.986	1.39–1	1.10	1.17–3	0
C_{sat}	37	27	0.988	1.37–1	1.11	2.69–3	–2

Temp. range K	A_1	A_2	A_3	A_4	Level of uncertainty
93.9–200.0	1.60047+1	–8.83788	4.41175	–6.55201–1	III
200.0–310.0	8.60837	2.25666	–1.13552	2.69344–1	IV
310.0–344.3	–3.19330+3	3.10087+3	–1.00069+3	1.07748+2	V
93.9–200.0	1.60814+1	–9.01118	4.53842	–6.85168–1	III
200.0–310.0	1.07454+1	–1.00719	5.36424–1	–1.81681–2	IV
310.0–344.3	–2.41314+3	2.34469+3	–7.56139+2	8.13448+1	V

TABLE 13.2.4. Recommended values of heat capacities

Temp. (K)	90	100	110	120	130	140	150
c_p (J $K^{-1}g^{-1}$)	2.202	2.158	2.124	2.098	2.081	2.071	2.067
C_p (J $K^{-1}mol^{-1}$)	92.68	90.82	89.37	88.30	87.57	87.14	87.00
c_{sat} (J $K^{-1}g^{-1}$)	2.203	2.158	2.124	2.098	2.081	2.071	2.067
C_{sat} (J $K^{-1}mol^{-1}$)	92.69	90.82	89.37	88.29	87.56	87.14	87.00
Temp. (K)	160	170	180	190	200	210	220
c_p (J $K^{-1}g^{-1}$)	2.070	2.077	2.088	2.103	2.121	2.141	2.163
C_p (J $K^{-1}mol^{-1}$)	87.09	87.40	87.88	88.51	89.25	90.08	91.00
c_{sat} (J $K^{-1}g^{-1}$)	2.070	2.077	2.088	2.103	2.120	2.139	2.160
C_{sat} (J $K^{-1}mol^{-1}$)	87.10	87.40	87.88	88.50	89.23	90.03	90.90
Temp. (K)	230	240	250	260	270	273.15	280
c_p (J $K^{-1}g^{-1}$)	2.187	2.214	2.245	2.279	2.317	2.329	2.359
C_p (J $K^{-1}mol^{-1}$)	92.03	93.18	94.47	95.90	97.49	98.02	99.25
c_{sat} (J $K^{-1}g^{-1}$)	2.182	2.206	2.232	2.259	2.288	2.297	2.318
C_{sat} (J $K^{-1}mol^{-1}$)	91.84	92.85	93.92	95.06	96.27	96.67	97.55
Temp. (K)	290	298.15	300	310	320	330	340
c_p (J $K^{-1}g^{-1}$)	2.405	2.446	2.456	2.512	2.60	2.81	3.29
C_p (J $K^{-1}mol^{-1}$)	101.2	102.9	103.4	105.7	109	118	138
c_{sat} (J $K^{-1}g^{-1}$)	2.350	2.377	2.383	2.418	2.47	2.62	2.96
C_{sat} (J $K^{-1}mol^{-1}$)	98.88	100.0	100.3	101.7	104	110	125
Temp. (K)	350						
c_p (J $K^{-1}g^{-1}$)	4.15						
C_p (J $K^{-1}mol^{-1}$)	175						
c_{sat} (J $K^{-1}g^{-1}$)	3.60						
C_{sat} (J $K^{-1}mol^{-1}$)	151						

TABLE 13.2.5. Parameters of quasi-polynomial equation

Heat capacity type	No. data points total	used	s_w	s/R	s_r %	s_b/R	+/–
C_p	37	20	1.663	2.81–1	2.04	–5.83–2	–4

Temp. range K	T_c K	A_1	A_2	A_3	A_4	Level of uncertainty
141.2–344.3	365.57	–2.88933–1	4.22652–1	9.55239	4.93799–2	V

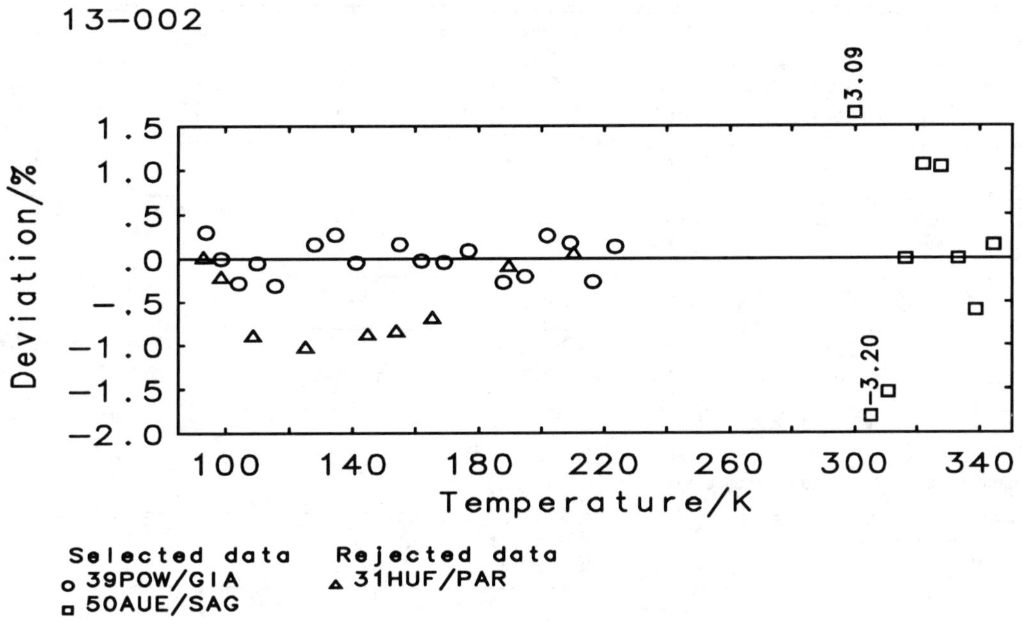

Name: 1,2–Butadiene
Formula: C_4H_6

CAS-RN: 590–19–2
Group No.: 13–003
Molar Mass: 54.09

TABLE 13.3.1. Experimental heat capacities

Reference		Temp. range K	No. pnts	Error %	Purity %	Purity method	Type capacity	Calorimeter Type	Calorimeter Reference
47AST/SZA	N	140.0–290.0	31S	0.20	99.89	melpt	C_p	BSAO	47AST/SZA

47AST/SZA errors below and above 200 K 0.1 % and 0.2 %, respectively

TABLE 13.3.3. Parameters of regression polynomial

Heat capacity type	No. data points total	No. data points used	s_w	s/R	s_r %	s_b/R	+/–
C_p	31	31	0.377	1.02–2	0.08	1.21–5	–2

Temp. range K	A_1	A_2	A_3	Level of uncertainty
140.0–290.0	1.61127+1	–3.51881	1.05399	III

TABLE 13.3.4. Recommended values of heat capacities

Temp. (K)	140	150	160	170	180	190	200
c_p ($J K^{-1} g^{-1}$)	2.037	2.030	2.026	2.025	2.028	2.034	2.043
C_p ($J K^{-1} mol^{-1}$)	110.2	109.8	109.6	109.6	109.7	110.0	110.5
Temp. (K)	210	220	230	240	250	260	270
c_p ($J K^{-1} g^{-1}$)	2.055	2.071	2.090	2.112	2.137	2.166	2.197
C_p ($J K^{-1} mol^{-1}$)	111.2	112.0	113.0	114.2	115.6	117.1	118.9
Temp. (K)	273.15	280	290	298.15			
c_p ($J K^{-1} g^{-1}$)	2.208	2.232	2.271	2.304			
C_p ($J K^{-1} mol^{-1}$)	119.4	120.8	122.8	124.6			

TABLE 13.3.5. Parameters of quasi-polynomial equation

Heat capacity type	No. data points total	used	s_w	s/R	s_r %	s_b/R	+/−
C_p	31	26	1.266	3.44–2	0.25	1.45–4	1

Temp. range K	T_c K	A_1	A_2	A_3	A_4	Level of uncertainty
165.0–290.0	452.20	7.27575	4.08823	8.80550	3.23713	III

Name: 1,3–Butadiene
Formula: C_4H_6

CAS-RN: 106–99–0
Group No.: 13–004
Molar Mass: 54.09

TABLE 13.4.1. Experimental heat capacities

Reference	Temp. range K	No. pnts	Error %	Purity %	Purity method	Type capacity	Calorimeter Type	Calorimeter Reference
45SCO/MEY	165.0–300.0	29S	0.10	99.95	melpt	C_{sat}	BSAO	45SCO/MEY

TABLE 13.4.3. Parameters of regression polynomial

Heat capacity type	No. data points total	used	s_w	s/R	s_r %	s_b/R	+/−
C_p	29	29	0.697	9.54–3	0.07	1.77–5	1
C_{sat}	29	29	0.658	8.77–3	0.07	1.19–5	1

Temp. range K	A_1	A_2	A_3	Level of uncertainty
165.0–300.0	1.57459+1	−4.15515	1.30027	II
165.0–300.0	1.54941+1	−3.90430	1.23869	II

TABLE 13.4.4. Recommended values of heat capacities

Temp. (K)	170	180	190	200	210	220	230
c_p ($J K^{-1}g^{-1}$)	1.912	1.918	1.928	1.942	1.960	1.983	2.009
C_p ($J K^{-1}mol^{-1}$)	103.4	103.8	104.3	105.1	106.0	107.2	108.6
c_{sat} ($J K^{-1}g^{-1}$)	1.912	1.918	1.929	1.943	1.961	1.983	2.009
C_{sat} ($J K^{-1}mol^{-1}$)	103.4	103.8	104.3	105.1	106.1	107.3	108.6
Temp. (K)	240	250	260	270	273.15	280	290
c_p ($J K^{-1}g^{-1}$)	2.039	2.073	2.111	2.153	2.167	2.199	2.249
C_p ($J K^{-1}mol^{-1}$)	110.3	112.1	114.2	116.5	117.2	118.9	121.7
c_{sat} ($J K^{-1}g^{-1}$)	2.038	2.071	2.108	2.149	2.163	2.194	2.243
C_{sat} ($J K^{-1}mol^{-1}$)	110.2	112.0	114.0	116.3	117.0	118.7	121.3
Temp. (K)	298.15	300					
c_p ($J K^{-1}g^{-1}$)	2.293	2.303					
C_p ($J K^{-1}mol^{-1}$)	124.0	124.6					
c_{sat} ($J K^{-1}g^{-1}$)	2.285	2.295					
C_{sat} ($J K^{-1}mol^{-1}$)	123.6	124.1					

TABLE 13.4.5. Parameters of quasi-polynomial equation

Heat capacity type	No. data points total	used	s_w	s/R	s_r %	s_b/R	+/–
C_p	29	29	2.842	3.73–2	0.28	1.71–4	2

Temp. range K	T_c K	A_1	A_2	A_3	A_4	Level of uncertainty
165.0–300.0	425.17	1.87702	2.24415	9.44921	3.92485–1	III

Name: 1–Butyne
Formula: C_4H_6

CAS-RN: 107–00–6
Group No.: 13–005
Molar Mass: 54.09

TABLE 13.5.1. Experimental heat capacities

Reference		Temp. range K	No. pnts	Error %	Purity %	method	Type capacity	Calorimeter Type	Reference
50AST/MAS	N	149.9–275.1	22	0.50	99.967	melpt	C_p	BSIO	36AST/MES

50AST/MAS errors below and above 200 K 0.3 % and 0.5 %, respectively

TABLE 13.5.3. Parameters of regression polynomial

Heat capacity type	No. data points total	used	s_w	s/R	s_r %	s_b/R	+/–
C_p	22	22	0.593	4.44–2	0.30	2.28–4	–2

Temp. range K	A_1	A_2	A_3	Level of uncertainty
149.9–275.1	1.65172+1	–3.73094	1.25624	III

TABLE 13.5.4. Recommended values of heat capacities

Temp. (K)	150	160	170	180	190	200	210
c_p (J K^{-1}g^{-1})	2.113	2.116	2.122	2.132	2.146	2.164	2.186
C_p (J K^{-1}mol^{-1})	114.3	114.4	114.8	115.3	116.1	117.1	118.3
Temp. (K)	220	230	240	250	260	270	273.15
c_p (J K^{-1}g^{-1})	2.212	2.241	2.275	2.312	2.353	2.398	2.413
C_p (J K^{-1}mol^{-1})	119.6	121.2	123.0	125.1	127.3	129.7	130.5
Temp. (K)	280						
c_p (J K^{-1}g^{-1})	2.447						
C_p (J K^{-1}mol^{-1})	132.4						

TABLE 13.5.5. Parameters of quasi-polynomial equation

Heat capacity type	No. data points total	used	s_w	s/R	s_r %	s_b/R	+/–
C_p	22	22	0.717	5.42–2	0.36	3.23–4	–2

Temp. range K	T_c K	A_1	A_2	A_3	A_4	Level of uncertainty
149.9–275.1	463.70	1.30790+1	7.34862	6.05548	5.81948	III

Name: 2–Butyne
Formula: C_4H_6

CAS-RN: 503–17–3
Group No.: 13–006
Molar Mass: 54.09

TABLE 13.6.1. Experimental heat capacities

Reference		Temp. range K	No. pnts	Error %	Purity %	Purity method	Type capacity	Calorimeter Type	Calorimeter Reference
41YOS/OSB	N	249.5–284.3	10	0.20	99.75	melpt	C_p	BSAO	41YOS/GAR
92BAO/CAC		255.1–290.1	8	1.00	99.5	estim	C_p	BDHT	92BAO/CAC

41YOS/OSB C_p value at 255.06 K misprinted, 29.57 should read 28.57 cal/(K.mol)

TABLE 13.6.2. Correlated heat capacities

Reference	Temp. range K	No. pnts used	$\sigma_r C$ %	d_w	d/R	d_r %	d_b/R	+/–
				Selected data				
41YOS/OSB	249.5–284.3	10	0.20	0.675	1.95–2	0.14	5.34–5	0
				Rejected data				
92BAO/CAC	(1.03–1, 0.72, –8.73–2, –6)							

TABLE 13.6.3. Parameters of regression polynomial

Heat capacity type	No. data points total	used	s_w	s/R	s_r %	s_b/R	+/–
C_p	18	10	0.755	2.18–2	0.15	5.34–5	0

Temp. range K	A_1	A_2	Level of uncertainty
249.5–284.3	1.06547+1	1.47410	III

TABLE 13.6.4. Recommended values of heat capacities

Temp. (K)	250	260	270	273.15	280	290	298.15
c_p ($J K^{-1}g^{-1}$)	2.204	2.227	2.250	2.257	2.272	2.295	2.313
C_p ($J K^{-1}mol^{-1}$)	119.2	120.5	121.7	122.1	122.9	124.1	125.1

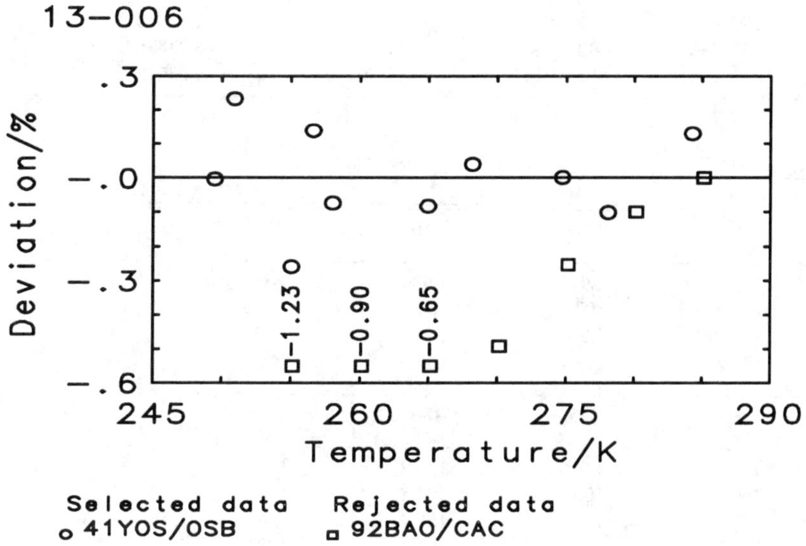

Name: 1–Butene
Formula: C_4H_8

CAS-RN: 106–98–9
Group No.: 13–007
Molar Mass: 56.11

TABLE 13.7.1. Experimental heat capacities

Reference		Temp. range K	No. pnts	Error %	Purity %	Purity method	Type capacity	Calorimeter Type	Calorimeter Reference
36TOD/PAR		81.3–253.4	24	0.50	99.9	estim	C_p	BSIO	25PAR
46AST/FIN	N	89.8–258.5	18	0.30	99.54	melpt	C_p	BSIO	28GIA/WIE1
49SCH/SAG		310.9–366.5	11	1.50	not specified		C_p	BSAO	39SAG/EVA

46AST/FIN errors below and above 200 K 0.3 % and 0.5 %, respectively

TABLE 13.7.2. Correlated heat capacities

Reference	Temp. range K	No. pnts used	$\sigma_r C$ %	d_w	d/R	d_r %	d_b/R	+/–
				Selected data				
36TOD/PAR	81.3253.4	24	0.50	0.470	3.17–2	0.23	1.27–2	8
46AST/FIN	89.8–258.5	18	0.30	0.729	2.90–2	0.22	–5.12–3	0
49SCH/SAG	310.9–366.5	11	1.50	0.774	2.20–1	1.16	–2.52–2	–5

TABLE 13.7.3. Parameters of cubic spline polynomials

Heat capacity type	No. data points total	used	s_w	s/R	s_r %	s_b/R	+/–
C_p	53	53	0.669	1.09–1	0.60	–1.23–3	3
C_{sat}	53	53	0.643	9.54–2	0.54	–9.01–5	5

Temp. range K	A_1	A_2	A_3	A_4	Level of uncertainty
81.3–230.0	1.75376+1	–8.18394	4.17454	–5.76677–1	III
230.0–366.5	–3.18046	1.88397+1	–7.57486	1.12613	V
81.3–230.0	1.75357+1	–8.18299	4.17677	–5.77886–1	III
230.0–366.5	–7.33269–1	1.56460+1	–6.18368	9.23629–1	V

TABLE 13.7.4. Recommended values of heat capacities

Temp. (K)	90	100	110	120	130	140	150
c_p (J K^{-1}g^{-1})	1.946	1.919	1.900	1.887	1.880	1.879	1.883
C_p (J K^{-1}mol^{-1})	109.2	107.7	106.6	105.9	105.5	105.4	105.7
c_{sat} (J K^{-1}g^{-1})	1.946	1.919	1.900	1.887	1.880	1.879	1.883
C_{sat} (J K^{-1}mol^{-1})	109.2	107.7	106.6	105.9	105.5	105.4	105.7
Temp. (K)	**160**	**170**	**180**	**190**	**200**	**210**	**220**
c_p (J K^{-1}g^{-1})	1.892	1.905	1.922	1.942	1.964	1.989	2.015
C_p (J K^{-1}mol^{-1})	106.2	106.9	107.8	108.9	110.2	111.6	113.1
c_{sat} (J K^{-1}g^{-1})	1.892	1.905	1.922	1.942	1.964	1.989	2.015
C_{sat} (J K^{-1}mol^{-1})	106.2	106.9	107.8	108.9	110.2	111.6	113.0
Temp. (K)	**230**	**240**	**250**	**260**	**270**	**273.15**	**280**
c_p (J K^{-1}g^{-1})	2.042	2.07	2.10	2.13	2.17	2.18	2.21
C_p (J K^{-1}mol^{-1})	114.6	116	118	120	122	122	124
c_{sat} (J K^{-1}g^{-1})	2.042	2.07	2.10	2.13	2.17	2.18	2.20
C_{sat} (J K^{-1}mol^{-1})	114.6	116	118	120	121	122	124
Temp. (K)	**290**	**298.15**	**300**	**310**	**320**	**330**	**340**
c_p (J K^{-1}g^{-1})	2.25	2.30	2.31	2.37	2.44	2.51	2.60
C_p (J K^{-1}mol^{-1})	127	129	129	133	137	141	146
c_{sat} (J K^{-1}g^{-1})	2.25	2.29	2.30	2.35	2.41	2.48	2.56
C_{sat} (J K^{-1}mol^{-1})	126	128	129	132	135	139	144
Temp. (K)	**350**	**360**	**370**				
c_p (J K^{-1}g^{-1})	2.70	2.82	2.94				
C_p (J K^{-1}mol^{-1})	152	158	165				
c_{sat} (J K^{-1}g^{-1})	2.65	2.75	2.86				
C_{sat} (J K^{-1}mol^{-1})	149	154	160				

TABLE 13.7.5. Parameters of quasi-polynomial equation

Heat capacity type	No. data points total	used	s_w	s/R	s_r %	s_b/R	+/-
C_{sat}	53	41	1.723	1.14–1	0.77	–1.76–2	–8

Temp. range K	T_c K	A_1	A_2	A_3	A_4	Level of uncertainty
124.6–366.5	419.95	–1.35127	6.90682–1	1.08104+1	6.60913–1	V

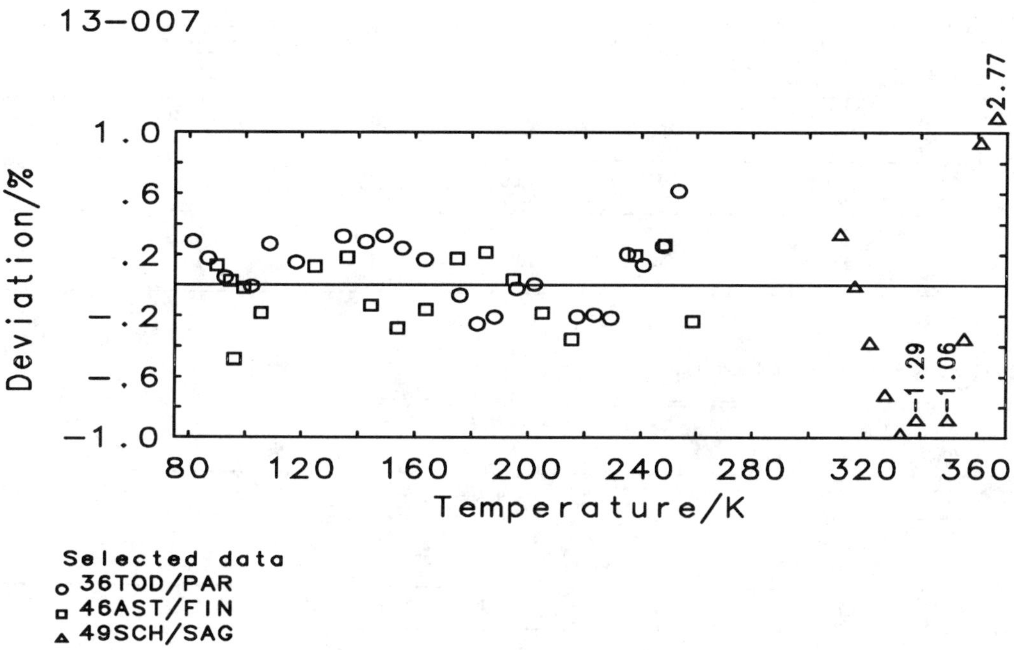

Name: (*E*)–2–Butene
Formula: C_4H_8

CAS-RN: 624–64–6
Group No.: 13–008
Molar Mass: 56.11

TABLE 13.8.1. Experimental heat capacities

Reference	Temp. range K	No. pnts	Error %	Purity %	Purity method	Type capacity	Calorimeter Type	Calorimeter Reference
36TOD/PAR	176.0–259.6	6	0.50	99.9	estim	C_p	BSIO	25PAR
45GUT/PIT	170.7–271.0	14	0.20	99.63	melpt	C_p	BSAO	45GUT/PIT

TABLE 13.8.2. Correlated heat capacities

Reference	Temp. range K	No. pnts used	$\sigma_r C$ %	d_w	d/R	d_r %	d_b/R	+/-
				Selected data				
45GUT/PIT	170.7–271.0	14	0.20	0.823	2.34–2	0.16	7.79–5	0
				Rejected data				
36TOD/PAR	(1.04–1, 0.72, 6.56–2, 2)							

TABLE 13.8.3. Parameters of regression polynomial

Heat capacity type	No. data points total	used	s_w	s/R	s_r %	s_b/R	+/–
C_p	20	14	0.974	2.76–2	0.19	7.79–5	0

Temp. range K	A_1	A_2	A_3	A_4	Level of uncertainty
170.7–271.0	2.25852+1	–1.35561+1	6.16090	–8.24713–1	II

TABLE 13.8.4. Recommended values of heat capacities

Temp. (K)	170	180	190	200	210	220	230
c_p (J K^{-1}g^{-1})	1.970	1.976	1.988	2.003	2.023	2.045	2.069
C_p (J K^{-1}mol^{-1})	110.5	110.9	111.5	112.4	113.5	114.7	116.1
Temp. (K)	240	250	260	270			
c_p (J K^{-1}g^{-1})	2.095	2.121	2.148	2.173			
C_p (J K^{-1}mol^{-1})	117.5	119.0	120.5	121.9			

TABLE 13.8.5. Parameters of quasi-polynomial equation

Heat capacity type	No. data points total	used	s_w	s/R	s_r %	s_b/R	+/–
C_p	20	14	1.445	4.07–2	0.29	1.83–4	4

Temp. range K	T_c K	A_1	A_2	A_3	A_4	Level of uncertainty
170.7–271.0	428.63	–3.73205–1	1.20479	1.10428+1	2.89016–2	III

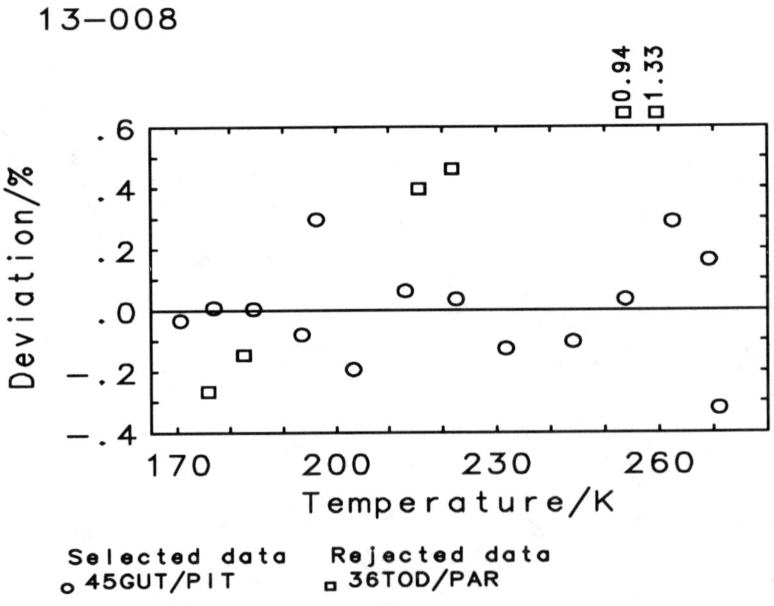

Name: (*Z*)–2–Butene
Formula: C_4H_8

CAS-RN: 590–18–1
Group No.: 13–009
Molar Mass: 56.11

TABLE 13.9.1. Experimental heat capacities

Reference	Temp. range K	No. pnts	Error %	Purity %	Purity method	Type capacity	Calorimeter Type	Calorimeter Reference
36TOD/PAR	138.2–266.6	18	0.50	99.9	estim	C_p	BSIO	25PAR
44SCO/FER	134.3–300.0	36S	0.10	99.943	melpt	C_{sat}	BSAO	33SOU/BRI
52SCH/SAG	299.8–366.5	13S	1.50	99.8	estim	C_p	BSAO	39SAG/EVA

TABLE 13.9.2. Correlated heat capacities

Reference	Temp. range K	No. pnts used	$\sigma_r C$ %	d_w	d/R	d_r %	d_b/R	+/–
				Selected data				
44SCO/FER	134.3–300.0	36	0.10	0.866	1.26–2	0.09	–3.88–4	0
52SCH/SAG	299.8–366.5	13	1.50	1.046	2.71–1	1.57	2.66–1	13
				Rejected data				
36TOD/PAR	(7.51–2, 0.56, –6.96–2, –18)							

TABLE 13.9.3. Parameters of cubic spline polynomials

Heat capacity type	No. data points total	No. data points used	s_w	s/R	s_r %	s_b/R	+/–
C_p	67	49	0.968	1.47–1	0.86	7.03–2	13
C_{sat}	67	49	0.936	1.43–1	0.84	6.82–2	16

Temp. range K	A_1	A_2	A_3	A_4	Level of uncertainty
134.3–250.0	1.95676+1	–7.48436	2.60326	–2.00614–1	II
250.0–366.5	–3.45238	2.01396+1	–8.44634	1.27267	V
134.3–250.0	1.96826+1	–7.68436	2.71714	–2.21850–1	II
250.0–366.5	–1.39561	1.76095+1	–7.40041	1.12716	V

TABLE 13.9.4. Recommended values of heat capacities

Temp. (K)	140	150	160	170	180	190	200
c_p (J K^{-1}g^{-1})	2.022	2.004	1.991	1.983	1.980	1.981	1.987
C_p (J K^{-1}mol^{-1})	113.4	112.4	111.7	111.3	111.1	111.2	111.5
c_{sat} (J K^{-1}g^{-1})	2.022	2.004	1.991	1.983	1.980	1.981	1.987
C_{sat} (J K^{-1}mol^{-1})	113.4	112.4	111.7	111.3	111.1	111.2	111.5
Temp. (K)	210	220	230	240	250	260	270
c_p (J K^{-1}g^{-1})	1.997	2.010	2.028	2.049	2.074	2.10	2.13
C_p (J K^{-1}mol^{-1})	112.0	112.8	113.8	115.0	116.3	118	120
c_{sat} (J K^{-1}g^{-1})	1.997	2.010	2.028	2.049	2.073	2.10	2.13
C_{sat} (J K^{-1}mol^{-1})	112.0	112.8	113.8	114.9	116.3	118	120
Temp. (K)	273.15	280	290	298.15	300	310	320
c_p (J K^{-1}g^{-1})	2.15	2.17	2.22	2.26	2.27	2.33	2.40
C_p (J K^{-1}mol^{-1})	120	122	124	127	127	131	135
c_{sat} (J K^{-1}g^{-1})	2.14	2.17	2.21	2.25	2.26	2.32	2.39
C_{sat} (J K^{-1}mol^{-1})	120	122	124	126	127	130	134
Temp. (K)	330	340	350	360	370		
c_p (J K^{-1}g^{-1})	2.48	2.58	2.69	2.81	2.95		
C_p (J K^{-1}mol^{-1})	139	145	151	158	165		
c_{sat} (J K^{-1}g^{-1})	2.46	2.55	2.65	2.77	2.90		
C_{sat} (J K^{-1}mol^{-1})	138	143	149	155	162		

TABLE 13.9.5. Parameters of quasi-polynomial equation

Heat capacity type	No. data points total	used	s_w	s/R	s_r %	s_b/R	+/–
C_p	67	43	2.984	8.23–2	0.51	2.03–2	9

Temp. range K	T_c K	A_1	A_2	A_3	A_4	A_5	A_6	Level of uncertainty
160.0–366.5	435.58	–1.11740	5.11459–3	1.12695+1	1.69324+1	–5.24214+1	4.90685+1	V

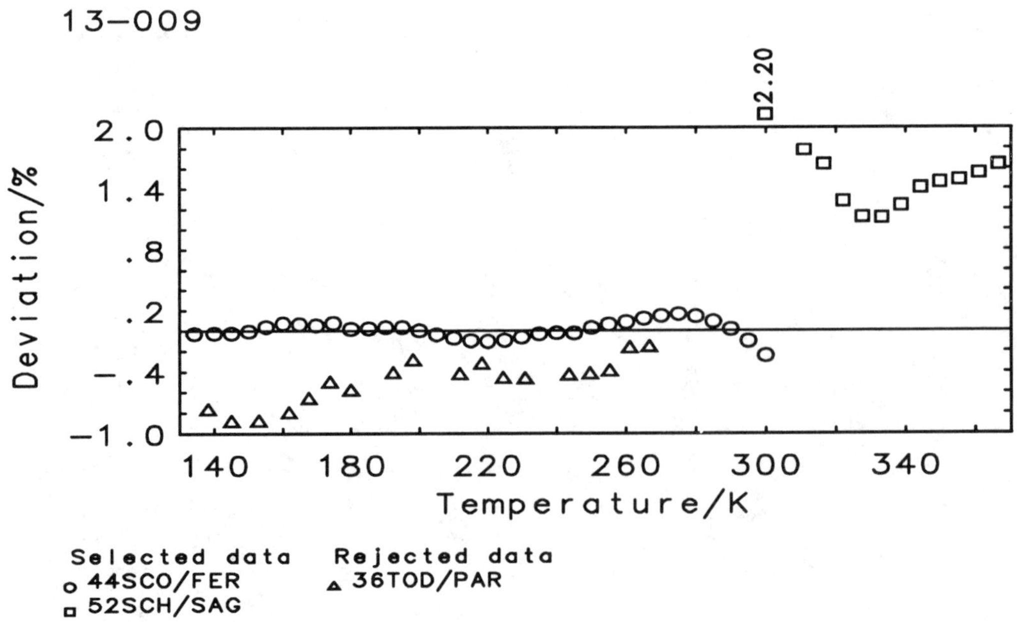

Name: 2–Methyl–1–propene
Formula: C_4H_8

CAS-RN: 115–11–7
Group No.: 13–010
Molar Mass: 56.11

TABLE 13.10.1. Experimental heat capacities

Reference	Temp. range K	No. pnts	Error %	Purity %	Purity method	Type capacity	Calorimeter Type	Calorimeter Reference
36TOD/PAR	139.2–253.1	15	0.50	99.9	estim	C_p	BSIO	25PAR

TABLE 13.10.3. Parameters of regression polynomial

Heat capacity type	No. data points total	No. data points used	s_w	s/R	s_r %	s_b/R	+/–
C_p	15	15	0.158	1.09–2	0.08	1.29–5	2

Temp. range K	A_1	A_2	A_3	A_4	Level of uncertainty
139.2–253.1	1.49551+1	–3.91890	2.00970	–2.04006–1	III

TABLE 13.10.4. Recommended values of heat capacities

Temp. (K)	140	150	160	170	180	190	200
c_p (J K^{-1}g^{-1})	1.904	1.913	1.926	1.941	1.959	1.981	2.004
C_p (J K^{-1}mol^{-1})	106.8	107.3	108.0	108.9	109.9	111.1	112.4
Temp. (K)	210	220	230	240	250		
c_p (J K^{-1}g^{-1})	2.030	2.058	2.088	2.120	2.153		
C_p (J K^{-1}mol^{-1})	113.9	115.5	117.2	118.9	120.8		

TABLE 13.10.5. Parameters of quasi–polynomial equation

Heat capacity type	No. data points total	No. data points used	s_w	s/R	s_r %	s_b/R	+/–
C_p	15	15	0.359	2.41–2	0.18	6.57–5	–1

Temp. range K	T_c K	A_1	A_2	A_3	A_4	Level of uncertainty
139.2–253.1	417.90	7.87193–1	2.14113	9.89321	7.23535–2	III

Name: 2–Methyl–1,3–butadiene
Formula: C_5H_8

CAS-RN: 78–79–5
Group No.: 13–011
Molar Mass: 68.12

TABLE 13.11.1. Experimental heat capacities

Reference		Temp. range K	No. pnts	Error %	Purity %	Purity method	Type capacity	Calorimeter Type	Calorimeter Reference
37BEK/WOO	N	130.2–298.5	35	0.20	not specified		C_p	BSAO	33SOU/BRI
70MES/TOD		130.3–303.9	21	0.20	99.83	melpt	C_{sat}	BSAO	47HUF

37BEK/WOO impurity 1.7 mole % estimated by 70MES/TOD from triple point temperature

TABLE 13.11.2. Correlated heat capacities

Reference	Temp. range K	No. pnts used	$\sigma_r C$ %	d_w	d/R	d_r %	d_b/R	+/–
				Selected data				
70MES/TOD	130.3–303.9	21	0.20#	0.112	3.59–3	0.02	4.54–8	–1
				Rejected data				
37BEK/WOO	(1.11–1, 0.66, 9.39–2, 32)							

TABLE 13.11.3. Parameters of regression polynomial

Heat capacity type	No. data points total	used	s_w	s/R	s_r %	s_b/R	+/–
C_{sat}	56	21	0.124	4.00–3	0.02	4.54–8	–1

Temp. range K	A_1	A_2	A_3	A_4	Level of uncertainty
130.3–303.9	1.92297+1	–6.84607	2.94929	–2.59048–1	II

TABLE 13.11.4. Recommended values of heat capacities

Temp. (K)	130	140	150	160	170	180	190
c_{sat} (J K^{-1}g^{-1})	1.800	1.796	1.797	1.802	1.812	1.825	1.842
C_{sat} (J K^{-1}mol^{-1})	122.6	122.3	122.4	122.8	123.4	124.3	125.5
Temp. (K)	200	210	220	230	240	250	260
c_{sat} (J K^{-1}g^{-1})	1.863	1.887	1.914	1.945	1.978	2.014	2.052
C_{sat} (J K^{-1}mol^{-1})	126.9	128.5	130.4	132.5	134.7	137.2	139.8
Temp. (K)	270	273.15	280	290	298.15	300	
c_{sat} (J K^{-1}g^{-1})	2.093	2.106	2.136	2.180	2.218	2.226	
C_{sat} (J K^{-1}mol^{-1})	142.6	143.5	145.5	148.5	151.1	151.7	

TABLE 13.11.5. Parameters of quasi–polynomial equation

Heat capacity type	No. data points total	used	s_w	s/R	s_r %	s_b/R	+/–
C_{sat}	56	21	0.702	2.20–2	0.14	4.03–5	–1

Temp. range K	T_c K	A_1	A_2	A_3	A_4	A_5	A_6	Level of uncertainty
130.3–303.9	475.70	1.88933+2	1.58383+1	–2.59707	1.92529+2	–1.03593+1	1.94658+2	II

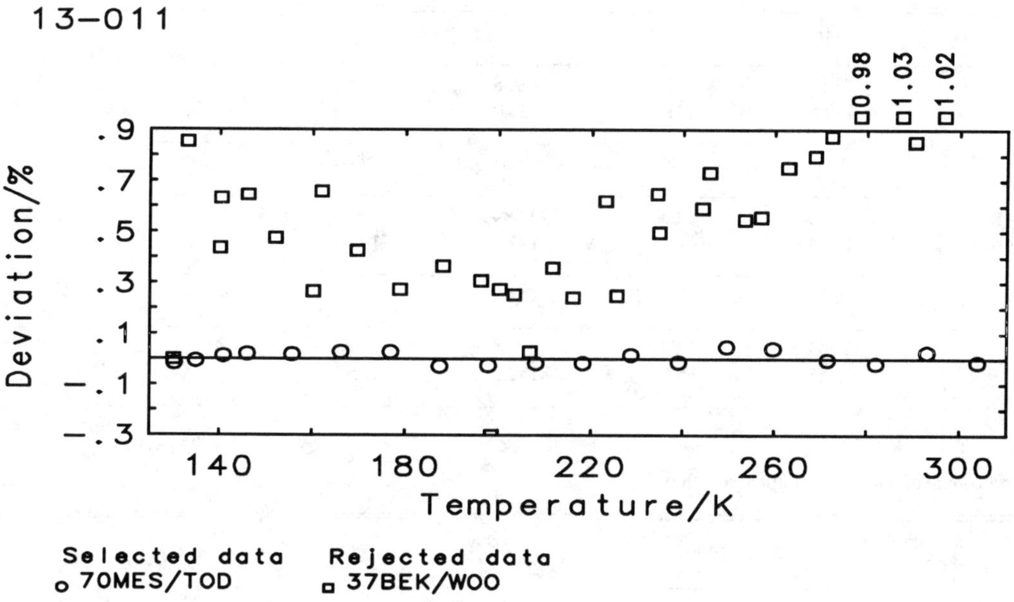

Name: 3–Methyl–1,2–butadiene
Formula: C_5H_8

CAS-RN: 598–25–4
Group No.: 13–012
Molar Mass: 68.12

TABLE 13.12.1. Experimental heat capacities

Reference	Temp. range K	No. pnts	Error %	Purity %	Purity method	Type capacity	Calorimeter Type	Calorimeter Reference
70MES/TOD	161.9–314.6	20	0.20	99.98	melpt	C_{sat}	BSAO	47HUF

TABLE 13.12.3. Parameters of regression polynomial

Heat capacity type	No. data points total	No. data points used	s_w	s/R	s_r %	s_b/R	+/–
C_{sat}	20	20	0.091	3.04–3	0.02	7.15–7	3

Temp. range K	A_1	A_2	A_3	A_4	Level of uncertainty
161.9–314.6	1.75969+1	–3.35718	1.52343	–1.05608–1	II

TABLE 13.12.4. Recommended values of heat capacities

Temp. (K)	160	170	180	190	200	210	220
c_{sat} (J K^{-1}g^{-1})	1.915	1.925	1.938	1.952	1.969	1.988	2.009
C_{sat} (J K^{-1}mol^{-1})	130.5	131.1	132.0	133.0	134.1	135.4	136.9
Temp. (K)	230	240	250	260	270	273.15	280
c_{sat} (J K^{-1}g^{-1})	2.032	2.057	2.084	2.113	2.143	2.153	2.175
C_{sat} (J K^{-1}mol^{-1})	138.4	140.1	142.0	143.9	146.0	146.7	148.2
Temp. (K)	290	298.15	300	310			
c_{sat} (J K^{-1}g^{-1})	2.209	2.237	2.244	2.281			
C_{sat} (J K^{-1}mol^{-1})	150.5	152.4	152.9	155.3			

TABLE 13.12.5. Parameters of quasi-polynomial equation

Heat capacity type	No. data points total	No. data points used	s_w	s/R	s_r %	s_b/R	+/-
C_{sat}	20	20	0.990	3.35–2	0.20	1.05–4	0

Temp. range K	T_c K	A_1	A_2	A_3	A_4	Level of uncertainty
161.9–314.6	489.20	-8.11580–1	2.05912	1.22293+1	7.99687–2	II

Name: 1,2–Pentadiene
Formula: C_5H_8

CAS-RN: 591–95–7
Group No.: 13–013
Molar Mass: 68.12

TABLE 13.13.1. Experimental heat capacities

Reference	Temp. range K	No. pnts	Error %	Purity %	Purity method	Type capacity	Calorimeter Type	Calorimeter Reference
70MES/TOD	142.3–307.5	21	0.20	99.92	melpt	C_{sat}	BSAO	47HUF

TABLE 13.13.3. Parameters of regression polynomial

Heat capacity type	No. data points total	No. data points used	s_w	s/R	s_r %	s_b/R	+/-
C_{sat}	21	21	0.077	2.48–3	0.02	1.59–6	–3

Temp. range K	A_1	A_2	A_3	A_4	Level of uncertainty
142.3–307.5	1.92139+1	-5.03246	2.06721	-1.67635–1	II

TABLE 13.13.4. Recommended values of heat capacities

Temp. (K)	140	150	160	170	180	190	200
c_{sat} (J K^{-1}g^{-1})	1.924	1.923	1.925	1.930	1.938	1.949	1.962
C_{sat} (J K^{-1}mol^{-1})	131.0	131.0	131.1	131.4	132.0	132.7	133.7
Temp. (K)	210	220	230	240	250	260	270
c_{sat} (J K^{-1}g^{-1})	1.979	1.997	2.018	2.042	2.067	2.094	2.123
C_{sat} (J K^{-1}mol^{-1})	134.8	136.0	137.5	139.1	140.8	142.7	144.6
Temp. (K)	273.15	280	290	298.15	300	310	
c_{sat} (J K^{-1}g^{-1})	2.133	2.154	2.187	2.215	2.221	2.256	
C_{sat} (J K^{-1}mol^{-1})	145.3	146.8	149.0	150.8	151.3	153.7	

TABLE 13.13.5. Parameters of quasi-polynomial equation

Heat capacity type	No. data points total	used	s_w	s/R	s_r %	s_b/R	+/−
C_{sat}	21	21	0.414	1.37−2	0.08	2.67−5	−4

Temp. range K	T_c K	A_1	A_2	A_3	A_4	A_5	A_6	Level of uncertainty
142.3–307.5	491.20	1.44883+2	1.21353+1	2.47926	1.47582+2	−7.78539	1.49292+2	II

Name: (E)–1,3–Pentadiene
Formula: C_5H_8

CAS-RN: 2004–70–8
Group No.: 13–014
Molar Mass: 68.12

TABLE 13.14.1. Experimental heat capacities

Reference	Temp. range K	No. pnts	Error %	Purity %	Purity method	Type capacity	Calorimeter Type	Calorimeter Reference
70MES/TOD	189.2–316.4	17	0.20	99.9	melpt	C_{sat}	BSAO	47HUF

TABLE 13.14.3. Parameters of regression polynomial

Heat capacity type	No. data points total	used	s_w	s/R	s_r %	s_b/R	+/−
C_{sat}	17	17	0.347	1.11−2	0.07	1.17−5	−1

Temp. range K	A_1	A_2	A_3	A_4	Level of uncertainty
189.2–316.4	2.58705+1	−1.28646+1	4.84097	−4.74840−1	II

TABLE 13.14.4. Recommended values of heat capacities

Temp. (K)	190	200	210	220	230	240	250
c_{sat} (J K^{-1}g^{-1})	1.910	1.917	1.929	1.946	1.967	1.991	2.020
C_{sat} (J K^{-1}mol^{-1})	130.1	130.6	131.4	132.6	134.0	135.7	137.6
Temp. (K)	260	270	273.15	280	290	298.15	300
c_{sat} (J K^{-1}g^{-1})	2.051	2.085	2.096	2.121	2.160	2.193	2.200
C_{sat} (J K^{-1}mol^{-1})	139.7	142.0	142.8	144.5	147.1	149.4	149.9
Temp. (K)	310						
c_{sat} (J K^{-1}g^{-1})	2.242						
C_{sat} (J K^{-1}mol^{-1})	152.7						

TABLE 13.14.5. Parameters of quasi-polynomial equation

Heat capacity type	No. data points total	used	s_w	s/R	s_r %	s_b/R	+/–
C_{sat}	17	17	0.552	1.78–2	0.11	4.31–5	–4

Temp. range K	T_c K	A_1	A_2	A_3	A_4	A_5	A_6	Level of uncertainty
189.2–316.4	487.20	1.65500+2	1.28633+1	–3.45211–1	1.85567+2	–5.42808+1	2.12088+2	II

Name: (*Z*)–1,3–Pentadiene
Formula: C_5H_8

CAS-RN: 1574–41–0
Group No.: 13–015
Molar Mass: 68.12

TABLE 13.15.1. Experimental heat capacities

Reference	Temp. range K	No. pnts	Error %	Purity %	Purity method	Type capacity	Calorimeter Type	Calorimeter Reference
70MES/TOD	138.2–316.1	22	0.20	99.91	melpt	C_{sat}	BSAO	47HUF

TABLE 13.15.3. Parameters of regression polynomial

Heat capacity type	No. data points total	used	s_w	s/R	s_r %	s_b/R	+/–
C_{sat}	22	22	0.079	2.50–3	0.02	–1.04–6	3

Temp. range K	A_1	A_2	A_3	A_4	Level of uncertainty
138.2–316.1	2.20720+1	–8.31741	2.99909	–2.37835–1	II

TABLE 13.15.4. Recommended values of heat capacities

Temp. (K)	140	150	160	170	180	190	200
c_{sat} (J K^{-1}g^{-1})	1.911	1.897	1.888	1.884	1.883	1.888	1.896
C_{sat} (J K^{-1}mol^{-1})	130.1	129.2	128.6	128.3	128.3	128.6	129.1
Temp. (K)	210	220	230	240	250	260	270
c_{sat} (J K^{-1}g^{-1})	1.908	1.923	1.942	1.965	1.990	2.019	2.050
C_{sat} (J K^{-1}mol^{-1})	129.9	131.0	132.3	133.8	135.6	137.5	139.7
Temp. (K)	273.15	280	290	298.15	300	310	
c_{sat} (J K^{-1}g^{-1})	2.061	2.084	2.121	2.152	2.159	2.200	
C_{sat} (J K^{-1}mol^{-1})	140.4	142.0	144.5	146.6	147.1	149.9	

TABLE 13.15.5. Parameters of quasi-polynomial equation

Heat capacity type	No. data points total	used	s_w	s/R	s_r %	s_b/R	+/–
C_{sat}	22	16	0.411	1.34–2	0.08	1.09–5	–3

Temp. range K	T_c K	A_1	A_2	A_3	A_4	A_5	A_6	Level of uncertainty
173.6–316.1	490.20	1.63995+2	1.27573+1	–5.33111–1	1.85335+2	–5.71994+1	2.12098+2	II

Name: 1,4–Pentadiene
Formula: C_5H_8

CAS-RN: 591–93–5
Group No.: 13–016
Molar Mass: 68.12

TABLE 13.16.1. Experimental heat capacities

Reference	Temp. range K	No. pnts	Error %	Purity %	Purity method	Type capacity	Calorimeter Type	Calorimeter Reference
36PAR/TOD2	125.4–292.5	23	0.50	99.9	melpt	C_p	BSIO	25PAR
70MES/TOD	128.0–300.9	22	0.20	99.98	melpt	C_{sat}	BSAO	47HUF

TABLE 13.16.2. Correlated heat capacities

Reference	Temp. range K	No. pnts used	$\sigma_r C$ %	d_w	d/R	d_r %	d_b/R	+/–
				Selected data				
70MES/TOD	128.0–300.9	22	0.20#	0.106	3.30–3	0.02	3.08–6	3
				Rejected data				
36PAR/TOD2	(6.45–2, 0.38, 1.98–2, 3)							

TABLE 13.16.3. Parameters of regression polynomial

Heat capacity type	No. data points total	used	s_w	s/R	s_r %	s_b/R	+/−
C_{sat}	45	22	0.117	3.65–3	0.02	3.08–6	3

Temp. range K	A_1	A_2	A_3	A_4	Level of uncertainty
128.0–300.9	1.96736+1	−6.11737	2.40621	−1.95032−1	II

TABLE 13.16.4. Recommended values of heat capacities

Temp. (K)	130	140	150	160	170	180	190
c_{sat} (J K^{-1}g^{-1})	1.875	1.866	1.862	1.861	1.864	1.870	1.880
C_{sat} (J K^{-1}mol^{-1})	127.7	127.1	126.8	126.8	127.0	127.4	128.0
Temp. (K)	200	210	220	230	240	250	260
c_{sat} (J K^{-1}g^{-1})	1.892	1.908	1.927	1.948	1.972	1.998	2.027
C_{sat} (J K^{-1}mol^{-1})	128.9	130.0	131.2	132.7	134.3	136.1	138.1
Temp. (K)	270	273.15	280	290	298.15	300	
c_{sat} (J K^{-1}g^{-1})	2.058	2.068	2.091	2.125	2.155	2.162	
C_{sat} (J K^{-1}mol^{-1})	140.2	140.9	142.4	144.8	146.8	147.3	

TABLE 13.16.5. Parameters of quasi-polynomial equation

Heat capacity type	No. data points total	used	s_w	s/R	s_r %	s_b/R	+/−
C_{sat}	45	16	0.325	1.04–2	0.07	1.07–5	−1

Temp. range K	T_c K	A_1	A_2	A_3	A_4	A_5	A_6	Level of uncertainty
154.1–300.9	462.20	8.60776+1	6.17988	7.68986	9.58082+1	−2.82775+1	1.18011+2	II

13–016

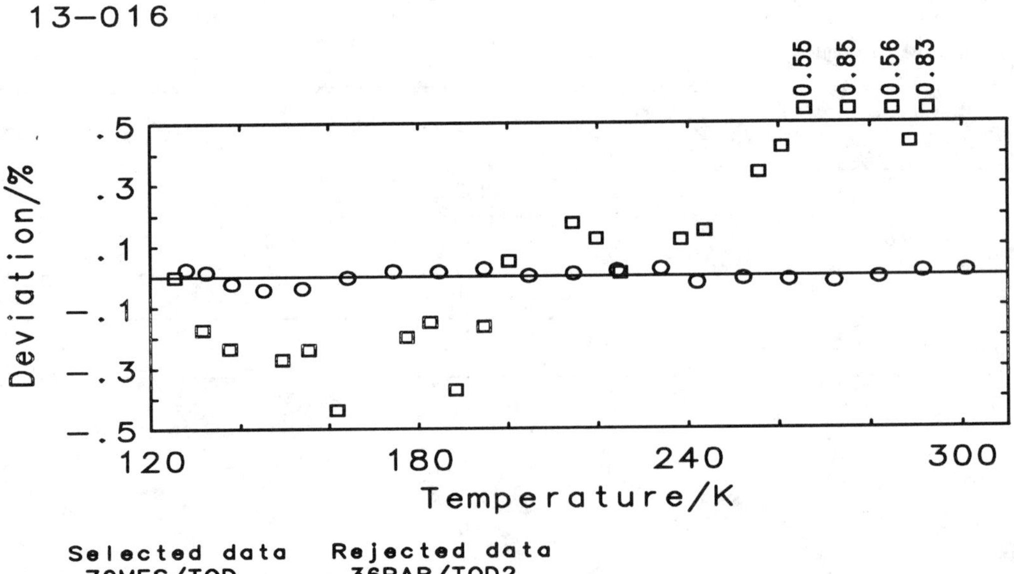

Name: 2,3–Pentadiene
Formula: C_5H_8

CAS-RN: 591–96–8
Group No.: 13–017
Molar Mass: 68.12

TABLE 13.17.1. Experimental heat capacities

Reference	Temp. range K	No. pnts	Error %	Purity %	Purity method	Type capacity	Calorimeter Type	Calorimeter Reference
70MES/TOD	150.7–315.1	21	0.20	99.88	melpt	C_{sat}	BSAO	47HUF

TABLE 13.17.3. Parameters of regression polynomial

Heat capacity type	No. data points total	used	s_w	s/R	s_r %	s_b/R	+/–
C_{sat}	21	21	0.078	2.60–3	0.02	8.17–7	–2

Temp. range K	A_1	A_2	A_3	A_4	Level of uncertainty
150.7–315.1	1.93375+1	–4.04968	1.49720	–8.48617–2	II

TABLE 13.17.4. Recommended values of heat capacities

Temp. (K)	150	160	170	180	190	200	210
c_{sat} (J K^{-1}g^{-1})	1.995	1.995	1.997	2.002	2.010	2.020	2.032
C_{sat} (J K^{-1}mol^{-1})	135.9	135.9	136.1	136.4	136.9	137.6	138.4
Temp. (K)	220	230	240	250	260	270	273.15
c_{sat} (J K^{-1}g^{-1})	2.047	2.064	2.083	2.105	2.128	2.154	2.163
C_{sat} (J K^{-1}mol^{-1})	139.4	140.6	141.9	143.4	145.0	146.7	147.3
Temp. (K)	280	290	298.15	300	310	320	
c_{sat} (J K^{-1}g^{-1})	2.182	2.211	2.237	2.242	2.276	2.310	
C_{sat} (J K^{-1}mol^{-1})	148.6	150.6	152.4	152.8	155.0	157.4	

TABLE 13.17.5. Parameters of quasi–polynomial equation

Heat capacity type	No. data points total	used	s_w	s/R	s_r %	s_b/R	+/–
C_{sat}	21	21	0.315	1.06–2	0.06	1.30–5	0

Temp. range K	T_c K	A_1	A_2	A_3	A_4	A_5	A_6	Level of uncertainty
150.7–315.1	497.20	9.34029+1	6.99568	8.27198	9.97942+1	–1.89302+1	1.16200+2	II

Name: 2–Methyl–1–butene
Formula: C_5H_{10}

CAS–RN: 563–46–2
Group No.: 13–018
Molar Mass: 70.13

TABLE 13.18.1. Experimental heat capacities

Reference	Temp. range K	No. pnts	Error %	Purity %	Purity method	Type capacity	Calorimeter Type	Calorimeter Reference
47TOD/OLI	140.1–292.9	23	0.20	99.95	melpt	C_{sat}	BSAO	47HUF

TABLE 13.18.3. Parameters of regression polynomial

Heat capacity type	No. data points total	No. data points used	s_w	s/R	s_r %	s_b/R	+/–
C_{sat}	23	23	0.126	4.20–3	0.03	2.65–6	–1

Temp. range K	A_1	A_2	A_3	A_4	Level of uncertainty
140.1–292.9	1.94579+1	–5.09285	2.08121	–1.45714–1	II

TABLE 13.18.4. Recommended values of heat capacities

Temp. (K)	140	150	160	170	180	190	200
c_{sat} (J K^{-1}g^{-1})	1.898	1.898	1.902	1.909	1.919	1.932	1.948
C_{sat} (J K^{-1}mol^{-1})	133.1	133.1	133.4	133.9	134.6	135.5	136.6
Temp. (K)	210	220	230	240	250	260	270
c_{sat} (J K^{-1}g^{-1})	1.967	1.989	2.013	2.040	2.069	2.101	2.135
C_{sat} (J K^{-1}mol^{-1})	138.0	139.5	141.2	143.1	145.1	147.4	149.8
Temp. (K)	273.15	280	290	298.15			
c_{sat} (J K^{-1}g^{-1})	2.146	2.171	2.210	2.242			
C_{sat} (J K^{-1}mol^{-1})	150.5	152.3	155.0	157.2			

TABLE 13.18.5. Parameters of quasi–polynomial equation

Heat capacity type	No. data points total	No. data points used	s_w	s/R	s_r %	s_b/R	+/–
C_{sat}	23	23	0.352	1.16–2	0.07	1.27–5	–4

Temp. range K	T_c K	A_1	A_2	A_3	A_4	A_5	A_6	Level of uncertainty
140.1–292.9	470.20	1.37454+2	1.10921+1	3.64526	1.42489+2	–1.46131+1	1.51531+2	II

Name: 2–Methyl–2–butene
Formula: C_5H_{10}

CAS-RN: 513–35–9
Group No.: 13–019
Molar Mass: 70.13

TABLE 13.19.1. Experimental heat capacities

Reference	Temp. range K	No. pnts	Error %	Purity %	Purity method	Type capacity	Calorimeter Type	Calorimeter Reference
30PAR/HUF1	143.9–293.9	15	1.00	not specified		C_p	BSIO	25PAR
47TOD/OLI	140.9–301.0	24	0.20	99.95	melpt	C_{sat}	BSAO	47HUF

TABLE 13.19.2. Correlated heat capacities

Reference	Temp. range K	No. pnts used	$\sigma_r C$ %	d_w	d/R	d_r %	d_b/R	+/–
				Selected data				
47TOD/OLI	140.9–301.0	24	0.20#	0.226	7.48–3	0.05	7.11–6	–1
				Rejected data				
30PAR/HUF1	(8.99–2, 0.52, –5.88–2, –10)							

TABLE 13.19.3. Parameters of regression polynomial

Heat capacity type	No. data points total	No. data points used	s_w	s/R	s_r %	s_b/R	+/–
C_{sat}	39	24	0.248	8.20–3	0.05	7.11–6	–1

Temp. range K	A_1	A_2	A_3	A_4	Level of uncertainty
140.9–301.0	1.90849+1	–4.42379	1.65433	–8.39777–2	II

TABLE 13.19.4. Recommended values of heat capacities

Temp. (K)	140	150	160	170	180	190	200
c_{sat} (J K^{-1}g^{-1})	1.885	1.884	1.885	1.889	1.896	1.906	1.918
C_{sat} (J K^{-1}mol^{-1})	132.2	132.1	132.2	132.5	133.0	133.7	134.6
Temp. (K)	210	220	230	240	250	260	270
c_{sat} (J K^{-1}g^{-1})	1.934	1.952	1.973	1.996	2.022	2.050	2.080
C_{sat} (J K^{-1}mol^{-1})	135.6	136.9	138.4	140.0	141.8	143.8	145.9
Temp. (K)	273.15	280	290	298.15	300		
c_{sat} (J K^{-1}g^{-1})	2.090	2.113	2.148	2.178	2.185		
C_{sat} (J K^{-1}mol^{-1})	146.6	148.2	150.7	152.8	153.3		

TABLE 13.19.5. Parameters of quasi-polynomial equation

Heat capacity type	No. data points total	used	s_w	s/R	s_r %	s_b/R	+/–
C_{sat}	39	24	0.460	1.52–2	0.09	3.26–5	0

Temp. range K	T_c K	A_1	A_2	A_3	A_4	A_5	A_6	Level of uncertainty
140.9–301.0	481.20	1.32943+2	1.05061+1	3.80543	1.41581+2	–2.45434+1	1.56121+2	II

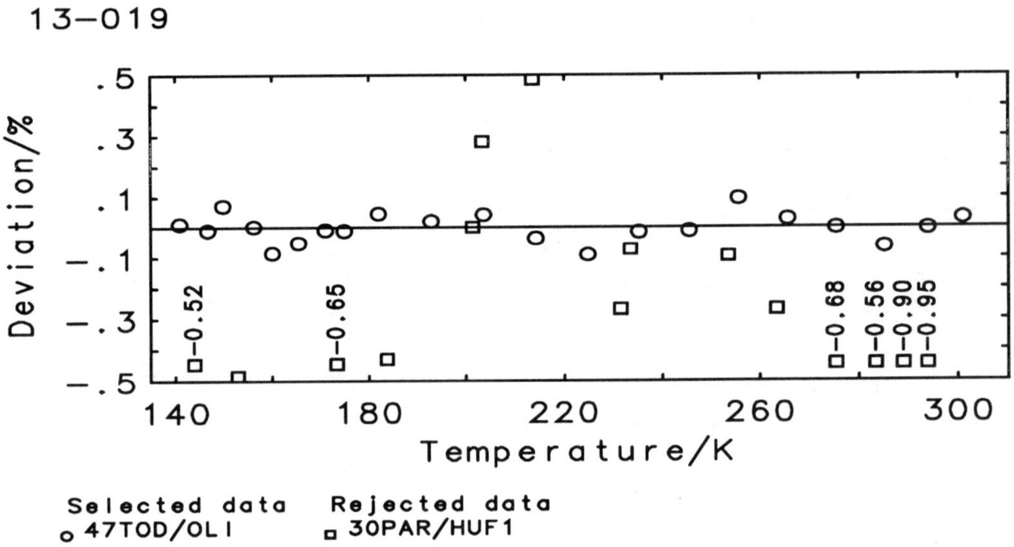

Name: 3–Methyl–1–butene
Formula: C_5H_{10}

CAS–RN: 563–45–1
Group No.: 13–020
Molar Mass: 70.13

TABLE 13.20.1. Experimental heat capacities

Reference	Temp. range K	No. pnts	Error %	Purity %	Purity method	Type capacity	Calorimeter Type	Calorimeter Reference
47TOD/OLI	110.6–298.4	24	0.20	99.995	melpt	C_{sat}	BSAO	47HUF

TABLE 13.20.3. Parameters of regression polynomial

Heat capacity type	No. data points total	used	s_w	s/R	s_r %	s_b/R	+/–
C_{sat}	24	24	0.426	1.37–2	0.09	2.06–5	–2

Temp. range K	A_1	A_2	A_3	A_4	Level of uncertainty
110.6–298.4	1.55001+1	–2.12757	1.26394	–6.20012–2	II

TABLE 13.20.4. Recommended values of heat capacities

Temp. (K)	110	120	130	140	150	160	170
c_{sat} (J K^{-1}g^{-1})	1.732	1.738	1.747	1.758	1.772	1.787	1.806
C_{sat} (J K^{-1}mol^{-1})	121.4	121.9	122.5	123.3	124.2	125.4	126.6
Temp. (K)	180	190	200	210	220	230	240
c_{sat} (J K^{-1}g^{-1})	1.826	1.849	1.874	1.901	1.930	1.961	1.994
C_{sat} (J K^{-1}mol^{-1})	128.1	129.7	131.4	133.3	135.3	137.5	139.8
Temp. (K)	250	260	270	273.15	280	290	298.15
c_{sat} (J K^{-1}g^{-1})	2.029	2.066	2.104	2.117	2.145	2.187	2.223
C_{sat} (J K^{-1}mol^{-1})	142.3	144.9	147.6	148.5	150.4	153.4	155.9
Temp. (K)	300						
c_{sat} (J K^{-1}g^{-1})	2.231						
C_{sat} (J K^{-1}mol^{-1})	156.5						

TABLE 13.20.5. Parameters of quasi-polynomial equation

Heat capacity type	No. data points total	used	s_w	s/R	s_r %	s_b/R	+/–
C_{sat}	24	24	0.620	1.98–2	0.12	4.08–5	–1

Temp. range K	T_c K	A_1	A_2	A_3	A_4	A_5	A_6	Level of uncertainty
110.6–298.4	453.20	2.22495+1	1.12826	1.32620+1	2.12359+1	7.05572	3.16854+1	II

Name: 1–Pentene
Formula: C_5H_{10}

CAS–RN: 109–67–1
Group No.: 13–021
Molar Mass: 70.13

TABLE 13.21.1. Experimental heat capacities

Reference	Temp. range K	No. pnts	Error %	Purity %	Purity method	Type capacity	Calorimeter Type	Calorimeter Reference
47TOD/OLI	125.7–295.3	24	0.20	not specified		C_{sat}	BSAO	47HUF
49SCH/SAG	310.9–366.5	11	1.50	not specified		C_p	BSAO	39SAG/EVA
90MES/TOD	111.0–308.4	25	0.10	99.913	melpt	C_{sat}	BSAO	47HUF

TABLE 13.21.2. Correlated heat capacities

Reference	Temp. range K	No. pnts used	$\sigma_r C$ %	d_w	d/R	d_r %	d_b/R	+/–
				Selected data				
47TOD/OLI	125.7–295.3	24	0.20	0.461	1.54–2	0.09	1.25–2	20
49SCH/SAG	310.9–366.5	11	1.50	0.572	1.70–1	0.86	–1.06–1	–7
90MES/TOD	111.0–308.4	25	0.10	0.535	9.51–3	0.05	–2.76–3	–13

TABLE 13.21.3. Parameters of cubic spline polynomials

Heat capacity type	No. data points total	used	s_w	s/R	s_r %	s_b/R	+/–
C_p	60	60	0.537	7.71–2	0.39	–1.57–2	0
C_{sat}	60	60	0.533	7.89–2	0.40	–1.62–2	2

Temp. range K	A_1	A_2	A_3	A_4	Level of uncertainty
111.0–300.0	1.81128+1	–4.16574	1.78245	–1.08500–1	I
300.0–366.5	1.17329+2	–1.03382+2	3.48544+1	–3.78316	V
111.0–300.0	1.82039+1	–4.32891	1.87602	–1.25713–1	I
300.0–366.5	1.26012+2	–1.12137+2	3.78119+1	–4.11859	V

TABLE 13.21.4. Recommended values of heat capacities

Temp. (K)	110	120	130	140	150	160	170
c_p (J K^{-1}g^{-1})	1.8426	1.8367	1.8341	1.8348	1.8386	1.8454	1.8552
C_p (J K^{-1}mol^{-1})	129.23	128.82	128.64	128.68	128.95	129.43	130.12
c_{sat} (J K^{-1}g^{-1})	1.8429	1.8368	1.8341	1.8346	1.8384	1.8453	1.8552
C_{sat} (J K^{-1}mol^{-1})	129.25	128.82	128.63	128.67	128.94	129.42	130.11
Temp. (K)	**180**	**190**	**200**	**210**	**220**	**230**	**240**
c_p (J K^{-1}g^{-1})	1.8680	1.8836	1.9019	1.9230	1.9466	1.9728	2.0014
C_p (J K^{-1}mol^{-1})	131.01	132.10	133.39	134.87	136.52	138.36	140.37
c_{sat} (J K^{-1}g^{-1})	1.8680	1.8837	1.9021	1.9232	1.9468	1.9729	2.0014
C_{sat} (J K^{-1}mol^{-1})	131.01	132.11	133.40	134.88	136.54	138.37	140.37
Temp. (K)	**250**	**260**	**270**	**273.15**	**280**	**290**	**298.15**
c_p (J K^{-1}g^{-1})	2.0324	2.0657	2.1012	2.1128	2.1388	2.1785	2.2124
C_p (J K^{-1}mol^{-1})	142.54	144.87	147.36	148.18	150.01	152.79	155.16
c_{sat} (J K^{-1}g^{-1})	2.0323	2.0653	2.1004	2.1119	2.1376	2.1768	2.2100
C_{sat} (J K^{-1}mol^{-1})	142.53	144.85	147.31	148.12	149.92	152.67	155.00
Temp. (K)	**300**	**310**	**320**	**330**	**340**	**350**	**360**
c_p (J K^{-1}g^{-1})	2.2202	2.26	2.31	2.34	2.38	2.40	2.41
C_p (J K^{-1}mol^{-1})	155.72	159	162	164	167	168	169
c_{sat} (J K^{-1}g^{-1})	2.2177	2.26	2.30	2.34	2.37	2.39	2.40
C_{sat} (J K^{-1}mol^{-1})	155.54	159	161	164	166	167	168

TABLE 13.21.5. Parameters of quasi-polynomial equation

Heat capacity type	No. data points total	used	s_w	s/R	s_r %	s_b/R	+/–
C_p	60	60	2.082	4.70–2	0.27	–5.95–3	–18

Temp. range K	T_c K	A_1	A_2	A_3	A_4	A_5	A_6	Level of uncertainty
111.0–366.5	464.78	5.22732+1	3.36498	1.17701+1	5.36613+1	–5.04617	7.09940+1	V

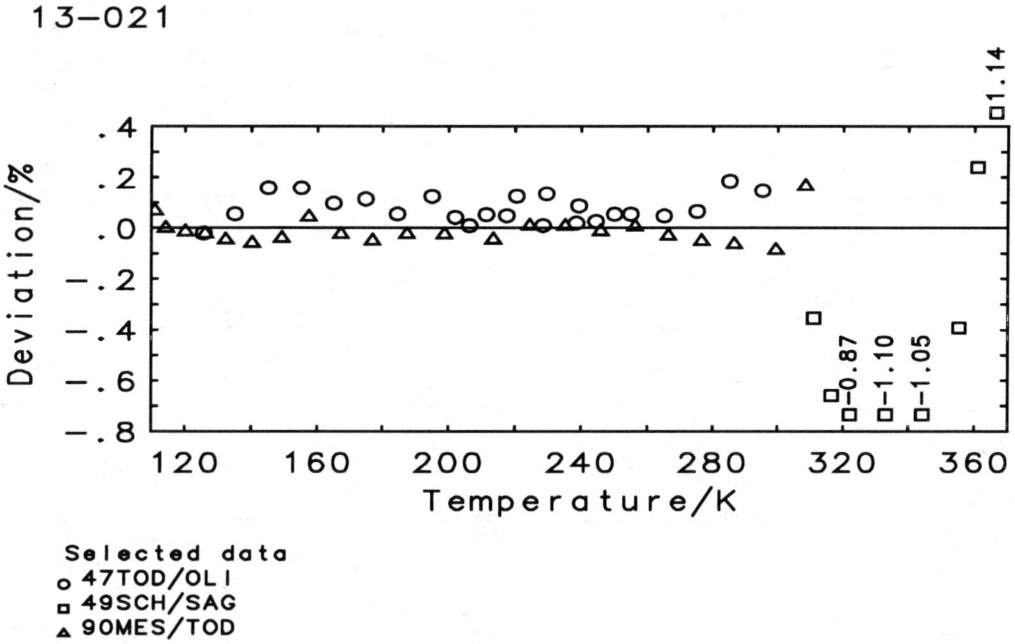

Name: 2–Pentene
Formula: C_5H_{10}

CAS–RN: 109–68–2
Group No.: 13–022
Molar Mass: 70.13

TABLE 13.22.1. Experimental heat capacities

Reference		Temp. range K	No. pnts	Error %	Purity %	Purity method	Type capacity	Calorimeter Type	Calorimeter Reference
*81VON	N	296.8–306.0	4S	nosp	not specified		C_{avg}	DSIO	*81VON
30PAR/HUF1		136.1–289.1	8	1.00	not specified		C_p	BSIO	25PAR

*81VON substance not clearly identified, name "Amylen" used; probably 2–Pentene (according to n.b.t.)

TABLE 13.22.2. Correlated heat capacities

Reference	Temp. range K	No. pnts used	$\sigma_r C$ %	d_w	d/R	d_r %	d_b/R	+/–
				Selected data				
*81VON	296.8–306.0	4	2.00#	0.593	2.26–1	1.19	1.93–1	4
30PAR/HUF1	136.1–289.1	8	1.00	0.385	6.75–2	0.38	–2.28–2	–4

TABLE 13.22.3. Parameters of regression polynomial

Heat capacity type	No. data points total	No. data points used	s_w	s/R	s_r %	s_b/R	+/–
C_p	12	12	0.536	1.63–1	0.87	4.91–2	0

Temp. range K	A_1	A_2	A_3	Level of uncertainty
136.1–306.0	1.65680+1	–2.15818	9.55251–1	V

TABLE 13.22.4. Recommended values of heat capacities

Temp. (K)	140	150	160	170	180	190	200
c_p (J K^{-1}g^{-1})	1.83	1.84	1.84	1.86	1.87	1.89	1.91
C_p (J K^{-1}mol^{-1})	128	129	129	130	131	132	134
Temp. (K)	210	220	230	240	250	260	270
c_p (J K^{-1}g^{-1})	1.93	1.95	1.97	2.00	2.03	2.06	2.10
C_p (J K^{-1}mol^{-1})	135	137	138	140	143	145	147
Temp. (K)	273.15	280	290	298.15	300	310	
c_p (J K^{-1}g^{-1})	2.11	2.14	2.17	2.21	2.22	2.26	
C_p (J K^{-1}mol^{-1})	148	150	153	155	155	158	

TABLE 13.22.5. Parameters of quasi–polynomial equation

Heat capacity type	No. data points total	used	s_w	s/R	s_r %	s_b/R	+/–
C_p	12	12	0.489	1.09–1	0.61	–8.52–3	–2

Temp. range K	T_c K	A_1	A_2	A_3	A_4	Level of uncertainty
136.1–306.0	475.00	4.77990	4.68347	1.01201+1	1.21958	V

Name: (E)–2–Pentene
Formula: C_5H_{10}

CAS–RN: 646–04–8
Group No.: 13–023
Molar Mass: 70.13

TABLE 13.23.1. Experimental heat capacities

Reference	Temp. range K	No. pnts	Error %	Purity %	Purity method	Type capacity	Calorimeter Type	Calorimeter Reference
47TOD/OLI	135.0–301.7	27	0.20	99.995	melpt	C_{sat}	BSAO	47HUF

TABLE 13.23.3. Parameters of regression polynomial

Heat capacity type	No. data points total	used	s_w	s/R	s_r %	s_b/R	+/–
C_{sat}	27	27	0.270	9.20–3	0.05	7.31–6	–1

Temp. range K	A_1	A_2	A_3	Level of uncertainty
135.0–301.7	1.59522+1	–1.61398	8.70985–1	II

TABLE 13.23.4. Recommended values of heat capacities

Temp. (K)	140	150	160	170	180	190	200
c_{sat} (J K^{-1}g^{-1})	1.826	1.836	1.849	1.864	1.881	1.900	1.922
C_{sat} (J K^{-1}mol^{-1})	128.0	128.8	129.7	130.8	131.9	133.3	134.8
Temp. (K)	210	220	230	240	250	260	270
c_{sat} (J K^{-1}g^{-1})	1.945	1.970	1.997	2.027	2.058	2.092	2.127
C_{sat} (J K^{-1}mol^{-1})	136.4	138.2	140.1	142.1	144.3	146.7	149.2
Temp. (K)	273.15	280	290	298.15	300		
c_{sat} (J K^{-1}g^{-1})	2.139	2.165	2.205	2.239	2.246		
C_{sat} (J K^{-1}mol^{-1})	150.0	151.8	154.6	157.0	157.6		

TABLE 13.23.5. Parameters of quasi-polynomial equation

Heat capacity type	No. data points total	used	s_w	s/R	s_r %	s_b/R	+/–
C_{sat}	27	27	1.131	3.86–2	0.23	1.39–4	0

Temp. range K	T_c K	A_1	A_2	A_3	A_4	Level of uncertainty
135.0–301.7	474.20	–8.11492–2	2.76166	1.14140+1	5.96126–4	III

Name: (Z)–2–Pentene
Formula: C_5H_{10}

CAS–RN: 627–20–3
Group No.: 13–024
Molar Mass: 70.13

TABLE 13.24.1. Experimental heat capacities

Reference	Temp. range K	No. pnts	Error %	Purity %	Purity method	Type capacity	Calorimeter Type	Calorimeter Reference
47TOD/OLI	128.7–295.3	37	0.20	99.995	melpt	C_{sat}	BSAO	47HUF

TABLE 13.24.3. Parameters of regression polynomial

Heat capacity type	No. data points total	used	s_w	s/R	s_r %	s_b/R	+/–
C_{sat}	37	37	0.396	1.36–2	0.08	1.82–5	6

Temp. range K	A_1	A_2	A_3	A_4	Level of uncertainty
128.7–295.3	2.13097+1	–7.11514	2.58379	–1.81446–1	II

TABLE 13.24.4. Recommended values of heat capacities

Temp. (K)	130	140	150	160	170	180	190
c_{sat} (J K^{-1}g^{-1})	1.900	1.887	1.878	1.873	1.872	1.875	1.882
C_{sat} (J K^{-1}mol^{-1})	133.3	132.3	131.7	131.3	131.3	131.5	132.0
Temp. (K)	200	210	220	230	240	250	260
c_{sat} (J K^{-1}g^{-1})	1.892	1.907	1.924	1.945	1.969	1.996	2.026
C_{sat} (J K^{-1}mol^{-1})	132.7	133.7	134.9	136.4	138.1	140.0	142.1
Temp. (K)	270	273.15	280	290	298.15	300	
c_{sat} (J K^{-1}g^{-1})	2.058	2.069	2.094	2.132	2.164	2.172	
C_{sat} (J K^{-1}mol^{-1})	144.4	145.1	146.8	149.5	151.8	152.3	

TABLE 13.24.5. Parameters of quasi-polynomial equation

Heat capacity type	No. data points total	used	s_w	s/R	s_r %	s_b/R	+/-
C_{sat}	37	29	0.466	1.62-2	0.09	1.66-5	-3

Temp. range K	T_c K	A_1	A_2	A_3	A_4	A_5	A_6	Level of uncertainty
163.7–295.3	475.20	2.19302+2	1.83175+1	-6.45813	2.42391+2	-5.98522+1	2.57149+2	II

Name: 1,5–Hexadiene
Formula: C_6H_{10}

CAS-RN: 592–42–7
Group No.: 13–025
Molar Mass: 82.15

TABLE 13.25.1. Experimental heat capacities

Reference	Temp. range K	No. pnts	Error %	Purity %	Purity method	Type capacity	Calorimeter Type	Calorimeter Reference
*81VON	297.0–309.7	4S	nosp	not specified		C_{avg}	DSIO	*81VON

TABLE 13.25.3. Parameters of regression polynomial

Heat capacity type	No. data points total	used	s_w	s/R	s_r %	s_b/R	+/-
C	4	4	0.003	1.38-3	0.01	-4.77-7	0

Temp. range K	A_1	A_2	Level of uncertainty
297.0–309.7	5.51252	5.21277	V

TABLE 13.25.4. Recommended values of heat capacities

Temp. (K)	300	310
c (J K^{-1}g^{-1})	2.14	2.19
C (J K^{-1}mol^{-1})	176	180

Name: 2,3–Dimethyl–2–butene
Formula: C_6H_{12}

CAS–RN: 563–79–1
Group No.: 13–026
Molar Mass: 84.16

TABLE 13.26.1. Experimental heat capacities

Reference	Temp. range K	No. pnts	Error %	Purity %	Purity method	Type capacity	Calorimeter Type	Calorimeter Reference
36PAR/TOD2	202.4–295.5	12	0.50	not specified		C_p	BSIO	25PAR
55SCO/FIN	204.3–318.1	18	0.20	99.95	melpt	C_{sat}	BSAO	43RUE/HUF

TABLE 13.26.2. Correlated heat capacities

Reference	Temp. range K	No. pnts used	$\sigma_r C$ %	d_w	d/R	d_r %	d_b/R	+/–
				Selected data				
55SCO/FIN	204.3–318.1	18	0.20	0.071	2.74–3	0.01	–5.30–7	1
				Rejected data				
36PAR/TOD2	(1.35–1, 0.66, 1.22–1, 11)							

TABLE 13.26.3. Parameters of regression polynomial

Heat capacity type	No. data points total	No. data points used	s_w	s/R	s_r %	s_b/R	+/–
C_{sat}	30	18	0.081	3.11–3	0.02	–5.30–7	1

Temp. range K	A_1	A_2	A_3	A_4	Level of uncertainty
204.3–318.1	2.55879+1	–9.70117	3.76391	–3.43721–1	II

TABLE 13.26.4. Recommended values of heat capacities

Temp. (K)	200	210	220	230	240	250	260
c_{sat} (J $K^{-1}g^{-1}$)	1.827	1.841	1.858	1.877	1.900	1.925	1.953
C_{sat} (J $K^{-1}mol^{-1}$)	153.7	154.9	156.3	158.0	159.9	162.0	164.4
Temp. (K)	270	273.15	280	290	298.15	300	310
c_{sat} (J $K^{-1}g^{-1}$)	1.983	1.992	2.014	2.048	2.076	2.082	2.119
C_{sat} (J $K^{-1}mol^{-1}$)	166.9	167.7	169.5	172.3	174.7	175.3	178.3
Temp. (K)	320						
c_{sat} (J $K^{-1}g^{-1}$)	2.156						
C_{sat} (J $K^{-1}mol^{-1}$)	181.5						

TABLE 13.26.5. Parameters of quasi-polynomial equation

Heat capacity type	No. data points total	used	s_w	s/R	s_r %	s_b/R	+/–
C_{sat}	30	18	0.664	2.65–2	0.13	5.50–5	2

Temp. range K	T_c K	A_1	A_2	A_3	A_4	Level of uncertainty
204.3–318.1	524.00	–3.13792–1	3.48304	1.26417+1	7.06749–3	II

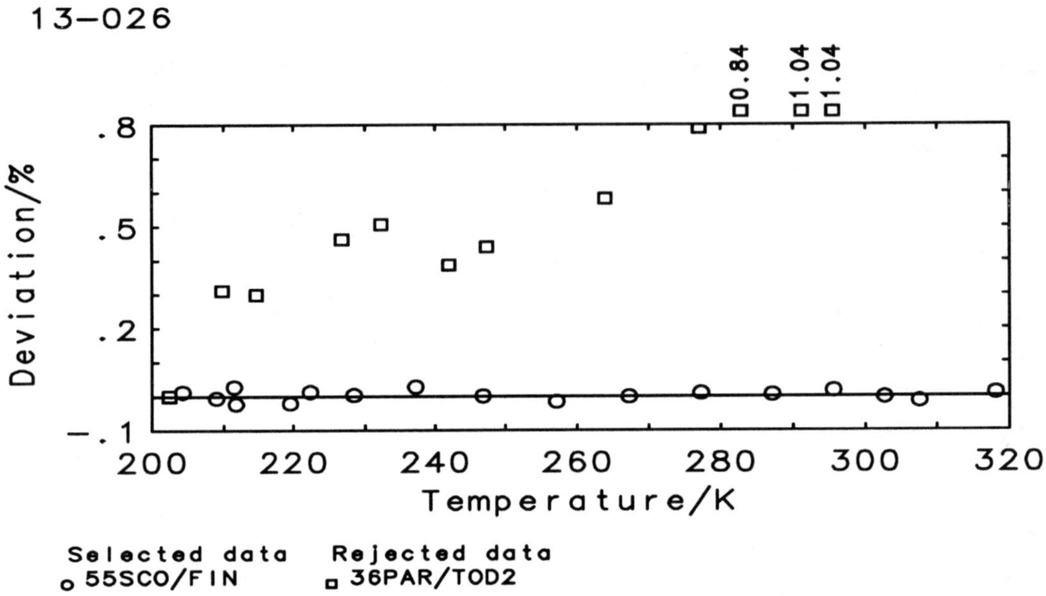

Name: 3,3–Dimethyl–1–butene
Formula: C_6H_{12}

CAS–RN: 558–37–2
Group No.: 13–027
Molar Mass: 84.16

TABLE 13.27.1. Experimental heat capacities

Reference	Temp. range K	No. pnts	Error %	Purity %	Purity method	Type capacity	Calorimeter Type	Calorimeter Reference
38KEN/SHO	165.9–295.9	20	0.70	99.8	estim	C_p	BSIO	25PAR

TABLE 13.27.3. Parameters of regression polynomial

Heat capacity type	No. data points total	used	s_w	s/R	s_r %	s_b/R	+/–
C_p	20	20	0.212	2.80–2	0.15	6.48–5	0

Temp. range K	A_1	A_2	A_3	A_4	Level of uncertainty
165.9–295.9	5.38184	1.31325+1	–4.96802	8.45690–1	IV

TABLE 13.27.4. Recommended values of heat capacities

Temp. (K)	170	180	190	200	210	220	230
c_p (J K^{-1}g^{-1})	1.729	1.764	1.798	1.832	1.866	1.900	1.936
C_p (J K^{-1}mol^{-1})	145.5	148.5	151.3	154.2	157.0	159.9	162.9
Temp. (K)	240	250	260	270	273.15	280	290
c_p (J K^{-1}g^{-1})	1.973	2.013	2.056	2.101	2.116	2.151	2.204
C_p (J K^{-1}mol^{-1})	166.1	169.4	173.0	176.8	178.1	181.0	185.5
Temp. (K)	298.15	300					
c_p (J K^{-1}g^{-1})	2.251	2.262					
C_p (J K^{-1}mol^{-1})	189.5	190.4					

TABLE 13.27.5. Parameters of quasi-polynomial equation

Heat capacity type	No. data points total	used	s_w	s/R	s_r %	s_b/R	+/–
C_p	20	20	0.296	3.99–2	0.21	1.42–4	1

Temp. range K	T_c K	A_1	A_2	A_3	A_4	Level of uncertainty
165.9–295.9	490.00	–4.91066	1.73956	1.15768+1	3.46561	IV

Name: 1–Hexene
Formula: C_6H_{12}

CAS–RN: 592–41–6
Group No.: 13–028
Molar Mass: 84.16

TABLE 13.28.1. Experimental heat capacities

Reference	Temp. range K	No. pnts	Error %	Purity %	Purity method	Type capacity	Calorimeter Type	Calorimeter Reference
57MCC/FIN1	136.2–308.2	23	0.20	99.96	melpt	C_{sat}	BSAO	47HUF
85KAL/WOY	186.6–301.1	31	nosp	99.72	chrom	C_p	BSAO	80KAL/JED

TABLE 13.28.2. Correlated heat capacities

Reference	Temp. range K	No. pnts used	$\sigma_r C$ %	d_w	d/R	d_r %	d_b/R	+/–
				Selected data				
57MCC/FIN1	136.2–308.2	23	0.20#	0.166	6.42–3	0.03	3.32–6	–1
				Rejected data				
85KAL/WOY	(1.31–1, 0.67, 8.59–2, 17)							

TABLE 13.28.3. Parameters of regression polynomial

Heat capacity type	No. data points total	used	s_w	s/R	s_r %	s_b/R	+/–
C_{sat}	54	23	0.183	7.06–3	0.04	3.32–6	–1

Temp. range K	A_1	A_2	A_3	A_4	Level of uncertainty
136.2–308.2	2.30342+1	–6.68116	2.80137	–2.25071–1	II

TABLE 13.28.4. Recommended values of heat capacities

Temp. (K)	140	150	160	170	180	190	200
c_{sat} (J K^{-1}g^{-1})	1.833	1.833	1.837	1.844	1.855	1.868	1.885
C_{sat} (J K^{-1}mol^{-1})	154.3	154.3	154.6	155.2	156.1	157.2	158.6
Temp. (K)	210	220	230	240	250	260	270
c_{sat} (J K^{-1}g^{-1})	1.904	1.926	1.951	1.978	2.008	2.040	2.073
C_{sat} (J K^{-1}mol^{-1})	160.2	162.1	164.2	166.5	169.0	171.6	174.5
Temp. (K)	273.15	280	290	298.15	300	310	
c_{sat} (J K^{-1}g^{-1})	2.084	2.109	2.147	2.179	2.186	2.227	
C_{sat} (J K^{-1}mol^{-1})	175.4	177.5	180.7	183.3	184.0	187.4	

TABLE 13.28.5. Parameters of quasi-polynomial equation

Heat capacity type	No. data points total	used	s_w	s/R	s_r %	s_b/R	+/–
C_{sat}	54	23	0.416	1.65–2	0.08	2.05–5	–1

Temp. range K	T_c K	A_1	A_2	A_3	A_4	A_5	A_6	Level of uncertainty
136.2–308.2	504.03	2.58690+2	2.29861+1	–5.88457	2.55772+2	8.39159	2.36986+2	II

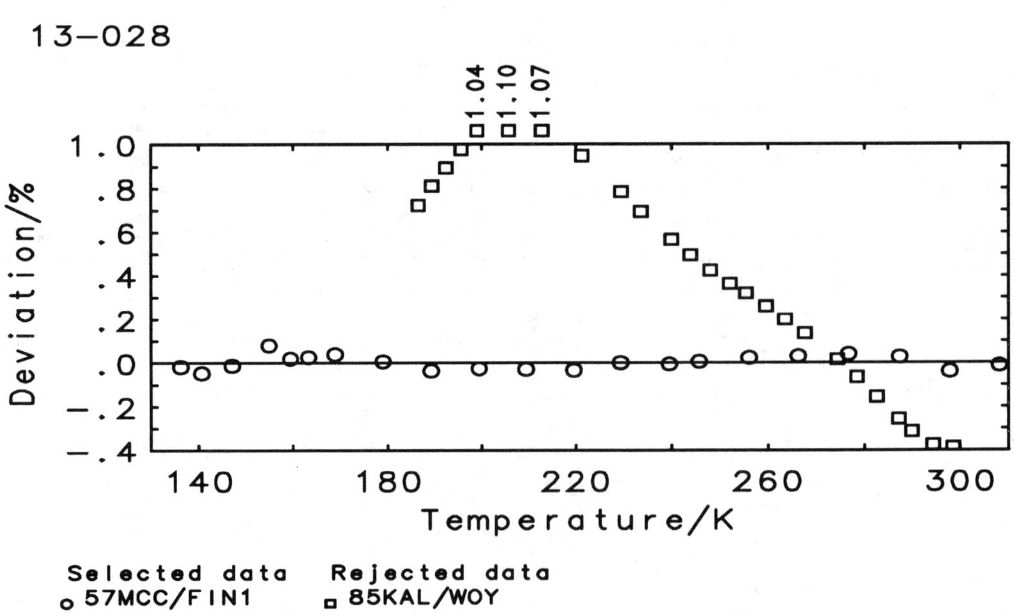

Name: (Z)–2–Hexene
Formula: C_6H_{12}

CAS–RN: 7688–21–3
Group No.: 13–029
Molar Mass: 84.16

TABLE 13.29.1. Experimental heat capacities

Reference	Temp. range K	No. pnts	Error %	Purity %	Purity method	Type capacity	Calorimeter Type	Calorimeter Reference
90MES/TOD	129.1–314.5	22	0.10	99.963	melpt	C_{sat}	BSAO	47HUF

TABLE 13.29.3. Parameters of regression polynomial

Heat capacity type	No. data points total	used	s_w	s/R	s_r %	s_b/R	+/–
C_{sat}	22	22	0.633	1.25–2	0.06	1.27–5	–1

Temp. range K	A_1	A_2	A_3	A_4	Level of uncertainty
129.1–314.5	2.37227+1	–7.34389	2.86313	–2.20334–1	II

TABLE 13.29.4. Recommended values of heat capacities

Temp. (K)	130	140	150	160	170	180	190
c_{sat} (J K^{-1}g^{-1})	1.831	1.823	1.818	1.818	1.821	1.827	1.837
C_{sat} (J K^{-1}mol^{-1})	154.1	153.4	153.0	153.0	153.2	153.8	154.6
Temp. (K)	200	210	220	230	240	250	260
c_{sat} (J K^{-1}g^{-1})	1.850	1.866	1.885	1.906	1.931	1.958	1.987
C_{sat} (J K^{-1}mol^{-1})	155.7	157.0	158.6	160.4	162.5	164.8	167.2
Temp. (K)	270	273.15	280	290	298.15	300	310
c_{sat} (J K^{-1}g^{-1})	2.018	2.029	2.052	2.088	2.118	2.125	2.164
C_{sat} (J K^{-1}mol^{-1})	169.9	170.7	172.7	175.7	178.3	178.8	182.1

TABLE 13.29.5. Parameters of quasi–polynomial equation

Heat capacity type	No. data points total	used	s_w	s/R	s_r %	s_b/R	+/–
C_{sat}	22	19	0.798	1.58–2	0.08	1.09–5	1

Temp. range K	T_c K	A_1	A_2	A_3	A_4	A_5	A_6	Level of uncertainty
153.3–314.5	512.00	2.35656+2	2.01229+1	–3.58592	2.39437+2	–1.07578+1	2.37094+2	II

Name: 1–Heptene
Formula: C_7H_{14}

CAS–RN: 592–76–7
Group No.: 13–030
Molar Mass: 98.19

TABLE 13.30.1. Experimental heat capacities

Reference	Temp. range K	No. pnts	Error %	Purity %	Purity method	Type capacity	Calorimeter Type	Calorimeter Reference
36PAR/TOD2	151.1–295.1	21	0.50	99.85	melpt	C_p	BSIO	25PAR
57MCC/FIN1	157.1–299.6	22	0.20	99.86	melpt	C_{sat}	BSAO	47HUF

TABLE 13.30.2. Correlated heat capacities

Reference	Temp. range K	No. pnts used	$\sigma_r C$ %	d_w	d/R	d_r %	d_b/R	+/–
				Selected data				
57MCC/FIN1	157.1–299.6	22	0.20#	0.110	5.10–3	0.02	6.94–7	0
				Rejected data				
36PAR/TOD2	(1.03–1, 0.43, 4.78–3, –1)							

TABLE 13.30.3. Parameters of regression polynomial

Heat capacity type	No. data points total	No. data points used	s_w	s/R	s_r %	s_b/R	+/–
C_{sat}	43	22	0.122	5.64–3	0.02	6.94–7	0

Temp. range K	A_1	A_2	A_3	A_4	Level of uncertainty
157.1–299.6	2.95049+1	–1.07928+1	4.41304	–4.18264–1	II

TABLE 13.30.4. Recommended values of heat capacities

Temp. (K)	160	170	180	190	200	210	220
c_{sat} ($J\,K^{-1}g^{-1}$)	1.848	1.851	1.858	1.868	1.882	1.899	1.919
C_{sat} ($J\,K^{-1}mol^{-1}$)	181.4	181.7	182.4	183.4	184.8	186.5	188.5
Temp. (K)	230	240	250	260	270	273.15	280
c_{sat} ($J\,K^{-1}g^{-1}$)	1.942	1.968	1.996	2.026	2.058	2.068	2.092
C_{sat} ($J\,K^{-1}mol^{-1}$)	190.7	193.2	196.0	198.9	202.1	203.1	205.4
Temp. (K)	290	298.15	300				
c_{sat} ($J\,K^{-1}g^{-1}$)	2.127	2.157	2.164				
C_{sat} ($J\,K^{-1}mol^{-1}$)	208.8	211.8	212.4				

TABLE 13.30.5. Parameters of quasi-polynomial equation

Heat capacity type	No. data points total	used	s_w	s/R	s_r %	s_b/R	+/–
C_{sat}	43	22	0.362	1.67–2	0.07	4.68–6	1

Temp. range K	T_c K	A_1	A_2	A_3	A_4	A_5	A_6	Level of uncertainty
157.1–299.6	537.30	6.82280+2	7.02369+1	–5.14378+1	6.52620+2	7.78060+1	4.99087+2	II

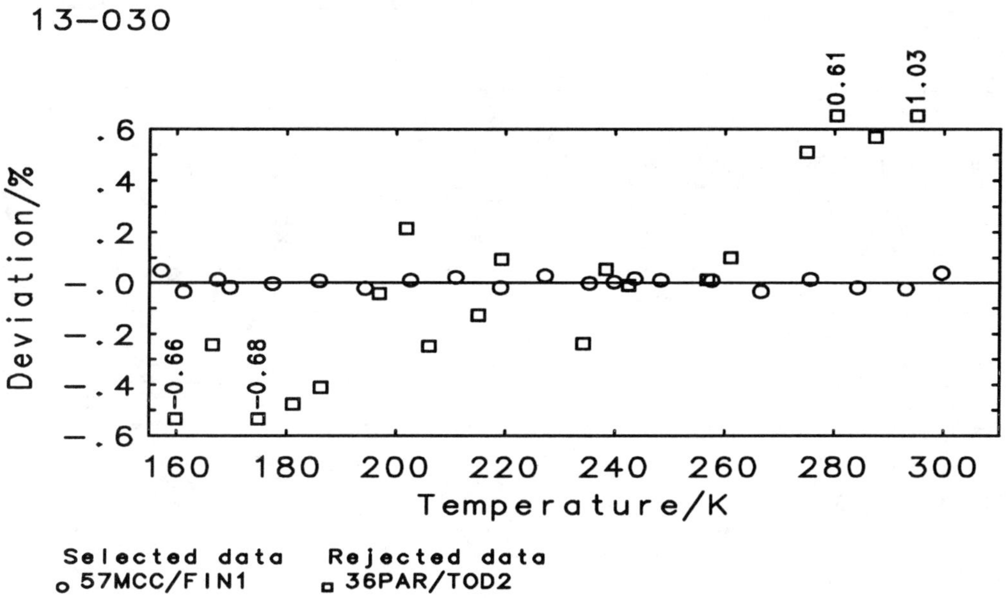

Name: 1–Octene
Formula: C_8H_{16}

CAS–RN: 111–66–0
Group No.: 13–031
Molar Mass: 112.22

TABLE 13.31.1. Experimental heat capacities

Reference	Temp. range K	No. pnts	Error %	Purity %	Purity method	Type capacity	Calorimeter Type	Calorimeter Reference
57MCC/FIN1	175.4–308.5	20	0.20	99.94	melpt	C_{sat}	BSAO	47HUF

TABLE 13.31.3. Parameters of regression polynomial

Heat capacity type	No. data points total	used	s_w	s/R	s_r %	s_b/R	+/–
C_{sat}	20	20	0.191	1.01–2	0.04	3.62–6	–2

Temp. range K	A_1	A_2	A_3	A_4	Level of uncertainty
175.4–308.5	4.53296+1	–2.51051+1	9.75750	–1.06384	II

TABLE 13.31.4. Recommended values of heat capacities

Temp. (K)	180	190	200	210	220	230	240
c_{sat} (J K^{-1}g^{-1})	1.893	1.894	1.900	1.911	1.926	1.946	1.969
C_{sat} (J K^{-1}mol^{-1})	212.4	212.5	213.2	214.4	216.2	218.4	221.0
Temp. (K)	250	260	270	273.15	280	290	298.15
c_{sat} (J K^{-1}g^{-1})	1.995	2.024	2.055	2.065	2.088	2.122	2.150
C_{sat} (J K^{-1}mol^{-1})	223.9	227.1	230.6	231.8	234.3	238.1	241.3
Temp. (K)	300	310					
c_{sat} (J K^{-1}g^{-1})	2.157	2.192					
C_{sat} (J K^{-1}mol^{-1})	242.0	245.9					

TABLE 13.31.5. Parameters of quasi-polynomial equation

Heat capacity type	No. data points total	used	s_w	s/R	s_r %	s_b/R	+/–
C_{sat}	20	20	0.609	3.21–2	0.12	7.95–5	–2

Temp. range K	T_c K	A_1	A_2	A_3	A_4	A_5	A_6	Level of uncertainty
175.4–308.5	566.65	2.79268+3	3.22992+2	–3.13143+2	2.65202+3	3.27256+2	1.78737+3	II

Name: 2,4,4–Trimethyl–1–pentene
Formula: C_8H_{16}

CAS–RN: 107–39–1
Group No.: 13–032
Molar Mass: 112.22

TABLE 13.32.1. Experimental heat capacities

Reference		Temp. range K	No. pnts	Error %	Purity %	method	Type capacity	Calorimeter Type	Reference
36PAR/TOD2	N	178.4–296.0	15	0.50	99.5	estim	C_p	BSIO	25PAR

36PAR/TOD2 identity of substance was not known exactly (estimated by authors)

TABLE 13.32.3. Parameters of regression polynomial

Heat capacity type	No. data points total	used	s_w	s/R	s_r %	s_b/R	+/–
C_p	15	15	0.177	2.22–2	0.09	2.82–5	1

Temp. range K	A_1	A_2	A_3	A_4	Level of uncertainty
178.4–296.0	1.32041+1	9.41313	–3.44770	6.72476–1	III

TABLE 13.32.4. Recommended values of heat capacities

Temp. (K)	180	190	200	210	220	230	240
c_p (J $K^{-1}g^{-1}$)	1.697	1.723	1.750	1.778	1.807	1.837	1.870
C_p (J $K^{-1}mol^{-1}$)	190.4	193.4	196.4	199.5	202.8	206.2	209.8
Temp. (K)	250	260	270	273.15	280	290	298.15
c_p (J $K^{-1}g^{-1}$)	1.904	1.941	1.980	1.993	2.022	2.068	2.108
C_p (J $K^{-1}mol^{-1}$)	213.7	217.8	222.2	223.6	226.9	232.0	236.5
Temp. (K)	300						
c_p (J $K^{-1}g^{-1}$)	2.117						
C_p (J $K^{-1}mol^{-1}$)	237.6						

TABLE 13.32.5. Parameters of quasi-polynomial equation

Heat capacity type	No. data points total	used	s_w	s/R	s_r %	s_b/R	+/–
C_p	15	15	0.194	2.40–2	0.10	4.02–5	1

Temp. range K	T_c K	A_1	A_2	A_3	A_4	Level of uncertainty
178.4–296.0	561.00	–1.65572	7.36809	1.13957+1	9.30166–2	III

Name: 2,4,4–Trimethyl–2–pentene
Formula: C_8H_{16}

CAS–RN: 107–40–4
Group No.: 13–033
Molar Mass: 112.22

TABLE 13.33.1. Experimental heat capacities

Reference		Temp. range K	No. pnts	Error %	Purity %	method	Type capacity	Calorimeter Type	Reference
30PAR/HUF1	N	183.0–296.0	8	1.00	not specified		C_p	BSIO	25PAR
36PAR/TOD2	N	170.0–298.6	15	0.50	98.	estim	C_p	BSIO	25PAR

30PAR/HUF1 unspecified isomer; probably mixture of isomers (according to n.m.t.)
36PAR/TOD2 identity of substance was not known exactly (estimated by authors)

TABLE 13.33.2. Correlated heat capacities

Reference	Temp. range K	No. pnts used	$\sigma_r C$ %	d_w	d/R	d_r %	d_b/R	+/–
				Selected data				
36PAR/TOD2	170.0–298.6	15	0.50	0.195	2.64–2	0.10	4.79–5	–5
				Rejected data				
30PAR/HUF1	(6.78–1, 2.68, –6.77–1, –8)							

TABLE 13.33.3. Parameters of regression polynomial

Heat capacity type	No. data points total	used	s_w	s/R	s_r %	s_b/R	+/–
C_p	23	15	0.228	3.08–2	0.11	4.79–5	–5

Temp. range K	A_1	A_2	A_3	A_4	Level of uncertainty
170.0–298.6	1.00471+1	1.37729+1	–5.04661	8.54456–1	III

TABLE 13.33.4. Recommended values of heat capacities

Temp. (K)	170	180	190	200	210	220	230
c_p (J K^{-1}g^{-1})	1.710	1.739	1.768	1.796	1.825	1.854	1.884
C_p (J K^{-1}mol^{-1})	191.9	195.1	198.4	201.6	204.8	208.0	211.4
Temp. (K)	240	250	260	270	273.15	280	290
c_p (J K^{-1}g^{-1})	1.915	1.948	1.983	2.020	2.032	2.060	2.103
C_p (J K^{-1}mol^{-1})	214.9	218.6	222.5	226.7	228.1	231.2	236.0
Temp. (K)	298.15	300					
c_p (J K^{-1}g^{-1})	2.141	2.150					
C_p (J K^{-1}mol^{-1})	240.3	241.3					

TABLE 13.33.5. Parameters of quasi–polynomial equation

Heat capacity type	No. data points total	used	s_w	s/R	s_r %	s_b/R	+/–
C_p	23	15	0.340	4.30–2	0.17	1.20–4	2

Temp. range K	T_c K	A_1	A_2	A_3	A_4	Level of uncertainty
170.0–298.6	566.00	–6.19620	4.05142	1.44028+1	2.36910	III

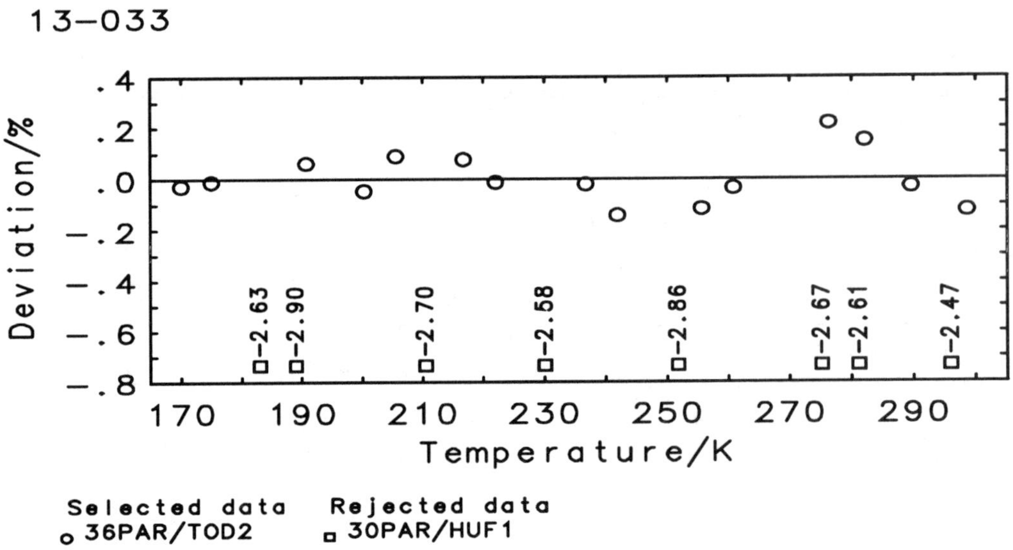

Name: 1–Nonene
Formula: C_9H_{18}

CAS–RN: 124–11–8
Group No.: 13–034
Molar Mass: 126.24

TABLE 13.34.1. Experimental heat capacities

Reference	Temp. range K	No. pnts	Error %	Purity %	Purity method	Type capacity	Calorimeter Type	Calorimeter Reference
90MES/TOD	199.4–380.3	25	0.10	99.979	melpt	C_{sat}	BSAO	47HUF

TABLE 13.34.3. Parameters of cubic spline polynomials

Heat capacity type	No. data points total	used	s_w	s/R	s_r %	s_b/R	+/–
C_{sat}	25	25	0.421	1.30–2	0.04	8.09–6	0

Temp. range K	A_1	A_2	A_3	A_4	Level of uncertainty
199.4–280.0	6.39017+1	–4.15822+1	1.56093+1	–1.74160	II
280.0–380.3	3.38566+1	–9.39106	4.11242	–3.72925–1	II

TABLE 13.34.4. Recommended values of heat capacities

Temp. (K)	200	210	220	230	240	250	260
c_{sat} (J K^{-1}g^{-1})	1.926	1.929	1.938	1.953	1.972	1.995	2.022
C_{sat} (J K^{-1}mol^{-1})	243.1	243.5	244.7	246.5	248.9	251.9	255.2
Temp. (K)	270	273.15	280	290	298.15	300	310
c_{sat} (J K^{-1}g^{-1})	2.051	2.061	2.082	2.115	2.142	2.149	2.184
C_{sat} (J K^{-1}mol^{-1})	258.9	260.1	262.9	267.0	270.5	271.3	275.7
Temp. (K)	320	330	340	350	360	370	380
c_{sat} (J K^{-1}g^{-1})	2.219	2.256	2.293	2.330	2.368	2.405	2.443
C_{sat} (J K^{-1}mol^{-1})	280.2	284.8	289.4	294.1	298.9	303.6	308.4

TABLE 13.34.5. Parameters of quasi–polynomial equation

Heat capacity type	No. data points total	used	s_w	s/R	s_r %	s_b/R	+/–
C_{sat}	25	25	1.850	5.74–2	0.19	1.66–4	0

Temp. range K	T_c K	A_1	A_2	A_3	A_4	A_5	A_6	Level of uncertainty
199.4–380.3	593.20	7.06944+2	6.71847+1	–4.22604+1	6.93342+2	3.70320+1	5.94951+2	II

Name: 1–Decene
Formula: $C_{10}H_{20}$

CAS-RN: 872–05–9
Group No.: 13–035
Molar Mass: 140.27

TABLE 13.35.1. Experimental heat capacities

Reference	Temp. range K	No. pnts	Error %	Purity %	Purity method	Type capacity	Calorimeter Type	Calorimeter Reference
57MCC/FIN1	211.4–355.8	19	0.20	99.9	melpt	C_{sat}	BSAO	47HUF

TABLE 13.35.3. Parameters of regression polynomial

Heat capacity type	No. data points total	No. data points used	s_w	s/R	s_r %	s_b/R	+/–
C_{sat}	19	19	0.558	3.87–2	0.11	5.14–5	–4

Temp. range K	A_1	A_2	A_3	A_4	Level of uncertainty
211.4–355.8	7.93412+1	–5.17038+1	1.81222+1	–1.89255	II

TABLE 13.35.4. Recommended values of heat capacities

Temp. (K)	210	220	230	240	250	260	270
c_{sat} (J K^{-1}g^{-1})	1.965	1.965	1.972	1.984	2.002	2.025	2.051
C_{sat} (J K^{-1}mol^{-1})	275.7	275.6	276.6	278.3	280.8	284.0	287.7
Temp. (K)	273.15	280	290	298.15	300	310	320
c_{sat} (J K^{-1}g^{-1})	2.060	2.081	2.113	2.141	2.148	2.183	2.220
C_{sat} (J K^{-1}mol^{-1})	289.0	291.9	296.4	300.3	301.2	306.2	311.3
Temp. (K)	330	340	350	360			
c_{sat} (J K^{-1}g^{-1})	2.256	2.291	2.326	2.358			
C_{sat} (J K^{-1}mol^{-1})	316.4	321.4	326.2	330.7			

TABLE 13.35.5. Parameters of quasi–polynomial equation

Heat capacity type	No. data points total	No. data points used	s_w	s/R	s_r %	s_b/R	+/–
C_{sat}	19	19	1.014	6.99–2	0.20	2.00–4	1

Temp. range K	T_c K	A_1	A_2	A_3	A_4	A_5	A_6	Level of uncertainty
211.4–355.8	616.70	4.15103+3	4.71647+2	–4.67816+2	4.00277+3	3.42953+2	2.81120+3	II

Name: 1–Undecene
Formula: $C_{11}H_{22}$

CAS-RN: 821–95–4
Group No.: 13–036
Molar Mass: 154.30

TABLE 13.36.1. Experimental heat capacities

Reference	Temp. range K	No. pnts	Error %	Purity %	Purity method	Type capacity	Calorimeter Type	Calorimeter Reference
57MCC/FIN1	227.2–311.4	12	0.20	99.96	melpt	C_{sat}	BSAO	47HUF

TABLE 13.36.3. Parameters of regression polynomial

Heat capacity type	No. data points total	used	s_w	s/R	s_r %	s_b/R	+/–
C_{sat}	12	12	0.403	3.06–2	0.08	3.05–5	1

Temp. range K	A_1	A_2	A_3	A_4	Level of uncertainty
227.2–311.4	1.44979+2	–1.18448+2	4.19396+1	–4.71375	II

TABLE 13.36.4. Recommended values of heat capacities

Temp. (K)	230	240	250	260	270	273.15	280
c_{sat} ($J K^{-1} g^{-1}$)	1.997	2.000	2.012	2.030	2.055	2.063	2.083
C_{sat} ($J K^{-1} mol^{-1}$)	308.1	308.6	310.4	313.3	317.0	318.3	321.4

Temp. (K)	290	298.15	300	310
c_{sat} ($J K^{-1} g^{-1}$)	2.114	2.140	2.146	2.177
C_{sat} ($J K^{-1} mol^{-1}$)	326.2	330.2	331.1	335.9

TABLE 13.36.5. Parameters of quasi-polynomial equation

Heat capacity type	No. data points total	used	s_w	s/R	s_r %	s_b/R	+/–
C_{sat}	12	12	0.680	5.13–2	0.14	–2.39–3	1

Temp. range K	T_c K	A_1	A_2	A_3	A_4	A_5	A_6	Level of uncertainty
227.2–311.4	637.90	5.18289+4	6.86719+3	–7.10688+3	4.78298+4	8.30768+3	2.67726+4	II

Name: 1–Dodecene
Formula: $C_{12}H_{24}$

CAS-RN: 112–41–4
Group No.: 13–037
Molar Mass: 168.32

TABLE 13.37.1. Experimental heat capacities

Reference	Temp. range K	No. pnts	Error %	Purity %	Purity method	Type capacity	Calorimeter Type	Calorimeter Reference
57MCC/FIN1	240.3–305.5	11	0.20	99.92	melpt	C_{sat}	BSAO	47HUF

TABLE 13.37.3. Parameters of regression polynomial

Heat capacity type	No. data points total	used	s_w	s/R	s_r %	s_b/R	+/–
C_{sat}	11	11	0.189	1.57–2	0.04	–7.98–6	1

Temp. range K	A_1	A_2	A_3	A_4	Level of uncertainty
240.3–305.5	2.16574+2	–1.88147+2	6.56951+1	–7.40332	II

TABLE 13.37.4. Recommended values of heat capacities

Temp. (K)	240	250	260	270	273.15	280	290
c_{sat} (J K^{-1}g^{-1})	2.029	2.031	2.043	2.063	2.071	2.089	2.118
C_{sat} (J K^{-1}mol^{-1})	341.6	341.9	344.0	347.3	348.6	351.6	356.5

Temp. (K)	298.15	300	310
c_{sat} (J K^{-1}g^{-1})	2.143	2.149	2.178
C_{sat} (J K^{-1}mol^{-1})	360.7	361.7	366.6

TABLE 13.37.5. Parameters of quasi–polynomial equation

Heat capacity type	No. data points total	used	s_w	s/R	s_r %	s_b/R	+/–
C_{sat}	11	11	0.390	3.24–2	0.08	–2.26–3	–3

Temp. range K	T_c K	A_1	A_2	A_3	A_4	A_5	A_6	Level of uncertainty
240.3–305.5	657.20	1.74601+5	2.40146+4	–2.47951+4	1.59051+5	3.15074+4	8.34620+4	II

Name: 1–Hexadecene
Formula: $C_{16}H_{32}$

CAS-RN: 629–73–2
Group No.: 13–038
Molar Mass: 224.43

TABLE 13.38.1. Experimental heat capacities

Reference	Temp. range K	No. pnts	Error %	Purity %	method	Type capacity	Calorimeter Type	Reference
57MCC/FIN1	279.4–303.8	14	0.20	98.7	melpt	C_{sat}	BSAO	47HUF
90MES/TOD	283.9–380.9	32	0.10	99.990	melpt	C_{sat}	BSAO	47HUF

TABLE 13.38.2. Correlated heat capacities

Reference	Temp. range K	No. pnts used	$\sigma_r C$ %	d_w	d/R	d_r %	d_b/R	+/–
				Selected data				
57MCC/FIN1	279.4–303.8	14	0.20#	0.805	9.35–2	0.16	–4.92–2	–4
90MES/TOD	283.9–380.9	32	0.10	0.708	4.31–2	0.07	5.48–3	0

TABLE 13.38.3. Parameters of regression polynomial

Heat capacity type	No. data points total	used	s_w	s/R	s_r %	s_b/R	+/–
C_{sat}	46	46	0.773	6.58–2	0.11	–1.12–2	–4

Temp. range K	A_1	A_2	A_3	A_4	Level of uncertainty
279.4–380.9	1.42512+2	–8.35976+1	2.54246+1	–2.29509	II

TABLE 13.38.4. Recommended values of heat capacities

Temp. (K)	280	290	298.15	300	310	320	330
c_{sat} (J $K^{-1}g^{-1}$)	2.126	2.146	2.165	2.170	2.198	2.228	2.261
C_{sat} (J $K^{-1}mol^{-1}$)	477.1	481.6	485.9	487.0	493.2	500.1	507.5
Temp. (K)	340	350	360	370	380		
c_{sat} (J $K^{-1}g^{-1}$)	2.296	2.333	2.370	2.408	2.446		
C_{sat} (J $K^{-1}mol^{-1}$)	515.4	523.6	532.0	540.5	549.1		

TABLE 13.38.5. Parameters of quasi-polynomial equation

Heat capacity type	No. data points total	used	s_w	s/R	s_r %	s_b/R	+/–
C_{sat}	46	46	0.865	6.96–2	0.12	6.69–3	5

Temp. range K	T_c K	A_1	A_2	A_3	A_4	A_5	A_6	Level of uncertainty
279.4–380.9	719.50	7.40744+3	8.59228+2	–8.41357+2	6.99954+3	9.54338+2	4.66373+3	II

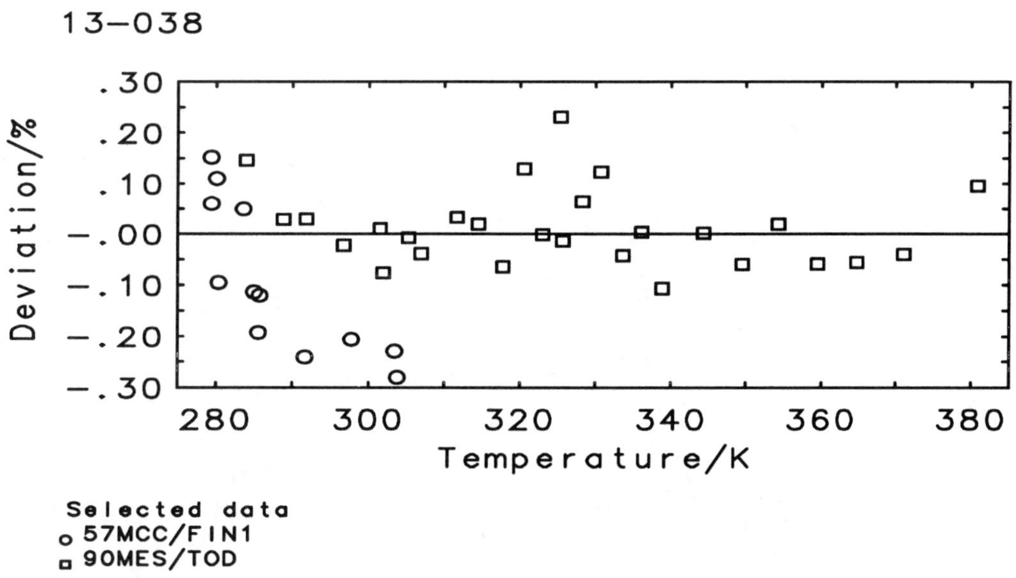

14. Aromatic and Unsaturated Cyclic Hydrocarbons

This family includes 97 compounds, 8 of which were measured at a single temperature only.

A total of 42 compounds of this family were measured at BMB (and later at NIPER) because of their relationships to petroleum, crude tar, and coal (44GUT/SPI, 48HUF/EAT, 48OLI/EAT, 49SCO/GRO, 56FIN/SCO, 57MCC/FIN2, 62SCO/GUT, 65MES/TOD2, 77FIN/MES, 79LEE/HOS, 81LEE/FIN, 82GAM/CAL, 88MES/FIN, 89CHI/HOS, 89CHI/KNI). The data were measured in a high precision adiabatic calorimeter (the reported error of 0.1 to 0.2 % seems to be quite realistic) using high purity samples and were used as a basis for recommendations in all cases. The last two sources included measurements for high-boiling compounds measured using DSC device up to the vicinity of the critical temperature with a reported error of 1 %.

Seven compounds measured at CIUG (67LEB/RAB1, 75LEB/TSV, 78LEB/LEB, 78LEB/TSV, 82LEB/BYK, 88LEB/KUL, 89LEB/SMI) with a reported error of 0.2 to 0.3 % were investigated as monomers which have potential uses in the synthesis of new polymeric materials.

Accurate measurements of five condensed aromatic and bicyclic compounds were carried out at UMMA (70AND/WES, 70GOU/GIR, 70WON/WES, 71WON/WES, 80WON/WES) mainly because of their unusual molecular structure.

Two older measurements for 1,4-dimethylbenzene and ethylbenzene (47COR/GIN, 45SCO/BRI) and two later measurements for two terphenyls (72CHA/BES, 83CHA) were carried out at NBSW.

All the above data have been included in the correlation with the stated uncertainty limits which was warranted by the good reputation of the laboratories and consistency between independent data sources.

Additional data used in the correlation and considered as reliable were obtained from the following laboratories: UOTO (73KIS/SUG, 77HAI/SUG1), UCB (43PIT/SCO, 46PIT/GUT), and RUH (55TAY/KIL, 57PUT/KIL) where three isomers of trimethylbenzene were studied.

When a compound was only measured by a single laboratory, then old measurements from JHUB (31SMI/AND1 1,1-diphenylethylene and 1,1-diphenylethane) and SUC (30HUF/PAR1 tert.butylbenzene and diphenylmethane; 31HUF/PAR 1-isopropyl-4-methylbenzene, 1,2,3,4- and 1,2,3,5-tetramethylbenzene; 32SPA/THO hexamethylbenzene; 33FER/THO 1,2,4,5-tetramethylbenzene and pentamethylbenzene) have been included whose uncertainty limit was about 1 %. Data obtained at PTIL (47KUR, 50KUR) and reported as average heat capacities are considered as only approximative with the error in our judgement exceeding 5 or even 10 % due to the impurities in compounds and the rudimentary experimental technique used.

A variety of compounds from this family were measured at WUSL; the data were presented in a form of an equation and published in two thesis (41SCH, 44EIB). The uncertainty stated by the authors does not seem to be quite realistic and is difficult to estimate what it should be due to the lack of good data from other sources; the expected error can be up to 3 %. In the final selection, the data for four compounds have been examined.

Measurements for 1,3- and 1,4-cyclohexadiene were carried out only at BAF (76GEI/WOL) with a reported error of 3 %. These samples contained considerable amounts of impurities in a form of similar compounds.

This family of compounds contains a large number of chemicals widely used in the chemical industry such as benzene, toluene, styrene, xylenes, and compounds serving as heat transport media in nuclear power stations (such as biphenyl and isopropyl biphenyl). Correspondingly, the frequency of measurements and number of laboratories involved in such measurements are large.

It is surprising that it was difficult to find consistent data for benzene which has been measured most frequently. Measurements from BMB (48OLI/EAT), together with high temperature data from GPI (75RAS/GRI) and MITC (75SAN) have been selected as a basis for the correlation. These data have been supplemented by results obtained at the University St Andrews, Dundee (56DUF/EVE) determined by non-calorimetric (piezometric) method with a reported error below 0.5 %. The piezoelectric data have been selected here due to the absence of reliable calorimetric data in the range between 330 to 353 K. Piezometric data from the same source have also been selected for diphenylmethane as relatively more reliable compared to four available calorimetric sets.

A substantially better situation exists for toluene where two major data sets (BMB measurements 62SCO/GUT, and GITA measurements 66HWA/ZIE, 75HOL/ZIE) agreed to within 0.1 %. At higher temperatures, the data have been supplemented by the measurements from GPI (75RAS/GRI) up to 462 K, (79AND/GRI) up to 373 K and (76SAN/MEL) up to 503 K.

An interesting situation exits in the case of naphthalene where several data sets agree with each other quite well. However, they all differed from the BMB measurements (57MCC/FIN2) which we feel are the most credible source of data. For that reason our recommendations have been based exclusively on the BMB values supplemented for extension of the temperature range by the consistent data generated from an equation given in reference (41SCH).

Different compounds suitable as heat transport media in the power industry were measured using commercial DSC instruments at EREL (83ORO/MRA), at NRLW (58WAL/BRO), and also at MCSL (56MCE/MAL). These measurements were of a routine character and the expected errors are likely to be between 1 and 5 %. An unspecified isomer of isopropylbiphenyl was investigated in fSU laboratories (63VAR/KOP, 64VUK/RAS) and was not fully identified in the literature and the results are subject to considerable uncertainty (over 3 %).

Reliable data for 1,4-dimethylbenzene were obtained at BYUP (79OTT/GOA), and for indane and indene at DCM (61STU/SIN).

Recent extensive measurements on benzene, biphenyl, terphenyl and condensed ring compounds at Laboratoire de Thermodynamique, INPL-ENSIC, Nancy, France (93DUR/AOU) were performed by DSC calorimetry over a wide temperature range. The results show a linear dependence with temperature, however, for those compounds where still other sources of accurate data were available, the slope of the above DSC data was often found to be significantly different. We have expected that the error could be 2–6 % and we included the data only when no other values were available.

Name: Cyclopentene
Formula: C_5H_8

CAS-RN: 142–29–0
Group No.: 14–001
Molar Mass: 68.12

TABLE 14.1.1. Experimental heat capacities

Reference	Temp. range K	No. pnts	Error %	Purity %	Purity method	Type capacity	Calorimeter Type	Calorimeter Reference
48HUF/EAT	140.9–300.8	25	0.20	99.996	melpt	C_{sat}	BSAO	43RUE/HUF

TABLE 14.1.3. Parameters of regression polynomial

Heat capacity type	No. data points total	No. data points used	s_w	s/R	s_r %	s_b/R	+/–
C_{sat}	25	25	0.184	4.75–3	0.04	4.08–6	–3

Temp. range K	A_1	A_2	A_3	A_4	Level of uncertainty
140.9–300.8	1.86400+1	–9.34139	3.76637	–3.60140–1	II

TABLE 14.1.4. Recommended values of heat capacities

Temp. (K)	140	150	160	170	180	190	200
c_{sat} (J K^{-1}g^{-1})	1.459	1.451	1.448	1.449	1.456	1.467	1.482
C_{sat} (J K^{-1}mol^{-1})	99.41	98.83	98.61	98.74	99.18	99.92	101.0
Temp. (K)	210	220	230	240	250	260	270
c_{sat} (J K^{-1}g^{-1})	1.501	1.524	1.550	1.579	1.611	1.646	1.683
C_{sat} (J K^{-1}mol^{-1})	102.2	103.8	105.6	107.6	109.7	112.1	114.6
Temp. (K)	273.15	280	290	298.15	300		
c_{sat} (J K^{-1}g^{-1})	1.695	1.722	1.763	1.797	1.805		
C_{sat} (J K^{-1}mol^{-1})	115.5	117.3	120.1	122.4	123.0		

TABLE 14.1.5. Parameters of quasi-polynomial equation

Heat capacity type	No. data points total	No. data points used	s_w	s/R	s_r %	s_b/R	+/–
C_{sat}	25	19	0.409	1.07–2	0.08	1.70–5	–2

Temp. range K	T_c K	A_1	A_2	A_3	A_4	A_5	A_6	Level of uncertainty
163.6–300.8	506.00	3.76498+2	3.57171+1	–2.67450+1	3.75992+2	1.33860	3.29834+2	II

Name: Benzene–d_6
Formula: C_6D_6

CAS-RN: 1076–43–3
Group No.: 14–002
Molar Mass: 84.15

TABLE 14.2.1. Experimental heat capacities

Reference		Temp. range K	No. pnts	Error %	Purity %	Purity method	Type capacity	Calorimeter Type	Calorimeter Reference
42ZIE/AND		283.5–322.6	9	2.00	99.8	melpt	C_p	BDHO	37STU
62RAB/NIK	N	283.1–303.1	5S	0.30	80.	anal	C_p	BSAO	47SKU

62RAB/NIK substance deuterated to 80 %; results extrapolated to 100 % deuteration

TABLE 14.2.2. Correlated heat capacities

Reference	Temp. range K	No. pnts used	$\sigma_r C$ %	d_w	d/R	d_r %	d_b/R	+/–
				Selected data				
62RAB/NIK	283.1–303.1	5	0.30	0.215	1.17–2	0.06	1.56–5	1
				Rejected data				
42ZIE/AND	(5.91–1, 3.39, –5.48–1, –4)							

TABLE 14.2.3. Parameters of regression polynomial

Heat capacity type	No. data points total	No. data points used	s_w	s/R	s_r %	s_b/R	+/–
C_p	14	5	0.277	1.52–2	0.08	1.56–5	1

Temp. range K	A_1	A_2	Level of uncertainty
283.1–303.1	1.24486+1	1.97130	III

TABLE 14.2.4. Recommended values of heat capacities

Temp. (K)	280	290	298.15	300
c_p (J K^{-1}g^{-1})	1.775	1.795	1.811	1.814
C_p (J K^{-1}mol^{-1})	149.4	151.0	152.4	152.7

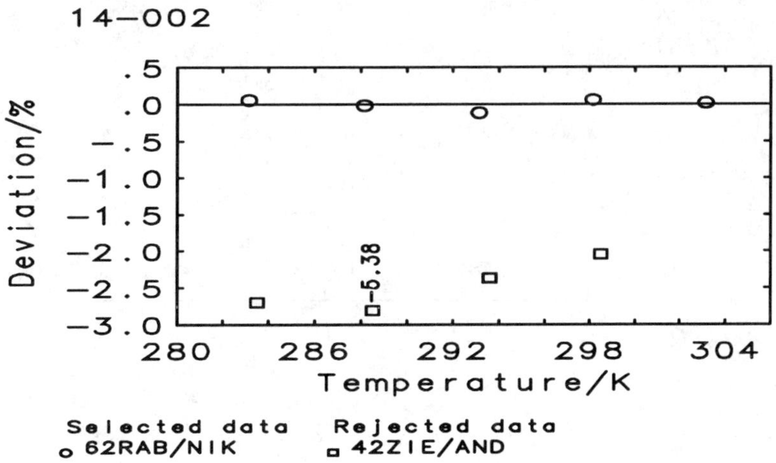

Name: Benzene
Formula: C_6H_6

CAS-RN: 71–43–2
Group No.: 14–003
Molar Mass: 78.11

TABLE 14.3.1. Experimental heat capacities

Reference		Temp. range K	No. pnts	Error %	Purity %	Purity method	Type capacity	Calorimeter Type	Calorimeter Reference
19DEJ		297.6–323.1	16	nosp	not specified		C_p	BSIO	19DEJ
21TRE		289.4–331.3	12	nosp	not specified		C_p	BSIO	49WEI
24WIL/DAN		305.0–335.0	eqn	nosp	not specified		C_p	BSAO	24WIL/DAN
25WIL/DAN		293.1–333.1	5S	nosp	not specified		C_p	BSAO	24WIL/DAN
26AND/LYN		293.0–383.0	eqn	nosp	not specified		C_p	DSIO	26AND/LYN
30HUF/PAR1		281.1–300.0	6	1.00	not specified		C_p	BSIO	25PAR
31FIO/GIN	N	328.1–378.1	6S	0.10	not specified		C_{sat}	BSIO	31FIO/GIN
33KOL/UDO	N	287.9	1	nosp	not specified		C_p	BSIT	34KOL/UDO2
37COH/BUI		283.1–295.1	7	nosp	not specified		C_p	BSAO	20COH/MOE
39PHI		301.4	1	nosp	not specified		C_p	BSIO	49WEI
41ZHD		281.4–318.8	3	nosp	not specified		C_p	BSIT	34KOL/UDO2
42ZIE/AND		283.5–322.6	9	2.00	99.99	melpt	C_p	BDHO	37STU
47KUR		290.0–321.8	5	nosp	not specified		C_{avg}	DSIO	47KUR
48OLI/EAT		286.9–336.9	9	0.20	99.967	melpt	C_{sat}	BSAO	43RUE/HUF
48TSC1		295.1	1	nosp	not specified		C_p	BSIO	48TSC1
48TSC5		295.1	1	nosp	not specified		C_p	BSIO	48TSC1
51SIE/CRU	N	293.1	1	nosp	not specified		C_p	not specified	
55STA/TUP		288.4–346.9	13	0.20	not specified		C_p	BSAO	55STA/TUP
56DUF/EVE	N	303.1–353.1	6S	0.50	not specified		C_p	not specified	
60SWI/ZIE	N	316.3–322.1	2	nosp	not specified		C_{avg}	DSIO	58SWI/ZIE1
62RAB/NIK		283.1–303.1	5S	0.30	not specified		C_p	BSAO	47SKU
65FIN/GRU		300.0	1	0.40	not specified		C_p	BDAO	65FIN/GRU
65KAU/BIT		293.1–349.1	8	1.00	not specified		C_p	FSIO	65KAU/BIT
67PAC		298.1	1	nosp	not specified		C_p	BDHT	79DU/COM
67RAS/GAN		293.1–353.1	4S	0.50	not specified		C_p	BSAO	67RAS/GAN
68REC1	N	298.0–313.0	eqn	nosp	not specified		C_p	BSAO	68REC1
71DES/BHA		298.1–318.1	3S	1.00	not specified		C_p	BSIO	56MUR/VAN
71KHA/SUB	N	298.1–313.1	2	nosp	not specified		C_p	BSIO	64MOE/THO
74RAJ/SUB		298.1–323.1	3	0.30	not specified		C_p	BSIO	64MOE/THO
75RAS/GRI	N	305.1–463.1	9	nosp	not specified		C_p	BDAO	75RAS/GRI
75SAN	N	433.1–493.1	6S	1.00	not specified		C_p	FSIO	75SAN
76FOR/BEN2		298.1	1	0.30	not specified		C_p	FSIT	71PIC/LED
77WIL/GRO		298.1	1	0.30	not specified		C_p	FSIT	71PIC/LED
78GRO/WIL		298.1	1	0.30	not specified		C_p	FSIT	71PIC/LED
81ATA/ELS		293.1	1	2.50	not specified		C_p	BDHO	81ATA/ELS
82GRO/ING		298.1	1	nosp	98.5	melpt	C_p	FSIT	71PIC/LED
82TAN		293.1–303.1	3S	0.30	99.99	melpt	C_p	FSIT	71PIC/LED
82WIL/FAR		298.1	1	nosp	98.5	melpt	C_p	FSIT	71PIC/LED
83GOR/SIM	N	283.8–348.5	12	0.25	99.9	anal	C_p	FSIO	83GOR/SIM
85OGA		298.1	1	0.10	not specified		C_p	FSIO	85OGA
85TAN		298.1	1	0.30	not specified		C_p	FSIT	71PIC/LED
86NAZ/BAS1		322.0–351.1	2	2.00	not specified		C_p	BDHO	86NAZ/BAS1
87KAL/KOH		293.1–313.1	2	nosp	not specified		C_p	FSIT	71PIC/LED
87TAN		293.1–303.1	3	nosp	not specified		C_p	FSIT	71PIC/LED

TABLE 14.3.1. Experimental heat capacities (continued)

Reference	Temp. range K	No. pnts	Error %	Purity %	Purity method	Type capacity	Calorimeter Type	Calorimeter Reference
88SHI/OGA1	298.1	1	nosp	99.9	chrom	C_p	FSIO	85OGA
89LAI/ROD	298.1	1	nosp	99.5	anal	C_p	FSIT	71PIC/LED
89PRA/RAJ	318.1–333.1	4	3.00	not specified		C_p	BDHT	89PRA/RAJ
90YAM/OGA	298.1	1	nosp	99.97	chrom	C_p	FSIO	85OGA
91CZA	298.5	1	2.00	not specified		C_p	BSIO	79CZA
91GRO/ROU	298.1	1	nosp	99.5	anal	C_p	FSIT	71PIC/LED
91TAN/ADA	298.1	1	nosp	not specified		C_p	FSIT	71PIC/LED
91WIL/JIM	298.1	1	nosp	99.	anal	C_p	FSIT	71PIC/LED
92MIY/TAM	298.1	1	nosp	99.99	chrom	C_p	FSIO	85OGA
93DUR/AOU	303.0–347.0	23	nosp	not specified		C_p	BDCT	86MER/BEN

31FIO/GIN water content below 0.01 %; data calculated using procedure by 85WIL/CHA
33KOL/UDO same datum in 34KOL/UDO2
51SIE/CRU heat of mixing calorimeter used
56DUF/EVE measured by a non–calorimetric method (piezo–thermometric)
60SWI/ZIE average values in temperature ranges 294–339 K and 294–350 K
68REC1 same data in 68REC2
71KHA/SUB reproducibility given as 0.3 %
75RAS/GRI data above 343 K measured at superambient pressures up to 1.33 MPa
75SAN same data in 76SAN/MEL; C_p at saturation curve extrapolated from high pressure measurements
83GOR/SIM same data in 82GOR/GRI

TABLE 14.3.2. Correlated heat capacities

Reference	Temp. range K	No. pnts used	$\sigma_r C$ %	d_w	d/R	d_r %	d_b/R	+/–
				Selected data				
48OLI/EAT	286.9–336.9	9	0.20	0.361	1.19–2	0.07	3.56–3	3
56DUF/EVE	303.1–353.1	6	0.50	0.527	4.46–2	0.26	6.12–3	0
65FIN/GRU	300.0	1	0.40	0.218	1.43–2	0.09	1.43–2	1
71KHA/SUB	298.1–313.1	2	0.50#	0.384	3.21–2	0.19	–2.20–2	–1
75RAS/GRI	305.1–463.1	9	0.80#	1.239	2.06–1	0.99	–5.40–2	–3
75SAN	433.1–493.1	6	1.00	0.580	1.34–1	0.58	8.40–2	4
76FOR/BEN2	298.1	1	0.30	0.315	1.54–2	0.09	–1.54–2	–1
77WIL/GRO	298.1	1	0.30	0.709	3.47–2	0.21	–3.47–2	–1
78GRO/WIL	298.1	1	0.30	0.684	3.35–2	0.21	–3.35–2	–1
82GRO/ING	298.1	1	0.50#	0.248	2.03–2	0.12	–2.03–2	–1
82TAN	293.1–303.1	3	0.30	0.433	2.12–2	0.13	–2.11–2	–3
85OGA	298.1	1	0.10	1.261	2.06–2	0.13	2.06–2	1
85TAN	298.1	1	0.30	0.414	2.03–2	0.12	–2.03–2	–1
87TAN	293.1–303.1	3	0.50#	0.322	2.63–2	0.16	–2.61–2	–3
89LAI/ROD	298.1	1	0.50#	0.396	3.23–2	0.20	–3.23–2	–1
91GRO/ROU	298.1	1	0.50#	0.366	2.99–2	0.18	–2.99–2	–1
91TAN/ADA	298.1	1	0.50#	0.307	2.51–2	0.15	–2.51–2	–1
92MIY/TAM	298.1	1	0.50#	0.076	6.20–3	0.04	6.20–3	1

TABLE 14.3.2. Correlated heat capacities (continued)

Reference	Temp. range K	No. pnts used	$\sigma_r C$ %	d_w	d/R	d_r %	d_b/R	+/–
				Rejected data				
19DEJ	(2.74–1, 1.61, 2.49–1, 16)			21TRE	(1.61–1, 0.96, 4.26–2, 4)			
24WIL/DAN	(2.90–1, 1.68, 3.93–2, 1)			25WIL/DAN	(3.38–1, 1.94, 1.60–3, –1)			
26AND/LYN	(3.30–1, 1.93, –1.33–3, 1)			30HUF/PAR1	(1.25–1, 0.77, –1.24–1, –3)			
31FIO/GIN	(1.32–1, 0.74, 1.30–1, 6)			33KOL/UDO	(3.05–1, 1.93, –3.05–1, –1)			
37COH/BUI	(2.31–1, 1.45, –2.30–1, –5)			39PHI	(7.37–2, 0.45, –7.37–2, –1)			
41ZHD	(3.53–2, 0.21, 2.42–2, 1)			42ZIE/AND	(1.62–1, 0.97, –6.86–2, –6)			
47KUR	(2.70–1, 1.70, –1.78–1, –3)			48TSC1	(2.31–1, 1.44, –2.31–1, –1)			
48TSC5	(1.53–1, 0.95, –1.53–1, –1)			51SIE/CRU	(2.09–1, 1.30, –2.09–1, –1)			
55STA/TUP	(1.30–1, 0.77, –1.19–1, –13)			60SWI/ZIE	(5.26–1, 3.23, –3.20–1, 0)			
62RAB/NIK	(1.10–1, 0.68, –1.10–1, –4)			65KAU/BIT	(3.54–1, 2.00, 1.05–1, 0)			
67PAC	(6.20–1, 3.94, –6.20–1, –1)			67RAS/GAN	(7.62–2, 0.44, 5.51–2, 2)			
68REC1	(5.90–2, 0.36, –5.88–2, –5)			71DES/BHA	(1.95–1, 1.19, –1.95–1, –3)			
74RAJ/SUB	(6.84–2, 0.41, –6.04–2, –3)			81ATA/ELS	(1.56–1, 0.97, –1.56–1, –1)			
82WIL/FAR	(3.47–2, 0.21, –3.47–2, –1)			83GOR/SIM	(8.78–2, 0.51, 7.55–2, 10)			
86NAZ/BAS1	(1.76–1, 1.01, –7.54–3, 0)			87KAL/KOH	(5.46–2, 0.33, –5.09–2, –2)			
88SHI/OGA1	(1.51–1, 0.93, –1.51–1, –1)			89PRA/RAJ	(4.57–1, 2.73, –4.39–1, –4)			
90YAM/OGA	(8.16–2, 0.50, –8.16–2, –1)			91CZA	(5.34–3, 0.03, –5.34–3, –1)			
91WIL/JIM	(3.47–2, 0.21, –3.47–2, –1)			93DUR/AOU	(2.78–1, 1.61, 2.76–1, 23)			

TABLE 14.3.3. Parameters of cubic spline polynomials

Heat capacity type	No. data points total	used	s_w	s/R	s_r %	s_b/R	+/–
C_p	226	49	0.726	1.08–1	0.52	–5.49–3	–8
C_{sat}	226	49	0.717	1.05–1	0.51	–5.70–3	–8

Temp. range K	A_1	A_2	A_3	A_4	Level of uncertainty
286.9–353.0	5.18892+1	–3.65279+1	1.16595+1	–1.14259	III
353.0–493.1	–5.36489+1	5.31646+1	–1.37492+1	1.25671	IV
286.9–353.0	4.90773+1	–3.39839+1	1.09030+1	–1.06896	III
353.0–493.1	–4.38720+1	4.50098+1	–1.14748+1	1.04415	IV

TABLE 14.3.4. Recommended values of heat capacities

Temp. (K)	290	298.15	300	310	320	330	340
c_p (J $K^{-1}g^{-1}$)	1.719	1.740	1.745	1.773	1.804	1.837	1.870
C_p (J $K^{-1}mol^{-1}$)	134.3	135.9	136.3	138.5	141.0	143.5	146.1
c_{sat} (J $K^{-1}g^{-1}$)	1.719	1.740	1.745	1.773	1.804	1.836	1.869
C_{sat} (J $K^{-1}mol^{-1}$)	134.3	135.9	136.3	138.5	140.9	143.4	146.0

Temp. (K)	350	360	370	380	390	400	410
c_p (J $K^{-1}g^{-1}$)	1.903	1.936	1.968	2.001	2.035	2.071	2.109
C_p (J $K^{-1}mol^{-1}$)	148.7	151.2	153.7	156.3	158.9	161.7	164.8
c_{sat} (J $K^{-1}g^{-1}$)	1.901	1.934	1.965	1.997	2.030	2.065	2.101
C_{sat} (J $K^{-1}mol^{-1}$)	148.5	151.0	153.5	156.0	158.6	161.3	164.1

Temp. (K)	420	430	440	450	460	470	480
c_p (J $K^{-1}g^{-1}$)	2.152	2.198	2.250	2.309	2.374	2.446	2.527
C_p (J $K^{-1}mol^{-1}$)	168.1	171.7	175.8	180.3	185.4	191.1	197.4
c_{sat} (J $K^{-1}g^{-1}$)	2.141	2.184	2.231	2.284	2.342	2.406	2.477
C_{sat} (J $K^{-1}mol^{-1}$)	167.2	170.6	174.3	178.4	182.9	187.9	193.5

Temp. (K)	490
c_p (J $K^{-1}g^{-1}$)	2.617
C_p (J $K^{-1}mol^{-1}$)	204.5
c_{sat} (J $K^{-1}g^{-1}$)	2.556
C_{sat} (J $K^{-1}mol^{-1}$)	199.6

TABLE 14.3.5. Parameters of quasi-polynomial equation

Heat capacity type	No. data points total	used	s_w	s/R	s_r %	s_b/R	+/–
C_p	226	49	0.795	1.11–1	0.53	–9.05–3	3

Temp. range K	T_c K	A_1	A_2	A	A_4	Level of uncertainty
286.9–493.1	562.16	–3.06796	3.49088–1	9.70407	6.74068	IV

14–003

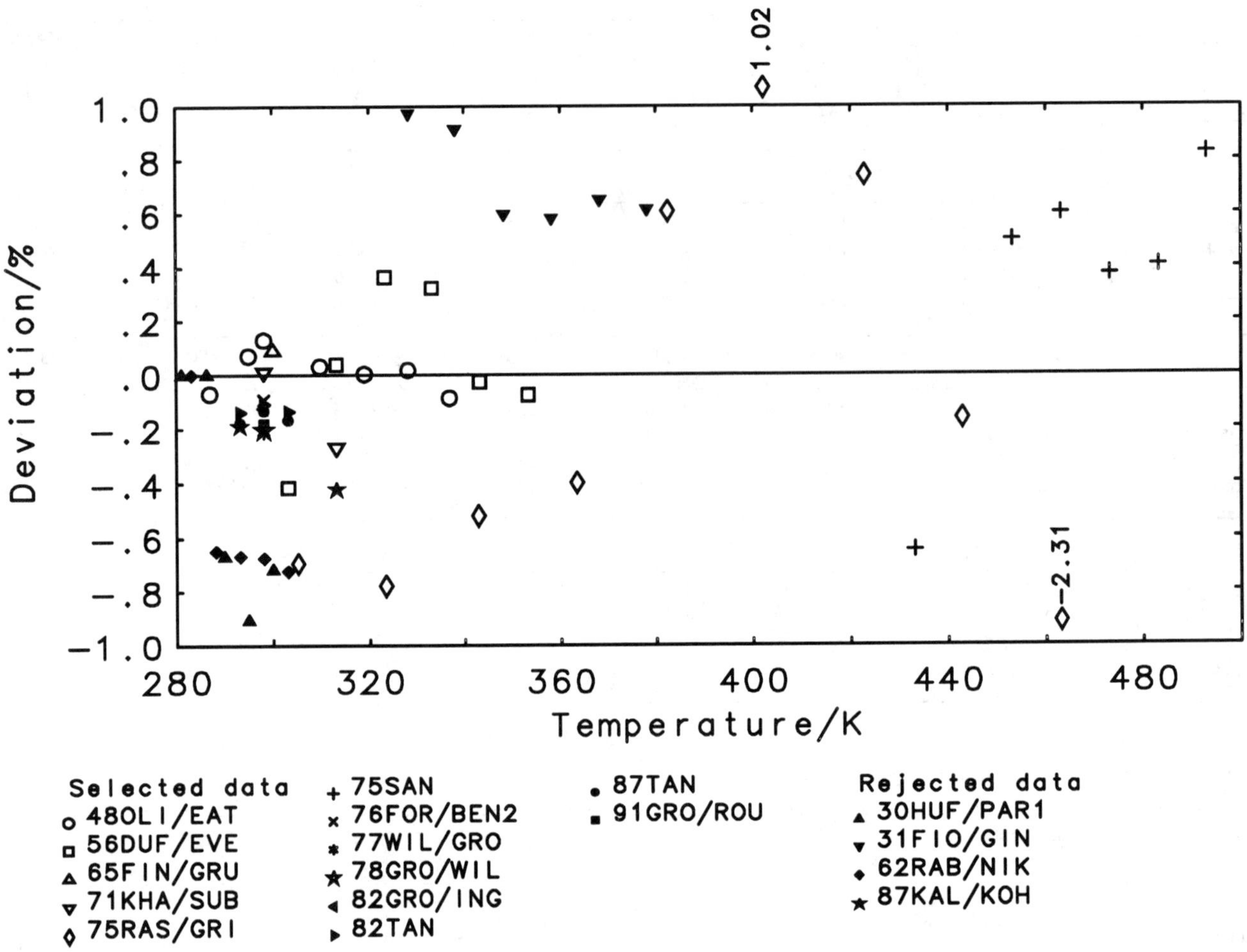

Name: 1,3–Cyclohexadiene
Formula: C_6H_8

CAS-RN: 592–57–4
Group No.: 14–004
Molar Mass: 80.13

TABLE 14.4.1. Experimental heat capacities

Reference		Temp. range K	No. pnts	Error %	Purity %	Purity method	Type capacity	Calorimeter Type	Calorimeter Reference
76GIE/WOL	N	170.0–300.0	14S	3.00	89.4	chrom	C_p	BSIO	76GIE/WOL
89STE/CHI3		298.1	1	1.00	99.95	chrom	C_p	BDHT	89KNI/ARC

76GIE/WOL sample contained 9 % 1,4–Cyclohexadiene, 0.8 % Benzene and 0.8 % Cyclohexene

TABLE 14.4.2. Correlated heat capacities

Reference	Temp. range K	No. pnts used	$\sigma_r C$ %	d_w	d/R	d_r %	d_b/R	+/–
				Selected data				
76GIE/WOL	170.0–300.0	14	3.00	0.230	1.17–1	0.69	4.92–2	0
89STE/CHI3	298.1	1	1.00	0.432	7.35–2	0.43	–7.35–2	–1

TABLE 14.4.3. Parameters of regression polynomial

Heat capacity type	No. data points total	used	s_w	s/R	s_r %	s_b/R	+/–
C_p	15	15	0.278	1.29–1	0.76	4.10–2	–1
Temp. range K		**A_1**	**A_2**	**A_3**			**Level of uncertainty**
170.0–300.0		1.29885+1	–4.76553–1	6.19334–1			V

TABLE 14.4.4. Recommended values of heat capacities

Temp. (K)	170	180	190	200	210	220	230
c_p (J K^{-1}g^{-1})	1.45	1.47	1.49	1.51	1.53	1.55	1.57
C_p (J K^{-1}mol^{-1})	116	118	119	121	122	124	126
Temp. (K)	**240**	**250**	**260**	**270**	**273.15**	**280**	**290**
c_p (J K^{-1}g^{-1})	1.60	1.63	1.65	1.68	1.69	1.71	1.74
C_p (J K^{-1}mol^{-1})	128	130	133	135	136	137	140
Temp. (K)	**298.15**	**300**					
c_p (J K^{-1}g^{-1})	1.77	1.78					
C_p (J K^{-1}mol^{-1})	142	142					

TABLE 14.4.5. Parameters of quasi-polynomial equation

Heat capacity type	No. data points total	used	s_w	s/R	s_r %	s_b/R	+/–
C_p	15	15	0.605	1.05–1	0.62	–1.77–2	2

Temp. range K	T_c K	A_1	A_2	A_3	A_4	Level of uncertainty
170.0–300.0	560.00	7.18648	8.35630	4.12380	1.54511	V

Name: 1,4–Cyclohexadiene
Formula: C_6H_8

CAS-RN: 628–41–1
Group No.: 14–005
Molar Mass: 80.13

TABLE 14.5.1. Experimental heat capacities

Reference		Temp. range K	No. pnts	Error %	Purity %	Purity method	Type capacity	Calorimeter Type	Calorimeter Reference
76GIE/WOL	N	230.0–300.0	8S	3.00	96.7	chrom	C_p	BSIO	76GIE/WOL
89STE/CHI3		298.1	1	1.00	99.9	chrom	C_p	BDHT	89KNI/ARC

76GIE/WOL sample contained 3.3 % Cyclohexene

TABLE 14.5.2. Correlated heat capacities

Reference	Temp. range K	No. pnts used	$\sigma_r C$ %	d_w	d/R	d_r %	d_b/R	+/−
				Selected data				
76GIE/WOL	230.0–300.0	8	3.00	0.324	1.70–1	0.97	7.71–2	0
89STE/CHI3	298.1	1	1.00	0.381	6.52–2	0.38	−6.52–2	−1

TABLE 14.5.3. Parameters of regression polynomial

Heat capacity type	No. data points total	used	s_w	s/R	s_r %	s_b/R	+/−
C_p	9	9	0.405	1.98–1	1.13	6.13–2	−1

Temp. range K	A_1	A_2	A_3	Level of uncertainty
230.0–300.0	9.54238	3.54110	−3.30214–1	V

TABLE 14.5.4. Recommended values of heat capacities

Temp. (K)	230	240	250	260	270	273.15	280
c_p ($J K^{-1} g^{-1}$)	1.65	1.67	1.69	1.71	1.73	1.74	1.75
C_p ($J K^{-1} mol^{-1}$)	133	134	136	137	139	139	140

Temp. (K)	290	298.15	300
c_p ($J K^{-1} g^{-1}$)	1.77	1.78	1.78
C_p ($J K^{-1} mol^{-1}$)	142	143	143

TABLE 14.5.5. Parameters of quasi-polynomial equation

Heat capacity type	No. data points total	used	s_w	s/R	s_r %	s_b/R	+/−
C_p	9	9	1.015	1.78–1	1.04	−4.14–2	1

Temp. range K	T_c K	A_1	A_2	A_3	A_4	Level of uncertainty
230.0–300.0	567.60	5.61346+1	2.41382+1	−8.64491	3.26359+1	V

Name: Cyclohexene
Formula: C_6H_{10}

CAS–RN: 110–83–8
Group No.: 14–006
Molar Mass: 82.15

TABLE 14.6.1. Experimental heat capacities

Reference	Temp. range K	No. pnts	Error %	Purity %	method	Type capacity	Calorimeter Type	Reference
30PAR/HUF1	175.2–293.2	12	1.00	not specified		C_p	BSIO	25PAR
48HUF/EAT	178.8–300.6	19	0.20	99.997	melpt	C_{sat}	BSAO	43RUE/HUF
77HAI/SUG1	171.2–293.4	42	nosp	99.96	melpt	C_{sat}	BSAO	65SUG/SEK
88KAL/WOY	183.0–298.1	26	nosp	not specified		C_p	BSAO	80KAL/JED

TABLE 14.6.2. Correlated heat capacities

Reference	Temp. range K	No. pnts used	$\sigma_r C$ %	d_w	d/R	d_r %	d_b/R	+/–
				Selected data				
48HUF/EAT	178.8–300.6	19	0.20#	1.703	5.39–2	0.34	5.34–2	19
77HAI/SUG1	171.2–293.4	42	0.20#	0.830	2.66–2	0.17	–2.39–2	–41
				Rejected data				
30PAR/HUF1	(1.33–1, 0.84, –1.23–1, –12)			88KAL/WOY	(2.19–1, 1.32, 1.99–1, 26)			

TABLE 14.6.3. Parameters of regression polynomial

Heat capacity type	No. data points total	used	s_w	s/R	s_r %	s_b/R	+/–
C_{sat}	99	61	1.214	3.86–2	0.24	1.75–4	–2 2

Temp. range K	A_1	A_2	A_3	A_4	Level of uncertainty
171.2–300.6	1.66241+1	–5.46846	2.71146	–2.46606–1	II

TABLE 14.6.4. Recommended values of heat capacities

Temp. (K)	170	180	190	200	210	220	230
c_{sat} (J K^{-1}g^{-1})	1.412	1.430	1.451	1.474	1.499	1.527	1.558
C_{sat} (J K^{-1}mol^{-1})	116.0	117.5	119.2	121.1	123.2	125.5	128.0
Temp. (K)	240	250	260	270	273.15	280	290
c_{sat} (J K^{-1}g^{-1})	1.590	1.624	1.660	1.698	1.710	1.737	1.777
C_{sat} (J K^{-1}mol^{-1})	130.6	133.4	136.4	139.5	140.4	142.7	146.0
Temp. (K)	298.15	300					
c_{sat} (J K^{-1}g^{-1})	1.810	1.818					
C_{sat} (J K^{-1}mol^{-1})	148.7	149.4					

TABLE 14.6.5. Parameters of quasi-polynomial equation

Heat capacity type	No. data points total	used	s_w	s/R	s_r %	s_b/R	+/–
C_{sat}	99	61	1.407	4.44–2	0.28	–2.89–4	–9

Temp. range K	T_c K	A_1	A_2	A_3	A_4	Level of uncertainty
171.2–300.6	560.50	3.29996–1	5.96405	5.45636	4.56474–3	III

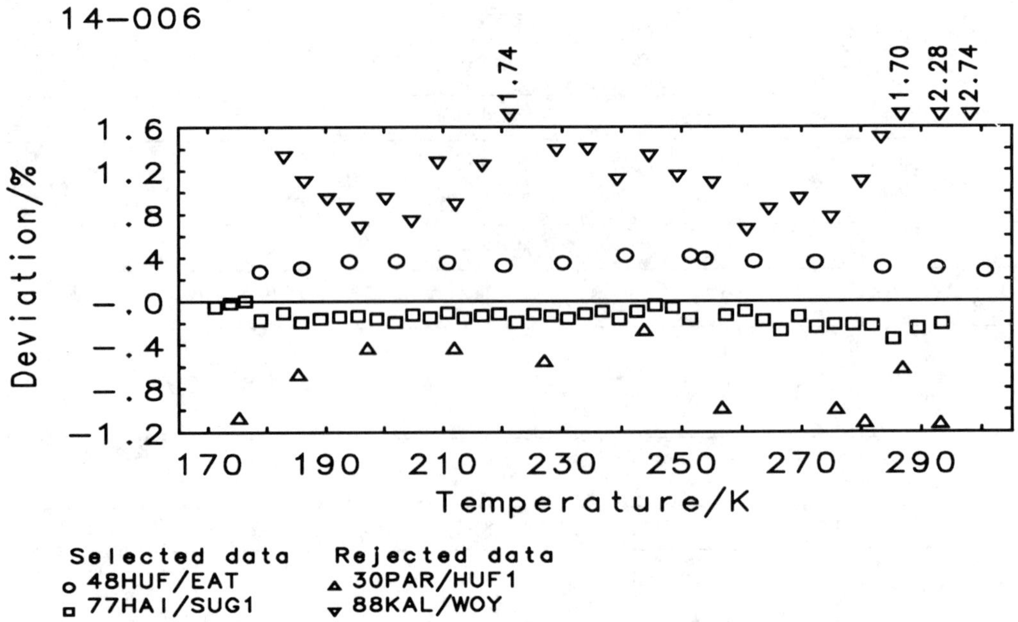

Name: 1–Methylcyclopentene
Formula: C_6H_{10}

CAS–RN: 693–89–0
Group No.: 14–007
Molar Mass: 82.15

TABLE 14.7.1. Experimental heat capacities

Reference	Temp. K	Capac. J/(K.g)	Error %	Purity %	Purity method	Type capacity	Calorimeter Type	Calorimeter Reference
79FUC/PEA	298.15	1.864	1.50	99.	chrom	C_p	BSIO	80FUC

Name: 3–Methylcyclopentene
Formula: C_6H_{10}

CAS–RN: 1120–62–3
Group No.: 14–008
Molar Mass: 82.15

TABLE 14.8.1. Experimental heat capacities

Reference	Temp. K	Capac. J/(K.g)	Error %	Purity %	Purity method	Type capacity	Calorimeter Type	Calorimeter Reference
79FUC/PEA	298.15	1.854	1.50	99.	chrom	C_p	BSIO	80FUC

Name: Bicyclo[2.2.1]hepta–2,5–diene
Formula: C_7H_8

CAS–RN: 121–46–0
Group No.: 14–009
Molar Mass: 92.14

TABLE 14.9.1. Experimental heat capacities

Reference	Temp. K	Capac. J/(K.g)	Error %	Purity %	Purity method	Type capacity	Calorimeter Type	Calorimeter Reference
73HAL/SMI	297.15	1.260	nosp	99.9	chrom	C_p	BDHT	71DU/COM

Name: 1,3,5–Cycloheptatriene
Formula: C_7H_8

CAS-RN: 544–25–2
Group No.: 14–010
Molar Mass: 92.14

TABLE 14.10.1. Experimental heat capacities

Reference	Temp. range K	No. pnts	Error %	Purity %	Purity method	Type capacity	Calorimeter Type	Calorimeter Reference
56FIN/SCO	201.5–312.2	19	0.20	99.986	melpt	C_{sat}	BSAO	43RUE/HUF

TABLE 14.10.3. Parameters of regression polynomial

Heat capacity type	No. data points total	No. data points used	s_w	s/R	s_r %	s_b/R	+/–
C_{sat}	19	19	0.104	3.81–3	0.02	1.91–6	–2

Temp. range K	A_1	A_2	A_3	A_4	Level of uncertainty
201.5–312.2	1.68356+1	–2.20006	1.26682	–7.38067–2	II

TABLE 14.10.4. Recommended values of heat capacities

Temp. (K)	200	210	220	230	240	250	260
c_{sat} ($J\,K^{-1}g^{-1}$)	1.526	1.545	1.565	1.586	1.609	1.633	1.659
C_{sat} ($J\,K^{-1}mol^{-1}$)	140.6	142.3	144.2	146.2	148.3	150.5	152.8
Temp. (K)	270	273.15	280	290	298.15	300	310
c_{sat} ($J\,K^{-1}g^{-1}$)	1.685	1.694	1.713	1.742	1.767	1.773	1.804
C_{sat} ($J\,K^{-1}mol^{-1}$)	155.3	156.1	157.9	160.5	162.8	163.3	166.2

TABLE 14.10.5. Parameters of quasi–polynomial equation

Heat capacity type	No. data points total	No. data points used	s_w	s/R	s_r %	s_b/R	+/–
C_{sat}	19	19	0.256	9.38–3	0.05	7.53–6	–2

Temp. range K	T_c K	A_1	A_2	A_3	A_4	Level of uncertainty
201.5–312.2	609.60	–2.31325	4.24880	9.55177	3.14862–1	II

Name: Methylbenzene
Formula: C_7H_8

CAS-RN: 108–88–3
Group No.: 14–011
Molar Mass: 92.14

TABLE 14.11.1. Experimental heat capacities

Reference		Temp. range K	No. pnts	Error %	Purity %	Purity method	Type capacity	Calorimeter Type	Calorimeter Reference
*81VON		304.8–341.3	7S	nosp	not specified		C_{avg}	DSIO	*81VON
*86SCH		309.3–329.0	11S	nosp	not specified		C_{avg}	DSIO	*86SCH
07BAT	N	181.1–248.1	6	nosp	not specified		C_p	BSIO	07BAT
24WIL/DAN		305.0–348.0	eqn	nosp	not specified		C_p	BSAO	24WIL/DAN
25WIL/DAN		293.1–333.1	5S	nosp	not specified		C_p	BSAO	24WIL/DAN
29KEL5	N	183.8–284.4	11	nosp	not specified		C_p	BSIO	29KEL1
30SOU/AND	N	183.8–244.8	7	0.50	not specified		C_p	BSAO	30SOU/AND
31SMI/AND1		184.4–298.5	9	nosp	not specified		C_p	DSIO	26AND/LYN
35AOY/KAN		195.2–227.8	2	nosp	not specified		C_p	BSAO	35AOY/KAN
37VOL		298.1	1	nosp	not specified		C_p	BSIO	37VOL
41ZHD		278.5–320.4	3	nosp	not specified		C_p	BSIT	34KOL/UDO2
47KUR		241.9–310.5	4	nosp	not specified		C_{avg}	DSIO	47KUR
48TSC1		295.1	1	nosp	not specified		C_p	BSIO	48TSC1
48TSC5		295.1	1	nosp	not specified		C_p	BSIO	48TSC1
56SCH/GOT		270.0–292.5	10	2.00	not specified		C_p	BSAO	33SOU/BRI
58SWI/ZIE2	N	323.9–333.2	2	nosp	not specified		C_{avg}	DSIO	58SWI/ZIE1
62SCO/GUT		183.2–371.0	23	0.20	99.999	melpt	C_{sat}	BSAO	47HUF
66HWA/ZIE		180.9–304.4	34	nosp	99.99	melpt	C_p	BSAO	45SCO/MEY
67RAS/GAN		293.1–373.1	5S	0.50	not specified		C_p	BSAO	67RAS/GAN
71DES/BHA		298.1–318.1	3S	1.00	not specified		C_p	BSIO	56MUR/VAN
73AKH/EKS		298.1	1	0.20	not specified		C_p	FSIO	59RIB/EGO
74RAJ/SUB		298.1–323.1	3	0.30	not specified		C_p	BSIO	64MOE/THO
75HOL/ZIE		162.0–312.0	eqn	0.20	not specified		C_p	BSAO	45SCO/MEY
75PED/KAY		300.6–347.3	19	nosp	not specified		C_p	BSIO	75PED/KAY
75RAS/GRI	N	303.0–462.7	9	1.00	not specified		C_p	BDAO	75RAS/GRI
76FOR/BEN1		298.1	1	0.30	not specified		C_p	FSIT	71PIC/LED
76SAN/MEL	N	393.1–503.1	8S	1.00	not specified		C_p	FSIO	75SAN
77FOR/BEN		298.1	1	nosp	not specified		C_p	FSIT	71PIC/LED
77WIL/GRO		298.1	1	0.30	not specified		C_p	FSIT	71PIC/LED
79AND/GRI	N	293.1–373.3	9	0.50	not specified		C_p	BSAO	67RAS/GAN
80NEF/FIL		300.0–520.0	12S	nosp	not specified		C_{sat}	BDHO	84FIL/LAU
81ATA/ELS		293.1	1	2.50	not specified		C_p	BDHO	81ATA/ELS
82GRO/ING		298.1	1	0.30	99.5	estim	C_p	FSIT	71PIC/LED
82WIL/FAR		298.1	1	0.30	99.5	estim	C_p	FSIT	71PIC/LED
84STE/OLS		266.1–318.1	11S	nosp	99.5	melpt	C_p	BDHT	69PER/COM
85COS/PAT6		298.2	1	nosp	99.	anal	C_p	FSIT	71PIC/LED
86ROU/GRO		298.1	1	nosp	99.	anal	C_p	FSIT	71PIC/LED
87OKH/RAZ		181.4–380.4	15	0.30	not specified		C_{sat}	BDAO	87OKH/RAZ
88SHI/OGA1		298.1	1	nosp	99.9	chrom	C_p	FSIO	85OGA
89PRA/RAJ		318.1–333.1	4	3.00	not specified		C_p	BDHT	89PRA/RAJ
90RAO/RAJ		318.1–333.1	4	4.00	not specified		C_p	BDHT	89PRA/RAJ
91COB/GAR		298.1–368.1	4	nosp	not specified		C_p	BDCT	83ROU/ROU
91TAN/ADA		298.1	1	nosp	not specified		C_p	FSIT	71PIC/LED
91WIL/JIM		298.1	1	nosp	99.	anal	C_p	FSIT	71PIC/LED

07BAT same data in 08BAT
29KEL5 high sample purity
30SOU/AND high sample purity
58SWI/ZIE2 average values in temperature ranges 294–354 K and 294–372 K
75RAS/GRI data above 364 K measured at superambient pressures up to 0.72 MPa
76SAN/MEL C_p at saturation curve extrapolated from high pressure measurements
79AND/GRI grade: pure

TABLE 14.11.2. Correlated heat capacities

Reference	Temp. range K	No. pnts used	$\sigma_r C$ %	d_w	d/R	d_r %	d_b/R	+/–
				Selected data				
62SCO/GUT	183.2–371.0	23	0.20	0.462	1.83–2	0.09	2.00–3	2
66HWA/ZIE	180.9–304.5	34	0.20#	0.426	1.50–2	0.09	2.43–3	–2
75HOL/ZIE	162.0–312.0	31	0.20	0.277	1.03–2	0.06	–6.72–3	–22
75RAS/GRI	303.0–462.7	9	1.00	0.624	1.50–1	0.62	–2.18–2	1
76FOR/BEN1	298.1	1	0.30	0.220	1.25–2	0.07	–1.25–2	–1
76SAN/MEL	393.1–503.1	8	1.00	0.427	1.12–1	0.43	7.29–2	6
77FOR/BEN	298.1	1	0.50#	0.062	5.87–3	0.03	–5.87–3	–1
77WIL/GRO	298.1	1	0.30	0.403	2.28–2	0.12	–2.28–2	–1
79AND/GRI	293.1–373.3	9	0.50	0.811	8.02–2	0.41	7.36–2	9
82GRO/ING	298.1	1	0.30	0.254	1.44–2	0.08	–1.44–2	–1
82WIL/FAR	298.1	1	0.30	0.043	2.42–3	0.01	2.42–3	1
86ROU/GRO	298.1	1	0.50#	0.063	5.99–3	0.03	–5.99–3	–1
91COB/GAR	298.1–368.1	4	0.50#	0.215	2.08–2	0.11	–1.65–2	–3
91TAN/ADA	298.1	1	0.50#	0.178	1.68–2	0.09	–1.68–2	–1
91WIL/JIM	298.1	1	0.50#	0.114	1.08–2	0.06	–1.08–2	–1

Rejected data

Reference		Reference	
*81VON	(3.03–1, 1.52, 2.66–1, 5)	*86SCH	(3.70–1, 1.83, 3.62–1, 11)
07BAT	(1.08–1, 0.63, 9.06–2, 6)	24WIL/DAN	(5.43–1, 2.82, –5.41–1, –5)
25WIL/DAN	(5.52–1, 2.93, –5.52–1, –5)	29KEL5	(2.11–1, 1.24, –1.81–1, –11)
30SOU/AND	(1.95–1, 1.17, –1.80–1, –7)	31SMI/AND1	(3.34–1, 1.92, –1.61–1, –5)
35AOY/KAN	(1.54–1, 0.90, 1.27–1, 2)	37VOL	(7.81–2, 0.41, –7.81–2, –1)
41ZHD	(2.61–1, 1.38, 1.88–2, 1)	47KUR	(6.39–1, 3.84, –2.84–1, 0)
48TSC1	(2.59–1, 1.36, 2.59–1, 1)	48TSC5	(4.90–1, 2.54, 4.90–1, 1)
56SCH/GOT	(4.74–1, 2.52, 4.71–1, 10)	58SWI/ZIE2	(2.87–1, 1.42, 2.87–1, 2)
67RAS/GAN	(2.24–1, 1.07, –1.15–1, –1)	71DES/BHA	(1.89–1, 0.98, –1.72–1, –3)
73AKH/EKS	(4.07–2, 0.21, 4.07–2, 1)	74RAJ/SUB	(1.43–1, 0.72, 9.72–2, 1)
75PED/KAY	(1.06–1, 0.53, 5.47–2, 4)	80NEF/FIL	(4.93–1, 1.85, –3.18–1, –9)
81ATA/ELS	(2.96–2, 0.16, 2.96–2, 1)	84STE/OLS	(1.67–1, 0.87, 1.55–1, 11)
85COS/PAT6	(1.87–1, 0.98, 1.87–1, 1)	87OKH/RAZ	(1.53–1, 0.77, 1.02–1, 10)
88SHI/OGA1	(1.41–1, 0.75, –1.41–1, –1)	89PRA/RAJ	(3.29–1, 1.62, 3.25–1, 4)
90RAO/RAJ	(7.97–1, 4.21, –7.81–1, –4)		

TABLE 14.11.3. Parameters of cubic spline polynomials

Heat capacity type	No. data points total	used	s_w	s/R	s_r %	s_b/R	+/–
C_p	291	126	0.455	5.64–2	0.24	6.49–3	–15
C_{sat}	291	126	0.459	5.67–2	0.24	6.58–3	–15

Temp. range K	A_1	A_2	A_3	A_4	Level of uncertainty
162.0–300.0	2.19038+1	–7.86604	3.13678	–2.80597–1	II
300.0–400.0	1.61637+1	–2.12595	1.22342	–6.80006–2	III
400.0–503.1	–9.49111+1	8.11801+1	–1.96031+1	1.66754	IV
162.0–300.0	2.19306+1	–7.90144	3.15203	–2.82743–1	II
300.0–400.0	1.68655+1	–2.83637	1.46367	–9.51468–2	III
400.0–503.1	–8.43158+1	7.30496+1	–1.75078+1	1.48581	IV

TABLE 14.11.4. Recommended values of heat capacities

Temp. (K)	170	180	190	200	210	220	230
c_p (J K^{-1}g^{-1})	1.463	1.468	1.476	1.487	1.500	1.515	1.533
C_p (J K^{-1}mol^{-1})	134.8	135.3	136.0	137.0	138.2	139.6	141.3
c_{sat} (J K^{-1}g^{-1})	1.464	1.468	1.476	1.487	1.500	1.515	1.533
C_{sat} (J K^{-1}mol^{-1})	134.8	135.3	136.0	137.0	138.2	139.6	141.3
Temp. (K)	**240**	**250**	**260**	**270**	**273.15**	**280**	**290**
c_p (J K^{-1}g^{-1})	1.553	1.575	1.599	1.625	1.634	1.652	1.681
C_p (J K^{-1}mol^{-1})	143.1	145.2	147.4	149.7	150.5	152.3	154.9
c_{sat} (J K^{-1}g^{-1})	1.553	1.575	1.599	1.625	1.634	1.652	1.681
C_{sat} (J K^{-1}mol^{-1})	143.1	145.2	147.4	149.7	150.5	152.3	154.9
Temp. (K)	**298.15**	**300**	**310**	**320**	**330**	**340**	**350**
c_p (J K^{-1}g^{-1})	1.705	1.711	1.742	1.774	1.807	1.841	1.876
C_p (J K^{-1}mol^{-1})	157.1	157.6	160.5	163.5	166.5	169.7	172.9
c_{sat} (J K^{-1}g^{-1})	1.705	1.711	1.742	1.774	1.807	1.841	1.876
C_{sat} (J K^{-1}mol^{-1})	157.1	157.6	160.5	163.5	166.5	169.6	172.8
Temp. (K)	**360**	**370**	**380**	**390**	**400**	**410**	**420**
c_p (J K^{-1}g^{-1})	1.912	1.949	1.987	2.026	2.065	2.105	2.147
C_p (J K^{-1}mol^{-1})	176.2	179.6	183.1	186.6	190.3	194.0	197.8
c_{sat} (J K^{-1}g^{-1})	1.912	1.948	1.985	2.023	2.062	2.101	2.142
C_{sat} (J K^{-1}mol^{-1})	176.1	179.5	182.9	186.4	190.0	193.6	197.3
Temp. (K)	**430**	**440**	**450**	**460**	**470**	**480**	**490**
c_p (J K^{-1}g^{-1})	2.191	2.239	2.291	2.349	2.412	2.483	2.561
C_p (J K^{-1}mol^{-1})	201.9	206.3	211.1	216.4	222.3	228.8	236.0
c_{sat} (J K^{-1}g^{-1})	2.185	2.231	2.280	2.334	2.394	2.460	2.533
C_{sat} (J K^{-1}mol^{-1})	201.3	205.5	210.1	215.1	220.6	226.7	233.4
Temp. (K)	**500**						
c_p (J K^{-1}g^{-1})	2.649						
C_p (J K^{-1}mol^{-1})	244.1						
c_{sat} (J K^{-1}g^{-1})	2.614						
C_{sat} (J K^{-1}mol^{-1})	240.8						

TABLE 14.11.5. Parameters of quasi-polynomial equation

Heat capacity type	No. data points total	used	s_w	s/R	s_r %	s_b/R	+/−
C_p	291	126	1.689	1.22−1	0.54	8.21−3	1 3

Temp. range K	T_c K	A_1	A_2	A_3	A_4	A_5	A_6	Level of uncertainty
162.0–503.1	591.79	2.96623	3.04607−2	1.51030+1	3.32066	6.12322	1.97779+1	IV

14–011

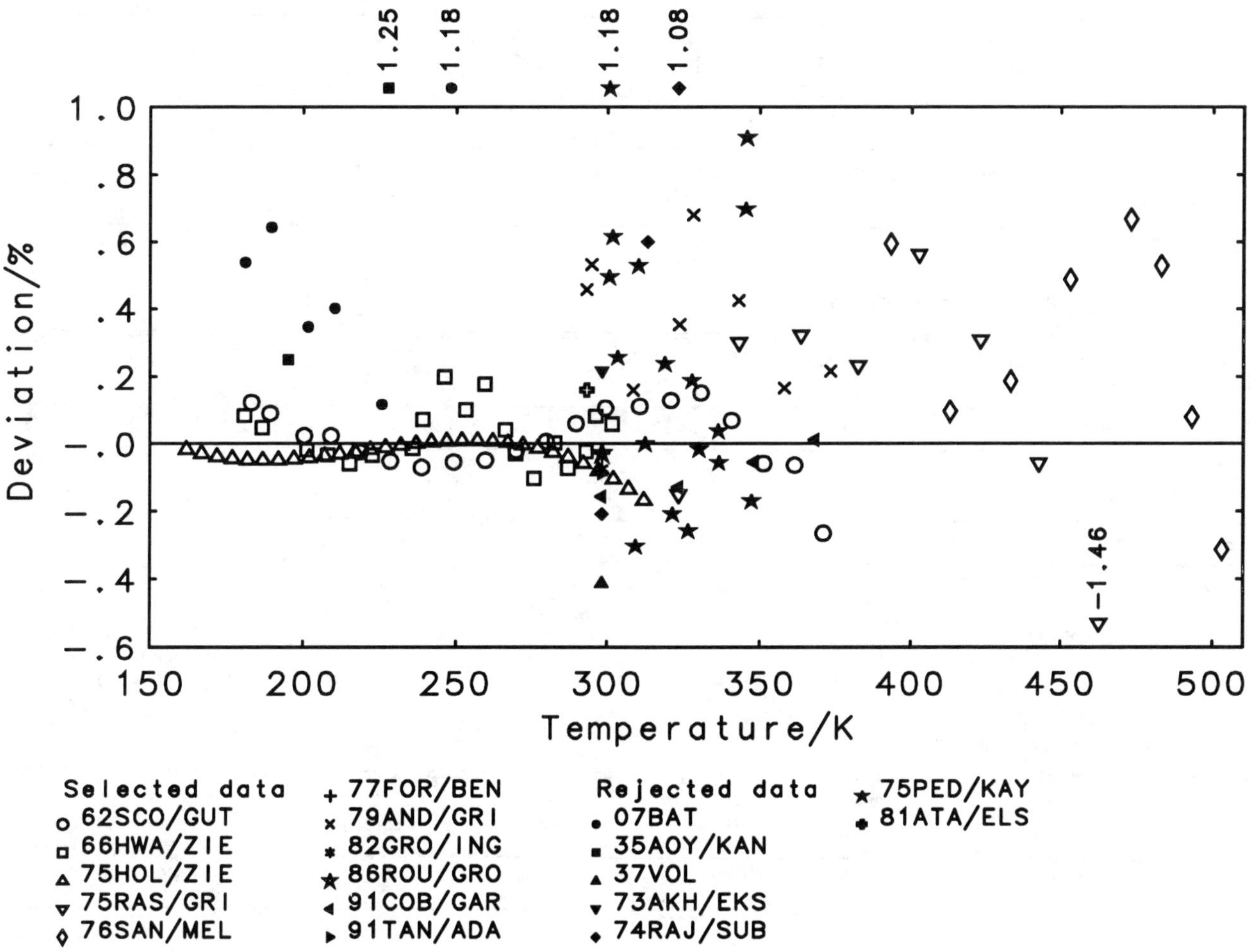

Name: (*Z*)–Cycloheptene
Formula: C_7H_{12}

CAS–RN: 45510–00–7
Group No.: 14–012
Molar Mass: 96.17

TABLE 14.12.1. Experimental heat capacities

Reference	Temp. range K	No. pnts	Error %	Purity %	Purity method	Type capacity	Calorimeter Type	Calorimeter Reference
89LEB/SMI	218.4–302.8	23	0.20	not specified		C_p	BSAO	76LEB/LIT

TABLE 14.12.3. Parameters of regression polynomial

Heat capacity type	No. data points total	No. data points used	s_w	s/R	s_r %	s_b/R	+/–
C_p	23	22	0.984	3.69–2	0.20	1.26–4	2

Temp. range K	A_1	A_2	A_3	Level of uncertainty
218.5–302.8	1.49129+1	–4.01879–1	7.82994–1	III

TABLE 14.12.4. Recommended values of heat capacities

Temp. (K)	220	230	240	250	260	270	273.15
c_p (J K^{-1}g^{-1})	1.540	1.567	1.596	1.626	1.657	1.689	1.699
C_p (J K^{-1}mol^{-1})	148.2	150.7	153.5	156.3	159.3	162.4	163.4
Temp. (K)	280	290	298.15	300			
c_p (J K^{-1}g^{-1})	1.723	1.758	1.787	1.794			
C_p (J K^{-1}mol^{-1})	165.7	169.1	171.9	172.6			

TABLE 14.12.5. Parameters of quasi–polynomial equation

Heat capacity type	No. data points total	used	s_w	s/R	s_r %	s_b/R	+/–
C_p	23	22	0.985	3.70–2	0.20	1.26–4	1

Temp. range K	T_c K	A_1	A_2	A_3	A_4	Level of uncertainty
218.5–302.8	600.00	–4.95633	3.67604	9.13518	1.67063	III

Name: 1–Ethylcyclopentene
Formula: C_7H_{12}

CAS–RN: 2146–38–5
Group No.: 14–013
Molar Mass: 96.17

TABLE 14.13.1. Experimental heat capacities

Reference	Temp. K	Capac. J/(K.g)	Error %	Purity %	Purity method	Type capacity	Calorimeter Type	Calorimeter Reference
79FUC/PEA	298.15	1.958	1.50	99.	chrom	C_p	BSIO	80FUC

Name: 4–Methylcyclohexene
Formula: C_7H_{12}

CAS–RN: 591–47–9
Group No.: 14–014
Molar Mass: 96.17

TABLE 14.14.1. Experimental heat capacities

Reference	Temp. range K	No. pnts	Error %	Purity %	Purity method	Type capacity	Calorimeter Type	Calorimeter Reference
88LEB/KUL	161.1–321.4	22	0.20	99.3	chrom	C_p	BSAO	76LEB/LIT

TABLE 14.14.3. Parameters of regression polynomial

Heat capacity type	No. data points total	used	s_w	s/R	s_r %	s_b/R	+/–
C_p	22	22	1.794	7.25–2	0.36	4.54–4	–4

Temp. range K	A_1	A_2	A_3	Level of uncertainty
161.1–321.4	1.61669+1	–8.75321–1	9.23000–1	III

TABLE 14.14.4. Recommended values of heat capacities

Temp. (K)	160	170	180	190	200	210	220
c_p (J K^{-1}g^{-1})	1.481	1.500	1.520	1.542	1.566	1.591	1.617
C_p (J K^{-1}mol^{-1})	142.4	144.2	146.2	148.3	150.6	153.0	155.6
Temp. (K)	230	240	250	260	270	273.15	280
c_p (J K^{-1}g^{-1})	1.646	1.676	1.707	1.740	1.775	1.786	1.811
C_p (J K^{-1}mol^{-1})	158.3	161.2	164.2	167.4	170.7	171.8	174.2
Temp. (K)	290	298.15	300	310	320		
c_p (J K^{-1}g^{-1})	1.849	1.881	1.889	1.930	1.973		
C_p (J K^{-1}mol^{-1})	177.9	180.9	181.7	185.6	189.7		

TABLE 14.14.5. Parameters of quasi-polynomial equation

Heat capacity type	No. data points total	used	s_w	s/R	s_r %	s_b/R	+/–
C_p	22	22	1.843	7.23–2	0.37	4.54–4	–4

Temp. range K	T_c K	A_1	A_2	A_3	A_4	Level of uncertainty
161.1–321.4	573.00	–3.04750	4.72273	9.40212	4.91626–1	III

Name: Ethynylbenzene
Formula: C_8H_6

CAS-RN: 536–74–3
Group No.: 14–015
Molar Mass: 102.14

TABLE 14.15.1. Experimental heat capacities

Reference	Temp. range K	No. pnts	Error %	Purity %	Purity method	Type capacity	Calorimeter Type	Calorimeter Reference
31SMI/AND1	231.7–298.5	6	nosp	99.0	estim	C_p	DSIO	26AND/LYN
82LEB/BYK	228.0–330.0	4S	0.30	98.4	melpt	C_p	BSAO	76LEB/LIT

TABLE 14.15.2. Correlated heat capacities

Reference	Temp. range K	No. pnts used	$\sigma_r C$ %	d_w	d/R	d_r %	d_b/R	+/–
				Selected data				
31SMI/AND1	231.7–298.5	6	1.20#	0.934	2.28–1	1.12	–2.05–1	–6
82LEB/BYK	228.0–330.0	4	0.30	0.523	3.36–2	0.16	1.98–2	2

TABLE 14.15.3. Parameters of regression polynomial

Heat capacity type	No. data points total	used	s_w	s/R	s_r %	s_b/R	+/–
C_p	10	10	0.951	2.12–1	1.04	–1.15–1	–4

Temp. range K	A_1	A_2	A_3	Level of uncertainty
228.0–330.0	1.89859+1	–1.61818	8.37050–1	IV

TABLE 14.15.4. Recommended values of heat capacities

Temp. (K)	230	240	250	260	270	273.15	280
c_p (J $K^{-1}g^{-1}$)	1.603	1.622	1.642	1.664	1.687	1.694	1.711
C_p (J $K^{-1}mol^{-1}$)	163.7	165.7	167.7	169.9	172.3	173.0	174.8
Temp. (K)	290	298.15	300	310	320	330	
c_p (J $K^{-1}g^{-1}$)	1.737	1.759	1.764	1.792	1.822	1.853	
C_p (J $K^{-1}mol^{-1}$)	177.4	179.6	180.1	183.0	186.1	189.2	

TABLE 14.15.5. Parameters of quasi–polynomial equation

Heat capacity type	No. data points total	used	s_w	s/R	s_r %	s_b/R	+/–
C_p	10	10	1.925	1.64–1	0.81	–2.97–3	0

Temp. range K	T_c K	A_1	A_2	A_3	A_4	Level of uncertainty
228.0–330.0	643.70	–6.46565–2	6.56850	9.25609	1.59110–4	IV

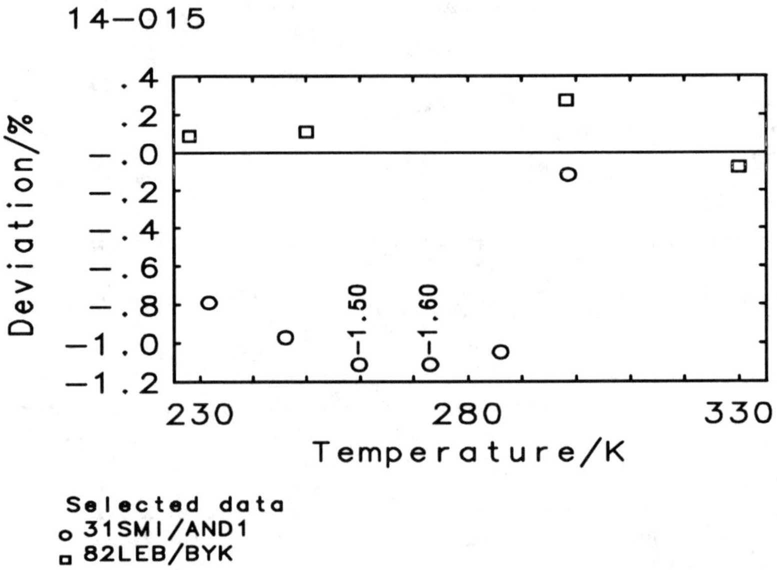

Name: 1,3,5,7–Cyclooctatetraene
Formula: C_8H_8

CAS-RN: 629–20–9
Group No.: 14–016
Molar Mass: 104.15

TABLE 14.16.1. Experimental heat capacities

Reference	Temp. range K	No. pnts	Error %	Purity %	Purity method	Type capacity	Calorimeter Type	Calorimeter Reference
49SCO/GRO	273.0–328.3	12	0.20	99.92	melpt	C_{sat}	BSAO	43RUE/HUF

TABLE 14.16.3. Parameters of regression polynomial

Heat capacity type	No. data points total	used	s_w	s/R	s_r %	s_b/R	+/−
C_{sat}	12	12	0.337	7.60–3	0.03	3.66–6	−1

Temp. range K	A_1	A_2	A_3	Level of uncertainty
273.0–328.3	1.66604+1	7.53020–1	3.78797–1	II

TABLE 14.16.4. Recommended values of heat capacities

Temp. (K)	280	290	298.15	300	310	320	330
c_{sat} ($J K^{-1}g^{-1}$)	1.735	1.759	1.778	1.783	1.807	1.832	1.858
C_{sat} ($J K^{-1}mol^{-1}$)	180.7	183.2	185.2	185.7	188.2	190.8	193.5

TABLE 14.16.5. Parameters of quasi–polynomial equation

Heat capacity type	No. data points total	used	s_w	s/R	s_r %	s_b/R	+/−
C_{sat}	12	12	0.344	7.76–3	0.03	3.97–6	0

Temp. range K	T_c K	A_1	A_2	A_3	A_4	Level of uncertainty
273.0–328.3	639.00	−4.97768	9.43203–1	1.43124+1	6.56732	II

Name: Ethenylbenzene
Formula: C_8H_8

CAS-RN: 100–42–5
Group No.: 14–017
Molar Mass: 104.15

TABLE 14.17.1. Experimental heat capacities

Reference	Temp. range K	No. pnts	Error %	Purity %	Purity method	Type capacity	Calorimeter Type	Calorimeter Reference
31SMI/AND1	246.0–298.5	5	nosp	99.0	estim	C_p	DSIO	26AND/LYN
46PIT/GUT	246.7–298.5	9	0.20	99.8	melpt	C_p	BSIO	28LAT/GRE
48TSC4	295.1	1	nosp	not specified		C_p	BSIO	48TSC1
48TSC6	294.1	1	nosp	not specified		C_p	BSIO	48TSC1
50KUR	308.3–353.3	5	nosp	not specified		C_{avg}	DSIO	47KUR

TABLE 14.17.2. Correlated heat capacities

Reference	Temp. range K	No. pnts used	$\sigma_r C$ %	d_w	d/R	d_r %	d_b/R	+/−
				Selected data				
46PIT/GUT	246.7–298.5	9	0.20	1.541	6.36–2	0.31	4.14–4	3
				Rejected data				
31SMI/AND1	(3.24–1, 1.54, −3.17–1, −4)			48TSC4	(3.81–1, 1.78, −3.81–1, −1)			
48TSC6	(4.53–1, 2.12, −4.53–1, −1)							

TABLE 14.17.3. Parameters of regression polynomial

Heat capacity type	No. data points total	used	s_w	s/R	s_r %	s_b/R	+/–
C_p	21	9	1.747	7.21–2	0.35	4.14–4	3

Temp. range K	A_1	A_2	Level of uncertainty
246.7–298.5	1.19343+1	3.34845	III

TABLE 14.17.4. Recommended values of heat capacities

Temp. (K)	250	260	270	273.15	280	290	298.15
c_p (J K^{-1}g^{-1})	1.621	1.648	1.674	1.683	1.701	1.728	1.750
C_p (J K^{-1}mol^{-1})	168.8	171.6	174.4	175.3	177.2	180.0	182.2

Temp. (K)	300
c_p (J K^{-1}g^{-1})	1.755
C_p (J K^{-1}mol^{-1})	182.8

TABLE 14.17.5. Parameters of quasi-polynomial equation

Heat capacity type	No. data points total	used	s_w	s/R	s_r %	s_b/R	+/–
C_p	21	9	1.579	6.55–2	0.32	2.77–4	1

Temp. range K	T_c K	A_1	A_2	A_3	A_4	Level of uncertainty
246.7–298.5	636.20	2.21355+1	1.73526+1	1.06718–2	7.05914	III

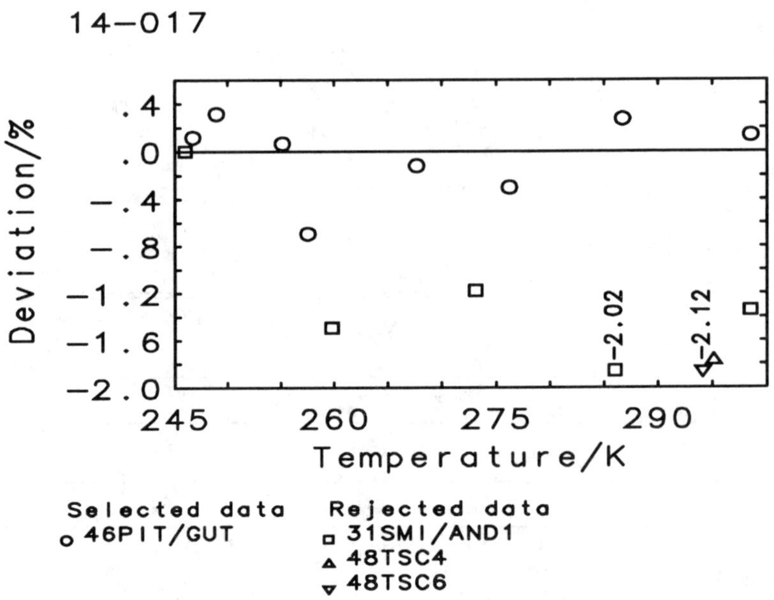

Name: 1,2–Dimethylbenzene
Formula: C_8H_{10}

CAS–RN: 95–47–6
Group No.: 14–018
Molar Mass: 106.17

TABLE 14.18.1. Experimental heat capacities

Reference		Temp. range K	No. pnts	Error %	Purity %	Purity method	Type capacity	Calorimeter Type	Calorimeter Reference
24WIL/DAN		303.0–348.0	eqn	nosp	not specified		C_p	BSAO	24WIL/DAN
30HUF/PAR1		253.3–295.1	8	1.00	not specified		C_p	BSIO	25PAR
43PIT/SCO		251.6–301.9	9	0.20	99.88	melpt	C_p	BSIO	28LAT/GRE
47KUR		308.8–346.6	3	nosp	not specified		C_{avg}	DSIO	47KUR
48TSC3		293.1	1	nosp	not specified		C_p	BSIO	48TSC1
48TSC6		294.1	1	nosp	not specified		C_p	BSIO	48TSC1
58SWI/ZIE2	N	347.3	1	nosp	not specified		C_{avg}	DSIO	58SWI/ZIE1
75RAS/GRI	N	303.2–463.1	10	1.00	not specified		C_p	BDAO	75RAS/GRI
77FOR/BEN		298.1	1	nosp	not specified		C_p	FSIT	71PIC/LED
79FOR/BEN		298.1	1	0.30	not specified		C_p	FSIT	71PIC/LED
91TAN/ADA		298.1	1	nosp	not specified		C_p	FSIT	71PIC/LED
93GAR/BAN		318.1–373.1	12	nosp	99.6	chrom	C_p	BDCT	91BAN/GAR

58SWI/ZIE2 average value in temperature range 333–373 K
75RAS/GRI data above 348 K measured at superambient pressures up to 0.51 MPa

TABLE 14.18.2. Correlated heat capacities

Reference	Temp. range K	No. pnts used	$\sigma_r C$ %	d_w	d/R	d_r %	d_b/R	+/–
				Selected data				
43PIT/SCO	251.6–301.9	9	0.20	1.263	5.64–2	0.25	7.83–3	2
75RAS/GRI	303.2–463.1	10	1.00	0.501	1.26–1	0.50	–2.22–2	–2
77FOR/BEN	298.1	1	0.50#	0.147	1.66–2	0.07	–1.66–2	–1
79FOR/BEN	298.1	1	0.30	0.123	8.32–3	0.04	–8.32–3	–1
91TAN/ADA	298.1	1	0.50#	0.162	1.83–2	0.08	–1.83–2	–1
93GAR/BAN	318.1–373.1	12	0.50#	0.418	4.91–2	0.21	–2.54–2	–4
				Rejected data				
24WIL/DAN	(7.05–1, 3.10, –7.02–1, –5)			30HUF/PAR1	(2.58–1, 1.19, –2.49–1, –8)			
47KUR	(2.09–1, 0.92, –1.37–1, –1)			48TSC3	(5.02–1, 2.29, –5.02–1, –1)			
48TSC6	(5.36–1, 2.45, –5.36–1, –1)			58SWI/ZIE2	(4.47–1, 1.80, 4.47–1, 1)			

TABLE 14.18.3. Parameters of cubic spline polynomials

Heat capacity type	No. data points total	No. data points used	s_w	s/R	s_r %	s_b/R	+/–
C_p	53	34	0.810	8.66–2	0.35	–1.47–2	–7
C_{sat}	53	34	0.809	8.65–2	0.35	–1.49–2	–7

Temp. range K	A_1	A_2	A_3	A_4	Level of uncertainty
251.6–350.0	–4.59064	2.10061+1	–6.07675	7.00164–1	III
350.0–463.1	4.77123+1	–2.38250+1	6.73215	–5.19731–1	IV
251.6–350.0	–4.39299	2.08056+1	–6.00945	6.92696–1	III
350.0–463.1	4.94168+1	–2.53171+1	7.16845	–5.62343–1	IV

TABLE 14.18.4. Recommended values of heat capacities

Temp. (K)	250	260	270	273.15	280	290	298.15
c_p (J K^{-1}g^{-1})	1.636	1.664	1.692	1.701	1.719	1.746	1.768
C_p (J K^{-1}mol^{-1})	173.6	176.7	179.7	180.6	182.5	185.4	187.7
c_{sat} (J K^{-1}g^{-1})	1.636	1.664	1.692	1.701	1.719	1.746	1.768
C_{sat} (J K^{-1}mol^{-1})	173.7	176.7	179.7	180.6	182.5	185.4	187.7
Temp. (K)	300	310	320	330	340	350	360
c_p (J K^{-1}g^{-1})	1.773	1.800	1.828	1.857	1.888	1.920	1.953
C_p (J K^{-1}mol^{-1})	188.3	191.1	194.1	197.2	200.4	203.8	207.4
c_{sat} (J K^{-1}g^{-1})	1.773	1.800	1.828	1.857	1.888	1.920	1.953
C_{sat} (J K^{-1}mol^{-1})	188.3	191.1	194.1	197.2	200.4	203.8	207.4
Temp. (K)	370	380	390	400	410	420	430
c_p (J K^{-1}g^{-1})	1.989	2.026	2.064	2.104	2.144	2.185	2.226
C_p (J K^{-1}mol^{-1})	211.2	215.1	219.2	223.4	227.6	231.9	236.3
c_{sat} (J K^{-1}g^{-1})	1.989	2.026	2.064	2.103	2.143	2.183	2.223
C_{sat} (J K^{-1}mol^{-1})	211.2	215.1	219.1	223.3	227.5	231.8	236.0
Temp. (K)	440	450	460				
c_p (J K^{-1}g^{-1})	2.267	2.308	2.348				
C_p (J K^{-1}mol^{-1})	240.7	245.0	249.3				
c_{sat} (J K^{-1}g^{-1})	2.263	2.303	2.342				
C_{sat} (J K^{-1}mol^{-1})	240.3	244.5	248.7				

TABLE 14.18.5. Parameters of quasi-polynomial equation

Heat capacity type	No. data points total	used	s_w	s/R	s_r %	s_b/R	+/–
C_p	53	34	0.863	1.14–1	0.44	–7.53–3	2

Temp. range K	T_c K	A_1	A_2	A_3	A_4	Level of uncertainty
251.6–463.1	630.30	–5.54618	8.70887–1	1.31753+1	8.83013	IV

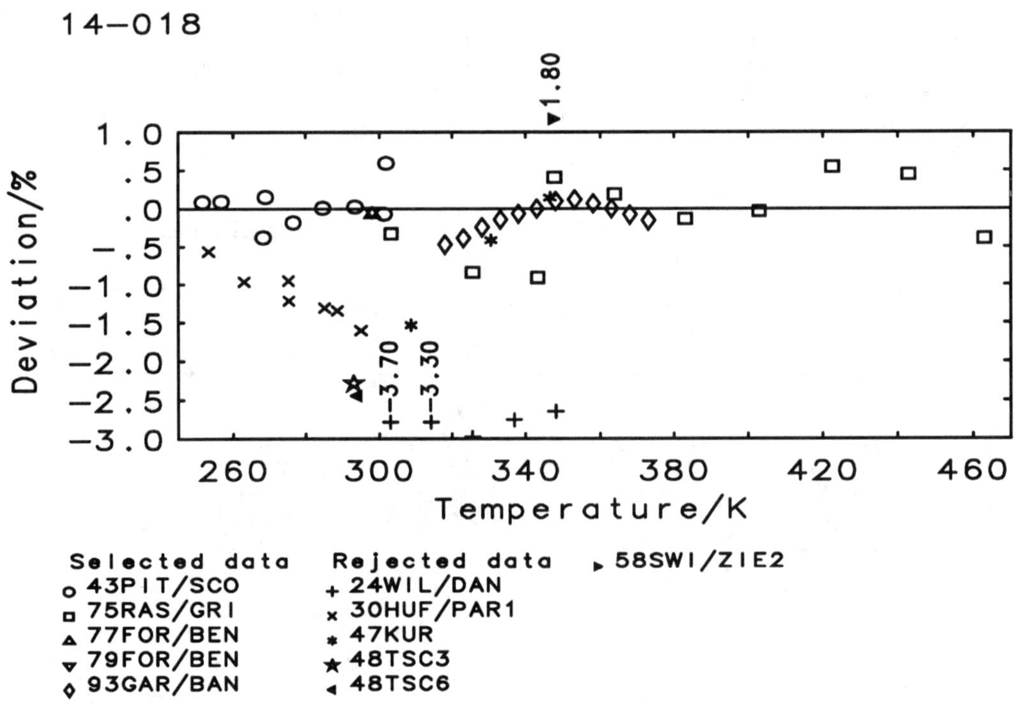

Name: 1,3–Dimethylbenzene
Formula: C_8H_{10}

CAS–RN: 108–38–3
Group No.: 14–019
Molar Mass: 106.17

TABLE 14.19.1. Experimental heat capacities

Reference		Temp. range K	No. pnts	Error %	Purity %	Purity method	Type capacity	Calorimeter Type	Calorimeter Reference
*86SCH		308.9–334.8	11S	nosp	not specified		C_{avg}	DSIO	*86SCH
24WIL/DAN		303.0–348.0	eqn	nosp	not specified		C_p	BSAO	24WIL/DAN
30HUF/PAR1	N	217.0–275.3	6	1.00	not specified		C_p	BSIO	25PAR
43PIT/SCO	N	231.4–318.2	11	0.20	97.66	melpt	C_p	BSIO	28LAT/GRE
47KUR		309.8–347.7	3	nosp	not specified		C_{avg}	DSIO	47KUR
48TSC3		293.1	1	nosp	not specified		C_p	BSIO	48TSC1
48TSC6		294.1	1	nosp	not specified		C_p	BSIO	48TSC1
58SWI/ZIE2	N	336.8	1	nosp	not specified		C_{avg}	DSIO	58SWI/ZIE1
75RAS/GRI	N	302.9–462.5	9	1.00	not specified		C_p	BDAO	75RAS/GRI
75SAN	N	411.1–540.2	8S	1.00	not specified		C_p	FSIO	75SAN
77FOR/BEN		298.1	1	nosp	not specified		C_p	FSIT	71PIC/LED
79FOR/BEN		298.1	1	0.30	not specified		C_p	FSIT	71PIC/LED
91TAN/ADA		298.1	1	nosp	not specified		C_p	FSIT	71PIC/LED
93GAR/BAN		318.1–373.1	12	nosp	99.3	chrom	C_p	BDCT	91BAN/GAR

30HUF/PAR1 low sample purity
43PIT/SCO corrected for content of 1,4–Dimethylbenzene
58SWI/ZIE2 average value in temperature range 333–373 K
75RAS/GRI data above 384 K measured at superambient pressures up to 0.45 MPa
75SAN same data in 76SAN/MEL; C_p at saturation line extrapolated from high pressure measurement

TABLE 14.19.2. Correlated heat capacities

Reference	Temp. range K	No. pnts used	$\sigma_r C$ %	d_w	d/R	d_r %	d_b/R	+/–
Selected data								
43PIT/SCO	231.4–318.2	11	0.20	1.597	6.72–2	0.32	1.24–2	3
75RAS/GRI	303.0–462.5	9	1.00	0.652	1.56–1	0.65	–5.34–2	–3
75SAN	411.1–540.2	8	1.00	0.723	2.28–1	0.72	8.12–2	4
93GAR/BAN	318.1–373.1	12	0.70#	0.847	1.41–1	0.59	–1.38–1	–12

Rejected data

Reference		Reference	
*86SCH	(3.41–1, 1.44, 3.06–1, 11)	24WIL/DAN	(6.85–1, 3.07, –6.84–1, –5)
30HUF/PAR1	(3.58–2, 0.17, 1.62–2, 0)	47KUR	(7.93–1, 3.24, 7.56–1, 3)
48TSC3	(1.25–1, 0.58, –1.25–1, –1)	48TSC6	(1.58–1, 0.72, 1.58–1, 1)
58SWI/ZIE2	(4.06–1, 1.70, 4.06–1, 1)	77FOR/BEN	(1.24–1, 0.57, –1.24–1, –1)
79FOR/BEN	(1.16–1, 0.53, –1.16–1, –1)	91TAN/ADA	(1.41–1, 0.65, –1.41–1, –1)

TABLE 14.19.3. Parameters of cubic spline polynomials

Heat capacity type	No. data points total	used	s_w	s/R	s_r %	s_b/R	+/–
C_p	71	40	1.130	1.62–1	0.62	–3.38–2	–8
C_{sat}	71	40	1.121	1.58–1	0.61	–3.40–2	–8

Temp. range K	A_1	A_2	A_3	A_4	Level of uncertainty
231.4–400.0	2.09114+1	–4.37338	2.00315	–1.40658–1	III
400.0–540.2	–1.66661+1	2.38098+1	–5.04264	4.46492–1	IV
231.4–400.0	2.10854+1	–4.56521	2.07307	–1.49080–1	III
400.0–540.2	–6.80272	1.63509+1	–3.15595	2.86672–1	IV

TABLE 14.19.4. Recommended values of heat capacities

Temp. (K)	230	240	250	260	270	273.15	280
c_p (J K^{-1}g^{-1})	1.546	1.567	1.590	1.614	1.640	1.648	1.667
C_p (J K^{-1}mol^{-1})	164.1	166.4	168.8	171.4	174.1	175.0	177.0
c_{sat} (J K^{-1}g^{-1})	1.546	1.567	1.590	1.614	1.640	1.648	1.667
C_{sat} (J K^{-1}mol^{-1})	164.1	166.4	168.8	171.4	174.1	175.0	177.0
Temp. (K)	**290**	**298.15**	**300**	**310**	**320**	**330**	**340**
c_p (J K^{-1}g^{-1})	1.695	1.719	1.725	1.755	1.787	1.820	1.854
C_p (J K^{-1}mol^{-1})	180.0	182.5	183.1	186.4	189.7	193.2	196.8
c_{sat} (J K^{-1}g^{-1})	1.695	1.719	1.725	1.755	1.787	1.820	1.854
C_{sat} (J K^{-1}mol^{-1})	180.0	182.5	183.1	186.4	189.7	193.2	196.8
Temp. (K)	**350**	**360**	**370**	**380**	**390**	**400**	**410**
c_p (J K^{-1}g^{-1})	1.888	1.924	1.960	1.997	2.035	2.073	2.111
C_p (J K^{-1}mol^{-1})	200.5	204.3	208.1	212.0	216.0	220.1	224.2
c_{sat} (J K^{-1}g^{-1})	1.888	1.924	1.960	1.996	2.034	2.072	2.110
C_{sat} (J K^{-1}mol^{-1})	200.5	204.2	208.1	212.0	215.9	219.9	224.0
Temp. (K)	**420**	**430**	**440**	**450**	**460**	**470**	**480**
c_p (J K^{-1}g^{-1})	2.151	2.191	2.232	2.275	2.319	2.365	2.413
C_p (J K^{-1}mol^{-1})	228.3	232.6	237.0	241.5	246.2	251.1	256.2
c_{sat} (J K^{-1}g^{-1})	2.149	2.189	2.229	2.270	2.313	2.357	2.402
C_{sat} (J K^{-1}mol^{-1})	228.1	232.3	236.6	241.0	245.6	250.2	255.0
Temp. (K)	**490**	**500**	**510**	**520**	**530**	**540**	
c_p (J K^{-1}g^{-1})	2.464	2.516	2.571	2.629	2.690	2.754	
C_p (J K^{-1}mol^{-1})	261.6	267.1	273.0	279.1	285.6	292.4	
c_{sat} (J K^{-1}g^{-1})	2.449	2.497	2.547	2.600	2.654	2.710	
C_{sat} (J K^{-1}mol^{-1})	260.0	265.1	270.5	276.0	281.7	287.7	

TABLE 14.19.5. Parameters of quasi-polynomial equation

Heat capacity type	No. data points total	used	s_w	s/R	s_r %	s_b/R	+/–
C_p	71	40	1.244	1.63–1	0.58	1.63–3	0

Temp. range K	T_c K	A_1	A_2	A_3	A_4	A_5	A_6	Level of uncertainty
231.4–540.2	617.05	8.32909–1	2.89196–3	1.59082+1	5.29671	1.20864+1	1.03672+1	IV

14–019

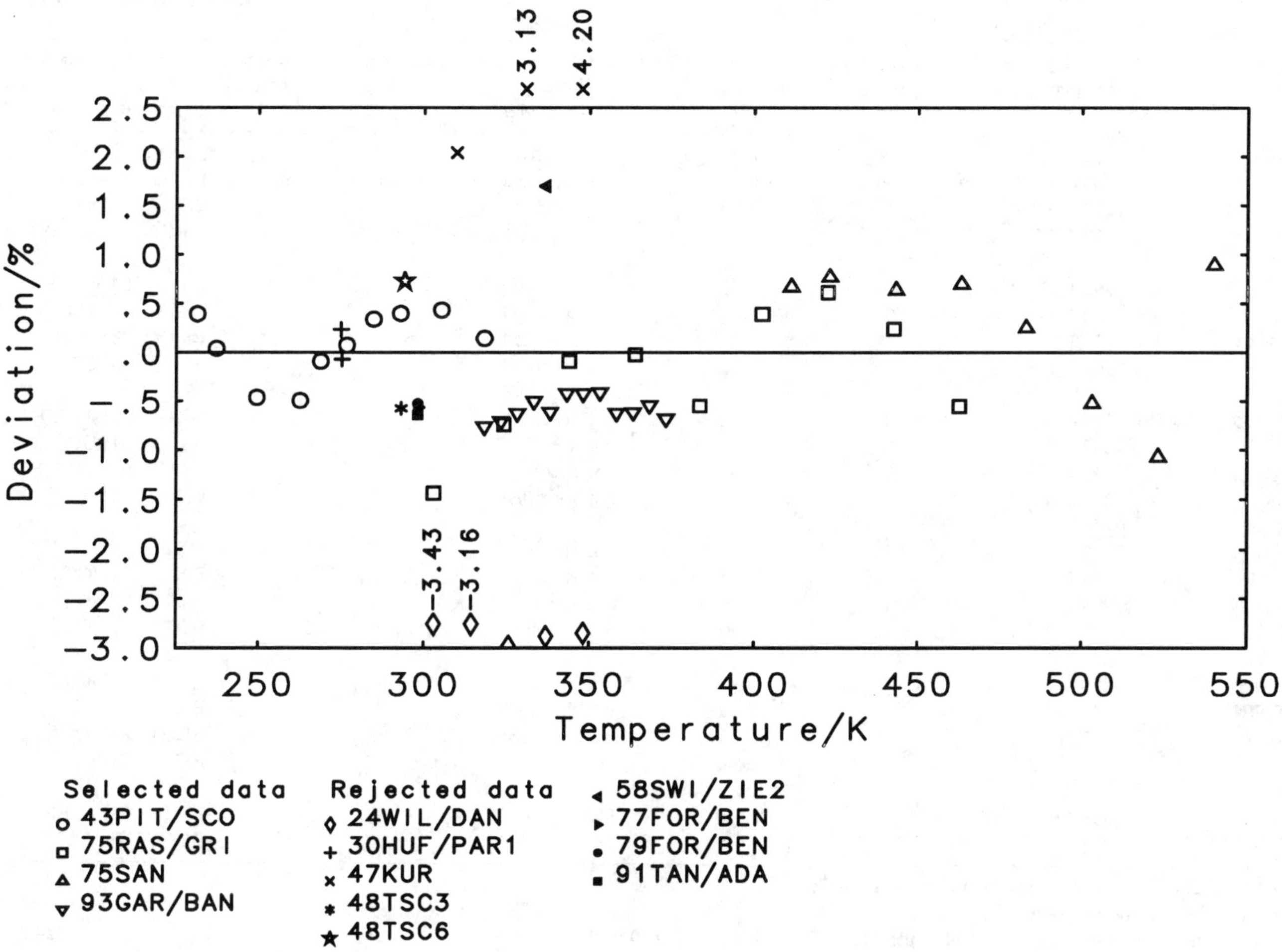
×3.13
×4.20
−3.43
−3.16
Deviation/%
Temperature/K
2.5
2.0
1.5
1.0
.5
.0
−.5
−1.0
−1.5
−2.0
−2.5
−3.0
250
300
350
400
450
500
550
Selected data
43PIT/SCO
75RAS/GRI
75SAN
93GAR/BAN
Rejected data
24WIL/DAN
30HUF/PAR1
47KUR
48TSC3
48TSC6
58SWI/ZIE2
77FOR/BEN
79FOR/BEN
91TAN/ADA

Name: 1,4–Dimethylbenzene
Formula: C_8H_{10}

CAS–RN: 106–42–3
Group No.: 14–020
Molar Mass: 106.17

TABLE 14.20.1. Experimental heat capacities

Reference		Temp. range K	No. pnts	Error %	Purity %	Purity method	Type capacity	Calorimeter Type	Calorimeter Reference
*86SCH		312.2–335.7	6S	nosp	not specified		C_{avg}	DSIO	*86SCH
24WIL/DAN		303.0–348.0	eqn	nosp	not specified		C_p	BSAO	24WIL/DAN
30HUF/PAR1		290.7–299.4	5	1.00	not specified		C_p	BSIO	25PAR
43PIT/SCO		292.0–354.6	10	0.20	99.98	melpt	C_p	BSIO	28LAT/GRE
47COR/GIN		293.1–573.2	15S	0.20	99.9	melpt	C_{sat}	DSTO	50GIN/DOU
47KUR		309.3–349.1	3	nosp	not specified		C_{avg}	DSIO	47KUR
48TSC3		293.1	1	nosp	not specified		C_p	BSIO	48TSC1
48TSC6		294.1	1	nosp	not specified		C_p	BSIO	48TSC1
58SWI/ZIE2	N	336.6–347.0	2	nosp	not specified		C_{avg}	DSIO	58SWI/ZIE1
61ROU		298.0–315.9	20	nosp	not specified		C_p	BSAO	61ROU
71KHA/SUB	N	298.1–313.1	2	nosp	not specified		C_p	BSIO	64MOE/THO
75RAS/GRI	N	303.6–462.9	9	1.00	not specified		C_p	BDAO	75RAS/GRI
77FOR/BEN		298.1	1	nosp	not specified		C_p	FSIT	71PIC/LED
77WIL/GRO		298.1	1	nosp	not specified		C_p	FSIT	71PIC/LED
79FOR/BEN		298.1	1	0.30	not specified		C_p	FSIT	71PIC/LED
79OTT/GOA		288.1–328.1	9S	0.30	99.9	melpt	C_p	FSIT	71PIC/LED
85COS/PAT7		298.2	1	nosp	99.	anal	C_p	FSIT	71PIC/LED
88MES/FIN		295.4–373.4	11	0.10	99.996	melpt	C_{sat}	BSAO	47HUF
89PRA/RAJ		318.1–333.1	4	3.00	not specified		C_p	BDHT	89PRA/RAJ
91TAN/ADA		298.1	1	nosp	not specified		C_p	FSIT	71PIC/LED
91WIL/JIM		298.1	1	nosp	99.	anal	C_p	FSIT	71PIC/LED
93GAR/BAN		318.1–373.1	12	nosp	99.7	chrom	C_p	BDCT	91BAN/GAR

58SWI/ZIE2 average values in temperature ranges 294–379 K and 295–399 K
71KHA/SUB reproducibility given as 0.3 %
75RAS/GRI data above 386 K measured at superambient pressures up to 0.49 MPa

TABLE 14.20.2. Correlated heat capacities

Reference	Temp. range K	No. pnts used	$\sigma_r C$ %	d_w	d/R	d_r %	d_b/R	+/–
				Selected data				
47COR/GIN	293.1–573.2	15	0.20	1.405	7.36–2	0.28	–3.41–2	–5
75RAS/GRI	303.6–462.9	9	1.00	0.588	1.60–1	0.59	7.07–2	3
77FOR/BEN	298.1	1	0.50#	0.174	1.90–2	0.09	–1.90–2	–1
79FOR/BEN	298.1	1	0.30	0.027	1.77–3	0.01	–1.77–3	0
79OTT/GOA	288.1–328.1	9	0.30	0.852	5.71–2	0.26	–6.37–3	–3
88MES/FIN	295.4–373.4	11	0.20#	1.307	6.31–2	0.26	5.69–2	11
91TAN/ADA	298.1	1	0.50#	0.277	3.03–2	0.14	–3.03–2	–1
91WIL/JIM	298.1	1	0.50#	0.448	4.91–2	0.22	4.91–2	1
93GAR/BAN	318.1–373.1	12	0.70#	0.642	1.06–1	0.45	–1.02–1	–12
				Rejected data				

*86SCH	(2.96–1, 1.26, 2.82–1, 6)	24WIL/DAN	(7.04–1, 3.19, –6.96–1, –5)
30HUF/PAR1	(2.06–1, 0.95, –1.99–1, –5)	43PIT/SCO	(4.51–1, 1.91, 1.97–1, 6)
47KUR	(9.00–1, 3.68, 8.57–1, 3)	48TSC3	(3.12–1, 1.46, –3.12–1, –1)
48TSC6	(2.99–1, 1.39, –2.99–1, –1)	58SWI/ZIE2	(3.52–1, 1.46, 3.52–1, 2)
61ROU	(8.31–1, 3.60, 8.27–1, 20)	71KHA/SUB	(7.98–2, 0.36, –6.95–2, –2)
77WIL/GRO	(4.83–2, 0.22, –4.83–2, –1)	85COS/PAT7	(2.02–1, 0.92, 2.02–1, 1)
89PRA/RAJ	(3.48–1, 1.47, 3.24–1, 4)		

TABLE 14.20.3. Parameters of cubic spline polynomials

Heat capacity type	No. data points total	used	s_w	s/R	s_r %	s_b/R	+/–
C_p	121	60	1.073	9.76–2	0.38	–8.94–3	–7
C_{sat}	121	60	1.020	9.40–2	0.38	–8.99–3	–7

Temp. range K	A_1	A_2	A_3	A_4	Level of uncertainty
288.1–420.0	1.78687+1	–2.57703	1.72250	–1.36338–1	II
420.0–573.2	–7.08471+1	6.07914+1	–1.33652+1	1.06110	IV
288.1–420.0	1.26505+1	1.77510	5.31803–1	–2.97523–2	II
420.0–573.2	–3.13461+1	3.32013+1	–6.95062	5.64090–1	IV

TABLE 14.20.4. Recommended values of heat capacities

Temp. (K)	290	298.15	300	310	320	330	340
c_p (J K^{-1}g^{-1})	1.688	1.714	1.720	1.752	1.785	1.819	1.853
C_p (J K^{-1}mol^{-1})	179.2	182.0	182.6	186.0	189.5	193.1	196.7
c_{sat} (J K^{-1}g^{-1})	1.687	1.714	1.720	1.753	1.786	1.819	1.853
C_{sat} (J K^{-1}mol^{-1})	179.1	181.9	182.6	186.1	189.6	193.2	196.8

Temp. (K)	350	360	370	380	390	400	410
c_p (J K^{-1}g^{-1})	1.888	1.923	1.959	1.994	2.031	2.067	2.104
C_p (J K^{-1}mol^{-1})	200.4	204.2	207.9	211.8	215.6	219.5	223.3
c_{sat} (J K^{-1}g^{-1})	1.888	1.922	1.957	1.993	2.028	2.064	2.100
C_{sat} (J K^{-1}mol^{-1})	200.4	204.1	207.8	211.5	215.3	219.1	223.0

Temp. (K)	420	430	440	450	460	470	480
c_p (J K^{-1}g^{-1})	2.140	2.177	2.214	2.252	2.292	2.334	2.378
C_p (J K^{-1}mol^{-1})	227.2	231.1	235.1	239.1	243.4	247.8	252.5
c_{sat} (J K^{-1}g^{-1})	2.137	2.173	2.211	2.249	2.288	2.328	2.370
C_{sat} (J K^{-1}mol^{-1})	226.8	230.7	234.7	238.7	242.9	247.2	251.6

Temp. (K)	490	500	510	520	530	540	550
c_p (J K^{-1}g^{-1})	2.425	2.476	2.531	2.590	2.654	2.724	2.800
C_p (J K^{-1}mol^{-1})	257.5	262.9	268.7	275.0	281.8	289.2	297.2
c_{sat} (J K^{-1}g^{-1})	2.414	2.460	2.508	2.559	2.612	2.669	2.730
C_{sat} (J K^{-1}mol^{-1})	256.3	261.1	266.2	271.6	277.4	283.4	289.8

Temp. (K)	560	570
c_p (J K^{-1}g^{-1})	2.882	2.971
C_p (J K^{-1}mol^{-1})	306.0	315.4
c_{sat} (J K^{-1}g^{-1})	2.794	2.862
C_{sat} (J K^{-1}mol^{-1})	296.6	303.8

TABLE 14.20.5. Parameters of quasi-polynomial equation

Heat capacity type	No. data points total	used	s_w	s/R	s_r %	s_b/R	+/–
C_p	121	60	1.165	8.81–2	0.35	–1.65–3	0

Temp. range K	T_c K	A_1	A_2	A_3	A_4	Level of uncertainty
288.1–573.2	616.20	–3.18089	1.32290–1	1.02433+1	1.91210+1	IV

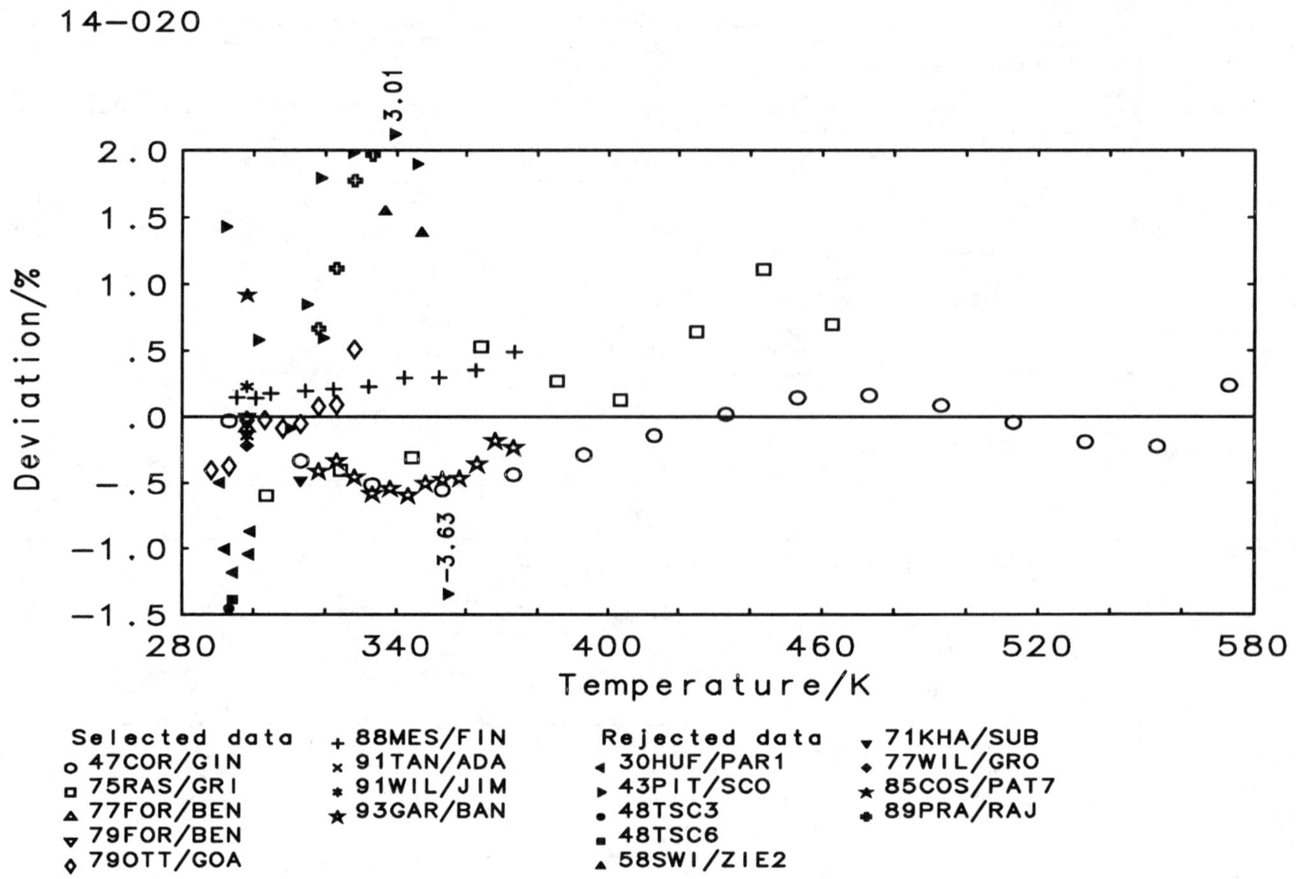

Name: Ethylbenzene
Formula: C_8H_{10}

CAS–RN: 100–41–4
Group No.: 14–021
Molar Mass: 106.17

TABLE 14.21.1. Experimental heat capacities

Reference		Temp. range K	No. pnts	Error %	Purity %	Purity method	Type capacity	Calorimeter Type	Calorimeter Reference
*81VON		304.8–362.1	8S	nosp	not specified		C_{avg}	DSIO	*81VON
*86SCH		309.0–335.7	11S	nosp	not specified		C_{avg}	DSIO	*86SCH
24WIL/DAN		303.0–343.0	eqn	nosp	not specified		C_p	BSAO	24WIL/DAN
30HUF/PAR1		185.0–304.9	16	1.00	not specified		C_p	BSIO	25PAR
31BLA/LEI		286.1–368.1	25	3.00	not specified		C_{sat}	BSIO	31BLA/LEI
31SMI/AND1		184.4–298.5	9	nosp	99.9	estim	C_p	DSIO	26AND/LYN
34KOL/UDO2	N	302.7	1	nosp	not specified		C_p	BSIT	34KOL/UDO2
44GUT/SPI		181.5–305.4	16	0.20	99.93	melpt	C_p	BSAO	43RUE/HUF
45SCO/BRI		178.2–300.0	27S	0.10	99.993	melpt	C_{sat}	BSAO	33SOU/BRI
47KUR		308.8–372.8	4	nosp	not specified		C_{avg}	DSIO	47KUR
48TSC1		295.1	1	nosp	not specified		C_p	BSIO	48TSC1
48TSC5		295.1	1	nosp	not specified		C_p	BSIO	48TSC1
76FOR/BEN2		298.1	1	0.30	not specified		C_p	FSIT	71PIC/LED
77FOR/BEN		298.1	1	nosp	not specified		C_p	FSIT	71PIC/LED
79AND/GRI	N	293.3–393.0	10	0.50	not specified		C_{sat}	BSAO	67RAS/GAN
79FOR/BEN		298.1	1	nosp	not specified		C_p	FSIT	71PIC/LED
83GRO/FAR1		298.1	1	nosp	not specified		C_p	FSIT	71PIC/LED
91TAN/ADA		298.1	1	nosp	not specified		C_p	FSIT	71PIC/LED
91WIL/JIM		298.1	1	nosp	99.	anal	C_p	FSIT	71PIC/LED
93GAR/BAN		318.1–373.1	12	nosp	99.5	chrom	C_p	BDCT	91BAN/GAR

34KOL/UDO2 same datum in 34KOL/UDO1
79AND/GRI error 0.5 % below 373 K and 0.8 % at 433 K

TABLE 14.21.2. Correlated heat capacities

Reference	Temp. range K	No. pnts used	$\sigma_r C$ %	d_w	d/R	d_r %	d_b/R	+/–
				Selected data				
44GUT/SPI	181.5–305.4	16	0.20#	0.322	1.36–2	0.06	–9.97–3	–11
45SCO/BRI	178.1–300.0	27	0.10	0.319	6.89–3	0.03	2.40–3	9
76FOR/BEN2	298.1	1	0.30	0.246	1.65–2	0.07	–1.65–2	–1
77FOR/BEN	298.1	1	0.50#	0.386	4.31–2	0.19	–4.31–2	–1
79AND/GRI	293.3–393.0	10	0.50	0.525	6.53–2	0.26	5.44–3	–2
83GRO/FAR1	298.1	1	0.50#	0.169	1.89–2	0.08	–1.89–2	–1
91WIL/JIM	298.1	1	0.50#	0.137	1.53–2	0.07	–1.53–2	–1
93GAR/BAN	318.1–373.1	12	0.70#	0.801	1.32–1	0.56	–8.76–2	–7

Rejected data

*81VON	(3.15–1, 1.30, –3.03–3, 2)	*86SCH	(3.40–1, 1.42, 3.22–1, 11)
24WIL/DAN	(6.17–1, 2.73, –6.12–1, –5)	30HUF/PAR1	(3.01–1, 1.40, –2.58–1, –16)
31BLA/LEI	(9.06–1, 3.56, 4.56–1, 15)	31SMI/AND1	(4.53–1, 2.35, –4.27–1, –9)
34KOL/UDO2	(1.04, 4.84, –1.04, –1)	47KUR	(8.43–1, 3.23, 7.21–1, 4)
48TSC1	(3.75–2, 0.17, 3.75–2, 1)	48TSC5	(1.44–1, 0.64, 1.44–1, 1)
79FOR/BEN	(4.15–2, 0.19, –4.15–2, –1)	91TAN/ADA	(3.82–2, 0.17, –3.82–2, –1)

TABLE 14.21.3. Parameters of regression polynomial

Heat capacity type	No. data points total	used	s_w	s/R	s_r %	s_b/R	+/–
C	152	69	0.482 6.	32–2	0.27	–1.72–2	–15

Temp. range K	A_1	A_2	A_3	A_4	Level of uncertainty
178.1–393.0	2.48043+1	–9.01410	3.87757	–3.78713–1	IV

TABLE 14.21.4. Recommended values of heat capacities

Temp. (K)	180	190	200	210	220	230	240
c (J K^{-1}g^{-1})	1.483	1.494	1.508	1.525	1.543	1.564	1.587
C (J K^{-1}mol^{-1})	157.4	158.6	160.1	161.9	163.9	166.1	168.5
Temp. (K)	250	260	270	273.15	280	290	298.15
c (J K^{-1}g^{-1})	1.612	1.639	1.667	1.676	1.696	1.726	1.751
C (J K^{-1}mol^{-1})	171.2	174.0	176.9	177.9	180.0	183.2	185.9
Temp. (K)	300	310	320	330	340	350	360
c (J K^{-1}g^{-1})	1.757	1.789	1.821	1.854	1.887	1.920	1.953
C (J K^{-1}mol^{-1})	186.5	189.9	193.4	196.8	200.3	203.9	207.3
Temp. (K)	370	380	390				
c (J K^{-1}g^{-1})	1.986	2.018	2.049				
C (J K^{-1}mol^{-1})	210.8	214.2	217.5				

TABLE 14.21.5. Parameters of quasi-polynomial equation

Heat capacity type	No. data points total	used	s_w	s/R	s_r %	s_b/R	+/–
C	152	69	0.718	7.88–2	0.33	–1.20–2	–7

Temp. range K	T_c K	A_1	A_2	A_3	A_4	A_5	A_6	Level of uncertainty
178.1–393.0	617.24	3.16141+2	3.04404+1	–1.23068+1	2.89465+2	8.96899+1	2.09550+2	IV

11–021

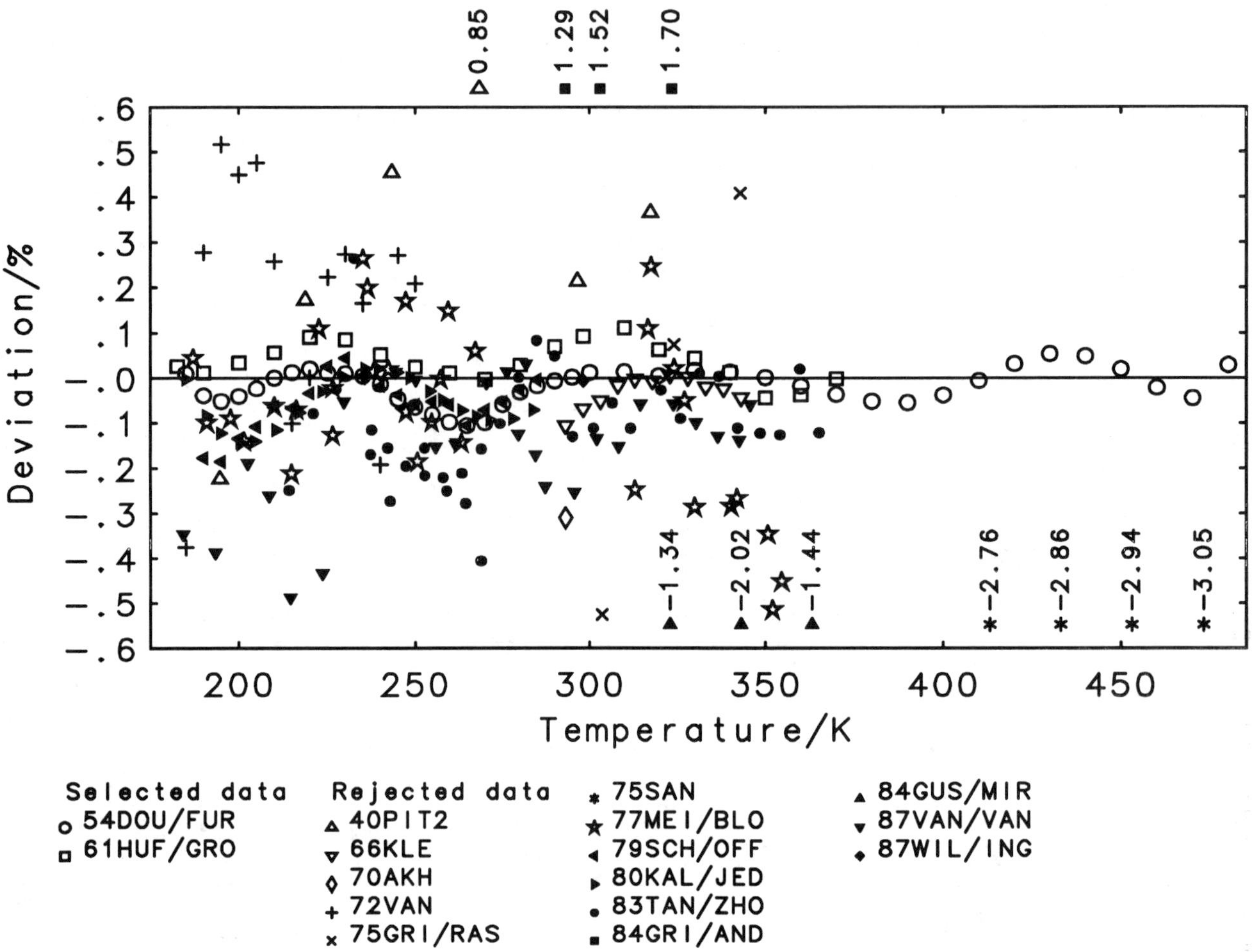

Name: Bicyclo[2.2.2]oct–2–ene
Formula: C_8H_{12}

CAS–RN: 931–64–6
Group No.: 14–022
Molar Mass: 108.18

TABLE 14.22.1. Experimental heat capacities

Reference	Temp. range K	No. pnts	Error %	Purity %	Purity method	Type capacity	Calorimeter Type	Calorimeter Reference
70WON/WES	392.5–408.3	7	nosp	99.82	melpt	C_{sat}	BSAO	68WES/FUR

TABLE 14.22.3. Parameters of regression polynomial

Heat capacity type	No. data points total	used	s_w	s/R	s_r %	s_b/R	+/–
C_{sat}	7	7	1.021	8.31–2	0.31	3.67–4	1

Temp. range K	A_1	A_2	Level of uncertainty
392.5–408.3	8.10580–1	6.60379	III

TABLE 14.22.4. Recommended values of heat capacities

Temp. (K)	390	400	410
c_{sat} (J K^{-1}g^{-1})	2.042	2.092	2.143
C_{sat} (J K^{-1}mol^{-1})	220.9	226.4	231.9

Name: 1,5–Cyclooctadiene (unspecified stereoisomer)
Formula: C_8H_{12}

CAS–RN: 111–78–4
Group No.: 14–023
Molar Mass: 108.18

TABLE 14.23.1. Experimental heat capacities

Reference		Temp. range K	No. pnts	Error %	Purity %	method	Type capacity	Calorimeter Type	Reference
75LEB/TSV	N	207.5–320.0	27	0.20	99.53	melpt	C_p	BSAO	66NIK/LEB

75LEB/TSV smoothed values in 75LEB/LEB

TABLE 14.23.3. Parameters of regression polynomial

Heat capacity type	No. data points total	used	s_w	s/R	s_r %	s_b/R	+/–
C_p	27	27	0.208	9.08–3	0.04	6.15–6	0

Temp. range K	A_1	A_2	A_3	A_4	Level of uncertainty
207.4–320.0	2.46266+1	–9.17333	4.28360	–4.30880–1	III

TABLE 14.23.4. Recommended values of heat capacities

Temp. (K)	210	220	230	240	250	260	270
c_p (J K^{-1}g^{-1})	1.557	1.582	1.610	1.639	1.670	1.703	1.737
C_p (J K^{-1}mol^{-1})	168.5	171.2	174.2	177.3	180.7	184.2	188.0
Temp. (K)	273.15	280	290	298.15	300	310	320
c_p (J K^{-1}g^{-1})	1.748	1.773	1.809	1.840	1.846	1.884	1.923
C_p (J K^{-1}mol^{-1})	189.1	191.8	195.7	199.0	199.8	203.9	208.0

TABLE 14.23.5. Parameters of quasi-polynomial equation

Heat capacity type	No. data points total	used	s_w	s/R	s_r %	s_b/R	+/−
C_p	27	27	0.500	2.22–2	0.10	3.72–5	0

Temp. range K	T_c K	A_1	A_2	A_3	A_4	Level of uncertainty
207.4–320.0	647.00	−5.54551	6.01692	8.74044	1.27776	III

Name: Cyclooctene (unspecified stereoisomer)
Formula: C_8H_{14}

CAS–RN: 931–88–4
Group No.: 14–024
Molar Mass: 110.20

TABLE 14.24.1. Experimental heat capacities

Reference		Temp. range K	No. pnts	Error %	Purity %	method	Type capacity	Calorimeter Type	Reference
78LEB/LEB	N	260.9–328.2	31	0.20	99.978	melpt	C_p	BSAO	76LEB/LIT

78LEB/LEB probably *cis*–isomer (according to n.m.t. and diferent stability of both stereoisomers)

TABLE 14.24.3. Parameters of regression polynomial

Heat capacity type	No. data points total	used	s_w	s/R	s_r %	s_b/R	+/−
C_p	31	30	1.442	7.22–2	0.29	3.84–4	4

Temp. range K	A_1	A_2	Level of uncertainty
262.2–328.2	8.50675	5.52350	III

TABLE 14.24.4. Recommended values of heat capacities

Temp. (K)	260	270	273.15	280	290	298.15	300
c_p (J K^{-1}g^{-1})	1.725	1.767	1.780	1.809	1.850	1.884	1.892
C_p (J K^{-1}mol^{-1})	190.1	194.7	196.2	199.3	203.9	207.7	208.5
Temp. (K)	310	320	330				
c_p (J K^{-1}g^{-1})	1.934	1.975	2.017				
C_p (J K^{-1}mol^{-1})	213.1	217.7	222.3				

TABLE 14.24.5. Parameters of quasi-polynomial equation

Heat capacity type	No. data points total	used	s_w	s/R	s_r %	s_b/R	+/−
C_p	31	30	1.467	7.36–2	0.29	3.86–4	4

Temp. range K	T_c K	A_1	A_2	A_3	A_4	Level of uncertainty
262.2–328.2	617.00	1.24672	1.06690–2	8.18025	3.64211+1	III

Name: 1*H*–Indene
Formula: C_9H_8

CAS-RN: 95–13–6
Group No.: 14–025
Molar Mass: 116.16

TABLE 14.25.1. Experimental heat capacities

Reference		Temp. range K	No. pnts	Error %	Purity %	Purity method	Type capacity	Calorimeter Type	Calorimeter Reference
61STU/SIN	N	280.0–300.0	3S	0.25	99.7	melpt	C_p	BSAO	58HIL/KRA
65ZIE	N	315.7–375.1	3	nosp	not specified		C_{avg}	DSIO	58SWI/ZIE1

61STU/SIN same data in 59STU/SIN
65ZIE average values in temperature ranges 296–454 K, 294–382 K and 294–338 K

TABLE 14.25.2. Correlated heat capacities

Reference	Temp. range K	No. pnts used	$\sigma_r C$ %	d_w	d/R	d_r %	d_b/R	+/–
				Selected data				
61STU/SIN	280.0–300.0	3	0.25	0.141	7.71–3	0.04	1.77–3	0
65ZIE	315.7–375.1	3	1.50#	0.877	3.24–1	1.32	–5.34–2	–1

TABLE 14.25.3. Parameters of regression polynomial

Heat capacity type	No. data points total	No. data points used	s_w	s/R	s_r %	s_b/R	+/–
C_p	6	6	0.770	2.80–1	1.14	–2.58–2	–1

Temp. range K	A_1	A_2	Level of uncertainty
280.0–375.1	1.03749+1	4.06333	V

TABLE 14.25.4. Recommended values of heat capacities

Temp. (K)	280	290	298.15	300	310	320	330
c_p ($J K^{-1} g^{-1}$)	1.56	1.59	1.61	1.62	1.64	1.67	1.70
C_p ($J K^{-1} mol^{-1}$)	181	184	187	188	191	194	198
Temp. (K)	340	350	360	370			
c_p ($J K^{-1} g^{-1}$)	1.73	1.76	1.79	1.82			
C_p ($J K^{-1} mol^{-1}$)	201	205	208	211			

Name: 2,3–Dihydroindene
Formula: C_9H_{10}

CAS-RN: 496–11–7
Group No.: 14–026
Molar Mass: 118.18

TABLE 14.26.1. Experimental heat capacities

Reference		Temp. range K	No. pnts	Error %	Purity %	Purity method	Type capacity	Calorimeter Type	Calorimeter Reference
61STU/SIN	N	240.0–320.0	6S	0.25	99.89	melpt	C_p	BSAO	58HIL/KRA

61STU/SIN same data in 59STU/SIN

TABLE 14.26.3. Parameters of regression polynomial

Heat capacity type	No. data points total	used	s_w	s/R	s_r %	s_b/R	+/–
C_p	6	6	0.533	3.02–2	0.13	4.58–5	2

Temp. range K	A_1	A_2	A_3	Level of uncertainty
240.0–320.0	2.13733+1	–3.96813	1.49978	II

TABLE 14.26.4. Recommended values of heat capacities

Temp. (K)	240	250	260	270	273.15	280	290
c_p (J K^{-1}g^{-1})	1.441	1.465	1.491	1.519	1.528	1.549	1.582
C_p (J K^{-1}mol^{-1})	170.4	173.2	176.2	179.5	180.6	183.1	186.9
Temp. (K)	298.15	300	310	320			
c_p (J K^{-1}g^{-1})	1.609	1.616	1.652	1.691			
C_p (J K^{-1}mol^{-1})	190.2	191.0	195.3	199.8			

TABLE 14.26.5. Parameters of quasi–polynomial equation

Heat capacity type	No. data points total	used	s_w	s/R	s_r %	s_b/R	+/–
C_p	6	6	0.450	2.56–2	0.11	2.80–5	2

Temp. range K	T_c K	A_1	A_2	A_3	A_4	Level of uncertainty
240.0–320.0	684.90	1.83258+1	1.98319+1	–3.62235	4.23351	II

Name: (1–Methylethenyl)benzene
Formula: C_9H_{10}

CAS–RN: 98–83–9
Group No.: 14–027
Molar Mass: 118.18

TABLE 14.27.1. Experimental heat capacities

Reference		Temp. range K	No. pnts	Error %	Purity %	Purity method	Type capacity	Calorimeter Type	Calorimeter Reference
67LEB/RAB1	N	250.8–300.0	6S	0.50	99.5	melpt	C_p	BSAO	66NIK/LEB

67LEB/RAB1 same data in 71LEB/RAB2

TABLE 14.27.3. Parameters of regression polynomial

Heat capacity type	No. data points total	used	s_w	s/R	s_r %	s_b/R	+/–
C_p	6	6	0.626	7.45–2	0.31	3.44–4	0

Temp. range K	A_1	A_2	Level of uncertainty
250.8–300.0	1.23867+1	3.94111	IV

TABLE 14.27.4. Recommended values of heat capacities

Temp. (K)	260	270	273.15	280	290	298.15	300
c_p (J $K^{-1}g^{-1}$)	1.592	1.620	1.629	1.648	1.676	1.698	1.703
C_p (J $K^{-1}mol^{-1}$)	188.2	191.5	192.5	194.7	198.0	200.7	201.3

Name: (1–Methylethyl)benzene
Formula: C_9H_{12}

CAS–RN: 98–82–8
Group No.: 14–028
Molar Mass: 120.19

TABLE 14.28.1. Experimental heat capacities

Reference		Temp. range K	No. pnts	Error %	Purity %	Purity method	Type capacity	Calorimeter Type	Calorimeter Reference
34KOL/UDO2	N	302.0	1	nosp	not specified		C_p	BSIT	34KOL/UDO2
47KUR		308.8–359.6	4	nosp	not specified		C_{avg}	DSIO	47KUR
52SCH/SAG		299.8–366.5	13S	1.00	99.8	estim	C_p	BSAO	39SAG/EVA
73KIS/SUG		179.9–313.1	45	nosp	99.93	melpt	C_{sat}	BSAO	65SUG/SEK
79AND/GRI	N	296.0–412.8	7	0.50	not specified		C_{sat}	BSAO	67RAS/GAN

34KOL/UDO2 same datum in 34KOL/UDO1
79AND/GRI error 0.5 % below 373 K and 0.8 % at 433 K

TABLE 14.28.2. Correlated heat capacities

Reference	Temp. range K	No. pnts used	$\sigma_r C$ %	d_w	d/R	d_r %	d_b/R	+/–
				Selected data				
73KIS/SUG	179.9–313.1	45	0.20#	0.362	1.78–2	0.07	1.65–3	0
79AND/GRI	296.0–412.8	7	0.50	0.728	1.03–1	0.36	–6.45–2	–3
				Rejected data				
34KOL/UDO2	(2.14, 8.96, –2.14, –1)	47KUR		(1.51–1, 0.55, –7.02–2, –2)				
52SCH/SAG	(4.93–1, 1.91, –3.84–1, –9)							

TABLE 14.28.3. Parameters of cubic spline polynomials

Heat capacity type	No. data points total	No. data points used	s_w	s/R	s_r %	s_b/R	+/–
C_{sat}	70	52	0.452	4.35–2	0.16	–7.25–3	–3

Temp. range K	A_1	A_2	A_3	A_4	Level of uncertainty
179.9–300.0	2.77820+1	–1.01809+1	4.62002	–4.75860–1	II
300.0–412.8	1.83532+1	–7.52100–1	1.47708	–1.26645–1	IV

TABLE 14.28.4. Recommended values of heat capacities

Temp. (K)	180	190	200	210	220	230	240
c_{sat} (J K^{-1}g^{-1})	1.498	1.512	1.528	1.547	1.569	1.592	1.617
C_{sat} (J K^{-1}mol^{-1})	180.0	181.7	183.7	186.0	188.6	191.4	194.4
Temp. (K)	250	260	270	273.15	280	290	298.15
c_{sat} (J K^{-1}g^{-1})	1.644	1.673	1.702	1.712	1.733	1.764	1.791
C_{sat} (J K^{-1}mol^{-1})	197.6	201.0	204.6	205.7	208.3	212.1	215.2
Temp. (K)	300	310	320	330	340	350	360
c_{sat} (J K^{-1}g^{-1})	1.797	1.829	1.862	1.896	1.930	1.964	1.998
C_{sat} (J K^{-1}mol^{-1})	215.9	219.9	223.8	227.9	231.9	236.0	240.1
Temp. (K)	370	380	390	400	410		
c_{sat} (J K^{-1}g^{-1})	2.032	2.067	2.101	2.136	2.170		
C_{sat} (J K^{-1}mol^{-1})	244.3	248.4	252.5	256.7	260.8		

TABLE 14.28.5. Parameters of quasi-polynomial equation

Heat capacity type	No. data points total	used	s_w	s/R		s_r %	s_b/R	+/–
C_{sat}	70	52	0.578	7.33–2		0.25	–9.51–3	1
Temp. range K	T_c K	A_1	A_2	A_3	A_4	A_5	A_6	Level of uncertainty
179.9–412.8	631.10	2.53596+2	2.31237+1	–2.73796	2.30950+2	9.16085+1	1.65041+2	III

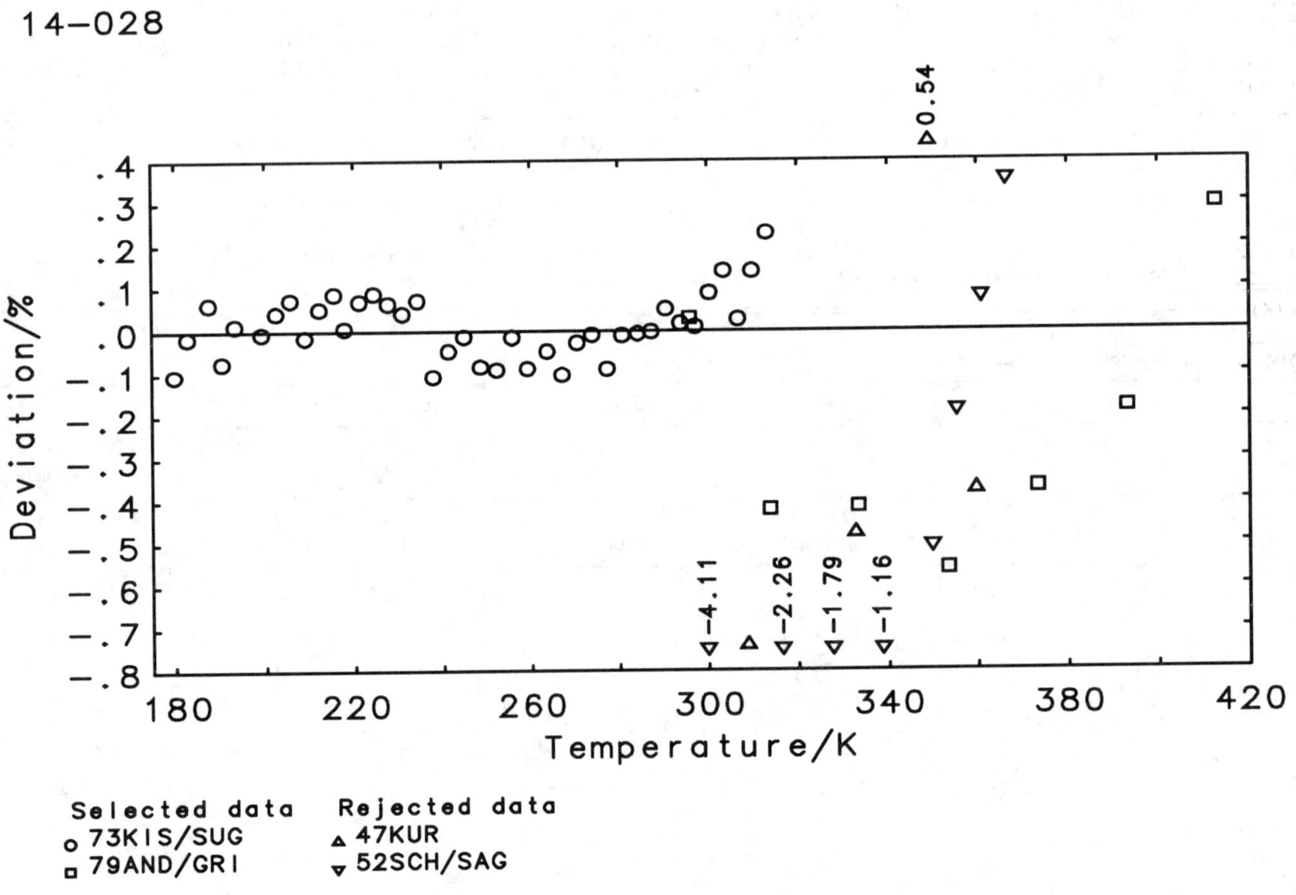

Name: Propylbenzene
Formula: C_9H_{12}

CAS–RN: 103–65–1
Group No.: 14–029
Molar Mass: 120.19

TABLE 14.29.1. Experimental heat capacities

Reference	Temp. range K	No. pnts	Error %	Purity %	Purity method	Type capacity	Calorimeter Type	Calorimeter Reference
*86SCH	308.6–346.9	9S	nosp	not specified		C_{avg}	DSIO	*86SCH
48TSC1	295.1	1	nosp	not specified		C_p	BSIO	48TSC1
48TSC5	294.1	1	nosp	not specified		C_p	BSIO	48TSC1
65MES/TOD2	180.9–370.5	23	0.20	99.97	melpt	C_{sat}	BSAO	47HUF
83GRO/FAR2	298.1	1	nosp	99.0	estim	C_p	FSIT	71PIC/LED

TABLE 14.29.2. Correlated heat capacities

Reference	Temp. range K	No. pnts used	$\sigma_r C$ %	d_w	d/R	d_r %	d_b/R	+/–
				Selected data				
65MES/TOD2	180.9–370.5	23	0.20#	0.116	6.07–3	0.02	1.82–6	4
				Rejected data				
*86SCH	(4.51–1, 1.58, 3.95–1, 9)			48TSC1	(3.45–2, 0.13, –3.45–2, –1)			
48TSC5	(4.90–2, 0.19, –4.90–2, –1)			83GRO/FAR2	(4.40–2, 0.17, –4.40–2, –1)			

TABLE 14.29.3. Parameters of regression polynomial

Heat capacity type	No. data points total	No. data points used	s_w	s/R	s_r %	s_b/R	+/–
C_{sat}	35	23	0.127	6.68–3	0.03	1.82–6	4

Temp. range K	A_1	A_2	A_3	A_4	Level of uncertainty
180.9–370.5	2.93134+1	–1.08270+1	4.52644	–4.32074–1	II

TABLE 14.29.4. Recommended values of heat capacities

Temp. (K)	180	190	200	210	220	230	240
c_{sat} (J K^{-1}g^{-1})	1.520	1.530	1.543	1.559	1.577	1.598	1.621
C_{sat} (J K^{-1}mol^{-1})	182.7	183.9	185.5	187.4	189.6	192.1	194.8
Temp. (K)	250	260	270	273.15	280	290	298.15
c_{sat} (J K^{-1}g^{-1})	1.645	1.672	1.700	1.709	1.729	1.760	1.786
C_{sat} (J K^{-1}mol^{-1})	197.8	200.9	204.3	205.4	207.9	211.6	214.7
Temp. (K)	300	310	320	330	340	350	360
c_{sat} (J K^{-1}g^{-1})	1.792	1.825	1.858	1.892	1.926	1.961	1.995
C_{sat} (J K^{-1}mol^{-1})	215.4	219.3	223.3	227.4	231.5	235.7	239.8
Temp. (K)	370						
c_{sat} (J K^{-1}g^{-1})	2.029						
C_{sat} (J K^{-1}mol^{-1})	243.9						

TABLE 14.29.5. Parameters of quasi-polynomial equation

Heat capacity type	No. data points total	used	s_w	s/R	s_r %	s_b/R	+/–
C_{sat}	35	23	0.343	1.72–2	0.07	2.86–5	–1

Temp. range K	T_c K	A_1	A_2	A_3	A_4	A_5	A_6	Level of uncertainty
180.9–370.5	638.30	4.95451+2	4.98904+1	–2.88255+1	4.52545+2	1.35495+2	3.13175+2	II

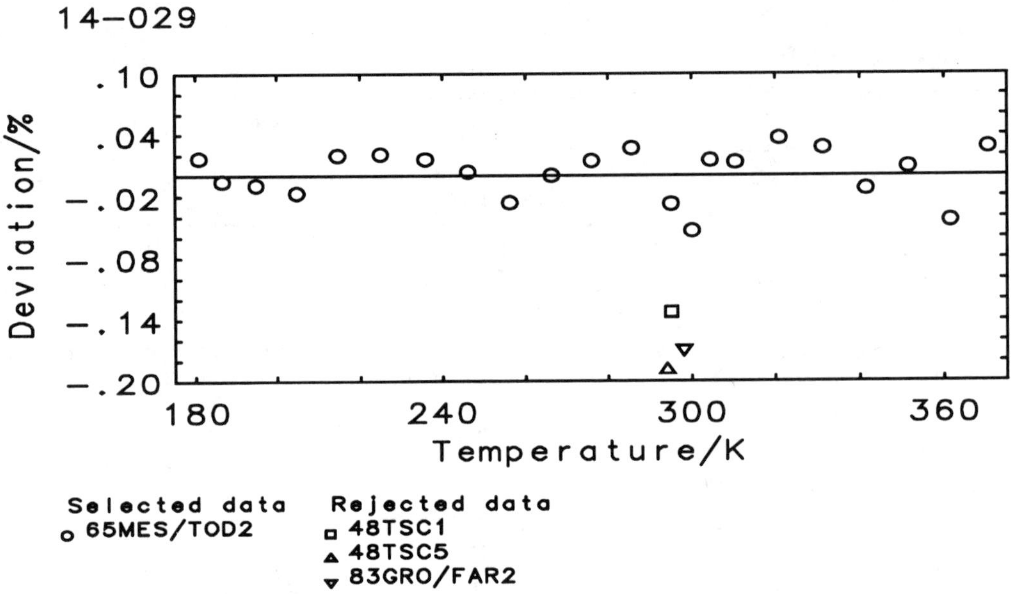

Name: 1,2,3–Trimethylbenzene
Formula: C_9H_{12}

CAS–RN: 526–73–8
Group No.: 14–030
Molar Mass: 120.19

TABLE 14.30.1. Experimental heat capacities

Reference	Temp. range K	No. pnts	Error %	Purity %	Purity method	Type capacity	Calorimeter Type	Calorimeter Reference
55TAY/JOH	250.0–301.0	27	0.30	99.99	melpt	C_p	BSIO	55TAY/JOH

TABLE 14.30.3. Parameters of regression polynomial

Heat capacity type	No. data points total	used	s_w	s/R	s_r %	s_b/R	+/–
C_p	27	27	1.111	8.32–2	0.33	5.24–4	8

Temp. range K	A_1	A_2	Level of uncertainty
250.0–301.0	1.44581+1	3.85259	III

TABLE 14.30.4. Recommended values of heat capacities

Temp. (K)	250	260	270	273.15	280	290	298.15
c_p (J K^{-1}g^{-1})	1.666	1.693	1.720	1.728	1.746	1.773	1.795
C_p (J K^{-1}mol^{-1})	200.3	203.5	206.7	207.7	209.9	213.1	215.7

Temp. (K)	300
c_p (J K^{-1}g^{-1})	1.800
C_p (J K^{-1}mol^{-1})	216.3

TABLE 14.30.5. Parameters of quasi-polynomial equation

Heat capacity type	No. data points total	used	s_w	s/R	s_r %	s_b/R	+/−
C_p	27	27	1.086	8.13–2	0.33	4.72–4	6

Temp. range K	T_c K	A_1	A_2	A_3	A_4	Level of uncertainty
250.0–301.0	664.47	5.99939	1.21025+1	7.27473	7.43496–1	III

Name: 1,2,4-Trimethylbenzene
Formula: C_9H_{12}

CAS-RN: 95–63–6
Group No.: 14–031
Molar Mass: 120.19

TABLE 14.31.1. Experimental heat capacities

Reference		Temp. range K	No. pnts	Error %	Purity %	method	Type capacity	Calorimeter Type	Reference
*86SCH		309.3–346.1	8S	nosp	not specified		C_{avg}	DSIO	*86SCH
31HUF/PAR		239.5–297.3	7	1.00	not specified		C_p	BSIO	25PAR
47KUR	N	308.6–365.3	5	nosp	not specified		C_{avg}	DSIO	47KUR
48TSC3		293.1	1	nosp	not specified		C_p	BSIO	48TSC1
48TSC6		294.1	1	nosp	not specified		C_p	BSIO	48TSC1
55HEL/HEI		294.3–377.6	16S	nosp	99.4	melpt	C_p	BSAO	39SAG/EVA
57PUT/KIL		230.2–301.0	39	0.30	99.7	melpt	C_p	BSIO	55TAY/JOH
79AND/GRI	N	293.1–429.9	19	0.50	not specified		C_{sat}	BSAO	67RAS/GAN
87WIL/ING		298.1	1	nosp	99.	anal	C_p	FSIT	71PIC/LED

47KUR estimated error about 5%
79AND/GRI error 0.5 % below 373 K and 0.8 % at 433 K

TABLE 14.31.2. Correlated heat capacities

Reference	Temp. range K	No. pnts used	$\sigma_r C$ %	d_w	d/R	d_r %	d_b/R	+/−
				Selected data				
57PUT/KIL	230.2–301.0	39	0.30	0.343	2.60–2	0.10	5.91–3	4
79AND/GRI	293.1–429.9	19	0.50	0.713	9.96–2	0.36	−3.27–2	−7
				Rejected data				

*86SCH	(3.47–1, 1.25, 1.70–1, 0)	31HUF/PAR	(1.31–1, 0.53, −1.19–1, −7)
47KUR	(4.43–1, 1.54, 3.55–1, 5)	48TSC3	(4.90–1, 1.95, −4.90–1, −1)
48TSC6	(6.51–1, 2.61, −6.51–1, −1)	55HEL/HEI	(3.82–1, 1.47, −2.96–1, −10)
87WIL/ING	(1.61–1, 0.63, −1.61–1, −1)		

TABLE 14.31.3. Parameters of cubic spline polynomials

Heat capacity type	No. data points total	used	s_w	s/R	s_r %	s_b/R	+/–
C	97	58	0.518	6.36–2	0.23	–6.74–3	–3

Temp. range K	A_1	A_2	A_3	A_4	Level of uncertainty
230.2–330.0	2.61593+1	–7.85653	3.79254	–4.02072–1	II
330.0–429.9	–2.66111+1	4.01166+1	–1.07448+1	1.06634	III

TABLE 14.31.4. Recommended values of heat capacities

Temp. (K)	230	240	250	260	270	273.15	280
c ($J K^{-1} g^{-1}$)	1.609	1.632	1.656	1.681	1.707	1.716	1.734
C ($J K^{-1} mol^{-1}$)	193.4	196.1	199.0	202.1	205.2	206.2	208.4
Temp. (K)	290	298.15	300	310	320	330	340
c ($J K^{-1} g^{-1}$)	1.762	1.784	1.789	1.817	1.846	1.874	1.901
C ($J K^{-1} mol^{-1}$)	211.7	214.4	215.1	218.4	221.8	225.2	228.5
Temp. (K)	350	360	370	380	390	400	410
c ($J K^{-1} g^{-1}$)	1.930	1.958	1.988	2.019	2.052	2.088	2.127
C ($J K^{-1} mol^{-1}$)	231.9	235.4	238.9	242.7	246.7	251.0	255.6
Temp. (K)	420	430					
c ($J K^{-1} g^{-1}$)	2.168	2.2137					
C ($J K^{-1} mol^{-1}$)	260.6	266.07					

TABLE 14.31.5. Parameters of quasi-polynomial equation

Heat capacity type	No. data points total	used	s_w	s/R	s_r %	s_b/R	+/–
C	97	58	0.580	6.65–2	0.24	–4.39–3	–5

Temp. range K	T_c K	A_1	A_2	A_3	A_4	Level of uncertainty
230.2–429.9	649.17	–6.20509	8.79443–1	1.52893+1	1.09453+1	III

14–031

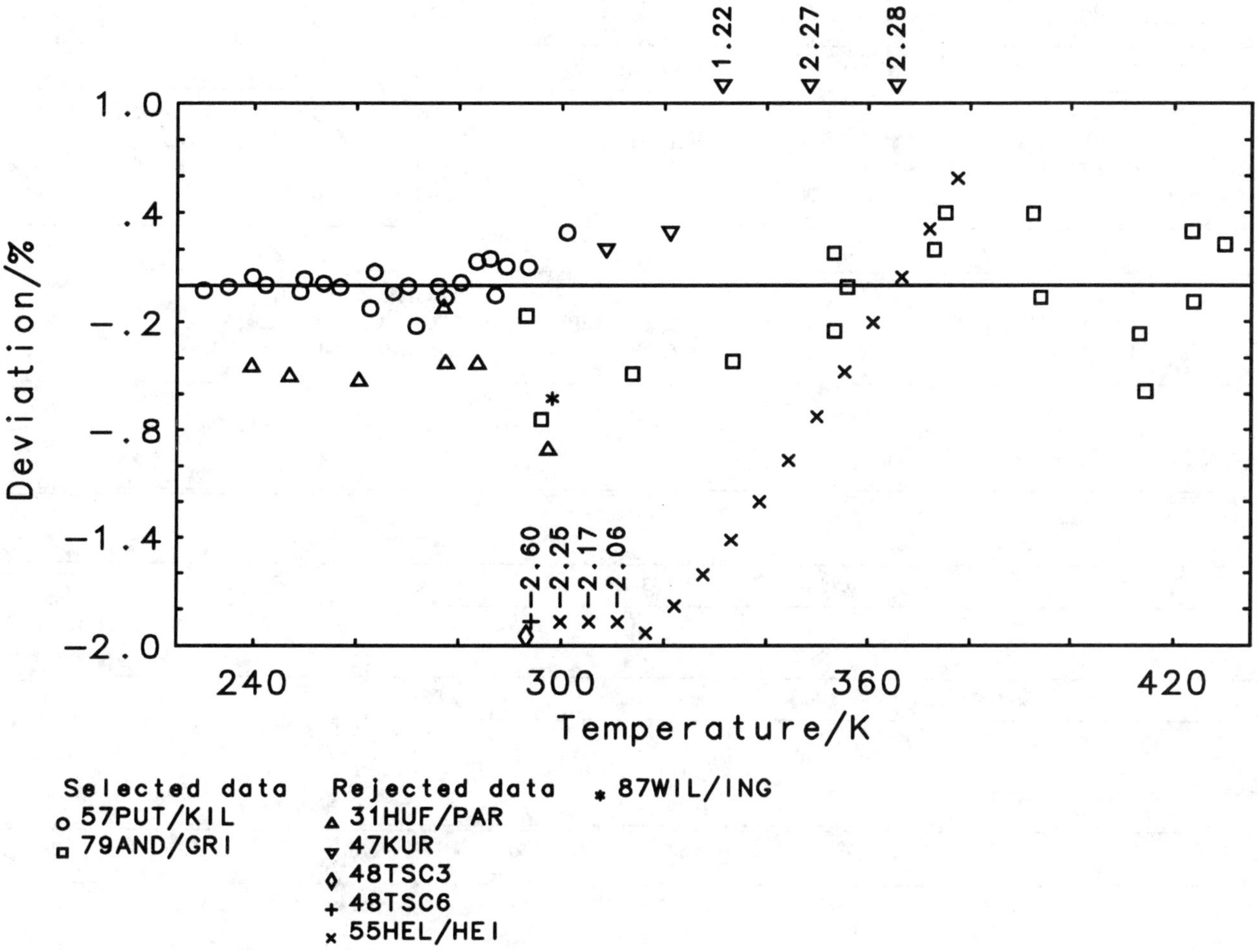

Name: 1,3,5–Trimethylbenzene
Formula: C_9H_{12}

CAS–RN: 108–67–8
Group No.: 14–032
Molar Mass: 120.19

TABLE 14.32.1. Experimental heat capacities

Reference		Temp. range K	No. pnts	Error %	Purity %	Purity method	Type capacity	Calorimeter Type	Calorimeter Reference
*81VON		303.9–348.8	4S	nosp	not specified		C_{avg}	DSIO	*81VON
*86SCH		320.5–349.4	8S	nosp	not specified		C_{avg}	DSIO	*86SCH
47KUR		309.9–359.3	5	nosp	not specified		C_{avg}	DSIO	47KUR
55HEL/HEI		294.3–377.6	16S	nosp	99.78	melpt	C_p	BSAO	39SAG/EVA
55TAY/KIL		230.0–304.5	20	0.30	99.95	melpt	C_p	BSIO	55TAY/JOH
65ZIE	N	315.6–366.3	3	nosp	not specified		C_{avg}	DSIO	58SWI/ZIE1
68REC1	N	298.0–313.0	eqn	nosp	not specified		C_p	BSAO	68REC1
77FOR/BEN		298.1	1	0.30	not specified		C_p	FSIT	71PIC/LED
79AND/GRI	N	295.0–423.6	8	0.50	not specified		C_{sat}	BSAO	67RAS/GAN
79WIL/FAR		298.1	1	0.30	99.0	estim	C_p	FSIT	71PIC/LED

65ZIE average values in temperature ranges 295–438 K and 294–338 K
68REC1 same data in 68REC2
79AND/GRI error 0.5 % below 373 K and 0.8 % at 433 K

TABLE 14.32.2. Correlated heat capacities

Reference	Temp. range K	No. pnts used	$\sigma_r C$ %	d_w	d/R	d_r %	d_b/R	+/–
				Selected data				
55TAY/KIL	230.0–304.5	20	0.30	0.644	4.72–2	0.19	1.52–2	1
77FOR/BEN	298.1	1	0.30	1.104	8.27–2	0.33	–8.27–2	–1
79AND/GRI	295.0–423.6	8	0.50	0.792	1.05–1	0.40	–4.44–2	0
79WIL/FAR	298.1	1	0.30	1.152	8.63–2	0.35	–8.63–2	–1
				Rejected data				
*81VON	(7.03–1, 2.60, 6.97–1, 4)			*86SCH	(8.94–1, 3.22, 8.73–1, 8)			
47KUR	(8.63–1, 3.11, 8.41–1, 5)			55HEL/HEI	(5.70–1, 2.22, –9.70–2, –2)			
65ZIE	(1.13–1, 0.42, –4.92–2, 1)			68REC1	(1.95–1, 0.78, –1.94–1, –5)			

TABLE 14.32.3. Parameters of regression polynomial

Heat capacity type	No. data points total	used	s_w	s/R	s_r %	s_b/R	+/–
C	71	30	0.781	7.53–2	0.29	–7.34–3	–1

Temp. range K	A_1	A_2	A_3	A_4	Level of uncertainty
230.0–423.6	7.97146	9.81903	–2.22329	2.85949–1	III

TABLE 14.32.4. Recommended values of heat capacities

Temp. (K)	230	240	250	260	270	273.15	280
c (J K^{-1}g^{-1})	1.541	1.569	1.597	1.625	1.654	1.662	1.682
C (J K^{-1}mol^{-1})	185.2	188.6	192.0	195.4	198.7	199.8	202.1
Temp. (K)	290	298.15	300	310	320	330	340
c (J K^{-1}g^{-1})	1.710	1.734	1.739	1.768	1.798	1.829	1.860
C (J K^{-1}mol^{-1})	205.6	208.4	209.0	212.5	216.1	219.8	223.6
Temp. (K)	350	360	370	380	390	400	410
c (J K^{-1}g^{-1})	1.893	1.926	1.961	1.997	2.035	2.074	2.114
C (J K^{-1}mol^{-1})	227.5	231.5	235.7	240.0	244.5	249.2	254.1
Temp. (K)	420						
c (J K^{-1}g^{-1})	2.157						
C (J K^{-1}mol^{-1})	259.2						

TABLE 14.32.5. Parameters of quasi-polynomial equation

Heat capacity type	No. data points total	used	s_w	s/R	s_r %	s_b/R	+/–
C	71	30	0.820	8.15–2	0.31	–1.48–2	–3

Temp. range K	T_c K	A_1	A_2	A_3	A_4	Level of uncertainty
230.0–423.6	637.25	–6.63746	9.50896–1	1.36631+1	1.15827+1	III

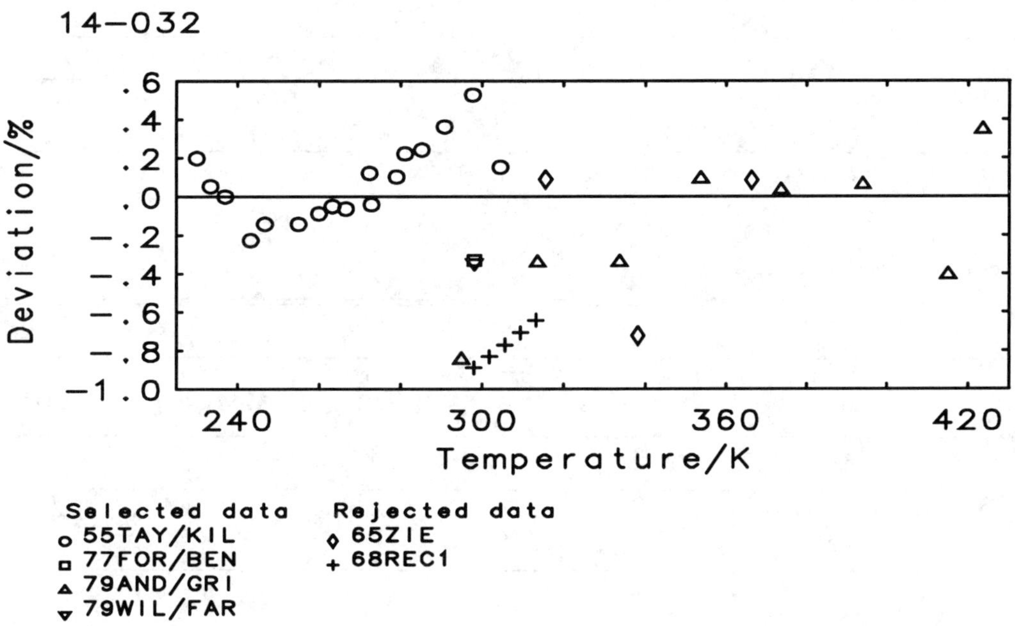

Name: Naphthalene
Formula: $C_{10}H_8$

CAS–RN: 91–20–3
Group No.: 14–033
Molar Mass: 128.17

TABLE 14.33.1. Experimental heat capacities

Reference		Temp. range K	No. pnts	Error %	Purity %	Purity method	Type capacity	Calorimeter Type	Calorimeter Reference
*90PIC		353.0–372.0	2	4.00	not specified		C_p	BDHO	*90PIC
26AND/LYN	N	354.0–473.0	eqn	nosp	not specified		C_p	DSIO	26AND/LYN
31BLA/LEI		382.0–465.8	17	3.00	not specified		C_{sat}	BSIO	31BLA/LEI
32SPA/THO		363.1–463.1	11S	1.00	not specified		C_p	BDHO	31THO/PAR
41SCH		353.0–473.0	eqn	nosp	not specified		C_p	not specified	
44EIB		353.0–473.0	eqn	nosp	not specified		C_p	not specified	
57MCC/FIN2		357.0–370.8	6	0.20	99.985	melpt	C_{sat}	BSAO	47HUF
64RAS/BAS		359.1	1	1.50	not specified		C_p	BSIO	64RAS/BAS
93DUR/AOU		373.0–473.0	102	nosp	not specified		C_p	BDCT	86MER/BEN

26AND/LYN calculated from temperature dependence of enthalpy by the compilers

TABLE 14.33.2. Correlated heat capacities

Reference	Temp. range K	No. pnts used	$\sigma_r C$ %	d_w	d/R	d_r %	d_b/R	+/–
				Selected data				
41SCH	353.0–472.7	10	0.50#	0.645	8.48–2	0.32	–4.45–2	–2
57MCC/FIN2	357.0–370.8	6	0.20	0.329	1.75–2	0.07	1.20–2	3
				Rejected data				
*90PIC	(4.59, 14.76, 4.57, 2)		26AND/LYN	(8.43–1, 2.76, 8.02–1, 9)				
31BLA/LEI	(5.63, 15.84, 5.58, 17)		32SPA/THO	(7.33–1, 2.49, 7.29–1, 11)				
44EIB	(4.50–1, 1.62, –4.42–1, –10)		64RAS/BAS	(2.44, 8.48, 2.44, 1)				
93DUR/AOU	(8.54–1, 2.84, 5.93–1, 56)							

TABLE 14.33.3. Parameters of regression polynomial

Heat capacity type	No. data points total	used	s_w	s/R	s_r %	s_b/R	+/–
C	169	16	0.608	7.54–2	0.29	–2.33–2	1

Temp. range K	A_1	A_2	A_3	Level of uncertainty
353.0–472.7	1.32901+1	2.51050	3.15992–1	III

TABLE 14.33.4. Recommended values of heat capacities

Temp. (K)	360	370	380	390	400	410	420
c $(J\,K^{-1}g^{-1})$	1.714	1.745	1.777	1.809	1.842	1.874	1.908
C $(J\,K^{-1}mol^{-1})$	219.7	223.7	227.8	231.9	236.0	240.2	244.5
Temp. (K)	430	440	450	460	470		
c $(J\,K^{-1}g^{-1})$	1.941	1.976	2.010	2.045	2.080		
C $(J\,K^{-1}mol^{-1})$	248.8	253.2	257.6	262.1	266.6		

TABLE 14.33.5. Parameters of quasi-polynomial equation

Heat capacity type	No. data points total	used	s_w	s/R	s_r %	s_b/R	+/–
C	169	16	0.883	6.58–2	0.25	1.15–3	3

Temp. range K	T_c K	A_1	A_2	A_3	A_4	Level of uncertainty
353.0–472.7	748.40	–3.43150	9.65806–2	9.26821	3.04803+1	III

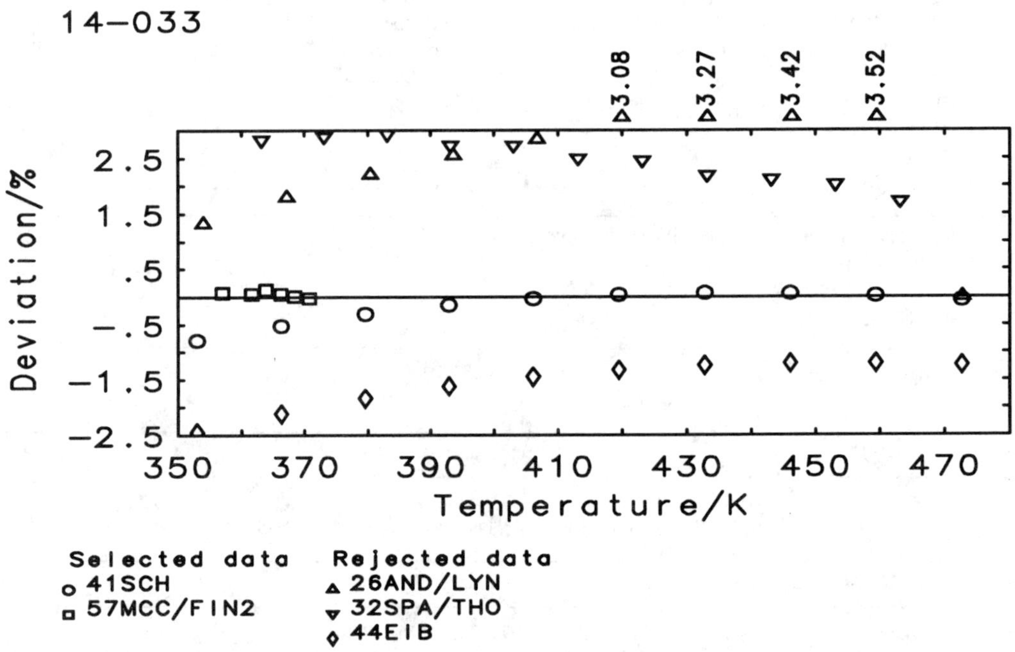

Name: Tricyclo[3.3.2.0^{2,8}]deca–3,6,9–triene
Formula: $C_{10}H_{10}$

CAS-RN: 1005–51–2
Group No.: 14–034
Molar Mass: 130.19

TABLE 14.34.1. Experimental heat capacities

Reference		Temp. range K	No. pnts	Error %	Purity %	Purity method	Type capacity	Calorimeter Type	Calorimeter Reference
80FAL	N	380.0–450.0	8S	nosp	99.87	melpt	C_{sat}	BDHT	69PER/COM

80FAL reproducibility given as 0.6–1.0 %

TABLE 14.34.3. Parameters of regression polynomial

Heat capacity type	No. data points total	No. data points used	s_w	s/R	s_r %	s_b/R	+/–
C_{sat}	8	8	0.925	4.30–1	1.39	9.24–3	6

Temp. range K	A_1	A_2	Level of uncertainty
380.0–450.0	1.39771+1	4.34494	V

TABLE 14.34.4. Recommended values of heat capacities

Temp. (K)	380	390	400	410	420	430	440
c_{sat} (J K^{-1}g^{-1})	1.95	1.97	2.00	2.03	2.06	2.09	2.11
C_{sat} (J K^{-1}mol^{-1})	253	257	261	264	268	272	275

Temp. (K)	450
c_{sat} (J K^{-1}g^{-1})	2.14
C_{sat} (J K^{-1}mol^{-1})	279

Name: 1,2,3,4–Tetrahydronaphthalene
Formula: $C_{10}H_{12}$

CAS-RN: 119–64–2
Group No.: 14–035
Molar Mass: 132.21

TABLE 14.35.1. Experimental heat capacities

Reference	Temp. range K	No. pnts	Error %	Purity %	Purity method	Type capacity	Calorimeter Type	Calorimeter Reference
22HER/SCH	289.6	1	nosp	not specified		C_p	DSIO	22HER/SCH
57MCC/FIN2	248.4–318.5	11	0.20	99.97	melpt	C_{sat}	BSAO	47HUF

TABLE 14.35.2. Correlated heat capacities

Reference	Temp. range K	No. pnts used	$\sigma_r C$ %	d_w	d/R	d_r %	d_b/R	+/–
				Selected data				
57MCC/FIN2	248.4–318.5	11	0.20	0.037	1.86–3	0.01	–1.04–6	–1
				Rejected data				
22HER/SCH	(1.14, 4.25, 1.14, 1)							

TABLE 14.35.3. Parameters of regression polynomial

Heat capacity type	No. data points total	used	s_w	s/R	s_r %	s_b/R	+/–
C_{sat}	12	11	0.047	2.33–3	0.01	–1.04–6	–1

Temp. range K	A_1	A_2	A_3	A_4	Level of uncertainty
248.4–318.5	2.44682+1	–8.73212	4.50995	–4.66721–1	II

TABLE 14.35.4. Recommended values of heat capacities

Temp. (K)	250	260	270	273.15	280	290	298.15
c_{sat} ($J K^{-1}g^{-1}$)	1.480	1.512	1.546	1.557	1.580	1.616	1.645
C_{sat} ($J K^{-1}mol^{-1}$)	195.7	200.0	204.4	205.8	209.0	213.6	217.5
Temp. (K)	300	310	320				
c_{sat} ($J K^{-1}g^{-1}$)	1.652	1.688	1.724				
C_{sat} ($J K^{-1}mol^{-1}$)	218.3	223.1	227.9				

TABLE 14.35.5. Parameters of quasi-polynomial equation

Heat capacity type	No. data points total	used	s_w	s/R	s_r %	s_b/R	+/–
C_{sat}	12	11	0.112	5.71–3	0.02	1.73–6	0

Temp. range K	T_c K	A_1	A_2	A_3	A_4	Level of uncertainty
248.4–318.5	719.00	–1.18383+1	2.75198	9.82621	1.27314+1	II

Name: Butylbenzene
Formula: $C_{10}H_{14}$

CAS-RN: 104–51–8
Group No.: 14–036
Molar Mass: 134.22

TABLE 14.36.1. Experimental heat capacities

Reference	Temp. range K	No. pnts	Error %	Purity %	Purity method	Type capacity	Calorimeter Type	Calorimeter Reference
31HUF/PAR	191.9–298.2	8	1.00	not specified		C_p	BSIO	25PAR
48TSC1	293.1	1	nosp	not specified		C_p	BSIO	48TSC1
48TSC5	295.1	1	nosp	not specified		C_p	BSIO	48TSC1
65MES/TOD2	193.8–370.0	19	nosp	99.92	melpt	C_{sat}	BSAO	47HUF
83GRO/FAR3	298.1	1	0.30	99.	chrom	C_p	FSIT	71PIC/LED

TABLE 14.36.2. Correlated heat capacities

Reference	Temp. range K	No. pnts used	$\sigma_r C$ %	d_w	d/R	d_r %	d_b/R	+/–
				Selected data				
65MES/TOD2	193.8–370.0	19	0.20#	0.063	3.61–3	0.01	–1.00–7	2
				Rejected data				
31HUF/PAR	(2.45–1, 0.87, –1.91–1, –7)			48TSC1	(3.72–1, 1.30, –3.72–1, –1)			
48TSC5	(3.42–1, 1.19, –3.42–1, –1)			83GRO/FAR3	(7.61–2, 0.26, –7.61–2, –1)			

TABLE 14.36.3. Parameters of regression polynomial

Heat capacity type	No. data points total	used	s_w	s/R	s_r %	s_b/R	+/–
C_{sat}	30	19	0.071	4.06–3	0.01	–1.00–7	2

Temp. range K	A_1	A_2	A_3	A_4	Level of uncertainty
193.8–370.0	3.30377+1	–1.24635+1	5.31326	–5.22035–1	II

TABLE 14.36.4. Recommended values of heat capacities

Temp. (K)	200	210	220	230	240	250	260
c_{sat} (J K^{-1}g^{-1})	1.560	1.577	1.597	1.618	1.642	1.668	1.696
C_{sat} (J K^{-1}mol^{-1})	209.4	211.7	214.3	217.2	220.4	223.9	227.6
Temp. (K)	270	273.15	280	290	298.15	300	
c_{sat} (J K^{-1}g^{-1})	1.725	1.734	1.755	1.787	1.813	1.819	
C_{sat} (J K^{-1}mol^{-1})	231.5	232.8	235.6	239.8	243.4	244.2	

TABLE 14.36.5. Parameters of quasi-polynomial equation

Heat capacity type	No. data points total	used	s_w	s/R	s_r %	s_b/R	+/–
C_{sat}	30	19	0.204	1.16–2	0.04	2.06–5	1

Temp. range K	T_c K	A_1	A_2	A_3	A_4	A_5	A_6	Level of uncertainty
193.8–370.0	660.50	6.56869+2	6.94704+1	–4.58059+1	5.92385+2	2.03210+2	3.69694+2	II

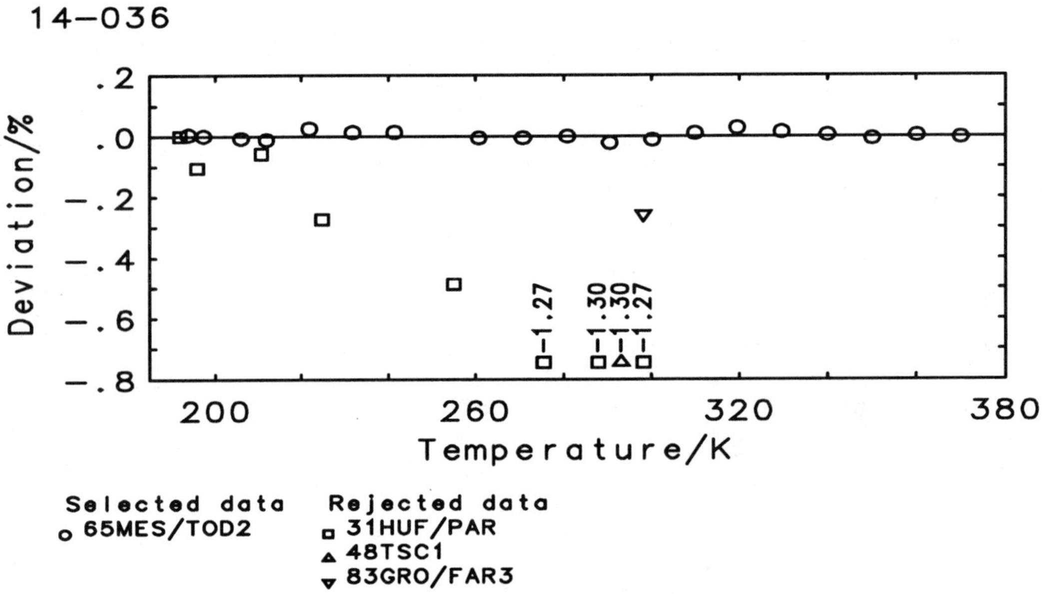

Name: (1,1–Dimethylethyl)benzene
Formula: $C_{10}H_{14}$

CAS-RN: 98–06–6
Group No.: 14–037
Molar Mass: 134.22

TABLE 14.37.1. Experimental heat capacities

Reference	Temp. range K	No. pnts	Error %	Purity %	Purity method	Type capacity	Calorimeter Type	Calorimeter Reference
30HUF/PAR1	220.4–294.3	8	1.00	not specified		C_p	BSIO	25PAR

TABLE 14.37.3. Parameters of regression polynomial

Heat capacity type	No. data points total	No. data points used	s_w	s/R	s_r %	s_b/R	+/–
C_p	8	8	0.195	5.23–2	0.19	1.34–4	2

Temp. range K	A_1	A_2	A_3	Level of uncertainty
220.4–294.3	2.55833+1	–3.80318	1.63987	IV

TABLE 14.37.4. Recommended values of heat capacities

Temp. (K)	220	230	240	250	260	270	273.15
c_p (J K^{-1}g^{-1})	1.558	1.580	1.604	1.631	1.659	1.689	1.699
C_p (J K^{-1}mol^{-1})	209.1	212.1	215.4	218.9	222.7	226.7	228.1

Temp. (K)	280	290	298.15
c_p (J K^{-1}g^{-1})	1.722	1.756	1.785
C_p (J K^{-1}mol^{-1})	231.1	235.7	239.6

TABLE 14.37.5. Parameters of quasi–polynomial equation

Heat capacity type	No. data points total	used	s_w	s/R	s_r %	s_b/R	+/–
C_p	8	8	0.178	4.79–2	0.18	1.08–4	2

Temp. range K	T_c K	A_1	A_2	A_3	A_4	Level of uncertainty
220.4–294.3	647.30	2.13126+1	2.17188+1	–6.76063–1	5.22848	IV

Name: 1–Methyl–4–(1–methylethyl)benzene
Formula: $C_{10}H_{14}$

CAS-RN: 99–87–6
Group No.: 14–038
Molar Mass: 134.22

TABLE 14.38.1. Experimental heat capacities

Reference		Temp. range K	No. pnts	Error %	Purity %	method	Type capacity	Calorimeter Type	Reference
*86SCH	N	320.0–349.8	7S	nosp	not specified		C_{avg}	DSIO	*86SCH
31HUF/PAR		210.8–297.1	8	1.00	not specified		C_p	BSIO	25PAR
47KUR		305.9–358.5	4	nosp	not specified		C_{avg}	DSIO	47KUR

*86SCH substance not clearly identified, name "Cymen" used

TABLE 14.38.2. Correlated heat capacities

Reference	Temp. range K	No. pnts used	$\sigma_r C$ %	d_w	d/R	d_r %	d_b/R	+/–
				Selected data				
*86SCH	320.0–349.8	7	1.00#	0.429	1.35–1	0.43	3.26–2	2
31HUF/PAR	210.8–297.1	8	1.00	0.271	7.49–2	0.27	–2.72–2	–2
				Rejected data				
47KUR	(3.28–1, 1.08, 1.83–2, –1)							

TABLE 14.38.3. Parameters of regression polynomial

Heat capacity type	No. data points total	used	s_w	s/R	s_r %	s_b/R	+/–
C_p	19	15	0.395	1.20–1	0.40	7.12–4	0

Temp. range K	A_1	A_2	A_3	Level of uncertainty
210.8–349.8	2.90266+1	–6.49744	2.13944	IV

TABLE 14.38.4. Recommended values of heat capacities

Temp. (K)	210	220	230	240	250	260	270
c_p (J $K^{-1}g^{-1}$)	1.537	1.554	1.573	1.595	1.620	1.648	1.678
C_p (J $K^{-1}mol^{-1}$)	206.3	208.6	211.2	214.1	217.5	221.1	225.2
Temp. (K)	273.15	280	290	298.15	300	310	320
c_p (J $K^{-1}g^{-1}$)	1.688	1.710	1.745	1.776	1.783	1.824	1.867
C_p (J $K^{-1}mol^{-1}$)	226.5	229.5	234.3	238.4	239.4	244.8	250.6
Temp. (K)	330	340	350				
c_p (J $K^{-1}g^{-1}$)	1.913	1.962	2.013				
C_p (J $K^{-1}mol^{-1}$)	256.8	263.3	270.2				

TABLE 14.38.5. Parameters of quasi-polynomial equation

Heat capacity type	No. data points total	used	s_w	s/R	s_r %	s_b/R	+/-
C_p	19	15	0.473	1.43–1	0.47	–5.74–3	1

Temp. range K	T_c K	A_1	A_2	A_3	A_4	Level of uncertainty
210.8–349.8	651.00	1.76135+1	1.99541+1	8.76667–1	3.88684	IV

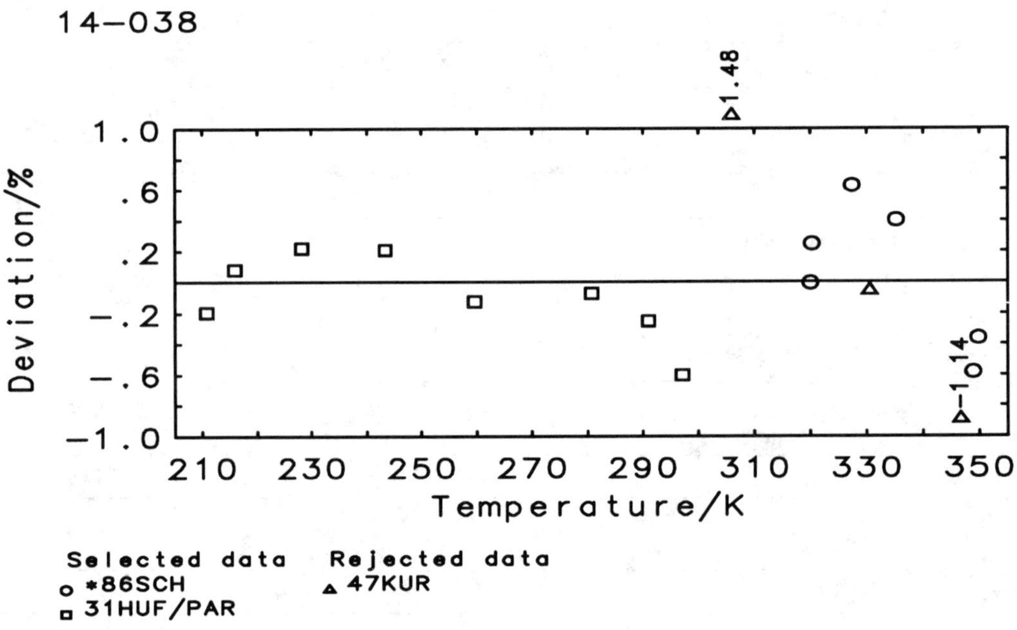

Name: (1–Methylpropyl)benzene
Formula: $C_{10}H_{14}$

CAS-RN: 135–98–8
Group No.: 14–039
Molar Mass: 134.22

TABLE 14.39.1. Experimental heat capacities

Reference		Temp. range K	No. pnts	Error %	Purity %	Purity method	Type capacity	Calorimeter Type	Calorimeter Reference
79AND/GRI	N	293.7–429.6	9	0.80	not specified		C_{sat}	BSAO	67RAS/GAN

79AND/GRI error 0.5 % below 373 K and 0.8 % at 433 K

TABLE 14.39.3. Parameters of regression polynomial

Heat capacity type	No. data points total	No. data points used	s_w	s/R	s_r %	s_b/R	+/–
C_{sat}	9	9	0.290	7.38–2	0.23	2.49–4	0

Temp. range K	A_1	A_2	A_3	Level of uncertainty
293.7–429.6	2.37272+1	–6.90785–1	8.58597–1	IV

TABLE 14.39.4. Recommended values of heat capacities

Temp. (K)	290	298.15	300	310	320	330	340
c_{sat} (J K^{-1}g^{-1})	1.793	1.815	1.820	1.848	1.878	1.908	1.939
C_{sat} (J K^{-1}mol^{-1})	240.7	243.6	244.3	248.1	252.0	256.1	260.3
Temp. (K)	350	360	370	380	390	400	410
c_{sat} (J K^{-1}g^{-1})	1.972	2.005	2.040	2.075	2.112	2.150	2.188
C_{sat} (J K^{-1}mol^{-1})	264.6	269.1	273.8	278.5	283.5	288.5	293.7
Temp. (K)	420	430					
c_{sat} (J K^{-1}g^{-1})	2.228	2.269					
C_{sat} (J K^{-1}mol^{-1})	299.1	304.6					

TABLE 14.39.5. Parameters of quasi–polynomial equation

Heat capacity type	No. data points total	No. data points used	s_w	s/R	s_r %	s_b/R	+/–
C_{sat}	9	9	0.256	6.53–2	0.21	1.77–4	–1

Temp. range K	T_c K	A_1	A_2	A_3	A_4	Level of uncertainty
293.7–429.6	649.00	–7.73412	1.40394	1.70464+1	1.06516+1	IV

Name: 1,2,3,4–Tetramethylbenzene
Formula: $C_{10}H_{14}$

CAS-RN: 488–23–3
Group No.: 14–040
Molar Mass: 134.22

TABLE 14.40.1. Experimental heat capacities

Reference	Temp. range K	No. pnts	Error %	Purity %	Purity method	Type capacity	Calorimeter Type	Calorimeter Reference
31HUF/PAR	276.5–291.9	4	1.00	not specified		C_p	BSIO	25PAR
47KUR	306.7–380.1	4	nosp	not specified		C_{avg}	DSIO	47KUR

TABLE 14.40.2. Correlated heat capacities

Reference	Temp. range K	No. pnts used	$\sigma_r C$ %	d_w	d/R	d_r %	d_b/R	+/–
				Selected data				
31HUF/PAR	276.5–291.9	4	1.00	1.077	3.04–1	1.08	–5.64–2	0
47KUR	306.7–380.1	4	2.00#	0.613	3.77–1	1.23	2.63–1	2

TABLE 14.40.3. Parameters of regression polynomial

Heat capacity type	No. data points total	No. data points used	s_w	s/R	s_r %	s_b/R	+/–
C_p	8	8	1.012	3.96–1	1.33	1.03–1	2

Temp. range K	A_1	A_2	Level of uncertainty
276.5–380.1	8.45928	6.97210	V

TABLE 14.40.4. Recommended values of heat capacities

Temp. (K)	280	290	298.15	300	310	320	330
c_p (J K^{-1}g^{-1})	1.73	1.78	1.81	1.82	1.86	1.91	1.95
C_p (J K^{-1}mol^{-1})	233	238	243	244	250	256	262
Temp. (K)	340	350	360	370	380		
c_p (J K^{-1}g^{-1})	1.99	2.04	2.08	2.12	2.17		
C_p (J K^{-1}mol^{-1})	267	273	279	285	291		

TABLE 14.40.5. Parameters of quasi–polynomial equation

Heat capacity type	No. data points total	No. data points used	s_w	s/R	s_r %	s_b/R	+/–
C_p	8	8	1.087	4.81–1	1.59	1.10–1	2

Temp. range K	T_c K	A_1	A_2	A_3	A_4	Level of uncertainty
276.5–380.1	689.70	–1.26889+1	2.91254	1.08890+1	1.38203+1	V

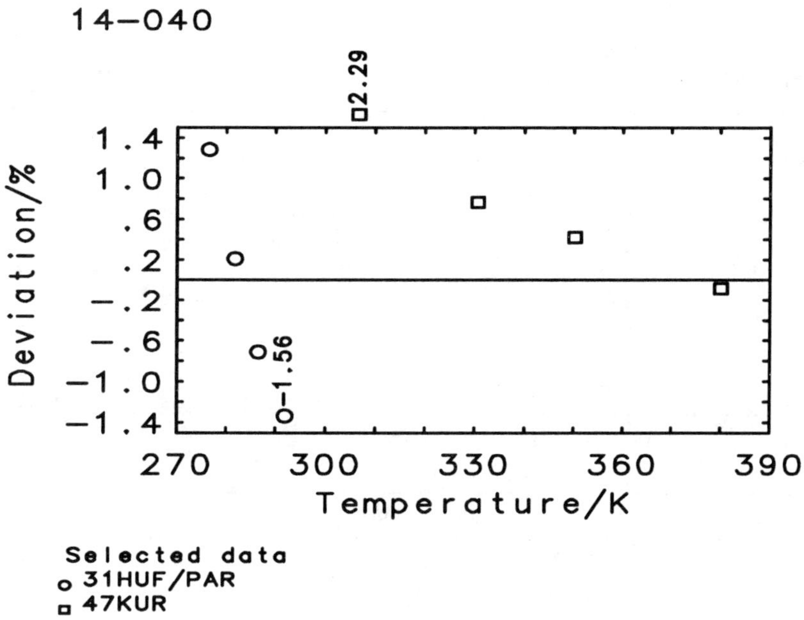

Name: 1,2,3,5–Tetramethylbenzene
Formula: $C_{10}H_{14}$

CAS-RN: 527–53–7
Group No.: 14–041
Molar Mass: 134.22

TABLE 14.41.1. Experimental heat capacities

Reference	Temp. range K	No. pnts	Error %	Purity %	Purity method	Type capacity	Calorimeter Type	Calorimeter Reference
31HUF/PAR	255.3–297.1	5	1.00	not specified		C_p	BSIO	25PAR

TABLE 14.41.3. Parameters of regression polynomial

Heat capacity type	No. data points total	No. data points used	s_w	s/R	s_r %	s_b/R	+/–
C_p	5	5	0.113	3.20–2	0.11	5.19–5	1

Temp. range K	A_1	A_2	Level of uncertainty
255.3–297.1	1.60468+1	4.31873	IV

TABLE 14.41.4. Recommended values of heat capacities

Temp. (K)	260	270	273.15	280	290	298.15	300
c_p (J K^{-1}g^{-1})	1.690	1.716	1.725	1.743	1.770	1.792	1.797
C_p (J K^{-1}mol^{-1})	226.8	230.4	231.5	234.0	237.6	240.5	241.1

Name: 1,2,4,5–Tetramethylbenzene
Formula: $C_{10}H_{14}$

CAS-RN: 95–93–2
Group No.: 14–042
Molar Mass: 134.22

TABLE 14.42.1. Experimental heat capacities

Reference	Temp. range K	No. pnts	Error %	Purity %	Purity method	Type capacity	Calorimeter Type	Calorimeter Reference
33FER/THO	363.1–393.1	4S	nosp	not specified		C_p	BDHO	31THO/PAR
44EIB	352.0–473.0	eqn	nosp	not specified		C_p	not specified	
47KUR	382.8–413.8	3	nosp	not specified		C_{avg}	DSIO	47KUR

TABLE 14.42.2. Correlated heat capacities

Reference	Temp. range K	No. pnts used	$\sigma_r C$ %	d_w	d/R	d_r %	d_b/R	+/–
				Selected data				
33FER/THO	363.1–393.1	4	1.50#	1.104	5.55–1	1.65	5.54–1	4
44EIB	352.0–473.2	13	1.00#	0.297	9.71–2	0.30	–7.28–2	–9
				Rejected data				
47KUR	(2.07, 5.78, 2.06, 3)							

TABLE 14.42.3. Parameters of regression polynomial

Heat capacity type	No. data points total	No. data points used	s_w	s/R	s_r %	s_b/R	+/–
C_p	20	17	0.634	3.01–1	0.90	7.48–2	–5

Temp. range K	A_1	A_2	Level of uncertainty
352.0–473.2	1.85460+1	3.80813	V

TABLE 14.42.4. Recommended values of heat capacities

Temp. (K)	360	370	380	390	400	410	420
c_p (J K^{-1}g^{-1})	2.00	2.02	2.05	2.07	2.09	2.12	2.14
C_p (J K^{-1}mol^{-1})	268	271	275	278	281	284	287
Temp. (K)	430	440	450	460	470		
c_p (J K^{-1}g^{-1})	2.16	2.19	2.21	2.23	2.26		
C_p (J K^{-1}mol^{-1})	290	294	297	300	303		

TABLE 14.42.5. Parameters of quasi–polynomial equation

Heat capacity type	No. data points total	No. data points used	s_w	s/R	s_r %	s_b/R	+/–
C_p	20	17	0.782	3.00–1	0.91	–3.04–2	–3

Temp. range K	T_c K	A_1	A_2	A_3	A_4	Level of uncertainty
352.0–473.2	675.00	–1.90983	4.86320–2	2.09303+1	1.87502+1	V

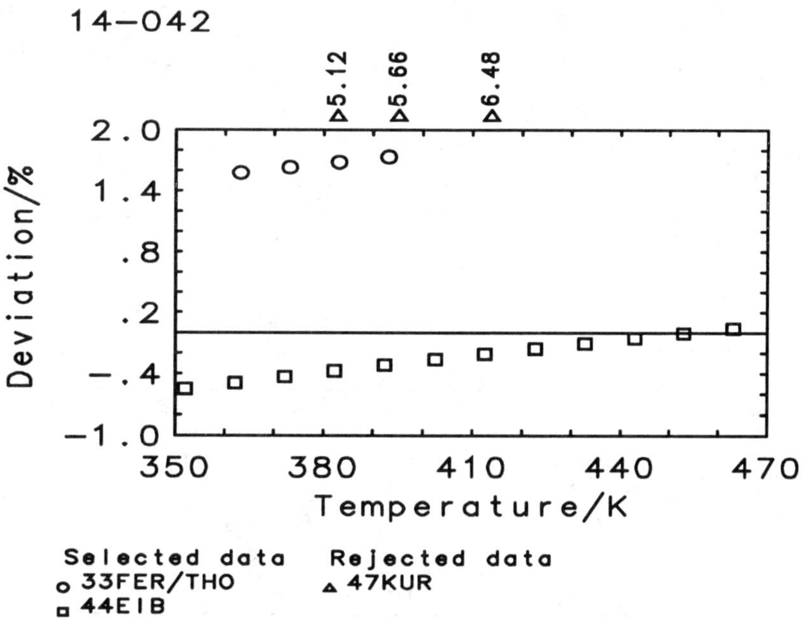

Name: 1–Methyl–4–(1–methylethenyl)cyclohexene
Formula: $C_{10}H_{16}$

CAS-RN: 138–86–3
Group No.: 14–043
Molar Mass: 136.24

TABLE 14.43.1. Experimental heat capacities

Reference		Temp. K	Capac. J/(K.g)	Error %	Purity %	Purity method	Type capacity	Calorimeter Type	Calorimeter Reference
33KOL/UDO	N	293.35	1.831	nosp	not specified		C_p	BSIT	34KOL/UDO2

33KOL/UDO same datum in 34KOL/UDO2

Name: 1–Methylnaphthalene
Formula: $C_{11}H_{10}$

CAS-RN: 90–12–0
Group No.: 14–044
Molar Mass: 142.20

TABLE 14.44.1. Experimental heat capacities

Reference	Temp. range K	No. pnts	Error %	Purity %	Purity method	Type capacity	Calorimeter Type	Calorimeter Reference
57MCC/FIN2	248.0–352.2	15	0.20	99.95	melpt	C_{sat}	BSAO	47HUF

TABLE 14.44.3. Parameters of regression polynomial

Heat capacity type	No. data points total	No. data points used	s_w	s/R	s_r %	s_b/R	+/–
C_{sat}	15	15	0.064	3.52–3	0.01	1.91–6	0

Temp. range K	A_1	A_2	A_3	A_4	Level of uncertainty
248.0–352.2	3.24344+1	–1.43371+1	6.05156	–6.22299–1	II

TABLE 14.44.4. Recommended values of heat capacities

Temp. (K)	250	260	270	273.15	280	290	298.15
c_{sat} ($J\,K^{-1}g^{-1}$)	1.444	1.469	1.496	1.505	1.525	1.554	1.578
C_{sat} ($J\,K^{-1}mol^{-1}$)	205.3	208.9	212.8	214.0	216.8	220.9	224.4
Temp. (K)	300	310	320	330	340	350	
c_{sat} ($J\,K^{-1}g^{-1}$)	1.584	1.614	1.645	1.676	1.706	1.737	
C_{sat} ($J\,K^{-1}mol^{-1}$)	225.2	229.5	233.9	238.3	242.7	247.0	

TABLE 14.44.5. Parameters of quasi-polynomial equation

Heat capacity type	No. data points total	used	s_w	s/R	s_r %	s_b/R	+/–
C_{sat}	15	15	0.354	1.91–2	0.07	2.20–5	0

Temp. range K	T_c K	A_1	A_2	A_3	A_4	Level of uncertainty
248.0–352.2	772.00	–1.16043+1	2.59766	1.20903+1	1.29597+1	II

Name: 2–Methylnaphthalene
Formula: $C_{11}H_{10}$

CAS-RN: 91–57–6
Group No.: 14–045
Molar Mass: 142.20

TABLE 14.45.1. Experimental heat capacities

Reference	Temp. range K	No. pnts	Error %	Purity %	Purity method	Type capacity	Calorimeter Type	Calorimeter Reference
31HUF/PAR	310.4	1	1.00	not specified		C_p	BSIO	25PAR
57MCC/FIN2	312.1–366.6	9	0.20	99.99	melpt	C_{sat}	BSAO	47HUF

TABLE 14.45.2. Correlated heat capacities

Reference	Temp. range K	No. pnts used	$\sigma_r C$ %	d_w	d/R	d_r %	d_b/R	+/–
				Selected data				
31HUF/PAR	310.4	1	1.00	0.336	9.21–2	0.34	–9.21–2	–1
57MCC/FIN2	312.1–366.6	9	0.20	0.165	9.23–3	0.03	4.22–4	–1

TABLE 14.45.3. Parameters of regression polynomial

Heat capacity type	No. data points total	used	s_w	s/R	s_r %	s_b/R	+/–
C	10	10	0.211	3.40–2	0.12	–8.83–3	–2

Temp. range K	A_1	A_2	Level of uncertainty
310.4–366.6	1.20537+1	4.97583	II

TABLE 14.45.4. Recommended values of heat capacities

Temp. (K)	310	320	330	340	350	360	370
c (J K^{-1}g^{-1})	1.607	1.636	1.665	1.694	1.723	1.752	1.781
C (J K^{-1}mol^{-1})	228.5	232.6	236.7	240.9	245.0	249.2	253.3

Name: 1,1–Dimethylindan
Formula: $C_{11}H_{14}$

CAS-RN: 4912–92–9
Group No.: 14–046
Molar Mass: 146.23

TABLE 14.46.1. Experimental heat capacities

Reference	Temp. range K	No. pnts	Error %	Purity %	Purity method	Type capacity	Calorimeter Type	Calorimeter Reference
81LEE/FIN	236.0–388.8	20	0.20	99.914	melpt	C_{sat}	BSAO	47HUF

TABLE 14.46.3. Parameters of regression polynomial

Heat capacity type	No. data points total	No. data points used	s_w	s/R	s_r %	s_b/R	+/–
C_{sat}	20	20	0.159	9.51–3	0.03	4.86–6	0

Temp. range K	A_1	A_2	A_3	A_4	Level of uncertainty
236.0–388.8	2.37687+1	–3.85559	2.66628	–2.25518–1	II

TABLE 14.46.4. Recommended values of heat capacities

Temp. (K)	240	250	260	270	273.15	280	290
c_{sat} (J K^{-1}g^{-1})	1.521	1.551	1.581	1.612	1.622	1.645	1.678
C_{sat} (J K^{-1}mol^{-1})	222.5	226.7	231.2	235.8	237.2	240.5	245.4
Temp. (K)	298.15	300	310	320	330	340	350
c_{sat} (J K^{-1}g^{-1})	1.706	1.712	1.747	1.782	1.818	1.855	1.891
C_{sat} (J K^{-1}mol^{-1})	249.4	250.3	255.4	260.6	265.9	271.2	276.6
Temp. (K)	360	370	380	390			
c_{sat} (J K^{-1}g^{-1})	1.929	1.966	2.004	2.042			
C_{sat} (J K^{-1}mol^{-1})	282.0	287.5	293.0	298.6			

Name: 4,6–Dimethylindan
Formula: $C_{11}H_{14}$

CAS-RN: 1685–82–1
Group No.: 14–047
Molar Mass: 146.23

TABLE 14.47.1. Experimental heat capacities

Reference	Temp. range K	No. pnts	Error %	Purity %	Purity method	Type capacity	Calorimeter Type	Calorimeter Reference
81LEE/FIN	262.0–370.5	12	0.20	99.992	melpt	C_{sat}	BSAO	47HUF

TABLE 14.47.3. Parameters of regression polynomial

Heat capacity type	No. data points total	used	s_w	s/R	s_r %	s_b/R	+/−
C_{sat}	12	12	0.395	2.30–2	0.08	2.04–5	−2

Temp. range K	A_1	A_2	A_3	Level of uncertainty
262.0–370.5	1.89140+1	1.16774	7.42060–1	II

TABLE 14.47.4. Recommended values of heat capacities

Temp. (K)	260	270	273.15	280	290	298.15	300
c_{sat} (J K^{-1}g^{-1})	1.533	1.562	1.572	1.592	1.623	1.648	1.654
C_{sat} (J K^{-1}mol^{-1})	224.2	228.5	229.8	232.8	237.3	241.1	241.9
Temp. (K)	310	320	330	340	350	360	370
c_{sat} (J K^{-1}g^{-1})	1.687	1.720	1.754	1.789	1.825	1.861	1.899
C_{sat} (J K^{-1}mol^{-1})	246.7	251.5	256.5	261.6	266.8	272.2	277.7

Name: 4,7–Dimethylindan
Formula: $C_{11}H_{14}$

CAS-RN: 6682–71–9
Group No.: 14–048
Molar Mass: 146.23

TABLE 14.48.1. Experimental heat capacities

Reference	Temp. range K	No. pnts	Error %	Purity %	Purity method	Type capacity	Calorimeter Type	Calorimeter Reference
81LEE/FIN	285.8–393.9	11	0.20	99.984	melpt	C_{sat}	BSAO	47HUF

TABLE 14.48.3. Parameters of regression polynomial

Heat capacity type	No. data points total	used	s_w	s/R	s_r %	s_b/R	+/−
C_{sat}	11	11	0.293	1.88–2	0.06	3.29–6	−2

Temp. range K	A_1	A_2	Level of uncertainty
285.8–393.9	1.14135+1	5.91154	II

TABLE 14.48.4. Recommended values of heat capacities

Temp. (K)	290	298.15	300	310	320	330	340
c_{sat} (J K^{-1}g^{-1})	1.624	1.651	1.657	1.691	1.725	1.758	1.792
C_{sat} (J K^{-1}mol^{-1})	237.4	241.4	242.4	247.3	252.2	257.1	262.0
Temp. (K)	350	360	370	380	390		
c_{sat} (J K^{-1}g^{-1})	1.825	1.859	1.893	1.926	1.960		
C_{sat} (J K^{-1}mol^{-1})	266.9	271.8	276.8	281.7	286.6		

Name: Pentamethylbenzene
Formula: $C_{11}H_{16}$

CAS-RN: 700–12–9
Group No.: 14–049
Molar Mass: 148.25

TABLE 14.49.1. Experimental heat capacities

Reference	Temp. range K	No. pnts	Error %	Purity %	Purity method	Type capacity	Calorimeter Type	Calorimeter Reference
33FER/THO	333.1–393.1	7S	nosp	not specified		C_p	BDHO	31THO/PAR
44EIB	328.0–473.0	eqn	nosp	not specified		C_p	not specified	

TABLE 14.49.2. Correlated heat capacities

Reference	Temp. range K	No. pnts used	$\sigma_r C$ %	d_w	d/R	d_r %	d_b/R	+/–
				Selected data				
44EIB	328.0–473.6	15	1.00#	0.000	2.41–6	0.00	5.09–7	0
				Rejected data				
33FER/THO	(8.34–1, 2.26, 6.53–1, 5)							

TABLE 14.49.3. Parameters of regression polynomial

Heat capacity type	No. data points total	No. data points used	s_w	s/R	s_r %	s_b/R	+/–
C_p	22	15	0.000	2.59–6	0.00	5.09–7	0

Temp. range K	A_1	A_2	Level of uncertainty
328.0–473.6	2.35673+1	3.17054	V

TABLE 14.49.4. Recommended values of heat capacities

Temp. (K)	330	340	350	360	370	380	390
c_p ($J K^{-1}g^{-1}$)	1.91	1.93	1.94	1.96	1.98	2.00	2.02
C_p ($J K^{-1}mol^{-1}$)	283	286	288	291	293	296	299
Temp. (K)	400	410	420	430	440	450	460
c_p ($J K^{-1}g^{-1}$)	2.03	2.05	2.07	2.09	2.10	2.12	2.14
C_p ($J K^{-1}mol^{-1}$)	301	304	307	309	312	315	317
Temp. (K)	470						
c_p ($J K^{-1}g^{-1}$)	2.16						
C_p ($J K^{-1}mol^{-1}$)	320						

TABLE 14.49.5. Parameters of quasi-polynomial equation

Heat capacity type	No. data points total	used	s_w	s/R	s_r %	s_b/R	+/–
C_p	22	15	0.000	2.91–6	0.00	–1.78–6	0

Temp. range K	T_c K	A_1	A_2	A_3	A_4	Level of uncertainty
328.0–473.6	719.00	–7.88201–5	6.8132–11	2.35673+1	2.27960+1	V

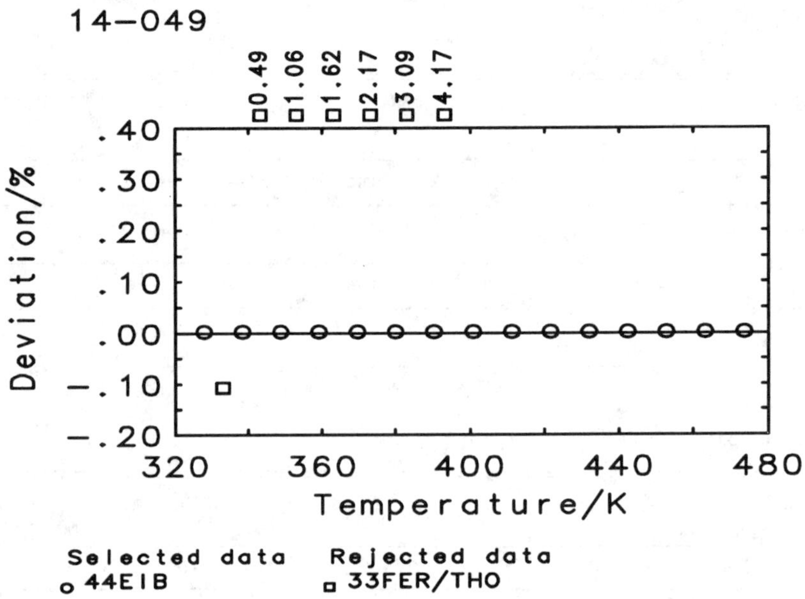

Name: Acenaphthylene
Formula: $C_{12}H_8$

CAS–RN: 208–96–8
Group No.: 14–050
Molar Mass: 152.20

TABLE 14.50.1. Experimental heat capacities

Reference		Temp. K	Capac. J/(K.g)	Error %	Purity %	Purity method	Type capacity	Calorimeter Type	Calorimeter Reference
69SAD/STE	N	393.15	1.093	nosp	not specified		C_{avg}	DSIO	69SAD/STE

69SAD/STE average value in temperature range 363–423 K; possible decomposition at higher temperatures

Name: Acenaphthene
Formula: $C_{12}H_{10}$

CAS–RN: 83–32–9
Group No.: 14–051
Molar Mass: 154.21

TABLE 14.51.1. Experimental heat capacities

Reference		Temp. range K	No. pnts	Error %	Purity %	Purity method	Type capacity	Calorimeter Type	Calorimeter Reference
35SKA	N	367.0–377.0	eqn	nosp	not specified		C_p	DSIO	26AND/LYN
44EIB		368.0–473.0	eqn	nosp	not specified		C_p	not specified	
69SAD/STE		366.0–473.0	eqn	nosp	not specified		C_p	DSIO	69SAD/STE
77FIN/MES		374.2–436.4	10	0.20	99.981	melpt	C_{sat}	BSAO	47HUF

35SKA temperature range not specified, recommended use near n.m.t.

TABLE 14.51.2. Correlated heat capacities

Reference	Temp. range K	No. pnts used	$\sigma_r C$ %	d_w	d/R	d_r %	d_b/R	+/–
				Selected data				
77FIN/MES	374.2–436.4	10	0.20	0.374	2.49–2	0.07	3.74–5	3
				Rejected data				
35SKA	(5.91–1, 1.80, 5.91–1, 1)			44EIB	(4.12, 10.93, 4.11, 5)			
69SAD/STE	(5.52, 13.68, 5.32, 9)							

TABLE 14.51.3. Parameters of regression polynomial

Heat capacity type	No. data points total	used	s_w	s/R	s_r %	s_b/R	+/–
C_{sat}	38	10	0.419	2.79–2	0.08	3.74–5	3

Temp. range K	A_1	A_2	Level of uncertainty
374.2–436.4	1.06600+1	5.70431	II

TABLE 14.51.4. Recommended values of heat capacities

Temp. (K)	380	390	400	410	420
c_{sat} (J K^{-1}g^{-1})	1.743	1.774	1.805	1.836	1.866
C_{sat} (J K^{-1}mol^{-1})	268.9	273.6	278.3	283.1	287.8

TABLE 14.51.5. Parameters of quasi-polynomial equation

Heat capacity type	No. data points total	used	s_w	s/R	s_r %	s_b/R	+/–
C_{sat}	38	10	0.447	2.98–2	0.09	3.78–5	3

Temp. range K	T_c K	A_1	A_2	A_3	A_4	Level of uncertainty
374.2–436.4	803.20	–3.26683–1	5.90845–4	1.07623+1	4.51563+1	II

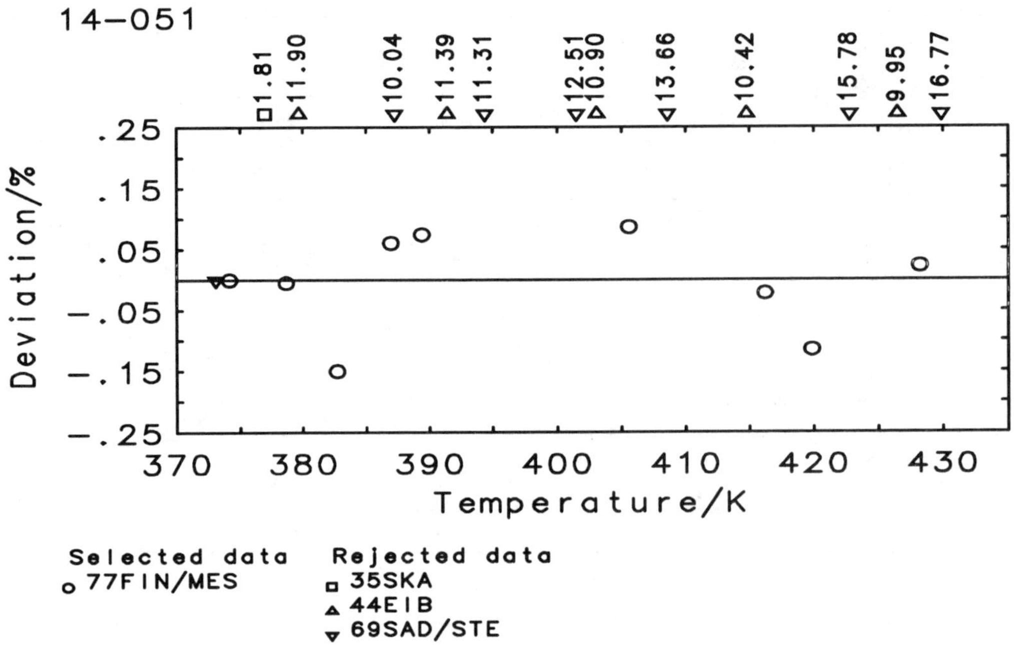

Name: 1,1'–Biphenyl
Formula: $C_{12}H_{10}$

CAS-RN: 92–52–4
Group No.: 14–052
Molar Mass: 154.21

TABLE 14.52.1. Experimental heat capacities

Reference		Temp. range K	No. pnts	Error %	Purity %	Purity method	Type capacity	Calorimeter Type	Calorimeter Reference
31FOR/BRU		350.8–469.6	22	2.00	not specified		C_p	BSIO	31FOR/BRU
31FOR/BRU		420.8–620.2	8	2.00	not specified		C_p	FSIO	31FOR/BRU
31NEW/KAU	N	344.0–573.0	eqn	1.00	not specified		C_p	not specified	
32SPA/THO		353.1–373.1	3S	nosp	not specified		C_p	BDHO	31THO/PAR
41SCH		342.0–473.0	eqn	nosp	not specified		C_p	not specified	
50KUR		386.4–492.8	10	nosp	not specified		C_{avg}	DSIO	47KUR
56MCE/MAL		422.0–588.7	4S	nosp	not specified		C_p	not specified	
58WAL/BRO		429.8–533.3	3	2.00	not specified		C_p	DSIO	56WAL/GRA
83ORO/MRA		350.0–480.0	24S	1.50	99.9	chrom	C_p	BDHT	69PER/COM
89CHI/KNI		349.1–440.3	11	0.10	99.983	melpt	C_{sat}	BSAO	47HUF
89CHI/KNI		360.0–700.0	18	1.00	99.983	melpt	C_{sat}	BDHT	89KNI/ARC
93DUR/AOU		373.0–473.0	51	nosp	not specified		C_p	BDCT	86MER/BEN

31NEW/KAU four different calorimeters used

TABLE 14.52.2. Correlated heat capacities

Reference	Temp. range K	No. pnts used	$\sigma_r C$ %	d_w	d/R	d_r %	d_b/R	+/–
				Selected data				
89CHI/KNI	349.1–440.3	11	0.10	0.190	6.56–3	0.02	–1.51–3	–3
89CHI/KNI	360.0–700.0	18	1.00#	0.680	2.87–1	0.68	9.64–2	4
				Rejected data				
31FOR/BRU	(1.41, 3.67, 3.33–1, –2)			31FOR/BRU	(4.11, 8.43, 3.54, 6)			
31NEW/KAU	(5.52–1, 1.36, –2.39–1, –3)			32SPA/THO	(1.64–1, 0.50, –1.41–1, –3)			
41SCH	(4.39–1, 1.20, –7.05–2, –1)			50KUR	(2.30, 5.86, 1.95, 10)			
56MCE/MAL	(9.51–1, 2.10, 6.21–1, 2)			58WAL/BRO	(1.98–1, 0.49, –2.98–2, –1)			
83ORO/MRA	(3.23–1, 0.89, –2.12–1, –12)			93DUR/AOU	(1.39, 3.85, –1.23, –51)			

TABLE 14.52.3. Parameters of regression polynomial

Heat capacity type	No. data points total	used	s_w	s/R	s_r %	s_b/R	+/–
C_p	174	29	0.591	2.43–1	0.58	5.92–2	1
C_{sat}	174	29	0.557	2.23–1	0.54	6.18–2	1

Temp. range K	A_1	A_2	A_3	A_4	Level of uncertainty
349.1–700.0	1.61319+1	4.17889	2.29446–1	–1.72138–2	IV
349.1–700.0	1.82988+1	2.60244	6.08539–1	–4.73438–2	IV

TABLE 14.52.4. Recommended values of heat capacities

Temp. (K)	350	360	370	380	390	400	410
c_p (J K^{-1}g^{-1})	1.770	1.798	1.826	1.854	1.882	1.910	1.938
C_p (J K^{-1}mol^{-1})	273.0	277.3	281.6	285.9	290.2	294.5	298.8
c_{sat} (J K^{-1}g^{-1})	1.770	1.798	1.826	1.854	1.881	1.909	1.938
C_{sat} (J K^{-1}mol^{-1})	273.0	277.3	281.5	285.8	290.1	294.5	298.8
Temp. (K)	420	430	440	450	460	470	480
c_p (J K^{-1}g^{-1})	1.966	1.994	2.022	2.050	2.078	2.106	2.134
C_p (J K^{-1}mol^{-1})	303.1	307.4	311.8	316.1	320.4	324.7	329.0
c_{sat} (J K^{-1}g^{-1})	1.966	1.994	2.022	2.050	2.078	2.106	2.134
C_{sat} (J K^{-1}mol^{-1})	303.1	307.4	311.8	316.1	320.4	324.7	329.1
Temp. (K)	490	500	510	520	530	540	550
c_p (J K^{-1}g^{-1})	2.162	2.190	2.218	2.245	2.273	2.301	2.329
C_p (J K^{-1}mol^{-1})	333.3	337.7	342.0	346.3	350.6	354.8	359.1
c_{sat} (J K^{-1}g^{-1})	2.162	2.189	2.217	2.245	2.272	2.299	2.326
C_{sat} (J K^{-1}mol^{-1})	333.3	337.6	341.9	346.1	350.4	354.5	358.7
Temp. (K)	560	570	580	590	600	610	620
c_p (J K^{-1}g^{-1})	2.356	2.384	2.412	2.439	2.467	2.494	2.521
C_p (J K^{-1}mol^{-1})	363.4	367.7	371.9	376.1	380.4	384.6	388.8
c_{sat} (J K^{-1}g^{-1})	2.353	2.380	2.406	2.432	2.458	2.484	2.509
C_{sat} (J K^{-1}mol^{-1})	362.9	367.0	371.1	375.1	379.1	383.1	387.0

TABLE 14.52.4. Recommended values of heat capacities (continued)

Temp. (K)	630	640	650	660	670	680	690
c_p (J K^{-1}g^{-1})	2.548	2.575	2.602	2.629	2.656	2.682	2.709
C_p (J K^{-1}mol^{-1})	393.0	397.1	401.3	405.4	409.5	413.6	417.7
c_{sat} (J K^{-1}g^{-1})	2.535	2.559	2.584	2.608	2.632	2.655	2.678
C_{sat} (J K^{-1}mol^{-1})	390.9	394.7	398.5	402.2	405.9	409.5	413.0
Temp. (K)	700						
c_p (J K^{-1}g^{-1})	2.735						
C_p (J K^{-1}mol^{-1})	421.7						
c_{sat} (J K^{-1}g^{-1})	2.701						
C_{sat} (J K^{-1}mol^{-1})	416.5						

TABLE 14.52.5. Parameters of quasi-polynomial equation

Heat capacity type	No. data points total	used	s_w	s/R	s_r %	s_b/R	+/–
C_p	174	29	1.253	2.42–1	0.54	–2.85–3	1

Temp. range K	T_c K	A_1	A_2	A_3	A_4	Level of uncertainty
349.1–700.0	780.00	1.33336	1.00099–2	1.36467+1	4.44026+1	IV

14–052

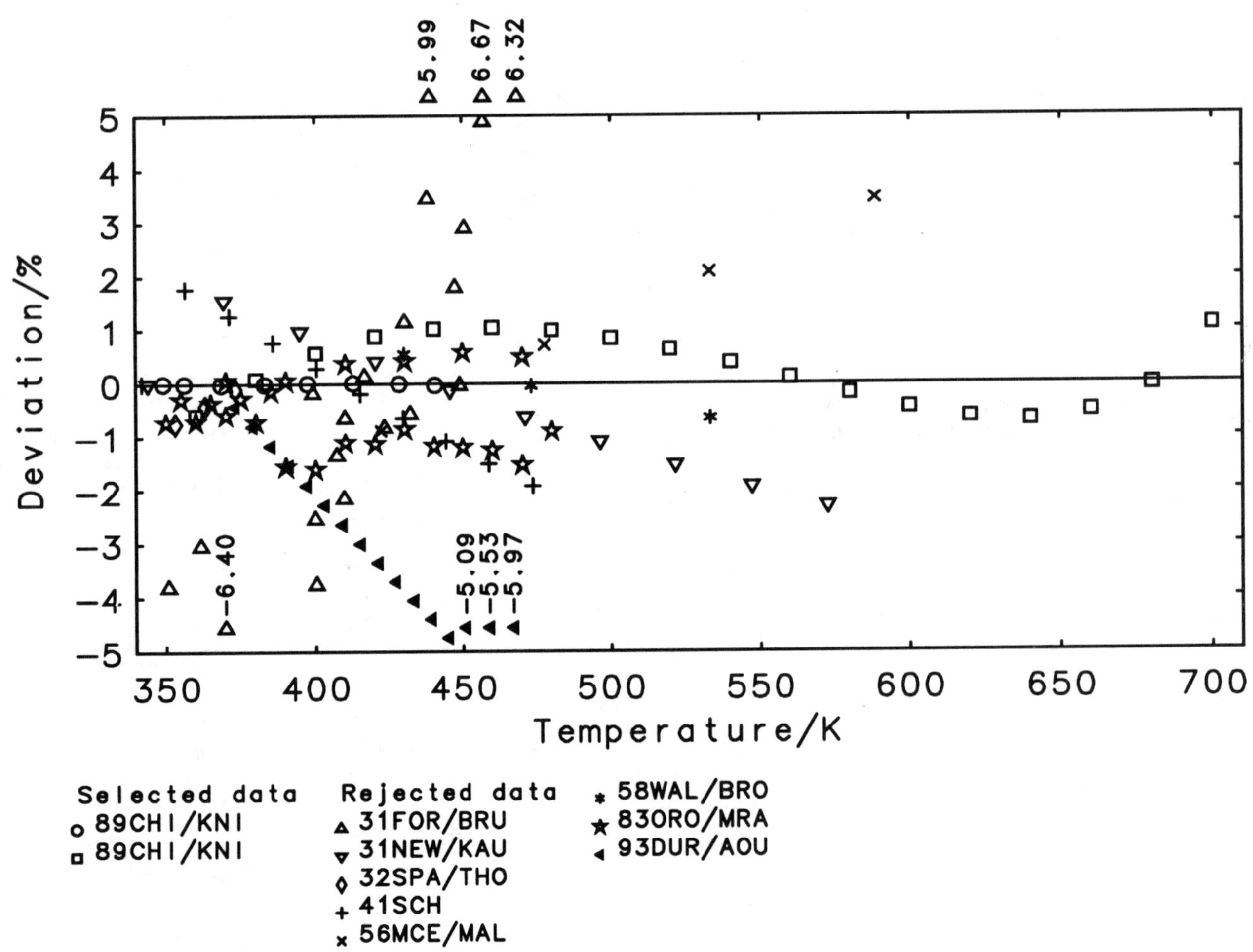

Name: 1,8–Dimethylnaphthalene
Formula: $C_{12}H_{12}$

CAS-RN: 569–41–5
Group No.: 14–053
Molar Mass: 156.23

TABLE 14.53.1. Experimental heat capacities

Reference	Temp. range K	No. pnts	Error %	Purity %	Purity method	Type capacity	Calorimeter Type	Calorimeter Reference
77FIN/MES	349.6–432.3	7	0.20	99.9955	melpt	C_{sat}	BSAO	47HUF

TABLE 14.53.3. Parameters of regression polynomial

Heat capacity type	No. data points total	No. data points used	s_w	s/R	s_r %	s_b/R	+/–
C_{sat}	7	7	0.265	1.91–2	0.05	–8.72–6	1

Temp. range K	A_1	A_2	Level of uncertainty
349.6–432.4	1.22890+1	6.09189	II

TABLE 14.53.4. Recommended values of heat capacities

Temp. (K)	350	360	370	380	390	400	410
c_{sat} (J K^{-1}g^{-1})	1.789	1.821	1.854	1.886	1.918	1.951	1.983
C_{sat} (J K^{-1}mol^{-1})	279.5	284.5	289.6	294.7	299.7	304.8	309.8

Temp. (K)	420	430
c_{sat} (J K^{-1}g^{-1})	2.016	2.048
C_{sat} (J K^{-1}mol^{-1})	314.9	320.0

TABLE 14.53.5. Parameters of quasi–polynomial equation

Heat capacity type	No. data points total	No. data points used	s_w	s/R	s_r %	s_b/R	+/–
C_{sat}	7	7	0.204	1.53–2	0.04	7.08–6	1

Temp. range K	T_c K	A_1	A_2	A_3	A_4	Level of uncertainty
349.6–432.4	792.00	6.38062	1.69089–1	1.04137+1	6.01934+1	II

Name: 2,3–Dimethylnaphthalene
Formula: $C_{12}H_{12}$

CAS-RN: 581–40–8
Group No.: 14–054
Molar Mass: 156.23

TABLE 14.54.1. Experimental heat capacities

Reference	Temp. range K	No. pnts	Error %	Purity %	Purity method	Type capacity	Calorimeter Type	Calorimeter Reference
88MES/FIN	384.5–415.1	8	0.10	99.935	melpt	C_{sat}	BSAO	47HUF

TABLE 14.54.3. Parameters of regression polynomial

Heat capacity type	No. data points total	used	s_w	s/R	s_r %	s_b/R	+/–
C_{sat}	8	8	0.706	2.58–2	0.07	2.43–5	–4

Temp. range K	A_1	A_2	Level of uncertainty
384.5–415.1	1.49680+1	5.41211	II

TABLE 14.54.4. Recommended values of heat capacities

Temp. (K)	380	390	400	410
c_{sat} (J K^{-1}g^{-1})	1.891	1.920	1.949	1.978
C_{sat} (J K^{-1}mol^{-1})	295.4	299.9	304.4	308.9

Name: 2,6–Dimethylnaphthalene
Formula: $C_{12}H_{12}$

CAS-RN: 581–42–0
Group No.: 14–055
Molar Mass: 156.23

TABLE 14.55.1. Experimental heat capacities

Reference	Temp. range K	No. pnts	Error %	Purity %	Purity method	Type capacity	Calorimeter Type	Calorimeter Reference
77FIN/MES	390.2–438.3	9	0.20	99.958	melpt	C_{sat}	BSAO	47HUF

TABLE 14.55.3. Parameters of regression polynomial

Heat capacity type	No. data points total	used	s_w	s/R	s_r %	s_b/R	+/–
C_{sat}	9	9	0.646	4.88–2	0.13	1.12–4	0

Temp. range K	A_1	A_2	Level of uncertainty
390.2–438.3	1.19193+1	6.00463	II

TABLE 14.55.4. Recommended values of heat capacities

Temp. (K)	390	400	410	420	430	440
c_{sat} (J K^{-1}g^{-1})	1.881	1.913	1.945	1.977	2.009	2.040
C_{sat} (J K^{-1}mol^{-1})	293.8	298.8	303.8	308.8	313.8	318.8

Name: 2,7–Dimethylnaphthalene
Formula: $C_{12}H_{12}$

CAS-RN: 582–16–1
Group No.: 14–056
Molar Mass: 156.23

TABLE 14.56.1. Experimental heat capacities

Reference	Temp. range K	No. pnts	Error %	Purity %	Purity method	Type capacity	Calorimeter Type	Calorimeter Reference
77FIN/MES	372.6–391.0	6	0.20	99.95	melpt	C_{sat}	BSAO	47HUF

TABLE 14.56.3. Parameters of regression polynomial

Heat capacity type	No. data points total	used	s_w	s/R	s_r %	s_b/R	+/−
C_{sat}	6	6	0.469	3.27–2	0.09	4.13–5	0

Temp. range K	A_1	A_2	Level of uncertainty
372.6–391.0	1.25511+1	5.87490	II

TABLE 14.56.4. Recommended values of heat capacities

Temp. (K)	370	380	390
c_{sat} ($J\,K^{-1}g^{-1}$)	1.825	1.856	1.887
C_{sat} ($J\,K^{-1}mol^{-1}$)	285.1	290.0	294.9

Name: Cyclohexylbenzene
Formula: $C_{12}H_{16}$

CAS-RN: 827–52–1
Group No.: 14–057
Molar Mass: 160.26

TABLE 14.57.1. Experimental heat capacities

Reference	Temp. range K	No. pnts	Error %	Purity %	Purity method	Type capacity	Calorimeter Type	Calorimeter Reference
83ORO/MRA	290.0–470.0	37S	1.00	99.9	chrom	C_p	BDHT	69PER/COM

TABLE 14.57.3. Parameters of regression polynomial

Heat capacity type	No. data points total	used	s_w	s/R	s_r %	s_b/R	+/−
C_p	37	37	0.905	3.34–1	0.91	5.13–3	−1 0

Temp. range K	A_1	A_2	Level of uncertainty
290.0–470.0	9.23013	7.50904	V

TABLE 14.57.4. Recommended values of heat capacities

Temp. (K)	290	298.15	300	310	320	330	340
c_p ($J\,K^{-1}g^{-1}$)	1.61	1.64	1.65	1.69	1.73	1.76	1.80
C_p ($J\,K^{-1}mol^{-1}$)	258	263	264	270	277	283	289
Temp. (K)	350	360	370	380	390	400	410
c_p ($J\,K^{-1}g^{-1}$)	1.84	1.88	1.92	1.96	2.00	2.04	2.08
C_p ($J\,K^{-1}mol^{-1}$)	295	302	308	314	320	326	333
Temp. (K)	420	430	440	450	460	470	
c_p ($J\,K^{-1}g^{-1}$)	2.12	2.15	2.19	2.23	2.27	2.31	
C_p ($J\,K^{-1}mol^{-1}$)	339	345	351	358	364	370	

TABLE 14.57.5. Parameters of quasi-polynomial equation

Heat capacity type	No. data points total	used	s_w	s/R	s_r %	s_b/R	+/-
C_p	37	37	0.908	3.35-1	0.91	5.50-3	-9

Temp. range K	T_c K	A_1	A_2	A_3	A_4	Level of uncertainty
290.0–470.0	742.00	6.03820	1.35392-1	7.40096	6.73228+1	V

Name: *E,E,Z*–1,5,9–Cyclododecatriene
Formula: $C_{12}H_{18}$

CAS-RN: 706–31–0
Group No.: 14–058
Molar Mass: 162.27

TABLE 14.58.1. Experimental heat capacities

Reference	Temp. K	Capac. J/(K.g)	Error %	Purity %	Purity method	Type capacity	Calorimeter Type	Calorimeter Reference
89LAI/ROD	298.15	1.773	nosp	99.	anal	C_p	FSIT	71PIC/LED

Name: Hexamethylbenzene
Formula: $C_{12}H_{18}$

CAS-RN: 87–85–4
Group No.: 14–059
Molar Mass: 162.27

TABLE 14.59.1. Experimental heat capacities

Reference	Temp. range K	No. pnts	Error %	Purity %	Purity method	Type capacity	Calorimeter Type	Calorimeter Reference
32SPA/THO	443.1–473.1	4S	1.00	not specified		C_p	BDHO	31THO/PAR
47KUR	470.3–492.6	2	nosp	not specified		C_{avg}	DSIO	47KUR

TABLE 14.59.2. Correlated heat capacities

Reference	Temp. range K	No. pnts used	$\sigma_r C$ %	d_w	d/R	d_r %	d_b/R	+/-
				Selected data				
32SPA/THO	443.1–473.1	4	1.00	0.039	1.83-2	0.04	1.24-5	0
				Rejected data				
47KUR	(1.97, 4.31, -1.97, -1)							

TABLE 14.59.3. Parameters of regression polynomial

Heat capacity type	No. data points total	used	s_w	s/R	s_r %	s_b/R	+/-
C_p	6	4	0.055	2.58-2	0.06	1.24-5	0

Temp. range K	A_1	A_2	Level of uncertainty
443.1–473.1	6.97009	8.65599	V

TABLE 14.59.4. Recommended values of heat capacities

Temp. (K)	440	450	460	470
c_p (J K^{-1}g^{-1})	2.31	2.35	2.40	2.44
C_p (J K^{-1}mol^{-1})	375	382	389	396

Name: Hexylbenzene
Formula: $C_{12}H_{18}$

CAS-RN: 1077–16–3
Group No.: 14–060
Molar Mass: 162.27

TABLE 14.60.1. Experimental heat capacities

Reference	Temp. range K	No. pnts	Error %	Purity %	Purity method	Type capacity	Calorimeter Type	Calorimeter Reference
48TSC1	293.1	1	nosp	not specified		C_p	BSIO	48TSC1
48TSC5	295.1	1	nosp	not specified		C_p	BSIO	48TSC1

TABLE 14.60.3. Parameters of regression polynomial

Heat capacity type	No. data points total	No. data points used	s_w	s/R	s_r %	s_b/R	+/–
C_p	2	2	0.000	0.00	0.00	0.00	0

Temp. range K	A_1	A_2	Level of uncertainty
293.1–295.1	5.92153+1	–8.16594	VI

TABLE 14.60.4. Recommended values of heat capacities

Temp. (K)	293	295
c_p (J K^{-1}g^{-1})	1.81	1.80
C_p (J K^{-1}mol^{-1})	293	292

Name: 9*H*–Fluorene
Formula: $C_{13}H_{10}$

CAS-RN: 86–73–7
Group No.: 14–061
Molar Mass: 166.22

TABLE 14.61.1. Experimental heat capacities

Reference	Temp. range K	No. pnts	Error %	Purity %	Purity method	Type capacity	Calorimeter Type	Calorimeter Reference
44EIB	387.0–473.0	eqn	nosp	not specified		C_p	not specified	
77FIN/MES	392.6–426.7	8	0.20	99.992	melpt	C_{sat}	BSAO	47HUF

TABLE 14.61.2. Correlated heat capacities

Reference	Temp. range K	No. pnts used	$\sigma_r C$ %	d_w	d/R	d_r %	d_b/R	+/–
				Selected data				
44EIB	387.0–472.0	18	1.00#	0.611	2.16–1	0.61	–1.16–1	–3
77FIN/MES	392.6–426.7	8	0.20	0.348	2.54–2	0.07	1.08–2	3

TABLE 14.61.3. Parameters of regression polynomial

Heat capacity type	No. data points total	used	s_w	s/R	s_r %	s_b/R	+/–
C	26	26	0.578	1.92–1	0.54	–7.74–2	0

Temp. range K	A_1	A_2	A_3	Level of uncertainty
387.0–472.0	3.10273+1	–3.23336	1.12314	III

TABLE 14.61.4. Recommended values of heat capacities

Temp. (K)	390	400	410	420	430	440	450
c (J K^{-1}g^{-1})	1.776	1.804	1.833	1.864	1.895	1.928	1.962
C (J K^{-1}mol^{-1})	295.2	299.9	304.7	309.8	315.0	320.5	326.1

Temp. (K)	460	470
c (J K^{-1}g^{-1})	1.997	2.033
C (J K^{-1}mol^{-1})	331.9	337.9

TABLE 14.61.5. Parameters of quasi–polynomial equation

Heat capacity type	No. data points total	used	s_w	s/R	s_r %	s_b/R	+/–
C	26	26	1.447	1.33–1	0.37	4.42–3	–4

Temp. range K	T_c K	A_1	A_2	A_3	A_4	Level of uncertainty
387.0–472.0	825.00	–1.04989+1	9.11211–1	1.24839+1	3.02420+1	III

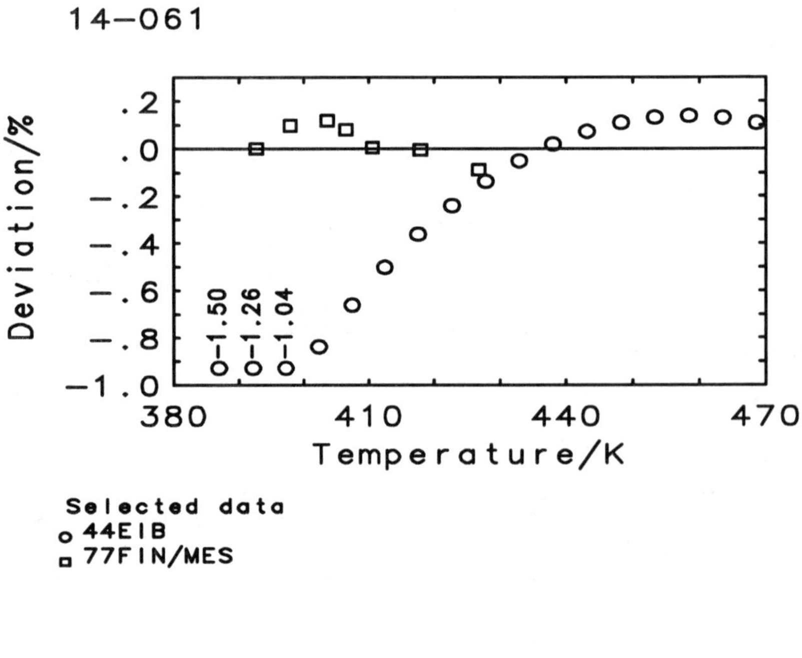

Name: 1,1'–Methylenebisbenzene
Formula: $C_{13}H_{12}$

CAS-RN: 101–81–5
Group No.: 14–062
Molar Mass: 168.24

TABLE 14.62.1. Experimental heat capacities

Reference		Temp. range K	No. pnts	Error %	Purity %	Purity method	Type capacity	Calorimeter Type	Calorimeter Reference
30HUF/PAR1		302.0–312.3	4	nosp	not specified		C_p	BSIO	25PAR
31BLA/LEI		308.1–368.1	13	nosp	not specified		C_p	BSIO	31BLA/LEI
31SMI/AND1		310.7–322.6	2	nosp	99.9	estim	C_p	DSIO	26AND/LYN
50KUR		313.8–414.4	8	nosp	not specified		C_{avg}	DSIO	47KUR
56DUF/EVE	N	303.1–353.1	6S	nosp	not specified		C_p	not specified	

56DUF/EVE measured by a non–calorimetric method (piezo–thermometric)

TABLE 14.62.2. Correlated heat capacities

Reference	Temp. range K	No. pnts used	$\sigma_r C$ %	d_w	d/R	d_r %	d_b/R	+/–
				Selected data				
30HUF/PAR1	302.0–312.3	4	1.00#	0.363	1.17–1	0.36	1.08–1	4
56DUF/EVE	303.1–353.1	6	0.50#	0.305	5.03–2	0.15	–1.78–2	–2
				Rejected data				
31BLA/LEI	(3.52, 9.43, 3.43, 10)		31SMI/AND1	(4.90–1, 1.48, 4.50–1, 2)				
50KUR	(2.41, 6.70, 2.40, 4)							

TABLE 14.62.3. Parameters of regression polynomial

Heat capacity type	No. data points total	No. data points used	s_w	s/R	s_r %	s_b/R	+/–
C_p	33	10	0.368	9.36–2	0.29	3.28–2	2

Temp. range K	A_1	A_2	Level of uncertainty
302.0–353.1	1.70065+1	4.94395	III

TABLE 14.62.4. Recommended values of heat capacities

Temp. (K)	300	310	320	330	340	350
c_p (J $K^{-1}g^{-1}$)	1.573	1.598	1.622	1.647	1.671	1.696
C_p (J $K^{-1}mol^{-1}$)	264.7	268.8	272.9	277.1	281.2	285.3

TABLE 14.62.5. Parameters of quasi–polynomial equation

Heat capacity type	No. data points total	No. data points used	s_w	s/R	s_r %	s_b/R	+/–
C_p	33	10	0.329	7.15–2	0.22	8.34–3	2

Temp. range K	T_c K	A_1	A_2	A_3	A_4	Level of uncertainty
302.0–353.1	770.00	6.04128+1	3.94115+1	–1.18128+1	2.31513+1	III

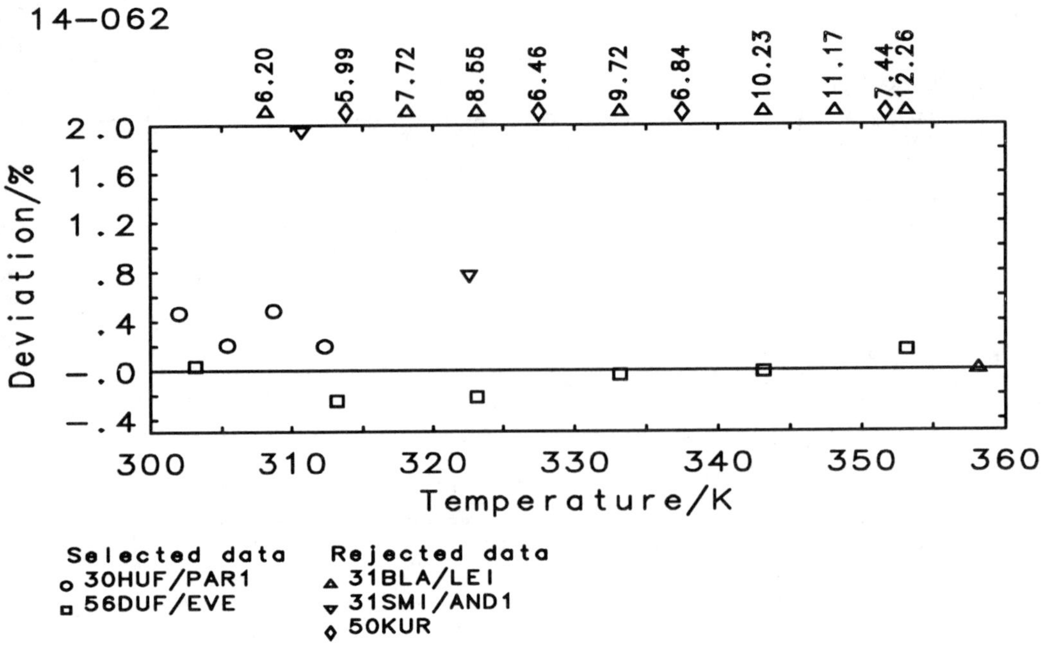

Name: 1,1,4,6–Tetramethylindan
Formula: $C_{13}H_{18}$

CAS-RN: 941–60–6
Group No.: 14–063
Molar Mass: 174.29

TABLE 14.63.1. Experimental heat capacities

Reference	Temp. range K	No. pnts	Error %	Purity %	Purity method	Type capacity	Calorimeter Type	Calorimeter Reference
81LEE/FIN	286.1–377.6	12	0.20	99.937	melpt	C_{sat}	BSAO	47HUF

TABLE 14.63.3. Parameters of regression polynomial

Heat capacity type	No. data points total	No. data points used	s_w	s/R	s_r %	s_b/R	+/–
C_{sat}	12	12	0.295	2.25–2	0.06	–5.40–6	2

Temp. range K	A_1	A_2	Level of uncertainty
286.1–377.6	1.35539+1	7.53843	II

TABLE 14.63.4. Recommended values of heat capacities

Temp. (K)	290	298.15	300	310	320	330	340
c_{sat} (J K^{-1}g^{-1})	1.690	1.719	1.725	1.761	1.797	1.833	1.869
C_{sat} (J K^{-1}mol^{-1})	294.5	299.6	300.7	307.0	313.3	319.5	325.8

Temp. (K)	350	360	370	380
c_{sat} (J K^{-1}g^{-1})	1.905	1.941	1.977	2.013
C_{sat} (J K^{-1}mol^{-1})	332.1	338.3	344.6	350.9

Name: 1,1,4,7–Tetramethylindan
Formula: $C_{13}H_{18}$

CAS-RN: 1078–04–2
Group No.: 14–064
Molar Mass: 174.29

TABLE 14.64.1. Experimental heat capacities

Reference	Temp. range K	No. pnts	Error %	Purity %	Purity method	Type capacity	Calorimeter Type	Calorimeter Reference
81LEE/FIN	170.1–376.4	20	0.20	99.67	melpt	C_{sat}	BSAO	47HUF

TABLE 14.64.3. Parameters of regression polynomial

Heat capacity type	No. data points total	No. data points used	s_w	s/R	s_r %	s_b/R	+/–
C_{sat}	20	20	0.263	1.74–2	0.05	1.55–5	–2

Temp. range K	A_1	A_2	A_3	A_4	Level of uncertainty
170.1–376.4	3.05143+1	–6.97221	4.25614	–4.21718–1	II

TABLE 14.64.4. Recommended values of heat capacities

Temp. (K)	170	180	190	200	210	220	230
c_{sat} ($J K^{-1} g^{-1}$)	1.378	1.398	1.419	1.442	1.466	1.492	1.520
C_{sat} ($J K^{-1} mol^{-1}$)	240.2	243.6	247.3	251.3	255.6	260.1	264.9
Temp. (K)	240	250	260	270	273.15	280	290
c_{sat} ($J K^{-1} g^{-1}$)	1.549	1.579	1.610	1.642	1.652	1.675	1.708
C_{sat} ($J K^{-1} mol^{-1}$)	269.9	275.2	280.6	286.2	287.9	291.9	297.7
Temp. (K)	298.15	300	310	320	330	340	350
c_{sat} ($J K^{-1} g^{-1}$)	1.736	1.742	1.777	1.811	1.846	1.881	1.916
C_{sat} ($J K^{-1} mol^{-1}$)	302.5	303.6	309.6	315.7	321.8	327.9	334.0
Temp. (K)	360	370	380				
c_{sat} ($J K^{-1} g^{-1}$)	1.951	1.986	2.020				
C_{sat} ($J K^{-1} mol^{-1}$)	340.0	346.1	352.0				

Name: Heptylbenzene
Formula: $C_{13}H_{20}$

CAS-RN: 1078–71–3
Group No.: 14–065
Molar Mass: 176.30

TABLE 14.65.1. Experimental heat capacities

Reference	Temp. range K	No. pnts	Error %	Purity %	Purity method	Type capacity	Calorimeter Type	Calorimeter Reference
48TSC1	294.1	1	nosp	not specified		C_p	BSIO	48TSC1
48TSC5	295.1	1	nosp	not specified		C_p	BSIO	48TSC1

TABLE 14.65.3. Parameters of regression polynomial

Heat capacity type	No. data points total	used	s_w	s/R	s_r %	s_b/R	+/−
C_p	2	2	0.000	0.00	0.00	0.00	0

Temp. range K	A_1	Level of uncertainty
294.1–295.1	3.87698+1	VI

TABLE 14.65.4. Recommended values of heat capacities

Temp. (K)	294	295
c_p (J K^{-1}g^{-1})	1.83	1.83
C_p (J K^{-1}mol^{-1})	322	322

Name: Anthracene
Formula: $C_{14}H_{10}$

CAS-RN: 120–12–7
Group No.: 14–066
Molar Mass: 178.23

TABLE 14.66.1. Experimental heat capacities

Reference		Temp. range K	No. pnts	Error %	Purity %	method	Type capacity	Calorimeter Type	Reference
17HIL/DUS	N	526.2–541.2	2	nosp	not specified		C_{avg}	DSIO	11LEW/RAN
70GOU/GIR	N	496.7–511.3	5	0.10	99.99	melpt	C_{sat}	BSAO	53WES/HAT
93DUR/AOU		492.0–592.0	51	nosp	not specified		C_p	BD*CT*	86MER/BEN

17HIL/DUS average values in temperature ranges 489–563 K and 489–593 K
70GOU/GIR smoothed value at 500 K in 68GOU/GIR

TABLE 14.66.2. Correlated heat capacities

Reference	Temp. range K	No. pnts used	$\sigma_r C$ %	d_w	d/R	d_r %	d_b/R	+/−
				Selected data				
17HIL/DUS	526.2–541.2	2	3.00#	0.229	3.17–1	0.69	1.80–1	0
70GOU/GIR	496.7–511.3	5	0.10	3.920	1.75–1	0.39	4.20–3	−1
93DUR/AOU	492.0–592.0	51	3.00#	0.403	5.34–1	1.21	−2.43–1	−7

TABLE 14.66.3. Parameters of regression polynomial

Heat capacity type	No. data points total	used	s_w	s/R	s_r %	s_b/R	+/−
C	58	58	1.245	5.21−1	1.18	−2.07−1	−8

Temp. range K	A_1	A_2	A_3	Level of uncertainty
492.0–592.0	1.74875+2	−5.28183+1	5.35102	V

TABLE 14.66.4. Recommended values of heat capacities

Temp. (K)	490	500	510	520	530	540	550
c (J $K^{-1}g^{-1}$)	2.08	2.08	2.08	2.10	2.11	2.13	2.16
C (J $K^{-1}mol^{-1}$)	370	370	372	373	376	380	384
Temp. (K)	560	570	580	590			
c (J $K^{-1}g^{-1}$)	2.19	2.22	2.26	2.31			
C (J $K^{-1}mol^{-1}$)	390	396	404	412			

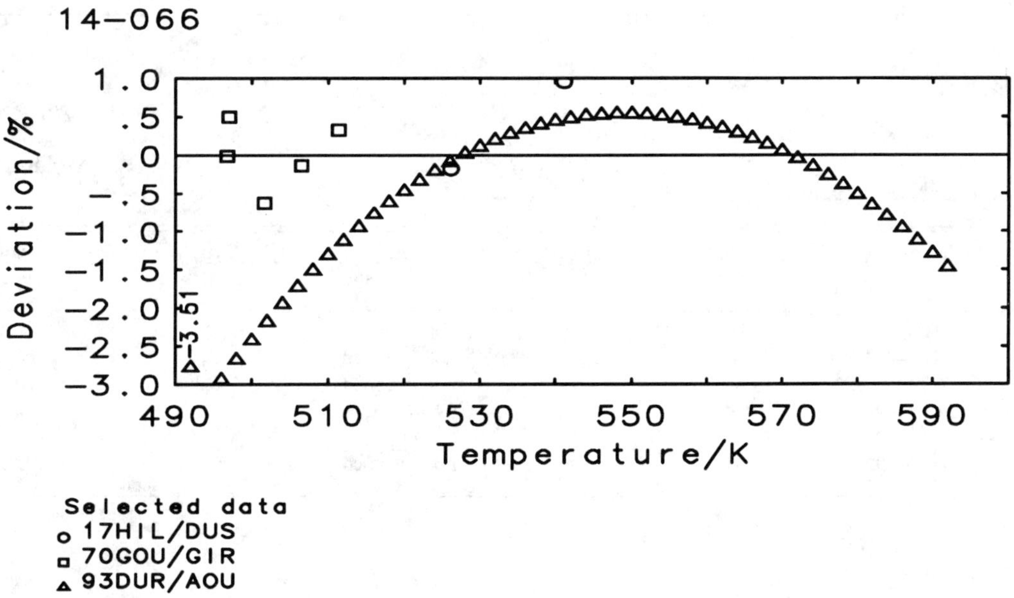

Name: 1,1'–(1,2–Ethynediyl)bisbenzene
Formula: $C_{14}H_{10}$

CAS-RN: 501–65–5
Group No.: 14–067
Molar Mass: 178.23

TABLE 14.67.1. Experimental heat capacities

Reference	Temp. range K	No. pnts	Error %	Purity %	Purity method	Type capacity	Calorimeter Type	Calorimeter Reference
50KUR	364.5–552.5	10	nosp	not specified		C_{avg}	DSIO	47KUR

TABLE 14.67.3. Parameters of regression polynomial

Heat capacity type	No. data points total	No. data points used	s_w	s/R	s_r %	s_b/R	+/–
C	10	10	0.265	4.65–1	1.06	8.84–3	4

Temp. range K	A_1	A_2	Level of uncertainty
364.5–552.5	1.23024+1	6.88427	VI

TABLE 14.67.4. Recommended values of heat capacities

Temp. (K)	370	380	390	400	410	420	430
c ($J K^{-1} g^{-1}$)	1.76	1.79	1.83	1.86	1.89	1.92	1.95
C ($J K^{-1} mol^{-1}$)	314	320	326	331	337	343	348
Temp. (K)	440	450	460	470	480	490	500
c ($J K^{-1} g^{-1}$)	1.99	2.02	2.05	2.08	2.12	2.15	2.18
C ($J K^{-1} mol^{-1}$)	354	360	366	371	377	383	388
Temp. (K)	510	520	530	540	550		
c ($J K^{-1} g^{-1}$)	2.21	2.24	2.28	2.31	2.34		
C ($J K^{-1} mol^{-1}$)	394	400	406	411	417		

Name: Phenanthrene
Formula: $C_{14}H_{10}$

CAS-RN: 85–01–8
Group No.: 14–068
Molar Mass: 178.23

TABLE 14.68.1. Experimental heat capacities

Reference	Temp. range K	No. pnts	Error %	Purity %	Purity method	Type capacity	Calorimeter Type	Calorimeter Reference
41SCH	371.0–473.0	eqn	nosp	not specified		C_p	not specified	
44EIB	371.0–473.0	eqn	nosp	not specified		C_p	not specified	
64RAS/BAS	379.1	1	1.50	not specified		C_p	BSIO	64RAS/BAS
77FIN/MES	383.3–408.6	4	0.20	99.987	melpt	C_{sat}	BSAO	47HUF
93DUR/AOU	392.0–492.0	51	nosp	not specified		C_p	BDCT	86MER/BEN

TABLE 14.68.2. Correlated heat capacities

Reference	Temp. range K	No. pnts used	$\sigma_r C$ %	d_w	d/R	d_r %	d_b/R	+/–
				Selected data				
41SCH	371.0–472.7	10	1.00#	0.758	3.07–1	0.76	9.97–3	–4
77FIN/MES	383.3–408.6	4	0.20	0.084	6.38–3	0.02	–4.16–3	–2
93DUR/AOU	392.0–492.0	51	2.00#	0.383	2.97–1	0.77	3.30–2	–13
				Rejected data				
44EIB	(1.04, 2.79, –3.55–1, –4)			64RAS/BAS	(1.55+1, 29.80, 1.55+1, 1)			

TABLE 14.68.3. Parameters of regression polynomial

Heat capacity type	No. data points total	No. data points used	s_w	s/R	s_r %	s_b/R	+/–
C	76	65	0.462	2.96–1	0.76	2.72–2	–19

Temp. range K	A_1	A_2	A_3	Level of uncertainty
371.0–492.0	–2.07752+1	2.30443+1	–2.09945	V

TABLE 14.68.4. Recommended values of heat capacities

Temp. (K)	380	390	400	410	420	430	440
c (J K^{-1}g^{-1})	1.70	1.73	1.76	1.79	1.82	1.84	1.86
C (J K^{-1}mol^{-1})	303	309	314	319	324	328	332
Temp. (K)	450	460	470	480	490		
c (J K^{-1}g^{-1})	1.89	1.90	1.92	1.93	1.95		
C (J K^{-1}mol^{-1})	336	339	342	345	347		

TABLE 14.68.5. Parameters of quasi-polynomial equation

Heat capacity type	No. data points total	used	s_w	s/R	s_r %	s_b/R	+/–
C	76	65	1.392	2.78–1	0.72	–4.99–2	–17

Temp. range K	T_c K	A_1	A_2	A_3	A_4	Level of uncertainty
371.0–492.0	873.00	2.69250+2	4.70347+1	–6.02175+1	3.85330+2	V

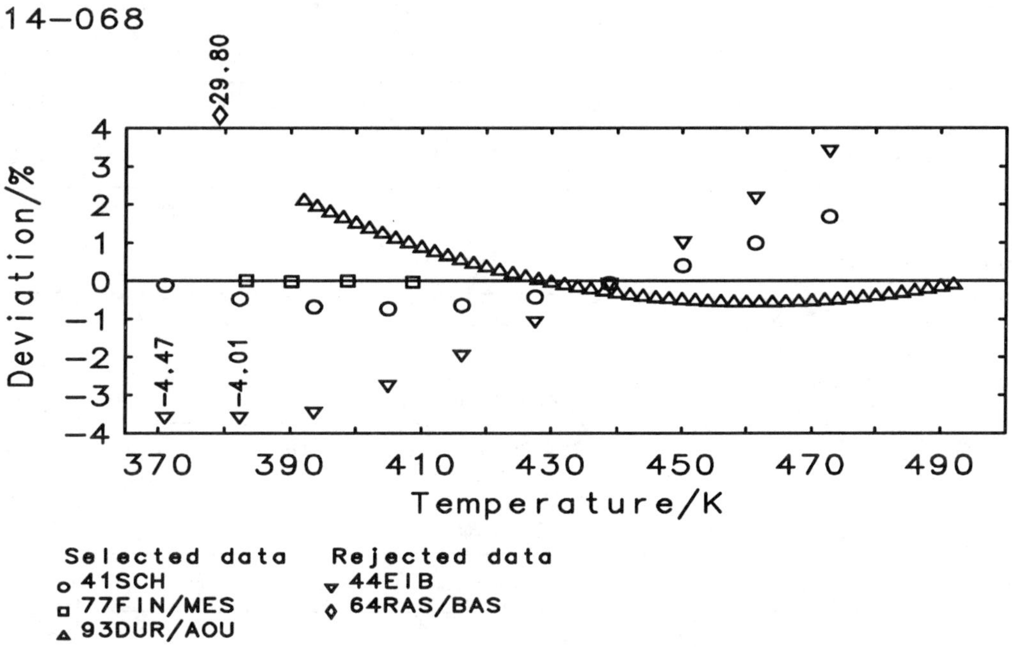

Name: 9,10–Dihydrophenanthrene
Formula: $C_{14}H_{12}$

CAS-RN: 776–35–2
Group No.: 14–069
Molar Mass: 180.25

TABLE 14.69.1. Experimental heat capacities

Reference	Temp. range K	No. pnts	Error %	Purity %	Purity method	Type capacity	Calorimeter Type	Calorimeter Reference
79LEE/HOS	310.2–349.5	34	0.20	99.75	melpt	C_{sat}	BSAO	47HUF

TABLE 14.69.3. Parameters of regression polynomial

Heat capacity type	No. data points total	used	s_w	s/R	s_r %	s_b/R	+/–
C_{sat}	34	34	0.968	6.75–2	0.19	2.50–4	1

Temp. range K	A_1	A_2	Level of uncertainty
310.2–349.5	1.44190+1	6.43817	II

TABLE 14.69.4. Recommended values of heat capacities

Temp. (K)	310	320	330	340	350
c_{sat} (J $K^{-1}g^{-1}$)	1.586	1.615	1.645	1.675	1.705
C_{sat} (J $K^{-1}mol^{-1}$)	285.8	291.2	296.5	301.9	307.2

Name: (*E*)–1,1'–(1,2–Ethenediyl)bisbenzene
Formula: $C_{14}H_{12}$

CAS-RN: 103–30–0
Group No.: 14–070
Molar Mass: 180.25

TABLE 14.70.1. Experimental heat capacities

Reference	Temp. range K	No. pnts	Error %	Purity %	method	Type capacity	Calorimeter Type	Reference
33FER/THO	403.1	1	nosp	not specified		C_p	BDHO	31THO/PAR
50KUR	449.3–552.9	6	nosp	not specified		C_{avg}	DSIO	47KUR
85BOU/DEL	401.6–410.6	4	nosp	99.95	melpt	C_p	BSAO	85BOU/DEL

TABLE 14.70.2. Correlated heat capacities

Reference	Temp. range K	No. pnts used	$\sigma_r C$ %	d_w	d/R	d_r %	d_b/R	+/–
				Selected data				
50KUR	449.3–552.9	6	3.00#	0.366	4.92–1	1.10	–2.38–1	–1
85BOU/DEL	401.6–410.6	4	0.30#	0.325	4.18–2	0.10	3.83–3	0
				Rejected data				
33FER/THO	(3.60–1, 0.85, –3.60–1, –1)							

TABLE 14.70.3. Parameters of regression polynomial

Heat capacity type	No. data points total	used	s_w	s/R	s_r %	s_b/R	+/–
C_p	11	10	0.418	4.57–1	1.02	–1.42–1	–1

Temp. range K	A_1	A_2	A_3	Level of uncertainty
401.6–552.9	3.57797+1	–9.26465–1	6.51152–1	V

TABLE 14.70.4. Recommended values of heat capacities

Temp. (K)	400	410	420	430	440	450	460
c_p (J K^{-1}g^{-1})	1.96	1.98	2.00	2.02	2.04	2.07	2.09
C_p (J K^{-1}mol^{-1})	353	357	361	364	368	372	377
Temp. (K)	470	480	490	500	510	520	530
c_p (J K^{-1}g^{-1})	2.11	2.14	2.16	2.19	2.21	2.24	2.27
C_p (J K^{-1}mol^{-1})	381	385	390	394	399	404	409
Temp. (K)	540	550					
c_p (J K^{-1}g^{-1})	2.30	2.32					
C_p (J K^{-1}mol^{-1})	414	419					

TABLE 14.70.5. Parameters of quasi-polynomial equation

Heat capacity type	No. data points total	used	s_w	s/R	s_r %	s_b/R	+/–
C_p	11	10	4.261	5.99–1	1.39	2.08–1	2

Temp. range K	T_c K	A_1	A_2	A_3	A_4	Level of uncertainty
401.6–552.9	819.00	-9.52009	1.25421	2.40722+1	1.80656+1	V

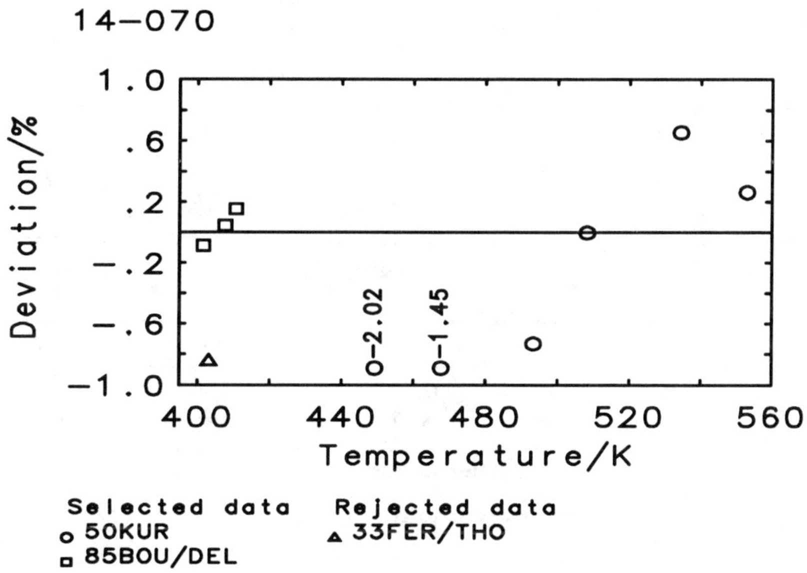

Name: 1,1'–Ethenylidenebisbenzene
Formula: $C_{14}H_{12}$

CAS-RN: 530–48–3
Group No.: 14–071
Molar Mass: 180.25

TABLE 14.71.1. Experimental heat capacities

Reference	Temp. range K	No. pnts	Error %	Purity %	Purity method	Type capacity	Calorimeter Type	Calorimeter Reference
31SMI/AND1	286.0–298.5	2	nosp	99.3	melpt	C_p	DSIO	26AND/LYN

TABLE 14.71.3. Parameters of regression polynomial

Heat capacity type	No. data points total	used	s_w	s/R	s_r %	s_b/R	+/–
C_p	2	2	0.000	0.00	0.00	0.00	0

Temp. range K	A_1	A_2	Level of uncertainty
286.0–298.5	1.91565+1	5.63602	V

TABLE 14.71.4. Recommended values of heat capacities

Temp. (K)	290	298.15	300
c_p (J K^{-1}g^{-1})	1.64	1.66	1.66
C_p (J K^{-1}mol^{-1})	295	299	300

Name: 1,1'–(1,2–Ethanediyl)bisbenzene
Formula: $C_{14}H_{14}$

CAS-RN: 103–29–7
Group No.: 14–072
Molar Mass: 182.27

TABLE 14.72.1. Experimental heat capacities

Reference		Temp. range K	No. pnts	Error %	Purity %	Purity method	Type capacity	Calorimeter Type	Calorimeter Reference
08BOG/WIN	N	348.1	1	nosp	not specified		C_{avg}	DSIO	08BOG/WIN
33FER/THO		333.1–343.1	2S	nosp	not specified		C_p	BDHO	31THO/PAR
41SCH		324.0–473.0	eqn	nosp	not specified		C_p	not specified	
50KUR		349.4–427.4	5	nosp	not specified		C_{avg}	DSIO	47KUR
88MES/FIN		335.3–372.3	15	0.10	99.961	melpt	C_{sat}	BSAO	47HUF

08BOG/WIN average value in temperature range 325–371 K

TABLE 14.72.2. Correlated heat capacities

Reference	Temp. range K	No. pnts used	$\sigma_r C$ %	d_w	d/R	d_r %	d_b/R	+/–
				Selected data				
41SCH	324.0–472.4	15	1.50#	0.528	3.16–1	0.79	1.10–2	–5
88MES/FIN	335.3–372.3	15	0.10	0.627	2.49–2	0.06	7.38–6	0
				Rejected data				
08BOG/WIN	(1.31, 3.23, 1.31, 1)		33FER/THO	(9.53–2, 0.25, 9.50–2, 2)				
50KUR	(9.08–1, 2.08, 8.78–1, 5)							

TABLE 14.72.3. Parameters of regression polynomial

Heat capacity type	No. data points total	used	s_w	s/R	s_r %	s_b/R	+/–
C	38	30	0.611	2.36–1	0.59	5.48–3	–5

Temp. range K	A_1	A_2	A_3	Level of uncertainty
324.0–472.4	3.42711	1.35622+1	–9.41607–1	III

TABLE 14.72.4. Recommended values of heat capacities

Temp. (K)	330	340	350	360	370	380	390
c (J K^{-1}g^{-1})	1.730	1.763	1.796	1.827	1.857	1.887	1.916
C (J K^{-1}mol^{-1})	315.4	321.4	327.3	333.0	338.5	343.9	349.2
Temp. (K)	400	410	420	430	440	450	460
c (J K^{-1}g^{-1})	1.944	1.971	1.997	2.022	2.047	2.071	2.093
C (J K^{-1}mol^{-1})	354.3	359.2	364.0	368.6	373.1	377.4	381.5
Temp. (K)	470						
c (J K^{-1}g^{-1})	2.115						
C (J K^{-1}mol^{-1})	385.5						

TABLE 14.72.5. Parameters of quasi-polynomial equation

Heat capacity type	No. data points total	used	s_w	s/R	s_r %	s_b/R	+/−
C	38	30	4.572	1.91−1	0.49	−7.50−2	−9

Temp. range K	T_c K	A_1	A_2	A_3	A_4	Level of uncertainty
324.0–472.4	785.00	8.62970	3.07069−1	1.70449+1	6.06311+1	III

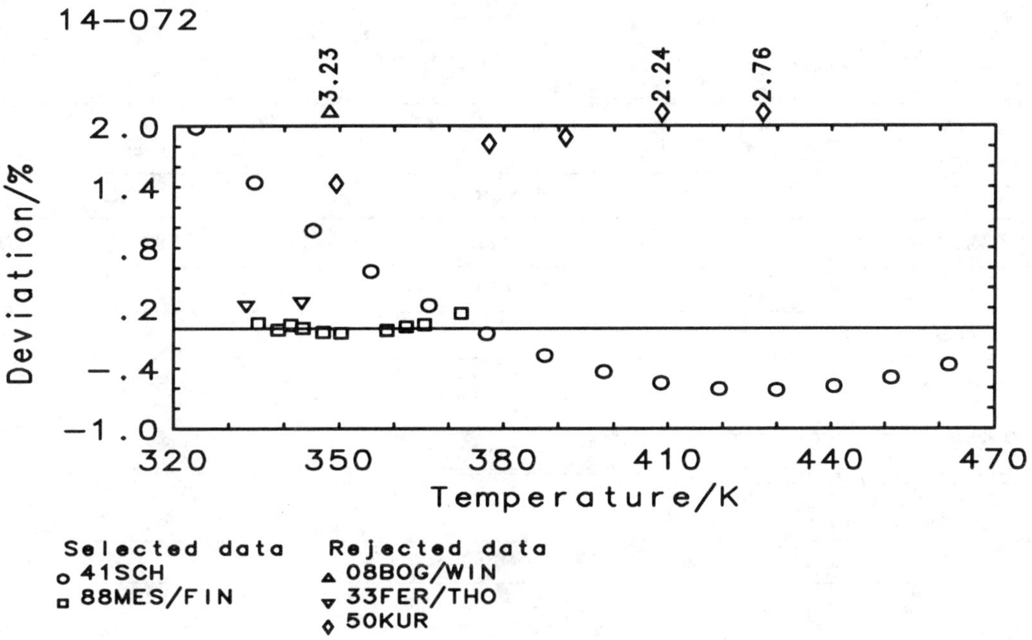

Name: 1,1'–Ethylidenebisbenzene
Formula: $C_{14}H_{14}$

CAS-RN: 612–00–0
Group No.: 14–073
Molar Mass: 182.27

TABLE 14.73.1. Experimental heat capacities

Reference	Temp. range K	No. pnts	Error %	Purity %	Purity method	Type capacity	Calorimeter Type	Calorimeter Reference
31SMI/AND1	259.8–298.5	4	nosp	not specified		C_p	DSIO	26AND/LYN

TABLE 14.73.3. Parameters of regression polynomial

Heat capacity type	No. data points total	used	s_w	s/R	s_r %	s_b/R	+/–
C_p	4	4	0.132	8.87–2	0.26	1.98–4	–2

Temp. range K	A_1	A_2	Level of uncertainty
259.8–298.5	1.43694+1	7.08295	V

TABLE 14.73.4. Recommended values of heat capacities

Temp. (K)	260	270	273.15	280	290	298.15	300
c_p (J K^{-1}g^{-1})	1.50	1.53	1.54	1.56	1.59	1.62	1.62
C_p (J K^{-1}mol^{-1})	273	278	280	284	290	295	296

Name: 1,2,3,4,5,6,7,8–Octahydroanthracene
Formula: $C_{14}H_{18}$

CAS-RN: 1079–71–6
Group No.: 14–074
Molar Mass: 186.30

TABLE 14.74.1. Experimental heat capacities

Reference	Temp. range K	No. pnts	Error %	Purity %	Purity method	Type capacity	Calorimeter Type	Calorimeter Reference
71BOY/SAN	372.0–390.0	2	nosp	99.9	melpt	C_p	BDHT	73PER/COM
82GAM/CAL	351.6–396.1	10	0.05	99.996	melpt	C_{sat}	BSAO	47HUF

TABLE 14.74.2. Correlated heat capacities

Reference	Temp. range K	No. pnts used	$\sigma_r C$ %	d_w	d/R	d_r %	d_b/R	+/–
				Selected data				
82GAM/CAL	351.6–396.1	10	0.05	0.241	5.10–3	0.01	1.53–6	0
				Rejected data				
71BOY/SAN	(3.08, 6.60, 3.05, 2)							

TABLE 14.74.3. Parameters of regression polynomial

Heat capacity type	No. data points total	used	s_w	s/R	s_r %	s_b/R	+/–
C_{sat}	12	10	0.289	6.10–3	0.01	1.53–6	0

Temp. range K	A_1	A_2	A_3	Level of uncertainty
351.6–396.1	–7.68131	1.87469+1	–1.40580	II

TABLE 14.74.4. Recommended values of heat capacities

Temp. (K)	350	360	370	380	390	400
c_{sat} (J K^{-1}g^{-1})	1.817	1.856	1.894	1.931	1.966	2.000
C_{sat} (J K^{-1}mol^{-1})	338.5	345.8	352.8	359.7	366.2	372.6

Name: Octylbenzene
Formula: $C_{14}H_{22}$

CAS-RN: 2189–60–8
Group No.: 14–075
Molar Mass: 190.33

TABLE 14.75.1. Experimental heat capacities

Reference	Temp. range K	No. pnts	Error %	Purity %	Purity method	Type capacity	Calorimeter Type	Calorimeter Reference
48TSC1	295.1	1	nosp	not specified		C_p	BSIO	48TSC1
48TSC5	295.1	1	nosp	not specified		C_p	BSIO	48TSC1

TABLE 14.75.3. Parameters of regression polynomial

Heat capacity type	No. data points total	No. data points used	s_w	s/R	s_r %	s_b/R	+/–
C_p	2	2	0.158	2.03–1	0.48	9.65–4	0

Temp. range K	A_1	Level of uncertainty
295.1–295.1	4.27633+1	V

TABLE 14.75.4. Recommended values of heat capacities

Temp. (K)	295.1
c_p (J K^{-1}g^{-1})	1.87
C_p (J K^{-1}mol^{-1})	356

Name: 4–Methylphenanthrene
Formula: $C_{15}H_{12}$

CAS-RN: 832–64–4
Group No.: 14–076
Molar Mass: 192.26

TABLE 14.76.1. Experimental heat capacities

Reference	Temp. range K	No. pnts	Error %	Purity %	Purity method	Type capacity	Calorimeter Type	Calorimeter Reference
89CHI/HOS	331.7–445.3	18	nosp	99.92	melpt	C_{sat}	BSAO	47HUF

TABLE 14.76.3. Parameters of regression polynomial

Heat capacity type	No. data points total	No. data points used	s_w	s/R	s_r %	s_b/R	+/–
C_{sat}	18	18	0.275	1.09–2	0.03	6.78–6	2

Temp. range K	A_1	A_2	A_3	Level of uncertainty
331.7–445.3	7.10016	1.00294+1	–3.43111–1	II

TABLE 14.76.4. Recommended values of heat capacities

Temp. (K)	330	340	350	360	370	380	390
c_{sat} (J $K^{-1}g^{-1}$)	1.577	1.610	1.643	1.676	1.709	1.741	1.773
C_{sat} (J $K^{-1}mol^{-1}$)	303.2	309.6	316.0	322.3	328.5	334.7	340.9
Temp. (K)	400	410	420	430	440		
c_{sat} (J $K^{-1}g^{-1}$)	1.805	1.836	1.867	1.898	1.928		
C_{sat} (J $K^{-1}mol^{-1}$)	346.9	353.0	358.9	364.9	370.7		

TABLE 14.76.5. Parameters of quasi-polynomial equation

Heat capacity type	No. data points total	used	s_w	s/R	s_r %	s_b/R	+/–
C_{sat}	18	18	0.264	1.04–2	0.03	5.51–6	2

Temp. range K	T_c K	A_1	A_2	A_3	A_4	Level of uncertainty
331.7–445.3	898.60	2.23454+1	1.22605	7.36100	1.01814+2	II

Name: (1–Methylethyl)–1,1'–biphenyl
Formula: $C_{15}H_{16}$

CAS-RN: 25640–78–2
Group No.: 14–077
Molar Mass: 196.29

TABLE 14.77.1. Experimental heat capacities

Reference		Temp. range K	No. pnts	Error %	Purity % method	Type capacity	Calorimeter Type	Reference
56MCE/MAL		422.0–588.7	4S	nosp	not specified	C_p	not specified	
63VAR/KOP	N	293.1–671.2	10	1.50	not specified	C_{sat}	BSAO	63VAR/KOP
64VUK/RAS		311.8–485.0	6	3.00	not specified	C_{sat}	BDHO	64VUK/RAS

63VAR/KOP maximum difference between reported values (measured at 1 MPa) for C_{sat} is 0.5 %

TABLE 14.77.2. Correlated heat capacities

Reference	Temp. range K	No. pnts used	$\sigma_r C$ %	d_w	d/R	d_r %	d_b/R	+/–
				Selected data				
56MCE/MAL	422.0–588.7	4	1.50#	0.916	8.84–1	1.37	–4.84–1	0
63VAR/KOP	293.1–671.2	10	1.50	0.591	4.96–1	0.89	3.08–1	6
64VUK/RAS	311.8–485.0	6	3.00	0.567	7.73–1	1.70	–5.97–1	–4

TABLE 14.77.3. Parameters of regression polynomial

Heat capacity type	No. data points total	used	s_w	s/R	s_r %	s_b/R	+/–
C	20	20	0.719	7.34–1	1.39	–1.22–1	2

Temp. range K	A_1	A_2	A_3	Level of uncertainty
293.1–671.2	3.08238+1	1.01804	8.63981–1	V

TABLE 14.77.4. Recommended values of heat capacities

Temp. (K)	290	298.15	300	310	320	330	340
c ($J K^{-1} g^{-1}$)	1.74	1.76	1.76	1.79	1.82	1.85	1.88
C ($J K^{-1} mol^{-1}$)	341	345	346	352	357	362	368
Temp. (K)	350	360	370	380	390	400	410
c ($J K^{-1} g^{-1}$)	1.90	1.94	1.97	2.00	2.03	2.06	2.10
C ($J K^{-1} mol^{-1}$)	374	380	386	392	399	405	412
Temp. (K)	420	430	440	450	460	470	480
c ($J K^{-1} g^{-1}$)	2.13	2.17	2.20	2.24	2.28	2.32	2.36
C ($J K^{-1} mol^{-1}$)	419	426	433	440	447	455	462
Temp. (K)	490	500	510	520	530	540	550
c ($J K^{-1} g^{-1}$)	2.40	2.44	2.48	2.52	2.56	2.61	2.65
C ($J K^{-1} mol^{-1}$)	470	478	486	495	503	511	520
Temp. (K)	560	570	580	590	600	610	620
c ($J K^{-1} g^{-1}$)	2.69	2.74	2.79	2.83	2.88	2.93	2.98
C ($J K^{-1} mol^{-1}$)	529	538	547	556	566	575	585
Temp. (K)	630	640	650	660	670		
c ($J K^{-1} g^{-1}$)	3.03	3.08	3.13	3.18	3.24		
C ($J K^{-1} mol^{-1}$)	595	605	615	625	635		

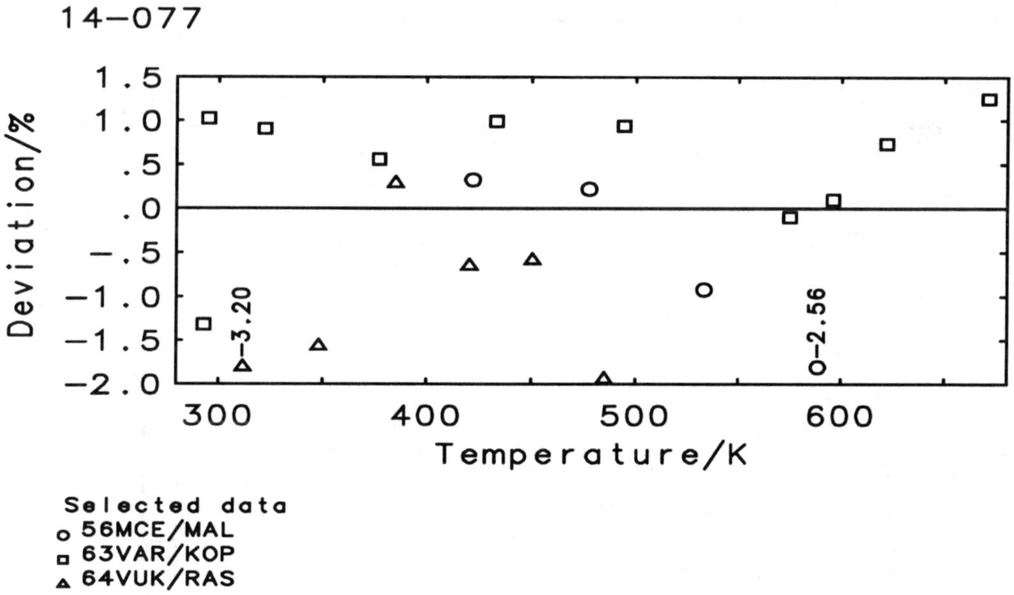

Name: Fluoranthene
Formula: $C_{16}H_{10}$

CAS-RN: 206–44–0
Group No.: 14–078
Molar Mass: 202.26

TABLE 14.78.1. Experimental heat capacities

Reference	Temp. range K	No. pnts	Error %	Purity %	Purity method	Type capacity	Calorimeter Type	Calorimeter Reference
71WON/WES	393.7–422.6	5	0.10	99.936	melpt	C_{sat}	BSAO	68WES/WES
71WON/WES	386.4–451.7	10	0.10	99.61	melpt	C_{sat}	BSAO	68WES/WES
93DUR/AOU	492.0–592.0	51	nosp	not specified		C_p	BDCT	86MER/BEN

TABLE 14.78.2. Correlated heat capacities

Reference	Temp. range K	No. pnts used	$\sigma_r C$ %	d_w	d/R	d_r %	d_b/R	+/–
				Selected data				
71WON/WES	393.7–422.6	5	0.10	3.557	1.53–1	0.36	9.50–2	4
71WON/WES	386.4–451.7	10	0.10	1.821	7.64–2	0.18	–4.68–2	–6

TABLE 14.78.3. Parameters of regression polynomial

Heat capacity type	No. data points total	No. data points used	s_w	s/R	s_r %	s_b/R	+/–
C_{sat}	66	15	2.724	1.16–1	0.27	5.05–4	–2

Temp. range K	A_1	A_2	Level of uncertainty
386.4–451.7	1.42876+1	6.76896	II

TABLE 14.78.4. Recommended values of heat capacities

Temp. (K)	390	400	410	420	430	440	450
c_{sat} (J K^{-1}g^{-1})	1.673	1.700	1.728	1.756	1.784	1.812	1.840
C_{sat} (J K^{-1}mol^{-1})	338.3	343.9	349.5	355.2	360.8	366.4	372.1

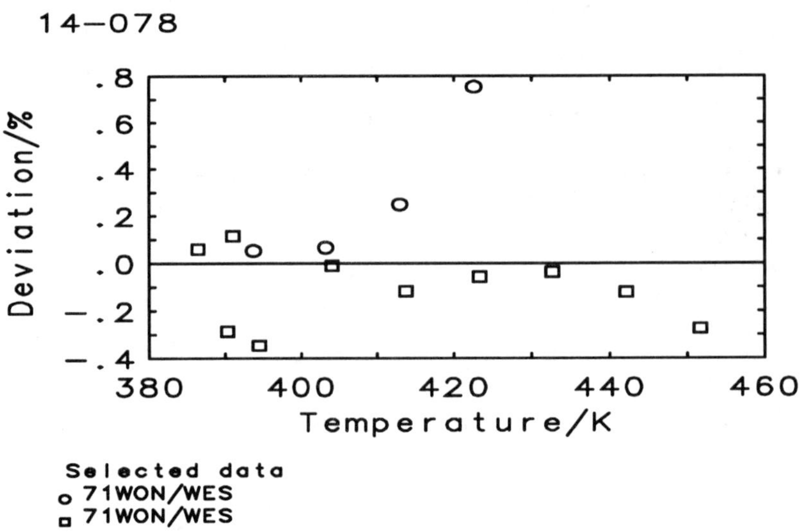

Name: Pyrene
Formula: $C_{16}H_{10}$

CAS-RN: 129–00–0
Group No.: 14–079
Molar Mass: 202.26

TABLE 14.79.1. Experimental heat capacities

Reference	Temp. range K	No. pnts	Error %	Purity %	Purity method	Type capacity	Calorimeter Type	Calorimeter Reference
71WON/WES	430.6–478.8	7	0.10	99.97	melpt	C_{sat}	BSAO	68WES/WES
93DUR/AOU	492.0–592.0	51	nosp	not specified		C_p	BDCT	86MER/BEN

TABLE 14.79.2. Correlated heat capacities

Reference	Temp. range K	No. pnts used	$\sigma_r C$ %	d_w	d/R	d_r %	d_b/R	+/–
				Selected data				
71WON/WES	430.6–478.8	7	0.10	0.689	3.02–2	0.07	–2.45–3	0
93DUR/AOU	492.0–592.0	51	3.00#	0.333	4.92–1	01.00	3.18–1	27

TABLE 14.79.3. Parameters of regression polynomial

Heat capacity type	No. data points total	No. data points used	s_w	s/R	s_r %	s_b/R	+/–
C	58	58	0.404	4.74–1	0.96	2.79–1	27

Temp. range K	A_1	A_2	A_3	Level of uncertainty
430.6–592.0	3.04330+1	–8.79318–1	8.49992–1	IV

TABLE 14.79.4. Recommended values of heat capacities

Temp. (K)	430	440	450	460	470	480	490
c (J K^{-1}g^{-1})	1.742	1.769	1.796	1.824	1.853	1.883	1.913
C (J K^{-1}mol^{-1})	352.3	357.7	363.2	368.9	374.8	380.8	386.9
Temp. (K)	500	510	520	530	540	550	560
c (J K^{-1}g^{-1})	1.944	1.976	2.008	2.041	2.075	2.109	2.144
C (J K^{-1}mol^{-1})	393.2	399.6	406.1	412.8	419.6	426.6	433.7
Temp. (K)	570	580	590				
c (J K^{-1}g^{-1})	2.180	2.217	2.254				
C (J K^{-1}mol^{-1})	441.0	448.4	455.9				

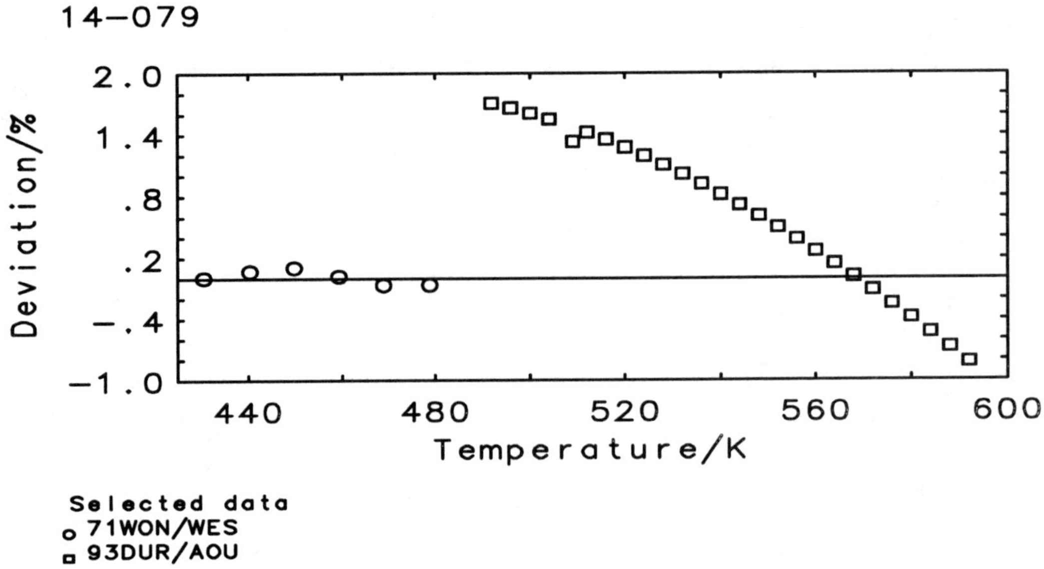

Name: Triphenylene
Formula: $C_{18}H_{12}$

CAS-RN: 217–59–4
Group No.: 14–080
Molar Mass: 228.29

TABLE 14.80.1. Experimental heat capacities

Reference	Temp. range K	No. pnts	Error %	Purity %	Purity method	Type capacity	Calorimeter Type	Calorimeter Reference
71WON/WES	481.0–509.5	4	0.10	99.962	melpt	C_{sat}	BSAO	68WES/WES

TABLE 14.80.3. Parameters of regression polynomial

Heat capacity type	No. data points total	No. data points used	s_w	s/R	s_r %	s_b/R	+/–
C_{sat}	4	4	0.567	3.09–2	0.06	5.72–6	0

Temp. range K	A_1	A_2	A_3	Level of uncertainty
481.0–509.5	2.35548+2	–7.99810+1	8.76376	II

TABLE 14.80.4. Recommended values of heat capacities

Temp. (K)	480	490	500	510
c_{sat} (J K^{-1}g^{-1})	1.951	1.969	1.994	2.025
C_{sat} (J K^{-1}mol^{-1})	445.3	449.5	455.1	462.2

Name: 1,1':2',1''–Terphenyl
Formula: $C_{18}H_{14}$

CAS-RN: 84–15–1
Group No.: 14–081
Molar Mass: 230.31

TABLE 14.81.1. Experimental heat capacities

Reference	Temp. range K	No. pnts	Error %	Purity %	Purity method	Type capacity	Calorimeter Type	Calorimeter Reference
72CHA/BES	330.1–355.6	14	0.10	99.995	melpt	C_p	BSAO	65STE/BLA

TABLE 14.81.3. Parameters of regression polynomial

Heat capacity type	No. data points total	No. data points used	s_w	s/R	s_r %	s_b/R	+/–
C_p	14	14	0.489	2.33–2	0.05	2.02–5	–1

Temp. range K	A_1	A_2	Level of uncertainty
330.1–355.6	2.03451+1	8.03673	II

TABLE 14.81.4. Recommended values of heat capacities

Temp. (K)	330	340	350
c_p (J K^{-1}g^{-1})	1.692	1.721	1.750
C_p (J K^{-1}mol^{-1})	389.7	396.4	403.0

Name: 1,1':3',1''–Terphenyl
Formula: $C_{18}H_{14}$

CAS-RN: 92–06–8
Group No.: 14–082
Molar Mass: 230.31

TABLE 14.82.1. Experimental heat capacities

Reference	Temp. range K	No. pnts	Error %	Purity %	Purity method	Type capacity	Calorimeter Type	Calorimeter Reference
58WAL/BRO	432.3–534.3	3	2.00	not specified		C_p	DSIO	56WAL/GRA

TABLE 14.82.3. Parameters of regression polynomial

Heat capacity type	No. data points total	No. data points used	s_w	s/R	s_r %	s_b/R	+/–
C_p	3	3	0.017	1.97–2	0.03	–1.53–5	–1

Temp. range K	A_1	A_2	Level of uncertainty
432.3–534.3	2.45197+1	6.93100	V

TABLE 14.82.4. Recommended values of heat capacities

Temp. (K)	440	450	460	470	480	490	500
c_p ($J K^{-1} g^{-1}$)	1.99	2.01	2.04	2.06	2.09	2.11	2.14
C_p ($J K^{-1} mol^{-1}$)	457	463	469	475	480	486	492
Temp. (K)	510	520	530				
c_p ($J K^{-1} g^{-1}$)	2.16	2.19	2.21				
C_p ($J K^{-1} mol^{-1}$)	498	504	509				

Name: 1,1':4',1''–Terphenyl
Formula: $C_{18}H_{14}$

CAS-RN: 92–94–4
Group No.: 14–083
Molar Mass: 230.31

TABLE 14.83.1. Experimental heat capacities

Reference	Temp. range K	No. pnts	Error %	Purity %	Purity method	Type capacity	Calorimeter Type	Calorimeter Reference
83CHA	487.0–580.0	6S	0.50	99.9	chrom	C_p	BDHT	69PER/COM
93DUR/AOU	492.0–592.0	102	nosp	not specified		C_p	BDCT	86MER/BEN

TABLE 14.83.2. Correlated heat capacities

Reference	Temp. range K	No. pnts used	$\sigma_r C$ %	d_w	d/R	d_r %	d_b/R	+/–
				Selected data				
83CHA	487.0–580.0	6	0.50	0.480	1.57–1	0.24	1.00–3	0
				Rejected data				
93DUR/AOU	(4.89, 8.35, –4.73, –90)							

TABLE 14.83.3. Parameters of regression polynomial

Heat capacity type	No. data points total	No. data points used	s_w	s/R	s_r %	s_b/R	+/–
C_p	108	6	0.588	1.92–1	0.29	1.00–3	0

Temp. range K	A_1	A_2	Level of uncertainty
487.0–580.0	1.15041+1	9.63474	III

TABLE 14.83.4. Recommended values of heat capacities

Temp. (K)	490	500	510	520	530	540	550
c_p ($J K^{-1} g^{-1}$)	2.120	2.154	2.189	2.224	2.259	2.294	2.328
C_p ($J K^{-1} mol^{-1}$)	488.2	496.2	504.2	512.2	520.2	528.2	536.2
Temp. (K)	560	570	580				
c_p ($J K^{-1} g^{-1}$)	2.363	2.398	2.433				
C_p ($J K^{-1} mol^{-1}$)	544.3	552.3	560.3				

Name: 1–Methyl–7–(1–methylethyl)phenanthrene
Formula: $C_{18}H_{18}$

CAS-RN: 483–65–8
Group No.: 14–084
Molar Mass: 234.34

TABLE 14.84.1. Experimental heat capacities

Reference	Temp. range K	No. pnts	Error %	Purity % method	Type capacity	Calorimeter Type Reference
44EIB	369.0–473.0	eqn	nosp	not specified	C_p	not specified

TABLE 14.84.3. Parameters of regression polynomial

Heat capacity type	No. data points total	used	s_w	s/R	s_r %	s_b/R	+/–
C_p	10	10	0.000	3.02–6	0.00	–3.82–7	0

Temp. range K	A_1	A_2	Level of uncertainty
369.0–473.4	2.16330+1	7.79479	V

TABLE 14.84.4. Recommended values of heat capacities

Temp. (K)	370	380	390	400	410	420	430
c_p (J K^{-1}g^{-1})	1.79	1.82	1.85	1.87	1.90	1.93	1.96
C_p (J K^{-1}mol^{-1})	420	426	433	439	446	452	459
Temp. (K)	440	450	460	470			
c_p (J K^{-1}g^{-1})	1.98	2.01	2.04	2.07			
C_p (J K^{-1}mol^{-1})	465	472	478	484			

Name: *ar*,*ar'*–Bis(1–methylethyl)–1,1'–biphenyl
Formula: $C_{18}H_{22}$

CAS-RN: 36876–13–8
Group No.: 14–085
Molar Mass: 238.37

TABLE 14.85.1. Experimental heat capacities

Reference		Temp. range K	No. pnts	Error %	Purity % method	Type capacity	Calorimeter Type Reference
56MCE/MAL	N	422.0–588.7	4S	nosp	not specified	C_p	not specified

56MCE/MAL probably mixture of isomers

TABLE 14.85.3. Parameters of regression polynomial

Heat capacity type	No. data points total	used	s_w	s/R	s_r %	s_b/R	+/–
C_p	4	4	0.034	4.68–2	0.07	–1.54–4	0

Temp. range K	A_1	A_2	Level of uncertainty
422.0–588.7	2.13374+1	9.78795	V

TABLE 14.85.4. Recommended values of heat capacities

Temp. (K)	420	430	440	450	460	470	480
c_p (J $K^{-1}g^{-1}$)	2.18	2.21	2.25	2.28	2.31	2.35	2.38
C_p (J $K^{-1}mol^{-1}$)	519	527	535	544	552	560	568
Temp. (K)	490	500	510	520	530	540	550
c_p (J $K^{-1}g^{-1}$)	2.42	2.45	2.49	2.52	2.55	2.59	2.62
C_p (J $K^{-1}mol^{-1}$)	576	584	592	601	609	617	625
Temp. (K)	560	570	580	590			
c_p (J $K^{-1}g^{-1}$)	2.66	2.69	2.72	2.76			
C_p (J $K^{-1}mol^{-1}$)	633	641	649	658			

Name: 1,1',1''–Methylidynetrisbenzene
Formula: $C_{19}H_{16}$

CAS-RN: 519–73–3
Group No.: 14–086
Molar Mass: 244.34

TABLE 14.86.1. Experimental heat capacities

Reference		Temp. range K	No. pnts	Error %	Purity %	Purity method	Type capacity	Calorimeter Type	Calorimeter Reference
17HIL/DUS	N	383.1	1	nosp	not specified		C_{avg}	DSIO	11LEW/RAN
32SPA/THO		373.1–383.1	2S	nosp	not specified		C_p	BDHO	31THO/PAR
44EIB		365.0–473.0	eqn	nosp	not specified		C_p	not specified	
50KUR		400.0–596.2	15	nosp	not specified		C_{avg}	DSIO	47KUR

17HIL/DUS average value in temperature range 369–408 K

TABLE 14.86.2. Correlated heat capacities

Reference	Temp. range K	No. pnts used	$\sigma_r C$ %	d_w	d/R	d_r %	d_b/R	+/–
				Selected data				
44EIB	365.0–465.0	11	1.00#	0.588	3.49–1	0.59	–2.57–1	–8
50KUR	400.0–596.2	15	3.00#	1.475	2.79	4.42	1.97	9
				Rejected data				
17HIL/DUS	(6.88, 11.69, 6.88, 1)		32SPA/THO	(3.33, 6.08, 3.33, 2)				

TABLE 14.86.3. Parameters of regression polynomial

Heat capacity type	No. data points total	No. data points used	s_w	s/R	s_r %	s_b/R	+/–
C_p	29	26	1.259	2.27	3.60	1.02	1

Temp. range K	A_1	A_2	A_3	Level of uncertainty
365.0–596.2	1.76016+1	6.68562	5.98972–1	V

TABLE 14.86.4. Recommended values of heat capacities

Temp. (K)	370	380	390	400	410	420	430
c_p (J K^{-1}g^{-1})	1.72	1.76	1.80	1.84	1.87	1.91	1.95
C_p (J K^{-1}mol^{-1})	420	429	439	448	458	468	477
Temp. (K)	440	450	460	470	480	490	500
c_p (J K^{-1}g^{-1})	1.99	2.04	2.08	2.12	2.16	2.20	2.25
C_p (J K^{-1}mol^{-1})	487	497	507	518	528	538	549
Temp. (K)	510	520	530	540	550	560	570
c_p (J K^{-1}g^{-1})	2.29	2.33	2.38	2.42	2.47	2.51	2.56
C_p (J K^{-1}mol^{-1})	559	570	581	592	603	614	625
Temp. (K)	580	590	600				
c_p (J K^{-1}g^{-1})	2.60	2.65	2.70				
C_p (J K^{-1}mol^{-1})	636	648	659				

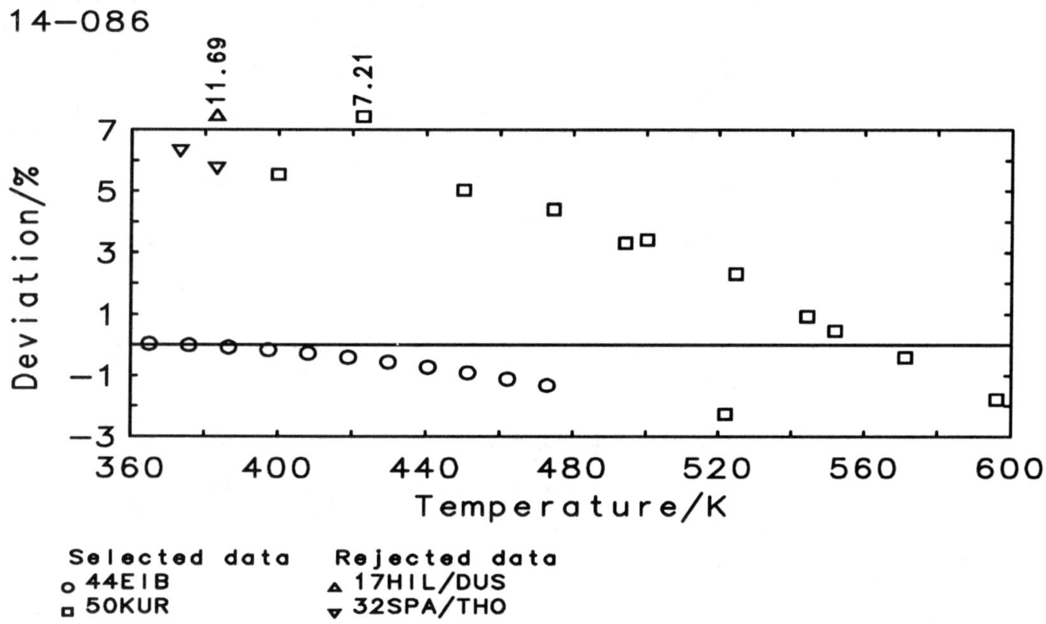

Name: Perylene
Formula: $C_{20}H_{12}$

CAS-RN: 198–55–0
Group No.: 14–087
Molar Mass: 252.32

TABLE 14.87.1. Experimental heat capacities

Reference	Temp. range K	No. pnts	Error %	Purity %	Purity method	Type capacity	Calorimeter Type	Calorimeter Reference
80WON/WES	554.4–573.0	5	0.10	99.97	melpt	C_p	BSAO	68WES/WES
93DUR/AOU	552.0–652.0	51	nosp	not specified		C_p	BDCT	86MER/BEN

TABLE 14.87.2. Correlated heat capacities

Reference	Temp. range K	No. pnts used	$\sigma_r C$ %	d_w	d/R	d_r %	d_b/R	+/–
				Selected data				
80WON/WES	554.4–573.0	5	0.30#	0.942	1.76–1	0.28	9.96–4	1
				Rejected data				
93DUR/AOU	(1.78, 2.92, –1.77, –9)							

TABLE 14.87.3. Parameters of regression polynomial

Heat capacity type	No. data points total	used	s_w	s/R	s_r %	s_b/R	+/–
C_p	56	5	1.216	2.27–1	0.36	9.96–4	1

Temp. range K	A_1	A_2	Level of uncertainty
554.4–573.0	3.75800+1	4.44166	III

TABLE 14.87.4. Recommended values of heat capacities

Temp. (K)	560	570	580
c_p (J K^{-1}g^{-1})	2.058	2.073	2.087
C_p (J K^{-1}mol^{-1})	519.3	523.0	526.7

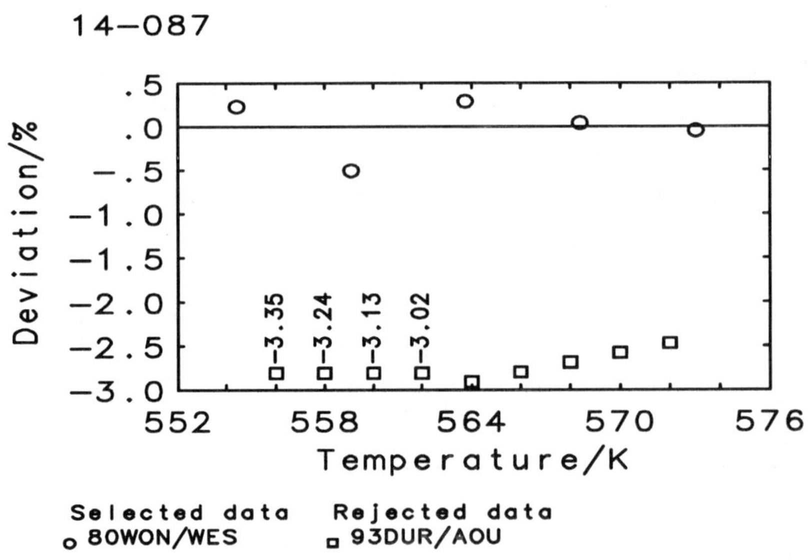

Name: 9,10–Dihydro–9,10[1',2']–benzenoanthracene
Formula: $C_{20}H_{14}$

CAS-RN: 477–75–8
Group No.: 14–088
Molar Mass: 254.33

TABLE 14.88.1. Experimental heat capacities

Reference	Temp. range K	No. pnts	Error %	Purity %	Purity method	Type capacity	Calorimeter Type	Calorimeter Reference
70AND/WES	532.5–548.8	7	nosp	99.9995	melpt	C_p	BSAO	68WES/WES

TABLE 14.88.3. Parameters of regression polynomial

Heat capacity type	No. data points total	No. data points used	s_w	s/R	s_r %	s_b/R	+/–
C_p	7	7	0.335	8.88–2	0.13	1.70–4	–1

Temp. range K	A_1	A_2	Level of uncertainty
532.5–548.8	3.09320+1	6.56449	III

TABLE 14.88.4. Recommended values of heat capacities

Temp. (K)	530	540	550
c_p (J K^{-1}g^{-1})	2.149	2.170	2.192
C_p (J K^{-1}mol^{-1})	546.5	551.9	557.4

Name: 1–(2–Naphthalenylmethyl)naphthalene
Formula: $C_{21}H_{16}$

CAS-RN: 611–48–3
Group No.: 14–089
Molar Mass: 268.36

TABLE 14.89.1. Experimental heat capacities

Reference	Temp. range K	No. pnts	Error %	Purity %	Purity method	Type capacity	Calorimeter Type	Calorimeter Reference
77FIN/MES	343.3–422.9	10	0.20	99.9	melpt	C_{sat}	BSAO	47HUF

TABLE 14.89.3. Parameters of regression polynomial

Heat capacity type	No. data points total	No. data points used	s_w	s/R	s_r %	s_b/R	+/–
C_{sat}	10	10	0.207	2.34–2	0.04	1.33–5	–4

Temp. range K	A_1	A_2	A_3	Level of uncertainty
343.3–422.9	9.86475	1.50534+1	–7.28695–1	II

TABLE 14.89.4. Recommended values of heat capacities

Temp. (K)	350	360	370	380	390	400	410
c_{sat} (J K^{-1}g^{-1})	1.661	1.692	1.722	1.752	1.781	1.810	1.838
C_{sat} (J K^{-1}mol^{-1})	445.9	454.1	462.2	470.1	478.0	485.7	493.3
Temp. (K)	420						
c_{sat} (J K^{-1}g^{-1})	1.866						
C_{sat} (J K^{-1}mol^{-1})	500.8						

Name: Pentacene
Formula: $C_{22}H_{14}$

CAS-RN: 135–48–8
Group No.: 14–090
Molar Mass: 278.35

TABLE 14.90.1. Experimental heat capacities

Reference	Temp. range K	No. pnts	Error %	Purity %	Purity method	Type capacity	Calorimeter Type	Calorimeter Reference
93DUR/AOU	546.0–594.0	25	nosp	not specified		C_p	BDCT	86MER/BEN

TABLE 14.90.3. Parameters of regression polynomial

Heat capacity type	No. data points total	No. data points used	s_w	s/R	s_r %	s_b/R	+/–
C_p	25	25	0.000	3.59–4	0.00	–9.15–7	0

Temp. range K	A_1	A_2	Level of uncertainty
546.0–594.0	1.05476+1	1.01792+1	VI

TABLE 14.90.4. Recommended values of heat capacities

Temp. (K)	550	560	570	580	590
c_p (J K^{-1}g^{-1})	1.99	2.02	2.05	2.08	2.11
C_p (J K^{-1}mol^{-1})	553	562	570	579	587

Name: 5'–Phenyl–1,1':3',1''–terphenyl
Formula: $C_{24}H_{18}$

CAS-RN: 612–71–5
Group No.: 14–091
Molar Mass: 306.41

TABLE 14.91.1. Experimental heat capacities

Reference	Temp. range K	No. pnts	Error %	Purity %	Purity method	Type capacity	Calorimeter Type	Calorimeter Reference
82LEB/BYK	446.0–480.0	2S	2.00	99.4	melpt	C_p	BDHT	74DSM/COM

TABLE 14.91.3. Parameters of regression polynomial

Heat capacity type	No. data points total	used	s_w	s/R	s_r %	s_b/R	+/–
C_p	2	2	0.000	0.00	0.00	0.00	0
Temp. range K		A_1	A_2				Level of uncertainty
446.0–480.0		1.49243+1	1.32652+1				V

TABLE 14.91.4. Recommended values of heat capacities

Temp. (K)	450	460	470	480
c_p (J K^{-1}g^{-1})	2.02	2.06	2.10	2.13
C_p (J K^{-1}mol^{-1})	620	631	642	653

Name: 1,1':3',1'':3'',1'''–Quaterphenyl
Formula: $C_{24}H_{18}$

CAS-RN: 1166–18–3
Group No.: 14–092
Molar Mass: 306.41

TABLE 14.92.1. Experimental heat capacities

Reference	Temp. range K	No. pnts	Error %	Purity %	method	Type capacity	Calorimeter Type	Reference
58WAL/BRO	431.5–534.0	3	2.00	not specified		C_p	DSIO	56WAL/GRA

TABLE 14.92.3. Parameters of regression polynomial

Heat capacity type	No. data points total	used	s_w	s/R	s_r %	s_b/R	+/–
C_p	3	3	0.009	1.40–2	0.02	1.27–5	–1
Temp. range K		A_1	A_2				Level of uncertainty
431.5–534.0		3.48860+1	8.56650				V

TABLE 14.92.4. Recommended values of heat capacities

Temp. (K)	430	440	450	460	470	480	490
c_p (J K^{-1}g^{-1})	1.95	1.97	1.99	2.02	2.04	2.06	2.09
C_p (J K^{-1}mol^{-1})	596	603	611	618	625	632	639
Temp. (K)	500	510	520	530			
c_p (J K^{-1}g^{-1})	2.11	2.13	2.16	2.18			
C_p (J K^{-1}mol^{-1})	646	653	660	668			

Name: 1,1–Diphenyldodecane
Formula: $C_{24}H_{34}$

CAS-RN: 1603–53–8
Group No.: 14–093
Molar Mass: 322.53

TABLE 14.93.1. Experimental heat capacities

Reference		Temp. range K	No. pnts	Error %	Purity %	Purity method	Type capacity	Calorimeter Type	Calorimeter Reference
60KAR/STR1	N	283.4–298.3	6	nosp	99.1	melpt	C_p	BSAO	54STR/ICK

60KAR/STR1 smoothed values in 60KAR/STR3

TABLE 14.93.3. Parameters of regression polynomial

Heat capacity type	No. data points total	No. data points used	s_w	s/R	s_r %	s_b/R	+/–
C_p	6	6	0.373	2.64–1	0.37	1.31–3	0

Temp. range K	A_1	A_2	Level of uncertainty
283.4–298.3	4.90212+1	7.49856	IV

TABLE 14.93.4. Recommended values of heat capacities

Temp. (K)	290	298.15	300
c_p ($J K^{-1}g^{-1}$)	1.824	1.840	1.844
C_p ($J K^{-1}mol^{-1}$)	588.4	593.5	594.6

Name: (1–Cyclohexyldodecyl)benzene
Formula: $C_{24}H_{40}$

CAS-RN: 62155–50–4
Group No.: 14–094
Molar Mass: 328.58

TABLE 14.94.1. Experimental heat capacities

Reference		Temp. range K	No. pnts	Error %	Purity %	Purity method	Type capacity	Calorimeter Type	Calorimeter Reference
60KAR/STR2	N	282.0–297.1	9	nosp	97.6	melpt	C_p	BSAO	54STR/ICK

60KAR/STR2 smoothed data in 60KAR/STR3

TABLE 14.94.3. Parameters of regression polynomial

Heat capacity type	No. data points total	No. data points used	s_w	s/R	s_r %	s_b/R	+/–
C_p	9	9	0.178	1.30–1	0.18	3.64–4	1

Temp. range K	A_1	A_2	Level of uncertainty
282.0–297.1	5.65882+1	5.63261	IV

TABLE 14.94.4. Recommended values of heat capacities

Temp. (K)	290	298.15	300
c_p ($J K^{-1}g^{-1}$)	1.845	1.857	1.860
C_p ($J K^{-1}mol^{-1}$)	606.3	610.1	611.0

Name: 9,10–Diphenylanthracene
Formula: $C_{26}H_{18}$

CAS-RN: 1499–10–1
Group No.: 14–095
Molar Mass: 330.43

TABLE 14.95.1. Experimental heat capacities

Reference	Temp. range K	No. pnts	Error %	Purity %	Purity method	Type capacity	Calorimeter Type	Calorimeter Reference
93DUR/AOU	522.0–622.0	51	nosp	not specified		C_p	BDCT	86MER/BEN

TABLE 14.95.3. Parameters of regression polynomial

Heat capacity type	No. data points total	No. data points used	s_w	s/R	s_r %	s_b/R	+/–
C_p	51	51	0.000	3.52–4	0.00	1.50–6	0

Temp. range K	A_1	A_2	Level of uncertainty
522.0–622.0	1.68489+1	1.19477+1	VI

TABLE 14.95.4. Recommended values of heat capacities

Temp. (K)	520	530	540	550	560	570	580
c_p (J $K^{-1}g^{-1}$)	1.99	2.02	2.05	2.08	2.11	2.14	2.17
C_p (J $K^{-1}mol^{-1}$)	657	667	677	686	696	706	716
Temp. (K)	590	600	610	620			
c_p (J $K^{-1}g^{-1}$)	2.20	2.23	2.26	2.29			
C_p (J $K^{-1}mol^{-1}$)	726	736	746	756			

Name: 11–Phenylheneicosane
Formula: $C_{27}H_{48}$

CAS-RN: 6703–80–6
Group No.: 14–096
Molar Mass: 372.68

TABLE 14.96.1. Experimental heat capacities

Reference	Temp. range K	No. pnts	Error %	Purity %	Purity method	Type capacity	Calorimeter Type	Calorimeter Reference
49PAR/MOO	300.00	2.054	nosp	not specified		C_p	BSIO	25PAR

Name: 1,1',1''–(1,3,5–Benzenetriyl)trisnaphthalene
Formula: $C_{36}H_{24}$

CAS-RN: 7059–70–3
Group No.: 14–097
Molar Mass: 456.59

TABLE 14.97.1. Experimental heat capacities

Reference	Temp. range K	No. pnts	Error %	Purity %	Purity method	Type capacity	Calorimeter Type	Calorimeter Reference
67MAG N	472.0–560.0	5	nosp	not specified		C_p	BDHT	66PER/COM

67MAG data from a graph only

TABLE 14.97.3. Parameters of regression polynomial

Heat capacity type	No. data points total	used	s_w	s/R	s_r %	s_b/R	+/–
C_p	5	5	0.160	3.53–1	0.32	1.12–3	–3

Temp. range K	A_1	A_2	Level of uncertainty
472.0–560.0	4.07841+1	1.35793+1	V

TABLE 14.97.4. Recommended values of heat capacities

Temp. (K)	480	490	500	510	520	530	540
c_p ($J\,K^{-1}g^{-1}$)	1.93	1.95	1.98	2.00	2.03	2.05	2.08
C_p ($J\,K^{-1}mol^{-1}$)	881	892	904	915	926	937	949
Temp. (K)	550	560					
c_p ($J\,K^{-1}g^{-1}$)	2.10	2.13					
C_p ($J\,K^{-1}mol^{-1}$)	960	971					

2. Compounds of Carbon, Hydrogen and Halogen

21. Fluorine Drivatives

The fluorine derivatives family contains 33 compounds, three of which were measured at one temperature only.

Most measurements were carried out in the 1960's and 1970's at NPLT where data on 6 fluorinated benzenes and toluenes (65COU/GRI, 68COU/HAL1, 68COU/HAL2 73AND/MAR1, 74AND/MAR) were obtained. The second laboratory which was involved in the study of these compounds, BMB (now called NIPER), performed measurements for four fluorinated benzenes and two fluorinated toluenes (56SCO/MCC1, 59SCO/DOU, 62SCO/MES, 63SCO/MES, 70MES/FIN). Data from both laboratories have uncertainties of 0.2 % and have been, therefore, included in the correlation. A similar situation holds for two data sets from NBSW (53FUR/MCC, 54FUR/MCC), two data sets from WRUC (67PAC/PLA, 69SMI/PAC), and one data set from UMMA (64TRO/WES2). On the other hand, the data contained in three articles originating from IICN (69PAU/LAV1, 71PAU, 75PAU1), have been rejected from the correlation because, in all cases, the data differed by 1 to 2 % from those of NPLT which we feel are more reliable. Exception has been made with the only available data for decafluorobiphenyl (71PAU/RAK).

Hexafluorobenzene has had its heat capacity measured most frequently. Accurate data (65COU/GRI, 70MES/FIN) have been supplemented by consistent data from IVTM (82GOR/SMI) with the reported error of 0.5 % for extending the upper temperature limit be several kelvins.

Low molar mass fluorinated alkanes and cycloalkanes are used as refrigerants and were thus measured repeatedly. Extensive measurements of temperature and pressure dependence far above the normal boiling temperature were carried out in the fSU (78SOL/SUK, 84STO/CHA) and data were published only as parameters of smoothing equations. In general, the data seem to have low accuracy. Also, for two substances, reliable low temperature data were available (measurements for trifluoromethane from UCB, 62VAL/BRO, and from CITP, 72COP/REA; measurements for octafluorocyclobutane from NBSW, 54FUR/MCC) which allowed one to extend the temperature range of recommendations considerably. Other Freons were measured at CIPT (44RUS/GOL, 72COP/REA) and at PSC (48PAC/AST).

Perfluorinated alkanes, C_5F_{12} and C_6F_{14}, were measured at the University of Strathclyde, Glasgow (74COC/NOR) using DSC and showed low accuracy over a limited temperature range. DSC measurements at UCCC (51OLI/GRI1) on C_7F_{16} were made and yielded results with high accuracy. Three differently hydrogenated and fluorinated naphthalenes were investigated at MSUM (81ZHO/KOS1) and results gave data having an estimated uncertainty of 0.5 %.

Name: Tetrafluoromethane
Formula: CF_4

CAS–RN: 75–73–0
Group No.: 21–001
Molar Mass: 88.00

TABLE 21.1.1. Experimental heat capacities

Reference	Temp. range K	No. pnts	Error %	Purity %	Purity method	Type capacity	Calorimeter Type	Calorimeter Reference
38EUC/SCH	100.0–145.0	6S	nosp	99.9	estim	C_p	BSAO	38EUC/SCH
69ENO/SHI2	90.4–91.7	2	nosp	99.985	melpt	C_p	BSAO	66SHI/ATA
69SMI/PAC	92.0–139.9	21	nosp	99.985	melpt	C_{sat}	BSAO	55PAC/PIE
72COP/REA	121.6–141.4	6	nosp	99.9	estim	C_p	BSAO	68WES/FUR

TABLE 21.1.2. Correlated heat capacities

Reference	Temp. range K	No. pnts used	$\sigma_r C$ %	d_w	d/R	d_r %	d_b/R	+/–
				Selected data				
69SMI/PAC	92.0–139.9	21	0.20#	0.672	1.26–2	0.13	3.42–5	–4
				Rejected data				
38EUC/SCH	(1.10–1, 1.16, 1.08–1, 4)			72COP/REA	(7.32–2, 0.77, 7.18–2, 5)			

TABLE 21.1.3. Parameters of regression polynomial

Heat capacity type	No. data points total	used	s_w	s/R	s_r %	s_b/R	+/–
C_{sat}	35	21	0.725	1.36–2	0.15	3.42–5	–4

Temp. range K	A_1	A_2	A_3	Level of uncertainty
92.0–139.9	1.28211+1	–6.39122	2.87755	II

TABLE 21.1.4. Recommended values of heat capacities

Temp. (K)	100	110	120	130	140
c_{sat} (J K^{-1}g^{-1})	0.879	0.876	0.878	0.886	0.899
C_{sat} (J K^{-1}mol^{-1})	77.39	77.10	77.29	77.95	79.10

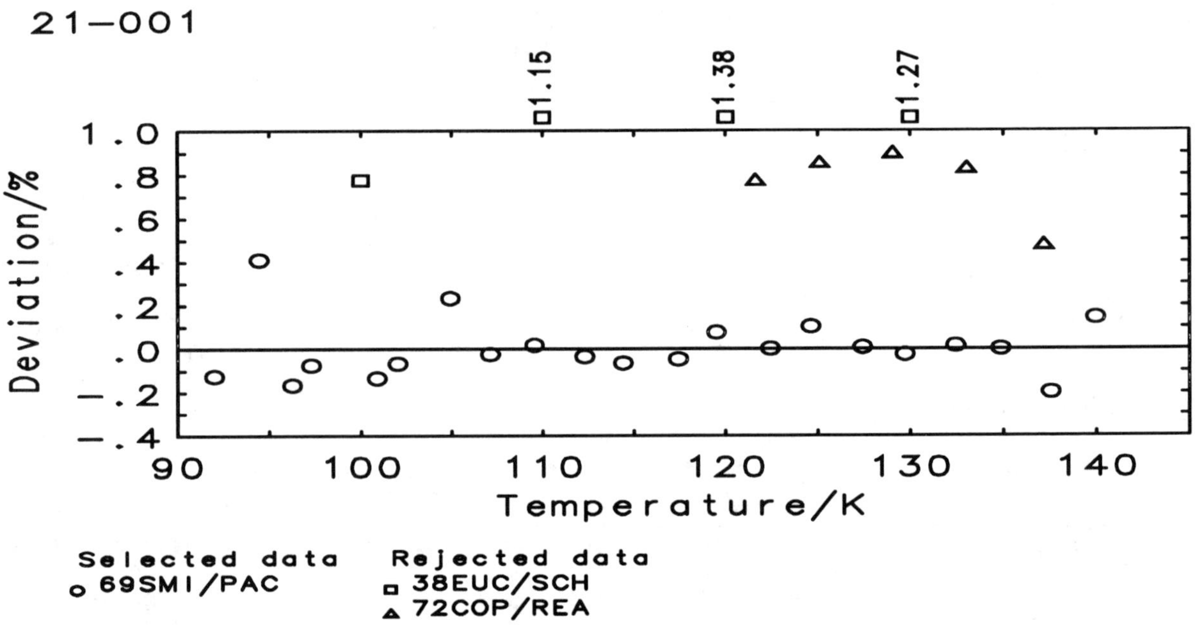

Name: Trifluoromethane
Formula: CHF_3

CAS–RN: 75–46–7
Group No.: 21–002
Molar Mass: 70.01

TABLE 21.2.1. Experimental heat capacities

Reference	Temp. range K	No. pnts	Error %	Purity %	Purity method	Type capacity	Calorimeter Type	Calorimeter Reference
62VAL/BRO	122.7–189.3	8	nosp	99.999	melpt	C_{sat}	BSIO	37GIA/EGA
72COP/REA	127.6–142.9	4	nosp	99.8	chrom	C_p	BSAO	68WES/FUR
78SOL/SUK	123.0–153.0	eqn	nosp	not specified		C_{sat}	BSAO	78SOL/SUK
78SOL/SUK	153.0–288.0	eqn	nosp	not specified		C_{sat}	BSAO	78SOL/SUK
84STO/CHA	173.0–191.0	eqn	0.30	99.98	anal	C_p	BSAO	78SOL/SUK

TABLE 21.2.2. Correlated heat capacities

Reference	Temp. range K	No. pnts used	$\sigma_r C$ %	d_w	d/R	d_r %	d_b/R	+/–
				Selected data				
62VAL/BRO	122.7–189.3	8	0.30#	0.508	1.56–2	0.15	1.54–3	2
72COP/REA	127.6–142.9	4	0.20#	0.639	1.29–2	0.13	–5.40–3	–1
78SOL/SUK	123.0–153.0	5	1.00#	0.041	4.17–3	0.04	–1.49–4	1
78SOL/SUK	153.0–279.0	15	1.00#	0.594	8.57–2	0.59	–8.17–4	–5
84STO/CHA	173.0–191.0	5	0.30	0.266	8.20–3	0.08	8.02–3	5

TABLE 21.2.3. Parameters of cubic spline polynomials

Heat capacity type	No. data points total	No. data points used	s_w	s/R	s_r %	s_b/R	+/–
C_p	38	37	0.541	5.95–2	0.42	4.82–4	2

Temp. range K	A_1	A_2	A_3	A_4	Level of uncertainty
122.7–220.0	1.42801+1	–5.98233	2.25110	–9.23258–2	III
220.0–279.0	–1.75344+2	2.52596+2	–1.15284+2	1.77161+1	IV

TABLE 21.2.4. Recommended values of heat capacities

Temp. (K)	120	130	140	150	160	170	180
c_p (J K^{-1}g^{-1})	1.209	1.200	1.195	1.195	1.199	1.207	1.219
C_p (J K^{-1}mol^{-1})	84.67	84.01	83.67	83.64	83.92	84.49	85.36
Temp. (K)	190	200	210	220	230	240	250
c_p (J K^{-1}g^{-1})	1.236	1.257	1.281	1.310	1.345	1.396	1.476
C_p (J K^{-1}mol^{-1})	86.53	87.98	89.71	91.72	94.15	97.74	103.4
Temp. (K)	260	270	273.15	280			
c_p (J K^{-1}g^{-1})	1.598	1.775	1.844	2.019			
C_p (J K^{-1}mol^{-1})	111.9	124.3	129.1	141.3			

TABLE 21.2.5. Parameters of quasi-polynomial equation

Heat capacity type	No. data points total	used	s_w		s/R		s_r %	s_b/R	+/–
C_p	38	37	0.966		4.20–2		0.40	9.71–3	1
Temp. range K	T_c K	A_1	A_2		A_3	A_4	A_5	A_6	Level of uncertainty
122.7–279.0	299.30	–2.64123	2.72549–1		8.96704	3.34691	–1.67958+1	1.02816+1	IV

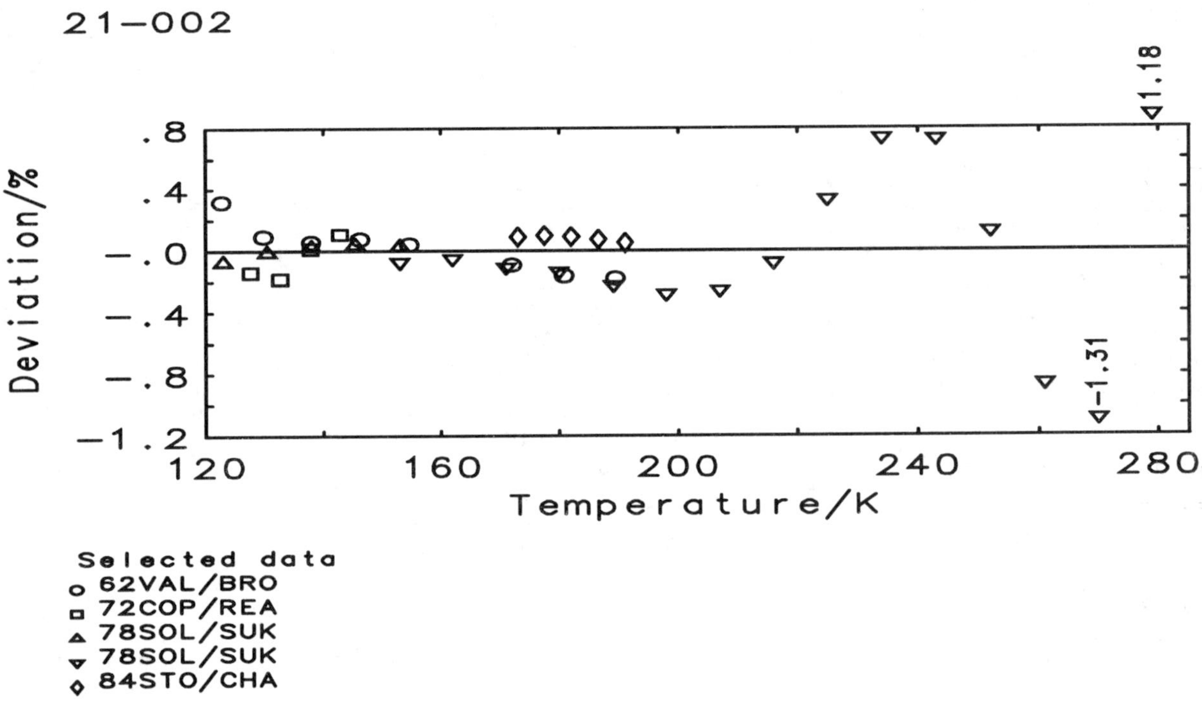

Name: Tetrafluoroethene
Formula: C_2F_4

CAS–RN: 116–14–3
Group No.: 21–003
Molar Mass: 100.02

TABLE 21.3.1. Experimental heat capacities

Reference	Temp. range K	No. pnts	Error %	Purity %	Purity method	Type capacity	Calorimeter Type	Calorimeter Reference
53FUR/MCC	145.0–210.0	15S	0.20	99.99	melpt	C_{sat}	BSAO	45SCO/MEY

TABLE 21.3.3. Parameters of regression polynomial

Heat capacity type	No. data points total	used	s_w	s/R	s_r %	s_b/R	+/–
C_{sat}	15	15	0.070	1.80–3	0.01	8.27–7	0
Temp. range K		A_1	A_2	A_3			Level of uncertainty
145.0–210.0		1.49227+1	–3.63698	1.47869			II

TABLE 21.3.4. Recommended values of heat capacities

Temp. (K)	150	160	170	180	190	200	210
c_{sat} (J K^{-1}g^{-1})	1.064	1.071	1.082	1.095	1.110	1.128	1.148
C_{sat} (J K^{-1}mol^{-1})	106.4	107.2	108.2	109.5	111.0	112.8	114.8

TABLE 21.3.5. Parameters of quasi-polynomial equation

Heat capacity type	No. data points total	used	s_w	s/R	s_r %	s_b/R	+/–
C_{sat}	15	15	0.356	9.34–3	0.07	1.04–5	0

Temp. range K	T_c K	A_1	A_2	A_3	A_4	Level of uncertainty
145.0–210.0	306.50	5.98733–1	1.06879	1.10593+1	8.38519–2	II

Name: Hexafluoroethane
Formula: C_2F_6

CAS–RN: 76–16–4
Group No.: 21–004
Molar Mass: 138.01

TABLE 21.4.1. Experimental heat capacities

Reference	Temp. range K	No. pnts	Error %	Purity %	Purity method	Type capacity	Calorimeter Type	Calorimeter Reference
48PAC/AST	174.9–190.7	10	0.50	99.84	melpt	C_p	BSIO	36AST/MES

TABLE 21.4.3. Parameters of regression polynomial

Heat capacity type	No. data points total	used	s_w	s/R	s_r %	s_b/R	+/–
C_p	10	10	0.211	1.59–2	0.11	2.69–5	–1

Temp. range K	A_1	A_2	A_3	Level of uncertainty
174.9–190.7	5.52188+1	–4.78117+1	1.41587+1	III

TABLE 21.4.4. Recommended values of heat capacities

Temp. (K)	180	185	190
c_p (J K^{-1}g^{-1})	0.906	0.917	0.933
C_p (J K^{-1}mol^{-1})	125.0	126.6	128.8

Name: 1,1,1–Trifluoroethane
Formula: $C_2H_3F_3$

CAS–RN: 420–46–2
Group No.: 21–005
Molar Mass: 84.04

TABLE 21.5.1. Experimental heat capacities

Reference	Temp. range K	No. pnts	Error %	Purity %	Purity method	Type capacity	Calorimeter Type	Calorimeter Reference
44RUS/GOL	165.4–220.7	11	0.10	99.95	melpt	C_p	BSAO	41YOS/GAR

TABLE 21.5.3. Parameters of regression polynomial

Heat capacity type	No. data points total	No. data points used	s_w	s/R	s_r %	s_b/R	+/–
C_p	11	11	0.866	1.10–2	0.09	1.38–5	–3

Temp. range K	A_1	A_2	A_3	Level of uncertainty
165.4–220.7	1.27400+1	–1.85308	9.34890–1	II

TABLE 21.5.4. Recommended values of heat capacities

Temp. (K)	170	180	190	200	210	220
c_p ($J K^{-1} g^{-1}$)	1.216	1.230	1.246	1.264	1.283	1.305
C_p ($J K^{-1} mol^{-1}$)	102.2	103.4	104.7	106.2	107.9	109.7

TABLE 21.5.5. Parameters of quasi-polynomial equation

Heat capacity type	No. data points total	No. data points used	s_w	s/R	s_r %	s_b/R	+/–
C_p	11	11	0.904	1.14–2	0.09	1.48–5	–2

Temp. range K	T_c K	A_1	A_2	A_3	A_4	Level of uncertainty
165.4–220.7	346.30	–8.09898–1	7.68275–1	1.01313+1	2.13444–1	II

Name: Octafluoropropane
Formula: C_3F_8

CAS–RN: 76–19–7
Group No.: 21–006
Molar Mass: 188.02

TABLE 21.6.1. Experimental heat capacities

Reference	Temp. range K	No. pnts	Error %	Purity %	Purity method	Type capacity	Calorimeter Type	Calorimeter Reference
67PAC/PLA	125.8–233.0	36	0.10	99.993	melpt	C_{sat}	BSAO	55PAC/PIE

TABLE 21.6.3. Parameters of regression polynomial

Heat capacity type	No. data points total	used	s_w	s/R	s_r %	s_b/R	+/–
C_{sat}	36	36	0.983	6.71–2	0.34	4.63–4	–1

Temp. range K	A_1	A_2	Level of uncertainty
125.8–233.0	1.28050+1	3.85835	III

TABLE 21.6.4. Recommended values of heat capacities

Temp. (K)	130	140	150	160	170	180	190
c_{sat} (J K^{-1}g^{-1})	0.788	0.805	0.822	0.839	0.856	0.873	0.890
C_{sat} (J K^{-1}mol^{-1})	148.2	151.4	154.6	157.8	161.0	164.2	167.4
Temp. (K)	200	210	220	230			
c_{sat} (J K^{-1}g^{-1})	0.907	0.925	0.942	0.959			
C_{sat} (J K^{-1}mol^{-1})	170.6	173.8	177.0	180.3			

TABLE 21.6.5. Parameters of quasi-polynomial equation

Heat capacity type	No. data points total	used	s_w	s/R	s_r %	s_b/R	+/–
C_{sat}	36	36	0.994	6.81–2	0.35	4.43–4	–2

Temp. range K	T_c K	A_1	A_2	A_3	A_4	Level of uncertainty
125.8–233.0	345.00	–2.84810–1	1.59589–3	1.29009+1	1.27071+1	III

Name: Octafluorocyclobutane
Formula: C_4F_8

CAS–RN: 115–25–3
Group No.: 21–007
Molar Mass: 200.03

TABLE 21.7.1. Experimental heat capacities

Reference	Temp. range K	No. pnts	Error %	Purity %	Purity method	Type capacity	Calorimeter Type	Calorimeter Reference
54FUR/MCC	234.7–268.5	9	0.20	99.98	melpt	C_{sat}	BSAO	45SCO/MEY
66VAS	237.1–316.9	17	0.80	99.8	melpt	C_{sat}	BDAO	64VAS2
84STO/CHA	233.0–267.0	eqn	0.30	99.98	anal	C_p	BSAO	78SOL/SUK

TABLE 21.7.2. Correlated heat capacities

Reference	Temp. range K	No. pnts used	$\sigma_r C$ %	d_w	d/R	d_r %	d_b/R	+/–
				Selected data				
54FUR/MCC	234.7–268.5	9	0.20	0.586	2.87–2	0.12	–1.27–2	–3
84STO/CHA	233.0–267.4	9	0.30	0.413	3.04–2	0.12	2.88–2	9
				Rejected data				
66VAS	(2.52, 11.44, –2.52, –9)							

TABLE 21.7.3. Parameters of regression polynomial

Heat capacity type	No. data points total	used	s_w	s/R	s_r %	s_b/R	+/–
C_p	35	18	0.538	3.14–2	0.13	8.07–3	6
C_{sat}	35	18	0.537	3.13–2	0.13	8.07–3	6

Temp. range K	A_1	A_2	Level of uncertainty
233.0–268.5	1.46654+1	3.95051	II
233.0–268.5	1.50684+1	3.77870	II

TABLE 21.7.4. Recommended values of heat capacities

Temp. (K)	240	250	260	270
c_p (J K^{-1}g^{-1})	1.004	1.020	1.037	1.053
C_p (J K^{-1}mol^{-1})	200.8	204.1	207.3	210.6
c_{sat} (J K^{-1}g^{-1})	1.003	1.019	1.035	1.050
C_{sat} (J K^{-1}mol^{-1})	200.7	203.8	207.0	210.1

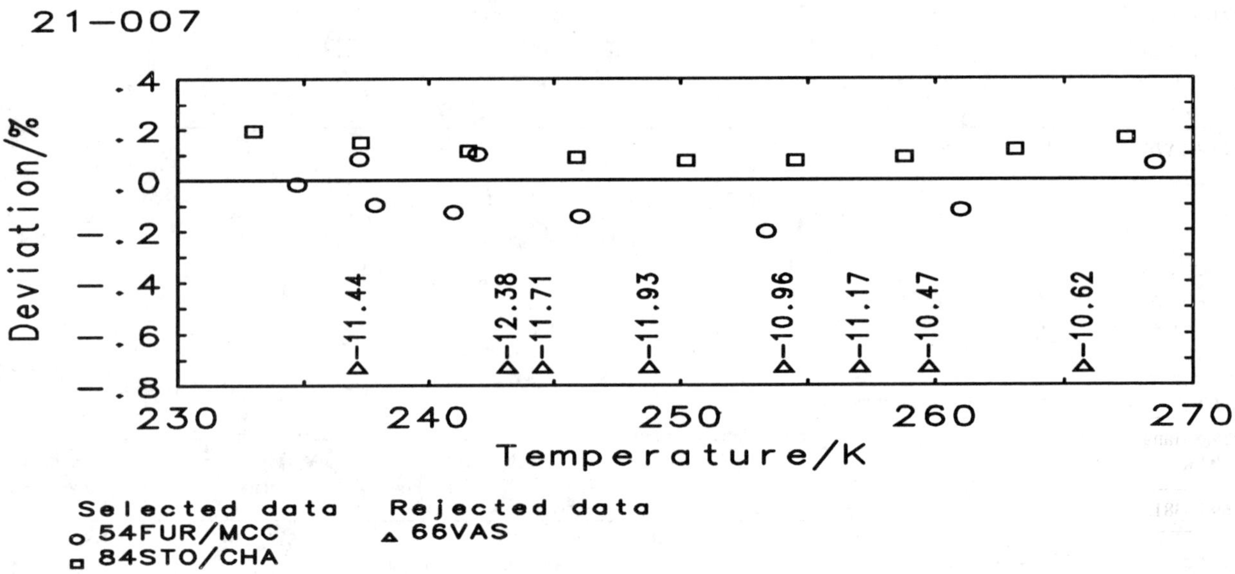

Name: Dodecafluoropentane
Formula: C_5F_{12}

CAS–RN: 678–26–2
Group No.: 21–008
Molar Mass: 288.04

TABLE 21.8.1. Experimental heat capacities

Reference		Temp. range K	No. pnts	Error %	Purity %	Purity method	Type capacity	Calorimeter Type	Calorimeter Reference
74COC/NOR	N	233.0–253.0	3S	nosp	90.9	chrom	C_p	BDHT	69PER/COM

74COC/NOR impurity of 9.1 % due to branched perfluorohydrocarbons

TABLE 21.8.3. Parameters of regression polynomial

Heat capacity type	No. data points total	used	s_w	s/R	s_r %	s_b/R	+/–
C_p	3	3	0.100	4.94–2	0.20	–5.72–5	–1

Temp. range K	A_1	A_2	Level of uncertainty
233.0–253.0	–3.25881+1	2.34911+1	V

TABLE 21.8.4. Recommended values of heat capacities

Temp. (K)	230	240	250
c_p (J K^{-1}g^{-1})	0.62	0.69	0.75
C_p (J K^{-1}mol^{-1})	178	198	217

Name: 1,3–Difluoro–2,2–bis(fluoromethyl)propane
Formula: $C_5H_8F_4$

CAS–RN: 338–23–8
Group No.: 21–009
Molar Mass: 144.11

TABLE 21.9.1. Experimental heat capacities

Reference	Temp. range K	No. pnts	Error %	Purity %	Purity method	Type capacity	Calorimeter Type	Calorimeter Reference
64TRO/WES2	369.2–381.0	9	0.10	99.82	melpt	C_{sat}	BSAO	53WES/HAT

TABLE 21.9.3. Parameters of regression polynomial

Heat capacity type	No. data points total	used	s_w	s/R	s_r %	s_b/R	+/–
C_{sat}	9	9	1.712	5.23–2	0.17	–1.89–2	–5

Temp. range K	A_1	A_2	A_3	A_4	Level of uncertainty
369.2–381.0	–1.05124+5	8.38439+4	–2.22836+4	1.97410+3	III

TABLE 21.9.4. Recommended values of heat capacities

Temp. (K)	370	375	380
c_{sat} (J K^{-1}g^{-1})	1.750	1.774	1.776
C_{sat} (J K^{-1}mol^{-1})	252.2	255.7	255.9

Name: Hexafluorobenzene
Formula: C_6F_6

CAS–RN: 392–56–3
Group No.: 21–010
Molar Mass: 186.06

TABLE 21.10.1. Experimental heat capacities

Reference		Temp. range K	No. pnts	Error %	Purity %	Purity method	Type capacity	Calorimeter Type	Calorimeter Reference
65COU/GRE		281.9–305.9	8	0.20	99.97	melpt	C_p	BSAO	63AND/COU1
70MES/FIN		284.8–342.7	8	0.20	99.93	melpt	C_{sat}	BSAO	47HUF
82GOR/SIM	N	284.0–349.7	18	0.50	99.74	melpt	C_p	FSIO	83GOR/SIM
87WIL/LAI		298.1	1	nosp	99.	anal	C_p	FSIT	71PIC/LED

82GOR/SIM same data in 82GOR/GRI

TABLE 21.10.2. Correlated heat capacities

Reference	Temp. range K	No. pnts used	$\sigma_r C$ %	d_w	d/R	d_r %	d_b/R	+/–
				Selected data				
65COU/GRE	281.9–305.9	8	0.20	0.683	3.61–2	0.14	–8.53–3	–2
70MES/FIN	284.8–342.7	8	0.20	0.572	3.09–2	0.11	–3.49–3	0
82GOR/SIM	284.0–349.7	18	0.50	0.542	7.46–2	0.27	3.43–2	7
				Rejected data				
87WIL/LAI	(4.16–1, 1.54, 4.16–1, 1)							

TABLE 21.10.3. Parameters of regression polynomial

Heat capacity type	No. data points total	No. data points used	s_w	s/R	s_r %	s_b/R	+/–
C	35	34	0.613	6.18–2	0.23	1.53–2	5

Temp. range K	A_1	A_2	A_3	Level of uncertainty
281.9–349.7	2.42710+1	–7.27364–1	5.11410–1	II

TABLE 21.10.4. Recommended values of heat capacities

Temp. (K)	280	290	298.15	300	310	320	330
c (J K^{-1}g^{-1})	1.173	1.183	1.191	1.193	1.203	1.215	1.226
C (J K^{-1}mol^{-1})	218.2	220.0	221.6	221.9	223.9	226.0	228.1
Temp. (K)	340	350					
c (J K^{-1}g^{-1})	1.238	1.251					
C (J K^{-1}mol^{-1})	230.4	232.7					

TABLE 21.10.5. Parameters of quasi-polynomial equation

Heat capacity type	No. data points total	No. data points used	s_w	s/R	s_r %	s_b/R	+/–
C	35	34	0.675	5.43–2	0.20	1.32–3	0

Temp. range K	T_c K	A_1	A_2	A_3	A_4	Level of uncertainty
281.9–349.7	516.73	–2.56982	6.34040–1	2.14362+1	2.60393	II

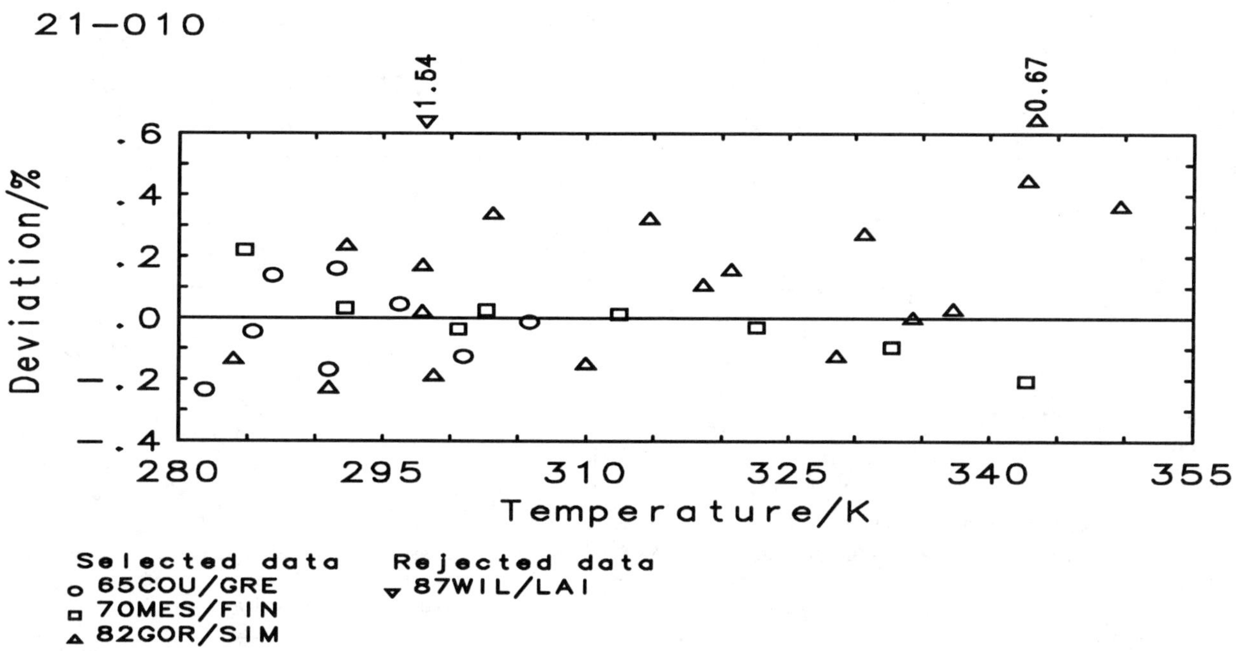

Name: Tetradecafluorohexane
Formula: C_6F_{14}

CAS–RN: 355–42–0
Group No.: 21–011
Molar Mass: 338.04

TABLE 21.11.1. Experimental heat capacities

Reference		Temp. range K	No. pnts	Error %	Purity %	Purity method	Type capacity	Calorimeter Type	Calorimeter Reference
74COC/NOR		233.0–273.0	5S	nosp	99.0	chrom	C_p	BDHT	69PER/COM
83CAM/DIA	N	293.0	1	nosp	not specified		C_p	BDHT	77LAG/PIE

83CAM/DIA a graph based on the same data in 82CAM/REY

TABLE 21.11.2. Correlated heat capacities

Reference	Temp. range K	No. pnts used	$\sigma_r C$ %	d_w	d/R	d_r %	d_b/R	+/–
				Selected data				
74COC/NOR	233.0–273.0	5	3.00#	1.288	1.04	3.87	3.05–1	1
83CAM/DIA	293.0	1	3.00#	1.168	1.08	3.50	–1.08	–1

TABLE 21.11.3. Parameters of regression polynomial

Heat capacity type	No. data points total	No. data points used	s_w	s/R	s_r %	s_b/R	+/–
C_p	6	6	1.554	1.28	4.66	7.50–2	0

Temp. range K	A_1	A_2	Level of uncertainty
233.0–293.0	–1.92412+1	1.74106+1	VI

TABLE 21.11.4. Recommended values of heat capacities

Temp. (K)	230	240	250	260	270	273.15	280
c_p (J K^{-1}g^{-1})	0.51	0.55	0.60	0.64	0.68	0.70	0.73
C_p (J K^{-1}mol^{-1})	173	187	202	216	231	235	245
Temp. (K)	290	298.15	300				
c_p (J K^{-1}g^{-1})	0.77	0.80	0.81				
C_p (J K^{-1}mol^{-1})	260	272	274				

Name: Pentafluorobenzene
Formula: C_6HF_5

CAS–RN: 363–72–4
Group No.: 21–012
Molar Mass: 168.07

TABLE 21.12.1. Experimental heat capacities

Reference	Temp. range K	No. pnts	Error %	Purity %	Purity method	Type capacity	Calorimeter Type	Calorimeter Reference
68COU/HAL2	230.7–324.1	21	nosp	99.87	melpt	C_p	BSAO	63AND/COU1
69PAU/LAV1	230.3–300.4	11	nosp	99.46	melpt	C_{sat}	BSAO	69PAU/LAV2

TABLE 21.12.2. Correlated heat capacities

Reference	Temp. range K	No. pnts used	$\sigma_r C$ %	d_w	d/R	d_r %	d_b/R	+/–
				Selected data				
68COU/HAL2	230.7–324.1	21	0.20#	0.241	1.17–2	0.05	1.18–5	2
				Rejected data				
69PAU/LAV1	(5.22–1, 2.13, 5.07–1, 10)							

TABLE 21.12.3. Parameters of regression polynomial

Heat capacity type	No. data points total	No. data points used	s_w	s/R	s_r %	s_b/R	+/–
C_p	32	21	0.260	1.26–2	0.05	1.18–5	2

Temp. range K	A_1	A_2	A_3	Level of uncertainty
230.7–324.1	1.95493+1	5.61347–1	3.82448–1	II

TABLE 21.12.4. Recommended values of heat capacities

Temp. (K)	230	240	250	260	270	273.15	280
c_p (J K^{-1}g^{-1})	1.131	1.143	1.155	1.167	1.180	1.184	1.193
C_p (J K^{-1}mol^{-1})	190.1	192.1	194.1	196.2	198.3	199.0	200.5
Temp. (K)	290	298.15	300	310	320	330	
c_p (J K^{-1}g^{-1})	1.207	1.218	1.221	1.235	1.250	1.265	
C_p (J K^{-1}mol^{-1})	202.8	204.7	205.2	207.6	210.0	212.6	

TABLE 21.12.5. Parameters of quasi-polynomial equation

Heat capacity type	No. data points total	used	s_w	s/R	s_r %	s_b/R	+/-
C_p	32	21	0.288	1.39–2	0.06	1.46–5	0

Temp. range K	T_c K	A_1	A_2	A_3	A_4	Level of uncertainty
230.7–324.1	532.00	−3.07933	3.92423–1	1.78115+1	6.04087	II

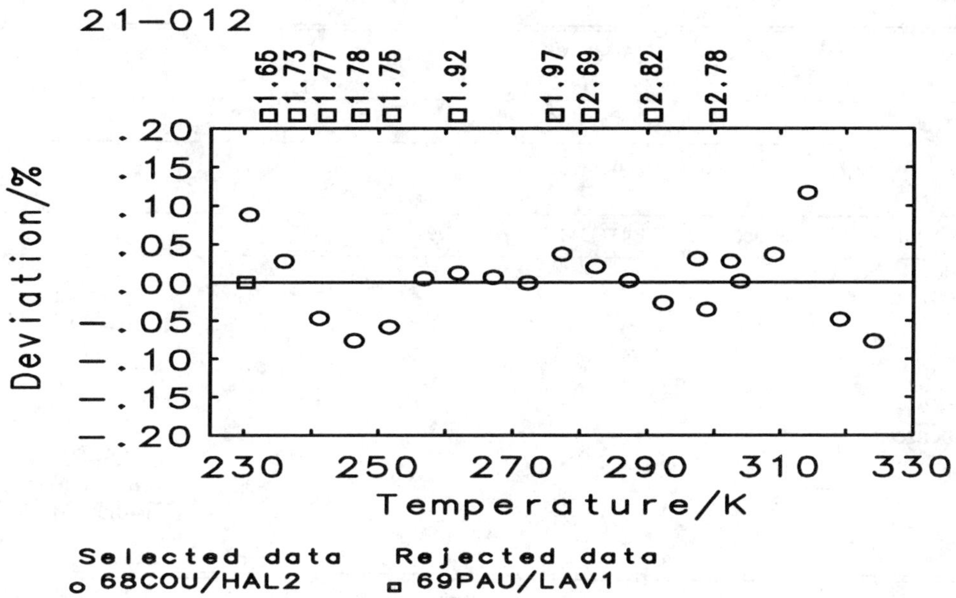

Name: 1,2,3,4–Tetrafluorobenzene
Formula: $C_6H_2F_4$

CAS–RN: 551–62–2
Group No.: 21–013
Molar Mass: 150.08

TABLE 21.13.1. Experimental heat capacities

Reference	Temp. range K	No. pnts	Error %	Purity %	Purity method	Type capacity	Calorimeter Type	Calorimeter Reference
73AND/MAR1	235.5–319.8	41	nosp	99.85	melpt	C_p	BSAO	63AND/COU1

TABLE 21.13.3. Parameters of regression polynomial

Heat capacity type	No. data points total	used	s_w	s/R	s_r %	s_b/R	+/-
C_p	41	41	0.296	1.30–2	0.06	1.37–5	−3

Temp. range K	A_1	A_2	A_3	A_4	Level of uncertainty
235.5–319.8	−2.62887	2.14433+1	−6.81722	8.35405–1	II

TABLE 21.13.4. Recommended values of heat capacities

Temp. (K)	240	250	260	270	273.15	280	290
c_p (J K^{-1}g^{-1})	1.170	1.187	1.203	1.220	1.225	1.236	1.252
C_p (J K^{-1}mol^{-1})	175.6	178.1	180.6	183.0	183.8	185.4	187.9
Temp. (K)	298.15	300	310	320			
c_p (J K^{-1}g^{-1})	1.266	1.269	1.286	1.305			
C_p (J K^{-1}mol^{-1})	189.9	190.4	193.1	195.9			

TABLE 21.13.5. Parameters of quasi-polynomial equation

Heat capacity type	No. data points total	used	s_w	s/R	s_r %	s_b/R	+/–
C_p	41	41	0.371	1.66–2	0.07	2.23–5	2
Temp. range K	T_c K	A_1	A_2	A_3	A_4		Level of uncertainty
235.5–319.8	550.80	–1.46170	3.98380–2	1.43733+1	1.34079+1		II

Name: 1,2,3,5–Tetrafluorobenzene
Formula: $C_6H_2F_4$

CAS–RN: 2367–82–0
Group No.: 21–014
Molar Mass: 150.08

TABLE 21.14.1. Experimental heat capacities

Reference	Temp. range K	No. pnts	Error %	Purity %	Purity method	Type capacity	Calorimeter Type	Calorimeter Reference
73AND/MAR1	229.3–311.2	25	nosp	99.99	melpt	C_p	BSAO	63AND/COU1

TABLE 21.14.3. Parameters of regression polynomial

Heat capacity type	No. data points total	used	s_w	s/R	s_r %	s_b/R	+/–
C_p	25	25	0.335	1.49–2	0.07	1.89–5	–3
Temp. range K		A_1	A_2	A_3			Level of uncertainty
229.3–311.2		1.94897+1	–1.02669	7.26879–1			II

TABLE 21.14.4. Recommended values of heat capacities

Temp. (K)	230	240	250	260	270	273.15	280
c_p (J K^{-1}g^{-1})	1.162	1.175	1.189	1.204	1.220	1.225	1.236
C_p (J K^{-1}mol^{-1})	174.4	176.4	178.5	180.7	183.1	183.8	185.5
Temp. (K)	290	298.15	300	310			
c_p (J K^{-1}g^{-1})	1.253	1.268	1.272	1.290			
C_p (J K^{-1}mol^{-1})	188.1	190.3	190.8	193.7			

TABLE 21.14.5. Parameters of quasi-polynomial equation

Heat capacity type	No. data points total	used	s_w	s/R	s_r %	s_b/R	+/–
C_p	25	25	0.309	1.37–2	0.06	1.49–5	–2

Temp. range K	T_c K	A_1	A_2	A_3	A_4	Level of uncertainty
229.3–311.2	535.20	–3.38317	1.68047	1.53904+1	1.70277	II

Name: 1,2,4,5–Tetrafluorobenzene
Formula: $C_6H_2F_4$

CAS–RN: 327–54–8
Group No.: 21–015
Molar Mass: 150.08

TABLE 21.15.1. Experimental heat capacities

Reference	Temp. range K	No. pnts	Error %	Purity %	Purity method	Type capacity	Calorimeter Type	Calorimeter Reference
73AND/MAR1	279.8–352.5	20	nosp	99.94	melpt	C_p	BSAO	63AND/COU1

TABLE 21.15.3. Parameters of regression polynomial

Heat capacity type	No. data points total	used	s_w	s/R	s_r %	s_b/R	+/–
C_p	20	20	0.284	1.33–2	0.06	1.27–5	1

Temp. range K	A_1	A_2	Level of uncertainty
279.8–352.5	1.41727+1	3.00019	II

TABLE 21.15.4. Recommended values of heat capacities

Temp. (K)	280	290	298.15	300	310	320	330
c_p (J K^{-1}g^{-1})	1.251	1.267	1.281	1.284	1.300	1.317	1.334
C_p (J K^{-1}mol^{-1})	187.7	190.2	192.2	192.7	195.2	197.7	200.2
Temp. (K)	340	350					
c_p (J K^{-1}g^{-1})	1.350	1.367					
C_p (J K^{-1}mol^{-1})	202.7	205.1					

TABLE 21.15.5. Parameters of quasi-polynomial equation

Heat capacity type	No. data points total	used	s_w	s/R	s_r %	s_b/R	+/–
C_p	20	20	0.292	1.37–2	0.06	1.31–5	1

Temp. range K	T_c K	A_1	A_2	A_3	A_4	Level of uncertainty
279.8–352.5	543.30	1.36161–1	2.78781–4	1.41018+1	1.66257+1	II

Name: 1,2–Difluorobenzene
Formula: $C_6H_4F_2$

CAS–RN: 367–11–3
Group No.: 21–016
Molar Mass: 114.09

TABLE 21.16.1. Experimental heat capacities

Reference	Temp. range K	No. pnts	Error %	Purity %	Purity method	Type capacity	Calorimeter Type	Calorimeter Reference
63SCO/MES	229.0–357.2	20	0.20	99.998	melpt	C_{sat}	BSAO	47HUF

TABLE 21.16.3. Parameters of regression polynomial

Heat capacity type	No. data points total	No. data points used	s_w	s/R	s_r %	s_b/R	+/–
C_{sat}	20	20	0.252	1.02–2	0.05	8.01–6	–2

Temp. range K	A_1	A_2	A_3	A_4	Level of uncertainty
229.0–357.2	2.32770+1	–1.04138+1	4.52483	–5.02627–1	II

TABLE 21.16.4. Recommended values of heat capacities

Temp. (K)	230	240	250	260	270	273.15	280
c_{sat} (J K^{-1}g^{-1})	1.250	1.268	1.288	1.308	1.330	1.337	1.352
C_{sat} (J K^{-1}mol^{-1})	142.6	144.7	146.9	149.3	151.8	152.6	154.3
Temp. (K)	290	298.15	300	310	320	330	340
c_{sat} (J K^{-1}g^{-1})	1.375	1.394	1.398	1.421	1.444	1.467	1.488
C_{sat} (J K^{-1}mol^{-1})	156.9	159.1	159.5	162.2	164.8	167.3	169.8
Temp. (K)	350	360					
c_{sat} (J K^{-1}g^{-1})	1.509	1.529					
C_{sat} (J K^{-1}mol^{-1})	172.2	174.4					

TABLE 21.16.5. Parameters of quasi-polynomial equation

Heat capacity type	No. data points total	No. data points used	s_w	s/R	s_r %	s_b/R	+/–
C_{sat}	20	20	0.735	2.85–2	0.15	7.70–5	0

Temp. range K	T_c K	A_1	A_2	A_3	A_4	Level of uncertainty
229.0–357.2	562.50	–1.11457	2.14668–2	1.05712+1	1.44674+1	II

Name: 1,3–Difluorobenzene
Formula: $C_6H_4F_2$

CAS–RN: 372–18–9
Group No.: 21–017
Molar Mass: 114.09

TABLE 21.17.1. Experimental heat capacities

Reference	Temp. range K	No. pnts	Error %	Purity %	Purity method	Type capacity	Calorimeter Type	Calorimeter Reference
70MES/FIN	214.3–354.9	15	0.20	99.999	melpt	C_{sat}	BSAO	47HUF

TABLE 21.17.3. Parameters of regression polynomial

Heat capacity type	No. data points total	used	s_w	s/R	s_r %	s_b/R	+/–
C_{sat}	15	15	0.090	3.16–3	0.02	1.91–6	–1

Temp. range K	A_1	A_2	A_3	A_4	Level of uncertainty
214.3–354.9	1.84883+1	–5.05609	2.51870	–2.51541–1	II

TABLE 21.17.4. Recommended values of heat capacities

Temp. (K)	210	220	230	240	250	260	270
c_{sat} (J $K^{-1}g^{-1}$)	1.213	1.230	1.248	1.267	1.287	1.308	1.330
C_{sat} (J $K^{-1}mol^{-1}$)	138.4	140.3	142.4	144.5	146.8	149.2	151.7
Temp. (K)	273.15	280	290	298.15	300	310	320
c_{sat} (J $K^{-1}g^{-1}$)	1.337	1.352	1.375	1.395	1.399	1.423	1.447
C_{sat} (J $K^{-1}mol^{-1}$)	152.5	154.3	156.9	159.1	159.6	162.3	165.1
Temp. (K)	330	340	350	360			
c_{sat} (J $K^{-1}g^{-1}$)	1.471	1.496	1.520	1.544			
C_{sat} (J $K^{-1}mol^{-1}$)	167.9	170.7	173.5	176.2			

TABLE 21.17.5. Parameters of quasi-polynomial equation

Heat capacity type	No. data points total	used	s_w	s/R	s_r %	s_b/R	+/–
C_{sat}	15	15	0.716	2.67–2	0.14	6.50–5	–1

Temp. range K	T_c K	A_1	A_2	A_3	A_4	Level of uncertainty
214.3–354.9	543.00	–3.37293	3.64435–1	1.13396+1	7.80431	II

Name: 1,4–Difluorobenzene
Formula: $C_6H_4F_2$

CAS–RN: 540–36–3
Group No.: 21–018
Molar Mass: 114.09

TABLE 21.18.1. Experimental heat capacities

Reference		Temp. range K	No. pnts	Error %	Purity %	Purity method	Type capacity	Calorimeter Type	Calorimeter Reference
91LIC	N	238.0–324.0	eqn	2.00	99.	anal	C_p	BDCT	89BRE/LIC

91LIC low temperature limit is below n.m.t.; undercooled liquid

TABLE 21.18.3. Parameters of regression polynomial

Heat capacity type	No. data points total	used	s_w	s/R	s_r %	s_b/R	+/–
C_p	10	10	0.000	0.00	0.00	0.00	0

Temp. range K	A_1	A_2	Level of uncertainty
238.0–324.4	1.34746+1	1.90029	V

TABLE 21.18.4. Recommended values of heat capacities

Temp. (K)	240	250	260	270	273.15	280	290
c_p (J K^{-1}g^{-1})	1.31	1.33	1.34	1.36	1.36	1.37	1.38
C_p (J K^{-1}mol^{-1})	150	152	153	155	155	156	158
Temp. (K)	298.15	300	310	320			
c_p (J K^{-1}g^{-1})	1.39	1.40	1.41	1.43			
C_p (J K^{-1}mol^{-1})	159	159	161	163			

Name: Fluorobenzene
Formula: C_6H_5F

CAS–RN: 462–06–6
Group No.: 21–019
Molar Mass: 96.10

TABLE 21.19.1. Experimental heat capacities

Reference	Temp. range K	No. pnts	Error %	Purity %	Purity method	Type capacity	Calorimeter Type	Calorimeter Reference
37STU	240.0–320.0	9S	nosp	not specified		C_p	BDHO	37STU
56SCO/MCC1	235.1–350.3	16	0.10	99.97	melpt	C_{sat}	BSAO	43RUE/HUF
84ROU/GRO	298.1	1	0.30	99.	estim	C_p	FSIT	71PIC/LED

TABLE 21.19.2. Correlated heat capacities

Reference	Temp. range K	No. pnts used	$\sigma_r C$ %	d_w	d/R	d_r %	d_b/R	+/–
				Selected data				
56SCO/MCC1	235.1–350.3	16	0.10	0.174	2.94–3	0.02	3.02–5	1
84ROU/GRO	298.1	1	0.30	0.080	4.20–3	0.02	–4.20–3	–1
				Rejected data				
37STU	(3.67–1, 2.15, 2.28–1, 3)							

TABLE 21.19.3. Parameters of regression polynomial

Heat capacity type	No. data points total	No. data points used	s_w	s/R	s_r %	s_b/R	+/–
C	26	17	0.194	3.46–3	0.02	–2.19–4	0

Temp. range K	A_1	A_2	A_3	A_4	Level of uncertainty
235.1–350.3	2.02402+1	–7.30473	3.00091	–2.84436–1	II

TABLE 21.19.4. Recommended values of heat capacities

Temp. (K)	240	250	260	270	273.15	280	290
c (J K^{-1}g^{-1})	1.390	1.409	1.431	1.453	1.460	1.477	1.502
C (J K^{-1}mol^{-1})	133.5	135.4	137.5	139.6	140.4	141.9	144.3
Temp. (K)	298.15	300	310	320	330	340	350
c (J K^{-1}g^{-1})	1.523	1.527	1.554	1.581	1.609	1.636	1.665
C (J K^{-1}mol^{-1})	146.3	146.8	149.3	151.9	154.6	157.3	160.0

TABLE 21.19.5. Parameters of quasi-polynomial equation

Heat capacity type	No. data points total	used	s_w	s/R	s_r %	s_b/R	+/–
C	26	17	1.047	1.93–2	0.10	–2.82–3	–2

Temp. range K	T_c K	A_1	A_2	A_3	A_4	Level of uncertainty
235.1–350.3	560.09	–3.82528	9.64460–1	1.06013+1	3.79301	II

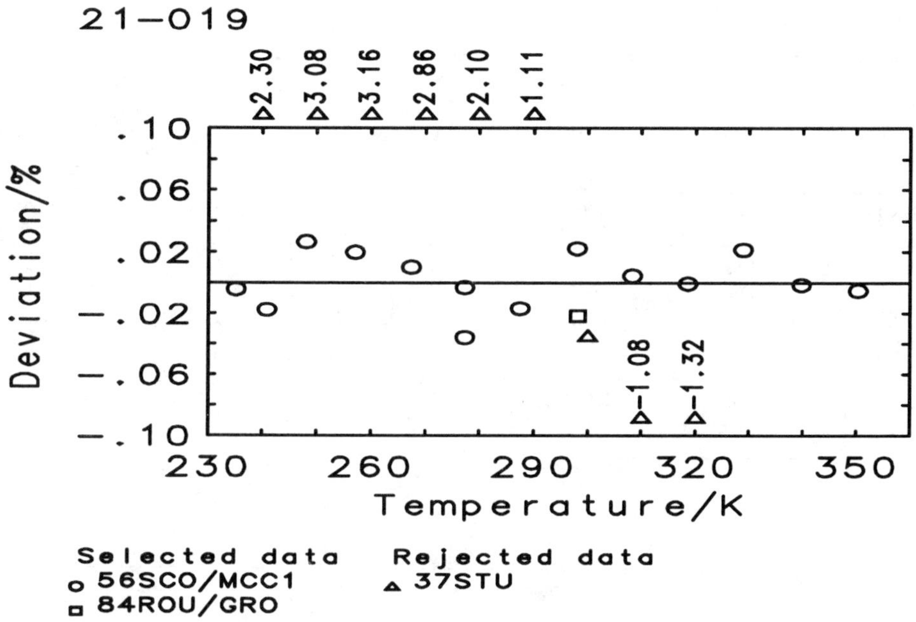

Name: Pentafluoro(trifluoromethyl)benzene
Formula: C_7F_8

CAS–RN: 434–64–0
Group No.: 21–020
Molar Mass: 236.06

TABLE 21.20.1. Experimental heat capacities

Reference	Temp. range K	No. pnts	Error %	Purity %	Purity method	Type capacity	Calorimeter Type	Calorimeter Reference
74AND/MAR	210.8–365.1	36	nosp	99.89	melpt	C_p	BSAO	63AND/COU1
75PAU1	210.8–298.8	23	nosp	99.47	melpt	C_{sat}	BSAO	69PAU/LAV2

TABLE 21.20.2. Correlated heat capacities

Reference	Temp. range K	No. pnts used	$\sigma_r C$ %	d_w	d/R	d_r %	d_b/R	+/–
				Selected data				
74AND/MAR	210.8–365.1	36	0.20#	0.193	1.19–2	0.04	7.79–6	–1
				Rejected data				
75PAU1	(4.12–1, 1.34, 4.04–1, 22)							

TABLE 21.20.3. Parameters of regression polynomial

Heat capacity type	No. data points total	used	s_w	s/R	s_r %	s_b/R	+/−
C_p	59	36	0.205	1.26−2	0.04	7.79−6	−1

Temp. range K	A_1	A_2	A_3	A_4	Level of uncertainty
210.8–365.1	3.12549+1	−5.68404	2.70101	−2.55401−1	II

TABLE 21.20.4. Recommended values of heat capacities

Temp. (K)	210	220	230	240	250	260	270
c_p (J K^{-1}g^{-1})	1.017	1.025	1.034	1.044	1.054	1.065	1.077
C_p (J K^{-1}mol^{-1})	240.0	242.0	244.1	246.4	248.9	251.5	254.2
Temp. (K)	273.15	280	290	298.15	300	310	320
c_p (J K^{-1}g^{-1})	1.080	1.089	1.101	1.111	1.114	1.126	1.140
C_p (J K^{-1}mol^{-1})	255.1	257.0	259.9	262.3	262.9	265.9	269.0
Temp. (K)	330	340	350	360	370		
c_p (J K^{-1}g^{-1})	1.153	1.166	1.180	1.193	1.207		
C_p (J K^{-1}mol^{-1})	272.2	275.3	278.5	281.7	284.9		

TABLE 21.20.5. Parameters of quasi-polynomial equation

Heat capacity type	No. data points total	used	s_w	s/R	s_r %	s_b/R	+/−
C_p	59	36	0.565	3.49−2	0.11	7.40−5	3

Temp. range K	T_c K	A_1	A_2	A_3	A_4	Level of uncertainty
210.8–365.1	534.47	−3.70808	4.36801−1	2.31329+1	7.86963	II

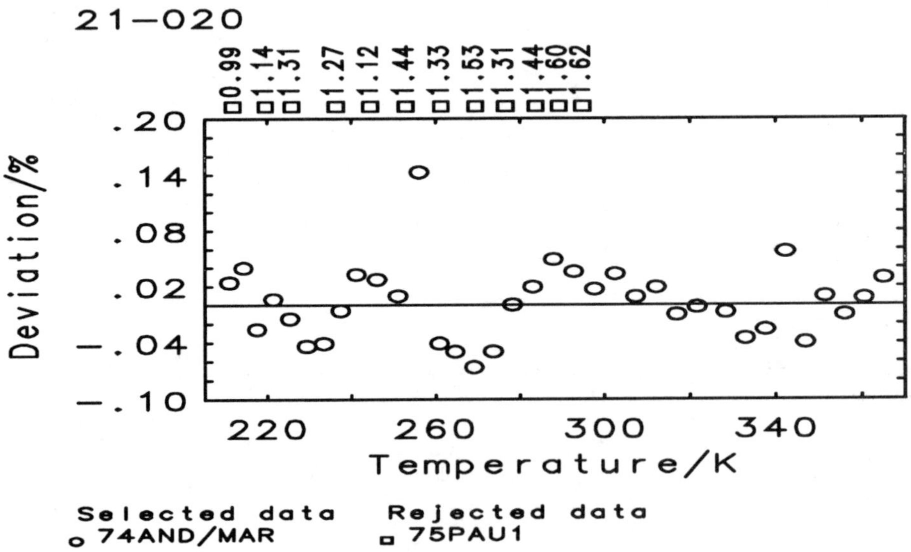

Name: (Trifluoromethyl)undecafluorocyclohexane
Formula: C_7F_{14}

CAS–RN: 355–02–2
Group No.: 21–021
Molar Mass: 350.05

TABLE 21.21.1. Experimental heat capacities

Reference	Temp. range K	No. pnts	Error %	Purity %	Purity method	Type capacity	Calorimeter Type	Calorimeter Reference
57YAR/KAY	303.0–335.0	eqn	1.10	not specified		C_p	BSIO	75PED/KAY

TABLE 21.21.3. Parameters of regression polynomial

Heat capacity type	No. data points total	No. data points used	s_w	s/R	s_r %	s_b/R	+/–
C_p	7	7	0.000	2.95–6	0.00	1.63–6	0

Temp. range K	A_1	A_2	Level of uncertainty
303.0–334.8	2.35457+1	6.34151	V

TABLE 21.21.4. Recommended values of heat capacities

Temp. (K)	300	310	320	330	340
c_p (J K^{-1}g^{-1})	1.01	1.03	1.04	1.06	1.07
C_p (J K^{-1}mol^{-1})	354	359	364	370	375

Name: Hexadecafluoroheptane
Formula: C_7F_{16}

CAS–RN: 335–57–9
Group No.: 21–022
Molar Mass: 388.05

TABLE 21.22.1. Experimental heat capacities

Reference	Temp. range K	No. pnts	Error %	Purity %	Purity method	Type capacity	Calorimeter Type	Calorimeter Reference
51OLI/GRI1	221.9–310.0	10S	0.20	99.94	melpt	C_{sat}	BSAO	43RUE/HUF
57YAR/KAY	295.0–345.0	eqn	1.10	not specified		C_p	BSIO	75PED/KAY

TABLE 21.22.2. Correlated heat capacities

Reference	Temp. range K	No. pnts used	$\sigma_r C$ %	d_w	d/R	d_r %	d_b/R	+/–
				Selected data				
51OLI/GRI1	221.9–310.0	10	0.20	0.154	1.43–2	0.03	–4.96–6	–1
				Rejected data				
57YAR/KAY	(8.55–1, 1.72, –8.53–1, –3)							

TABLE 21.22.3. Parameters of regression polynomial

Heat capacity type	No. data points total	used	s_w	s/R	s_r %	s_b/R	+/–
C_{sat}	19	10	0.172	1.60–2	0.03	–4.96–6	–1

Temp. range K	A_1	A_2	Level of uncertainty
221.9–310.0	3.01875+1	6.75732	II

TABLE 21.22.4. Recommended values of heat capacities

Temp. (K)	220	230	240	250	260	270	273.15
c_{sat} (J K^{-1}g^{-1})	0.965	0.980	0.994	1.009	1.023	1.038	1.042
C_{sat} (J K^{-1}mol^{-1})	374.6	380.2	385.8	391.5	397.1	402.7	404.5
Temp. (K)	280	290	298.15	300	310		
c_{sat} (J K^{-1}g^{-1})	1.052	1.067	1.078	1.081	1.096		
C_{sat} (J K^{-1}mol^{-1})	408.3	413.9	418.5	419.5	425.2		

TABLE 21.22.5. Parameters of quasi-polynomial equation

Heat capacity type	No. data points total	used	s_w	s/R	s_r %	s_b/R	+/–
C_{sat}	19	10	0.146	1.39–2	0.03	3.82–6	–1

Temp. range K	T_c K	A_1	A_2	A_3	A_4	Level of uncertainty
221.9–310.0	474.80	1.14138	9.40148–3	2.96808+1	3.46421+1	II

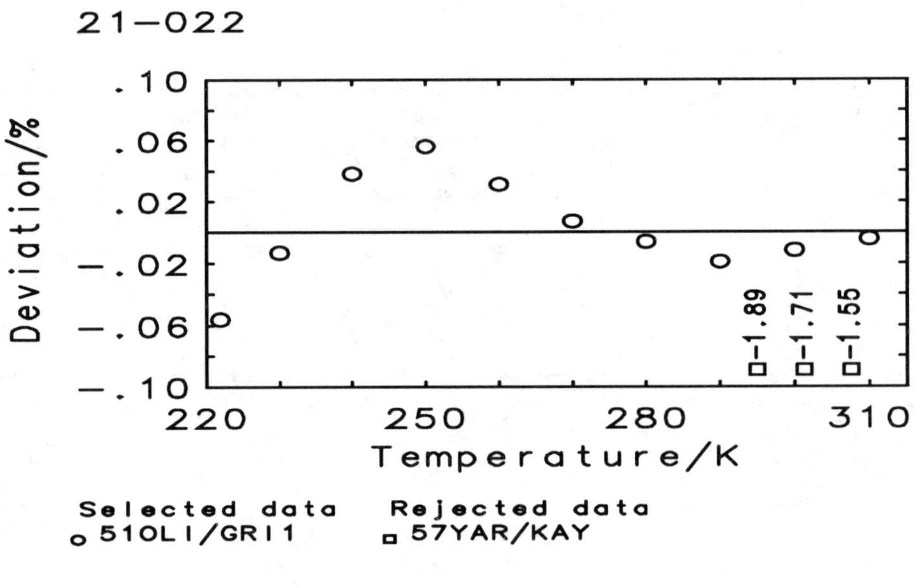

Name: Pentafluoromethylbenzene
Formula: $C_7H_3F_5$

CAS–RN: 771–56–2
Group No.: 21–023
Molar Mass: 182.09

TABLE 21.23.1. Experimental heat capacities

Reference		Temp. range K	No. pnts	Error %	Purity %	Purity method	Type capacity	Calorimeter Type	Calorimeter Reference
68COU/HAL1		252.8–376.5	30	nosp	99.95	melpt	C_p	BSAO	63AND/COU1
71PAU2	N	243.7–305.0	9S	nosp	not specified		C_{sat}	BSAO	69PAU/LAV2

71PAU2 measured data deposited in VINITI No 2538–71

TABLE 21.23.2. Correlated heat capacities

Reference	Temp. range K	No. pnts used	$\sigma_r C$ %	d_w	d/R	d_r %	d_b/R	+/–
				Selected data				
68COU/HAL1	252.8–376.5	30	0.20#	0.222	1.23–2	0.04	1.20–5	1
				Rejected data				
71PAU2	(5.51–1, 2.01, 5.45–1, 7)							

TABLE 21.23.3. Parameters of regression polynomial

Heat capacity type	No. data points total	No. data points used	s_w	s/R	s_r %	s_b/R	+/–
C_p	39	30	0.234	1.29–2	0.05	1.20–5	1

Temp. range K	A_1	A_2	A_3	Level of uncertainty
252.8–376.5	1.89055+1	2.03596	2.46068–1	II

TABLE 21.23.4. Recommended values of heat capacities

Temp. (K)	250	260	270	273.15	280	290	298.15
c_p (J K^{-1}g^{-1})	1.166	1.181	1.196	1.201	1.212	1.227	1.240
C_p (J K^{-1}mol^{-1})	212.3	215.0	217.8	218.7	220.6	223.5	225.8
Temp. (K)	300	310	320	330	340	350	360
c_p (J K^{-1}g^{-1})	1.243	1.259	1.276	1.292	1.309	1.326	1.344
C_p (J K^{-1}mol^{-1})	226.4	229.3	232.3	235.3	238.4	241.5	244.6
Temp. (K)	370	380					
c_p (J K^{-1}g^{-1})	1.361	1.379					
C_p (J K^{-1}mol^{-1})	247.8	251.1					

TABLE 21.23.5. Parameters of quasi-polynomial equation

Heat capacity type	No. data points total	No. data points used	s_w	s/R	s_r %	s_b/R	+/–
C_p	39	30	0.236	1.31–2	0.05	1.04–5	2

Temp. range K	T_c K	A_1	A_2	A_3	A_4	Level of uncertainty
252.8–376.5	566.50	–2.52279	1.14009–1	1.76947+1	1.39561+1	II

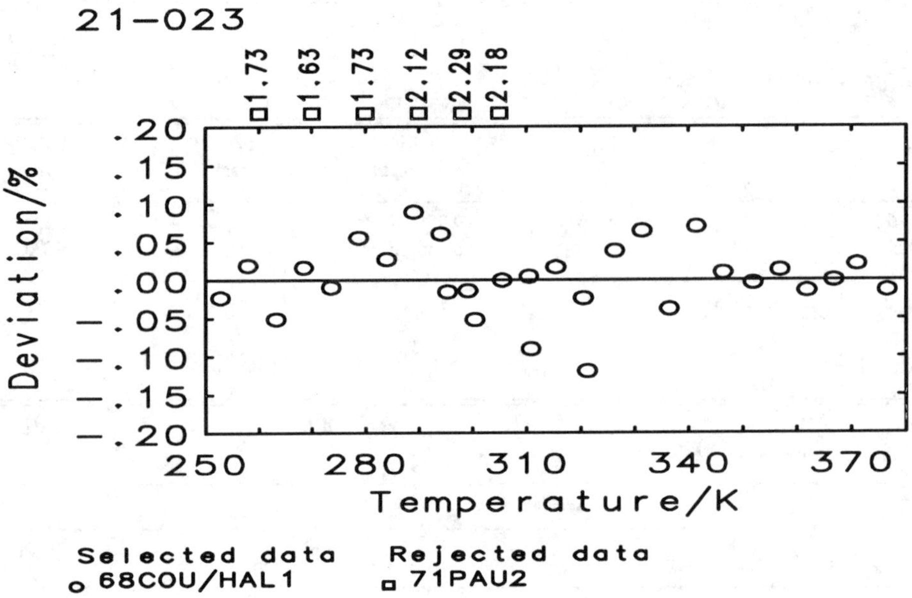

Name: (Trifluoromethyl)benzene
Formula: $C_7H_5F_3$

CAS–RN: 98–08–8
Group No.: 21–024
Molar Mass: 146.11

TABLE 21.24.1. Experimental heat capacities

Reference	Temp. range K	No. pnts	Error %	Purity %	Purity method	Type capacity	Calorimeter Type	Calorimeter Reference
59SCO/DOU	252.5–364.7	13	0.20	99.9987	melpt	C_{sat}	BSAO	47HUF
89JAD/FRA	325.0–365.0	5	nosp	99.	anal	C_p	BDCT	86MER/BEN

TABLE 21.24.2. Correlated heat capacities

Reference	Temp. range K	No. pnts used	$\sigma_r C$ %	d_w	d/R	d_r %	d_b/R	+/–
				Selected data				
59SCO/DOU	252.5–364.7	13	0.20	0.153	6.94–3	0.03	4.26–6	3
				Rejected data				
89JAD/FRA	(5.37–1, 2.09, 4.29–1, 4)							

TABLE 21.24.3. Parameters of regression polynomial

Heat capacity type	No. data points total	No. data points used	s_w	s/R	s_r %	s_b/R	+/–
C_{sat}	18	13	0.174	7.91–3	0.03	4.26–6	3

Temp. range K	A_1	A_2	A_3	Level of uncertainty
252.5–364.7	1.30944+1	2.32085	2.98263–1	II

TABLE 21.24.4. Recommended values of heat capacities

Temp. (K)	250	260	270	273.15	280	290	298.15
c_{sat} (J K^{-1}g^{-1})	1.181	1.203	1.225	1.233	1.248	1.271	1.290
C_{sat} (J K^{-1}mol^{-1})	172.6	175.8	179.1	180.1	182.3	185.7	188.5
Temp. (K)	300	310	320	330	340	350	360
c_{sat} (J K^{-1}g^{-1})	1.294	1.318	1.342	1.366	1.390	1.415	1.441
C_{sat} (J K^{-1}mol^{-1})	189.1	192.5	196.0	199.6	203.1	206.8	210.5
Temp. (K)	370						
c_{sat} (J K^{-1}g^{-1})	1.466						
C_{sat} (J K^{-1}mol^{-1})	214.2						

TABLE 21.24.5. Parameters of quasi-polynomial equation

Heat capacity type	No. data points total	used	s_w	s/R		s_r %	s_b/R	+/-
C_{sat}	18	13	0.213	9.78–3		0.04	6.31–6	2
Temp. range K	T_c K	A_1	A_2	A_3	A_4			Level of uncertainty
252.5–364.7	562.60	–3.07158	1.49560–1	1.16726+1	1.57706+1			II

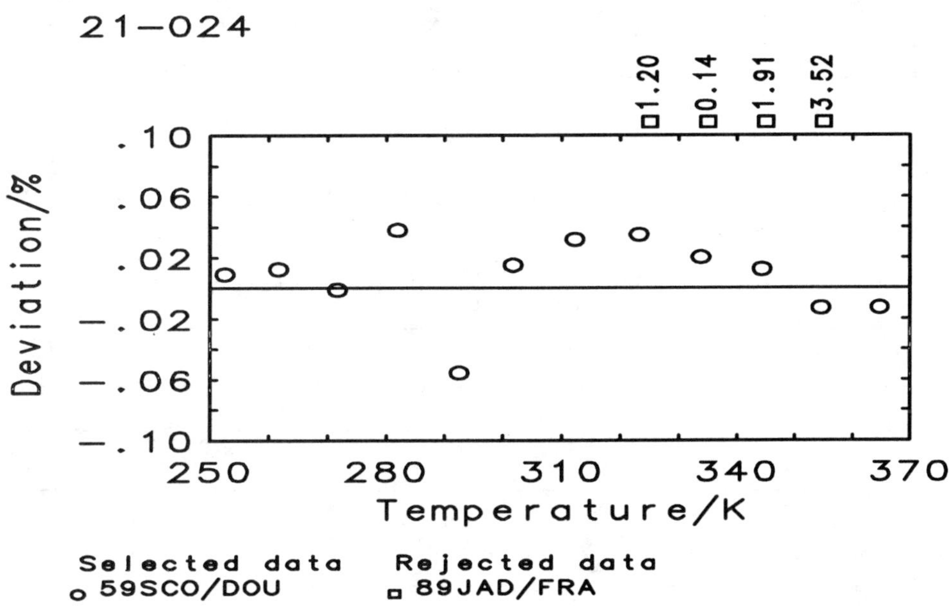

Name: 1–Fluoro–2–methylbenzene
Formula: C_7H_7F

CAS–RN: 95–52–3
Group No.: 21–025
Molar Mass: 110.13

TABLE 21.25.1. Experimental heat capacities

Reference	Temp. K	Capac. J/(K.g)	Error %	Purity %	Purity method	Type capacity	Calorimeter Type	Calorimeter Reference
90MEV/LIC	210.65	1.298	3.00	99.	anal	C_p	BDCT	89BRE/LIC

Name: 1–Fluoro–3–methylbenzene
Formula: C_7H_7F

CAS–RN: 352–70–5
Group No.: 21–026
Molar Mass: 110.13

TABLE 21.26.1. Experimental heat capacities

Reference	Temp. K	Capac. J/(K.g)	Error %	Purity %	Purity method	Type capacity	Calorimeter Type	Calorimeter Reference
90MEV/LIC	183.95	1.217	3.00	99.	anal	C_p	BDCT	89BRE/LIC

Name: 1–Fluoro–4–methylbenzene
Formula: C_7H_7F

CAS–RN: 352–32–9
Group No.: 21–027
Molar Mass: 110.13

TABLE 21.27.1. Experimental heat capacities

Reference		Temp. range K	No. pnts	Error %	Purity %	Purity method	Type capacity	Calorimeter Type	Calorimeter Reference
62SCO/MES		221.4–360.5	19	0.20	99.92	melpt	C_{sat}	BSAO	47HUF
90MEV/LIC	N	216.0–298.0	eqn	3.00	99.	anal	C_p	BDCT	89BRE/LIC

90MEV/LIC same equation in 91LIC

TABLE 21.27.2. Correlated heat capacities

Reference	Temp. range K	No. pnts used	$\sigma_r C$ %	d_w	d/R	d_r %	d_b/R	+/–
				Selected data				
62SCO/MES	221.4–360.5	19	0.20	0.171	7.35–3	0.03	1.81–6	–2
				Rejected data				
90MEV/LIC	(2.25–1, 1.14, 5.15–2, 0)							

TABLE 21.27.3. Parameters of regression polynomial

Heat capacity type	No. data points total	No. data points used	s_w	s/R	s_r %	s_b/R	+/–
C_{sat}	24	19	0.192	8.27–3	0.04	1.81–6	–2

Temp. range K	A_1	A_2	A_3	A_4	Level of uncertainty
221.4–360.5	2.54744+1	–1.15407+1	4.89196	–5.27311–1	II

TABLE 21.27.4. Recommended values of heat capacities

Temp. (K)	220	230	240	250	260	270	273.15
c_{sat} (J K^{-1}g^{-1})	1.370	1.389	1.409	1.431	1.455	1.480	1.488
C_{sat} (J K^{-1}mol^{-1})	150.9	152.9	155.2	157.6	160.2	162.9	163.8
Temp. (K)	280	290	298.15	300	310	320	330
c_{sat} (J K^{-1}g^{-1})	1.505	1.532	1.553	1.558	1.585	1.613	1.639
C_{sat} (J K^{-1}mol^{-1})	165.8	168.7	171.1	171.6	174.6	177.6	180.5
Temp. (K)	340	350	360				
c_{sat} (J K^{-1}g^{-1})	1.666	1.691	1.716				
C_{sat} (J K^{-1}mol^{-1})	183.4	186.2	189.0				

TABLE 21.27.5. Parameters of quasi-polynomial equation

Heat capacity type	No. data points total	used	s_w	s/R	s_r %	s_b/R	+/–
C_{sat}	24	19	0.959	3.81–2	0.19	1.28–4	–1

Temp. range K	T_c K	A_1	A_2	A_3	A_4	Level of uncertainty
221.4–360.5	589.20	–3.80900	3.34663–1	1.17122+1	1.08381+1	II

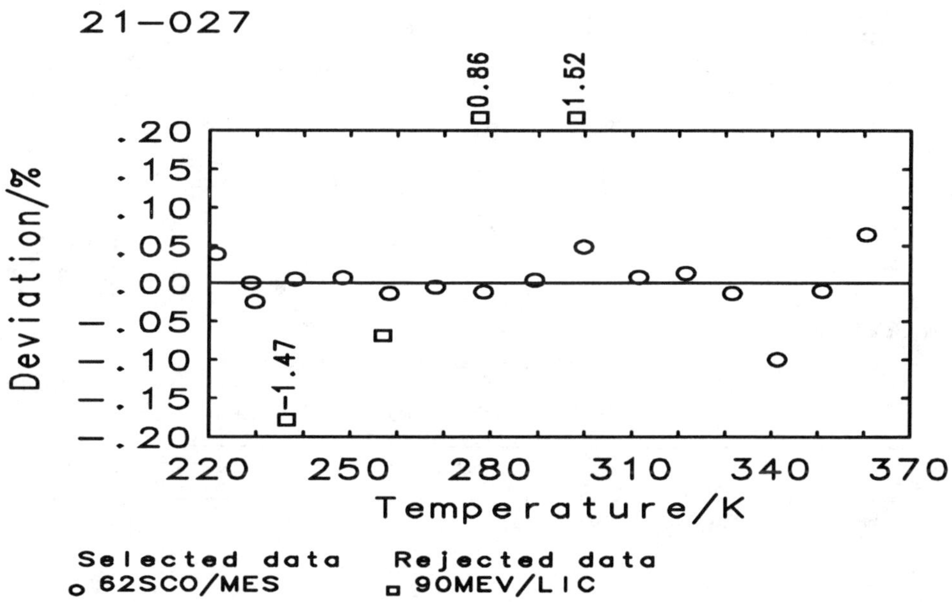

Name: Decafluorobis(trifluoromethyl)cyclohexane
Formula: C_8F_{16}

CAS–RN: 26637–68–3
Group No.: 21–028
Molar Mass: 400.06

TABLE 21.28.1. Experimental heat capacities

Reference		Temp. range K	No. pnts	Error %	Purity %	Purity method	Type capacity	Calorimeter Type	Calorimeter Reference
57YAR/KAY	N	304.0–360.0	eqn	1.10	not specified		C_p	BSIO	75PED/KAY

57YAR/KAY probably mixture of isomers (according to n.b.t.)

TABLE 21.28.3. Parameters of regression polynomial

Heat capacity type	No. data points total	No. data points used	s_w	s/R	s_r %	s_b/R	+/–
C_p	6	6	0.000	1.91–6	0.00	6.36–7	0

Temp. range K	A_1	A_2	Level of uncertainty
304.0–360.0	2.95116+1	6.44217	V

TABLE 21.28.4. Recommended values of heat capacities

Temp. (K)	300	310	320	330	340	350	360
c_p (J K^{-1}g^{-1})	1.02	1.03	1.04	1.06	1.07	1.08	1.10
C_p (J K^{-1}mol^{-1})	406	411	417	422	427	433	438

Temp. (K)	370
c_p (J K^{-1}g^{-1})	1.11
C_p (J K^{-1}mol^{-1})	444

Name: Octadecafluorooctane
Formula: C_8F_{18}

CAS–RN: 307–34–6
Group No.: 21–029
Molar Mass: 438.06

TABLE 21.29.1. Experimental heat capacities

Reference		Temp. K	Capac. J/(K.g)	Error %	Purity %	Purity method	Type capacity	Calorimeter Type	Calorimeter Reference
83CAM/DIA	N	293.00	0.8883	nosp	not specified		C_p	BDHT	77LAG/PIE

83CAM/DIA a graph based on the same data in 82CAM/REY

Name: Hexadecafluoro–1,2,3,4,5,6,7,8–octahydronaphthalene
Formula: $C_{10}F_{16}$

CAS–RN: 54939–04–7
Group No.: 21–030
Molar Mass: 424.08

TABLE 21.30.1. Experimental heat capacities

Reference	Temp. range K	No. pnts	Error %	Purity %	Purity method	Type capacity	Calorimeter Type	Calorimeter Reference
81ZHO/KOS2	266.8–300.8	7	nosp	99.59	melpt	C_{sat}	BSAO	79ZHO/KOS

TABLE 21.30.3. Parameters of regression polynomial

Heat capacity type	No. data points total	used	s_w	s/R	s_r %	s_b/R	+/–
C_{sat}	7	7	0.136	3.41–2	0.07	3.76–5	–1

Temp. range K	A_1	A_2	Level of uncertainty
266.8–300.8	2.64594+1	8.42922	IV

TABLE 21.30.4. Recommended values of heat capacities

Temp. (K)	270	273.15	280	290	298.15	300
c_{sat} (J K^{-1}g^{-1})	0.965	0.970	0.981	0.998	1.011	1.015
C_{sat} (J K^{-1}mol^{-1})	409.2	411.4	416.2	423.2	429.0	430.3

Name: *cis*–Octadecafluorodecahydronaphthalene
Formula: $C_{10}F_{18}$

CAS–RN: 60433–11–6
Group No.: 21–031
Molar Mass: 462.08

TABLE 21.31.1. Experimental heat capacities

Reference	Temp. range K	No. pnts	Error %	Purity %	method	Type capacity	Calorimeter Type	Reference
81ZHO/KOS1	269.5–300.6	6	nosp	99.57	melpt	C_{sat}	BSAO	79ZHO/KOS

TABLE 21.31.3. Parameters of regression polynomial

Heat capacity type	No. data points total	used	s_w	s/R	s_r %	s_b/R	+/–
C_{sat}	6	6	0.157	4.17–2	0.08	4.83–5	1

Temp. range K	A_1	A_2	Level of uncertainty
269.5–300.6	2.82565+1	8.65756	IV

TABLE 21.31.4. Recommended values of heat capacities

Temp. (K)	270	273.15	280	290	298.15	300
c_{sat} (J K^{-1}g^{-1})	0.929	0.934	0.945	0.960	0.973	0.976
C_{sat} (J K^{-1}mol^{-1})	429.3	431.6	436.5	443.7	449.6	450.9

Name: *trans*–Octadecafluorodecahydronaphthalene
Formula: $C_{10}F_{18}$

CAS–RN: 60433–12–7
Group No.: 21–032
Molar Mass: 462.08

TABLE 21.32.1. Experimental heat capacities

Reference	Temp. range K	No. pnts	Error %	Purity %	method	Type capacity	Calorimeter Type	Reference
81ZHO/KOS1	296.6–306.1	5	nosp	99.46	melpt	C_{sat}	BSAO	79ZHO/KOS

TABLE 21.32.3. Parameters of regression polynomial

Heat capacity type	No. data points total	used	s_w	s/R	s_r %	s_b/R	+/–
C_{sat}	5	5	0.241	6.55–2	0.12	9.61–5	1
Temp. range K	A_1		A_2				Level of uncertainty
296.6–306.1	2.06769+1		1.10978+1				IV

TABLE 21.32.4. Recommended values of heat capacities

Temp. (K)	295	298.15	300	305
c_{sat} (J K^{-1}g^{-1})	0.961	0.967	0.971	0.981
C_{sat} (J K^{-1}mol^{-1})	444.1	447.0	448.7	453.3

Name: 2,2',3,3',4,4',5,5',6,6'–Decafluoro–1,1'–biphenyl
Formula: $C_{12}F_{10}$

CAS–RN: 434–90–2
Group No.: 21–033
Molar Mass: 334.12

TABLE 21.33.1. Experimental heat capacities

Reference	Temp. range K	No. pnts	Error %	Purity %	Purity method	Type capacity	Calorimeter Type	Calorimeter Reference
71PAU/RAK	347.5–349.8	3	nosp	99.93	melpt	C_p	BSAO	69PAU/LAV2

TABLE 21.33.3. Parameters of regression polynomial

Heat capacity type	No. data points total	used	s_w	s/R	s_r %	s_b/R	+/–
C_p	3	3	0.001	4.36–4	0.00	2.54–6	0
Temp. range K	A_1		A_2				Level of uncertainty
347.5–349.8	4.15903+1		2.14082				IV

TABLE 21.33.4. Recommended values of heat capacities

Temp. (K)	348	350
c_p (J K^{-1}g^{-1})	1.220	1.221
C_p (J K^{-1}mol^{-1})	407.7	408.1

22. Chlorine Derivatives

The chlorine family contains 49 compounds, 8 of which have data at only one temperature.

No single laboratory specifically focused its attention on the heat capacities of chlorinated compounds. The largest number of chlorinated organic compounds for which heat capacity data were reported (15 substances) were those of Kurbatov (48KUR) from PTIL, however, these data appear to be of low reliability due to the limits of the calorimeter used and due to the uncertain quality of samples studied. Therefore, the data of Kurbatov have been included only in those cases when no other more reliable data were available or when it allowed extension of the temperature interval of the recommended data. The same applied to the data of the German laboratories (*81VON, *87SCH) measured in the late 1800's.

Three sets of data from UCB (40PIT, 48GOR/GIA, 56LI/PIT) and three sets from CIUG (67LEB/RAB2, 78MOS/RAB, 86NOV/RAB) were considered to be high quality data. In the range of higher temperatures, up to about 350 K, the values used were reported from GPI (67RAS/GAN) to have an error of 0.7 %. For two compounds, the data from UCE (57HAR/MOE) with a reported error of 0.5 % have been selected. Also, three compounds measured at ICLO (55STA/TUP) having an uncertainty of 1.0 % were included in the final selection.

Various low molar mass chlorinated hydrocarbons are used as common solvents. It is surprising that despite a large number of measurements, it is difficult to find reliable and consistent data. As an example, for tetrachloromethane, where data from 34 different sources were collected, it is difficult to deduce what is the real dependence of $C_p = f(T)$. Several authors (36LOR/BLA, 44HIC/HOO, 55STA/TUP, 89PET/PES) have found a minimum in the C_p versus temperature curve but each of them reported a different temperature (in the range between 254 to 304 K). There is a real possibility for the existence of a minimum as a similar variation of C_p with temperature was observed for other chlorinated methanes and tetrafluoromethane. A detailed discussion regarding minima on heat capacity curves can be found in the reference (93ZAB/BUR). In the hope of obtaining meaningful recommended values from inconsistent measurements for tetrachloromethane, the data from five sources (and several additional one point measurements) have been simultaneously correlated. The situation was not much better for trichloromethane where the data from the references (55STA/TUP, 57HAR/MOE, 67RAS/GAN) are consistent only to about 1.5 % and the minimum on the C_p curve near 278 K was reported only in the reference (86ALP/PES) from work at IKNR.

For dichloromethane, two data sets were obtained from work at KITH (37PER, 40RIE) both with an error estimated by us to be 1.5 %. They have been, however, discarded and the preference has been given to more recent measurements from CIUG (78MOS/RAB) with a reported error 0.3 %.

No heat capacity data for chloromethane were published after 1940. Both accurate data from PSC (40MES/AST),less reliable data from the Engineering Committee of the Board, London (24SHO), and data from NPLT (40AWB/GRI) have been combined and included to give recommended values over the largest temperature interval possible.

An error of 0.3 % was claimed for the data on tetrachloroethene and reported by CIUG (86NOV/RAB) over the interval from 253 to 300 K; however, at 298 K the C_p data obtained from excess heat capacity measurements (82GRO/ING, 89WIL/LAI) differ by more than 7 %. The data for 1,1-dichloroethene from DCM (59HIL/MCD) are considered to be highly reliable as well as the data for 1,1,1-trichloroethane from NPLT (73AND/CON). The data obtained from the simultaneous measurements of C_p and dielectric constant at PUP (50CRO/SMI) were omitted because of their uncertain reliability for 1,1,1-trichloroethane, but had to be considered for 1,1,2-trichloroethane because it was the only available data source. We expect the error to be between 1 and 2 %.

1,2-Dichloroethane is a frequently measured compound where the data from three sources (40PIT1, 55STA/TUP, 67RAS/GAN) have been supplemented by data for pure components obtained as part of excess heat capacity measurements at the Technological University, Delft, Netherlands (55RUI2) and at UWA (69WIL/SCH).

Most heat capacity measurements for chlorinated benzenes come from the old literature with the exception of the source (74PET/TER) which reports data for 1,2,4-trichlorobenzene; however, the reliability and accuracy appear to be low. Several data sources are available for chlorobenzene at 298 K from measurements on pure components as part of excess heat capacity studies. The temperature dependence for chlorobenzene was reported in the references (37STU) from JHUB and (90RAO/RUJ) from IISB. The latter source has not been considered in the final selection of recommended values because the real uncertainty of the data seems to be much higher than the claimed error 4 %.

Name: Tetrachloromethane
Formula: CCl_4

CAS–RN: 56–23–5
Group No.: 22–001
Molar Mass: 153.82

TABLE 22.1.1. Experimental heat capacities

Reference		Temp. range K	No. pnts	Error %	Purity %	Purity method	Type capacity	Calorimeter Type	Calorimeter Reference
12SCH2		293.1–323.1	3	nosp	not specified		C_p	BSIO	12SCH1
22LAT		253.8–290.0	3	2.00	not specified		C_p	BSIO	20GIB/LAT
24WIL/DAN		303.0–331.0	eqn	nosp	not specified		C_p	BSAO	24WIL/DAN
25WIL/DAN		293.1–333.1	5S	nosp	not specified		C_p	BSAO	24WIL/DAN
33KOL/UDO	N	288.3	1	nosp	not specified		C_p	BSIT	34KOL/UDO2
36LOR/BLA		260.0–330.0	8S	1.50	not specified		C_p	DSIO	26AND/LYN
37STU		260.0–320.0	7S	nosp	not specified		C_p	BDHO	37STU
37VOL		298.1	1	nosp	not specified		C_p	BSIO	37VOL
39PHI		301.1	1	nosp	not specified		C_p	BSIO	49WEI
41ZHD		278.6–319.2	3	nosp	not specified		C_p	BSIT	34KOL/UDO2
44HIC/HOO		253.8–298.5	13	nosp	99.98	melpt	C_p	BSIO	38HIC
48KUR		299.1	1	nosp	not specified		C_{avg}	DSIO	47KUR
49TSC/RIC3		298.1	1	nosp	not specified		C_p	BSIO	49TSC/RIC1
55STA/TUP		295.9–338.8	7	1.00	not specified		C_p	BSAO	55STA/TUP
57HAR/MOE		254.4–303.3	6	0.50	not specified		C_p	BSIO	57HAR/MOE
67GRA		308.1–338.1	7	1.00	not specified		C_p	BSIO	67GRA
67RAS/GAN		293.1–333.1	3S	0.50	not specified		C_p	BSAO	67RAS/GAN
71DES/BHA		298.1–318.1	3S	nosp	not specified		C_p	BSIO	56MUR/VAN
71VAN		255.0–300.0	10S	nosp	99.97	chrom	C_{sat}	BSAO	72VAN
72ARE/VAN		252.6–256.1	3	nosp	99.97	melpt	C_p	BSAO	72VAN
72REC/SAD		298.1–298.1	2	0.30	not specified		C_p	BSIO	70REC
73SUB/RAJ		298.1–323.1	3	0.30	not specified		C_p	BSIO	64MOE/THO
74WIL/ZET		273.1–323.1	6	nosp	not specified		C_p	BSAO	64ARN1
75GRO/BEN		298.1	1	0.30	not specified		C_p	FSIT	71PIC/LED
76FOR/BEN1		298.1	1	0.30	not specified		C_p	FSIT	71PIC/LED
76FOR/BEN2		298.1	1	0.30	not specified		C_p	FSIT	71PIC/LED
78GRO/WIL		298.1	1	0.30	not specified		C_p	FSIT	71PIC/LED
79GRO/HAM		298.1	1	0.30	not specified		C_p	FSIT	71PIC/LED
79WIL/FAR		298.1	1	0.30	99.5	chrom	C_p	FSIT	71PIC/LED
81ATA/ELS		293.1	1	2.50	not specified		C_p	BDHO	81ATA/ELS
82TAN		293.1–303.1	3	nosp	99.99	chrom	C_p	FSIT	71PIC/LED
85NKI/CHA		298.0	1	0.30	not specified		C_p	FSIT	71PIC/LED
89LAI/ROD		298.1	1	nosp	99.5	anal	C_p	FSIT	71PIC/LED
89PET/PES		258.1–318.1	4	nosp	not specified		C_p	BSAO	83KUK/KOR

33KOL/UDO same datum in 34KOL/UDO2

TABLE 22.1.2. Correlated heat capacities

Reference	Temp. range K	No. pnts used	$\sigma_r C$ %	d_w	d/R	d_r %	d_b/R	+/–
+/–								
				Selected data				
44HIC/HOO	253.8–298.5	13	0.50#	0.919	7.24–2	0.46	5.49–3	1
55STA/TUP	295.9–338.8	7	1.00	0.790	1.27–1	0.79	1.22–1	7
67RAS/GAN	293.1–333.1	3	0.50	0.833	6.68–2	0.42	5.93–2	3
73SUB/RAJ	298.1–323.1	3	0.30	1.494	7.07–2	0.45	–6.89–2	–3
74WIL/ZET	273.1–323.1	6	0.30#	0.607	2.88–2	0.18	2.50–2	5
75GRO/BEN	298.1	1	0.30	0.863	4.11–2	0.26	4.11–2	1
76FOR/BEN1	298.1	1	0.30	0.428	2.03–2	0.13	–2.03–2	–1
76FOR/BEN2	298.1	1	0.30	0.504	2.39–2	0.15	–2.39–2	–1
78GRO/WIL	298.1	1	0.30	0.029	1.38–3	0.01	1.38–3	0
79GRO/HAM	298.1	1	0.30	0.029	1.38–3	0.01	1.38–3	0
79WIL/FAR	298.1	1	0.30	0.504	2.39–2	0.15	–2.39–2	–1
82TAN	293.1–303.1	3	0.30#	0.546	2.59–2	0.16	–2.58–2	–3

Rejected data

12SCH2	(1.00, 6.07, 2.94–1, 1)	22LAT	(1.60–1, 1.01, 1.60–1, 2)
24WIL/DAN	(3.57–1, 2.23, 2.86–2, 1)	25WIL/DAN	(2.77–1, 1.74, –4.64–2, –3)
33KOL/UDO	(5.78–1, 3.80, –5.78–1, –1)	36LOR/BLA	(2.78–1, 1.74, 4.65–2, 2)
37STU	(1.12–1, 0.71, 4.45–2, 3)	37VOL	(1.69–1, 1.06, 1.69–1, 1)
39PHI	(1.86–1, 1.16, 1.86–1, 1)	41ZHD	(1.95–1, 1.22, 6.62–2, 1)
48KUR	(3.28–1, 2.12, –3.28–1, –1)	49TSC/RIC3	(3.55–1, 2.19, 3.55–1, 1)
57HAR/MOE	(1.00–1, 0.64, –9.39–2, –6)	67GRA	(1.91–1, 1.18, 1.88–1, 7)
71DES/BHA	(1.22–1, 0.76, 6.39–2, 1)	71VAN	(5.53–1, 3.64, –5.52–1, –10)
72ARE/VAN	(2.12–1, 1.34, 2.11–1, 2)	72REC/SAD	(3.13–1, 2.03, –2.98–1, –2)
81ATA/ELS	(1.86–1, 1.19, –1.86–1, –1)	85NKI/CHA	(1.62–1, 1.01, 1.62–1, 1)
89LAI/ROD	(2.15–1, 1.34, 2.15–1, 1)	89PET/PES	(4.54–1, 2.78, 4.11–1, 4)

TABLE 22.1.3. Parameters of regression polynomial

Heat capacity type	No. data points total	used	s_w	s/R	s_r %	s_b/R	+/–
C_p	116	41	0.869	7.60–2	0.48	2.31–2	8

Temp. range K	A_1	A_2	A_3	Level of uncertainty
253.8–338.8	1.62691+1	–7.50981–1	2.01666–1	IV

TABLE 22.1.4. Recommended values of heat capacities

Temp. (K)	260	270	273.15	280	290	298.15	300
c_p (J K^{-1}g^{-1})	0.848	0.849	0.850	0.851	0.853	0.855	0.856
C_p (J K^{-1}mol^{-1})	130.4	130.6	130.7	130.9	131.3	131.6	131.6

Temp. (K)	310	320	330	340
c_p (J K^{-1}g^{-1})	0.858	0.861	0.864	0.867
C_p (J K^{-1}mol^{-1})	132.0	132.5	132.9	133.4

TABLE 22.1.5. Parameters of quasi-polynomial equation

Heat capacity type	No. data points total	used	s_w	s/R	s_r %	s_b/R	+/−
C_p	116	41	0.951	6.74–2	0.43	4.91–3	1

Temp. range K	T_c K	A_1	A_2	A_3	A_4	Level of uncertainty
253.8–338.8	556.40	7.24295	3.12242	1.24136+1	4.20029	IV

22–001

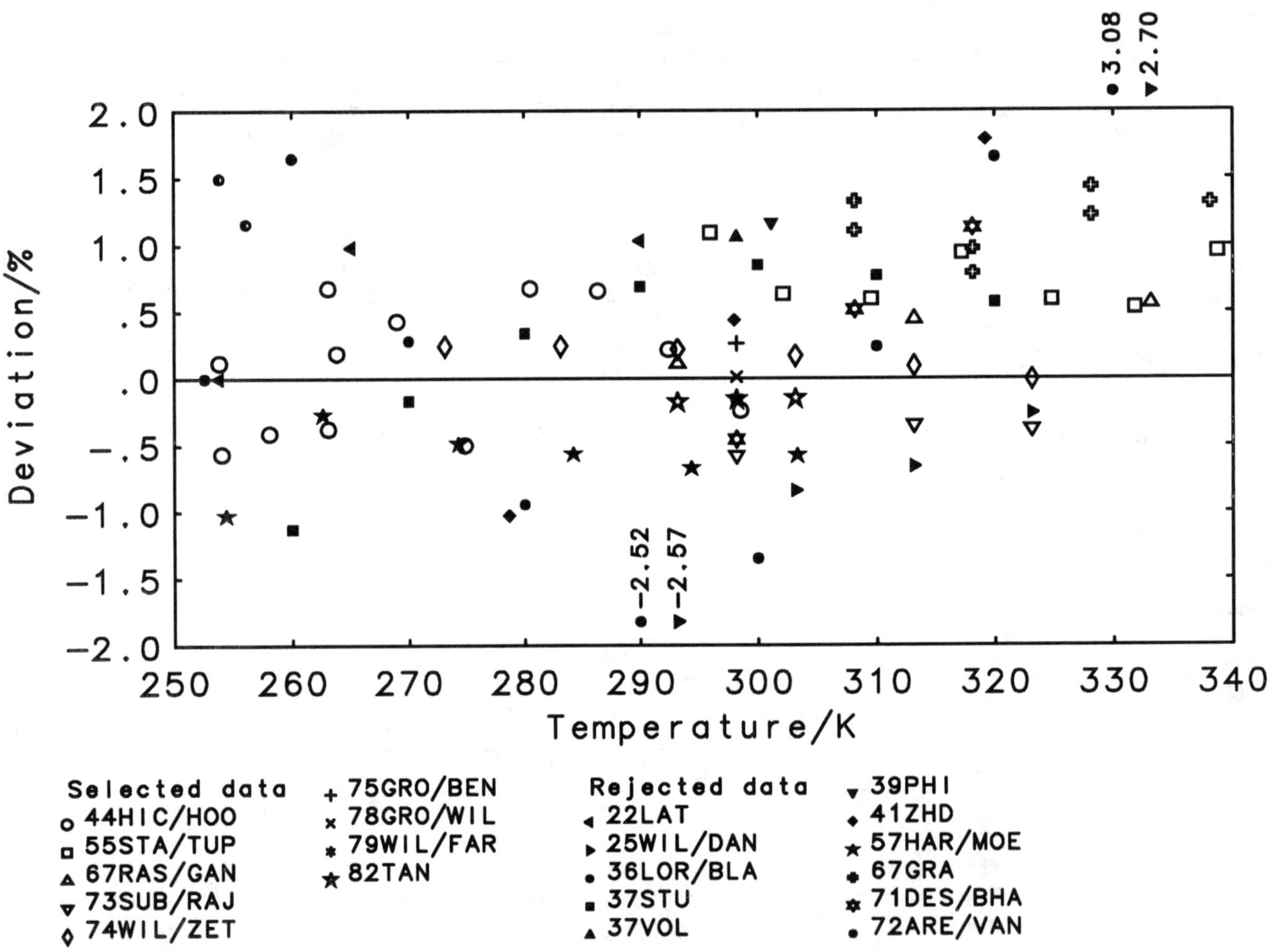

Name: Trichloromethane
Formula: $CHCl_3$

CAS-RN: 67–66–3
Group No.: 22–002
Molar Mass: 119.38

TABLE 22.2.1. Experimental heat capacities

Reference		Temp. range K	No. pnts	Error %	Purity %	Purity method	Type capacity	Calorimeter Type	Calorimeter Reference
12SCH1		223.1–328.1	9	nosp	not specified		C_p	BSIO	12SCH1
21TRE		295.5–321.8	7	nosp	not specified		C_p	BSIO	49WEI
24WIL/DAN		295.0–318.0	eqn	nosp	not specified		C_p	BSAO	24WIL/DAN
25WIL/DAN		293.1–323.1	4S	nosp	not specified		C_p	BSAO	24WIL/DAN
39PHI		303.5	1	nosp	not specified		C_p	BSIO	49WEI
47SKU		293.1	1	0.30	not specified		C_p	BSAO	47SKU
48KUR		252.9–305.8	4	nosp	not specified		C_{avg}	DSIO	47KUR
49TSC/RIC3		293.1	1	nosp	not specified		C_p	BSIO	49TSC/RIC1
55STA/TUP		284.0–329.0	11	1.00	not specified		C_p	BSAO	55STA/TUP
57HAR/MOE		245.9–303.2	8	0.50	not specified		C_p	BSIO	57HAR/MOE
67RAS/GAN		293.1–333.1	3S	0.50	not specified		C_p	BSAO	67RAS/GAN
85HEP/KOO		298.1	1	nosp	not specified		C_p	FSIT	71PIC/LED
86ALP/PES	N	258.1–318.1	4	0.20	not specified		C_p	BSAO	83KUK/KOR
89BAR/KOO1		298.1	1	nosp	not specified		C_p	FSIT	71PIC/LED
89BAR/KOO2		298.1	1	nosp	99.5	anal	C_p	FSIT	71PIC/LED
89ING		298.1	1	nosp	99.5	anal	C_p	FSIT	71PIC/LED
90YAM/OGA		298.1	1	nosp	99.97	chrom	C_p	FSIO	85OGA
91GRO/ROU		298.1	1	nosp	99.	anal	C_p	FSIT	71PIC/LED

86ALP/PES same data in 89PET/PES and 90ALP/PES

TABLE 22.2.2. Correlated heat capacities

Reference	Temp. range K	No. pnts used	$\sigma_r C$ %	d_w	d/R	d_r %	d_b/R	+/–
				Selected data				
55STA/TUP	284.0–329.0	11	1.00	0.930	1.29–1	0.93	–8.90–2	–5
57HAR/MOE	245.9–303.2	8	0.50	0.973	6.60–2	0.49	–3.27–2	–4
67RAS/GAN	293.1–333.1	3	0.50	3.201	2.25–1	1.60	2.05–1	3
89BAR/KOO1	298.1	1	0.50#	0.163	1.12–2	0.08	1.12–2	1
89BAR/KOO2	298.1	1	0.50#	0.143	9.81–3	0.07	9.81–3	1
89ING	298.1	1	0.50#	0.025	1.70–3	0.01	1.70–3	1
90YAM/OGA	298.1	1	0.50#	0.713	4.88–2	0.36	–4.88–2	–1
91GRO/ROU	298.1	1	0.50#	0.713	4.88–2	0.36	–4.88–2	–1

Rejected data

Reference		Reference	
12SCH1	(4.10–1, 2.94, 3.65–1, 6)	21TRE	(1.27–1, 0.91, 5.58–2, 3)
24WIL/DAN	(2.53–1, 1.79, 2.52–1, 3)	25WIL/DAN	(3.28–1, 2.28, 3.16–1, 4)
39PHI	(3.00, 17.84, 3.00, 1)	47SKU	(3.85–1, 2.74, 3.85–1, 1)
48KUR	(3.72–1, 2.74, 3.95–2, 0)	49TSC/RIC3	(5.65–1, 3.97, 5.65–1, 1)
85HEP/KOO	(6.20–2, 0.45, –6.20–2, –1)	86ALP/PES	(3.91–1, 2.80, 3.02–1, 4)

TABLE 22.2.3. Parameters of regression polynomial

Heat capacity type	No. data points total	No. data points used	s_w	s/R	s_r %	s_b/R	+/–
C_p	62	27	1.427	1.25–1	0.90	–2.59–2	–5

Temp. range K	A_1	A_2	A_3	Level of uncertainty
245.9–333.1	1.78697+1	–4.18738	9.39942–1	V

TABLE 22.2.4. Recommended values of heat capacities

Temp. (K)	240	250	260	270	273.15	280	290
c_p (J K^{-1}g^{-1})	0.92	0.92	0.93	0.93	0.94	0.94	0.95
C_p (J K^{-1}mol^{-1})	110	110	111	112	112	112	113
Temp. (K)	298.15	300	310	320	330		
c_p (J K^{-1}g^{-1})	0.96	0.96	0.97	0.98	1.00		
C_p (J K^{-1}mol^{-1})	114	114	116	117	119		

TABLE 22.2.5. Parameters of quasi-polynomial equation

Heat capacity type	No. data points total	used	s_w	s/R	s_r %	s_b/R	+/–
C_p	62	27	1.499	1.17–1	0.84	4.41–4	–5

Temp. range K	T_c K	A_1	A_2	A_3	A_4	Level of uncertainty
245.9–333.1	536.40	2.94568	2.39978	1.02086+1	9.03939–1	V

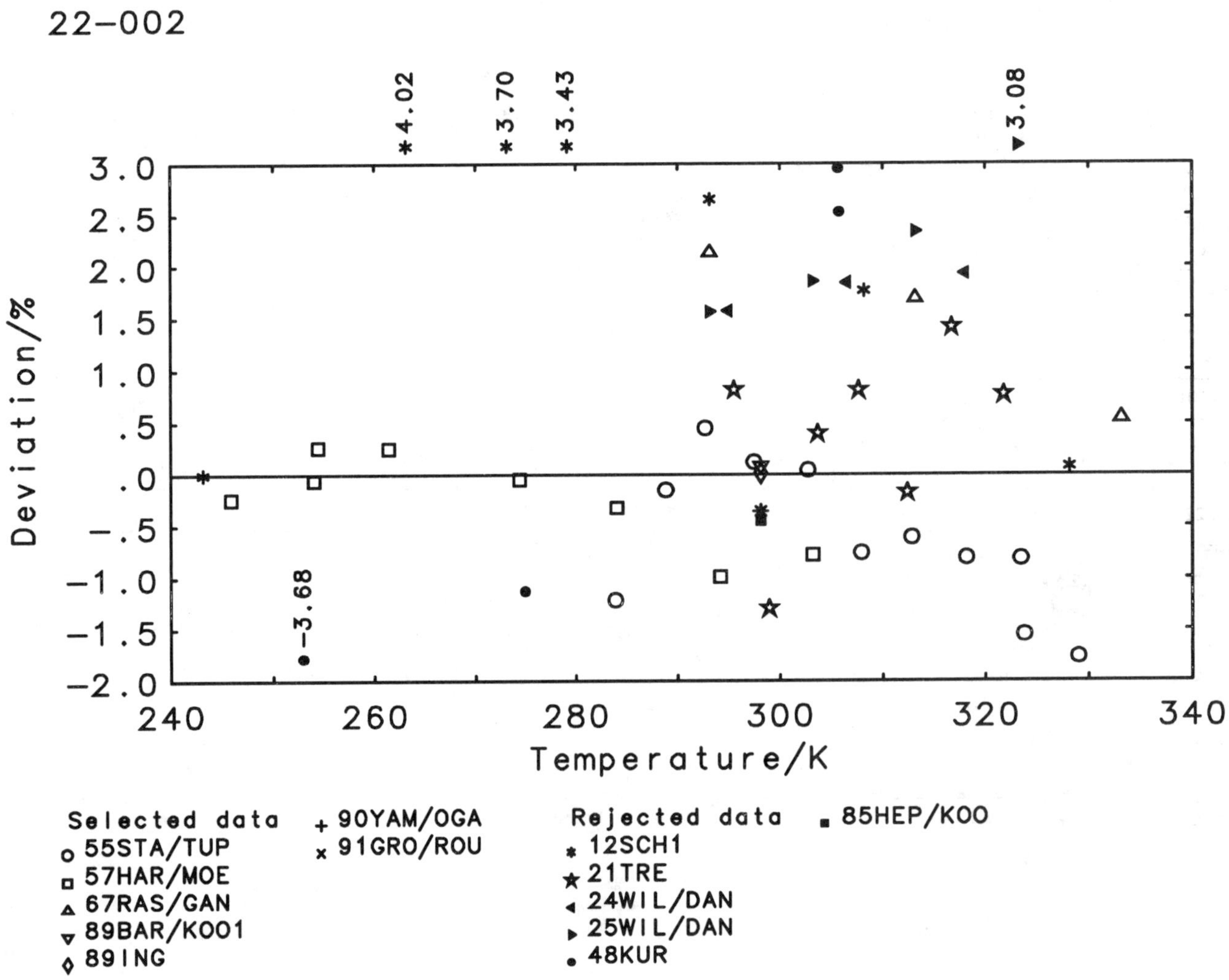

Name: Dichloromethane
Formula: CH_2Cl_2

CAS–RN: 75–09–2
Group No.: 22–003
Molar Mass: 84.93

TABLE 22.3.1. Experimental heat capacities

Reference		Temp. range K	No. pnts	Error %	Purity %	Purity method	Type capacity	Calorimeter Type	Calorimeter Reference
37PER1	N	215.1–292.4	15	0.40	not specified		C_p	BSIO	37PER1
40RIE	N	226.3–313.9	12	nosp	not specified		C_{sat}	BSIO	39RIE3
48KUR		241.0–299.9	4	nosp	not specified		C_{avg}	DSIO	47KUR
78MOS/RAB		182.6–301.0	30	0.30	99.97	melpt	C_p	BSAO	76LEB/LIT

37PER1 same data in 37PER2
40RIE same data in 41RIE1

TABLE 22 3.2. Correlated heat capacities

Reference	Temp. range K	No. pnts used	$\sigma_r C$ %	d_w	d/R	d_r %	d_b/R	+/–
				Selected data				
78MOS/RAB	182.6–301.0	29	0.30	0.580	2.02–2	0.17	7.08–5	2
				Rejected data				
37PER1	(1.94–1, 1.64, 1.64–1, 13)			40RIE	(2.11–1, 1.78, 1.06–1, 5)			
48KUR	(2.45, 17.00, 2.41, 4)							

TABLE 22.3.3. Parameters of regression polynomial

Heat capacity type	No. data points total	No. data points used	s_w	s/R	s_r %	s_b/R	+/–
C_p	61	29	0.625	2.17–2	0.19	7.08–5	2

Temp. range K	A_1	A_2	A_3	A_4	Level of uncertainty
182.6–301.0	8.44513	6.28476	–3.75910	6.98459–1	III

TABLE 22.3.4. Recommended values of heat capacities

Temp. (K)	180	190	200	210	220	230	240
c_p ($J K^{-1} g^{-1}$)	1.141	1.136	1.132	1.129	1.127	1.127	1.129
C_p ($J K^{-1} mol^{-1}$)	96.88	96.50	96.16	95.90	95.74	95.72	95.88
Temp. (K)	250	260	270	273.15	280	290	298.15
c_p ($J K^{-1} g^{-1}$)	1.133	1.140	1.151	1.155	1.165	1.184	1.202
C_p ($J K^{-1} mol^{-1}$)	96.25	96.86	97.76	98.11	98.97	100.5	102.1
Temp. (K)	300						
c_p ($J K^{-1} g^{-1}$)	1.207						
C_p ($J K^{-1} mol^{-1}$)	102.5						

TABLE 22.3.5. Parameters of quasi-polynomial equation

Heat capacity type	No. data points total	used	s_w	s/R	s_r %	s_b/R	+/–
C_p	61	17	0.821	2.84–2	0.25	1.15–4	4

Temp. range K	T_c K	A_1	A_2	A_3	A_4	Level of uncertainty
220.6–301.0	510.00	6.64132+1	2.09089+1	–1.05361+1	5.27374+1	III

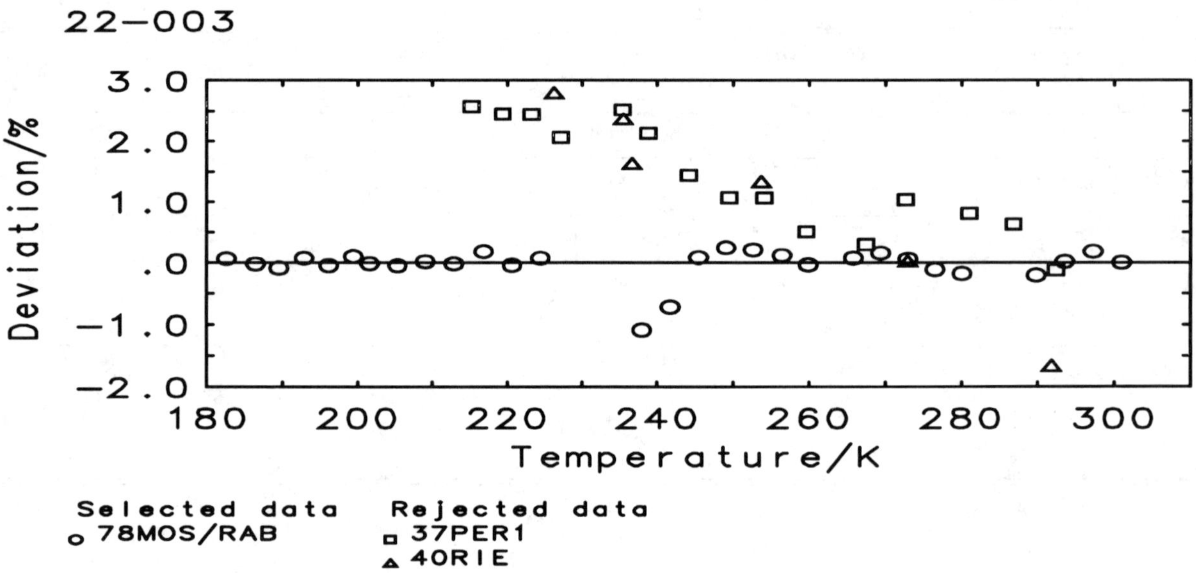

Name: Chloromethane
Formula: CH_3Cl

CAS–RN: 74–87–3
Group No.: 22–004
Molar Mass: 50.49

TABLE 22.4.1. Experimental heat capacities

Reference	Temp. range K	No. pnts	Error %	Purity %	method	Type capacity	Calorimeter Type	Reference
24SHO	243.1–313.1	8	nosp	not specified		C_{sat}	BDHO	24SHO
28EUC/HAU	190.0–240.0	6S	nosp	not specified		C_{sat}	BSIO	28EUC/HAU
40AWB/GRI	243.1–303.1	7S	1.00	99.9998	anal	C_{sat}	BSAO	40AWB/GRI
40MES/AST	182.2–249.7	10	0.50	not specified		C_p	BSIO	36AST/MES

TABLE 22.4.2. Correlated heat capacities

Reference	Temp. range K	No. pnts used	$\sigma_r C$ %	d_w	d/R	d_r %	d_b/R	+/–
				Selected data				
24SHO	243.1–313.1	8	1.00#	0.789	7.53–2	0.79	3.66–2	4
40AWB/GRI	243.1–303.1	7	1.00	0.395	3.73–2	0.40	1.67–2	3
40MES/AST	182.2–249.7	10	0.50	0.521	2.35–2	0.26	–9.84–3	0
				Rejected data				
28EUC/HAU	(9.36–2, 1.05, 1.03–2, 2)							

TABLE 22.4.3. Parameters of regression polynomial

Heat capacity type	No. data points total	used	s_w	s/R	s_r %	s_b/R	+/–
C_p	31	25	0.632	5.25–2	0.55	1.25–2	7
C_{sat}	31	25	0.662	5.54–2	0.59	1.32–2	7

Temp. range K	A_1	A_2	A_3	Level of uncertainty
182.2–313.1	1.30804+1	–4.10065	1.01114	IV
182.2–313.1	1.26596+1	–3.69710	9.14966–1	IV

TABLE 22.4.4. Recommended values of heat capacities

Temp. (K)	180	190	200	210	220	230	240
c_p (J K^{-1}g^{-1})	1.478	1.472	1.470	1.470	1.474	1.482	1.493
C_p (J K^{-1}mol^{-1})	74.63	74.33	74.20	74.23	74.44	74.81	75.35
c_{sat} (J K^{-1}g^{-1})	1.477	1.472	1.470	1.471	1.475	1.482	1.492
C_{sat} (J K^{-1}mol^{-1})	74.58	74.32	74.21	74.25	74.45	74.80	75.30
Temp. (K)	250	260	270	273.15	280	290	298.15
c_p (J K^{-1}g^{-1})	1.507	1.524	1.545	1.552	1.569	1.596	1.621
C_p (J K^{-1}mol^{-1})	76.06	76.94	77.99	78.35	79.20	80.59	81.84
c_{sat} (J K^{-1}g^{-1})	1.504	1.520	1.539	1.546	1.561	1.586	1.609
C_{sat} (J K^{-1}mol^{-1})	75.96	76.76	77.72	78.05	78.83	80.09	81.23
Temp. (K)	300	310					
c_p (J K^{-1}g^{-1})	1.627	1.661					
C_p (J K^{-1}mol^{-1})	82.14	83.85					
c_{sat} (J K^{-1}g^{-1})	1.614	1.645					
C_{sat} (J K^{-1}mol^{-1})	81.51	83.07					

TABLE 22.4.5. Parameters of quasi-polynomial equation

Heat capacity type	No. data points total	used	s_w	s/R	s_r %	s_b/R	+/–
C_p	31	22	1.043	5.80–2	0.62	–6.15–3	2

Temp. range K	T_c K	A_1	A_2	A_3	A_4	Level of uncertainty
204.9–313.1	416.25	–5.89496–1	3.84573–1	7.56995	2.25904–1	IV

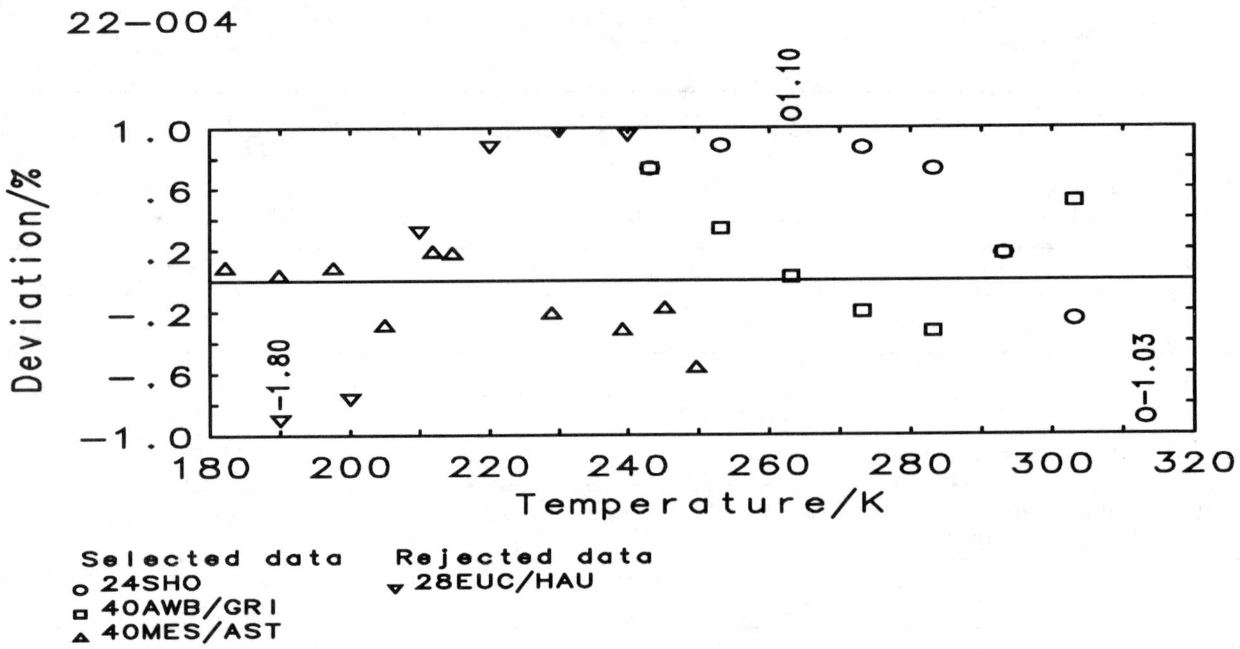

Name: Tetrachloroethene
Formula: C_2Cl_4

CAS–RN: 127–18–4
Group No.: 22–005
Molar Mass: 165.83

TABLE 22.5.1. Experimental heat capacities

Reference	Temp. range K	No. pnts	Error %	Purity %	Purity method	Type capacity	Calorimeter Type	Calorimeter Reference
*81VON	317.7–346.1	3S	nosp	not specified		C_{avg}	DSIO	*81VON
13HER/RAT	293.1	1	nosp	not specified		C_p	DSIO	22HER/SCH
48KUR	319.3–340.9	2	nosp	not specified		C_{avg}	DSIO	47KUR
66SAV	293.0–333.0	eqn	1.20	not specified		C_p	BSIO	66SAV
82GRO/ING	298.1	1	0.30	99.	estim	C_p	FSIT	71PIC/LED
86NOV/RAB	253.1–300.0	14	0.30	99.77	melpt	C_p	BSAO	76LEB/LIT
89WIL/LAI	298.1	1	nosp	99.7	anal	C_p	FSIT	71PIC/LED

TABLE 22.5.2. Correlated heat capacities

Reference	Temp. range K	No. pnts used	$\sigma_r C$ %	d_w	d/R	d_r %	d_b/R	+/–
				Selected data				
86NOV/RAB	253.1–300.0	14	0.30	0.221	1.20–2	0.07	1.20–5	–1
				Rejected data				
13HER/RAT	(7.17–1, 3.98, –7.17–1, –1)			66SAV	(2.83, 17.81, –2.83, –1)			
82GRO/ING	(1.36, 7.73, –1.36, –1)			89WIL/LAI	(1.28, 7.24, –1.28, –1)			

TABLE 22.5.3. Parameters of regression polynomial

Heat capacity type	No. data points total	used	s_w	s/R	s_r %	s_b/R	+/–
C_p	25	14	0.262	1.42–2	0.08	1.20–5	–1

Temp. range K	A_1	A_2	A_3	A_4	Level of uncertainty
253.1–300.0	4.71766+2	–4.72380+2	1.62500+2	–1.84466+1	IV

TABLE 22.5.4. Recommended values of heat capacities

Temp. (K)	260	270	273.15	280	290	298.15	300
c_p (J K^{-1}g^{-1})	0.895	0.896	0.900	0.911	0.932	0.952	0.956
C_p (J K^{-1}mol^{-1})	148.5	148.7	149.2	151.0	154.6	157.8	158.5

TABLE 22.5.5. Parameters of quasi-polynomial equation

Heat capacity type	No. data points total	used	s_w	s/R	s_r %	s_b/R	+/–
C_p	25	11	0.751	4.10–2	0.23	9.17–5	–3

Temp. range K	T_c K	A_1	A_2	A_3	A_4	Level of uncertainty
260.3–300.0	620.20	1.05500+3	3.32726+2	–3.32429+2	8.36292+2	IV

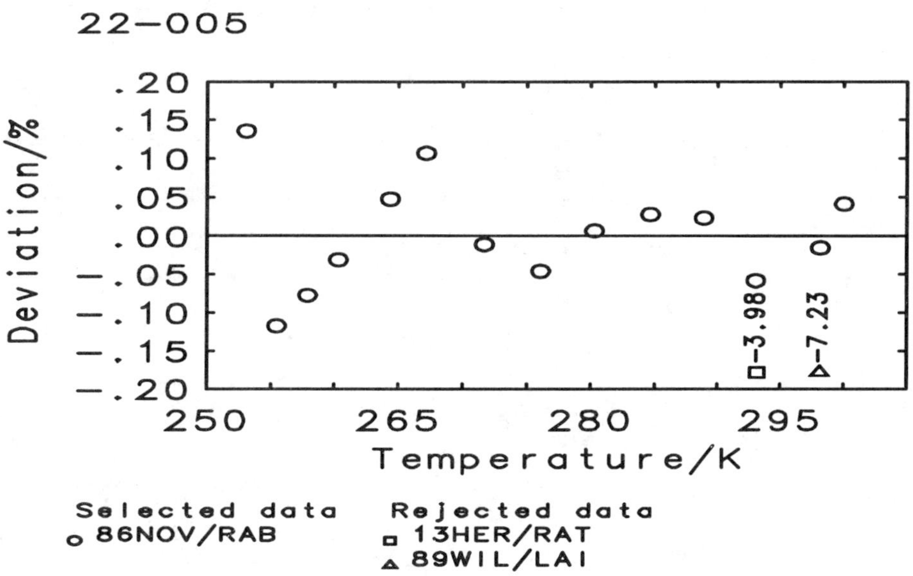

Name: Trichloroethene
Formula: C_2HCl_3

CAS–RN: 79–01–6
Group No.: 22–006
Molar Mass: 131.39

TABLE 22.6.1. Experimental heat capacities

Reference	Temp. range K	No. pnts	Error %	Purity %	Purity method	Type capacity	Calorimeter Type	Calorimeter Reference
13HER/RAT	293.1	1	nosp	not specified		C_p	DSIO	22HER/SCH
33TRE/WAT	298.1	1	nosp	not specified		C_p	BSIO	49WEI
48KUR	298.9–320.9	3	nosp	not specified		C_{avg}	DSIO	47KUR
49TSC/RIC3	298.1	1	nosp	not specified		C_p	BSIO	49TSC/RIC1

TABLE 22.6.2. Correlated heat capacities

Reference	Temp. range K	No. pnts used	$\sigma_r C$ %	d_w	d/R	d_r %	d_b/R	+/–
				Selected data				
13HER/RAT	293.1	1	2.00#	0.000	9.54–7	0.00	9.54–7	0
33TRE/WAT	298.1	1	2.00#	0.137	4.12–2	0.27	4.12–2	1
49TSC/RIC3	298.1	1	3.00#	0.204	9.10–2	0.61	–9.10–2	–1

TABLE 22.6.3. Parameters of regression polynomial

Heat capacity type	No. data points total	No. data points used	s_w	s/R	s_r %	s_b/R	+/–
C_p	6	3	0.246	9.99–2	0.67	–1.66–2	0

Temp. range K	A_1	A_2	Level of uncertainty
293.1–298.1	1.65317	4.46558	VI

TABLE 22.6.4. Recommended values of heat capacities

Temp. (K)	290	298.15	300
c_p (J $K^{-1}g^{-1}$)	0.92	0.95	0.95
C_p (J $K^{-1}mol^{-1}$)	121	124	125

Name: Pentachloroethane
Formula: C_2HCl_5

CAS–RN: 76–01–7
Group No.: 22–007
Molar Mass: 202.29

TABLE 22.7.1. Experimental heat capacities

Reference	Temp. range K	No. pnts	Error %	Purity %	Purity method	Type capacity	Calorimeter Type	Calorimeter Reference
13HER/RAT	293.1	1	nosp	not specified		C_p	DSIO	22HER/SCH
48KUR	331.2–358.7	3	nosp	not specified		C_{avg}	DSIO	47KUR

TABLE 22.7.2. Correlated heat capacities

Reference	Temp. range K	No. pnts used	$\sigma_r C$ %	d_w	d/R	d_r %	d_b/R	+/–
				Selected data				
13HER/RAT	293.1	1	3.00#	1.127	9.16–1	3.38	9.16–1	1
48KUR	331.2–358.7	3	5.00#	0.692	8.66–1	3.46	–7.30–1	–3

TABLE 22.7.3. Parameters of regression polynomial

Heat capacity type	No. data points total	used	s_w	s/R	s_r %	s_b/R	+/–
C_p	4	4	0.950	1.01	3.97	–3.19–1	–2

Temp. range K	A_1	Level of uncertainty
293.1–358.7	2.61625+1	VI

TABLE 22.7.4. Recommended values of heat capacities

Temp. (K)	300	320	340	360
c_p (J K^{-1}g^{-1})	1.08	1.08	1.08	1.08
C_p (J K^{-1}mol^{-1})	218	218	218	218

Name: 1,1–Dichloroethene
Formula: $C_2H_2Cl_2$

CAS–RN: 75–35–4
Group No.: 22–008
Molar Mass: 96.94

TABLE 22.8.1. Experimental heat capacities

Reference		Temp. range K	No. pnts	Error %	Purity %	Purity method	Type capacity	Calorimeter Type	Calorimeter Reference
*81BER/OGI	N	304.6	1	nosp	not specified		C_{avg}	DSIO	*79BER
59HIL/MCD		157.4–290.9	37	0.25	99.97	melpt	C_p	BSAO	58HIL/KRA

*81BER/OGI average value in temperature range 286–323 K

TABLE 22.8.2. Correlated heat capacities

Reference	Temp. range K	No. pnts used	$\sigma_r C$ %	d_w	d/R	d_r %	d_b/R	+/–
				Selected data				
59HIL/MCD	157.4–290.9	37	0.25	0.918	2.86–2	0.23	1.30–4	–3

TABLE 22.8.3. Parameters of regression polynomial

Heat capacity type	No. data points total	used	s_w	s/R	s_r %	s_b/R	+/–
C_p	38	37	0.957	2.98–2	0.24	1.30–4	–3

Temp. range K	A_1	A_2	A_3	Level of uncertainty
157.4–290.9	1.38528+1	–2.12017	6.53036–1	II

TABLE 22.8.4. Recommended values of heat capacities

Temp. (K)	160	170	180	190	200	210	220
c_p (J K^{-1}g^{-1})	1.041	1.041	1.042	1.045	1.048	1.053	1.059
C_p (J K^{-1}mol^{-1})	100.9	100.9	101.0	101.3	101.6	102.1	102.7
Temp. (K)	230	240	250	260	270	273.15	280
c_p (J K^{-1}g^{-1})	1.066	1.074	1.084	1.094	1.105	1.109	1.118
C_p (J K^{-1}mol^{-1})	103.4	104.1	105.0	106.1	107.2	107.5	108.4
Temp. (K)	290	298.15					
c_p (J K^{-1}g^{-1})	1.132	1.144					
C_p (J K^{-1}mol^{-1})	109.7	110.9					

TABLE 22.8.5. Parameters of quasi-polynomial equation

Heat capacity type	No. data points total	used	s_w	s/R	s_r %	s_b/R	+/–
C_p	38	37	1.143	3.53–2	0.29	1.82–4	–5

Temp. range K	T_c K	A_1	A_2	A_3	A_4	Level of uncertainty
157.4–290.9	491.30	1.04006+1	5.12849	6.88139	5.27314	III

Name: 1,2–Dichloroethene
Formula: $C_2H_2Cl_2$

CAS–RN: 540–59–0
Group No.: 22–009
Molar Mass: 96.94

TABLE 22.9.1. Experimental heat capacities

Reference		Temp. range K	No. pnts	Error %	Purity %	Purity method	Type capacity	Calorimeter Type	Calorimeter Reference
48KUR	N	265.2–306.5	3	nosp	not specified		C_{avg}	DSIO	47KUR

48KUR mixture of stereoisomers (according to n.b.t.)

TABLE 22.9.3. Parameters of regression polynomial

Heat capacity type	No. data points total	used	s_w	s/R	s_r %	s_b/R	+/–
C	3	3	0.112	7.50–2	0.56	3.03–4	1

Temp. range K	A_1	A_2	Level of uncertainty
265.2–306.5	5.52070	2.63049	VI

TABLE 22.9.4. Recommended values of heat capacities

Temp. (K)	270	273.15	280	290	298.15	300	310
c (J K^{-1}g^{-1})	1.08	1.09	1.11	1.13	1.15	1.15	1.17
C (J K^{-1}mol^{-1})	105	106	107	109	111	112	114

Name: (E)–1,2–Dichloroethene
Formula: $C_2H_2Cl_2$

CAS–RN: 156–60–5
Group No.: 22–010
Molar Mass: 96.94

TABLE 22.10.1. Experimental heat capacities

Reference		Temp. K	Capac. J/(K.g)	Error %	Purity %	Purity method	Type capacity	Calorimeter Type	Calorimeter Reference
34MEH2	N	288.15	1.163	1.50	not specified		C_p	BSIO	49WEI

34MEH2 isomer not specified, n.b.t. given for Z–isomer

Name: (Z)–1,2–Dichloroethene
Formula: $C_2H_2Cl_2$

CAS–RN: 156–59–2
Group No.: 22–011
Molar Mass: 96.94

TABLE 22.11.1. Experimental heat capacities

Reference		Temp. K	Capac. J/(K.g)	Error %	Purity %	Purity method	Type capacity	Calorimeter Type	Calorimeter Reference
34MEH2	N	288.15	1.176	1.50	not specified		C_p	BSIO	49WEI

34MEH2 isomer not specified, n.b.t. given for E–isomer

Name: 1,1,1,2–Tetrachloroethane
Formula: $C_2H_2Cl_4$

CAS–RN: 630–20–6
Group No.: 22–012
Molar Mass: 167.85

TABLE 22.12.1. Experimental heat capacities

Reference		Temp. K	Capac. J/(K.g)	Error %	Purity %	Purity method	Type capacity	Calorimeter Type	Calorimeter Reference
78GUN/HEA	N	295.00	1.120	nosp	99.96	chrom	C_p	not specified	

78GUN/HEA the origin of data unclear

Name: 1,1,2,2–Tetrachloroethane
Formula: $C_2H_2Cl_4$

CAS–RN: 79–34–5
Group No.: 22–013
Molar Mass: 167.85

TABLE 22.13.1. Experimental heat capacities

Reference	Temp. range K	No. pnts	Error %	Purity %	Purity method	Type capacity	Calorimeter Type	Calorimeter Reference
13HER/RAT	293.1	1	nosp	not specified		C_p	DSIO	22HER/SCH
34MEH1	289.1	1	1.50	not specified		C_p	BSIO	49WEI
48KUR	308.6–353.6	6	nosp	not specified		C_{avg}	DSIO	47KUR
49TSC/RIC3	298.1	1	nosp	not specified		C_p	BSIO	49TSC/RIC1
82KOS/KOL	233.6–296.5	12	nosp	99.4	melpt	C_{sat}	BSAO	79ZHO/KOS
89PRA/RAJ	318.1–333.1	4	3.00	not specified		C_p	BDHT	89PRA/RAJ
89WIL/LAI	298.1	1	nosp	not specified		C_p	FSIT	71PIC/LED

TABLE 22.13.2. Correlated heat capacities

Reference	Temp. range K	No. pnts used	$\sigma_r C$ %	d_w	d/R	d_r %	d_b/R	+/–
				Selected data				
48KUR	308.6–353.6	6	5.00#	0.635	6.21–1	3.18	5.16–2	2
82KOS/KOL	233.6–296.5	12	0.40#	0.484	3.85–2	0.19	–1.03–2	0
89WIL/LAI	298.1	1	0.50#	1.973	1.99–1	0.99	1.99–1	1
				Rejected data				
13HER/RAT	(2.68, 11.84, 2.68, 1)			34MEH1	(2.55, 11.33, 2.55, 1)			
49TSC/RIC3	(2.29, 10.28, 2.29, 1)			89PRA/RAJ	(3.80, 22.96, –3.80, –4)			

TABLE 22.13.3. Parameters of regression polynomial

Heat capacity type	No. data points total	No. data points used	s_w	s/R	s_r %	s_b/R	+/–
C	26	19	0.755	3.85–1	1.97	2.03–2	3

Temp. range K	A_1	A_2	A_3	Level of uncertainty
233.6–353.6	2.66394+1	–5.49537	1.09697	V

TABLE 22.13.4. Recommended values of heat capacities

Temp. (K)	240	250	260	270	273.15	280	290
c (J K^{-1}g^{-1})	0.98	0.98	0.98	0.98	0.98	0.98	0.99
C (J K^{-1}mol^{-1})	164	164	164	165	165	165	166

Temp. (K)	298.15	300	310	320	330	340	350
c (J K^{-1}g^{-1})	0.99	0.99	1.00	1.00	1.01	1.02	1.03
C (J K^{-1}mol^{-1})	166	167	168	169	170	172	173

TABLE 22.13.5. Parameters of quasi-polynomial equation

Heat capacity type	No. data points total	used	s_w	s/R	s_r %	s_b/R	+/–
C	26	19	0.841	3.63–1	1.86	9.86–3	11

Temp. range K	T_c K	A_1	A_2	A_3	A_4	Level of uncertainty
233.6–353.6	626.00	1.67164+2	5.31104+1	–3.59982+1	1.31537+2	V

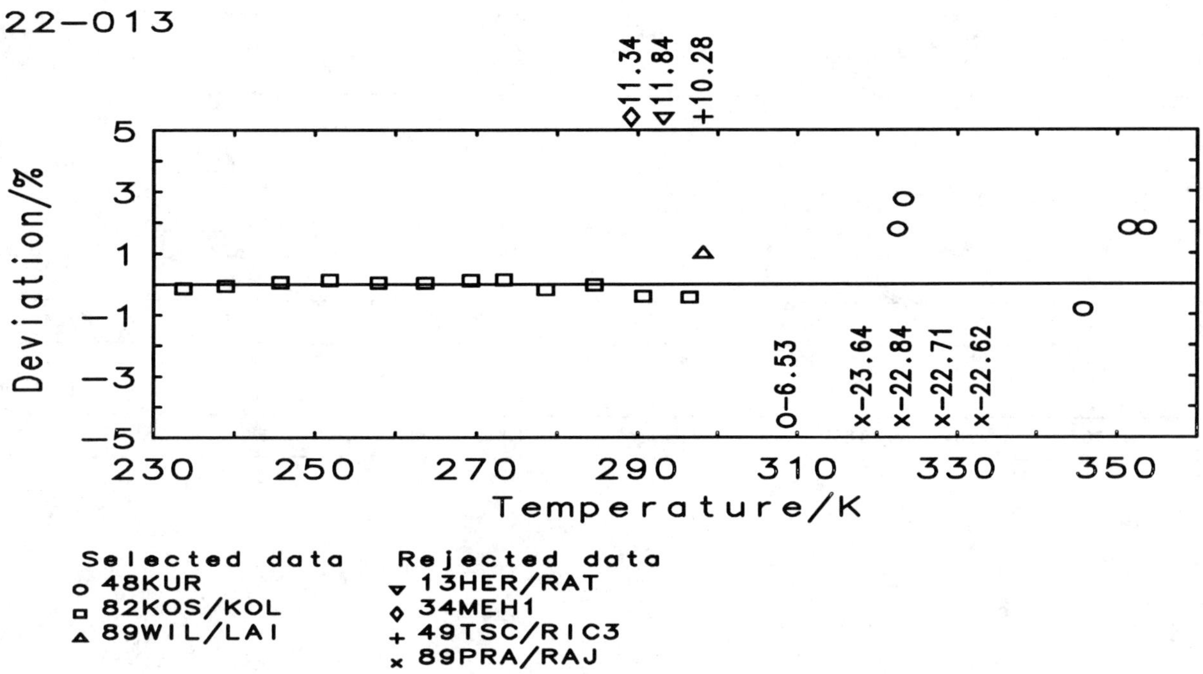

Name: Chloroethene
Formula: C_2H_3Cl

CAS–RN: 75–01–4
Group No.: 22–014
Molar Mass: 62.50

TABLE 22.14.1. Experimental heat capacities

Reference	Temp. range K	No. pnts	Error %	Purity %	Purity method	Type capacity	Calorimeter Type	Calorimeter Reference
67LEB/RAB2	119.2–300.0	10S	0.50	99.7	melpt	C_{sat}	BSAO	56POP/KOL

TABLE 22.14.3. Parameters of cubic spline polynomials

Heat capacity type	No. data points total	used	s_w	s/R	s_r %	s_b/R	+/–
C_{sat}	10	10	0.147	7.58–3	0.07	5.63–6	0

Temp. range K	A_1	A_2	A_3	A_4	Level of uncertainty
119.2–240.0	9.72177	8.87961–1	–5.12663–1	9.77939–2	IV
240.0–300.0	–1.53611+1	3.22416+1	–1.35767+1	1.91224	IV

TABLE 22.14.4. Recommended values of heat capacities

Temp. (K)	120	130	140	150	160	170	180
c_{sat} (J K^{-1}g^{-1})	1.359	1.360	1.361	1.361	1.361	1.361	1.361
C_{sat} (J K^{-1}mol^{-1})	84.96	85.01	85.04	85.06	85.06	85.06	85.05
Temp. (K)	190	200	210	220	230	240	250
c_{sat} (J K^{-1}g^{-1})	1.361	1.361	1.361	1.362	1.363	1.364	1.366
C_{sat} (J K^{-1}mol^{-1})	85.05	85.05	85.07	85.10	85.16	85.24	85.37
Temp. (K)	260	270	273.15	280	290	298.15	300
c_{sat} (J K^{-1}g^{-1})	1.370	1.378	1.381	1.390	1.410	1.432	1.437
C_{sat} (J K^{-1}mol^{-1})	85.62	86.10	86.32	86.90	88.11	89.47	89.83

TABLE 22.14.5. Parameters of quasi-polynomial equation

Heat capacity type	No. data points total	used	s_w	s/R	s_r %	s_b/R	+/–
C_{sat}	10	10	0.421	2.17–2	0.21	–2.61–3	–1

Temp. range K	T_c K	A_1	A_2	A_3	A_4	Level of uncertainty
119.2–300.0	429.70	1.37504+1	3.86185	5.92523	1.22398+1	IV

Name: 1,1,1–Trichloroethane
Formula: $C_2H_3Cl_3$

CAS–RN: 71–55–6
Group No.: 22–015
Molar Mass: 133.40

TABLE 22.15.1. Experimental heat capacities

Reference	Temp. range K	No. pnts	Error %	Purity %	method	Type capacity	Calorimeter Type	Reference
44RUB/LEV	245.4–299.6	10	2.00	99.7	melpt	C_p	BSAO	41YOS/GAR
50CRO/SMY2	243.1–257.1	6	0.50	not specified		C_p	BDHO	50KUS/CRO
73AND/COU	243.1–310.0	10S	nosp	99.99	melpt	C_p	BSAO	63AND/COU1
88ING	298.1	1	nosp	99.	anal	C_p	FSIT	71PIC/LED
89PRA/RAJ	318.1–333.1	4	3.00	not specified		C_p	BDHT	89PRA/RAJ

TABLE 22.15.2. Correlated heat capacities

Reference	Temp. range K	No. pnts used	$\sigma_r C$ %	d_w	d/R	d_r %	d_b/R	+/–
				Selected data				
73AND/COU	243.1–310.0	10	0.20#	0.348	1.19–2	0.07	1.47–5	1
				Rejected data				
44RUB/LEV	(5.69–2, 0.34, –4.37–2, –10)			50CRO/SMY2	(2.10–1, 1.26, –2.08–1, –6)			
88ING	(1.48–1, 0.84, 1.48–1, 1)							

TABLE 22.15.3. Parameters of regression polynomial

Heat capacity type	No. data points total	used	s_w	s/R	s_r %	s_b/R	+/–
C_p	31	10	0.389	1.33–2	0.08	1.47–5	1

Temp. range K	A_1	A_2	Level of uncertainty
243.1–310.0	1.41433+1	1.08096	II

TABLE 22.15.4. Recommended values of heat capacities

Temp. (K)	250	260	270	273.15	280	290	298.15
c_p (J K^{-1}g^{-1})	1.050	1.057	1.063	1.066	1.070	1.077	1.082
C_p (J K^{-1}mol^{-1})	140.1	141.0	141.9	142.1	142.8	143.7	144.4

Temp. (K)	300	310
c_p (J K^{-1}g^{-1})	1.084	1.090
C_p (J K^{-1}mol^{-1})	144.6	145.5

TABLE 22.15.5. Parameters of quasi-polynomial equation

Heat capacity type	No. data points total	used	s_w	s/R	s_r %	s_b/R	+/–
C_p	31	10	0.114	3.91–3	0.02	1.72–6	1

Temp. range K	T_c K	A_1	A_2	A_3	A_4	Level of uncertainty
243.1–310.0	545.00	7.01313	7.15455–1	1.19382+1	1.71863+1	II

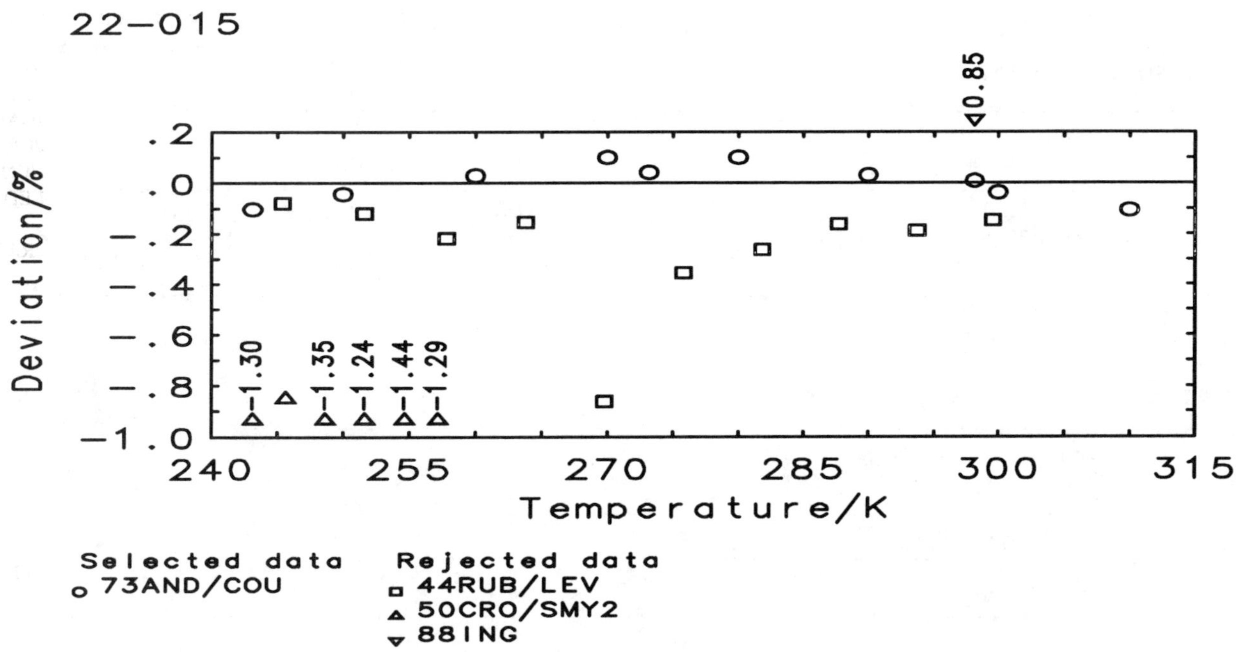

Name: 1,1,2–Trichloroethane
Formula: $C_2H_3Cl_3$

CAS–RN: 79–00–5
Group No.: 22–016
Molar Mass: 133.40

TABLE 22.16.1. Experimental heat capacities

Reference	Temp. range K	No. pnts	Error %	Purity %	Purity method	Type capacity	Calorimeter Type	Calorimeter Reference
50CRO/SMY2	240.1–251.6	4	1.50	not specified		C_p	BDHO	50KUS/CRO

TABLE 22.16.3. Parameters of regression polynomial

Heat capacity type	No. data points total	No. data points used	s_w	s/R	s_r %	s_b/R	+/–
C_p	4	4	0.611	1.57–1	0.92	1.44–3	0

Temp. range K	A_1	A_2	Level of uncertainty
240.1–251.6	1.13833+1	2.35057	V

TABLE 22.16.4. Recommended values of heat capacities

Temp. (K)	240	250
c_p (J $K^{-1}g^{-1}$)	1.06	1.08
C_p (J $K^{-1}mol^{-1}$)	142	144

Name: 1,1–Dichloroethane
Formula: $C_2H_4Cl_2$

CAS–RN: 75–34–3
Group No.: 22–017
Molar Mass: 98.96

TABLE 22.17.1. Experimental heat capacities

Reference	Temp. range K	No. pnts	Error %	Purity %	Purity method	Type capacity	Calorimeter Type	Calorimeter Reference
*81VON	297.9–312.3	3S	nosp	not specified		C_{avg}	DSIO	*81VON
48KUR	253.0–309.6	4	nosp	not specified		C_{avg}	DSIO	47KUR
56LI/PIT	177.1–294.3	18	0.20	99.87	melpt	C_p	BSIO	37GIA/EGA

TABLE 22.17.2. Correlated heat capacities

Reference	Temp. range K	No. pnts used	$\sigma_r C$ %	d_w	d/R	d_r %	d_b/R	+/–
				Selected data				
56LI/PIT	177.1–294.3	18	0.20	0.595	1.75–2	0.12	4.18–5	–3
				Rejected data				
48KUR	(6.19–1, 4.37, –6.19–1, –1)							

TABLE 22.17.3. Parameters of regression polynomial

Heat capacity type	No. data points total	used	s_w	s/R	s_r %	s_b/R	+/–
C_p	25	18	0.652	1.91–2	0.13	4.18–5	–3

Temp. range K	A_1	A_2	A_3	Level of uncertainty
177.1–294.3	1.48825+1	–8.54817–1	3.22034–1	II

TABLE 22.17.4. Recommended values of heat capacities

Temp. (K)	180	190	200	210	220	230	240
c_p (J K^{-1}g^{-1})	1.209	1.212	1.215	1.219	1.223	1.228	1.234
C_p (J K^{-1}mol^{-1})	119.6	119.9	120.2	120.6	121.1	121.6	122.1
Temp. (K)	250	260	270	273.15	280	290	298.15
c_p (J K^{-1}g^{-1})	1.240	1.247	1.254	1.256	1.261	1.270	1.277
C_p (J K^{-1}mol^{-1})	122.7	123.4	124.1	124.3	124.8	125.6	126.4
Temp. (K)	300						
c_p (J K^{-1}g^{-1})	1.278						
C_p (J K^{-1}mol^{-1})	126.5						

TABLE 22.17.5. Parameters of quasi-polynomial equation

Heat capacity type	No. data points total	used	s_w	s/R	s_r %	s_b/R	+/–
C_p	25	18	0.685	2.02–2	0.14	4.58–5	–2

Temp. range K	T_c K	A_1	A_2	A_3	A_4	Level of uncertainty
177.1–294.3	523.00	1.81454	1.86630	1.21488+1	4.41056–1	II

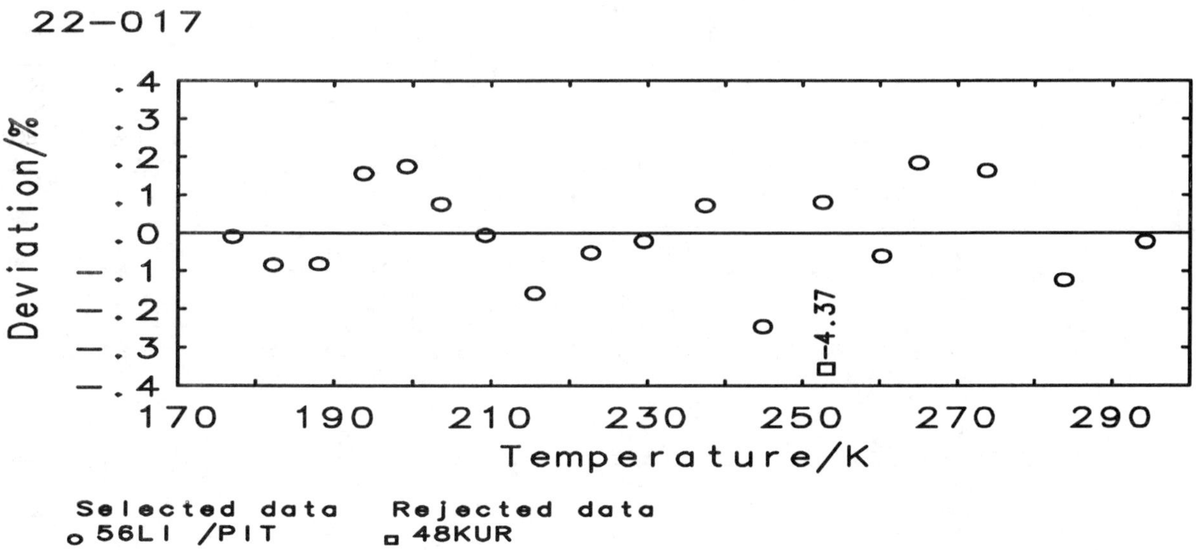

Name: 1,2–Dichloroethane
Formula: $C_2H_4Cl_2$

CAS–RN: 107–06–2
Group No.: 22–018
Molar Mass: 98.96

TABLE 22.18.1. Experimental heat capacities

Reference		Temp. range K	No. pnts	Error %	Purity %	Purity method	Type capacity	Calorimeter Type	Calorimeter Reference
12SCH1		293.1–323.1	3	nosp	not specified		C_p	BSIO	12SCH1
39RAI	N	240.0–322.0	15	2.00	99.96	melpt	C_p	BDHO	37STU
40PIT1		238.7–307.5	6	0.20	99.942	melpt	C_p	BSIO	28LAT/GRE
48KUR		270.5–322.1	4	nosp	not specified		C_{avg}	DSIO	47KUR
48TSC5		294.1	1	nosp	not specified		C_p	BSIO	48TSC1
48TSC6		294.1	1	nosp	not specified		C_p	BSIO	48TSC1
49TSC/RIC3		298.1	1	nosp	not specified		C_p	BSIO	49TSC/RIC1
51SIE/CRU	N	293.1	1	nosp	not specified		C_p	not specified	
55RUI2		280.8–323.7	4	nosp	not specified		C_p	BSIO	55RUI1
55STA/TUP		284.1–348.0	14	1.00	not specified		C_p	BSAO	55STA/TUP
67RAS/GAN		293.1–353.1	4S	0.50	not specified		C_p	BSAO	67RAS/GAN
69WIL/SCH		293.1–313.1	3	0.40	not specified		C_p	BDAO	65FIN/GRU
77WIL/GRO		298.1	1	0.30	not specified		C_p	FSIT	71PIC/LED
79WIL/FAR		298.1	1	0.30	99.5	melpt	C_p	FSIT	71PIC/LED
79WIL/GRO		298.1	1	0.30	not specified		C_p	FSIT	71PIC/LED
89PRA/RAJ		318.1–333.1	4	3.00	not specified		C_p	BDHT	89PRA/RAJ
93HAL		298.1	1	0.15	99.9	chrom	C_p	DDCT	74SUU/WAD

39RAI data from a graph only
51SIE/CRU heat of mixing calorimeter used

TABLE 22.18.2. Correlated heat capacities

Reference	Temp. range K	No. pnts used	$\sigma_r C$ %	d_w	d/R	d_r %	d_b/R	+/–
				Selected data				
40PIT1	238.7–307.5	6	0.20	1.688	5.12–2	0.34	–7.29–3	–1
55RUI2	280.8–323.7	4	0.50#	0.914	7.18–2	0.46	6.09–2	4
55STA/TUP	284.1–348.0	14	1.00	0.696	1.12–1	0.70	9.21–2	10
67RAS/GAN	293.1–353.1	4	0.50	1.377	1.10–1	0.69	2.84–3	1
69WIL/SCH	293.1–313.1	3	0.40	1.433	8.92–2	0.57	–7.29–2	–3
77WIL/GRO	298.1	1	0.30	0.214	9.96–3	0.06	9.96–3	1
79WIL/FAR	298.1	1	0.30	0.214	9.96–3	0.06	9.96–3	1
79WIL/GRO	298.1	1	0.30	0.111	5.14–3	0.03	5.14–3	1

Rejected data

12SCH1	(3.47–1, 2.29, –3.28–1, –3)	39RAI	(3.13–1, 2.08, 3.43–2, 3)
48KUR	(7.56–1, 5.32, –6.46–1, –4)	48TSC5	(1.70–1, 1.11, –1.70–1, –1)
48TSC6	(1.45–1, 0.94, –1.45–1, –1)	49TSC/RIC3	(5.31–1, 3.31, 5.31–1, 1)
51SIE/CRU	(5.11–1, 3.42, –5.11–1, –1)	89PRA/RAJ	(8.90–1, 5.93, –8.89–1, –4)
93HAL	(5.93–2, 0.38, 5.93–2, 1)		

TABLE 22.18.3. Parameters of regression polynomial

Heat capacity type	No. data points total	used	s_w	s/R	s_r %	s_b/R	+/–
C_p	65	34	1.151	9.58–2	0.60	3.84–2	14

Temp. range K	A_1	A_2	A_3	Level of uncertainty
238.7–353.1	1.99782+1	–4.19575	9.03923–1	IV

TABLE 22.18.4. Recommended values of heat capacities

Temp. (K)	240	250	260	270	273.15	280	290
c_p (J K^{-1}g^{-1})	1.270	1.272	1.275	1.280	1.282	1.287	1.295
C_p (J K^{-1}mol^{-1})	125.7	125.9	126.2	126.7	126.9	127.4	128.1
Temp. (K)	298.15	300	310	320	330	340	350
c_p (J K^{-1}g^{-1})	1.303	1.305	1.316	1.328	1.342	1.358	1.375
C_p (J K^{-1}mol^{-1})	128.9	129.1	130.2	131.4	132.8	134.4	136.1

TABLE 22.18.5. Parameters of quasi-polynomial equation

Heat capacity type	No. data points total	used	s_w	s/R	s_r %	s_b/R	+/–
C_p	65	34	1.280	9.27–2	0.58	2.15–2	6

Temp. range K	T_c K	A_1	A_2	A_3	A_4	Level of uncertainty
238.7–353.1	561.60	1.34623+1	6.11509	8.73738	7.40930	IV

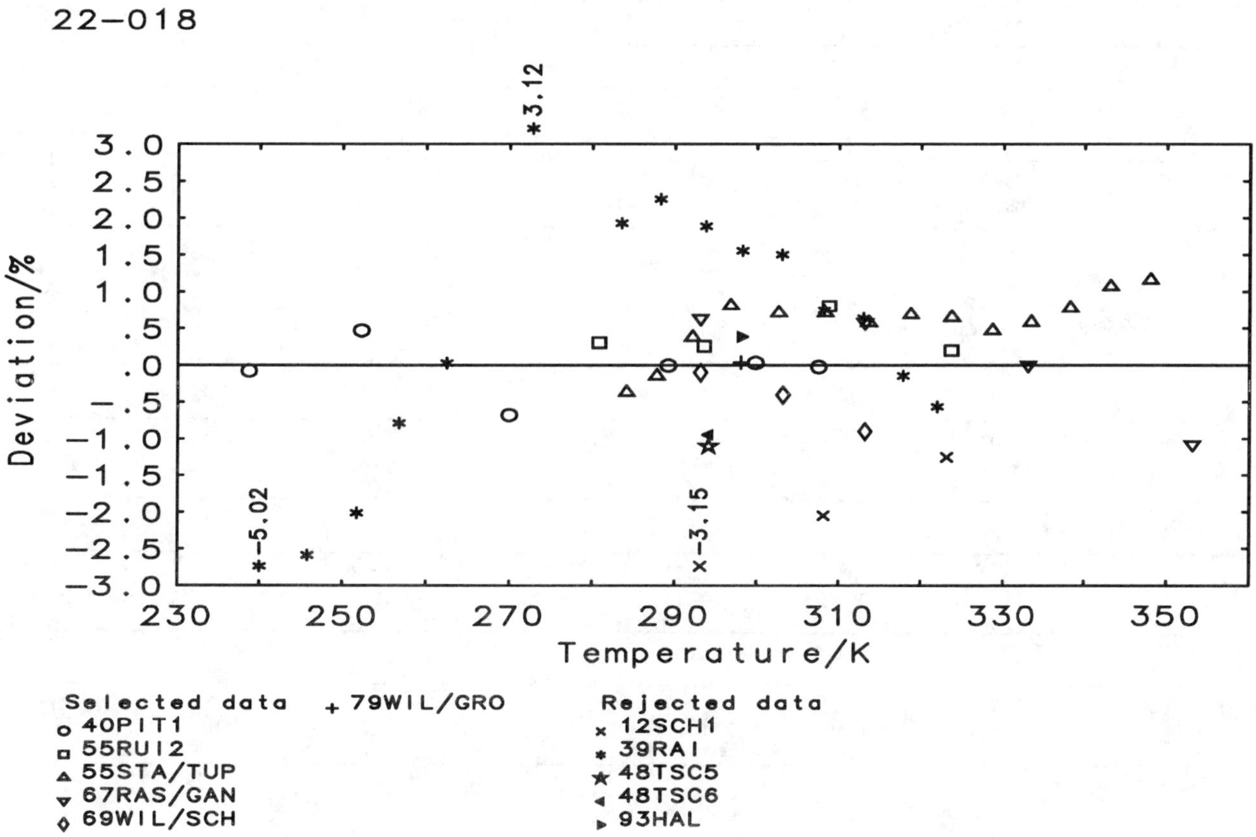

Name: Chloroethane
Formula: C_2H_5Cl

CAS–RN: 75–00–3
Group No.: 22–019
Molar Mass: 64.51

TABLE 22.19.1. Experimental heat capacities

Reference		Temp. range K	No. pnts	Error %	Purity %	Purity method	Type capacity	Calorimeter Type	Calorimeter Reference
23JEN/SHO	N	243.1–313.1	14S	nosp	not specified		C_{sat}	FSIO	23JEN/SHO
40RIE	N	224.8–318.1	12	nosp	not specified		C_{sat}	BSIO	39RIE3
48GOR/GIA		139.8–284.5	26	nosp	99.98	melpt	C_p	BSIO	28GIA/WIE1
48KUR		247.1–277.3	3	nosp	not specified		C_{avg}	DSIO	47KUR

23JEN/SHO values calculated from temperature dependence of enthalpy by the authors
40RIE same data in 41RIE1

TABLE 22.19.2. Correlated heat capacities

Reference	Temp. range K	No. pnts used	$\sigma_r C$ %	d_w	d/R	d_r %	d_b/R	+/–
				Selected data				
40RIE	224.8–318.1	12	1.50#	1.169	2.22–1	1.75	1.62–1	8
48GOR/GIA	139.8–284.5	26	0.40#	0.466	2.23–2	0.19	–4.99–3	–7
				Rejected data				
23JEN/SHO	(2.28–1, 1.95, –4.36–2, 2)			48KUR		(4.47–1, 3.57, 4.42–1, 3)		

TABLE 22.19.3. Parameters of cubic spline polynomials

Heat capacity type	No. data points total	No. data points used	s_w	s/R	s_r %	s_b/R	+/–
C_p	55	38	0.817	1.36–1	1.07	4.79–2	1

Temp. range K	A_1	A_2	A_3	A_4	Level of uncertainty
139.8–260.0	1.46953+1	–3.66454	1.13262	–4.72340–2	III
260.0–318.1	–5.08597+1	7.19759+1	–2.79598+1	3.68257	V

TABLE 22.19.4. Recommended values of heat capacities

Temp. (K)	140	150	160	170	180	190	200
c_p (J K^{-1}g^{-1})	1.502	1.493	1.487	1.483	1.481	1.482	1.485
C_p (J K^{-1}mol^{-1})	96.91	96.34	95.93	95.67	95.56	95.60	95.77
Temp. (K)	210	220	230	240	250	260	270
c_p (J K^{-1}g^{-1})	1.489	1.497	1.506	1.517	1.530	1.546	1.56
C_p (J K^{-1}mol^{-1})	96.09	96.55	97.14	97.87	98.73	99.72	101
Temp. (K)	273.15	280	290	298.15	300	310	320
c_p (J K^{-1}g^{-1})	1.57	1.59	1.62	1.65	1.66	1.71	1.78
C_p (J K^{-1}mol^{-1})	101	102	104	106	107	110	115

TABLE 22.19.5. Parameters of quasi-polynomial equation

Heat capacity type	No. data points total	used	s_w	s/R	s_r %	s_b/R	+/–
C_p	55	32	1.682	1.22–1	0.98	1.53–2	–2

Temp. range K	T_c K	A_1	A_2	A_3	A_4	Level of uncertainty
174.1–318.1	460.40	8.59220	4.29204	6.93412	4.30017	V

22–019

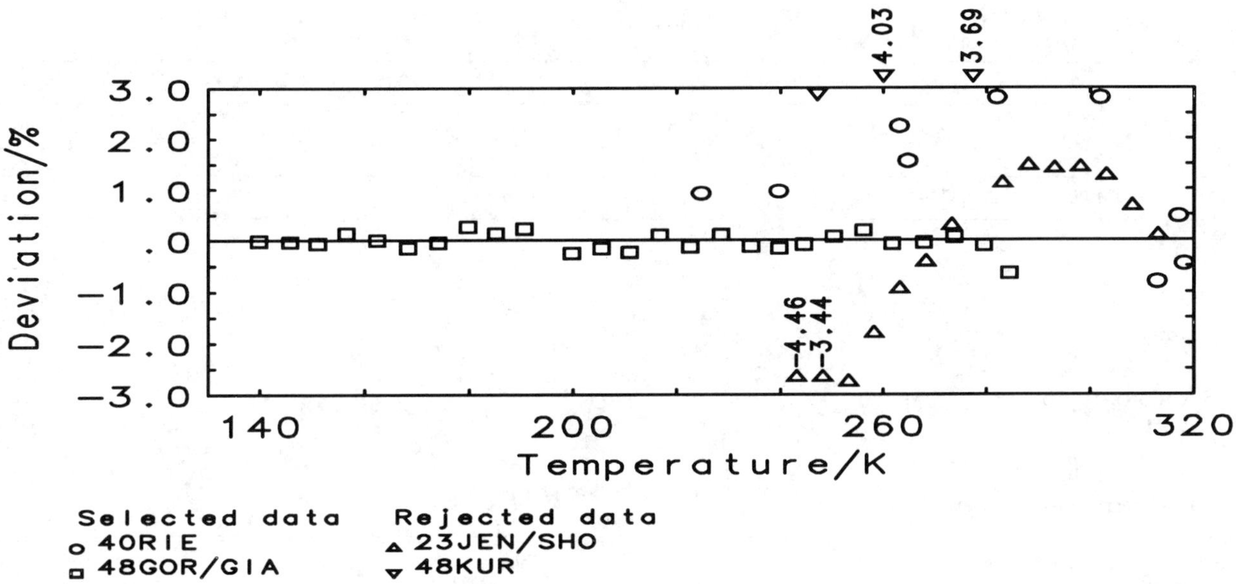

Name: 1,1,1,3–Tetrachloropropane
Formula: $C_3H_4Cl_4$

CAS–RN: 1070–78–6
Group No.: 22–020
Molar Mass: 181.88

TABLE 22.20.1. Experimental heat capacities

Reference	Temp. range K	No. pnts	Error %	Purity %	Purity method	Type capacity	Calorimeter Type	Calorimeter Reference
74KOL/VOR	239.6–288.5	11	nosp	99.82	melpt	C_{sat}	BSAO	62KOL/SER

TABLE 22.20.3. Parameters of regression polynomial

Heat capacity type	No. data points total	used	s_w	s/R	s_r %	s_b/R	+/–
C_{sat}	11	11	0.098	6.83–3	0.03	3.12–6	–1

Temp. range K	A_1	A_2	A_3	Level of uncertainty
239.6–288.5	2.59535+1	–3.41993	8.85508–1	III

TABLE 22.20.4. Recommended values of heat capacities

Temp. (K)	240	250	260	270	273.15	280	290
c_{sat} (J K^{-1}g^{-1})	1.044	1.049	1.054	1.059	1.061	1.066	1.074
C_{sat} (J K^{-1}mol^{-1})	190.0	190.7	191.6	192.7	193.1	193.9	195.2

Temp. (K)	298.15
c_{sat} (J K^{-1}g^{-1})	1.080
C_{sat} (J K^{-1}mol^{-1})	196.5

TABLE 22.20.5. Parameters of quasi-polynomial equation

Heat capacity type	No. data points total	used	s_w	s/R	s_r %	s_b/R	+/–
C_{sat}	11	11	0.091	6.34–3	0.03	2.08–6	–2

Temp. range K	T_c K	A_1	A_2	A_3	A_4	Level of uncertainty
239.6–288.5	571.00	2.00474+1	9.88188	1.24560+1	1.01676+1	III

Name: 3–Chloro–1–propene
Formula: C_3H_5Cl

CAS–RN: 107–05–1
Group No.: 22–021
Molar Mass: 76.53

TABLE 22.21.1. Experimental heat capacities

Reference	Temp. range K	No. pnts	Error %	Purity %	Purity method	Type capacity	Calorimeter Type	Calorimeter Reference
*81VON	299.3–312.1	4S	nosp	not specified		C_{avg}	DSIO	*81VON

TABLE 22.21.3. Parameters of regression polynomial

Heat capacity type	No. data points total	used	s_w	s/R	s_r %	s_b/R	+/–
C	4	4	0.002	9.52–4	0.01	2.38–7	0

Temp. range K	A_1	A_2	Level of uncertainty
299.3–312.1	1.03308+1	1.56868	V

TABLE 22.21.4. Recommended values of heat capacities

Temp. (K)	300	310
c (J K^{-1}g^{-1})	1.63	1.65
C (J K^{-1}mol^{-1})	125	126

Name: 1,2,3–Trichloropropane
Formula: $C_3H_5Cl_3$

CAS–RN: 96–18–4
Group No.: 22–022
Molar Mass: 147.43

TABLE 22.22.1. Experimental heat capacities

Reference	Temp. range K	No. pnts	Error %	Purity %	Purity method	Type capacity	Calorimeter Type	Calorimeter Reference
41NEL/NEW	273.0–333.0	eqn	0.50	not	specified	C_p	BSIO	41NEL/NEW
48KUR	320.1–358.9	3	nosp	not	specified	C_{avg}	DSIO	47KUR

TABLE 22.22.2. Correlated heat capacities

Reference	Temp. range K	No. pnts used	$\sigma_r C$ %	d_w	d/R	d_r %	d_b/R	+/–
				Selected data				
41NEL/NEW	273.0–333.2	8	0.50	0.000	1.51–6	0.00	1.19–6	0
				Rejected data				
48KUR	(7.75–1, 3.58, –7.75–1, –1)							

TABLE 22.22.3. Parameters of regression polynomial

Heat capacity type	No. data points total	No. data points used	s_w	s/R	s_r %	s_b/R	+/–
C_p	11	8	0.000	1.74–6	0.00	1.19–6	0

Temp. range K	A_1	A_2	Level of uncertainty
273.0–333.2	1.72456+1	1.61733	V

TABLE 22.22.4. Recommended values of heat capacities

Temp. (K)	270	273.15	280	290	298.15	300	310
c_p ($J\,K^{-1}g^{-1}$)	1.22	1.22	1.23	1.24	1.24	1.25	1.26
C_p ($J\,K^{-1}mol^{-1}$)	180	180	181	182	183	184	185

Temp. (K)	320	330
c_p ($J\,K^{-1}g^{-1}$)	1.26	1.27
C_p ($J\,K^{-1}mol^{-1}$)	186	188

Name: 1,2–Dichloropropane
Formula: $C_3H_6Cl_2$

CAS–RN: 78–87–5
Group No.: 22–023
Molar Mass: 112.99

TABLE 22.23.1. Experimental heat capacities

Reference	Temp. range K	No. pnts	Error %	Purity %	Purity method	Type capacity	Calorimeter Type	Calorimeter Reference
48KUR	305.9–359.3	3	nosp	not	specified	C_{avg}	DSIO	47KUR

TABLE 22.23.3. Parameters of regression polynomial

Heat capacity type	No. data points total	used	s_w	s/R	s_r %	s_b/R	+/–
C	3	3	0.111	1.10–1	0.55	3.03–4	–1

Temp. range K	A_1	A_2	Level of uncertainty
306.0–359.3	5.30569	4.34878	VI

TABLE 22.23.4. Recommended values of heat capacities

Temp. (K)	310	320	330	340	350	360
c ($J\,K^{-1}g^{-1}$)	1.38	1.41	1.45	1.48	1.51	1.54
C ($J\,K^{-1}mol^{-1}$)	156	160	163	167	171	174

Name: 1,3–Dichloropropane
Formula: $C_3H_6Cl_2$

CAS–RN: 142–28–9
Group No.: 22–024
Molar Mass: 112.99

TABLE 22.24.1. Experimental heat capacities

Reference	Temp. K	Capac. J/(K.g)	Error %	Purity %	Purity method	Type capacity	Calorimeter Type	Calorimeter Reference
93HAL	298.15	1.390	0.15	99.9	chrom	C_p	DDCT	74SUU/WAD

Name: 2,2–Dichloropropane
Formula: $C_3H_6Cl_2$

CAS–RN: 594–20–7
Group No.: 22–025
Molar Mass: 112.99

TABLE 22.25.1. Experimental heat capacities

Reference	Temp. range K	No. pnts	Error %	Purity %	Purity method	Type capacity	Calorimeter Type	Calorimeter Reference
72VAN	240.5–266.7	8	0.50	99.72	melpt	C_p	BSAO	72VAN

TABLE 22.25.3. Parameters of regression polynomial

Heat capacity type	No. data points total	used	s_w	s/R	s_r %	s_b/R	+/–
C_p	8	8	0.354	3.20–2	0.18	8.51–5	–2

Temp. range K	A_1	A_2	Level of uncertainty
240.5–266.7	1.54043+1	1.03091	III

TABLE 22.25.4. Recommended values of heat capacities

Temp. (K)	240	250	260	270
c_p ($J\,K^{-1}g^{-1}$)	1.316	1.323	1.331	1.338
C_p ($J\,K^{-1}mol^{-1}$)	148.7	149.5	150.4	151.2

Name: 1–Chloropropane
Formula: C_3H_7Cl

CAS–RN: 540–54–5
Group No.: 22–026
Molar Mass: 78.54

TABLE 22.26.1. Experimental heat capacities

Reference		Temp. range K	No. pnts	Error %	Purity % method	Type capacity	Calorimeter Type Reference
*81VON		299.1–311.6	4S	nosp	not specified	C_{avg}	DSIO *81VON
48EUC	N	200.0–293.0	4S	nosp	not specified	C_p	not specified
48KUR		261.9–300.8	2	nosp	not specified	C_{avg}	DSIO 47KUR

48EUC unpublished data measured by A.Landsberg

TABLE 22.26.2. Correlated heat capacities

Reference	Temp. range K	No. pnts used	$\sigma_r C$ %	d_w	d/R	d_r %	d_b/R	+/–
				Selected data				
*81VON	299.1–311.6	4	2.00#	0.484	1.55–1	0.97	–4.90–2	0
48EUC	200.0–293.0	4	1.50#	0.534	1.18–1	0.80	3.09–2	2
				Rejected data				
48KUR	(3.33–1, 2.23, 2.86–1, 2)							

TABLE 22.26.3. Parameters of regression polynomial

Heat capacity type	No. data points total	used	s_w	s/R	s_r %	s_b/R	+/–
C_p	10	8	0.644	1.74–1	1.12	–9.04–3	2

Temp. range K	A_1	A_2	A_3	Level of uncertainty
200.0–311.6	1.61110+1	–4.40676	1.43637	V

TABLE 22.26.4. Recommended values of heat capacities

Temp. (K)	200	210	220	230	240	250	260
c_p (J K^{-1}g^{-1})	1.38	1.40	1.42	1.44	1.46	1.49	1.52
C_p (J K^{-1}mol^{-1})	108	110	111	113	115	117	119
Temp. (K)	270	273.15	280	290	298.15	300	310
c_p (J K^{-1}g^{-1})	1.55	1.57	1.59	1.63	1.67	1.67	1.72
C_p (J K^{-1}mol^{-1})	122	123	125	128	131	132	135

TABLE 22.26.5. Parameters of quasi-polynomial equation

Heat capacity type	No. data points total	used	s_w	s/R	s_r %	s_b/R	+/–
C_p	10	8	0.814	1.89–1	1.30	–1.35–2	0

Temp. range K	T_c K	A_1	A_2	A_3	A_4	Level of uncertainty
200.0–311.6	503.00	–3.40854	1.24416	8.24083	2.33452	V

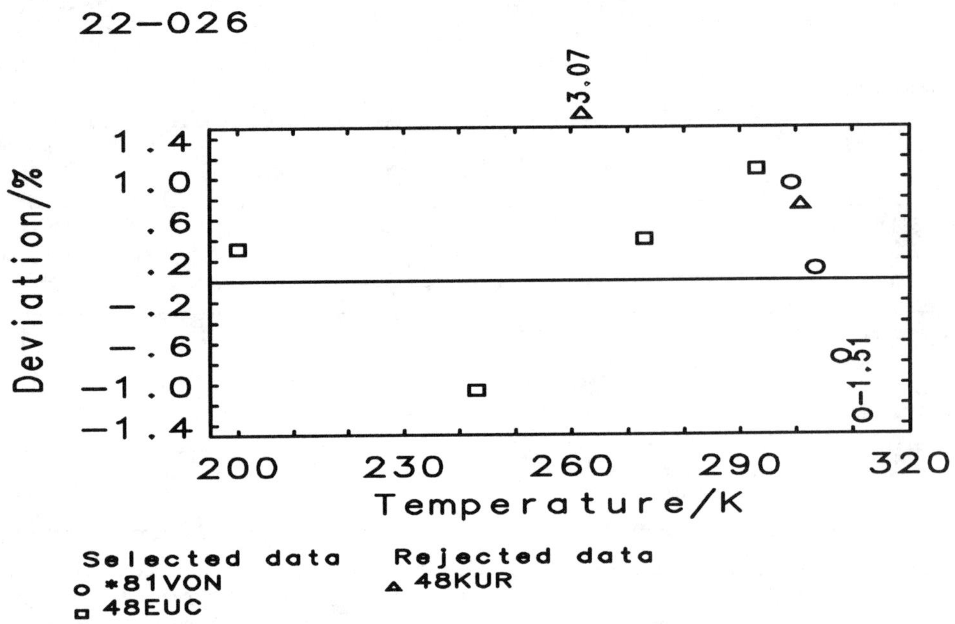

Name: 2–Chloro–1,3–butadiene
Formula: C_4H_5Cl

CAS–RN: 126–99–8
Group No.: 22–027
Molar Mass: 88.54

TABLE 22.27.1. Experimental heat capacities

Reference		Temp. K	Capac. J/(K.g)	Error %	Purity %	Purity method	Type capacity	Calorimeter Type	Calorimeter Reference
60JOH	N	293.15	1.314	nosp	not specified		C_p	not specified	

60JOH the origin of data is unclear

Name: 1,4–Dichlorobutane
Formula: $C_4H_8Cl_2$

CAS–RN: 110–56–5
Group No.: 22–028
Molar Mass: 127.01

TABLE 22.28.1. Experimental heat capacities

Reference	Temp. range K	No. pnts	Error %	Purity %	Purity method	Type capacity	Calorimeter Type	Calorimeter Reference
85LAI/WIL	298.1	1	0.30	99.5	estim	C_p	FSIT	71PIC/LED
93HAL	298.1	1	0.15	99.9	chrom	C_p	DDCT	74SUU/WAD

TABLE 22.28.2. Correlated heat capacities

Reference	Temp. range K	No. pnts used	$\sigma_r C$ %	d_w	d/R	d_r %	d_b/R	+/–
				Selected data				
85LAI/WIL	298.1	1	0.30	0.595	3.94–2	0.18	–3.94–2	–1
93HAL	298.1	1	0.15	0.298	9.90–3	0.04	9.90–3	1

TABLE 22.28.3. Parameters of regression polynomial

Heat capacity type	No. data points total	used	s_w	s/R	s_r %	s_b/R	+/–
C_p	2	2	0.665	4.06–2	0.18	–1.48–2	0

Temp. range K	A_1	Level of uncertainty
298.1–298.1	2.21201+1	III

TABLE 22.28.4. Recommended values of heat capacities

Temp. (K)	298.15
c_p (J K^{-1}g^{-1})	1.448
C_p (J K^{-1}mol^{-1})	183.9

Name: 1–Chlorobutane
Formula: C_4H_9Cl

CAS–RN: 109–69–3
Group No.: 22–029
Molar Mass: 92.57

TABLE 22.29.1. Experimental heat capacities

Reference	Temp. range K	No. pnts	Error %	Purity %	method	Type capacity	Calorimeter Type	Reference
61ROU	300.9–309.0	12	nosp	not specified		C_p	BSAO	61ROU
85LAI/WIL	298.1	1	0.30	99.5	estim	C_p	FSIT	71PIC/LED

TABLE 22.29.2. Correlated heat capacities

Reference	Temp. range K	No. pnts used	$\sigma_r C$ %	d_w	d/R	d_r %	d_b/R	+/–
				Selected data				
61ROU	300.9–309.0	12	1.00#	0.597	1.17–1	0.60	3.64–2	2
85LAI/WIL	298.1	1	0.30	0.655	3.77–2	0.20	–3.77–2	–1

TABLE 22.29.3. Parameters of regression polynomial

Heat capacity type	No. data points total	used	s_w	s/R	s_r %	s_b/R	+/–
C_p	13	13	0.654	1.23–1	0.63	3.07–2	1

Temp. range K	A_1	A_2	Level of uncertainty
298.1–309.0	1.23357	6.03867	IV

TABLE 22.29.4. Recommended values of heat capacities

Temp. (K)	300	310
c_p (J K^{-1}g^{-1})	1.738	1.792
C_p (J K^{-1}mol^{-1})	160.9	165.9

Name: 1–Chloro–2–methylpropane
Formula: C_4H_9Cl

CAS–RN: 513–36–0
Group No.: 22–030
Molar Mass: 92.57

TABLE 22.30.1. Experimental heat capacities

Reference	Temp. range K	No. pnts	Error %	Purity %	Purity method	Type capacity	Calorimeter Type	Calorimeter Reference
48KUR	292.1–309.6	2	nosp	not specified		C_{avg}	DSIO	47KUR

TABLE 22.30.3. Parameters of regression polynomial

Heat capacity type	No. data points total	No. data points used	s_w	s/R	s_r %	s_b/R	+/–
C	2	2	0.000	0.00	0.00	–3.82–6	0

Temp. range K	A_1	A_2	Level of uncertainty
292.1–309.6	–3.40121+1	1.72752+1	VI

able 22.30.4. Recommended values of heat capacities

Temp. (K)	290	298.15	300	310
c (J $K^{-1}g^{-1}$)	1.44	1.57	1.60	1.76
C (J $K^{-1}mol^{-1}$)	134	145	148	162

Name: 2–Chloro–2–methylpropane
Formula: C_4H_9Cl

CAS–RN: 507–20–0
Group No.: 22–031
Molar Mass: 92.57

TABLE 22.31.1. Experimental heat capacities

Reference	Temp. range K	No. pnts	Error %	Purity %	Purity method	Type capacity	Calorimeter Type	Calorimeter Reference
50KUS/CRO	251.6–259.6	4	2.00	not specified		C_p	BDHO	50KUS/CRO
66DWO/GUI	252.4–272.7	8	2.00	99.75	melpt	C_p	BSAO	66DWO/GUI
92KAL/KOH	293.1–313.1	2	1.00	98.8	chrom	C_p	FSIT	71PIC/LED

TABLE 22.31.2. Correlated heat capacities

Reference	Temp. range K	No. pnts used	$\sigma_r C$ %	d_w	d/R	d_r %	d_b/R	+/–
				Selected data				
50KUS/CRO	251.6–259.6	4	2.00	0.274	1.01–1	0.55	3.64–2	–1
92KAL/KOH	293.1–313.1	2	1.00	0.589	1.17–1	0.59	–1.61–2	0
				Rejected data				
66DWO/GUI	(1.84, 9.00, 1.83, 8)							

TABLE 22.31.3. Parameters of regression polynomial

Heat capacity type	No. data points total	used	s_w	s/R	s_r %	s_b/R	+/–
C_p	14	6	0.499	1.30–1	0.69	1.89–2	–1

Temp. range K	A_1	A_2	Level of uncertainty
251.6–313.1	9.10350	3.58306	V

TABLE 22.31.4. Recommended values of heat capacities

Temp. (K)	250	260	270	273.15	280	290	298.15
c_p ($J K^{-1} g^{-1}$)	1.62	1.65	1.69	1.70	1.72	1.75	1.78
C_p ($J K^{-1} mol^{-1}$)	150	153	156	157	159	162	165

Temp. (K)	300	310
c_p ($J K^{-1} g^{-1}$)	1.78	1.82
C_p ($J K^{-1} mol^{-1}$)	165	168

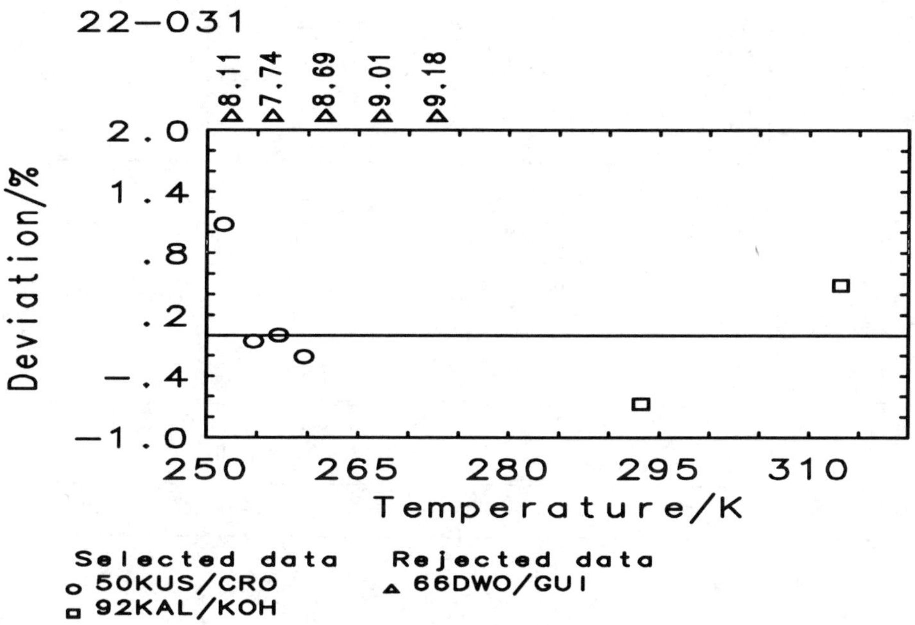

Name: 1,3–Dichloro–2,2–bis(chloromethyl)propane
Formula: $C_5H_8Cl_4$

CAS–RN: 3228–99–7
Group No.: 22–032
Molar Mass: 209.93

TABLE 22.32.1. Experimental heat capacities

Reference	Temp. range K	No. pnts	Error %	Purity %	method	Type capacity	Calorimeter Type	Reference
65CLE/WON	372.8–397.6	8	0.15	99.07	melpt	C_{sat}	BSAO	53WES/HAT

TABLE 22.32.3. Parameters of regression polynomial

Heat capacity type	No. data points total	used	s_w	s/R	s_r %	s_b/R	+/–
C_{sat}	8	8	0.614	3.08–2	0.09	4.20–5	–1

Temp. range K	A_1	A_2	Level of uncertainty
372.8–397.6	1.88473+1	3.81495	III

TABLE 22.32.4. Recommended values of heat capacities

Temp. (K)	370	380	390	400
c_{sat} (J K^{-1}g^{-1})	1.306	1.321	1.336	1.351
C_{sat} (J K^{-1}mol^{-1})	274.1	277.2	280.4	283.6

Name: 1,5–Dichloropentane
Formula: $C_5H_{10}Cl_2$

CAS–RN: 628–76–2
Group No.: 22–033
Molar Mass: 141.04

TABLE 22.33.1. Experimental heat capacities

Reference	Temp. K	Capac. J/(K.g)	Error %	Purity %	Purity method	Type capacity	Calorimeter Type	Calorimeter Reference
93HAL	298.15	1.513	0.15	99.9	chrom	C_p	DDCT	74SUU/WAD

Name: 1–Chloro–3–methylbutane
Formula: $C_5H_{11}Cl$

CAS–RN: 107–84–6
Group No.: 22–034
Molar Mass: 106.60

TABLE 22.34.1. Experimental heat capacities

Reference	Temp. range K	No. pnts	Error %	Purity %	Purity method	Type capacity	Calorimeter Type	Calorimeter Reference
48KUR	307.3–329.1	2	nosp	not specified		C_{avg}	DSIO	47KUR

TABLE 22.34.3. Parameters of regression polynomial

Heat capacity type	No. data points total	used	s_w	s/R	s_r %	s_b/R	+/–
C	2	2	0.000	0.00	0.00	0.00	0

Temp. range K	A_1	A_2	Level of uncertainty
307.3–329.1	1.16622+1	3.18413	VI

TABLE 22.34.4. Recommended values of heat capacities

Temp. (K)	310	320	330
c (J K^{-1}g^{-1})	1.68	1.70	1.73
C (J K^{-1}mol^{-1})	179	182	184

Name: 1,2,4–Trichlorobenzene
Formula: $C_6H_3Cl_3$

CAS–RN: 120–82–1
Group No.: 22–035
Molar Mass: 181.45

TABLE 22.35.1. Experimental heat capacities

Reference	Temp. range K	No. pnts	Error %	Purity %	Purity method	Type capacity	Calorimeter Type	Calorimeter Reference
69WIL/ROT	293.1–303.1	2	0.40	not specified		C_p	BDAO	65FIN/GRU
74PET/TER	298.0–469.2	15	nosp	98.	melpt	C_p	BDCT	74PET/TER
82WIL/ING	298.1	1	0.30	not specified		C_p	FSIT	71PIC/LED
86WIL/LAI	298.1	1	nosp	not specified		C_p	FSIT	71PIC/LED

TABLE 22.35.2. Correlated heat capacities

Reference	Temp. range K	No. pnts used	$\sigma_r C$ %	d_w	d/R	d_r %	d_b/R	+/–
				Selected data				
69WIL/ROT	293.1–303.1	2	0.40	3.229	2.95–1	1.29	–2.94–1	–2
74PET/TER	298.0–459.2	13	3.00#	0.750	5.69–1	2.25	1.68–2	1
82WIL/ING	298.1	1	0.30	3.491	2.45–1	1.05	2.45–1	1
86WIL/LAI	298.1	1	0.50#	2.450	2.87–1	1.23	2.87–1	1

TABLE 22.35.3. Parameters of regression polynomial

Heat capacity type	No. data points total	No. data points used	s_w	s/R	s_r %	s_b/R	+/–
C_p	19	17	1.758	5.49–1	2.19	9.55–3	1

Temp. range K	A_1	A_2	Level of uncertainty
293.1–459.2	1.54985+1	2.56761	V

TABLE 22.35.4. Recommended values of heat capacities

Temp. (K)	290	298.15	300	310	320	330	340
c_p (J $K^{-1}g^{-1}$)	1.05	1.06	1.06	1.07	1.09	1.10	1.11
C_p (J $K^{-1}mol^{-1}$)	191	193	193	195	197	199	201
Temp. (K)	350	360	370	380	390	400	410
c_p (J $K^{-1}g^{-1}$)	1.12	1.13	1.15	1.16	1.17	1.18	1.19
C_p (J $K^{-1}mol^{-1}$)	204	206	208	210	212	214	216
Temp. (K)	420	430	440	450	460		
c_p (J $K^{-1}g^{-1}$)	1.20	1.22	1.23	1.24	1.25		
C_p (J $K^{-1}mol^{-1}$)	219	221	223	225	227		

TABLE 22.35.5. Parameters of quasi-polynomial equation

Heat capacity type	No. data points total	No. data points used	s_w	s/R	s_r %	s_b/R	+/–
C_p	19	17	2.825	5.22–1	2.11	1.42–2	1

Temp. range K	T_c K	A_1	A_2	A_3	A_4	Level of uncertainty
293.1–459.2	734.90	8.91423+1	1.37867+1	–1.24739+1	1.44094+2	V

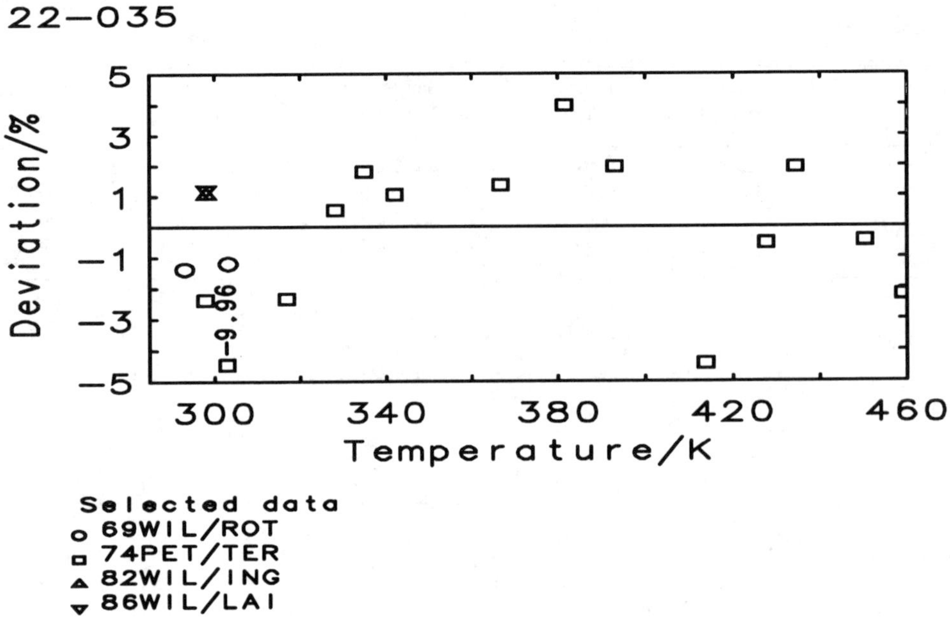

Name: 1,2–Dichlorobenzene
Formula: $C_6H_4Cl_2$

CAS–RN: 95–50–1
Group No.: 22–036
Molar Mass: 147.00

TABLE 22.36.1. Experimental heat capacities

Reference	Temp. range K	No. pnts	Error %	Purity %	Purity method	Type capacity	Calorimeter Type	Calorimeter Reference
18NAR	291.1–326.1	4S	nosp	not specified		C_{avg}	DSIO	18NAR

TABLE 22.36.3. Parameters of regression polynomial

Heat capacity type	No. data points total	No. data points used	s_w	s/R	s_r %	s_b/R	+/–
C	4	4	0.081	8.38–2	0.40	3.45–4	0

Temp. range K	A_1	A_2	Level of uncertainty
291.1–326.1	1.32591+1	2.42863	VI

TABLE 22.36.4. Recommended values of heat capacities

Temp. (K)	290	298.15	300	310	320	330
c $(\mathrm{J\,K^{-1}g^{-1}})$	1.15	1.16	1.16	1.18	1.19	1.20
C $(\mathrm{J\,K^{-1}mol^{-1}})$	169	170	171	173	175	177

Name: 1,3–Dichlorobenzene
Formula: $C_6H_4Cl_2$

CAS–RN: 541–73–1
Group No.: 22–037
Molar Mass: 147.00

TABLE 22.37.1. Experimental heat capacities

Reference	Temp. range K	No. pnts	Error %	Purity %	Purity method	Type capacity	Calorimeter Type	Calorimeter Reference
18NAR	253.9–327.5	6S	nosp	not specified		C_{avg}	DSIO	18NAR

TABLE 22.37.3. Parameters of regression polynomial

Heat capacity type	No. data points total	No. data points used	s_w	s/R	s_r %	s_b/R	+/–
C	6	6	0.071	7.38–2	0.36	3.82–4	–2

Temp. range K	A_1	A_2	Level of uncertainty
253.9–327.5	1.40629+1	2.17998	VI

TABLE 22.37.4. Recommended values of heat capacities

Temp. (K)	260	270	273.15	280	290	298.15	300
c ($J\,K^{-1}g^{-1}$)	1.12	1.13	1.13	1.14	1.15	1.16	1.17
C ($J\,K^{-1}mol^{-1}$)	164	166	166	168	169	171	171

Temp. (K)	310	320	330
c ($J\,K^{-1}g^{-1}$)	1.18	1.19	1.20
C ($J\,K^{-1}mol^{-1}$)	173	175	177

Name: 1,4–Dichlorobenzene
Formula: $C_6H_4Cl_2$

CAS–RN: 106–46–7
Group No.: 22–038
Molar Mass: 147.00

TABLE 22.38.1. Experimental heat capacities

Reference		Temp. range K	No. pnts	Error %	Purity %	Purity method	Type capacity	Calorimeter Type	Calorimeter Reference
18NAR	N	349.1	1	nosp	not specified		C_{avg}	DSIO	18NAR
76DWO/FIG		328.2–329.3	2	nosp	99.93	melpt	C_{sat}	BSAO	76DWO/FIG

18NAR average value in temperature range 326–372 K

TABLE 22.38.2. Correlated heat capacities

Reference	Temp. range K	No. pnts used	$\sigma_r C$ %	d_w	d/R	d_r %	d_b/R	+/–
				Selected data				
18NAR	349.1	1	5.00#	0.043	4.73–2	0.21	–4.73–2	–1
76DWO/FIG	328.2–329.3	2	2.00#	0.316	1.38–1	0.63	5.55–3	0

TABLE 22.38.3. Parameters of regression polynomial

Heat capacity type	No. data points total	used	s_w	s/R	s_r %	s_b/R	+/−
C_{sat}	3	3	0.449	2.01−1	0.92	−1.21−2	−1

Temp. range K	A_1	A_2	Level of uncertainty
328.2–349.1	1.75778+1	1.29282	VI

TABLE 22.38.4. Recommended values of heat capacities

Temp. (K)	330	340	350
c_{sat} (J K^{-1}g^{-1})	1.24	1.24	1.25
C_{sat} (J K^{-1}mol^{-1})	182	183	184

Name: Chlorobenzene
Formula: C_6H_5Cl

CAS–RN: 108–90–7
Group No.: 22–039
Molar Mass: 112.56

TABLE 22.39.1. Experimental heat capacities

Reference		Temp. range K	No. pnts	Error %	Purity %	Purity method	Type capacity	Calorimeter Type	Calorimeter Reference
*81VON		343.8–355.3	3S	nosp	not specified		C_{avg}	DSIO	*81VON
*87SCH		308.6–332.9	6S	nosp	not specified		C_{avg}	DSIO	*86SCH
25WIL/DAN	N	293.1–353.1	4S	nosp	not specified		C_p	BSAO	24WIL/DAN
37STU		230.0–320.0	10S	nosp	not specified		C_p	BDHO	37STU
39PHI		305.5	1	nosp	not specified		C_p	BSIO	49WEI
49TSC/RIC3		298.1	1	nosp	not specified		C_p	BSIO	49TSC/RIC1
61ROU	N	296.1–318.1	2	nosp	not specified		C_p	BSAO	61ROU
71DES/BHA		298.1–318.1	3S	nosp	not specified		C_p	BSIO	56MUR/VAN
77FOR/BEN		298.1	1	nosp	not specified		C_p	FSIT	71PIC/LED
88PER/AIC		298.1	1	nosp	99.	anal	C_p	FSIT	71PIC/LED
90RAO/RAJ		318.1–333.1	4	4.00	not specified		C_p	BDHT	89PRA/RAJ
91TAN/ADA		298.1	1	nosp	not specified		C_p	FSIT	71PIC/LED
92KAL/KOH		293.1–313.1	2	1.00	99.98	chrom	C_p	FSIT	71PIC/LED

25WIL/DAN origin of data unclear
61ROU constant value in temperature range 296–318 K obtained by the author

TABLE 22.39.2. Correlated heat capacities

Reference	Temp. range K	No. pnts used	$\sigma_r C$ %	d_w	d/R	d_r %	d_b/R	+/–
				Selected data				
*81VON	343.8–355.3	3	2.00#	0.282	1.10–1	0.56	8.73–2	1
37STU	230.0–320.0	10	1.50#	0.836	2.18–1	1.25	2.36–2	2
77FOR/BEN	298.1	1	0.50#	0.215	1.94–2	0.11	–1.94–2	–1
91TAN/ADA	298.1	1	0.50#	0.264	2.39–2	0.13	–2.39–2	–1
92KAL/KOH	293.1–313.1	2	1.00	1.127	2.07–1	1.13	1.91–2	0
				Rejected data				
*87SCH	(3.10–1, 1.60, 2.59–1, 6)			25WIL/DAN	(4.83–1, 2.69, –4.49–1, –4)			
39PHI	(6.04–1, 3.19, 6.04–1, 1)			49TSC/RIC3	(5.93–1, 3.16, 5.93–1, 1)			
61ROU	(2.87–1, 1.55, 1.56–1, 0)			71DES/BHA	(3.04–1, 1.70, –2.91–1, –3)			
88PER/AIC	(3.40–1, 1.84, 3.40–1, 1)			90RAO/RAJ	(9.25–1, 4.67, 8.98–1, 4)			

TABLE 22.39.3. Parameters of regression polynomial

Heat capacity type	No. data points total	used	s_w	s/R	s_r %	s_b/R	+/–
C_p	39	17	0.840	2.07–1	1.17	2.90–2	1
Temp. range K		A_1	A_2	A_3			Level of uncertainty
230.0–355.3		1.52484+1	–1.73449–1	3.85133–1			V

TABLE 22.39.4. Recommended values of heat capacities

Temp. (K)	230	240	250	260	270	273.15	280
c_p (J K^{-1}g^{-1})	1.25	1.26	1.27	1.29	1.30	1.30	1.31
C_p (J K^{-1}mol^{-1})	140	142	143	145	146	147	148
Temp. (K)	290	298.15	300	310	320	330	340
c_p (J K^{-1}g^{-1})	1.33	1.34	1.34	1.36	1.38	1.39	1.41
C_p (J K^{-1}mol^{-1})	150	151	151	153	155	157	159
Temp. (K)	350	360					
c_p (J K^{-1}g^{-1})	1.43	1.45					
C_p (J K^{-1}mol^{-1})	161	163					

TABLE 22.39.5. Parameters of quasi-polynomial equation

Heat capacity type	No. data points total	used	s_w	s/R	s_r %	s_b/R	+/–
C_p	39	17	0.936	2.21–1	1.29	4.93–3	3

Temp. range K	T_c K	A_1	A_2	A_3	A_4	Level of uncertainty
230.0–355.3	632.40	1.30912+1	8.99227	7.18891	4.76461	V

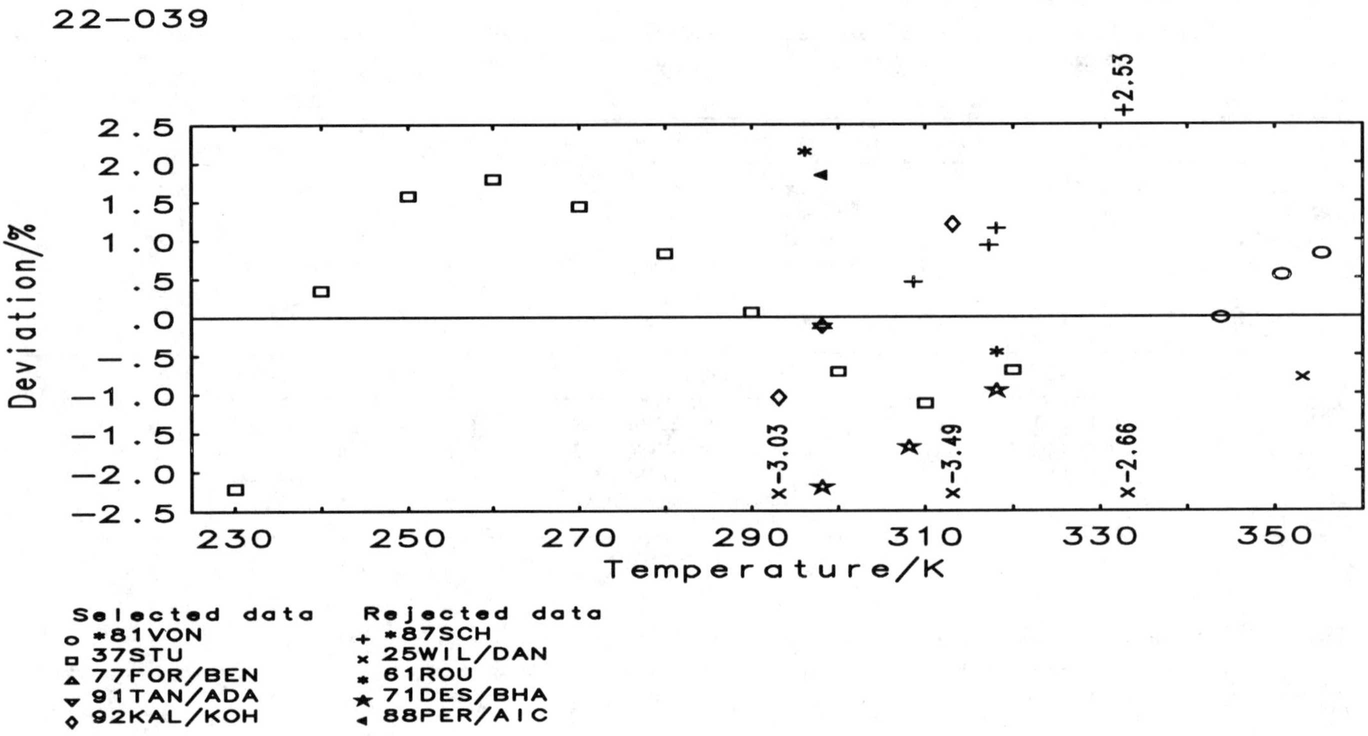

Name: 1,6–Dichlorohexane
Formula: $C_6H_{12}Cl_2$

CAS–RN: 2163–00–0
Group No.: 22–040
Molar Mass: 155.07

TABLE 22.40.1. Experimental heat capacities

Reference	Temp. K	Capac. J/(K.g)	Error %	Purity %	Purity method	Type capacity	Calorimeter Type	Calorimeter Reference
85LAI/GRO	298.15	1.545	nosp	not specified		C_p	FSIT	71PIC/LED

Name: (Chloromethyl)benzene
Formula: C_7H_7Cl

CAS–RN: 100–44–7
Group No.: 22–041
Molar Mass: 126.59

TABLE 22.41.1. Experimental heat capacities

Reference	Temp. range K	No. pnts	Error %	Purity %	Purity method	Type capacity	Calorimeter Type	Calorimeter Reference
*87SCH	317.5–346.5	6S	nosp	not specified		C_p	DSIO	*86SCH
31SMI/AND1	246.0–298.5	5	nosp	99.5	estim	C_p	DSIO	26AND/LYN

TABLE 22.41.2. Correlated heat capacities

Reference	Temp. range K	No. pnts used	$\sigma_r C$ %	d_w	d/R	d_r %	d_b/R	+/–
				Selected data				
*87SCH	317.5–346.5	6	2.00#	0.061	2.82–2	0.12	1.88–3	1
31SMI/AND1	246.0–298.5	5	1.50#	0.205	6.58–2	0.31	–7.87–4	1

TABLE 22.41.3. Parameters of regression polynomial

Heat capacity type	No. data points total	used	s_w	s/R	s_r %	s_b/R	+/–
C_p	11	11	0.171	5.75–2	0.27	6.69–4	2
Temp. range K	A_1	A_2	A_3				**Level of uncertainty**
246.0–346.5	2.95624+1	–8.17737	1.89337				V

TABLE 22.41.4. Recommended values of heat capacities

Temp. (K)	250	260	270	273.15	280	290	298.15
c_p (J K^{-1}g^{-1})	1.38	1.39	1.40	1.40	1.41	1.43	1.45
C_p (J K^{-1}mol^{-1})	174	175	177	178	179	181	183
Temp. (K)	300	310	320	330	340	350	
c_p (J K^{-1}g^{-1})	1.45	1.47	1.50	1.52	1.55	1.59	
C_p (J K^{-1}mol^{-1})	184	186	189	193	197	201	

TABLE 22.41.5. Parameters of quasi-polynomial equation

Heat capacity type	No. data points total	used	s_w	s/R	s_r %	s_b/R	+/–
C_p	11	11	0.147	5.05–2	0.23	–3.45–3	–1

Temp. range K	T_c K	A_1	A_2	A_3	A_4	Level of uncertainty
246.0–346.5	660.00	6.04442+1	2.98666+1	–9.94535	3.05818+1	V

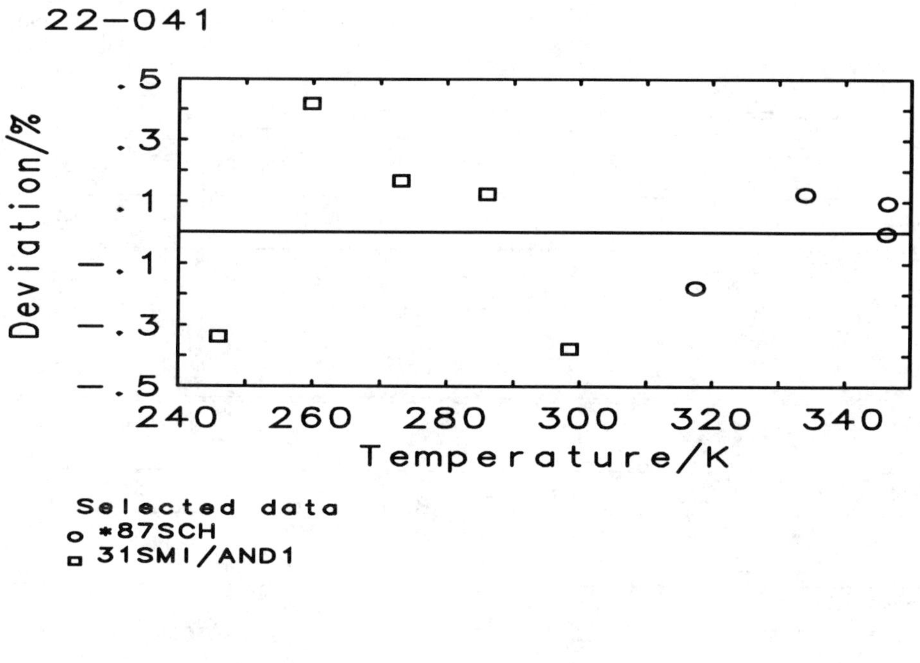

Name: 1–Chloro–2–methylbenzene
Formula: C_7H_7Cl

CAS–RN: 95–49–8
Group No.: 22–042
Molar Mass: 126.59

TABLE 22.42.1. Experimental heat capacities

Reference		Temp. range K	No. pnts	Error %	Purity %	Purity method	Type capacity	Calorimeter Type	Calorimeter Reference
*87SCH	N	316.6–345.8	6S	nosp	not specified		C_{avg}	DSIO	*86SCH

*87SCH unspecified isomer; probably 1-Chloro-2-methylbenzene (according to n.b.t.)

TABLE 22.42.3. Parameters of regression polynomial

Heat capacity type	No. data points total	used	s_w	s/R	s_r %	s_b/R	+/–
C	6	6	0.024	1.09–2	0.05	3.82–6	–2

Temp. range K	A_1	A_2	Level of uncertainty
316.6–345.8	7.02596	4.78215	V

TABLE 22.42.4. Recommended values of heat capacities

Temp. (K)	320	330	340	350
c $(J\,K^{-1}g^{-1})$	1.47	1.50	1.53	1.56
C $(J\,K^{-1}mol^{-1})$	186	190	194	198

Name: 3–Chloroheptane
Formula: $C_7H_{15}Cl$

CAS–RN: 999–52–0
Group No.: 22–043
Molar Mass: 134.65

TABLE 22.43.1. Experimental heat capacities

Reference	Temp. range K	No. pnts	Error %	Purity %	Purity method	Type capacity	Calorimeter Type	Calorimeter Reference
88MEL/VER	233.1–573.2	11	2.50	not specified		C_{sat}	not specified	

TABLE 22.43.3. Parameters of regression polynomial

Heat capacity type	No. data points total	used	s_w	s/R	s_r %	s_b/R	+/–
C_{sat}	11	11	0.443	6.05–1	1.11	6.58–3	–1

Temp. range K	A_1	A_2	A_3	A_4	Level of uncertainty
233.1–573.2	1.29397+1	1.22924+1	–3.56564	5.18093–1	V

TABLE 22.43.4. Recommended values of heat capacities

Temp. (K)	230	240	250	260	270	273.15	280
c_{sat} (J K^{-1}g^{-1})	1.77	1.79	1.82	1.85	1.87	1.88	1.90
C_{sat} (J K^{-1}mol^{-1})	238	242	245	249	252	253	256
Temp. (K)	290	298.15	300	310	320	330	340
c_{sat} (J K^{-1}g^{-1})	1.93	1.95	1.96	1.99	2.02	2.06	2.09
C_{sat} (J K^{-1}mol^{-1})	260	263	264	268	272	277	282
Temp. (K)	350	360	370	380	390	400	410
c_{sat} (J K^{-1}g^{-1})	2.13	2.17	2.21	2.26	2.31	2.36	2.41
C_{sat} (J K^{-1}mol^{-1})	287	292	298	304	311	318	325
Temp. (K)	420	430	440	450	460	470	480
c_{sat} (J K^{-1}g^{-1})	2.47	2.54	2.60	2.67	2.75	2.82	2.91
C_{sat} (J K^{-1}mol^{-1})	333	341	350	360	370	380	392
Temp. (K)	490	500	510	520	530	540	550
c_{sat} (J K^{-1}g^{-1})	3.00	3.09	3.19	3.29	3.40	3.52	3.64
C_{sat} (J K^{-1}mol^{-1})	403	416	429	443	458	473	490
Temp. (K)	560	570					
c_{sat} (J K^{-1}g^{-1})	3.76	3.90					
C_{sat} (J K^{-1}mol^{-1})	507	525					

Name: 1–Chloronaphthalene
Formula: $C_{10}H_7Cl$

CAS–RN: 90–13–1
Group No.: 22–044
Molar Mass: 162.62

TABLE 22.44.1. Experimental heat capacities

Reference	Temp. range K	No. pnts	Error %	Purity %	Purity method	Type capacity	Calorimeter Type	Calorimeter Reference
81GRO/ING	298.1	1	nosp	99.0	chrom	C_p	FSIT	71PIC/LED
86WIL/LAI	298.1	1	nosp	99.	anal	C_p	FSIT	71PIC/LED
88COS/VAN	298.1	1	nosp	99.	anal	C_p	FSIT	71PIC/LED

TABLE 22.44.2. Correlated heat capacities

Reference	Temp. range K	No. pnts used	$\sigma_r C$ %	d_w	d/R	d_r %	d_b/R	+/–
				Selected data				
81GRO/ING	298.1	1	0.50#	0.043	5.42–3	0.02	5.42–3	1
86WIL/LAI	298.1	1	0.50#	0.043	5.41–3	0.02	–5.41–3	–1
				Rejected data				
88COS/VAN	(1.53–1, 0.60, 1.53–1, 1)							

TABLE 22.44.3. Parameters of regression polynomial

Heat capacity type	No. data points total	used	s_w	s/R	s_r %	s_b/R	+/–
C_p	3	2	0.060	7.65–3	0.03	2.86–6	0

Temp. range K	A_1	Level of uncertainty
298.1–298.1	2.54164+1	III

TABLE 22.44.4. Recommended values of heat capacities

Temp. (K)	298.15
c_p (J $K^{-1}g^{-1}$)	1.300
C_p (J $K^{-1}mol^{-1}$)	211.3

Name: 2–Chloronaphthalene
Formula: $C_{10}H_7Cl$

CAS–RN: 91–58–7
Group No.: 22–045
Molar Mass: 162.62

TABLE 22.45.1. Experimental heat capacities

Reference	Temp. range K	No. pnts	Error %	Purity %	Purity method	Type capacity	Calorimeter Type	Calorimeter Reference
91VAN/VER	333.5–367.5	34	nosp	99.91	melpt	C_p	BSAO	87VAN/VAN

TABLE 22.45.3. Parameters of regression polynomial

Heat capacity type	No. data points total	used	s_w	s/R	s_r %	s_b/R	+/–
C_p	34	34	0.204	1.69–2	0.06	2.02–5	11

Temp. range K	A_1	A_2	Level of uncertainty
333.5–367.5	1.19413+1	4.58432	III

TABLE 22.45.4. Recommended values of heat capacities

Temp. (K)	340	350	360	370
c_p (J $K^{-1}g^{-1}$)	1.407	1.431	1.454	1.478
C_p (J $K^{-1}mol^{-1}$)	228.9	232.7	236.5	240.3

Name: 2–Chloro–1,1'–biphenyl
Formula: $C_{12}H_9Cl$

CAS–RN: 2051–60–7
Group No.: 22–046
Molar Mass: 188.66

TABLE 22.46.1. Experimental heat capacities

Reference	Temp. range K	No. pnts	Error %	Purity %	Purity method	Type capacity	Calorimeter Type	Calorimeter Reference
74GEI/DZH2	304.8–330.0	4S	nosp	99.66	melpt	C_p	BSAO	54STR/ICK

TABLE 22.46.3. Parameters of regression polynomial

Heat capacity type	No. data points total	used	s_w	s/R	s_r %	s_b/R	+/–
C_p	4	4	0.006	1.78–3	0.01	–4.77–7	0

Temp. range K	A_1	A_2	Level of uncertainty
304.8–330.0	1.80393+1	4.46051	IV

TABLE 22.46.4. Recommended values of heat capacities

Temp. (K)	310	320	330
c_p (J K^{-1}g^{-1})	1.404	1.424	1.444
C_p (J K^{-1}mol^{-1})	265.0	268.7	272.4

Name: 4–Chloro–1,1'–biphenyl
Formula: $C_{12}H_9Cl$

CAS–RN: 2051–62–9
Group No.: 22–047
Molar Mass: 188.66

TABLE 22.47.1. Experimental heat capacities

Reference	Temp. range K	No. pnts	Error %	Purity %	Purity method	Type capacity	Calorimeter Type	Calorimeter Reference
75GEI/DZH	347.9–370.0	2S	0.30	99.1	melpt	C_p	BSAO	54STR/ICK

TABLE 22.47.3. Parameters of regression polynomial

Heat capacity type	No. data points total	used	s_w	s/R	s_r %	s_b/R	+/–
C_p	2	2	0.000	0.00	0.00	0.00	0

Temp. range K	A_1	A_2	Level of uncertainty
347.9–370.0	2.48100+1	2.28571	IV

TABLE 22.47.4. Recommended values of heat capacities

Temp. (K)	350	360	370
c_p (J K^{-1}g^{-1})	1.446	1.456	1.466
C_p (J K^{-1}mol^{-1})	272.8	274.7	276.6

Name: 1,1'–(Chloromethylene)bisbenzene
Formula: $C_{13}H_{11}Cl$

CAS–RN: 90–99–3
Group No.: 22–048
Molar Mass: 202.68

TABLE 22.48.1. Experimental heat capacities

Reference	Temp. range K	No. pnts	Error %	Purity %	Purity method	Type capacity	Calorimeter Type	Calorimeter Reference
31SMI/AND1	298.5–310.7	2	nosp	98.5	estim	C_p	DSIO	26AND/LYN

TABLE 22.48.3. Parameters of regression polynomial

Heat capacity type	No. data points total	No. data points used	s_w	s/R	s_r %	s_b/R	+/–
C_p	2	2	0.000	0.00	0.00	0.00	0

Temp. range K	A_1	A_2	Level of uncertainty
298.5–310.7	1.89174+1	5.36208	V

TABLE 22.48.4. Recommended values of heat capacities

Temp. (K)	300	310
c_p (J K^{-1}g^{-1})	1.44	1.46
C_p (J K^{-1}mol^{-1})	291	295

Name: 1–Chlorooctadecane
Formula: $C_{18}H_{37}Cl$

CAS–RN: 3386–33–2
Group No.: 22–049
Molar Mass: 288.94

TABLE 22.49.1. Experimental heat capacities

Reference		Temp. K	Capac. J/(K.g)	Error %	Purity %	Purity method	Type capacity	Calorimeter Type	Calorimeter Reference
75STR/SUN	N	301.00	2.100	3.00	not specified		C_p	BDHT	69PER/COM

75STR/SUN content of Octadecane below 0.02 mass %

23. Bromine Derivatives

The bromine family contains 40 compounds, of which 8 were measured only at one temperature.

Some laboratories produced values for a large number of bromine derivatives but in most cases the data pertain to the older literature (such as measurements from THA, *81VON) and/or sources reporting poor quality data (such as results for 14 compounds from PTIL, 48KUR). These sources had to be included only because of the absence of better quality data or for extending data over a temperature interval. The remaining articles containing more than one measured bromine compound are those from JHUB (31DEE) and from PUP (50KUS/CRO, 50CRO/SMI2); the uncertainty of these measurements is according to our judgement between 1.5 to 2 %.

With some exceptions, only data for bromoalkanes were reported. Measurements over a wide temperature range were performed for bromomethane at UCB (38EGA/KEM) and for dibromomethane at UCE (57HAR/MOE). The most frequently measured compound in the bromine family is 1,2-dibromoethane, but most of the data are from older sources. Since the two more reliable measurements from UCB (40PIT1) and UWA (69WIL/SCH) were carried out over a narrow temperature interval, some of the less reliable measurements had to be taken account (33LEB/MOE, 39RAI) in the correlation to obtain the recommended values. The data measured for 3-bromohexane at KGPI (88MEL/VER) gave results with a reported error of 2.5 %, and those for 1-bromodocosane and 1-bromotriacontane from General Electric Research Laboratory Schenectady, N.Y. (53HOF/DEC) seem also to be of poor quality with a reported error of 4 % for sample purity of only 90 %.

The most accurate measurements in this family appear to be the data for bromobenzene from NBSW (75MAS/SCO) and those for pentaerythrityl bromide from UMAA (65CLE/WON). The former data have been supplemented by the older measurements of (*81VON).

Unusual studies were carried out on 10 brominated and deuterated ethanes at one temperature in Belgium (49WUY/JUN, 49DHO/JUN). As for some of these compounds, any new information has not been published since this time, hence, their CASRN's are not assigned or available.

Name: Tetrabromomethane
Formula: CBr_4

CAS–RN: 558–13–4
Group No.: 23–001
Molar Mass: 331.63

TABLE 23.1.1. Experimental heat capacities

Reference		Temp. range K	No. pnts	Error %	Purity %	Purity method	Type capacity	Calorimeter Type	Calorimeter Reference
39FRE/HIL	N	363.3–423.1	2	nosp	not specified		C_{avg}	DSIO	38FRE/HIL1
48KUR		403.9–446.8	3	nosp	not specified		C_{avg}	DSIO	47KUR

39FRE/HIL constant value calculated from temperature dependence of enthalpy by authors

TABLE 23.1.2. Correlated heat capacities

Reference	Temp. range K	No. pnts used	$\sigma_r C$ %	d_w	d/R	d_r %	d_b/R	+/–
				Selected data				
39FRE/HIL	363.3	1	3.00#	0.154	8.56–2	0.46	–8.56–2	–1
48KUR	403.9–446.8	3	5.00#	0.323	3.41–1	1.61	8.94–2	1

TABLE 23.1.3. Parameters of regression polynomial

Heat capacity type	No. data points total	No. data points used	s_w	s/R	s_r %	s_b/R	+/–
C	5	4	0.410	4.22–1	2.00	4.57–2	0

Temp. range K	A_1	A_2	Level of uncertainty
363.3–446.8	2.77717	4.34315	VI

TABLE 23.1.4. Recommended values of heat capacities

Temp. (K)	360	370	380	390	400	410	420
c (J $K^{-1}g^{-1}$)	0.46	0.47	.48	0.49	0.51	0.52	0.53
C (J $K^{-1}mol^{-1}$)	153	157	160	164	168	171	175
Temp. (K)	430	440	450				
c (J $K^{-1}g^{-1}$)	0.54	0.55	0.56				
C (J $K^{-1}mol^{-1}$)	178	182	186				

Name: Tribromomethane
Formula: $CHBr_3$

CAS–RN: 75–25–2
Group No.: 23–002
Molar Mass: 252.73

TABLE 23.2.1. Experimental heat capacities

Reference	Temp. range K	No. pnts	Error %	Purity %	Purity method	Type capacity	Calorimeter Type	Calorimeter Reference
22COH/MOE	292.1–292.1	2	nosp	not	specified	C_p	BSAO	20COH/MOE
32TRE	298.0	1	nosp	not	specified	C_p	BSIO	49WEI
48KUR	305.2–356.9	4	nosp	not	specified	C_{avg}	DSIO	47KUR
49TSC/RIC3	298.1	1	nosp	not	specified	C_p	BSIO	49TSC/RIC1

TABLE 23.2.2. Correlated heat capacities

Reference	Temp. range K	No. pnts used	$\sigma_r C$ %	d_w	d/R	d_r %	d_b/R	+/–
				Selected data				
22COH/MOE	292.1	2	0.70#	0.069	7.80–3	0.05	–4.51–3	0
32TRE	298.0	1	2.00#	0.447	1.45–1	0.89	1.45–1	1
48KUR	305.2–356.9	4	5.00#	0.285	2.28–1	1.43	–1.13–1	0
49TSC/RIC3	298.1	1	3.00#	0.035	1.70–2	0.11	1.70–2	1

TABLE 23.2.3. Parameters of regression polynomial

Heat capacity type	No. data points total	No. data points used	s_w	s/R	s_r %	s_b/R	+/–
C_p	8	8	0.299	1.96–1	1.22	–3.74–2	2

Temp. range K	A_1	A_2	Level of uncertainty
292.1–356.9	1.37276+1	8.07330–1	VI

TABLE 23.2.4. Recommended values of heat capacities

Temp. (K)	290	298.15	300	310	320	330	340
c_p (J $K^{-1}g^{-1}$)	0.53	0.53	0.53	0.53	0.54	0.54	0.54
C_p (J $K^{-1}mol^{-1}$)	134	134	134	135	136	136	137
Temp. (K)	350	360					
c_p (J $K^{-1}g^{-1}$)	0.54	0.55					
C_p (J $K^{-1}mol^{-1}$)	138	138					

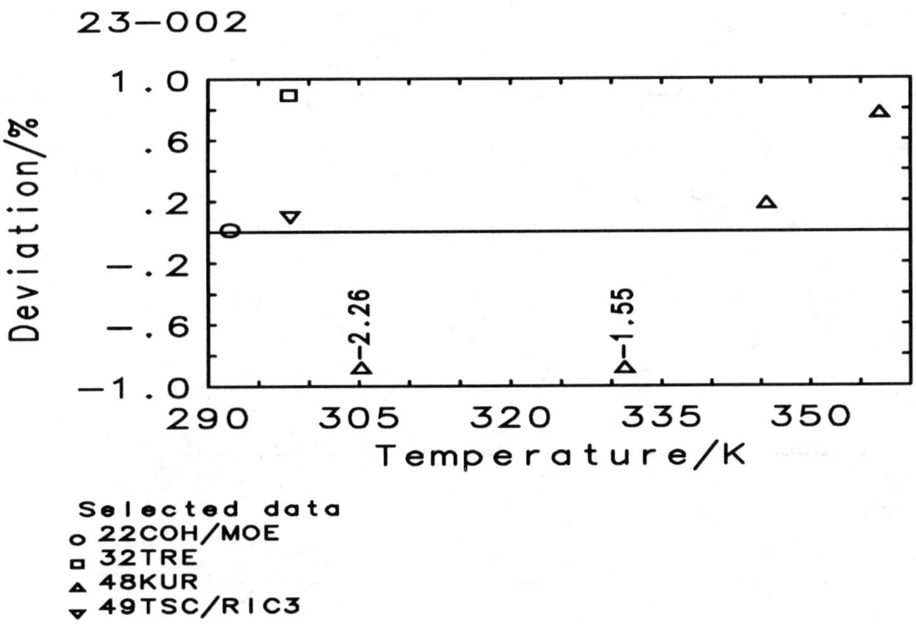

Name: Dibromomethane
Formula: CH_2Br_2

CAS-RN: 74–95–3
Group No.: 23–003
Molar Mass: 173.83

TABLE 23.3.1. Experimental heat capacities

Reference	Temp. range K	No. pnts	Error %	Purity %	Purity method	Type capacity	Calorimeter Type	Calorimeter Reference
48KUR	272.1–329.9	4	nosp	not	specified	C_{avg}	DSIO	47KUR
57HAR/MOE	244.9–303.2	7	0.50	not	specified	C_p	BSIO	57HAR/MOE

TABLE 23.3.2. Correlated heat capacities

Reference	Temp. range K	No. pnts used	$\sigma_r C$ %	d_w	d/R	d_r %	d_b/R	+/–
				Selected data				
57HAR/MOE	244.9–303.2	7	0.50	1.241	7.77–2	0.62	9.66–4	3
				Rejected data				
48KUR	(2.59, 17.04, 2.58, 3)							

TABLE 23.3.3. Parameters of regression polynomial

Heat capacity type	No. data points total	used	s_w	s/R	s_r %	s_b/R	+/–
C_p	11	7	1.468	9.19–2	0.73	9.66–4	3

Temp. range K	A_1	A_2	Level of uncertainty
244.9–303.2	1.25861+1	–4.53477–3	IV

TABLE 23.3.4. Recommended values of heat capacities

Temp. (K)	240	250	260	270	273.15	280	290
c_p (J K^{-1}g^{-1})	0.601	0.601	0.601	0.601	0.601	0.601	0.601
C_p (J K^{-1}mol^{-1})	104.6	104.6	104.5	104.5	104.5	104.5	104.5
Temp. (K)	298.15	300					
c_p (J K^{-1}g^{-1})	0.601	0.601					
C_p (J K^{-1}mol^{-1})	104.5	104.5					

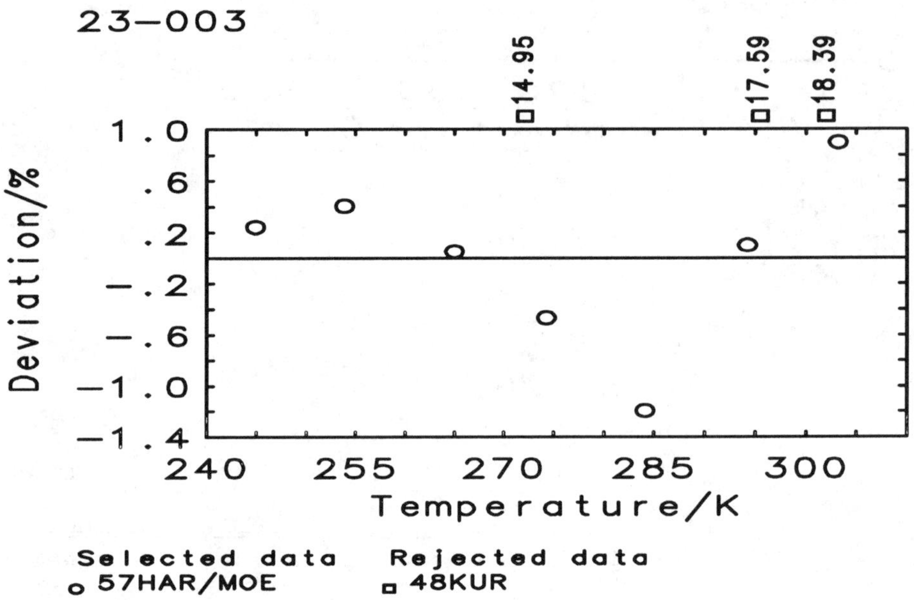

Name: Bromomethane
Formula: CH_3Br

CAS–RN: 74–83–9
Group No.: 23–004
Molar Mass: 94.94

TABLE 23.4.1. Experimental heat capacities

Reference		Temp. range K	No. pnts	Error %	Purity %	Purity method	Type capacity	Calorimeter Type	Calorimeter Reference
38EGA/KEM		184.5–275.6	16	nosp	99.999	melpt	C_p	BSIO	28GIA/WIE1
48KUR	N	244.1–266.5	3	nosp	not specified		C_{avg}	DSIO	47KUR

48KUR suspect values

TABLE 23.4.2. Correlated heat capacities

Reference	Temp. range K	No. pnts used	$\sigma_r C$ %	d_w	d/R	d_r %	d_b/R	+/–
				Selected data				
38EGA/KEM	184.5–275.6	16	0.30#	0.534	1.50–2	0.16	4.79–5	0
				Rejected data				
48KUR	(3.87, 29.26, 3.87, 3)							

TABLE 23.4.3. Parameters of regression polynomial

Heat capacity type	No. data points total	used	s_w	s/R	s_r %	s_b/R	+/–
C_p	19	16	0.593	1.66–2	0.18	4.79–5	0

Temp. range K	A_1	A_2	A_3	Level of uncertainty
184.5–275.6	1.27783+1	–3.14160	7.04852–1	III

TABLE 23.4.4. Recommended values of heat capacities

Temp. (K)	190	200	210	220	230	240	250
c_p (J K^{-1}g^{-1})	0.819	0.816	0.814	0.813	0.813	0.814	0.817
C_p (J K^{-1}mol^{-1})	77.77	77.45	77.24	77.14	77.17	77.31	77.57

Temp. (K)	260	270
c_p (J K^{-1}g^{-1})	0.821	0.826
C_p (J K^{-1}mol^{-1})	77.95	78.44

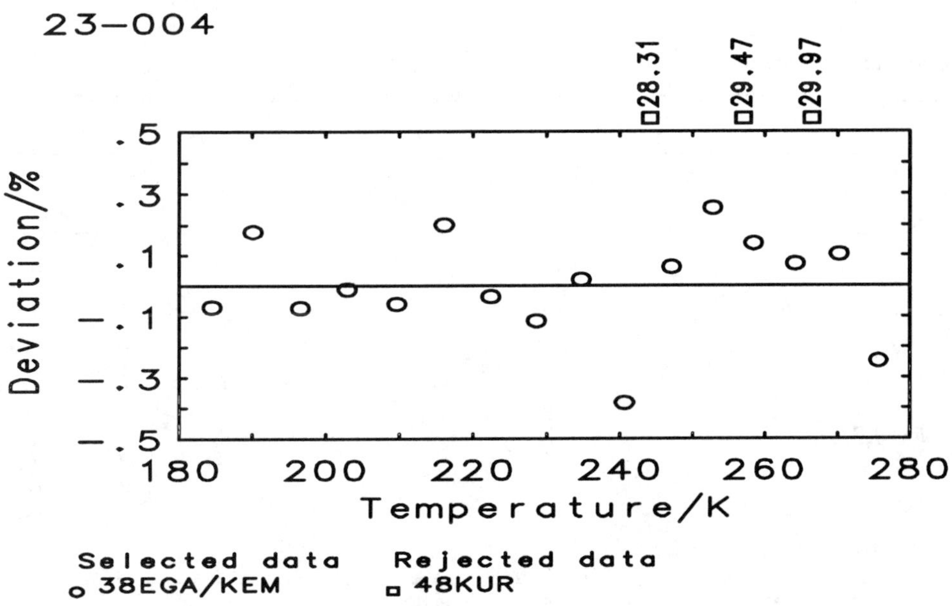

Name: 1,2–Dibromoethane–1,1,2,2–d_4
Formula: $C_2Br_2D_4$

CAS–RN: 22581–63–1
Group No.: 23–005
Molar Mass: 191.89

TABLE 23.5.1. Experimental heat capacities

Reference		Temp. range K	No. pnts	Error %	Purity %	Purity method	Type capacity	Calorimeter Type	Calorimeter Reference
49DHO/JUN	N	293.1	1	nosp	not specified		C_p	BDHO	49WUY/JUN
49WUY/JUN	N	310.1	1	nosp	not specified		C_p	BDHO	49WUY/JUN

49DHO/JUN correction for incomplete deuteration (3.5 % of isotope D_3H)
49WUY/JUN correction for incomplete deuteration (3.5 % of isotope D_3H)

TABLE 23.5.2. Correlated heat capacities

Reference	Temp. range K	No. pnts used	$\sigma_r C$ %	d_w	d/R	d_r %	d_b/R	+/–
				Selected data				
49DHO/JUN	293.1	1	3.00#	0.000	0.00	0.00	0.00	0
49WUY/JUN	310.1	1	3.00#	0.000	0.00	0.00	0.00	0

TABLE 23.5.3. Parameters of regression polynomial

Heat capacity type	No. data points total	No. data points used	s_w	s/R	s_r %	s_b/R	+/–
C_p	2	2	0.000	0.00	0.00	0.00	0

Temp. range K	A_1	A_2	Level of uncertainty
293.1–310.1	1.47018+1	1.04235	V

TABLE 23.5.4. Recommended values of heat capacities

Temp. (K)	290	298.15	300	310
c_p (J K^{-1}g^{-1})	0.77	0.77	0.77	0.78
C_p (J K^{-1}mol^{-1})	147	148	148	149

Name: 1,1,2–Tribromoethane–1,2,2–d_3
Formula: $C_2Br_3D_3$

CAS–RN: unknown
Group No.: 23–006
Molar Mass: 269.78

TABLE 23.6.1. Experimental heat capacities

Reference		Temp. K	Capac. J/(K.g)	Error %	Purity %	Purity method	Type capacity	Calorimeter Type	Calorimeter Reference
49DHO/JUN	N	293.15	0.6627	nosp	not specified		C_p	BDHO	49WUY/JUN

49DHO/JUN correction for incomplete deuteration (1.0 % of isotope D_2H)

Name: 1,2–Dibromoethane–1,1,2–d_3
Formula: $C_2HBr_2D_3$

CAS–RN: 117164–17–7
Group No.: 23–007
Molar Mass: 190.88

TABLE 23.7.1. Experimental heat capacities

Reference		Temp. K	Capac. J/(K.g)	Error %	Purity %	Purity method	Type capacity	Calorimeter Type	Calorimeter Reference
49WUY/JUN	N	310.15	0.7604	nosp	not specified		C_p	BDHO	49WUY/JUN

49WUY/JUN correction for incomplete deuteration (7.0 % of isotope D_2H_2)

Name: 1,1,2–Tribromoethane–2,2–d_2
Formula: $C_2HBr_3D_2$

CAS–RN: unknown
Group No.: 23–008
Molar Mass: 268.77

TABLE 23.8.1. Experimental heat capacities

Reference		Temp. K	Capac. J/(K.g)	Error %	Purity %	Purity method	Type capacity	Calorimeter Type	Calorimeter Reference
49DHO/JUN	N	293.15	0.6489	nosp	not specified		C_p	BDHO	49WUY/JUN

49DHO/JUN correction for incomplete deuteration (6.0 % of isotope HDH)

Name: 1,2–Dibromoethane–1,1–d_2
Formula: $C_2H_2Br_2D_2$

CAS–RN: unknown
Group No.: 23–009
Molar Mass: 189.87

TABLE 23.9.1. Experimental heat capacities

Reference		Temp. range K	No. pnts	Error %	Purity %	Purity method	Type capacity	Calorimeter Type	Calorimeter Reference
49DHO/JUN	N	293.1	1	nosp	not specified		C_p	BDHO	49WUY/JUN
49WUY/JUN	N	310.1	1	nosp	not specified		C_p	BDHO	49WUY/JUN

49DHO/JUN correction for incomplete deuteration (8.0 % of isotope H_3D)
49WUY/JUN correction for incomplete deuteration (8.0 % of isotope H_3D)

TABLE 23.9.2. Correlated heat capacities

Reference	Temp. range K	No. pnts used	$\sigma_r C$ %	d_w	d/R	d_r %	d_b/R	+/–
				Selected data				
49DHO/JUN	293.1	1	3.00#	0.000	0.00	0.00	0.00	0
49WUY/JUN	310.1	1	3.00#	0.000	0.00	0.00	0.00	0

TABLE 23.9.3. Parameters of regression polynomial

Heat capacity type	No. data points total	used	s_w	s/R	s_r %	s_b/R	+/–
C_p	2	2	0.000	0.00	0.00	0.00	0

Temp. range K	A_1	A_2	Level of uncertainty
293.1–310.1	1.51763+1	6.76794–1	V

TABLE 23.9.4. Recommended values of heat capacities

Temp. (K)	290	298.15	300	310
c_p (J $K^{-1}g^{-1}$)	0.75	0.75	0.75	0.76
C_p (J $K^{-1}mol^{-1}$)	143	143	143	144

Name: 1,2–Dibromoethane–1,2–d_2
Formula: $C_2H_2Br_2D_2$

CAS–RN: unknown
Group No.: 23–010
Molar Mass: 189.87

TABLE 23.10.1. Experimental heat capacities

Reference		Temp. range K	No. pnts	Error %	Purity % method	Type capacity	Calorimeter Type	Calorimeter Reference
49DHO/JUN	N	293.1	1	nosp	not specified	C_p	BDHO	49WUY/JUN
49WUY/JUN	N	310.1	1	nosp	not specified	C_p	BDHO	49WUY/JUN

49DHO/JUN correction for incomplete deuteration (2.0 % of isotope H_3D)
49WUY/JUN correction for incomplete deuteration (2.0 % of isotope H_3D)

TABLE 23.10.2. Correlated heat capacities

Reference	Temp. range K	No. pnts used	$\sigma_r C$ %	d_w	d/R	d_r %	d_b/R	+/–
				Selected data				
49DHO/JUN	293.1	1	3.00#	0.000	0.00	0.00	0.00	0
49WUY/JUN	310.1	1	3.00#	0.000	0.00	0.00	0.00	0

TABLE 23.10.3. Parameters of regression polynomial

Heat capacity type	No. data points total	used	s_w	s/R	s_r %	s_b/R	+/–
C_p	2	2	0.000	0.00	0.00	0.00	0

Temp. range K	A_1	A_2	Level of uncertainty
293.1–310.1	1.60341+1	3.41819–1	V

TABLE 23.10.4. Recommended values of heat capacities

Temp. (K)	290	298.15	300	310
c_p (J $K^{-1}g^{-1}$)	0.75	0.75	0.75	0.75
C_p (J $K^{-1}mol^{-1}$)	142	142	142	142

Name: 1,1,2–Tribromoethane–1–*d*
Formula: $C_2H_2Br_3D$

CAS–RN: unknown
Group No.: 23–011
Molar Mass: 267.76

TABLE 23.11.1. Experimental heat capacities

Reference		Temp. K	Capac. J/(K.g)	Error %	Purity % method	Type capacity	Calorimeter Type Reference
49DHO/JUN	N	293.15	0.6339	nosp	not specified	C_p	BDHO 49WUY/JUN

49DHO/JUN correction for incomplete deuteration (8.0 % of isotope H_3)

Name: 1,1,2–Tribromoethane–2–*d*
Formula: $C_2H_2Br_3D$

CAS–RN: unknown
Group No.: 23–012
Molar Mass: 267.76

TABLE 23.12.1. Experimental heat capacities

Reference		Temp. K	Capac. J/(K.g)	Error %	Purity % method	Type capacity	Calorimeter Type Reference
49DHO/JUN	N	293.15	0.6326	nosp	not specified	C_p	BDHO 49WUY/JUN

49DHO/JUN correction for incomplete deuteration (3.0 % of isotope H_3)

Name: 1,1,2,2–Tetrabromoethane
Formula: $C_2H_2Br_4$

CAS–RN: 79–27–6
Group No.: 23–013
Molar Mass: 345.65

TABLE 23.13.1. Experimental heat capacities

Reference	Temp. range K	No. pnts	Error %	Purity % method	Type capacity	Calorimeter Type Reference
48KUR	306.5–346.8	3	nosp	not specified	C_{avg}	DSIO 47KUR

TABLE 23.13.3. Parameters of regression polynomial

Heat capacity type	No. data points total	used	s_w	s/R	s_r %	s_b/R	+/–
C	3	3	0.000	4.20–4	0.00	0.00	0

Temp. range K	A_1	A_2	Level of uncertainty
306.5–346.8	7.26458	4.24055	VI

TABLE 23.13.4. Recommended values of heat capacities

Temp. (K)	310	320	330	340	350
c (J K^{-1}g^{-1})	0.49	0.50	0.51	0.52	0.53
C (J K^{-1}mol^{-1})	170	173	177	180	184

Name: Bromoethene
Formula: C_2H_3Br

CAS–RN: 593–60–2
Group No.: 23–014
Molar Mass: 106.95

TABLE 23.14.1. Experimental heat capacities

Reference	Temp. K	Capac. J/(K.g)	Error %	Purity %	Purity method	Type capacity	Calorimeter Type	Calorimeter Reference
34MEH2	288.15	1.004	1.50	not specified		C_p	BSIO	49WEI

Name: 1,2–Dibromoethane–*d*
Formula: $C_2H_3Br_2D$

CAS–RN: unknown
Group No.: 23–015
Molar Mass: 188.87

TABLE 23.15.1. Experimental heat capacities

Reference		Temp. range K	No. pnts	Error %	Purity %	Purity method	Type capacity	Calorimeter Type	Calorimeter Reference
49DHO/JUN	N	293.1	1	nosp	not specified		C_p	BDHO	49WUY/JUN
49WUY/JUN	N	310.1	1	nosp	not specified		C_p	BDHO	49WUY/JUN

49DHO/JUN correction for incomplete deuteration (6.5 % of isotope H_4)
49WUY/JUN correction for incomplete deuteration (6.5 % of isotope H_4)

TABLE 23.15.2. Correlated heat capacities

Reference	Temp. range K	No. pnts used	$\sigma_r C$ %	d_w	d/R	d_r %	d_b/R	+/–
				Selected data				
49DHO/JUN	293.1	1	3.00#	0.000	0.00	0.00	0.00	0
49WUY/JUN	310.1	1	3.00#	0.000	0.00	0.00	0.00	0

TABLE 23.15.3. Parameters of regression polynomial

Heat capacity type	No. data points total	No. data points used	s_w	s/R	s_r %	s_b/R	+/–
C_p	2	2	0.000	0.00	0.00	0.00	0

Temp. range K	A_1	A_2	Level of uncertainty
293.1–310.1	1.52613+1	4.41708–1	V

TABLE 23.15.4. Recommended values of heat capacities

Temp. (K)	290	298.15	300	310
c_p (J K^{-1}g^{-1})	0.73	0.73	0.73	0.73
C_p (J K^{-1}mol^{-1})	138	138	138	138

Name: 1,1,2–Tribromoethane
Formula: $C_2H_3Br_3$

CAS–RN: 78–74–0
Group No.: 23–016
Molar Mass: 266.76

TABLE 23.16.1. Experimental heat capacities

Reference	Temp. K	Capac. J/(K.g)	Error %	Purity %	Purity method	Type capacity	Calorimeter Type	Calorimeter Reference
49DHO/JUN	293.15	0.6209	nosp	not specified		C_p	BDHO	49WUY/JUN

Name: 1,2–Dibromoethane
Formula: $C_2H_4Br_2$

CAS–RN: 106–93–4
Group No.: 23–017
Molar Mass: 187.86

TABLE 23.17.1. Experimental heat capacities

Reference		Temp. range K	No. pnts	Error %	Purity %	Purity method	Type capacity	Calorimeter Type	Calorimeter Reference
33LEB/MOE		283.3–298.5	5	nosp	not specified		C_p	BSIO	33LEB/MOE
39RAI	N	283.3–321.7	9	2.00	99.92	melpt	C_p	BDHO	37STU
40PIT1		286.5–318.0	4	0.20	99.74	melpt	C_p	BSIO	28LAT/GRE
48KUR		309.1–345.7	3	nosp	not specified		C_{avg}	DSIO	47KUR
49DHO/JUN		293.1	1	nosp	not specified		C_p	BDHO	49WUY/JUN
49TSC/RIC3		298.1	1	nosp	not specified		C_p	BSIO	49TSC/RIC1
49WUY/JUN		310.1	1	nosp	not specified		C_p	BDHO	49WUY/JUN
65FIN/GRU		300.0	1	0.40	not specified		C_p	BDAO	65FIN/GRU
69WIL/SCH		293.1–313.1	3	0.40	not specified		C_p	BDAO	65FIN/GRU

39RAI data from a graph only

TABLE 23.17.2. Correlated heat capacities

Reference	Temp. range K	No. pnts used	$\sigma_r C$ %	d_w	d/R	d_r %	d_b/R	+/–
				Selected data				
33LEB/MOE	283.3–298.5	5	1.00#	1.137	1.89–1	1.14	7.02–2	1
39RAI	283.3–321.7	9	2.00	0.723	2.32–1	1.45	–1.75–1	–7
40PIT1	286.5–318.0	4	0.20	1.101	3.60–2	0.22	1.97–2	2
49TSC/RIC3	298.1	1	3.00#	0.093	4.55–2	0.28	–4.55–2	–1
65FIN/GRU	300.0	1	0.40	0.552	3.60–2	0.22	–3.60–2	–1
69WIL/SCH	293.1–313.1	3	0.40	1.346	8.75–2	0.54	–8.67–2	–3
				Rejected data				
48KUR	(1.79–1, 1.08, 1.79–1, 1)			49DHO/JUN	(9.48–2, 0.58, –9.48–2, –1)			
49WUY/JUN	(1.10–1, 0.67, –1.10–1, –1)							

TABLE 23.17.3. Parameters of regression polynomial

Heat capacity type	No. data points total	used	s_w	s/R	s_r %	s_b/R	+/–
C_p	28	23	1.043	1.86–1	1.15	–6.47–2	–9

Temp. range K	A_1	A_2	A_3	Level of uncertainty
283.3–321.7	4.32852+1	–1.82118+1	3.07322	IV

TABLE 23.17.4. Recommended values of heat capacities

Temp. (K)	290	298.15	300	310	320
c_p ($J K^{-1}g^{-1}$)	0.722	0.722	0.722	0.724	0.729
C_p ($J K^{-1}mol^{-1}$)	135.7	135.6	135.6	136.0	137.0

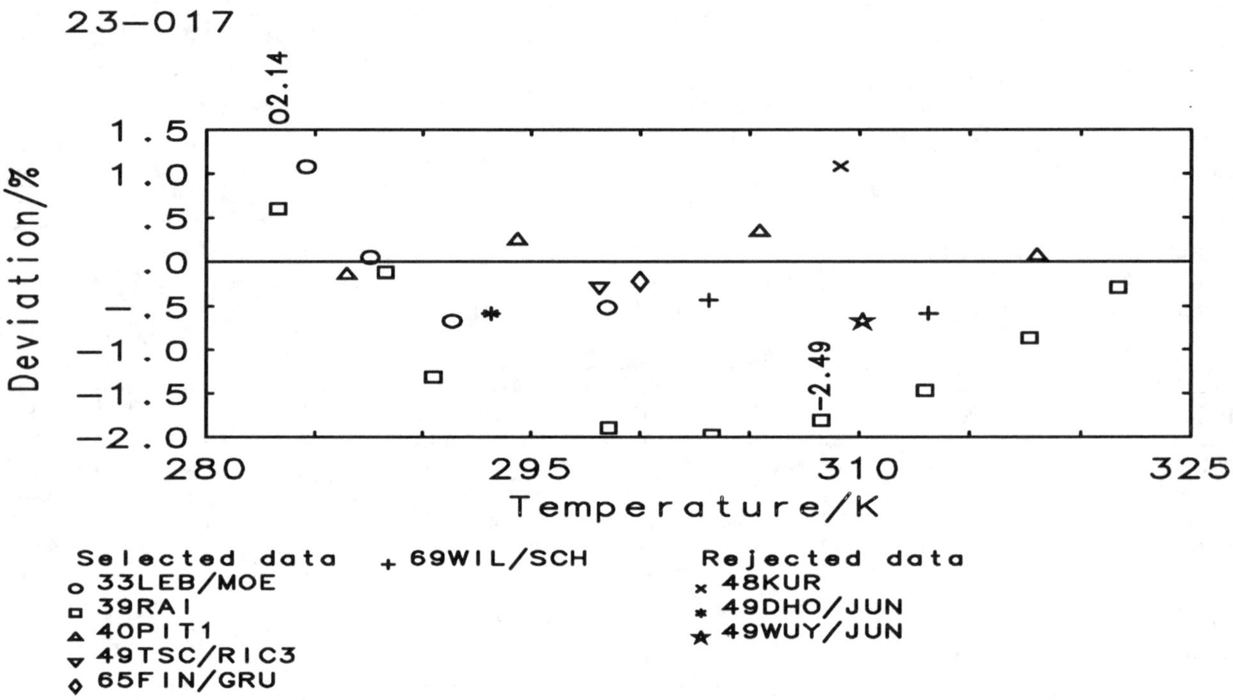

Name: Bromoethane
Formula: C_2H_5Br

CAS–RN: 74–96–4
Group No.: 23–018
Molar Mass: 108.97

TABLE 23.18.1. Experimental heat capacities

Reference		Temp. range K	No. pnts	Error %	Purity %	Purity method	Type capacity	Calorimeter Type	Calorimeter Reference
07BAT	N	167.7–244.3	5	nosp	not	specified	C_p	BSIO	07BAT
12SCH2		280.6–290.6	3	nosp	not	specified	C_p	BSIO	12SCH1
48KUR		256.5–300.0	5	nosp	not	specified	C_{avg}	DSIO	47KUR
49TSC/RIC3		298.1	1	nosp	not	specified	C_p	BSIO	49TSC/RIC1

07BAT same data in 08BAT

TABLE 23.18.2. Correlated heat capacities

Reference	Temp. range K	No. pnts used	$\sigma_r C$ %	d_w	d/R	d_r %	d_b/R	+/–
				Selected data				
07BAT	167.7–244.3	5	2.00#	0.130	2.86–2	0.26	1.14–2	–1
12SCH2	280.6–290.6	3	3.00#	0.313	1.11–1	0.94	1.05–2	–1
48KUR	256.5–300.0	5	5.00#	0.363	2.04–1	1.81	–7.67–2	–1
				Rejected data				
49TSC/RIC3	(3.90, 24.54, 3.90, 1)							

TABLE 23.18.3. Parameters of regression polynomial

Heat capacity type	No. data points total	used	s_w	s/R	s_r %	s_b/R	+/–
C_p	14	13	0.322	1.58–1	1.39	–2.27–2	–3

Temp. range K	A_1	A_2	A_3	Level of uncertainty
167.7–300.0	1.22805+1	–2.02028	6.46068–1	VI

TABLE 23.18.4. Recommended values of heat capacities

Temp. (K)	170	180	190	200	210	220	230
c_p (J K^{-1}g^{-1})	0.82	0.82	0.82	0.83	0.83	0.84	0.84
C_p (J K^{-1}mol^{-1})	89.1	89.3	89.6	90.0	90.5	91.2	91.9
Temp. (K)	240	250	260	270	273.15	280	290
c_p (J K^{-1}g^{-1})	0.85	0.86	0.87	0.88	0.88	0.89	0.90
C_p (J K^{-1}mol^{-1})	92.7	93.7	94.7	95.9	96.3	97.2	98.6
Temp. (K)	298.15	300					
c_p (J K^{-1}g^{-1})	0.92	0.92					
C_p (J K^{-1}mol^{-1})	99.8	100					

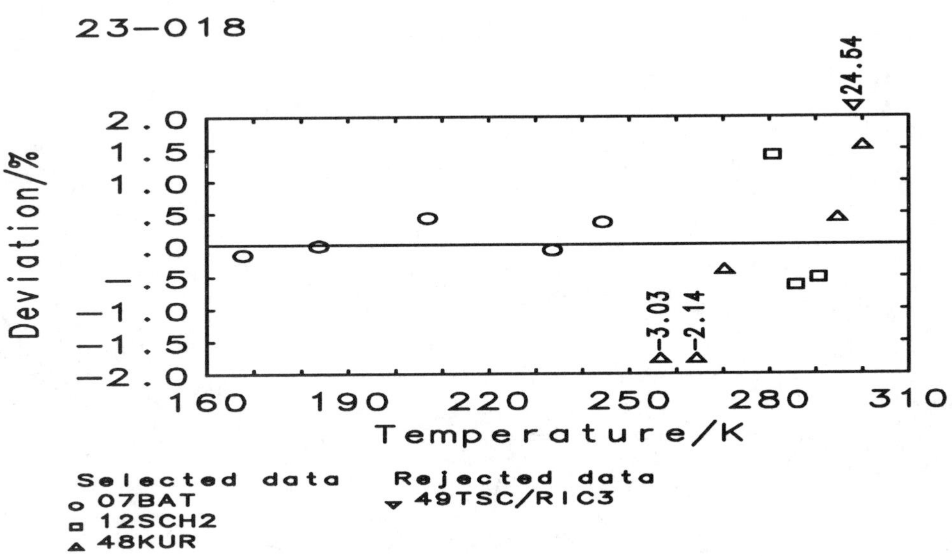

Name: 3–Bromo–1–propene
Formula: C_3H_5Br

CAS–RN: 106–95–6
Group No.: 23–019
Molar Mass: 120.98

TABLE 23.19.1. Experimental heat capacities

Reference	Temp. K	Capac. J/(K.g)	Error %	Purity %	Purity method	Type capacity	Calorimeter Type	Calorimeter Reference
49TSC/RIC3	298.15	1.088	nosp	not specified		C_p	BSIO	49TSC/RIC1

Name: 1,2,3–Tribromopropane
Formula: $C_3H_5Br_3$

CAS–RN: 96–11–7
Group No.: 23–020
Molar Mass: 280.78

TABLE 23.20.1. Experimental heat capacities

Reference	Temp. range K	No. pnts	Error %	Purity %	Purity method	Type capacity	Calorimeter Type	Calorimeter Reference
48KUR	320.2–390.6	6	nosp	not specified		C_{avg}	DSIO	47KUR

TABLE 23.20.3. Parameters of regression polynomial

Heat capacity type	No. data points total	No. data points used	s_w	s/R	s_r %	s_b/R	+/–
C	6	6	0.392	4.95–1	1.96	1.35–2	–2

Temp. range K	A_1	A_2	Level of uncertainty
320.2–390.6	4.78323	5.32870	VI

TABLE 23.20.4. Recommended values of heat capacities

Temp. (K)	320	330	340	350	360	370	380
c ($J\,K^{-1}g^{-1}$)	0.65	0.66	0.68	0.69	0.71	0.73	0.74
C ($J\,K^{-1}mol^{-1}$)	182	186	190	195	199	204	208

Temp. (K)	390
c ($J\,K^{-1}g^{-1}$)	0.76
C ($J\,K^{-1}mol^{-1}$)	213

Name: 1,2–Dibromopropane
Formula: $C_3H_6Br_2$

CAS–RN: 78–75–1
Group No.: 23–021
Molar Mass: 201.89

TABLE 23.21.1. Experimental heat capacities

Reference	Temp. range K	No. pnts	Error %	Purity %	Purity method	Type capacity	Calorimeter Type	Calorimeter Reference
48KUR	305.4–346.6	3	nosp	not specified		C_{avg}	DSIO	47KUR

TABLE 23.21.3. Parameters of regression polynomial

Heat capacity type	No. data points total	used	s_w	s/R	s_r %	s_b/R	+/−
C	3	3	0.002	2.01–3	0.01	−6.36–7	0

Temp. range K	A_1	A_2	Level of uncertainty
305.4–346.6	8.07941	3.74501	VI

TABLE 23.21.4. Recommended values of heat capacities

Temp. (K)	310	320	330	340	350
c (J K^{-1}g^{-1})	0.81	0.83	0.84	0.86	0.87
C (J K^{-1}mol^{-1})	164	167	170	173	176

Name: 1,3–Dibromopropane
Formula: $C_3H_6Br_2$

CAS–RN: 109–64–8
Group No.: 23–022
Molar Mass: 201.89

TABLE 23.22.1. Experimental heat capacities

Reference	Temp. range K	No. pnts	Error %	Purity %	Purity method	Type capacity	Calorimeter Type	Calorimeter Reference
48KUR	332.3–358.3	3	nosp	not specified		C_{avg}	DSIO	47KUR
50CRO/SMY2	243.1–245.6	2	1.40	not specified		C_p	BDHO	50KUS/CRO

TABLE 23.22.2. Correlated heat capacities

Reference	Temp. range K	No. pnts used	$\sigma_r C$ %	d_w	d/R	d_r %	d_b/R	+/−
				Selected data				
48KUR	332.3–358.4	3	5.00#	0.224	2.32–1	1.12	−1.69–2	−1
50CRO/SMY2	243.1–245.6	2	1.40	0.200	5.24–2	0.28	2.91–3	0

TABLE 23.22.3. Parameters of regression polynomial

Heat capacity type	No. data points total	used	s_w	s/R	s_r %	s_b/R	+/−
C_p	5	5	0.277	2.36–1	1.14	−9.00–3	−1

Temp. range K	A_1	A_2	Level of uncertainty
243.1–358.4	1.41627+1	1.85308	VI

TABLE 23.22.4. Recommended values of heat capacities

Temp. (K)	240	250	260	270	273.15	280	290
c_p (J K^{-1}g^{-1})	0.77	0.77	0.78	0.79	0.79	0.80	0.80
C_p (J K^{-1}mol^{-1})	155	156	158	159	160	161	162
Temp. (K)	298.15	300	310	320	330	340	350
c_p (J K^{-1}g^{-1})	0.81	0.81	0.82	0.83	0.84	0.84	0.85
C_p (J K^{-1}mol^{-1})	164	164	166	167	169	170	172
Temp. (K)	360						
c_p (J K^{-1}g^{-1})	0.86						
C_p (J K^{-1}mol^{-1})	173						

Name: 1–Bromopropane
Formula: C_3H_7Br

CAS–RN: 106–94–5
Group No.: 23–023
Molar Mass: 122.99

TABLE 23.23.1. Experimental heat capacities

Reference	Temp. range K	No. pnts	Error %	Purity %	Purity method	Type capacity	Calorimeter Type	Calorimeter Reference
*81VON	305.1–323.3	3S	nosp	not specified		C_{avg}	DSIO	*81VON
48KUR	268.1–312.9	3	nosp	not specified		C_{avg}	DSIO	47KUR
49TSC/RIC3	298.1	1	nosp	not specified		C_p	BSIO	49TSC/RIC1

TABLE 23.23.2. Correlated heat capacities

Reference	Temp. range K	No. pnts used	$\sigma_r C$ %	d_w	d/R	d_r %	d_b/R	+/–
				Selected data				
*81VON	305.1–323.3	3	3.00#	0.585	2.79–1	1.75	–2.26–1	–3
48KUR	268.1–312.9	3	5.00#	0.791	6.78–1	3.96	4.46–1	1
49TSC/RIC3	298.1	1	3.00#	0.592	2.95–1	1.78	2.95–1	1

TABLE 23.23.3. Parameters of regression polynomial

Heat capacity type	No. data points total	No. data points used	s_w	s/R	s_r %	s_b/R	+/–
C_p	7	7	0.807	5.83–1	3.45	1.36–1	–1

Temp. range K	A_1	A_2	Level of uncertainty
268.1–323.3	1.45860+1	5.72349–1	VI

TABLE 23.23.4. Recommended values of heat capacities

Temp. (K)	270	273.15	280	290	298.15	300	310
c_p (J K^{-1}g^{-1})	1.09	1.09	1.09	1.10	1.10	1.10	1.11
C_p (J K^{-1}mol^{-1})	134	134	135	135	135	136	136

Temp. (K)	320
c_p (J K^{-1}g^{-1})	1.11
C_p (J K^{-1}mol^{-1})	137

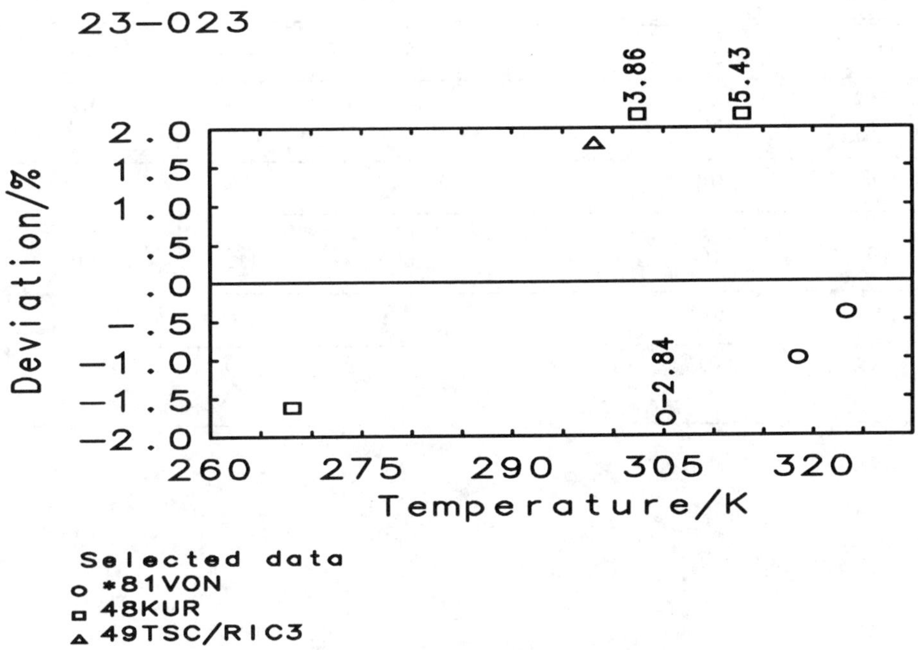

Name: 2–Bromopropane
Formula: C_3H_7Br

CAS–RN: 75–26–3
Group No.: 23–024
Molar Mass: 122.99

TABLE 23.24.1. Experimental heat capacities

Reference	Temp. range K	No. pnts	Error %	Purity %	Purity method	Type capacity	Calorimeter Type	Calorimeter Reference
*81VON	298.0–317.7	4S	nosp	not specified		C_{avg}	DSIO	*81VON
50KUS/CRO	190.1–209.6	7	1.70	not specified		C_p	BDHO	50KUS/CRO

TABLE 23.24.2. Correlated heat capacities

Reference	Temp. range K	No. pnts used	$\sigma_r C$ %	d_w	d/R	d_r %	d_b/R	+/–
				Selected data				
*81VON	298.0–317.7	4	2.00#	0.110	3.51–2	0.22	–4.75–3	0
50KUS/CRO	190.1–209.6	7	1.70	0.156	4.01–2	0.27	2.24–3	2

TABLE 23.24.3. Parameters of regression polynomial

Heat capacity type	No. data points total	used	s_w	s/R	s_r %	s_b/R	+/−
C_p	11	11	0.156	4.24–2	0.28	−3.04–4	2

Temp. range K	A_1	A_2	Level of uncertainty
190.1–317.7	1.34755+1	8.34816–1	V

TABLE 23.24.4. Recommended values of heat capacities

Temp. (K)	190	200	210	220	230	240	250
c_p (J K^{-1}g^{-1})	1.02	1.02	1.03	1.04	1.04	1.05	1.05
C_p (J K^{-1}mol^{-1})	125	126	127	127	128	129	129
Temp. (K)	260	270	273.15	280	290	298.15	300
c_p (J K^{-1}g^{-1})	1.06	1.06	1.07	1.07	1.07	1.08	1.08
C_p (J K^{-1}mol^{-1})	130	131	131	131	132	133	133
Temp. (K)	310	320					
c_p (J K^{-1}g^{-1})	1.09	1.09					
C_p (J K^{-1}mol^{-1})	134	134					

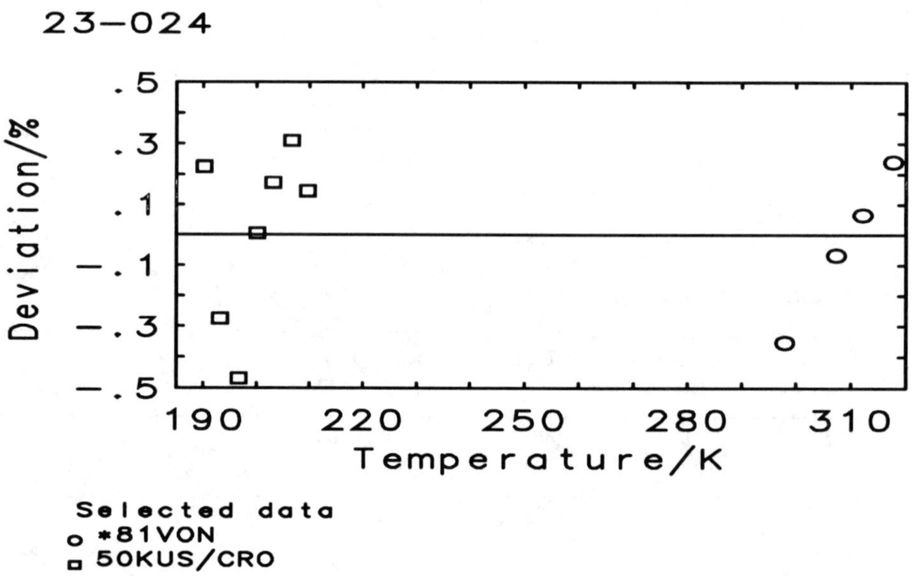

Name: 1–Bromobutane
Formula: C_4H_9Br

CAS–RN: 109–65–9
Group No.: 23–025
Molar Mass: 137.02

TABLE 23.25.1. Experimental heat capacities

Reference	Temp. range K	No. pnts	Error %	Purity %	Purity method	Type capacity	Calorimeter Type	Calorimeter Reference
31DEE	176.9–292.3	27	nosp	not	specified	C_p	BDAO	31DEE
48KUR	307.6–331.1	2	nosp	not	specified	C_{avg}	DSIO	47KUR
49TSC/RIC3	298.1	1	nosp	not	specified	C_p	BSIO	49TSC/RIC1

TABLE 23.25.2. Correlated heat capacities

Reference	Temp. range K	No. pnts used	$\sigma_r C$ %	d_w	d/R	d_r %	d_b/R	+/–
				Selected data				
31DEE	176.9–292.3	27	2.00#	0.830	2.80–1	1.66	–1.77–3	1
48KUR	307.6–331.1	2	5.00#	1.027	1.07	5.13	1.05	2
				Rejected data				
49TSC/RIC3	(1.85, 8.92, 1.85, 1)							

TABLE 23.25.3. Parameters of regression polynomial

Heat capacity type	No. data points total	No. data points used	s_w	s/R	s_r %	s_b/R	+/–
C_p	30	29	0.893	4.12–1	2.21	7.11–2	3

Temp. range K	A_1	A_2	A_3	Level of uncertainty
176.9–331.1	2.67701+1	–9.71229	2.37252	V

TABLE 23.25.4. Recommended values of heat capacities

Temp. (K)	180	190	200	210	220	230	240
c_p (J K^{-1}g^{-1})	1.03	1.02	1.02	1.02	1.02	1.03	1.04
C_p (J K^{-1}mol^{-1})	141	140	140	140	140	141	142
Temp. (K)	250	260	270	273.15	280	290	298.15
c_p (J K^{-1}g^{-1})	1.05	1.07	1.08	1.09	1.10	1.13	1.15
C_p (J K^{-1}mol^{-1})	144	146	148	149	151	154	157
Temp. (K)	300	310	320	330			
c_p (J K^{-1}g^{-1})	1.15	1.18	1.21	1.25			
C_p (J K^{-1}mol^{-1})	158	162	166	171			

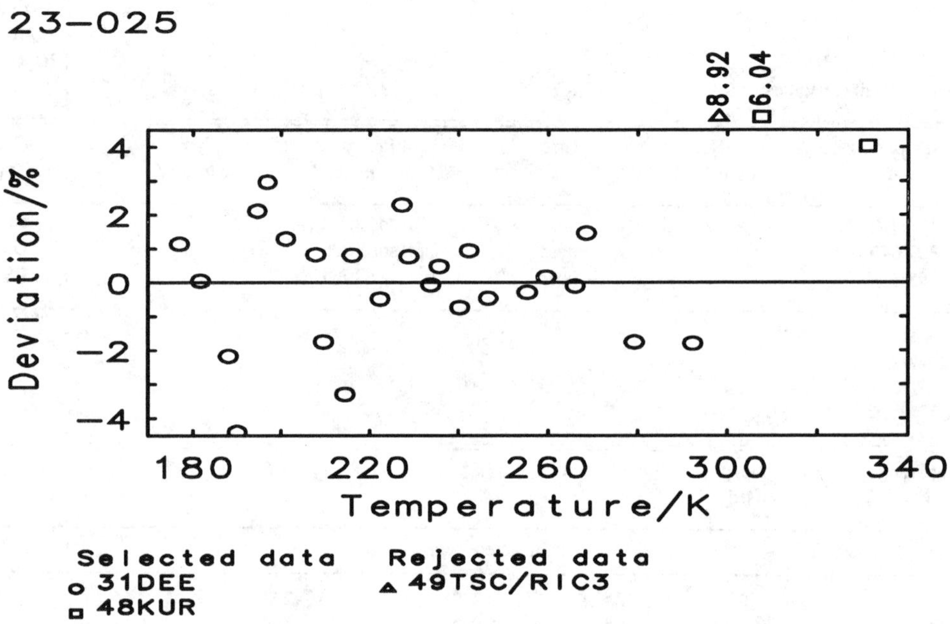

Name: 1–Bromo–2–methylpropane
Formula: C_4H_9Br

CAS–RN: 78–77–3
Group No.: 23–026
Molar Mass: 137.02

TABLE 23.26.1. Experimental heat capacities

Reference	Temp. range K	No. pnts	Error %	Purity %	Purity method	Type capacity	Calorimeter Type	Calorimeter Reference
48KUR	306.1–319.1	2	nosp	not specified		C_{avg}	DSIO	47KUR

TABLE 23.26.3. Parameters of regression polynomial

Heat capacity type	No. data points total	No. data points used	s_w	s/R	s_r %	s_b/R	+/–
C	2	2	0.000	0.00	0.00	0.00	0

Temp. range K	A_1	A_2	Level of uncertainty
306.1–319.1	7.64004	3.65968	VI

TABLE 23.26.4. Recommended values of heat capacities

Temp. (K)	310	320
c ($J K^{-1} g^{-1}$)	1.15	1.17
C ($J K^{-1} mol^{-1}$)	158	161

Name: 2–Bromo–2–methylpropane
Formula: C_4H_9Br

CAS–RN: 507–19–7
Group No.: 23–027
Molar Mass: 137.02

TABLE 23.27.1. Experimental heat capacities

Reference	Temp. range K	No. pnts	Error %	Purity %	Purity method	Type capacity	Calorimeter Type	Calorimeter Reference
50KUS/CRO	259.6–265.1	3	2.00	not specified		C_p	BDHO	50KUS/CRO

TABLE 23.27.3. Parameters of regression polynomial

Heat capacity type	No. data points total	No. data points used	s_w	s/R	s_r %	s_b/R	+/–
C_p	3	3	0.062	2.24–2	0.12	1.84–5	1

Temp. range K	A_1	A_2	Level of uncertainty
259.6–265.1	1.58127+1	8.83802–1	V

TABLE 23.27.4. Recommended values of heat capacities

Temp. (K)	260	265
c_p (J K^{-1}g^{-1})	1.10	1.10
C_p (J K^{-1}mol^{-1})	151	151

Name: 1,3–Dibromo–2,2–bis(bromomethyl)propane
Formula: $C_5H_8Br_4$

CAS–RN: 3229–00–3
Group No.: 23–028
Molar Mass: 387.73

TABLE 23.28.1. Experimental heat capacities

Reference	Temp. range K	No. pnts	Error %	Purity %	Purity method	Type capacity	Calorimeter Type	Calorimeter Reference
65CLE/WON	440.3–459.8	5	0.15	99.67	melpt	C_{sat}	BSAO	53WES/HAT

TABLE 23.28.3. Parameters of regression polynomial

Heat capacity type	No. data points total	No. data points used	s_w	s/R	s_r %	s_b/R	+/–
C_{sat}	5	5	0.296	1.67–2	0.04	–1.53–6	1

Temp. range K	A_1	A_2	A_3	Level of uncertainty
440.3–459.8	2.66935+2	–1.05536+2	1.21291+1	III

TABLE 23.28.4. Recommended values of heat capacities

Temp. (K)	440	450	460
c_{sat} (J K^{-1}g^{-1})	0.802	0.807	0.817
C_{sat} (J K^{-1}mol^{-1})	310.9	312.9	316.9

Name: 1–Bromo–3–methylbutane
Formula: $C_5H_{11}Br$

CAS–RN: 107–82–4
Group No.: 23–029
Molar Mass: 151.05

TABLE 23.29.1. Experimental heat capacities

Reference	Temp. range K	No. pnts	Error %	Purity %	Purity method	Type capacity	Calorimeter Type	Calorimeter Reference
48KUR	306.9–330.3	2	nosp	not specified		C_{avg}	DSIO	47KUR

TABLE 23.29.3. Parameters of regression polynomial

Heat capacity type	No. data points total	No. data points used	s_w	s/R	s_r %	s_b/R	+/–
C	2	2	0.000	0.00	0.00	0.00	0

Temp. range K	A_1	A_2	Level of uncertainty
306.9–330.3	7.74454	4.90486	VI

TABLE 23.29.4. Recommended values of heat capacities

Temp. (K)	310	320	330
c ($J K^{-1} g^{-1}$)	1.26	1.29	1.32
C ($J K^{-1} mol^{-1}$)	191	195	199

Name: 1–Bromopentane
Formula: $C_5H_{11}Br$

CAS–RN: 110–53–2
Group No.: 23–030
Molar Mass: 151.05

TABLE 23.30.1. Experimental heat capacities

Reference	Temp. range K	No. pnts	Error %	Purity %	Purity method	Type capacity	Calorimeter Type	Calorimeter Reference
31DEE	195.8–290.7	10	nosp	not specified		C_p	BDAO	31DEE
50KUS/CRO	190.1–206.6	6	1.50	not specified		C_p	BDHO	50KUS/CRO

TABLE 23.30.2. Correlated heat capacities

Reference	Temp. range K	No. pnts used	$\sigma_r C$ %	d_w	d/R	d_r %	d_b/R	+/–
				Selected data				
31DEE	195.8–290.7	10	2.00#	1.075	4.25–1	2.15	–2.28–1	–4
50KUS/CRO	190.1–206.6	6	1.50	1.165	3.66–1	1.75	2.44–1	2

TABLE 23.30.3. Parameters of regression polynomial

Heat capacity type	No. data points total	used	s_w	s/R	s_r %	s_b/R	+/–
C_p	16	16	1.231	4.48–1	2.23	–5.12–2	–2

Temp. range K	A_1	A_2	A_3	Level of uncertainty
190.1–290.7	3.46947+1	–1.22464+1	2.56531	V

TABLE 23.30.4. Recommended values of heat capacities

Temp. (K)	190	200	210	220	230	240	250
c_p (J K^{-1}g^{-1})	1.14	1.13	1.12	1.11	1.11	1.11	1.11
C_p (J K^{-1}mol^{-1})	172	170	169	168	167	167	167

Temp. (K)	260	270	273.15	280	290	298.15
c_p (J K^{-1}g^{-1})	1.11	1.12	1.12	1.13	1.14	1.16
C_p (J K^{-1}mol^{-1})	168	169	169	171	173	174

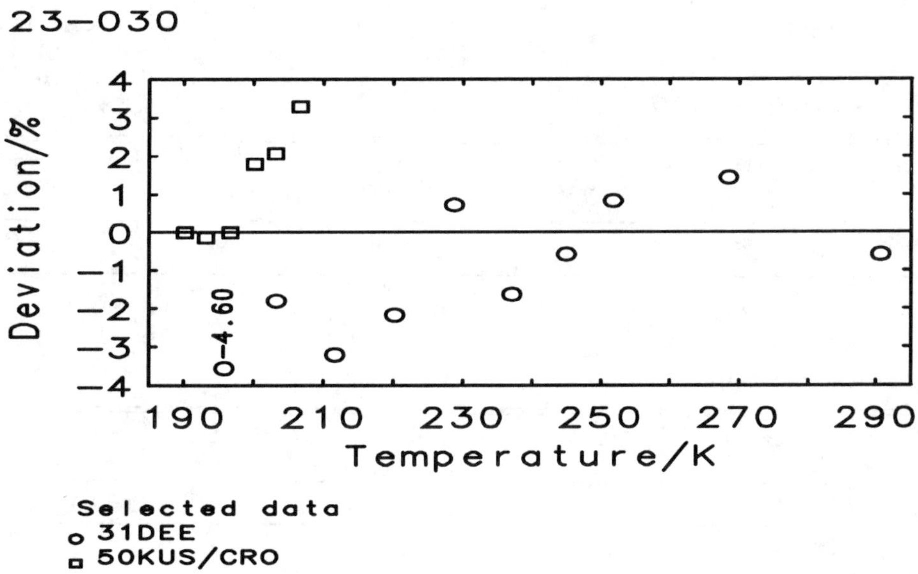

Name: 1,2–Dibromobenzene
Formula: $C_6H_4Br_2$

CAS–RN: 583–53–9
Group No.: 23–031
Molar Mass: 235.91

TABLE 23.31.1. Experimental heat capacities

Reference	Temp. range K	No. pnts	Error %	Purity %	Purity method	Type capacity	Calorimeter Type	Calorimeter Reference
18NAR	291.5–326.5	4S	nosp	not specified		C_{avg}	DSIO	18NAR

TABLE 23.31.3. Parameters of regression polynomial

Heat capacity type	No. data points total	used	s_w	s/R	s_r %	s_b/R	+/–
C	4	4	0.035	3.97–2	0.18	6.77–5	0

Temp. range K	A_1	A_2	Level of uncertainty
291.5–326.5	1.41109+1	2.64264	VI

TABLE 23.31.4. Recommended values of heat capacities

Temp. (K)	290	298.15	300	310	320	330
c ($J\,K^{-1}g^{-1}$)	0.77	0.78	0.78	0.79	0.80	0.80
C ($J\,K^{-1}mol^{-1}$)	181	183	183	185	188	190

Name: 1,3–Dibromobenzene
Formula: $C_6H_4Br_2$

CAS–RN: 108–36–1
Group No.: 23–032
Molar Mass: 235.91

TABLE 23.32.1. Experimental heat capacities

Reference	Temp. range K	No. pnts	Error %	Purity %	Purity method	Type capacity	Calorimeter Type	Calorimeter Reference
18NAR	290.9–326.4	4S	nosp	not specified		C_{avg}	DSIO	18NAR

TABLE 23.32.3. Parameters of regression polynomial

Heat capacity type	No. data points total	used	s_w	s/R	s_r %	s_b/R	+/–
C	4	4	0.105	1.16–1	0.52	6.10–4	0

Temp. range K	A_1	A_2	Level of uncertainty
290.9–326.4	1.43928+1	2.38844	VI

TABLE 23.32.4. Recommended values of heat capacities

Temp. (K)	290	298.15	300	310	320	330
c ($J\,K^{-1}g^{-1}$)	0.75	0.76	0.76	0.77	0.78	0.79
C ($J\,K^{-1}mol^{-1}$)	177	179	179	181	183	185

Name: Bromobenzene
Formula: C_6H_5Br

CAS–RN: 108–86–1
Group No.: 23–033
Molar Mass: 157.01

TABLE 23.33.1. Experimental heat capacities

Reference		Temp. range K	No. pnts	Error %	Purity %	Purity method	Type capacity	Calorimeter Type	Calorimeter Reference
*81VON		323.0–369.1	5S	nosp	not specified		C_{avg}	DSIO	*81VON
25WIL/DAN		293.1–353.1	4S	nosp	not specified		C_p	BSAO	24WIL/DAN
34KOL/UDO2	N	302.6	1	nosp	not specified		C_p	BSIT	34KOL/UDO2
37STU		250.0–320.0	8S	nosp	not specified		C_p	BDHO	37STU
49TSC/RIC3		298.1	1	nosp	not specified		C_p	BSIO	49TSC/RIC1
75MAS/SCO		243.3–296.4	17	nosp	99.998	melpt	C_{sat}	BSAO	45SCO/MEY

34KOL/UDO2 same datum in 34KOL/UDO1

TABLE 23.33.2. Correlated heat capacities

Reference	Temp. range K	No. pnts used	$\sigma_r C$ %	d_w	d/R	d_r %	d_b/R	+/–
				Selected data				
*81VON	323.0–369.1	5	2.00#	0.090	3.48–2	0.18	–1.82–2	–1
75MAS/SCO	243.3–296.4	17	0.20#	0.324	1.15–2	0.06	6.92–5	–1
				Rejected data				
25WIL/DAN	(4.98–1, 2.64, –4.72–1, –4)			34KOL/UDO2	(1.15, 6.60, –1.15, –1)			
37STU	(3.03–1, 1.65, 2.37–1, 6)			49TSC/RIC3	(4.99–1, 2.62, 4.99–1, 1)			

TABLE 23.33.3. Parameters of regression polynomial

Heat capacity type	No. data points total	No. data points used	s_w	s/R	s_r %	s_b/R	+/–
C	36	22	0.319	2.15–2	0.11	–4.09–3	–2

Temp. range K	A_1	A_2	A_3	A_4	Level of uncertainty
243.3–369.1	2.61761+1	–1.16331+1	4.40349	–4.56322–1	IV

TABLE 23.33.4. Recommended values of heat capacities

Temp. (K)	250	260	270	273.15	280	290	298.15
c (J K^{-1}g^{-1})	0.926	0.936	0.947	0.951	0.959	0.971	0.982
C (J K^{-1}mol^{-1})	145.4	147.0	148.7	149.3	150.6	152.5	154.2
Temp. (K)	300	310	320	330	340	350	360
c (J K^{-1}g^{-1})	0.984	0.998	1.011	1.024	1.038	1.051	1.063
C (J K^{-1}mol^{-1})	154.5	156.6	158.7	160.8	162.9	164.9	166.9
Temp. (K)	370						
c (J K^{-1}g^{-1})	1.075						
C (J K^{-1}mol^{-1})	168.8						

TABLE 23.33.5. Parameters of quasi-polynomial equation

Heat capacity type	No. data points total	used	s_w	s/R	s_r %	s_b/R	+/–
C	36	22	0.536	1.95–2	0.11	6.01–3	5

Temp. range K	T_c K	A_1	A_2	A_3	A_4	Level of uncertainty
243.3–369.1	670.00	–4.09271	7.36012–1	1.22557+1	5.68953	IV

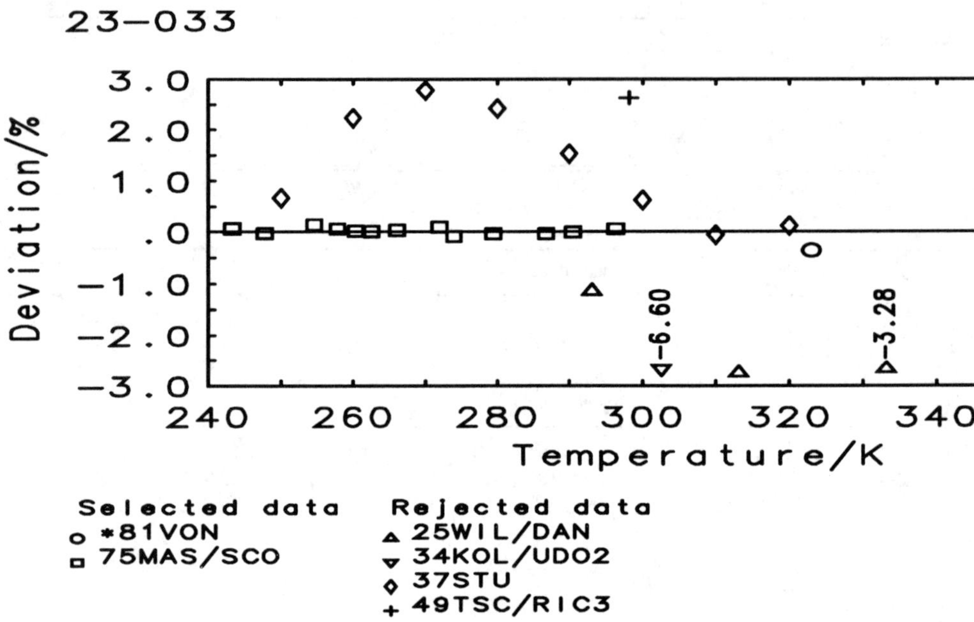

Name: 1–Bromohexane
Formula: $C_6H_{13}Br$

CAS–RN: 111–25–1
Group No.: 23–034
Molar Mass: 165.07

TABLE 23.34.1. Experimental heat capacities

Reference	Temp. range K	No. pnts	Error %	Purity %	Purity method	Type capacity	Calorimeter Type	Calorimeter Reference
31DEE	214.0–289.8	12	nosp	not specified		C_p	BDAO	31DEE
49TSC/RIC3	298.1	1	nosp	not specified		C_p	BSIO	49TSC/RIC1

TABLE 23.34.2. Correlated heat capacities

Reference	Temp. range K	No. pnts used	$\sigma_r C$ %	d_w	d/R	d_r %	d_b/R	+/–
				Selected data				
31DEE	214.0–289.8	12	2.00#	0.470	2.21–1	0.94	4.15–3	2

TABLE 23.34.3. Parameters of regression polynomial

Heat capacity type	No. data points total	used	s_w	s/R	s_r %	s_b/R	+/–
C_p	13	12	0.515	2.42–1	1.03	4.15–3	2

Temp. range K	A_1	A_2	Level of uncertainty
214.0–289.8	1.95961+1	1.46585	V

TABLE 23.34.4. Recommended values of heat capacities

Temp. (K)	210	220	230	240	250	260	270
c_p (J $K^{-1}g^{-1}$)	1.14	1.15	1.16	1.16	1.17	1.18	1.19
C_p (J $K^{-1}mol^{-1}$)	189	190	191	192	193	195	196

Temp. (K)	273.15	280	290	298.15
c_p (J $K^{-1}g^{-1}$)	1.19	1.19	1.20	1.21
C_p (J $K^{-1}mol^{-1}$)	196	197	198	199

TABLE 23.34.5. Parameters of quasi-polynomial equation

Heat capacity type	No. data points total	used	s_w	s/R	s_r %	s_b/R	+/–
C_p	13	12	0.543	2.55–1	1.09	4.15–3	2

Temp. range K	T_c K	A_1	A_2	A_3	A_4	Level of uncertainty
214.0–289.8	629.00	–7.12945–1	1.59151–2	1.96982+1	7.98442	V

Name: 3–Bromohexane
Formula: $C_6H_{13}Br$

CAS-RN: 3377–87–5
Group No.: 23–035
Molar Mass: 165.07

TABLE 23.35.1. Experimental heat capacities

Reference	Temp. range K	No. pnts	Error %	Purity %	Purity method	Type capacity	Calorimeter Type	Calorimeter Reference
88MEL/VER	213.1–523.2	10	2.50	not specified		C_{sat}	not specified	

TABLE 23.35.3. Parameters of regression polynomial

Heat capacity type	No. data points total	used	s_w	s/R	s_r %	s_b/R	+/–
C_{sat}	10	10	0.330	2.59–1	0.82	3.42–3	0

Temp. range K	A_1	A_2	A_3	A_4	Level of uncertainty
213.1–523.2	1.90435+1	4.36607	–1.80177	3.55138–1	V

TABLE 23.35.4. Recommended values of heat capacities

Temp. (K)	210	220	230	240	250	260	270
c_{sat} (J K^{-1}g^{-1})	1.19	1.19	1.20	1.21	1.22	1.23	1.24
C_{sat} (J K^{-1}mol^{-1})	196	197	199	200	202	203	205
Temp. (K)	273.15	280	290	298.15	300	310	320
c_{sat} (J K^{-1}g^{-1})	1.25	1.26	1.27	1.28	1.29	1.30	1.32
C_{sat} (J K^{-1}mol^{-1})	206	207	210	212	212	215	218
Temp. (K)	330	340	350	360	370	380	390
c_{sat} (J K^{-1}g^{-1})	1.34	1.36	1.38	1.41	1.44	1.47	1.50
C_{sat} (J K^{-1}mol^{-1})	221	225	228	233	237	242	247
Temp. (K)	400	410	420	430	440	450	460
c_{sat} (J K^{-1}g^{-1})	1.53	1.57	1.61	1.65	1.69	1.74	1.79
C_{sat} (J K^{-1}mol^{-1})	253	259	265	272	280	287	296
Temp. (K)	470	480	490	500	510	520	
c_{sat} (J K^{-1}g^{-1})	1.85	1.90	1.96	2.03	2.09	2.16	
C_{sat} (J K^{-1}mol^{-1})	305	314	324	334	346	357	

TABLE 23.35.5. Parameters of quasi-polynomial equation

Heat capacity type	No. data points total	used	s_w	s/R	s_r %	s_b/R	+/–
C_{sat}	10	10	0.552	3.88–1	1.38	6.16–3	0
Temp. range K	T_c K	A_1	A_2	A_3	A_4		Level of uncertainty
213.1–523.2	608.00	–5.18058	1.98483	1.67436+1	3.38044		V

Name: 1–Bromoheptane
Formula: $C_7H_{15}Br$

CAS–RN: 629–04–9
Group No.: 23–036
Molar Mass: 179.10

TABLE 23.36.1. Experimental heat capacities

Reference	Temp. range K	No. pnts	Error %	Purity %	Purity method	Type capacity	Calorimeter Type	Calorimeter Reference
50CRO/SMY1	222.1–231.1	4	1.00	not specified		C_p	BDHO	50KUS/CRO

TABLE 23.36.3. Parameters of regression polynomial

Heat capacity type	No. data points total	used	s_w	s/R	s_r %	s_b/R	+/–
C_p	4	4	0.228	5.40–2	0.23	1.23–4	0
Temp. range K		A_1	A_2				Level of uncertainty
222.1–231.1		1.86821+1	2.18682				IV

TABLE 23.36.4. Recommended values of heat capacities

Temp. (K)	220	230
c_p (J $K^{-1}g^{-1}$)	1.091	1.101
C_p (J $K^{-1}mol^{-1}$)	195.3	197.2

Name: 1–Bromooctane
Formula: $C_8H_{17}Br$

CAS–RN: 111–83–1
Group No.: 23–037
Molar Mass: 193.13

TABLE 23.37.1. Experimental heat capacities

Reference	Temp. range K	No. pnts	Error %	Purity %	Purity method	Type capacity	Calorimeter Type	Calorimeter Reference
50CRO/SMY1	222.1–234.1	3	1.00	not specified		C_p	BDHO	50KUS/CRO

TABLE 23.37.3. Parameters of regression polynomial

Heat capacity type	No. data points total	used	s_w	s/R	s_r %	s_b/R	+/–
C_p	3	3	0.093	2.91–2	0.09	3.50–5	–1

Temp. range K	A_1	Level of uncertainty
222.1–234.1	3.13671+1	IV

TABLE 23.37.4. Recommended values of heat capacities

Temp. (K)	220	230
c_p (J $K^{-1}g^{-1}$)	1.350	1.350
C_p (J $K^{-1}mol^{-1}$)	260.8	260.8

Name: 1–Bromononane
Formula: $C_9H_{19}Br$

CAS–RN: 693–58–3
Group No.: 23–038
Molar Mass: 207.15

TABLE 23.38.1. Experimental heat capacities

Reference	Temp. range K	No. pnts	Error %	Purity %	Purity method	Type capacity	Calorimeter Type	Calorimeter Reference
50CRO/SMY1	248.6–257.1	4	1.00	not specified		C_p	BDHO	50KUS/CRO

TABLE 23.38.3. Parameters of regression polynomial

Heat capacity type	No. data points total	used	s_w	s/R	s_r %	s_b/R	+/–
C_p	4	4	0.300	1.04–1	0.30	4.67–4	0

Temp. range K	A_1	Level of uncertainty
248.6–257.1	3.45579+1	IV

TABLE 23.38.4. Recommended values of heat capacities

Temp. (K)	250	260
c_p (J $K^{-1}g^{-1}$)	1.387	1.387
C_p (J $K^{-1}mol^{-1}$)	287.3	287.3

Name: 1–Bromodocosane
Formula: $C_{22}H_{45}Br$

CAS–RN: 6938–66–5
Group No.: 23–039
Molar Mass: 389.50

TABLE 23.39.1. Experimental heat capacities

Reference	Temp. range K	No. pnts	Error %	Purity %	Purity method	Type capacity	Calorimeter Type	Calorimeter Reference
53HOF/DEC	318.1–333.1	3	4.00	90.	estim	C_p	BDHO	52HOF

TABLE 23.39.3. Parameters of regression polynomial

Heat capacity type	No. data points total	No. data points used	s_w	s/R	s_r %	s_b/R	+/–
C_p	3	3	0.752	3.00	3.01	1.19–1	–1

Temp. range K	A_1	Level of uncertainty
318.1–333.1	9.85362+1	VI

TABLE 23.39.4. Recommended values of heat capacities

Temp. (K)	320	330
c_p (J $K^{-1}g^{-1}$)	2.10	2.10
C_p (J $K^{-1}mol^{-1}$)	819	819

Name: 1–Bromotriacontane
Formula: $C_{30}H_{61}Br$

CAS–RN: 4209–22–7
Group No.: 23–040
Molar Mass: 501.72

TABLE 23.40.1. Experimental heat capacities

Reference	Temp. range K	No. pnts	Error %	Purity %	Purity method	Type capacity	Calorimeter Type	Calorimeter Reference
53HOF/DEC	343.1–353.1	2	4.00	90.	estim	C_p	BDHO	52HOF

TABLE 23.40.3. Parameters of regression polynomial

Heat capacity type	No. data points total	used	s_w	s/R	s_r %	s_b/R	+/−
C_p	2	2	0.000	0.00	0.00	0.00	0
Temp. range K		A_1	A_2				Level of uncertainty
343.1–353.1		4.71744+1	2.52474+1				VI

TABLE 23.40.4. Recommended values of heat capacities

Temp. (K)	340	350
c_p (J K^{-1}g^{-1})	2.20	2.25
C_p (J K^{-1}mol^{-1})	1110	1130

24. Iodine Derivatives

As iodine derivatives have not been studied with much frequency, there are only 9 such compounds in this family for which C_p data were reported at several temperatures.

For most compounds, the recommended data have been derived from old measurements (*81BER/OGI, *81VON, 18NAR) performed in German laboratories or they relate to studies at PTIL (48KUR) which are of limited and uncertain value, however, for many iodine compounds no other sources were available. The only extensive measurements with good accuracy (reported error of 0.5 %) were carried out for iodomethane at UCE (57HAR/MOE, 61LOW/MOE) and with somewhat lower accuracy for iodobenzene at JHUB (37STU, 61ROU) showing a reported error between 1 and 2 %.

Name: Diiodomethane
Formula: CH_2I_2

CAS-RN: 75–11–6
Group No.: 24–001
Molar Mass: 267.84

TABLE 24.1.1. Experimental heat capacities

Reference	Temp. range K	No. pnts	Error %	Purity %	Purity method	Type capacity	Calorimeter Type	Calorimeter Reference
48KUR	307.1–363.0	3	nosp	not specified		C_{avg}	DSIO	47KUR

TABLE 24.1.3. Parameters of regression polynomial

Heat capacity type	No. data points total	No. data points used	s_w	s/R	s_r %	s_b/R	+/–
C	3	3	0.062	5.20–2	0.31	9.35–5	–1

Temp. range K	A_1	A_2	Level of uncertainty
307.1–363.0	1.02351+1	1.94967	VI

TABLE 24.1.4. Recommended values of heat capacities

Temp. (K)	310	320	330	340	350	360
c (J K^{-1}g^{-1})	0.51	0.51	0.52	0.52	0.53	0.54
C (J K^{-1}mol^{-1})	135	137	139	140	142	143

Name: Iodomethane
Formula: CH_3I

CAS-RN: 74–88–4
Group No.: 24–002
Molar Mass: 141.94

TABLE 24.2.1. Experimental heat capacities

Reference	Temp. range K	No. pnts	Error %	Purity %	Purity method	Type capacity	Calorimeter Type	Calorimeter Reference
48KUR	254.9–299.6	5	nosp	not specified		C_{avg}	DSIO	47KUR
57HAR/MOE	243.4–303.2	8	0.50	not specified		C_p	BSIO	57HAR/MOE
62LOW/MOE	253.2–308.2	7S	0.50	not specified		C_p	BSIO	57HAR/MOE

TABLE 24.2.2. Correlated heat capacities

Reference	Temp. range K	No. pnts used	$\sigma_r C$ %	d_w	d/R	d_r %	d_b/R	+/–
				Selected data				
57HAR/MOE	243.4–303.2	8	0.50	0.735	3.59–2	0.37	–8.09–3	–2
62LOW/MOE	253.2–308.2	7	0.50	0.446	2.19–2	0.22	9.65–3	3
				Rejected data				
48KUR	(4.98, 33.66, 4.97, 5)							

TABLE 24.2.3. Parameters of regression polynomial

Heat capacity type	No. data points total	used	s_w	s/R	s_r %	s_b/R	+/–
C_p	20	15	0.457	2.24–2	0.23	6.74–5	4
Temp. range K	A_1		A_2	A_3	A_4		Level of uncertainty
243.4–308.2	–5.08573+1		7.25074+1	–2.87128+1	3.76706		IV

TABLE 24.2.4. Recommended values of heat capacities

Temp. (K)	240	250	260	270	273.15	280	290
c_p (J K^{-1}g^{-1})	0.577	0.575	0.573	0.571	0.571	0.571	0.575
C_p (J K^{-1}mol^{-1})	81.90	81.62	81.26	81.01	80.99	81.06	81.60
Temp. (K)	298.15	300	310				
c_p (J K^{-1}g^{-1})	0.581	0.583	0.598				
C_p (J K^{-1}mol^{-1})	82.53	82.82	84.89				

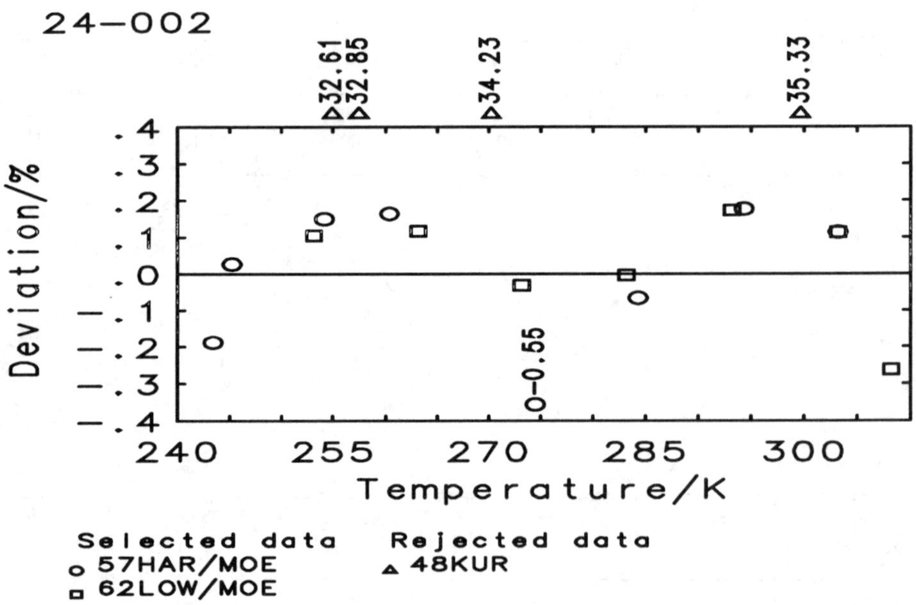

Name: Iodoethane
Formula: C_2H_5I

CAS–RN: 75–03–6
Group No.: 24–003
Molar Mass: 155.97

TABLE 24.3.1. Experimental heat capacities

Reference	Temp. range K	No. pnts	Error %	Purity %	Purity method	Type capacity	Calorimeter Type	Calorimeter Reference
*81BER/OGI	243.1–333.1	4S	nosp	not specified		C_{avg}	DSIO	*79BER
48KUR	263.4–316.6	3	nosp	not specified		C_{avg}	DSIO	47KUR

TABLE 24.3.2. Correlated heat capacities

Reference	Temp. range K	No. pnts used	$\sigma_r C$ %	d_w	d/R	d_r %	d_b/R	+/–
				Selected data				
*81BER/OGI	243.1–333.1	4	5.00#	0.003	1.75–3	0.01	4.77–7	0
				Rejected data				
48KUR	(7.86–1, 5.61, 7.29–1, 3)							

TABLE 24.3.3. Parameters of regression polynomial

Heat capacity type	No. data points total	No. data points used	s_w	s/R	s_r %	s_b/R	+/–
C	7	4	0.004	2.48–3	0.02	4.77–7	0

Temp. range K	A_1	A_2	Level of uncertainty
243.1–333.1	9.15537	1.29237	VI

TABLE 24.3.4. Recommended values of heat capacities

Temp. (K)	240	250	260	270	273.15	280	290
c ($J K^{-1} g^{-1}$)	0.65	0.66	0.67	0.67	0.68	0.68	0.69
C ($J K^{-1} mol^{-1}$)	102	103	104	105	105	106	107

Temp. (K)	298.15	300	310	320	330
c ($J K^{-1} g^{-1}$)	0.69	0.69	0.70	0.71	0.72
C ($J K^{-1} mol^{-1}$)	108	108	109	111	112

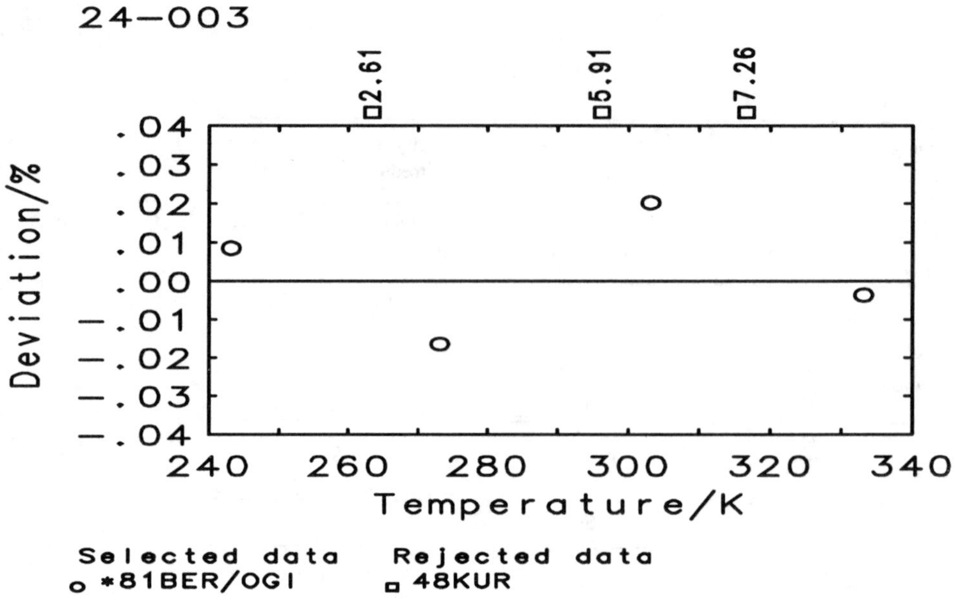

Name: 1–Iodopropane
Formula: C_3H_7I

CAS–RN: 107–08–4
Group No.: 24–004
Molar Mass: 169.99

TABLE 24.4.1. Experimental heat capacities

Reference	Temp. range K	No. pnts	Error %	Purity %	Purity method	Type capacity	Calorimeter Type	Calorimeter Reference
*81VON	317.3–339.1	4S	nosp	not specified		C_{avg}	DSIO	*81VON

TABLE 24.4.3. Parameters of regression polynomial

Heat capacity type	No. data points total	No. data points used	s_w	s/R	s_r %	s_b/R	+/–
C	4	4	0.183	6.17–2	0.37	2.21–4	0

Temp. range K	A_1	A_2	Level of uncertainty
317.3–339.1	9.06545	2.40243	VI

TABLE 24.4.4. Recommended values of heat capacities

Temp. (K)	320	330	340
c (J $K^{-1}g^{-1}$)	0.82	0.83	0.84
C (J $K^{-1}mol^{-1}$)	139	141	143

Name: 1–Iodo–2–methylpropane
Formula: C_4H_9I

CAS–RN: 513–38–2
Group No.: 24–005
Molar Mass: 184.02

TABLE 24.5.1. Experimental heat capacities

Reference	Temp. range K	No. pnts	Error %	Purity %	Purity method	Type capacity	Calorimeter Type	Calorimeter Reference
*81VON	315.0–344.5	3S	nosp	not specified		C_{avg}	DSIO	*81VON

TABLE 24.5.3. Parameters of regression polynomial

Heat capacity type	No. data points total	No. data points used	s_w	s/R	s_r %	s_b/R	+/–
C	3	3	0.000	0.00	0.00	0.00	0

Temp. range K	A_1	Level of uncertainty
315.0–344.5	2.05021+1	VI

TABLE 24.5.4. Recommended values of heat capacities

Temp. (K)	320	330	340
c ($J K^{-1} g^{-1}$)	0.93	0.93	0.93
C ($J K^{-1} mol^{-1}$)	170	170	170

Name: 1–Iodo–3–methylbutane
Formula: $C_5H_{11}I$

CAS–RN: 541–28–6
Group No.: 24–006
Molar Mass: 198.05

TABLE 24.6.1. Experimental heat capacities

Reference	Temp. range K	No. pnts	Error %	Purity %	Purity method	Type capacity	Calorimeter Type	Calorimeter Reference
48KUR	306.6–349.6	3	nosp	not specified		C_{avg}	DSIO	47KUR

TABLE 24.6.3. Parameters of regression polynomial

Heat capacity type	No. data points total	No. data points used	s_w	s/R	s_r %	s_b/R	+/–
C	3	3	0.003	3.78–3	0.02	–1.91–6	1

Temp. range K	A_1	A_2	Level of uncertainty
306.6–349.6	8.53242	4.49787	VI

TABLE 24.6.4. Recommended values of heat capacities

Temp. (K)	310	320	330	340	350
c ($J K^{-1} g^{-1}$)	0.94	0.96	0.98	1.00	1.02
C ($J K^{-1} mol^{-1}$)	187	191	194	198	202

Name: 1,2–Diiodobenzene
Formula: $C_6H_4I_2$

CAS–RN: 615–42–9
Group No.: 24–007
Molar Mass: 329.91

TABLE 24.7.1. Experimental heat capacities

Reference	Temp. range K	No. pnts	Error %	Purity %	Purity method	Type capacity	Calorimeter Type	Calorimeter Reference
18NAR	297.0–325.2	3S	nosp	not specified		C_{avg}	DSIO	18NAR

TABLE 24.7.3. Parameters of regression polynomial

Heat capacity type	No. data points total	No. data points used	s_w	s/R	s_r %	s_b/R	+/–
C	3	3	0.009	1.07–2	0.05	3.82–6	1

Temp. range K	A_1	A_2	Level of uncertainty
297.0–325.2	1.80372+1	1.59961	VI

TABLE 24.7.4. Recommended values of heat capacities

Temp. (K)	300	310	320	330
c (J $K^{-1}g^{-1}$)	0.58	0.58	0.58	0.59
C (J $K^{-1}mol^{-1}$)	190	191	193	194

Name: 1,3–Diiodobenzene
Formula: $C_6H_4I_2$

CAS–RN: 626–00–6
Group No.: 24–008
Molar Mass: 329.91

TABLE 24.8.1. Experimental heat capacities

Reference	Temp. range K	No. pnts	Error %	Purity %	Purity method	Type capacity	Calorimeter Type	Calorimeter Reference
18NAR	321.9–340.0	2	nosp	not specified		C_{avg}	DSIO	18NAR

TABLE 24.8.3. Parameters of regression polynomial

Heat capacity type	No. data points total	No. data points used	s_w	s/R	s_r %	s_b/R	+/–
C	2	2	0.000	0.00	0.00	0.00	0

Temp. range K	A_1	A_2	Level of uncertainty
321.9–340.0	1.39365+1	2.73652	VI

TABLE 24.8.4. Recommended values of heat capacities

Temp. (K)	320	330	340
c (J $K^{-1}g^{-1}$)	0.57	0.58	0.59
C (J $K^{-1}mol^{-1}$)	189	191	193

Name: Iodobenzene
Formula: C_6H_5I

CAS–RN: 591–50–4
Group No.: 24–009
Molar Mass: 204.01

TABLE 24.9.1. Experimental heat capacities

Reference	Temp. range K	No. pnts	Error %	Purity %	Purity method	Type capacity	Calorimeter Type	Calorimeter Reference
37STU	250.0–320.0	8S	nosp	not specified		C_p	BDHO	37STU
61ROU	301.3–316.2	7	nosp	not specified		C_p	BSAO	61ROU

TABLE 24.9.2. Correlated heat capacities

Reference	Temp. range K	No. pnts used	$\sigma_r C$ %	d_w	d/R	d_r %	d_b/R	+/–
Selected data								
37STU	250.0–320.0	8	2.00#	0.646	2.48–1	1.29	–1.48–1	–2
61ROU	301.3–316.2	7	1.00#	0.448	8.84–2	0.45	4.50–2	3

TABLE 24.9.3. Parameters of regression polynomial

Heat capacity type	No. data points total	No. data points used	s_w	s/R	s_r %	s_b/R	+/–
C_p	15	15	0.604	2.05–1	1.07	–5.81–2	1

Temp. range K	A_1	A_2	Level of uncertainty
250.0–320.0	1.42074+1	1.72619	V

TABLE 24.9.4. Recommended values of heat capacities

Temp. (K)	250	260	270	273.15	280	290	298.15
c_p (J K^{-1}g^{-1})	0.75	0.76	0.77	0.77	0.78	0.78	0.79
C_p (J K^{-1}mol^{-1})	154	155	157	157	158	160	161

Temp. (K)	300	310	320
c_p (J K^{-1}g^{-1})	0.79	0.80	0.80
C_p (J K^{-1}mol^{-1})	161	163	164

TABLE 24.9.5. Parameters of quasi-polynomial equation

Heat capacity type	No. data points total	No. data points used	s_w	s/R	s_r %	s_b/R	+/–
C_p	15	15	0.703	1.95–1	1.01	–1.54–2	3

Temp. range K	T_c K	A_1	A_2	A_3	A_4	Level of uncertainty
250.0–320.0	721.20	6.14866–1	8.06828–3	1.47785+1	1.17144+1	V

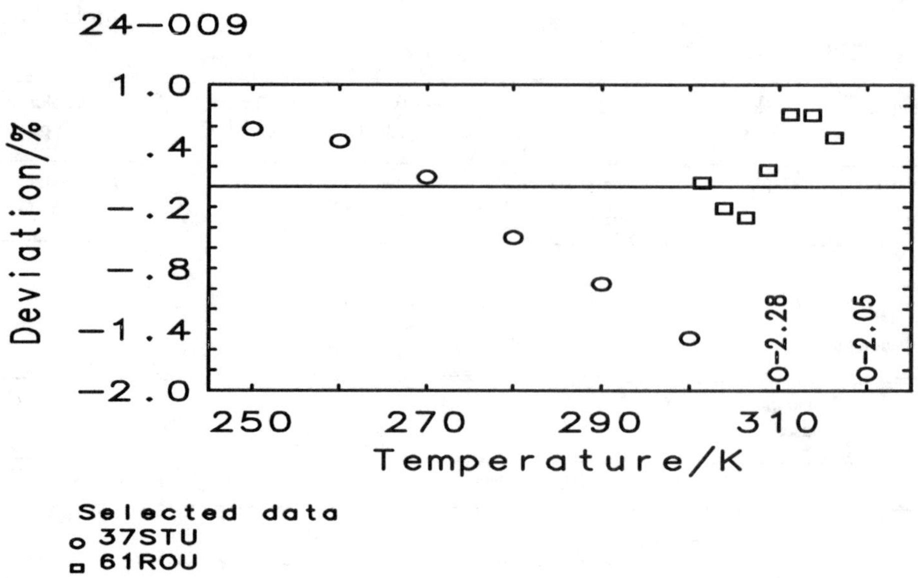
24–009
Deviation/%
1.0
.4
–.2
–.8
–1.4
–2.0
250
270
290
310
Temperature/K
O–2.28
O–2.05
Selected data
37STU
61ROU

25. Mixed Halogen Derivatives

This mixed halogen family consists of 35 compounds containing at least two different halogen atoms of which two were measured at one temperature only.

Some of these mixed halogen compounds belong to the Freon refrigerants. A large number of measurements over a wide temperature and pressure range reflects the wide use and importance of Freons. The recommendations were made previously by Makita (77MAK) as parameters of polynomial expressing the temperature dependence for the recommended data of 14 Freons. The equations are based on a data compilation containing both experimental and estimated data.

The most extensive measurements (8 compounds, with a reported error 0.2 to 0.4 %) were performed at MSUM (71KOL/VOR, 72KOL/VOR, 74KOL/VOR, 74VOR/KOL, 78KOS/KOL, 81KOL/KOS, 82KOS/ZHO) over a wide temperature range and have been all included in the final correlation (with one exception 78KOS/KOL). Another laboratory reporting data on several mixed halogen derivatives is ICTP where 7 compounds were measured between 298 and 318 K with a reported error below 0.5 % (88SVO/VES, 88VES/ZAB).

Regarding substituted methanes, the interest was focused on $CBrF_3$ and $CClF_3$ (both refrigerants) which were investigated over a wide temperature and pressure range in an insufficiently identified Russian laboratory (84STO/CHA); the data were published as parameters of a correlating equation only. The compound, $CClF_3$, was also investigated at KITH (41RIE2). The reliability of both previous sources seems to be substantially lower than the claimed accuracy. Another compound investigated at KITH was CCl_2F_2 (39RIE3) and three other halogenated methane derivatives were measured up to above the normal boiling temperature at KCIW (40BEN/MCH), resulting in data with apparently low reliability. Other substituted methanes were measured at the following institutions: apparently high quality data for CCl_3F at CITP (41OSB/GAR); good quality data for $CHClF_2$ at OSU (57NEI/WHI) with a claimed error 0.5 % which seems to be realistic as well as data for $CHCl_2F$ along the saturation curve up to 0.65 MPa at ITN (74GRU/SHE). The latter data were reported as isobaric heat capacities and were converted to C_{sat} before the correlation procedure. From the last source, data were also obtained for 1,2-dibromo-1,1,2,2-tetrafluoroethane.

Beside the laboratories mentioned above, the halogenated ethanes were measured with good accuracy at PSC (55AST/WIL) for C_2ClF_5 and at UOTO (78KIS/SUG) for 1,1,2-trichloro-1,2,2-trifluoroethane over a narrow temperature range . The latter compound was measured and yielded results with lower accuracy but extended to high temperatures at KCIW (40BEN/MCH). Two additional ethane derivatives used as refrigerants were measured at KITH: unspecified isomer of $C_2Cl_2F_4$ (37PER1) and 1-chloro-1,1-difluoroethane (37PER1, 41RIE3). The data from 1941 have been discarded as better data were available in the same temperature range for the latter substance from LISI (64VAS2) with a reported error of 0.5 %. Two less common refrigerants R122 (1,1,2-trichloro-1,2-difluoroethane) and R132 (1,1-dichloro-1,2-difluoroethane) have been recently investigated over a wide temperature interval at CIUG (92LEB/KUL).

Regarding the derivatives of aromatic hydrocarbons, data are available for C_6BrF_5, C_6ClF_5 and 1,3,5-trichloro-2,4,6-trifluorobenzene measured at IICN (69PAU/GLU2, 75PAU2). In the case of the latter two compounds, more reliable data from NPLT (68AND/COU1, 73AND/MAR2) have been preferred as they differed substantially from the data obtained at IICN (by 1 and 3 % resp.) in the above reference.

Name: Bromotrichloromethane
Formula: $CBrCl_3$

CAS–RN: 75–62–7
Group No.: 25–001
Molar Mass: 198.27

TABLE 25.1.1. Experimental heat capacities

Reference	Temp. K	Capac. J/(K.g)	Error %	Purity %	Purity method	Type capacity	Calorimeter Type	Calorimeter Reference
59BEN/THO	298.00	0.7531	nosp	not specified		C_{avg}	BDHT	59BEN/THO

Name: Bromotrifluoromethane
Formula: $CBrF_3$

CAS–RN: 75–63–8
Group No.: 25–002
Molar Mass: 148.91

TABLE 25.2.1. Experimental heat capacities

Reference	Temp. range K	No. pnts	Error %	Purity %	Purity method	Type capacity	Calorimeter Type	Calorimeter Reference
84STO/CHA	173.0–214.0	eqn	0.30	99.98	anal	C_p	BSAO	78SOL/SUK

TABLE 25.2.3. Parameters of regression polynomial

Heat capacity type	No. data points total	No. data points used	s_w	s/R	s_r %	s_b/R	+/–
C_p	15	15	0.002	7.14–5	0.00	–5.09–7	0

Temp. range K	A_1	A_2	A_3	A_4	Level of uncertainty
173.0–213.6	7.36402	4.71839	–2.32916	5.28139–1	IV

TABLE 25.2.4. Recommended values of heat capacities

Temp. (K)	170	180	190	200	210
c_p (J K^{-1}g^{-1})	0.628	0.636	0.645	0.654	0.664
C_p (J K^{-1}mol^{-1})	93.53	94.71	95.98	97.36	98.88

Name: Chlorotrifluoromethane
Formula: $CClF_3$

CAS–RN: 75–72–9
Group No.: 25–003
Molar Mass: 104.46

TABLE 25.3.1. Experimental heat capacities

Reference	Temp. range K	No. pnts	Error %	Purity %	Purity method	Type capacity	Calorimeter Type	Calorimeter Reference
41RIE2	184.5–275.8	8	5.00	not specified		C_{sat}	BSIO	39RIE3
84STO/CHA	163.0–192.0	eqn	0.30	99.98	anal	C_p	BSAO	78SOL/SUK

TABLE 25.3.2. Correlated heat capacities

Reference	Temp. range K	No. pnts used	$\sigma_r C$ %	d_w	d/R	d_r %	d_b/R	+/–
				Selected data				
41RIE2	184.6–275.8	8	5.00	0.612	3.23–1	3.06	–2.08–1	–4
84STO/CHA	163.0–192.0	6	0.30	0.081	2.66–3	0.02	1.10–3	1

TABLE 25.3.3. Parameters of regression polynomial

Heat capacity type	No. data points total	used	s_w	s/R	s_r %	s_b/R	+/–
C_p	14	14	0.551	2.89–1	2.74	–1.19–1	–3
C_{sat}	14	14	0.556	2.95–1	2.77	–1.11–1	–3

Temp. range K	A_1	A_2	A_3	A_4	Level of uncertainty
163.0–275.8	–6.02362	2.62709+1	–1.46529+1	2.93097	V
163.0–275.8	5.20781–1	1.53838+1	–8.59481	1.80149	V

TABLE 25.3.4. Recommended values of heat capacities

Temp. (K)	160	170	180	190	200	210	220
c_p (J $K^{-1}g^{-1}$)	0.84	0.85	0.87	0.88	0.90	0.93	0.96
C_p (J $K^{-1}mol^{-1}$)	87.3	88.9	90.5	92.3	94.4	97.0	100
c_{sat} (J $K^{-1}g^{-1}$)	0.84	0.85	0.87	0.88	0.90	0.92	0.95
C_{sat} (J $K^{-1}mol^{-1}$)	87.4	88.8	90.4	92.1	94.1	96.5	99.3
Temp. (K)	**230**	**240**	**250**	**260**	**270**	**273.15**	**280**
c_p (J $K^{-1}g^{-1}$)	1.00	1.05	1.10	1.17	1.26	1.28	1.35
C_p (J $K^{-1}mol^{-1}$)	104	109	115	123	131	134	141
c_{sat} (J $K^{-1}g^{-1}$)	0.98	1.02	1.07	1.12	1.18	1.20	1.25
C_{sat} (J $K^{-1}mol^{-1}$)	103	107	112	117	124	126	131

TABLE 25.3.5. Parameters of quasi-polynomial equation

Heat capacity type	No. data points total	used	s_w	s/R	s_r %	s_b/R	+/–
C_p	14	14	10.330	3.45–1	3.19	1.63–1	4

Temp. range K	T_c K	A_1	A_2	A_3	A_4	A_5	A_6	Level of uncertainty
163.0–275.8	302.00	–2.25339–1	2.90917–4	8.61948	1.06802+1	–3.29462+1	3.28366+1	V

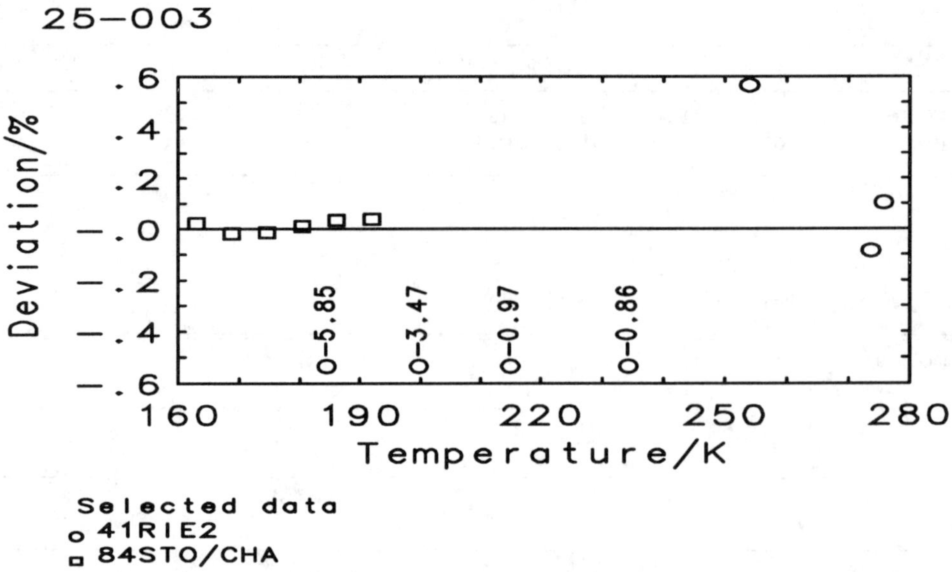

Name: Dichlorodifluoromethane
Formula: CCl_2F_2

CAS–RN: 75–71–8
Group No.: 25–004
Molar Mass: 120.91

TABLE 25.4.1. Experimental heat capacities

Reference		Temp. range K	No. pnts	Error %	Purity %	Purity method	Type capacity	Calorimeter Type	Calorimeter Reference
31BUF/FLE	N	230.1–290.1	2	4.00	not specified		C_{avg}	not specified	
39RIE3		194.4–293.5	12	1.00	not specified		C_{sat}	BSIO	39RIE3

31BUF/FLE the first datum from drop calorimetry; the second datum from isoperibol calorimetry

TABLE 25.4.2. Correlated heat capacities

Reference	Temp. range K	No. pnts used	$\sigma_r C$ %	d_w	d/R	d_r %	d_b/R	+/−
				Selected data				
39RIE3	194.4–293.5	12	1.00	0.456	5.95–2	0.46	5.54–4	2
				Rejected data				
31BUF/FLE	(9.39–1, 6.10, 6.53–1, 0)							

TABLE 25.4.3. Parameters of regression polynomial

Heat capacity type	No. data points total	No. data points used	s_w	s/R	s_r %	s_b/R	+/−
C_p	14	12	0.526	6.86–2	0.53	5.54–4	2
C_{sat}	14	12	0.511	6.65–2	0.51	5.17–4	2

Temp. range K	A_1	A_2	A_3	Level of uncertainty
194.4–293.5	1.47834+1	−3.20041	1.01807	V
194.4–293.5	1.38976+1	−2.35789	8.18319–1	V

TABLE 25.4.4. Recommended values of heat capacities

Temp. (K)	190	200	210	220	230	240	250
c_p (J K^{-1}g^{-1})	0.85	0.86	0.86	0.87	0.88	0.89	0.90
C_p (J K^{-1}mol^{-1})	103	104	104	105	106	108	109
c_{sat} (J K^{-1}g^{-1})	0.85	0.86	0.86	0.87	0.88	0.89	0.90
C_{sat} (J K^{-1}mol^{-1})	103	104	104	105	106	108	109
Temp. (K)	260	270	273.15	280	290	298.15	
c_p (J K^{-1}g^{-1})	0.92	0.93	0.94	0.95	0.97	0.98	
C_p (J K^{-1}mol^{-1})	111	113	113	115	117	119	
c_{sat} (J K^{-1}g^{-1})	0.91	0.93	0.93	0.94	0.96	0.97	
C_{sat} (J K^{-1}mol^{-1})	111	112	113	114	116	118	

TABLE 25.4.5. Parameters of quasi-polynomial equation

Heat capacity type	No. data points total	used	s_w	s/R	s_r %	s_b/R	+/−
C_p	14	12	0.493	6.42–2	0.49	4.76–4	0

Temp. range K	T_c K	A_1	A_2	A_3	A_4	Level of uncertainty
194.4–293.5	384.95	−7.73430–1	5.23607–1	1.06495+1	2.85612–1	V

Name: Trichlorofluoromethane
Formula: CCl_3F

CAS–RN: 75–69–4
Group No.: 25–005
Molar Mass: 137.37

TABLE 25.5.1. Experimental heat capacities

Reference	Temp. range K	No. pnts	Error %	Purity %	Purity method	Type capacity	Calorimeter Type	Calorimeter Reference
40BEN/MCH	261.4–347.5	4	nosp	not specified		C_{sat}	BSIO	40BEN/MCH
41OSB/GAR	170.6–287.8	13	0.10	99.95	melpt	C_{sat}	BSAO	41YOS/GAR

TABLE 25.5.2. Correlated heat capacities

Reference	Temp. range K	No. pnts used	$\sigma_r C$ %	d_w	d/R	d_r %	d_b/R	+/−
				Selected data				
40BEN/MCH	261.5–347.5	4	1.50#	0.882	1.92–1	1.32	1.10–1	0
41OSB/GAR	170.6–287.8	13	0.10	0.260	3.63–3	0.03	−1.41–4	0

TABLE 25.5.3. Parameters of regression polynomial

Heat capacity type	No. data points total	used	s_w	s/R	s_r %	s_b/R	+/−
C_p	17	17	0.554	1.07–1	0.73	2.57–2	0
C_{sat}	17	17	0.548	1.07–1	0.74	2.56–2	0

Temp. range K	A_1	A_2	A_3	A_4	Level of uncertainty
170.6–347.5	1.09354+1	1.96990	−5.70283–1	1.09750–1	IV
170.6–347.5	1.12836+1	1.46261	−3.26291–1	7.10062–2	IV

TABLE 25.5.4. Recommended values of heat capacities

Temp. (K)	170	180	190	200	210	220	230
c_p (J K^{-1}g^{-1})	0.797	0.803	0.809	0.815	0.822	0.828	0.834
C_p (J K^{-1}mol^{-1})	109.5	110.4	111.2	112.0	112.9	113.7	114.6
c_{sat} (J K^{-1}g^{-1})	0.798	0.803	0.809	0.815	0.822	0.828	0.834
C_{sat} (J K^{-1}mol^{-1})	109.6	110.4	111.2	112.0	112.9	113.7	114.6
Temp. (K)	240	250	260	270	273.15	280	290
c_p (J K^{-1}g^{-1})	0.841	0.848	0.855	0.863	0.865	0.871	0.879
C_p (J K^{-1}mol^{-1})	115.5	116.5	117.5	118.5	118.9	119.6	120.8
c_{sat} (J K^{-1}g^{-1})	0.841	0.848	0.855	0.863	0.865	0.870	0.878
C_{sat} (J K^{-1}mol^{-1})	115.5	116.5	117.5	118.5	118.8	119.6	120.7
Temp. (K)	298.15	300	310	320	330	340	350
c_p (J K^{-1}g^{-1})	0.887	0.888	0.898	0.908	0.918	0.929	0.941
C_p (J K^{-1}mol^{-1})	121.8	122.0	123.3	124.7	126.1	127.7	129.3
c_{sat} (J K^{-1}g^{-1})	0.885	0.887	0.896	0.905	0.914	0.925	0.935
C_{sat} (J K^{-1}mol^{-1})	121.6	121.8	123.0	124.3	125.6	127.0	128.5

TABLE 25.5.5. Parameters of quasi-polynomial equation

Heat capacity type	No. data points total	used	s_w	s/R	s_r %	s_b/R	+/−
C_p	17	17	4.809	9.84–2	0.70	2.01–2	–1

Temp. range K	T_c K	A_1	A_2	A_3	A_4	Level of uncertainty
170.6–347.5	471.20	–5.35921–1	1.44721–2	1.09985+1	4.96145	IV

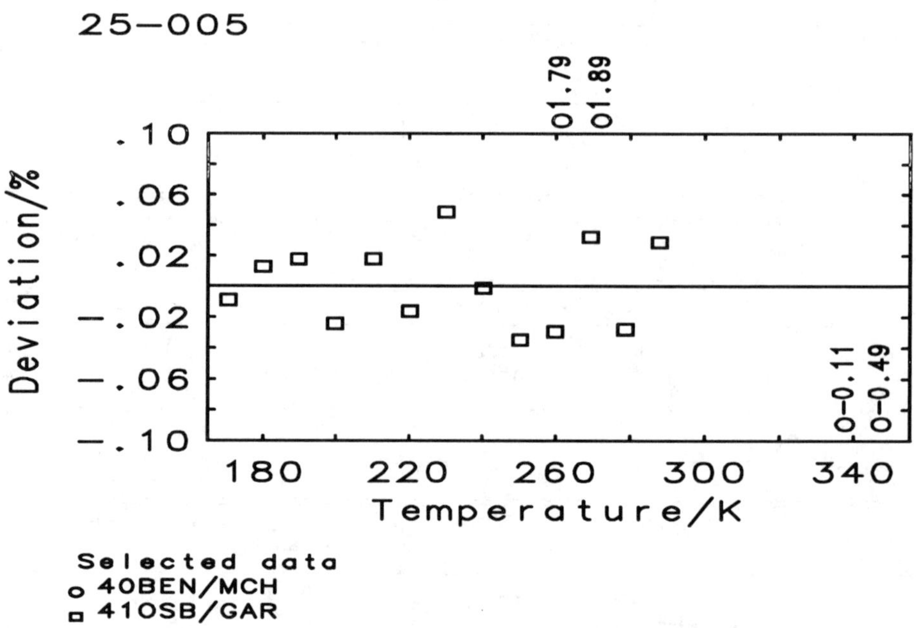

Name: Chlorodifluoromethane
Formula: $CHClF_2$

CAS-RN: 75–45–6
Group No.: 25–006
Molar Mass: 86.47

TABLE 25.6.1. Experimental heat capacities

Reference	Temp. range K	No. pnts	Error %	Purity %	Purity method	Type capacity	Calorimeter Type	Calorimeter Reference
40BEN/MCH	255.8–328.3	4	nosp	not specified		C_{sat}	BSIO	40BEN/MCH
57NEI/WHI	121.8–226.1	12	0.50	99.98	melpt	C_p	BSIO	50JOH/CLA

TABLE 25.6.2. Correlated heat capacities

Reference	Temp. range K	No. pnts used	$\sigma_r C$ %	d_w	d/R	d_r %	d_b/R	+/–
				Selected data				
40BEN/MCH	255.9–328.3	4	1.50#	0.537	1.11–1	0.81	4.72–2	2
57NEI/WHI	121.8–226.1	12	0.50	0.905	5.02–2	0.45	–1.19–3	0

TABLE 25.6.3. Parameters of regression polynomial

Heat capacity type	No. data points total	No. data points used	s_w	s/R	s_r %	s_b/R	+/–
C_p	16	16	0.957	8.14–2	0.65	1.09–2	2
C_{sat}	16	16	0.909	7.54–2	0.64	1.43–2	2

Temp. range K	A_1	A_2	A_3	A_4	Level of uncertainty
121.8–328.3	8.15765	5.79995	–3.78720	8.07200–1	V
121.8–328.3	9.17639	3.96913	–2.71469	6.01885–1	V

TABLE 25.6.4. Recommended values of heat capacities

Temp. (K)	120	130	140	150	160	170	180
c_p (J $K^{-1}g^{-1}$)	1.06	1.06	1.06	1.06	1.06	1.06	1.06
C_p (J $K^{-1}mol^{-1}$)	91.9	92.0	92.0	92.0	91.9	91.8	91.7
c_{sat} (J $K^{-1}g^{-1}$)	1.06	1.06	1.06	1.06	1.06	1.06	1.06
C_{sat} (J $K^{-1}mol^{-1}$)	92.0	92.0	92.0	91.9	91.8	91.8	91.8

Temp. (K)	190	200	210	220	230	240	250
c_p (J $K^{-1}g^{-1}$)	1.06	1.06	1.07	1.08	1.09	1.10	1.12
C_p (J $K^{-1}mol^{-1}$)	91.8	92.0	92.4	93.0	93.8	95.0	96.4
c_{sat} (J $K^{-1}g^{-1}$)	1.06	1.06	1.07	1.07	1.08	1.09	1.11
C_{sat} (J $K^{-1}mol^{-1}$)	91.8	92.0	92.4	92.9	93.7	94.7	95.9

Temp. (K)	260	270	273.15	280	290	298.15	300
c_p (J $K^{-1}g^{-1}$)	1.14	1.16	1.17	1.19	1.23	1.27	1.28
C_p (J $K^{-1}mol^{-1}$)	98.3	101	101	103	107	110	110
c_{sat} (J $K^{-1}g^{-1}$)	1.13	1.15	1.16	1.17	1.21	1.23	1.24
C_{sat} (J $K^{-1}mol^{-1}$)	97.5	99.4	100	102	104	107	107

Temp. (K)	310	320	330
c_p (J $K^{-1}g^{-1}$)	1.33	1.38	1.45
C_p (J $K^{-1}mol^{-1}$)	115	120	125
c_{sat} (J $K^{-1}g^{-1}$)	1.28	1.33	1.38
C_{sat} (J $K^{-1}mol^{-1}$)	111	115	119

TABLE 25.6.5. Parameters of quasi-polynomial equation

Heat capacity type	No. data points total	used	s_w	s/R	s_r %	s_b/R	+/–
C_p	16	16	1.184	7.68–2	0.67	1.40–2	6

Temp. range K	T_c K	A_1	A_2	A_3	A_4	A_5	A_6	Level of uncertainty
121.8–328.3	369.30	-2.31206	7.97997–2	9.76315	6.63517	-2.42557+1	1.81513+1	V

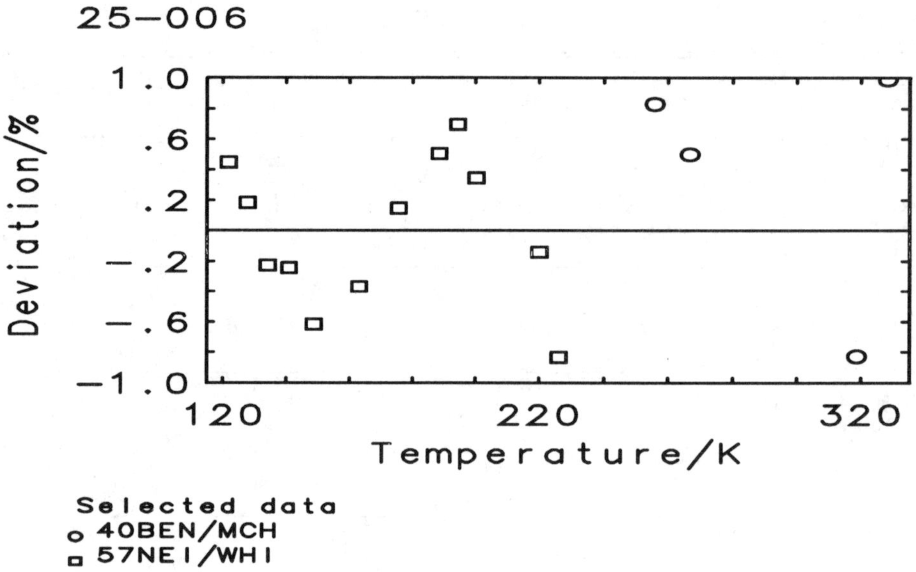

Name: Dichlorofluoromethane
Formula: $CHCl_2F$

CAS–RN: 75–43–4
Group No.: 25–007
Molar Mass: 102.92

TABLE 25.7.1. Experimental heat capacities

Reference		Temp. range K	No. pnts	Error %	Purity %	method	Type capacity	Calorimeter Type	Reference
40BEN/MCH		261.0–337.8	4	nosp	not specified		C_{sat}	BSIO	40BEN/MCH
74GRU/SHE	N	307.1–363.1	7S	1.00	99.77	chrom	C_p	BDHO	40TUR/BAR

74GRU/SHE C_p at saturation line (max. pressure 0.65 MPa)

TABLE 25.7.2. Correlated heat capacities

Reference	Temp. range K	No. pnts used	$\sigma_r C$ %	d_w	d/R	d_r %	d_b/R	+/–
				Selected data				
40BEN/MCH	261.0–337.9	4	1.50#	0.120	2.42–2	0.18	–6.40–3	0
74GRU/SHE	307.1–363.1	7	1.00	0.090	1.25–2	0.09	1.67–3	1

TABLE 25.7.3. Parameters of regression polynomial

Heat capacity type	No. data points total	used	s_w	s/R	s_r %	s_b/R	+/−
C_p	11	11	0.119	2.07–2	0.15	−1.26–3	1
C_{sat}	11	11	0.093	1.70–2	0.13	−6.19–4	−1

Temp. range K	A_1	A_2	A_3	Level of uncertainty
261.0–363.1	1.94294+1	−5.41461	1.08982	V
261.0–363.1	1.71561+1	−3.77772	7.94319–1	V

TABLE 25.7.4. Recommended values of heat capacities

Temp. (K)	260	270	273.15	280	290	298.15	300
c_p (J K^{-1}g^{-1})	1.03	1.03	1.03	1.04	1.04	1.05	1.05
C_p (J K^{-1}mol^{-1})	106	106	106	107	107	108	108
c_{sat} (J K^{-1}g^{-1})	1.03	1.03	1.03	1.03	1.04	1.05	1.05
C_{sat} (J K^{-1}mol^{-1})	106	106	106	106	107	108	108

Temp. (K)	310	320	330	340	350	360
c_p (J K^{-1}g^{-1})	1.06	1.07	1.08	1.10	1.12	1.14
C_p (J K^{-1}mol^{-1})	109	110	112	113	115	117
c_{sat} (J K^{-1}g^{-1})	1.06	1.07	1.08	1.09	1.10	1.12
C_{sat} (J K^{-1}mol^{-1})	109	110	111	112	114	115

TABLE 25.7.5. Parameters of quasi-polynomial equation

Heat capacity type	No. data points total	used	s_w	s/R	s_r %	s_b/R	+/−
C_p	11	11	0.101	1.71–2	0.13	−1.74–3	2

Temp. range K	T_c K	A_1	A_2	A_3	A_4	Level of uncertainty
261.0–363.1	451.60	−7.51806–2	5.13431–1	1.14156+1	2.75213–3	V

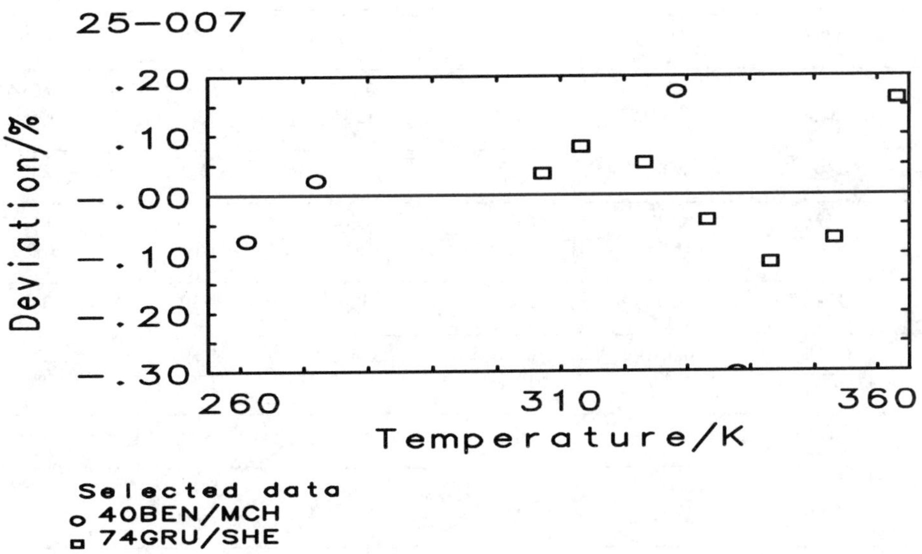

Name: 1,2–Dibromo–1,1,2,2–tetrafluoroethane
Formula: $C_2Br_2F_4$

CAS–RN: 124–73–2
Group No.: 25–008
Molar Mass: 259.82

TABLE 25.8.1. Experimental heat capacities

Reference		Temp. range K	No. pnts	Error %	Purity %	Purity method	Type capacity	Calorimeter Type	Calorimeter Reference
74GRU/SHE	N	323.1–413.1	10S	1.00	99.995	chrom	C_p	BDHO	40TUR/BAR
82KOS/ZHO		165.7–297.0	25	nosp	99.46	melpt	C_{sat}	BSAO	79ZHO/KOS
88VES/ZAB		298.1–318.1	5	0.30	100.0	chrom	C_p	BSAO	79VES/ZAB

74GRU/SHE C_p at saturation line (max. pressure 1 MPa)

TABLE 25.8.2. Correlated heat capacities

Reference	Temp. range K	No. pnts used	$\sigma_r C$ %	d_w	d/R	d_r %	d_b/R	+/–
				Selected data				
74GRU/SHE	323.1–413.1	10	1.00	1.520	3.41–1	1.52	1.35–1	2
82KOS/ZHO	165.7–297.0	25	0.40#	1.157	9.25–2	0.46	–3.05–2	–3
88VES/ZAB	298.1–318.1	5	0.30	1.424	8.96–2	0.43	6.63–2	3

TABLE 25.8.3. Parameters of regression polynomial

Heat capacity type	No. data points total	No. data points used	s_w	s/R	s_r %	s_b/R	+/–
C	40	40	1.361	1.99–1	0.90	2.31–2	2

Temp. range K	A_1	A_2	A_3	A_4	Level of uncertainty
165.7–413.1	1.39894+1	3.97258	–1.18635	2.06692–1	IV

TABLE 25.8.4. Recommended values of heat capacities

Temp. (K)	170	180	190	200	210	220	230
c (J K^{-1}g^{-1})	0.587	0.592	0.598	0.603	0.608	0.614	0.620
C (J K^{-1}mol^{-1})	152.4	153.8	155.3	156.7	158.1	159.5	161.0
Temp. (K)	240	250	260	270	273.15	280	290
c (J K^{-1}g^{-1})	0.626	0.632	0.638	0.644	0.646	0.651	0.658
C (J K^{-1}mol^{-1})	162.5	164.1	165.7	167.4	168.0	169.2	171.1
Temp. (K)	298.15	300	310	320	330	340	350
c (J K^{-1}g^{-1})	0.665	0.666	0.674	0.682	0.691	0.701	0.711
C (J K^{-1}mol^{-1})	172.7	173.0	175.1	177.3	179.7	182.1	184.8
Temp. (K)	360	370	380	390	400	410	
c (J K^{-1}g^{-1})	0.722	0.733	0.745	0.758	0.772	0.787	
C (J K^{-1}mol^{-1})	187.6	190.5	193.7	197.0	200.6	204.4	

TABLE 25.8.5. Parameters of quasi-polynomial equation

Heat capacity type	No. data points total	used	s_w	s/R	s_r %	s_b/R	+/–
C	40	40	2.049	2.09–1	1.01	4.22–2	6

Temp. range K	T_c K	A_1	A_2	A_3	A_4	Level of uncertainty
165.7–413.1	487.80	–1.81087	1.28482–1	1.49397+1	6.38073	V

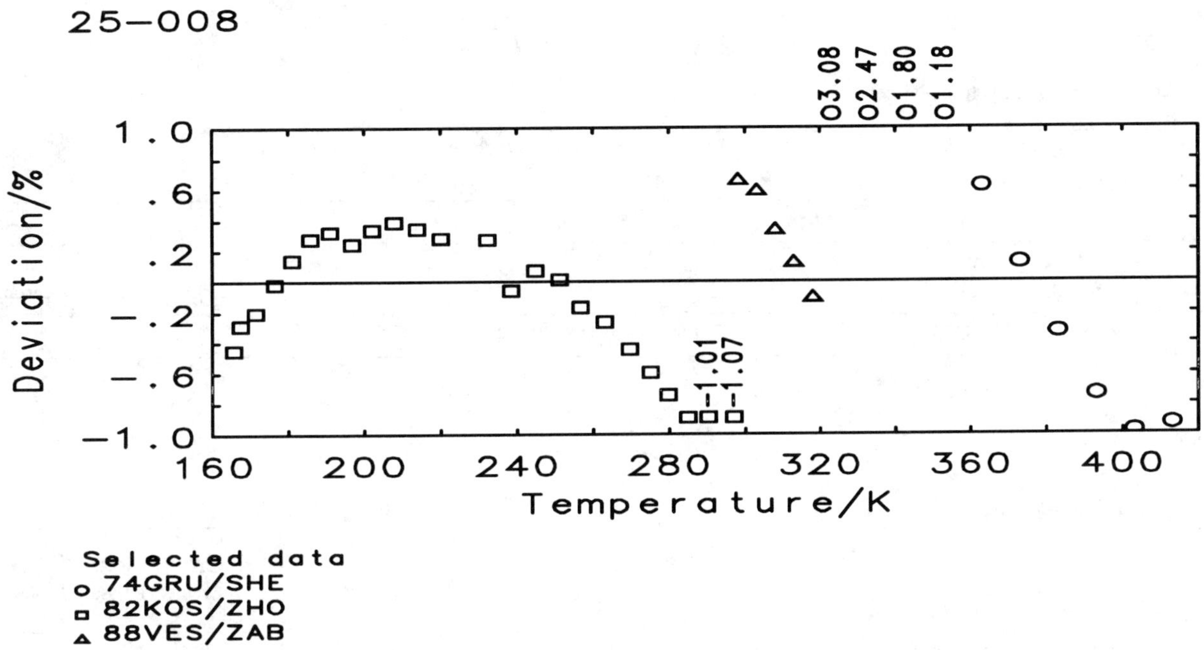

Name: Chlorotrifluoroethene
Formula: C_2ClF_3

CAS–RN: 79–38–9
Group No.: 25–009
Molar Mass: 116.47

TABLE 25.9.1. Experimental heat capacities

Reference	Temp. range K	No. pnts	Error %	Purity %	Purity method	Type capacity	Calorimeter Type	Calorimeter Reference
51OLI/GRI2	120.0–240.0	11S	0.20	99.84	melpt	C_{sat}	BSAO	43RUE/HUF

TABLE 25.9.3. Parameters of regression polynomial

Heat capacity type	No. data points total	used	s_w	s/R	s_r %	s_b/R	+/–
C_{sat}	11	11	0.247	6.75–3	0.05	3.03–6	–2

Temp. range K	A_1	A_2	A_3	A_4	Level of uncertainty
120.0–240.0	1.81031+1	–7.41371	3.67702	–4.94745–1	II

TABLE 25.9.4. Recommended values of heat capacities

Temp. (K)	120	130	140	150	160	170	180
c_{sat} (J K^{-1}g^{-1})	0.974	0.970	0.969	0.970	0.973	0.978	0.984
C_{sat} (J K^{-1}mol^{-1})	113.5	113.0	112.9	113.0	113.3	113.9	114.6
Temp. (K)	190	200	210	220	230	240	
c_{sat} (J K^{-1}g^{-1})	0.992	1.001	1.011	1.022	1.034	1.046	
C_{sat} (J K^{-1}mol^{-1})	115.6	116.6	117.8	119.1	120.4	121.8	

TABLE 25.9.5. Parameters of quasi-polynomial equation

Heat capacity type	No. data points total	used	s_w	s/R	s_r %	s_b/R	+/–
C_{sat}	11	9	1.058	2.95–2	0.21	8.13–5	–3

Temp. range K	T_c K	A_1	A_2	A_3	A_4	Level of uncertainty
140.0–240.0	379.00	2.07355	1.84817	1.13435+1	5.81604–1	II

Name: Chloropentafluoroethane
Formula: C_2ClF_5

CAS–RN: 76–15–3
Group No.: 25–010
Molar Mass: 154.47

TABLE 25.10.1. Experimental heat capacities

Reference		Temp. range K	No. pnts	Error %	Purity %	Purity method	Type capacity	Calorimeter Type	Calorimeter Reference
55AST/WIL		175.0–230.0	12S	0.20	99.992	melpt	C_p	BSIO	36AST/MES
56PLA	N	303.1	1	nosp	not specified		C_{sat}	not specified	

56PLA original datum in Kinetic Chemicals Inc.Bulletin B–2, 1950

TABLE 25.10.2. Correlated heat capacities

Reference	Temp. range K	No. pnts used	$\sigma_r C$ %	d_w	d/R	d_r %	d_b/R	+/–
				Selected data				
55AST/WIL	175.0–230.0	12	0.20	0.530	1.76–2	0.11	–8.65–5	–1
56PLA	303.1	1	1.00#	0.169	3.81–2	0.17	3.81–2	1

TABLE 25.10.3. Parameters of regression polynomial

Heat capacity type	No. data points total	used	s_w	s/R	s_r %	s_b/R	+/–
C	13	13	0.615	2.39–2	0.13	2.85–3	0

Temp. range K	A_1	A_2	A_3	A_4	Level of uncertainty
175.0–303.1	8.84825	9.09189	–4.95390	1.13497	III

TABLE 25.10.4. Recommended values of heat capacities

Temp. (K)	180	190	200	210	220	230	240
c (J $K^{-1}g^{-1}$)	0.850	0.863	0.877	0.894	0.913	0.935	0.959
C (J $K^{-1}mol^{-1}$)	131.2	133.2	135.5	138.1	141.0	144.4	148.2
Temp. (K)	250	260	270	273.15	280	290	298.15
c (J $K^{-1}g^{-1}$)	0.988	1.020	1.056	1.069	1.097	1.143	1.184
C (J $K^{-1}mol^{-1}$)	152.6	157.5	163.1	165.1	169.5	176.5	182.9
Temp. (K)	300						
c (J $K^{-1}g^{-1}$)	1.194						
C (J $K^{-1}mol^{-1}$)	184.4						

Name: Dichlorotetrafluoroethane (unspecified isomer)
Formula: $C_2Cl_2F_4$

CAS–RN: 1320–37–2
Group No.: 25–011
Molar Mass: 170.92

TABLE 25.11.1. Experimental heat capacities

Reference		Temp. range K	No. pnts	Error %	Purity %	Purity method	Type capacity	Calorimeter Type	Calorimeter Reference
37PER1	N	183.1–293.3	26	nosp	not specified		C_{sat}	BSIO	37PER1

37PER1 mixture of two isomers, commercial product by Du Pont Co.; same data in 37PER2

TABLE 25.11.3. Parameters of regression polynomial

Heat capacity type	No. data points total	No. data points used	s_w	s/R	s_r %	s_b/R	+/–
C_{sat}	26	26	0.642	1.23–1	0.64	1.44–3	6

Temp. range K	A_1	A_2	Level of uncertainty
183.1–293.3	1.26750+1	2.69731	V

TABLE 25.11.4. Recommended values of heat capacities

Temp. (K)	180	190	200	210	220	230	240
c_{sat} (J $K^{-1}g^{-1}$)	0.85	0.87	0.88	0.89	0.91	0.92	0.93
C_{sat} (J $K^{-1}mol^{-1}$)	146	148	150	152	155	157	159
Temp. (K)	250	260	270	273.15	280	290	298.15
c_{sat} (J $K^{-1}g^{-1}$)	0.94	0.96	0.97	0.97	0.98	1.00	1.01
C_{sat} (J $K^{-1}mol^{-1}$)	161	164	166	167	168	170	172

Name: 1,2–Dichloro–1,1,2,2–tetrafluoroethane
Formula: $C_2Cl_2F_4$

CAS–RN: 76–14–2
Group No.: 25–012
Molar Mass: 170.92

TABLE 25.12.1. Experimental heat capacities

Reference		Temp. range K	No. pnts	Error %	Purity %	Purity method	Type capacity	Calorimeter Type	Calorimeter Reference
60MAR	N	237.1–275.9	2	nosp	not specified		C_{sat}	not specified	
81KOL/KOS		182.0–295.9	21	0.10	99.79	melpt	C_{sat}	BSAO	62KOL/SER

60MAR unpublished data: J.G.Aston, private communication (1955) and R.C.McHarness, private communication (1951)

TABLE 25.12.2. Correlated heat capacities

Reference	Temp. range K	No. pnts used	$\sigma_r C$ %	d_w	d/R	d_r %	d_b/R	+/–
				Selected data				
81KOL/KOS	182.0–295.9	21	0.10	1.388	2.51–2	0.14	7.14–5	–2
				Rejected data				
60MAR	(3.79–1, 1.95, 3.50–1, 2)							

TABLE 25.12.3. Parameters of regression polynomial

Heat capacity type	No. data points total	No. data points used	s_w	s/R	s_r %	s_b/R	+/–
C_p	23	21	1.499	2.71–2	0.15	7.14–5	–2
C_{sat}	23	21	1.431	2.58–2	0.14	6.42–5	–1

Temp. range K	A_1	A_2	A_3	Level of uncertainty
182.0–295.9	1.44940+1	8.20599–1	3.17576–1	III
182.0–295.9	1.41292+1	1.16680	2.36227–1	III

TABLE 25.12.4. Recommended values of heat capacities

Temp. (K)	180	190	200	210	220	230	240
c_p (J K^{-1}g^{-1})	0.827	0.837	0.847	0.857	0.868	0.879	0.890
C_p (J K^{-1}mol^{-1})	141.3	143.0	144.7	146.5	148.3	150.2	152.1
c_{sat} (J K^{-1}g^{-1})	0.827	0.837	0.847	0.857	0.868	0.879	0.890
C_{sat} (J K^{-1}mol^{-1})	141.3	143.0	144.7	146.5	148.3	150.2	152.1

Temp. (K)	250	260	270	273.15	280	290	298.15
c_p (J K^{-1}g^{-1})	0.901	0.913	0.925	0.929	0.938	0.951	0.961
C_p (J K^{-1}mol^{-1})	154.1	156.1	158.2	158.8	160.3	162.5	164.3
c_{sat} (J K^{-1}g^{-1})	0.901	0.913	0.924	0.928	0.936	0.949	0.959
C_{sat} (J K^{-1}mol^{-1})	154.0	156.0	158.0	158.6	160.0	162.1	163.9

Temp. (K)	300
c_p (J K^{-1}g^{-1})	0.964
C_p (J K^{-1}mol^{-1})	164.7
c_{sat} (J K^{-1}g^{-1})	0.961
C_{sat} (J K^{-1}mol^{-1})	164.3

TABLE 25.12.5. Parameters of quasi-polynomial equation

Heat capacity type	No. data points total	used	s_w	s/R	s_r %	s_b/R	+/–
C_p	23	21	1.377	2.47–2	0.14	5.68–5	–2

Temp. range K	T_c K	A_1	A_2	A_3	A_4	Level of uncertainty
182.0–295.9	418.78	–1.53323	1.06437–1	1.35726+1	5.52154	III

25–012

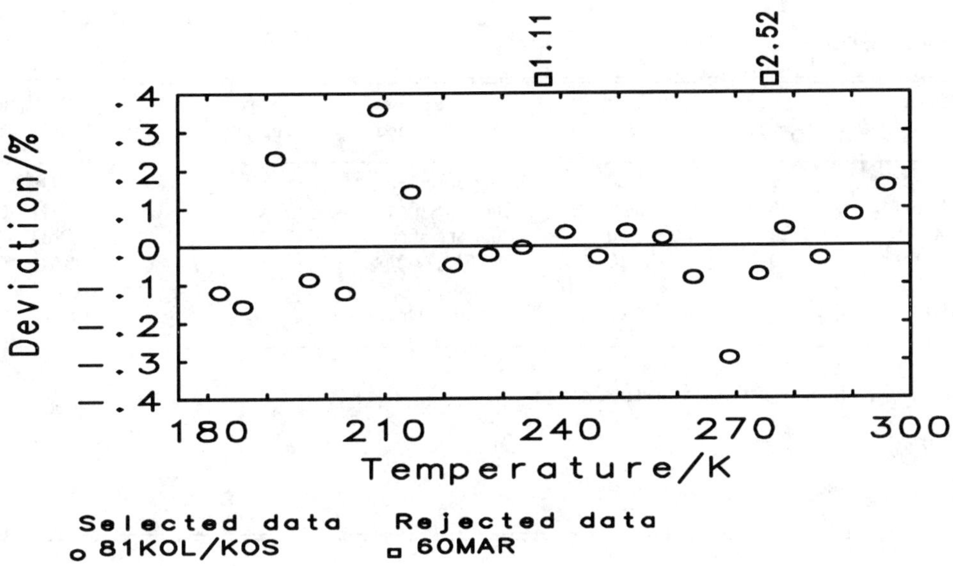

Name: 1,1,1–Trichloro–2,2,2–trifluoroethane
Formula: $C_2Cl_3F_3$

CAS–RN: 354–58–5
Group No.: 25–013
Molar Mass: 187.38

TABLE 25.13.1. Experimental heat capacities

Reference	Temp. range K	No. pnts	Error %	Purity %	Purity method	Type capacity	Calorimeter Type	Calorimeter Reference
88SVO/VES	298.1–318.1	5S	0.30	99.5	chrom	C_p	BSAO	79VES/ZAB

TABLE 25.13.3. Parameters of regression polynomial

Heat capacity type	No. data points total	used	s_w	s/R	s_r %	s_b/R	+/–
C_p	5	5	0.000	0.00	0.00	0.00	0

Temp. range K	A_1	A_2	Level of uncertainty
298.1–318.1	1.31060+1	2.40543	III

TABLE 25.13.4. Recommended values of heat capacities

Temp. (K)	290	298.15	300	310	320
c_p (J $K^{-1}g^{-1}$)	0.891	0.900	0.902	0.912	0.923
C_p (J $K^{-1}mol^{-1}$)	167.0	168.6	169.0	171.0	173.0

Name: 1,1,2–Trichloro–1,2,2–trifluoroethane
Formula: $C_2Cl_3F_3$

CAS–RN: 76–13–1
Group No.: 25–014
Molar Mass: 187.38

TABLE 25.14.1. Experimental heat capacities

Reference		Temp. range K	No. pnts	Error %	Purity %	Purity method	Type capacity	Calorimeter Type	Calorimeter Reference
38RIE	N	242.9–333.9	19	0.50	not specified		C_{sat}	BSIO	39RIE3
40BEN/MCH		261.6–336.4	6	nosp	not specified		C_{sat}	BSIO	40BEN/MCH
54BUB	N	253.1–458.1	22S	3.30	not specified		C_p	BDAO	75RAS/GRI
81KOL/KOS		240.5–298.9	10	0.10	99.81	melpt	C_{sat}	BSAO	62KOL/SER
88VES/ZAB		298.1–318.1	5	0.30	100.0	chrom	C_p	BSAO	79VES/ZAB

38RIE same data in 39RIE1
54BUB C_p at saturation line extrapolated from high pressure measurements

TABLE 25.14.2. Correlated heat capacities

Reference	Temp. range K	No. pnts used	$\sigma_r C$ %	d_w	d/R	d_r %	d_b/R	+/–
				Selected data				
40BEN/MCH	261.6–336.5	6	1.50#	0.539	1.70–1	0.81	–2.05–2	2
54BUB	253.1–458.1	22	3.30	0.792	5.92–1	2.62	4.28–1	14
81KOL/KOS	240.5–298.9	10	0.10	1.019	2.08–2	0.10	1.47–3	0
88VES/ZAB	298.1–318.1	5	0.30	1.119	7.10–2	0.3	4–3.86–2	–1
				Rejected data				
38RIE	(5.66–1, 2.60, 5.18–1, 19)							

TABLE 25.14.3. Parameters of cubic spline polynomials

Heat capacity type	No. data points total	No. data points used	s_w	s/R	s_r %	s_b/R	+/–
C_p	62	43	0.920	4.57–1	2.02	2.12–1	15
C_{sat}	62	43	0.928	4.47–1	2.01	2.23–1	15

Temp. range K	A_1	A_2	A_3	A_4	Level of uncertainty
240.5–330.0	3.12424+1	–1.50403+1	5.63489	–5.92727–1	III
330.0–458.1	–9.79129+1	1.02374+2	–2.99451+1	3.00121	V
240.5–330.0	3.02747+1	–1.42054+1	5.42177	–5.79060–1	III
330.0–458.1	–7.62112+1	8.26000+1	–2.39132+1	2.38407	V

TABLE 25.14.4. Recommended values of heat capacities

Temp. (K)	240	250	260	270	273.15	280	290
c_p (J K^{-1}g^{-1})	0.861	0.870	0.879	0.889	0.893	0.901	0.912
C_p (J K^{-1}mol^{-1})	161.4	163.0	164.7	166.7	167.3	168.7	170.9
c_{sat} (J K^{-1}g^{-1})	0.861	0.870	0.879	0.890	0.893	0.901	0.912
C_{sat} (J K^{-1}mol^{-1})	161.4	163.0	164.7	166.7	167.3	168.7	170.9
Temp. (K)	298.15	300	310	320	330	340	350
c_p (J K^{-1}g^{-1})	0.922	0.924	0.937	0.949	0.962	0.97	0.99
C_p (J K^{-1}mol^{-1})	172.8	173.2	175.5	177.9	180.2	183	185
c_{sat} (J K^{-1}g^{-1})	0.922	0.924	0.936	0.948	0.960	0.97	0.98
C_{sat} (J K^{-1}mol^{-1})	172.7	173.1	175.4	177.6	179.8	182	184
Temp. (K)	360	370	380	390	400	410	420
c_p (J K^{-1}g^{-1})	1.00	1.02	1.04	1.06	1.09	1.12	1.16
C_p (J K^{-1}mol^{-1})	188	191	194	199	204	210	218
c_{sat} (J K^{-1}g^{-1})	1.00	1.01	1.03	1.05	1.07	1.10	1.13
C_{sat} (J K^{-1}mol^{-1})	187	190	193	196	201	206	212
Temp. (K)	430	440	450	460			
c_p (J K^{-1}g^{-1})	1.21	1.26	1.33	1.40			
C_p (J K^{-1}mol^{-1})	226	237	248	262			
c_{sat} (J K^{-1}g^{-1})	1.17	1.21	1.26	1.32			
C_{sat} (J K^{-1}mol^{-1})	219	227	237	248			

TABLE 25.14.5. Parameters of quasi-polynomial equation

Heat capacity type	No. data points total	used	s_w	s/R	s_r %	s_b/R	+/–
C_p	62	43	7.769	3.32–1	1.54	–7.02–2	–11

Temp. range K	T_c K	A_1	A_2	A_3	A_4	Level of uncertainty
240.5–458.1	487.30	–2.42293	3.01084–1	1.50316+1	4.87454	V

25–014

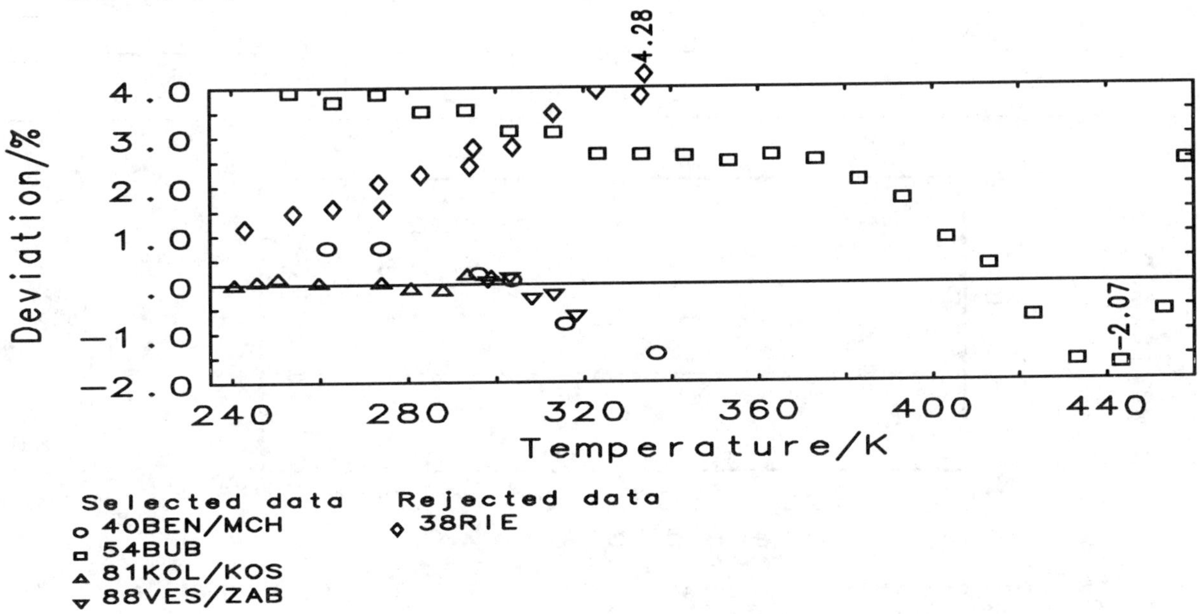

Name: 1,1,2,2–Tetrachloro–1,2–difluoroethane
Formula: $C_2Cl_4F_2$

CAS–RN: 76–12–0
Group No.: 25–015
Molar Mass: 203.83

TABLE 25.15.1. Experimental heat capacities

Reference	Temp. range K	No. pnts	Error %	Purity %	Purity method	Type capacity	Calorimeter Type	Calorimeter Reference
57YAR/KAY	305.0–353.0	eqn	1.10	not specified		C_p	BSIO	75PED/KAY
78KIS/SUG	299.0–310.3	5	0.30	99.93	melpt	C_p	BSAO	65SUG/SEK
78KOS/KOL	300.9–309.4	10	nosp	99.27	melpt	C_{sat}	BSAO	62KOL/SER

TABLE 25.15.2. Correlated heat capacities

Reference	Temp. range K	No. pnts used	$\sigma_r C$ %	d_w	d/R	d_r %	d_b/R	+/–
				Selected data				
78KIS/SUG	299.0–310.3	5	0.30	0.048	3.09–3	0.01	0.00	0
				Rejected data				
57YAR/KAY	(5.03–1, 2.39, –5.03–1, –1)			78KOS/KOL	(1.93–1, 0.88, 1.68–1, 10)			

TABLE 25.15.3. Parameters of regression polynomial

Heat capacity type	No. data points total	No. data points used	s_w	s/R	s_r %	s_b/R	+/–
C_p	21	5	0.062	3.98–3	0.02	0.00	0

Temp. range K	A_1	A_2	Level of uncertainty
299.0–310.3	1.60637+1	1.80535	III

TABLE 25.15.4. Recommended values of heat capacities

Temp. (K)	300	310
c_p (J K^{-1}g^{-1})	0.876	0.884
C_p (J K^{-1}mol^{-1})	178.6	180.1

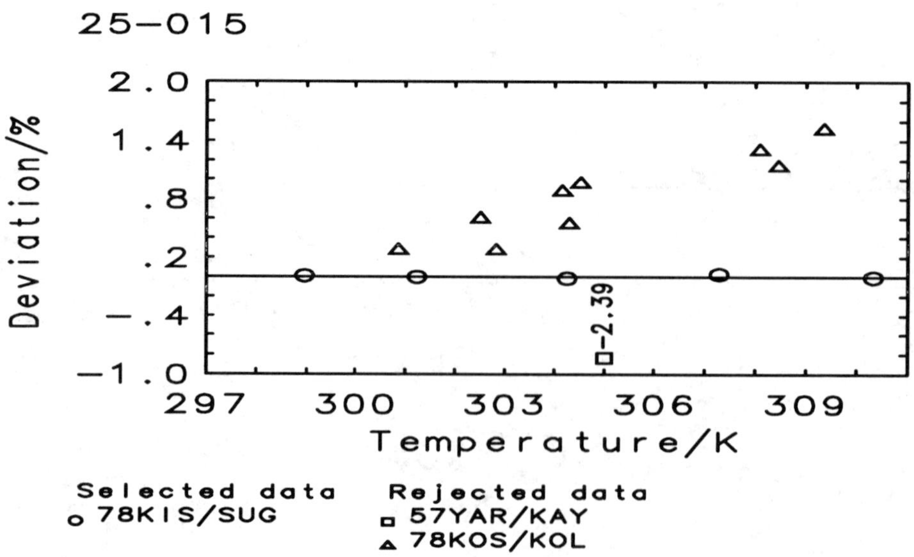

Name: 1–Bromo–2–chloro–1,1,2–trifluoroethane
Formula: $C_2HBrClF_3$

CAS–RN: 354–06–3
Group No.: 25–016
Molar Mass: 197.38

TABLE 25.16.1. Experimental heat capacities

Reference	Temp. range K	No. pnts	Error %	Purity %	Purity method	Type capacity	Calorimeter Type	Calorimeter Reference
88VES/ZAB	298.1–318.1	5	0.30	99.0	chrom	C_p	BSAO	79VES/ZAB

TABLE 25.16.3. Parameters of regression polynomial

Heat capacity type	No. data points total	No. data points used	s_w	s/R	s_r %	s_b/R	+/–
C_p	5	5	0.222	1.30–2	0.07	1.03–5	1

Temp. range K	A_1	A_2	Level of uncertainty
298.1–318.1	1.39079+1	1.80563	III

TABLE 25.16.4. Recommended values of heat capacities

Temp. (K)	290	298.15	300	310	320
c_p (J K^{-1}g^{-1})	0.806	0.813	0.814	0.822	0.829
C_p (J K^{-1}mol^{-1})	159.2	160.4	160.7	162.2	163.7

Name: 2–Bromo–2–chloro–1,1,1–trifluoroethane
Formula: $C_2HBrClF_3$

CAS–RN: 151–67–7
Group No.: 25–017
Molar Mass: 197.38

TABLE 25.17.1. Experimental heat capacities

Reference	Temp. range K	No. pnts	Error %	Purity %	Purity method	Type capacity	Calorimeter Type	Calorimeter Reference
88VES/ZAB	298.1–318.1	5	0.30	99.0	chrom	C_p	BSAO	79VES/ZAB

TABLE 25.17.3. Parameters of regression polynomial

Heat capacity type	No. data points total	No. data points used	s_w	s/R	s_r %	s_b/R	+/–
C_p	5	5	0.255	1.46–2	0.08	1.30–5	2

Temp. range K	A_1	A_2	Level of uncertainty
298.1–318.1	1.29511+1	1.97094	III

TABLE 25.17.4. Recommended values of heat capacities

Temp. (K)	290	298.15	300	310	320
c_p (J K^{-1}g^{-1})	0.786	0.793	0.795	0.803	0.811
C_p (J K^{-1}mol^{-1})	155.2	156.5	156.8	158.5	160.1

Name: 1,1,2–Trichloro–1,2–difluoroethane
Formula: $C_2HCl_3F_2$

CAS–RN: 354–15–4
Group No.: 25–018
Molar Mass: 169.38

TABLE 25.18.1. Experimental heat capacities

Reference	Temp. range K	No. pnts	Error %	Purity %	Purity method	Type capacity	Calorimeter Type	Calorimeter Reference
92LEB/KUL1	104.8–328.1	82	0.20	98.9	chrom	C_p	BSAO	76LEB/LIT

TABLE 25.18.3. Parameters of regression polynomial

Heat capacity type	No. data points total	No. data points used	s_w	s/R	s_r %	s_b/R	+/–
C_p	82	82	2.630	1.21–2	0.53	1.22–4	4

Temp. range K	A_1	A_2	A_3	A_4	Level of uncertainty
104.8–328.1	2.68703	–7.19962–1	3.11068–1	–3.26272–2	III

TABLE 25.18.4. Recommended values of heat capacities

Temp. (K)	100	110	120	130	140	150	160
c_p ($J K^{-1}g^{-1}$)	0.110	0.109	0.109	0.108	0.108	0.108	0.108
C_p ($J K^{-1}mol^{-1}$)	18.67	18.53	18.41	18.33	18.29	18.27	18.27
Temp. (K)	170	180	190	200	210	220	230
c_p ($J K^{-1}g^{-1}$)	0.108	0.108	0.109	0.109	0.110	0.111	0.112
C_p ($J K^{-1}mol^{-1}$)	18.31	18.36	18.44	18.54	18.66	18.80	18.95
Temp. (K)	240	250	260	270	273.15	280	290
c_p ($J K^{-1}g^{-1}$)	0.113	0.114	0.115	0.116	0.117	0.117	0.119
C_p ($J K^{-1}mol^{-1}$)	19.12	19.30	19.49	19.69	19.76	19.90	20.12
Temp. (K)	298.15	300	310	320	330		
c_p ($J K^{-1}g^{-1}$)	0.120	0.120	0.121	0.123	0.124		
C_p ($J K^{-1}mol^{-1}$)	20.29	20.34	20.56	20.78	21.00		

TABLE 25.18.5. Parameters of quasi-polynomial equation

Heat capacity type	No. data points total	No. data points used	s_w	s/R	s_r %	s_b/R	+/–
C_p	82	82	4.535	2.09–2	0.91	3.54–4	1

Temp. range K	T_c K	A_1	A_2	A_3	A_4	Level of uncertainty
104.8–328.1	525.10	1.32588	8.00097–1	1.38006	5.49299–1	IV

Name: 1,1–Dichloro–1,2–difluoroethane
Formula: $C_2H_2Cl_2F_2$

CAS–RN: 1842–05–3
Group No.: 25–019
Molar Mass: 134.94

TABLE 25.19.1. Experimental heat capacities

Reference	Temp. range K	No. pnts	Error %	Purity %	Purity method	Type capacity	Calorimeter Type	Calorimeter Reference
92LEB/KUL1	173.5–313.4	27	0.20	99.6	melpt	C_p	BSAO	76LEB/LIT

TABLE 25.19.3. Parameters of regression polynomial

Heat capacity type	No. data points total	No. data points used	s_w	s/R	s_r %	s_b/R	+/–
C_p	27	26	1.453	6.15–3	0.29	3.15–5	2

Temp. range K	A_1	A_2	A_3	Level of uncertainty
173.5–307.6	2.09557	–1.31798–1	5.58456–2	III

TABLE 25.19.4. Recommended values of heat capacities

Temp. (K)	170	180	190	200	210	220	230
c_p (J K^{-1}g^{-1})	0.125	0.126	0.126	0.127	0.127	0.128	0.129
C_p (J K^{-1}mol^{-1})	16.90	16.96	17.02	17.09	17.17	17.26	17.36
Temp. (K)	240	250	260	270	273.15	280	290
c_p (J K^{-1}g^{-1})	0.129	0.130	0.131	0.132	0.133	0.133	0.135
C_p (J K^{-1}mol^{-1})	17.47	17.59	17.71	17.85	17.89	18.00	18.15
Temp. (K)	298.15	300	310				
c_p (J K^{-1}g^{-1})	0.135	0.136	0.137				
C_p (J K^{-1}mol^{-1})	18.28	18.32	18.49				

Name: 1–Chloro–1,1–difluoroethane
Formula: $C_2H_3ClF_2$

CAS–RN: 75–68–3
Group No.: 25–020
Molar Mass: 100.50

TABLE 25.20.1. Experimental heat capacities

Reference		Temp. range K	No. pnts	Error %	Purity %	Purity method	Type capacity	Calorimeter Type	Calorimeter Reference
37PER1	N	145.9–291.5	28	0.40	not specified		C_p	BSIO	37PER1
41RIE3	N	212.0–294.8	10	1.00	not specified		C_{sat}	BSIO	39RIE3
58CHE	N	245.1–370.1	15	nosp	not specified		C_{sat}	BSIO	58CHE
64VAS2		229.2–302.9	12	0.50	not specified		C_{sat}	BDAO	64VAS2

37PER1 same data in 37PER2
41RIE3 same data in 42RIE2
58CHE reproducibility given as 1–2 %

TABLE 25.20.2. Correlated heat capacities

Reference	Temp. range K	No. pnts used	$\sigma_r C$ %	d_w	d/R	d_r %	d_b/R	+/–
				Selected data				
37PER1	145.9–291.5	28	1.50#	0.818	1.83–1	1.23	1.23–1	18
41RIE3	212.1–294.8	10	1.00	0.836	1.23–1	0.84	–7.26–2	–6
64VAS2	229.2–303.0	12	0.70#	0.546	5.69–2	0.38	–2.92–2	–4
				Rejected data				
58CHE	(8.16–1, 5.74, –8.00–1, –8)							

TABLE 25.20.3. Parameters of regression polynomial

Heat capacity type	No. data points total	used	s_w	s/R	s_r %	s_b/R	+/–
C_p	65	50	0.790	1.55–1	1.04	4.73–2	8

Temp. range K	A_1	A_2	A_3	Level of uncertainty
145.9–303.0	1.34379+1	–8.18832–1	5.28473–1	V

TABLE 25.20.4. Recommended values of heat capacities

Temp. (K)	150	160	170	180	190	200	210
c_p (J K^{-1}g^{-1})	1.11	1.12	1.12	1.13	1.14	1.15	1.16
C_p (J K^{-1}mol^{-1})	111	112	113	114	115	116	117
Temp. (K)	220	230	240	250	260	270	273.15
c_p (J K^{-1}g^{-1})	1.17	1.19	1.20	1.22	1.23	1.25	1.25
C_p (J K^{-1}mol^{-1})	118	119	121	122	124	125	126
Temp. (K)	280	290	298.15	300			
c_p (J K^{-1}g^{-1})	1.26	1.28	1.30	1.30			
C_p (J K^{-1}mol^{-1})	127	129	130	131			

TABLE 25.20.5. Parameters of quasi-polynomial equation

Heat capacity type	No. data points total	used	s_w	s/R	s_r %	s_b/R	+/–
C_p	65	50	0.850	1.44–1	0.97	5.86–3	0

Temp. range K	T_c K	A_1	A_2	A_3	A_4	Level of uncertainty
145.9–303.0	409.60	–1.39843	2.58966–1	1.16073+1	1.88789	V

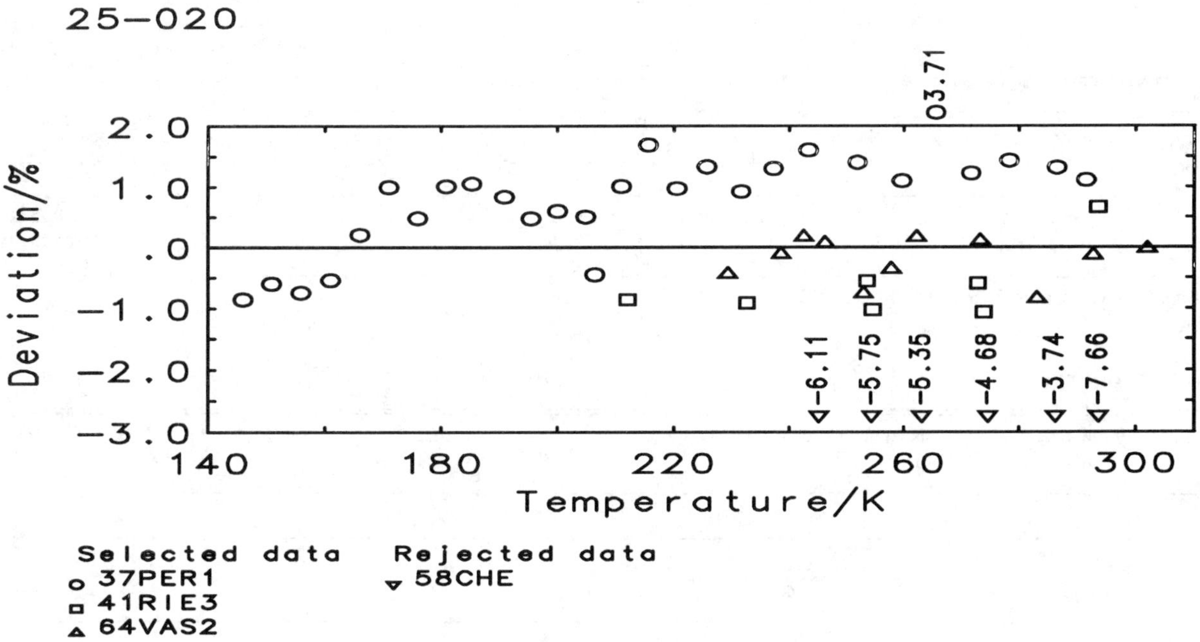

Name: 1–Bromo–2–chloroethane
Formula: C_2H_4BrCl

CAS–RN: 107–04–0
Group No.: 25–021
Molar Mass: 143.41

TABLE 25.21.1. Experimental heat capacities

Reference		Temp. range K	No. pnts	Error %	Purity %	Purity method	Type capacity	Calorimeter Type	Calorimeter Reference
39RAI	N	262.5–321.5	11	2.00	99.8	melpt	C_p	BDHO	37STU

39RAI data from a graph only

TABLE 25.21.3. Parameters of regression polynomial

Heat capacity type	No. data points total	No. data points used	s_w	s/R	s_r %	s_b/R	+/–
C_p	11	11	0.306	9.44–2	0.61	9.43–4	–1

Temp. range K	A_1	A_2	Level of uncertainty
262.5–321.5	1.26695+1	9.70011–1	V

TABLE 25.21.4. Recommended values of heat capacities

Temp. (K)	260	270	273.15	280	290	298.15	300
c_p ($J K^{-1}g^{-1}$)	0.88	0.89	0.89	0.89	0.90	0.90	0.90
C_p ($J K^{-1}mol^{-1}$)	126	127	127	128	129	129	130

Temp. (K)	310	320
c_p ($J K^{-1}g^{-1}$)	0.91	0.91
C_p ($J K^{-1}mol^{-1}$)	130	131

Name: 1–Chloro–1,1,3,3,3–pentafluoropropane
Formula: $C_3H_2ClF_5$

CAS–RN: 460–92–4
Group No.: 25–022
Molar Mass: 168.49

TABLE 25.22.1. Experimental heat capacities

Reference		Temp. range K	No. pnts	Error %	Purity %	Purity method	Type capacity	Calorimeter Type	Calorimeter Reference
73KOL/VOR		298.1	1	nosp	99.82	melpt	C_p	BSAO	62KOL/SER
74VOR/KOL	N	167.3–286.4	30	nosp	99.82	melpt	C_{sat}	BSAO	62KOL/SER

74VOR/KOL smoothed value at 298.15 K in 73KOL/VOR

TABLE 25.22.2. Correlated heat capacities

Reference	Temp. range K	No. pnts used	$\sigma_r C$ %	d_w	d/R	d_r %	d_b/R	+/–
				Selected data				
73KOL/VOR	298.1	1	0.40#	0.043	4.10–3	0.02	–4.10–3	–1
74VOR/KOL	167.3–286.4	30	0.40#	0.220	1.83–2	0.09	1.70–4	–4

TABLE 25.22.3. Parameters of regression polynomial

Heat capacity type	No. data points total	No. data points used	s_w	s/R	s_r %	s_b/R	+/–
C	31	31	0.232	1.93–2	0.09	3.23–5	–5

Temp. range K	A_1	A_2	A_3	A_4	Level of uncertainty
167.3–298.1	2.23369+1	–4.94861	2.35063	–1.82728–1	III

TABLE 25.22.4. Recommended values of heat capacities

Temp. (K)	170	180	190	200	210	220	230
c (J K^{-1}g^{-1})	0.978	0.986	0.995	1.006	1.017	1.030	1.044
C (J K^{-1}mol^{-1})	164.8	166.1	167.7	169.5	171.4	173.6	176.0
Temp. (K)	240	250	260	270	273.15	280	290
c (J K^{-1}g^{-1})	1.060	1.076	1.093	1.111	1.117	1.130	1.150
C (J K^{-1}mol^{-1})	178.5	181.3	184.2	187.2	188.2	190.4	193.7
Temp. (K)	298.15	300					
c (J K^{-1}g^{-1})	1.166	1.170					
C (J K^{-1}mol^{-1})	196.5	197.2					

Name: 1,1,1–Trichloro–3,3,3–trifluoropropane
Formula: $C_3H_2Cl_3F_3$

CAS–RN: 7125–84–0
Group No.: 25–023
Molar Mass: 201.40

TABLE 25.23.1. Experimental heat capacities

Reference	Temp. range K	No. pnts	Error %	Purity %	Purity method	Type capacity	Calorimeter Type	Calorimeter Reference
71KOL/VOR	246.3–296.8	6	0.20	99.59	melpt	C_{sat}	BSAO	62KOL/SER

TABLE 25.23.3. Parameters of regression polynomial

Heat capacity type	No. data points total	used	s_w	s/R	s_r %	s_b/R	+/–
C_{sat}	6	6	0.364	1.70–2	0.07	1.72–5	0

Temp. range K	A_1	A_2	Level of uncertainty
246.4–296.8	1.74468+1	2.21533	III

TABLE 25.23.4. Recommended values of heat capacities

Temp. (K)	250	260	270	273.15	280	290	298.15
c_{sat} ($J K^{-1} g^{-1}$)	0.949	0.958	0.967	0.970	0.976	0.985	0.993
C_{sat} ($J K^{-1} mol^{-1}$)	191.1	193.0	194.8	195.4	196.6	198.5	200.0

Temp. (K)	300
c_{sat} ($J K^{-1} g^{-1}$)	0.995
C_{sat} ($J K^{-1} mol^{-1}$)	200.3

Name: 3,3–Dichloro–1,1,1–trifluoropropane
Formula: $C_3H_3Cl_2F_3$

CAS–RN: 460–69–5
Group No.: 25–024
Molar Mass: 166.96

TABLE 25.24.1. Experimental heat capacities

Reference		Temp. range K	No. pnts	Error %	Purity %	method	Type capacity	Calorimeter Type	Reference
72VOR/KOL	N	184.8–293.6	20	nosp	99.16	melpt	C_{sat}	BSAO	56POP/KOL

72VOR/KOL smoothed values in 72KOL/VOR

TABLE 25.24.3. Parameters of regression polynomial

Heat capacity type	No. data points total	used	s_w	s/R	s_r %	s_b/R	+/–
C_{sat}	20	20	0.341	3.05–2	0.14	7.07–5	–6

Temp. range K	A_1	A_2	A_3	Level of uncertainty
184.8–293.6	2.20955+1	–1.92133	7.47765–1	III

TABLE 25.24.4. Recommended values of heat capacities

Temp. (K)	190	200	210	220	230	240	250
c_{sat} ($J K^{-1} g^{-1}$)	1.053	1.058	1.064	1.070	1.077	1.085	1.094
C_{sat} ($J K^{-1} mol^{-1}$)	175.8	176.6	177.6	178.7	179.9	181.2	182.6

Temp. (K)	260	270	273.15	280	290
c_{sat} ($J K^{-1} g^{-1}$)	1.103	1.113	1.117	1.124	1.136
C_{sat} ($J K^{-1} mol^{-1}$)	184.2	185.9	186.5	187.7	189.7

Name: 3–Chloro–1,1,1–trifluoropropane
Formula: $C_3H_4ClF_3$

CAS–RN: 460–35–5
Group No.: 25–025
Molar Mass: 132.51

TABLE 25.25.1. Experimental heat capacities

Reference		Temp. range K	No. pnts	Error %	Purity %	Purity method	Type capacity	Calorimeter Type	Calorimeter Reference
74KOL/VOR	N	184.4–299.0	32	nosp	99.8	melpt	C_{sat}	BSAO	62KOL/SER

74KOL/VOR smoothed values in 72KOL/VOR

TABLE 25.25.3. Parameters of regression polynomial

Heat capacity type	No. data points total	No. data points used	s_w	s/R	s_r %	s_b/R	+/–
C_{sat}	32	32	0.628	3.71–2	0.19	1.27–4	–4

Temp. range K	A_1	A_2	A_3	Level of uncertainty
184.4–299.0	1.83387+1	–8.04779–1	5.19408–1	III

TABLE 25.25.4. Recommended values of heat capacities

Temp. (K)	180	190	200	210	220	230	240
c_{sat} ($J\,K^{-1}g^{-1}$)	1.165	1.172	1.180	1.188	1.197	1.207	1.217
C_{sat} ($J\,K^{-1}mol^{-1}$)	154.4	155.4	156.4	157.5	158.7	159.9	161.3
Temp. (K)	250	260	270	273.15	280	290	298.15
c_{sat} ($J\,K^{-1}g^{-1}$)	1.228	1.240	1.252	1.256	1.265	1.278	1.290
C_{sat} ($J\,K^{-1}mol^{-1}$)	162.7	164.3	165.9	166.4	167.6	169.4	170.9
Temp. (K)	300						
c_{sat} ($J\,K^{-1}g^{-1}$)	1.292						
C_{sat} ($J\,K^{-1}mol^{-1}$)	171.3						

Name: 1,4–Dibromo–2,3–dichloro–1,1,2,3,4,4–hexafluorobutane
Formula: $C_4Br_2Cl_2F_6$

CAS–RN: 375–42–8
Group No.: 25–026
Molar Mass: 392.75

TABLE 25.26.1. Experimental heat capacities

Reference	Temp. range K	No. pnts	Error %	Purity %	Purity method	Type capacity	Calorimeter Type	Calorimeter Reference
88SVO/VES	298.1–318.1	5S	0.30	99.5	chrom	C_p	BSAO	79VES/ZAB

TABLE 25.26.3. Parameters of regression polynomial

Heat capacity type	No. data points total	No. data points used	s_w	s/R	s_r %	s_b/R	+/–
C_p	5	5	0.067	7.28–3	0.02	2.29–6	0

Temp. range K	A_1	A_2	Level of uncertainty
298.1–318.1	2.59282+1	3.34333	III

TABLE 25.26.4. Recommended values of heat capacities

Temp. (K)	290	298.15	300	310	320
c_p (J $K^{-1}g^{-1}$)	0.754	0.760	0.761	0.768	0.775
C_p (J $K^{-1}mol^{-1}$)	296.2	298.5	299.0	301.8	304.5

Name: Bromopentafluorobenzene
Formula: C_6BrF_5

CAS–RN: 344–04–7
Group No.: 25–027
Molar Mass: 246.96

TABLE 25.27.1. Experimental heat capacities

Reference	Temp. range K	No. pnts	Error %	Purity %	Purity method	Type capacity	Calorimeter Type	Calorimeter Reference
75PAU2	245.9–281.5	6	nosp	99.9	melpt	C_{sat}	BSAO	69PAU/LAV2

TABLE 25.27.3. Parameters of regression polynomial

Heat capacity type	No. data points total	No. data points used	s_w	s/R	s_r %	s_b/R	+/–
C_{sat}	6	6	0.117	1.52–2	0.06	6.99–6	0

Temp. range K	A_1	A_2	A_3	Level of uncertainty
245.9–281.5	3.32317+1	–7.85791	1.92818	IV

TABLE 25.27.4. Recommended values of heat capacities

Temp. (K)	250	260	270	273.15	280
c_{sat} (J $K^{-1}g^{-1}$)	0.863	0.870	0.878	0.881	0.887
C_{sat} (J $K^{-1}mol^{-1}$)	213.2	214.8	216.8	217.5	219.1

Name: 1,6–Dibromo–2,3,5–trichloro–1,1,2,3,4,4,5,6,6–nonafluorohexane
Formula: $C_6Br_2Cl_3F_9$

CAS–RN: 85131–86–8
Group No.: 25–028
Molar Mass: 509.22

TABLE 25.28.1. Experimental heat capacities

Reference	Temp. range K	No. pnts	Error %	Purity %	Purity method	Type capacity	Calorimeter Type	Calorimeter Reference
88SVO/VES	298.1–318.1	5S	0.30	99.5	chrom	C_p	BSAO	79VES/ZAB

TABLE 25.28.3. Parameters of regression polynomial

Heat capacity type	No. data points total	used	s_w	s/R	s_r %	s_b/R	+/–
C_p	5	5	0.075	1.14–2	0.02	3.82–6	1

Temp. range K	A_1	A_2	Level of uncertainty
298.1–318.1	4.03701+1	3.34342	III

TABLE 25.28.4. Recommended values of heat capacities

Temp. (K)	290	298.15	300	310	320
c_p (J $K^{-1}g^{-1}$)	0.817	0.822	0.823	0.828	0.834
C_p (J $K^{-1}mol^{-1}$)	416.3	418.5	419.1	421.8	424.6

Name: Chloropentafluorobenzene
Formula: C_6ClF_5

CAS–RN: 344–07–0
Group No.: 25–029
Molar Mass: 202.51

TABLE 25.29.1. Experimental heat capacities

Reference	Temp. range K	No. pnts	Error %	Purity %	Purity method	Type capacity	Calorimeter Type	Calorimeter Reference
68AND/COU1	263.0–394.8	35	nosp	99.94	melpt	C_p	BSAO	63AND/COU1
69PAU/GLU2	263.8–302.7	14	nosp	98.89	melpt	C_p	BSAO	54STR/ICK

TABLE 25.29.2. Correlated heat capacities

Reference	Temp. range K	No. pnts used	$\sigma_r C$ %	d_w	d/R	d_r %	d_b/R	+/–
				Selected data				
68AND/COU1	263.0–394.8	35	0.20#	0.227	1.23–2	0.05	1.13–5	5
				Rejected data				
69PAU/GLU2	(2.54–1, 0.96, 2.33–1, 12)							

TABLE 25.29.3. Parameters of regression polynomial

Heat capacity type	No. data points total	used	s_w	s/R	s_r %	s_b/R	+/–
C_p	49	35	0.237	1.28–2	0.05	1.13–5	5

Temp. range K	A_1	A_2	A_3	Level of uncertainty
263.0–394.8	2.11722+1	1.17587	2.19901–1	II

TABLE 25.29.4. Recommended values of heat capacities

Temp. (K)	260	270	273.15	280	290	298.15	300
c_p (J K^{-1}g^{-1})	1.056	1.065	1.069	1.075	1.085	1.093	1.095
C_p (J K^{-1}mol^{-1})	213.8	215.8	216.4	217.7	219.8	221.4	221.8
Temp. (K)	310	320	330	340	350	360	370
c_p (J K^{-1}g^{-1})	1.106	1.116	1.127	1.138	1.149	1.160	1.171
C_p (J K^{-1}mol^{-1})	223.9	226.0	228.2	230.4	232.7	234.9	237.2
Temp. (K)	380	390					
c_p (J K^{-1}g^{-1})	1.183	1.195					
C_p (J K^{-1}mol^{-1})	239.6	242.0					

TABLE 25.29.5. Parameters of quasi-polynomial equation

Heat capacity type	No. data points total	used	s_w	s/R		s_r %	s_b/R	+/–
C_p	49	35	0.238	1.28–2		0.05	1.18–5	5
Temp. range K	T_c K	A_1	A_2	A_3	A_4			Level of uncertainty
263.0–394.8	558.00	−1.82501	8.70882–2	1.99453+1	9.56122			II

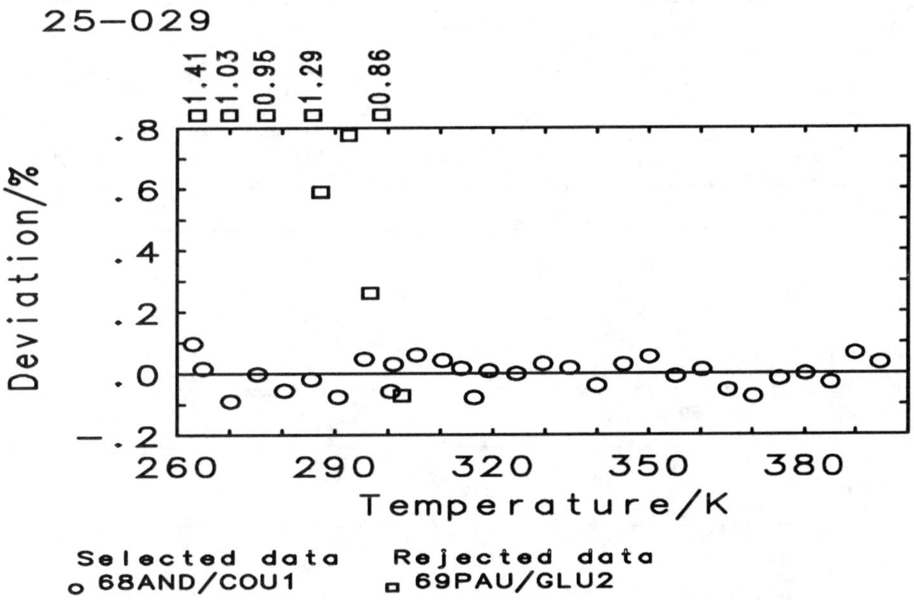

Name: 1,3,5–Trichloro–2,4,6–trifluorobenzene
Formula: $C_6Cl_3F_3$

CAS–RN: 319–88–0
Group No.: 25–030
Molar Mass: 235.42

TABLE 25.30.1. Experimental heat capacities

Reference	Temp. range K	No. pnts	Error %	Purity %	Purity method	Type capacity	Calorimeter Type	Calorimeter Reference
69PAU/GLU1	341.0–354.1	7	0.10	99.12	melpt	C_p	BSAO	54STR/ICK
73AND/MAR2	339.1–347.5	3	nosp	99.8	melpt	C_p	BSAO	63AND/COU1

TABLE 25.30.2. Correlated heat capacities

Reference	Temp. range K	No. pnts used	$\sigma_r C$ %	d_w	d/R	d_r %	d_b/R	+/–
				Selected data				
73AND/MAR2	339.1–347.5	3	0.20#	0.141	8.07–3	0.03	5.09–6	–1
				Rejected data				
69PAU/GLU1	(1.09, 3.68, 1.09, 4)							

TABLE 25.30.3. Parameters of regression polynomial

Heat capacity type	No. data points total	No. data points used	s_w	s/R	s_r %	s_b/R	+/–
C_p	10	3	0.245	1.40–2	0.05	5.09–6	–1

Temp. range K	A_1	A_2	Level of uncertainty
339.1–347.5	1.87546+1	2.85573	II

TABLE 25.30.4. Recommended values of heat capacities

Temp. (K)	340	350
c_p (J K^{-1}g^{-1})	1.005	1.015
C_p (J K^{-1}mol^{-1})	236.7	239.0

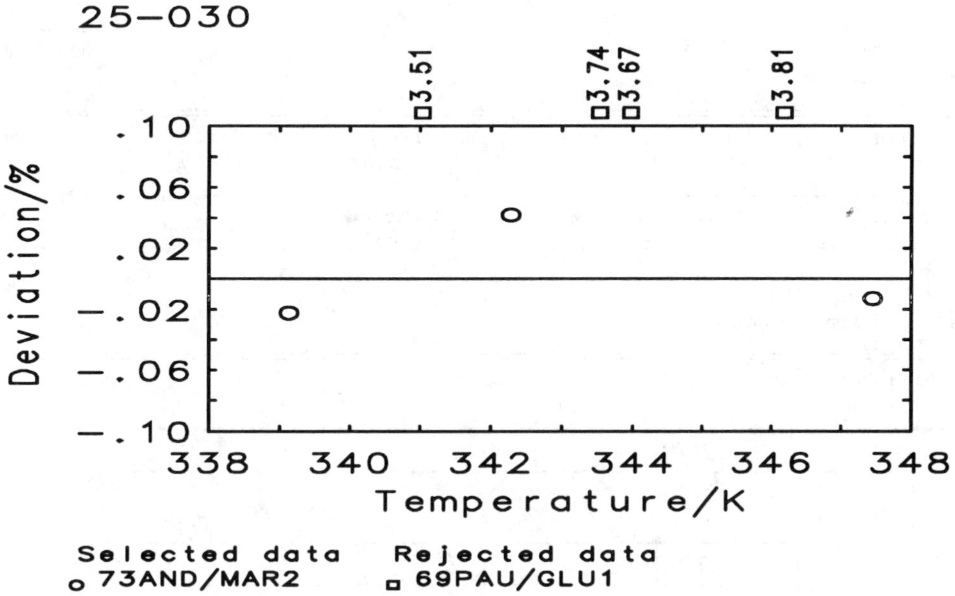

Name: 1–Bromo–2–chlorobenzene
Formula: C_6H_4BrCl

CAS–RN: 694–80–4
Group No.: 25–031
Molar Mass: 191.45

TABLE 25.31.1. Experimental heat capacities

Reference	Temp. range K	No. pnts	Error %	Purity %	Purity method	Type capacity	Calorimeter Type	Calorimeter Reference
18NAR	306.0–326.3	4S	nosp	not specified		C_{avg}	DSIO	18NAR

TABLE 25.31.3. Parameters of regression polynomial

Heat capacity type	No. data points total	No. data points used	s_w	s/R	s_r %	s_b/R	+/–
C	4	4	0.042	4.47–2	0.21	9.25–5	0

Temp. range K	A_1	A_2	Level of uncertainty
306.0–326.3	1.45410+1	2.25090	VI

TABLE 25.31.4. Recommended values of heat capacities

Temp. (K)	310	320	330
c ($J K^{-1} g^{-1}$)	0.93	0.94	0.95
C ($J K^{-1} mol^{-1}$)	179	181	183

Name: 1–Bromo–3–chlorobenzene
Formula: C_6H_4BrCl

CAS–RN: 108–37–2
Group No.: 25–032
Molar Mass: 191.45

TABLE 25.32.1. Experimental heat capacities

Reference	Temp. range K	No. pnts	Error %	Purity %	Purity method	Type capacity	Calorimeter Type	Calorimeter Reference
18NAR	254.3–326.5	4S	nosp	not specified		C_{avg}	DSIO	18NAR

TABLE 25.32.3. Parameters of regression polynomial

Heat capacity type	No. data points total	No. data points used	s_w	s/R	s_r %	s_b/R	+/–
C	4	4	0.071	7.63–2	0.35	2.83–4	0

Temp. range K	A_1	A_2	Level of uncertainty
254.3–326.5	1.44468+1	2.20624	VI

TABLE 25.32.4. Recommended values of heat capacities

Temp. (K)	260	270	273.15	280	290	298.15	300
c $(J\,K^{-1}g^{-1})$	0.88	0.89	0.89	0.90	0.91	0.91	0.91
C $(J\,K^{-1}mol^{-1})$	168	170	170	171	173	175	175

Temp. (K)	310	320	330
c $(J\,K^{-1}g^{-1})$	0.92	0.93	0.94
C $(J\,K^{-1}mol^{-1})$	177	179	181

Name: 1–Bromo–4–chlorobenzene
Formula: C_6H_4BrCl

CAS–RN: 106–39–8
Group No.: 25–033
Molar Mass: 191.45

TABLE 25.33.1. Experimental heat capacities

Reference	Temp. K	Capac. J/(K.g)	Error %	Purity %	Purity method	Type capacity	Calorimeter Type	Calorimeter Reference
18NAR	355.35	1.0000	nosp	not specified		C_{avg}	DSIO	18NAR

Name: 1–Bromo–2–iodobenzene
Formula: C_6H_4BrI

CAS–RN: 583–55–1
Group No.: 25–034
Molar Mass: 282.91

TABLE 25.34.1. Experimental heat capacities

Reference	Temp. range K	No. pnts	Error %	Purity %	Purity method	Type capacity	Calorimeter Type	Calorimeter Reference
18NAR	290.9–325.5	3S	nosp	not specified		C_{avg}	DSIO	18NAR

TABLE 25.34.3. Parameters of regression polynomial

Heat capacity type	No. data points total	used	s_w	s/R	s_r %	s_b/R	+/–
C	3	3	0.041	4.60–2	0.21	6.55–5	1

Temp. range K	A_1	A_2	Level of uncertainty
290.9–325.5	1.72090+1	1.69233	VI

TABLE 25.34.4. Recommended values of heat capacities

Temp. (K)	290	298.15	300	310	320	330
c (J $K^{-1}g^{-1}$)	0.65	0.65	0.65	0.66	0.66	0.67
C (J $K^{-1}mol^{-1}$)	184	185	185	187	188	190

Name: 1–Bromo–3–iodobenzene
Formula: C_6H_4BrI

CAS–RN: 591–18–4
Group No.: 25–035
Molar Mass: 282.91

TABLE 25.35.1. Experimental heat capacities

Reference	Temp. range K	No. pnts	Error %	Purity %	Purity method	Type capacity	Calorimeter Type	Calorimeter Reference
18NAR	290.9–325.5	3S	nosp	not specified		C_{avg}	DSIO	18NAR

TABLE 25.35.3. Parameters of regression polynomial

Heat capacity type	No. data points total	used	s_w	s/R	s_r %	s_b/R	+/–
C	3	3	0.023	2.53–2	0.11	1.97–5	1

Temp. range K	A_1	A_2	Level of uncertainty
290.9–325.5	1.63780+1	1.89294	VI

TABLE 25.35.4. Recommended values of heat capacities

Temp. (K)	290	298.15	300	310	320	330
c (J $K^{-1}g^{-1}$)	0.64	0.65	0.65	0.65	0.66	0.66
C (J $K^{-1}mol^{-1}$)	182	183	183	185	187	188